T0134772

Lecture Notes in Business Information Processing 417

Series Editors

More information about this series at http://www.springer.com/series/7911

Joaquim Filipe · Michał Śmiałek ·
Alexander Brodsky · Slimane Hammoudi (Eds.)

Enterprise Information Systems

22nd International Conference, ICEIS 2020 Virtual Event, May 5–7, 2020 Revised Selected Papers

Editors
Joaquim Filipe
Polytechnic Institute of Setúbal/INSTICC
Setúbal, Portugal

Michał Śmiałek
Warsaw University of Technology
Warsaw, Poland

Alexander Brodsky
George Mason University
Fairfax, VA, USA

Slimane Hammoudi
MODESTE/ESEO
Angers, France

ISSN 1865-1348 ISSN 1865-1356 (electronic)
Lecture Notes in Business Information Processing
ISBN 978-3-030-75417-4 ISBN 978-3-030-75418-1 (eBook)
https://doi.org/10.1007/978-3-030-75418-1

This Springer imprint is published by the registered company Springer Nature Switzerland AG
The registered company address is: Gewerbestrasse 11, 6330 Cham, Switzerland

Preface

The present book includes extended and revised versions of a set of selected papers from the 22nd International Conference on Enterprise Information Systems (ICEIS 2020), exceptionally held as a web-based event, due to the COVID-19 pandemic, during May 5–7, 2020.

ICEIS 2020 received 255 paper submissions from 49 countries, of which 16% were included in this book. The papers were selected by the event chairs and their selection is based on a number of criteria that include classifications and comments provided by Program Committee members, session chairs' assessments and also the program chairs' global view of all the papers included in the technical program. The authors of selected papers were then invited to submit revised and extended versions of their papers having at least 30% innovative material.

The purpose of the 22nd International Conference on Enterprise Information Systems (ICEIS 2020) was to bring together researchers, engineers, and practitioners interested in advances and business applications of information systems. Six simultaneous tracks were held, covering different aspects of Enterprise Information Systems Applications, including Enterprise Database Technology, Systems Integration, Artificial Intelligence, Decision Support Systems, Information Systems Analysis and Specification, Internet Computing, Electronic Commerce, Human Factors, and Enterprise Architecture.

We are confident that the papers included in this book will strongly contribute to the understanding of some current research trends in Enterprise Information Systems. Such systems require diverse approaches to answer the challenges of contemporary enterprises. Thus, this book covers such diverse but complementary areas as: Data Science and Databases, Ontologies, Social Networks, Knowledge Management, Software Development, Human-Computer Interaction and Multimedia.

We would like to thank all the authors for their contributions and the reviewers for their hard work which has helped ensuring the quality of this publication.

May 2020

Joaquim Filipe
Michal Smialek
Alexander Brodsky
Slimane Hammoudi

Preface

Organization

Conference Co-chairs

Alexander Brodsky	George Mason University, USA
Slimane Hammoudi	ESEO, France

Program Co-chairs

Joaquim Filipe	Polytechnic Institute of Setúbal/INSTICC, Portugal
Michal Smialek	Warsaw University of Technology, Poland

Program Committee

Amna Abidi	Altran Research, France
Adeel Ahmad	Laboratoire d'Informatique Signal et Image de la Côte d'Opale, France
Zahid Akhtar	State University of New York Polytechnic Institute, USA
Patrick Albers	ESEO, France
Javier Albusac	Universidad de Castilla-La Mancha, Spain
Eduardo Alchieri	Universidade de Brasilia, Brazil
Julien Aligon	IRIT, France
Mohammad Al-Shamri	Ibb University, Yemen
Luis Álvarez Sabucedo	University of Vigo, Spain
Omar Alvarez-Xochihua	Universidad Autónoma de Baja California, Mexico
Andreas Andreou	Cyprus University of Technology, Cyprus
Leandro Antonelli	Universidad Nacional de La Plata, Argentina
Olatz Arbelaitz Gallego	Universidad del Pais Vasco, Spain
Youcef Baghdadi	Sultan Qaboos University, Oman
José Banares Bañares	Universidad de Zaragoza, Spain
Veena Bansal	Indian Institute of Technology Kanpur, India
Cecilia Baranauskas	State University of Campinas, Brazil
Ken Barker	University of Calgary, Canada
Jean-Paul Barthes	Université de Technologie de Compiègne, France
Rémi Bastide	University of Toulouse, France
Smaranda Belciug	University of Craiova, Romania
Marta Beltrán	Universidad Rey Juan Carlos, Spain
Jorge Bernardino	Polytechnic Institute of Coimbra, Portugal
Edward Bernroider	Vienna University of Economics and Business, Austria
Huseyin Bicen	Near East University, Cyprus
Ilia Bider	Stockholm University, Sweden
Frederique Biennier	INSA Lyon, France

Tales Bogoni	Universidade do Estado de Mato Grosso, Brazil
Zita Bošnjak	University of Novi Sad, Serbia
Jean-Louis Boulanger	CERTIFER, France
Grégory Bourguin	Université du Littoral Côte d'Opale, France
Andrés Boza	Universitat Politècnica de València, Spain
Maria-Eugenia Cabello-Espinosa	University of Colima, Mexico
David Cabrero Souto	University of A Coruña, Spain
Daniel Callegari	Pontificia Universidade Católica do Rio Grande do Sul, Brazil
Luis Camarinha-Matos	New University of Lisbon, Portugal
Roy Campbell	University of Illinois at Urbana-Champaign, USA
Manuel Capel-Tuñón	University of Granada, Spain
João Luís Cardoso de Moraes	Federal University of São Carlos, Brazil
Glauco Carneiro	Universidade Salvador, Brazil
Angélica Caro	University of Bio-Bio, Chile
Diego Carvalho	Federal Centre of Engineering Studies and Technological Education, Brazil
Marco Casanova	Pontificia Universidade Católica do Rio de Janeiro, Brazil
Laura M. Castro	Universidade da Coruña, Spain
Salvatore Cavalieri	University of Catania, Italy
Luca Cernuzzi	Universidad Católica "Nuestra Señora de la Asunción", Paraguay
Olena Chebanyuk	National Aviation University, Ukraine
Shiping Chen	CSIRO, Australia
Max Chevalier	IRIT, France
Nan-Hsing Chiu	Chien Hsin University of Science and Technology, Taiwan
Witold Chmielarz	Warsaw University, Poland
Betim Cico	Epoka University, Albania
Daniela Claro	Universidade Federal da Bahia, Brazil
Tadeu Classe	Universidade Federal do Estado do Rio de Janeiro, Brazil
Pedro Coelho	State University of Rio de Janeiro, Brazil
Ticiana Coelho da Silva	Universidade Federal do Ceará, Brazil
Cesar Collazos	Universidad del Cauca, Colombia
Antonio Corral	University of Almeria, Spain
Jean-Valère Cossu	My Local Influence - MyLI, France
Henrique Cota de Freitas	Pontifícia Universidade Católica de Minas Gerais, Brazil
Karl Cox	University of Brighton, UK
Sharon Cox	Birmingham City University, UK
Maria Cruz	Escola Politécnica de Pernambuco, Brazil
Beata Czarnacka-Chrobot	Warsaw School of Economics, Poland

Pedro Damián Reyes	Universidad de Colima, Mexico
José de Almeida Amazonas	University of São Paulo, Spain
Antonio De Nicola	ENEA, Italy
Marcos de Oliveira	Universidade Federal do Ceará, Brazil
Lucélia de Souza	UNICENTRO, Brazil
Rosario Delgado de la Torre	University Autonomous of Barcelone, Spain
Bruna Diirr	Federal University of the State of Rio de Janeiro, Brazil
Kamil Dimililer	Near East University, Cyprus
Aleksandar Dimov	Sofia University "St. Kliment Ohridski", Bulgaria
Dulce Domingos	Universidade de Lisboa, Portugal
César Domínguez	Universidad de La Rioja, Spain
António Dourado	University of Coimbra, Portugal
Helena Dudycz	Wroclaw University of Economics, Poland
Sophie Ebersold	IRIT, France
El-Sayed A. El-Dahshan	Egyptian E-Learning University, Cairo, Egypt
Fabrício Enembreck	Pontifical Catholic University of Paraná, Brazil
Indrit Enesi	Polytechnic University of Tirana, Albania
Marcelo Fantinato	University of São Paulo, Brazil
João Faria	University of Porto, Portugal
Antonio Fariña	University of A Coruña, Spain
Fausto Fasano	University of Molise, Italy
Edilson Ferneda	Catholic University of Brasilia, Brazil
Maria Ferreira	Universidade Portucalense, Portugal
Paulo Ferreira	INESC-ID/IST, Portugal
Luis Ferreira Pires	University of Twente, Netherlands
Jiri Feuerlicht	Unicorn College, Czech Republic
Gustavo Figueroa	Instituto Nacional de Electricidad y Energías Limpias, Mexico
Adriano Fiorese	Santa Catarina State University, Brazil
Francesco Folino	ICAR-CNR, Italy
Kevin Foltz	IDA, USA
Anna Formica	CNR-IASI, Italy
Lixin Fu	University of North Carolina, Greensboro, USA
Mariagrazia Fugini	Politecnico di Milano, Italy
Leonid Galchynsky	National Technical University of Ukraine "Igor Sikorsky Kyiv Polytechnic Institute", Ukraine
Cristian García Bauza	National University of the Center of the Buenos Aires Province/PLADEMA/CONICET, Argentina
Alejandra Garrido	Universidad Nacional de La Plata/CONICET, Argentina
Mouzhi Ge	Deggendorf Institute of Technology, Germany
Dejan Gjorgjevikj	Ss. Cyril and Methodius University of Skopje, North Macedonia
Danilo Gligoroski	Norwegian University of Science and Technology, Norway
Manuel Gómez Olmedo	University of Granada, Spain

Enyo Gonçalves	Universidade Federal do Ceará, Brazil
Amr Goneid	The American University in Cairo, Egypt
Pascual Gonzalez	University of Castilla-La Mancha, Spain
Virginie Govaere	INRS, France
Janis Grabis	Riga Technical University, Latvia
Rogéria Gratão de Souza	São Paulo State University, Brazil
Luca Greco	University of Salerno, Italy
Andreas Gregoriades	Cyprus University of Technology, Cyprus
Olivera Grljevic	University of Novi Sad, Serbia
Wieslawa Gryncewicz	Wroclaw University of Economics, Poland
Dirk Habich	Technische Universität Dresden, Germany
Wahab Hamou-Lhadj	Concordia University, Canada
Karin Harbusch	Universität Koblenz-Landau, Germany
Mohammad Hasan	Staffordshire University, UK
Celso Hirata	Instituto Tecnológico de Aeronáutica, Brazil
Violeta Holmes	University of Huddersfield, UK
Wladyslaw Homenda	Warsaw University of Technology, Poland
Wei-Chiang Hong	Oriental Institute of Technology, Taiwan
Miguel J. Hornos	University of Granada, Spain
Fu-Shiung Hsieh	Chaoyang University of Technology, Taiwan
Chun-Che Huang	National Chi Nan University, Taiwan
Kai-I Huang	Tunghai University, Taiwan
Miroslav Hudec	VSB - Technical University of Ostrava, Czech Republic
Nantia Iakovidou	Aristotle University, Greece
George Ioannidis	University of Patras, Greece
Arturo Jaime	Universidad de La Rioja, Spain
Agnieszka Jastrzebska	Warsaw University of Technology, Poland
Dorota Jelonek	Czestochowa University of Technology, Poland
Luis Jiménez Linares	University of de Castilla-La Mancha, Spain
Paul Johannesson	Royal Institute of Technology, Sweden
Olga Kalimullina	The Bonch-Bruevich St. Petersburg State University of Telecommunications, Russian Federation
Christos Kalloniatis	University of the Aegean, Greece
Dimitrios Katsaros	University of Thessaly, Greece
Euclid Keramopoulos	Alexander Technological Education Institute of Thessaloniki, Greece
Nina Khairova	National Technical University "Kharkiv Polytechnic Institute", Ukraine
Angeliki Kitsiou	University of the Aegean, Greece
Alexander Knapp	Universität Augsburg, Germany
Fotios Kokkoras	University of Thessaly, Greece
Christophe Kolski	Université Polytechnique Hauts-de-France, France
Natalija Kozmina	University of Latvia, Latvia
Rob Kusters	Open Universiteit Nederland, Netherlands

Jan Kwiatkowski	Wroclaw University of Science and Technology, Poland
Ramon Lawrence	University of British Columbia Okanagan, Canada
Vanya Lazarova	University of National and World Economy, Bulgaria
Jintae Lee	Leeds School of Business, USA
Carlos León de Mora	University of Seville, Spain
Da-Yin Liao	Straight & Up Intelligent Innovations Group Co., USA
Therese Libourel	University of Montpellier, France
Stephane Loiseau	University of Angers, France
Maria Lopes	Universidade Portucalense Infante D. Henrique, Portugal
Katrinna MacFarlane	University of Sunderland, UK
Rita Suzana Maciel	Federal University of Bahia, Brazil
Regis Magalhães	Federal University of Ceara, Brazil
Hani Mahdi	Ain Shams University, Egypt
Antonio Martí Campoy	Universitat Politècnica de València, Spain
Ivo Martinik	VSB – Technical University of Ostrava, Czech Republic
David Martins de Matos	L2F/INESC-ID Lisboa/Instituto Superior Técnico, Universidade de Lisboa, Portugal
Riccardo Martoglia	University of Modena and Reggio Emilia, Italy
Wolfgang Mayer	University of South Australia, Australia
Rafael Mayo-García	CIEMAT, Spain
Marcin Michalak	Silesian University of Technology, Poland
Jerzy Michnik	University of Economics in Katowice, Poland
Alfredo Milani	University of Perugia, Italy
Michele Missikoff	ISTC-CNR, Italy
Hiroyuki Mitsuhara	Tokushima University, Japan
Ghodrat Moghadampour	Vaasa University of Applied Sciences, Finland
Lars Mönch	FernUniversität in Hagen, Germany
Francisco Montero	University of Castilla-La Mancha, Spain
Fernando Moreira	Universidade Portucalense, Portugal
Edward Moreno	Federal University of Sergipe, Brazil
Javier Muguerza	University of the Basque Country, Spain
Hamid Mukhtar	National University of Sciences and Technology, Pakistan
Pietro Murano	Oslo Metropolitan University, Norway
Omnia Neffati	King Khalid University, Saudi Arabia/RIADI, Tunisia
Leandro Neves	São Paulo State University, Brazil
Vincent Ng	The Hong Kong Polytechnic University, Hong Kong
Ovidiu Noran	Griffith University, Australia
Joshua Nwokeji	Gannon University, USA
Edson Oliveira Jr.	State University of Maringá, Brazil
Ermelinda Oro	National Research Council, Italy
Wendy Osborn	University of Lethbridge, Canada
Pavels Osipovs	Riga Technical University, Latvia

Ozay Ozaydin	Dogus University, Turkey
Claus Pahl	Free University of Bozen-Bolzano, Italy
Malgorzata Pankowska	University of Economics in Katowice, Poland
Rafael Parpinelli	Universidade do Estado de Santa Catarina, Brazil
Silvia Parusheva	University of Economics - Varna, Bulgaria
Rafael Pasquini	Universidade Federal de Uberlândia, Brazil
Ignazio Passero	University of Salerno, Italy
Maria Penadés Gramaje	Universitat Politècnica de València, Spain
Taoxin Peng	Edinburgh Napier University, UK
Valéria Pequeno	Universidade Autónoma de Lisboa, Portugal
Ricardo Pérez-Castillo	University of Castilla-La Mancha, Spain
Dana Petcu	West University of Timisoara, Romania
Dessislava Petrova-Antonova	Sofia University "St. Kl. Ohidski", Bulgaria
Anna Piotrowska	Nicolaus Copernicus University in Torun, Poland
Matus Pleva	Technical University of Kosice, Slovak Republic
Placide Poba-Nzaou	Université du Québec à Montréal, Canada
Alexandra Pomares-Quimbaya	Pontificia Universidad Javeriana, Colombia
Luigi Pontieri	National Research Council, Italy
Filipe Portela	University of Minho, Portugal
Rodica Potolea	Techincal University of Cluj-Napoca, Romania
Naveen Prakash	IIITD, India
Maryam Radgui	National Institute of Statistics and Applied Economics, Morocco
Daniele Radicioni	University of Turin, Italy
T. Ramayah	Universiti Sains Malaysia, Malaysia
Carlos Arturo Raymundo Ibañez	Universidad Peruana de Ciencias Aplicadas, Peru
Francisco Regateiro	Instituto Superior Técnico, Portugal
Ulrich Reimer	Eastern Switzerland University of Applied Sciences, Switzerland
Yassine Rhazali	Moulay Ismail University of Meknes, Morocco
Nuno Ribeiro	Universidade Fernando Pessoa, Portugal
Ricardo Ribeiro	ISCTE-IUL/INESC-ID Lisboa, Portugal
Michele Risi	University of Salerno, Italy
Sonja Ristic	University of Novi Sad, Serbia
Sashko Ristov	University of Innsbruck, Austria
Jose Rivero	National University of La Plata, Argentina
Carlos Roberto Valêncio	São Paulo State University, Brazil
Leonardo Rocha	Federal University of São João Del Rei, Brazil
Alfonso Rodriguez	University of Bio-Bio, Chile
Daniel Rodriguez	University of Alcalá, Spain
Luciana Romani	Embrapa Agricultural Informatics, Brazil
Jose Raul Romero	University of Cordoba, Spain
Philippe Roose	IUT de Bayonne et du Pays Basque/UPPA, France

Michael Rosemann	Queensland University of Technology, Australia
Leon Rothkrantz	Delft University of Technology, Netherlands
Ounsa Roudiès	Ecole Mohammadia d'Ingénieurs, Morocco
Antonio Juan Rubio-Montero	Complutense University of Madrid, Spain
Rafael Sachetto	Federal University of São João Del Rei, Brazil
Oumaima Saidani	Princess Nora Bint Abdul Rahman University, Saudi Arabia
George Sammour	Princess Sumaya University for Technology, Jordan
Luis Enrique Sánchez	Universidad de Castilla-La Mancha, Spain
Jurek Sasiadek	Carleton University, Canada
Lucile Sautot	AgroParisTech, France
Snezana Savoska	"St Kliment Ohridski" University - Bitola, North Macedonia
Manuel Serrano	University of Castilla-La Mancha, Spain
Isabel Seruca	Universidade Portucalense, Portugal
Ali Fawaz Shareef	Cyryx College, Maldives
Markus Siepermann	TU Dortmund, Germany
Alberto Silva	IST/INESC-ID, Portugal
Rodrigo Silva	São Paulo State Technological College, Brazil
Stavros Simou	University of the Aegean, Greece
Seppo Sirkemaa	University of Turku, Finland
Michel Soares	Federal University of Sergipe, Brazil
Chantal Soule-Dupuy	Toulouse 1 Capitole University, France
Damires Souza	Federal Institute of Education, Science and Technology of Paraiba, Brazil
Patricia Souza	UFMT, Brazil
Marco Spohn	Federal University of Fronteira Sul, Brazil
Clare Stanier	Staffordshire University, UK
Chris Stary	Johannes Kepler University of Linz, Austria
Hans-Peter Steinbacher	University of Applied Science Kufstein, Austria
Hiroki Suguri	Miyagi University, Japan
Sagar Sunkle	Tata Consultancy Services, India
Marcos Sunye	Federal University of Parana, Brazil
Reima Suomi	University of Turku, Finland
Nestori Syynimaa	University of Jyväskylä, Finland
Zoltán Szabó	Corvinus University of Budapest, Hungary
Ryszard Tadeusiewicz	Agh University Science Technology, Poland
Mohan Tanniru	Oakland University, USA
Sotirios Terzis	University of Strathclyde, UK
Lucineia Heloisa Thom	Universidade Federal do Rio Grande do Sul, Brazil
Claudio Toledo	São Paulo University, Brazil
Ricardo Torres	Norwegian University of Science and Technology, Norway
Dimitar Trajanov	Ss. Cyril and Methodius University of Skopje, North Macedonia

Denis Trcek	University of Ljubljana, Slovenia
Mario Vacca	Italian Ministry of Education, Italy
David Vallejo	University of Castilla-La Mancha, Spain
Michael Vassilakopoulos	University of Thessaly, Greece
Jose Vazquez-Poletti	Universidad Complutense de Madrid, Spain
Belen Vela Sanchez	Rey Juan Carlos University, Spain
Melina Vidoni	RMIT University, Australia
M. A. Vila	University of Granada, Spain
Gualtiero Volpe	Università degli Studi di Genova, Italy
Vasiliki Vrana	International Hellenic University, Greece
Boris Vrdoljak	University of Zagreb, Croatia
Miljan Vucetic	VLATACOM Institute of High Technologies, Serbia
Dariusz Wawrzyniak	Wroclaw University of Economics, Poland
Hans Weigand	Tilburg University, Netherlands
Janusz Wielki	Opole University of Technology, Poland
Daniel Wilusz	Poznan University of Economics and Business, Poland
Adam Wójtowicz	Poznan University of Economics and Business, Poland
Stanislaw Wrycza	University of Gdansk, Poland
Mudasser Wyne	National University, USA
Shuxiang Xu	University of Tasmania, Australia
Muhammed Younas	Oxford Brookes University, UK
Geraldo Zafalon	São Paulo State University, Brazil
Brahmi Zaki	RIADI, Tunisia
Ying Zhang	Peking University, China
Yifeng Zhou	Southeast University, China
Eugenio Zimeo	University of Sannio, Italy

Additional Reviewers

S. Assoul	Morocco
Anthony Atkins	Staffordshire University, UK
Andre Cordeiro	State University of Maringa, Brazil
Raquel Fialho	UFU, Brazil
Paloma García de Marina	Rey Juan Carlos University, Spain
Ricardo Geraldi	State University of Maringá, Brazil
Rodrique Kafando	INRAE, France
Aurora Ramirez	University of Córdoba, Spain

Invited Speakers

Stanislaw Jarzabek	Bialystok University of Technology, Poland
Alon Halevy	Facebook AI, USA
Kecheng Liu	University of Reading, UK
Fred Phillips	University of New Mexico, USA

Contents

Information Systems Analysis and Specification

Databases and Information Systems Integration

Anonimisation, Impacts and Challenges into Big Data: A Case Studies

Artur Potiguara Carvalho[1(✉)], Edna Dias Canedo[1,2(✉)], Fernanda Potiguara Carvalho[3(✉)], and Pedro Henrique Potiguara Carvalho[1(✉)]

[1] Electrical Engineering Department (ENE), University of Brasília (UnB), P.O. Box 4466, Brasília, DF, Brazil
artur.carvalho@redes.unb.br, ednacanedo@unb.br

[2] Department of Computer Science, University of Brasília (UnB), P.O. Box 4466, Brasília, DF, Brazil

[3] Law School (FD), University of Brasília (UnB), Brasília, DF, Brazil
carvalho.fernanda@aluno.unb.br

Abstract. In a context in which privacy is increasingly demanded by citizens and by various institutions, reflected in protection laws, anonymity emerges as an essential tool. Both the General Data Protection Regulation (GDPR) in the EU and the Brazilian General Data Protection Law (LGPD) provide a softer regulation for anonymised data, compared to personal data. Despite the legal advantages in their use, anonymisation tools have limits that should be considered, especially when it comes to massive data contexts. The work seeks to analyze whether anonymisation techniques can satisfactorily ensure privacy in big data environments, without taking other measures in favor of privacy. Based on two hypothetical cases, we realized that the anonymisation techniques, although well implemented, must be associated with governance techniques to avoid latent breaches of privacy. Besides that, we point out some guidelines identified in the case studies for the use of anonymous data in Big Data.

Keywords: Anonymisation · Big Data · Privacy · Governance · Compliance

1 Introduction

The growing concern about privacy calls for a new era of big data systems. Sometimes even for nefarious purposes, the indiscriminate use of data can cause personal information to leak, and it has led several countries around the world to develop specific protection laws.

One of the measures for protecting personal data is the efforts to reduce the data submitted to treatment. Because of this, "data minimisation" has been presented as a guiding principle of the regulation of such software. In the European

J. Filipe et al. (Eds.): ICEIS 2020, LNBIP 417, pp. 3–23, 2021.
https://doi.org/10.1007/978-3-030-75418-1_1

context, that expression is mentioned in the General Data Protection Regulation (GDPR), Article 5(1)(c) [19], which posits personal data shall be: "adequate, relevant and limited to what is necessary in relation to the purposes for which they are processed" ('data minimisation'). Also, the Brazilian General Data Protection Law (LGPD), contains similar wording in its article 6 [21], that providing the "principle of necessity". Under this principle, data processing should be limited to the minimum necessary to achieve its purposes, using only data that is relevant, proportionate and not excessive to the purposes of the processing.

Underlying these legal structures, Data Governance (DG) has been used to foster standardization of and quality control in internal data management. To accomplish this, "data minimisation" has been proposed as a way to rationalize otherwise costly and expansive DG [8,24]. However, minimizing data usage is particularly sensitive when it comes to big data. It is because Big Data (BD) is a massive data processing technology [6], which often includes all kinds of personal data, while not providing clear guidelines on how to store them. This aggregation method impacts data protection directly [2,8,12,16,17]. In this context, several countries have sought strategies to preserve privacy and guide Big Data use.

The challenge is to conciliate Personal Data Regulations (PDR) and BD mechanisms, mitigating friction between companies and governments. In this paper, we investigate an important tool for the compliance of BD mechanisms with PDR: anonymisation[1] techniques. These are important because once anonymised, these data are exempt from the requirements of Personal Data Regulations (PDR), including the principle of "data minimisation" [19].

To guide this work, we present the Background exploring the limits of expectations placed upon this tool. The question is whether anonymisation used exclusively can meet the demands of privacy and demands of the BD. The justification for choosing the problem in focus is specified by pointing the tool limits, since difficulties of conceptualizing the term to the problems arise from the practical application. In this moment, an overview of the academy's work in the area is presented. We strive to counterbalance the advantages and risks of using anonymisation as a form of compliance to Personal Data Regulation (PDR). We raised the hypothesis that, although anonymisation is an important tool to increase data protection, it needs to be used with assistance from other mechanisms developed by compliance-oriented governance to ensure privacy.

The main goal is to present anonymisation risks in order to promote better use of this tool to privacy protections and BD demands. In section Related Work, we raise the main bibliographical references for the subject. We point out as a research method the literature review and the study of two hypothetical cases. In section Related Work, we bring the results obtained so far, which been compared, when we bring a brief discussion about areas prominence and limitations of this work.

[1] The term is spelled with two variants: "anonymisation", used in the European context; or "anonymization" used in the US context. We adopt in this article the European variant because the work uses the GDPR [19] as reference.

We conclude that it is not possible to complete BD compliance with Personal Data Regulation (PDR) and privacy protection exclusively by anonymisation tools [1,4,22,24], [7]. To solve this problem we aim to conduct future research about frameworks that can promote good practices that, associated with anonymisation mechanisms, can secure data protection in BD environments.

This work is an expansion of the article published in the ICEIS 2020 conference (22th International Conference on Enterprise Information Systems) [3] and presents a new case study applying the concepts presented in the previous article in a real and potential use context. We present some anonimisation techniques applied in the real applications and their vulnerabilities and risks when dissociated from a good governance practices.

2 Background

It is important to emphasize that anonymization through BD is an excellent starting point for solving data protection and privacy problems but is not a solution in itself. Organizations need to keep in mind that a more technical study is required on the subject once data protection regulations can directly impact their data protection processes and policies. [1,4,5,18,22]. For this, in this work we investigate the following research question:

RQ.1 Is anonymisation sufficient to conciliate Big Data compliance with Personal Data Regulations and data privacy at large?

Before answering the RQ.1, it is essential to comment on the concept of anonymization, its mechanisms, and legal treatment. The text preceding the articles of Regulation, pertaining to the European Economic Area, guides the anonymisation is point 26 [19]. It states that the "principles of data protection should apply to any information concerning an identified or identifiable natural person". Therefore, since the anonymous data refers to a non-identifiable data source, the principles previously mentioned do not apply to it, namely, "to personal data rendered anonymous in such a manner that the data subject is not or no longer identifiable".

In Article 12 of the LGPD [21], it is possible to observe a similar exclusion. Regulators conclude that, once anonymous, information do not violate privacy, because data can no longer be linked to an identified or identifiable person. However, this premise implies some challenges. First of all, even if it is not possible to know the person's name to whom the data refer, it still can be considered personal. It happens because, even with the name's lack, it is still possible to re-identify a person by the data associated with them. The name is just a way of identifying a person, and its absence does not make it impossible to re-identify the data.

Second, in a BD context, connecting information becomes extremely easy, even when it comes to metadata or data fragments. It happens precisely because BD deals with massive data. Thus, some easy anonymisation techniques, such as masking, can be effective in closed and smaller databases, but not in BD.

When it comes to BD context, it needs more complex data manipulation techniques to gets real effective anonimysation solutions [17].

Also, inference techniques are more readily applicable in BD contexts, allowing the re-identification of apparently anonymous data. Inference is one of the techniques where information, although not explicit, can be assumed through the available data. Analyzing the propositional logic, we can say that there is inference when three propositions A, B and C respect the following equations:

$$A \Rightarrow B \tag{1}$$

$$B \Rightarrow C \tag{2}$$

so,

$$A \Rightarrow C \tag{3}$$

In a example (using a Shapiro's CarPool World [23]) if A is "Betty is a driver", B is "Tom is a passenger" and C is "Betty drives Tom", we can say by inference that every time that Betty is a driver and Tom is the passenger, Betty drives Tom (Eq. 3).

Based on that analysis, it is possible to assume that anonymisation projects in a BD context becomes even more vulnerable when techniques such as inference are considered, turning data into auxiliary information to indicate personal data.

In consequence, PDR specifies assumptions about what is considered identified or identifiable data. The introduction to GDPR states, in point 26 [19], that to determine whether an individual is identified or identifiable, all reasonable means must be taken into account, namely, "all objective factors such as costs and the amount of time required must be considered for identification, taking into account the technology available at the time of processing and technological developments". Since the text assumes that, in massive contexts, data identification is possible depending on the effort employed to re-identify it, we could highlight this classification as a third point of concern.

The fourth and last one point of concern would be the difficulties of determining the anonymity of a particular data witch the identification will depend on criteria that, although specified by law, will change according to technical advances or even by the specific analysis conditions. This makes the data process anonymisation in this case always uncertain.

These four concerns converge to the point that it is not possible to sustain the unexamined belief in anonymisation as a surefire way of ensuring privacy in BD contexts, which leads us to the hypothesis of the present paper. As noted, anonymisation in BD involves risks, especially to user privacy. Therefore, we argue that anonymisation should be used with the assistance of other privacy mechanisms, so as to better manage this data. Considering this, we can define the hypothesis of this research as follows:

HP.1 BD compliance with the PDR and data privacy cannot be achieved solely by anonymisation tools.

This does not mean that anonymity is a useless tool. Instead, it is an excellent ally when using BD platforms as it is one of the most powerful privacy protection techniques. The point, however, is that it is not possible to rely solely on this technique, leaving aside the use of privacy-oriented governance. This means that to some extent BD must also adapt to privacy, either by increasing data capture criteria in the sense of minimization or by strengthening the governance of such data, even if anonymised.

In this paper, we intend to present anonymisation risks and promote the better use of this tool for BD and the necessary privacy protection. Our main goal is to expose privacy threats related to anonymisation process as an alternative to PDR enforcement. Thus, to foster privacy, we point out that anonymity tools should follow the protection guidelines.

We used literature review as a research method. We aimed to explore the evolution of the concept, classification, demands, improvements about anonymity. To demonstrate the weaknesses of the tool we present a hypothetical case study. Random anonymous data were organized within a BD platform and analyzed with basic data from a specific database structure. Thus, although the data were anonymised in relation to the platform itself, they could be re-identified when exposed to external data. The result of the hypothetical case were analyzed in light of current legislation, and will be discuss in the following sections.

To guide the consultation of PDR, in particular, the GDPR and LGPD, follows a comparative Table 1 of regulations, which will be used throughout the paper.

Table 1. Comparative table of regulations [3].

Concepts	GDPR	LGPD
"Data minimization" concept	Article 5(1)(c)	Article 6(III)
Anonimisation concept	Text preceding the articles of GDPR, point 26; Article 4(5)	Article 5(XI)
Personal data concept	Article 4(1)	Article 5(I)
Exclusion of anonymous data from personal data classification	Text preceding the articles of GDPR, point 26	Article 12
Processing concept	Article 4(2)	Article 5(X)
Sensitive data concept	Text preceding the articles of GDPR, point 51	Article 5(II), Article 11

2.1 Related Work

Back in 2015, H. Liu had already announced the challenge of managing legal frameworks, privacy protection, individual autonomy, and data applications.

[14]. In 2016, Mehmood et al. detailed a group of methods and techniques that provides encryption and protection to data inside BD [15]. In the same year, Dalla Farvera and da Silva discuss veiled threats to data privacy in the BD era [5]. Still in 2016, Lin et al. presents a model considering differential privacy (varying by datasets privacy loss) [13].

The vision of the organization governance model to address the new protection data regulations (PDR) issues was added by Ryan and Brinkley in 2017 [22]. In that year, many other authors discussed the same subject [4,11,18]. In 2018, Ventura and Coeli introduce the concept of the right to information in the context of personal data protection and governance [24], while Brasher [1] criticize the current process of anonymisation in BD.

Domingo-Ferrer [7] summarizes the Brashers review in 2019, mainly in BD platforms, presenting the issues of anonymisation and its specificities. Jensen et al. [10] discuss how to realize value with BD projects and the best practices to measure and control it. Mustafa et al. [16] indicate a framework about privacy protection for application in the health field. They present the threats of privacy involving medical data in the light of regulation (GDPR).

As stated by Brasher [1], "anonymisation protects data subjects privacy by reducing the link-ability of the data to its subjects", which is similar to the concept outlined by PDR. In line with this definition, it is possible to highlight two types of data: Personally Identifiable Information ("PII"), which may include the quasi-identifiers and contains security liabilities concerning personal data, and Auxiliary Data ("AD"), which can reveal the subjects referenced. These two types of data must be handled separately by anonymisation, according to the risks inherent to each.

About quasi-identifiers, Brasher [1] describes: "non-facially identifiable data that can be linked to auxiliary information to re-identify data subjects". Mehmood et al. [15] complements: "The attributes that cannot uniquely identify a record by themselves, but if linked with some external dataset may be able to re-identify the records." To exemplify that description, Mehmood et al. [15] show an example (Fig. 1) of link quasi-identifiers from records of medical application and movie reviews application:

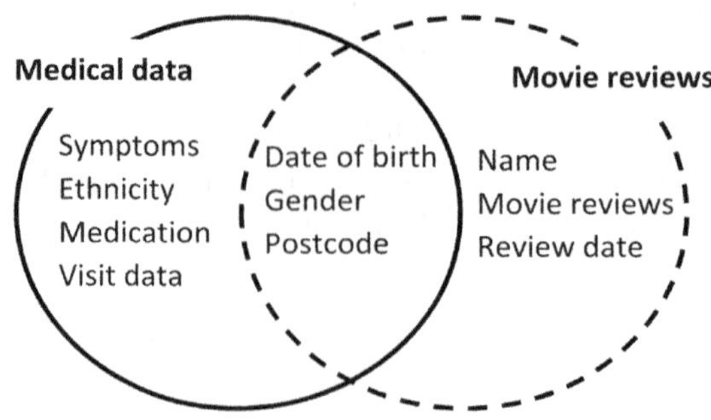

Fig. 1. Quasi-identifiers and linking records [15].

Brasher's work [1] presents the five most common anonymisation techniques: (1) **Suppression**, (2) **Generalization**, (3) **Aggregation**, (4) **Noise Addition**, and (5) **Substitution**, as shown in Fig. 2.

1) **Suppression** is the process that excludes any PII from the base.
2) **Generalization** shuffles PII identifiers, without excluding any information, reducing their link-ability.
3) In **Aggregation**, both data types (PII and AD) go through some reducing treatment that maintains some properties of data (average, statistical distribution, or any others property) and also reduces their link-ability.
4) **Noise addition** adds some non-productive data to confuse the link between PII/AD and their subjects.
5) Finally, **Substitution** is similar to Generalization, while it differs in that: it shuffles not the identifier, but the value of the data itself, replacing the original dataset with other parameters. This process can be applied to both PII and AD [1].

According to Mehmood et al. [15], the privacy protection by anonymisation is also divided into five different operations: (1) **Suppression**, (2) **Generalization**, (3) **Permutation**, (4) **Perturbation**, and (5) **Anatomization**, all of which correspond to the strategies presented by [1]. This can be seen in Fig. 2.

The **Suppression** (strategy 1 in Fig. 2) strategy is criticized by Domingo-Ferrer [7]. Anonymising data in BD is not enough because re-linking the deleted identifiers becomes trivial in this massive context, especially if external data is factored into the analysis. According to the author, concerns about the social impact of this insufficient protection are as great as to have surfaced on mainstream media [7].

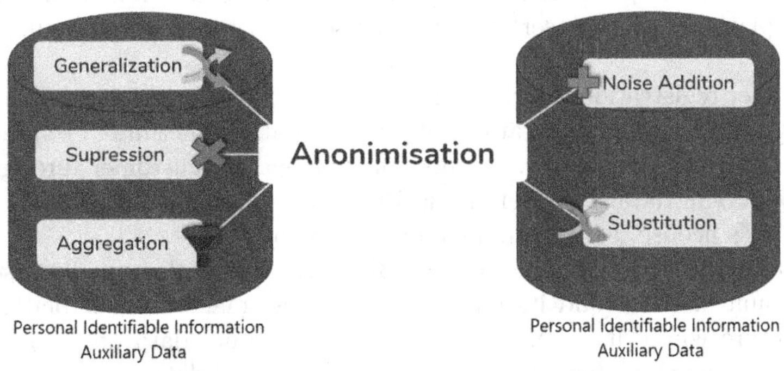

Fig. 2. Anonymisation Techniques, adapted from ([1,3,15]).

The author goes on to explain that efficient privacy protection must consider balancing these two aspects: utility loss and privacy gain of PII-based data. Supposed privacy gains occur at the expense of utility loss. When a suppressed piece of data is discarded less valuable information can be extracted [7].

BD anonymisation is still limited [7]. Domingo-Ferrer presents three main limitations to current big data anonymisation processes:

1) Trust in data controllers, granted by PDR, is undermined by lack of actionable management criteria for the treatment of confidentiality;
2) The weakness of the anonymisation methods, which satisfy an insufficiently broad set of Statistic Disclosure Controls (SDC);
3) and the utility cost of the process of data anonymisation which may incur the difficulty of merging and exploring anonymised data.

Mehmood et al. [15] and Domingo-Ferrer [7] agree about the trade-off between privacy by anonymisation and utility, and its negative relation mainly in the BD context. Applying some anonymisation strategy as the only action regarding data privacy leads to the decrease of potential insights on PII and AD.

Quoting the weakness of the anonymisation methods, Lin et al. [13] apply differential privacy to body sensor network using sensitive BD. In their work, Lin et al. [13] combine strategies 3 and 4 (Fig. 2) to hardening the privacy of a given dataset. But as shown, the scheme adopted by Lin only considers the information given by the internal dataset, ignoring possible attacks using other ADs available on the Internet, for example. Lin et al. [13] also discuss the risk of data loss through the anonymisation process.

3 Partial Results

3.1 Hypothetical Case 1

To demonstrate a hypothetical example 1, we will use a data repository proposal on a BD platform whose inserted data represents customers of a financial institution.

Customer registration information (usually not just for financial companies) represents significant concentrations of personal data, sometimes, even sensitive. Besides, in the financial sector, it is possible to identify a customer through other non-conventional data (considered quasi-identifiers) such as identity, social register, driver's license, bank account number, among others.

The hypothetical example will use BD because, as already discussed, the large amount of data (and its intrinsic challenges) make the BD platform the infrastructure where it is easier to re-identify personal data once to treat its countless relationships (explicit or implicit) of personal data can be an arduous and expensive task in terms of processing.

Consider a certain data structure in a BD platform according to Fig. 3:

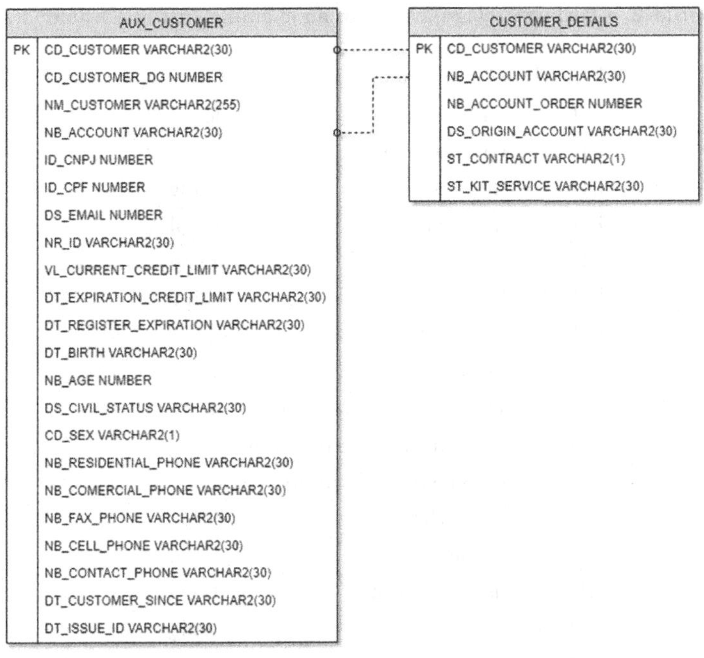

Fig. 3. Hypothetical structure data model [3].

This structure is implemented on a BD platform, to enable the analysis of the customer (current or potential) characteristics of a certain financial company. This analysis would contain personal data filters such as age, sex or relationship time with the company and will support several departments in this organization. Also datasets AUX_CUSTOMER and CUSTOMER_DETAIL were considered and classified according to Tables 2 and 3.

Now, we must consider the anonymisation applied by combining the strategies 1–5 described before, according to the showing:

1) Using strategy 1 (**Suppression**): The registers with CD_CUSTOMER lower than 100500 were excluded from this table (from 100000 to 100500).
2) Using strategy 2 (**Generalization**): From 100500 to 100800, the identification was weakened by shuffling the CD_CUSTOMER.
3) Using strategy 3 (**Aggregation**): The register with the same ID_CPF (22464662100) was converted to a unique register (CD_CUSTOMER = 603093, 603094, 603095 and 603096) by the sum of attribute value VL_CURRENT_CREDIT_LIMIT and the max operation over attribute values DT_EXPIRA-TION_ CREDIT_LIMIT, DT_REGISTER_EXPIRATION, NB_AGE and the min operation over attribute values DT _BIRTH, DT_CUSTOMER_SINCE and DT_ISSUE_ID.
4) Using strategy 4 (**Noise Addition**): Was included the register identified by the CD_CUSTOMER 100623 with random information.

Table 2. Attributes/classification of an example customer table [3].

PII/AD	Column name	Data type
PII	cd_customer	double
PII	cd_customer_dg	double
PII	nm_customer	string
PII	nb_account	double
PII	id_cnpj	string
PII	id_cpf	string
PII	ds_email	string
PII	nb_id	string
AD	vl_current_credit_limit	double
AD	dt_expiration_credit_limit	string
AD	dt_register_expiration	string
AD	dt_birth	string
AD	nb_age	double
AD	ds_civil_status	string
AD	cd_sex	string
AD	nb_residential_phone	string
AD	nb_comercial_phone	string
AD	nb_fax_phone	string
AD	nb_cell_phone	string
AD	nb_contact_phone	string
AD	dt_customer_since	string
AD	dt_issue_id	string

Table 3. Attributes/classification of a example customer details table [3].

PII/AD	Column name	Data type
PII	cd_customer	double
PII	nb_account	double
PII	nb_account_order	double
AD	ds_origin_account	string
AD	st_contract	string
AD	st_kit_service	string

5) Using strategy 5 (**Substitution**): Was divided two groups of registers (G1 - from CD_CUSTOMER 100800 to 101100 and G2 - from CD_CUSTOMER 1013000 to 101600) and the AD attributes were shuffled between these two groups, preserving the original characteristics.

Based on the difficulty of transforming data privacy governance concepts into operational data protection actions (as described by Ventura and Coeli [24]), suppose that only part of the data in the structure shown by Fig. 3 has been classified as identifiable of the respective subject. Only the data contained in the dataset AX_CUSTOMER will be anonymised, excluding the data present in the dataset CUSTOMER _DETAILS.

In the actual production environment, several reasons could lead to the BD information not being taken into consideration while in providing anonymisation, such as data governance process failures, misinterpretation of regulation, mistakes in internal concept of sensitive personal data, difficult to manage large amounts of data or many different datasets, among others.

Using another dataset (concerning customer details) from the same schema that was extracted from the previous customer table, it is possible to undo or disturb the anonymisation (weakening the privacy protection) according to the shown:

1) Concerning strategy 1 (**Suppression**): The registers excluded were identified (provide that the application of the anonymisation method was known) by the referential integrity (not explicit) with the table CUSTOMER_DETAIL by the attribute CD_CUSTOMER. Besides, exclusion is the most aggressive strategy, and produces the greatest loss of utility.
2) Concerning strategy 2 (**Generalization**): Using the attribute NB_ACCOUNT (not search index, but personal data), it was possible to identify the shuffling, since this attribute can identify an individual.
3) Concerning strategy 3 (**Aggregation**): The presence of the register with the CD_ CUSTOMER = 603093, 603094, 603095 and 603096 in the table CUSTOMER_ DETAIL denouncement that these registers were manipulated in the original table.
4) Concerning strategy 4 (**Noise Addition**): The absence of the register with CD_ CUSTOMER = 100623 indicates that this register was added to the original table.
5) Concerning strategy 5 (**Substitution**): Combining the CD_CUSTOMER and the NB_ACCOUNT from these two tables its possible to identify the manipulation of these data, even if it is hard to define what exactly was modified.

Note that the data used to undo/detect the anonymisation process belonged to the same data schema as the original base. In the context of BD, it would be common that in the large universe of data there would be replications of PII or quasi-identifiers like the shown in the example.

Thus, it is possible that within the DB database there are reliable data to guide the conclusions against anonymisation. Besides, the data used to re-identify can be accessed by internet, social network, another BD or any other external data repository. Both present themselves as weaknesses in BD platforms, as they will provide insight into the anonymisation methods used.

Once the anonymisation method is detected, it is simpler to look for mechanisms to complete missing information or even rearrange and restructure information that has been merged or added noises. For this, public databases can be an effective source for obtaining specific information.

Also, knowing which data has been anonymised greatly weakens database protection. This is because data that has not undergone the anonymisation process, for example, or data that is reorganized within the platform, will constitute a remnant base that maintains its integrity. Thus, unchanged data is known to be intact and can be used to obtain relevant information.

Finally, we clarified that all scripts used for create/populate the examples data structures are available [3]:

```
CREATE TABLE
AUX_CUSTOMER
    (
    CD_CUSTOMER VARCHAR2(30)
    PRIMARY KEY,
    CD_CUSTOMER_DG NUMBER,
    NM_CUSTOMER VARCHAR2(255),
    NB_ACCOUNT VARCHAR2(30),
    ID_CNPJ NUMBER,
    ID_CPF NUMBER,
    DS_EMAIL NUMBER,
    NR_ID VARCHAR2(30),
    VL_CURRENT_CREDIT_LIMIT
    VARCHAR2(30),
    DT_EXPIRATION_CREDIT_LIMIT
    VARCHAR2(30),
    DT_REGISTER_EXPIRATION
    VARCHAR2(30),
    DT_BIRTH VARCHAR2(30),
    NB_AGE NUMBER,
    DS_CIVIL_STATUS
    VARCHAR2(30),
    CD_SEX VARCHAR2(1),
    NB_RESIDENTIAL_PHONE
    VARCHAR2(30),
    NB_COMERCIAL_PHONE
    VARCHAR2(30),
    NB_FAX_PHONE VARCHAR2(30),
    NB_CELL_PHONE VARCHAR2(30),
    NB_CONTACT_PHONE VARCHAR2(30),
    DT_CUSTOMER_SINCE VARCHAR2(30),
    DT_ISSUE_ID VARCHAR2(30)
    );
CREATE TABLE
```

```
CUSTOMER_DETAILS
  (
  CD_CUSTOMER VARCHAR2(30)
  PRIMARY KEY,
  NB_ACCOUNT VARCHAR2(30),
  NB_ACCOUNT_ORDER NUMBER,
  DS_ORIGIN_ACCOUNT VARCHAR2(30),
  ST_CONTRACT VARCHAR2(1),
  ST_KIT_SERVICE VARCHAR2(30)
  ) ;
```

3.2 The Hipothetical Case 2

Now, consider that in the data model (Fig. 3) presented in hypothetical case 1 (3.1), the data controller decides to exclude all personal information contained in the database, except the name of each individual and some digits of their "id_cnpj/id_cpf".

Initially, it is essential to state that, in some cases, the name can be considered a "quasi-identifier". For example, this happens with homonyms, which are relatively common in the Brazilian context when it comes to popular names. In such cases, disclosure of the name may not affect the privacy of the personal data represented. On the other hand, the "CPF" represents the identification number of the Brazilian citizen, which consists of eleven digits in total. It is a unique number whose purpose is precisely to identify the citizen, even in the case of homonyms. However, in the example, the CPF will be incomplete and "unidentifiable".

Further, CNPJ is the identification number of companies registered in Brazil. Therefore, because it refers directly to legal entities, we will not cover in our hypothetical case, which will be restricted to customers who are individuals.

Thus, strategy number 1, Suppression, was applied in all personal information (and in some digits of the "id_cnpj/id_cpf" information). In this case, a large part of the personal information was abdicated by the controller to guarantee the privacy and protection of personal data. Was left identifiable only the name of the person represented (with their namesakes and possible data quality issues).

To detail the method applied for the anonymisation of "id_cnpj/id_cpf" information, keep in mind the character of this attribute. Specifically, in the case of the CPF field, it has nine digits plus two check digits indicated after a dash. Therefore, as mentioned, a CPF has 11 digits. The CPF number is written in the form of ABC.DEF.GHI-JK or directly as ABCDEFGHIJK, where the digits cannot all be the same.

The J is called the 1st check digit of the CPF number.

The K is called the 2nd check digit of the CPF number.

First Check Digit: To obtain multiple multipliers A, B, C, D, E, F, G, H and I, by the following constants (Table 4):

Table 4. Formation of CPF.

A	B	C	D	E	F	G	H	I
x10	x9	x8	x7	x6	x5	x4	x3	x2

The result of the sum,

$$10A + 9B + 8C + 7D + 6E + 5F + 4G + 3H + 2I, \tag{4}$$

is divided by 11.

We analyze then or REMAINDER [R] of that division:

$$J = \begin{cases} 0, & \text{if } R \text{ in } [0,1] \\ 11 - R, & \text{if } 2 \leq \text{R} \leq 10 \end{cases} \tag{5}$$

Second Check Digit: We already have J. To obtain K multipliers A, B, C, D, E, F, G, H, I and J by the following constants (Table 5):

Table 5. Formation of CPF (continuation).

A	B	C	D	E	F	G	H	I	J
x11	x10	x9	x8	x7	x6	x5	x4	x3	x2

The result of the sum,

$$11A + 10B + 9C + 8D + 7E + 6F + 5G + 4H + 3I + 2J, \tag{6}$$

is divided by 11. We then verify the REMAINDER [R] of that division:

$$K = \begin{cases} 0, & \text{if } R \text{ in } [0,1] \\ 11 - R, & \text{if } 2 \leq \text{R} \leq 10 \end{cases} \tag{7}$$

The method of anonymising information (which did not take into account the open databases available on the internet) replaced the digits of the two central groups of the attribute with the letter X, making the information non-identifiable. For example, for the CPF number: 123.456.789-00, the attribute would be described as follows: 123.XXX.XXX-00. In fact, this method is already used in some information like the one in the public source: http://www.in.gov.br/web/dou/-/edital-n-28-de-3-de-abril-de-2020-252134150.

As mentioned in the hypothetical case 1 (3.1), the internet is a vast source of data for breaking anonymisation procedures. It is possible to explore various information on the network, specially those available on the internet, because of government policies of greater transparency and data access. This information is particularly reliable since it is made possible by the state managers themselves to notify the population.

In the Brazilian context, we had the enactment of the law on access to information- LAI [20], which imposes on the State the publication of data involving finance and public management.

Through the disclosures imposed by LAI cite leiLAI, it is possible to find public databases relating various information to the citizens' CPF. Usually, for the disclosure of these data, CPF anonymisation mechanisms are used by suppressing some digits. For example, the database available at http://www.portaltransparencia.gov.br/pessoa-fisica/busca/lista?pagina=1&tamanhoPagina=10, anonymised the CPF by hiding the group digits at the ends from Camp. In this reference, the CPF used in the example becomes ***.456.789-**.

Suppose we compare the anonymisation carried out in our hypothetical case and the anonymisation carried out by the state database mentioned above. In that case, it is easy to see the complementarity between these bases. Its combination is very likely to restore the original CPF number, re-identifying the information. Only in cases of homonymous people would re-identification encounter obstacles.

In addition, the anonymisation of the CPF also faces internal challenges. This is because the CPF number has a formation rule, which facilitates the validation of re-identification (applying the Formulas 4, 5, 6, 7 and comparing its results) (Table 6).

Table 6. Combining two anonimised CPF.

src1	1	2	3	x	x	x	x	x	x	0	0
src2	*	*	*	4	5	6	7	8	9	*	*
res	1	2	3	4	5	6	7	8	9	0	0

In this way, we can test the number of CPF generated by comparing the suggested bases to verify its validity.

So, in this simple example, we can see that using anonymisation without considering the databases available on the networks can lead to a grotesquely flawed procedure.

Besides, the more data made available by globally insufficient anonymisation, so the higher the risks to privacy. The information obtained so far (name and CPF) is sufficient in most sites that provide personal information, and enable individuals to identify themselves as the subject of this information.

For example, the Brazilian IRS website requires only the CPF number and the birth date for the consultation of the registration status of this citizen. In this site, it is possible to acquire information such as invalid CPF and registration pending issues.

And, similarly, with the name, it is possible to obtain the date of birth of a certain individual, through public databases available on the internet. For example, the base present in http://www.idecan.org.br/concursos/146/40.pdf.

Sometimes, the subject of the information themselves expose data such as phone number, address, or date of birth on their social networks, making them not only public but also indexable by internet search engines (such as Google).

All of this information available on the network can weaken anonymisation, even if the procedure is reliable, considering only the anonymised basis. We discuss these fragilities more in these next topics.

4 Threats and Validation

4.1 The Hypothetical Case 1

Partial results have given us a perspective about the threats involved anonymisation. In a few cases, as when attributes have been shuffled, comparative analysis with the table CUSTOMER_DETAILS makes it possible to re-identify and rearrange the information [3].

But, in general, it was possible to conclude at least the existence of data processing. For example, when deleting data, comparison with the CUSTOMER_DETAILS table reveals that information has been suppressed.

It means that the use of anonymisation is clear from a simple comparison with a table within the same database. This is true even with suppression, which is the most aggressive anonymisation technique.

This reveals which data has been modified, deleted or shuffled and provides a remnant base that maintains its integrity and can be used.

Also, it provides information to complete or organize all bases through external reinforcement, as with public base, as mentioned.

This requires that the comparison be based on information whose integrity is assured. Obtaining such secure information is not only possible but is likely, mainly in the context of BD, that to take into account large databases, that are stored without effective governance. Also, it is likely that exist database there is unfeasible anonymisation due to the need to link information to users, as in the case of personalized services, within the same database [3].

Therefore, these anonymised data still present risks when they are indiscriminately shared on different bases, marketed or made available.

Lack of management increases the likelihood of leakage of this data, which could cause information to be easily obtained.

Therefore, anonymisation, taken in isolation, while providing a sense of security, contains factors that make its misuse extremely risky.

4.2 The Hypothetical Case 2

In the example giving in hypothetical case 1, the most aggressive strategy (as already mentioned) was applied, which should raise the level of data protection.

Despite the strength of the anonymization technique employed and its correct application, the base remained vulnerable in the face of a simple comparison with a complimentary public base.

The example confirms the idea that anonymisation without effective data governance and a risk management strategy is not enough to protect privacy. By neglecting public bases that store part of the treated personal information, it becomes viable to reidentify individuals and obtain their data.

Although hypothetical, in our experience with official Brazilian data sites, we can observe situations very similar to the one presented. Such cases, for reasons of ethics, were reported to the base managers.

Such cases serve to conclude that anonymisation that does not observe the data available in public can be doomed to failure, in even odd ways, as presented.

This example used some pages of information access portals (including those of the Brazilian Government's transparency portal). In this case, it was possible to demonstrate the technique's fragility for the entire anonymised database by comparing corresponding data.

However, a supposed data privacy attacker may not be limited, making use of any public bases available to him. Thus, in a targeted attack, it would be even more difficult to keep the data anonymous.

4.3 Validation

Considering the risks presented in this paper, and also based on the criticisms raised by [7], we highlight some discussion points and propose, for each of them, guidelines to the use of anonymous data in the context of BD mechanisms.

1) Is data anonymised by comparing the entire company database or the public database?
   As we discussed in the Hypothetical case 2, data is usually considered anonymous within their own platforms. However, anonymisation cannot neglect that, in our BD age, a large amount of data is available through other sources. We suggest that, as a minimum requirement for anonymisation to be considered effective, it should analyze its own database and - at least - other databases that are public, organized, and freely accessible.
2) Acknowledging the loss of utility caused by the anonymisation process, which data can and cannot be anonymised?
   This is an important issue because, depending on the company's activity, anonymisation may be a technique that will render data unusable for certain purposes. If the company deals, for example, with personalized services, knowing which data relates to each customer becomes essential. Thus, companies need to choose which data to store, reducing the cost of maintaining large anonymised databases.
   This is justified by the fact that maintaining anonymity requires continuous readjustment according to the evolution of the technique, as highlighted earlier. Besides, keeping smaller databases minimizes the risk of leakage, which increases as more data is stored. Finally, better choosing which data to anonymise forestalls the need for an anonymous database not to be re-identified in order to meet business demands.
3) Is anonymisation a type of "processing of personal data"? While some researchers argue differently [9], we argue that anonymisation is a form of "processing of personal data".
   Once anonymised, the data can be used, even for purposes other than originally stated when it was collected as personal data, as we can see in article 6(4)(e) [19] and the point 26 of the GDPR introduction text. But, for

anonymisation to be considered a lawful processing method, it is necessary to follow the requirements outlined in the GDPR, Article 6 (1) [19], such as the subject's consent or vital interest. However, we highlight some criticisms of article 6 (1) (f) [19], which will be explained in the next point.

4) Can anonymisation be applied by legitimate interests?
   Article 6 (1)(f) [19] stresses that data may be used for "legitimate interests" pursued by the controllers. On this point, the introduction of GDPR states in point 47 [19] that:

   > "The legitimate interests of a controller, including those of a controller to which the personal data may be disclosed, or of a third party, may provide a legal basis for processing, provided that the interests or the fundamental rights and freedoms of the data subject are not overriding, taking into consideration the reasonable expectations of data subjects based on their relationship with the controller".

   Thus, legitimate interest is an abstract term that may be used to create a means of escaping regulation, rather than data protection [25]. Furthermore, the same point 47 highlights that "At any rate the existence of a legitimate interest would need careful assessment including whether a data subject can reasonably expect at the time and in the context of the collection of the personal data that processing for that purpose may take place".
   As seen, the article 6(4)(e) [19] and point 26 of the GDPR introductory text state that the anonymous data is not a personal data, and, therefore, processing of this data need not respect the original purpose of the data. Due to this, it is not possible to initially predict which purposes the data will serve after anonymisation.
   In a nutshell, legitimate interest definition involves a high degree of abstraction, and besides that, once anonymised, data can be used even for purposes other than the original. Thus, we argue that legitimate interests can not be considered sufficient to make anonymisation processing legal. On the other hand, we consider that due to risk, the best way to enact lawful processing of anonymisation is through given consent, without excluding remaining case applications described in Article 6 (1) [19].

Importantly, based on the hypothetical case and the observations already exposed, we adhere to the position described by [7] about the three main limitations to anonymisation. Also, we add the following observations:

1) Are data controllers trusty? Although granted by PDR, it is undermined by a lack of actionable management criteria for the treatment of confidentiality. Therefore, especially for anonymous data, it is likely that bad data processing will be detected only with data leakage. This is why we support stricter regulation of anonymity, as a way to increasing care with this data and promoting good governance practices for its management. This is a tool to consider objective factors for measuring trust in data controllers.
2) As with [7] discourse, the many anonymisation methods proposed and its privacy models satisfies a specific SDC. A schema designed with focus in

the decentralized process would fortify the data protection, approaching its issues holistically, especially because the BD platform requires the scalability property, both in terms of performance and infrastructure.

3) The utility cost of the data anonymisation process that can result in the difficulty of merging and exploiting anonymised data.
As indicated, anonymisation requires continuous improvement of its processing, considering the evolution of the techniques. So keep data anonymous on these platforms implies expenditure of maintenance resources.
Anonymisation also implies a reduction in the utility of data. It difficult its use in businesses using personalized services, as mentioned. These make in some cases unfeasible to use anonymisation tools. On the other hand, it is possible to overcome these hurdles in the future. One example is homomorphic encryption, which allows personalized services with anonymous data analysis without re-identifying this information. It seems to be an alternative to this type of database analysis and to reduce risk. However, this tool needs to be refined to analyze the DB. Currently, homomorphic encryption requires an unreasonable amount of time to perform on DB platforms.

5 Conclusions

As concluded by Carvalho et al. [3], its a loss of opportunity excluding a data governance review based exclusively on the GDPR [19] and LGPD [21] expectations about anonymisation techniques and its safe use of personal data insurance. This tool, according to the examples presented, must be reconsidered, in view of the better (and conscious) exploitation of its use, limitations and combinations with the other tools available to increase the security of data privacy.

This loss of opportunity is even greater when it comes to the Big Data context, when considering that even anonymous data is still easily de-anonymisable. Drawing from the case studies presented, as well as the Big Data tool and its internal aspects, below are some points of observation relevant to the context of anonymous data, and possible problems in distinguishing them from personal data. They are:

1. Ease of identifying an subject, given a context or subgroup of characteristics;
2. Big Data is a conducive environment for restoring information from patterns, metadata or data fragments;
3. Protection of data privacy can be improved if combined with other data security mechanisms, mainly because the re-identification of subjects in the massive data environment is possible, even for the relevant legislation, assuming the use of the necessary effort;
4. The dependence on measuring the anonymity of criteria that vary with specific conditions of analysis or technical contexts.

Carvalho et al. [3] proved that it is possible to discover the anonymisation technique used if you have a public personal database to compare with, and, in some cases, also re-identify subject data. Therefore, it is clear that anonymisation is not sufficient to reconcile Big Data compliance with Personal Data Regulations and data privacy. This does not mean that anonymisation is a useless tool, but it needs to be applied with the assistance of other mechanisms developed by compliance-oriented governance.

Overall, the anonymization process by itself cannot guarantee the privacy and security attributes simultaneously in the context of BD. Besides, anonymisation needs to consider some other factors listed, such as interference from external data, such as public databases, which can make feasible data re-identification; the need to comply with legal requirements for the processing personal data to promote anonymisation; the recognition of the loss of utility that this technique involves.

Also as concluded by Carvalho et al. [3], not all guidelines consider the data anonymisation process, or the data itself after such a process, as included in data management activities (and therefore subject to the application of certain principles such as "data minimization"). From this conclusion, it follows that a Big Data-driven framework is necessary in order to recommend good practices combined with the application of anonymisation tools for data protection in BD environments at the same time that legal compliance problems are addressed. We expect to investigate and present this proposal in future research.

Acknowledgments. This research work has the support of the Research Support Foundation of the Federal District (FAPDF) research grant 05/2018.

References

1. Brasher, E.A.: Addressing the failure of anonymization: guidance from the European union's general data protection regulation. Colum. Bus. L. Rev. p. 209 (2018)
2. Brkan, M.: Do algorithms rule the world? Algorithmic decision-making and data protection in the framework of the GDPR and beyond. Int. J. Law Inf. Technol. **27**(2), 91–121 (2019)
3. Carvalho, A.P., Canedo, E.D., Carvalho, F.P., Carvalho, P.H.P.: Anonymisation and compliance to protection data: Impacts and challenges into big data. In: 22th International Conference on Enterprise Information Systems, pp. 31–41 (2020)
4. Casanovas, P., De Koker, L., Mendelson, D., Watts, D.: Regulation of big data: perspectives on strategy, policy, law and privacy. Health Technol. **7**(4), 335–349 (2017)
5. Dalla Favera, R.B., da Silva, R.L.: Cibersegurança na união europeia e no mercosul: big data e surveillance versus privacidade e proteção de dados na internet. Revista de Direito, Governança e Novas Tecnologias **2**(2), 112–134 (2016)
6. De Mauro, A., Greco, M., Grimaldi, M.: A formal definition of big data based on its essential features. Lib. Rev. **65**(3), 122–135 (2016)
7. Domingo-Ferrer, J.: Personal big data, GDPR and anonymization. In: Cuzzocrea, A., Greco, S., Larsen, H.L., Saccà, D., Andreasen, T., Christiansen, H. (eds.) FQAS 2019. LNCS (LNAI), vol. 11529, pp. 7–10. Springer, Cham (2019). https://doi.org/10.1007/978-3-030-27629-4_2
8. Fothergill, D.B., Knight, W., Stahl, B.C., Ulnicane, I.: Responsible data governance of neuroscience big data. Front. Neuroinf. **13**, 28 (2019)

9. Hintze, M., El Emam, K.: Comparing the benefits of pseudonymisation and anonymisation under the GDPR. J. Data Prot. Priv. **2**(2), 145–158 (2018)
10. Jensen, M.H., Nielsen, P.A., Persson, J.S.: Managing big data analytics projects: the challenges of realizing value (2018)
11. Joyce, D.: Data associations and the protection of reputation online in Australia. Big Data Soc. **4**(1), 2053951717709829 (2017)
12. Koutli, M., et al.: Secure IoT e-health applications using vicinity framework and GDPR guidelines. In: 2019 15th International Conference on Distributed Computing in Sensor Systems (DCOSS), pp. 263–270. IEEE (2019)
13. Lin, C., Wang, P., Song, H., Zhou, Y., Liu, Q., Wu, G.: A differential privacy protection scheme for sensitive big data in body sensor networks. Ann. Telecommun. 465–475 (2016). https://doi.org/10.1007/s12243-016-0498-7
14. Liu, H.: Visions of big data and the risk of privacy protection: a case study from the Taiwan health databank project. Ann. Global Health **1**(81), 77–78 (2015)
15. Mehmood, A., Natgunanathan, I., Xiang, Y., Hua, G., Guo, S.: Protection of big data privacy. IEEE Access **4**, 1821–1834 (2016)
16. Mustafa, U., Pflugel, E., Philip, N.: A novel privacy framework for secure m-health applications: the case of the GDPR. In: 2019 IEEE 12th International Conference on Global Security, Safety and Sustainability (ICGS3), pp. 1–9. IEEE (2019)
17. Pomares-Quimbaya, A., Sierra-Múnera, A., Mendoza-Mendoza, J., Malaver-Moreno, J., Carvajal, H., Moncayo, V.: Anonylitics: from a small data to a big data anonymization system for analytical projects. In: 21st International Conference on Enterprise Information Systems, pp. 61–71 (2019)
18. Popovich, C., Jeanson, F., Behan, B., Lefaivre, S., Shukla, A.: Big data, big responsibility! building best-practice privacy strategies into a large-scale neuroinformatics platform. Int. J. Popul. Data Sci. **1**(1), 1–343 (2017)
19. Regulation, G.D.P.: EU data protection rules. European Commission, Accessed in October 9, 2019 (2018). https://ec.europa.eu/commission/priorities/justice-and-fundamental-rights/data-protection/2018-reform-eu-data-protection-rules_en
20. da República, P.: Lei geral de acesso a informação (lai). Secretaria-Geral (2011). http://www.mpf.mp.br/atuacao-tematica/sci/normas-e-legislacao/legislacao/legislacao-em-ingles/law-12.527. Accessed 4 Feb 2020
21. da República, P.: Lei geral de proteção de dados pessoais (LGPD). Secretaria-Geral (2018). https://www.pnm.adv.br/wp-content/uploads/2018/08/Brazilian-General-Data-Protection-Law.pdf. Accessed 19 Nov 2019
22. Ryan, M., Brinkley, M.: Navigating privacy in a sea of change: new data protection regulations require thoughtful analysis and incorporation into the organization's governance model. Int. Auditor **74**(3), 61–63 (2017)
23. Shapiro, S.: Propositional, first-order and higher-order logics: basic definitions, rules of inference, examples. Natural Language Processing and Knowledge Representation: Language for Knowledge and Knowledge for Language. AAAI Press/The MIT Press, Menlo Park, CA (1995)
24. Ventura, M., Coeli, C.M.: Para além da privacidade: direito à informação na saúde, proteção de dados pessoais e governança. Cadernos de Saúde Pública **34**, e00106818 (2018)
25. Zikopoulos, P., Eaton, C., et al.: Understanding big data: analytics for enterprise class hadoop and streaming data. McGraw-Hill Osborne Media (2011)

Streaming Set Similarity Joins

Lucas Pacífico and Leonardo Andrade Ribeiro(✉)

Instituto de Informática, Universidade Federal de Goiás, Goiânia, Brazil
{lucasopacifico,laribeiro}@inf.ufg.br

Abstract. We consider the problem of efficiently answering set similarity joins over streams. This problem is challenging both in terms of CPU cost, because similarity matching is computationally much more expensive than equality comparisons, and memory requirements, due to the unbounded nature of streams. This article presents SSTR, a novel similarity join algorithm for streams of sets. We adopt the concept of temporal similarity and exploit its properties to improve efficiency and reduce memory usage. Furthermore, we propose a sampling-based technique for ordering set elements that increases the pruning power of SSTR and, thus, reduce even further the number of similarity comparisons and memory consumption. We provide an extensive experimental study on several synthetic as well as real-world datasets. Our results show that the techniques we proposed significantly reduce memory consumption, improve scalability, and lead to substantial performance gains over the baseline approach.

Keywords: Advanced query processing · Data cleaning and Integration · Data streams · Set similarity join

1 Introduction

In the current Big Data era, large volumes of data are continuously generated over time at high speeds. Very often, there is a need for immediate processing of such stream of data to deliver analysis results in a timely fashion. Examples of such application scenarios abound, including social networks, Internet of Things, sensor networks, and a wide variety of log processing systems. Over the years, several stream processing systems have emerged seeking to meet this demand [1,9].

However, the requirements for stream processing systems are often conflicting. Many applications demand comparisons between historical and live data, together with the requirements for instantaneously processing and fast response times (see Rules 5 and 8 in [33]). To deliver results in real-time, it is imperative to avoid extreme latencies caused by disk accesses. However, maintaining all data in the main memory is impractical for unbounded data streams.

The problem becomes even more challenging in the presence of stream imperfections, which must be handled without causing significant delays in operations

J. Filipe et al. (Eds.): ICEIS 2020, LNBIP 417, pp. 24–42, 2021.
https://doi.org/10.1007/978-3-030-75418-1_2

Table 1. Messages from distinct sources about a football match [23].

Source	Time	Message
X	270	Great chance missed within the penalty area
Y	275	Shooting chance missed within the penalty area
Z	420	Great chance missed within the penalty area

(Rule 3 in [33]). In the case of streams coming from different sources, such imperfections may include the so-called *fuzzy duplicates*, i.e., multiple and non-identical representations of the same information. The presence of this type of redundancy degrades the quality of the data delivered to application programs, thereby leading to a myriad of problems, including inaccurate data mining models, the inability of correlating information related to the same entity or event, and unnecessarily repeated operations.

The identification of fuzzy duplicates requires similarity comparisons, which are computationally much more expensive than simple equality comparisons. Furthermore, data stream has an intrinsic temporal nature. A timestamp is typically associated with each data object recording, for example, the time of its arrival. This temporal attribute represents important semantic information and, thus, can affect a given notion of similarity. Therefore, it is intuitive to consider that the similarity between two data objects decreases with their temporal distance.

As a concrete example, consider a web site providing live scores and commentary about sporting events, which aggregates streams from different sources. Because an event can be covered by more than one source, multiple arriving messages can be actually describing a same moment. Posting such redundant messages are likely to annoy users and degrade their experience. This issue can be addressed by performing a similarity (self-)join over the incoming streams — a similarity join returns all data objects whose similarity is not less than a specified threshold. Thus, a new message is only posted if there are no previous ones that are similar to it. In this context, temporal information is crucial for similarity assessment because two textually similar messages might be considered as distinct if the difference in their arrival time is large. For example, Table 1 shows three messages about a soccer match from different sources. All messages are very similar to one another. However, considering the time of arrival of each message, one can conclude that, while the first two messages refer to the same moment of the match, the third message, despite being identical to the first one, actually is related to a different moment.

Morales and Gionis introduced the concept of temporal similarity for streams [22]. Besides expressing the notion of time-dependent similarity, this concept is directly used to design efficient similarity join algorithms for streams of vectors. The best-performing algorithm exploits temporal similarity to reduce the number of comparisons. Moreover, such time-dependent similarity allows to establish an "aging factor": after some time, a given data object cannot be similar to any new data arriving in the stream and, thus, can be safely discarded to reduce memory consumption.

This article presents an algorithm for similarity joins over set streams. There is a vast literature on set similarity joins for static data [11,20,24,27,29,30,34, 36,37]; however, to the best of our knowledge, there is no prior work on this type of similarity join for stream data. Here, we adapt the notion of temporal similarity to sets and exploit its properties to reduce both comparison and memory spaces. We provide an extensive experimental study on several synthetic as well as real-world datasets. Our results show that the techniques we proposed significantly improve scalability and lead to substantial performance gains over the baseline approach.

This article is an extended and revised version of our earlier conference paper [23]. Here, we present a sampling-based approach to order set elements according to their estimated frequency in the stream. This ordering avoids considering high-frequency elements in the earlier stages of the algorithm and, thus, further reduces the number of similarity comparisons. We significantly extended our experimental study to include an in-depth analysis of the performance impact of this new proposal.

The remainder of the article is organized as follows. Section 2 provides background material. Section 3 introduces the SSTR algorithm and its underlying techniques. Section 4 presents our proposal for ordering set elements. Section 5 describes the experimental study and analyzes its results. Section 6 reviews relevant related work. Section 7 summarizes the article and discusses future research.

2 Background

In this section, we define the notions of set and temporal similarity, together with essential optimizations derived from these definitions. Then, we formally state the problem considered in this article.

2.1 Set Similarity

This work focus on streams of data objects represented as sets. Intuitively, the similarity between two sets is determined by their intersection. Representing data objects as sets for similarity assessment is a widely used approach for string data [11].

Strings can be mapped to sets in several ways. For example, the string *"Great chance missed within the penalty area"* can be mapped to the set of words {*'Great', 'chance', 'missed', 'within', 'the', 'penalty', 'area'*}. Another well-known method is based on the concept of q-grams, i.e., substrings of length q obtained by "sliding" a window over the characters of the input string. For example, the string *"similar"* can be mapped to the set of *3-grams* {*'sim', 'imi', 'mil', 'ila', 'lar'*}. Henceforth, we generically refer to a set element as a *token*.

A *similarity function* returns a value in the interval [0, 1] quantifying the underlying notion of similarity between two sets; greater values indicate higher similarity. In this article, we focus on the well-known Jaccard similarity; nevertheless, all techniques described here apply to other similarity functions such as *Dice* e *Cosine* [37].

Definition 1 (Jaccard Similarity). *Given two sets x and y, the Jaccard similarity between them is defined as* $J(x, y) = \frac{|x \cap y|}{|x \cup y|}$.

Example 1. Consider the sets x and y below, derived from the two first messages in Table 1 (sources X and Y):

$$x = \{\text{'}Great\text{'}, \text{'}chance\text{'}, \text{'}missed\text{'}\ \text{'}within\text{'}, \text{'}the\text{'}, \text{'}penalty\text{'}, \text{'}area\text{'}\},$$
$$y = \{\text{'}Shooting\text{'}, \text{'}chance\text{'}, \text{'}missed\text{'}, \text{'}within\text{'}, \text{'}the\text{'}, \text{'}penalty\text{'}, \text{'}area\text{'}\}.$$

Then, we have $J(x, y) = \frac{6}{7+7-6} = 0.75$.

A fundamental property of the Jaccard similarity is that any predicate of the form $J(x, y) \geq \gamma$, where γ is a threshold, can be equivalently rewritten in terms of an *overlap bound.*

Lemma 1 (Overlap Bound [11]). *Given two sets, r and s, and a similarity threshold γ, let $\mathcal{O}(x, y, \gamma)$ denote the corresponding overlap bound, for which the following holds:*

$$J(x, y) \geq \gamma \iff |x \cap y| \geq \mathcal{O}(x, y, \gamma) = \frac{\gamma}{1+\gamma} \times (|x| + |y|).$$

Overlap bound provides the basis for several filtering techniques. Arguably, the most popular and effective techniques are *size-based filter* [30], *prefix filter* [11], and *positional filter* [37], which we review in the following.

2.2 Optimization Techniques

Intuitively, the difference in size between two similar sets cannot be too large. Thus, one can quickly discard set pairs whose sizes differ enough.

Lemma 2 (Size-based Filter [30]). *For any two sets x and y, and a similarity threshold γ, the following holds:*

$$J(x, y) \geq \gamma \implies \gamma \leq \frac{|x|}{|y|} \leq \frac{1}{\gamma}.$$

Prefix filter allows discarding candidate set pairs by only inspecting a fraction of them. To this end, we first fix a total order on the universe $\mathcal{U}$ from which all tokens are drawn; we defer the discussion about a specific token order to Sect. 4.

Lemma 3 (Prefix Filter [11]**).** *Given a set r and a similarity threshold γ, let $pref(x, \gamma) \subseteq x$ denote the subset of x containing its first $|x| - \lceil |x| \times \gamma \rceil + 1$ tokens. For any two sets x e y, and a similarity threshold γ, the following holds:*

$$J(x, y) \geq \gamma \implies pref(x, \gamma) \cap pref(y, \gamma) \neq \varnothing.$$

The positional filter also exploits token ordering for pruning. This technique filters dissimilar set pairs using the position of matching tokens.

Lemma 4 (Positional Filter [37]**).** *Given a set x, let $w = x[i]$ be a token of x at position i, which divides x into two partitions, $x_l(w) = x[1, .., (i-1)]$ and $x_r(w) = x[i, .., |x|]$. Thus, for any two sets x e y, and a similarity threshold γ, the following holds:*

$$J(x, y) \geq \gamma \implies |x_l(w) \cap y_l(w)| + min(|x_r(w)|, |y_r(w)|) \geq \mathcal{O}(x, y, \gamma).$$

2.3 Temporal Similarity

Each set x is associated with a timestamp, denoted by $t(x)$, which indicates its generation or arrival time. We assume that the timestamp is monotonically increasing. Formally, the input stream is denoted by $\mathcal{S} = \langle ..., (x_i, t(x_i)), (x_{i+1}, t(x_{i+1}), ...\rangle$, where $t(x_i) \leq t(x_{i+1})$.

The concept of temporal similarity captures the intuition that the similarity between two sets diminishes with their temporal distance. To this end, the difference in the arrival time is incorporated into the similarity function.

Definition 2 (Temporal Similarity [22]**).** *Given two sets x e y, let $\Delta t_{xy} = |t(x) - t(y)|$ be the difference in their arrival time. Therefore, the temporal similarity between x and y is defined as*

$$J_{\Delta t}(x, y) = J(x, y) \times e^{-\lambda \times \Delta t_{xy}},$$

where λ is a time-decay parameter.

Example 2. Consider again the sets x and y from Example 1, obtained from sources X and Y, respectively, in Table 1, and $\lambda = 0.01$. We have $\Delta t_{xy} = 5$ and, thus, $J_{\Delta t}(x, y) = 0.75 \times e^{-\lambda \times 5} \approx 0.71$. Consider now set z obtained from source Z. Despite of sharing all tokens, x and z have a relatively large temporal distance, i.e., $\Delta t_{xz} = 150$. As a result, we have $J_{\Delta t}(x, z) = 1 \times e^{-\lambda \times 150} \approx 0.22$.

Note that $J_{\Delta t}(x, y) = J(x, y)$ when $\Delta t_{xy} = 0$ or $\lambda = 0$, and its limit is 0 as Δt_{xy} approaches infinity, at an exponential rate modulated by λ. The time-decay factor, together with the similarity threshold, allows defining a *time filter*: given a set x, after a certain period, called *time horizon*, no newly arriving set can be similar to x.

Lemma 5 (Time Filter [22]**).** *Given a time-decay factor λ, let $\tau = \frac{1}{\lambda} \times ln\frac{1}{\gamma}$ be the time horizon. Thus, for any two sets x e y, the following holds:*

$$J_{\Delta t}(x, y) \geq \gamma \implies \Delta t_{xy} < \tau.$$

Note that the time horizon establishes a temporal window of fixed size, which slides as a new set arrives. This behavior resembles the traditional sliding window model [5]. Also note that the number of sets can vary widely across different temporal windows.

2.4 Problem Statement

We are now ready to formally define the problem considered in this article.

Definition 3 (Similarity Join over Set Streams) . *Given a stream of timestamped sets $\mathcal{S}$, a similarity threshold γ, and a time-decay factor λ, a similarity join over $\mathcal{S}$ returns all set pairs (x, y) in $\mathcal{S}$ such that $J_{\Delta t}(x, y) \geq \gamma$.*

3 Similarity Join over Set Streams

In this section, we present our proposal to solve the problem of efficiently answering similarity joins over set streams. We first describe a baseline approach based on a straightforward adaptation of an existing set similarity join algorithm for static data. Then, we present the main contribution of this article, a new algorithm deeply integrating characteristics of temporal similarity to both improve runtime performance and reduce memory consumption.

3.1 Baseline Approach

Most state-of-the-art set similarity join algorithms follow a filtering-verification framework [20]. In this framework, the input set collection is scanned sequentially, and each set goes through the filtering and verification phases. In the filtering phase, tokens of the current set (called henceforth probe set) are used to find potentially similar sets that have already been processed (called henceforth candidate sets). The filters discussed in the previous section are then applied to reduce the number of candidates. This phase is supported by an inverted index, which is incrementally built as the sets are processed. In the verification phase, the similarity between the probe set and each of the surviving candidates is fully calculated, and those pairs satisfying the similarity predicate are sent to the output.

A naive way to perform set similarity join in a stream setting is to simply carry out the filtering and verification phases of an existing algorithm on each incoming set. The temporal decay is then applied to the similarity of the pairs returned by the verification in a post-processing phase, before sending results to the output.

Algorithm 1 describes this naive approach for PPJoin [37], one of the best performing algorithms in a recent empirical evaluation [20]. The algorithm continuously processes sets from the input stream as they arrive. The filtering phase uses prefix tokens (Line 5) to probe the inverted index (Line 7). Each set found in the associated inverted list is considered a candidate and checked against conditions using the size-based filter (Line 8) and the positional filter (Lines 10–11).

Algorithm 1. The PPJoin algorithm over set streams.

```
   Input: Set stream S, threshold γ, decay λ
   Output: All pairs (x, y) ∈ S s.t. J_Δt(x, y) ≥ γ
1  I_i ← ∅ (1 ≤ i ≤ |U|)
2  while true do
3      x ← read (S)
4      M ← empty map from set id to int
5      for i ← 1 to |pref (x, γ)| do
6          k ← x[i]
7          foreach (y, j) ∈ I_k do
8              if |y| < |x| × γ then
9                  continue
10             ubound ← 1 + min (|x| − i, |y| − j)
11             if M[y] + ubound ≥ O(x, y, γ) then
12                 M[y] ← M[y] + 1
13             else
14                 M[y] ← −∞
15         I_k ← I_k ∪ (x, i)
16     R_tmp ← Verify (x, M, γ)
17     R ← ApplyDecay (R_tmp, γ, λ)
18     Emit (R)
```

A reference to the probe set is appended to the inverted list associated with each prefix token (Line 15). Not shown in the algorithm, the verification phase (Line 16) can be highly optimized by exploiting the token ordering in a merge-like fashion and the overlap bound to define early stopping conditions [27]. Finally, the temporal decay is applied, and a last check against the threshold is performed to produce an output (Line 17).

Clearly, the above approach has two serious drawbacks. First, space consumption of the inverted index can be exorbitant and quickly exceed the available memory. Even worse, a large part of the index can be stale entries, i.e., entries referencing sets that will not be similar to any set arriving in the future. Second, temporal decay is applied only after the verification phase. Therefore, much computation in the verification is wasted on set pairs that cannot be similar owing to the difference in their arrival times.

3.2 The SSTR Algorithm

We now present our proposed algorithm called SSTR for similarity joins over set streams. SSTR exploits properties of the temporal similarity definition to avoid the pitfalls of the naive approach. First, SSTR dynamically removes old entries from the inverted lists that are outside the window induced by the probe set and the time horizon. Second, it uses the temporal decay to derive a new similarity threshold between the probe set and each candidate set. This new threshold is

Algorithm 2. The SSTR algorithm.

```
   Input: Set stream S, threshold γ, decay λ
   Output: All pairs (x, y) ∈ F s.t. J_Δt(x, y) ≥ γ
1  τ = 1/λ × ln(1/γ)
2  I_i ← ∅ (1 ≤ i ≤ |U|)
3  while true do
4      x ← read(S)
5      M ← empty map from set id to int
6      for i ← 1 to |pref(x, γ)| do
7          k ← x[i]
8          Remove all (y, j) from I_k s.t. Δt_xy > τ
9          foreach (y, j) ∈ I_k do
10             γ' ← γ / e^(−λ×Δt_xy)
11             if |y| < |x| × γ' then
12                 M[y] ← −∞
13                 continue
14             ubound ← 1 + min(|x| − i, |y| − j)
15             if M[y].s + ubound ≥ O(x, y, γ') then
16                 M[y].s ← M[y].s + 1
17             else
18                 M[y] ← −∞
19         I_k ← I_k ∪ (x, i)
20     R ← Verify(x, M, γ, λ)
21     Emit(R)
```

greater than the original, which increases the effectiveness of the size-based and positional filters.

The steps of STTR are formalized in Algorithm 2. References to sets whose difference in arrival time with the probe set is greater than the time horizon is removed as the inverted lists are scanned (Line 8). Note that the entries in the inverted lists are sorted in increasing timestamp order. Thus, all stale entries are grouped at the beginning of the lists. For each candidate set, a new threshold value is calculated (Line 10), which is used in the size-based filter and to calculate the overlap bound (Lines 11 and 15, respectively). In the same way, such increased, candidate-specific threshold is also used in the verification phase to obtain greater overlap bounds and, thus, improve the effectiveness of the early-stop conditions. For this reason, the time-decay parameter is passed to the *Verify* procedure (Line 20), which now directly produces output pairs[1].

Even with the removal of stale entries from the inverted lists, SSTR still can incurs into high memory consumption issues for temporal windows containing too many sets. This situation can happen due to very small time-decay param-

[1] In our implementation, we avoid repeated calculations of candidate-specific thresholds and overlap bounds by storing them in the map M.

eters leading to large windows or at peak data stream rate leading to "dense" windows. In such cases, sacrificing timeliness by resorting to some approximation method, such as batch processing [5], is inevitable. Nevertheless, considering a practical scenario with a pre-defined memory budget, the SSTR algorithm can dramatically reduce the frequency of such batch processing modes in comparison to the baseline approach, as we empirically show next.

4 The Partial Token-Frequency Table

Recall from Sect. 2 that prefix filter requires all sets to be sorted according to a total order over the universe of tokens; positional filter also requires such ordering. While any token ordering ensures the correctness of the filters, the choice of a specific ordering has performance implications [11]. The standard approach is to order the sets by increasing frequency in the dataset. In this way, rare tokens are placed in the prefix positions. As a result, inverted lists are shorter and there is less prefix overlap between dissimilar sets, thereby decreasing the number of generated candidates.

Unfortunately, token frequency order cannot be directly used in a stream setting because the frequency of tokens is continuously updated as new sets arrive. Therefore, one has to pick an ordering that does not depend on the frequency, such as the lexicographical ordering. The problem is that high-frequency tokens may end up in the prefix positions, thereby increasing the number of candidates.

We now propose a new approach to avoid the presence of high-frequency tokens in the prefixes. In a pre-processing phase, we build a token-frequency table and use this table to order the tokens during similarity join processing. Figure 1 illustrates this idea. The objective of the table is to capture the relative frequency order of the tokens. We can use a random sampling algorithm to collect the tokens of the set stream in the pre-processing phase, such as the well-known reservoir sampling algorithm [35].

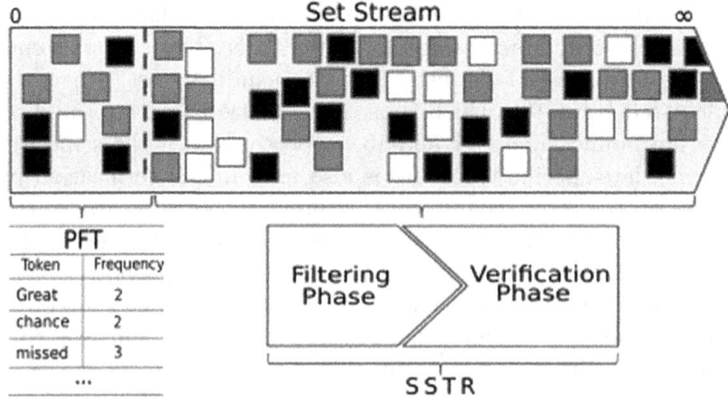

Fig. 1. Partial frequency table scheme.

Obviously, the token-frequency table is partial, i.e., the frequencies collected will change afterward and part of the universe of tokens will be absent. A question that arises is how to order tokens with no entry in the table. Here, we assign the minimum frequency among all sampled tokens, typically a very low value. The rationale behind this decision is that word frequency distributions tend to obey power laws and, thus, are characterized by a large number of rare words [4].

Finally, we note that, alternatively, other data stream mining techniques can be used to identify high-frequency tokens. In particular, there is a wealth of available algorithms for finding frequent items in streams [13]. The evaluation of such algorithms in our context is left for future work.

5 Experiments

We now present an experimental study of the techniques proposed in this paper. The goals of our experiments are as follows.

- **G1:** Evaluate the effectiveness of our proposed techniques for reducing comparison space and memory consumption. To this end, we compare our SSTR algorithm with the baseline approach.
- **G2:** Evaluate the effect on performance of the sample size used to build the partial token-frequency table.
- **G3:** Compare the runtime performance of SSTR on sets sorted in lexicographic order and sets sorted in increasing token frequency order based on values obtained from the partial token-frequency table.

Orthogonally to the three goals above, we also evaluate the effect on performance of the parameters γ and λ.

5.1 Datasets and Setup

We used four datasets: DBLP[2], containing information about computer science publications; WIKI[3], an encyclopedia containing generalized information about different topics; TWITTER[4], geocoded tweets collected during the Brazilian election of 2018; and REDDIT, a social news aggregation, web content rating, and discussion website [6]. For DBLP and WIKI, we started by randomly selecting 70k and 200k article titles, respectively. Then, we generated four fuzzy duplicates from each string by performing transformations on string attributes, such as characters insertions, deletions, or substitutions. We end up with 350k and 1M strings for DBLP and WIKI, respectively. Finally, we assigned artificial timestamps to each string in these datasets, sampled from Poisson (DBLP) and Uniform (WIKI) distribution functions. For this reason, we call DBLP and

---

[2] dblp.uni-trier.de/xml.

[3] https://en.wikipedia.org/wiki/Wikipedia:Database_download.

[4] https://developer.twitter.com/en/products/tweets.

Table 2. Datasets statistics [23].

Name	Population	Avg. set size	Timestamp
DBLP	350000	76	Poisson
WIKI	1000000	53	Uniform
TWITTER	2824998	90	Publishing date
REDDIT	19456493	53	Publishing date

WIKI (semi)synthetic datasets. For TWITTER and REDDIT, we used the complete dataset available without applying any modification, where the publication time available for each item was used as timestamp. For this reason, we call TWITTER and REDDIT real-world datasets. The datasets are heterogeneous, exhibiting different characteristics, as summarized in Table 2.

For the similarity threshold γ, we explore a range of values in $[0.5, 0.95]$, while for the time-decay factor λ we use exponentially increasing values in the range $[10^{-3}, 10^{-1}]$. We built the partial token-frequency table using random sample sizes corresponding to 1%, 5%, and 10% of the original dataset. For all datasets, we tokenized the strings into sets of *3*-grams and hashed the tokens into four byte values.

In the experimental charts, the baseline approach is abbreviated to SPPJ (streaming PPJoin). The lexicographic token ordering is abbreviated to Lex, whereas the ordering based on the partial token-frequency table is abbreviated to PTF.

We conduct our experiments on an Intel E5-2620 @ 2.10 GHz with 15 MB of cache, 16 GB of RAM, running Ubuntu 16.04 LTS. We report the average runtime over repeated runs. All algorithms were implemented using Java SDK 11.

We set a timeout of 3 h for each execution. Some parameter configurations were very troublesome to execute in our hardware environment, both in terms of runtime and memory; this is particularly the case for SPPJ on the largest datasets. As a result, we were unable to finish the execution of the algorithms in some settings. Table 3 shows a summary of the outcomes of the evaluations under different configurations. SPPJ failed to finish due to timeout or memory requirements, whereas SSTR only due to timeout.

5.2 Results

We now report our experimental results addressing the three aforementioned goals, namely, comparison between SSTR and SPPJ, evaluation of different sample sizes for building the partial token-frequency table, and comparison between the variants of SSTR based on Lex and PTF token orderings.

G1: SSTR vs. SPPJ. Figure 2 shows the results of SSTR and SPPJ. We used Lex token ordering for both algorithms. We first analyze the results on the syn-

Table 3. The fraction of the algorithm configurations that successfully terminated.

SPPJ		SSTR			
Dataset	Lex	Lex	PTF-1%	PTF-5%	PTF-10%
DBLP	0.22	0.55	1.0	1.0	1.0
WIKI	0.33	1.0	1.0	1.0	1.0
Twitter	0	1.0	1.0	1.0	1.0
Reddit	0	1.0	1.0	1.0	1.0

thetic datasets shown in Fig. 2(a). As expected, SPPJ was only able to finish its execution for very high threshold values. In contrast, SSTR successfully terminated in all settings on WIKI. Higher threshold values increase the effectiveness of the prefix filter, which benefits both SPPJ and SSTR. Yet, in most cases where SPPJ was able to terminate, SSTR was up to three orders of magnitude faster. These results highlight the effectiveness of our techniques in drastically reducing the number of similarity comparisons as well as memory usage.

On the DBLP dataset, SSTR terminates within the time limit for all threshold values only for $\lambda = 0.1$. The reason is that the Poisson distribution generates some very dense temporal windows, with set objects temporally very close to each other. For small time-decay values, temporal windows are large and more sets have to be kept in the inverted lists and compared in the verification phase. Conversely, greater time-decay values translate into a smaller time horizon and, thus, narrower temporal windows. As a result, the time filter is more effective for pruning stale entries from the inverted lists. Moreover, time-decay values lead to greater candidate-specific thresholds, which, in turn, improve the pruning power of the size-based and positional filters.

We now analyze the results on the real-world datasets. Figure 2(b) plots the runtimes for SSTR on TWITTER and REDDIT. We do not show the results for SPPJ because it failed due to lack of memory on these datasets in all settings. Obviously, as SPPJ does not prune stale entries from the index, it cannot directly handle the largest datasets in our experimental setting. Note that we can always reconstruct the inverted index, for example after having reached some space limit. However, this strategy sacrifices timeliness, accuracy, or both. While resorting to such batch processing mode is inevitable in stressful scenarios, the results show that the SSTR algorithm can nevertheless sustain continuous stream processing much longer than SPPJ.

Another important observation is that, overall, SSTR successfully terminates in all settings on real-world datasets; the only exception is on REDDIT for the smallest λ value. Moreover, even though those datasets are larger than DBLP and WIKI, SSTR is up to two orders of magnitude faster on them. The explanation lies in the timestamp distribution of the real-world datasets, which exhibit more "gaps" as compared to the synthetic ones. Hence, the induced temporal windows are sparser on those datasets. Which is effectively exploited by the time filter to dynamically maintain the length of the inverted lists reduced to a minimum. The

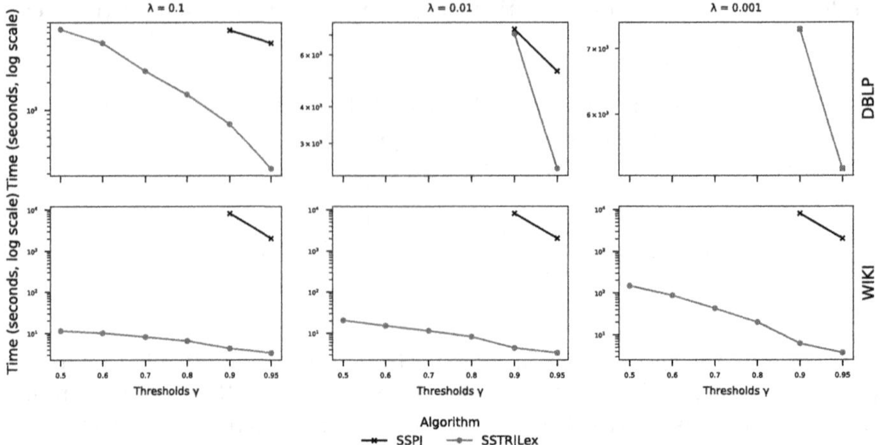

(a) Runtime results of the algorithms SSTR and SPPJ on synthetic datasets.

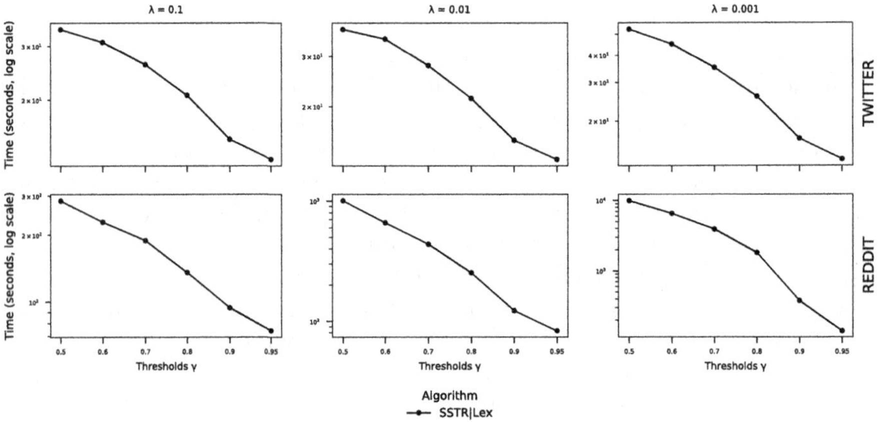

(b) Runtime results of the algorithms SSTR and SPPJ on real-world datasets. datasets.

Fig. 2. Experimental results comparing SSTR and SPPJ [23].

other trends remain the same: execution times increase and decrease as similarity thresholds and time-decay parameters decrease and increase, respectively.

G2: Sample Sizes. We now evaluate the effect of the sample size used to build the token-frequency table. Table 4 shows the results for SSTR on all datasets along varying thresholds and for $\lambda = 0.01$; results for other values of λ followed the same trends.

We observe that there is no significant performance gain as we increase the sample size. At first glance, this result can be surprising, because using more sample data is expected to result in a more accurate PTF. However, recall that we only need to capture the relative frequency order of the tokens or, at least,

Table 4. Time in seconds for all PTF percentage variations and with $\lambda = 0.01$.

Dataset	%/γ	0.5	0.6	0.7	0.8	0.9	0.95
DBLP	1%	5574.07	2765.71	1266.63	395.45	45.51	10.29
	5%	5555.59	2741.99	1255.5	385.92	42.95	9.21
	10%	5587.67	2725.12	1238.41	383.92	42.41	9.12
WIKI	1%	13.54	11.41	9.67	8.17	7.3	6.53
	5%	13.22	11.32	9.48	8.28	7.06	6.61
	10%	13.27	10.9	9.87	8.18	7.03	6.72
TWITTER	1%	41.87	35.8	33.35	29.69	24.3	19.26
	5%	41.91	37.37	32.9	28.62	22.51	19.39
	10%	41.85	36.62	33.63	29.66	23.72	19.64
REDDIT	1%	429.47	322.17	230.99	176.19	139.49	121.65
	5%	433.23	320.87	233.7	178.37	140.83	120.53
	10%	427.72	323.41	232.69	176.53	142.48	122.26

separate high-frequency tokens from lower-frequency ones. We conclude from the results that we can achieve these goals with relatively small sample sizes. In the following, we only report results for sample size corresponding to 1% of the original dataset.

G3: Lex vs. PTF. We now evaluate the effectiveness of the partial table by comparing the runtime results of SSTR on sets that are sorted according to Lex and PTF orderings. Figure 3 shows the timings on all datasets and configuration parameters. Our first observation is that PTF successfully terminates in all settings. On the other hand, Lex did not terminate owing to timeout on DBLP and Reddit (as already shown in Fig. 2); only for the smallest value of γ in the latter. Under the Lex ordering, the presence of high-frequent tokens in the prefix generated a large number of candidates, which were not pruned by the time filter due to low value of λ.

PTF obtain better results in most settings. Nonetheless, Lex is even faster than PTF in some cases. In general, the performance advantage of PTF is inversely proportional to the effectiveness of the time filter. When the time filter performs better, it tends to overshadow the gains of the prefix filter, which benefits most of the token frequency order. Conversely, when time filter performs poorer, reduction of the comparison space is obtained mostly from the prefix filter; thus, the increase of its pruning power owing to PTF is decisive for the overall performance.

In fact, PTF tends to be faster than Lex for lower values of λ, which imply in larger temporal windows. In such cases, much more sets are within a temporal distance of each other that is smaller than the time horizon. In contrast, for greater time-decay values, the vast majority of candidates are pruned by the time filter, thereby rendering prefix filter less important for overall performance.

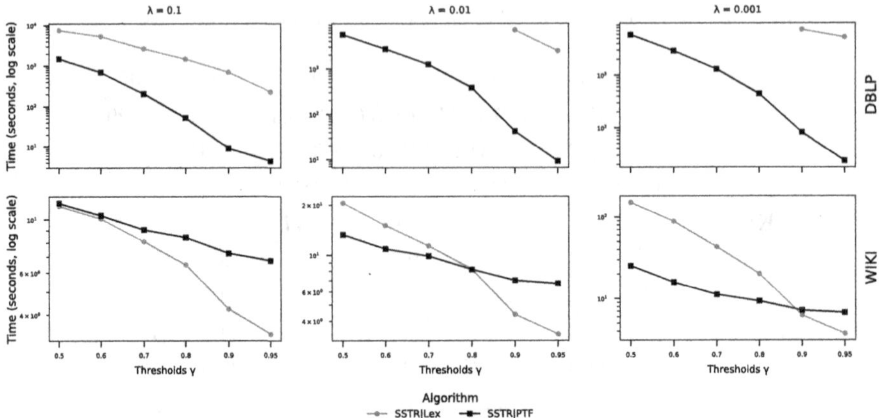

(a) Runtime results of SSTR using Lex and PTF on synthetic datasets.

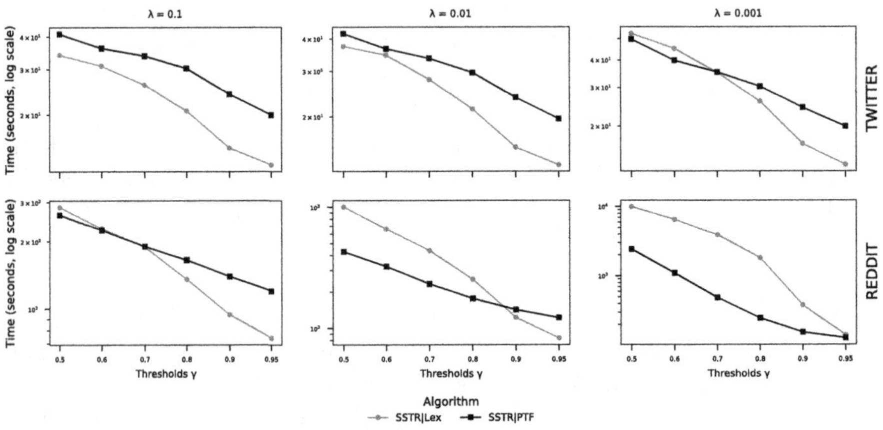

(b) Runtime results of SSTR using Lex and PTF on real datasets.

Fig. 3. Experimental results comparing Lex and PTF token orderings.

DBLP is the only dataset where PTF outperforms Lex in all settings. As already mentioned, the reason is that the Poisson distribution generates some very dense temporal windows, with set objects temporally very close to each other. Again, time filter is not efficient in this scenario and, as a result, the performance of prefix filter is crucial for reducing the comparison space. In the other datasets, the underlying temporal distribution is much sparser, which greatly benefits the time filter.

Finally, we observe that, while the pruning power of all filters decreases with the threshold value, performance degradation is more noticeable for time filter. Indeed, PTF tends to be faster than Lex for lower threshold values.

We conclude that PTF is the variant of choice for settings involving lower time-decay values, dense temporal distributions, and lower threshold values. It turns out that such settings are the most stressful in terms of processing and memory requirements and. thus, PTF is essential to avoid having to resort to batch processing in these contexts.

6 Related Work

There is a long line of research on efficiently answering set similarity joins [11,20,24,27,29,30,34,36,37]. Popular optimizations, such as size-based filtering, prefix filtering, and positional filtering, were incorporated into our algorithm. Recently, reference [36] exploited set relations to improve performance — the key insight is that similar sets produce similar results. However, one of the underlying techniques, the so-called index-level skipping, relies on building the whole inverted index before start processing and, thus, cannot be used in our context where new sets are continuously arriving.

Further, set similarity join has been addressed in a wide variety of settings, including: distributed platforms [10,34]; many-core architectures [24,29]; relational DBMS, either declaratively in SQL [28] or within the query engine as a physical operator [11]; cloud environments [32]; integrated into clustering algorithms [25,26]; and probabilistic, either for increasing performance (at the expense of missing some valid results) [8,12] or modeling uncertain data [18]. However, none of these previous studies considered similarity join over set streams.

Previous work on similarity join over streams focused on data objects represented as vectors, where the similarity between two using vectors is measured using Euclidean distance [17,19] or cosine [22]. Lian and Chen [17] proposed an adaptive approach based on a formal cost model for multi-way similarity join over streams. The same authors later addressed similarity joins over uncertain streams [19].

Morales and Gionis [22] introduced the notion of time-dependent similarity. The authors then adapted existing similarity join algorithms for vectors, namely AllPairs [7] and L2AP [3], to incorporate this notion and exploit its properties to reduce the number of candidate pairs and dynamically remove stale entries from the inverted index. We follow a similar approach here, but the details of these optimizations are not directly applicable to our context, as we focus on a stream of data objects represented as sets. Moreover, we proposed a technique for token ordering based on a partial frequency table generated through sampling.

Processing the entire data of possibly unbounded streams is clearly infeasible. Therefore, some method has to be used to limit the portion of stream history processed at each query evaluation. The sliding window model is popularly used in streaming similarity processing [17,19,31], where only recent data from the streams are evaluated [5]. As already mentioned, our approach resembles this model as the temporal similarity induces a sliding temporal window (i.e., the time horizon).

Streaming similarity search finds all data objects that are similar to a given query [16]. To some extent, similarity join can be viewed as a sequence of searches using each arriving object as a query object. A fundamental difference in this context is that the threshold is fixed for joins, while it can vary along distinct queries for searches.

Top-k queries have also been studied in the streaming setting [2,31]. Focusing on streams of vectors, Shen et al. [31] proposed a framework supporting queries with different similarity functions and window sizes. Amagata et al. [2] presented an algorithm for kNN self-join, a type top-k query that finds the k most similar objects for each object. This work assumes objects represented as sets, however the dynamic scenario considered is very different: instead of a stream of sets, the focus is on a stream of updates continuously inserting and deleting elements of existing sets.

Finally, duplicate detection in streams is a well-studied problem [14,15,21]. A common approach to dealing with unbounded streams is to employ space-preserving, probabilistic data structures, such as Bloom Filters and Quotient Filters together with window models. However, these proposals aim at detecting exact duplicates and, therefore, similarity matching is not addressed.

7 Conclusions and Future Work

In this paper, we considered the problem of efficiently answering set similarity joins over streams. To the best of our knowledge, set similarity join has not been previously investigated in a streaming setting. We presented a new algorithm called SSTR, which adopts the concept of temporal similarity and exploits its properties to reduce processing cost and memory usage. We also proposed a sampling-based approach to build a partial token-frequency table, which is used to order set elements according to their estimated frequency in the stream. This ordering increased the pruning power of SSTR, leading to further improvements in performance and memory consumption. We reported an extensive experimental study on synthetic and real-world datasets. Our results showed that SSTR is efficient and outperforms the baseline approach by several orders of magnitude. Future work is mainly oriented towards designing a parallel version of SSTR and an algorithmic framework for seamless integration with batch processing models.

References

1. Abadi, D.J., et al.: The design of the borealis stream processing engine. In: Proceedings of the Conference on Innovative Data Systems Research, pp. 277–289 (2005)
2. Amagata, D., Hara, T., Xiao, C.: Dynamic Set kNN Self-Join. In: Proceedings of the IEEE International Conference on Data Engineering, pp. 818–829 (2019)
3. Anastasiu, D.C., Karypis, G.: L2AP: fast cosine similarity search with prefix L-2 norm bounds. In: Proceedings of the IEEE International Conference on Data Engineering, pp. 784–795 (2014)
4. Baayen, R.H.: Word Frequency Distributions, Text, Speech and Language Technology, vol. 18. Kluwer Academic Publishers (2001)

5. Babcock, B., Babu, S., Datar, M., Motwani, R., Widom, J.: Models and issues in data stream systems. In: Proceedings of the ACM Symposium on Principles of Database Systems, pp. 1–16 (2002)
6. Baumgartner, J.: Reddit May 2019 submissions. Harv. Dataverse (2019). https://doi.org/10.7910/DVN/JVI8CT
7. Bayardo, R.J., Ma, Y., Srikant, R.: Scaling up all pairs similarity search. In: Proceedings of the International World Wide Web Conferences, pp. 131–140. ACM (2007)
8. Broder, A.Z., Charikar, M., Frieze, A.M., Mitzenmacher, M.: Min-wise independent permutations (extended abstract). In: Proceedings of the ACM SIGACT Symposium on Theory of Computing, pp. 327–336. ACM (1998)
9. Carbone, P., Katsifodimos, A., Ewen, S., Markl, V., Haridi, S., Tzoumas, K.: Apache Flink^{TM}: stream and batch processing in a single engine. IEEE Data Eng. Bull. **38**(4), 28–38 (2015)
10. do Carmo Oliveira, D.J., Borges, F.F., Ribeiro, L.A., Cuzzocrea, A.: Set similarity joins with complex expressions on distributed platforms. In: Proceedings of the Symposium on Advances in Databases and Information Systems, pp. 216–230 (2018)
11. Chaudhuri, S., Ganti, V., Kaushik, R.: A primitive operator for similarity joins in data cleaning. In: Proceedings of the IEEE International Conference on Data Engineering, p. 5. IEEE Computer Society (2006)
12. Christiani, T., Pagh, R., Sivertsen, J.: Scalable and robust set similarity join. In: Proceedings of the IEEE International Conference on Data Engineering, pp. 1240–1243. IEEE Computer Society (2018)
13. Cormode, G., Hadjieleftheriou, M.: Finding frequent items in data streams. Proc. VLDB Endow. **1**(2), 1530–1541 (2008)
14. Deng, F., Rafiei, D.: Approximately detecting duplicates for streaming data using stable bloom filters. In: Proceedings of the ACM SIGMOD International Conference on Management of Data, pp. 25–36 (2006)
15. Dutta, S., Narang, A., Bera, S.K.: Streaming quotient filter: a near optimal approximate duplicate detection approach for data streams. Proc. VLDB Endow. **6**(8), 589–600 (2013)
16. Kraus, N., Carmel, D., Keidar, I.: Fishing in the stream: similarity search over endless data. In: bigdata, pp. 964–969 (2017)
17. Lian, X., Chen, L.: Efficient similarity join over multiple stream time series. IEEE Trans. Knowl. Data Eng. **21**(11), 1544–1558 (2009)
18. Lian, X., Chen, L.: Set similarity join on probabilistic data. Proc. VLDB Endow. **3**(1), 650–659 (2010)
19. Lian, X., Chen, L.: Similarity join processing on uncertain data streams. IEEE Trans. Knowl. Data Eng. **23**(11), 1718–1734 (2011)
20. Mann, W., Augsten, N., Bouros, P.: An empirical evaluation of set similarity join techniques. PVLDB **9**(9), 636–647 (2016)
21. Metwally, A., Agrawal, D., El Abbadi, A.: Duplicate detection in click streams. In: Proceedings of the International World Wide Web Conferences, pp. 12–21 (2005)
22. Morales, G.D.F., Gionis, A.: Streaming similarity self-join. Proc. VLDB Endow. **9**(10), 792–803 (2016)
23. Pacífico, L., Ribeiro, L.A.: SSTR: set similarity join over stream data. In: International Conference on Enterprise Information Systems, pp. 52–60. SCITEPRESS (2020)

24. Quirino, R.D., Ribeiro-Júnior, S., Ribeiro, L.A., Martins, W.S.: fgssjoin: A GPU-based algorithm for set similarity joins. In: International Conference on Enterprise Information Systems, pp. 152–161. SCITEPRESS (2017)
25. Ribeiro, L.A., Cuzzocrea, A., Bezerra, K.A.A., do Nascimento, B.H.B.: SJClust: towards a framework for integrating similarity join algorithms and clustering. In: International Conference on Enterprise Information Systems, pp. 75–80. SCITEPRESS (2016)
26. Ribeiro, L.A., Cuzzocrea, A., Bezerra, K.A.A., do Nascimento, B.H.B.: SjClust: a framework for incorporating clustering into set similarity join algorithms. LNCS Trans. Large Scale Data Knowl. Center. Syst. **38**, 89–118 (2018)
27. Ribeiro, L.A., Härder, T.: Generalizing prefix filtering to improve set similarity joins. Inf. Syst. **36**(1), 62–78 (2011)
28. Ribeiro, L.A., Schneider, N.C., de Souza Inácio, A., Wagner, H.M., von Wangenheim, A.: Bridging database applications and declarative similarity matching. J. Inf. Data Manage. **7**(3), 217–232 (2016)
29. Ribeiro-Júnior, S., Quirino, R.D., Ribeiro, L.A., Martins, W.S.: Fast parallel set similarity joins on many-core architectures. J. Inf. Data Manage. **8**(3), 255–270 (2017)
30. Sarawagi, S., Kirpal, A.: Efficient set joins on similarity predicates. In: Proceedings of the ACM SIGMOD International Conference on Management of Data, pp. 743–754 (2004)
31. Shen, Z., Cheema, M.A., Lin, X., Zhang, W., Wang, H.: A generic framework for top-k pairs and top-k objects queries over sliding windows. IEEE Trans. Knowl. Data Eng. **26**(6), 1349–1366 (2014)
32. Sidney, C.F., Mendes, D.S., Ribeiro, L.A., Härder, T.: Performance prediction for set similarity joins. In: Proceedings of the ACM Symposium on Applied Computing, pp. 967–972 (2015)
33. Stonebraker, M., Çetintemel, U., Zdonik, S.B.: The 8 requirements of real-time stream processing. SIGMOD Rec. **34**(4), 42–47 (2005)
34. Vernica, R., Carey, M.J., Li, C.: Efficient parallel set-similarity joins using MapReduce. In: Proceedings of the ACM SIGMOD International Conference on Management of Data, pp. 495–506. ACM (2010)
35. Vitter, J.S.: Random sampling with a reservoir. ACM Trans. Math. Softw. **11**(1), 37–57 (1985)
36. Wang, X., Qin, L., Lin, X., Zhang, Y., Chang, L.: Leveraging set relations in exact set similarity join. Proc. VLDB Endow. **10**(9), 925–936 (2017)
37. Xiao, C., Wang, W., Lin, X., Yu, J.X., Wang, G.: Efficient similarity joins for near-duplicate detection. ACM Trans. Database Syst. **36**(3), 15:1–15:41 (2011)

Flexible OPC UA Data Load Optimizations on the Edge of Production

Johannes Lipp[1,2(✉)], Michael Rath[3], Maximilian Rudack[4], Uwe Vroomen[4], and Andreas Bührig-Polaczek[4]

[1] Fraunhofer Institute for Applied Information Technology FIT, Schloss Birlinghoven, Sankt Augustin, Germany
johannes.lipp@fit.fraunhofer.de

[2] Chair of Databases and Information Systems, RWTH Aachen University, Aachen, Germany
johannes.lipp@rwth-aachen.de

[3] Institute of Information Management in Mechanical Engineering, RWTH Aachen University, Aachen, Germany
michael.rath@ima.rwth-aachen.de

[4] Chair for Comprehensive Foundry Science and Foundry Institute, RWTH Aachen University, Aachen, Germany
{m.rudack,u.vroomen,sekretariat}@gi.rwth-aachen.de

https://fit.fraunhofer.de/, https://cybernetics-lab.de/, https://www.gi.rwth-aachen.de/

Abstract. Recent trends like the (Industrial) Internet of Things and Industry 4.0 lead to highly integrated machines and thus to greater challenges in dealing with data, mostly with respect to its volume and velocity. It is impossible to collect all data available, both at maximum breadth (number of values) and maximum depth (frequency and precision). The goal is to achieve an optimal trade-off between bandwidth utilization versus information transmitted. This requires optimized data collection strategies, which can extensively profit from involving the domain expert's knowledge about the process.

In this paper, we build on our previously presented optimized data load methods, that leverage process-driven data collection. These enable data providers (i) to split their production process into phases, (ii) for each phase to precisely define what data to collect and how and (iiii) to model transitions between phases via a data-driven method. This paper extends the previous approach in both breadth and depth and focuses especially on making its benefits, like the demonstrated 39% savings in bandwidth, to domain experts. We propose a novel, user-friendly assistant that enables domain experts to define, deploy and maintain a flexible data integration pipeline from the edge of production to the cloud.

Keywords: Internet of Production · Industry 4.0 · OPC UA · Data collection

Funded by the Deutsche Forschungsgemeinschaft (DFG, German Research Foundation) under Germany's Excellence Strategy – EXC-2023 Internet of Production – 390621612.

J. Filipe et al. (Eds.): ICEIS 2020, LNBIP 417, pp. 43–61, 2021.
https://doi.org/10.1007/978-3-030-75418-1_3

1 Introduction

The current trends in Industry 4.0, Internet of Production [13] and adjacent research areas are leading to massive connectivity of machines, sensors and systems. Modern protocols such as MQTT [19] and OPC UA [1], which are standardized and openly accessible, allow flexible data acquisition independent of manufacturer specifications. Production machines usually offer numerous nodes via OPC UA, each of which maps a machine value in real-time or even an entire array of them (cf. Sect. 5). Common data collection methods provide two pieces of information, namely what information should be collected and how. For the former, all required values are listed, and for the latter, either fixed time intervals or subscriptions to specific data changes are specified.

The issue with these solutions is that both details are defined a-priori and stay constant throughout the entire data collection. The integration of more and more production machines leads to an uncontrollable increase in network traffic, since detailed analyses of observed processes require a fine granularity and thus a high frequency and precision of the collected data values. A higher collection frequency linearly increases the data load and quickly reaches the limits of available network and processing resources. However, the need for high-frequency data rarely remains valid throughout an entire process cycle, but mostly in short phases only. This means that values are recorded in a high frequency over a long period of time, even though they are only of minor importance in some process phases. To the best of our knowledge, there are no suitable methods that flexibly adjust the data acquisition rates based on the production process currently observed.

A major challenge in practice is, even if such methods did exist, to make these available for domain experts. Both, easy-to-use tool support and intuitive user interfaces are missing. After demonstrating a data saving of 39%, we do not focus on pushing this value, but on bringing them into application and leverage their benefits. In this paper, we build on previously presented concept [4] and implementation of a process-driven data collection, that allows machine operators (i) to split the digital shadow of their production process into phases, (ii) For each phase to precisely define what data to collect and how, (iii) To model transitions between phases via a data-driven scheme and (iv) to integrate the declaration of process phases into common, user-friendly workflows. This data collection scheme considers the actual need of information, instead of static rules and thus reduces the amount of unnecessarily captured data significantly.

This extension of our previous work [4] includes the following major items:

- A more sophisticated network infrastructure setup,
- more details about our implementation and architecture,
- highly important steps to transfer our approach into application, including a Node-RED embedding.

The remainder of this paper is structured as follows. We describe a real-world use case that requires flexible data collection in Sect. 2 and highlight related work in Sect. 3. Section 4 presents our concept for process phases and transitions between them, followed by an implementation with technical results in Sect. 5. The completely new section Sect. 6 proposes a user-friendly integration for domain experts, before we conclude our work in Sect. 7.

2 Problem Statement

This section motivates the two-folded problem statement this paper originates from. The first challenge is the actual use-case, namely high pressure die casting, which we in detail focused in our previous work [4]. Given a the proof-of-concept solution we developed previously, we motivate in Sect. 2.2 the crucial need for usability raised by domain experts, to ensure that deployed solutions will even be used in practice.

2.1 High Pressure Die Casting

The following section describes the HPDC (High Pressure Die Casting) process, that serves as a use-case for the presented work. It is an adaption from our previous paper [4].

The HPDC process is used to manufacture high volumes of geometrically intricate light metal components for various applications, such as automotive structural parts or power train components. It operates in a discontinuous, iterative cycle, where one casting is usually produced in 40–60 s depending on the specific machine and tool combination. For the cold chamber variety of the HPDC process, molten metal is injected into a cavity with a plunger system, solidified under high pressure and consecutively ejected from the machine.

A HPDC production cell does not only include the HPDC machine itself, but also additional auxiliary systems, which are used to enable a highly automated production system that requires an adequate information network [17]. A typical machine consists of the shot end, which injects the metal into the cavity, as well as the locking unit which holds the mold shut during injection. External systems enable the machine to automatically produce parts, such as a molten metal holding furnace, a dosing unit to transfer the metal to the machine, multiple oil- or water-based thermal regulation units, that deliver coolant to the die, a vacuum system that evacuates the cavity shortly before injection, a plunger lubrication system and a spray head to apply a parting agent to the die surface after part ejection. These auxiliary systems are used at different points in time or sub-cycle phases of the overall cycle, when they are needed.

Most sensors provide values continuously throughout the cycle. One could gather all these values by assigning a constant sampling rate independently of the machine's current sub-cycle state. However, the usage of constant sampling rates does impose limitations on data acquisition. The Te duration of the overall cycle and its sub-cycle phases varies either due to adjustments to the process control by the operator or production interruptions caused by unstable process conditions. Furthermore, the different sub-cycles require different sampling rates to capture all important information in the sensor signal. For this reason it is beneficial to detect the transition from one sub-cycle to the consecutive cycle phase, as well as having the capability to set sampling rates depending on the machines current state. For instance, the plunger injection phase can be divided into the slow shot, that typically lasts 750–1500 ms and the fast shot which is concluded after only 25–100 ms. The Programmable Logic Controls (PLC) of real-time controlled cold chamber HPDC machines are designed to replicate the programmed process parameters with minimal fluctuations. Due to wear, operator, or environmentally induced changes of the state of the machine however, irregularities and fluctuations can be introduced. Such condition changes are indirectly reflected in the process data, albeit with a certain degree of ambiguity. By collecting an information density, that is adequate for the machine's state, one can enable the domain expert or mathematical models to have a reasonable chance of properly detecting the root cause of an unstable process.

This can only be realized by transitioning from the constant sampling rate, to a more sophisticated, data-driven approach, that takes sub-cycle phases into account.

2.2 Usability for Domain Experts

Clear separations between the digital and the physical world become increasingly difficult to make, since many of our physical systems are controlled by software. This is especially true within the domain of production technology. However, engineers are typically not familliar with both, software engineering and metallurgy. The rare ones that are, are very sought after and therefore, difficult to hire. Not all problems that can by tackled by deploying software solutions on existing machinery do necessarily require a background in computer scinece. Most engineers are trained to use proprietary software solutions that enable simplified programming of a PLC or acquisition of measurement values of a technical system.

Measurement systems for laboratory use are often highly reconfigurable in terms of what sensor value shall be acquired, e.g.: force, pressure, temperature or velocity. Accordingly, the laboratory software for such systems often is high reconfigurable and based on a Graphical User Interfaces (GUI), that allows to connect logic blocks, which executes a certain action, e.g.: write a value from a measurement channel at a defined frequency. This kind of system most engineers have experience with or can adopt to very quickly. The necessity of writing many lines of computer code to access a measurement value may be a very frustrating experience for the domain expert, who is more interested in dealing with parameters of physical nature, such as the velocity of a piston.

The authors consider the use of open source GUI based visual programming tools a possible solution for this dilemma. These tools lower the barrier of entry for domain experts, that lack sophisticated programming skills, to create data processing solutions. With help of these tools, a user can create "flows" by sticking logic blocks together, that fulfill certain functions. These logic block hide the hard coding work and release the production domain expert from the burden of being an avid user of the command line.

Since modern production machines increasingly adapt open, standardized protocols such as OPC Unified Architecture (OPC UA) compatible blocks for reading OPC UA servers already exist and the domain expert can make use of them with little effort. We therefore identified the necessity to transition the data collection system based upon our custom OPC UA client to such a visual programming solution, to enable wider adoption.

3 Related Work

This related work sections extends the state of the art for OPC UA data collection we presented in [4] with the completely new aspect of usability for domain experts. First of all, we revisit both the environment and the requirements for data collection again, followed by a section dedicated to investigations on visual programming environments in order to increase the usability for users.

3.1 Technical Environment

The technical context of this paper is shown in Fig. 1, which depicts the traditional extract-transform-load (ETL), as well as the more modern extract-load-transform (ELT) process. Solutions with relational data models, like data warehouses, first transform extracted data into their designated schema and then load it [15,16,22]. In contrast, data lakes and other NoSQL systems make data quickly available by first loading it and then in-place transforming it later in time (ELT). The data collection approach we propose however, tackles the OPC UA data collection within the extraction phase and thus is independent of whether an ETL or ELT architecture was chosen, because either one starts with the extraction (E) [4]. Therefore, we focus on related work on the extraction phase, applied on OPC UA in particular, and skip the not directly related load and transform steps in this section.

3.2 Data Collection via OPC UA

Classical data collection using OPC UA [1] either regularly reads data via so-called polling, or uses subscription functionalities, that push changes to the client on demand. Polling follows a simple approach, where the client defines a time interval, that is mostly fixed. Whenever that timer fires, the client actively contacts the server and reads the specified data. An example for this is to collect a certain temperature value every 2000 ms. Depending on the choice for the interval, this approach however, can be very inefficient, because of the following reasons: It can miss value changes, if the interval is too long, e.g. if the polling interval fires before an update and after the next update. For example, a polling at $t = 0$ and $t = 2000$ would be lead to bad results if the value of interest changes at $t = 300$ and returns to its initial value at $t = 1800$. The opposite could happen if the poll frequency is chosen too high and thus repetitively reads a value that stays constant over a long period of time. In the above example, a

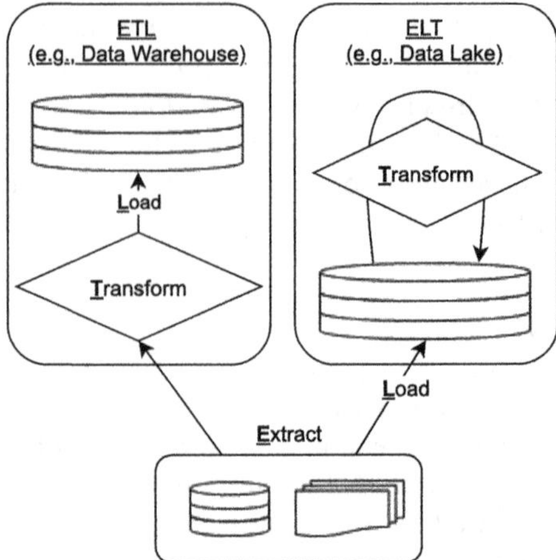

Fig. 1. Common data integration processes as we presented in [4]. Data warehouses typically implement relational models and transform extracted data before loading it (ETL). NoSQL approaches like data lakes load data directly after extraction and later perform in-place transformations (ELT). Note that our approach improves the extraction phase and thus is independent from the ETL/ELT process.

data read every 2000 ms would be create lots of unnecessary data if the value only changes once a month. The latter issue can be tackled by decreasing the polling frequency, which worsens the former, and vice versa.

Besides polling, OPC UA provides a more sophisticated functionality, defined in OPC UA Part 1 [10], which is called subscriptions. A client registers a subscription to OPC UA nodes of interest on the server, which will then monitor these locally. Whenever observed value changes exceed a given threshold (absolute or relative), the server transmits these updates to the client, which solves both disadvantages mentioned above. This functionality of controlling which value changes should be reported is called filters and was introduced in OPC UA Part 4 [11].

OPC OA Part 14 [12] introduced *PubSub* as an even more sophisticated approach, which is short for publish subscribe. Its design builds on a pattern, which allows any publisher distribute data to any interested subscriber via a central broker. In both the previous and this paper, we unfortunately could not evaluate this OPC UA PubSub feature in our real-world production use-case, because the available machine does not have that OPC UA standard implemented yet. Since this is a strongly growing area in industry, we will surely focus on this in future work, as soon as that implementation is available on our machine.

Please note that the PubSub mechanism available in OPC UA properly controls, when to transmit updates to clients, but does not support flexible adoptions on demand. Keeping track of the current state of the production process and updating the subscription rules based on this still to be done by the client. This leaves the task of flexibly minimizing the data load up to the clients and thus does not fully leverage its potential.

3.3 Visual Programming Environments

Section 2.2 demonstrates the importance of visual toosl to support bridging the gap between software technologies and domain experts in production. This section briefly evaluates the state of the art for visual programming environments, which are required to achieve a broad acceptance and application of the process-driven data collection we propose previously in [4]. Dataflow programming started as a topic of research in the 1970s and Johnston et al. in [3] reviewed its evolution and subsequently discuss the trend towards visual programming languages. Note that this paper appeared in 2004, and multiple approaches and implementations were developed in the meantime. The high popularity among people other than pure software developers remains high and might even increase due to the recent trends in Industry 4.0 and Internet of Production, which require domain experts to handle data in databases and Digital Shadows [13]. We present some popular frameworks and tools from this area in the following and will subsequently select the one that suits best for further design and implementation in the scope of this paper.

N8n.io. The free and open workflow automation tool n8n.io (nodemation) [7] was developed in 2019 by the n8n GmbH. Its goal is to provide an easy-to-use and highly customizable way to connect APIs though a intuitive web interface. This tool is written in Typescript.

Microsoft Power Automate. This proprietary tool started as Microsoft Flow in 2016 and was renamed to Microsoft Power Automate in November 2019 [6]. It supports users to automate their rule-based business and office processes.

OPC Router. Developed and marketed by inray Industriesoftware GmbH, "the OPC Router is a middle-software, that connects databases and different sources and thus drives the idea of IoT and Industry 4.0." [2] It is a visual tool that enables control of data flow between components like OPC UA, databases, Manufacturing Execution Systems (MES), Supervisory Control and Data Acquisition (SCADA) Systems and other parts in industrial automation.

Apache SteamPipes. This Java-based project was developed by the Forschungszentrum Informatik in Karlsruhe, Germany. It is currently listed as an incubating project at the Apache Software Foundation [20] and focues on real time data collection in Industrial Internet of Things (IIoT) systems. It is designed to integrate well with industry standards like OPC UA, as well as with commonly used toolboxes of the Apache ecosystem.

Node-RED. Node-RED [9] is a browser-based visual programming environment, initially developed by IBM, which is now maintained as an open-source tool by the JS Foundation. As a visual programming tool, data processing in Node-RED is modelled by dragging and dropping of components onto a canvas.

Node-RED is flow based. Data sources, data processing units and data sinks are represented by nodes that have input and output connectors. A node's connectors are connected by drawing lines, so that messages are passed from the output of one node to the input of another node. Within nodes, messages containing data can be arbitrarily processed. A large variety of predefined nodes, as well as complete flows, are available within the Node-RED software package and user generated nodes are available through the Node-RED Library.

Node-RED has previously been used in multiple Industrial Internet of Things use-cases. Toc et al. [21] applied a Node-RED driven middleware to the development of a Modbus-OPC UA Wrapper for the use in water industries. To analyze machine health metrics, Newman et al. [8] used Node-RED on embedded linux computer to monitor OPC UA devices, while Sosa et al. [18] studied the performance of offloading on Raspberry Pis running Node-RED.

Tool Selection. We decided for Node-RED as our tool of choice, because of its mature state of development, the extensive community and ease of use. The availability of a detailed documentation makes Node-RED a well suited visual development environment.

4 Process Phases Design

This section revisits the design we developed and presented in [4], without adding major updates to it. Our concept includes modeling process phases with transitions, data collection rules, and start/end conditions of phases. Listing 1.1 shows an extract of the general structure, which we briefly present below. A process phase describes one or more conditions for it to start and end, transitions to and from other process phases, and what data the client should collect while the process is in that phase. A production process as whole can start with one or more phases, and can be cyclic if the start condition for the first phase includes the end of the last phase, respectively.

```
"phases":[
  {
    "name":"Phase 1",
    "whatToCollect":[...],
    "howToCollect":[...],
    "startConditions":[...],
    "endConditions":[...]
  },
  [...]
]
```

Listing 1.1. JSON extract that shows a process phase on a high-level view. Each phase has a name, what to collect and how during that phase, and start/end conditions that model transmissions between different phases. This design is a result of our previous paper [4].

4.1 Key Elements of Process Phases

As presented in [4], process phases in our context represent phases of a real-world production process, and multiple ones can even run in parallel. The level of detail of the process phases depends on the view of the data scientist, and thus can be specified accordingly. We defined a process phase as a collection of:

1. A name and/or description,
2. A list with data of interest (what),
3. Rules for acquiring the data (how),
4. A start condition, and
5. An end condition.

Start and end conditions in our current implementation cover the observation of values (boolean, integer, double etc.) via polling or subscriptions.

4.2 Process Phases and Transitions

Phase-driven data collection, as introduced in [4], defines a strategy of what data to acquire while a particular phase is active and how to do it. Please note that we focus on OPC UA in both the previous and this paper, but our design can be extended to also support other data sources like databases or arbitrary sensors. The supported data collection strategies, which can be applied to each phase, include:

1. Interval-based (polling): Read values based on a fixed time interval, e.g. 500 ms,
2. Subscription-based (absolute): The current value is fixed and data collection is triggered whenever the current value exceeds the given absolute deadband value, e.g. when the temperature changes by more than 2 °C,
3. Subscription-based (relative): Same rule as for absolute, but the change calculation is relative instead of absolute.

These rules have been implemented in [4] and are ready for further extensions such as more complex or flexible applications.

The previous paper included start and end conditions based on values as well as the state of other phases, i.e. "start a phase once another phase ends". Combining these rules allows to traverse through all phases linearly, in parallel, or both. We also motivated possible extensions like more complex rules (cf. to a process diagram or a Petri net [14]) to model rules like "start phase k once phases l,m and n have ended" in [4].

5 Technical Results

This section presents preliminary results of our approach. We model process phases including the above-mentioned required details for a use-case at the Foundry Institute of the RWTH Aachen University. This HPDC use-case is described in detail in Sect. 2 and the machine offers data via an OPC UA server, which we collect in the process phases as depicted in Fig. 3 using our approach proposed in Sect. 4 and finally discuss preliminary benefits.

5.1 Architecture Setup

We use a horizontal cold chamber HPDC machine (cf. Sect. 2) with an embedded OPC UA server, which we access via an Ethernet cable. A Java-based client installed on an edge device connects to the machine using OPC UA, and is capable of collecting, processing and storing collected data. We create a config for process-driven data collection (cf. Sect. 4) and load it on the edge device. At the edge device, we perform precise measurements during one work shift to compare our approach with a fixed collection baseline.

The HPDC machine comes with an embedded OPC UA server that offers indirect access to values form the PLC as well as additional sensors. A Raspberry Pi 4 serves as edge device. It is equipped with a Broadcom quad core 64-bit SoC @ 1.5 GHz, 4 GB SDRAM, Ethernet connectivity, and a 32 GB SD card. This edge device in embedded in the network architecture depicted in Fig. 2, which we present in the following.

As shown in Fig. 2, our network architecture consists of three zones. The production network resides in the manufacturing hall itself and is securely connected to the office network through a firewall. The office network allows user access from workstations, and has an uplink to our Kubernetes cluster located in a self-managed cloud infrastructure. We explain the three individual network zones and their purpose in detail in the following.

The heart of the production network is the HPDC machine itself, together with its components. It consists of the PLC that controls the machine and has all important data values for that machine. We can access these valuable information via two ways. First, the human machine interaction (HMI) panel, which provides exploratory access for pre-configured values or setups. This is mainly used by the machine operator for supervision purposes, and it keeps few to no historical values for later use. The embedded OPC UA server provides a second way to access the data, and can be freely configured in collaboration with the machine manufacturer. We access these values through a signed and encrypted connection, which includes username+password authentication as well as signed X.509 certificates. A firewall additionally secures any access from within the office network.

The office network depicted on top of Fig. 2 bridges control and data flow from the production network to the data sink, which in our case is a Kubernetes cluster located in a self-managed cloud. It consists of two main entities, a standalone

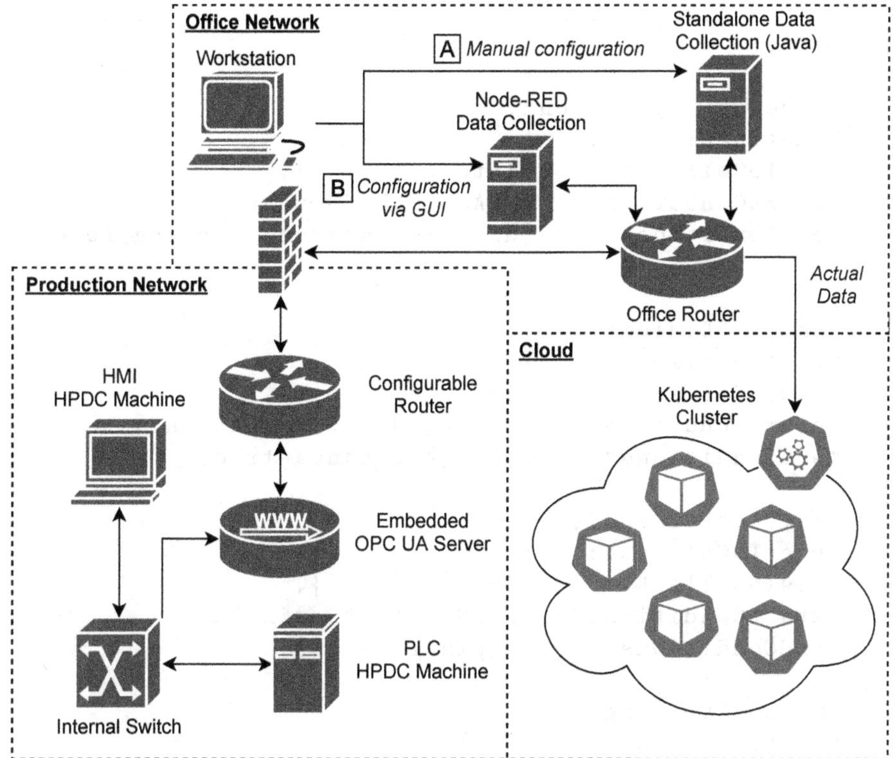

Fig. 2. The HPDC machine is located in a Production Network. The data collection devices access this machine's data form a Office Network and forward selected data to the Cloud.

data collection (A) as we presented in our previous work [4] as well as a novel data collection (B) configurable via Node-RED.

Both options are controlled from a workstation by a domain expert. Our first version (A) requires a manual configuration, where the domain expert constructs JSON configuration files and copies these to the edge deice, cf. Sect. 4 and Sect. 5.2. In this paper, we introduce the novel option (B), which offers a user-friendly configuration via a GUI for the Node-RED framework, cf. Sect. 2.2, Sect. 3.3, and particularly our solution in Sect. 6.

5.2 HPDC Process Phases

The HPDC process was analyzed with regard to the necessary requirements for data extraction during multiple cycle phases. This serves the purpose of only acquiring data when it is relevant. Process phases that occur during the HPDC cycle, including transitions are depicted in Fig. 3, and seperate the HPDC process in five distinct process phases. The collection intervals and other practical steps are presented as a JSON config in Listing 1.2. This section is an adaption from our previous paper [4].

```
"phases":[
  {
    "name":"Preparation",
    "whatToCollect":    [...],
    "howToCollect":     {interval:4000},
    "startConditions": [PHASE_END:"Removal"],
    "endConditions":    [die_closed=true, locking_force>
      400]
  },
    "name":"Dosing",
    "whatToCollect":    [...],
    "howToCollect":     {interval:200},
    "startConditions": [PHASE_END:"Preparation"],
    "endConditions":    [shot_clearance=true]
  },
    "name":"Filling",
    "whatToCollect":    [...],
    "howToCollect":     {interval:200},
    "startConditions": [PHASE_END:"Dosing"],
    "endConditions":    [x_plunger>500]
  },
    "name":"Cooling",
    "whatToCollect":    [...],
    "howToCollect":     {interval:200},
    "startConditions": [PHASE_END:"Filling"],
    "endConditions":    [cooling_completed=true]
  },
    "name":"Removal",
    "whatToCollect":    [...],
    "howToCollect":     {interval:4000},
    "startConditions": [PHASE_END:"Cooling"],
    "endConditions":    [ejected=true]
  },
]
```

Listing 1.2. Reformatted JSON Config file from the results presented in [4]. It defines the five process phases shown in Fig. 3 and their transitions. In the use-case we tackled previously, the collection intervals differ by a factor 20 from 200 to 4000 ms. This config snippet avoids listing any `whatToCollect` OPC UA node ids due to better readability.

Preparation of the machine is the initial process phase, in which mainly the spraying unit and the ejection system are active while most other systems are on hold until completion of preparation. A data collection frequency of 4000 ms for selected temperature values is sufficient (cf. Listing 1.2). Temperature values are read from the OPC UA server's nodes, the OPC UA identifiers were omitted

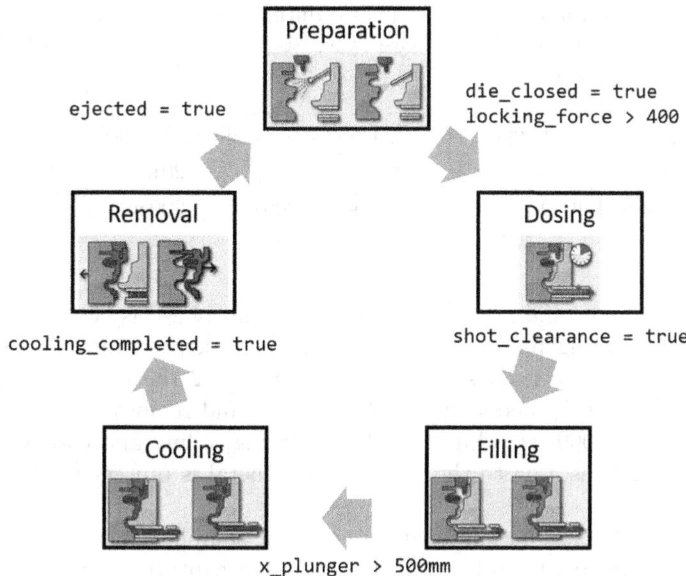

Fig. 3. The five process phases of the HPDC process and their transition conditions as developed with the HPDC process owner in [4]. The process starts with preparation of the machine, which is followed by dosing and filling. After cooling, the workpiece is ejected and the process starts over. Images are extracted from Magmasoft 5.4 [5].

for improved clearness of the chosen configuration. In this case the end of the preparation phase is detected by two conditions which are fulfilled, the die is closed and the locking force is greater than 400 metric tons. The collection interval during the dosing phase is significantly reduced since the temperature in the shot sleeve is acquired. The phase ends when the dosing unit clears the the machine to let the shot commence. This high data collection frequency is held throughout the next two phases filling and cooling. A transition from filling to cooling happens as soon as the plunger closes in on its terminal position after cavity filling (`x_plunger` $>$ 500). The HPDC machine sets a flag, the time is defined by the operator, when the cooling is completed, which triggers the die to open and eject the part. This concludes the HPDC cycle and the process starts over with the preparation cycle. We achieve this functionality by adding an additional start condition to the preparation phase (cf. Listing 1.2).

5.3 Results

The results from the experimentally acquired process values stem from the machine trials. The amount of data that has to be passed through the edge device is investigated, OPC UA control messages are ignored due to their infrequentness. The focus is laid on the data flow itself. Table 1 gives the

Table 1. Average duration of process phases during the experiments as measured in [4]. The sampling rates for that use-case were also suggested by the HPDC process owner in that pervious paper.

	(1) Preparation	(2) Dosing	(3) Filling	(4) Cooling	(5) Removal
Average duration	15 s	6.5 s	1.2 s	20 s	12 s
Collection interval	4000 ms	200 ms	200 ms	200 ms	4000 ms

duration of each specific process phase. The filling phase is concluded after a mere 1.2 s, the dosing phase takes 6.5 s and removal, preparation and cooling require two digit time frames in seconds. In Sect. 5.2 the collection intervalls were specified by the process owner. Preparation and removal only require one data value every 4000 ms, while the phases dosing, filling and cooling demand an interval of 200 ms, due to the fact that the metal is present in the machine during these times.

The data collection baseline with a constant sampling interval reduces the flexibility with regard to individualizing the data capturing during the different cycle phases. Due to quick changes of temperature during doing, filling and solidification at the measurement point a high sampling rate of one value per 200 ms is desired.

The cycle time for the used HPDC process is the sum of all averages in Table 1, which is around 54.7 s for a regular cycle. Setting a constant non adjustable interval of 200 ms leads to 238 (rounded up) data messages for one full production cycle for that specific thermocouple.

The proposed process-driven data collection flexibly changes the collection interval in order to realize a sufficient information density for the cycle phases. That leads to (all rounded up) 4 messages for the preparation phase, 33 for dosing, 6 for filling, 100 for cooling and 3 for removal. This is a total of 146 messages for the complete process cycle, which is about 61% of the baseline and thus saves about 39% bandwidth.

6 User-Friendly Integration for Domain Experts

The main intent of using a process phase based data transmission is to integrate process domain knowledge into data acquisition to optimize the amount of data to be transferred. Domain experts, having deep insight into the technical processes to be monitored, usually lack the knowledge about how to program and maintain the data collection devices, that they need to configure according to the processes of interest. To facilitate the configuration of the data collection device, we implemented the domain driven data collection as a custom node within the visual programming environment Node-RED and plan to make it available to the public after a thorough testing phase.

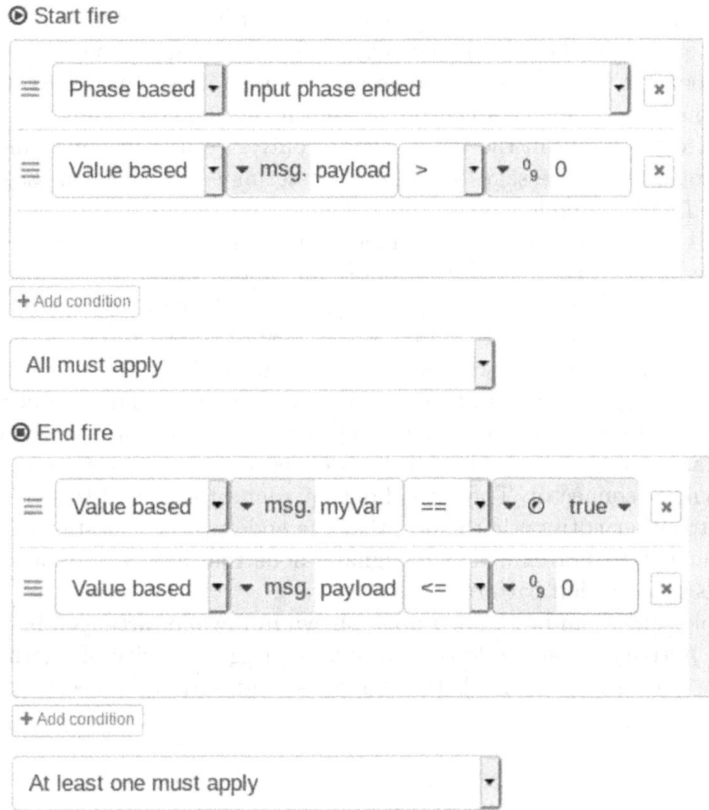

Fig. 4. Configuration window of the start and end conditions of a process phase. In this example, the phase starts as soon as the previous phase has sent an end-signal and the attribute named `payload` of the incoming message is greater than zero. It ends when the variable `myVar` equals `true` and the `payload` of the incoming message is less than or equal to zero.

Fire type Interval

Interval [ms] 4000

Fig. 5. Configuration window of the fire type. In this example, the phase node forwards a message every 4 s.

6.1 The Process Phase Node

To make the process phase-based data acquisition accessible to the domain experts, we realized it's functionality as a custom node within the Node-RED framework. One node represents hereby a single process phase, that acts as a forwarding mechanism on incoming messages according to defined rules. Those rules can be configured by the domain expert through an input mask. The expert defines, when the node should become active and start sending data, when to stop sending it and how (frequently) to send it. In contrast, what needs to be sent is not defined within the node, since it only forwards incoming messages. The content of these messages is therefore determined by those messages that are injected into the node.

Figure 4 depicts the before mentioned input masks in which an arbitrary number of rules can be defined, that determine under which conditions a phase starts and under which it ends, respectively. In accordance with the definition of process phases in Sect. 4, a user also chooses how to collect the data: in a fixed interval or as one of two subscription variants (cf. Fig. 5).

Each node, representing a process phase, has one input and two output connectors. The input connector receives messages from the previous node, the messages are processed according to the defined rules and are then output to the first output connector. The second output connector is used to send specific messages to either notify other nodes, that the node's phase has started (i.e. the node became active) or ended, or to signal a node that acts as a data source to change it's subscription pattern.

Multiple Nodes can be chained up as shown in Fig. 6 or arranged in complex networks. Activity of one node is then able to trigger activity of another node by sending start and end signals through its second output connector.

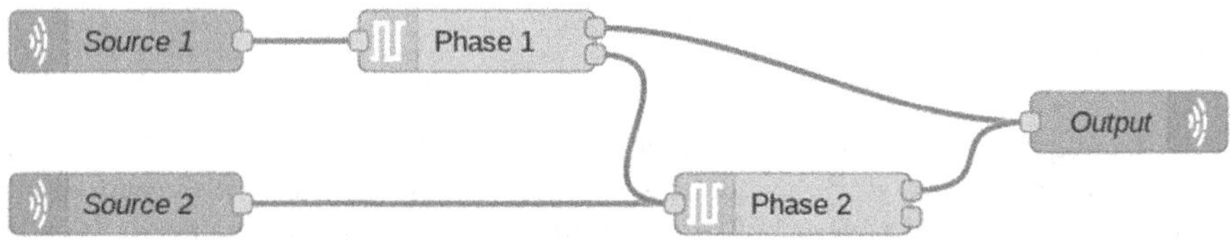

Fig. 6. Two nodes named Phase 1 and Phase 2 represent process phases. They forward incoming messages from data sources according to predefined rules. The link between Phase 1 and Phase 2 can be used to trigger activity of Phase 2 depending on the behavior of Phase 1 and Source 2.

6.2 The HPDC Flow

A data pipeline that realizes data collection of the HPDC process using the visual programming environment Node-RED consists of three parts: A OPC UA client as a data source, our process phase node and a data sink of choice.

Publicly available nodes, that implement the functionality of OPC UA clients[1], are used as data sources. These nodes connect and subscribe to the HPDC Machine's OPC UA server, generate messages containing machine data

[1] https://flows.nodered.org/node/node-red-contrib-opcua.

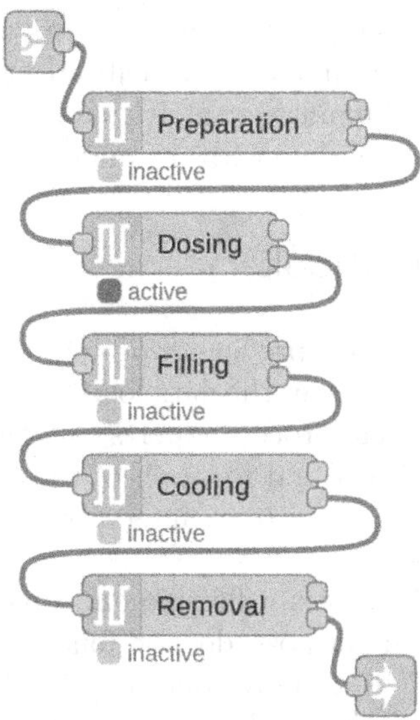

Fig. 7. The HPDC process modeled as a Node-RED flow. The two nodes labeled with grey arrows are "Link in/out" nodes, which are connected to each other to close the circle. The actual data connections to OPC UA sources and outputs are neglected for clarity in this image.

and inject those messages into Node-RED's flow. Which OPC UA endpoints to subscribe to or at what frequency to query them, can be modified by the messages injected into the OPC UA node.

Using our process phase node defined in the previous section, we modeled the HPDC process with it's five phases: preparation, dosing, filling, cooling and removal. Each phase is represented by a separate node. To model the cyclical nature of the process, output two (the one sending activity information) of a node is connected to the input of the node that represents its subsequent phase. The resulting network of nodes is depicted in Fig. 7. Messages generated by OPC UA nodes are injected into the respective process phase node, where they are either blocked or forwarded, depending on the activity of the phase.

As the third component of our flow, the data sink, we use an AMQP node[2]. This node is connected to all process phase nodes. Its purpose is to send incoming messages to the cloud, where it is stored in databases running on an on-premises Kubernetes cluster.

The whole flow consisting of sources, phase nodes and sinks is deployed to an installation of Node-RED running on a Rasperry Pi 4. In Fig. 2, this device is indicated as "Node-RED data collection". The flow can be viewed and modified by the domain expert through a web interface from anywhere within the office network, without any additional network configuration or programming efforts.

[2] https://flows.nodered.org/node/node-red-contrib-amqp.

This presented approach supports domain experts with setup and configuration of machine data collection pipelines. It facilitates the definition of processes phases, their deployment and maintenance. It enables engineers to easily modify the pipelines to keep them up to date with the underlying machine configuration.

7 Conclusion and Outlook

This paper is an extension to our previous work [4], which presented a process-driven data collection that optimizes the data load by flexibly adapting collection frequencies. With our approach, process experts can split the real-world production process into process phases and for each of these define (i) What data to collect, (ii) How to collect the data (e.g., interval), and (iii) Transitions to and from other phases.

The work we presented enables a flexible and precise control of what data to collect and when. We on purpose do not follow data lake paradigms like "collect first, then analyze", but apply sophisticated data saving techniques on the edge, as we found that bandwidth bottlenecks commonly occur in production networks.

We added notable contributions in both depth and breadth, which we group in the three areas infrastructure, implementation, and user-friendly application. We extended our network infrastructure by new components, and presented its sophisticated setup in detail. Our presentation of both problem statement and concrete implementation steps contains new insights and is more comprehensive.

Novel contents include a user-friendly approach to actually bring our process-driven data collection into practice. We motivated why user-friendly tools are especially required for non-experts and integrated both design and implementation of a comfortable Node-RED setup into our approach. This allows us to actually transform our bandwidth saving achievements of around 39% from experimental test into real-world applications.

Our work is an important step towards handling the ever growing number of data sources in the Internet of Production, where simple "collect, then analyze" solutions are not sufficient. It is particularly strong in scenarios where even small changes during the process have extensive effects. We collaborate with process experts in order to focus on important phases of the production process and to reduce the limited data load. It is important that they are able to define, adjust, and maintain the defined data flows even without us technical experts. Therefore, we proposed a user-friendly solution via Node-RED that allows a graphical configuration and thus enables it for practical use by domain experts.

Future work includes the implementation of more complex use-cases, extended evaluations, user-experience studies, and publish our implementation open-source to motivate community collaboration. For this concrete use-case in the research project *Internet of Production*, we also plan an even more comprehensive data collection by integration three additional OPC UA servers, which would include more data about the PLC, an extended PLC with additional sensors, and details about the dedicated dosing unit.

References

1. Hannelius, T., Salmenpera, M., Kuikka, S.: Roadmap to adopting OPC UA. In: 2008 6th IEEE International Conference on Industrial Informatics, pp. 756–761. IEEE (2008)
2. inray Industriesoftware GmbH: OPC Router. https://opcfoundation.org/products/view/opc-router. Accessed 12 Aug 2020
3. Johnston, W.M., Hanna, J.P., Millar, R.J.: Advances in dataflow programming languages. ACM Comput. Surv. (CSUR) **36**(1), 1–34 (2004)
4. Lipp, J., Rudack, M., Vroomen, U., Bührig-Polaczek, A.: When to collect what? Optimizing data load via process-driven data collection. In: Proceedings of the 22nd International Conference on Enterprise Information Systems - Volume 1: ICEIS, pp. 220–225. INSTICC, SciTePress (2020). https://doi.org/10.5220/0009439502200225
5. MAGMA Gießereitechnologie GmbH: MAGMASOFT autonomous engineering (2019). https://www.magmasoft.com/en/solutions/magmasoft/. Accessed 02 Feb 2020
6. Microsoft: microsoft power automate (2016/2019). https://flow.microsoft.com/. Accessed 13 Aug 2020
7. n8n GmbH: n8n.io - free and open workflow automation tool (2019). https://n8n.io/. Accessed 13 Aug 2020
8. Newman, D., Parto, M., Saleeby, K., Kurfess, T., Dugenske, A.: Development of a digital architecture for distributed CNC machine health monitoring (2019)
9. O'Leary, N., Conway-Jones, D.: IBM emerging technologies, now part of the JS foundation: node-RED (2013). https://nodered.org/. Accessed 24 Jul 2020
10. OPC foundation: OPC UA Part 1: overview and concepts (2017). https://reference.opcfoundation.org/v104/Core/docs/Part1/. Accessed 06 Dec 2019
11. OPC foundation: OPC UA part 4: services (2017). https://reference.opcfoundation.org/v104/Core/docs/Part4/. Accessed 06 Dec 2019
12. OPC foundation: OPC UA part 14: PubSub (2018). https://reference.opcfoundation.org/v104/Core/docs/Part14/. Accessed 06 Dec 2019
13. Pennekamp, J., et al.: Towards an infrastructure enabling the internet of production. In: 2019 IEEE International Conference on Industrial Cyber Physical Systems (ICPS), pp. 31–37, May 2019. https://doi.org/10.1109/ICPHYS.2019.8780276
14. Peterson, J.L.: Petri nets. ACM Comput. Surv. (CSUR) **9**(3), 223–252 (1977)
15. Quix, C.: Metadata management for quality-oriented information logistics in data warehouse systems. Ph.D. thesis, RWTH Aachen University (2003)
16. Quix, C., Hai, R., Vatov, I.: Metadata extraction and management in data lakes with gemms. Complex Syst. Inf. Model. Quart. **9**(9), 67–83 (2016)
17. Rix, M., Kujat, B., Meisen, T., Jeschke, S.: An agile information processing framework for high pressure die casting applications in modern manufacturing systems. Procedia CIRP **41**, 1084–1089 (2016)
18. Sosa, R., Kiraly, C., Rodriguez, J.D.P.: Offloading execution from edge to cloud: a dynamic node-red based approach. In: 2018 IEEE International Conference on Cloud Computing Technology and Science (CloudCom), pp. 149–152. IEEE (2018)
19. Stanford-Clark, A., Hunkeler, U.: MQ telemetry transport (MQTT) (1999). http://mqtt.org/. Accessed 06 Dec 2019
20. The apache software foundation: apache StreamPipes (incubating). https://streampipes.apache.org/. Accessed 17 Aug 2020
21. Toc, S.I., Korodi, A.: Modbus-OPC UA wrapper using node-red and IoT-2040 with application in the water industry. In: 2018 IEEE 16th International Symposium on Intelligent Systems and Informatics (SISY), pp. 000099–000104. IEEE (2018)
22. Vassiliadis, P., Simitsis, A.: Extraction, transformation, and loading. Encycl. Database Syst. **10** (2009)

Using Image Mining Techniques from a Business Process Perspective

Myriel Fichtner[1(✉)], Stefan Schönig[2(✉)], and Stefan Jablonski[1(✉)]

[1] University of Bayreuth, Bayreuth, Germany
{myriel.fichtner,stefan.jablonski}@uni-bayreuth.de
[2] University of Regensburg, Regensburg, Germany
stefan.schoenig@ur.de

Abstract. Business process modeling is an established method to improve business procedures and to provide more insights into internal workflows. Once the process is visualized in a business process model, future process executions correspond to the workflow prescribed by the process model. Process details like input specifications or the order of internal sub-steps are only considered during process execution if contained in the process model. These details may be decisive since they can have an impact on the success of the overall process. In some cases, such important process details are not modeled due to different aspects, like modeling with a high degree of abstraction to preserve the traceability. Nevertheless, it is necessary to identify missing but essential process details that reduce the success rate of a process. In this paper, we present a conceptual approach to use image mining techniques in order to analyze and extract process details from image data recorded during process executions. We propose to redesign business process models considering the analysis results to ensure successful process executions. We discuss different requirements regarding the image analysis output and present an exemplary prototype.

Keywords: Image mining · Process model enhancement · Process model extension · Process detail identification · Quality control · Recommendation system · Process redesign

1 Introduction

1.1 Motivation

Business process models are an established and useful method to visualize workflows in industrial as well as in small and medium-sized enterprises. Today there exist various process modeling languages meeting different requirements and leading to different process models. But in complex cases, all of them might lead to large and hardly traceable process models, containing hundreds of modeling elements [19]. Besides, though there exists modeling recommendations regarding general aspects like correctness or comparability [2], there is no concrete rule or

J. Filipe et al. (Eds.): ICEIS 2020, LNBIP 417, pp. 62–83, 2021.
https://doi.org/10.1007/978-3-030-75418-1_4

guidance how detailed a process has to be modelled. Thus, it is quite obvious that processes are often modelled in an abstract way to keep aspects like clarity and traceability. This procedure is described as process model abstraction and is an important subject in the discipline of business process management [26]. Although this approach contributes to an improvement of large process model understanding, this high level of abstraction leads to the fact that detailed information of process steps are omitted [3,20]. In other cases, process modellers are prevented to add more detailed (sub-)steps due to missing knowledge about process details. We want to give two examples that explain these thoughts.

In a first case, diverse liquids have to be filled into a casting mold. In the process model the process step is described as "add all ingredients". The description of this task is kept abstract because the process modelers want to keep it simple or because they have no further knowledge about this process step. Looking into process details reveals that three ingredients have to be added. So the process could also be refined into three process (sub-)steps "add blue liquid", "add red liquid", and "add green liquid". Since these details are not explicity modeled, process participants are free to choose the execution order of these low-level process steps. This leads to an excessive flexibility in the execution step. In some other cases, process modelers are not aware of the optimal execution sequence of process steps. For instance, they allow that two steps are executed independent from each other, i.e. in arbitrary order. However, the execution sequence does have an impact on process performance, regarding the execution time and/or the quality of process outcomes. Following the example above, the addition of the red liquid as second step instead of third step results in a mixture with reduced binding capability, i.e. this execution sequence is finally not desired. This affects the overall process success since the result of any subsequent task depends on the quality of the mixture.

In a second case, machine parts have to be disposed on a pallet. Also in this example, we assume that due to missing knowledge or due to keeping the process description simple, the process designers give no further information considering details of the placing task. Furthermore, some tasks are hard to describe or important details can not be described with established process modeling languages. In this palletisation scenario this may include (i) information considering the position of objects in the product environment, e.g. exact numerical data (ii) information that is related to behavioral patterns, e.g. special movements. If even expressible by a modeling language, such information can be integrated in a process model inefficiently. For example through a large number of process modeling elements. Especially if multiple options or branchings are allowed, the number of modeling elements strongly increases what contradicts the overall goal to keep business process models traceable through preserving a certain level of abstraction. Although there are approaches that deal with that challenge, e.g. [12,19,29], they assume that knowledge about the correlation between this details and the process success already exists.

To enhance process executions and to ensure process success, necessary details of activities that are not yet contained in the business process model have to be identified. Although quality control and process monitoring are popular topics in research, existing approaches (e.g. [13,21]) focus on the identification of deviations in process results but are not able to discover missing process model details.

1.2 Problem Statements

We conclude the observations and partially already adressed problems above in a list of problem statements. Each point highlights an aspect why relevant process details could miss in process models.

Problem Statement 1: To avoid overload and preserve readability, a certain degree of abstraction is kept in process models, e.g. [2,3,19,20]. As consequence, process details are omitted and possibly important information which ensures process success is missing.

Problem Statement 2: Established process modeling language are restricted in their expressiveness. Although there are requirements that process modeling languages should meet, like providing different perspectives (cf. [9]), only a few approaches enable the integration of complex information content, e.g. [29]. Some types of process details can not be considered by existing modeling approaches or it is not defined how to integrate them.

Problem Statement 3: The optimal process and important process details are not always known. Especially in complex processes, even a capable business analyst is not able to identify all improvement levers [16]. In a worst case, the optimal process is never identified and the overall process success (cost, time, quality) is reduced permanently. One possibility to tackle this problem provides the work of [23] which enables process model design with parameter uncertainty. A more straight approach would be to analyze optimal processes automatically by considering additional data sources.

Problem Statement 4: Existing business process optimization approaches which could identify missing details are restricted. The central question of these methods is to optimize the process model according to different aspects, e.g. performance [1,10] or readability [3,26]. Therefore the approaches often rely on existing process models, static models or artificially generated data. To achieve an overall process optimization, it is necessary to more focus on the analysis of the execution of a task when to analyze the existing process model. In most cases, the success inhibiting details of a process are not contained in the process model but have to be identified first from process executions.

If important process details are missing in a process model due to one of the problems presented above, enterprises could make the following observation: The process results of process model executions differ from each other although the same unchanged process model was executed in each case. At this point, it is obvious that important process details are missing and that analysis techniques to identify these details are necessary. Furthermore, additional data sources which contain also complex information about the process have to be taken into account.

In a previous paper [7], we suggest to apply image mining techniques in the process management context to enhance business process models and to ensure process success. To be more detailed, our concept shows a new approach how image data recorded during activity executions can be anaylzed to reach process success.

As extension to our previous work, this paper gives more insights to the image analysis step which is the most challenging part of our conceptual approach. We explain more detailed how image mining techniques can be used in our concept. For this purpose we define requirements the image analysis results have to meet to be integrated in an existing process model. We present different types of output data and discuss their differences regarding the requirements. Furthermore, we present an implementation which serves as small protoype in order to show an example of how image mining techniques can be used in a simple use case scenario to reach overall process improvement. Concluding we discuss how our concept tackles the newly defined problems presented above in Sect. 4.

2 Background and Related Work

An appropriate overview that points out important background necessary for our overall concept is already given in our previous paper [7]. We reproduce these insights to provide a self-contained work.

In general, images can be understood as complex data collection. Depending on the context in which they are created, the knowledge about this context and other associations, images may contain meaningful information if analyzed and interpreted correctly. How to achieve an effective extraction of this information is the research question in work that is related to image mining. According to [31], *image mining deals with the extraction of implicit knowledge, image data relationship, or other patterns not explicitly stored in the image databases.* Among others, different methods from computer vision, data mining and machine learning are used to process low-level pixel representations contained in raw images or image sequences in order to identify high-level spatial objects and relationships. The overall image mining process is well described in [15]. Summarized, the process can be divided into three parts:

Pre-processing: In order to reduce the cost of the analysis step which can be high in time and space, images have to be preprocessed. Therefore unnecessary or unrelated data is cleaned up and quality reductions due to noise are eliminated through filtering operations. This step may include image-thresholding, border-tracing an wavelet-based segmentation.

Feature Extraction: Algorithms are used to detect features such as shapes, edges or other basic elements in the images. Therefore the image content is reduced while unimportant features can easily be discarded. A promising feature extraction approach can be found in [8], where a combination of the features color, edge and texture is suggested.

Image Mining Technology: Image mining techniques are used on the extracted feature vectors to reveal, evaluate and explain high-level knowledge. Several methods have been developped which realize this procedure in different ways: Image classification, clustering, indexing and retrieval, object recognition, association rule mining and approaches that work with neural networks.

The techniques are used in many different real-world applications, like for example the analysis of paths and trends of forest fires over years in satellite images in order to enable firefighters to fight fire more effectively [32]. Another work uses images gathered from the Web for learning of a generic image classification system and enables a Web image mining approach for generic image classification [30]. Image mining was introduced by [17] as new approach for data mining. The fundamental concepts of discovering knowledge from data stored in relational databases are transfered to image databases.

Related to this idea and the fact that process mining builds on data mining, also the association of image mining with process mining techniques or rather the application of image mining in the context of business process management is reasonable. However, the application of image mining techniques or in general the integration of images in this field is not yet fully explored. The work of [25,29] suggest two different approaches to introduce images and related mining techniques in business process management. In the thesis of [29], the business process modeling language BPMN is extended to a more intuitive modeling language which allows to annotate tasks with multimedial content like images or videos. This approach enables to add non-formalized descriptions to a process task and enriches the process model with additional information. During the execution of this task, process participants can follow the referenced execution in the video. The work of [25] confirms the potential of image mining for business process management. The process management lifecycle according to [6] is presented and the application and intregation of images and suitable image mining techniques for each phase are discussed. This approach focuses on image data which is created in each phase. The authors differentiate between documents, drawings and pictures, while documents contain textual information and are analyzed with optical character recognition methods. Drawings and pictures are analyzed by using one of the image mining techniques as described above. Furthermore, the authors present a protoype for object recognition of business process models which detects modeling elements like gateways, activities etc. from images.

Altogether, both authors explain how images contribute to support the overall process. Particulary interesting is the suggestion of [25] to analyze pictures, taken with phones or tablets, which contain information of the production environment. Covering the process monitoring and controlling step, these images are analyzed subsequently in order to find possible process improvements. This proposal corresponds with our approach, while we share the idea that any issues that may reduce the overall process success could revealed through monitoring the process execution.

In contrast to the work of [25], we suggest a concrete approach that aims to process improvement through an overall system that solely bases on the image data that is produced in the process execution step (cf. Sect. 3.1). Our system is restricted to images or videos which contain real information of the production environment or capture actual states of a product. We go further by restoring the analyzed improvements in the existing process model. For this we propose to translate them in the considered process modeling language or to use media annotations like suggested by [29]. Compared to classical machine vision approaches that control the quality of a product like [14,18] or [24], our concept starts one stage earlier and identifies the causes of defects in products if they are related to human task executions. However, such systems can easily be integrated in our overall concept while the implemented techniques can be used in the image analysis step of our concept.

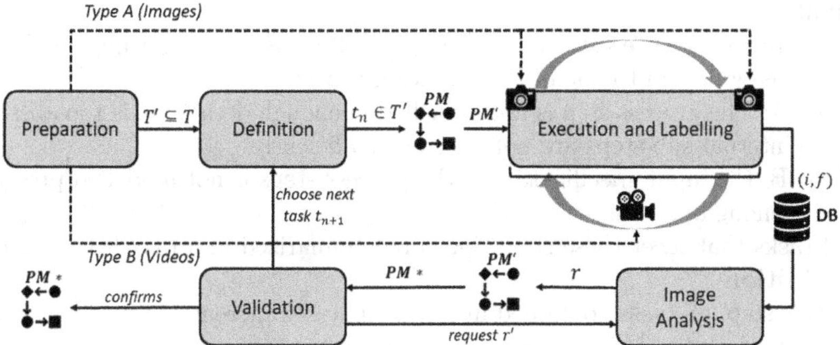

Fig. 1. The 5 high-level steps of the overall concept according to [7].

3 Using Image Mining Techniques in the Context of Process Management

3.1 Concept

The overall concept published in [7] consists of 5 steps: Preparation, Definition, Execution and Labelling, Image Analysis and Validation. In the following we explain these steps in detail and define their inputs and outputs. In the next section we present an example to point out the purpose of each step.

Figure 1 gives an overview of all steps and their dependencies. As introduction, we briefly summarize the procedure. The conceptual process starts with the Preparation step and ends if the Validation step confirms the extended process model $PM*$. In one cycle, task t_n of all process tasks T is taken into account. Then, the existing business process model PM is adapted to PM' to enable the consideration of t_n. Database DB serves as storage for pairs of recorded image data (images and videos) and related feedback of each process execution (i, f).

All entries of the DB are analyzed and reference data r is computed. To validate this data, r is integrated in PM resulting in $PM*$. Based on the validation result either $PM*$ is confirmed, new reference data r' is evaluated or $t_{n+1} \in T$ is taken into account.

1. **Preparation**
   According to [22], we define a given process model by a tuple as follows $PM = (N, E, EC, NT)$. The set N is defined as a set of process nodes, while the set of edges of PM, which are the connections between process nodes, is defined as $E \subset N \times N$. Furthermore, the function EC assigns optionally transition condition to control edges and the function NT assigns to each process node a node type provided by the process modeling language, e.g. activities or start nodes. For the sake of simplicity we define a set $T \in N$ containing all activities which need to be executed by process participants manually or which contain parts that need to be executed manually, e.g. arranging the input of a task.
   The Preparation step starts by examining all tasks in T regarding their ability to assign them to one of the following types:
   Type A: The process step is not modelled in enough detail, i.e. is too abstract since internal sub-steps are not specified sufficiently.
   Type B: the input specification of the process steps is not modelled prescriptively enough.
   All tasks that meet these conditions are summarized in a task list $T' \subseteq T$.
2. **Definition**
   In this step, all tasks contained in T' are analyzed incrementally. That means that any task $t \in T'$ is selected and prepared for further examination. Therefore PM has to be redesigned to PM' while the definition of t has to be adapted as follows. If t belongs to *Type A*, the new definition of this task has to contain the information that its execution has to be monitored by recording videos of the execution. If t belongs to *Type B*, the task has to be redesigned so that an image has to be taken. Depending on the context, the image has to be taken at the beginning, before starting the execution or after finishing the execution of t. Furthermore, a camera system has to be provided and installed in order to enable the recording of image data.
3. **Execution and Labelling**
   In this step, the redefined process model PM' is executed and image data (images and videos) is generated as specified in the model. At the end of each process execution, outcome data is labelled, i.e. process experts evaluate the process success through giving feedback considering different criteria. Traditional criteria that are related to process success are the production time, cost and quality of the result [5]. It is necessary to give the feedback at the end of the overall process execution in order to identify the influence of the considered task on subsequent tasks in the process model. Depending on the use case, feedback that is given directly after the execution of the considered task possibly does not reveal correlations within the process like we will see later in the example. However, if the feedback is given directly after the execution, the

considered task is analyzed related to classical machine vision approaches but provides more flexibility. For both alternatives it should be ensured that all feedback refers to the same criterion in order to enable comparability between them. Any feedback f and the related image data i that has been recorded during the execution step are stored as pair (i, f) in a database DB. Each set of image data referring to one execution is hence associated with an evaluation of success. This procedure is repeated until DB contains enough data to serve as basis for a meaningful analysis. The size of a sufficient database depends on the use case as well as on other aspects, e.g. how often the process model is executed in a real environment.

4. **Image (and video) Analysis**
   In this step, all entries in DB are analyzed by using image mining techniques including all 3 parts of the image mining process (cf. Sect. 2). Depending on the type of i that is stored in DB, different objectives are defined and therefore different methods have to be applied. In all cases, the overall goal is to analyze the differences between all entries and to relate them with the given feedback in order to identify which differences cause good or insufficient process results. Based on the analysis results, reference data r is created which serves as input for the validation step and in some cases also for the camera system. This data can be understood as guidance for further executions. In Sect. 3.3 we present different types of this data and suggest possibilties how this data can be presented in the process model.
   If DB contains images (i.e. i is of Type B), it holds entries capturing the same state of each process execution. To somehow compare these images and to find relevant features that differ them, image retrieval techniques as well as image classification and image clustering are suitable methods. But also the application of neural networks is quite promising, especially in finding adequate reference data.
   If DB contains videos (i.e. i is of Type A), it holds several recordings of the full execution of the same task. As described above these executions may differ in the order of underlying subtasks. Therefore all subtasks has to be identified in a first step and then the order of them has to be analyzed and compared in a second step. These requirements could be reformulated as challenge to extract event logs and therefore process models from videos and to find techniques to compare them. The order of the subtasks that has the most success is finally stored as r.
   The image analysis is the most challenging step of the overall approach regarding its technical and conceptual realization. The difficulty lies in the automated extraction of features and the identification of exactly those that have an impact on the overall process success. Further aspects are discussed in Sect. 3.3.
5. **Validation**
   The reference data computed in the previous image analysis step is incorporated into PM' resulting in $PM*$, while t is modified by r. Therefore PM might be extended by additional modeling elements to include necessary subtasks or other detailed information. Afterwards, the redesigned process

model $PM*$ is further executed. Reobservations and the repeated measuring of process success determines whether this new process version is accepted or further investigation are necessary. In the latter case, the whole process improvement procedure has to start from its beginning.
In the Validation step, t is still taken into account and image data is recorded while $PM*$ is executed. In contrast to the previous execution step related to PM or PM', process participants now execute an estimated more successful version of t. Furthermore, the camera systems can be used to recognize deviations, leading to an interruption of the execution if the recorded image data differs from r. If it does, the execution has to be adapted until it matches r including a predefined threshold. In this step, again image mining techniques are used to identify this deviations. Just as in the previous execution step, pairs of image data and feedback are collected and analyzed in order to validate if the restriction to executions that relate to r really lead to process sucess. If the validation fails and DB still contains negative feedback, either r has to be recomputed based on a larger number of entries in DB or t was not the task that led to the unsatisfying process success and the procedure has to be repeated for the next task $t' \in T*$ while $T* = T' \setminus \{t\}$. If the definition of r as guideline leads to more successful process executions, the modified process model $PM*$ replaces PM permanently.

Concluding, the presented approach enables the identification of process model execution details that have an impact on the overall success of the process. The existing process model is only extended by information, which is identified as necessary in the analysis step. Therefore the predefined abstraction of the process model is preserved but details which improve the overall process are included.

Since the concept considers the overall success of process execution, it is possible to reveal complex interrelations and dependencies between tasks that are not contained in the process model but influence the process result. Quality assurance approaches only evaluate the result of a single (mostly fully-automized, machine-based) task execution. Therefore they are only able to analyze the direct correlation of a task execution with its output. In contrast, the overall process success is considered and all tasks which have either no fixed input or where multiple executions are possible are analyzed. The proposed concept therefore enables to analyze the effect of a task on a subsequent task which leads to a reduction of the process success. In most cases, all single tasks were executed according to their rules defined in the process model while a direct evaluation of the task execution result would confirm process success. In contrast, the result of the overall process might not be satisfying. At this point, existing approaches reach their limitations since they are not able to identify the cause of the reduced process success sourcing in the dependency between tasks. Like described above, the concept tackles this issue and provides a more general process execution analysis.

The approach is based on the analysis of image data while therefore an innovative application of image mining techniques in the process management context is introduced. Image data serves as valuable source for complex analysis since it has high information content, provides several features and supports the flexibility of the presented approach. This flexibility reflects in the possibility to deal with static as well as dynamic information related to the process environment.

Concluding, the system does not only extract relevant features from images but identifies process execution dependencies affecting the process success and suggests reference data related to optimal executions. The intregration of this information in the existing process model is important to ensure process success in future executions but to avoid overload due to unnecessary information at the same time.

3.2 Example

The following example serves to illustrate the presented conceptual approach by a simple use case. For this example, we think of an use case from the manufacturing sector. We assume that a certain product that has to be manufactured according to a given process model PM. We further assume that PM contains all tasks T of the process necessary to create that product. Although all (most) process executions were successful, i.e. the manufacturing processes were completed correctly, the final products were revealing quite different quality. Since the process model is already in place and process executions did not show errors, the assumption was that not all process steps were modelled and thus finally performed optimally.

By examining the process model, the process experts identify all tasks of Type A and Type B as candidate steps that might lead to disparate production results (Preparation). Although, at this point in time, the process experts still do not know whether this assumption is correct and in case it is correct how the solutions would look like. We assume that the experts identified exactly two tasks t_1 ("add all ingredients to the glass") matching *Type A* and t_2 ("place all machine parts on the palette") matching *Type B* with $t_1, t_2 \in T$ out of overall n tasks ($|T| = n$) leading to $T' = \{t_1, t_2\}$. The process experts perceive that task t_1 is composite. However, they are not sure what the sub-tasks are at all. So, they should monitor the upcoming executions of this process step by videoing it. Additionally, the process experts see that task t_2 requires a complex input configuration. Thus, it is proposed that process experts take pictures of the input configuration of upcoming process executions. Due to the presumption of the process experts that t_2 might have more impact on the overall process sucess, they decide to start with the analysis of t_2. Therefore PM is redefined in the way that an image is taken after placing all machine parts on the palette (Definition).

The process model PM' is executed 100 times under the assumption that the process experts know that 100 executions provide a sufficient data basis for analysis. This means that 100 images are recorded, which show the resulting palette after an employee placed all machine parts on it. At the end of each execution, the process experts rate the overall process success regarding any aspect, e.g. the product quality. Together with the associated feedback 100 pairs (i, f) are stored in DB (Execution and Labelling). Hence, all 100 entries in DB

are analyzed, while it turns out that the positioning of the machine parts on the palette seems to have an impact on the overall process sucess (Image Analaysis). The image data which refers to an ideal execution is stored in reference data r. By considering r, the process model PM' is extended to $PM*$. The process is executed based on the modified process model $PM*$ and the executions are reanalyzed with regards to the process success (Validation). We assume that the validation step fails, providing the process experts 2 possibilties. Either, they analyze the other task in T' or they try to compute further reference data related to t_2. The decision depends on knowledge about the process context. In some case there might be further well-founded assumptions that one task causes reduced process success. In this example, the process experts decide to further examine t_2 instead of taking t_1 into account. Therefore they initiate to perform further process model executions in order to reach a larger data set for DB what enables the identification of a more meaningful reference image r'.

3.3 Image Analysis

In this section we further describe details and characteristics of the image analysis step from our overall concept. The small prototype in the last subsection (cf. Sect. 3.3) is used to make realization suggestions and to get more insights in our idea.

From a first view, the image analysis step can be described as a black box, which gets an amount of labelled image data as input and calculates so-called reference data as output. Like already outlined in Sect. 3.1, the reference data is used as a guideline or reference in the sense of a positive example for future process model executions. It is included in the process model leading to a modification of the original model. The goal of the modification is to enhance the process model and consequently improve the overall process success. To reach this objective, there are several questions that have to be answered. We want to outline the most important ones that have to be considered by the image analysis step:

1. Which contents of the image data recorded during process execution are relevant for the process?
2. How can this process relevant content be extracted and analyzed?
3. How is process relevant content represented?
4. How can process relevant content be evaluated with regrards to process enhancement?
5. How can process relevant content be compared?
6. How is the existing process model extended by process relevant content, how is this content represented in the process model?

By taking the overall image mining process (cf. Sect. 2) into account, the first two questions are related to the challenge to find an appropriate feature extraction technique. The others are more assigned to the image mining technology part while high-level knowledge has to be evaluated and explained. To realize the proposed overall concept, all of the questions above have to be covered by the participating image analysis system. Nevertheless, in the next two subsections we want to focus on the last question. From our view, this aspect is the most important one when it comes to process management since it leads to changes in the existing process model.

Output Requirements. Like stated above, the output of the image analysis procedure has to meet some requirements. In the following we specify some general aspects which should be considered when an existing process model is extended by process relevant analysis results.

Interpretability: The most important requirement deals with comprehension or interpretability. Since our overall concept focusses on human or manually executed tasks, also the calculated reference data referring to such a task will relate to human executions. Numeric data or low-level, machine-oriented information are not always convenient for integrating them in process models. The complexity of the reference data should not exceed a certain level in order to enable its integration in the process model while preserving its readability. This means that possibly the output information of the image analysis step has to be reprocessed to a more abstract level to stick to process modeling guidelines and to keep the traceability of process models. Another aspect that comes along with this point is that the reference data should only contain the information that is absolutely necessary for improving the respective task. All other information has to be neglected. This reduces the amount of information a priori and therefore avoids information overload in the process model.

Robustness: The output of the image analysis step has to be robust against failures or should minimize them. Like described in Sect. 2, one substep of the image mining process is the extraction of features and the handling of their specifications in an image. Consider, for instance, the differentiation between two objects by using classification techniques based on a common feature, e.g. the color. If failures like misclassifications caused by outliers occur, the reference data is also defective. Consequently the process model would not be modified based on an ideal reference data but on a negative example. At this point it has to be discussed which level of quality reference data should have to cause a valuable improvement. This strongly depends on the use case and the number of input samples. Nevertheless, image mining techniques suitable for specific applications must be selected to avoid computation failures. But to ensure process enhancement in the case of failures, we suggest to focus on the choice of the reference data's representation type.

Scalability: The reference data needs to contain all information that is necessary to improve the considered task in the process model. In other words, it may not be restricted to one single aspect, if the analysis confirms that multiple aspects have an impact on the process success. In some cases there might be dependencies between two resources while they both have to be considered in the process execution.

Extensibility: In order to enable the inclusion of new or additional information content, the reference data has to be extensible. As we will see in the next section, the reference data can be represented in different ways. Extensions within one representations type as well as flexible combinations have to be possible.

Types of Reference Data. The analyzed aspects of a task that influence the overall process success can be of different types. In the business process management context, this could be any aspect related to one of the 5 process perspectives which are often considered, e.g. [4,27,28]: The functional, data and dataflow, organizational, operational and the behavioral perspective. These aspects could differ in their information content. We want to give a few examples to highlight their differences. For example from the data perspective, the analyzed information could possibly be any detail considering the position of an object, the number of objects, the type of objects and so forth. From a behavioral perspective, the analyzed aspects could rather deal with the ordering of subprocesses or special movements.

Every type of information has different requirements when it comes to their integration in the process model. This means that depending on the information content, we have to consider different types of process model modification which can be more or less suitable. Allover we suggest 3 possibilities on how the analyzed information can be integrated into the existing process model.

Media Annotations: Like already described in the first part of this paper, the work of Wiedmann [29] suggests to use media annotations to add further information to a process task. The concept allows for adding images, videos, audio files and other data like documents to process tasks in the model. According to our work, the author points out that the annotations can be used to support process participants in the execution of manual tasks. The reference data could for example be a raw image of an ideal state in the execution or a video of an optimal execution sequence.
Discussing the presented requirements above, we focus on image and video annotations. An obvious advantage of these two annotation types is that the interpretion of information is left to the process participants. No additional techniques have to be considered. If you think for example of image annotations, humans are able to understand even multi-dimensional or complex information. This fits under the assumptions that the information content is restricted and that the image focusses on required information so that human interpretation only differs in details. But as long as the image contains that information, no additional information extraction processes are necessary. Concluding we state that the interpretability depends on the amount of information visualized in an image and the cognitive skills of the interpreter. The criteron robustness strongly depends on the content that is visualized. We want to clarify this by means of two examples: In a first example, an image is annotated that only contains a state, which should be reached during execution. We assume that the output of the image analysis step delivers

bad reference data due to a miscomputation caused by insufficient samples. Consequently, the depicted state is related to a bad example and the overall process leads to reduced process success. Since this image is marked as reference, process participants would try to reproduce this state why future executions would affect the process. In a second example, an image uniting an amount of images of the same execution state from previous executions is added. In this image, good executions that led to overall process sucess are visually distinguished from unsuccessful executions. In this example we assume that few images were labelled wrong. In the case that the images are related to process success but marked as unsuccessful, the process participant will ignore them and focus on the good labels. Thus the failure will have no impact and we can indicate the reference data as robust. Conversely, if two samples are misclassified, the process participants includes them in their set of possible executions. But since the reference data contains multiple samples, process participants will unlikely choose these outliers as reference because they intuitively consider aspects like density.

The requirements scalability and extensibility are met by this representation type. Images or videos may contain multi-dimensional information and can be easily extended by further information, also by including different data formats, e.g. text or audio.

Structural Modifications: Another approach how the analyzed information can be integrated in the existing process model, is by performing a structural process model modification. In this case, the reference data can be understood as instruction how to remodel a particular part of the process model. In most cases this will lead to an increase of the number of process modeling elements, since we want to add missing information to the process. For example this could be the decomposition of a task in multiple sequential subtasks. Another possible restructuring is to add further modeling elements like gateways to define options or conditions. Both variants lead to a decomposition into small subprocesses which were implicitly modelled by one task before. Although this modifications raise the complexity of the process model in terms of the number of model elements, too abstract design descisions are concretized.

In contrast to the media annotation approach, the raw output data of the image analysis step has to be interpreted technically. The extracted information has to be transformed into modeling instructions. This might require the use of different and complex techniques. For this reason the interpretability depends on the use case and the selected analysis methods. But if once integrated in the process model, all process steps are clear during execution.

Failures in the interpretation or information transformation step will lead to wrong process models. In such cases the whole process model structure can be affected. To recorrect modeling errors, an analysis of the whole process model would be necessary. For this reason this representation type does not meet the robustness requirement.

The aspects scability and extensibility are limited when using structural modifications. In fact, a structural modification is mainly related to the behavioral process perspective. Other perspectives like the organizational are possible, but limited. In contrast to media annotations, which extend the expressiveness of process modeling languages, structural modifications can only be done by using defined elements of a process modeling language. Consequently, structural modifications are restricted by the expressiveness of the process modeling language and their rules. It cannot be guaranteed that the modifications that a process modeling language permits reflect the modifications that are actually necessary.

Descriptive Modifications: As third type of reference data, we suggest descriptive modifications. This approach modifies the textual description of the considered process task. In other words, the instruction process participants follow is changed. Just like for the structural modifications, the amount of information visualized by the process model is increased but in contrast, the overall structure is not changed. A descriptive modification is suitable if simple but necessary instructions are included somehow in the process model. We illustrate this by the examplary task "place machine part on the palette". We assume that the image analysis step identifies that executions where the machine part is placed in the upper part of the palette increase the overall process success. Thus the description of the task should include this information through modifying it as follows: "place machine part in the upper area of the palette".

For this approach, the output of the image analysis has to be transformed into a textual description. The interpretability of this textual description depends on the degree of abstraction. Too concret, low level task descriptions, e.g. "place machine part between coordinates (x_1, y_1) and (x_2, y_2) would be difficult to realize by process participants. Therefore, the output of the image analysis step should be transformed in instructions that are easy to understand by using appropriate techniques that are able to transform formal expressions in natural language instructions. In some cases, low-level instructions might also be useful but in general such details are not part of process models. In general, natural language instructions are easy to understand, while on the other hand, they are intentionally fuzzy. Depending on the use case they therefore lead to more or less clear interpretations.

Due to the fuzziness of this type of reference data, failures have minor impact on the overall process success than in structural modifications. Similar to the media annotation approach, it strongly depends on the number of input samples the analysis was based on. Another difference to a structural modification is that the process model does not change its overall structure while textual corrections are easy to realize.

A clear advantage of a descriptive modification is that it is not limited with respect to the aspects scalability and extensibility. Any information can be expressed by text while multiple dimensions are considerable. Although any textual instruction can be added to a task, the information should be formulated efficiently in order to keep the traceability of a process task.

We summarized all representation types of reference data and their evaluation regarding the defined requirements in Table 1. As shown in the table, the robustness aspect is the most critical one. Especially when it comes to a process enhancement tool, failures in the analysis step counter the objective of our approach. We tackle this issue by integrating a Validation Step after the image analysis to ensure and double-check that the identified aspects leads to process success.

Table 1. Evaluation of the different types of reference data regarding the four requirements Interpretability, Robustness, Scalability and Extensibility.

	Media annotations	Structural modifications	Descriptive modifications
Interpretability	Depending	Fulfilled	Depending
Robustness	Depending	Not fulfilled	Depending
Scalability	Fulfilled	Limited	Fulfilled
Extensibility	Fulfilled	Limited	Limited

Implementation. In order to improve the comprehension of our concept and to make the image analysis step more traceable, we provide an exemplary implementation. This prototype shows how necessary information can be extracted from a collection of images. Furthermore, we show how this information is transformed in reference data to prepare it for integration in a process model.

In real life environments, complex images or photographies of real working environments would be generated. In contrast, we want to give a first idea on how an implementation could look like. For this reason we restrict to simple image data, i.e. noisefree images depicting plain scenes with uncomplex objects. For this purpose we programmatically generated 100 images based on one photography, containing scenes with one object clearly distinguishable from the background by a feature, like the color. Our prototype further focuses on the analysis of process steps referring to Type *B* or tasks where the input specification is not modelled precisely enough, e.g. "place a machine part on the palette" (cf. Sect. 3.1). Since the input of the image analysis Step is labelled image data, we label each generated image with a number between 0 and 100. We classify the labels in two groups representative for the quality of the process result: bad (0–69) and good (70–100). The labels as well as their numerical assignments are variable. To stick to our task example, we assume that the position of the objects in the scene is strongly related to the type of label. Allover we set up a small python-based implementation including functions for input data generation, object recognition and image analysis.

To illustrate our approach, we consider a set of image data depicting the same scene with only one red object. The position of the object varies while its orientation stays the same in all generated images (cf. Fig. 2). For this example we assume that the color is a good criteria for feature extraction and that the position of the objects plays a key role. When it comes to real data, these aspects have to be investigated first. Each image is labelled according to the object position while we assume that the further the object is positioned in the upper part of the image, the higher is the label value. This analytical information is of course not known at this point in time when it comes to real data. But according to our assumption, the labels of the three images in Fig. 2 could be 47 (left), 83 (middle) and 22 (right). At this point we want to underline that these steps are usually not part of the image analysis step but serve as its input and are generated during previous process model executions.

The analytic process of our prototypal implementation succeeds incrementally. For each input image, the brick is recognized by its color and its position is extracted. The label value referring to the respective image determines its visualization afterwards. Depending on the image label, either the brick or its centroid will be colored green, if the image relates to good process results or red if not. All assignments are brought together resulting in an output image that gives an overview of good and bad positions of the object. This can be realized by either representing the bricks at their respective positions (cf. Fig. 3) or marking only their centroids (cf. Fig. 4). Both variants of output images can be used as reference data by adding them as media anntotations to the respective task in the process model. In this case the interpretation of the images would be left to the process participant. If interpreted correctly, the process participants would place the machine parts in the upper area of the table during future executions. The image analysis step would be completed and the Validation Step would be considered next. If another respresentation type of the reference data shall be computed, further analysis steps are necessary. We show these steps by computing the appropriate descriptive modification. For this purpose we assume that the two analyzed classes related to "good" and "bad" label types are linear separable. In order to identify the relationship between these two classes, linear classification techniques can be used. The purpose of linear classification is to find a hyperplane that separates multiple classes. An overview of different

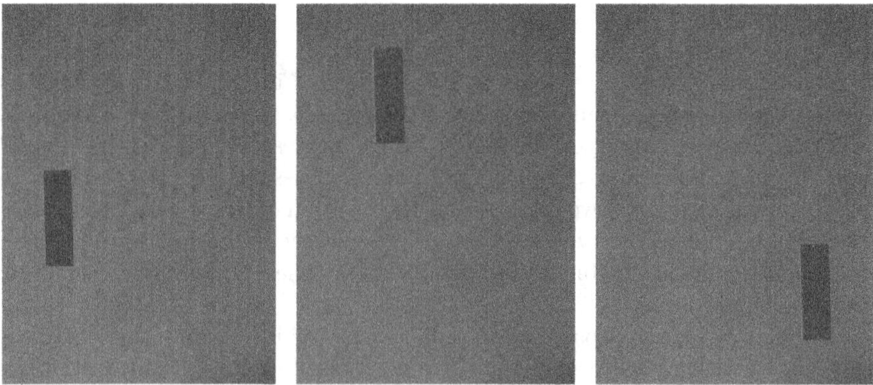

Fig. 2. Three exemplary and generated scenes: A red brick is placed at different positions on the table. (Color figure online)

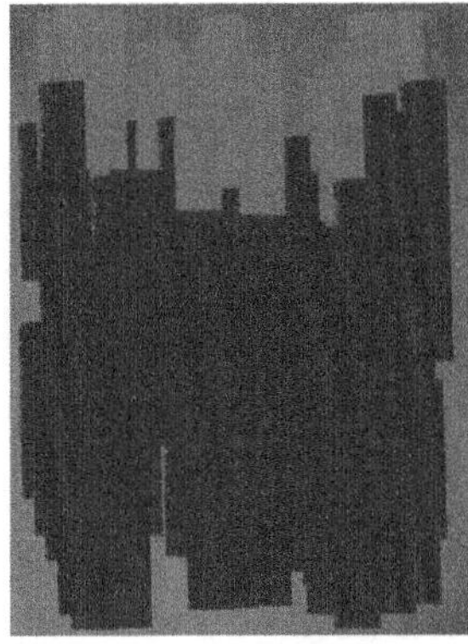

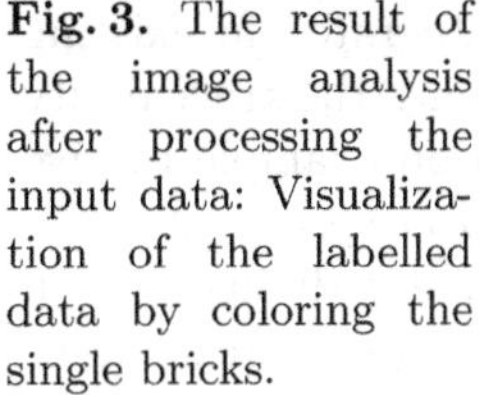

Fig. 3. The result of the image analysis after processing the input data: Visualization of the labelled data by coloring the single bricks.

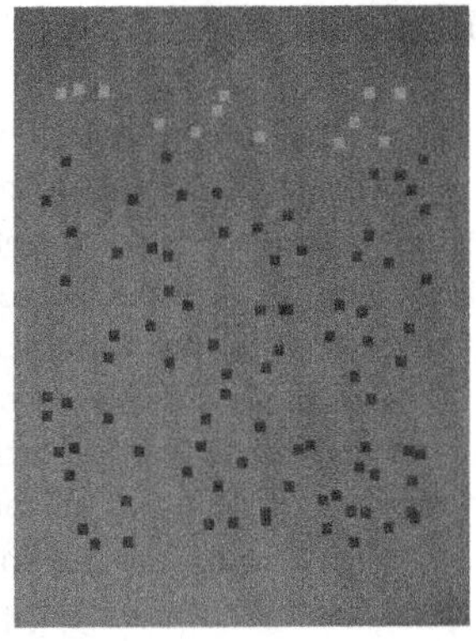

Fig. 4. The result of the image analysis after processing the input data: Visualization of the labelled data by coloring the centroid of the bricks.

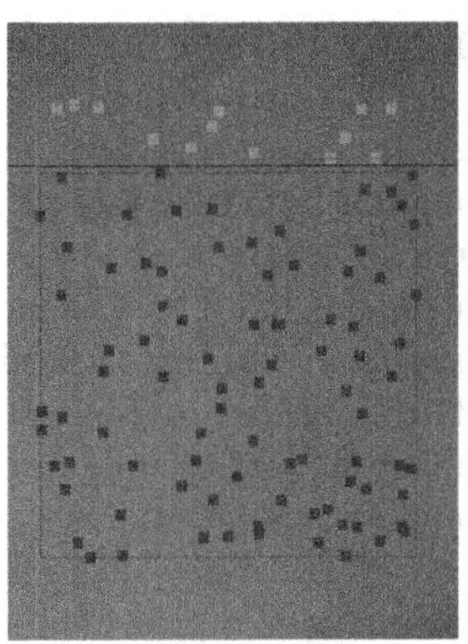

Fig. 5. The linear function (black) that separates the two minimal bounding boxes of each data set related to the classes "good" (green) and "bad" (red). (Color figure online)

classification methods that are applicable for our example is given in [11]. We compute this hyperplane by following two steps. First, we determine the minimal bounding box of each set of data points related to one class. In our case, the minimal bounding box is the smallest rectangle within which all samples of one class are located. Figure 5 shows the minimal bounding box of the "good" labelled data points in green and the minimal bounding box of the "bad" labelled data points in red. Second, we identify the separating hyperplane by computing the middle between the two bounding boxes shown as black line in Fig. 5. The line can be understood as decision point where the overall process success is increased (if objects are placed above the line) or decreased (if object are placed below the line). This information can be transformed in a descriptive modification of the process model successively by using formal-to-informal translation approaches. We think of a possible transformation process as follow. In a first step, a more concrete description of the needed information is derived, like "if $y < 1350$: 1, else: 0". While y is related with the height of the image and the two classes are distinguished by 1 (green) and 0 (red). By considering the image height, we can further derive that values smaller than 1350 lie in the upper 25% of the image. Or by transforming this in a more natural language expression, in the upper quarter of the image. Finally this information have to be added to the process model resulting in the process task description "place machine part on the palette in the upper quarter".

4 Conclusions and Discussion

In this paper we consider the problem of enhancing process models by adding missing but necessary process details. The presented concept tackles this problem by integrating image mining techniques in the context of business process management. Process model executions are monitored while image data, i.e. images and videos related to the process environment are collected. This image data is labelled according to the overall process success and is analyzed by using image mining techniques. Thus, execution-specific features that have an impact on the process success are identified. The analysis output is interpreted and transformed into reference data which contains the information about missing but necessary process details. We present different types of reference data and discuss them regarding four aspects which we newly define as requirements for the image analysis output. We explain how they can be integrated in existing process models to enhance future executions without reducing the traceability of process models. Finally, we provide a small implementation example to give more insights in the realization from a more technical perspective.

Camera systems are cost-efficient sensors which are often already installed in small- and medium-sized enterprises and industrial working areas. In most cases, such systems are used for different kind of monitoring or surveillance purposes. Nevertheless, the recorded image data created in such process environments contains valuable information which is often not fully analyzed. The approach presented in this paper is able to further analyze such data automatically and can be easily integrated in existing working environments. This aspect is not only a benefit but is necessary regarding the defined problem statements in Sect. 1.2. Process details might be absent in the process model for different reasons.

In the case that a detail is not modelled because the optimal process is not known or the process modeler is not conscious of this aspect, an automatic analysis is needed. The presented approach meets this requirement by using adequate analysis techniques on image data which contains multi-dimensional information. Image data can either contain static information, like the input or output of a task or dynamic information, like the task processing itself. Both may contain necessary information for process enhancement. We therefore dinstinguish between the analysis of images and videos. Images are related to snap-shots of the execution while videos capture the whole execution of the task including related subtasks.

We tackle the problem of omitting details in process modeling due to traceability and size limitation reasons by integrating only necessary information. If the goal of process modelers is to keep the process model small, they tend to model in an abstract way and do not distinguish between different types of details. In contrast, the analyzed reference data by which the process model is extended contains only the aspects that have an impact on the overall process success. Other detected deviations or differences in process executions are neglected. Therefore the presented concept does not only identify missing details but also only integrates the necessary ones resulting in a preservation of the traceability of the process model.

The fact that some details cannot be modelled or that it is not possible to express them by using modeling languages is also covered by our approach. We present different possibilities how the analyzed reference data can be integrated in an existing process model. For example media annotations are able to complement the expressiveness of existing process modeling languages.

In contrast to other process optimization methods, the presented approach focuses on the process execution step. Recorded image data that contains real information of the process environment is used as base for analysis. This enables the identification of details that were not considered before and are not yet contained in the existing process model.

The aspects above point out the great potential of using image mining techniques as suggested in this paper. The review of existing work further confirms that the application of image mining techniques in the process management context is useful and that more research in this area is needed. Concluding, although the overall conceptual approach is quite promising, we are aware that the technical realization is challenging. Further distinctions between different image mining techniques and their implementation costs are necessary. For this reason, we aim at considering the whole concept from a more technical perspective in a next step. Whereby all questions that were defined for the image analysis step should be addressed. We plan to investigate which use cases require what kind of image mining techniques. These insights will be used in order to extend the presented approach to evaluate further exemplary data and to prepare a system which is able to analyze simple data of real working environments.

References

1. Ahmadikatouli, A., Aboutalebi, M.: New evolutionary approach to business process model optimization. Int. MultiConf. Eng. Comput. Sci. **2**, 1119–1122 (2011)
2. Becker, J., Rosemann, M., von Uthmann, C.: Guidelines of business process modeling. In: van der Aalst, W., Desel, J., Oberweis, A. (eds.) Business Process Management. LNCS, vol. 1806, pp. 30–49. Springer, Heidelberg (2000). https://doi.org/10.1007/3-540-45594-9_3
3. Bobrik, R., Reichert, M., Bauer, T.: View-based process visualization. In: Alonso, G., Dadam, P., Rosemann, M. (eds.) BPM 2007. LNCS, vol. 4714, pp. 88–95. Springer, Heidelberg (2007). https://doi.org/10.1007/978-3-540-75183-0_7
4. Burattin, A., Maggi, F.M., Sperduti, A.: Conformance checking based on multi-perspective declarative process models. Expert Syst. Appl. **65**, 194–211 (2016)
5. Collins, A., Baccarini, D.: Project success - a survey. J. Const. Res. **5**(2), 211–231 (2004)
6. Dumas, M., La Rosa, M., Mendling, J., Reijers, H.A.: Process-aware information systems. Fundamentals of Business Process Management, pp. 341–369. Springer, Heidelberg (2018). https://doi.org/10.1007/978-3-662-56509-4_9
7. Fichtner, M., Schönig, S., Jablonski, S.: Process management enhancement by using image mining techniques: a position paper. In: ICEIS 2020, vol. 1, pp. 249–255 (2020)
8. Foschi, P. G., Kolippakkam, D., Liu, H.: Feature extraction for image mining. In: Multimedia Information Systems, pp. 103–109 (2002)

9. Giaglis, G.: A taxonomy of business process modeling and information systems modeling techniques. Int. J. Flex. Manufact. Syst. **13**(2), 209–228 (2001)
10. Gounaris, A.: Towards automated performance optimization of BPMN business processes. Commun. Comput. Inf. Sci. **637**, 19–28 (2016)
11. Kumar, R., Verma, R.: Classification algorithms for data mining: a survey. Int. J. Innov. Eng. Technol. (IJIET) **1**(2), 7–14 (2012)
12. La Rosa, M., Ter Hofstede, A.H., Wohed, P.: Process model abstraction: managing process model complexity via concrete syntax modifications. IEEE Trans. Indus. Inf. **7**(2), 255–265 (2011)
13. Lee, D.E., Hwang, I., Valente, C.: Precision manufacturing process monitoring with acoustic emission. In: Wang, L., Gao, R.X. (eds.) Condition Monitoring and Control for Intelligent Manufacturing, pp. 33–54. Springer, London (2006). https://doi.org/10.1007/1-84628-269-1_2
14. Manigel, J., Leonhard, W.: Vehicle control by computer vision. IEEE Trans. Indus. Electron. **39**(3), 181–188 (1992)
15. Min, H., Shuangyuan, Y.: Overview of image mining research. In: ICCSE 2010–5th International Conference on Computer Science and Education, Final Program and Book of Abstracts, pp. 1868–1970 (2010)
16. Niedermann, F., Radeschütz S., Mitschang, B.: Deep business optimization: a platform for automated process optimization. In: INFORMATIK 2010 - Business Process and Service Science, Proceedings of ISSS and BPSC, pp. 168–180 (2010)
17. Ordonez, C., Omiecinski, E.R.: Image mining: a new approach for data mining. Georgia Institute of Technology (1998)
18. Paulo, H., Davies, R., Correia, B.: A machine vision quality control system for industrial acrylic fibre production. EURASIP J. Adv. Signal Process. **2002**(7), 1–8 (2002)
19. Polyvyanyy, A., Smirnov, S., Weske, M.: Reducing complexity of large EPCs. In: Modellierung betrieblicher Informationssysteme (MobIS 2008), Gesellschaft für Informatik e.V., Bonn, pp. 195–207 (2008)
20. Polyvyanyy, A., Smirnov, S., Weske, M.: Process model abstraction: a slider approach. In: 2008 12th International IEEE Enterprise Distributed Object Computing Conference, pp. 325–331. IEEE (2008)
21. Prykäri, T., Czajkowski, J., Alarousu, E.: Optical coherence tomography as an accurate inspection and quality evaluation technique in paper industry. Opt. Rev. **17**(3), 218–222 (2010)
22. Reichert, M., Kolb, J., Bobrik, R.: Enabling personalized visualization of large business processes through parameterizable views. In: Proceedings of the 27th Annual ACM Symposium on Applied Computing - SAC 2012, pp. 1653–1660 (2012)
23. Rooney, W., Biegler, L.: Optimal process design with model parameter uncertainty and process variability. AIChE J. **49**, 438–449 (2003)
24. Saldaña, E., Siche, R., Luján, M.: Computer vision applied to the inspection and quality control of fruits and vegetables. Braz. J. Food Technol. **16**(4), 254–272 (2013)
25. Schmidt, R., Möhring, M., Zimmermann, A., Härting, R.-C., Keller, B.: Potentials of image mining for business process management. In: Czarnowski, I., Caballero, A.M., Howlett, R.J., Jain, L.C. (eds.) Intelligent Decision Technologies 2016. SIST, vol. 57, pp. 429–440. Springer, Cham (2016). https://doi.org/10.1007/978-3-319-39627-9_38
26. Smirnov, S., Reijers, H.A., Weske, M.: Business process model abstraction: a definition, catalog, and survey. Distrib. Parallel Databases **30**(1), 63–99 (2012)

27. Van der Aaalst, W.M.P., Adriansyah, A., De Medeiros, A.K.A.: Process mining manifesto. In: International Conference on Business Process Management, pp. 169–194 (2011)
28. Van der Aaalst, W.M.P.: Process Mining - Data Science in Action, 2nd edn. Springer, Berlin, Heidelberg (2016)
29. Wiedmann, P.: Agiles Geschäftsprozessmanagement auf Basis gebrauchssprachlicher Modellierung. Doctoral dissertation, University of Bayreuth (2017)
30. Yanai, K.: Web image mining toward generic image recognition. In: Poster Proceedings. 12th International World Wide Web Conference, pp. 1–6 (2003)
31. Zhang, J., Hsu, W., Lee, M.: Image mining: issues, frameworks and techniques. In: Proceedings of the 2nd ACM SIGKDD International Workshop on Multimedia Data Mining (MDM/KDD 2001), pp. 1–7 (2016)
32. Zhang, J., Hsu, W., Lee, M.: Image mining: trends and developments. J. Intell. Inf. Syst. **19**, 7–23 (2002)

Artificial Intelligence and Decision Support Systems

A Machine Learning Based Framework for Enterprise Document Classification

Juris Rāts[1](✉), Inguna Pede[1], Tatjana Rubina[2], and Gatis Vītols[2]

[1] RIX Technologies, Blaumana 5a-3, Riga 1011, Latvia
{juris.rats,inguna.pede}@rixtech.lv

[2] Faculty of Information Technologies, Latvia University of Life Sciences and Technologies, 2 Liela Street, Jelgava 3001, Latvia
{tatjana.rubina,gatis.vitols}@llu.lv

Abstract. Enterprise Content Management (ECM) systems store large amounts of documents that have to be conveniently labelled for easy managed and searching. The classification rules behind the labelling process are informal and tend to change that complicates the labelling even more. We propose a machine learning based document classification framework (Framework) that allows for continuous retraining of the classification bots, for easy analysis of the bot training results and for tuning of the further training runs to address this challenge. Documents and metadata fields typical for ECM systems are used in the research. The Framework comprises selection of classification and vectorization methods, configuration of the methods hyperparameters and general learning related parameters. The model provides user with visual tools to analyze the classification performance and to tune the further steps of the learning. A couple of challenges are addressed – as handling informal and eventually changing criteria for document classification, and dealing with imbalanced data sets. A prototype of the proposed Framework is developed and short analysis of the prototype performance is presented in the article.

Keywords: Document classification · Machine learning · Supervised learning · Enterprise Content Management · Python · Elasticsearch

1 Introduction

Enterprise Content Management (ECM) systems store large amounts of documents. The documents have to be conveniently classified (labelled) to be easily managed and searched for. Some kind of analysis and evaluation effort has to be performed to store them in appropriate folders, assign appropriate types, attribute to convenient cases and projects, route to appropriate user for handling.

Organizations usually hire staff (clerks) to handle labelling of documents received from other organizations or individuals (received document) to manage this appropriately. Received documents are routed to a clerk who reads the document, understands the content and attaches appropriate labels. Usually organizations have some guidelines or

J. Filipe et al. (Eds.): ICEIS 2020, LNBIP 417, pp. 87–99, 2021.
https://doi.org/10.1007/978-3-030-75418-1_5

rules to guide this process, still the proper labelling depends significantly on the person's experience and knowledge.

A convenient method for handling the cases where no formal algorithms can be easily specified is to use machine learning methods. We use supervised machine learning as ECM normally has a set of properly labelled samples - documents already in the repository.

Supervised machine learning is an intensely explored domain with a rich set of high performing algorithms ready for handling diverse kinds of tasks – including document and text classification (see below the Related work section for discussion). The aim of our research is to create a flexible Framework on top of the available algorithms that would provide users with:

- classification bots that would learn to classify the captured documents;
- advanced visual tools for analysis of the results of document classification
- configuration tools allowing to tune the next steps of learning of the classification bots.

Although we employ in our Framework the means for rule handling, the rules are used to handle exceptions only. This allows to reduce importantly the manual work to create classification rules used in rule-based classification systems.

This paper is a revised and extended version of the ICEIS conference paper [1]. The text is adjusted to match the new findings of the research, the Proof of concept Chapter is included, the Performance Evaluation chapter has been reworked to cover the latest measurements.

2 Related Work

Text classification methods are researched in a number of fields, as:

- sentiment analysis [2–5],
- news classification [6–9],
- web page classification [10, 11].

Although close to our domain of research, sentiment analysis, news classification and web page classification do not cover fully the specifics of the ECM document classification (e.g., ECM repositories span large numbers of voluminous documents).

Commercial mainly rule-based solutions are created for similar tasks like Xtracta [12], Serimag [13], ABBY FlexiCapture [14]. They mainly focus though on OCR technologies to streamline document scanning. We aim to address the process of classification of digital documents instead.

Machine learning algorithms are used for wide range of domains, including computer vision, speech understanding and synthesis, and human language translation. Each of the domains has its list of best performing algorithms. Some of the favorites for the document/text classification are Naïve Bayes classifier, K-Nearest Neighbors, Support Vector Machine and Deep learning models [15].

Naive Bayes (NB) [2, 16] shows good results in document classification. Kadhim [17] argues NB showed the second-best accuracy out of the 5 methods tested.

Support Vector Machine (SVM) is one of the most efficient machine learning algorithms [18], applied in particular to text categorization. Several drawbacks exist though as the lack of transparency in results caused by a high number of dimensions

Deep learning networks, e.g., Recurrent neural networks, Convolutional neural network, Hierarchical Attention networks, are applied for text classification [19–22], as well as combined RCNN - Convolutional neural with Recurrent neural network [9].

Basic methods as Naïve Bayes show good results with smaller data sets, while Convolutional neural network shows superior performance with larger data sets [23].

Text documents have to be translated to numeric vectors (i.e. vectorized) because machine learning methods operate numeric data. One of the core approaches here is the Bag of Words (BoW) model that has been successfully applied in a number of domains including the medical domain with automating medical image annotation [24], biomedical concept extraction [25], and recommender systems for medical events [26].

Several extensions of the BoW model are developed to overcome the shortcoming that BoW ignores the semantic information. Bag of meta-words (BoMW) [4], e.g., uses meta-words instead of words as building blocks. Several embedding methods (Word2Vec, GloVe, FastText) are found capable to improve accuracy (e.g. for large data volumes) and make the learning curve steeper [27, 28]. Another research [2] shows Doc2Vec has better accuracy than tf-idf for most cases.

The feature sets of the ECM documents extracted for machine learning tend to have large dimensionality because ECM systems normally handle voluminous documents. Dimensionality reduction methods like Principal Component Analysis (PCA), Linear Discriminant Analysis (LDA), and non-negative matrix factorization (NMF) allow for more efficient use of machine learning algorithms because of the time and memory complexity reduction [15, 17, 24].

According to the results of research projects mentioned here we selected classification algorithms for implementation of our Framework (see Table 1) as well as decided on vectorization methods to use. A number of other design decisions were made as well including what hyperparameter values for the classification algorithms to test.

3 Machine Learning Features Used

Machine learning based document classification models generally comprise four phases: feature and label extraction, dimension reduction, selection (and applying) of the classification algorithm, and evaluations (adapted from [15]).

Feature extraction is one of the most important factors in successful machine learning projects [29]. Feature extraction converts the sample text to a set of (numeric) features usable for machine learning algorithms.

Dimension reduction aims to reduce the dimensionality of the feature set used for training.

Evaluation. The performance of the trained model is evaluated on the test data prior to using it for prediction. Evaluation results are used to select the best performing classification algorithms (or to decide if the model has been trained sufficiently).

In most cases it is not possible to say in advance which of the classification algorithms will perform better for a domain in question. Several algorithms therefore are used for training and the best is selected, or ensemble methods are used to combine the predictions of individual algorithms. Our model uses the first option – selects the algorithm performing best in the test phase.

A couple of Machine Learning related aspects important for our model are discussed in the sections below.

3.1 Label Extraction

ECM document classification usually involves determining values of metadata fields such as document type, folder, assignee. Two approaches may be considered to relate labels to the metadata fields. The label may be related:

- to each of the fields separately;
- to a combination of all metadata fields (e.g. a separate label would be created for a combination of document type, folder and assignee); the document is classified against the combined label.

The measurements we performed show the second approach is preferable as it takes less processing time (i.e. processing of each separate label takes as much time as the processing of the combined label) and the predicting accuracy for both cases is comparable.

To handle the combined label our model must provide for conversion from metadata fields to labels and vice versa. Document metadata must be converted to labels when preparing the learning data set. Labels predicted by the classification bot have to be converted back to metadata fields.

3.2 Handling Imbalanced Data

The top five most popular labels (combined label document type + document folder + case) for the data set we research account for 94.76% of all samples while the rest (43) - for only 5.24%. A number of other ECM data sets we explored show similar trend. This means we are dealing with highly imbalanced data sets.

Imbalanced data sets can cause problems for classification algorithms for domains where rare cases tend to be more significant than the frequent ones (e.g. fraud detection and disease diagnosis) [30]. This does not happen with ECM documents – we can assume that a penalty for an incorrect prediction does not depend on how frequent the cases class is. In other words – all classes are of equal importance for our case.

There is a problem with the rare cases though - they have less training data. We propose the method of Balancing of Major Classes (BMC) to address this problem.

BMC consists of four consecutive steps.

1. The set of N label classes L is ordered descending into ordered set L^{ord} by appearance count in samples (popularity).

2. L^{ord} is split into two sets – major classes (or Majors - M_n, comprising first n classes from L^{ord}

$$M_n = \left(L_1^{ord}, L_2^{ord}, \ldots, L_n^{ord}\right) \tag{1}$$

and minor classes (Minors)

$$m_n = \left(L_{n+1}^{ord}, L_{n+2}^{ord}, \ldots\right) \tag{2}$$

with total popularity

$$P(m_n) = \sum_{i=n+1}^{N} L_i^{ord} \tag{3}$$

not more than some a priori set *Minors Threshold* (Tm, $P(m_n) <= Tm$), such as $P(m_{n-1}) > Tm$.
3. Majors set is supplemented with Others class (O) to represent all minor classes

$$M_n = \left(L_1^{ord}, L_2^{ord}, \ldots, L_n^{ord}, O\right) \tag{4}$$

4. The training sample set is created for each Major class.

It should be noted that predicting class O means that the algorithm is not able to determine the meta data fields for the case. Thus, in our model the classification algorithm has 3 possible prediction outcomes – accurate prediction, false prediction and no prediction.

The proposed BMC method allows to produce training data for highly imbalanced data sets by creating samples for the classes that have data available and to merge rare classes where there are no sufficient data anyway.

3.3 Continuous Training

Machine learning models usually are evaluated on static data sets. The data set is split into learning and test subsets, the model is trained on the former and tested on the latter, the performance gets evaluated and the model is ready for the prediction.

In an ECM repository new data is captured constantly therefore classification models have to be trained repeatedly as new data appears. The means have to be put in place to control a number of parameters, like frequency to retrain the model, volume of training samples to use, classification algorithm to incorporate, hyperparameter values of the classification algorithms and others.

4 Document Classification Framework

The document classification Framework we propose comprises three processes (sample retrieval, training and predicting) and three data stores (Fig. 1).

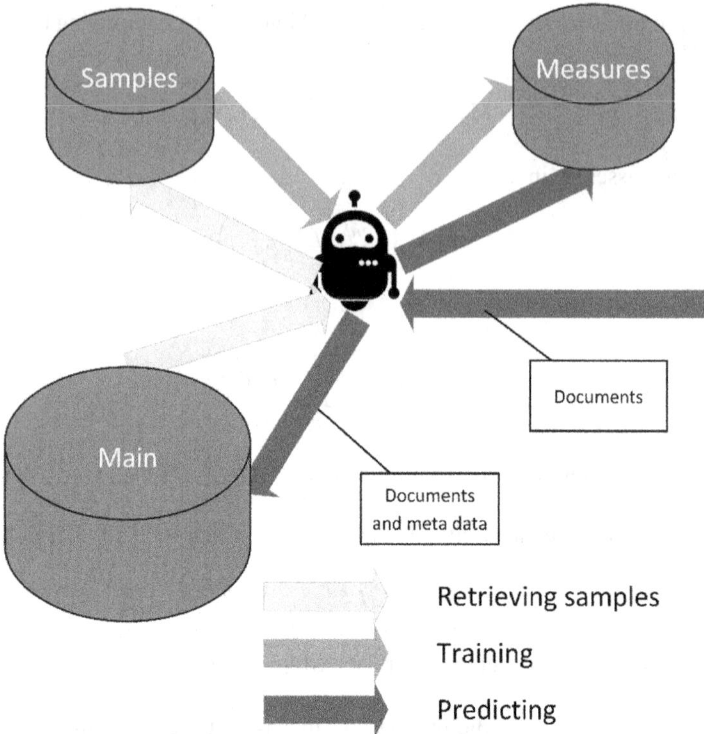

Fig. 1. Document classification Framework [1].

Sample retrieval process retrieves periodically the new (and modified) samples (documents and meta data) from the main store (ECM repository) and saves them into the Samples store.

Training process retrieves the samples for each bot, trains the bots and evaluates them. Bot training and evaluation statistics are saved in the Measures store.

Trained bots listen for their channels and *predict* the metadata values, results are saved in the Measures store.

Some of the important parameters supported by the Framework are listed in Table 1.

The main metric to measure the performance of the classification against is the prediction accuracy on all samples in total (as noted in Sect. 3.2 - all classes are equally important). According to the learning model (Chapter 3) the goal of the document classification process may be either to maximize true predictions or to minimize false predictions. The final decision should be made by the user according to her preferences.

Other features of the proposed document classification Framework are:

- the Framework handles continuous learning process as described in the previous chapter; the bots are retrained periodically on refreshed data; the user is empowered with tools to analyze the samples classified incorrectly by the bots and to eventually

Table 1. Framework parameters.

Parameter	Comments
Vectorization methods	Tf-idf, hashing, word2vec, fastText
Classification algorithms	Stochastic gradient descent, Multinomial naïve Bayes, Passive-aggressive classificator, Logistic regression, Support vector classifier, Linear support vector classifier, Convolutional neuron networks
Sample volume	Count of samples (training plus testing)
Minors threshold	Determines the relative part of the samples related to the label classes not included in the Majors set (e.g. for Minors threshold 0.1 not more than 10% of all samples should relate to Minor classes)
Test size	Determines the part of all samples hold out for testing

change the configuration for the next training cycles or to remove noise data from the training sets;

- document capturing via channels and classification bots dedicated for each of the channels;
- user defined rules for conversion between meta fields and labels;
- a broad range of parameters, including bot retraining frequency, sample volumes, feature extraction and classification methods (includes selection of methods and configuration of methods parameters), Minors Threshold etc.

Channels and user defined rules are outlined below (Sects. 4.1 and 4.2).

4.1 Channels

A document may be captured by the ECM system through one or more capturing channels (or simply *channels*). Sample channel examples are a particular e-mail address of the organization, interface with an external application or interaction with a particular group of users who add documents manually. Organizations normally have several channels (several e-mail addresses, several streams of documents from external applications etc.) dedicated to capture different kinds of documents (e.g. separate channels for invoices and for customer complaints). Thus, the document classification Framework should provide several bots – each for its own channel. The bot must be trained on documents and labels captured earlier through the same channel.

The bots should profit from the rules existing in the organization linking the channel to specific document metadata values. For example – documents captured through a particular e-mail address might be saved in a specific folder meaning the bot must not predict the folder.

4.2 User Defined Rules

Document classification rules may change. E.g. the documents previously saved to the folder A are switched to folder B. These changes if unhandled may deteriorate the performance of the classification bots. To address this, we incorporate into the Framework a means of user defined rules. For a sample above the rule may be introduced that instructs to use folder B instead of the folder A starting the specified date D. When creating train/test sets after the date D in samples dated before the date D folder A will be substituted than with the folder B.

5 Proof of Concept

Python, Java and R were considered as candidate platforms for the implementation of the Framework. We decided in favor of Python as the most popular for the practical implementations of machine learning tasks [31].

A prototype solution DOBO was created to assess the proposed Framework. The prototype solution implements all important aspects of the Framework, including the retrieval of the sample date from the ECM repository, preparing the train/test sample sets, configuration of the hyperparameter values, browsing and evaluation of the test results.

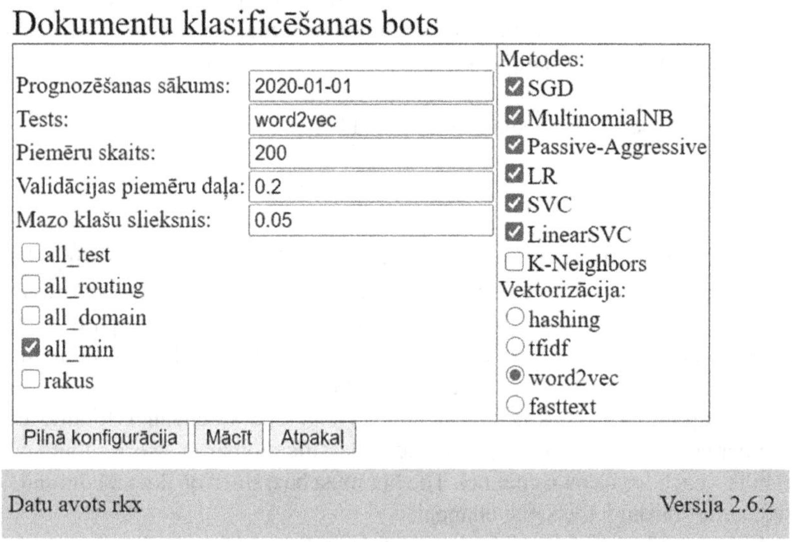

Dokumentu klasificēšanas bots
Prognozēšanas sākums: 2020-01-01
Tests: word2vec
Piemēru skaits: 200
Validācijas piemēru daļa: 0.2
Mazo klašu slieksnis: 0.05
all_test
all_routing
all_domain
all_min
rakus
Metodes:
SGD
MultinomialNB
Passive-Aggressive
LR
SVC
LinearSVC
K-Neighbors
Vektorizācija:
hashing
tfidf
word2vec
fasttext
Pilnā konfigurācija
Mācīt
Atpakaļ
Datu avots rkx
Versija 2.6.2

Fig. 2. Bot training configuration and launch page.

Figure 2 shows the bot training configuration/launch page. The page allows to set (left side top to bottom) the start date of the prediction period (the train/test samples are taken the newest ones before that date), the test name, the number of samples (for training and testing in total), part of the samples (0.2 or 20%) used for testing, Minors

Threshold, channels used for the test. Classification algorithms may be checked for use in the test and vectorization method may be selected on the right side.

The pushbuttons below are for accessing the advanced configuration (allows to configure a full set of hyperparameters, for launching the bot training, and for returning to the DOBO main page without launching the training.

DOBO features a number of pages for browsing and analyzing the bot training results.

Elasticsearch is used to implement the data persistence layer of the Framework, Kibana provides additional visualization means for data and performance analysis. Python (including machine learning libraries sklearn and keras) is used to implement the core functionality.

6 Performance Evaluation

The performance of the document classification models created and configured with the Framework was evaluated on a data set containing more than 160 thousand of documents. Combined label (document type + document folder + case) is used for the classification.

Seven classification methods and four vectorization methods (hashing, tf-idf, word2vec and fastText) were tested for a number of hyperparameter values. Other parameters explored are volume of the training set, number of features selected for text representation, analysis of bigrams and trigrams, Minors Threshold etc.

Figure 3 demonstrates the influence of Minors Threshold. Rightmost bar (value 0) represents the case when a bot attempts to predict values of all label classes while the leftmost (value 0.2) – when a bot ignores (does not predict) the minority classes accounting in total for 20% of all samples.

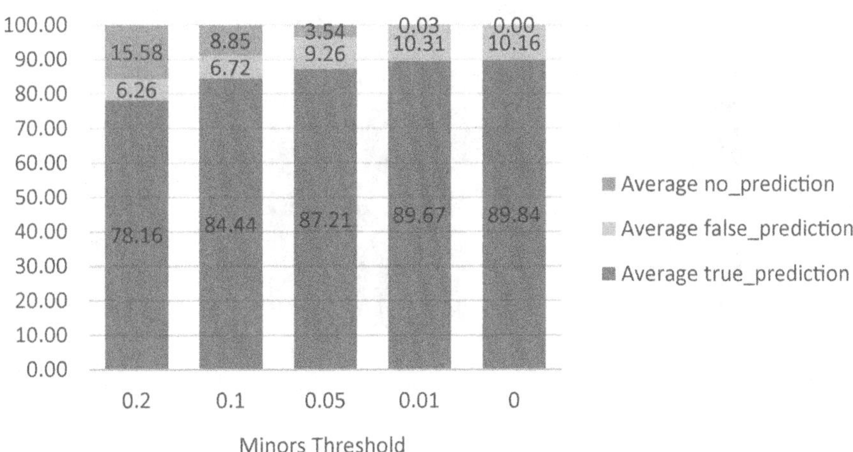

Fig. 3. Accuracy by Minors Threshold [1].

Figure 4 shows the bot performance depending on the number of features used for the training. The number of features in case of BoW model is the number of most important words of the joint sample dictionary used to represent the samples. One can see here that increasing the feature count above 5000 does not improve significantly the accuracy.

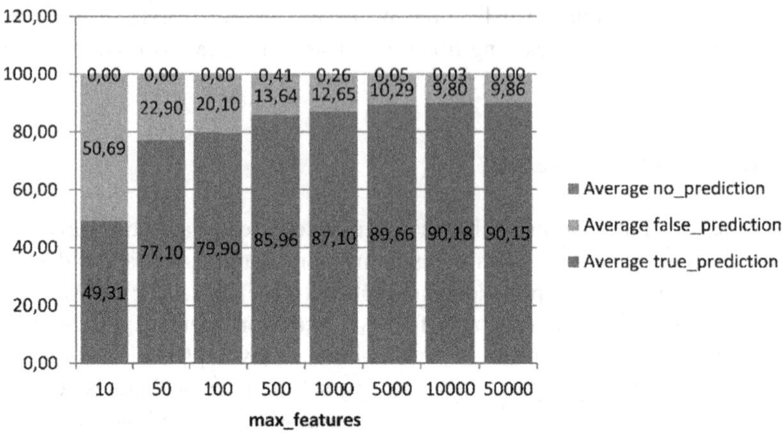

Fig. 4. Accuracy by feature count.

An interesting conclusion from the test measurements is that the accuracy of the document classification bots trained with different classification methods do not differ significantly (see Fig. 5). Multinomial Naïve Bayes (MultinomialNB) has lower accuracy

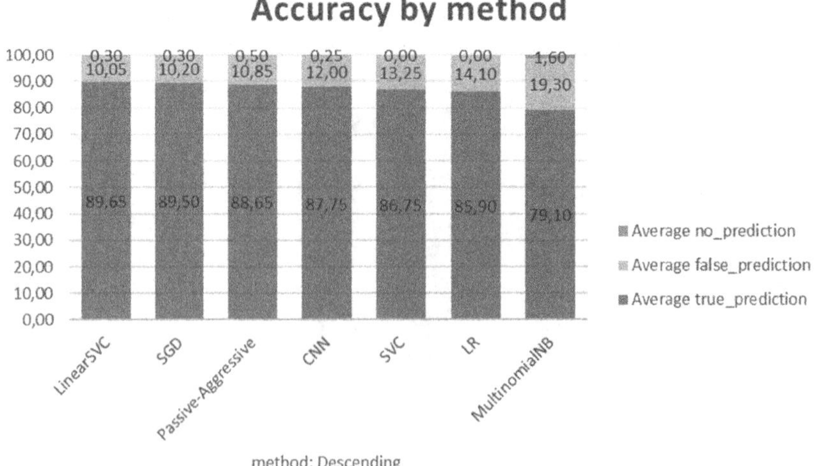

Fig. 5. Accuracy by method [1].

here but this may change with new documents captured. Every next learning cycle may have its own best methods.

It should be noted that the deep learning methods like multi-layered Convolution Neuron Network (CNN) did not show better results than basic methods like Stochastic Gradient Descent (SGD), Support Vector Classifier (SVC), Linear SVC, Passive-Aggressive Classifier and Logistic Regression (LR).

7 Conclusions

The aim of this research is to create a flexible Framework on top of the available text classification algorithms that would provide users with:

- classification bots that would learn to classify the captured documents;
- advanced visual tools for analysis of the results of document classification
- configuration tools allowing to tune the next steps of learning of the classification bots.

The proposed document classification Framework is based on the following main observations from the ECM domain:

- documents are captured through channels; each channel handles its own specific set of documents;
- label class distribution of captured documents is highly imbalanced;
- the classification rules are informal and tend to change.

Document classification models have to be retrained periodically on newly captured documents to adapt to the possible changes of document vocabulary and wording. Results of the bot training and testing may be used to spot possible disruptive changes in document classification rules.

An important requirement for the ECM document classification is the ability to handle highly imbalanced data. We propose to merge all rare classes into one for the bot training and to interpret the cases when bots predict that merged class as – no prediction made. Thus, our model avoids to make a prediction for cases where there is not enough data anyway.

Machine learning based document classification process has to be monitored and configured periodically to work properly. The configuration may include a selection of classification and vectorization methods, tuning of hyperparameters of the methods, a configuration of a number of general parameters of the training process. The measurements we run did not reveal any significant advantages of any particular classification method. It is rather important to tune the process to the actual data and actual classification rules.

These discoveries are the main reason why the document classification Framework we propose focuses on providing tools both for analysis of training/predicting performance and for periodic fine tuning.

Acknowledgements. The research accounted for in this paper is co-funded by the European Regional Development Fund (ERDF) (project No. 1.2.1.1/18/A/003).

References

1. Rāts, J., Pede, I., Rubina, T., Vītols, G.: A flexible model for enterprise document capturing automation. In: Proceedings of the 22nd International Conference on Enterprise Information Systems, pp. 297–304. SCITEPRESS - Science and Technology Publications (2020). https://doi.org/10.5220/0009034802970304
2. Avinash, M., Sivasankar, E.: A study of feature extraction techniques for sentiment analysis, pp. 1–12 (2019)
3. Pahwa, B., Taruna, S., Kasliwal, N.: Sentiment analysis- strategy for text pre-processing. Int. J. Comput. Appl. **180**, 15–18 (2018). https://doi.org/10.5120/ijca2018916865
4. Fu, M., Qu, H., Huang, L., Lu, L.: Bag of meta-words: a novel method to represent document for the sentiment classification. Expert Syst. Appl. **113**, 33–43 (2018). https://doi.org/10.1016/J.ESWA.2018.06.052
5. Tam Hoang, D.: Sentiment Analysis: Polarity Dataset. Charles University in Prague (2014)
6. Dilrukshi, I., De Zoysa, K., Caldera, A.: Twitter news classification using SVM. In: Proceedings of the 8th International Conference on Computer Science and Education, ICCSE 2013, pp. 287–291 (2013). https://doi.org/10.1109/ICCSE.2013.6553926
7. Liliana, D.Y., Hardianto, A., Ridok, M.: Indonesian news classification using support vector machine. Eng. Technol. World Acad. Sci. **57**, 767–770 (2011). https://doi.org/10.5281/zenodo.1074439
8. Sreedevi, J., Rama Bai, M., Reddy, C.: Newspaper article classification using machine learning techniques. Int. J. Innov. Technol. Explor. Eng. 2278–3075 (2020). https://doi.org/10.35940/ijitee.E2753.039520
9. Lin, R., Fu, C., Mao, C., Wei, J., Li, J.: Academic news text classification model based on attention mechanism and RCNN. Presented at the (2019). https://doi.org/10.1007/978-981-13-3044-5_38
10. Yu, H., Han, J., Chang, K.C.-C.: PEBL. In: Proceedings of the Eighth ACM SIGKDD International Conference on Knowledge Discovery and Data Mining - KDD 2002, p. 239. ACM Press, New York (2002). https://doi.org/10.1145/775047.775083
11. Shawon, A., Zuhori, S.T., Mahmud, F., Rahman, J.: Website classification using word based multiple n-gram models and random search oriented feature parameters. In: 2018 21st International Conference of Computer and Information Technology (ICCIT), pp. 1–6. IEEE (2018). https://doi.org/10.1109/ICCITECHN.2018.8631907
12. Xtracta: automated data entry software powered by AI
13. Serimag - artificial intelligence for document automation
14. Intelligent document processing platform - ABBYY FlexiCapture
15. Kowsari, K., Meimandi, K.J., Heidarysafa, M., Mendu, S., Barnes, L.E., Brown, D.E.: Text classification algorithms: a survey. Inf. 10 (2019). https://doi.org/10.3390/info10040150
16. Porter, M.F.: An algorithm for suffix stripping. Program **40**, 211–218 (2006). https://doi.org/10.1108/00330330610681286
17. Kadhim, A.I.: Survey on supervised machine learning techniques for automatic text classification. Artif. Intell. Rev. **52**, 273–292 (2019). https://doi.org/10.1007/s10462-018-09677-1

18. Karamizadeh, S., Abdullah, S.M., Halimi, M., Shayan, J., Rajabi, M.J.: Advantage and drawback of support vector machine functionality. In: I4CT 2014 - 1st International Conference on Computer, Communications, and Control Technology, Proceedings, pp. 63–65. Institute of Electrical and Electronics Engineers Inc. (2014). https://doi.org/10.1109/I4CT.2014.6914146
19. LeCun, Y., Bengio, Y., Hinton, G.: Deep learning. Nature **521**, 436–444 (2015). https://doi.org/10.1038/nature14539
20. Johnson, R., Zhang, T.: Effective use of word order for text categorization with convolutional neural networks. In: Proceedings of the 2015 Conference of the North American Chapter of the Association for Computational Linguistics: Human Language Technologies, pp. 103–112. Association for Computational Linguistics, Stroudsburg, PA, USA (2015). https://doi.org/10.3115/v1/N15-1011
21. Jacovi, A., Sar Shalom, O., Goldberg, Y.: Understanding convolutional neural networks for text classification, pp. 56–65 (2019). https://doi.org/10.18653/v1/w18-5408
22. Yang, Z., Yang, D., Dyer, C., He, X., Smola, A., Hovy, E.: Hierarchical attention networks for document classification. In: Proceedings of the 2016 Conference of the North American Chapter of the Association for Computational Linguistics: Human Language Technologies, pp. 1480–1489. Association for Computational Linguistics, Stroudsburg, PA, USA (2016). https://doi.org/10.18653/v1/N16-1174
23. Wei, F., Qin, H., Ye, S., Zhao, H.: Empirical study of deep learning for text classification in legal document review. In: 2018 IEEE International Conference on Big Data (Big Data), pp. 3317–3320. IEEE (2018). https://doi.org/10.1109/BigData.2018.8622157
24. Lauren, P., Qu, G., Zhang, F., Lendasse, A.: Discriminant document embeddings with an extreme learning machine for classifying clinical narratives. Neurocomputing. **277**, 129–138 (2018). https://doi.org/10.1016/J.NEUCOM.2017.01.117
25. Dinh, D., Tamine, L.: Towards a context sensitive approach to searching information based on domain specific knowledge sources. J. Web Semant. **12–13**, 41–52 (2012). https://doi.org/10.1016/J.WEBSEM.2011.11.009
26. Bayyapu, K.R., Dolog, P.: Tag and neighbour based recommender system for medical events. In: Proceedings of the First International Workshop on Web Science and Information Exchange in the Medical Web, MedEx 2010, pp. 14–24. APA (2010)
27. Kadriu, A., Abazi, L., Abazi, H.: Albanian text classification: bag of words model and word analogies. Bus. Syst. Res. J. **10**, 74–87 (2019). https://doi.org/10.2478/bsrj-2019-0006
28. Stein, R.A., Jaques, P.A., Valiati, J.F.: An analysis of hierarchical text classification using word embeddings (2018). https://doi.org/10.1016/j.ins.2018.09.001
29. Faggella, D.: What is machine learning? https://emerj.com/ai-glossary-terms/what-is-machine-learning/. Accessed 10 Oct 2019
30. Wong, A., Kamel, M.S., Sun, Y., Wong, A.K.C.: Classification of imbalanced data: a review pattern-directed aligned pattern clustering view project pattern discovery in gene expression data view project classification of imbalanced data: a review. Artic. Int. J. Pattern Recognit. Artif. Intell. 23 (2011). https://doi.org/10.1142/S0218001409007326
31. Ciapetti, A., Di Florio, R., Lomasto, L., Miscione, G., Ruggiero, G., Toti, D.: NETHIC: a system for automatic text classification using neural networks and hierarchical taxonomies. In: Proceedings of the 21st International Conference on Enterprise Information Systems, pp. 296–306. SCITEPRESS - Science and Technology Publications (2019). https://doi.org/10.5220/0007709702960306

Improving Corporate Support by Predicting Customer e-Mail Response Time: Experimental Evaluation and a Practical Use Case

Anton Borg[1(✉)], Jim Ahlstrand[2], and Martin Boldt[1]

[1] Blekinge Institute of Technology, 37179 Karlskrona, Sweden
{anton.borg,martin.boldt}@bth.se
[2] Telenor AB, Karlskrona, Sweden
jim.ahlstrand@telenor.se

Abstract. Customer satisfaction is an important aspect for any corporations customer support process. One important factor keeping the time customers' wait for a reply at acceptable levels.

By utilizing learning models based on the Random Forest Algorithm, the extent to which it is possible to predict e-Mail time-to-respond is investigated. This is investigated both for customers, but also for customer support agents. The former focusing on how long until customers reply, and the latter focusing on how long until a customer receives an answer.

The models are trained on a data set consisting of 51,682 customer support e-Mails. The e-Mails covers various topics from a large telecom operator. The models are able to predict the time-to-respond for customer support agents with an AUC of 0.90, and for customers with an AUC of 0.85. These results indicate that it is possible to predict the TTR for both groups. The approach were also implemented in an initial trial in a live environment.

How the predictions can be applied to improve communication efficiency, e.g. by anticipating the staff needs in customer support, is discussed in more detail in the paper. Further, insights gained from an initial implementation are provided.

Keywords: e-Mail time-to-respond · Prediction · Random forest · machine learning · Decision support

1 Introduction

An important element in any corporation is to maintain high-quality and cost-efficient interaction with the customers. This is especially important for interactions between the organization and customer via customer support, since failing

This research work is supported by the Knowledge Foundation in Sweden through project "Scalable resource-efficient systems for big data analytics" (grant: 20140032).

J. Filipe et al. (Eds.): ICEIS 2020, LNBIP 417, pp. 100–121, 2021.
https://doi.org/10.1007/978-3-030-75418-1_6

to resolve customers issues satisfactorily risk negatively affecting the customers view of the organization. Further, in a prolongation this might affect the overall reputation of the organization. In highly competitive markets, a single negative customer service experience can deter potential new customers from a company or increase the risk of existing customers to drop out [15], both negatively affecting the volume of business.

For many customers E-mails still account for an important means of communication due to both its ease and widespread use within almost all age groups [19]. As such, implementing efficient customer service processes that target customer E-mail communication is a necessity for corporations as they receive large numbers of such customer service E-mails each day. Furthermore, the customers expect short response times to digital messages, which further complicates the customer service process [10].

This paper is an invited journal extension of a previously published conference paper [7]. In this study, and the original, we investigate the possibility to use supervised machine learning in order to predict when an E-mail response will be received, time-to-respond (TTR) or responsiveness. The semi-automated customer service E-mail management system studied exists within one of the bigger telecom operators in Europe with over 200 million customers worldwide, and some 2.5 million in Sweden. When these customers experience problems they often turn to E-mail as their means of communication with the company, by submitting an E-mail to a generic customer service E-mail address. Under-staffing might impact the efficiency of customer support, negatively impacting customer relations. However, while over-staffing might produce quick responses, it might also result in customer support agents being idle. Consequently, it is important to be able to predict the customer support workload in order to successfully schedule personnel and improve communication efficiency [28]. This extension investigates the previous results in more detail, as well as an implementation of the proposed system in a production environment. Further, approaches to utilizing the suggested system is discussed, with arguments for and against.

Customer service E-mails, provided by the telecom company, contains support errands with different topics. Each customer service E-mail might contain different topics, and the importance of each topic might be of varying importance, depending on the customer. Different topics require different actions by customers, and thus would require varying time before a response can be expected. The content of an E-mail within a topic, e.g. invoice, might also affect time-to-respond, as certain actions are more complicated than others. Further, a customer service E-mail might contain two paragraphs of text, one detailing a technical issue, and the other one an order errand. As such, the E-mail topic would be sorted as *Invoice*, *TechicalIssue*, and *Order*. This would further affect response time.

1.1 Aims and Objectives

In this study we investigate the possibility to predict the time-to-respond for received E-mails based on its content. If successful, it would be possible to adjust the schedules for customer support personnel in order to improve efficiency. The two main questions investigated in this work are as follows. First, to what extent it is possible to predict the time required by customer support agents to respond to E-mails. Second, to what extent it is possible to predict the time it takes customers to respond to E-mails from customer support personnel.

1.2 Scope and Limitations

The scope of this study is within a Swedish setting, involving E-mail messages written in Swedish sent to the customer service branch of the studied telecom company. However, the problem studied is general enough to be of interest for other organizations as well. In this study, E-mails where no reply exists have been excluded, as it has been suggested to be a separate classification task [16]. Further, time-to-respond (TTR) is investigated independent of the workload of agents, and the content of the E-mails.

2 Related Work

Time-to-respond, or responsiveness, can affect the perceived relationship between people both positively and negatively [2,3,10].

Investigations into mobile instant messaging (e.g. SMS) indicates that it is possible to predict whether a user will read a message within a few minutes of receiving it (70.6% accuracy) [23]. This can be predicted based on only seven features, e.g. screen activity, or ringer mode.

Responsiveness to IM has been investigated, and been predicted successfully (90% accuracy) [3]. The paper where limited to messages initiating new sessions, but the model where capable of predicting whether an initiated session would get a response within 30 s, 1, 2, 5, or 10 min. Predicting the response time when interacting with chatbots using IM have also been investigated, within four time intervals <10 s, 10–30 s, 30–300 s, and >300 s (Accuracy of 0.89), but also whether a message will receive a response [16].

Similarly to IM, response time in chat-rooms have also been investigated, with one study finding that the cognitive and emotional load affect response time within and between customer support agents [24]. In a customer support setting, the cognitive load denotes e.g. the number of words or amount of information that must be processed. TTR predictions have also been investigated in chat rooms (AUC 0.971), intending to detect short or long response times [18].

However, it seems that there is little research that have investigated predicting the TTR of E-mails in a customer support setting. This presents a research gap as it has been argued that E-mails are a distinct type of text compared to types of text [4]. Research indicates that it is possible to estimate the time for an E-mail response to arrive, within the time intervals of <25 min, 25–245 min, or >245 min [28]. Similarly, research has been conducted on personal E-mail (i.e. non-corporate) [19]. In addition to this, prior research has mainly focused on assigning broad topical topics to E-mail messages. However, the work by

Wang et al. from 2019 indicate that sentence-level intent identification could be achieved using both machine learning approaches [27]. Which could open up for investigating different predicted response times for different subsets of sentences in an E-mail.

The identified gap in existing research, that the present article is based upon, it first that there is a quite limited amount of prior research about reply-time prediction of E-mails in a corporate setting. Secondly, the research that do exists investigates quite small TTRs which, although suitable for employee E-mails, might not conform to the customer support setting according to domain experts. Further, the workload estimation of customer support agents work resolution benefits from an increased resolution, i.e. more bins.

3 Data

The data set consists of 51,682 E-mails from the customer service department from a Swedish branch of a major telecom corporation. Each E-mail consists of the:

- subject line,
- send-to address,
- sent time, and
- e-Mail body text content.

Each e-Mail is also labeled with at least one label. In total there exists 36 distinct topic labels, each independent from the others, where several of these might be present in any given E-mail. The topics have been set by a rule-based system that was manually developed, configured and fine-tuned over several years by domain expertise within the corporation.

Table 1. Description of the features extracted or calculated from the data set [7].

Feature name	Type	Value range	Description
Text sentiment	Float	$[-1, +1]$	Text sentiment of an E-mail ranging negative to positive
Customer escalated	Boolean	$\{0, 1\}$	Whether customer changed between messages in thread
Agent escalated	Boolean	$\{0, 1\}$	Whether agent changed between messages in thread
Old	Boolean	$\{0, 1\}$	Whether a message is older than 48 h, or not
Text complexity	Float	$[0, 100]$	Indication of the text complexity
Sender	Categorical	Text	The E-mail address of the sender
Message length	Integer	≥ 1	Number of characters in each E-mail message

A *DoNotUnderstand* topic label acts as the last resort for any E-mail that the current labeling system is unable to classify. Those E-mails have been excluded from the data set and each E-mail has been anonymized. Further, ends of threads have been excluded from the data set (i.e. E-mails where no reply exists), as that has been suggested to be a separate classification task [16].

The E-mails are grouped into conversation threads, and for each E-mail the date and time sent is available, enabling the construction of a timeline for each thread. Further, it is possible to shift the time-date information in each thread by one step, so that the future sent time is available for each E-mail in the thread. As such, this data can be considered the TTR. In order to adjust the resolution of the TTR, the date-time where binned into groups [3]. The bins were decided by consulting with the telecom company and thus using their domain knowledge. Six bins where utilized: response within 2 h, between 2–4 h, between 4–8 h, between 8–24 h, between 24–48 h, and more than 48 h. The bins are considered as the class labels.

The data set is divided into subsets, by topic and sender. The topics *Credit* ($n = 6,239$), *Order* ($n = 2,221$), and *ChangeUser* ($n = 1,398$) are used to investigate this problem. Further, similar to the work by Yang et al. each topic is divided into one set for E-mails sent from the telecom corporation and another set for E-mails sent by the customer [28]. In this case, the agents can be considered a more homogeneous group (similar training and experience), whereas the customers could be regarded as a heterogeneous group (different background and experiences). As such, six data sets have been created.

The class distribution in the data set is exemplified by ChangeUser topic in Fig. 1 for agents and Fig. 2 for customers. A majority of the messages have a TTR within two hours, followed by a TTR longer than 48 h, 8–24 h. A minority of messages have a TTR between 2–4, 4–8, or 24–48 h. Consequently, it would seem that messages get responses "immediately", the next day, or after 2 days.

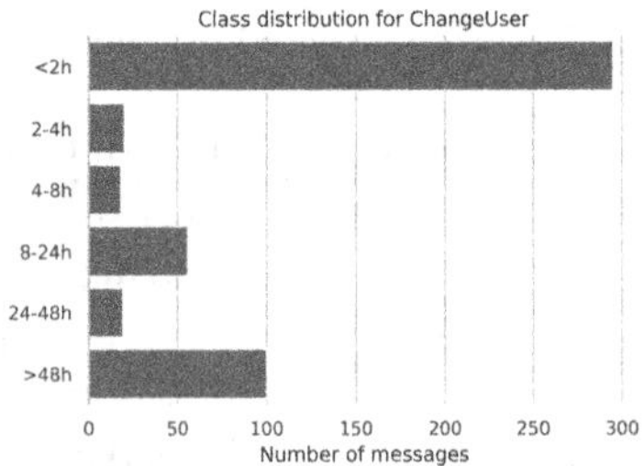

Fig. 1. ChangeUser agent TTR class distribution [7].

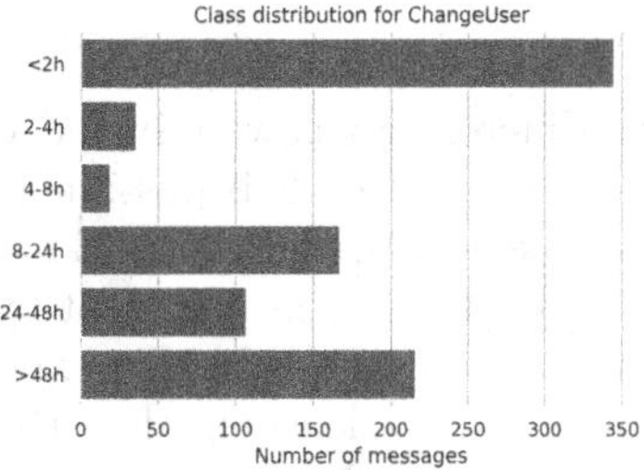

Fig. 2. ChangeUser customer TTR class distribution [7].

3.1 Feature Extraction

For each E-mail in the data set, seven features are calculated or extracted. A summary of all features used in the study are shown in Table 1. First, Vader sentiment is used to calculate the *Text sentiment* for each E-mail [17,24]. As the primary language in this data set is Swedish, a list of Swedish stop-words was used[1]. However, the Swedish stop-words were extended by English stop-words, as a fair amount of English also occurs due to the corporate environment.

Second, for each message it was calculated whether the customer or support agent responding participating in the conversation had changed over the thread timeline, denoted by the Boolean variables *Customer escalated* and *Agent escalated* respectively. A change in e.g. customer support agent indicates the involvement of an agent experienced in the current support errand. However, related work indicates that as the number of participants increase, so do the time to respond [28]. The variable *Old* denotes if the message has not received a response for 48 h or more, as per internal rules at the company. A *Text complexity* factor for the text is also calculated as per

$$CF = \frac{|\{x\}|}{|x|} \times 100, \qquad (1)$$

where x is the E-mail content [1]. Consequently, Eq. 1 is the number of unique words in the E-mail divided by the number of words in the E-mail. A higher score indicates a higher complexity in the text, which can affect the TTR [24]. It should be noted that there exist different readability scores for the English language, e.g. *Flesch–Kincaid* score [11]. However, the applicability of these on Swedish text is unknown. Finally, the *Sender* and *Length* of the E-mail is also included as variables.

4 Method

This section describes the experimental approach, which includes for instance the design and chosen evaluation metrics.

[1] https://gist.github.com/peterdalle/8865eb918a824a475b7ac5561f2f88e9.

4.1 Experiment Design

Two experiments with two different goals were included in this study. The first experiment aimed to investigate whether it is possible to predict the time a customer support agent would take to respond to the E-mail received. As such, the experiment used the data set containing E-mails sent by the customer and tried to predict when the agent would respond. In this experiment the independent variable was the models trained by the learning algorithms described in Sect. 4.2. The dependent variables were the evaluation metrics described in Sect. 4.4, of which the AUC metric was chosen as primary.

The second experiment is similar to the first one, but instead uses the data sets containing E-mails sent by the customer support agents, thus aiming to predict when the customer will respond. As such both the independent and the dependent variables were the same as in the first experiment.

Evaluation of the classification performance was handled using a 10-times 10-fold cross-validation approach in order to train and evaluate the models [13]. Each model's performance was measured using the metrics presented in Sect. 4.4.

4.2 Included Learning Algorithms

Random Forest [9] was selected as the learning algorithm to investigate in this study. It is a suitable algorithm as the data contains both Boolean, categorical, and continuous variables. Initially a SVM model [13] was also evaluated, but since Random Forest significantly outperformed the SVM models, they were excluded from the study. The reason to why the SVM model showed inferior performance is not clear. Although it is in line with the "No free lunch" theorem, stating that no single model is best in every situation. Thus, models' performance varies when evaluated over different problems.

The models trained by the Random Forest algorithm were compared against a Random Guesser classifier using a uniform random guesser as baseline [22,28].

4.3 Class-Balance

In order to deal with the class imbalance of the different bins, a multi-class oversampling strategy was used that relied on SMOTE and cleaned by removing instances which are considered Tomek links [5,20]. Using only oversampling can lead to over-fitting of the classifiers as majority class examples might overlap the minority class space, and the artificial minority class examples might be sampled too deep into the majority class space [5].

4.4 Evaluation Metrics

The models predictive performance in the experiments were evaluated using standard evaluation metrics calculated based on the True Positives (TP), False Positives (FP), True Negatives (TN), and False Negatives (FN). The evaluation metrics consists of the F_1*-score* (micro average), *Accuracy*, and *Area under ROC-curve* (AUC) (micro average).

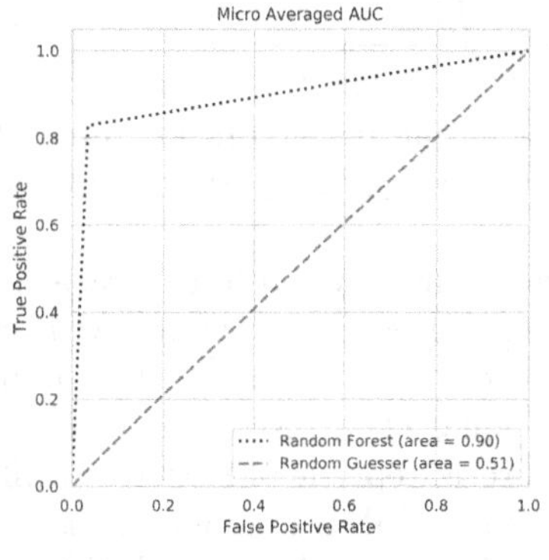

(a) Agent TTR predictions.

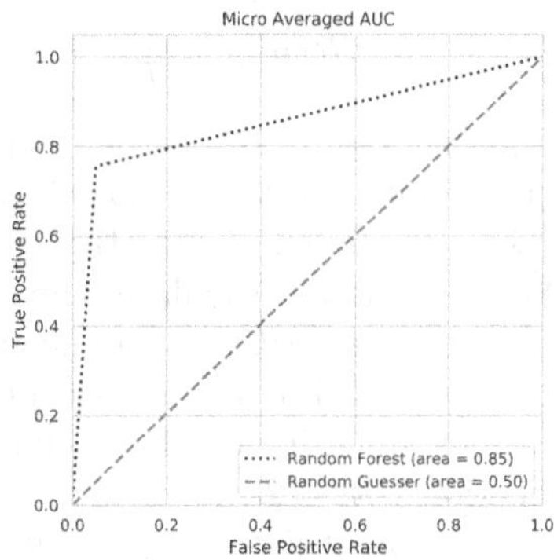

(b) Customer TTR Predictions.

Fig. 3. Micro averaged AUC for agent predictions (3a) and customer predictions (3b) over the different topics [7].

The theoretical ground for these metrics are explained by Flach [13]. The first metric is the traditional Accuracy that is defined as in Eq. 2 [29]:

$$Acc = \frac{TP + TN}{TP + TN + FP + FN} \tag{2}$$

It is a measurement of how well the model is capable of predicting TP and TN compared to the total number of instances. For the multi-class case, the accuracy is equivalent to the Jaccard index. Accuracy ranges between 0.0–1.0, where 1.0 is a perfect score.

However, in cases where there is a high number of negatives, e.g. in a multi-class setting, accuracy is not representative. In these cases the F_1-score is often used as an alternative, as it doesn't take true negatives into account [13]. Similar to the Accuracy, the F_1-score ranges between 0.0–1.0, where 1.0 is a perfect score. It is calculated as described in Eq. 5, based on Eq. 3 and Eq. 4. In this study micro-averaging was used as the number of labels might vary between classes [29].

$$Precision = \frac{TP}{TP + FP} \tag{3}$$

$$Recall = \frac{TP}{TP + FN} \tag{4}$$

$$F_1 = 2 * \frac{Precison * Recall}{Precision + Recall} \tag{5}$$

For micro-averaging, precision and recall are calculated according to Eq. 6 and Eq. 7 respectively, where n is the number of classes.

$$Precision_{\mu} = \frac{TP_1 + ... + TP_n}{TP_1 + ... + TP_n + FP_1 + ... + FP_n} \tag{6}$$

$$Recall_{\mu} = \frac{TP_1 + ... + TP_n}{TP_1 + ... + TP_n + FN_1 + ... + FN_n} \tag{7}$$

Hamming loss measures the fraction of labels that are incorrect compared to the total number of labels [26]. A score of 0.0 represents a perfect score as no labels were predicted incorrectly.

The AUC metric calculates the area under a curve, which in this case is the ROC. Hence, the AUC is also known as the Area under ROC curve (AUROC). AUC is often used as a standard performance measure in various data mining applications since it does not depend on an equal class distribution and misclassification cost [12]. A perfect AUC measure is represented by 1.0, while a measure of 0.5 is the worst possible score since it equals a random guesser.

To investigate how much the predictions deviate from the actual TTR, three regression metrics are used: Mean Absolute Error (MAE), Median Absolute Error (MedAE), and Mean Squared Error (MSE) [22]. Mean absolute Error is defined in Eq. 8 [22].

$$MAE(y, \hat{y}) = \frac{1}{n} \sum_{i=0}^{n-1} |y_i - \hat{y}_i| \tag{8}$$

Similar to MAE is the MedAE, included due to its resilience to outliers [22]. It is defined as in Eq. 9.

$$MedAE(y, \hat{y} = median(|y_i - \hat{y}_i|, ..., |y_n - \hat{y}_n| \tag{9}$$

MSE is defined in Eq. 10, and measures the average squared error [22].

$$MSE(y, \hat{y}) = \frac{1}{n} \sum_{i=0}^{n-1} (y_i - \hat{y})^2 \tag{10}$$

5 Results

The results are divided into two subsections, one for each of the experiments described in Sect. 4.1.

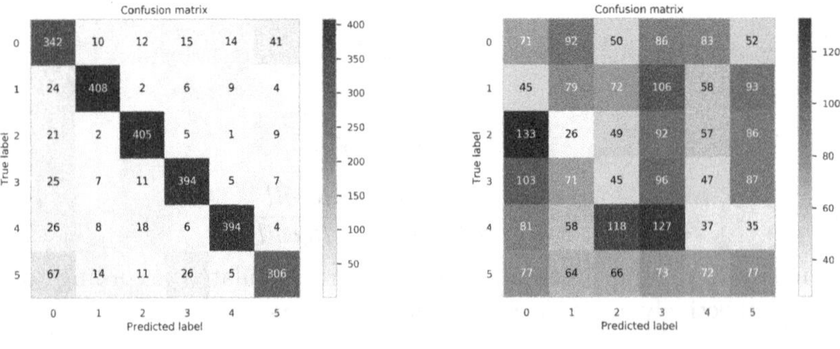

(a) Random Forest. **(b)** Random Guesser baseline.

Fig. 4. Agent TTR prediction performance per class for Random Forest (4a) and Random Guesser (4a) for the customer support topic Order [7].

Table 2. Agent time-to-reply (TTR) classification results [7].

Topic	Model	Accuracy (std)	AUC (std)	F_1-score (std)	Hamming (std)
ChangeUser	Random Forest	0.8720 (0.0214)	0.9232 (0.0128)	0.8720 (0.0214)	0.1279 (0.0214)
	Baseline	0.1758 (0.0324)	0.5055 (0.0194)	0.1758 (0.0324)	0.8241 (0.0324)
Credit	Random Forest	0.8149 (0.0096)	0.8889 (0.0057)	0.8149 (0.0096)	0.1850 (0.0096)
	Baseline	0.1822 (0.0082)	0.5093 (0.0049)	0.1822 (0.0082)	0.8177 (0.0082)
Order	Random Forest	0.8442 (0.0310)	0.9065 (0.0186)	0.8442 (0.0310)	0.1557 (0.0310)
	Baseline	0.1535 (0.0114)	0.4921 (0.0068)	0.1535 (0.0114)	0.8464 (0.0114)

Table 3. Agent time-to-reply (TTR) deviations. Indicates how many steps from the correct class the predictions are. The values are mean values and standard deviation over the folds.

Topic	Model	MAE (std)	MedAE (std)	MSE (std)
ChangeUser	Random Forest	0.5652 (0.0863)	0.0 (0.0000)	1.8353 (0.3685)
	Baseline	1.9227 (0.1190)	1.7 (0.4830)	5.8345 (0.5370)
Credit	Random Forest	0.6912 (0.0484)	0.0 (0.0000)	2.4379 (0.2240)
	Baseline	1.9230 (0.0288)	2.0 (0.0000)	5.7048 (0.1425)
Order	Random Forest	0.6372 (0.0955)	0.0 (0.0000)	2.1949 (0.4540)
	Baseline	1.9282 (0.0818)	2.0 (0.0000)	5.5868 (0.3628)

5.1 Experiment 1: Customer Agent Response Prediction

Figure 3a shows the micro-averaged AUC over the different topics for Random Forest and the random guesser when predicting support agents' E-mail response times. As expected, the random guesser models have a worst-case AUC metric of 0.51. While the models trained by the Random Forest algorithm show interesting predictive results with an overall AUC metric of 0.90, which significantly outperforms random chance.

Figure 4 shows the absolute confusion matrix for the predicted agent response times vs. the true agent response times for the Random Forest algorithms (Fig. 4a) and the Random Guesser baseline (Fig. 4b). The matrix shows the aggregated results over the different test folds showing that the Random Guesser baseline classifier randomly appoints the classes. In contrast the Random Forest model has a clear diagonal score that indicates significantly better prediction performance compared to the random baseline.

This is supported by the results shown in Table 2 in which the Random Forest models have AUC scores slightly above or below 0.9. In fact the 95 % confidence interval of the AUC metric for each of the three class labels *ChangeUser*, *Credit* and *Order* were 0.92 ± 0.026, 0.89 ± 0.011 and 0.91 ± 0.037 respectively. This indicates the models have a good ability to predict the TTR over various class labels. This is further strengthened when evaluating the predictive performance in terms of accuracy or F_1-scores instead. Although using these metrics, the

Random Forest models still performs well above 0.80, whereas the Random Guesser baseline models are associated with useless scores at 0.18, or worse. Figure 4a further indicates that the models have a slightly higher misclassification for label 0, and 5 compared to the other labels. This indicate that TTR prediction within 2 h and beyond 48 h are slightly more difficult to predict.

Table 3 details how the predicted TTR deviates from the actual TTR. This is shown using three regression metrics: Mean Absolute Error (MAE), Median Absolute Error (MedAE), and Mean Squared Error (MSE). The Random Forest model deviates, on average, 0, 4526 bins from the actual value. The Random Guesser, however, on average deviates 1, 9209 bins from the actual value. Similarly, the MedAE for the random forest is 0, and for the Random Guesser it is on average 1.9. Given that the the actual TTR is between 4–8 h, the Random Forest classifier will likely predict that value, whereas the Random Guesser is likely to suggest TTR values as either <2 h or 24–48 h.

5.2 Experiment 2: Customer Response Prediction

Similar to the previous experiment, Fig. 3b shows that the Random Guesser baseline models have an AUC metric of 0.50 while the Random Forest models significantly outperforms that with a metric of 0.85 when predicting customers' E-mail response times. Similar to the results in Sect. 5.1, Fig. 5 shows the absolute confusion matrix for the predicted agent response times vs. the true agent response times for the Random Forest models (Fig. 5a) and the Random Guesser baseline models (Fig. 5b).

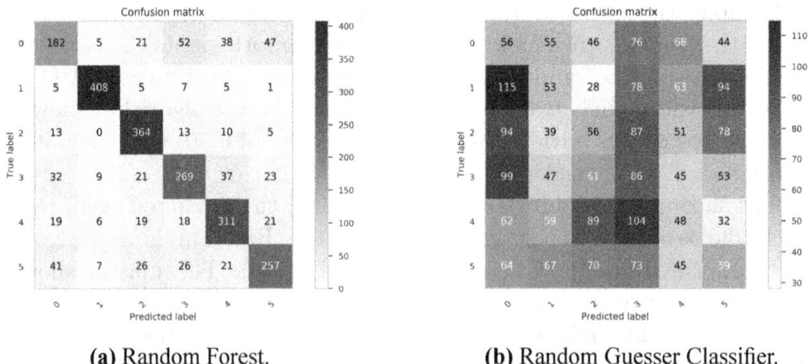

(a) Random Forest.

(b) Random Guesser Classifier.

Fig. 5. Customer TTR prediction performance per class for Random Forest (5a) and Random Guesser (5b) for the customer support topic Order [7].

The matrix shows the aggregated results over the different test folds that clearly show the increased predictability performance for the Random Forest models.

Table 4. Customer time-to-reply (TTR) classification results [7].

Topic	Model	Accuracy (std)	AUC (std)	F_1-score (std)	Hamming (std)
ChangeUser	Random Forest	0.7760 (0.0231)	0.8656 (0.0138)	0.7760 (0.0231)	0.2239 (0.0231)
	Baseline	0.1592 (0.0223)	0.4955 (0.0134)	0.1592 (0.0223)	0.8407 (0.0223)
Credit	Random Forest	0.7486 (0.0088)	0.8491 (0.0053)	0.7486 (0.0088)	0.2513 (0.0088)
	Baseline	0.1798 (0.0055)	0.5079 (0.0033)	0.1798 (0.0055)	0.8201 (0.0055)
Order	Random Forest	0.7640 (0.0154)	0.8584 (0.0092)	0.7640 (0.0154)	0.2359 (0.0154)
	Baseline	0.1527 (0.0216)	0.4916 (0.0130)	0.1527 (0.0216)	0.8472 (0.0216)

Table 5. Customer time-to-reply (TTR) deviations. Indicates how many steps from the correct class the predictions are.

Topic	Model	MAE (std)	MedAE (std)	MSE (std)
ChangeUser	Random Forest	0.3742 (0.1307)	0.0 (0.0)	1.3538 (0.6225)
	Baseline	1.9678 (0.0705)	2.0 (0.0)	5.8679 (0.3808)
Credit	Random Forest	0.5181 (0.0616)	0.0 (0.0)	1.8310 (0.2864)
	Baseline	1.8968 (0.0251)	2.0 (0.0)	5.5927 (0.1449)
Order	Random Forest	0.4657 (0.0926)	0.0 (0.0)	1.7425 (0.3723)
	Baseline	1.8982 (0.0525)	2.0 (0.0)	5.5500 (0.2747)

This is supported by the results shown in Table 4 where the Random Forest models have F_1-scores around 0.75, whereas the Random Guesser baseline have F_1-scores around than 0.15. Further, the Random Forest model have mean AUC scores between 0.85 and 0.87. In fact the 95 % confidence interval of the AUC metric for each of the three class labels *ChangeUser*, *Credit* and *Order* were 0.87 ± 0.028, 0.85 ± 0.011 and 0.86 ± 0.018 respectively. Although the metrics are slightly lower compared to the results from the first experiment, this similarly indicates the potential in predicting TTR for E-mails.

Similar to Fig. 4a, Fig. 5a indicates that the model have a higher misclassification for label 0 and 5 than the other labels, indicating that a TTR within 2 h and beyond 48 h are more difficult to predict. Two things are different from Fig. 4a. First, label 3 is also more difficult to predict. Second, the misclassifications are slightly worse than for Fig. 4a.

Tabel 5 details how the predicted TTR deviates from the actual TTR. The Random Forest model deviates, on average, 0, 4436 bins from the actual value. The Random Guesser, however, on average deviates 1, 9209 bins from the actual value. Similarly, the MedAE for the random forest is 0, and for the Random Guesser it is on average 2. Given that the actual TTR is between 4–8 h, the Random Forest classifier will likely predict that value, whereas the Random Guesser is likely to suggest TTR values as either <2 h or 24–48 h. These results are similar to the results in Experiment 1.

6 Practical Use Case: An Initial Implementation

The suggested approach for predicting characteristics in support E-mails has been implemented in the customer support branch of a large telecom corporation. In this setting the implemented system has been evaluated for a small number of E-mail groups over 3 months time. During this time it has predicted the response time for 3, 158 E-mails. The goal with this practical use case is to shed light on whether or not it is possible to predict the customer response time in a practical customer support setting.

The training data was based on E-mails from March 2019 through April 2020, consisting of 78,000 threads of a minimum of 3 E-mails in each thread. There was no overlap between this data and the data used in experiment 1 & 2. The following E-mail topics, considered e.g. internal, were removed: *ThanksReply*, *DoNotUnderstand*, *Spam* and *Undeliverable*. The topic within those E-mails can be considered to have no customer contact, unknown content and no response could be expected.

Compared to the experimental evaluation described in previous sections there were several methodological changes made. Most importantly a single model was trained for all 36 topics, instead of one model-per-topic as used in the experiments, due to resource issues. Further, the bins where changed from within 2 h, between 2–4 h, between 4–8 h, between 8–24 h, between 24–48 h, and more than 48 h. Instead the bins used are within a couple of minutes, between a couple of minutes to hours, between a few hours, between a couple of hours, between hours to days, and more than 2 days. As a consequence of those modifications, it is challenging to compare the results from the practical use case implementation to the experiments. However, the results from this use case implementation are still interesting as they provide practical experience in using the proposed approach.

The evaluation metrics used in the practical use case are the same as for the experiments, i.e. Accuracy, F_1-score, Jaccard index, Hamming loss, as well as MAE and MedAE. The mean and median absolute error is measured on the bins, i.e. how many bins away from the correct TTR do the model predict.

The F_1-score and Accuracy are both 0.2222, Jaccard index is 0.1250, and the Hamming loss is 0.7777, while the mean absolute error is 1.6375 and the median absolute error is 2.0. It is clear that the results are quite poor compared to the experimental evaluation. This is to some extent due to that the current data is much more heterogeneous compared to the data used in the experiments, and maybe also due to the changed bin-division. Although, the most significant reason for the poor results is likely that the practical use case uses a single model rather than one individual model-per-topic that were used in the experiments. The reason for using a single model solution was since that requires considerably less time to train and setup, compared to individually trained models per topic. Finally, it should be noted that the baseline of a uniform random guesser have an accuracy of 0.1666, thus the results are still slightly better than chance (Fig. 6).

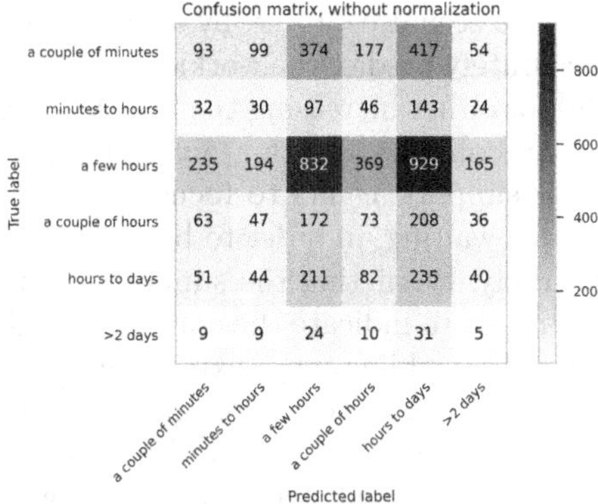

Fig. 6. Agent TTR prediction performance per class for Random Forest in the use case.

The main conclusion that can be drawn from the practical use case is that a single model approach is not preferable to a model-per-topic approach. This can potentially be explained by 1) the heterogeneous data used, and 2) the class imbalance of the data (as shown in Fig. 7), e.g. the topic DoNotUnderstand contains E-mails that the topic classifier [8] is unable to categorize. Consequently, it might contain vastly different E-mails and, as such, the predictability of the time-to-reply should be quite low, i.e. out of 811 E-mails, only 178 where correctly classified. The topic have a mean and median absolute error of 1.6991 and 2.0 respectively, as well as an mean squared error of 4.5425. Given the result of the use case, combined with the experimental results, the model-per-topic approach should be implemented for a more extensive trial.

7 Analysis and Discussion

The results presented in Sect. 5 indicates that it is possible to predict the response time for customer support agents, as well as the response time for when customers will respond to E-mails received. Organizations can benefit from these

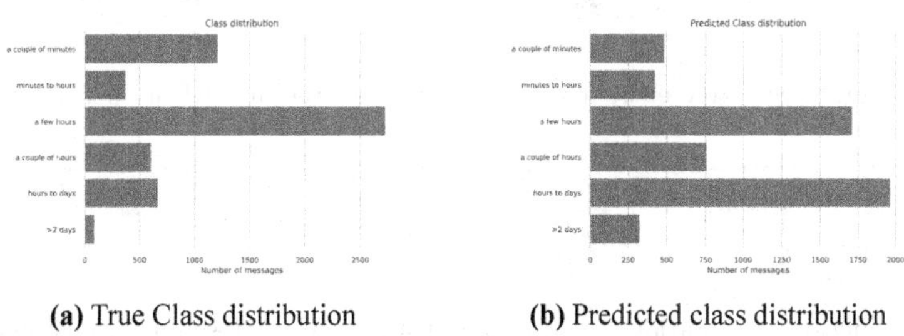

(a) True Class distribution **(b)** Predicted class distribution

Fig. 7. Customer TTR distribution per class for the ground truth (7a) and predicted data (7a) in use case test data.

conclusions in (at least) two scenarios. First, by aggregating the customer TTR, it is possible to more accurately predict the workload of agents the next couple of days. This can be useful for either increasing the workforce, or shifting personnel working on other topics. Secondly, given that a predicted customer TTR is low, it might be advisable for support agents to focus on other E-mails with a low predicted agent TTR while waiting, in order to be able to respond quickly when the customers eventual reply. Predicting the support agents TTR is also useful since it can be used as a proxy to indicate the emotional and cognitive load associated with each E-mail [24], enabling more experienced agents to handle them, planning the agents' workload (e.g. several low agent TTR E-mails, or a few long agent TTR E-mails). Further, predicting agent TTR allow for customers to be alerted when a support errand is predicted to take a longer time than usual.

Even though the feature set of this experiment is based partly on related work and partly on domain expertise, it is important to investigate that the models have not learnt trivial solutions. For this reason, a random tree in the Random Forest model has been extracted and visualized using the Graphviz framework [14], of which a sub-tree can be seen in Fig. 8. This tree indicates that the model has indeed not learnt a trivial solution when predicting TTR.

As a way to further investigate the internals of the models trained by Random Forest, the ELI5 model interpretation framework was used to estimate the features' impact on the model's class assignment. Table 6 shows the relative impact each feature has on the predictive result in the Random Forest model.

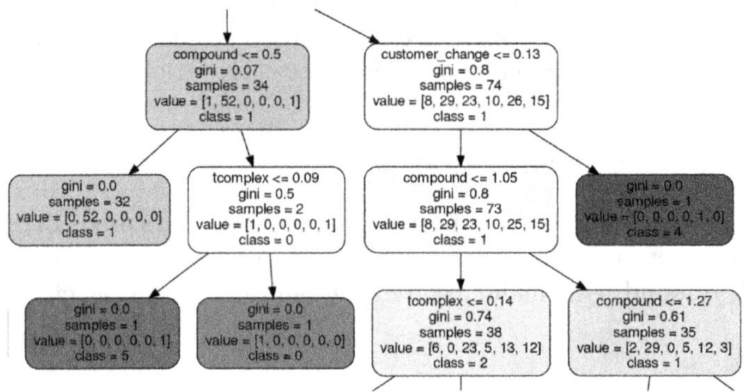

Fig. 8. Sub-tree extracted from a random tree in a RF model [7].

y=<2h (probability 0.007) top features		y=2-4h (probability 0.001) top features		y=4-8h (probability 0.980) top features		y=8-24h (probability 0.002) top features		y=24-48h (probability 0.002) top features		y=>48h (probability 0.008) top features	
Contribution	Feature	Contribution	Feature	Contribution	Feature	Contribution	Feature	Contribution	Feature	Contribution	Feature
+0.143	<BIAS>	+0.185	<BIAS>	+0.324	Sender	+0.164	<BIAS>	+0.173	<BIAS>	+0.162	<BIAS>
+0.006	Sentiment Score	+0.021	Sender	+0.173	<BIAS>	+0.000	Old	-0.010	Customer Escalated	+0.016	Sentiment Score
-0.001	Customer Escalated	+0.004	Old	+0.170	Sentiment Score	-0.000	Customer Escalated	-0.012	Old	-0.006	Customer Escalated
-0.005	Old	-0.004	Customer Escalated	+0.138	Message Length	-0.002	Sentiment Score	-0.021	Text Complexity Factor	-0.009	Old
-0.031	Text Complexity Factor	-0.015	Text Complexity Factor	+0.131	Text Complexity Factor	-0.025	Message Length	-0.022	Message Length	-0.028	Message Length
-0.031	Message Length	-0.033	Message Length	+0.021	Old	-0.031	Text Complexity Factor	-0.034	Sentiment Score	-0.033	Text Complexity Factor
-0.075	Sender	-0.156	Sentiment Score	+0.021	Customer Escalated	-0.104	Sender	-0.071	Sender	-0.095	Sender

Fig. 9. Prediction explanation for a random instance in the test set for customer TTR prediction. The instance is a true positive, where an response were sent between 4–8 h from receiving the E-mail [7].

The most highly ranked feature is *Sender* followed by the *Message length* and *Text complexity*, which seem reasonable.

Table 6. Feature weights for a model predicting customer TTR [7].

Weight	Feature
0.2988 ± 0.0899	Sender
0.2094 ± 0.0770	Message length
0.2092 ± 0.0772	Text complexity factor
0.1685 ± 0.0724	Text sentiment
0.0598 ± 0.0365	Old
0.0544 ± 0.0322	Customer escalated
0 ± 0.0000	Escalated

An example of an instance being predicted using a Random Forest model can be seen in Fig. 9, where the probability and feature impact is shown for each possible response time bin. This particular instance, is a true positive as it is correctly assigned to the *4–8* bin with an accuracy of 0.98%. The most significant feature in favor for this decision is *Sender* that is assigned a weight of +0.324. The feature *BIAS* is the expected average score based on the distribution of the data[2]. In this case, the BIAS is quite similar between the classes as the data, after the preprocessing, are balanced between the classes. Overall, this analysis of feature impacts indicate that the model has picked up patterns in the E-mails that are relevant for predicting TTR. Thus, it can be concluded that the models have not learned useless patterns from artefacts in the data sets.

The results from both experiments in this study suggests that the models' performance are in line with results from related research in the problem domain of instant messaging: 0.71 accuracy [23], 0.89 accuracy [16], 0.90 accuracy [3]. These results compare well to the results in this study. See Table 2 and Table 4 that have accuracy scores between 0.81–0.87 and 0.74–0.77 respectively. Thus, compared to the state-of-the-art, the results presented in this study indicates improved performance ($AUC \approx 0.85$, $F_1 \approx 0.84$, $accuracy \approx 0.82$ in mean performance for agent TTR), compared to the best performing model among related work that was ADABoost ($AUC = 0.72$, $F_1 = 0.45$, $accuracy = 0.46$) [28].

While the MAE in reported in Table 5 indicates that there is a slight difference between the true and predicted TTR for the Random Forest, the MedAE for all three topics is 0.0. This indicates that a large amount of instances are correctly classified. In comparison, the MedAE for the random guesser is 2.0, i.e. a majority of the predictions are two steps from their correct class. The results reported in Table 3 are similar.

[2] https://stackoverflow.com/questions/49402701/eli5-explaining-prediction-xgboost-model, accessed: 2020-02-15.

The problem was investigated separately for support agents and for customers. The results suggest that it is easier to predict the time to respond for agents, than it is for customers (c.f. Fig. 4a and Fig. 5a). This supports the prior statement that in this setting the agents can be considered a more homogeneous group due to similar training and experience, whereas the customers could be regarded as a more heterogeneous group of persons with different background and experiences.

Finally, it should also be noted that a high TTR doesn't indicate a lower quality of customer support agents works. Rather, complex errands can take longer time. In this case weekends and non-working hours haven't been considered by the preprocessing. Consequently, an E-mail received outside of working hours will not be addressed until the next day, giving it a possible TTR of 12 h.

7.1 Enhanced Prioritization of Customer Support E-Mails

The ability to predict customers time-to-reply allows for more efficient prioritization of incoming E-mail messages. This work has mainly focused on predicting support agents predicted response time and customers predicted time to respond to agents' E-mails. However, there exists other features that could be useful for such a prioritization scheme. An initial list of interesting aspects to consider when prioritizing E-mail messages are:

- *Privilege*: a topic's privilege within the system, e.g. whether an ordinary topic or a topic of increased importance,
- *Time-to-reply*: how quickly the customer it likely to reply to a response from the customer support,
- *Sentiment*: whether a message is generally positive, negative or neutral,
- *Complexity*: the complexity of the E-mail content,
- *Escalated*: whether an E-mail has been escalated to a more senior support agent, and
- *Old*: whether an E-mail message remain unanswered after a certain amount of time, e.g. 48 h.

Making use of any of these aspects individually, or in combinations, open for interesting E-mail prioritization opportunities. As an example, let's consider the following three different hypothetical cases.

Simple Prioritization Setup: all incoming E-mail messages are handled by any of the support agents as to their own choice, with one exception. E-mail messages that are highlighted as being "old" (e.g. not answered within 48 h) must be addressed by the support agents before attending any other support errands. A positive aspect with this prioritization case is both that it is simple and thus easy to teach the support agents, but also that it makes sure all E-mails are handled within the predefined time frame. However, there are several negative concerns with this prioritization, e.g. that support errands concerning privileged topics or from upset customers for a quick response to their problems.

Moderate Prioritization Setup: incoming E-mails from privileged topics should be considered first, after that E-mails that are marked as old, and finally the remaining E-mails. While this setup it slightly more refined compared to the prior case, it is still quite simple to explain. Further, the setup addresses the problem with privileged topics not getting a timely response, but unfortunately fails to identify negative or upset customers.

Enhanced Prioritization Setup: E-mails with a negative sentiment that are also predicted to have a quick time-to-reply are prioritized first. Secondly, E-mails from privileged topics with a predicted quick time-to-reply. Thirdly, and E-mails that are marked as old. Finally, any remaining E-mail messages should be attended. The main positive aspect with this setup is that it addresses the customers with negative sentiment first, e.g. mentioning that they have a problem or maybe even that they are upset, since those customers also are the ones that potentially could cancel the service unless their issues are resolved in a timely fashion. Secondly, privileged topics that are important to handle in a timely fashion (i.e. those with a predicted quick time-to-reply). Note: E-mails concerning privileged topics with a negative sentiment are already addressed in the first step, while those that have a longer estimated time-to-reply are put on hold for a bit longer, e.g. if they just send a final positive thank you message to inform the their issue has been solved. One negative aspect with this setup is its complexity, since it would be more challenging to teach support personnel to adopt compared to the prior cases, or even harder to obey at all times. As a consequence, such prioritization setups instead could be implemented into decision-support modules within the E-mail client the personnel is using, i.e. so that the E-mails automatically are ranked based on their prioritization scores so that the personnel simply need to pick E-mail from the top to the list.

As shown in the three cases above, the more advanced the prioritization becomes the more business value is added. E.g. since privileged topics might get quicker responses, as well as negative customers that otherwise could risk canceling their subscriptions. Unfortunately the more advanced the prioritization gets, the harder it is for humans to implement without technical assistance in the form of decision-support systems.

However, another challenge could be to design the prioritization schemes in a suitable manner, especially if they consists of several parameters. For instance, imagine the enhanced prioritization setup described above, but add two additional parameters: E-mail complexity and whether an E-mail has been escalated to a more senior support agent. The schemes quickly become complex and hard to manually configure, especially if specific configurations are needed for different topics. A solution to this problem could be to make use of machine learning approaches that could train a suitable prioritization model based on the available parameters if some measure of customer satisfaction is available, which sometimes are collected already, e.g. after closed support errands. By using machine learning it would be possible to train a prioritization model that is implemented by the decision-support system. It could even be possible to use an online learning approach so that the prioritization model is automatically updated as the stream of new customer satisfaction measures reach the system. Thus, constantly keeping the system up to date.

7.2 e-Mail Forecasting

A another important aspect is that predicting TTR enables another approach to predicting the workload of agents the next couple of days. There are several methods available for forecasting [21]. Knowledge of historical trends can be used to estimate the number of new E-mails arriving each hour. This can be combined with the predicted TTR to give an more detailed prognosis. In this case forecasting models can give an idea of the number of new E-mails arriving each hour. Although evaluating the possible forecasting of new E-mails is outside of the scope of this paper, suggesting a way to incorporate the predicted TTR is not.

Traditional forecasting can predict x E-mails arriving at hour h, but the suggested approach for predicting TTR doesn't provide such an exact resolution. Rather the predicted TTR is a time span, which in some cases might have been possible to apply to hour h. However, in this case it might not be possible for two reasons. First, the predicted TTR is from when the customer support agents sent the E-mail to the customer, i.e. a rolling window approach and might not fit into hour h. Secondly, the predicted time spans that have been used are of varying size. E.g. for experiment 1 sizes of 2 h, 4 h, 16 h, 24 h, and more than 48 h are utilized. Fitting such predictions into e.g. 2 h slots might be difficult.

Utilizing time spans is not a new phenomenon, e.g. in criminology for residential burglaries is this quite common. In that case law enforcement often know that the burglary has taken place during a certain time span, but no further time information is available [25]. A suitable temporal analysis method when the data contains a time interval rather than an exact point in time is the aoristic method [25]. This method suits the predicted time-to-reply intervals in the present study, and could be used for analyzing the overall E-mail load on customer support. The methods works by choosing a structured time or temporal resolution, e.g. minutes or hours, and then assign each E-mail a point of 1.0. This point is then proportionally divided among the units within the time interval it spans [6]. The division is done regardless whether the time slots are fully or partially covered as shown in Eq. 11 in which n represents the number of slots covered.

$$aoristic = \frac{1.0}{n} \tag{11}$$

E.g. given an e-Mail that is predicted to arrive between 08:00–10:30 (i.e. a time span of 2.5 h) and a temporal resolution of 30 min, the time span point is divided by 5. Each 30 min. slot that the predicted times span overlaps consequently gets 0.2 points. A second E-mail is predicted to arrive between 09:00–10:00 (i.e. a time span of 1 h), and it's point is thus divided by 2. Each 30 min slot that the predicted times span overlaps consequently gets 0.5 points. This procedure is then repeated for every E-mail investigated. Finally, the points of every temporal unit are then summarized and used to indicate high profile time periods, see example in Fig. 10.

The aoristic score can be visualized in a bar plot, as in Fig. 10, or in a heatmap where columns and rows can indicate e.g. hours and weekday respectively.

Consequently, the method provides an intuitive way of visualizing when customers predicted TTR.

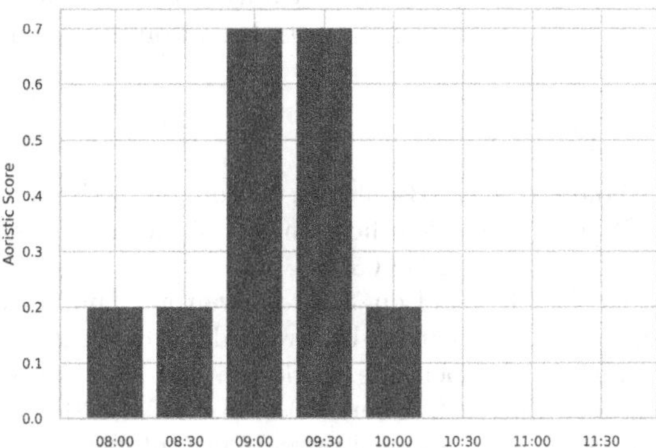

Fig. 10. The Aoristic score for two example e-Mails which predicted time-to-reply fall within the time span of 08:00–11:30.

8 Conclusion and Future Work

This study discusses a method for prioritizing E-mails within a corporate customer support setting though the use of predictive models from the machine learning field, which includes a practical trial implementation. Additionally, experiments are used to evaluate the predictive performance of such models. The method for prioritizing incoming customer support E-mails can make use of several E-mail properties when determining how time-critical a specific E-mail is to attend by the support personnel. Those E-mail properties could include for instance the sentiment of the content (whether positive, neutral or negative), the text complexity of the content, or how quickly the customer is likely to reply to a response. This study discusses different prioritization setups based on different combinations of the E-mail properties.

In addition, the predictive performance of various machine learning models for predicting future response-times for E-mails are investigated using experimental evaluations. The experimental results show that a predictive model that uses seven features from the E-mail message can predict both the time it take for support personnel to reply to an incoming E-mail (AUC 0.90), as well as the time-to-reply for customers to respond to E-mails sent by the support personnel (AUC 0.85). Overall, this study indicates that it is possible to better prioritize incoming customer support E-mail. So that the customer support process efficiency can be increased and so that time-critical E-mails are attended to in a more timely fashion.

Whilst not reaching similar performance, the trial implementation has still provided valuable insight into implementation requirements and how to utilize the predicted TTR. As future work, it would be interesting to implement one of the prioritization setups discussed in this work in a practical setting, and then evaluate the performance, e.g. to what extent it can effectively predict the workload of the customer agents. It would further be interesting to investigate whether such an implementation would actually improve efficiency to map cognitive and emotional load to different agents based on experience.

References

1. Abdallah, E., Abdallah, A.E., Bsoul, M., Otoom, A., Al Daoud, E.: Simplified features for email authorship identification. Int. J. Secur. Netw. **8**, 72–81 (2013). https://doi.org/10.1504/IJSN.2013.055941
2. Avrahami, D., Fussell, S.R., Hudson, S.E.: Im waiting: Timing and responsiveness in semi-synchronous communication. In: Proceedings of the 2008 ACM Conference on Computer Supported Cooperative Work, pp. 285–294. CSCW '08, ACM, New York, NY, USA (2008). https://doi.org/10.1145/1460563.1460610
3. Avrahami, D., Hudson, S.E.: Responsiveness in instant messaging: Predictive models supporting inter-personal communication. In: Proceedings of the SIGCHI Conference on Human Factors in Computing Systems, pp. 731–740. CHI '06, ACM, New York, NY, USA (2006). https://doi.org/10.1145/1124772.1124881
4. Baron, N.S.: Letters by phone or speech by other means: the linguistics of email. Lang. Commun. **18**(2), 133–170 (1998)
5. Batista, G.E., Bazzan, A.L., Monard, M.C.: Balancing training data for automated annotation of keywords: a case study. In: WOB, pp. 10–18 (2003)
6. Boldt, M., Borg, A.: Evaluating temporal analysis methods using residential burglary data. ISPRS Int. J. Geo-Inform. **5**, 148 (2016)
7. Borg, A., Ahlstrand, J., Boldt, M.: Predicting e-mail response time incorporate customer support. In: Proceedings of the 22nd International Conference on Enterprise Information Systems - Volume 1: ICEIS. vol. 1, pp. 305–314 (2020)
8. Borg, A., Boldt, M., Rosander, O., Ahlstrand, J.: E-mail classification with machine learning and word embeddings for improved customer support. Neural Comput. Appl. **36**, 1–22 (2020)
9. Breiman, L.: Random forests. Mach. Learn. **45**(1), 5–32 (2001). https://doi.org/10.1023/A:1010933404324
10. Church, K., de Oliveira, R.: What's up with whatsapp?: Comparing mobile instant messaging behaviors with traditional sms. In: Proceedings of the 15th International Conference on Human-computer Interaction with Mobile Devices and Services, pp. 352–361. MobileHCI '13, ACM, New York, NY, USA (2013). https://doi.org/10.1145/2493190.2493225
11. Farr, J.N., Jenkins, J.J., Paterson, D.G.: Simplification of flesch reading ease formula. J. Appl. Psychol. **35**(5), 333 (1951)
12. Fawcett, T.: Roc graphs: notes and practical considerations for researchers. Mach. Learn. **31**(1), 1–38 (2004)
13. Flach, P.: Machine Learning: The Art and Science of Algorithms That Make Sense of Data. Cambridge University Press, Cambridge (2012)
14. Gansner, E.R., North, S.C.: An open graph visualization system and its applications to software engineering. Softw. - Pract. Exp. **30**(11), 1203–1233 (2000)
15. Halpin, N.: The customer service report: why great customer service matters even more in the age of e-commerce and the channels that perform best (2016). http://www.businessinsider.com/customer-service-experiences-are-more-important-than-ever-in-the-age-of-e-commerce-2016-3?r=US&IR=T&IR=T

16. Huang, C., Ku, L.: Emotionpush: Emotion and response time prediction towards human-like chatbots. In: 2018 IEEE Global Communications Conference (GLOBECOM), pp. 206–212 (2018). https://doi.org/10.1109/GLOCOM.2018.8647331
17. Hutto, C., Gilbert, E.: Vader: a parsimonious rule-based model for sentiment analysis of social media text (2015)
18. Ikoro, G.O., Mondragon, R.J., White, G.: Predicting response waiting time in a chat room. In: 2017 Computing Conference, pp. 127–130 (2017). https://doi.org/10.1109/SAI.2017.8252092
19. Kooti, F., Aiello, L.M., Grbovic, M., Lerman, K., Mantrach, A.: Evolution of conversations in the age of email overload. In: Proceedings of the 24th International Conference on World Wide Web, pp. 603–613. WWW '15, International World Wide Web Conferences Steering Committee, Republic and Canton of Geneva, Switzerland (2015). https://doi.org/10.1145/2736277.2741130
20. Lemaître, G., Nogueira, F., Aridas, C.K.: Imbalanced-learn: A python toolbox to tackle the curse of imbalanced datasets in machine learning. J. Mach. Learni. Res. **18**(17), 1–5 (2017). http://jmlr.org/papers/v18/16-365.html
21. Makridakis, S., Spiliotis, E., Assimakopoulos, V.: Statistical and machine learning forecasting methods: concerns and ways forward. PLOS ONE **13**(3), 1–26 (2018). https://doi.org/10.1371/journal.pone.0194889
22. Pedregosa, F., Varoquaux, G., Gramfort, A., Michel, V., Thirion, B., Grisel, O., Blondel, M., Prettenhofer, P., Weiss, R., Dubourg, V., Vanderplas, J., Passos, A., Cournapeau, D., Brucher, M., Perrot, M., Duchesnay, E.: Scikit-learn: machine learning in Python. J. Mach. Learn. Res. **12**, 2825–2830 (2011)
23. Pielot, M., de Oliveira, R., Kwak, H., Oliver, N.: Didn't you see my message?: Predicting attentiveness to mobile instant messages. In: Proceedings of the 32nd Annual ACM Conference on Human Factors in Computing Systems, pp. 3319–3328. CHI '14, ACM, New York, NY, USA (2014). https://doi.org/10.1145/2556288.2556973
24. Rafaeli, A., Altman, D., Yom-Tov, G.: Cognitive and emotional load influence response time of service agents: a large scale analysis of chat service conversations. In: Proceedings of the 52nd Hawaii International Conference on System Sciences (2019)
25. Ratcliffe, J.: Aoristic analysis: the spatial interpretation of unspecific temporal events. Int. J. Geograph. Inform. Sci. **14**, 669–679 (2000). https://doi.org/10.1080/136588100424963
26. Tsoumakas, G., Katakis, I., Vlahavas, I.: Mining Multi-label Data, pp. 667–685. Springer, US, Boston, MA (2010)
27. Wang, W., Hosseini, S., Awadallah, A.H., Bennett, P.N., Quirk, C.: Context-aware intent identification in email conversations. In: Proceedings of the 42nd International ACM SIGIR Conference on Research and Development in Information Retrieval, p. 585–594. SIGIR'19, Association for Computing Machinery, New York, NY, USA (2019). https://doi.org/10.1145/3331184.3331260
28. Yang, L., Dumais, S.T., Bennett, P.N., Awadallah, A.H.: Characterizing and predicting enterprise email reply behavior. In: Proceedings of the 40th International ACM SIGIR Conference on Research and Development in Information Retrieval, pp. 235–244. SIGIR '17, ACM, New York, NY, USA (2017). https://doi.org/10.1145/3077136.3080782
29. Yang, Y.: An evaluation of statistical approaches to text categorization. Inf. Retrieval **1**(1), 69–90 (1999). https://doi.org/10.1023/A:1009982220290

A Mixed Approach for Pallet Building Problem with Practical Constraints

Manuel Iori[1], Marco Locatelli[2], Mayron C. O. Moreira[3], and Tiago Silveira[2](✉)

[1] Department of Sciences and Methods for Engineering, University of Modena and Reggio Emilia, Reggio Emilia, Italy
manuel.iori@unimore.it

[2] Department of Engineering and Architecture, University of Parma, Parma, Italy
{marco.locatelli,tiago.silveira}@unipr.it

[3] Department of Computer Science, Federal University of Lavras, Lavras, Brazil
mayron.moreira@ufla.br

Abstract. We study a pallet building problem that originates from a case study in a company that produces robotized systems for freight transportation and logistics. We generalize the problem by including the concept of family of items, which allows us to consider specific constraints such as visibility and contiguity. We solve the problem with an algorithm based on a two-step strategy: an Extreme Points heuristic is used to group items into horizontal layers and an exact method is invoked to stack layers one over the other to form pallets. The performance of the algorithm is assessed through extensive computational tests on real-world instances. The results show that the exact model considerably increases the solution quality, creating very compact packings with a limited computational effort.

Keywords: Pallet building problem · Practical constraints · Two-step heuristic · Exact model · Real-world instances

1 Introduction

Cutting and Packing (C&P) is one of the most widely studied fields of Operations Research, and involves several interesting practical problems [8,12,23,25,40]. Although conceptually different, the C&P problems have a similar structure: in cutting problems, a set of stock units has to be cut to produce smaller items, while in packing problems, a set of small items has to be packed into one or more containers. Due to their simple concept and vast applicability, these problems have attracted researchers' attention both for their practical and theoretical interest. Some practical applications involve the production of materials that come in panels (such as wood or glass), the optimization of layouts (as in industry or newspaper paging), and the loading and subsequent transportation of items employing containers, among others.

Supported by University of Parma, and by University of Modena and Reggio Emilia under grant FAR 2018.

J. Filipe et al. (Eds.): ICEIS 2020, LNBIP 417, pp. 122–139, 2021.
https://doi.org/10.1007/978-3-030-75418-1_7

Several surveys and books have been published in recent years to try to review the fast-growing C&P literature [12,25,27,35,40,42]. A typology of C&P problems that was widely used has been proposed in [46]. Later, Bortfeldt and Wäscher [8] extended the paper [46], formalizing the main concepts of the container loading problems and the related constraints. They noted that "the space available for packing above a pallet might be interpreted as a container, too", so they discussed constraints that can arise either when loading a container or a pallet.

Generally, C&P problems are associated with mass-production operations. Thus, improvements in techniques to reduce material/space wasted are directly related to the application of efficient methods, as automated (robotized) systems increasingly make C&P tasks instead of skilled experts who performed manual activities. The work that we face in this paper is an example of an automated-packing system.

The problem that we study is a packing problem named Pallet Building Problem (PBP). We have to load a given set of items in the minimum number of pallets, meeting general and specific constraints. This objective is similar to the objective of the Bin Packing Problem, but the specific constraints of the PBP can make it much harder to find an optimized solution.

The specific PBP that we address here originates from a real-world robotized application and is thus subject to some non-trivial operational constraints. Items should be packed into layers, that must then be piled one over the other while respecting given stackability rules. Also, items are grouped into families, and to facilitate loading/unloading operations, items from the same family packed into the same layer should be contiguous one with the other (as detailed in Sect. 3). Typical items include, e.g., food packaging, soft drink bottles, and cans. That said, the loading of items and, therefore, the creation of pallets for this problem is a fundamental area for logistics, as it impacts not only the company's costs but also the customer's final price (e.g., with the impact of freight logistics).

PBP is NP-hard because it is a generalization of the Bin Packing Problem, which is known to be NP-hard [16]. Besides that, PBP is part of the operational level in the everyday working activity of robotized packing systems in companies. For these reasons, we found necessary to adopt heuristic solution algorithms, mixing them with efficient exact algorithms to increase the quality of the solution. We derived part of the proposed algorithms from the most successful and recent C&P studies, by embedding in them in a tailored way the additional operational constraints of the problem at hand. In a preliminary work, which was presented in [26], we developed a greedy heuristic and performed some preliminary computational tests. In this new research, we formalize part of the problem with a mathematical model, improving, in this way, the original technique. Besides, we propose a more comprehensive experimental evaluation of the algorithms, which allows us to find a more adapted parametric configuration.

The main contributions of this paper can be sketched as follows:

(i) a real-world industrial application that addresses the concept of family, contiguity, and visibility is presented;
(ii) a new heuristic to solve this problem is proposed, made up by a constructive heuristic and a mathematical model;
(iii) extensive computational tests on instances derived from the real-world case study are given.

The remainder of the paper is organized as follows. Section 2 reviews the related literature. Section 3 provides a formal description of the problem. Section 4 presents the full algorithm that we implemented. Section 5 gives the outcome of extensive computational tests and a discussion about their main highlights. Conclusions and some future research directions are given in Sect. 6.

2 Literature Review

Approaches to handle C&P problems started around 1960, with the adoption of simple mathematical models for 2D versions of the problem [20]. However, advanced researches with practical utilization addressing 3D models started only years later, with the utilization of appropriate solutions for real cases by Tsai [44], who proposed a robotic 3D pallet loading with boxes of several sizes.

The PBP emerges as a variant of the *Container Loading Problem* (CLP), which has received a good amount of attention in the last years. In [8], a comprehensive survey of the main constraints used in the literature is presented. The authors verified that heuristic approaches are more frequently used than exact and approximation-guaranteed algorithms. Indeed, apart from container-related constraints (such as weight limit and distribution), they also discussed item-related constraints (as loading priorities, orientation, and stacking) that can be useful for pallets. The authors also considered cargo-related constraints, positioning constraints, and other load-related constraints that may appear during transportation (see also [27,28]), such as vertical and horizontal stability. In [42], the *Pallet Loading Problem* (PLP) is considered. In this problem, a set of two-dimensional rectangular items needs to be packed without overlapping and by allowing a 90° rotation into a two-dimensional bin. The authors proposed a broad analysis of the solution methods and some aspects concerning computational complexity, upper bounds, and data sets most used in numerical experiments. In [12], a survey is presented about 2D and 3D Orthogonal Packing Problems, focusing on data structures for the packing representation and the item-positioning rules. Concerning criteria to place items, we highlight the *extreme points* in [11], that increase the number of feasible regions on the partial packing. Recently, Iori et al. [25] proposed a survey on variants of 2D packing problems, considering techniques to represent and handle items, relaxation methods, as well as exact and heuristic approaches. We also refer to [46] for a categorized typology of 2D packing problems, besides [17] for a state-of-the-art computational analysis.

The PBP can be separated into two subsequent decisions: the first one consists in creating 2D layers, while the second one involves stacking layers to form pallets and thus considers the 3D characteristics. In the following, we discuss some relevant approaches for 2D and 3D, respectively.

For what concerns heuristics for 2D problems, Chazelle [10] described an efficient way to implement the famous bottom-left heuristic, which packs the items, one at a time in a given order, in the lowest-most and left-most position. In [9], a new placement heuristic, called best-fit, is presented for a 2D cutting problem, allowing rotations of 90 degrees. This technique uses a dynamic search based on the "niches", which are the available bottom-most gaps for an item in the partial packing. In terms of metaheuristics, in [5], a GRASP approach is developed for the 2D Strip Packing Problem

(2SPP), which is the problem of packing items into a strip of a given width by minimizing the used height. In the constructive phase, the items are placed into rectangles following specific criteria. A new rule attempts to foresee the future effect of the tallest object in the final solution to avoid spikes. The local search iteratively destroys and rebuilds portions of the current solution. Extensive computational experiments attested to the effectiveness of the proposed strategy. Imahori and Yagiura [24] improved the technique proposed in [9] by presenting a quicker implementation based on a balanced binary search tree and a priority heap with a doubly-linked list. In [34], a complete set of techniques to deal with the 2SPP is presented. The authors use the so-called "skyline" approach in conjunction with greedy local search, a simulated annealing metaheuristic, and a multi-start diversification strategy. In terms of exact algorithms, the 2SPP has been solved by Côté, Dell'Amico, and Iori [13], by means of a combinatorial Benders decomposition.

Regarding 3D problems, Haessler and Talbot [22] addressed a real CLP involving some practical constraints. They proposed an integer programming formulation and a heuristic algorithm. In [6], two heuristics for the CLP are presented: the first one produces loading patterns with a high degree of stability; the second one considers a multi-drop situation in which a Last-In-First-Out constraint is imposed on the cargo. Terno et al. [43] proposed the parallel generalized layer-wise loading approach (PGL-approach) for the CLP. The constraints they consider are: (i) weight capacity; (ii) placement (some items cannot be inserted over the others); (iii) splitting (items of the same type should be loaded in a minimum number of containers); (iv) connectivity; and (v) stability. Using a complex branch-and-bound algorithm, the authors show a certain degree of competitiveness compared with classical solutions reported in the literature. Bortfeldt and Gehring [7] proposed a hybrid genetic algorithm to solve the CLP with a single container and a strongly heterogeneous set of boxes, considering orientation, stability, stacking, weight, and balance. The results showed a better efficiency with more significant heterogeneity of the box sets. Egeblad et al. [18] addressed the CLP for one container (in the knapsack version), using irregular shaped items, and taking a stability constraint into account. They performed tests on randomly generated instances and on real-world instances deriving from a prominent European furniture producer. Józefowska et al. [32] study the CLP considering rotation, stackability, and stability constraints. They considered a case study arising from a household equipment factory, proposing a best fit heuristic based on the idea of wall-building over available space. Kurpel et al. [33] presented techniques to obtain bounds and exact approaches for solving input minimization and output maximization versions of the multiple CLP with rectangular boxes. They adapted and evaluated four discretization techniques from the literature, in addition to considering mathematical formulations to practical constraints such as rotation, vertical stability, and separation of the boxes. Da Silva et al. [14] proposed a matheuristic framework to solve two CLP named Three-dimensional Single Large Object Placement Problem and Single Stock-Size Cutting Stock Problem (according to the typology in [46]), considering seven practical constraints (orientation, load balance, loading priorities, positioning, stacking, stability and weight limit). The method consists of a wall-building approach by means of box arrangements: first, the solution is created by dividing the original problem into two smaller problems (Knapsack Problem and one-dimensional Packing Problem), which are solved through two mathematical formulations; then, a greedy improvement is carried out, using a heuristic that removes the walls with the lowest quality, recreates them, and re-applies the previous mathematical formulations to try to improve the solution. In [21], an extension of the paper by Elhedhli et al. [19] is proposed to solve the 3D Bin Packing Problem. The authors proposed a layer-based column generation approach through a column generation framework, which provides a pool of layers to construct bins. The bins are created one at a time by means of a heuristic that selects layers ordered in descending order of density, meeting several constraints such as vertical support, load-bearing, planogram sequencing, and bin weight limits. About load-bearing, the algorithm ensures that the items of the top layer have sufficient support from below of, at least, 70%, and, in addition, it uses an efficient graph representation to update the total weight to which an item is subjected. Other approaches based on mathematical models can be found in [4,36,37,39,47]. For what concerns the use of 3D packing problems arising in freight transportation, we refer to the surveys in [27,41,45].

The works that most resemble ours are the ones in [3,38], which, however, do not address the concept of family, and contiguity and visibility constraints. Ranck Júnior et al. [38] addressed a real problem of a beverage company for packing boxes into a multi-compartment container and delivering them over a predefined route, meeting practical constraints of orientation, stability, load-bearing strength, and load balancing. Alonso et al. [3] considered practical constraints through a real example originating from a logistics company required to load products into pallets (pallet building) and then load the created pallets into trucks (truck loading), by considering several practical constraints. For the pallet building, they incorporated orientation, support, priority, and stackability constraints. Regarding the truck loading, they adopted restrictions due to priority among pallets, stability, and stackability. They proposed a GRASP algorithm using a constructive phase, a randomized strategy to diversify the solutions, and an improvement phase. The efficiency of their GRASP was analyzed by comparing it with lower bounding procedures, showing good results. The study was later extended in [1,2], by focusing on the development of mathematical models for the case of multiple container loading, addressing a number of additional practical constraints such as vertical and horizontal stability, multi-drop, and load balance. It is worth mentioning in relation to both works that, although they deal with several practical constraints and present similarities to our problem, their main idea is optimizing the occupation in bins, i.e., maximizing the fill factor. In this sense, our proposal goes further by additionally requiring other practical constraints (contiguity for items, visibility for families, layer disposition) to solve the problem.

3 Problem Description

We are given a set R of identical pallets. Each pallet has a two-dimensional loading surface of width $W \in \mathbb{Z}_+^*$ and length $L \in \mathbb{Z}_+^*$, which can be used to load items up to a maximum height $H \in \mathbb{Z}_+^*$. We are also given a set $I = \{1, 2, \ldots, n\}$ of 3D rectangular item types, where each item type $i \in I$ contains b_i identical items, each having width $w_i \in \mathbb{Z}_+^*$, length $l_i \in \mathbb{Z}_+^*$, and height $h_i \in \mathbb{Z}_+^*$, such that $w_i \leq W$, $l_i \leq L$, and $h_i \leq H$. In addition, this problem makes use of the concept of family, which is a generalization of item type, covering at least one characteristic, e.g., geometric dimensions, material type, or the purpose of use. As we are mainly interested in the packing problem, it is enough to take only geometric characteristics into account.

Considering the previous concepts, item types are partitioned into a set F of families as follows. Each item type i belongs to a given family $f \in F$, which, in the real instances we have addressed, is defined as a set of item types with similar height and weight. Note, however, that in other applications, families could also be defined in different ways, e.g., each family could be made up by products of the same company. Items belonging to the same family can be used to form layers. Each layer is a 2D packing of items whose total width does not exceed W, and whose total length does not exceed L. In general, we consider three types of layers:

- *single-item type layers* are formed by a unique type of items;
- *single-family layers* are formed by different item types, but all belong to the same family;
- *residual layers* are formed by a combination of items of different families or items belonging to the same family but with occupation lower than a pre-defined threshold.

Let us call a *group* a set of items having the same item type and being loaded in the same layer. Packings of items in a layer should fulfill two operational constraints that concern groups and are aimed at easing unloading operations when the pallet is delivered. These constraints are named *contiguity* and *visibility* constraints, and are defined by the following concepts:

- *contiguity:* two items of the same type cannot be placed far apart in a layer because of the contiguity constraint. For this purpose, we establish the maximum Euclidean distance that can separate the two items without violating the constraint as follows. Let ℓ be the smallest edge length among all items in I. Let also $G \subseteq I$ be a generic subset of the same type of items packed into a layer. Then, two items $i, j \in G$ satisfy the contiguity constraint if the smallest Euclidean distance between the edges of i and the edges of j is strictly lower than ℓ. Roughly speaking, in this way we guarantee that no other item can fit between i and j. Therefore, the contiguity constraint for a generic layer r is met if, for each group G packed in r, any item in G meets the minimum required Euclidean distance from at least another item in G, and there are no separated sub-groups of G (i.e., subgroups whose distance is ℓ or more).
- *visibility:* similar to what we stated for the contiguity, we say that a group is visible from the outside if, for at least one item in the group, the Euclidean distance between its edges and the borders of the layer is strictly lower than the smallest edge length ℓ.

Single-item type layers and single-family layers can be used in a 3D packing to support other layers that are packed on top of them. For this purpose, we establish two additional conditions:

- a *stackability constraint* imposes that resistant items cannot be on top of fragile ones. Formally, each family $f \in F$ is assigned with a level of resistance ρ_f, with small values indicating fragile items and large values indicating resistant ones. Items belonging to family f cannot be put below items belonging to family g if $\rho_f < \rho_g$;

- a layer can be used to support other layers only if its total area loaded with items reaches a minimum fraction of α of the total loading surface WL. Parameter $0 < \alpha < 1$ is called *fill-factor*. A layer with loaded area lower than αWL can still be used to build a pallet, but can only be the topmost layer. We call this the *minimum supporting area constraint*.

We note that the stackability constraint above introduces a simplification of the real weight that an item has to bear in a load. The simplification is widely adopted in the literature as it works well in practice, see, e.g., [8]. For a more elaborate formulation of load-bearing constraints, we refer to, e.g., [15].

The PBP that we face aims to load all items into the minimum number of pallets, by ensuring that the following constraints are satisfied:

- all items should be packed in layers by satisfying packing, contiguity and visibility constraints;
- at most one residual layer can be used per pallet, and, in such a case, it must be placed at the top of the pallet;
- single-item layers can be used to support any type of layers on top of them, as well as single-family layers can be used to support single-family and residual layers, as long as support and fill-factor constraints are satisfied;
- the total height of the layers in any pallet should not exceed H.

4 Solution Algorithm

This section presents the full technique that we developed to solve the PBP with visibility and contiguity constraints, which we call *Mixed Extreme Points Modified Heuristic* (MEPMH). MEPMH produces feasible solutions by using a two-step heuristic: first, it creates layers to deal with individual items separately using a heuristic method (Sect. 4.1 – *Creating Layers*), forcing the presence of packing, fill factor, visibility, and contiguity constraints; second, the pallet generation step tries to minimize the number of pallets using an exact method (Sect. 4.1 – *Creating Pallets*), forcing the presence of stackability and layers sequence constraints.

4.1 Two-Step Heuristic

Creating Layers. The first step is related to the creation of the 2D layers. All of them are created considering only the dimensions l_i and w_i of item $i \in I$. In this case, MEPMH consists of the following parts.

Item Positioning. To find the position of an item in a layer, we use an adaptation of the Extreme Points Heuristic (EPH) proposed by [11]. Originally, EPH was described for 3D packings. As we need to deal with just a 2D layer, we limit our description here to this simpler case. To this aim, let us consider that the 2D layer under construction is placed in the positive quadrant of the cartesian system, with width parallel to the x-axis, length parallel to the y-axis, and bottom-left corner located in the origin of the axes.

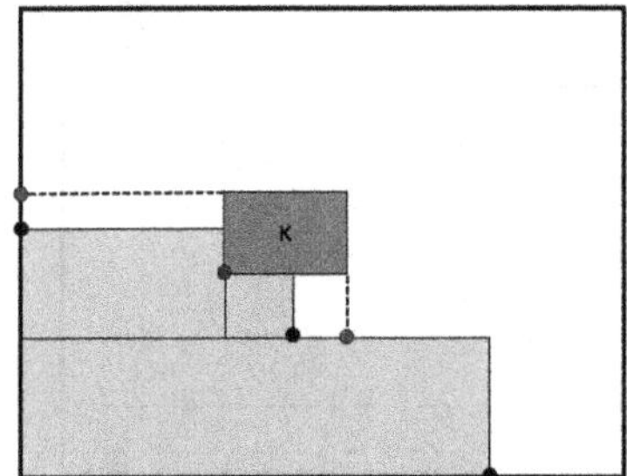

Fig. 1. Extreme points: black and green points form the current set E; the green point is chosen to pack item k (and is then removed); blue points are the new extreme points added to E after packing k (from Iori et al. [26]). (Color figure online)

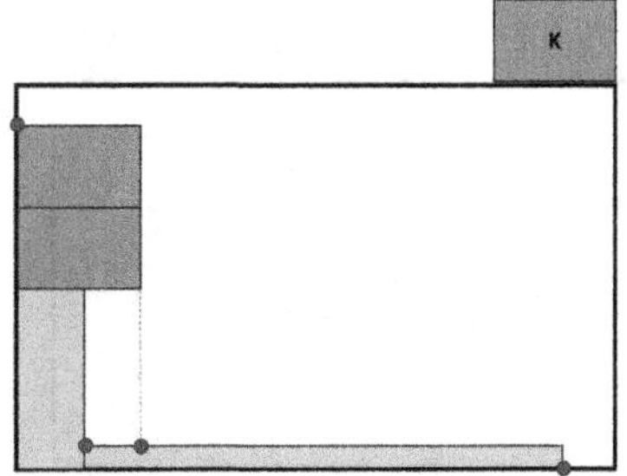

Fig. 2. Case where the EPH generates only extreme points that are infeasible for the PBP with visibility and contiguity constraints. Packing the next item k is impossible, so the current layer is closed and a new one is opened (from Iori et al. [26]).

EPH works with the concept of extreme points. An extreme point e is a point in the 2D space, where an item $k \in I$ can be packed by taking into account a partial packing solution built so far. For the sake of clarity, packing an item k in an extreme point e means packing the bottom-left corner of k in e.

In EPH, the items are packed one at a time in the layer, considering a set E of available extreme points. The set is initialized with the origin point $(0, 0)$. Each time a new item is packed, the set is updated by removing the point used for the packing of the item, and possibly inserting new extreme points. These new extreme points are obtained by computing the projection of k over the partial packing solution under construction, considering the two axes.

For the x-axis, EPH horizontally projects the top edge of item k to its left, until the projection touches a previously packed item or the left border of the layer (i.e., the y-axis). This is the first extreme point that is possibly created. For the y-axis, instead, EPH considers the right edge of item k and vertically projects it towards the bottom until the projection reaches a previously packed item or the bottom border of the layer. This is the second extreme point possibly created. These two points are added to E if they were not already included in it. Figure 1 depicts an example with a set E formed by green and black points. The green point is selected for the packing of item k and is thus removed from E. The blue points are the new extreme points added to E after the packing of k. An extreme point e can be either feasible or infeasible for the packing of the next item k, depending on the fact that such packing meets all required constraints.

Note that, besides the projections from k, it is mandatory to verify all projections from $i \in I$ that lie on k. This step that we name past projections is needed to find new extreme points not yet available. We execute it right after having computed the projections from k.

The original EPH solves pure 2D packing problems, so we need to include a set of changes to be able to solve the more complex PBP with visibility and contiguity constraints.

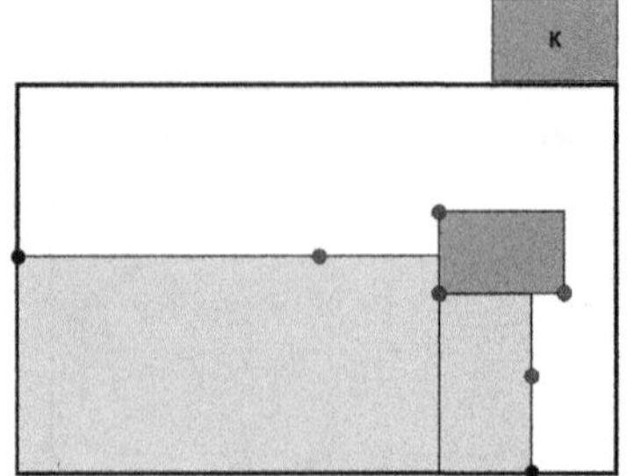

Fig. 3. New extreme points created for contiguity: after packing k in the green point (which is then removed from the current set E), four extreme points (represented in blue) are added (from Iori et al. [26]). (Color figure online)

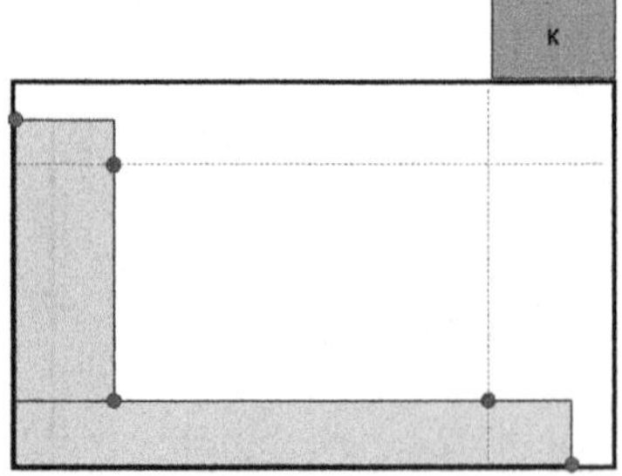

Fig. 4. New extreme points created for visibility: previous extreme points that do not meet visibility and the packing constraints are shown in red; new feasible extreme points that meet these constraints are shown in blue (from Iori et al. [26]). (Color figure online)

The first modification we apply consists of increasing the search space inside a layer. The EPH creates only two new extreme points at a time, whereas visibility and contiguity constraints narrow the search space. As a consequence, a layer could be closed even when its occupation is low because set E does not contain any feasible extreme point. Figure 2 shows an example where this situation occurs since none of the current feasible extreme points fulfills the contiguity constraints.

To overcome this limitation, we included some new extreme points, considering two cases: *contiguity* and *visibility*. In the first case, the extreme points are included around item k (the last item packed), to allow the contiguity with the next items of the same type. Basically, four new points are added: the k top-left and bottom-right corners allow the top and right connections, respectively, and the points on the left and below k enable the left and bottom connections, respectively. Figure 3 shows the new extreme points.

Extreme points regarding the visibility case are included close to the top and right layer borders, when item k has a different type from the previously packed items, and there is no point in E that allows k to meet the contiguity with the layer and the packing constraints. In this case, two new points are added: we proceed from right to the left in the layer, and we take the first feasible point that meets the contiguity on the top layer border. Moreover, going from top to bottom in the layer, we consider the first feasible point that meets the contiguity on the right layer border. Figure 4 shows the new extreme points for this case.

The next modification is related to performance. Current extreme points are directly linked to visibility and contiguity constraints. Note that these points can be enabled according to item k and its current item group in the partial solution. Thus, we work with two extreme points sets: *feasible extreme points*, E_f, and *infeasible extreme points*, E_i, representing, respectively, points that meet or do not meet the visibility and contiguity constraints for item k. Therefore, before adding k to the partial solution, we first update these two sets, and only after this is done, we check if k overlaps with previously packed

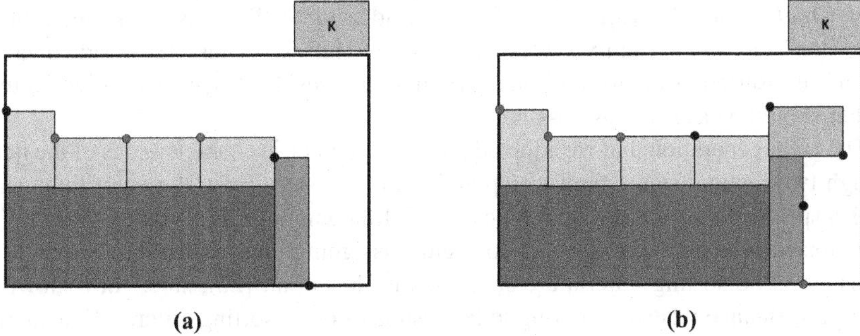

Fig. 5. Partial solutions and their feasible (black) and infeasible (red) extreme points for the next item k: representation of the extreme points that meet the boundary (5a), and contiguity (5b) constraints (from Iori et al. [26]). (Color figure online)

items. Note that these two sets are relative to the current configuration, and after the addition of the new items, infeasible points may become feasible and vice-versa. The benefit of using these sets is that we avoid checking the overlapping constraint for all extreme points. An example of sets E_f and E_i is provided in Fig. 5.

The latest modification increases the search space by adding randomness to the process. In this case, we include the possibility of selecting extreme points that are not the best ones. This strategy depends on the evaluation functions explanation (Sect. 4.1 – *Evaluation Functions*), and it is detailed in Sect. 4.1 – *Layer Creation and Classification*.

Evaluation Function. Given the next item k and its set E_f, the point $e \in E_f$ that results in the best packing has to be chosen. For that purpose, the algorithm uses an fitness evaluation function d for each e to aggregate k into a group over the partial solution. The function d that we adopted is the Bounding Box function, as suggested by Iori et al. [26], since this function showed a better performance when comparing to the results obtained by other fitness evaluation functions. To sum up, this function calculates the bounding box area – minimum rectangle area covering a set of items – with k and finds the extreme point e that minimizes the area among all bounding boxes generated when positioning the reference point of k over e.

Item Grouping. A group is a subset of items of the same type over the partial packing. Given an evaluation function d, the fitness of k is calculated from a group or the partial packing as follows. When k is the first item in the group, its fitness is calculated considering all items of the partial packing in the layer; when there is at least one item of the same type of k, the fitness is calculated by considering the group formed by items of this specific type.

Layer Creation and Classification. The main idea of MEPMH is to create as many single-item and single-family layers as possible. This is because the residual layers have to be inserted at the top of pallets, at most one residual layer per pallet, so their number should be as small as possible.

The layer generation that we adopted analyzes geometric characteristics of the items through two distinct steps: family step and item type step. First, the algorithm establishes a specific order for the item types $i \in I$, following a *random* order for families, and a non-increasing order of *width* for items, assigning them to the data structure S. In this case, the sorting criteria are suggested by Iori et al. [26], since they showed a better performance when comparing to the results of other sorting criteria. Considering the selection process over S, we define the selectable range based on the parameter ϵ, which stands for the percentage of elements analyzed on a random choice. Let F_S be the sorted set of families which represents the order of the families in S, and $G_f \subseteq S$ the sorted set of item types $s \in S$, which represents the order of the item types of a family $f \in F_S$. In this case, a family f is selected randomly among the first $\epsilon|F_S|$ families, and then $s \in G_f$ is selected among the first $\epsilon|G_f|$ item types. For example, if we set $\epsilon = 0.15$, this means that 15% of the first $|F_S|$ families and 15% of the first $|G_f|$ item types can be chosen by this method. After the item type s has been chosen, MEPMH continues the layer creation process.

In the same line, we describe how to select an extreme point. Given an item of type $s \in S$ to be packed and a fitness evaluation function $d_s \in D$, we should determine where to place it in the partial packing. In this case, we propose to tailor the EPH method (deterministic choices that select the best point according to a specific fitness evaluation function) to use a parameter that includes the possibility of random choices to select the next extreme point through parameter ϵ, in the following way: given s and its respective E_f, the algorithm creates the list T by sorting all extreme points of E_f in a non-decreasing order of the values calculated by d_s. The heuristic selects through a uniform distribution an extreme point among the first $\epsilon|T|$ extreme points of T, i.e., the parameter ϵ stands for the percentage of the best extreme points in T that can be chosen randomly.

The classification of a layer is made after it has been closed, analyzing its occupation, the families and the item types that are included in it.

Creating Pallets. After all items have been packed into layers, we need to put layers together into the minimum number of pallets. Let us introduce the following notation:

P : set of used pallets;

R : set of residual layers;

I : set of single-item layers;

F : set of single-family layers;

$M : I \cup F$;

$L : R \cup M$;

H : height of the pallet;

h_l : height of layer l;

S_l : stackability of layer l;

x_{lp} : it is equal to 1 if layer $l \in L$ is packed in pallet $p \in P$, 0 otherwise;

y_p : it is equal to 1 if pallet $p \in P$ is used, 0 otherwise.

We propose the following mathematical model:

$$\text{Minimize} \sum_{p \in P} y_p \tag{1}$$

$$\text{Subject to} \sum_{l \in L} h_l x_{lp} \leq H y_p \qquad p \in P \tag{2}$$

$$\sum_{p \in P} x_{lp} = 1 \qquad l \in L \tag{3}$$

$$\sum_{l \in R} x_{lp} \leq y_p \qquad p \in P \tag{4}$$

$$\sum_{l' \in M : S_{l'} < S_l} x_{l'p} \leq |M|(1 - x_{lp}) \qquad l \in R,\ p \in P \tag{5}$$

$$\sum_{l' \in I : S_{l'} < S_l} x_{l'p} \leq |I|(1 - x_{lp}) \qquad l \in F,\ p \in P \tag{6}$$

$$x_{lp} \in \{0, 1\} \qquad l \in L,\ p \in P \tag{7}$$

$$y_p \in \{0, 1\} \qquad p \in P \tag{8}$$

The objective function (1) aims at minimizing the number of used pallets. Constraints (2) ensure a layer $l \in L$ can be inserted in pallet $p \in P$ when the sum of the heights of all layers in p is not higher than H. Constraints (3) ensure all layers $l \in L$ are packed. Constraints (4) ensure at most a residual layer $l \in R$ can be packed in a pallet $p \in P$. Constraints (5) ensure that, for each layer $l \in R$, the layers $l' \in M$ with stackability lower than l will not be used when l is in the pallet. Constraints (6) are similar to the previous ones, but we have the correspondence between single-item and single-family layers in the pallet, by requiring all stackability values of the single-item layers be greater than the stackability values of the single-family layers. Constraints (7) and (8) are binary conditions for the variables.

We highlight that the height of each layer is the height of the highest item in the layer. Similarly, the stackability of a layer is computed as the maximum stackability value among the items in the layer. This exact algorithm concludes the 2-step heuristic and allows us to have a feasible solution.

5 Computational Results

All experiments have been conducted on a Virtual Machine VMware®, Intel® Xeon® CPU E5-2640 v2 2.00 GHz, 16 GB RAM, Linux Ubuntu Server 18.04 OS. The algorithms have been implemented in Java and ran using Oracle® JDK 11. We used the solver IBM ILOG CPLEX 12.10 to solve the mathematical model.

We solved the proposed instances with the MEPMH algorithm (Sect. 4). For each instance, we ran the algorithm 15 times. The time limit for the solver was set to 30 s. We used the instances presented in [26], that are separated into four groups, each containing six instances characterized by different intervals in the number of distinct items. Table 1 summarizes the details of the instances, reporting an instance field and its respective value by row: ID number, number of item types, number of families, and total number of items. For all instances, the container dimensions were set equal to 1500, 1250, and 1050 for height, width, and length, respectively. We tested them, by allowing rotation of 90 degrees of the items.

We carried out a detailed test for selecting the value of both fill factor and parameter ϵ. About the fill factor, we have tested values $\{0.55, 0.65, 0.75\}$; these values define a threshold that represents the minimum value of occupation for defining a residual layer. In this way, as long as the final layers present a fill factor that is able to create stable pallets, a proper adjustment of this parameter might allow us to generate more single-item and single-family layers, increasing the quality of the final solution as a result. About parameter ϵ, we have chosen the value set $\{0.00, 0.05, 0.15, 0.25\}$. In this case, the larger the value is, the more random the process is.

5.1 Experimental Evaluation

We report the average results for MEPMH to analyze its performance, considering all the previous configurations. Besides that, we show the average results obtained by heuristic EPMH, which uses the greedy algorithm to create pallets, as described in [26]. We have conducted these experiments over the same instances to compare the performance of the two strategies consistently.

The average results for both MEPHM and EPHM are summarized in Table 2. In this table, each row provides average results for a given algorithm on the 360 solutions obtained, 15 for each of the 24 attempted instances. The information about the algorithm is contained in the first column, followed by the minimum fill factor to create a layer, the choice factor (value of ϵ), the total number of pallets in the solution and the total number of layers created. In the six successive columns, we report, respectively, the minimum, maximum, and average pallet utilization and the minimum, maximum, and

Table 1. Instances settings (from Iori et al. [26]).

Field	Value																							
ID	01	02	03	04	05	06	07	08	09	10	11	12	13	14	15	16	17	18	19	20	21	22	23	24
n	20		23		29		34		37		48		59		64		61		75		79		66	
N. of families	4	12	4	13	6	15	7	17	6	17	9	18	10	19	11	19	10	19	16	20	17	21	13	19
Sum b_i	877	269	388	115	438	2103	698	1322	300	449	395	748	455	633	658	683	300	797	790	829	855	944	738	767

average fill factor, considering the 2D area of all layers. Then, the average number of single-item, single-family and residual layers, and the computational time (represented by seconds.milliseconds). About the EPMH heuristic, we reported the best results that met rotation, contiguity and visibility constraints (all constraints used in this work), using the Bounding Box function.

The first comment on Table 2 is about the number of created layers. As EPMH and MEPMH use the same algorithm to create layers, the 2D fill factor and the number of layers are very similar, showing only slight differences due to the randomness of the process provided by the choice factor. Besides that, the distribution among layers (single-item, single-family and residual) presents the same similarities.

Although the creation of layers is similar for both algorithms, the analysis of the parameters of minimum fill factor and choice factor is critical. Let us focus first on the analysis of the minimum fill factor. Here, the lower is this value, the higher the improvement in the quality of the solution is. Therefore, we notice an interesting situation: even though the algorithm uses the smallest value for the minimum fill factor (0.55), the final average and minimum 2D fill factor remain quite similar to the values obtained when we use the highest value for this parameter (0.75). This is an important consideration, because, notwithstanding a company needs a minimum occupation of a layer and even when we allow the algorithm to use a value for this parameter that is lower than this threshold, the algorithm still creates better solutions maintaining the stability of the pallets.

Considering the choice factor, this parameter presents a general behavior: a more significant improvement in the quality of the solution made by the configurations 0.05

Table 2. Computational results (individual average).

Algorithm	Min. fill factor	Choice factor (ϵ)	N. of pallets	N. of layers	Pallet filling			2D fill factor			Single-item layers	Single-family layers	Resid. layers (RL)	Time (sec.mil)
					Min	Max	Avg	Min	Max	Avg				
EPMH	0.75	0.00	12.84	40.93	0.16	0.85	0.53	0.41	0.96	0.86	21.79	11.53	7.61	0.016
		0.05	12.79	40.92	0.17	0.85	0.53	0.42	0.96	0.86	21.79	11.53	7.59	0.016
		0.15	13.81	42.29	0.13	0.82	0.49	0.41	0.95	0.83	21.36	12.20	8.73	0.020
		0.25	18.49	42.93	0.09	0.80	0.37	0.39	0.94	0.82	18.13	10.46	14.35	0.023
	0.65	0.00	11.41	41.55	0.22	0.84	0.59	0.36	0.96	0.84	22.79	14.00	4.76	0.015
		0.05	11.40	41.57	0.22	0.84	0.59	0.37	0.96	0.84	22.79	14.03	4.74	0.015
		0.15	11.91	42.96	0.16	0.83	0.57	0.40	0.95	0.82	23.05	15.11	4.80	0.017
		0.25	13.06	44.46	0.12	0.80	0.52	0.39	0.94	0.79	23.72	14.25	6.49	0.018
	0.55	0.00	11.26	41.81	0.18	0.85	0.59	0.41	0.96	0.84	23.13	15.00	3.68	0.014
		0.05	11.30	41.79	0.18	0.85	0.59	0.41	0.96	0.84	23.13	14.99	3.68	0.015
		0.15	11.48	43.25	0.20	0.83	0.59	0.41	0.95	0.81	23.42	16.20	3.64	0.016
		0.25	11.94	45.43	0.20	0.80	0.56	0.35	0.94	0.78	26.15	15.44	3.84	0.017
MEPMH	0.75	0.00	12.36	40.94	0.16	0.85	0.55	0.41	0.96	0.86	21.79	11.52	7.63	12.114
		0.05	12.29	40.91	0.17	0.84	0.56	0.43	0.96	0.86	21.79	11.51	7.61	11.133
		0.15	13.36	42.28	0.15	0.81	0.51	0.41	0.95	0.83	21.33	12.11	8.84	13.731
		0.25	17.96	42.98	0.10	0.79	0.39	0.38	0.95	0.82	18.26	10.44	14.28	13.796
	0.65	0.00	10.75	41.56	0.27	0.86	0.63	0.36	0.96	0.84	22.79	14.00	4.76	9.157
		0.05	10.74	41.57	0.28	0.86	0.63	0.38	0.96	0.84	22.79	14.03	4.74	9.562
		0.15	11.16	42.98	0.23	0.83	0.61	0.42	0.95	0.82	23.17	15.00	4.81	10.672
		0.25	12.39	44.45	0.17	0.80	0.55	0.40	0.94	0.79	23.44	14.53	6.48	12.723
	0.55	0.00	10.36	41.80	0.30	0.86	0.66	0.41	0.96	0.84	23.13	15.00	3.68	8.188
		0.05	10.38	41.77	0.30	0.85	0.65	0.43	0.96	0.84	23.13	14.99	3.65	8.719
		0.15	10.81	43.35	0.27	0.83	0.63	0.37	0.94	0.81	23.44	16.19	3.72	9.426
		0.25	11.24	45.38	0.28	0.79	0.60	0.37	0.94	0.78	26.08	15.45	3.85	9.417

and 0.15. It is important to highlight that the higher the values for this parameter, the larger the algorithm's randomness. Therefore, analyzing the results, we notice that high randomness brings a broad exploration, but the algorithm is not able to find good-quality solutions.

Last but not least, let us consider the part of the algorithm to create pallets. In this scenario, we can notice substantial differences. The first one is about the execution time. In this case, the execution time of MEPMH is greater than the execution time of EPMH, since the mathematical model in MEPMH is more complex to solve than the respective heuristic algorithm in EPMH. However, the final execution time of MEPMH leveled out between 8.188 and 13.796 s on average, thus showing a high performance, that can be easily accepted for an industrial software application.

The most important analysis when considering the final solution is about the comparison of the final number of pallets between the two algorithms. In this case, we can notice the improvement that MEPMH adds to the algorithm. To understand this comparison, we highlight the respective values in Table 3.

More in detail, Table 3 shows the minimum fill factor to create a layer, the choice factor (value of the parameter ϵ), and the percentage of improvement that MEPMH presented over EPMH. In this table, we can notice the improvement leveling out between 2.87% and 8.14%. Therefore, in the same line as in the previous analysis, the greatest improvements occur when the algorithm uses the lowest value for the minimum fill factor (0.55) and low values for the choice factor (0.05 and 0.15).

Table 3. Improvement in solution when using MEPMH.

Min. fill factor	Choice factor (ϵ)	Improvement (%)
0.75	0.00	3.74
	0.05	3.91
	0.15	3.26
	0.25	2.87
0.65	0.00	5.78
	0.05	5.79
	0.15	6.30
	0.25	5.13
0.55	0.00	7.99
	0.05	8.14
	0.15	5.84
	0.25	5.86

6 Conclusions

This paper addressed a pallet building problem from a case study in a company by considering practical constraints, such as rotation of the items, fill factor, stackability, visibility, and contiguity constraints. We proposed an extension of the work by [26], using a modification of a 2-step heuristic that includes a mathematical model to create pallets. We use the efficient Extreme Points heuristic for creating 2D layers, including in the algorithm the exact method to increase the quality of the solution to create 3D pallets. To analyze the effectiveness of the proposed heuristic, we carried out extensive computational experiments on real-world instances, testing a set of parameter values for the minimum fill factor of a layer, as well as for the random choice factor used to choose the next item to be packed. Despite the complexity of the real-world instances, good final results are reported in a short execution time, thus showing that the inclusion of the mathematical model in the algorithm is an efficient strategy to create compact solutions for this type of problems.

As future work, we intend to replace the Extreme Point heuristic with an exact mathematical model to solve the layer creation to try to enhance the quality of the solutions generated. Therefore, it will be necessary to express the concept of contiguity and visibility of items in a consistent way, as this is still a gap in the existing literature nowadays. Besides that, we are interested in including metaheuristics to improve the effectiveness of the algorithm, and in adding learning mechanisms such as Support Vector Machine, fixed set search metaheuristic [29–31], etc. We are also interested in extending the concept of family to address characteristics beyond the geometric ones, in order to make it possible to combine different problems, e.g., packing and delivery, to try to improve the logistic of the whole process.

References

1. Alonso, M.T., Alvarez-Valdes, R., Iori, M., Parreño, F.: Mathematical models for multi container loading problems with practical constraints. Comput. Ind. Eng. **127**, 722–733 (2019)
2. Alonso, M.T., Alvarez-Valdes, R., Iori, M., Parreño, F., Tamarit, J.M.: Mathematical models for multi container loading problems. OMEGA **66**, 106–117 (2017)
3. Alonso, M.T., Alvarez-Valdes, R., Parreño, F., Tamarit, J.M.: Algorithms for pallet building and truck loading in an interdepot transportation problem. Math. Probl. Eng. **2016**, 1–11 (2016)
4. Alvarez-Valdes, R., Parreño, F., Tamarit, J.M.: A branch-and-cut algorithm for the pallet loading problem. Comput. Oper. Res. **32**, 3007–3029 (2005)
5. Alvarez-Valdes, R., Parreño, F., Tamarit, J.M.: Reactive GRASP for the strip-packing problem. Comput. Oper. Res. **35**, 1065–1083 (2008)
6. Bischoff, E.E., Ratcliff, M.S.W.: Issues in the development of approaches to container loading. Omega **23**, 377–390 (1995)
7. Bortfeldt, A., Gehring, H.: A hybrid genetic algorithm for the container loading problem. Eur. J. Oper. Res. **131**, 143–161 (2001)
8. Bortfeldt, A., Wäscher, G.: Constraints in container loading - a state-of-the-art review. Eur. J. Oper. Res. **229**, 1–20 (2013)
9. Burke, E.K., Kendall, G., Whitwell, G.: A new placement heuristic for the orthogonal stock-cutting problem. Oper. Res. **52**, 655–671 (2004)
10. Chazelle, B.: The bottomn-left bin-packing heuristic: an efficient implementation. IEEE Trans. Comput. **C-32**, 697–707 (1983)
11. Crainic, T.G., Perboli, G., Tadei, R.: Extreme point-based heuristics for three-dimensional bin packing. INFORMS J. Comput. **20**, 368–384 (2008)

12. Crainic, T.G., Perboli, G., Tadei, R.: Recent advances in multi-dimensional packing problems. In: New Technologies, chap. 5. IntechOpen (2012)
13. Côté, J.F., Dell'Amico, M., Iori, M.: Combinatorial benders' cuts for the strip packing problem. Oper. Res. **62**, 643–661 (2014)
14. da Silva, E.F., Leão, A.A.S., Toledo, F.M.B., Wauters, T.: A matheuristic framework for the three-dimensional single large object placement problem with practical constraints. Comput. Oper. Res. **124**, 105058 (2020)
15. de Queiroz, T.A., Miyazawa, F.K.: Two-dimensional strip packing problem with load balancing, load bearing and multi-drop constraints. Int. J. Prod. Econ. **145**, 511–530 (2013)
16. Delorme, M., Iori, M., Martello, S.: Bin packing and cutting stock problems: mathematical models and exact algorithms. Eur. J. Oper. Res. **255**, 1–20 (2016)
17. Delorme, M., Iori, M., Martello, S.: Logic based benders decomposition for orthogonal stock cutting problems. Comput. Oper. Res. **78**, 290–298 (2017)
18. Egeblad, J., Garavelli, C., Lisi, S., Pisinger, D.: Heuristics for container loading of furniture.Eur. J. Oper. Res. **200**, 881–892 (2010)
19. Elhedhli, S., Gzara, F., Yildiz, B.: Three-dimensional bin packing and mixed-case palletization. INFORMS J. Optim. **1**(4), 323–352 (2019)
20. Gilmore, P.C., Gomory, R.E.: Multistage cutting stock problems of two or more dimensions. Oper. Res. **13**, 94–120 (1965)
21. Gzara, F., Elhedhli, S., Yildiz, B.C.: The pallet loading problem: three-dimensional bin packing with practical constraints. Eur. J. Oper. Res. **287**(3), 1062–1074 (2020)
22. Haessler, R.W., Talbot, F.B.: Load planning for shipments of low density products. Eur. J. Oper. Res. **44**, 289–299 (1990)
23. Hopper, E., Turton, B.: Application of genetic algorithms to packing problems - a review. In: Soft Computing in Engineering Design and Manufacturing, pp. 279–288 (1998)
24. Imahori, S., Yagiura, M.: The best-fit heuristic for the rectangular strip packing problem: an efficient implementation and the worst-case approximation ratio. Comput. Oper. Res. **37**, 325–333 (2010)
25. Iori, M., de Lima, V.L., Martello, S., Miyazawa, F.K., Monaci, M.: Two-dimensional cutting and packing: Problems and solution techniques. Eur. J. Oper. Res. (2020). (forthcoming)
26. Iori, M., Locatelli, M., Moreira, M.C.O., Silveira, T.: Solution of a practical pallet building problem with visibility and contiguity constraints. In: International Conference on Enterprise Information Systems, vol. 1, pp. 327–338. SciTePress (2020)
27. Iori, M., Martello, S.: Routing problems with loading constraints. TOP **18**, 4–27 (2010)
28. Iori, M., Martello, S.: An annotated bibliography of combined routing and loading problems. Yugoslav J. Oper. Res. **23**, 311–326 (2013)
29. Jovanovic, R., Tuba, M., Voß, S.: Fixed set search applied to the traveling salesman problem, pp. 63–77. International Workshop on Hybrid Metaheuristics (2019)
30. Jovanovic, R., Voß, S.: Fixed set search applied to the minimum weighted vertex cover problem. In: Kotsireas, I., Pardalos, P., Parsopoulos, K.E., Souravlias, D., Tsokas, A. (eds.) SEA 2019. LNCS, vol. 11544, pp. 490–504. Springer, Cham (2019). https://doi.org/10.1007/978-3-030-34029-2_31
31. Jovanovic, R., Voß, S.: The fixed set search applied to the power dominating set problem. Expert Systems, p. e12559 (2020)
32. Józefowska, J., Pawlak, G., Pesch, E., Morze, M., Kowalski, D.: Fast truck-packing of 3D boxes. Eng. Manage. Prod. Serv. **10**, 29–40 (2018)
33. Kurpel, D.V., Scarpin, C.T., Pécora Junior, J.E., Schenekemberg, C.M., Coelho, L.C.: The exact solutions of several types of container loading problems. Eur. J. Oper. Res. **284**, 87–107 (2020)
34. Leung, S.C.H., Zhang, D., Sim, K.M.: A two-stage intelligent search algorithm for the two-dimensional strip packing problem. Eur. J. Oper. Res. **215**, 57–69 (2011)

35. Lodi, A., Martello, S., Monaci, M., Vigo, D.: Two-Dimensional Bin Packing Problems, pp. 107–129. John Wiley & Sons, Ltd (2014)
36. Martins, G.H.A., Dell, R.F.: Solving the pallet loading problem. Eur. J. Oper. Res. **184**, 429–440 (2008)
37. Neliβen, J.: How to use structural constraints to compute an upper bound for the pallet loading problem. Eur. J. Oper. Res. **84**, 662–680 (1995)
38. Ranck Júnior, R., Yanasse, H.H., Morabito, R., Junqueira, L.: A hybrid approach for a multi-compartment container loading problem. Expert Syst. Appl. **137**, 471–492 (2019)
39. Ribeiro, G.M., Lorena, L.A.N.: Lagrangean relaxation with clusters and column generation for the manufacturers pallet loading problem. Comput. Oper. Res. **34**, 2695–2708 (2007)
40. Scheithauer, G.: Introduction to Cutting and Packing Optimization. Springer International Publishing (2018)
41. Schmid, V., Doerner, K.F., Laporte, G.: Rich routing problems arising in supply chain management. Eur. J. Oper. Res. **224**, 435–448 (2013)
42. Silva, E., Oliveira, J.F., Wäscher, G.: The pallet loading problem: a review of solution methods and computational experiments. Int. Trans. Oper. Res. **23**, 147–172 (2016)
43. Terno, J.J., Scheithauer, G., Sommerweiβ, U., Riehme, J.: An efficient approach for the multi-pallet loading problem. J. Eur. J. Oper. Res. **123**, 372–381 (2000)
44. Tsai, D.: Modeling and analysis of three-dimensional robotic palletizing systems for mixed carton sizes. Ph.D. thesis, Iowa State University (1987)
45. Vidal, T., Crainic, T.G., Gendreau, M., Prins, C.: Heuristics for multi-attribute vehicle routing problems: a survey and synthesis. Eur. J. Oper. Res. **231**, 1–21 (2013)
46. Wäscher, G., Hauβner, H., Schumann, H.: An improved typology of cutting and packing problems. Eur. J. Oper. Res. **183**, 1109–1130 (2007)
47. Wu, K.C., Ting, C.J.: A two-phase algorithm for the manufacturer's pallet loading problem. In: IEEE International Conference on Industrial Engineering and Engineering Management, pp. 1574–1578 (2007)

A Reference Process and Domain Model for Machine Learning Based Production Fault Analysis

Christian Seiffer(✉), Alexander Gerling, Ulf Schreier, and Holger Ziekow

Furtwangen University, 78120 Furtwangen, Germany
{christian.seiffer,alexander.gerling,ulf.schreier, holger.ziekow}@hs-furtwangen.de

Abstract. Early detection of errors in production processes is of crucial importance for manufacturing companies. With the advent of machine learning (ML) methods, the interaction of ML and human expertise offers the opportunity to develop a targeted understanding of error causes and thus to proactively avoid errors. The power of such a model can only be used if relevant domain knowledge is taken into account and applied correctly. When using ML methods, a systematic failure analysis without the need of deeper ML knowledge is crucial for an efficient quality management. Focusing on these two aspects, we develop an updated holistic solution by expanding and detailing our previously proposed approach to support production quality.

Keywords: Reference model · Machine learning · Manufacturing

1 Introduction

The use of machine learning (ML) methods in the context of optimizing production processes has already proven to be useful in the past [4,6,16]. Experts assume that ML will play an increasingly important role in the future [1,8,12]. The possibilities to benefit from machine learning methods are manifold [18]. Due to technological progress, large amounts of data can be stored, processed and utilized in manufacturing [10]. In particular, the focus is on data that is recorded during production processes to test and assure production quality. For example, it is common practice to record, store and evaluate the results of individual processing steps in a production line. Depending on the characteristics and complexity of the product, different features have to be checked during the individual production steps to ensure quality. Of particular interest is the extent to which the measurement results of such a test step can be used to draw conclusions about subsequent results or even the resulting product quality. Due to complex product properties, such correlations are usually hard to recognize by humans in the real world, which is why it is difficult for quality engineers to define e.g. threshold values that define the passing of individual tests. Here, the

J. Filipe et al. (Eds.): ICEIS 2020, LNBIP 417, pp. 140–157, 2021.
https://doi.org/10.1007/978-3-030-75418-1_8

use of machine learning is suitable to get insights into the complexity of recorded data and to detect errors and defects during the production process as early as possible. This is the focus of the PREFERML (Proactive Error Avoidance in Production through Machine Learning) project [19], which is researching the challenges in this context and developing holistic systems solutions.

The integration of ML into the production process changes existing roles and tasks. The decisive and at the same time challenging aspect is that the hoped-for benefit does not arise from replacing the quality engineer but rather from the interaction with the ML system. This ML system should be designed in a way that the quality engineer can work with it, without needing additional knowledge and ML expertise for regular use. Furthermore, transferability to other production processes is necessary to provide a generalizable solution.

A reference model for a corresponding solution was proposed in our basic work [3]. In the context of a real industry case, requirements for a ML system were validated to facilitate an effective implementation. Here, domain knowledge of the process participants is of great importance. From a workflow for troubleshooting and error correction in the classical sense without ML, a corresponding workflow including the support of an ML system was developed. Both processes were compared and the implications, benefits and challenges of using machine learning were discussed. Based on this, in this work we want to focus more on how domain knowledge can be used and integrated into the developed solution. Furthermore, we want to extend and discuss in more detail the process step of error identification and derivation of decisions and the role of machine learning. For this purpose, Sect. 2 describes the domain, which is used to describe the problem and task. Section 3 deals with the necessary methodical basics. Section 4 introduces the overall concept of the solution. Section 5 introduces the underlying process for error handling. The focus on the integration of domain knowledge is set in Sect. 6 and Sect. 7 describes the extension of the To-Be process with regard to error identification with the help of ML. Subsequently, relevant research is referred to in Sect. 8 and the work is concluded in Sect. 9.

2 Domain

Reducing errors in production processes is a difficult task. Typical challenges in the manufacturing environment are heterogeneous product families, diverse error types, imbalanced data distributions, non-linear production processes, self-loops and concept drifts [16]. Gerling et al. [3] illustrate a structured procedure that reflects the how quality engineers identify and correct errors. Quality engineers are main actors in this process, supported by a test system. The first steps are to select the correct data and to evaluate the data quality. This includes information regarding the production environment, such as weekday or temperature. If there are already findings from similar products, these can be used to obtain additional information. Afterwards, the cause of the error is identified, sometimes with the help of classic statistics tools. If necessary, the test specification can be adapted and a new test run can be performed based on this adaptation using the test system. Once the cause of the error has been determined, corrective measures can be taken.

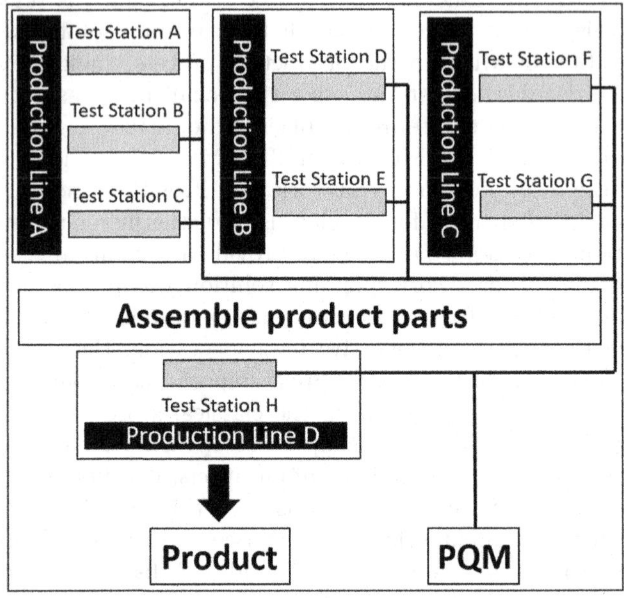

Fig. 1. Typical production setup [3].

In practice, there is a large number of different production lines, but they usually have important characteristics in common. Figure 1 shows a typical production setup. It consists of different production lines, which in turn can include several test stations. Their task is to measure the various product properties at different points in time within the production process. The measurements are stored in a Product Quality Management system (PQM). The product or product component is only passed on, if the measurement is within a desired tolerance range. Defects can also occur if corresponding measurements at previous stations were good. The defective part is then repaired separately. If an adequate repair is no longer possible, the product is discarded.

To detect as many defects as early as possible, it is very important to understand the production environment and to model relevant aspects. Only in this way can ML approaches deliver reliable and meaningful results. This includes the exact design of each production line, relevant product features and characteristics of test procedures. Often products have many variants, are divided into groups and individual rules can apply to the corresponding production steps.

This work is based on the cooperation with SICK AG, a manufacturer of intelligent sensors and sensor solutions for factories, logistics and process automation. SICK AG uses state of the art assembly lines to produce a variety of products.

3 Fundamentals

In this section we review established procedures for quality management and data analysis as basis for our subsequent discussion. We refer to the reviewed quality management procedures of Subsect. 3.1 later on and compare them with our approach. Subsect. 3.2 serves as introduction to ML methods that we integrated in our concept in Sect. 7.

3.1 Quality Management Procedures

A standardized methodology in the context of process improvement is Six Sigma [13]. Especially in the field of quality management, it is used to address performance problems by providing a framework to reduce or even avoid the number of failures, e.g. in a production process. In this context, sigma represents variability in terms of the number of defects.

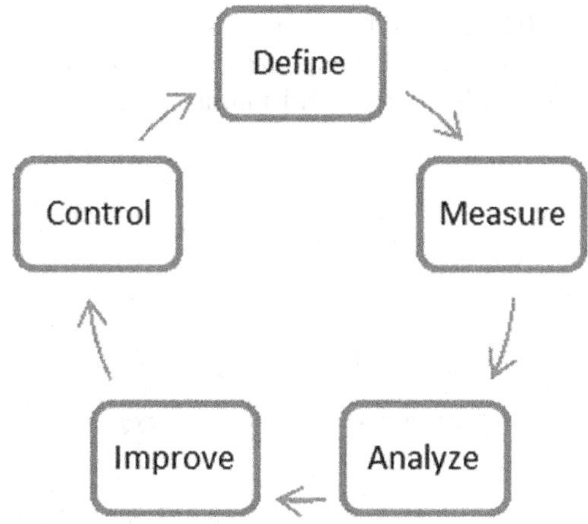

Fig. 2. The closed-loop improvement model DMAIC.

In order to make an effective contribution to quality management, a structured improvement procedure known by the acronym DMAIC, which stands for define, measure, analyze, improve and control, has been introduced [7]. Its closed-loop nature is illustrated in Fig. 2. At the beginning, objectives of the performance improvement are recorded. The next steps include the measurement of the actual state. This provides a quantitative reference for monitoring throughout the process. The analysis of the system is necessary to determine any discrepancies between the previously defined goal and the actual state and aims to find ways to fill these gaps. An improvement can then be made. The final step involves maintaining the improvement in order to achieve better performance in the long term.

One of the well-established data mining approaches is called Cross-Industry Standard Process for Data Mining, known as CRISP-DM [13]. Its six steps serve as framework for a systematic implementation of data science projects. The sequential nature of the workflow, as can be seen from Fig. 3, requires by implication thoughtful outcomes in each stage, as subsequent phases depend highly on the preceding ones. However, the need to adapt earlier steps cannot usually be avoided, so that an iterative procedure is common.

The first phase of CRISP-DM aims at understanding the business (1). Specifying the problem and identifying objectives and requirements are crucial for the success of the project. Subsequently, relevant datasets have to be determined (2), accompanying data familiarization. Data preparation (3) consists mainly of selecting, cleaning and transforming real-world data. Once well-structured data is available, modeling techniques can be selected and applied (4). In order to identify the most appropriate model, candidate models have to be adjusted and assessed. In Phase 5, the extent to which the selected models correspond to the objectives derived in step 1, is examined. Due to different perspective and abilities, collaboration between decision makers, who defined the underlying objectives, and data scientists, who built the models, is essential. Finally, the chosen model can be deployed (6), which comprises making results understandable to the end user as well as implementing and maintaining a repeating data mining process.

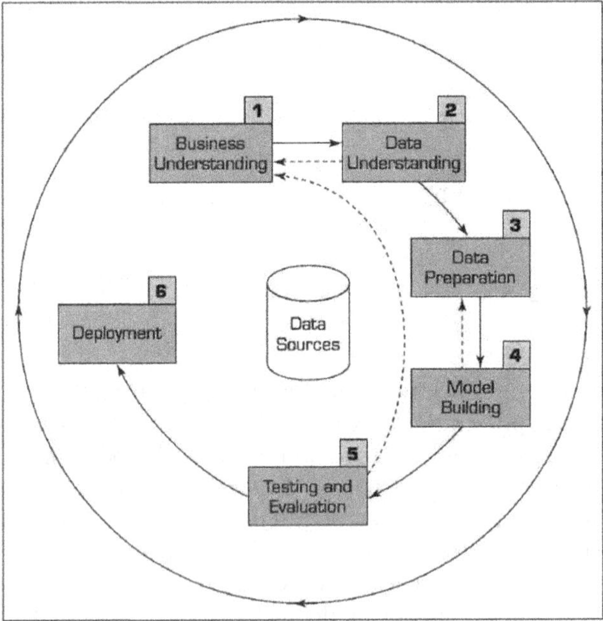

Fig. 3. CRISP-DM. Reprinted from business intelligence and analytics: systems for decision support (p. 205), by R. Sharda, 2015, Pearson Education.

3.2 Data Analysis Process

In practice, data mining projects often face the challenge that the underlying data an derived models are highly complex. In the context of recorded manufacturing data, this is due to several reasons. The complexity is expressed in the large number of product instances examined, but also in the large number of characteristics for which measurements were taken at the respective test station. Therefore, an analysis based on the complete data set is usually not targetoriented and cannot be interpreted intuitively. Machine learning methods can capture complex relations in complex data sets. However, there resulting models are complex as well and hard to interpret by humans. Hence, there is a need to reduce complexity and improve interpretability of the machine learning results. Various methods offer a remedy for this, which can be referred to as explainable ML. Some of these approaches are the following [2]:

Feature Importance
The ML algorithm calculates the importance of the features with which an error can be predicted for each correlation found. Both the ranking and the importance of the features provide information about which features are particularly important.

Shapley Values
The importance of the respective features is calculated at instance level. It computes the fair contribution of the feature to the model output. This allows for an interpretation which features were particularly important for the respective error instance. The importance can also be added or averaged over several instances, so that an interpretation over several instances is possible.

Surrogate Models
The idea is to approximate the complex prediction models of an ML algorithm and to allow simple interpretations. The original black-box-characteristic of the ML approach is turned into a transparent model which can often interpreted intuitively with the help of visualizations. In the case of a classification problem, for example, one may build a decision tree. Usually the depth of a tree is limited to a few levels for the sake of interpretability. The structure of the tree allows easy interpretation by humans as well as the derivation of decision rules.

Anchors
The so-called anchors also refer to the instance level. Taken relevant features into account, subspaces are defined that include at least one instance. Instances within this space, for example, have a high failure probability in common so that for each subspace a decision rule based on the space-defining features and their characteristics can be explained, which explains the respective prediction. This means that changes to non-relevant features do not affect the prediction whereby the prediction gets anchored.

4 Overall Solution

Early error detection to proactive error avoidance is a key competitive advantage for many companies in the manufacturing industry. We designed a system that aims to achieve financial benefits through reliable error predictions and the associated reduction in rejects [19]. Taking into account the increasingly dynamic processes, a high degree of automation of the used methods is of great importance in order to keep manual expenditure of time low. The proposed approach, as shown in Fig. 4, provides for the close interlinking of machine learning processes, big data technologies and knowledge modeling (see [19]). This way, in particular the following properties should be fulfilled for a scalable application:

- predict production errors with high prediction quality
- automatically create a variety of prediction models
- automatically maintain these models

One of the central ideas of the approach is to make use of the domain experts' existing knowledge of products and production processes. The modeling of relevant relationships on the basis of this knowledge enables and supports the automated creation of models, e.g. by considering only potentially relevant features for error prediction. In a similar way, this domain knowledge can be used to improve and maintain the developed models. The concrete modeling can be based on existing approaches such as ontologies, which can be adapted to the necessary requirements.

The automated creation of predictive models is another important feature of the concept. With the help of these ML models, both classification and regression

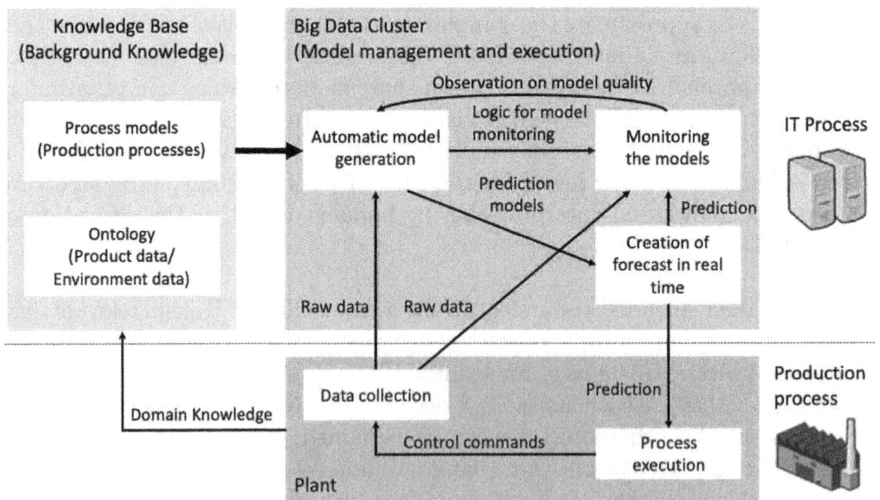

Fig. 4. Conception overview of PREFERML [19].

methods can be implemented. For example, regression predicts certain product features and a comparison with the objective allows an early proactive intervention in the production process.

A major challenge in the application of ML approaches is the handling of large amounts of complex data. The use of AutoML further increases complexity. The use of parallel and highly scalable data processing methods is therefore essential. These big data technologies enable fast processing of complex calculations in order to find optimized solutions for specific cases.

As a rule, production processes are subject to constant dynamics, so that relationships between errors and their causes are often not stable over a long period of time. Such a concept drift leads to the loss of the predictive power of a trained ML model. In order to prevent this, methods should be developed that detect relevant changes early on. The functionality required for this can be implemented as part of the automated model generation and aims to automatically adapt the underlying monitoring logic if necessary.

Figure 4 illustrates the automated improvement cycle for machine learning, how data and knowledge is used to create and maintain ML models. The next sections show how quality engineers are involved and interact with the PREFERML system. Data scientists are only needed for supervision.

5 To-Be-Process

In this section we introduce a process to leverage machine learning in manufacturing quality management. Here we expand and detail earlier work presented in [3]. Specifically we explain the investigation of errors using ML support in more detail and refer to analogies of the quality management procedures presented in Subsect. 3.1.

Integrating machine learning approaches into error detection methods in the context of manufacturing processes increases its potential performance on the one hand. On the other hand, it alters roles and requirements with respect to competences and responsibilities of the participating actors [3].

A **quality engineer** commands all relevant product and production information, which we are referring to as domain knowledge. Therefore, a quality engineer's task is to monitor a large amount of products within the manufacturing process and to document individual stages. Their corresponding product measurements are of special interest, as he or she is responsible for the specification of these tests. A quality engineer's knowledge is necessary to create an optimized ML model, e.g. he or she can estimate monetary gains or losses of interventions in the process. The application of the ML model includes primarily specifying the configuration of each experiment. For example, a quality engineer knows which choice of errors, groups of errors or considered period is most reasonable. His or her knowledge should be sufficient to create the ML model and to interpret the results. The ML model shall show previously undiscovered connections in the data. It increases the quality engineer's understanding of the data, like most important features as identified error sources or simply anomalies in the data. Furthermore, the ML model helps to reveal hints if the underlying concept changes over time, known as concept drift. The gained knowledge may support the quality engineer in adapting ML model, improving production processes, and ultimately the product or product part as well.

The support by machine learning requires a constant supervision of the ML models in use. A **data scientist** has appropriate knowledge to take on these controlling tasks. For instance, they may check performance measures to secure that the ML model is acceptable.

Based on the described competencies of the actors, a process for error identification and correction can be designed, which is supported by machine learning. As shown in Fig. 5, this process covers the entire product life cycle and is repeated until the product is withdrawn from the market. At the beginning of the process, the relevant domain knowledge of the quality engineer with regards to a specific product is recorded in a structured manner (T1). The ML model will later obtain all the necessary data from this source. This step corresponds to the define phase of the DMAIC model and can be seen as combination of business and data understanding of the CRISP-DM process. Data preparation is then carried out directly as part of the automated procedure. In a second step, the quality engineer creates such a model for a specific product (T2) and the data scientist supervises it (T3). For this purpose, performance measures and visualizations are provided and assessed by the data scientist (T3). Referring to the DMAIC model, T2 and T3 replace the measure phase. If the quality of the ML model is ensured, the quality engineer can start troubleshooting and check the plausibility of the results (T4). Indications of relations in the data given by the ML model can provide the quality engineer with important insights (T5). The systematic investigation of the errors on the basis of the results serves to discover previously unknown relationships between the features and the examined errors (T6). Section 7 deals with the detailed procedure. The knowledge gained in this step helps to identify and understand a particular error and its causes (T7). In the further course there are two options. On the one hand, the knowledge gained can be used, after consultation with the production team, to take precautions to correct the error in the production process (A11). Further optimizations can then be carried out by repeating the steps from T2. Alternatively, the developed model can be actively integrated into the production process (A21+A22), which corresponds to the deployment phase of the CRISP-DM Process. After the quality engineer has adapted the corresponding configuration, the test system can continuously use the model and record its results in a database. These recordings also make it possible to determine changes over time in the relationships between errors and their causes (A23). The quality engineer can then intervene and create an adapted ML model for certain errors (T2). Regarding the CRISP-DM process this adds a transition from deployment to model building or data preparation respectively, giving the approach a closed-loop nature.

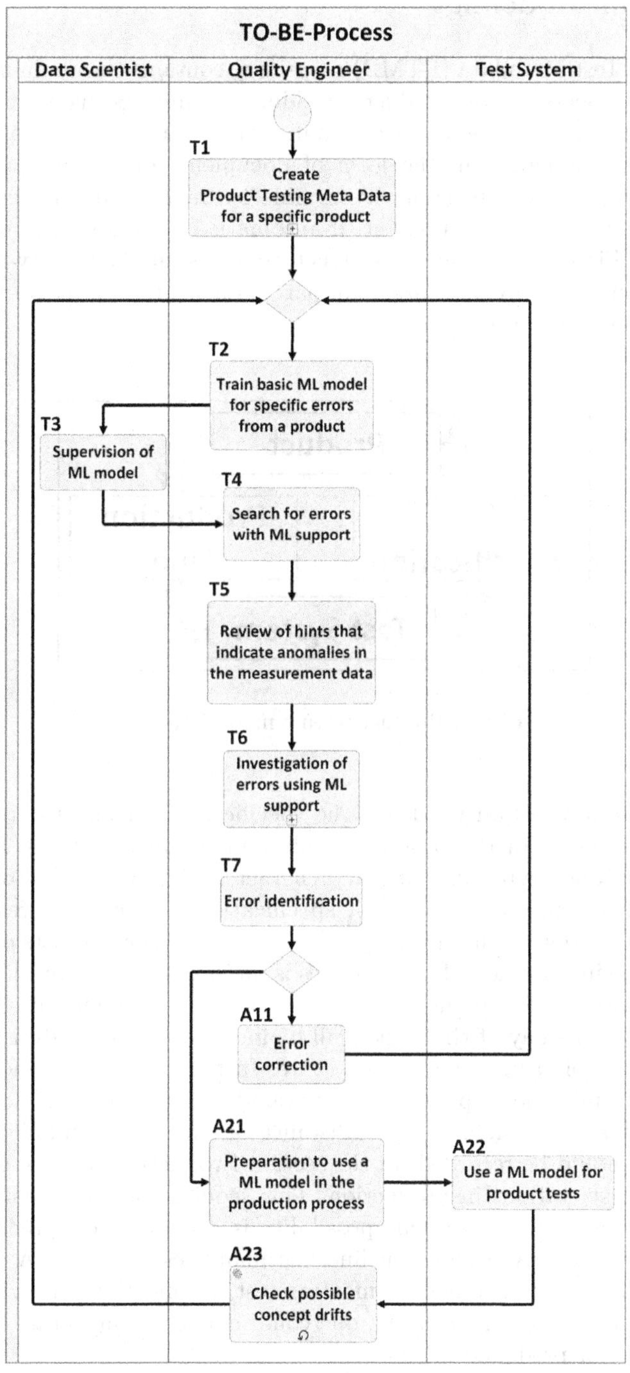

Fig. 5. To-Be-Process [3].

6 Domain Modeling

The Product Test Metadata (PTMD) repository contains the domain knowledge of quality engineers about products, production and testing that guides the work of PREFERML. It is a special configuration and selection of knowledge that exists in companies in the form of documents or database entries, but needs tailoring in order to support machine learning. A quality engineer can view this knowledge and can edit it, if automatic extraction from databases is not possible. Figure 6 illustrates the structure as a simplified entity-relationship representation. It is divided into four parts. Each of these parts represent a separate information section.

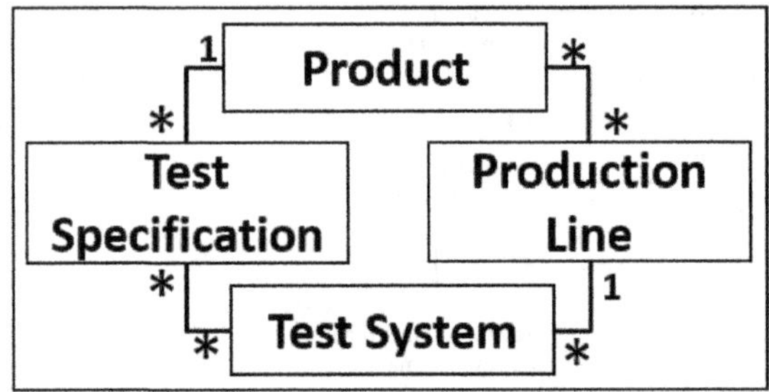

Fig. 6. Product testing meta data [3].

In the product section we define the specific product and the relationships with other members of the product family and product variants. Information about the product type and product characteristics must be stored in this section. Each product has many test specifications, since many test steps are installed after major production steps. A test specification consists of many test features. Type information of test features is important for the machine learning process. A distinction of numerical, categorical and time information is needed. For example, if the day of the week shall be included in the analysis, one might define this variable as a categorical feature. In the case of numerical information, physical unit and upper and lower bounds are provided (giving tolerance intervals for tests). As an example, if distances are measured in millimeters, this information should be recorded. In the section Production Line we describe the sequence of test stations in production. This sequence is important to analyze and identify product defects. One possibility is, to state each predecessor of a certain test station. A production line has many test systems. A test system can load many test programs implementing test specifications and vice versa. In order to consider economic aspects, one could define the monetary benefit of a correctly as error predicted instance.

7 Fault Analysis Process with Explainable ML

One of the key points during the suggested To-Be-Process (5) is the investigation of errors using ML support (T6). It is crucial to efficiently identify relevant errors which also includes ignoring insignificant errors. However, the benefit of the whole process highly depends on the validity of the drawn conclusions. Therefore, interpretation of ML results has to be made carefully and human experts must be enabled to judge the results. In order to prevent misinterpretation, this subprocess is examined in more detail (Fig. 7).

The starting point is that the search for errors using ML support has been completed. Subsequently, the errors identified up to this point must be checked for their relevance (T6.1 in Fig. 5). This decision is based on the metadata of the ML model in relation to the respective error message. In the case of the considered classification problem, there are a number of statistics available. Dependent on the context, an automated procedure is also conceivable for this step, saving additional work for the quality manager.

On the basis of selected key figures, a decision can be made as to whether an error message is relevant and should be investigated further. The key figures used and any threshold values on which this decision is based are context-dependent. Quality metrics such as the area under the ROC Curve (AUC) may be used. However, the target domain usually has heavily inbalanced data [16]. One should take this into account by using alternatives such as the Matthews correlation coefficient or a cost-sensitive metric. Furthermore, you could use the number of error instances within the time period examined as a decision criterion and only continue to look at those errors that occur more frequently than the specified threshold value.

These values should be selected in the automated procedure in such a way that at least clearly excludable cases are not selected. A too strict procedure can be disadvantageous under certain circumstances, since for example rare errors could still be relevant due to their high costs. Again, a cost-sensitive metric can provide a remedy by assigning a monetary weighting to the different cases of correctly and incorrect classification.

Once the relevant errors have been selected, the quality engineer can take a closer look at the evaluation of the ML model (T6.2). For this purpose, a confusion matrix is primarily used, which assigns both the instances classified as errors and the instances classified as correct to their true classes for the optimal configuration of the ML model. Various key figures such as accuracy or recall further aid the quantification of the classification quality. Yet, for the manufacturing context we argue that is advisable to derive a monetary benefit, as described above.

In addition, descriptive statistics help the quality manager to get a first impression of the respective error message. These include, for example, the number of errors in the period under investigation or the occurrence of errors over time. Periodic fluctuations as well as seasonal differences can thus be easily identified.

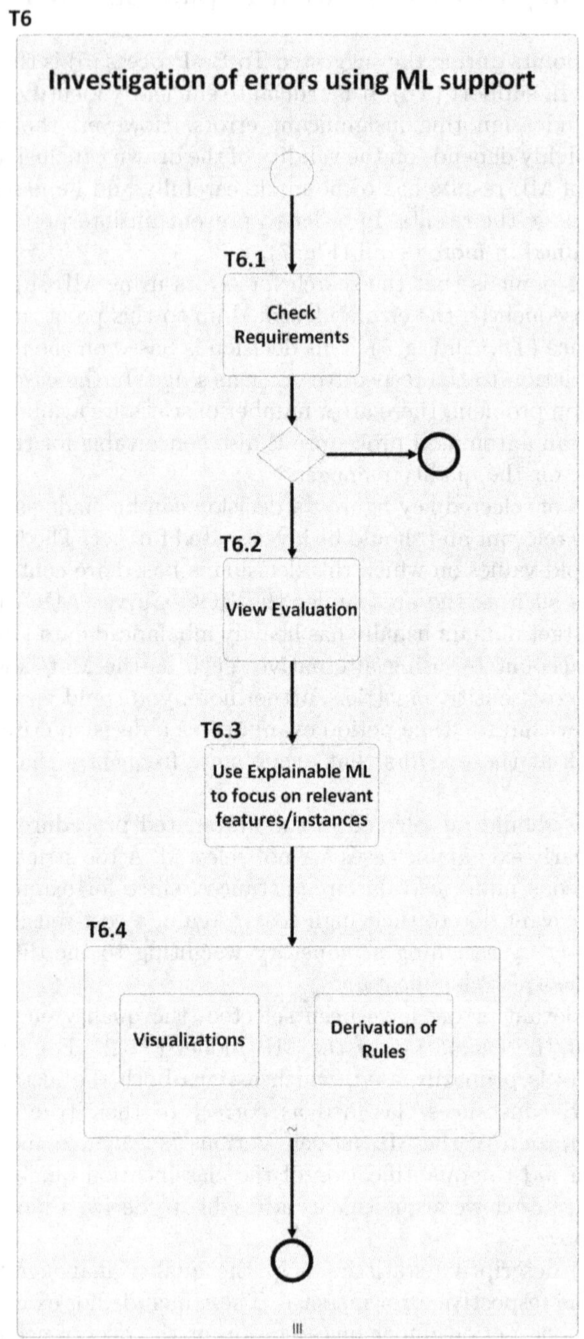

Fig. 7. Investigation of errors using ML support - detailed process step.

The underlying relationships between the errors that occur and their causes are usually very complex. Both the number of instances examined and the number of measurements and thus the measurement characteristics are so high that simplifications are needed to describe the relevant relationships more or less well. In practice, it is common that not all features considered are relevant for the explanation of errors, and likewise these correlations cannot be generalized for all instances. Accordingly, a targeted reduction to certain characteristics and certain instances is required in order to identify meaningful and yet comprehensible correlations (T6.3). Here the explainable ML methods are suitable.

One of the most essential methods in the context of error identification in manufacturing processes is the ranking of features according to their importance. The ML model indicates, based on certain metrics, which features are of crucial importance for the classification, called feature importance. Due to this weighting, the quality engineer can exclude a large number of supposedly unimportant features when investigating a certain defect and focus on a few features when searching for relevant correlations. Often even single features offer the possibility to recognize correlations well. Through visualization of such features, quality mangers can get hints about error causes (T6.4).

A possible visualization is the expression of an important feature over time, as in Fig. 8, based on artificial data that resembles real data a real case. The ordinate of the diagram reflects the order of the instances examined, which is equivalent to a temporal progression. The respective instance is colored here depending on the actual classification, red represents a faulty instance, blue a correct test piece. Since the number of correct test pieces is usually much larger,

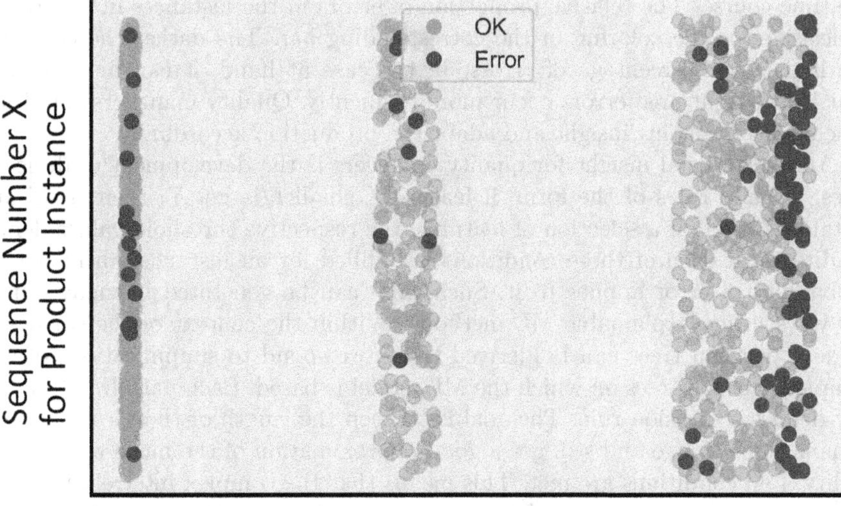

Fig. 8. Value range of important feature over time (synthetic data).

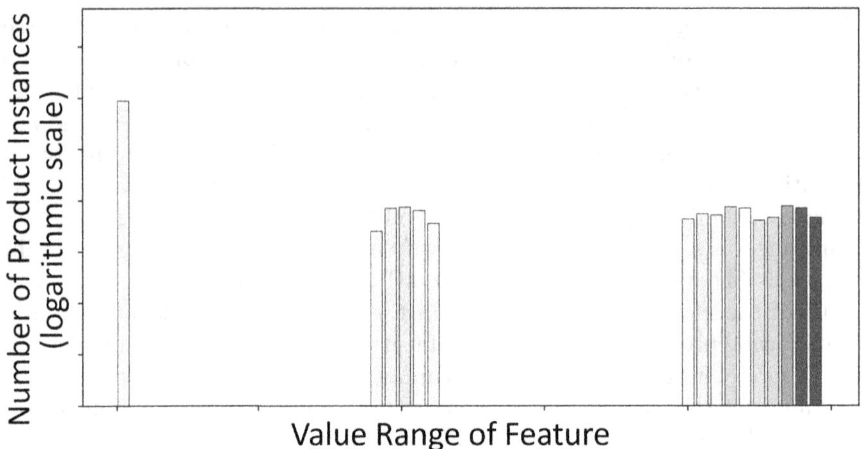

Fig. 9. Natural logarithm of number of product instances over value range of important feature (synthetic data). Coloring of bars reflects portion of error instances within the given range.

only 2000 randomly selected instances were plotted. Here, one can clearly see in which value ranges the characteristic occurs and where errors are particularly frequent. This gives first hints which value ranges of a characteristic could be critical.

This becomes even more clear in Fig. 9. Here the frequency of the instances in the respective range is shown on a logarithmic scale of the basis 10 instead of the time course. The relative proportion of errors in the instances in this range is decisive for the coloring of the corresponding bar. The darker the coloring, the higher the percentage of errors. In the case at hand, it can be seen well that in the right area errors occur more frequently. Quality managers can draw conclusions from this insight and adapt the production accordingly.

Another crucial insight for quality engineers is the development of decision rules. Usually rules of the form 'if feature X smaller/larger Y, then error' are suitable. Based on a selection of features, the respective threshold values define conditions. If each of these conditions is fulfilled for an instance, an increased probability of error applies to it. Such rules can be generated in many different ways using Explainable ML methods. Within the context of the surrogate models, decision trees can be derived that correspond to simplifications of the complex decision trees on which the ML model is based. Each path from root to leaf defines a decision rule. The goal is to keep the length of these rules within a manageable range and still get a good approximation of the mean error probability if all conditions are met. This means that the complex interrelationships are brought to a comprehensible and meaningful representation. Another way to generate such rules is offered by the anchors. Starting from the entire space, which the features span, small subspaces are formed. The value ranges defining the space form the conditions contained in the decision rule. All instances, whose characteristic values lie in such a subspace, show an increased probability of error. Here, too, the quality engineer gets an easy-to-interpret representation of the relationship between feature values and the occurrence of errors.

Once the correlations have been worked out, errors can be identified according to step T7 in Fig. 5 and the process can be continued.

8 Related Work

The potential of machine learning methods in manufacturing was recognized early on. Monostori et al. [9] discuss several approaches in their work and evaluate different methods regarding their applicability. Depending on their field of application they describe advantages and disadvantages as well as challenges. They emphasize that a harmonious interaction between humans and machine learning is the key to successful projects. Also Wang et al. [16] also investigate machine learning approaches that aim to make manufacturing smarter. They focus on the so-called deep learning approaches and compare existing methods. An advanced analytic framework serves to integrate deep learning approaches into the manufacturing processes. With regard to their possible applications, the authors evaluate the advantages and disadvantages of different methods. Among other things, they examine applications in the areas of diagnostic analysis for fault assessment and predictive analytics for defect prognosis. They also describe central challenges, including the quality of the data on which the deep learning approaches are based and the need to make the analytics solution comprehensible to the people involved in the manufacturing process. However, they do not mention in detail the requirements necessary for implementing an ML system for production. Wu et al. [17] develop a prognostic method in the context of smart manufacturing and compare the performance of their approach with other methods. In describing their experiment, however, they refrain from giving an exact description of the requirements necessary for the implementation. A structured overview of applications, advantages and challenges of ML methods in the context of manufacturing is given by Wuest et al. [18]. They explain why ML methods can be helpful for application purposes in manufacturing processes, and with regard to the challenges they see in this context, they go into more detail about the approaches ML methods offer to exploit different potentials. The authors do not go into detail about the conditions necessary to integrate an ML system into an existing manufacturing environment. Focusing on automatic production lines, Stanisavljevic and Spitzer [14] review some relevant papers on machine learning applications in this area. In particular, referring to Pham and Afify [11], they describe requirements that machine learning processes must meet in order to be applicable in the context of manufacturing. For example, an ML algorithm must be able to handle different data types, such as numbers, text or images and must be able to take into account noise and outlier. Usually huge datasets have to be processed, in some cases in real time. Susto et al. [15] mention four challenges when modeling semiconductor manufacturing processes, one of them is high dimensionality due to a large number of input variables.

We recognize such challenges and propose counter measures in our process by automatically removing unimportant data and taking only the essential information into account to create the ML model.

Explainable ML methods in the context of manufacturing have hardly been considered in research so far. Kharal [5] in particular uses the concept of feature importance and rule derivation to generate human-interpretable insights in processes of steal plate manufacturing. He proposes to reintegrated knowledge gained by the explainable ML methods directly into the model development process to iteratively achieve improved performance.

9 Conclusion

Integration of ML systems into product quality management promises improvements for production failure analysis approaches. However, certain requirements have to be met in order to obtain the benefits. Only when the data are given a meaning in the context of the production process, the ML approaches can provide results that are useful in terms of quality management. This can be achieved by structured modeling of domain knowledge, which includes, for example, the sequence of production steps or certain product characteristics. Whenever an ML method needs background knowledge, it is made available. We developed a procedure for error detection in manufacturing processes supported by machine learning using the integrated domain knowledge. In this paper we describe in more detail how a quality engineer can use this support during troubleshooting. The results of an ML algorithm are highly complex and usually not intuitively interpretable for humans. A reduction of the complexity by means of so-called explainable ML methods is the basis of our systematic approach to fault cause research in production. Based on simple decision criteria, quality assurance staff without special ML knowledge can use the results of the ML model to derive recommendations for action to proactively avoid errors in the future. In addition, changes in the context are recognized and remodeling can be initiated. Our developed procedure differs from the established techniques mentioned in Sect. 3. Because of context specific properties we feed backward at several stages. We use the CRISP-DM process as a closed loop procedure by providing a detailed adoption to ML based quality management in manufacturing. The steps of the DMAIC process were modified for our purposes so that they fit the given use case.

References

1. Dashchenko, A.I.: Manufacturing Technologies for Machines of the Future: 21st Century Technologies. Springer Science & Business Media, Berlin (2012)
2. Elshawi, R., Al-Mallah, M.H., Sakr, S.: On the interpretability of machine learning-based model for predicting hypertension. BMC Med. Inform. Decis. Mak. **19**(1), 146 (2019). https://doi.org/10.1186/s12911-019-0874-0

3. Gerling, A., Schreier, U., Hess, A., Saleh, A., Ziekow, H., Abdeslam, D.: A reference process model for machine learning aided production quality management, pp. 515–523. In: Proceedings of the 22nd International Conference on Enterprise Information Systems, 5–7 May 2020, vol. 1 (2020). https://doi.org/10.5220/0009379705150523
4. Hirsch, V., Reimann, P., Mitschang, B.: Data-driven fault diagnosis in end-of-line testing of complex products. In: 2019 IEEE International Conference on Data Science and Advanced Analytics (DSAA), pp. 492–503. IEEE (2019)
5. Kharal, A.: Explainable artificial intelligence based fault diagnosis and insight harvesting for steel plates manufacturing. arXiv preprint arXiv:2008.04448 (2020)
6. Kim, D., Kang, P., Cho, S., Lee, H.J., Doh, S.: Machine learning-based novelty detection for faulty wafer detection in semiconductor manufacturing. Expert Syst. Appl. **39**(4), 4075–4083 (2012)
7. de Mast, J., Lokkerbol, J.: An analysis of the six sigma DMAIC method from the perspective of problem solving. Int. J. Prod. Econ. **139**(2), 604–614 (2012). https://doi.org/10.1016/j.ijpe.2012.05.035
8. Mehta, D., Hamke, A.K.: In-depth. Artificial Intelligence (2019)
9. Monostori, L., Márkus, A., Van Brussel, H., Westkämpfer, E.: Machine learning approaches to manufacturing. CIRP Ann. **45**(2), 675–712 (1996)
10. O'Donovan, P., Leahy, K., Bruton, K., O'Sullivan, D.T.: Big data in manufacturing: a systematic mapping study. J. Big Data **2**(1), 20 (2015)
11. Pham, D.T., Afify, A.A.: Machine-learning techniques and their applications in manufacturing. Proc. Inst. Mech. Eng. Part B J. Eng. Manuf. **219**(5), 395–412 (2005). https://doi.org/10.1243/095440505X32274
12. Qu, Y., Ming, X., Liu, Z., Zhang, X., Hou, Z.: Smart manufacturing systems: state of the art and future trends. Int. J. Adv. Manuf. Technol. **103**(9–12), 3751–3768 (2019)
13. Sharda, R., Delen, D., Turban, E., Aronson, J.E., Liang, T.P., King, D.: Business Intelligence and Analytics: Systems for Decision Support, 10th edn. Pearson, Boston (2015)
14. Stanisavljevic, D., Spitzer, M.: A review of related work on machine learning in semiconductor manufacturing and assembly lines (Oct 2016)
15. Susto, G.A., Pampuri, S., Schirru, A., De Nicolao, G., McLoone, S.F., Beghi, A.: Automatic control and machine learning for semiconductor manufacturing: review and challenges. In: Proceedings of the 10th European Workshop on Advanced Control and Diagnosis (ACD 2012) (2012)
16. Wang, J., Ma, Y., Zhang, L., Gao, R.X., Wu, D.: Deep learning for smart manufacturing: methods and applications. J. Manuf. Syst. **48**, 144–156 (2018)
17. Wu, D., Jennings, C., Terpenny, J., Gao, R.X., Kumara, S.: A comparative study on machine learning algorithms for smart manufacturing: tool wear prediction using random forests. J. Manuf. Sci. Eng. **139**(7), 071018 (2017). https://doi.org/10.1115/1.4036350
18. Wuest, T., Weimer, D., Irgens, C., Thoben, K.D.: Machine learning in manufacturing: advantages, challenges, and applications. Prod. Manuf. Res. **4**(1), 23–45 (2016)
19. Ziekow, H., et al.: Proactive error prevention in manufacturing based on an adaptable machine learning environment. In: Artificial Intelligence: From Research to Application: The Upper-Rhine Artificial Intelligence Symposium UR-AI 2019, 13 Mar 2019, Offenburg, Germany, pp. 113–117. Hochschule Karlsruhe - Technik und Wirtschaft, Karlsruhe (2019)

An Investigation of Problem Instance Difficulty for Case-Based Reasoning and Heuristic Search

Hermann Kaindl[1(✉)], Ralph Hoch[1], Roman Popp[2], Thomas Rathfux[3], and Franz Lukasch[4]

[1] Institute of Computer Technology, TU Wien, Vienna, Austria
{hermann.kaindl,ralph.hoch}@tuwien.ac.at
[2] Purgstall an der Erlauf, Austria
[3] Mödling, Austria
[4] Robert Bosch AG, Göllnergasse 15-17, Vienna, Austria
franz.lukasch@at.bosch.com

Abstract. For managing the ever increasing variability of hardware/software interfaces (HSIs), e.g., in automotive systems, there is a need for the reuse of already existing HSIs. This reuse should be automated, and we (meta-)modeled the HSI domain for *design space exploration*. These models together with additionally defined transformation rules that lead from a model of one specific HSI to another one facilitate automatic adaptations of HSI instances in these models and, hence, both *case-based reasoning* (CBR) and *(heuristic) search*. Using these approaches for solving concrete problem instances, estimating their difficulty really matters, but there is not much theory available.

This work compares different approaches to *estimating* problem instance difficulty (similarity metrics, heuristic functions). It also shows that even *measuring* problem instance difficulty depends on the ground truth available and used. In order to avoid finding only domain-specific insights, we also employed sliding-tile puzzles for our experiments. The experimental results in both domains show how different approaches statistically correlate. Overall, this paper investigates problem instance difficulty for CBR and heuristic search. This investigation led to the insight that admissible functions guiding heuristic search may also be used for retrieving cases for CBR.

Keywords: Case-based reasoning · Similarity metric · Heuristic search · Admissible heuristic · Problem difficulty · Model-driven design · Hardware-Software Interfaces

---

Roman Popp did this work when he was with the Institute of Computer Technology at TU Wien.
Thomas Rathfux did this work when he was with the Institute of Computer Technology at TU Wien.

J. Filipe et al. (Eds.): ICEIS 2020, LNBIP 417, pp. 158–183, 2021.
https://doi.org/10.1007/978-3-030-75418-1_9

Notation

s, t	Start node and goal node, respectively
$g^*(n)$	Cost of an optimal path from s to n
$h^*(n)$	Cost of an optimal path from n to t
$g(n), h(n)$	Estimates of $g^*(n)$ and $h^*(n)$, respectively
$f(n)$	Static evaluation function: $g(n) + h(n)$
C^*	Cost of an optimal path from s to t
$N\#$	Number of nodes generated

1 Introduction

This research was motivated by real-world problems in the automotive domain. For reusing hardware/software interfaces (HSIs) in automotive systems, a stored most similar one is to be selected and retrieved, and its configuration to be transformed to another one that fulfills all the new requirements. The number of transformation steps is desired to be minimal in order to avoid having to create new hardware. We prepared a meta-model of this domain and transformation rules according to model-driven design, in order to facilitate automated design space exploration [26].

Hence, the desired reuse is addressed here through *case-based reasoning* (CBR) (see, e.g., [13]). Stored cases consist of previous problems together with their solutions (in the sense of specified HSI configurations). One of these cases is selected and the stored HSI configuration to be transformed to a solution of the problem posed by the new requirements (the problem formulation of the case at hand).

Our approach finds such a transformation automatically through *heuristic search*. More precisely, this search is performed by A* [14] with an *admissible* heuristic, so that the solutions have guaranteed minimal cost (actually, a minimum number of steps). In order to avoid potential misunderstandings with respect to the key notion "solution", note that a solution found by A* is a sequence of steps. These steps transform the selected HSI configuration to one that satisfies the given requirements, a solution in the sense of CBR. In the remainder of this paper, the given context disambiguates the notion of a solution in question.

In CBR, estimating the effort for adapting some stored solution of a previously solved problem instance to a solution of a new problem instance is usually addressed by retrieving the nearest neighbor determined through *similarity metrics*. The underlying assumption is that the relative effort for solution adaptation of stored instances (cases) correlates with the similarity of these instances with the given problem instance. In heuristic search for optimal solutions using A*, estimating the cost from a particular problem instance n to a goal is done through admissible heuristic functions $h(n)$.

Apart from estimating in two different ways, these approaches actually estimate two different measures:

- The effort of finding a solution vs.
- The cost of the solution.

In fact, these two measures indicate two *different views of problem difficulty*. The former view is the one of a problem solver that needs to come up with a solution within a reasonable period of time. The latter view is the one of an entity executing a solution, e.g., a robot moving along a path defined by it.

With regard to problem difficulty, there is a lot of theory on *complexity classes* of problems, e.g., the famous issue of **P** vs. **NP**, but not much on the difficulty of solving concrete problem *instances* of the same problem *class* (e.g., the well-known Fifteen Puzzle, see http://kociemba.org/themen/fifteen/fifteensolver.html). For solving concrete problem instances, e.g., through *heuristic search* [11,25], however, estimating their difficulty really matters.

Both views of problem *instance* difficulty actually depend on each other. The number of nodes generated and the time needed for finding a solution are good measures for the difficulty in terms of the effort for finding a solution, while C^* (the cost of an optimal path from the start node s to a goal node t) is a good measure of problem difficulty in terms of the cost of a solution. In fact, there is an excellent study of the relation between the two for IDA* and optimal solutions [23]. For finding error-bounded solutions, there is a trade-off (see, e.g., [20]), and finding any solution usually requires much less effort, of course.

For studying different estimation techniques (in the sense of heuristic functions and similarity metrics) more generally, we also employed the well-known Fifteen Puzzle. In contrast to the real-world domain of HSIs, it is widely known and was often used in the literature on heuristic search. All the results can be easily replicated publicly. This version of the sliding-tile puzzle is only used here as a vehicle for comparing such estimates (including simpler and less precise ones) and *not* for presenting a new search algorithm that would be faster than any other before (by using the best heuristic estimate known), when the Twenty-four Puzzle would have to be the choice today, such as in [4].

For the Fifteen Puzzle, usually the instances are ordered according to the lengths of their minimal solution paths C^* (due to *unit costs*, the length is the same here as the cost of an optimal path). But how well do they correlate with the numbers of nodes explored by some algorithm for solving them (also in the HSI domain)? And how do the numbers of nodes explored by, e.g., A* correlate with those of IDA* [22]?

Hence, this paper actually compares different approaches to both estimating and measuring problem instance difficulty with respect to CBR and heuristic search. For estimating, we employ both various similarity metrics and different h functions. For measuring problem instance difficulty, we employ minimum solution length (cost), the number of nodes generated for solving them by some algorithm, and two new measures very recently defined by us in [27].

An interesting question was then, whether to use a heuristic h instead of similarity metrics for selecting a stored HSI. Especially when using heuristic search for automatic adaptation, there is the additional possibility to estimate the adaptation effort using its heuristic function.

The remainder of this paper is organized in the following manner. First, we present some background material and discuss related work, in order to make this paper self-contained. Then we explain the HSI domain and our (meta-)model that we developed for it, as well as the transformation rules, the goal condition defined by formally specified requirements and the admissible heuristic function that we defined for this domain. After that, we explain several approaches to measuring problem instance difficulty, followed by explaining several approaches to estimating it. Based on all that, we present our experimental results and interpret them. Finally, we provide a conclusion based on our experimental results.

2 Background and Related Work

2.1 CBR in a Nutshell

Case-based Reasoning (CBR) is an approach to solve problem instances based on previously solved problem instances [21]. It is assumed to be closely related to the way a person solves problems: by recalling past experiences and applying that knowledge to the new problem, or in other words, mapping the new problem to previous experiences. For the cognitive science foundations of CBR see [24].

The first CBR systems have been developed in the early 1980s and were based on the work of [29]. Essentially, CBR is a cyclic process for solving problems [1] and consists of four major process steps. First, a fitting problem instance for a new problem instance is *retrieved* from the repository. This step is typically performed using *similarity metrics*. Next, the solution of the retrieved problem instance is *reused* by adapting it to the new problem instance. Note, that his step may be supported by an automatic adaption of the given problem instance to the new one by using, e.g., heuristic search. The resulting solution is then tested and, if necessary, *revised*. Finally, the resulting solution is stored, together with the new problem instance, in the repository, i.e., the new knowledge is *retained* in the repository.

CBR has been applied in a variety of domains, see, e.g., [5,6,17]. Search techniques have been used to help optimize a CBR system for predicting software project effort in [19]. An early combination of CBR with minimax search for game-playing can be found in [28].

2.2 Similarity Metrics

Similarity metrics [7] are used to determine how similar two objects are to each other, e.g., in CBR. It is an approach commonly used in information retrieval systems, e.g., to compare text documents, or for clustering of data in data mining applications [2]. For each object, a similarity value can be calculated in relation to another object, i.e., each pair of objects has a similarity score assigned. Typically, similarity scores are normalized values in the interval $[0, 1]$, where 0 indicates no similarity at all and 1 indicates maximum similarity. A popular similarity metric, among others, is the *Cosine Similarity* [30].

2.3 A* and IDA*

The traditional *best-first* search algorithm A* [14] maintains a list of so-called *open* nodes that have been generated but not yet expanded, i.e., the frontier nodes. It always selects a node from this list with minimum estimated cost, one of those it considers "best". This node is expanded by generating all its successor nodes and removed from this list. A* specifically estimates the cost of some node n with an evaluation function of the form $f(n) = g(n) + h(n)$, where $g(n)$ is the (sum) cost of a path found from s to n, and $h(n)$ is a heuristic estimate of the cost of reaching a goal from n, i.e., the cost of an optimal path from s to some goal t. If $h(n)$ never overestimates this cost for all nodes n (it is said to be *admissible*) and if a solution exists, then A* is guaranteed to return an *optimal* solution with minimum cost C^* (it is also said to be *admissible*). Under certain conditions, A* is optimal over admissible unidirectional heuristic search algorithms using the same information, in the sense that it never expands more nodes than any of these [9]. The major limitation of A* is its memory requirement, which is proportional to the number of nodes stored and, therefore, in most practical cases exponential.

IDA* [22] was designed to address this memory problem, while using the same heuristic evaluation function $f(n)$ as A*. IDA* performs iterations of *depth-first* searches. Consequently, it has linear-space requirements only. IDA*'s depth-first searches are guided by a threshold that is initially set to the estimated cost of s; the threshold for each succeeding iteration is the minimum f-value that exceeded the threshold on the previous iteration. While IDA* shows its best performance on trees, one of its major problems is that in its pure form it cannot utilize duplicate nodes in the sense of *transpositions*. A transposition arises, when several paths lead to the same node, and such a search space is represented by a *directed acyclic graph* (DAG). IDA* "treeifies" DAGs, and this disadvantage of IDA* relates to its advantage of requiring only linear space.

In principle, we could have used for this research other search algorithms as well, e.g., with limited memory [16] or for bidirectional search [15], but experiments using A* and IDA* were sufficient.

2.4 VIATRA2

VIATRA2 is a model-driven framework for design space exploration. Based on such a meta-model, this framework supports defining search strategies for traversing the design space, starting from an initial model by applying *transformation rules*. VIATRA2 also allows defining such rules based on the meta-model used. They are applied throughout the search to explore the design space. Goals are defined as conditions that must be satisfied for a solution.

VIATRA2 as used in our work presented here, is actually just one tool of a set of tools developed over time, where different tools provide different techniques for design space exploration [3].

3 HSI Domain

Parts of our findings are based on the real-world domain of Electronic Control Units (ECUs) and Hardware-Software Interfaces (HSIs) in automotive vehicles. One characteristic of this domain is its variability of possible configurations. In this section, we first present the structure of ECUs and HSIs. Based on this structure, we provide a detailed description of our modeling approach using a meta-model and transformation rules used during search. Finally, we specify goal conditions for search and show how we systematically derived an admissible heuristic function.

3.1 ECUs and HSIs

ECUs provide functionality of hardware components and software functions for external systems through an HSI. An ECU may contain different building blocks, but commonly includes a microcontroller (μC), which is used to run the software, and its internal resources alongside with its pins (μC-pin), hardware components for pre-processing signals and ECU pins. Internally, these building blocks are (potentially) connected to others through the wiring of the ECU. The wiring of the ECU also defines which μC-pins are connected to which ECU pins. External hardware components connect through ECU pins and, therefore, indirectly to a number of μC-pins. Figure 1 schematically illustrates the structure of an ECU with its building blocks and connections. Note, that a real ECU is both larger and more complex, but for the purposes of the explanations in this paper, this schematic illustration should be sufficient.

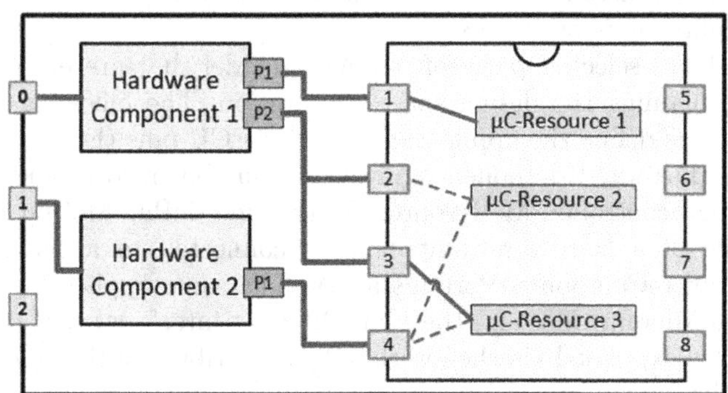

Fig. 1. Schematic illustration of an ECU and its HSI [26].

The hardware components represent any type of hardware processing input or output signals (e.g., power output states, low- or high-pass filters) and support a number of *interface types*. They are connected to ECU pins and to one or more μC-pin. The latter connections are realized through ports (e.g., P1, P2).

The microcontroller contains internal resources, e.g., analogue digital converter (ADC), timer input module (TIM), which can be made available on some of its pins. The possible connections of internal resources and μC-pins are pre-defined and can be active or not. Figure 1 illustrates potentially available connections as dashed lines, and currently active connections as bold solid lines. For example, in Fig. 1, μC-Resource 2 has potentially available connections to μC-pin 2 and μC-pin 4, while there is an active connection between μC-Resource 1 and μC-pin 1.

Hardware components may require specific μC resources to provide their interface type properly, e.g., analogue measurement of input signals may require an ADC. Hence, the μC resources have to be configured on μC-pins accordingly. The variability in configurations is mainly represented by the possible connections of the μC resources. All interface types of hardware components together with their connected μC resource configurations define an HSI. In essence, an HSI specifies which interface types are available on which ECU pins.

3.2 Meta-model

We formalized the application domain sketched above by means of meta-modeling in Ecore and the Eclipse Modeling Framework (EMF) [10,31], which is directly supported by the design space exploration tool VIATRA2. We used VIATRA2 for searching HSI configurations that fulfill given requirements. Our meta-model covers the structure of an ECU, the possible variations for activating connections to μC resources, and the currently selected configuration representing the HSI. A formal specification of the searched configuration, i.e., the goal conditions, is given by *Requirements*, which are also generically included in the meta-model. Each instance of the meta-model represents a specific ECU model (and the Requirements on its HSI).

Figure 2 shows selected parts of our meta-model that are essential for the purpose of explaining the design space exploration. The following classes and their associations define the application domain: ECU pins (EcuPin), hardware components (HardwareComponent, HardwareComponentPort) and the microcontroller (Microcontroller, McResource, McHwPin, McPin, McPinConnection). Each instance of a HardwareComponent is connected to a McPin instance through HardwareComponentPort instances. The same applies for the McResources of the Microntroller and the Ecu. Different interface types on hardware components are expressed via the interfaceType attribute of the HardwareComponent class [26].

The possible configurations, i.e., the variations, are modeled through instances of McPinConnection, which connects instances of McResources to instances of McPins and, consequently, to HardwareComponentPorts. As described above, each hardware component needs specific resources to function properly. In the meta-model, this is represented by the classes AssignmentConstraintSet and AssignmentConstraint. Each HardwareComponentPort may contain several AssignmentConstraintSets with AssignmentConstraints and one specific AssignmentConstraint specifies which (type of) McResource is needed.

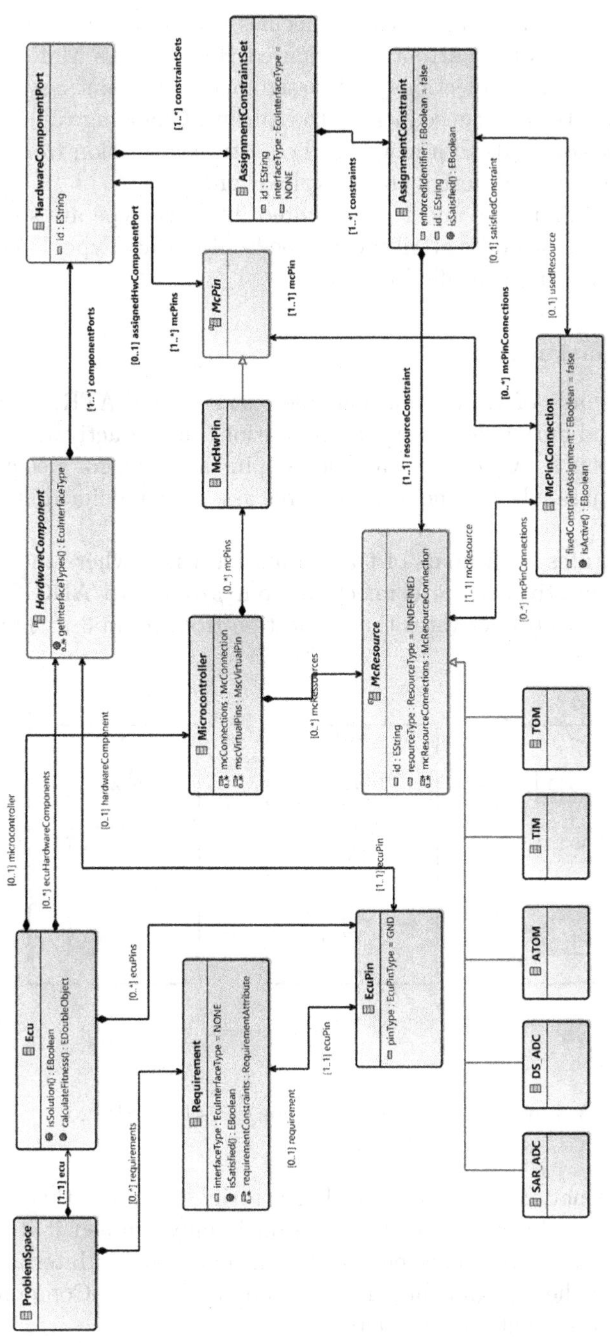

Fig. 2. Selected parts of our meta-model [26].

This information is processed to determine the transformation steps during design space exploration.

The association *usedResource*, which specifies if a connection to a resource is active or not, defines the currently configured HSI. It links McResources on McPins through McPinConnections and AssignmentConstraint(Set)s to HardwareComponentPorts, and, consequently, to EcuPins. Changing values on a usedResource association leads to a new HSI, i.e., a transformation from one state to another in the course of design space exploration.

Finally, the requirements and their required interface type are specified via the Requirements class and its attribute of type EcuInterfaceType. Each requirement is associated with a specific EcuPin.

3.3 Transformation Rules

We defined two kinds of transformation rules using the VIATRA2 query language to be applied during design space exploration: one for activating a specific connection between a μC-resource and a μC-pin, and one for deactivating a connection, i.e., manipulating the usedResource association defined in the meta-model.

Figure 3 illustrates both kinds of transformation rules, where the transition from (a) to (b) deactivates the connection from μ-pin 4 to ADC 2, and the transition from (b) to (c) activates the connection from μ-pin 3 to TIM 1.

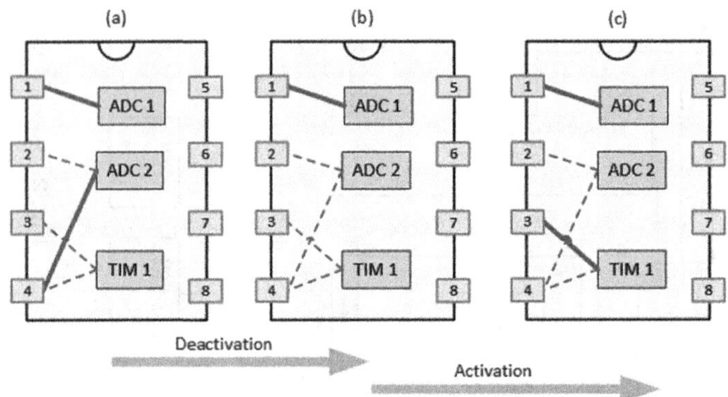

Fig. 3. Illustration of transformation rules for HSIs [26].

These transformations are not applied arbitrarily, but are rather selected in a goal-oriented way. That is, a transformation is only applied if it assists in satisfying (parts) of a given requirement. To this end, the EcuInterfaceType of Requirements and the AssignmentConstraintSets of HardwareComponentPorts defined in the meta-model are considered.

Technically, such a transformation rule as defined in VIATRA2 consists of three parts: pre-condition, transformation and post-condition. The pre-condition specifies the applicability of a transformation rule, in our application in terms of the constraints of HardwareComponentPorts. The transformation specifies the changes in the model, i.e., the instance of the meta-model that the transformation rule is applied to. The post-condition specifies the result of the rule application as a condition defined according to the meta-model.

3.4 Goal Condition

In the HSI domain sketched above, a goal condition is given by a set of requirements that need to be fulfilled. Each requirement is specified by a specific interface type that is required to be available on a specific ECU pin. We use first order logic to formally specify these goal conditions. As each requirement is specified as a specific interface type $IF(if)$ at a specific ECU-pin p, we introduce a requirement predicate $REQ(p, IF(if))$.

Each hardware component may support several interface types on its ports. Hence, we introduce a predicate $HW(p, IF(if), port)$, which defines a connection of an ECU-pin to the hardware component and the supported interface at a specific port. At each port, there may be a variety of supported interface types. If a specific interface type is supported, then an *assignment constraint set* has to be defined for it, i.e., $ACS(port,\ IF(if))$. These assignment constraint sets are or-connected:

$$\bigvee_{port_i \in Port} (ACS(port_i, IF(if))) \tag{1}$$

One assignment constraint set is specified by and-connected *assignment constraints*, i.e., $AC(RType(RT), SR(R))$, where RType denotes the requested resource type and SR specifies if a resource has been selected:

$$ACS(port, IF(if)) = \bigwedge_{rt \in REQ\_RES_{if}} AC(RType(RT_{rt}), SR(R_{rt})) \tag{2}$$

A port supports an interface type if there exists at least one ACS that fulfills its requirements. Hence, the interface types of all ports of a hardware component are defined as:

$$\bigvee_{port_i \in Port} (ACS(port_i, IF(if)) \tag{3}$$

Therefore, a specific interface type at a port of a hardware component is defined as:

$$HW(p, IF(if), port) = \bigvee_{acs \in port} (ACS_{acs}(port, IF(if)) \tag{4}$$

Using these formulas, we can formally specify that a requirement is fulfilled if a connected hardware component exists that provides the interface type on a port:

$$REQ(p, IF(if)) = HW(p, IF(if), port) \tag{5}$$

Based on this formula, our goal condition is a conjunction of all requirements and all ports on connected hardware components for a specific interface type.

$$\bigwedge_{if \in Req\_IF} \bigwedge_{p \in HWP_{if}} (\bigvee_{port_i \in Port_p} ACS(port_i, IF(if))) \tag{6}$$

3.5 Admissible Heuristic Function

For A*, *admissibility* of the heuristic function is important for guaranteeing the optimality of solutions found. For evaluating a configuration in such a function with respect to its goal achievement, the number of not (yet) fulfilled conjunctively related goal conditions is counted. In case of disjunctively related conditions, the minimum is taken. The resulting number can be used as the heuristic value, since each condition needs at least one application of a transformation rule. In fact, these can only be activation rules. Deactivation rules may additionally be necessary, in order to deactivate some connection so that another one needed can be activated at this particular pin. Consequently, this number is less than or equal to the number of minimal steps to achieve the goal condition, i.e., this is an admissible heuristic.

This can also be explained more theoretically based on the *meta-heuristic* of problem relaxation, see [25]. A relaxed problem would only need activation rules for its solution, i.e., the number calculated by our heuristic function.

Let us illustrate this in detail using a specific example. We elaborate on the calculation of our heuristic function for the case of an analogue signal delivered by a sensor. Figure 4 illustrates this example through a requirement condition R1 (at the left) and its possible fulfillments (in the middle) based on a partial configuration (at the right).

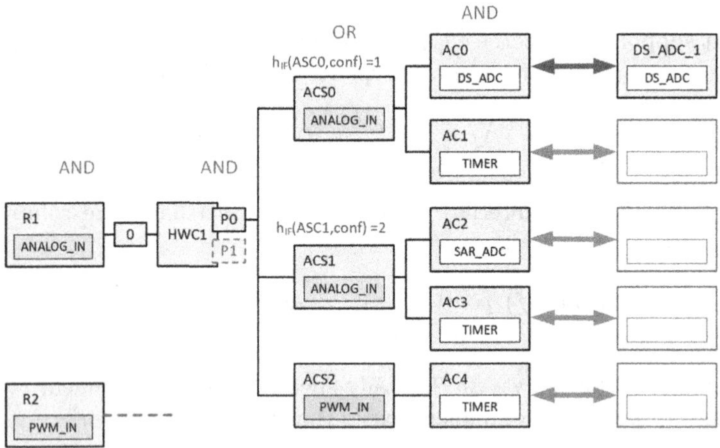

Fig. 4. A requirement and its possible fulfillments in a partial configuration [27].

The heuristic function optimistically estimates the distance from a given configuration to a goal by calculating the minimal number of transformation steps necessary to reach a configuration that fulfills the goal condition. We explain this using a part of a goal condition, where the clause of its conjunction corresponding to R1 (as shown in Fig. 4) is defined for this example as

$$REQ(0, IF(ANALOG\_IN))$$

Formally, this means a disjunction of $ACS(P0, IF(\_))$ instantiated for ACS0, ACS1 and ACS2. However, ACS2 has to be discarded, since its interface type PWM_IN is different from the required ANALOG_IN. Of course, only assignment constraint sets with matching interface types are taken into account.

The remaining formula can be written as follows, according to the concrete example shown in Fig. 4:

$$\begin{aligned} &(AC(RType(DS\_ADC), SR(DS\_ADC\_1)) \wedge AC(RType(TIMER), SR(Empty))) \\ &\vee \\ &(AC(RType(SAR\_ADC), SR(Empty)) \wedge AC(RType(TIMER), SR(Empty))) \end{aligned} \tag{7}$$

$SR(DS\_ADC\_1)$ means that the concrete resource DS_ADC_1 is selected for the assignment constraint AC0 as shown in the figure. Since the type of the selected resource matches the required resource type of the assignment constraint, the predicate *AC* evaluates to true for this example. *Empty* as given for the other assignment constraints in this example means that no resource is selected (yet) for them. Therefore, the predicate *AC* for these examples evaluates to false. This defines the current partial configuration in the course of the search that is evaluated heuristically.

For such a partial configuration, the corresponding formula evaluates to false. In order to fulfill the goal condition, all the remaining and-connected parts need to evaluate to true as well. This requires certain transformation steps, and for calculating the admissible heuristic, we determine the minimum number of such steps.

In fact, for changing a single *AC* from false to true, at least a single transformation step is necessary. Note, that a single step cannot make more than one *AC* true according to the definition of the transformation rules based on the meta-model.

We define a heuristic function h_{IF} of a specific interface type for a given configuration $(h_{IF}(ASCS_{IF}, CONF))$ in such a way, that it never over-estimates the minimum number of steps for fulfilling the requirements of the interface type. Hence, for or-connected parts, it is necessary to take the minimum, so that the heuristic estimate is optimistic. In the example, therefore, $h_{IF}(ASC0, conf) = 1$ must be taken for port $P0$.

For calculating the complete heuristic function for a given configuration $conf$, all requirements and associated ports must be taken into account. A configuration is specified by all currently selected resources. Therefore, the heuristic function can be calculated as follows:

$$h_{HSI}(conf) = \sum_{r \in Req} \sum_{p \in Ports(r)} \min_{acs \in ACS(p, IF(r))} \{h_{IF}(acs, conf)\} \tag{8}$$

4 Measuring and Estimating Problem Instance Difficulty

We provide here the definitions of measures and estimates used in our experiment. Two of them are actually hybrids in the sense that a measure is combined with an estimate.

4.1 Measuring

For measuring the difficulty of some problem instance in some domain (as compared to the difficulty of other problem instances), the question arises, what "problem instance difficulty" really means. Another question is what ground truth is available and used for measuring it. Of course, all of these measures are only available after the fact, so that *estimating* is necessary in practice (see below). However, our experiment had them available, of course. For getting a better theoretical understanding, it is still useful to study relations between measures and estimates, see also [23] for the number of nodes explored by IDA* using a given heuristic h, based on knowledge about its distribution over the problem space.

According to one view, the one of a problem solver that needs to come up with a solution within a reasonable period of time, the difficulty of a problem instance is characterized through the time it takes to solve it (and indirectly also the size of memory used). This time is usually proportional to the number of nodes explored for a given algorithm like IDA* (with only linear memory requirements). This depends on the algorithm used, however! In particular, this is different for A*, even though IDA* is conceptually derived from A*. However, the memory need of A* grows with the number of nodes explored, and maintaining the stored nodes for fast access incurs additional overhead.

According to the other view, which is independent from a specific algorithm, the difficulty of a problem instance is characterized through the cost of the solution, where we consider C^*, the cost of an *optimal* path from s to t (for unit costs, its length). This is the view of an entity executing a solution, e.g., a robot moving along a path defined by it.

An assumption is that C^* of a given problem instance correlates with the time it takes for solving it. How strong is this correlation actually?

4.2 Estimating

For *estimating* problem instance difficulty, let us distinguish here admissible heuristics from similarity metrics, where the former are employed for heuristic search and the latter for CBR.

Admissible Heuristics. Such heuristics fully depend, of course, on the problem domain. For the real-world HSI domain, we had to develop an admissible heuristic on our own before, which is defined above and denoted as h_{HSI}. Since we followed the constraint relaxation meta-heuristic (given already in [25]) for its development, h_{HSI} is actually *consistent*, but this is only relevant for our purposes here for implying admissibility of h_{HSI}.

For the Fifteen Puzzle domain previously studied in-depth, all the heuristics we include in our experiment of estimating problem instance difficulty are well-known, published and their admissibility is proven. In fact, already in [25] the constraint relaxation meta-heuristic was illustrated for two such heuristic functions:

- h_{TR}, simply counting the number of tiles that are positioned "right", i.e., in the position as defined by the target configuration t,
- h_M, the so-called *Manhattan Distance* heuristic.

Both of them obviously return estimated numbers of steps that are at least needed to arrive at the target node t. While h_{TR} provides only very rough estimates, h_M is obviously more accurate. Both, however, cannot compare with specifically pre-compiled *pattern databases*, see [12]. Essentially, these cache heuristic values from breadth-first searches from the target node t backwards. In contrast to computing heuristic values on demand like h_M, pattern databases are pre-computed for later use in heuristic search for some given problem instance. They may largely differ in their sizes, e.g., a static 5-5-5 partitioning version requires about 3MB, 6-6-3 partitioning about 33MB, and 7–8 partitioning more than half a GB. The heuristic function using these small, medium-sized and large versions are denoted below as h_{P555}, h_{P663} and h_{P78}, respectively.

Similarity Metrics. Typically, calculating similarity metrics is done using vectors encoding information. We studied different variations on how to create such vectors.

In the HSI domain, we made heavy use of binary vectors [8], where each entry in such a vector is either true (1) or false (0). We constructed the vectors using the requirements given for each pin. A vector has $n \times i$ entries, where n is the number of ECU-Pins and i is the number of interface types. Each entry in the vector corresponds to interface type it on ECU-Pin p and is 1 if this particular interface type is required on this pin. Using these vectors, the heuristic value for s_{H_BIN} can be calculated by counting the entries that are the same in both the original vector and the vector to be compared to. s_{RR_BIN} is defined analogously, but instead of counting the matching entries the differing entries are counted. Hence, this is actually a *dissimilarity* metric.

Additionally, we constructed vectors using the interface types of requirements at each ECU-Pin. In contrast to the vectors described above, we used the interface types directly (as a string representation) in the vector. Hence, these vectors have a maximum of n entries, where n is the number of ECU-Pins. Using such vectors, we defined a similarity metric s_R where we counted the conformance of interface types on ECU-Pins of two HSIs. We normalized these similarity values by dividing the result by the number of rows of the larger vector.

A more common approach to calculating similarity metrics is using *Cosine* similarity, which essentially calculates the angle between two vectors. Two vectors that have the same orientation have a similarity score of 1, where two vectors that are orthogonal to each other have a similarity score of 0. This similarity metric is not dependent on the domain and, thus, we employed it in the HSI as well as the Puzzle domain. The calculation of the cosine similarity is shown in Eq. 9.

$$s_{Cos}(s) = cos(\theta) = \frac{A \times B}{\|A\| \cdot \|B\|} = \frac{\sum_{i=1}^{N} A_i \cdot B_i}{\sqrt{\sum_{i=1}^{N} A_i^2} \cdot \sqrt{\sum_{i=1}^{N} B_i^2}} \tag{9}$$

In the HSI domain, we defined the cosine similarity s_{COS_BIN} using the same binary vectors as for s_{H_BIN}. Additionally, we created two similarity metrics s_{IT} and s_{ITR} using the cosine similarity. The vectors of these metrics are constructed by counting the number of times a specific interface type occurs in a requirement. In s_{ITR}, the total number of requirements is added as an additional entry in the vector.

For the puzzle domain, we defined the cosine similarity s_{COS} as well. We used the values of the tiles and their positions, i.e., which number is located where, for the construction of the vector and compared it with the vector of the goal state.

Additionally we defined a metric $s_h(s)$, where we used the count of all tiles that are already positioned correctly, i.e., are in the same position they have to be in the target state t, and divided it by the number of tiles, making it normalized in the interval $[0, 1]$. This metric is shown in Eq. 10.

$$s_h(s) = \frac{\sum_{n=0}^{15} TilePosCorrect_n}{15} \tag{10}$$

The same metric can also be defined using the heuristic h_{TR} from above as shown in Eq. 11.

$$s_h(s) = \frac{h_{tr}}{15} \tag{11}$$

Finally, we defined a Russel-Rao Dissimilarity s_{RR}, where the tiles that are not in the right position are used for its calculation.

4.3 Hybrids of Measuring and Estimating

Since "blind" algorithms like Dijkstra's famous shortest-path algorithm cannot solve "difficult" problem instances, algorithms like A* and IDA* employ heuristic estimates h and still guarantee that a solution found is optimal, if h is *admissible* (see also above). The time they take for that heavily depends on the quality of h used (see, e.g., [25]). Hence, we found it interesting to combine h with a measure into a hybrid, for getting yet another view.

In fact, it is shown in [23] that the effect of a heuristic function is to reduce the effective depth of search by a constant, relative to a brute-force search, rather than reducing the effective branching factor. And especially for distinguishing "easy" problems in the HSI domain from the others, d_{abs} turned out to be very useful in [27]:

$$d_{abs}(s) = C^* - h(s) \tag{12}$$

d_{rel} provides yet another view, since d_{abs} is most likely not very useful when comparing solutions with a large difference in cost:

$$d_{rel}(s) = \frac{d_{abs}(s)}{C^*} \tag{13}$$

Below, we use these formulas in the category of measures, since by including C^* they cannot be used for estimating in the course of a search for a newly given problem instance.

5 Experiment

The experiment is designed to explore how these different approaches to estimating problem instance difficulty relate to measuring problem instance difficulty, where also for the latter different approaches are compared as listed above. Results on this relationship are presented both using Pearson correlation coefficients and by numbers of selection errors that the different estimates make, as counted with respect to the various measuring approaches. All this has been done in both domains, the proprietary HSIs and the widely known Fifteen Puzzle.

5.1 Experiment Design

First, we created a repository of problem instances for both domains by running A* and for the Fifteen Puzzle also IDA*, and with a variety of admissible heuristics to find optimal solutions. We stored for each problem instance the values C^*, d_{abs}, d_{rel} and $N\#$ (the number of nodes generated by A* and IDA*, respectively). Since $N\#$, d_{abs}, d_{rel} also depend on the heuristic used, we stored these values for each of them, and the values calculated by the various heuristic functions separately as well. In addition, we calculated the values of a variety of similarity metrics (as explained above) and stored them.

When using A* to generate $N\#$ we utilized tie-breakers to select the next nodes to expand among all the nodes with the same minimal f-value. For the HSI domain, we used the smallest h-value as a primary tie-breaker and then applied a LIFO tie-breaking policy as a second-level criterion [27]. In the puzzle domain, breaking ties was done by a LIFO policy only.

For studying the correlation between the values of heuristic functions, similarity metrics and the different measures of the ground truth, we calculated *Pearson correlation coefficients*:

$$r = \frac{\sum_{i=1}^{n}(x_i - \bar{x}) \cdot (y_i - \bar{y})}{\sqrt{\sum_{i=1}^{n}(x_i - \bar{x})^2} \cdot \sqrt{\sum_{i=1}^{n}(y_i - \bar{y})^2} \tag{14}$$

If the absolute value of a correlation coefficient is higher than the absolute value of another one, the first one indicates a stronger relationship. In fact, we actually calculated these coefficients between all these values, since we were interested in the correlations among the estimates and also among the measures as well.

However, Pearson correlation coefficients are defined for *linear* correlations. h-values as estimates of C^*, however, are usually exponentially related to the number of explored nodes in searches by A*, IDA* and the like, for finding optimal solutions. That is why we determined Pearson correlation coefficients between h-values and the *logarithm* of the number of nodes.

In addition, for each Pearson correlation coefficient we determined its *significance* as the probability p that it could be the result of random fluctuation. All the calculations of Pearson correlation coefficients and their significance were run using Matlab. For each pair of estimates (heuristic functions/similarity metrics) and measures of problem instance difficulty, there are the following *null hypotheses*, which are correlation hypotheses. Let us formulate them generically here for all the cases that we investigated, rather than writing down each of them separately:

$NH_{e,m}$: *There is no statistically significant correlation (p-value = 0.05) between the estimate e and the measure m.*

For example, this formulation includes the null hypothesis NH_{h_M,C^*} for the estimate h_M and the measure C^*.

We were also interested in the respective numbers of *selection errors* that each of the estimates would make when being used for selecting a case from a case base. An error means to select a case that is worse than another one with respect to the problem instance difficulty. Since we have different measures for that, we were interested in the numbers of errors for each of them. However, executing an experiment directly with a case base and then finding optimal solutions for each of its cases stored and the given goal (state or condition, respectively) would be very expensive.

Instead, we used the repository initially created as follows. For each of its entries, we defined all the tuples $(est, diff)$ where for each *est* through a heuristic function or similarity metric, respectively, and for each measure of problem instance difficulty $diff$, one of the latter is assigned to the former. For each tuple, we calculated how many *errors* it induces. To calculate these errors, we ordered all tuples by their similarity or heuristic values, respectively, in a vector, and then checked it against the vector of the corresponding measure. The assumption is that a higher similarity value or a lower heuristic value, respectively, indicates a case with lower problem instance difficulty than another case with lower similarity or higher heuristic value, respectively. Hence, the order defined by the respective measure should be the same, e.g., the number of nodes generated $N\#$. Given that, an error is defined to occur for a given tuple, if the orders in the corresponding vectors are different.

As an example, let us consider two entries `(15,4)` and `(16,3)`, which are ordered in the sequence <`(15,4), (16,3)`> based on their estimate. Since the second tuple `(16,3)` has a higher estimate but lower difficulty measure, e.g., C^*, than the previous one, then it is counted as an error.

As some values of the tuples in a given vector may have the same similarity or heuristic value, respectively, we shuffled all tuples and then ordered them, introducing a different order of tuples with the same estimate each time. We ran these process 100 times for each measure and calculated the mean and median values of the numbers of errors. Of course, we included for both domains problem instances with varying difficulty. The C^* values varied from 3 to 17 for the HSI domain, and from 41 to 66 for the Fifteen Puzzle. Since $N\#$ depends on the algorithm used, we attempted to perform the searches with both A* and IDA*, in order to investigate the correlation between their respective $N\#$ values. Unfortunately, using IDA* is infeasible for the more difficult HSI problem instances due to the very high number of DAGs in this search space. (There are known means to address this, but then it would become some variant of IDA*.) For the Fifteen Puzzle, it was possible to solve all the 100 random instances listed in [22] running A* on a state-of-the-art laptop with 32 GB of memory.

More precisely, we executed the experiment runs on a standard Windows laptop computer with an Intel Core i7-8750H Processor (9 MB Cache, up to 4.1 GHz, 6 Cores). It has a DDR4-2666 MHz memory of 32 GB. The disk does not matter, since all the experimental data were gathered using the internal memory only.

5.2 Experimental Results

After having run the experiment as designed above for both the HSI domain and the Fifteen Puzzle, we got the results as presented below.

HSI Domain. In the HSI domain, we used more than 1,000 HSI specifications for our experiment. These specifications have been randomly adapted from ten base configurations and are spread across three categories: Base5, Base7 and Base9. These categories indicate the number of requirements that are already configured and fulfilled by the HSI.

Each of the 1,000 cases has between three and seventeen randomly selected requirements. For each of these cases, their h_{HSI} and their similarity values (using a variety of metrics as defined above) were calculated with regard to fitting base models, e.g., models using the same ECU. An automatic adaptation from each of the 1,000 HSI specifications to ones that satisfy the new requirements (as goal specifications) has been performed through heuristic search using A^* with our admissible heuristic function h_{HSI}. This guarantees an optimal solution C^* each, i.e., one with a minimal number of adaptation steps.

For these data, all correlation coefficients between the measures of the ground truth and the heuristic functions and similarity metrics, respectively, are given in Table 1. All the data in this table are highly statistically significant even with $p < 0.01$, that is, for all those pairs the null hypotheses generically defined above can be rejected. A higher (absolute) value indicates a stronger correlation with C^* and thus a better estimate. Some of these correlation coefficients are extremely high, such as the one between h_{HSI} and C^*. That is, this heuristic function is very good at estimating the number of steps minimally required for solution adaptation. The reason is that the vast majority of problem instances is "easy" in the sense that they require no reconfigurations, and only a few problem instances actually require a few reconfigurations, see [27]. Apart from that, h_{HSI} captures the knowledge for estimation very well. It is actually because of this distribution of problem instances that also some of the similarity metrics correlate with C^* so strongly. The difference in the knowledge involved only matters for very few problem instances here. For the absolute and the relative error measures of the ground truth, however, the correlations are very low. The reason is that both $d_{abs} = d_{rel} = 0$ for the majority of problem instances here.

Table 1. Correlation coefficients—HSI (adapted from [18, Table 1]).

	h_{HSI}	S_{H_BIN}	S_{RR_BIN}	S_R	S_{COS_BIN}	S_{IT}	S_{ITR}
C*	0.99	0.99	0.99	0.99	0.89	0.32	0.31
C*-h_{HSI}	0.02	0.02	0.02	0.02	0.02	0.08	0.08
(C*-h_{HSI})/C*	0.04	0.04	0.04	0.04	0.05	0.08	0.08
N#A*	0.63	0.63	0.63	0.63	0.52	0.20	0.19

Even for the numbers of nodes generated by A*, more precisely $ln(N\#\text{A})^*$, the correlations with some of the estimates are fairly high. For the given problem instances, where only a few problem instances actually require a few reconfigurations, these numbers can be predicted well.

We also calculated the numbers of errors for the heuristic function and each similarity metric. Table 2 shows the numbers of selection errors in the HSI domain. (More precisely these are the mean values as explained above, and we omit the median values since they are consistent with the mean values, i.e., outliers do not play a role here.) In fact, the numbers of errors of predicting $N\#\text{A}^*$ are very high. The selection results of predicting the other measures of the ground truth are much better.

Table 2. Numbers of selection errors—HSI (adapted from [18, Table 2]).

	h_{HSI}	S_{H_BIN}	S_{RR_BIN}	S_R	S_{COS_BIN}	S_{ITR}	S_{IT}
C*	16.39	16.42	16.47	16.33	65.93	328.82	338.14
C*-h_{HSI}	16.77	16.63	16.68	16.75	16.65	16.17	16
(C*-h_{HSI})/C*	16.63	16.62	16.67	16.68	16.66	16.23	16
N#A*	495.46	496.57	496.76	496.44	495.26	504.97	498.28

Fifteen Puzzle. As indicated above, we ran the set of 100 random Fifteen Puzzle instances published in [22] for gathering their data on the different heuristics h_{TR}, h_M, h_{P555}, h_{P663} and h_{P78}, as well as the C^*, d_{abs} and d_{rel} data (for the latter with all these heuristics). Since both IDA* and A* were able to solve all those instances even when using h_M (but not with h_{TR}), we were also able to get the data on $N\#$IDA* and $N\#$A*, more precisely the various numbers when using the different heuristic functions (except h_{TR}).

For these data, all correlation coefficients between the measures of the ground truth and the heuristic functions and similarity metrics, respectively, are given in Table 3. Note again, that we determined Pearson correlation coefficients between each h-value and the logarithm of the number of nodes. As a base, we took the value 2.1304 from [23], since it was determined there as the asymptotic branching factor for the Fifteen Puzzle.

Table 3. Correlation coefficients—Fifteen Puzzle (adapted from [18, Table 3]).

	h_M	h_{P555}	h_{P663}	h_{P78}	h_{TR}	s_{cos}	s_H	s_{RR}
C*	0.77	0.84	0.86	0.91	0.23	0.52	0.24	0.24
N#IDA*h_M	0.30	0.49	0.50	0.57	0.02	0.21	0.01	0.01
N#IDA*h_{P555}	0.41	0.49	0.51	0.58	0.05	0.30	0.05	0.05
N#IDA*h_{P663}	0.37	0.47	0.46	0.55	0.01	0.27	0.02	0.02
N#IDA*h_{P78}	0.31	0.40	0.41	0.44	0.01	0.24	0.00	0.00
N#A*h_M	0.40	0.57	0.58	0.66	0.02	0.29	0.03	0.03
N#A*h_{P555}	0.49	0.56	0.59	0.66	0.10	0.37	0.10	0.10
N#A*h_{P663}	0.40	0.49	0.48	0.59	0.04	0.33	0.04	0.04
N#A*h_{P78}	0.35	0.45	0.45	0.48	0.02	0.25	0.03	0.03
C*-h_M	0.22	0.07	0.07	0.20	0.38	0.10	0.38	0.38
(C*-h_M)/C*	0.64	0.34	0.35	0.24	0.54	0.36	0.54	0.54
C*-h_{P555}	0.18	0.10	0.23	0.38	0.06	0.17	0.04	0.04
(C*-h_{P555})/C*	0.19	0.31	0.16	0.01	0.05	0.06	0.08	0.08
C*-h_{P663}	0.11	0.17	0.09	0.32	0.02	0.14	0.04	0.04
(C*-h_{P663})/C*	0.25	0.21	0.31	0.07	0.14	0.09	0.17	0.17
C*-h_{P78}	0.09	0.17	0.16	0.12	0.02	0.00	0.03	0.03
(C*-h_{P78})/C*	0.19	0.14	0.15	0.22	0.11	0.19	0.12	0.12

The null hypothesis $NH_{h_{TR},N\#}$ for the estimate h_{TR} and the measure $N\#$ (for both $N\#$IDA* and $N\#$A*) cannot be rejected ($p = 0.412$). We attribute this to the fact that h_{TR} is a very weak heuristic, which we included here for illustration purposes only and which is not normally used.

However, most of the remaining data in Table 3 are highly statistically significant with $p < 0.01$, while especially those related to $N\#$IDA* are only statistically significant with $p < 0.05$, which is due to the fluctuation of the $N\#$IDA* data. Still, all the related null hypotheses apart from $NH_{h_{TR},N\#}$ can be rejected.

For the heuristic functions used here, their respective domain knowledge involved and their corresponding accuracy are known. In fact, it is exactly reflected here in their correlation with C^*. Estimates with more knowledge correlate more strongly with C^* than the ones with less knowledge. The similarity metric corresponding to h_{TR} in this regard, *Hamming*, has more or less the same correlation coefficient. In contrast, *Cosine* corresponds much better with C^* than *Hamming.* Still, its (absolute) correlation coefficient is much lower than the ones of the heuristic functions incorporating more domain knowledge. Neither $NH_{s_H,N\#}$ (for both $N\#$IDA* and $N\#$A*) nor $NH_{s_{RR},N\#}$ (for both $N\#$IDA* snd $N\#$A*) can be rejected because of $p = 0.412$ for both of them. Hence, these results may be due to chance fluctuation.

For the absolute and the relative error measures of the ground truth, however, this pattern is less clearly pronounced. d_{abs} is very hard to estimate for similarity metrics, which makes sense, since a heuristic function is actually part of its calculation. However, also for heuristic functions estimating d_{abs} is difficult. The same applies to d_{rel}. All these data are statistically significant with $p < 0.013$, that is, for all those pairs the null hypotheses can be rejected.

For the numbers of nodes generated by IDA* and A*, respectively, more precisely their logarithm, the correlations are lower than in the HSI domain. This can be explained by the fact that the problem instances in the HSI domain are much less difficult, both with respect to their solutions lengths and the effort to determine optimal solutions (in terms of the numbers of nodes explored). In addition, IDA* shows stronger fluctuations of these numbers than A*, when both use the same admissible heuristic. This can also be seen in Fig. 5 (generated by a clustering algorithm), where all the correlations are visualized as distances between nodes. If the distance between two nodes is small then this indicates a strong correlation and, vice versa, a larger distance indicates a weak correlation. The distances in the picture between C^* and the data points corresponding to the numbers of nodes generated are not smaller than those between C^* and most of the heuristic functions. The distance from C^* to the cluster of similarity metrics, however, is clearly larger, and this corresponds well with the observation made above, of course. In fact, all the heuristic functions attempt to estimate C^* somehow, especially the ones based on pattern databases and to a lesser extent h_M.

Overall, Fig. 5 actually illustrates clusters. There are two distinct clusters among the ground truth values. The first one contains all $N\#$ values. Although these values vary strongly in the number of nodes, they correlate well with each other. The second cluster contains the variations of d_{rel} and d_{abs}. They do not correlate well with estimating functions, but with each other.

Fig. 5. Two-dimensional clustering of correlations—Fifteen Puzzle (taken from [18, Figure 2]).

Finally, we also checked the number of errors during selection, see Table 4. The somewhat surprising result is, that a higher correlation coefficient does not necessarily carry over to a reduction of the number of errors. As pointed out above, $N\#$ values are hard to estimate and even higher correlation coefficients do not yield lower error values when selecting. The best selections results are accomplished in terms of C^*. In this case, a higher correlation coefficient also means fewer errors during selection. Also, heuristics with more knowledge provide better results, and the heuristic functions lead to fewer selection errors than the similarity metrics in this regard. Overall, it seems as though the domain knowledge included matters most.

Table 4. Numbers of selection errors—Fifteen Puzzle (adapted from [18, Table 4]).

	h_M	h_{P555}	h_{P663}	h_{P78}	h_{TR}	s_{COS}	s_H	s_{RR}
C*	41.76	37.32	38.82	36.83	46.79	45.21	46.73	46.53
N#IDA*h_M	48.73				49.95	49.54	49.85	49.25
N#IDA*h_{P555}		50.29			49.58	48.53	49.52	49.52
N#IDA*h_{P663}			48.39		49.17	49.77	49.66	49.68
N#IDA*h_{P78}				49.59	49.81	53.24	49.42	49.74
N#A*h_M	49.28				49.02	50.86	49.81	49.93
N#A*h_{P555}		49.39			49.68	49.36	49.59	49.93
N#A*h_{P663}			48.62		49.12	49.77	49.77	49.98
N#A*h_{P78}				49.7	50.11	48.49	49.45	49.86
C*-h_M	42.83				41.61	42.22	41.83	42.18
(C*-h_M)/C*	44.87				49.01	49.15	49.31	48.76
C*-h_{P555}		38.31			41.09	38.73	41.13	41.13
(C*-h_{P555})/C*		41.52			49.06	47.71	48.69	48.69
C*-h_{P663}			39.14		40.18	39.96	40.67	39.89
(C*-h_{P663})/C*			41.88		49.14	49.54	49.26	49.02
C*-h_{P78}				38.27	37.81	38.6	37.26	37.73
(C*-h_{P78})/C*				43.08	48.4	50.76	48.17	48.86

6 Conclusion

Our results indicate the feasibility of this approach for reusing realistic HSI designs. A* can search deep enough with the heuristic knowledge that we provided in the form of an admissible heuristic. Hence, the transformations from a retrieved model of one specific HSI to another one satisfying newly given requirements can be automated in CBR by using heuristic search.

In CBR, an underlying assumption is that the relative effort for solution adaptation of stored instances (cases) correlates with the similarity of these instances with the given problem instance. Since heuristic search is used for automatic solution adaptation in our approach, we conjectured that admissible heuristic functions h guiding search through estimating minimal costs to a given goal state or condition, may be used for retrieving cases for CBR as well.

From the results of our experiment, we conclude that the numbers of selection errors in the context of CBR can be reduced by using such heuristic functions, if they have more knowledge incorporated than the similarity metrics, and if the problem instances are difficult (as shown for the Fifteen Puzzle instances).

Unfortunately, however, both for the HSI and the Fifteen Puzzle instances, the optimal solution length does not correlate that strongly with the actual effort needed for finding the solutions, neither for IDA* nor for A*. Such heuristic functions are created, of course, for making these algorithms finding solutions efficiently, but they do not estimate this effort well. For future work it may be interesting to create estimators for that as well, in order to achieve a better understanding of the relationship between the two views of problem instance difficulty.

Acknowledgments. The InteReUse project (No. 855399) has been funded by the Austrian Federal Ministry of Transport, Innovation and Technology (BMVIT) under the program "ICT of the Future" between September 2016 and August 2019. More information can be found at https://iktderzukunft.at/en/.

The VIATRA team provided us with their VIATRA2 tool. Our implementations of the Fifteen Puzzle are based on the very efficient C code of IDA* and A* made available by Richard Korf and an efficient hashing schema by Jonathan Shaeffer. Ariel Felner and Shahaf Shperberg provided us with hints about the availability of code for the Fifteen Puzzle pattern databases.

Last but not least, Alexander Seiler and Lukas Schröer helped us with getting all the C code running under Windows for our Fifteen Puzzle experiment.

References

1. Aamodt, A., Plaza, E.: Case-based reasoning: foundational issues, methodological variations, and system approaches. AI Commun. **7**(1), 39–59 (1994). http://dl.acm.org/citation.cfm?id=196108.196115
2. Bandyopadhyay, S., Saha, S.: Unsupervised Classification: Similarity Measures, Classical and Metaheuristic Approaches, and Applications. Springer, Heidelberg (2012). https://doi.org/10.1007/978-3-642-32451-2
3. Bergmann, G., et al.: VIATRA 3: a reactive model transformation platform. In: Kolovos, D., Wimmer, M. (eds.) ICMT 2015. LNCS, vol. 9152, pp. 101–110. Springer, Cham (2015). https://doi.org/10.1007/978-3-319-21155-8_8
4. Bu, Z., Korf, R.E.: A*+IDA*: A simple hybrid search algorithm. In: Proceedings of the Twenty-Eighth International Joint Conference on Artificial Intelligence, IJCAI-2019, pp. 1206–1212. International Joint Conferences on Artificial Intelligence Organization, July 2019. https://doi.org/10.24963/ijcai.2019/168
5. Bulitko, V., Björnsson, Y., Lawrence, R.: Case-based subgoaling in real-time heuristic search for video game pathfinding. J. Artif. Int. Res. **39**(1), 269–300 (2010). http://dl.acm.org/citation.cfm?id=1946417.1946423
6. Burke, E.K., Petrovic, S., Qu, R.: Case-based heuristic selection for timetabling problems. J. Sched. **9**(2), 115–132 (2006). https://doi.org/10.1007/s10951-006-6775-y
7. Cha, S.H.: Comprehensive survey on distance/similarity measures between probability density functions. Int. J. Math. Models Methods Appl. Sci. **1**(4), 300–307 (2007). http://www.gly.fsu.edu/~parker/geostats/Cha.pdf
8. Choi, S., Cha, S., Tappert, C.C.: A survey of binary similarity and distance measures. J. Syst. Cybern. Inform. **8**(1), 43–48 (2010)
9. Dechter, R., Pearl, J.: Generalized best-first strategies and the optimality of a*. J. ACM **32**(3), 505–536 (1985)
10. Eclipse: Package org.eclipse.emf.ecore (EMF JavaDoc) (2017). http://download.eclipse.org/modeling/emf/emf/javadoc/2.9.0/
11. Edelkamp, S., Schroedl, S.: Heuristic Search: Theory and Applications. Morgan Kaufmann, Waltham (2012)
12. Felner, A., Korf, R.E., Hanan, S.: Additive pattern database heuristics. J. Artif. Int. Res. **22**(1), 279–318 (2004). http://dl.acm.org/citation.cfm?id=1622487.1622496
13. Goel, A.K., Diaz-Agudo, B.: What's hot in case-based reasoning. In: Proceedings of Thirty-First AAAI Conference on Artificial Intelligence (AAAI-2017), pp. 5067–5069. AAAI Press/The MIT Press, Menlo Park (2017)

14. Hart, P., Nilsson, N., Raphael, B.: A formal basis for the heuristic determination of minimum cost paths. IEEE Trans. Syst. Sci. Cybern. (SSC) **SSC-4**(2), 100–107 (1968)
15. Kaindl, H., Kainz, G.: Bidirectional heuristic search reconsidered. J. Artif. Intell. Res. (JAIR) **7**, 283–317 (1997)
16. Kaindl, H., Kainz, G., Leeb, A., Smetana, H.: How to use limited memory in heuristic search. In: Proceedings of Fourteenth International Joint Conference on Artificial Intelligence (IJCAI-1995), pp. 236–242. Morgan Kaufmann Publishers, San Francisco (1995)
17. Kaindl, H., Smialek, M., Nowakowski, W.: Case-based reuse with partial requirements specifications. In: Proceedings of the 18th IEEE International Requirements Engineering Conference (RE 2010), pp. 399–400, September 2010. https://doi.org/10.1109/RE.2010.57
18. Kaindl, H., Hoch, R., Popp, R.: Estimating problem instance difficulty. In: Filipe, J., Smialek, M., Brodsky, A., Hammoudi, S. (eds.) Proceedings of the 22nd International Conference on Enterprise Information Systems, ICEIS 2020, Prague, Czech Republic, 5–7 May 2020, vol. 1, pp. 359–369. SCITEPRESS (2020). https://doi.org/10.5220/0009390003590369
19. Kirsopp, C., Shepperd, M., Hart, J.: Search heuristics, case-based reasoning and software project effort prediction. In: Proceedings of the 4th Annual Conference on Genetic and Evolutionary Computation, GECCO 2002, pp. 1367–1374. Morgan Kaufmann Publishers Inc., San Francisco (2002). http://dl.acm.org/citation.cfm?id=2955491.2955741
20. Köll, A., Kaindl, H.: Bidirectional best-first search with bounded error: summary of results. In: Proceedings of Thirteenth International Joint Conference on Artificial Intelligence (IJCAI-1993), pp. 217–223. Morgan Kaufmann Publishers, San Francisco (1993)
21. Kolodner, J.: Case-Based Reasoning. Morgan Kaufmann Publishers Inc., San Francisco (1993)
22. Korf, R.: Depth-first iterative deepening: an optimal admissible tree search. Artif. Intell. **27**(1), 97–109 (1985)
23. Korf, R.E., Reid, M., Edelkamp, S.: Time complexity of iterative-deepening-A*. Artif. Intell. **129**(1), 199–218 (2001). https://doi.org/10.1016/S0004-3702(01)00094-7
24. Lopez de Mantaras, R., et al.: Retrieval, reuse, revision and retention in case-based reasoning. Knowl. Eng. Rev. **20**(3), 215–240 (2005). https://doi.org/10.1017/S0269888906000646
25. Pearl, J.: Heuristics: Intelligent Search Strategies for Computer Problem Solving. Addison-Wesley, Reading (1984)
26. Rathfux, T., Kaindl, H., Hoch, R., Lukasch, F.: An experimental evaluation of design space exploration of hardware/software interfaces. In: Proceedings of the 14th International Conference on Evaluation of Novel Approaches to Software Engineering, ENASE 2019, pp. 289–296. INSTICC, SciTePress (2019). https://doi.org/10.5220/0007689002890296
27. Rathfux, T., Kaindl, H., Hoch, R., Lukasch, F.: Efficiently finding optimal solutions to easy problems in design space exploration: A* tie-breaking. In: van Sinderen, M., Maciaszek, L.A. (eds.) Proceedings of the 14th International Conference on Software Technologies, ICSOFT 2019, Prague, Czech Republic, 26–28 July 2019, pp. 595–604. SciTePress (2019). https://doi.org/10.5220/0008119405950604

28. Reiser, C., Kaindl, H.: Case-based reasoning for multi-step problems and its integration with heuristic search. In: Haton, J.-P., Keane, M., Manago, M. (eds.) EWCBR 1994. LNCS, vol. 984, pp. 113–125. Springer, Heidelberg (1995). https://doi.org/10.1007/3-540-60364-6_31
29. Schank, R.C.: Dynamic Memory: A Theory of Reminding and Learning in Computers and People. Cambridge University Press, New York (1983)
30. Sohangir, S., Wang, D.: Improved sqrt-cosine similarity measurement. J. Big Data **4**(1), 25 (2017). https://doi.org/10.1186/s40537-017-0083-6
31. Steinberg, D., Budinsky, F., Paternostro, M., Merks, E.: EMF: Eclipse Modeling Framework 2.0, 2nd edn. Addison-Wesley Professional, Boston (2009)

Opti-Soft: Decision Guidance on Software Release Scheduling to Minimize the Cost of Business Processes

Fernando Boccanera(✉) and Alexander Brodsky

George Mason University, Fairfax, VA 22030, USA
{fboccane,brodsky}@gmu.edu

Abstract. Many approaches have been developed to increase the return on a software investment, but each one has drawbacks due to subjectivity or imprecision. This paper proposes a novel approach that addresses this problem for a particular class of information systems that improve business processes. The approach, called Opti-Soft, models a software development project as a mixed integer linear programming problem, where the objective function maximizes the return on investment and the corresponding decision variables produce an optimal release schedule. The uniqueness of Opti-Soft is in accurately modeling the business process and its improvement due to new software features, which leads to cost reduction. The approach includes a formal model, a methodology and a decision-guidance system.

Keywords: Decision guidance system · Decision support system · Software release schedule · Optimization · Mixed-integer linear programming

1 Introduction

Software development projects that are completed successfully and deliver a product that returns considerable value to the business are not the norm. Pucciarelli and Wiklund [12], in a highly publicized report by the International Data Corporation (IDC), estimate that 25% of projects fail and another 25% do not provide any Return on Investment (ROI). The Standish Group, in their 2015 Chaos Report [13] had similar findings: 29% of projects were successful, 52% were challenged and 19% failed. Standish also found that of the projects that do not fail, 45% of the functionality is never used, resulting in zero business value. Many approaches have been developed to address this business benefit gap but each one has pitfalls caused by imprecision, difficulty in providing the data necessary to conduct the analysis or lack of completeness.

This paper addresses the business benefit gap for a class of software projects that implement information systems that improve a business process. It proposes a new approach and a decision guidance system that uses the savings in the business processes caused by delivered functionality as a proxy for business value and then recommends a particular software delivery sequence that optimizes the business benefit.

J. Filipe et al. (Eds.): ICEIS 2020, LNBIP 417, pp. 184–216, 2021.
https://doi.org/10.1007/978-3-030-75418-1_10

One reason that business benefits are not realized is because often the project execution team is not driven by them. There is a disconnect between decisions made on software functionality during project execution and the business case, which creates a benefit gap. In the past fifteen years, there has been considerable research to address the benefit gap, by embedding business value considerations into the software development life cycle. These value-based approaches to software engineering evaluate benefits not at the level of a software product but at the level of functionality, providing greater granularity, which in turn makes estimations easier.

Value-oriented approaches are best suited to projects that deliver software in increments called releases so that the business value can be harvested after each release. Several value-oriented methods have been proposed and the ones based on financial aspects are the most popular. They draw from the field of finance and usually use Net Present Value (NPV) as a proxy for measuring benefits of the investment in software because NPV is widely used in capital budgeting to analyze the profitability of potential investments. It is the difference between the Present Value of cash inflows and the Present Value of cash outflows over a period of time. Present Value is the current worth of a future sum of money, given a specified rate of return. The rate used in capital budgeting is called the hurdle rate, which is defined as the minimum rate that an entity expects to earn from the investment.

A pre-requisite to the calculation of NPV is the estimation of cash flows over a number of periods within a time horizon. The cash flow estimation can be conducted at many levels of granularity, with the lowest level being at the requirement level.

The most popular and highly influential financial-based approach is the Incremental Funding Methodology (IFM) proposed by Denne and Cleland-Huang [4–6]. IFM associates cash flows (revenue and cost) to each software feature but it does not provide any guidance on how to estimate the cash flows, it simply assumes that they are given. A second financial-based approach is FEVOLVE*, proposed by Ruhe et al. [9]. FEVOLVE* is an iterative and evolutionary approach that facilitates the involvement of stakeholders to achieve increments (releases) that result in the highest degree of satisfaction among different stakeholders. The approach provides a decision support for the generation and selection of release plan alternatives. It requires that the savings revenue as well as the cost of the software be given as a single combined cash flow. A third approach by Akker et al. [14] applies integer linear programming to maximize the revenue.

A major drawback of all financial-based approaches, including IFM and FEVOLVE is the need to estimate cash flows for both revenue and software cost. Such estimation is not a trivial task and for a software investment is even more challenging due to the difficulty of drawing a direct correlation between a particular business benefit, like an increase in revenue, and a specific piece of software. Some researchers have acknowledged this difficulty, e.g., Devaraj and Kohli [7] noted that "the principal issue encountered is whether we can isolate the effect of IT on firm performance. It does not have an easy answer, because it means disentangling the effect of IT from various other factors such as competition, economic cycle, capacity utilization, and many other context-specific issues."

The challenge in the financial-based approach is that estimations of cash flow usually produce inaccurate results due to the difficulty in isolating the impact of software features in an organization's financial metrics. Another issue is that every dollar of revenue has to be mapped to one and only one software feature.

We believe that the accuracy can be increased for a class of software projects where the benefit, as measured by the cash flows, can be isolated from other factors that influence the financial performance of an organization. Such an isolation is possible in information systems that improve a business process. Intuitively, a business process is like a factory, where raw materials are incrementally transformed into a finished product. As an example, one business process of the US Patent Office transforms patent applications into approved patents, or various response letters. More generally, a business process consumes inputs and produces a product as an output, and the cost of a unit of product is a function of the labor rates of the workstation workers and the time that each intermediate workstation takes to produce intermediate products. This means that the cost, or cash outflow, of a business process can be accurately computed.

Taking this concept one step further, the financial benefit of a software feature that improves a business process can be determined by calculating the cost of the business process before and after the software implementation. This novel approach addresses the challenges described in Sect. 1.3 and is the focus of this paper. As far as I know, there is no method that accurately calculates the financial benefit of software features, they all assume that the benefit is given.

The focus of this paper is addressing the limitations of the existing value-based release scheduling approaches for the class of software projects that improve a Business Process Network (BPN). We address the limitations by proposing a decision-guidance framework that is more precise than existing approaches because it is based on a formal model of the BPN and its evolution following the implementation of software features.

The uniqueness of our approach is in leveraging the idea that the implementation of software features allows improvements in the BPN, which lead to a reduction in cost. As a consequence of this idea, the business benefit is not attributed to individual features in silos like in the current approaches, but rather to the synergetic effect of multiple interrelated features on the reduction of the overall cost of the BPN. *Opti-Soft* moves the benefit estimation from a guesswork to a systematic model-based methodology, which, we believe, will result in considerably higher return on software investment. To achieve that, we (1) develop a formal optimization model and solution based on a reusable library of analytical component models; (2) develop a decision guidance system and methodology for software release scheduling; and (3) demonstrate the methodology using an example from the U.S. Patent and Trademark Office.

More specifically, the first contribution of the formal model captures the entire space of alternatives for BPN configurations which produce some outputs from inputs (e.g., documents, requests, approvals, etc.). The formal model decides on (1) which interdependent software features are to be implemented and in which software release, and (2) which specific alternatives of the BPN are to be activated for each software release over the investment horizon. To be activated, atomic processes in the BPN may require new inter-dependent software features to be implemented. Improvements in the BPN are measured as cash flows and their associated Net Present Value (NPV). Cash flows are calculated to represent the ongoing costs of the BPN, as well as software development. Each potential software release schedule impacts the cash flow and results in a different NPV.

The formal optimization problem is to maximize the NPV of the combined cash flow of the BPN plus the software cost, while satisfying the constraints of (1) feature-to-release allocation, (2) dependencies among features, and (3) business processes activation.

As a second contribution, we develop a Decision Guidance System (DGS) and methodology that are centered around solving the optimization model and producing an optimal release sequence. The DGS is based on the formal model and is implemented in the Decision Guidance Analytics Language (DGAL) [3] within Unity [10, 11], a generic platform for the creation and execution of decision guidance systems.

Finally, to demonstrate the approach, we show a simplified example from the United States Patent and Trademark Office, with all the essential components.

This paper is organized as follows: Sect. 2 discusses related work, Sect. 3 explains the financial and software cost model and intuitively explains *Opti-Soft* through an example; Sect. 4 describes the formal model; Sect. 5 shows a full example, Sect. 6 introduces the decision guidance system; Sect. 7 discusses the methodology, and Sect. 8 provides concluding remarks.

2 Related Work

Agile Release Planning. Agile software development is an approach where self-organizing teams collaborate with the customer to conduct analysis, design, development and implementation in short, timeboxed and iterative periods where working software is delivered constantly. The approach has the potential to deliver software that will actually be useful to the customer because the customer is an integral member of the development team.

The term was coined and popularized in the Agile Manifesto [8], which codifies four values and twelve principles. Principle number one states that "Our highest priority is to satisfy the customer through early and continuous delivery of valuable software". This principle applies to our proposed approach, called *Opti-Soft*, because the main goal of the approach is to optimize the business value of the delivered software.

In Agile, requirements are placed in a product backlog. Through a planning process, product backlog items are assigned to timeboxed iterations. Some approaches like Scrum have two types of timeboxes: sprints and releases. A sprint has a fixed number of weeks (usually 1 to 4) and a release contains a fixed number of sprints. There are at least two types of items in the product backlog: features and user stories. Features are at the granularity of the business, that is, they are a self-contained pieces of functionality that are identifiable by the customer. Features are broken down into User Stories, which are small units that can be developed in a few days.

Features and User Stories have their size estimated, usually in points. An Agile point is a proxy for effort, the higher the effort, the higher the number of points. For each timebox (Sprint or Release), the capacity of the team is estimated in number of points. The team's capacity is a function of the duration of the timebox, the number of developers and the average productivity of the developers. Release Planning is the process of assigning features to releases and is constrained by the release capacity, there is, no release can have more feature points than the capacity of the release.

F-EVOLVE. In [9], Ruhe and Ngo-The propose EVOLVE*, an iterative and evolutionary approach that facilitates the involvement of stakeholders to achieve increments (releases) that result in the highest degree of satisfaction among different stakeholders. The authors extend EVOLVE* by proposing F-EVOLVE*. "F-EVOLVE* provides decision support in the generation and selection of release plan alternatives. The model aids in aligning the software development process with the goals of the business, to "achieve greater profit in the shortest time possible." The approach models the problem formally with the goal of offering the most profitable sequence of features. Although the approach has some similarity to ours, it does not address the calculation of cash flows, it just assumes that they are provided, that is, they are based on subjective estimations of both software cost and business savings.

Incremental Funding Method. The most popular and highly influential financial-based approach is the Incremental Funding Methodology (IFM) proposed by Denne and Cleland-Huang [4–6]. IFM's approach is to deliver software features as early as possible in order to maximize their business value. It assumes a software development life cycle that delivers software continuously and iteratively, consequently it can be utilized within the framework of modern Agile methodologies like Scrum.

In [6], the authors provide an example to demonstrate how the early delivery of software features increases the positive cash flow. In a traditional waterfall SDLC were the software is delivered at the end, the resulting profit in the IFM example is $200K and the ROI is 10%. Applying the method produces an NPV of $1.1M.

The IFM innovated because it combines the revenue and the cost of software development at the feature level and then provides an algorithm for selecting a sequence. The pitfalls of the IFM are:

1. It is based on inaccurate estimations of cash flow.
2. The scheduling algorithms are approximations, heuristics, they do not produce an optimal result.
3. It does not take into consideration the effort to develop each feature. The example assumes that each quarter produces one single feature.

3 Release Scheduling Approach

Our proposed release scheduling approach exhibits three properties: 1) accurately determine the financial benefit of software features, 2) accurately determine the cost of the software and 3) recommend a release schedule that optimizes the financial benefit.

3.1 Financial Benefit Model: Business Process

This section addresses property #1 of *Opti-Soft*, which is the need to accurately determine the financial benefit of software features. The class of problems that we will concentrate on are software projects that implement information systems that improve a business process. Figure 1 shows that the implementation of software features allows an initial business process to be improved and the amount of improvement is the financial benefit or business value. Consider a business process that takes as input a paper application and produces as output an application decision. The process is driven by the input and there is a one-to-one relationship between the input and the output. The unity cost, or cost outflow, is easily computed by multiplying the labor rate by the time it takes to process a single application. This means that the labor cost of a busines process can be accurately computed. The total benefit, or business value (BV), is the delta between the labor cost of the initial business process and the improved process after the software is released. Our approach uses this delta, which is precisely calculated, as the proxy for benefit, as opposed to previous approaches that assume the BV is given.

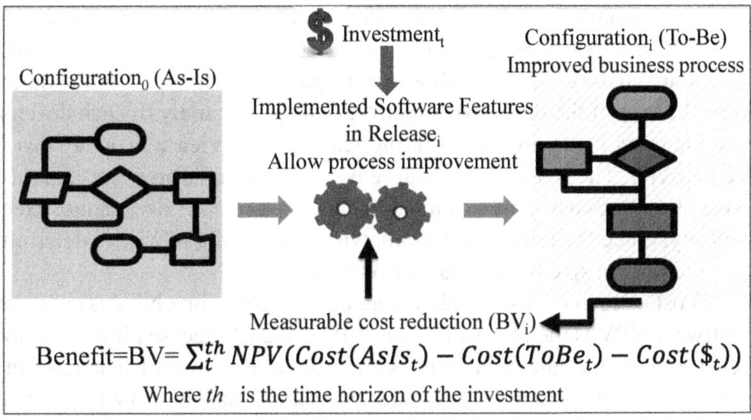

Fig. 1. Business value of the investment in software features.

Now that we know how to accurately calculate the cost outflow of a process, let's examine the link between a business process and the implemented software features. To leverage the labor reduction, the initial process, the As-Is, needs to change at the end of each release to take advantage of the software features just implemented. In an incremental SDLC, like in Fig. 1, each software feature *i* is assigned to a specific release *i* and each release *i* delivers functionality that allows the business process to transition to configuration *i*, which is benefited by BV_i. The business process will go through several configurations until it reaches the final, To-Be configuration *n*.

We adopt Net Present Value (NPV) as the proxy for BV. NPV is the discounted present value of the labor cost at each point in time. The total benefit, or BV, is the NPV of the labor cost of the As-Is, minus the labor cost of the To-Be, minus the cost of each software release, over the time of the investment.

The above insight, that the implementation of software features allows the adoption of more efficient business process networks (BPN) is key to our approach, because each new BPN configuration can be modeled and its cost measured. Note that in this approach, there is no need to estimate the cost of each individual feature, a feature is just a device that triggers a change in the BPN configuration, while cost is precisely calculated at the level of the BPN.

3.2 Financial Benefit Model: SDLC

This section addresses property #2 of *Opti-Soft*, which is the need to accurately determine the cost of the software.

Opti-Soft requires a SDLC that incrementally delivers functionality in releases and where release planning is conducted at the feature level. Such approach is well documented in the Agile practice literature and is called feature-driven. Release planning is conducted at the feature level, that is, features are removed from the product backlog and assigned to releases. Usually some kind of prioritization is utilized to drive release planning; we use the NPV of the labor cost of the BPN configuration as the prioritization technique; the higher the NPV, the higher the priority.

Features are customarily estimated in points, which are based on the perceived effort to implement the features. In Scrum, capacity is used to determine how many points can fit in a particular timebox (Sprint or Release). Capacity is a function of the duration of the timebox, the size of the developer's team and the productivity of each developer.

Release Planning is the process of fitting features in a release in a way that it does not exceed capacity. The size of each feature is estimated in effort points consequently the total size of features assigned to a release, called the release size, cannot exceed the release capacity. Once the release size is known, the cost of a release is determined by multiplying the release size by the cost per point.

Once the cost of each release is calculated, the corresponding NPV is computed and the total software NPV is then the sum of the NPVs of each release plus the period after the last release. Note that the NPV is not calculated at the level of features, but at the level of a release. Previous approaches required that the cash outflow be estimated for each feature, which assumes that each dollar spent must be assigned to one and only one feature. In *Opti-Soft*, the cash outflows and NPV are more precise because they are calculated at the release level and based on the change in labor cost of the BPN configuration. On top of this key advantage, our approach has the following benefits: 1) the calculations are simpler because there is no need to estimate the cost of every single feature, and 2) it eliminates the assumption that every dollar be allocated to a single feature because a release might implement many features.

3.3 Release Scheduling Example

Consider an organization, like the United States Patent Office, which processes applications for patents. Consider a simplified and partial version of the BPN, depicted in Fig. 2. The process starts with Application Intake (A), which takes a User Application and either accepts it by producing a Compliant Application or rejects it by producing a Non-compliance Notice. Compliant applications go through Adjudication (B) and then Adjudication Review (C), which produces an Adjudicated Application Letter.

Let us assume that initially, processes A, B and C are manual, and the Patent Office is considering implementing a software system to automate these processes to save cost. To reason about possible alternatives for automation, the Office creates the diagram shown in Fig. 3. In it, process A has three alternatives; AA (Manual Application Intake), AB (Electronic Application Intake) and AC (Self-service Application Intake), where AA is the initial manual process and AB and AC are increasingly automated alternatives of A.

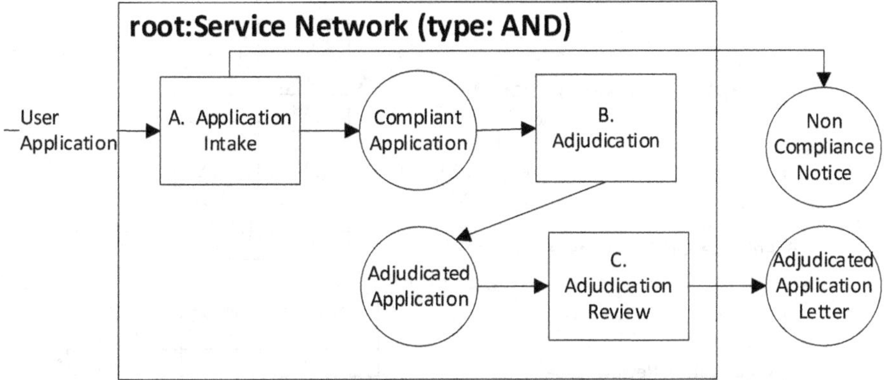

Fig. 2. Simplified patent adjudication BPN. Reprinted from [1].

Similarly, for B, BA is the initial manual process while BB is its automated alternative and for C, CA is manual while CB is its automated alternative. In essence, Figs. 2 and 3 show all possible configurations of the BPN, composed of a combination of alternatives to processes A, B and C.

Initially, in Configuration 0, there is no software system, consequently the BPN is made of manual processes AA (Manual Application Intake), BA (Manual Adjudication) and CA (Manual Adjudication Review). As the software system is implemented, the BPN configuration changes to take advantage of more efficient processes; AA transitions to AB (Electronic Application Intake), BA transitions to BB (Electronic Adjudication), etc.

We model the BPN as a Business Service Network (BSN) [2], which is a "network of service-oriented components that are linked together to produce products". Figure 2 depicts the root BSN, while Fig. 3 details its subservices A, B and C. Because the root service requires subservices A, B and C, we call this an AND-type service. Whereas, because service A requires only one of subservices AA, AB or AC, service A is an OR-type. Services B and C are also OR-types while all the other subservices are atomic.

The transition from a process alternative to another within the same composite process, requires the implementation of specific software functionality called features. For example, process alternative AB (Electronic Application Intake) requires business feature BF1 which is the capability to create and edit an electronic application. Assuming that feature BF1 is implemented in release 1, Fig. 3 shows that after release 1 is completed, the initial Config_0 AA, BA, CA transitions to Config_1 AB, BA, CA. The reason is that AB, which has a lower labor cost than AA, requires feature BF1, that is, BF1 'activates' AB.

3.4 Optimal Release Schedule

This section addresses property #3 of *Opti-Soft*, which is the recommendation of a release schedule that maximizes the financial benefit.

The result of release planning is a schedule, which is a table of all releases and the features planned to be implemented in each release. Every software release schedule, such as the one depicted in Fig. 3, has a total cost which is the cash outflow of the

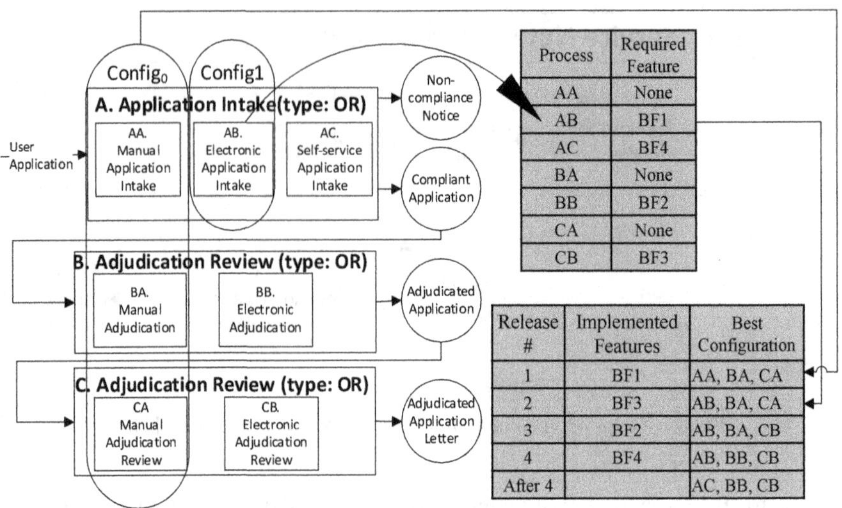

Fig. 3. BPN composite processes A, B and C.

BPN plus the cost of the SDLC. We approach the problem of determining the release schedule as an optimization problem. We maximize the overall NPV after all releases are implemented, subject to constraints such as the space of process alternatives, the required software features, the interdependencies among features, the one-to-one mapping between features and releases and the capacity of the development team. In our optimization problem, the main decision variables are:

1. *On(s,r)* – a Boolean indicating whether process *s* is activated in the BPN configuration associated with release *r*.
2. *Implemented(f,r)*–a Boolean indicating whether feature*f* is implemented in release *r.*
3. *InputThru(s,i,r)*– an integer indication the number of inputs of type *i* that go through process *s* in the BPN associated with release *r.*

4 Release Scheduling Formal Optimization Model [1]

The release scheduling problem formulation in mixed-integer linear programming is:

$$Max_{DV} O(P, DV)$$

$$s.st. C(P, DV)$$

Where:

P is a set of parameters,

DV is a set of decision variables,

C(P, DV) is a predicate, expressed as a function of parameters *P* and decision variables *DV*, that need to be satisfied, and.

$O(P, DV)$ is the NPV metric, expressed as a function of P sand DV.

The components of the optimization problem are described using the **ReleaseScheduling** formalization, which is a tuple ⟨Parameters, DecisionVariables, Computation, Constraints, InterfaceMetrics⟩, detailed in Sect. 4.1. At a high level, the **ReleaseScheduling** formalization is described in Fig. 4 as a hierarchy of components.

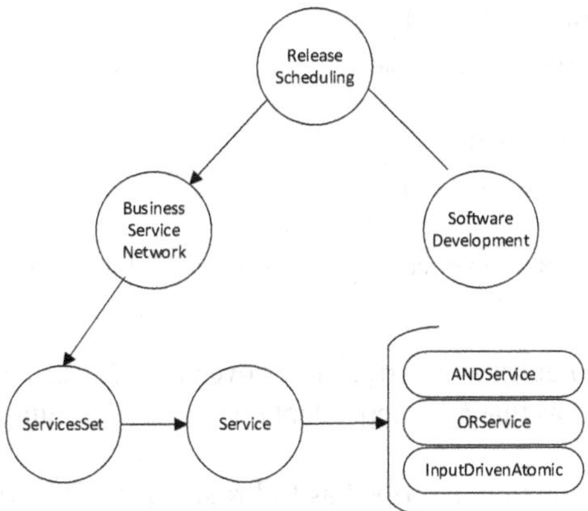

Fig. 4. Hierarchy of the formalizations of the release scheduling model. Reprinted from [1].

The hierarchy in Fig. 4 establishes a parent-child relationship where the child inherits all formalizations from the parent and the parent has access to all the formalizations of the child. For example, ReleaseScheduling is the parent of Business Service Network (BSN), consequently the BSN tuple is available to ReleaseScheduling and BSN inherits *Parameters, DecisionVariables, Computations* and *InterfaceMetrics* from ReleaseScheduling. In the next sections we describe the components of the Release Scheduling formalization hierarchy in details.

Figure 5 shows the hierarchy of the calculations. At the lowest level, the cost per day for each *InputDrivenAtomic* service is calculated. The result is rolled up recursively all the way up to the highest level.

4.1 Release Scheduling Formalization

ReleaseScheduling (RSch) formalization is a tuple ⟨*Parameters, DecisionVariables, Computation, Constraints, InterfaceMetrics*⟩

where:

Parameters, also denoted **Parm,** is a tuple ⟨*Features, TH, DiscountRate, ReleaseInfo, BSN.Parameters, SWD.Parameters*⟩

Where *Features* is a tuple ⟨*BF, TF, DG, FS*⟩ where:

- ***BF*** is a set of business features
- ***TF*** is a set of technical features, such that

$$BF \cap TF = \varnothing$$

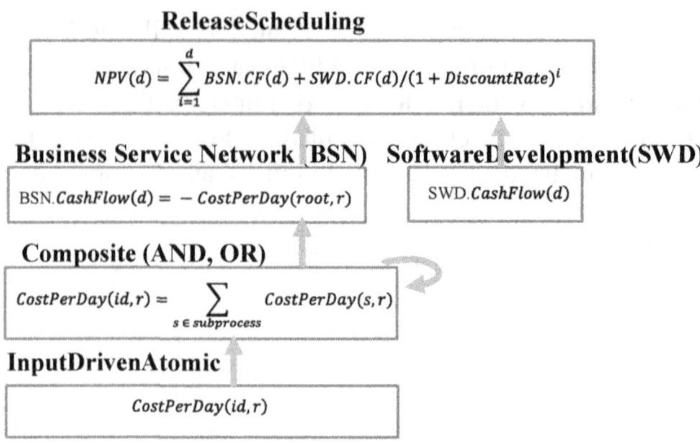

Fig. 5. Roll up of calculations from the bottom to the top.

- ***DG***, (Dependency Graph), is a partial order over $F = BF \cup TF$, $(f_1, f_2) \in DG$ also denoted $f_1 \prec f_2$, means that f_2 is dependent on f_1, that is, feature f_1 is a pre-requisite for feature f_2.
- ***FS*** : $F \rightarrow \mathbb{R}^+$ is a function described as follows: $(\forall f \in F)$, $FS(f)$ gives the size, in effort point, of each feature f.
- ***TH*** is the time horizon for analysis in days
- ***DiscountRate*** is the daily rate to discount cash flows.
- ***ReleaseInfo*** is a tuple $\langle NR, RD \rangle$, where:

  - ***NR*** is the number or releases
  - ***RD*** : $[1..NR] \rightarrow \mathbb{R}^+$, is a function described as follows: $(\forall r \in [1..NR])$, $RD(r)$ gives the maximum duration in days for release r.

- ***BSN.Parameters*** is defined in Sect. 4.2
- ***SWD.Parameters*** is defined in Sect. 4.7

DecisionVariables, also denoted **DV,** is a tuple $\langle IBF, ITF, BSN.DecisionVariables, SWD.DecisionVariables \rangle$.
where:

- ***IBF*** : $[1..NR] \rightarrow 2^{BF}$ is a function described as follows: $(\forall r \in [1..NR])$, $IBF(r)$ gives a set of business features planned to be implemented in release r.
- ***ITF*** : $[1..NR] \rightarrow 2^{BF}$ is a function described as follows: $(\forall r \in [1..NR])$, $ITF(r)$ gives a set of technical features planned to be implemented in release r.
- ***BSN.DecisionVariables*** is defined in Sect. 4.2.
- ***SWD.DecisionVariables*** is defined in Sect. 4.7.

Computation

1. Let *SoFarIBF*: $[1..NR+1] \rightarrow 2^{BF}$ be a function described as follows: $(\forall r \in [1..NR+1])$, $SoFarIBF(r)$ gives the set of all business features implemented up to release r or the period after the last release, computed as follows:

$$SoFarIBF(r) = \bigcup_{i=1}^{r-1} IBF(i)$$

2. Let *CombinedCashFlow* : $[1..TH] \rightarrow \mathbb{R}$ be a function described as follows: $(\forall d \in [1..TH])$, $CombinedCashFlow(d)$ gives the combined income/expenditure of both the Business Service Network and the Software Development, $(\forall d \in [1..TH])$, computed as follows:

$$CombinedCashFlow(d) = BSN.IM.Cashflow(d) + SWD.IM.CashFlow(d)$$

where:

- ***BSN.IM.CashFlow*** is defined in section *BSN.InterfaceMetrics* of Sect. 4.2
- ***SWD.IM.CashFlow*** is defined in section *Software.InterfaceMetrics* of Sect. 4.7.

Note that a negative cash flow means that it is a cash outflow.

3. Let *TimeWindowNPV* : $[1..TH] \rightarrow \mathbb{R}$ be a function described as follows: $(\forall d \in [1..TH])$, $TimeWindowNPV(d)$ gives the Net Present Value (NPV) of the *CombinedCashFlow* for the time investment window $[1..d]$, computed as follows:

$$TimeWindowNPV(d) = \sum_{i=1}^{d} \frac{CombinedCashFlow(i)}{(1 + DiscountRate)^i}$$

4. Let $F = BF \cup TF$
5. Let $IF(r) = IBF(r) \cup ITF(r)$, $(\forall r \in [1..NR])$
6. *FeatureSetsForReleasesArePairwiseDisjoint* constraint is:

$$(\forall i, j, \in [1..NR], i \neq j), IF(i) \cap IF(j) = \varnothing$$

7. *DependencyGraphIsSatisfied* constraint is:

$$(\forall r \in [1 \ldots NR])(\forall f_1, f_2 \in F),$$

$$(f_1 \prec f_2 \wedge f_2 \in IF(r)) \rightarrow (f_1 \in \bigcup_{i=1}^{r} IF(i))$$

Constraints

1. ***FeatureSetsForReleasesArePairwiseDisjoint*** is defined in computation #6 above.
2. ***DependencyGraphIsSatisfied*** *is defined in computation #7 above.*
3. ***BSN.Constraints*** is defined in Sect. 4.2.
4. ***SWD.Constraints*** is defined in Sect. 4.7.

InterfaceMetrics, also denoted **IM,** is a tuple.
⟨*SoFarIBF*, *CombinedCashFlow*, *TimeWindowNPV*, *BSN.InterfaceMetrics*, *SWD.InterfaceMetrics*⟩,
where:

- ***CombinedCashFlow*** is defined in computation #2 above.
- ***TimeWindowNPV*** is defined in computation #3 above.
- ***BSN.InterfaceMetrics*** is defined in Sect. 4.2
- ***SWD.InterfaceMetrics*** is defined in Sect. 4.7

4.2 Business Service Network Formalization

BusinessServiceNetwork formalization, also denoted **BSN,** is a tuple ⟨Parameters, DecisionVariables, Computation, Constraints, InterfaceMetrics⟩, where:

Parameters, also denoted **Parm,** is a tuple ⟨LaborRates, LaborPaySched, BSNDemand, ServicesSet, rootID⟩,
where:

- ***LaborRates*** is a tuple ⟨*LR, Rate*⟩ where:
- ***LR*** is a set of labor roles
- ***Rate*** : $LR \to \mathbb{R}^+$ is a function described as follows: $(\forall l \in LR)$, $Rate(l)$ gives the daily rate for labor role l.
- ***LaborPaySched***, the labor cost payment schedule, is a tuple ⟨*NLP*, *LaborPayDays*⟩, where:
  - ***NLP*** $\in \mathbb{R}^+$ is the number of labor payments over the entire time horizon
  - ***LaborPayDay*** : $[1..NLP] \to [1..TH]$ is a function described as follows: $(\forall p \in [1..NLP])$, $LaborPayDay(p)$ gives the day, relative to the first day of the time horizon, on which a payment p is made.
- ***BSNDemand***, is a tuple ⟨*BSNI*, *BSNO*, *Demand*⟩,
  where:
  - ***BSNI*** is a set of input items ids that have to be processed by the Service Network.
  - ***BSNO*** is a set of output items ids that have to be produced by the Service Network.
  - ***Demand*** : $BSNI \cup BSNO \to \mathbb{R}^+$ is a function described as follows: $(\forall j \in BSNI \bigcup BSNO)$, $Demand(j)$ gives for every item j, the required processing throughput per hour.

- ***ServicesSet*** is the set of all services in the Service Network, defined in Sect. 4.3.
- ***rootID*** is the *id* of the Service, in the *ServicesSet,* which is designated to be the "root". The definition of a Service is given in Sect. 4.2.

DecisionVariables is the set $\{s.DecisionVariables | s \in ServicesSet\}$. See Sect. 4.3.

Computation

1. Let *root* be a Service in *ServicesSet* with $id = rootid$
2. *BSNDemandIsSatisfied* constraint:

$$(\forall i \in BSNI)(\forall r \in [1..NR+1]),$$

$$Service.IM.InputThru(rootID,\ i,\ r) \geq Demand(i)$$

- $Service.IM.InputThru(rootID,\ r)$ is defined in Sect. 4.3

3. *BSNSupplyIsSatisfied* constraint:

$$(\forall o \in BSNO)(\forall r \in [1..NR+1]),$$

$$Service.IM.OutputThru(rootID,\ o,\ r) \geq Demand(o)$$

- $Service.IM.OutputThru(rootID,\ r)$ is defined in Sect. 4.3

4. Let $BSNCostForDay : [1..TH] \rightarrow \mathbb{R}^+$ be a function described as follows: $(\forall d \in [1..TH])$, $BSNCostForDay(d)$ gives the service network labor cost accrued for day d computed as:

$$BSNCostForDay(d) = Service.IM.CostPerDay(rootID,\ r)$$

Where:

- r is the release period (or period after the last release) where day d appears, i.e.,

$$SWD.IM.firstDay(r) \leq d \leq SWD.IM.lastDay(r)$$

- $Service.IM.CostPerDay(rootID,\ r)$ is defined in Sect. 4.3
- $SWD.IM.firstDay$ and $SWD.IM.lastDay$ are defined in Sect. 4.7.

5. Let *BSNPayment* : [1..*NLP*] $\to$ $\mathbb{R}$ be a function described as follows: ($\forall p \in [1..NLP]$), $BSNPayment(p)$ gives the service network labor payment in dollars, for each scheduled payment *p*, computed as:

$$BSNPayment(p) = \begin{cases} \sum_{d=1}^{LaborPayDay(p)} BSNCostForDay(d) \ \forall p = 1 \\ \sum_{d=LaborPayDay(p-1)+1}^{LaborPayDay(p)} BSNCostForDay(d) \ \forall p = [2..NLP] \end{cases}$$

6. Let *CashFlow* : $[1..TH] \to \mathbb{R}^+$ be a function described as follows: ($\forall d \in [1..TH]$), $CashFlow(d)$ gives the cash flow for the entire Business Service Network for day d, computed as follows:
if $d = LaborPayDay(p)$ *for some payment p*
Then $CashFlow(d) = -BSNPayment(p)$
Otherwise $CashFlow(d) = 0$

Constraints.

1. ***BSNDemandIsSatisfied*** (see Computation #2)
2. ***BSNSupplyIsSatisfied*** (see Computation #3)
3. ***Service.IM(rootID, r).Constraints***
(See Sect. 4.3)

InterfaceMetrics, also denoted **IM,** is a tuple $\langle CashFlow \rangle$, where:

- ***CashFlow*** is defined in computation #6 above.

4.3 Service Formalization

ServicesSet formalization is a set of **Service**, where:

Service is a tuple ⟨Parameters, DecisionVariables, Computation, Constraints, InterfaceMetrics⟩, defined separately for each *ServiceType* ∈ {*ANDservice*, *ORservice*, *InputDrivenAtomicService*}.

- Every service has an ***id*** and a ***ServiceType***. We denote by $service(id)$ the service with identifier *id*.
- ***ANDservice*** type is defined in Sect. 4.4.
- ***ORservice*** type is defined in Sect. 4.5.
- ***InputDrivenAtomicService*** type is defined in Sect. 4.6.

4.4 ANDservice Formalization

Intuitively, an ANDservice is a composite service, that is, an aggregation of sub-services such that all sub-services are activated.

ANDservice formalization is a tuple ⟨ *Parameters, DecisionVariables, Computation, Constraints, InterfaceMetrics*⟩

where:

Parameters, also denoted **Parm,** is a tuple ⟨*id, ServiceType(id),I(id),O(id), Subservices(id)*⟩

where:

- ***id*** is the Service id, which must be unique across all services in the *ServicesSet.*
- ***I(id)*** is a set of inputs
- ***O(id)*** is a set of outputs
- ***Subservices(id)*** is a set of the ids of the sub-services.
- ***ServiceType(id)*** is *ANDservice.*

DecisionVariables, also denoted **DV**, is a tuple ⟨$On(id)$, $InputThru(id)$, $OutputThru(id)$⟩

where:

- $\boldsymbol{On(id)} : [1..NR+1] \rightarrow \{0, 1\}$ is a function that determines whether the Service *id* is activated or not, for a particular release, i.e., $(\forall r \in [1..NR+1])$, $On(id)(r)$, also denoted by *On(id,r)* is as follows:

$$On(id, r) = \begin{cases} 1 \ \textit{if service id is activated in release r} \\ 0 \ \textit{otherwise} \end{cases}$$

- $\boldsymbol{InputThru}(id) : I(id) \times [1..NR+1] \rightarrow \mathbb{R}^+$ is a function described as follows: $(\forall i \in I(id), \forall r \in [1..NR+1])$, $InputThru(id)(i, r)$, also denoted $InputThru(id, i, r)$, gives the throughput of *i* (or quantity per day) during release *r* or the period after the last release.
- $\boldsymbol{OuputThru}(id) : O(id) \times [1..NR+1] \rightarrow \mathbb{R}^+$ is a function described as follows: $(\forall o \in O(id), \forall r \in [1..NR+1])$, $OutputThru(id)(o, r)$, also denoted $OutputThru(id, o, r)$, gives the throughput of *o* (or quantity per day) during release *r* or the period after the last release.

Computation

1. *AllSubservicesAreActivated* constraint:

   Let *n* be the cardinality of *Subservices(id).* Then the constraint is:

$$\sum_{i \in Subservices(id)} On(i, r) = n{*}On(id, r), \forall r \in [1..NR+1]$$

2. Let $SetAllIO(id)$ be a set of inputs and outputs, computed as follows:

$$SetAllIO(id) = I(id) \cup O(id) \cup \left(\bigcup_{i \in Subservices(id)} I(i) \right) \cup \left(\bigcup_{i \in Subservices(id)} O(i) \right)$$

3. Let $ServiceItemSupply(id) : SetAllIO \times [1..NR + 1] \rightarrow \mathbb{R}$ be a function described as follows: $(\forall j \in SetAllIO(id), \forall r \in [1..NR + 1])$, $ServiceItemSupply(id)(j, r)$, also denoted $ServiceSupply(id, j, r)$, gives the total supply of item j during release r (and the period after the last release), computed as follows:

$$ServiceItemSupply(id, j, r) = InputItemThru(id, j, r) + \sum_{s \in Subservices(id)} OutputItemThru(s, j, r)$$

Where:

$$InputItemThru(id, j, r) = \begin{cases} InputThru(id, j, r) & if\ j \in I(id) \\ 0 & otherwise \end{cases}$$

$$OutputItemThru(s, j, r) = \begin{cases} OutputThru(s, j, r) & if\ j \in O(s) \\ 0 & otherwise \end{cases}$$

4. Let $ServiceItemDemand(id) : SetAllIO \times [1..NR + 1] \rightarrow \mathbb{R}$ be a function described as follows:

$(\forall j \in SetAllIO(id), \forall r \in [1..NR + 1])$, $ServiceItem(id)(j, r)$, also denoted $ServiceConsumption(id, j, r)$, gives the total demand of item j during release r (and the period after the last release), computed as follows:

$$ServiceItemDemand(id, j, r) = OutputItemThru(id, j, r) + \sum_{s \in Subservices(id)} InputItemThru(s, j, r)$$

Where:

$$InputItemThru(s, j, r) = \begin{cases} InputThru(s, j, r) & if\ j \in I(id) \\ 0 & otherwise \end{cases}$$

$$OutputItemThru(id, j, r) = \begin{cases} OutputThru(id, j, r) & if\ j \in O(s) \\ 0 & otherwise \end{cases}$$

5. *SupplyItemMatchesDemandItem* constraint is:

$$\forall j \in SetAllIO(id), \forall r \in [1..NR + 1],$$

$$ServiceItemSupply(id, j, r) = ServiceItemDemand(id, j, r)$$

6. Let $CostPerDay(id) : [1..NR + 1] \rightarrow \mathbb{R}$ be a function described as follows: $(\forall r \in [1..NR + 1])$, $CostPerDay(id)(r)$, also denoted $CostperDay(id, r)$, gives the total dollar cost per day during period r and the period after the last period, computed as:

$$CostPerDay(id, r) = \sum_{i \in Subservices(id)} Service.IM.CostPerDay(i, r)$$

Constraints are as follows:

1. ***AllSubservicesAreActivated*** (see computation #1)
2. ***SupplyItemMatchesDemand*** (see computation #5)

InterfaceMetrics, also denoted **IM,** is a tuple ⟨ *CostPerDay(id), InputThru(id), OutputThru(id)*⟩
where:

- ***CostPerDay***(id) is defined in computation #6 above.
- ***InputThru***(id) is defined in DecisionVariables above.
- ***OutputThru***(id) is defined in DecisionVariables above.

4.5 ORservice Formalization

Intuitively, an ORservice is a composite service, that is, an aggregation of sub-services such that only one sub-services is activated.

ORservice formalization is a tuple ⟨ *Parameters, DecisionVariables, Computation, Constraints, InterfaceMetrics* ⟩
where:

Parameters, also denoted **Parm,** is a tuple $\langle id, ServiceType(id), I(id), O(id) O(id), Subservices(id)\rangle$
where:

- ***id*** is the Service id, which must be unique across all services in the *ServicesSet*.
- ***I(id)*** is a set of inputs
- ***O(id)*** is a set of outputs
- ***Subservices****(id)* is a set of the ids of the sub-services.
- ***ServiceType****(id)* is *ORservice.*

DecisionVariables, also denoted **DV,** is a tuple $\langle On(id), InputThru(id), OutputThru(id)\rangle$
where:

- ***On***$(id) : [1..NR + 1] \rightarrow \{0, 1\}$ is a function that determines whether the Service *id* is activated or not, for a particular release, i.e., $(\forall r \in [1..NR + 1])$, $On(id)(r)$, also denoted by *On(id,r)* is as follows:

$$On(id, r) = \begin{cases} 1 \text{ } if \text{ } service \text{ } id \text{ } is \text{ } activated \text{ } in \text{ } release \text{ } r \\ 0 \text{ } otherwise \end{cases}$$

- ***InputThru***$(id) : I(id) \times [1..NR + 1] \rightarrow \mathbb{R}^+$ is a function described as follows: $(\forall i \in I(id), \forall r \in [1..NR + 1])$, $InputThru(id)(i, r)$, also denoted $InputThru(id, i, r)$, gives the throughput of i (or quantity per day) during release r or the period after the last release.
- ***OuputThru***$(id) : O(id) \times [1..NR + 1] \rightarrow \mathbb{R}^+$ is a function described as follows: $(\forall o \in O(id), \forall r \in [1..NR + 1])$, $OutputThru(id)(o, r)$, also denoted $OutputThru(id, o, r)$, gives the throughput of o (or quantity per day) during release r or the period after the last release.

Computation

1. *OnlyOneServiceIsActivated* constraint:

$$\sum_{i \in Subservices(id)} On(i, r) = On(id, r), \forall r \in [1..NR + 1]$$

2. Same as ANDservice computation #2
3. Sam as ANDservice computation #4
4. Same as ANDservice computation #5
5. *SupplyItemMatchesDemandItem* constraint: same as ANDservice computation #5
6. CostPerDay computation: Same as ANDservice computation #6

Constraints are as follows:

1. ***OnlyOneServiceIsActivated*** (see computation #1)
2. ***SupplyItemMatchesDemand*** (see computation #5)

InterfaceMetrics, also denoted **IM,** is a tuple ⟨ *CostPerDay(id), InputThru(id), OutputThru(id)*⟩
where:

- CostPerDay(id) is defined in #6 above.
- InputThru(id) is defined in DecisionVariables.
- OutputThru(id) is defined in DecisionVariables.

4.6 InputDrivenAtomicService Formalization

Intuitively, an Input DrivenAtomicService is an indivisible service which's throughput is driven by the number of inputs that it needs to consume, for example, a process that receives applications and adjudicates them.

InputDrivenAtomicService formalization is a tuple ⟨*Parameters, DecisionVariables, Computation, Constraints, InterfaceMetrics*⟩

Parameters, also denoted **Parm,** is a tuple ⟨id, $ServiceType(id)$, $I(id)$, $O(id)$, $RBF(id)$, $ServiceRoles(id)$, $IOthruRatio(id)$, $RoleTimePerIO(id)$⟩
where:

- ***id*** is the Service id.
- ***I(id)****is a set of inputs*
- ***O(id)****is a set of outputs*
- ***RBF***$(id) \subseteq$ *ReleaseScheduling.Parm.BF* is a set of business features required by Service *id*
- ***ServiceRoles***$(id) \subseteq$ *BSN.Parm.LR* is a set of roles involved in the business service

- ***IOthruRatio***(id): $I(id) \times O(id) \rightarrow \mathbb{R}^+$ is a function described as follows: $(\forall i \in I(id))$, $(\forall o \in O(id))$, $IOthruRatio(id)(i, o)$ also denoted as $IOthruRatio(id, i, o)$, gives for input i and output o, the ratio of input throughput toward the output throughput.

- ***RoleTimePerIO***(id) : $ServiceRoles(id) \times (I(id) \bigcup O(id)) \rightarrow \mathbb{R}^+$ is a function described as follows: $\left(\forall l \in ServiceRoles(id), \forall j \in I(id) \bigcup O(id)\right)$, $RoleTimePerIO(id)(l, j)$, also denoted as $RoleTimePerIO(id, l, j)$, gives the amount of time in hours that role l spends per item j.
- ***ServiceType****(id)* is *InputDrivenAtomicService*

DecisionVariables, also denoted **DV,** is a tuple $\langle On(id), InputThru(id), OutputThru(id)\rangle$
where:

- ***On***(id) : $[1..NR + 1] \rightarrow \{0, 1\}$ is a function that determines whether the Service *id* is activated or not, for a particular release, i.e., $(\forall r \in [1..NR + 1])$, $On(id)(r)$, also denoted by *On(id,r)* is as follows:

$$On(id, r) = \begin{cases} 1 \textit{ if service id is activated in release r} \\ 0 \textit{ otherwise} \end{cases}$$

- ***InputThru***(id) : $I(id) \times [1..NR + 1] \rightarrow \mathbb{R}^+$ is a function described as follows: $(\forall i \in I(id), \forall r \in [1..NR + 1])$, $InputThru(id)(i, r)$, also denoted $InputThru(id, i, r)$, gives the throughput of i (or quantity per day) during release r or the period after the last release.

Computation

1. *FeatureDependencyIsSatisfied* constraint:

$$\forall r \in [1..NR + 1],$$

$$On(id, r) = 1 \rightarrow$$

$$RBF(id) \subseteq RSch.IM.SoFarIBF(r)$$

2. *DeactivatedServicesIsSatisfied* constraint:

$$\forall i \in I(id), \forall r \in [1..NR + 1],$$

$$On(id, r) = 0 \rightarrow InputThru(id, i, r) = 0$$

3. Let $OuputThru(id) : O(id) \times [1..NR + 1] \rightarrow \mathbb{R}^{+}$ be a function described as follows: $(\forall o \in O(id), \forall r \in [1..NR + 1])$, $OutputThru(id)(o, r)$, also denoted $OutputThru(id, o, r)$, gives the throughput of o (or quantity per day) during release r or the period after the last release, computed as

$$\forall o \in O(id), \forall r \in [1..NR + 1],$$

$$OutputThru(id, o, r) = \sum_{i \in I(id)} IOthruRatio(id, i, o) \times InputThru(id, i, r)$$

4. Let
$TimePerDay(id) : [1..NR + 1] \times ServiceRoles(id) \rightarrow \mathbb{R}^{+}$ be a function described as follows: $(\forall l \in ServiceRoles(id), r \in [1..NR + 1])$, $TimePerDay(id)(l, r)$, also denoted $TimePerDay(id, l, r)$, gives the total duration per day for role l and release r (and the period after the last release), computed as:

$$TimePerDay(id, l, r) = \sum_{j \in I(id)} (RoleTimePerIO(id, l, j) \times InputThru(id, j, r)) + \sum_{j \in O(id)} (RoleTimePerIO(id, l, j) \times OutputThru(id, j, r))$$

5. Let $CostPerDay(id) : [1..NR + 1] \rightarrow \mathbb{R}$ be a function described as follows:$(\forall r \in [1..NR + 1])$, $CostPerDay(id)(r)$, also denoted $CostperDay(id, r)$, gives the total dollar cost per day during period r (and the period after the last period), computed as:

$$CostPerDay(id, r) = \sum_{l \in ServiceRoles} (BSN.Parm.Rate(l) \times TimePerDay(id, l, r))$$

Constraints are as follows:

1. ***FeatureDependencyIsSatisfied*** (see computation #1)
2. ***DeactivatedServicesIsSatisfied*** (see computation #2)

InterfaceMetrics, also denoted **IM,** is a tuple ⟨*CostPerDay(id), InputThru(id), OutputThru(id)*⟩
where:

- ***CostPerDay***(id) is defined in computation #5.
- ***InputThru***(id) is defined in DecisionVariables.
- ***OutputThru***(id) is defined in computation #3.

4.7 Software Development Formalization

SoftwareDevelopment formalization, also denoted **SWD**, is a tuple ⟨*Parameters, DecisionVariables, Computation, Constraints, InterfaceMetrics*⟩
where:
Parameters, also denoted **Parm**, is a tuple ⟨*TS, DP, DC, OC, SS, SWPaySched*⟩,
where:

- ***TS*** : $[1 \ldots NR] \rightarrow \mathbb{R}^+$**is** a function that gives the team size, in full time equivalents, for each release.
- ***DP*** : $[1..NR] \rightarrow \mathbb{R}^+$ is a function that gives the developer productivity for each release in effort points per day.
- ***DC*** $\in \mathbb{R}^+$ is the developer cost in dollars per effort point.
- ***OC*** $\in \mathbb{R}^+$ is the operations cost in dollars per effort point per day.
- ***SS*** $\in \mathbb{R}^+$ is the size, in effort points, of the As-Is system (prior to development).
- ***SWPaySched***, the software cost payment schedule, is a tuple ⟨*NSP*, *SWPayDays*⟩, where:

    - *NSP* $\in R^+$ is the number of payments to the software team over the entire time horizon.
    - *SWPayDay* : $[1..NSP] \rightarrow [1..TH]$ is a function, i.e. $(\forall p \in [1..NSP])$, *SWPayDay*(p) gives the day (relative to the first day of the software development project) where payment p is made.

DecisionVariables, also denoted **DV**, is an empty tuple.

Computation:

1. Let $RC : [1..NR] \rightarrow \mathbb{R}^+$ be a function described as follows: $(\forall r \in [1..NR])$, $RC(r)$ gives the maximum capacity, in effort points, for release r computed as:

$$RC(r) = TS(r) \times DP(r) \times RSch.Parm.RD(r)$$

2. Let $RS : [1..NR] \rightarrow \mathbb{R}^+$ be a function described as follows: $(\forall r \in [1..NR])$, $RS(r)$ gives the actual size, in effort points, of release r, once features are assigned to it. The computation is as follows:

$$RS(r) = \left(\sum_{j \in RSch.DV.IBF(r)} RSch.Parm.FS(j) + \sum_{j \in RSch.DV.ITF(r)} RSch.Parm.FS(j) \right)$$

3. *ReleaseSizeCannotExceedCapacity* constraint:

$$0 \leq RS(r) \leq RC(r) \forall\ r \in [1..NR]$$

4. Let $firstDay : [1..NR+1] \rightarrow \mathbb{R}^+$ be a function described as follows: $(\forall r \in [1..NR+1])$, $firstDay(r)$ gives the day when release r actually starts, computed as:

$$firstDay(r) = \begin{cases} 1 \quad r = 1 \\ Rsch.Parm.RD(r) + firstDay(r-1) \forall r = [2..NR+1] \end{cases}$$

5. Let $lastDay : [1..NR+1] \rightarrow \mathbb{R}^+$ be a function described as follows: $(\forall r \in [1..NR+1])$, $lastDay(r)$ gives the day when release r ends, computed as:

$$lastDay(r) = \begin{cases} firstDay(r+1) - 1 \quad r = [1..NR] \\ RSch.TH \quad (r = NR+1) \end{cases}$$

6. Let $devCostPerDay : [1..NR+1] \rightarrow \mathbb{R}^+$ be a function described as follows: $(\forall r \in [1..NR+1])$, $devCostPerDay(r)$ gives the dollar cost of development per day for release$r$, computed as:

$$\forall r = [1..NR+1],$$

$$devCostPerDay(r) = \begin{cases} \left(\frac{RC(r)}{RSch.Parm.RD(r)} \times DC \quad (\forall r = [1..NR]) \right) \\ 0 \quad (r = NR+1) \end{cases}$$

7. Let $opsCostPerDay : [1..NR+1] \rightarrow \mathbb{R}^+$ be a function described as follows: $(\forall r \in [1..NR+1])$, $opsCostPerDay(r)$ gives the dollar cost of operations per day for release$r$, and the period after the last release, computed as:

$$opsCostPerDay(r) = \begin{cases} (SS \times OC) \quad r = 1 \\ \left(\left(\sum_{i=1}^{r-1} RC(i) \right) + SS \right) \times OC \quad \forall r = [2..NR+1] \end{cases}$$

8. Let $SWCostForDay : [1..TH] \rightarrow \mathbb{R}^+$ be a function described as follows: ($\forall d \in [1..TH]$), $SWCostForDay(d)$ gives the software cost accrued for each day d in the time horizon, computed as:

$$SWCostForDay(d) = devCostPerDay(r) + opsCostPerDay(r)$$

where r is the release period (or period after the last release), where day d appears, i.e.,

$$firstDay(r) \leq d \leq lastDay(r)$$

9. Let $SWPayment : [1..NSP] \rightarrow \mathbb{R}$ be a function described as follows: ($\forall p \in [1..NSP]$), $SWPayment(d)$ gives the software payment in dollars, for each scheduled payment p, computed as follows:

$$SWPayment(p) = \begin{cases} \sum_{d=1}^{SWPayDay(p)} SWCostForDay(d) & p = 1 \\ \sum_{d=SWPayDay(p-1)+1}^{SWPayDay(p)} SWCostForDay(d) & p = [2.NSP] \end{cases}$$

10. Let$CashFlow : [1..TH] \rightarrow \mathbb{R}^+$, be a function described as follows: ($\forall d \in [1..TH]$), $CashFlow(d)$ gives the cash flow of software cost for day d, is computed as:
if $d = SWPayDay(p)$ *for some payment* p
then $CashFlow(d) = -SWPayment(p)$
else $CashFlow(d) = 0$

Constraints.

1. ***ReleaseSizeCannotExceedCapacity*** (defined fin computation #3)

InterfaceMetrics, also denoted **IM,** is a tuple $\langle CashFlow, firstDay, lastDay \rangle$, where:

- ***CashFlow***(d) is defined in computation #10.
- ***firstDay***(r) is defined in computation #4.
- ***lastDay***(r) is defined in computation #5.

4.8 Optimization Formulation

The formalizations in the previous sections are building blocks; we now use them to formulate the optimization of the NPV of the final BPN configuration, called the To-Be. Given the top-level formal optimization model

$$RSch\langle Parameters,\ DecisionVariables,\ Computation,\ Constraints,\ IM\rangle,$$

the optimal NPV for the To-Be BPN, for a time horizon of *th* days, is:

$$\begin{aligned} NPV_{ToBe} = &\ Max\ RSch.IM.TimeWindowNPV(th) \\ &\ s.t.\ RSch.Constraints \end{aligned}$$

5 Release Scheduling Example

We now present a full solution to the example introduced in Sect. 3.3. The formal model in Sect. 4 describes seven components: *ReleaseScheduling, BusinessServiceNetwork, Service, ANDservice, ORservice, InputDrivenAtomicService* and *SoftwareDevelopment.* For the example in Sect. 3.3, we show the parameters for all the components in Tables 1, 2, 3, 4, 5, 6, and 7. We assume the inputs in Table 8, and use the IOThruRatio from Table 5 to calculate the outputs.

Table 1. *ReleaseScheduling.Parameters.* Reprinted from [1].

Parameter	Value
BF (business features)	{BF1, BF2, BF2, BF4}
TF (technical features)	{TF1}
DG (dependency graph)	{(TF1, BF1), (BF1, BF2), (TF1, BF3), (BF1, BF4)}
FS (feature size)	{(TF1,140), (BF1,140), (BF2,280), (BF3,280), (BF4,280)
TH (time horizon)	520
DiscountRate (daily)	0.01923076923%
NR (number releases)	4
RD (release duration)	60

The first calculation is the duration of each service; for example, the total duration of service BA is $70 \times 0.042 + 70 \times 0.208 = 17.50$. The cost per day for BA is the duration, multiplied by the labor rate of AO, which is 17.50×400, resulting in $7,000.00. The results, in Table 9, show that the greatest savings are in the services AB, CB, BB and AC in this order.

Table 10 shows a possible release sequence and the cost per day of the entire BSN. The cash outflow is the cost per day times the duration, which is 60 days for releases 1

Table 2. *BusinessServiceNetwork.Parameters.* Reprinted from [1].

Parameter	Value
LR (Labor Roles)	{IO, AO, A, S}
Rate	{(IO, 160), (AO, 400), (A, 0), (S, 0)}
NLP (Number Labor Payments)	5
LaborPayDay	[60,120,180,240,520]
BSNI (root input)	{User Application}
BSNO (root output)	{}
Demand(User Application) – # root inputs	100
ServicesSet	{Adj, A, B, C, AA, AB, AC, BA, BB, CA, CB}
rootID (id of root process)	Adj

Table 3. *AndService.Parameters, OrService.Parameters.*

id	Type	Input	Output	Sub services
Adj	AND	User Application (UA)	Non-compliance Notice (NCN) Adjudicated Application Letter (AAL)	A, B, C
A	OR	User Application (UA)	Compliant Application (CA) Non-compliance Notice (NCN)	AA, AB, AC
B	OR	Compliant Application (CA)	Adjudicated Application (AA)	BA, BB
C	OR	Adjudicated Application (AA)	Adjudicated Application Letter (AAL)	CA, CB

Table 4. *InputDrivenAtomicService.Parameters* part 1.

id	Type	Input	Output	RBF
AA	Atomic	UA	CA,NCN	None
AB	Atomic	UA	CA,NCN	BF1
AC	Atomic	UA	CA,NCN	BF4
BA	Atomic	CA	AA	None
BB	Atomic	CA	AA	BF2
CA	Atomic	AA	AAL	None
CB	Atomic	AA	AAL	BF3

Table 5. *InputDrivenAtomicService.Parameters* part 2. Reprinted from [1].

id	Input	Output	IO Thru Ratio
AA	UA	CA	70%
AA	UA	NCN	30%
AB	UA	CA	70%
AB	UA	NCN	30%
AC	UA	CA	70%
AC	UA	NCN	30%
BA	CA	AA	100%
BB	CA	AA	100%
CA	AA	AAL	100%
CB	AA	AAL	100%

Table 6. *InputDrivenAtomicService.Parameters* part 3.

id	Role	In-put	RoleTime PerIO
AA	IO	UA	0.25
AB	IO	UA	0.145
AC	A	UA	0.063
BA	AO	CA	0.042
BB	AO	CA	0.021
CA	AO	AA	0.021
CB	AO	AA	0.017

id	Role	Out-put	RoleTime PerIO
AA	IO	CA	0.125
AA	IO	NCN	0.219
AB	S	CA	0
AB	S	NCN	0
AC	S	CA	0
AC	S	NCN	0
BA	AO	AA	0.208
BB	AO	AA	0.143
CA	AO	AL	0.167
CB	AO	AL	0.083

to 4 and 280 for the subsequent period. The resulting cost per day is also shown in Table 10.

Next computation is the software cost. The daily cost is the team size, times the productivity, times the cost of each point, or $5 \times 1 \times \$1{,}040 = \$5{,}200$. From release 2 on, there is a small cost to maintain the software just released, so the cost increases as the software grows. Table 11 shows the results of the software cost calculations.

Once the BSN and the software cash outflow are known, we can calculate the NPV, which is the sum of the cash flows of the BSN plus the software, discounted at 5% per year. Table 12 shows the result, **−$6,411,432.73.**

Once the NPV of the To-Be is determined, we can calculate the NPV of the As-Is, which results in −$9,085,928.80. The NPV of the savings, which is the NPV of the To-Be minus the NPV of the As-Is, is $2,674,496.07. This means that investing in the

Table 7. *SoftwareDevelopment.Parameters.* Reprinted from [1].

Parameter	Value	Unit
TS (Team Size)	5	
DP (Dev Productivity)	1	(points/day)
DC (Dev Cost)	1,040	(US$/point)
OC (Operations Cost)	0.25	(US$/point/day)
SS (System Size)	0	(points)
NSP (# Soft Payments)	5	
SWPayDay	[60, 120, 180, 240, 520]	

Table 8. Number of inputs and outputs through the service network.

	InputThru	OutputThru
UA	100	
NCN		30
CA	70	70
AA	70	70
AAL		70

Table 9. Cost per day of atomic services.

id	Role	Time	Rate	CostPerDay	Savings
AA	IO	40.32	$160.00	$6,451.20	
AB	IO	14.50	$160.00	$2,320.00	$4,131.20
AC	A	6.30	$0.00	$0.00	$2,320.00
BA	AO	17.50	$400.00	$7,000.00	
BB	AO	11.48	$400.00	$4,592.00	$2,408.00
CA	AO	13.16	$400.00	$5,264.00	
CB	AO	7.00	$400.00	$2,800.00	$2,464.00

software release schedule as depicted in Table 10, produces a significant business value over 2 years.

Table 10. Optimal Release Sequence and BSN Cost Per Day.

Rel	IBF, ITF	BSN	BSN CostPerDay	BSN CashFlow
1	TF1, BF1	AA, BA, CA	$18,715.20	−1,122,912.00
2	BF3	AB, BA, CA	$14,584.00	−875,040.00
3	BF2	AB, BA, CB	$12,120.00	−727,200.00
4	BF4	AB, BB, CB	$9,712.00	−559,200.00
After 4		AC, BB, CB	$7,392.00	−1,960,000.00

Table 11. Software Cost.

Rel	# days	devCostPerDay	opsCostPerDay	SWD CashFlow
1	60	$5,200	$0	−$312,000
2	60	$5,200	$75	−$316,500
3	60	$5,200	$150	−$321,000
4	60	$5,200	$225	−$325,500
After 4	280	0	$300	−$84,000

Table 12. Combined cash outflow and NPV. Reprinted from [1].

Rel	BSN CashFlow	SWD CashFlow	NPV
1	−1,122,912.00	−312,000.00	−1,418,452.05
2	−875,040.00	−316,500.00	−1,164,360.35
3	−727,200.00	−321,000.00	−1,012,540.33
4	−582,720.00	−325,500.00	−867,258.63
After 4	−2,069,760.00	−84,000.00	−1,948,821.37
Total:	−5,377,632.00	−1,359,000.00	**−6,411,432.73**

6 Decision Guidance System

Decision Guidance Systems (DGSs) are an advance class of Decision Support Systems (DSS) that are designed to provide "actionable recommendations, typically based on formal analytical models and techniques" [3]. We use Unity [10, 11], a platform for building DGSs from reusable Analytical Models (AMs). Unity exposes an algebra of operators and provides an unified, high-level language called Decision Guidance Analytics Language (DGAL) [3].

The optimization model formalized in Sect. 4 is implemented, in *Opti-Soft*, as a Decision Guidance System (DGS). It uses the *Parameters* in the input file to maximize the NPV, subject to the *Constraints*. During the maximization, *Opti-Soft* performs the *Computation* and chooses the optimal *DecisionVariables*. The *InterfaceMetrics* are implemented by making them available to other components of the formalization hierarchy. The *Opti-Soft* DGS architecture is shown in Fig. 6.

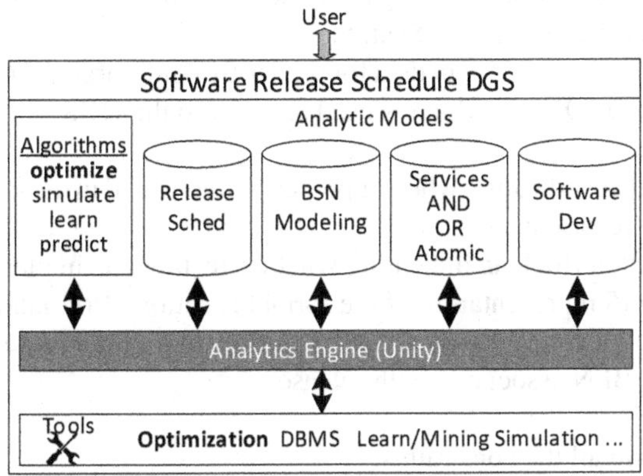

Fig. 6. *Opti-Soft* DGS Architecture. Reprinted from [1].

7 Methodology

The *Opti-Soft* DGS is an integral part of the Release Scheduling Methodology [1], which contains the following key steps:

1. Gather the ReleaseScheduling parameters
2. Model the BPN
3. Gather the SoftwareDevelopment parameters
4. Setup the parameter file for the To-Be BPN configuration. The file contains constant values as well as unresolved values representing the decision variables.
5. Run the DGS with the parameter file from step 4, to produce another parameter file with the decision variables optimally instantiated to maximize the NPV
6. Run the DGS with the instantiated parameter file produced in step 5, to calculate the optimal NPV for the To-Be configuration
7. Setup the DGS parameter file for the As-Is BPN configuration
8. Run the DGS to produce the NPV for the As-Is configuration
9. Calculate the total NPV of the savings

We now follow the methodology. In the first step, we gather the *ReleaseScheduling* parameters which include software features, their sizes and dependencies. For the example in Sect. 5, the parameters are shown in Table 1.

In step 2, the BSN is modelled. Each process is assigned a type of atomic, AND or OR, their inputs, outputs, labor rates and the features that enable them are set and the top-level or root process is defined. The daily number of inputs to the BSN, called *Demand,* is set, as well as the payment schedule. The *BusinessServiceNetwork.Parameters,* for the Sect. 5 example is shown in Table 2 and *Service.Parameters* are shown in Tables 3, 4, 5 and 6. Note that the parameters are a numerical codification of the BSN diagrams in Figs. 1 and 2.

In step 3 of the methodology, we gather the *SoftwareDevelopment* parameters as shown in Table 7 for the example in Sect. 5.

In step 4, we setup the DGS input data for the To-Be configuration. All the parameters above are coded in a JSON file which is used as input to the DGS.

1. *On(s,r)* – a Boolean indicating whether process*s* is activated in the BPN configuration associated with release *r*.
2. *Implemented(f,r)*- a Boolean indicating whether feature *f* is implemented in release *r*. This is the DGS representation of the formal functions *IBF* and *ITF*.
3. *InputThru(s,i,r)* – an integer indicating the number of inputs of type *i* that go through process *s* in the BPN associated with release *r.*

Below we explain all the constraints:

1. *FeatureSetsForReleasesArePairwiseDisjoint* – guarantees that a feature is implemented in one and only one release
2. *DependencyGraphIsSatisfied*– guarantees that features that are pre-requisites are implemented before the dependent ones. For example, the graph in Table 1 shows that TF1 and BF1 are pre-requisites for the other features BF2, BF3 and BF4.
3. *BSNDemandIsSatisfied*– This constraint preserves inputs. For the example in Sect. 5, the demand for input 'User Application (UA)', shown in Table 2, is 100 inputs, consequently the *InputThru* for the root process, "Adj" has to be at least 100.
4. *BSNSupplyIsSatisfied* – This constraint preservers outputs. In the Sect. 5 example, Table 2, *Demand(output)* is empty, consequently this constraint is always satisfied.
5. *AllSubservicesAreActivated* – Applies only to services of type AND. It forces all subservices to be activated. In the Sect. 5 example in Tables 2 and 3, service 'Adj' is the root, consequently *On(Adj,r) = 1*. Because 'Adj' has subservices A, B and C and its type is AND, the constraint forces *On(A,r) = 1, On(B,r) = 1, On(C,r) = 1* for every release *r*.
6. *OnlyOneServiceIsActivated* – Applies only to services of type OR. Because in the Sect. 5 example A, B and C have type OR, they can only have one subservice. This constraint, in conjunction with the previous one, guarantees that the configuration has one and only one subservice for A, B and C. This means that the configuration AA, BA, CA is valid but AA, AB, CA is not.
7. *SupplyItemMatchesDemand* – Applies only to services of type AND or OR, called composites. A composite service is made of subservices and this constraint forces the supply (number of inputs through the service) to be equal to the demand (number of outputs through the service).

8. *FeatureDependencyIsSatisfied* - Applies only to services of type ATOMIC and guarantees that the feature dependencies as shown in Table 1 are satisfied. Services AA, BA and CA don't require any feature consequently they force the BSN configuration prior to the implementation of release 1 to be AA, BA, CA, that is, *On(AA,1) = 1, On(BA,1) = 1, On(CA,1) = 1.*
9. *DeactivatedServicesIsSatisfied* - Applies only to services of type ATOMIC and guarantees that if a service is not activated, then no inputs are processed by it. For example, we know that AB is not activated in release 1, that is, *On(AB,1) = 0,* consequently *InputThru(AB,UA,1) = 0.*
10. *ReleaseSizeCannotExceedCapacity* – Guarantees that the sum of the points of features implemented in any release does not exceed the capacity of the team. Capacity is the size of the team, times the productivity, times the release duration. For the example in Table 7, the result is 300 points ($5 \times 1 \times 60$). Because the size of features BF2 and BF3 are 280, they both cannot be implemented in the same release.

Now that we understand the parameters, the decision variables and the constraints, we can formulate the decision problem as follows:

Given $ReleaseSchedule(Parameters, DecisionVariables, Computation, Constraints, IM)$

The optimal NPV for the To-Be BPN, for a time horizon of *th* days, is:

$$NPV_{ToBe} = MaxReleaseScheduling.InterfaceMetrics.TimeWindowNPV(th)$$

$$s.t.\ ReleaseScheduling.Constraints$$

Where the $TimeWindowNPV(th)$, or just $NPV(th)$,is the sum of the NPV for each day up to *th*.

Still in step 5, the DGS solver produces an output file with all decision variables instantiated so that the NPV is optimized. In step 6, we run the DGS with the instantiated parameter file produced in step 5, to calculate the optimal NPV for the To-Be configuration, which in the Sect. 5 example, results in −\$6,411,432.73.

In step 7, we setup the DGS parameter file with the data for the As-Is and in step 8, we run the DGS to produce the NPV for the As-Is, which in the Sect. 5 example, results in −\$9,085,928.80. In step 9, we compute the NPV of the savings, which is the NPV of the To-Be minus the NPV of the As-Is, resulting in \$2,674,496.07.

8 Conclusion

In this paper we focused on a class of problems where software is developed in order to reduce the cost of a business process. For that class of problems, we propose a software release scheduling approach that is more precise than existing approaches and guarantees that that result is an optimal solution. The precision is due to the fact that 1) the business value of the investment in the software can be measured by the reduction in labor cost due to productivity gains, and 2) there is no need to assign a business value to individual software features, instead, software features are just a mechanism for the business process to evolve into more efficient configurations.

The solution is optimal because it is based on a formal model which is then converted, through a Decision Guidance System, into a mixed-integer optimization model.

References

1. Boccanera, F., Brodsky, A.: Decision guidance on software feature selection to maximize the benefit to organizational processes. In: ICEIS, vol. 1, pp. 381–395 (2020)
2. Brodsky, A., Krishnamoorthy, M., Nachawati, M.O., Bernstein, W.Z., Menascé, D.A.: Manufacturing and contract service networks: composition, optimization and tradeoff analysis based on a reusable repository of performance models. In: 2017 IEEE International Conference on Big Data (Big Data), 1716–1725 (2017)
3. Brodsky, A., Luo, J.: Decision Guidance Analytics Language (DGAL)-toward reusable knowledge base centric modeling. In: 17th International Conference on Enterprise Information Systems (ICEIS), pp 67–78 (2015)
4. Cleland-Huang, J., Denne, M.: Financially informed requirements prioritization. In: 2005 Proceedings of the 27th International Conference on Software Engineering, ICSE 2005, p. 710 (2005)
5. Denne, M., Cleland-Huang, J.: The incremental funding method: data-driven software development. IEEE Softw. **21**(3), 39–47 (2004)
6. Denne, M., Cleland-Huang, J.: Software by Numbers: Low-Risk, High-Return Development. Prentice Hall (2003)
7. Devaraj, S., Kohli, R.: The IT payoff: measuring the business value of information technology investments. FT Press (2002)
8. Fowler, M., et al.: The agile manifesto. Software Development Magazine
9. Maurice, S., Ruhe, G., Saliu, O., Ngo-The, A.: Decision support for value-based software release planning. In: Value-Based Software Engineering pp. 247–261. Springer, Heidelberg (2006)
10. Nachawati, M.O., Brodsky, A., Luo, J.: Unity: a NoSQL-based platform for building decision guidance systems from reusable analytics models. In: Technical Report GMU-CS-TR-2016-4. George Mason University (2016)
11. Nachawati, M.O., Brodsky, A., Luo, J.: Unity decision guidance management system: analytics engine and reusable model repository. ICEIS **1**, 312–323 (2017)
12. Pucciarelli, J., Wiklund, D.: Improving IT project outcomes by systematically managing and hedging risk. IDC Report (2009)
13. The Standish Group: CHAOS Manifesto (2014)
14. Van den Akker, M., Brinkkemper, S., Diepen, G., Versendaal, J.: Determination of the Next Release of a Software Product: An Approach using Integer Linear Programming. CAiSE Short Paper Proceedings (2005)

Fast and Efficient Parallel Execution of SARIMA Prediction Model

Tiago Batista da Silveira(✉), Felipe Augusto Lara Soares, and Henrique Cota de Freitas

Graduate Program in Informatics, Pontifícia Universidade Católica de Minas Gerais (PUC Minas), Belo Horizonte, MG, Brazil

Abstract. Mathematical models for predicting values in time series are powerfultools for the process of knowledge discovery and decision making in several areas. However, the choice of the predictive model and its configuration are not trivial tasks, requiring a long processing time to obtain the results due to the high complexity of the models and the uncertainty of the value of the best parameters. Calculations performed by these approaches use sampling from the dataset, which can present discrepancies and variations that can directly impact the final result. Therefore, this work presents a new approach based on the SARIMA model for the prediction of values in time series. The proposal aims at predictive calculation from multiple executions of SARIMA in parallel, configured with predefined order and seasonal order parameters and applied to values already known in a time series. Thus, from the results obtained in past observations, it is possible to determine the percentage of precision that each parameter obtained, and, in this way, to determine the parameters that are more likely to obtain more accurate values in future observations, thus, eliminating the need to use specific algorithms to estimate them. The proposed approach is capable of achieving results with greater precision and performance compared to the traditional SARIMA execution, achieving results with greater assertiveness, reaching up to 10.77% of better accuracy and with better processing times, without the need for validation and parameter adjustments required by the settings obtained by functions, such as ACF and PACF.

Keywords: SARIMA · Forecasting · Time series · Parameter estimation · Parallel execution

1 Introduction

The current computing landscape is adapting to new trends in processing and information analysis. Exploring methods of knowledge discovery and more efficiently data analyzing is important to deal with the growing volume of information [9].

Obtaining knowledge of this mass of data is very important for decision making, however, the process of extracting the relevant information is delicate

J. Filipe et al. (Eds.): ICEIS 2020, LNBIP 417, pp. 217–241, 2021.
https://doi.org/10.1007/978-3-030-75418-1_11

and laborious. Often, this information is implicit in this data, requiring methods to extract, filter and prepare it.

With the advancement in the processes digitization, the collection and storage of data have become essential processes for control, business analysis and decision making. Each day, the speed and amount of data generated and stored increase. Thus, methods for extracting knowledge need to keep up with this growing trend.

The enrichment of these data is an extremely important step, since relevant information may be implicit in the data [4]. Mathematical and statistical methods can be used to enrich the dataset with valuable information [8], revealing characteristics and patterns that can be explored in the knowledge discovery process. In many cases, these stored data give rise to time series. These, in turn, are collections of data sequentially arranged over a period of time, presenting patterns and trends in their data.

Given this scenario, the use of artificial intelligence, machine learning and mathematical methods, which use formism to express relationships between variables [12], have led to advances in the knowledge discovery process in several fields [7]. These approaches are based on the analysis of large datasets, which in turn has the role of training the algorithms for decision making, pattern detection and predictions of future behavior.

Although the use of algorithms for predicting values in time series enables decision making based on the future behavior of the data, its use must be carried out with caution, since non-normalized, erroneous and dissonant data generate interference in the results, changing the percentage of accuracy and assertiveness of these algorithms [20].

Mathematical models for predicting values in time series based on the auto regression model and moving average such as the Autoregressive Integrated Moving Average (ARIMA)and Seasonal Autoregressive Integrated Moving Average (SARIMA) are models that achieve good predictive results in time series and are widely used. These approaches require a set of configuration parameters and these values define how auto-regression, moving average and differentiation will be applied in the time series, as well as, seasonality and the trend in the values of the analyzed series. These parameters can be calculated from the Autocorrelation Function (ACF), and Partial Autocorrelation Function (PACF), which directly impact the final predictions.

The execution of these algorithms presents two main problems: i) high processing time to obtain the results due to the estimation, application and comparison of the parameters to the time series, ii) and the uncertainty of the result, since it is obtained from only a set of seasonal order and order, without taking into account other possible combinations. Therefore, our main goal is to propose and evaluate a parallel SARIMA execution for performing predictive calculations. Our new approach presents the following contribution to the state-of-the-art, as well as, to our previous work [19]: the predictions are made in a pre-determined set of combinations of parameters of order and seasonal order in a parallel way, thus eliminating the need to use external algorithms, such as,

ACF and PACF, for analysis, comparison and estimation of configuration values, thus reducing processing time.

The proposed approach consists of using the predictive results achieved by a predetermined set of parameters in past observations of the analyzed series, thus, a coefficient is calculated, indicating which combinations of order presented the best results. The predictions are then made with the chosen combinations and the final result is obtained. Thus, an assertiveness gain is expected, since the results are now calculated from a group of parameters that obtained the best results in past observations of the time series, taking into account multiple autocorrelation and partial autocorrelation values. Tests and results presented a speedup of up to 229 times in execution time and up to 10.77% in accuracy compared to the traditional SARIMA approach.

2 Main Concepts Overview

The Partial Autocorrelation Function (PACF) and Autocorrelation Function (ACF) [1], [18] are mathematical tools that make possible to estimate parameters for the autoregressive moving average (ARMA) and derivatives [3] models, being responsible respectively for determining the order of Auto-regression and Moving Average, or how many past observations should be made for each model in a series. By definition, Autocorrelation is the process to which a correlation is calculated for values of the same ordered variable over time, in the case of a time series.

To apply the ACF and PACF functions, the time series must be stationary or its values over time must be around a constant average. Otherwise, it is necessary to apply differentiation functions to stabilize the trend.

The PACF is the correlation between the value of a variable and its occurrences in past observations, since it uses the residuals or, values whose variation already calculated in the series has been removed, thus disregarding the influence of values prior to the compared one. In this way, the resulting value of this function is used in ARMA models (auto-regression and moving average) and derivatives to determine the AR order (auto-regression) of the model since it is based on the correlation between two values.

The ACF is the correlation between the value of a variable and its occurrences in past observations. It determines how correlated these values are. Thus, it is possible to state that the result of the autocorrelation function shows how much the existence of one value is affected by the existence of another. In time series analysis, ACF has the role of providing the moving average value for the predictive model ARMA and its derivatives.

The application of the PACF and ACF functions in a time series generates a set of values that indicate the degree of correlation and partial correlation between the elements of a time series. Only values above a confidence interval are taken into account for determining the values of auto-regression and moving average, the confidence interval indicates an interval where with a percentage of chance the real value of the average of the set will be.

2.1 Transformation Functions

Data transformation is used in time series in order to improve the predictive result from noise reduction and trend attenuation, making the series as stationary as possible, or that the mean, variance and auto-correlation structure vary little over time.

Often, the time series has non-stationary characteristics, with the trend on average being the most common. These characteristics make the work of mathematical models such as SARIMA more difficult, since the trend and seasonality show an increase or decrease throughout the series, making predictive operations difficult.

The application of these transformations in the data is used to perform the monotonic transformation of a set, that is, to transform a set of numbers into another set that preserves the order of the original set. For this, a function is used to map real numbers to real numbers, which satisfies the property: $x > y$, então $f(x) > f(y)$, thus, stabilizing the variance, making the data more similar to the normal distribution and improving the validity of the association measures.

One of the types of transformations that can be applied to time series data is the Box-Cox transformation [16]. This function is a configurable type capable of evaluating data from the application of different transformations, automatically selecting the best approach for transforming the data.

2.2 ARIMA and SARIMA Mathematical Models

Mathematical models for the analysis and prediction of values in time series are the object of study in several areas. Models, such as ARIMA and SARIMA, are examples derived from the method proposed by [1]. The models based on this method have favorable characteristics that make them flexible in different situations, allowing the analysis of the time series only by the auto-regressive part or moving average, thus, better adapting the analyzed time series. These statistical analysis models are able to process and present information to better understand the data or to calculate the probability of a future value being between two specific limits.

It is also possible to affirm that this method consists of proposing and adjusting linear models to an observed time series. Thus, the configuration of the model is based on the analysis of the data itself, as proposed by [1], where the PACF and ACF based on the work of [22,23] made possible estimation of parameters for the ARMA model.

ARIMA is a model of statistical analysis, which from series containing temporal data is able to process and present information to better understand the data. However, this model presents a problem, it does not take into account patterns in the series or seasonality, which are repetition cycles in the data. To correct the problem, a method has been proposed to take seasonality into account, SARIMA is an extension of the ARIMA model that now takes into account the seasonal element of the series. Both ARIMA and SARIMA are members of the same family of predictive algorithms, being composed of two main parts: Autoregressive (AR) e Moving Average (MA).

The AR determines a dependency relationship between an observation and a number of outdated observations in the time series, or number of lags. It allows to incorporate the effect of values passed in the model. In the ARMA (p, q) approach, AR is directly related to the p parameter, which determines how many delayed observations should be made. From the auto-regression model, it is possible to attest that the next value in a series is likely to be close to its predecessors, if their value has remained constant previously. This determination is maintained for derivations of the AR model and can be defined by the Eq. 1:

$$X_t = \varepsilon_t + \sum_{i=1}^{p} \varphi_i X_{t-i} \tag{1}$$

where $\varphi_1 ... \varphi_p$ are the parameters resulting from the auto-correlation between the terms and ε_t is the noise error value.

a) Current values of AR (p) depend on their own p-previous values;
b) p is the order of AR processes.

The MA uses the dependency between the observed value and the mean of the set, or residual error of previous observations (lags). It allows the definition of the model error as a linear combination of the previous values. In the ARMA (p, q) model, MA is directly related to the q parameter, which determines the size of the window containing values for the moving average. The moving average can be determined from the Eq. 2:

$$X_t = \varepsilon_t + \sum_{i=1}^{q} \theta_i \varepsilon_{t-i} \tag{2}$$

where $\theta_t ... \theta_{t-1}$ represents the parameters of the moving average between the terms and ε_t and ε_{t-1} are noise error values.

a) The current deviation from the mean depends on MA (q) previous deviations;
b) q is the order of MA processes.

Based on improvements in the ARMA model, the I element was added, which represents the integrated part of the model and the differentiation in the series, that is, normalizes the series, eliminating seasonality and trend, in addition to stabilizing the average.

In the ARIMA (p, d, q) approach, I is directly related to the d parameter, which determines the number of differentiation required to make the series stationary. This part indicates that the data values have been replaced with the difference between their values and the previous values of the same variable. Thus, it is possible to say that the next value in a series is likely to remain close to its predecessors, if the difference between the values in previous observations is very small, according to Eq. 3:

$$y'_t = y_t - y_{t-1} \tag{3}$$

where the value of y'_t is equal to the difference between its value and a past observation of it.

The SARIMA method is a derivation of the previous models, where the seasonal part of the time series is taken into account in the computation. Seasonality is the definition given to a time series in which it is possible to determine a constant repetition pattern, that is, phenomena that occur throughout the series tend to be repeated at regular intervals.

For the configuration of the SARIMA model, the seasonal order values represented by the (P, D, Q, m) variables are added in addition to the (p, d, q) order parameters present in the ARIMA model. Seasonal values are responsible for performing Auto-Regression, Differentiation and Moving Average taking into account time intervals called the seasonal pattern, represented by the variable m. The determination of the seasonal variable is obtained from the seasonal decomposition of the time series. In this step, the series is divided into its seasonal component, trend and remaining or residual value. The interval between repetitions in the pattern is assigned to m.

2.3 Root-Mean-Square Deviation

Root Mean Square Deviation ($RMSD$) is one of the best general model performance measures to compare the results found with the expected values, as its error value is presented in the same dimensions as the analyzed variable [21]. Therefore, the measure ($RMSD$) presents itself as an important metric to compare different models and define which one obtained the best results. The measure is given by Eq. 4:

$$RMSD = \sqrt{\frac{1}{n}\sum_{i=1}^{n}(P_i - O_i)^2} \tag{4}$$

where P_i is the value resulting from the simulation and O_i is the observed value. Results of $RMSD$ close to 0 indicate greater assertiveness of the algorithm, and this measure can be used when comparing the assertiveness of the prediction of algorithms in the same dataset.

3 Related Work

This section presents works that aim to improve the processes of parameters estimation of the ARMA model and derivatives. In addition, the authors also attesting to the importance of this step in the modeling of these algorithms in time series.

Hyndman [10] proposes the use of autocovariance estimators presented by [11] to obtain new plots of ACF and PACF. The author states that ACF standards are not reliable for large lags and that the graphs of the generated autocorrelation functions make the determination process more complicated than necessary. Tests and comparisons were also carried out between the proposed model and

AR and MA models. However, the author states that the presented method obtained worse results than simple algorithms.

Mohammadi et al. [14] use two approaches for the identification of parameters for the ARMA model, maximum likelihood, widely used and error minimization, using Goal Programming (GP) language. The authors state that the second method showed better results and the advantage of being able to calibrate the model for the whole series simultaneously. Mohammadi et al. also state that the disadvantage of estimating using GP is the high computational cost and that this can be overcome by optimizing the algorithms.

Ding et al. [5] propose an equivalence model based on least squares iterative algorithm (ME-LSI) for the determination of parameters for stochastic dynamical systems with ARMA noise based on the works of Meng et al. [13] and Ding et al. [6]. The authors indicate a gain in the accuracy of the estimation of parameters being confirmed by tests and simulations performed.

Sheta et al. [17] affirm in their work that parameter estimation for the ARMA model is fundamental to obtain good predictive results, and when applied to businesses it has the ability to reduce costs, increase productivity and profit. In their work, the authors propose the use of nature-inspired Meta-heuristic algorithms and that this approach is capable of achieving an optimal global solution and can successfully be used to tune the ARMA model parameters with minimum modeling error.

Our work presents a new approach to optimize the parameter estimation process in ARMA and derivative models. The proposed approach differs from the others based on the use of parallel computing concepts to perform predictive calculations and the total elimination of the parameter estimation methods to configure the models. Thus, it is possible to eliminate the use of algorithms for estimating model configuration parameters, such as those based on the autocorrelation and partial autocorrelation functions, widely used for the ARMA configuration. So, it is possible to reduce the time needed for predictions, eliminate the need to validate the results and increase the assertiveness of the method, since the predictive results are now calculated from predefined parameters that presented good results in past predictions in the analyzed series. This way, we can summarize our contribution and findings as follows: i) SARIMA's parallel execution approach to eliminate the execution of additional algorithms for parameter estimation and validation; ii) time reduction required to obtain predictions; iii) higher accuracy in predictive results.

4 Methodology

This section presents our design and evaluation methodology divided into materials and method.

4.1 Materials

The experiments were performed on an Intel Core i7 8700 3.20 GHz, 6 cores and 12 threads, 64-bit Ubuntu Operating System and 16 GB of shared RAM.

We selected two datasets to perform our experiments. The first[1] contains data about the flights tracked in Brazil over a period of 3 years and 8 months, extracted from the open platform of the National Aviation Agency. The second[2] contains data on the occurrence of crimes in the city of São Paulo. Table 1 presents information on these datasets used.

Table 1. Dataset description.

Dataset	Description	Months
Set 1	Flights in Brazil	January 2015 to August 2017
Set 2	Crimes in São Paulo	January 2007 to December 2016

Dataset 1 provides the following information: flights, airline, departure (planned and actual), arrival (planned and actual), flight status, justification code, airport (origin and destination), city (origin and destination), state (origin and destination), country of destination and information related to latitude and longitude, totaling 21 columns and more than 2,000,000 rows. To test the proposed method, only columns with expected time data (expected departure and expected arrival) were used. These columns have data in the format datetime, containing year, month, day, hour and minutes. This set is not in the time series format, therefore, it is necessary to apply the time calculations, as well as the use of strategies for the removal and treatment of missing and discrepant data.

Dataset 2 presents data on the occurrence of crimes in the city of São Paulo from January 2007 to December 2016. The data are divided into latitude and longitude, date, nature of the crime and others. The data were subdivided into 15 groups based on the geolocation of the occurrence according to [19]. Subsequently, the data for each group were grouped by day, giving rise to 15 time series. The purpose of this grouping is to create time series according to [19], so that it is possible to compare the results.

4.2 Method

This section discusses how to work with the datasets, when our proposal is used and how to evaluate it.

Step 1 - Dataset Selection: In this stage, the data contained in the dataset are analyzed in order to detect inconsistencies. In this way, the information is adjusted and standardized, thus avoiding conversion problems during the time calculation step.

[1] https://www.kaggle.com/ramirobentes/flights-in-brazil. Accessed August 6, 2020.

[2] https://www.kaggle.com/inquisitivecrow/crime-data-in-brazil. Accessed August 6, 2020.

Step 2 - Preprocessing: Several datasets place the necessary information for time series creation, generally these data are arranged in the form of occurrences, or the date and time of a given event. Thus, this step becomes necessary for the processing and grouping of occurrences present in the dataset, giving rise to time series. In this stage, the granularity of the series is also defined, or the interval in which the occurrences will be grouped.

Step 3 - Outliers: After the previous steps, the data are analyzed in order to remove outliers from the dataset. This approach softens the trend and seasonality of the data, removing values that are entered incorrectly or that differ greatly from the time series pattern. For this approach, the IQR method was used to find the different elements of the dataset, this method is used to measure the degree of dispersion of the data and can be used to find the most disparate elements of the data set. These are found by subtracting the highest quartile (Q3) from the base by the lowest quartile (Q1). To remove the discrepant data from the set, the values of Q1 and Q3 were respectively 0.15 and 0.85.

Step 4 - Division of the Training Dataset: After the previous steps, the data are already prepared for the application of the proposed model. The first step is to define the training base. This dataset aims to enable the determination of an efficient, represented by δ, which indicates the precision of each predetermined configuration parameters of the SARIMA model in the training set predictions, or, a value that indicates which combinations of order and seasonal order achieved the best results in the elements of the training dataset.

In this stage, it is determined how many values of the time series will be part of the training dataset. This dataset must consist of the last m records belonging to the analyzed time series, whose equivalent values have already been observed. Thus, the training sample is closer to the prediction horizon, thus avoiding the use of obsolete values. The number of elements used for training the model directly impacts the final precision. Thus, it is advisable that the number of elements of the training dataset be equal to or multiples of the seasonality of the analyzed series, thus including the periodicity present in the data as a whole.

Still at this stage, the two datasets used in our tests gave rise to their respective training datasets, both time series present granularity in days and seasonal period of 7 days.

For tests and comparisons, the last 10 elements of the dataset have been removed, of these, the first 7 are used to determine the values of the δ error coefficient, and the last 3 are used for tests and model validations.

Step 5 - Application of the Method for Predictions in the Training Dataset and Determination of the Percentage of Error: After defining the training dataset according to the established rules, predictions are made to determine the values of the δ error coefficient. In this step, n predictions are made for m values, n being the number of predefined parameters of (P, D, Q) (p, d, q) and m the amount of values present in the training dataset. Thus, from the obtained and expected results it is possible to calculate the error coefficient of each combination of (P, D, Q) (p, d, q), that is, a value that indicates how close the value obtained in the prediction reached the real value.

The proposed approach eliminates the use of algorithms for parameter estimation of the SARIMA model such as those based on ACF and PACF. Unlike traditional methods, where these algorithms are used at intervals in the dataset in order to determine the best order value and seasonal order (p, d, q) (P, D, Q) for making predictions. The proposed approach performs the computation of several predefined parameters in parallel, including the possible combinations of parameters that would be obtained by executing the ACF and PACF functions. The δ error determination process happens as follows:

1. The number of elements in the training dataset, m, is defined according to Sect. 4.2, the m elements are removed regressively from the last element in the original dataset.
2. All combinations of predefined parameters are used to perform the prediction of the next value of the original set, or, first value of the training set, including the combinations possibly obtained by parameter estimation algorithms based on the ACF and PACF functions.
3. After performing the prediction of the value by all predefined combinations of order and seasonal order, the value is removed from the training dataset and inserted again in the original dataset.
4. The process is repeated until there are no more elements in the training dataset.

Thus, a matrix is generated, as illustrated in Fig. 1 containing all the order combinations used for each expected value in the training series, the values obtained in the predictions and their error coefficient. From the results, it is possible to combine the error value of each prediction in order to estimate the best parameters for the prediction of the next value in the complete series.

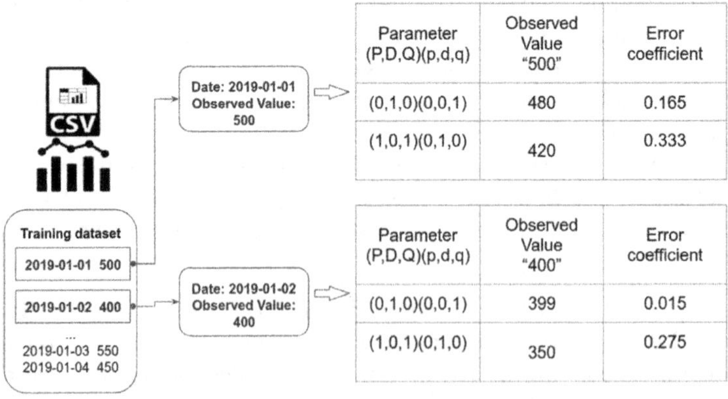

Fig. 1. Training dataset predictions.

Figure 1 illustrates the calculation in the values of a training dataset considering two predefined combinations of parameters. Where, the first expected value, 500, is compared with the result obtained by the two different parameters. Then, the process is repeated for the next value in the series, 400, and the result is stored.

The process is repeated until there are no more elements in the training time series, at the end of this step, the sum of the error values δ of each element of the training set is obtained and used to determine the best parameters to be used for the prediction of the next value. In this example, the order (0,1,0) (0,0,1) obtained the lowest error values, therefore, being a possible parameter for the prediction of the next value.

The prediction process is repeated several times, suffering a direct impact according to the number of elements in the training base and the number of predefined parameters. In a scenario where 64 combinations of order and seasonal order are predefined, the addition of an element in the training base represents another 64 executions of the SARIMA model, thus, a parallelism strategy was proposed. The approach is based on data parallelism, and the load, or the number of executions of the model, is divided equally among the cores of the system, thereby reducing the total processing time required.

In this step, the SARIMA method is used to make the prediction in the training dataset. However, unlike traditional approaches, where the configuration values of the mathematical model are defined by estimation algorithms based on the Autocorrelation functions, in the proposed approach, the parameter values are predetermined.

The values of parameters used in this step are the possible combinations of order and seasonal order parameters, where each element belonging to (P, D, Q) and (p, d, q) represents an integer value. However, it is not recommended to use values greater than 3 for $P + D + Q$ and $p + d + q$, since the predictive results achieved with these did not show significant gains.

It is worth noting that the values determined for the model configuration variables can assume any integer value, being limited to 0 and 1 in this work only to reduce the time of parameter estimation by the estimation algorithms based on the ACF and PACF for comparison purposes and also to respect $P + D + Q$ and $p + d + q$ less than or equal to 3, as proposed by [2]. The use of values greater than 1 for the variables of order and seasonal order represents only an increase in processing time, due to the need to make more predictions, thus, not adding extra complexity to the execution of the method, since the calculations are performed by the SARIMA model already determined.

All combinations of parameters are applied to the elements of the training dataset. The Eq. 5 is used to determine the processing time for this step, where n represents the total number of parameter combinations, m, the number of elements, t, the time required for making a prediction and tp the total time for calculations in the training dataset.

$$Tp = t * n * m \tag{5}$$

Thus, for the first element of the training dataset, n predictions are made with different combinations of the order and seasonal order parameters for the model. The calculations are repeated until all training base values have been calculated.

a) Predictions - In this stage, the predictions are made in the training dataset. All results of the n parameters are calculated for the first value. They are stored, the value in question is then removed from the training set and returned to the original dataset. The process is repeated for all elements of the training dataset until it is empty.
b) Determination of the Error Coefficient - The values obtained in the predictive calculations performed in the previous step are compared with the real values of the dataset. From the results of predictions and expected values, it is possible to determine an error coefficient, using the Eq. 6:

$$\delta_i = \sum_{i=1}^{q} (|y_i - x_i|) / \sum_{i=1}^{q} y_i \tag{6}$$

where δ represents the total error of each predetermined combination of parameters used in the training dataset elements. It represents a real number directly related to the success of the parameter used in the predictions of the training set values.
Table 2 illustrates the process for two distinct parameters. In this example, the training dataset consists of five elements, represented by the months from December to April, as the objective is to find the best parameters for the May prediction. The Eq. 6 is applied to the values obtained in the predictions, and the value of δ is calculated. Thus, it is possible to observe that the parameter (0,0,0) (0,1,0) obtained a delta value of 0.05, which corresponds to an assertiveness of 94.62%. This value is related to all the predictions made with this parameter in the training dataset. Therefore, it is expected that this parameter will obtain a high percentage of assertiveness in the prediction of the next month, different from the parameter (0,0,1) (1,0,1), which obtained a higher δ value with low assertiveness in the training set values. Thus, a low percentage of assertiveness is expected for the next month.
c) Selection of parameters and prediction of future values. From the results obtained in the previous step, all parameters have a value of δ for the training dataset. This value is directly linked to the success in predicting each of the tested parameters.

Table 2. Error coefficient.

	December	January	February	March	April	δ	Accuracy
Actual Value	63	59	57	66	71		
(0,0,0) (0,1,0)	65	63	59	57	66	0.05	94.62%
(0,0,1) (1,0,1)	28	33.8	34.5	31.5	31.7	0.37	62.93%

All months used in the calculation of δ already had their respective values registered in the time series. Thus, the month of May represents the month to be predicted for obtaining knowledge. The SARIMA model is then used to calculate the month value for each combination of (p, d, q) (P, D, Q) in the predefined parameters.

Table 3 illustrates the selection process in a group of parameters. The values in the δ column were calculated by the prediction and comparison of the previous four months, December to April. Table 2 illustrates two parameters used. The column *May* illustrates the results obtained by the predictions, where the real value 70 was removed, thus simulating a value to be predicted. In sequence, for better understanding, the delta value is transformed into a percentage that indicates the precision of the parameter in the previous predictions.

Table 3. Parameter selection.

Parameter	May	δ	Accuracy
...	...	...	...
(0,0,0) (1,0,0)	68.0	0.065	93.45%
(0,0,1) (0,1,0)	71.0	0.069	93.04%
(0,0,1) (1,0,0)	66.9	0.073	92.68%
(0,0,0) (1,0,1)	52.5	0.165	83.42%
(0,0,0) (1,1,0)	63.7	0.106	89.38%
(0,0,1) (0,1,1)	52.6	0.166	83.32%
(0,0,0) (1,1,1)	53.15	0.167	83.23%
(0,0,1) (0,0,1)	52.0	0.333	66.65%
(0,0,1) (0,0,0)	42.19	0.495	50.49%
...	...	...	...

The next step is to determine the precision cut. In this step, the cutoff values must be configured according to the manager's needs. For presentation, the following cutoff groups were determined: values δ greater than 90%, between 75% and 90% and values less than 75%. Table 3 illustrates the separation of parameters.

a) $> 90\%$ - For this grouping, three parameters were selected. The average percentage of δ was 93.05%. The average of the predicted values was 68.63, representing 93.05% of the expected value.
b) $< 90\%$ e $> 75\%$ - For this grouping, four parameters were selected. The average percentage of δ was 84.83%. The average of the predicted values was 55.48 representing 79.25% of the expected value.
c) $< 75\%$ - For this grouping, two parameters were selected. The average percentage of δ was 58.57% and the predictive average was 47.09, representing 67.21% of the expected value.

Therefore, when determining the prediction value from the average of the desired group, it is possible to obtain greater precision in the prediction of the desired value as illustrated in the Table 4.

Table 4. Comparison between the best parameter and the proposed approach.

Parameter	Value	Accuracy
Best parameter (0,0,0)(1,0,0)	68	97.14%
>90% Group average	68.63	97.57%

Step 6 - Proposal for Improving Predictions: At this stage, the two datasets used in the preparation of this work are ready for the proposed model application. Thus, both are in time series format, that is, two columns, the first containing the date and the second containing the occurrences. The data present in the datasets have already had their outliers removed and have undergone data transformation, attenuating the trend and making the distribution more linear. With the data prepared, the bases are subjected to the calculations of the SARIMA model using parallel computing concepts.

64 Configuration parameters were determined for the tests. This value was determined from the number of possible combinations for (p, d, q) (P, D, Q) when each variable takes the value of 0 or 1. The 64 parameters were divided equally between 8 processor cores, thus, each one will receive 8 parameters for the execution. Table 5 illustrates the range of parameters used by two of the eight processor cores.

Table 5. Data range.

Parameter	Core	Parameter	Core
(0,0,0) (0,0,0)	Core 0	(0,0,1) (0,1,0)	Core 1
(0,0,0) (0,0,1)	Core 0	(0,0,1) (0,1,1)	Core 1
(0,0,0) (0,1,1)	Core 0	(0,0,1) (1,0,0)	Core 1
(0,0,0) (1,0,0)	Core 0	(0,0,1) (1,0,1)	Core 1
(0,0,0) (1,1,0)	Core 0	(0,0,1) (1,1,0)	Core 1
(0,0,0) (1,1,1)	Core 0	(0,0,1) (1,1,1)	Core 1
(0,0,1) (0,0,0)	Core 0	(0,1,0) (0,0,0)	Core 1
(0,0,1) (0,0,1)	Core 0	(0,1,0) (0,0,1)	Core 1

The execution time of each core for its eight parameters on average is 2.56 s. It is worth noting that the eight executions of threads for applying the model occur in parallel. Thus, the total execution time of 5.77 s is measured from the beginning of the execution until the end of the execution, also taking into account the approximate time of 3.20 s required for the application of the proposed approach (Sect. 4.2, Eq. 6), and generation of data frames with partial (training) and final (prediction) results.

For the computation, the last 10 elements of the time series have been removed, 7 belonging to the test dataset and 3 as the validation values, or the data whose value is not yet registered in the dataset, or value to be predicted to obtain knowledge. In this way, 7 executions of the model were performed, one for each value of the training dataset, where in each execution the SARIMA model was calculated 64 times, with this load being divided equally among the system cores. Table 6 illustrates the results obtained by core 0 of the system.

Table 6. Data range.

Parameter	Expected value - 39
(0,0,0) (0,0,1)	Value obtained - 29
(0,0,0) (0,1,1)	Value obtained - 47
(0,0,0) (1,0,0)	Value obtained - 44
(0,0,0) (1,1,0)	Value obtained - 45
(0,0,0) (1,1,1)	Value obtained - 45
(0,0,1) (0,0,0)	Value obtained - 58
(0,0,1) (0,0,1)	Value obtained - 50

From the results obtained in this step, the values of δ for each parameter were calculated, taking into account the 7 elements belonging to the training dataset. Thus, for the predictions of the next value, or the value for which knowledge is to be obtained, the model is run again for the 64 parameters. From the results of δ, it is possible to determine the percentage of assertiveness of each parameter. Therefore, for the final result of the prediction, all parameters are selected with a percentage of precision greater than or equal to the desired cut, as exemplified in Sect. 4.2.

4.3 Parallel Processing Proposal

Our proposal also aims to reduce the processing time of the SARIMA model as part of the parameter elimination approach. For this reason, we distribute a fixed load among the cores available in the processor, i.e. Strong Scaling. Thus, as an example, all combinations of Order and Seasonal Order were used, respecting that the values of the configuration variables (p, d, q) (P, D, Q) can only assume values 0 and 1. Therefore, 64 combinations are possible, of which the combination (0,0,0) (0,0,0) was disregarded, as it does not represent a valid ARMA model. Thus, with the established parameter range, the load is divided equally for all the cores available in the system.

Figure 2 illustrates an example of the parallel processing in 8 cores. The predictive processing is performed from the computation made based on the predetermined order and seasonal order values. We execute multiple parameters in parallel, thus taking advantage of the performance gain obtained by multiple threads execution simultaneously. The list containing the parameters is divided equally among the available cores. In this way, each core receives a portion to be processed, reducing the total execution time.

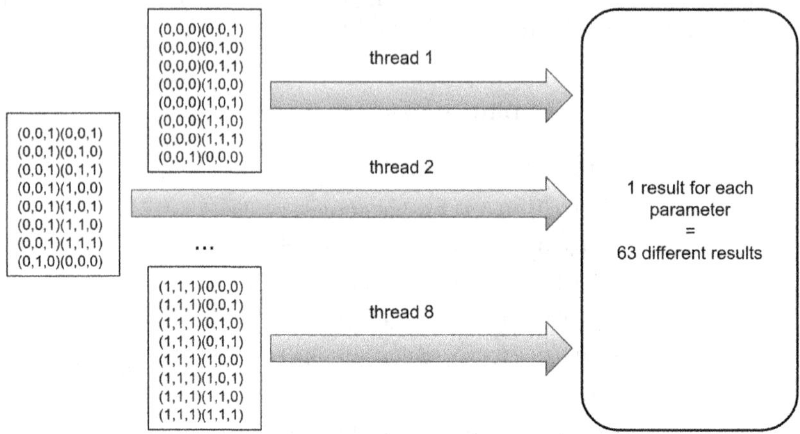

Fig. 2. Example of parallel processing.

5 Results

For calculations and comparisons regarding the performance of the approach, an average of the processing times was performed. Five model runs were performed for each of the 15 clusters proposed by [19] in relation to the Dataset 2. For the calculations of parameters based on ACF and PACF, a parallel approach was also used, where, an execution by set was made. Table 7 illustrates the values obtained, mean and standard deviation in seconds obtained from the execution of the proposed approach, and Table 8 shows the values obtained from the execution of ACF and PACF.

Table 7. Run time in seconds - (8 threads).

Group	Exc 1	Exc 2	Exc 3	Exc 4	Exc 5	Mean	std
C1	7.4269	6.3802	7.1031	6.6653	7.0087	6.9168	0.4045
C2	5.9480	6.2079	6.4606	6.8063	6.3751	6.3596	0.3172
C3	5.8912	5.7939	6.3039	6.1224	5.7509	5.9724	0.2345
C4	6.0756	5.8352	6.7272	6.2170	6.5416	6.2793	0.3578
C5	6.4145	7.2413	6.4321	6.9247	6.5408	6.7107	0.3611
C6	8.0747	7.4359	7.2580	7.6100	7.2085	7.5174	0.3494
C7	5.8024	6.1722	6.5525	5.7710	6.0984	6.0793	0.3181
C8	6.2669	6.2983	6.1723	6.5665	7.0862	6.4780	0.3701
C9	5.8886	6.6745	6.1020	6.5455	5.9297	6.2281	0.3606
C10	7.1994	8.0680	7.4295	8.0466	7.4700	7.6427	0.3924
C11	6.6726	6.1737	6.2438	6.2581	6.5657	6.3828	0.2214
C12	6.7456	6.5624	6.7685	6.7260	7.1362	6.7877	0.2110
C13	6.1633	6.2318	6.2973	6.3100	7.0274	6.4060	0.3523
C14	6.1388	6.1143	6.4104	6.2051	6.8081	6.3353	0.2888
C15	6.2623	6.3887	6.9734	6.7152	7.2660	6.7211	0.4128
Overall average	6.4647	6.5052	6.6156	6.6326	6.7209	6.5878	0.3301

Table 8. Run time in seconds - (8 threads).

Group	Exc 1
C1	1195.8847
C2	1215.3856
C3	1258.1969
C4	1210.9434
C5	1285.2525
C6	1247.6898
C7	1249.4468
C8	1233.5547
C9	1272.5777
C10	1199.5446
C11	1247.2221
C12	1218.2248
C13	1211.1544
C14	1262.4544
C15	1205.1055
Mean	1233.4729
Std	28.0512

Table 9. Dataset 1 - predictions and δ.

Observed value =	285	323	311	309	301	319	313	δ	δ(%)
(0,0,1) (0,0,0)	232	272	283	257	279	248	297	0.1416	85.8%
(0,0,0) (0,0,1)	260	278	277	273	271	272	266	0.1257	87.4%
(0,0,1) (0,0,1)	261	292	296	283	285	283	295	0.0784	92.2%
(0,1,0) (0,0,0)	314	285	323	311	309	301	319	0.0614	93.9%
(1,0,0) (0,0,0)	312	284	321	309	307	299	317	0.0585	94.1%
(1,1,0) (0,0,0)	316	294	312	315	310	304	314	0.0516	94.8%
(1,1,1) (0,0,0)	309	307	308	309	309	308	309	0.0361	96.4%
(1,0,0) (0,0,1)	298	300	317	310	304	307	315	0.0336	96.6%
(1,1,0) (0,0,1)	302	304	312	313	306	309	313	0.0314	96.8%
(1,0,1) (0,0,1)	300	311	308	308	305	311	310	0.0246	97.5%
(1,1,0) (1,1,1)	285	320	313	306	304	323	312	0.0087	99.1%
(1,1,0) (1,0,0)	287	317	311	310	303	319	314	0.0047	99.5%

5.1 Proposed Approach

After the necessary steps to adapt the datasets, they were submitted to our proposed approach. Thus, 1280 predictions were divided into Datasets 1 and 2. The results obtained were compared with the results of conventional approaches for setting parameters of the SARIMA model. All the predictions made in the training base for Dataset 1 were used.

Table 9 illustrates 12 of 64 predictive results achieved in the 7 intervals belonging to training set of Dataset 1. This table shows that the parameter (0,0,1) (0,0,0) has the highest value of δ. It represents the value with the lowest chances of obtaining a satisfactory predictive result for the prediction of the next time interval. In the same way, the parameter (1,1,0) (1,0,0) obtained the lowest value of δ. Consequently, the chances of this parameter obtaining results close to the actual observation in the prediction of the next interval are greater.

Although the value of δ presents the parameters with the best chance of achieving predictive results closer to the real value, for better precision in the process, multiple parameters are used. Thus, an average is performed in order to obtain the prediction value.

Table 11 illustrates the result obtained from the aggregate of parameters according to the cut percentage used as described in Sect. 4.2. The expected value for this prediction is 313. For the results obtained in Dataset 1, three precision intervals were determined, values with a percentage of δ, less than 95%, between 95% and 99% and values greater than 99%. All predictions made in the training set for Dataset 1 were used.

Besides, according to Table 11, the parameter with the lowest value of δ shows the results closest to the actual values. Table 10 illustrates the values and their respective precision percentages.

Thus, it is possible to attest that our proposed approach eliminates the need of the Auto-correlation Functions for the estimation of parameters. The results achieved by the calculations are already validated by the Eq. 6, which takes into account past results in order to determine the best combination of order and seasonal order for the SARIMA model.

Table 10. Dataset 1 - predictions and δ.

Expected =	285	323	311	309	301	319	313
(1,1,0) (1,0,0)	287	317	311	310	303	319	314
Accuracy	99.3%	98.14%	100%	99.67%	99.33%	100%	99.68%

5.2 Comparisons Between Approaches

In order to compare the proposed approach accuracy, tests were performed using Dataset 1 and the results obtained compared. Table 12 illustrates the comparison between the result obtained from the use of an algorithm for estimating parameters based on ACF and PACF, and the result obtained from our proposed approach. The test was performed on Dataset 1 for two different dates. The result of the proposed approach was obtained from the average of the parameters that obtained lower values of δ for Dataset 1.

From the results it is possible to notice a gain of 5.94% and 10.77% in accuracy of the proposed approach in comparison to the results achieved by the algorithms based on ACF and PACF in relation to the expected value.

Another way to check the best parameters is to use the Box-Jenkins [2] method, composed of three steps: model identification, parameter estimation and diagnostics. The selection of the possible best parameters is made using

Table 11. Dataset 1 - prediction of the next value.

Cut	Parameter qty	Result	Accuracy
Less than 95%	6	299	95.5%
In-between 95–99%	9	309	98.7%
Greater than 99%	49	311	99.3%

Table 12. Comparative - ACF/PACF and proposed approach Dataset 1 - accuracy.

Date	Actual value	(P)ACF	(P)ACF (%)	Proposal	Proposal (%)	Gain
2017/07/30	320	301	94.06%	320	100%	5.94%
2017/07/31	311	256	88.88%	310	99.65%	10.77%

ACF and PACF. However, the values found do not necessarily present the best results for the dataset, so it is important to perform tests on the model and adapt it [15]. In view of the need to test several parameters in order to find the ideal and the uncertainty that it awakens in the results found, the proposed model has advantages over these points.

5.3 Prediction of Crime Rates in the City of São Paulo

A case study was carried out between the results achieved by a hybrid approach using the conventional SARIMA model, proposed in [19], and the proposed approach to improve the SARIMA model presented in this work.

The tests using the proposed approach were performed on the same dataset, crimes that occurred in the city of São Paulo (Dataset 2), for the same period of time, December 2016, using the same subdivision proposed by the authors, 15 groups defined by the geolocation of the occurrences. Table 13 [19] illustrates the average precision in each of the 15 groups, as well as the average RMSD achieved by using three different approaches, the conventional SARIMA model, an Artificial Neural Network (ANN), and a hybrid model (SARIMA + ANN). The table also shows the predictive average and RMSD of all groups.

Table 13. Predictive results - December 2016 [19].

Group	δ _SARIMA	RMSD _SARIMA	δ _ANN	RMSD _ANN	δ _SARIMA+ANN	RMSD _SARIMA+ANN
C1	70.18%	1.96	75.71%	1.58	75.31%	1.67
C2	76.34%	2.79	76.67%	2.43	81.75%	2.47
C3	76.16%	2.06	67.74%	2.92	80.10%	1.98
C4	68.81%	1.69	65.07%	1.94	73.27%	1.61
C5	69.43%	2.78	61.54%	3.36	70.75%	2.65
C6	74.27%	3.45	70.80%	3.70	77.78%	3.33
C7	77.04%	1.35	70.25%	1.73	79.50%	1.37
C8	71.87%	1.75	66.32%	2.17	73.67%	1.81
C9	68.88%	3.46	67.66%	3.62	74.02%	3.18
C10	71.08%	2.36	63.83%	3.10	74.13%	2.30
C11	65.10%	2.25	64.15%	2.37	69.49%	2.13
C12	70.53%	1.26	62.78%	1.49	73.90%	1.25
C13	80.96%	1.59	79.24%	1.73	85.41%	1.49
C14	76.53%	1.29	74.64%	1.46	79.45%	1.30
C15	77.39%	4.02	74.70%	4.24	81.62%	3.83
Mean	72.97%	2.27	69.41%	2.52	76.68%	2.16

Table 14. Predictive results - December 2016 - proposed approach.

Group	δ _SARIMA	RMSD _SARIMA	δ _ANN	RMSD _ANN	δ _SARIMA+ANN	RMSD _SARIMA+ANN
C1	70.62 %	1.93	75.71%	1.58	76.00 %	1.63
C2	76.70 %	2.75	76.67%	2.43	82.18 %	2.44
C3	78.05 %	1.92	67.74%	2.92	81.84 %	1.85
C4	67.39%	1.70	65.07%	1.94	72.20%	1.60
C5	68.77%	2.84	61.54%	3.36	70.01%	2.71
C6	74.07%	3.45	70.80%	3.70	77.60%	3.32
C7	77.28 %	1.34	70.25%	1.73	80.28 %	1.34
C8	73.82 %	1.63	66.32%	2.17	75.74 %	1.68
C9	69.15 %	3.44	67.66%	3.62	74.37 %	3.16
C10	72.25 %	2.21	63.83%	3.10	75.23 %	2.2
C11	65.23 %	2.14	64.15%	2.37	69.87 %	2.07
C12	72.34 %	1.12	62.78%	1.49	75.74 %	1.14
C13	81.67 %	1.55	79.24%	1.73	85.88 %	1.48
C14	76.77 %	1.27	74.64%	1.46	79.65 %	1.27
C15	78.39 %	3.32	74.70%	4.24	82.78%	3.07
Mean	73.52%	2.18	69.41%	2.52	77.25%	2.06

Table 14 illustrates the results obtained from the SARIMA execution according our proposed approach. The Artificial Neural Network predictive results presented in Table 13 was maintained.

Table 15 illustrates the accuracy gain obtained from the proposed SARIMA approach. The results indicated improvement in 12 of the 15 groups analyzed, with the greatest gains in the C8, C3 and C12 groups, with an accuracy gain of 1.94%, 1.89% and 1.81%, respectively.

An evaluation of the execution time was also carried out in order to verify performance gains, using the C1 cluster as a basis. The tests were performed on the same system with the same number of cores applied to the solution. For conventional model parameters estimation, an estimation algorithm based on ACF and PACF functions were performed in parallel, with a processing time of 1184.40 s. In contrast, the calculation time for the proposed approach was 5.616 s, showing a speedup of 210.897 times.

5.4 Proposed Approach Performance

According to the objectives proposed in this work, the reduction of processing time is essential to adapt the proposed approach to trends in data analysis and processing. Thus, this section aims to measure and compare the processing time of the proposed approach and the other methods for estimating parameters.

Table 15. Comparisons between the proposed approach and conventional SARIMA.

Group	Proposed approach	SARIMA	**Accuracy gain**
C1	70.62%	70.18%	0.44%
C2	76.70%	76.34%	0.36%
C3	78.08%	76.16%	1.89%
C4	67.39%	68.81%	−1.42%
C5	68.77%	69.43%	−0.66%
C6	74.07%	74.27%	−0.19%
C7	77.28%	77.04%	0.24 %
C8	73.82%	71.87%	1.94%
C9	69.15%	68.88%	0.28%
C10	72.25%	71.08%	1.51%
C11	65.23%	65.10%	0.14%
C12	72.34%	70.53%	1.81%
C13	81.67%	80.96%	0.71%
C14	76.77%	76.53%	0.24%
C15	78.39%	77.39%	1.01%
Average	73.53%	72.97%	0.55%

The Eq. 7 shows the theoretical processing time for our proposed approach:

$$t * g * q + r * t + c \tag{7}$$

where t represents the execution time of a single parameter, g the total number of parameters defined in the interval, q the number of delayed views to be considered, r the number of parameters with results greater than or equal cut value. The constant c represents the approximate time to perform all the necessary operations after the predictive calculations, such as, data reading and writing.

The values of q and g represent a major impact on the model's total execution time, since their values multiply the total value of the parameter range. In the tests performed, 64 parameters were used in the interval g, 7 past observations q and the execution time for each one was 2.56 s (t), as shown in Sect. 4.2.

Table 16 shows the execution time of the prediction method proposed in this work applied to Dataset 2. The execution time for an interval of 64 predefined parameters in the SARIMA model was 38.13 s, while the parallel version in 2 cores was 22.23 s, 4 cores 14.01 s, 8 cores 9.51 s, 10 cores 7.89 s and 12 cores 6.84 s. Thus, it is possible to notice a reduction in the total execution time for each core used.

The tests were performed on Dataset 1 in the same system, with a total of 12 threads applied in both approaches. The ACF/PACF calculations obtained a processing time of 1595.41 s. This value was added to the normal SARIMA

Table 16. Time, gain and efficiency of model parallelization.

Threads	Time (s)	Gain	Efficiency
1	38.13	-	-
2	22.23	1.71	85.76%
4	14.01	2.72	68.04%
8	9.51	4.00	50.12%
10	7.89	4.83	48.33%
12	6.84	5.57	46.45%

execution time, totaling 1596.77 s, while the proposed parallel approach was 6.84 s, achieving a speedup of 233.4 times.

Table 16 illustrates the time and performance gain in relation to the number of threads being executed. The time decreases as the number of threads increases. Thus, execution on 12 threads obtained less processing time. The speedup increases according to the number of threads used. Thus, running on 12 threads achieved the highest performance.

6 Conclusion

The use of the SARIMA model for prediction has the ability to present important data for the knowledge discovery and decision making. The definition of order parameters and seasonal order for the model are essential to achieve satisfactory results in predictions.

Conventional methods for the definition of parameters bring uncertainty to the values selected by the model, requiring validation and adjustments. The use of our proposed approach offers the greatest assertiveness in the predictions. From the selection and aggregation of the parameters that most closely approximated the expected values in past predictions, the model is able to calculate the error coefficient of all possible parameter arrangements in a parallel fashion. This way, there is a reduction on the total execution time by indicating the possible configurations with the greatest chance of obtaining good results in the next prediction.

From the selected values, the predictive results of the model have an accuracy of more than 99% in some cases. The presented method is capable of calculating predictive results with a higher accuracy percentage when compared to the traditionally SARIMA model, reaching up to 10.77% more precision in certain cases and eliminating the need for parameter validation. In addition, our proposed approach is capable of performing the necessary calculations with a reduction of up to 90% in the total execution time in relation to the traditional methods of estimating parameters, such algorithms based on ACF and PACF.

Moreover, it was possible to identify the best performance of the proposed approach compared to the traditional use of the SARIMA model, extending the results presented in our previous paper [19] up to 1.91% in accuracy and 229x in execution performance.

In future works, we suggest to refactor and adapt the proposed codes in order to increase the performance of the application. Also, we suggest the execution of the proposal with a greater number of cores and the exploration of other parallelism strategies, such as tasks, programming models by shared variable and message passing, as well as distributed memory architectures, and the use of graphics processing units. It is also suggested to develop a method for defining the best number of parameters to be selected within the cut percentage, in order to increase the accuracy of the model.

Acknowledgements. The present work was carried out with the support of the Coordenação de Aperfeiçoamento de Pessoal de Nível Superior - Brazil (CAPES) - Financing Code 001. The authors thank CNPq, FAPEMIG, PUC Minas and REVEX for the partial support in the execution of this work.

References

1. Box, G.E., Jenkins, G.M.: Time Series Analysis: Forecasting and Control Holden-day, p. 498, San Francisco (1970)
2. Box, G.E., Jenkins, G.M., Reinsel, G.C., Ljung, G.M.: Time Series Analysis: Forecasting and Control. Wiley, Hoboken (2015)
3. Brockwell, P.J., Davis, R.A., Fienberg, S.E.: Time Series: Theory and methods. Springer, New York (1991). https://doi.org/10.1007/978-1-4419-0320-4
4. Castellanos, M.: Semantic enrichment of interoperable databases. In: Proceedings RIDE-IMS 1993: Third International Workshop on Research Issues in Data Engineering: Interoperability in Multidatabase Systems, pp. 126–129, April 1993. https://doi.org/10.1109/RIDE.1993.281934
5. Ding, F., Meng, D., Dai, J., Li, Q., Alsaedi, A., Hayat, T.: Least squares based iterative parameter estimation algorithm for stochastic dynamical systems with ARMA noise using the model equivalence. Int. J. Control Autom. Syst. **16**(2), 630–639 (2018)
6. Ding, F., Meng, D., Wang, Q.: The model equivalence based parameter estimation methods for Box-Jenkins systems. J. Franklin Inst. **352**(12), 5473–5485 (2015)
7. He, T., Zheng, S.: Time series analysis and forcast based on active learning artificial neural network. In: 2009 Second International Symposium on Knowledge Acquisition and Modeling, vol. 1, pp. 84–87, November 2009. https://doi.org/10.1109/KAM.2009.303
8. Holley, K., Sivakumar, G., Kannan, K.: Enrichment patterns for big data. In: 2014 IEEE International Congress on Big Data, pp. 796–799, June 2014. https://doi.org/10.1109/BigData.Congress.2014.127
9. HongJu, X., Fei, W., FenMei, W., XiuZhen, W.: Some key problems of data management in army data engineering based on big data. In: 2017 IEEE 2nd International Conference on Big Data Analysis (ICBDA), pp. 149–152, March 2017. https://doi.org/10.1109/ICBDA.2017.8078796
10. Hyndman, R.: Better ACF and PACF plots, but no optimal linear prediction. Electron. J. Stat. [E] **8**(2), 2296–2300 (2014)
11. McMurry, T.L., Politis, D.N.: Banded and tapered estimates for autocovariance matrices and the linear process bootstrap. J. Time Ser. Anal. **31**(6), 471–482 (2010)

12. Meerschaert, M.M.: Mathematical Modeling. Elsevier, Amsterdam (2007)
13. Meng, D., Ding, F.: Model equivalence-based identification algorithm for equation-error systems with colored noise. Algorithms **8**(2), 280–291 (2015)
14. Mohammadi, K., Eslami, H., Kahawita, R.: Parameter estimation of an ARMA model for river flow forecasting using goal programming. J. Hydrol. **331**(1–2), 293–299 (2006)
15. Olsson, M., Soder, L.: Modeling real-time balancing power market prices using combined SARIMA and Markov processes. IEEE Trans. Power Syst. **23**(2), 443–450 (2008). https://doi.org/10.1109/TPWRS.2008.920046
16. Sakia, R.M.: The box-cox transformation technique: a review. J. Roy. Stat. Soc. Ser. D (Stat.) **41**(2), 169–178 (1992)
17. Sheta, A.F., Faris, H., Aljarah, I.: Estimating ARMA model parameters of an industrial process using meta-heuristic search algorithms. Int. J. Eng. Technol. **7**(3.10), 187–194 (2018)
18. Shumway, R.H., Stoffer, D.S.: Time Series Analysis and Its Applications: With R Examples. Springer, Heidelberg (2017). https://doi.org/10.1007/978-3-319-52452-8
19. Soares., F.A.L., Silveira., T.B., Freitas., H.C.: Hybrid approach based on SARIMA and artificial neural networks for knowledge discovery applied to crime rates prediction. In: Proceedings of the 22nd International Conference on Enterprise Information Systems: ICEIS, vol. 1, pp. 407–415. INSTICC, SciTePress (2020). https://doi.org/10.5220/0009412704070415
20. Susanti, S.P., Azizah, F.N.: Imputation of missing value using dynamic Bayesian network for multivariate time series data. In: 2017 International Conference on Data and Software Engineering (ICoDSE), pp. 1–5, November 2017. https://doi.org/10.1109/ICODSE.2017.8285864
21. Willmott, C.J.: Some comments on the evaluation of model performance. Bull. Am. Meteor. Soc. **63**(11), 1309–1313 (1982)
22. Wold, H.: A study in the analysis of stationary time series. Ph.D. thesis, Almqvist & Wiksell (1938)
23. Yule, G.U.: Why do we sometimes get nonsense-correlations between time-series? - A study in sampling and the nature of time-series. J. Roy. Stat. Soc. **89**(1), 1–63 (1926)

An Approach to Intelligent Control Public Transportation System Using a Multi-agent System

Nabil Morri[1,3(✉)], Sameh Hadouaj[2,3], and Lamjed Ben Said[3]

[1] Emirates College of Technology, Baniyas Tower B, Abu Dhabi, United Arab Emirates
[2] Higher Colleges of Technology, PO Box 25035, Abu Dhabi, United Arab Emirates
[3] ISG, Université de Tunis, 41 Avenue de la Liberté, 2000 Bouchoucha, Bardo, Tunisia
lamjed.bensaid@isg.rnu.tn

Abstract. Traffic congestion has increased globally during the last decade representing an undoubted menace to the quality of urban life. A significant contribution can be made by the public transport system in reducing the problem intensity if it provides high-quality service. However, public transportation systems are highly complex because of the modes involved, the multitude of origins and destinations, and the amount and variety of traffic. They have to cope with dynamic environments where many complex and random phenomena appear and disturb the traffic network. To ensure good service quality, a control system should be used in order to maintain the public transport scheduled timetable. The quality service should be measured in terms of public transport key performance indicators (KPIs) for the wider urban transport system and issues. In fact, in the absence of a set of widely accepted performance measures and transferable methodologies, it is very difficult for public transport to objectively assess the effects of specific regulation system and to make use of lessons learned from other public transport systems. Moreover, vehicle traffic control tasks are distributed geographically and functionally, and disturbances might influence on many itineraries and occur simultaneously. Unfortunately, most existing traffic control systems consider only a part of the performance criteria and propose a solution without man-aging its influence on neighboring areas of the network. This paper sets the context of performance measurement in the field of public traffic management and presents the regulation support system of public transportation (RSSPT). The aim of this regulation support system is (i) to detect the traffic perturbation by distinguishing a critical performance variation of the current traffic, (ii) and to find the regulation action by optimizing the performance of the quality service of the public transportation. We adopt a multi-agent approach to model the system, as their distributed nature, allows managing several disturbances concurrently. The validation of our model is based on the data of an entire journey of the New York City transport system in which two perturbation scenarios occur. This net-work has the nation's largest bus fleet and more subway and commuter rail cars than all other U.S. transit systems combined. The obtained results show the efficiency of our system especially in case many performance indicators are needed to regulate a disturbance situation. It demonstrates the advantage as well of the multiagent approach and shows how the agents of different neighboring zones on which the disturbance has an impact, coordinate and adapt their plans and solve the issue.

J. Filipe et al. (Eds.): ICEIS 2020, LNBIP 417, pp. 242–267, 2021.
https://doi.org/10.1007/978-3-030-75418-1_12

Keywords: Multi-agent systems · Public transportation · Regulation system · Optimization · Key performance indicators

1 Introduction

Traffic congestion is a highly complex problem that has been increasing in much of the world, developed or not, representing an undoubted menace to the quality of urban life. Urban public transport is an essential transportation mode that plays a major role in overcoming this issue especially if the public transport system provides effective service. Nowadays, as the demands for public transport in urban areas are growing, many technological efforts are being made to improve the services offered by urban transport networks. Indeed, Public transportation systems are complex and dynamic networks that are constantly exposed to disturbances. This complexity is due to the difficulty of respecting the scheduled timetable of vehicle passage and the emergence of random phenomena that disturbs the network traffic of vehicle. Moreover, vehicle traffic control tasks are distributed geographically and functionally, and many disturbances might occur simultaneously and influence several vehicles on different itineraries. Thus, to improve the quality service of public transport, we have to design and build a regulation support system that detects disturbances and regulates the traffic of public transport, it should also be able to handle the dynamic of the traffic by managing the impact of the chosen regulation on the neighboring areas. The quality service should be measured in terms of KPIs.

Against this perspective, modeling and simulating such systems show real problems because the current methods and simulation tools do not combine various KPI measures into a single performance value, potentially covering multiple dimensions or goal categories and consequently propose regulation in a very limited context that essentially applies to very specific criteria with precise constraints. Some of them propose a centralized resolution as well in a single module and therefore only allow managing one perturbation at a time on a restricted itinerary. Moreover, there is no clear knowledge of the notion of KPIs and no common framework standard of quality service in public transport.

This paper describes particularly the public traffic control challenges, introduces in detail the existing public traffic control systems, and classify them into mathematical models and multi agent models, and then it discusses their limits. Extending [1], this paper sets the context of performance measurement in the field of public traffic management and presents the RSSPT. The aim of this regulation support system is to detect the traffic perturbation by verifying the adequacy between the planned and the current performance measures and find the most appropriate regulation action by optimizing the performance of the quality service of the public transportation. We adopt a multi-agent approach to model the system. Unlike the experimentations presented in [1] that were based on a simple situation, where only operational efficiency KPIs (punctuality, regularity, and correspondence) were relevant. This paper, tackle more complex case study where the frequency of trips is higher, and disturbance might influence on many itineraries simultaneously. Furthermore, beside the operational efficiency KPIs, we take into consideration mobility and reliability KPIs.

This paper is organized as follows. Section 2 describes the public transport control challenges, Sect. 3 introduces the existing control systems and discusses their limits. Section 4 discusses Key performance indicators for traffic management and their measures. Section 5 defines the optimization problem. Section 6 describes the regulation process and the Multi-Agents System design. Section 7 validates our model by providing experimentation and result of real scenarios occurred in New York City transport system. In Sect. 8, we conclude and discuss some future works.

2 Public Road Traffic Regulation: Issues and Challenges

From an economic, environmental, and political perspective, public transport has many advantages. It diminishes fuel costs for private cars both financially and environmentally, reduces congestion on roads and in parking areas, and is as well safer than private transport. Public transport has a wide variety of resources. It has different transport modes and infrastructure including multi-ride lines, stations, roads, etc. These resources have to be well exploited to ensure a good quality of service, especially to manage the perturbations that may occur during the day. With the evolution of today's transport systems and the advanced technologies they use, there is a major transformation in the systems of transport regulation. Transportation managers are using intelligent systems such as automatic vehicle location systems (e.g. [2, 3]) and new communication technologies to better manage perturbations. In the field of public transport, a perturbation is an event that suddenly appears and changes the traffic state of the network into an unsatisfactory situation in terms of quality of service. The complexity of the road network may result in several perturbations that can occur at the same time. Sometimes, one perturbation can generate others. In addition, the data from public transport systems are very dynamic and voluminous, hence the great need for intelligent systems to make better use of transport system resources and provide the best performance. The transport system must adapt to changing traffic conditions to ensure the required quality of service. Consequently, public transport control systems must quickly detect disturbances and adapt to new situations with the aim of improving the quality of service through performance measurements. These measures represent Key Performance Indicators (KPIs).

Performance measures represent a quantitative measure or index that numerically expresses a specific quality. In line with the European Commission's strategy for the future of transport presented in the 2001 and 2011 White Papers [4], four themes for strategic measures in urban traffic management are addressed: traffic efficiency; road safety; pollution reduction; and social inclusion.

In this study context, reference is made to KPIs, that provide sufficient understanding of the quality of public transport service from a passenger perspective and relative comparisons in perturbation management. These KPIs concern only to the traffic efficiency theme of transit traffic management. This theme includes indicators related to the objectives: mobility, reliability, and operational efficiency. The mobility KPI mainly considers travel times. The Reliability KPI, expresses the period of traffic congestion. The operational efficiency KPI focuses on three indicators, which are (i) scheduled departure time in stations for punctuality, (ii) time intervals between vehicles on the same route for regularity, and (iii) the time taken to put passengers in the transfer station for correspondence. The objective "system condition" does not fall within the scope of our study because they are related to the infrastructure and the traffic demand related to regulations for a long-term strategic action.

Consequently, the task of transport control consists mainly in optimizing the performance indicators. The control system must provide comparative information that allows to identify performance gaps and set targets and measures to fill them. In the event of a perturbation, the control system needs to know what quality of service is expected, and then proceeds to optimize KPIs and regulate transport traffic towards these targets. Therefore, a good control system needs to take into account the KPIs for public transport traffic management in order to detect and identify optimal control action. It must also be able to manage several perturbations at the same time. An efficient optimization method and accurate knowledge of the network condition improves public transport traffic management in case of perturbations.

3 Related Works

We distinguish two classes of control system modelling: mathematical and multi-agent. In mathematical modelling, the model is a set of equations, and the execution consists in evaluating them. Whereas, in multi-agent modelling, the model consists of a set of agents that encapsulate the behaviors of the different participants in the system, and the execution consists in emulating these behaviors.

3.1 Mathematical Models

Newell and Potts [5] first studied the management of bus disturbances. Using a simplified model based on mathematical equations on route rescheduling summarizing the general state of traffic, the authors proved the instability of a line by bringing buses together. They developed self-control algorithms to monitor the progress of consecutive buses and control the regularity of the vehicles. After stating the problem of bus reconciliation, many research works have been devoted to corrective strategies to reduce bus reconciliation. They are generally based on mathematical models using schedule-based control. They try to reduce the number of buses by adjusting schedules based on an analysis of route irregularities at the terminus and at pairs of consecutive stops [6–9], and its impact on traffic based on a timetable dealing with punctuality. They developed a framework to translate the massive data from AVL (Automatic Vehicle Location) and APC (automatic passenger counter) into valuable measurements. They used sequence-mining algorithms, called PrefixSpan, to find sequences of bus stops where systematic deviation events of regularities cause the creation of bus closures. The objective of these types of algorithms is to find interesting events available in time series.

[7–12] have studied the main causes of the disturbance and have recommended to switch from static regulation based on TMT to dynamic regulation based on the frequency of buses on the same line for a more efficient and fast regularity punctuality. They propose stochastic models describing the evolution of the probability distribution of delays when a bus follows a route with respect to the travel time between successive stops. The disadvantage of these studies is that they always ignore other KPIs such as correspondence and total travel time (mobility) and traffic fluidity (reliability). An important problem with this approach is to find a trade-off between the arrival of buses at regular intervals at stations and the optimal frequency of service. In the same context (solving the problem of bus grouping) [13] proposes a holding strategy based on cooperative control of two buses. He compares the performance of three different scenarios: before the control strategy, under the strategy for a single bus line and under the coordinated strategy for several bus lines. The results show that the cooperative strategy performs better on the regularity and punctuality criteria.

[14, 15] develop mathematical models based on dynamic exploitation of boarding times and limitations of stopping times (Dwell) dealing only with punctuality. They propose solutions that use self-balancing techniques for the natural progression of the system. It is an optimization using linear scheduling of timetable for reliability to punctuality criteria.

Other works, such as that of [16], propose a technique based on "Support Vector Machine" (SVM) and ant colony algorithms using a mathematical tool derived from static learning theory. The control action acts only on the observed delays of the vehicles without considering other objectives such as correspondence and regularity. The proposed solution is a reconfiguration of new schedules or races according to the new traffic conditions.

In other works, the authors take adequate regulation measures for each incident. For example, in [17], he treats the problem of on-line regulation as an optimization problem and provides the regulator with effective decisions evaluated and ranked by taking into account the criteria of regularity, correspondence and punctuality. The proposed module is based on a hybrid approach using a fuzzy evaluation method and evolutionary algorithms.

A set of works such as [18–21] in which the control is done between stations on speed and bus overloads in the form of numerical analyses in mathematical models. These techniques are called non-linear control algorithms. They ensure only punctuality and regularity with a gain in resolution time.

However, the author of [22] describes how the dynamic evolution of road traffic can be predicted at the mesoscopic level by Triangular Petri Lots Networks (Triangular RoLoTs). In this work, the author deals only with the mobility of public transport traffic. Tan disque, [23] with mathematical equations only treats the problem of regularity by rearranging crew schedules in his work on delay management in public transport. The authors describe a taboo search procedure for online vehicle scheduling optimizing service regularity and column generation approach for crew rescheduling.

The author of [24] finds an interesting solution to minimize passenger waiting time at processing stations while trying to maintain efficient regularity. He presents new mathematical formulations of mixed-integration nonlinear programming for the problem of arrival schedules using adaptive particle swarm algorithms.

In the same class, [25] calculates the performance based on the progress made, with the difference between the achieved regularity and the expected regularity at each station, which is then weighted by the passenger boarding capacity. However, there is no direct link between the calculated values and their impact on the modification of vehicle kinematic values (e.g. position, speed, acceleration). In this work, the author establishes the technique of linear mathematical models characterizing the arrival of vehicles at stations with its stopping time to control the regularity and punctuality of the runs.

3.2 Multi-agent Models

The first multi-agent approaches to regulation such as [26] propose a representation on the "property-based coordination principle" (PbC). Their objective is to solve three recurrent problems in the design of solutions related to knowledge, the space-time dimension, and the dynamics of the real environment. The tests show the importance of multi-agent representation. However, it only covers the criterion of punctuality. Other approaches propose to solve only the correspondence problem using a limited number of actions. As in [27], the author presents the process of control and anticipation by determining the roles of the agents.

Other works such as that of [28] which presents a regulation model detailing the cognitive activities of the agents in the process based on the "Basic and Combined Actions" (BCA) technique. The decision is integrated in an interactive environment, but it is based only on reliability and punctuality. Other authors [29, 30] use the multi-agent approach and treat a technique based respectively on swarm intelligence and a predictive control approach by MPC model (Model Predectif Control). These techniques are very fast and pragmatic but provide only acceptable and non-optimal actions. It is difficult to respect all KPIs with their exact weightings in the domain of the space-time problem. They develop artificial systems to solve complex traffic problems by dealing only with the two criteria punctuality and regularity.

[31] use multi-agent simulation. It is based on the static exploitation of schedules and minimum and maximum waiting time at stations. He discusses two methods that attempt to achieve equal progress, called minimum and maximum, by dealing only with regularity without considering other performance criteria.

Some other approaches use intelligent agents as a technique to model traffic signal priority mechanisms for public transit. One example is the work of [32, 33] which summarizes the use of various evolutionary techniques for traffic management and congestion avoidance in intelligent transportation. It proposes a real-time centralized evolutionary optimization technique, based on several intelligent agents, for urban traffic management in the field of traffic signal control. This scheme uses an evolutionary strategy for traffic signal control. In the same framework, [34–36] implement agents at different levels of abstraction (macroscopic and microscopic) using the linear quadratic equation to give priority to public transport. These types of regulations are only interested in the fluidity (reliability) of traffic without specifying which KPIs to consider.

The work of [37] is based on the design of a regulation system for multimodal transport networks. It provides effective real-time solutions to the problem of road congestion to solve only the mobility objective. The proposed system is a hybrid approach between network graph modelling and a multi-agent system and considers only the mobility criterion while neglecting the other performance criteria.

In the same context, we cite the work of [38]; its objective is to regulate a disturbed multimodal transport network producing the most efficient decision through heterogeneous agents. These agents are able to detect perturbations, analyze, propose and evaluate solutions using the integral Choquet. But this work only deals with the two criteria punctuality and regularity.

3.3 Discussion

Based on the above literature review, we can conclude that the mathematical models have a centralized resolution in a single module and therefore only allow to manage one perturbation at a time and without considering its impact on the neighbor areas. In contrast, multi-agent models, as a result of their distributed nature, have the advantage of managing simultaneously several perturbations. Indeed, vehicle traffic control tasks are distributed geographically and functionally. The public transport system contains elements dispersed throughout the network (stations, vehicles, sensors, etc.). Vehicles are moving as well in separate geographic areas controlled and regulated by autonomous regulators. Thus, the different system components used in road traffic control can become more autonomous and therefore more adaptive if they are modelled as agents. Data for each zone can be captured locally by regional agents that monitor the performance of the zone's network. If a detected disturbance has consequences on neighboring zones, the corresponding agents could be informed so that they can adapt and coordinate their plan accordingly.

However, most of the existing mathematical or multi-agent models do not take into account all the KPIs and consequently propose regulation in a very limited context that essentially applies to very specific criteria with precise constraints. To overcome this gap, designing a multi-agent control support system that manages several perturbations at the same time and produces an optimal KPI-based control action for public transport is a promising solution.

4 Key Performance Indicators for Traffic Management (KPI)

4.1 Performance Measurement: Literature Review

Several research works provide quantitative measures of performance in the area of transport service. [39] evaluates the performance by the excess waiting time (EWT: Excess wait time). This indicator is defined as the difference between the actual waiting time (AWT) and the scheduled waiting time (SWT). Moreover, in [40, 41], this performance is defined by the average waiting time expected by passengers. This indicator calculates the perceived regularity that measures the average additional waiting time of passengers. In fact, the low EWT means that the performance of the service is fairly regular.

[42] defines the performance by the deviations of the time intervals observed between the trips of the same line with respect to the regular frequency of the vehicles during a given period. This indicator is calculated as a standard deviation between the observed frequency and the programmed frequency. In addition, in other specific projects, to give more meaning to the evolution of the performance during abrupt changes in the transport traffic state, [40–42] provide another complementary definition for performance. This performance is defined as a percentage of deviations that no longer deviate from a quantity in absolute minutes. It represents the coefficient of variation.

[43, 44] describe the Gini index as another indicator by regularity index. Economists and sociologists use the Gini ratio to measure the degree of income inequality within groups of people. By analogy in the field of public transport, the authors measure by this ration the degree of inequality of performance within a group of trips of the same line to quickly detect the abnormal phenomena that disturb the traffic. [45] describes regularity as an index based on vehicle entries at stations. This indicator is specific to a line. Its formula is expressed as a percentage of unpunctual vehicle entries in relation to the total number of planned entries at the stations.

Other projects define the punctuality as another indicator that determines the performance. [46] defines the punctuality as a comparison of the actual departure times and scheduled departure times at the station. In [47] the authors distinguish three types of punctuality measures: the Punctuality Index based on Routes (PIR), the Deviation Index based on Stops (DIS) and the Evenness Index based on Stops (EIS). The PIR is defined as the probability that a bus will arrive at the terminals during a given period. The DIS is the ability to maintain distances and minimize the typical waiting time of a passenger at the stop, while the EIS is the ability to determine the consistency and balance of the distance between the vehicles. However, in [48], the author considers the punctuality index for a race, $P = 0$ if the bus arrives on time in all the stations of its trip and $P = 1$ if the bus does not arrive on time at all stations. For convenience, the punctuality index, P can be converted to percent for as in $P\ (\%) = (1 - P) \times 100$ to define the proportion of the trip that was punctual.

In [48], three alternative performance measures are proposed: Earliness Index (EI), Width Index (WI), and Second-Order Stochastic Dominance Index (SSD). These indices are used in two forms to capture the characteristics of the unreliability of bus service: (i) the distribution of the time interval deviations of trips for frequent services, (ii) the distribution of delays for non-frequent services.

[49] adds the transfer time as another indicator. This indicator covers the time spent when the passenger is waiting for the vehicle in changing the line at a connecting station. Other authors add the running time (time needed to change stop by walking in the transfer station) in the calculation of the transfer time.

[50] details and explains the formula of the Headway Buffer Time. This indicator indicates the additional travel time required to allow passengers to arrive on time. It can be used to capture the additional unreliability caused by an incident.

The authors of [51–53] examine another indicator called "Dwell" which is the bus downtime at stations including terminuses. This indicator refers to the time a vehicle, such as a bus or a train, goes to a stop without moving. In general, this time is spent on boarding or on embarking passengers, but it can also be used to wait for traffic to be restored [54, 55].

4.2 Discussion

According to the literature review presented above, there is no standard significance of the key performance indicators. The challenge in defining KPIs is to select the right ones that will give a sufficient accepting of overall performance on public transport.

Moreover, goals and objectives should be clear, concise, and achievable, in order to model the good performance formula for the regulation process. Indeed, the performance of public transport is an abstract term. In order to include performance considerations in a detailed engineering public transportation design and to evaluate the differences between existing and suggested service alternatives, it is necessary to describe it in mathematical terms. In fact, with a mathematical function, we can apply an optimization approach to the performance formula to reach the target. In addition, the main drawback of possible real-time performance regulation actions is the lack of prudent modeling and software that can activate automatically or semi-automatically these actions. Hence, build a regulation system to optimize the service performance based on KPIs in case of perturbation becomes an absolute necessity.

5 Optimization Resolution

5.1 Linear Programming Optimization

This section introduces the notion of optimization, using operations research (OR) principles and methods. Optimization usually means finding the best solution to some problem from a set of alternatives respecting constraints. Formally, an optimization problem can be described by a set U of potential solutions, a set $L \subset U$ of feasible solutions, and an objective function $F: L \rightarrow IR$. In the regulation problem, we are looking for regulation maneuver $x* \in L$ that minimizes the value of the objective function F overall feasible solutions. There is a list of feasible regulation actions that can be used to handle public transport traffic. This list should respect the constraints of the optimization problem.

In an existing problem, *F* is arbitrary, and the question is whether the set of feasible solutions is nonempty. The optimization should consider all KPIs and constraints fixed by experts of the traffic. We present a method that can be applied to regulate different traffic perturbations. This method is inspired by the work of [56].

In the RSSPT, a linear program with *n* criteria (KPIs) and *m* constraints is a minimization problem defined on a vector $x = (x_1,\ldots,x_n)$ of real-valued KPIs. The objective function is a linear function *F* of *x*, i.e.

$$F:\ IR^n \rightarrow IR \text{ with } F(x) = c * x \tag{1}$$

Where $c = (c_1,\ldots, c_n)$ is called cost vector. It is relative to the importance of different KPIs. E.g. punctuality criteria for buses of low-frequency lines (large headways) is more important than regularity while regularity for buses of high-frequency lines (large headways) is more important than punctuality, against keeping good transfer time criteria is more interesting for lines presented transfer stations with a high passengers' crowding. The variables are constrained by m linear constraints of the form:

$$a_i * x \bowtie_i b_i, \text{ Where } \bowtie_i \in \{\leq, \geq, =\},\ a_i = (a_{i1},\ldots,a_{in}) \in IR^n, \text{ and } b_i \in IR \text{ for } i \in 1..m \tag{2}$$

Consequently, the vector of criteria values of the feasible solutions is given by:

$$L = \{x \in IR^n: \forall\, i \in 1..m \text{ and } j \in 1..n: x_j \geq 0 \wedge a_i * x \bowtie_i b_i\} \tag{3}$$

5.2 Optimization Formulas

Notation List. The following data notations are used in the formulations of the optimization function and of the constraints:

- n: the number of vehicles of the same line arriving at station in a defined period.
- $\overline{h}$: $\frac{1}{n-1}\sum_{2}^{n-1}(t_i - t_{i-1})$ the average headway for n vehicles.
- t_i: the actual arrival time of the i-th vehicle.
- t_t: the scheduled arrival time of the i-th vehicle.
- t_{ij}: $t_j - t_i$ time between the departure time t_j of station j and the departure time t_i of station i. i and j represent respectively the two successive stations of the link l_{ij}.
- T_{ci}: estimated total travel time i.
- T_{ct}: scheduled total travel time i.
- h_i: $t_i - t_{i-1}$ ($i = 2,\ldots,I$), the current headway of the i-th vehicle.
- h_t: the scheduled headway of the i-th vehicle.
- H_{min_i}: minimum headway in the i station.
- H_{max_i}: maximum headway in station i.
- c_i: the current correspondence of the i-th vehicle.
- c_t: the scheduled correspondence of the i-th vehicle.
- $\overline{c}$: the average of the correspondence for the n vehicles.
- N_i: number of performed trips in station i.
- V_{PUN_i}: punctuality value in station i.
- $V_{PUN_{max}}$: permitted punctuality max value in station i.
- V_{REG_i}: regularity value in station i.
- $V_{REG_{max}}$: permitted regularity max value in station i.
- I_{MOB}: Mobility indicator
- I_{REL}: Reliability indicator
- I_{OPE}: Operational efficiency indicator
- $|C|$: describes the number of trips in the period of the journey
- *c:* describes the current trip
- ATT^c: describes the estimated travel time for the trip *c*.
- $\overline{MOB}$: the mobility average for n vehicles.
- MOB_i: the real mobility of the *i-th* vehicle.
- MOB_t: the theoretical (scheduled) mobility of the *i-th* vehicle.
- *L*: all links to the current trip.
- CT^l: the total duration of congestion on link l.
- w_l: the relative importance of the link l.
- T_{w_l}: the period in which congestion is monitored with the importance w_l.
- $\overline{REL}$: the reliability average for n vehicles.
- REL_i: the real reliability of the i-th vehicle.
- REL_t: the theoretical (scheduled) reliability of the i-th vehicle.

Formulation of the Optimization Function. The list of KPIs selected to measure the quality of performance as mentioned in Sect. 2 are based on the objectives of the traffic efficiency theme: mobility, reliability, and operational efficiency. These KPIs require an operational definition in one unit of measurement. The following sections summarize the measures and create the basis for defining the KPIs related to each objective. The formulas for these measures are drawn from [10] and some other research work such as [46, 49, 56, 57]. They represent a standard that can assess the overall performance of public transit in terms of traffic efficiency in transportation engineering and ITS.

$$F = W_{MOB}.I_{MOB} + W_{REL}.I_{REL} + W_{OPE}.I_{OPE} \tag{4}$$

Here, the W_{MOB}, W_{REL} and W_{OPE} represent the weight (cost) of the KPIs in the calculation of the performance value. It is necessary that: $W_{MOB} + W_{REL} + W_{OPE} = 1$. To calculate the weights, an experimental method is suggested capable of achieving a twofold objective: (i) to provide a methodology for constructing a measure of performance that can be adapted to any plan or transport program, and (ii) providing a methodology that can be transferred between projects. The technique chosen by the experts is the Delphi method [57].

Mobility. It defines the travel time distribution of the line trip I [58]. Its formula is:

$$MOB = \frac{1}{|C|}\sum\nolimits_{c\in C}^{|C|}\frac{ATT^c}{D_c} \tag{5}$$

The formula for the mobility indicator I_{MOB} is:

$$I_{MOB} = \frac{S_1^2}{\overline{MOB}^2} \tag{6}$$

$$S_1^2 = \frac{1}{n}\sum\nolimits_{i=1}^{n}(MOB_i - MOB_t)^2 \tag{7}$$

$$\overline{MOB} = \frac{1}{n}\sum\nolimits_{i=1}^{n-1}(MOB_i - MOB_{i-1}) \tag{8}$$

The unit of MOB is the "Travel time per km".

Reliability. It is defined as follows:

$$REL = 1 - \sum\nolimits_{l\in L}^{|L|} w_l.\frac{CT^l}{T_{w_l}} \tag{9}$$

To compute the estimated total duration of congestion, we need to calculate the speed performance index (SPI) as an indicator to evaluate the traffic state of the link [56]. The weight w_l is defined according to the length, the type (primary or secondary road), and the season or the period of the journey. The formula for the reliability indicator I_{REL} is:

$$I_{REL} = \frac{S_2^2}{\overline{REL}^2} \tag{10}$$

$$S_2^2 = \frac{1}{n}\sum\nolimits_{i=1}^{n}(REL_i - REL_t)^2 \tag{11}$$

$$\overline{REL} = \frac{1}{n}\sum\nolimits_{i=1}^{n-1}(REL_i - REL_{i-1}) \tag{12}$$

Operational Efficiency. This KPI corresponds to the vehicle at the station. According to [58], it is composed of three criteria: punctuality, regularity, and correspondence. The formula is as follows:

$$I_{OPE}(i) = W_{PUN}.I_{PUN} + W_{REG}.I_{REG} + W_{COR}.I_{COR} \tag{13}$$

Here, the W_{PUN}, W_{REG} and W_{COR} represent the importance of the criteria in the calculation of the operational efficiency and system condition KPI. E.g. the punctuality for buses of lines characterized by low-frequency services plays the most significant role; on the other hand, the regularity becomes more important for lines with high frequency [41]. It is necessary that: $W_{PUN} + W_{REG} + W_{COR} = 1$.

Punctuality Criteria. Punctuality is defined in [46] as a comparison of actual departure times with expected departure times at the station. Its formula is:

$$V_{PUN} = \frac{S_3^2}{\bar{h}^2} \text{ with } S_3^2 = \frac{1}{n}\sum_{i=1}^{n} (t_i - t_t)^2 \tag{14}$$

Regularity Criteria. It Measures the Differences in the Time Intervals Observed Between Successive Vehicles of the Same Line with Respect to the Scheduled Headway. Its Formula is:

$$V_{REG} = \frac{S_4^2}{\bar{h}^2} \text{ with } S_4^2 = \frac{1}{n-1}\sum_{i=2}^{n} (h_i - h_t)^2 \tag{15}$$

Correspondence Criteria. The correspondence criterion signifies the differences between the observed correspondence values with those of the scheduled correspondence. His formula is as follows:

$$V_{COR} = \frac{S_5^2}{\bar{c}^2} \text{ with } S_5^2 = \frac{1}{n}\sum_{i=1}^{n} (c_i - c_t)^2 \tag{16}$$

The current correspondence value 'c_i' (or the scheduled 'c_t') of the i-th vehicle is the sum of the waiting time between the vehicle 'I' and all coming vehicles to the transfer station. It is equal to:

$$C_i = \sum_{j=1}^{n} f_j(\Delta_{ij}) \tag{17}$$

f_j determines the importance factor of the vehicle 'j' which is in connection with the vehicle 'I'. This factor is calculated by experts according to the passengers waiting time of in the connection station for the vehicle in connection "j" [41]. It is necessary that:

$$\sum_{j \in n}^{|n|} f_j = 1 \tag{18}$$

And Δ_{ij} represents the gap time in relation to the scheduled waiting time of i-th connecting vehicle. It is equal to:

$$\Delta_{ij} = t_i - t_j \tag{19}$$

t_i is the current arrival time for the vehicle 'I', while t_j is the current departure time for the vehicle in connection 'j'.

Formulation of the Constraints. Based on [49], the Problem is Feasible Under the Following Constraints:

$$I_{REG_i} \leq I_{REG_{max}} \tag{20}$$

$$I_{PUN_i} \leq \min(I_{PUN_{max}}, I_{REG_{max}}) \tag{21}$$

$$t_i \leq N_i . H_{max_i} \tag{22}$$

$$t_i \geq (N_i - 1) . H_{min_i} \tag{23}$$

$$T_{ci} \leq T_{cmax} \text{ with } T_{cmax} = T_{ct} + \left(n * I_{REG_{max}}\right) \tag{24}$$

$$t_i \in \left[0, I_{REG_{max}}\right] \tag{25}$$

These constraints are mandatory in order to verify the following:

- not to exceed the maximum regularity value permissible limit (Eq. 20).
- the next trip does not catch up with the regulated trip (Eq. 21).
- the departure time at each station i does not exceed the maximum hour allowed during a regulation (Eq. 22).
- respect the minimum regularity between the vehicles of the same line (Eq. 23).
- not to exceed the maximum time allowed for a given trip (Eq. 24).
- not to have a conjunction of two consecutive trips in the starting station (Eq. 25).

6 The Regulation System of Public Transport

6.1 Regulation Process

To model our regulation system, the transit control problem is formulated as an optimization problem including. It divided to two phases: detection and decision.

- The mathematical model developed on the basis (Eq. 4).
- The constraints (Eq. 20, 21, 22, 23, 24 and 25).
- The purpose of the system mentioned above: Detecting the disturbance and making the decision.

The figure below (see Fig. 1) describes the regulation process of our system. This process involves five steps:

- Operations supervision: In this step, trips are monitored to see if everything is going as planned. The same thing happens for boarding passengers. The surveillance is done on route and at station after gartering information from (i) the Automatic vehicle location Module (AVLM) that is the GPS vehicle tracking system that continuously records

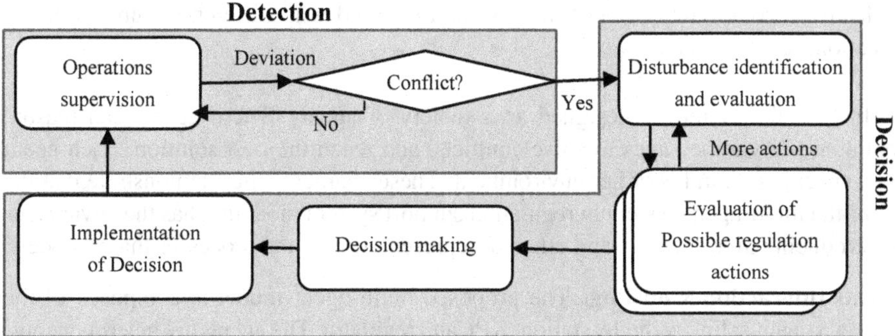

Fig. 1. Regulation process in our RSSPT.

and automatically transmits the geographic location and the speed of a vehicle, and (ii) detectors or loops to provide the properties of roads (length, speed max, density max, and current density) and station (passenger embarking and passenger boarding flow).

- Disturbance identification and evaluation: If an event occurs that produces a considerable deviation of the objective function F, for example, if an accident or works take place on a road, a rapid assessment is performed to determine if a regulation action is required. Otherwise, monitoring continues. The detection is based on the impact of the performance variability of the KPIs.
- Evaluation of possible regulation actions: the system selects the possible regulation actions from the existing list by using a classification method. This list is defined and updated by experts. An example of this list can be found in [59].
- Decision making: After filtering out possible solutions, a decision must be made by using an optimization resolution then, the system chooses the adequate action.
- Implementation of Decision: After choosing the decision, it must be applied to the environment with the update of the operational plan.

6.2 Multi-agent Design

Multi Agents System for Regulation Support System Modeling. Multi-agent modeling can give a suitable solution to public transport network activities where autonomous entities, called agents, interact with each other in a distributed, open, heterogeneous, and dynamic environment. We note that multi-agent systems are increasingly present in the field of traffic regulation. The following is a short description of the main characteristics for public transport regulation system:

- Distributed: the information is geographically dispersed over the network requires distributed agents.
- Dynamic: there is a daily change of information, for example, a vehicle can move forward, slow down, accelerate and communicate its passage with other agents like stations. As well, perturbations change the state of the traffic.
- Open: the vehicles can enter to or exit from the traffic network.

- Heterogeneous: The actors of the system are varied with different natures: vehicle, station, regulator, etc.

In fact, the system is designed as a society of agents. These agents can reason, communicate via messages to solve conflicts, and reach the best solution. Each agent has a specific role in his or her environment. These characteristics demonstrate that the use of the multi-agent system in regulation support system modeling has the advantages of introducing more flexible and efficient representation in the processes that it models.

Agents Interaction Modeling. The proposed multi-agent model is composed of the following agents: link, vehicle, station, KPI, and regulator. The agents are heterogeneous, and their interactions reflect the state of the transit system. We describe the behavior and the interactions between agents in the following sequence diagram of messages (Fig. 2).

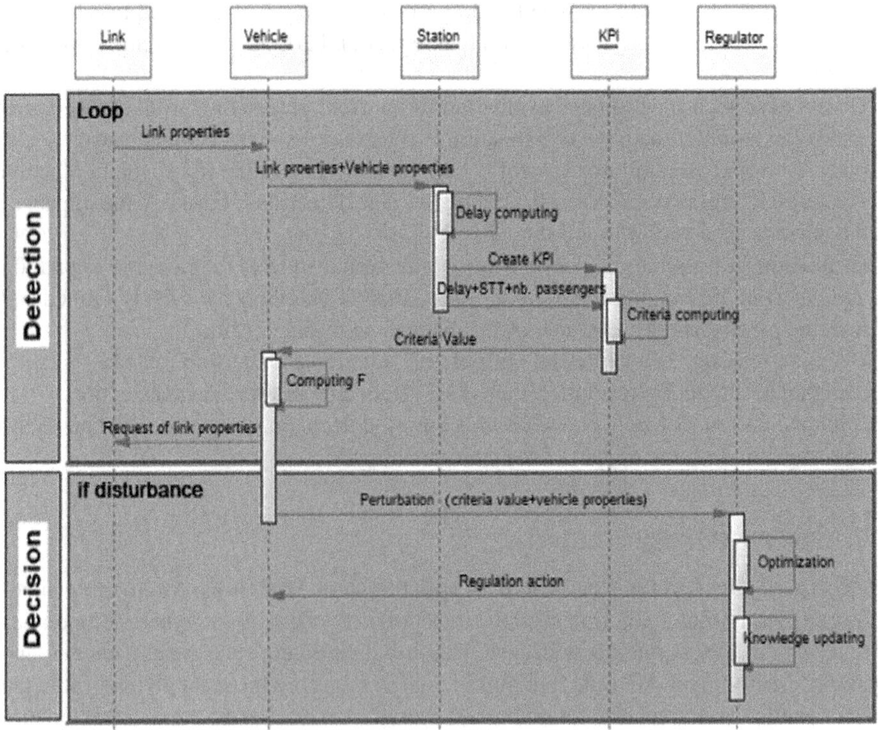

Fig. 2. Sequence diagram of messages of RSSPT.

The agents are described as follows:

- Vehicle: Vehicle agent memorizes all its properties such as position, type (bus, metro, and tramway), speed, capacity, number of passengers, line, mission (school bus, special, and passenger), driver, the properties of the current link and the values of the

KPIs involved. It calculates the overall performance and detects the disturbance. In the case of a disturbance, it transmits a message to the regulator concerned in order to trigger the optimization resolution to evaluate the possible regulation actions and make the appropriate decision. Also, each vehicle agent continuously provides its properties with those of the current link to the next station agent in order to estimate the remaining time.

- Station: It represents a departure or arrival of one or more links. It memorizes the passenger alighting and boarding flows, as well as the scheduled timetable and the current passage time of vehicle. It calculates the delayed time for the arrived vehicle. Then it creates the necessaries KPI agents for each coming vehicle according to the optimization problem and sends to them the calculated delayed time and the waiting passenger number to calculate the key performance criteria value.
- Link: It represents the connection between two consecutive stations. It should be related at least to one line. It memorizes two types of information: static properties (length, speed max, and density max) and dynamic properties (average of vehicle speeds and current density). These data are sent to the vehicle agent. The link agent used to analyze and detect link congestion by calculating the SPI as an indicator for evaluating the traffic congestion of the link. This indicator is transmitted to the KPI agent through vehicle agents to calculate the reliability indicator.
- KPI: It calculates the value of the Key Performance Indicator and sends it to the regulator agent the concerned vehicle agent. In our system KPIs are classified by objectives.
- Regulator: Each regulator agent is responsible for a geographical area of the network. It receives the KPI values of each disturbed vehicle. Then it defines the perturbation cause (vehicle breakdown, exceed parking time, driver uneasiness, an accident on the road, etc.) and follows an optimization process to find the regulation action.

6.3 Regulation Algorithm

The regulation process begins after the detection of perturbation. In following algorithm, we have to check that the constraints of our optimization model are respected. Practically, the verification is performed at the time of the choice of the regulation action to test the respect of the rules before they are applied.

```
ALGORITHM 1. Regulation
Input
Actual status of traffic network: scheduled and current
timetable of public transport.
Output:
Regulation action
begin
repeat //detection process
   - Link Agent calculates the SPI value and diffuses it
     with its properties to the corresponding vehicle
     agents. //vehicles running on the link
   - Vehicle Agent sends its properties and those of the
     current link to the next connected station agent.
   - Station Agent calculates the estimated arrival time
     of each coming vehicle and sends it to the KPI
     agents.
   - Each KPI Agent receives the estimated time of its
     corresponding vehicle and calculates its criteria
     value
   - Each Vehicle Agent receives all corresponding cri-
     teria value and calculates the variability of the
     performance function "F".
until "F" falls down in the critical area //Disturbance
   - Regulator agent receives All KIPs values and the
     properties of the disturbed vehicle.
   - The Regulator Agent fixes the incident /*vehicle
     breakdown, congestion, driver malaise*/.
   - Optimization Module
End
```

In optimization module, the system uses the classification method to search the list of possible solutions. If this list presents satisfactory actions, i.e. their performance variability values are adequate, it will choose the most optimal action that has the smallest variability value. We describe the optimization module in the following algorithm:

```
ALGORITHM 2. Optimization module
Input:
 -Vehicle with their properties and incident
 -Population ←vector (Xi)//set of regulation actions
Output:
Regulation action X //rule to apply
Begin
    -  Possible control actions of size n ←search (deci-
       sion tree (vector (Xi)) /*classification by mode of
       transport, days of the week, public holidays...*/
    -  Population ←(X1, X2, ..., Xn)/* set of possible
       actions*/
    -  L←(X1,X2,...,Xp) /* with p≤n; The regulator agent
       checks his knowledge base and chooses the possible
       solutions by applying the forward chaining of the
       inference engine: perturbation expressed by known
       facts and basic rules.*/
    -  Ls←∅ ////initialize satisfactory solutions list
   if (not empty (L)) then
    For each (solution Xi) /*Construction of "Ls" with
    i∈{1,2,…p}*/
    - The regulator agent calculates the variation value
    "F".
      if (F(Xi)∈[Vmin,Vmax]) /*the solution Xi presents
      an adequate variation*/
      - Ls←insert(Ls,Xi)
      end if
     end for
    end if
    -  X←optimization(F,Ls)/* X is the optimal regulation
       action obtained following optimization of function
       "F" on the list of actions "Ls" * /
End
```

7 Experimentation and Result

7.1 Description

We test our model using the data of a real traffic network of New York City and based on the KPIs for traffic management and ITS related to the objectives: reliability, mobility, and operational efficiency. The data was collected from the Metropolitan Transportation Authority (MTA) network. The MTA network has the nation's largest bus fleet and more subway and commuter rail cars than all other U.S. transit systems combined. It provides over 2.6 billion trips each year, accounting for about one-third of the nation's mass transit users.

The resolution is expressed in terms of an optimization problem with the objective function F using linear programming presented above. We used AnyLogic to simulate traffic scenarios and estimate measures needed to calculate the KPIs values. AnyLogic is a program for computer-aided transport planning, which determines the impacts of existing or planned supply that can encompass public transportation by simulating traffic scenarios (https://www.anylogic.com/). In addition, AnyLogic combines a dynamic simulation engine for animation and analytical tools for optimization. We present in the Fig. 3 the real traffic network of the itinerary M1with perturbation (Alert) on 18 September 2019. The data was imported to the AnyLogic as GTFS files to model the public transportation map data like lines, links, stations, vehicles.

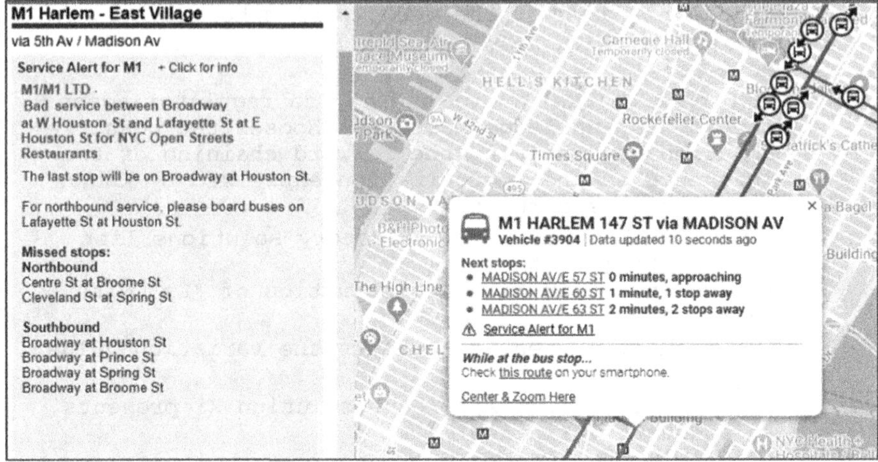

Fig. 3. Traffic network of the itinerary M1with perturbation (Alert) on 18 September 2019.

The study itinerary M1 contains 111 trips going from "Harlem W 146 St/Mlcolm X Blvd" to "Soho Grand St/Centre St". In fact, the journey starts with the trip 1 (04:48 am to 5:28 am) and finish with the trip 111 (0:05 am of the next day to 0:38 am).

We present, in the Fig. 4 the observed delays on each station in all trips run into the line M1. We distinguish two perturbations. The first one is due to the inclement weather conditions. It occurred in the morning on the trip 3. This perturbation is fully disappeared on the trip 25. The second perturbation is due to the United Nations General Assembly, it occurred in the evening between the trip 65 at and the trip 80.

In order to validate our regulation system, we simulate the scenario using our RSSPT. Then we compare the obtained results with the observed ones and demonstrate how it can manage the influence of the perturbations on many itineraries at the same time (see Fig. 5).

For verification, we use the variation of the performance F for each vehicle (trip) during the entire journey. We assume that all impact factor *fi* are equal for each connecting bus of each station. We assume, as well, that the distribution of weights W_{REG}, W_{PUN} and W_{COR} gives more importance to the regularity criteria because the itinerary M1 is characterized by high frequency (111 trips during the journey). All weights are given in the following table (see Table 1).

To detect perturbation, the system checks the value of the performance variation F on each station for each vehicle. When it exceeds the critic value (we suppose that this value is fixed to 0.15 by the experts of the traffic) the regulator receives the speed

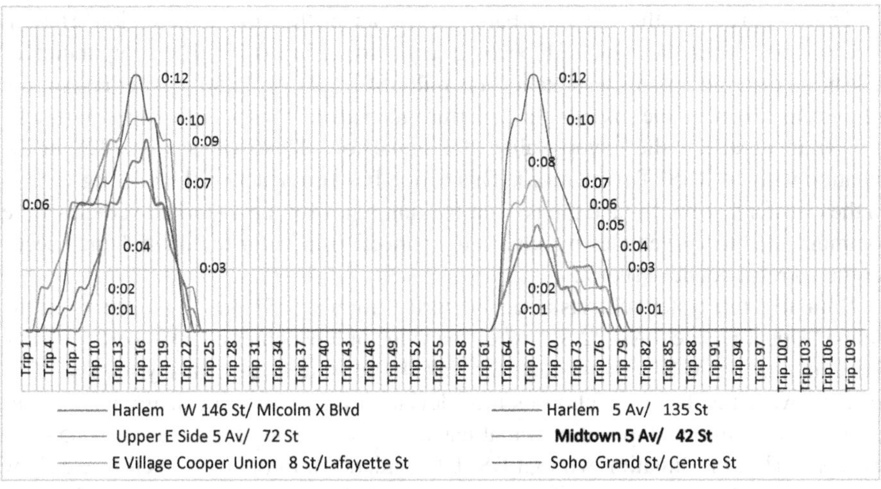

Fig. 4. Observed vehicle delays of M1 service from Harlem to Solo on 18 Sept 2019.

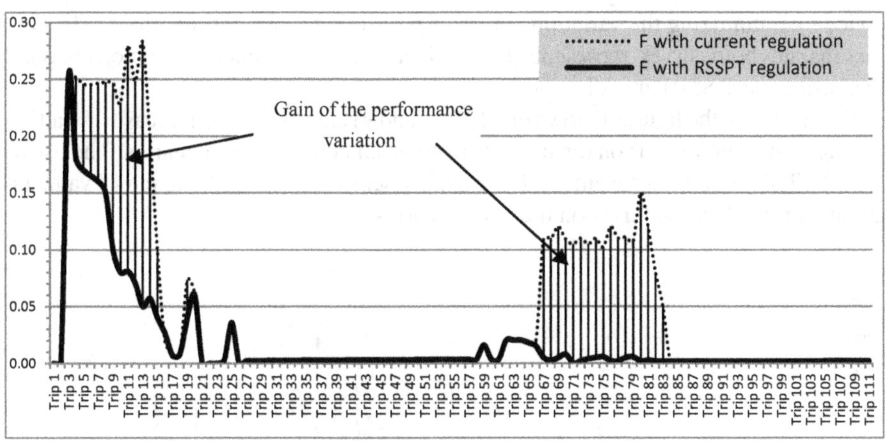

Fig. 5. The observed and the RSSPT performance F evolution of the itinerary M1.

Table 1. Weight KPIs distribution of the itinerary M1.

W_{MOB}	W_{REL}	W_{OPE}	W_{REG}	W_{PUN}	W_{COR}
0.25	0.25	0.5	0.5	0.25	0.25

performance index and starts the resolution process to find the best regulation maneuver. The regulator extracts the list of the feasible regulation actions and chooses the one offering the lowest value of F. This regulation actions list is classified into two groups: at terminus and online [60].

The system detects the first perturbation at 5:25 am on the trip 3 at the station "Harlem 5 Av/ 135 St", when the value of F becomes equal to 0.25 (>0.15). The computation of SPI shows that, the link traffic state level is mild (SPI = 37 ∈ (25,50]). After resolution, the system chooses "the inserting a departure maneuver" (insert new trip at terminus at 5:30 am) that has the lowest value of F equal to 0.02845 (it was 0.07428 in the situation without RSSPT regulation). This action has an impact on the coming vehicle on the transfer station. In fact, the new embarking times for the disturbed vehicles were communicated to all vehicles which are in connection in order to adjust theirs speeds for the correspondence criteria.

Similarly, the system detects a second perturbation at 3:33 pm on the trip 65 at the station "Upper E Side 5 Av/ 72 St", when the value of F becomes equal to 0.18 (>0.15). The computation of SPI shows that, the link traffic state level is heavy (SPI = 12 ∈ [0, 25]). After resolution, it chooses the "deviation maneuver" for disturbed vehicles (change road without skip stations) that has the lowest value of F equal to 0.00546 (it was 0.10107 in the situation without RSSPT regulation). In the figure below, we show the evolution of the performance variation F for each trip of the observed traffic with the current regulation and of those after applying our RSSPT for the disturbed itinerary M1 during the entire journey. The obtained results show an improvement of the quality service by minimizing the variation values of F during the two perturbations. The area that separates both curves, represents the gain in terms of performance variation obtained when using our RSSPT model.

We present in the figures below (Fig. 6, 7, 8, and 9) the percentage increase of waiting passengers per station (PI) on disturbed trips for each perturbation scenarios. The use of the RSSPT shows an improvement of the quality service by minimizing the PI value on stations for the disturbed trips on the two scenarios.

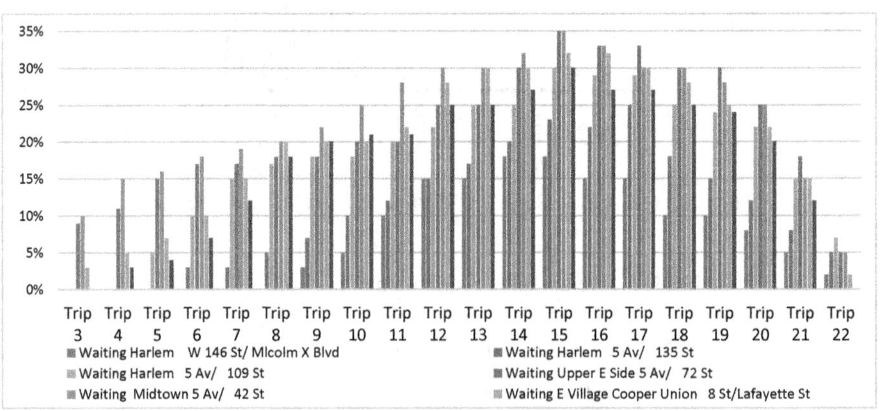

Fig. 6. PI per station on disturbed trips for scenario 1 without RSSPT.

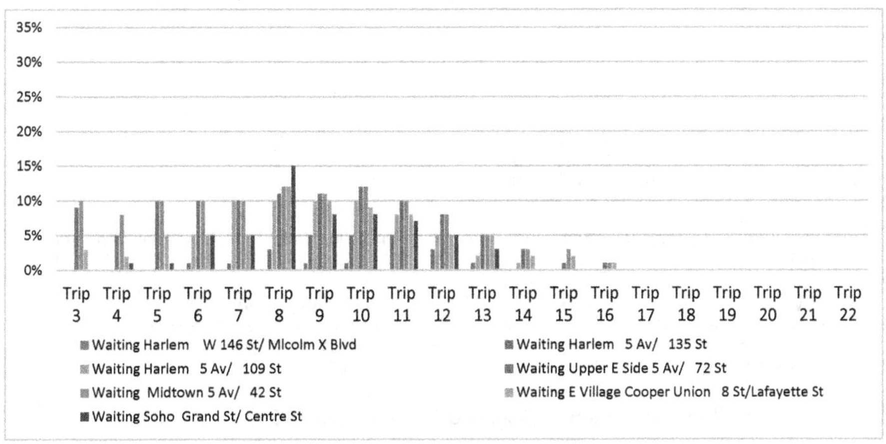

Fig. 7. PI per station on disturbed trips for scenario 1 using RSSPT.

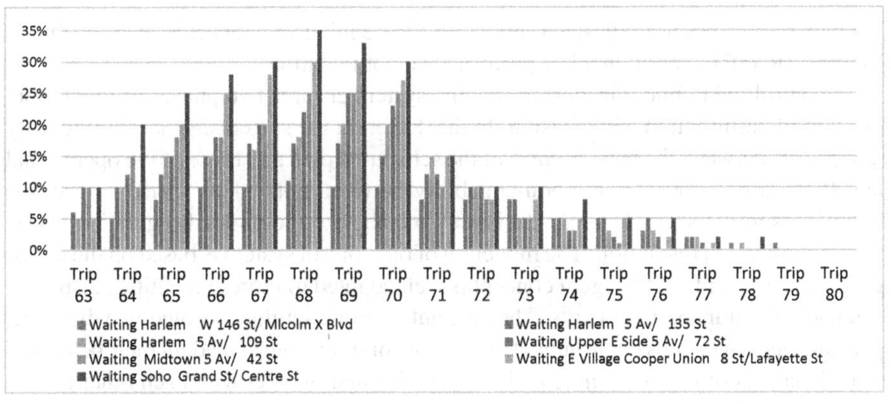

Fig. 8. PI per station on disturbed trips for scenario 2 Without RSSPT.

We observe that when we apply our RSSPT, the resolution of the disturbance is more rapid: In fact, the disturbance was fully resolved on the trip 17 instead of trip 23 on the scenario 1 and on the trip 73 instead of trip 79 on the scenario 2. Furthermore, the number of waiting passengers on each disturbed trip are considerably reduced with RSSPT on the two perturbation scenarios.

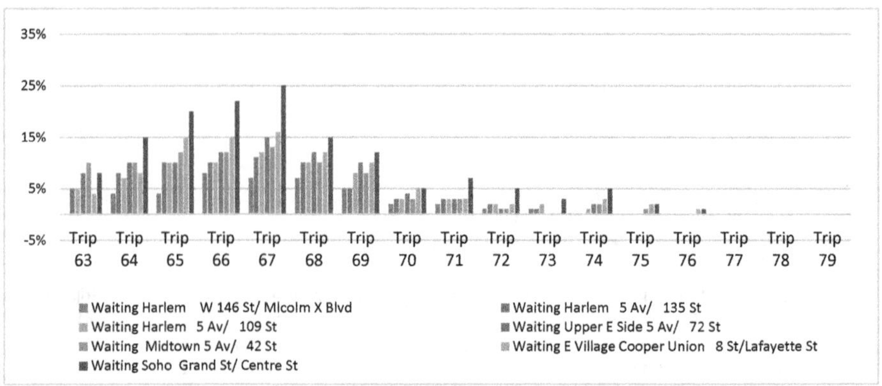

Fig. 9. PI per station on disturbed trips for scenario 2 using RSSPT.

8 Conclusion and Future Works

This paper describes a multi-agent modeling for Regulation Support System of Public Transport (RSSPT) based on a key performance optimization.

To control the public transport traffic, the system ensures two phases of regulation: detection of perturbation and decision. In the detection the system checks continuously the variation between the performance of the scheduled plan and those of the operational plan. If there is a considerable non-equability between both plans' performance, the system goes to the second phase "decision" to elaborate the optimal regulation action via an optimization resolution. The modeling of our control system is based on the multi-agent approach. In fact, the agent concept is well adapted to represent entities dispersed throughout the transport networks. They are autonomous entities, located in a dynamic, open, and heterogeneous environment. They can communicate and adapt their behaviors to the dynamics of the environment. They perceive and interact with each other to deal with different disturbances simultaneously.

We tested our model using the data of a real traffic network of New York City. The studied itinerary contains 111 trips going from "Harlem W 146 St/ Mlcolm X Blvd" to "Soho Grand St/ Centre St" and involves many perturbations. The obtained results show an improvement of the quality service. The use of our RSSPT model allows a faster regulation of the traffic perturbations.

A future work direction consists in providing an evolutionary approach to the optimization problem so that the system through the regulator agent can remember the results for future situations. Therefore, when there is a new situation (unknown disturbance, new traffic parameter, etc.), our model should suggest a neighboring solution as a future action with new experiments using learning process. In this case, the regulator should be responsible for improving its behavior by updating its knowledge base.

References

1. Morri, N., Hadouaj, S., Ben Said, L.: Intelligent regulation system to optimize the service performance of the public transport. In: Proceedings of the 22nd International Conference on Enterprise Information Systems - Volume 1: ICEIS, pp. 416–427 (2020). ISBN 978-989-758-423-7. https://doi.org/10.5220/0009416104160427
2. Barbeau, S., Georggi, N., Winters, P.: Global positioning system integrated with personalized real-time transit information from automatic vehicle location. Transp. Res. Rec. J. Transp. Res. Board **2143**, 168–176 (2010)
3. Yan, Y.: Bus transit travel time reliability evaluation based on automatic vehicle location data. J. Southeast Univ. **28**(1), 100–105 (2012)
4. European Commission, "White paper – Roadmap to a single European Transport Area" - Towards a competitive and resource efficient transport system (2011)
5. Newell, G.F., Potts, R.B.: Maintaining a bus schedule. In: 1964 2nd Conference of the Australian Road Research Board (ARRB), Melbourne (1964)
6. Feng, W., Figliozzi, M. using archived AVL/APC bus data to identify spatial-temporal causes of bus bunching. Compendium of Papers of 90th Transportation Research Board, Annual Meeting, Washington DC (2011)
7. Feng, W., Figliozzi, M.: Empirical findings of bus bunching distributions and attributes using archived avl/apc bus data. In: Proceedings of the 11th International Conference of Chinese Transportation Professionals (ICCTP). ASCE Reston, VA (2011)
8. Moreira-Matias, L., Ferreira, C., Gama, J., Mendes-Moreira, J., Sousa, J.: Bus bunching detection by mining sequences of headway deviations. In: Perner, P.. (ed.) Advances in Data Mining. Applications and Theoretical Aspects. Lecture Notes in Computer Science (Lecture Notes in Artificial Intelligence), vol. 7377, pp. 77–91. Springer, Heidelberg (2012)
9. Verbich, D., Diab, E., El-Geneidy, A.: Have they bunched yet? An exploratory study of the impacts of bus bunching on dwell and running times. Pub. Transp. **8**(2), 225–242 (2016). https://doi.org/10.1007/s12469-016-0126-y
10. Newell, G.: Unstable Brownian motion of a bus trip. In: Statistical Mechanics and Statistical Methods in Theory and Application, pp. pp. 645–667. Springer, Heidelberg (1977)
11. Zhao, J., Dessouky, M., Bukkapatnam, S.: Optimal slack time for schedule-based transit operations. Transp. Sci. **40**(4), 529–539 (2006)
12. Bellei, G., Gkoumas, K.: Transit vehicles headway distribution and service irregularity. Pub. Transp. **2**, 269–289 (2010). https://doi.org/10.1007/s12469-010-0024-7
13. Zhou, X., Wang, Y., Ji, X., Cottrill, C.: Coordinated control strategy for multi-line bus bunching in common corridors. Sustainability **11**, 6221 (2019). https://doi.org/10.3390/su1122 6221
14. Zolfaghari, S., Azizi, N., Jaber, M.Y.: A model for holding strategy in public transit systems with real-time information. Int. J. Transp. Manage. **2**(2), 99–110 (2004)
15. Bartholdi, J.J., Eisenstein, D.D.: A self-coördinating bus route to resist bus bunching. Transp. Res. Part B **46**(4), 481–491 (2012)
16. Zidi, S., Maouche, S., Hammadi, S.: Real-time route planning of the public transportation system. In: 2006 IEEE Intelligent Transportation Systems Conference, Toronto, Ont., Canada (2006). https://doi.org/10.1109/ITSC.2006.1706718
17. Ould Sidi, M., Hayat, S., Hammadi, S., Borne, P.: A novel approach to developing and evaluating regulation strategies for urban transport disrupted networks. Int. J. Comput. Integ. Manuf. **21**(4), 480–493 (2008).
18. Pilachowski, J.M.: An Approach to Reducing Bus Bunching. University of California Transportation Center (2009)

19. Daganzo, C.F., Pilachowski, J.: Reducing bunching with bus-to-bus cooperation. Transp. Res. Part B **45**(1), 267–277 (2011)
20. He, S.-X.: An anti-bunching strategy to improve bus schedule and headway reliability by making use of the available accurate information. Comput. Ind. Eng. **85**, 17–32 (2015)
21. Schmöcker, J., Sun, W., Fonzone, A., Liu, R.: Bus bunching along a corridor served by two lines. Transp. Res. Part B Meth. **93**, 300–317 (2016)
22. Gaddouri, R., Brenner, L., Demongodin, I.: Modélisation et simulation du trafic routier par réseaux de petri lots triangulaires. In: 10e Conférence Internationale de MOdélisation, Optimisation et Simulation - MOSIM'14 - Nancy – France (2014)
23. Carosi, S., Gualandi, S., Malucelli, F., Tresoldi, E.: Delay management in public transportation: service regularity issues and crew re-scheduling. In: 18th Euro Working Group on Transportation, EWGT, Delft, The Netherlands (2015)
24. Hassannayebi, E., Zegordi, S.H., Amin-Naseri, M.R., Yaghini, M.: Optimizing headways for urban rail transit services using adaptive particle swarm algorithms. Pub. Transp. **10**(1), 23–62 (2017). https://doi.org/10.1007/s12469-016-0147-6
25. Li, Y., Tan, H.: En-route headway-based bus reliability with real-time data at network scale. Transp. Saf. **2**, 236–245 (2020). https://doi.org/10.1093/tse/tdaa017
26. Fabien, B., Scemama G.: Modélisation d'une perturbation sur un réseau de transport: le modèle Incident. In: 16–19 mai 2000, IAE - 15, Quai Claude Bernard, Lyon, 69007 Inrets - Gretia, Institut National de Recherche sur les Transports et leur Sécurité (2000)
27. Laichour, H.: Thesis, Modélisation Multi-agent et aide à la décision: application à la régulation des correspondances dans les réseaux de transport urbain. Université des sciences et technologies de Lille (2002)
28. Bouamrane, K., Liazid, A., Amrani, F., Hamdadou, D.: Modelling and cognitive simulation in dynamic situation: decision-making for regulation of an urban transportation system, 5–7 Avril 2006 (2006). ISBN 9957-8592-0-X
29. Teodorovic, D.: Transport modeling by Multi-agent systems: a swarm intelligence approach. Transp. Plan. Technol. **26**, 289–312 (2003)
30. de Oliveira, L.B., Camponogara, E.: Multi-agent model predictive control of signaling split in urban traffic networks. Trans. Res. Part C **18**, 120–139 (2010)
31. Gershenson, C., Pineda, L.A.: Why does public transport not arrive on time? The pervasiveness of equal headway instability. PLOS ONE **4**(10), e7292 (2009)
32. Balaji, P.G., Sachdeva, G., Srinivasan, D.: Multi-agent system based urban traffic management. In: 2007 IEEE Congress on Evolutionary Computation (2007)
33. Kachroudi. Thesis, Commande et optimisation pour la régulation du trafic urbain multimodale sur de grands réseaux urbains. Université d'Evry Val d'Essone (2010)
34. Fayech. Thesis, Régulation des réseaux de transport multimodal: Systèmes multi-agents et algorithmes évolutionnistes. Université de Lille1- Sciences et Technologies (2003)
35. Bhouri, N., Balbo, F., Pinson, S., Tlig, M.: IEEE/WIC/ACM International Conference on Web Intelligence and Intelligent Agent Technology, Lyon (2011). ISBN 978-0-7695-4513-4
36. Albright, E., Figliozzi, M.A.: Analysis of the impacts of transit signal priority on bus bunching and performance. In: Proceedings of the Conference on Advanced Systems for Public Transport (CASPT), Santiago, Chile (2012)
37. Mejri, H.: Thesis, Un système d'aide à la régulation d'un réseau de transport multimodal perturbé: réponse au problème de congestion. École Centrale de Lille (2012)
38. Ben Rabah, N., Hammadi, S., Tahon, C.: Intelligent regulation support system for multimodal traffic. In: IEEE/WIC/ACM International Joint Conferences on Web Intelligence (WI) and Intelligent Agent Technologies (IAT) (2014).https://doi.org/10.1109/WI-IAT.2014.191
39. Tromp, M., Liu, X., Graham, D.J.: Development of key performance indicator to compare regularity of service between urban bus operators. Transp. Res. Rec. J. Transp. Res. Board **2216**(1), 33–41 (2011)

40. Napiah, M., Kamaruddin, I., Suwardo: Punctuality index and expected average waiting time of stage buses in mixed traffic. WIT Trans. Built Environ. **116**, 215–226 (2015). WIT Press. ISSN 1743–3509 (online)
41. Trompet, M.: The Development of a Performance Indicator to Compare Regularity of Service between Urban Bus Operators. Imperial College London, Skempton (2010)
42. Cats, O., Burghout, W., Toledo, T., Koutsopoulos, H.N.: Mesoscopic modeling of bus public transportation. Transp. Res. Board Nat. Acad. Washington D.C. **2188**, 9–18 (2010)
43. Bhouri, N., Aron, M., Scemama, G.: Gini index for evaluating bus reliability performances for operators and riders. Transportation Research Board, Washington, United States. Transportation Research Board, 13 p. (2016)
44. Henderson, G., Kwong, P., Adkins, H.: Regularity indices for evaluating transit performance. Transp. Res. Rec. **1297**, 3–9 (1991)
45. Carosi, S.., Gualandi, S., Malucelli, F., Tresoldi, E.: Delay management in public transportation: service regularity issues and crew re-scheduling. In: 18th Euro Working Group on Transportation, EWGT, Delft, The Netherlands, July 2015 (2015)
46. Yaakub, N., Napiah, M.: public transport: punctuality index for bus operation. World Acad. Sci. Eng. Technol. Int. J. Civ. Environ. Eng. **5**(12), 857–862 (2011)
47. Chen, X., Yu, L., Zhang, Y., Guo, J.: Analyzing urban bus service reliability at the stop, route, and network levels. Transp. Res. Part A **43**, 722–734 (2009)
48. Saberi, M., Ali Zockaie, K.: Definition and properties of alternative bus service reliability measures at the stop level. J. Pub. Transp. **16**(1), 97–122 (2013)
49. Ceder, A. A.: Public Transit Planning and Operation: Theory, Modelling and Practice. Elsevier Ltd. (2007)
50. Ma, Z., Luis, F., Mahmoud, M.: A framework for the development of bus service reliability measures. In: Australasian Transport Research Forum, Australia (2013)
51. Dueker, K.J., Kimpel, T.J., Strathman, J.G.: Determinants of bus dwell time. J. Pub. Transp. **7**(1), 21–40 (2014). https://doi.org/10.5038/2375-0901.7.1.2
52. TRT: Dwell time - Transportation Research Thesaurus (TRT). trt.trb.org (2017)
53. Levinson, H.: Analyzing transit travel time performance. Transp. Res. Rec. **915**, 1–6 (1983)
54. Tran, V.T., Eklund, P., Cook, C.: Toward real-time decision making for bus service reliability. In: International Symposium on Communications and Information Technologies (ISCIT) (2012). https://doi.org/10.1109/ISCIT.2012.6380856
55. Cats, O., Larijani, A.N., Koutsopoulos, H.N., Burghout, W.: Impacts of holding control strategies on transit performance: a bus simulation model analysis. Transp. Res. Rec. J. Transp. Res. Board **2216**, 51–58 (2011)
56. Yan, X.Y., Crookes, R.J.: Reduction potentials of energy demand and GHG emissions in China's road transport sector. Energy Policy **37**, 658–668 (2009)
57. Linstone, H.A., Turrof, M.: The Delphi Method - Techniques and Applications. Addison-Wesley Publishing Company (1975)
58. Cambridge Systematics Inc., PB Consult Inc., and System Metrics Group: Analytical tools for asset management. 545, NCHRP report (2005)
59. Foloff, E., Rizi, M., Sapiroto, A.: Bases et pratiques de regulation, RATP Direction du reseau routier, RC/MSE (1989)
60. Morri, N., Hadouaj, S., Ben Said, L.: Multi-agent optimization model for multi-criteria regulation of multi-modal public transport. NNGT Int. J. Artif. Intell. **3** (2015). https://doi.org/03.IJAI.2015.1.8

Comparative Evaluation of the Supervised Machine Learning Classification Methods and the Concept Drift Detection Methods in the Financial Business Problems

Victor Ulisses Pugliese(✉), Renato Duarte Costa(✉), and Celso Massaki Hirata(✉)

Instituto Tecnológico de Aeronáutica, Praça Marechal Eduardo Gomes 50, São José dos Campos, Brazil
{pugliese,rdcosta,hirata}@ita.br

Abstract. Machine Learning methods are key tools for aiding in the decision making of financial business problems, such as risk analysis, fraud detection, and credit-granting evaluations, reducing the time and effort and increasing accuracy. Supervised machine learning classification methods learn patterns in data to improve prediction. In the long term, the data patterns may change in a process known as concept drift, with the changes requesting retraining the classification methods to maintain their accuracies. We conducted a comparative study using twelve classification methods and seven concept drift detection methods. The evaluated methods are Gaussian and Incremental Naïve Bayes, Logistic Regression, Support Vector Classifier, k-Nearest Neighbors, Decision Tree, Random Forest, Gradient Boosting, XGBoost, Multilayer Perceptron, Stochastic Gradient Descent, and Hoeffding Tree. The analyzed concept drift detection methods are ADWIN, DDM, EDDM, HDDMa, HDDMw, KSWIN, and Page Hinkley. We used the next-generation hyperparameter optimization framework Optuna and applied the non-parametric Friedman test to infer hypotheses and Nemeyni as a posthoc test to validate the results. We used five datasets in the financial domain. With the performance metrics of F1 and AUROC scores for classification, XGBoost outperformed other methods in the classification experiments. In the data stream experiments with concept drift, using accuracy as performance metrics, Hoeffding Tree and XGBoost showed the best results with the HDDMw, KSWIN, and ADWIN concept drift detection methods. We conclude that XGBoost with HDDMw is the recommended combination when financial datasets that exhibit concept drift.

Keywords: Supervised learning · Concept drift · Ranking methods

1 Introduction

Business success and failure have been extensively studied. Most of the studies try to identify the various determinants that can affect business existence [40].

J. Filipe et al. (Eds.): ICEIS 2020, LNBIP 417, pp. 268–292, 2021.
https://doi.org/10.1007/978-3-030-75418-1_13

Business operations are conducted based on how companies make financial decisions and depend on models to support them. Inadequate models can lead to business failure [10].

Decisions are based on the prediction of classification for problems such as credit granting, credit card fraud detection, bankruptcy risk analysis, and are commonly treated as binary classification problems [21,40].

This paper is a substantial extension of the publication [29] in which the authors presented a comparative study of nine classification methods and found that XGBoost is the recommended classification method for financial domain applications. In this paper, we include three additional supervised classification methods and introduce the evaluation of seven supervised concept drift detection methods. We employ the concept drift detection methods in order to provide long term accuracy sustainability.

In this paper, we conduct a comparative study of twelve classification methods and seven concept drift detection methods in order to identify which supervised methods perform best on problems of analyzing risk, detecting the occurrence of fraud, and deciding about credit granting. The three new classification methods are introduced due to their incremental implementation. We also made an incremental implementation of the XGBoost method in order to allow its use with the concept drift methods. The motivation is to identify methods that have the best performance considering the possibility of concept drift occurring in the aforementioned problems.

This paper is organized as follows. The next section briefly describes the twelve predictive methods and the seven concept drift detection methods. Section 2 also describes the performance metrics and the Optuna hyperparameter optimization technique. Section 3 describes the related work. We describe the evaluation of the predictive methods with five datasets of the financial domain in Sect. 4. The results are shown in Sect. 5. Section 6 describes the evaluations of the twelve predictive methods for three datasets in other than financial domains. Section 7 describes the evaluation of concept drift detection methods for financial datasets. The main findings, discussions, and conclusions are presented in Sect. 8.

2 Background

This section briefly describes the twelve methods for financial forecasting, the performance metrics, Friedman Tests, the Optuna hyperparameter optimization technique, and the seven methods to detect concept drift in data.

2.1 Classification Methods

The predictive methods used are Gaussian (GNB) and Incremental (INB) Naive Bayes, Logistic Regression, Support Vector Classifier, k-Nearest Neighbors, Decision Tree, Random Forest, Gradient Boosting, XGBoost, Multilayer Perceptron, Stochastic Gradient Descent, and Hoeffding Tree.

Naïve Bayes classifier is based on applying Bayes's theorem with strong (naïve) independence assumptions [33].

$$p(X|Y) = \prod_{i=1}^{n} p(X_i|Y) \tag{1}$$

where p is a probability, $X(X_1, \ldots, X_n)$ is a feature vector and Y is a class. The theorem establishes that the class Y given the feature X, the posterior probability, $p(Y|X)$, can be calculated by the class prior probability, $p(Y)$, multiplied by the observed feature probability, $p(X|Y)$, or likelihood, divided by the total feature probability, $p(X)$, which is constant for all classes [27].

$$p(Y|X) = \frac{p(Y) * p(X|Y)}{p(X)} \tag{2}$$

Although the independence between features is a condition not fully sustained in most cases, the Naïve Bayes method has proved its strength in practical situations with comparable performance to Neural Network and Decision Tree classifiers [18].

Logistic Regression (LR) is a classification method used to predict the probability of a categorical dependent variable assigning observations to a discrete set of classes (yes or no, success or failure). Unlike linear regression which outputs continuous number values, logistic regression transforms its output using the logistic sigmoid function (Eq. 3) to return a probability value, which can then be mapped to discrete classes. The logistic sigmoid function maps any real value into another value between 0 and 1. A decision threshold classifies values into classes 0 or 1.

$$S(z) = \frac{1}{1 + e^{-z}} \tag{3}$$

The Logistic Regression is binary if the dependent variable is a binary variable (pass or fail), multinomial if the dependent variable is categorical as type of animal or flower, and ordinal for ordered classes like Low, Medium or High. Ng and Jordan [25] present a comparison between Naïve Bayes and Logistic Regression classifier algorithms.

k-Nearest Neighbor (kNN) is a non-parametric method for classification and regression tasks. It is one of the most fundamental and simplest methods, being the first choice method for classification when there is little or no prior knowledge about the distribution of the data [28]. Examples are classified based on the class of their nearest neighbors. It is usually used to identify more than one neighbor, where k is a referee for determining class number.

Support Vector Classifier (SVC) is a statistical learning method that is suitable for binary classification [41]. The objective of the Support Vector Classifier is to find a hyper-plane in n-dimensional space, where n is the number of features, that distinctly classifies the data [36].

Decision Tree (DT) is a flow chart like tree structure, where each internal node denotes a test on an attribute, each branch represents an outcome of the test, and each leaf node holds a class label [19]. Decision tree classifiers are commonly used in credit card, automobile insurance, and corporate fraud problems.

Random Forest (RandFC) is proposed as an additional layer of randomness bagging tree [7,20]. The Random Forest collects data and searches a random selection of features for the best division on each node, regardless of previous trees. In the end, a simple majority vote is made for prediction.

Random Forest performs very well compared to many other classifiers, including discriminant analysis, support vector classifier and neural networks, being robust against overfitting [20].

Gradient Boosting (GradB) is based on a different constructive strategy of ensemble set like Random Forest. Boosting's main idea is to add new models to the ensemble sequentially [24]. Boosting fits the "weak" tree classifiers to different observation weights in a dataset [32]. In the end, a weighted vote is made for prediction [20].

XGBoost (XGB) is a scalable machine learning system for optimized tree boosting. The method is available as an open source package. Its impact has been widely recognized in a number of machine learning and data mining challenges [8]. XGBoost has several features such as parallel computation with OpenMP. It is generally over 10 times faster than Gradient Boosting. It supports customized objective function and evaluation function. It has better performance on several different datasets.

Multilayer Perceptron (NN) is a feed-forward artificial neural network model for supervised learning, composed by a series of layers of nodes or neurons with full interconnection between adjacent layer nodes. The feature vector X is presented to the input layer. Its nodes output values are fully connected to the next layer neurons through weighted synapses. The connections repeat until the output layer, responsible to present the results of the network. The learning of NN is made by the back-propagation algorithm. The training is made layer by layer, adjusting the synaptic weights from the last to the first layer, minimizing the error. Error metrics such as minimum square error is used. The algorithm repeats the training process several times. Each iteration is called epoch. On each epoch, the configuration that presents the best results is used as the seed for the next interaction, until some criterion as accuracy or number of iterations is reached.

Stochastic Gradient Descent (SGD) is an optimization method that stochastically estimates the gradient. Typically, the gradient methods use the entire training set to calculate the gradients at each stage. SGD takes a random instance of the training set at each stage and calculates the gradients based on that single instance only. SGD is faster than other gradient methods, using fewer resources, being, on the other hand, more unstable. The available hyperparameters are regularization, the number of iterations, and selection of loss function [5,14].

Hoeffding Tree (HT) is an incremental decision tree based on the theoretical guarantees of the Hoeffding bound, which is capable of learning from large data streams. The node is expanded as soon as there is sufficient statistical evidence that an optimal splitting feature exists. Theoretically, the Hoeffding tree's model is asymptotically nearly identical to the one built by a non-incremental learner, if the number of training instances is large enough [17]. In [4], they included ADWIN to monitor performance of branches on the tree and to replace them with new branches when their accuracy decreases if the new branches are more accurate.

To measure the performance of the predictive methods employed in this study, we used the following metrics: *Accuracy, Precision, Recall, F1, AUROC, and Mean Square Error (MSE).*

Accuracy is simply a ratio of the correct predicted observation to the total observations.

Precision is the number of correct positive results divided by the number of positive results predicted.

Recall is the number of correct positive results divided by the number of all samples that should have been identified as positive.

F1 Score is the harmonic mean of *Precision* and *Recall.* F1 score reaches its best value at 1 (perfect precision and recall) and worst at 0 (Eq. 4).

$$F1\, Score = 2 * \frac{Precision * Recall}{Precision + Recall} \tag{4}$$

AUROC (Area under the Receiver Operating Characteristic) is a usual metric for the goodness of a predictor in a binary classification task.

Mean Square Error (MSE) is used to observe how close a set of instances is to the regression or objective line. It considers the square of the distance between the instance to the regression line (the distance is the "error"). It also gives greater weight to larger differences. It is called the mean square error because it finds the mean of a set of errors.

To evaluate the methods with the datasets, we employ *Friedman* and *Nemenyi* tests.

The *Friedman* test is a nonparametric equivalent of repeated measure analysis of variance (ANOVA). The purpose of the test is to determine if one can conclude from a sample of results that there is a difference between the treatment effect [13].

The *Nemenyi* test is a post-hoc test of Friedman applied when all possible pairwise comparisons need to be performed. It assumes that the value of the significance level α is adjusted in a single step by dividing it merely by the number of comparisons performed.

Hyperparameter optimization is one of the essential steps in training machine learning models. With many parameters to optimize, long training time, and multiple folds to limit information leak, hyperparameter adjustment is a cumbersome endeavor. There are a few methods to deal with the issue: grid search, random search, and Bayesian methods. Optuna is an implementation of the last one.

Optuna is a next-generation Hyperparameter Optimization Framework [1]. It has the following features: define-by-run API that allows users to construct the parameter search space dynamically; efficient implementation of both searching and pruning strategies; and easy-to-setup, versatile architecture.

2.2 Concept Drift Detection Methods

Concept drift is the phenomenon in which the statistical properties of the target variable, which the supervised classification method is trying to predict, tend to change over time [38]. As a result, predictions of models trained in the past may become less accurate over time. Thus, learning models need to have mechanisms for continuous performance diagnostics and be able to adapt to changes in the data over time. These mechanisms include the concept drift detection methods.

Python Scikit-Multiflow is a library specialized in anomaly detection methods, data streaming methods, and concept drift detection methods [22]. The concept drift detection methods listed in the library are: ADWIN, DDM, EDDM, HDDMa, HDDMw, KSWIN, and Page Hinkley.

Adaptive Windowing (ADWIN) uses sliding windows of varying sizes, recalculated online according to the rate of change that is observed from the data in those windows. If there is no apparent change in the context, then the window is increased, otherwise, when a change is detected, the window is reduced [15]. To detect changes in the statistical distribution of the data, the algorithm checks for averages in two sequential sub-windows. If the averages differ above a detection level, a concept drift is detected.

Drift Detection Method (DDM) considers the base learner data classification error rate. An indication of concept drift is given after an increase in error rate measurement [15].

Early Drift Detection Method (EDDM) uses the distance between classification errors of the base learner to identify whether a drift occurs. This metric computes the rate of classification errors. If it increases above a certain level, then the distance between errors is short and a concept drift is detected [15].

There are two drift detection methods based on Hoeffding's bound: HDDMa and HDDMw [12]. The bound is set up to define an upper limit of the level of divergence between the averages. The HDDMa uses the moving average as an estimator for detecting drifts, while the HDDMw uses the Exponentially Weighted Moving Average (EWMA) statistic as an estimator [34].

Kolmogorov-Smirnov Windowing (KSWIN) is based on the Kolmogorov-Smirnov (KS) test. This statistical test does not assume the distribution of the underlying data, but it compares the distance of the cumulative data distribution. Whenever the distance is greater than a limit, KSWIN detects a drift [31].

Page Hinkley considers the base learner accuracy measures and their average. Over time, a base learner fails to correctly classify the instances received and the actual accuracy decreases. The method analyzes the cumulative difference between the values and the minimum difference. Higher values indicate that the observed values differ considerably from their previous values. When the difference is above a specified limit, a change in distribution is detected.

3 Related Work

There are two systematic literature reviews [6,35] that describe the works on data mining techniques applied in financial frauds, healthcare insurance frauds, and automobile insurance frauds. With respect to concept drift, Webb et al. characterize it whereas Zliobaite et al. [38] describe its applications [42].

Moro et al. [23] propose a data mining technique approach for the selection of bank marketing clients. They compare four models: Logistic Regression, Decision Tree, Neural Networks, Support Vector Machines, using the performance metrics AUROC and LIFT. For both metrics, the best results were obtained by Neural Network. Moro et al. do not use bagging or boosting tree.

Zareapoor and Shamsolmoali [41] apply five predictive methods: Naive Bayes, k-Nearest Neighbors, Support Vector Classifier, Decision Tree, and Bagging Tree to credit card's dataset. They report that Bagging Tree shows better results than others. Zareapoor and Shamsolmoali do not use Boosting Tree, Neural Network and Logistic Regression as we do. Their survey does not have a nonparametric test.

Wang et al. [37] explore credit scoring with three bank credit datasets: Australian, German, and Chinese. They made a comparative assessment of performance of three ensemble methods, Bagging, Boosting, and Stacking based on four base learners, Logistic Regression, Decision Tree, Neural Network, and Support Vector Machine. They found that Bagging performs better than Boosting across all credit datasets. Wang et al. do not use k-Nearest Neighbors and XGBoost.

Gonçalves et al. [15] propose a comparative study of the concept drift detectors ADWIN, DDM, EDDM, Page Hinkley, Paired Learners (PL), ECDD, STEPD, and DOF. They employed the detectors in twelve datasets: four synthetic data stream with abrupt concept drift, four with gradual concept drift, and four synthetic data stream real datasets. They found the best hyperparameters settings for each method and noted that DDM performs better than others. The authors neither use HDDMw, HDDMa, and KSWIN nor employ Stochastic Gradient Descent and XGBoost as base learners, over the financial domain.

Barros and Santos [3] compare fourteen concept drifts methods using synthetic datasets with gradual concept drift. Their study shows that the best drift detection methods are HDDMa, FTDD, WSTD, FHDDM, and HDDMw. In opposite, the worst methods are DDM, EDDM, ADWIN, ECDD, STEPD, and SEED. They neither employ KSWIN nor apply Stochastic Gradient Descent and XGBoost as base learners.

4 Evaluation of the Methods with Datasets of the Financial Area

We use five datasets of the financial domain in the evaluations. They are briefly described as follows.

The *Bank Marketing* dataset is about direct marketing campaigns (phone calls) of a Portuguese banking institution. It contains personal information and banking transaction data of clients. The classification goal is to predict if a client will subscribe to a term deposit. The dataset is multivariate with 41,188 instances (4,640 subscriptions), 21 attributes (5 real, 5 integer and 11 object), and no missing values. It is available at https://archive.ics.uci.edu/ml/datasets/Bank+Marketing [23].

The *Default of Credit Card Clients* dataset contains information of default payments, demographic factors, credit data, history of payment, and bill statements of credit card clients in Taiwan from April to September 2005. The classification goal is to predict if the client is credible. The dataset is multivariate with 30,000 instances (6,636 creditation), 24 integer attributes, and no missing values. The dataset is available at https://archive.ics.uci.edu/ml/datasets/default+of+credit+card+clients [2].

The *Kaggle Credit Card* dataset is a modified version of Default of Credit Card Clients, with data in the same period. Both datasets have the same classification goal: predict if the client is credible. However, Kaggle Credit Card has more features, with 31 numerical attributes and a lower number of positive credible client instances. The dataset has 284,807 instances, with 492 positive credible client instances, which accounts for 0.173%. We note that the dataset is highly unbalanced. It is available at https://www.kaggle.com/uciml/default-of-credit-card-clients-dataset [9].

The *Statlog German Credit* dataset contains categorical and symbolic attributes. It contains credit history, purpose, personal client data, nationality, and other information. The goal is to classify clients using a set of attributes as good or bad for credit risk. We used an alternative dataset provided by Strathclyde University. The file was edited and several indicator variables were added to make it suitable for algorithms that cannot cope with categorical variables. Several attributes that are ordered categorically were coded as integer. The dataset is multivariate with 1,000 instances, 300 instances are classified as bad, 24 integer attributes, and no missing values. The dataset is available at https://archive.ics.uci.edu/ml/datasets/statlog+(german+credit+data) [16].

The *Statlog Australian Credit Approval* dataset is used for analysis of credit card operations. All attribute names and values were anonymized to protect data privacy. The dataset is multivariate with 690 instances (307 instances are labeled as bad), 14 attributes (3 real, 11 integer), and no missing values. It is available at http://archive.ics.uci.edu/ml/datasets/statlog+(australian+credit+approval) [30].

Initially, for each dataset, we preprocessed the attributes, sampled the data, and divided the data into 90% for training and 10% for testing. After splitting the dataset, we employed cross-validation with ten stratified k-folds, fifteen seeds (55, 67, 200, 245, 256, 302, 327, 336, 385, 407, 423, 456, 489, 515, 537), and twelve predictive methods. Firstly, the methods used the scikit-learn default hyperparameters. The *F1 Score and AUROC* metrics were measured. Tests were performed on the measured metrics to rank statistic differences over methods.

Finally, we employed Optuna to optimize the hyperparameters and used the classification methods again.

The main scikit-learn default hyperparameters were used to test the Gaussian Naive Bayes, Logistic Regression, kNN, SVC, Decision Tree, Random Forest, Gradient Boosting, XGBoost, and Multilayer Perceptron classification methods. For this work, we upgraded the Scikit-Learn to '0.23.2' version, and the XGBoost to '1.1.1' version, keeping similar hyperparameters as presented in our last paper [29]. Stochastic Gradient Descent uses only the default of Scikit-Learn, applied over Hinge regularization.

The main scikit-multiflow default hyperparameters were used to test the Incremental Naive Bayes and Hoeffding Tree methods. The Incremental Naive Bayes does not have relevant hyperparameters. The default hyperparameters for Hoeffding Tree method are:

binary_split = False, bootstrap_sampling = True,
grace_period = 200, leaf_prediction = 'nba', max_byte_size = 33554432,
memory_estimate_period = 1000000, nb_threshold = 0, no_preprune = False,
nominal_attributes = None, random_state = None, remove_poor_atts = False,
split_confidence = 1e−07, split_criterion = 'info_gain',
stop_mem_management = False, tie_threshold = 0.05.

We used Optuna to optimize the hyperparameters of the methods, running one study with 100 iterations, using the same following ranges used in [29]. We did not optimize SVC, Stochastic Gradient Descent, and Naive Bayes. For the new methods, only the Hoeffding tree has parameters to optimize. They are:

split_criterion: 0 or 1, grace_period: 100 to 300, and
tie_ threshold: 0.01 to 0.10.

For checking the concept drift detection on streaming data, one cannot shuffle or balance the data. It is necessary to keep the sequential statistic characteristics of the data. This fact requires a dataset with enough data for training that is not unbalanced. The Bank Marketing and Default of Credit Card Clients datasets fulfill these requirements and were used for the concept drift detection method evaluation.

We divided the datasets into 20% for initial training and 80% for concept drift detection testing, employing incremental methods, including XGBoost. Mean and minimum accuracy and mean square error were used as measure of performance for the seven concept drift detection methods: ADWIN, DDM, EDDM, HDDMa, HDDMw, KSWIN, and Page Hinkley. Furthermore, we used as base reference level for the tests: no retraining and fixed interval retraining conditions. We added the following parameters to control the algorithms: maximum number of detected instances; minimum distance between detected instances; and length of retraining.

5 Prediction Results with Datasets of the Financial Domain

In this section, we present the results of the classification methods for the five datasets in the financial domain.

For the *Bank Marketing* dataset, we transformed categorical data with One-Hot-Encoding. Afterward, we applied to the undersample to balance the dataset. Undersampling is an algorithm to deal with class-imbalance problems. It uses only a subset of the majority class for efficiency. Subsequently, we used cross-validation methods to evaluate them, and finally, we employed Optuna over the methods in the dataset. The results are shown in Table 1.

Table 1. Bank Marketing dataset experiments.

1a - Cross-validation					1b - Optimization with Optuna				
Method	F1		AUROC		Method	F1		AUROC	
	Mean	Std	Mean	Std		Mean	Std	Mean	Std
XGB	88.91	0.00	88.62	0.00	XGB	89.59	0.00	89.25	0.00
GradB	88.77	0.01	88.46	0.01	GradB	89.40	0.03	89.12	0.03
RandFC	88.77	0.13	88.42	0.13	RandFC	88.98	0.05	88.50	0.03
KNN	85.80	0.00	85.61	0.00	DT	87.60	0.02	87.35	0.02
LR	85.74	0.00	85.82	0.00	NN	86.79	0.47	86.73	0.29
HT	83.55	0.55	82.98	0.83	KNN	86.13	0.00	86.02	0.00
DT	83.28	0.17	83.49	0.11	LR	85.82	0.00	85.89	0.00
SVC	82.77	0.00	83.33	0.00	HT	84.96	0.60	84.30	0.58
NN	80.67	2.46	81.62	2.25	SVC	82.77	0.00	83.33	0.00
SGD	70.04	4.04	70.38	3.89	SGD	70.04	4.04	70.38	3.89
GNB	65.89	0.00	71.94	0.00	GNB	65.89	0.00	71.94	0.00
INB	65.70	0.00	71.83	0.00	INB	65.70	0.00	71.83	0.00

As one can observe in Table 1, the lowest values were from Naive Bayes (GNB/INB). The best results were achieved by XGBoost, with 0% of standard deviation for both metrics, using these Optuna hyperparameters: 'booster': 'gbtree', 'lambda': 7.28024139 6829422e−08, 'alpha': 1.3345521568506561e−05, 'max_depth': 6, 'eta': 0.06702444912 759584, 'gamma': 0.056863049059355034, 'grow_policy': 'lossguide'.

Figure 1 illustrates the Critical Difference diagram constructed using Nemenyi test for F1 Score and AUROC in the Bank Marketing cross-validation.

As one can see in Fig. 1 that XGBoost, Gradient Boosting, Random Forest, Decision Tree, and Multilayer Perceptron are the best, but no significant statistical difference could be observed among them. Thus, the methods can be used, with similar efficiency, to classify clients for a term deposit.

For the *Default of Credit Card Clients dataset*, we applied to the undersample to balance the dataset. Afterwards, we employed the methods. Subsequently, we used cross-validation methods to evaluate them, and finally, we employed Optuna over the methods in the dataset. The results are shown in Table 2.

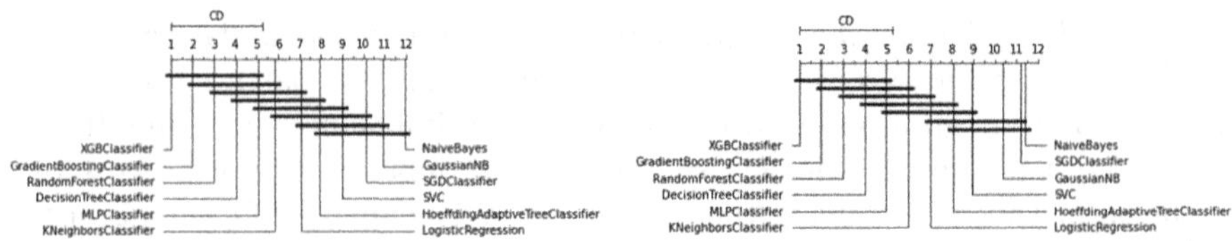

Fig. 1. Critical difference diagram over F1 and AUROC measures obtained by Optuna experiment in the Bank Marketing dataset.

Table 2. Default of Credit Card Clients dataset experiments.

2a - Cross-validation					2b - Optimization with Optuna				
Method	F1		AUROC		Method	F1		AUROC	
	Mean	Std	Mean	Std		Mean	Std	Mean	Std
GradB	69.03	0.00	71.44	0.01	XGB	69.22	0.00	71.37	0.00
XGB	68.97	0.00	70.06	0.00	RandFC	69.18	0.04	71.66	0.08
RandFC	68.32	0.19	70.49	0.21	GradB	68.21	0.17	69.37	0.12
INB	67.95	0.00	57.64	0.00	INB	67.95	0.00	57.64	0.00
GNB	67.43	0.00	54.17	0.00	GNB	67.43	0.00	54.17	0.00
HT	64.34	1.51	67.00	0.77	DT	66.30	0.03	67.33	0.03
SVC	62.91	0.00	60.85	0.00	HT	65.03	0.92	67.84	0.49
DT	62.59	0.23	62.27	0.20	KNN	64.53	0.00	60.85	0.00
LR	61.23	0.00	60.31	0.00	SVC	62.91	0.00	60.85	0.00
KNN	59.09	0.00	58.26	0.00	LR	61.29	0.00	60.42	0.00
NN	56.85	3.91	57.78	0.73	NN	57.86	2.31	57.63	1.01
SGD	51.06	4.13	53.41	1.33	SGD	51.06	4.13	53.41	1.33

As one can observe in Table 2, the lowest values were obtained by Stochastic Gradient Descent. The best values were achieved by XGBoost, with 0% of standard deviation for both metrics, using these Optuna hyperparameters: 'booster': 'gbtree', 'lambda': 2.8480032815406225e−07, 'alpha': 7.313771237085843e−06, 'max_depth': 3, 'eta': 0.13530501122715882, 'gamma': 0.0007703150869627277, 'grow_policy': 'lossguide'.

Figure 2 shows the Critical difference diagram constructed using Nemenyi test for F1 Score and AUROC in the Default of Credit Card Clients dataset cross-validation.

As one can see in Fig. 2, XGBoost, Random Forest, Gradient Boosting, Incremental and Gaussian Naive Bayes are the best, but no significant statistical difference could be observed among them. Thus, the above methods can be used with similar efficiency for classifying a credible client.

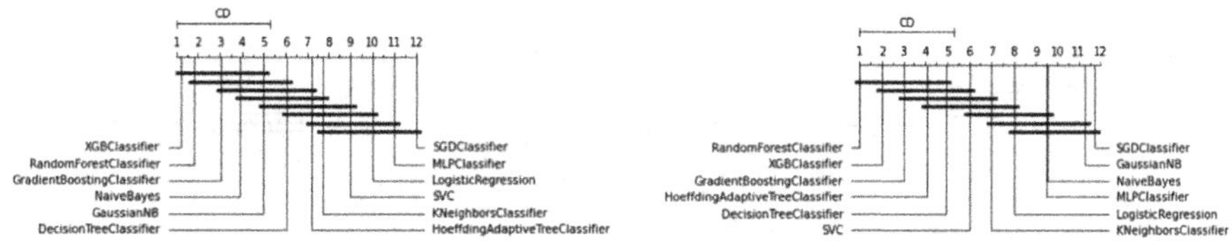

Fig. 2. Critical difference diagram over F1 and AUROC measures obtained by Optuna experiment in the Default of Credit Card Clients dataset.

For the *Kaggle Credit Card dataset*, we applied to the undersample for balancing the dataset. Subsequently, we used cross-validation methods to evaluate them, and finally, we employed Optuna over the methods in the dataset. The results are shown in Table 3.

Table 3. Kaggle Credit Card dataset experiments.

3a - Cross-validation					3b - Optimization with Optuna				
Method	F1		AUROC		Method	F1		AUROC	
	Mean	Std	Mean	Std		Mean	Std	Mean	Std
GradB	93.72	0.12	93.86	0.12	XGB	94.26	0.00	94.43	0.00
XGB	93.67	0.00	93.87	0.00	GradB	93.87	0.11	93.94	0.15
RandFC	93.02	0.28	93.29	0.20	RandFC	93.14	0.26	93.35	0.21
LR	92.14	0.00	92.40	0.00	LR	91.87	0.00	92.17	0.00
DT	90.21	0.36	90.32	0.31	DT	90.57	0.42	90.75	0.31
INB	89.51	0.00	90.31	0.00	INB	89.51	0.00	90.31	0.00
HT	89.44	0.20	90.24	0.20	HT	89.46	0.21	90.22	0.15
GNB	81.75	0.00	84.37	0.00	GNB	81.75	0.00	84.37	0.00
KNN	60.58	0.00	61.89	0.00	KNN	63.83	0.00	64.39	0.00
NN	57.59	4.87	57.38	4.80	NN	58.59	4.84	58.10	2.66
SVC	51.86	0.00	57.51	0.00	SVC	51.86	0.00	57.51	0.00
SGD	49.15	8.23	50.14	0.50	SGD	49.15	8.23	50.14	0.50

As one can observe in Table 3, the lowest values were obtained by Stochastic Gradient Descent. The best values were achieved by XGBoost, with respect 0% of standard deviation for both metrics, using these Optuna hyperparameters: 'booster': 'gbtree', 'lambda': 3.516629984966917e−05, 'alpha': 0.00041663981021167635, 'max_depth': 4, 'eta': 0.0562730158760733, 'gamma': 1.6266386520410115e−06, 'grow_policy': 'lossguide'.

Figure 3 brings the Critical Difference Diagram constructed using Nemenyi test for F1 Score and AUROC in the Kaggle Credit Card dataset cross-validation.

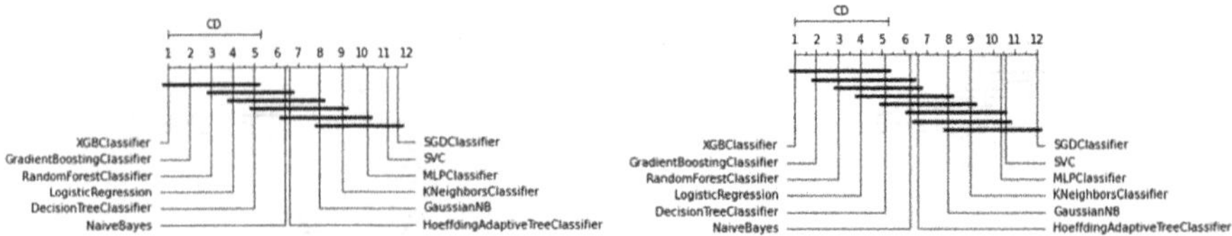

Fig. 3. Critical difference diagram over F1 and AUROC measures obtained by Optuna experiment in the Kaggle Credit Card dataset.

As one can note in Fig. 3, XGBoost, Gradient Boosting, Random Forest and Logistic Regression obtained the best results, but no significant statistical difference could be observed among them. Thus, the above methods can be used with similar efficiency for classifying creditable client.

For Statlog German Credit dataset, we applied SMOTE algorithm to balance the dataset. In SMOTE, the minority class is oversampled by duplicating samples. Depending on the oversampling required, numbers of nearest neighbors are randomly chosen. Subsequently, we used cross-validation methods to evaluate them, and finally, we employed Optuna over the methods in the dataset. The results are shown in Table 4.

Table 4. Statlog German Credit experiments.

4a - Cross-validation					4b - Optimization with Optuna				
Method	F1		AUROC		Method	F1		AUROC	
	Mean	Std	Mean	Std		Mean	Std	Mean	Std
XGB	83.72	0.00	83.28	0.00	XGB	84.54	0.00	84.34	0.00
RandFC	83.45	0.43	83.05	0.45	GradB	84.47	0.31	83.96	0.37
GradB	82.51	0.00	81.83	0.00	RandFC	83.81	0.31	83.25	0.24
NN	81.86	0.63	81.53	0.56	KNN	83.08	0.00	80.82	0.00
LR	77.61	0.00	77.34	0.00	NN	82.59	0.04	82.27	0.46
KNN	77.51	0.00	74.20	0.00	LR	79.24	0.00	78.93	0.00
INB	77.17	0.00	74.39	0.00	DT	78.87	0.04	76.60	0.04
GNB	77.08	0.00	74.32	0.00	INB	77.17	0.00	74.39	0.00
DT	75.94	0.50	74.86	0.58	GNB	77.08	0.00	74.32	0.00
HT	75.75	0.92	73.86	0.56	HT	73.41	1.60	71.99	1.60
SVC	73.11	0.00	71.21	0.00	SVC	73.11	0.00	71.21	0.00
SGD	71.02	2.27	67.31	2.69	SGD	71.02	2.27	67.31	2.69

As one can observe in Table 4, the lowest values were obtained by Stochastic Gradient Descent. The best results were achieved by XGBoost, with respect 0% of standard deviation for both metrics, using these Optuna hyperparameters: 'booster': 'gbtree', 'lambda': 4.390180375617503e−05, 'alpha': 2.7519932161346493e−05, 'max_depth': 5, 'eta': 0.22105660872526461, 'gamma': 0.06059710183782625, 'grow_policy': 'depthwise'.

Figure 4 shows the Critical difference diagram constructed using Nemenyi test for F1 Score and AUROC in the Statlog German Credit dataset.

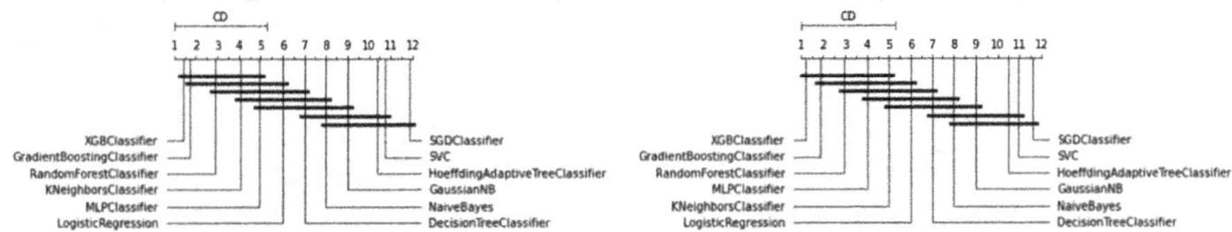

Fig. 4. Critical difference diagram over F1 and AUROC measures obtained by Optuna experiment in the Statlog German Credit dataset

As one can note in Fig. 4, XGBoost, Gradient Boosting, Random Forest, kNN, and Multilayer Perceptron obtained the best results, but no significant statistical difference could be observed among them. Thus, the aforementioned methods can be employed with similar efficiency for classifying who is credible client.

For the Statlog Australian Credit Approval dataset, we just used cross-validation methods for evaluation and employed Optuna over the methods in the dataset. The results are shown in Table 5.

Table 5. Statlog Australian Credit dataset experiments.

5a - Cross-validation					5b - Optimization with Optuna				
Method	F1		AUROC		Method	F1		AUROC	
	Mean	Std	Mean	Std		Mean	Std	Mean	Std
XGB	86.64	0.00	87.95	0.00	XGB	87.38	0.00	88.60	0.00
RandFC	85.81	0.70	87.10	0.48	GradB	86.50	0.00	88.00	0.00
HT	84.66	0.64	86.06	0.36	RandFC	86.20	0.18	87.48	0.31
GradB	84.41	0.20	85.97	0.16	DT	85.19	0.00	86.34	0.00
LR	79.31	0.00	81.83	0.00	HT	84.72	0.50	85.76	0.63
DT	78.03	0.81	80.78	0.60	NN	81.22	0.89	83.27	0.59
INB	76.43	0.00	79.73	0.00	LR	80.53	0.00	82.84	0.00
GNB	75.50	0.00	79.11	0.00	INB	76.43	0.00	79.73	0.00
NN	72.43	2.39	75.86	1.91	GNB	75.50	0.00	79.11	0.00
KNN	61.23	0.00	67.28	0.00	SGD	61.67	2.87	62.75	2.24
SGD	61.67	2.87	62.75	2.24	KNN	61.23	0.00	67.28	0.00
SVC	50.12	0.00	64.51	0.00	SVC	50.12	0.00	64.51	0.00

As one can observe in Table 5, the worst values were obtained by Support Vector Classifier. The best values were achieved by XGBoost, using these Optuna hyperparameters: 'booster': 'gbtree', 'lambda': 1.7500758050198892e−06, 'alpha': 4.58329922951 2845e−06, 'max_depth': 5, 'eta': 0.568694520947508, 'gamma': 6.081656560005183e−05, 'grow_policy': 'depthwise'.

Figure 5 shows the Critical Difference Diagram constructed using Nemenyi Test for F1 Score and AUROC in the Statlog German Credit dataset.

Fig. 5. Critical difference diagram over F1 and AUROC measures obtained by Optuna experiment in the Statlog Australian Credit dataset.

As one can see in Fig. 5, XGBoost, Gradient Boosting, Decision Tree, and Hoeffding Tree were considered the best, but no significant statistical difference can be observed among them. Thus, the methods can be used with similar efficiency for classifying who is a credible client.

6 Evaluations of the Methods with Datasets of Domains Other Than Finance

In this section, we show the results of the methods with datasets of domains other than finance. We used three datasets in the healthcare and ionosphere domains to verify the performance of the classification methods.

The *Heart Disease* dataset contains information on patient's heart exams. The complete dataset has 76 attributes. Published experiences refer to the use of a subset with no missing values and 14 numerical attributes, such as client personal data and cardiac test results. We used the Cleveland database. Its purpose is to classify who has or not a heart disease. It is available at https://archive.ics.uci.edu/ml/datasets/Heart+Disease [11].

For the Heart Disease dataset, we just employed the predictive methods without preprocessing the data. The results are shown in Table 6.

As one can observe in Table 6, the worst values were achieved by Support Vector Classifier and Stochastic Gradient Descent. The best values were achieved by XGBoost, with respect 0% of standard deviation for both metrics, using the Optuna hyperparameters.

Figure 6 shows the Critical Difference Diagram constructed using Nemenyi Test between F1 Score and AUROC of Heart Disease dataset.

As one can note, Logistic Regression, Gaussian and Incremental Naive Bayes, Hoeffding Tree, and Random Forest are considered the best, but no significant

Table 6. Heart Disease dataset experiments.

6a - Cross-validation					6b - Optimization with Optuna				
Method	F1		AUROC		Method	F1		AUROC	
	Mean	Std	Mean	Std		Mean	Std	Mean	Std
HT	80.18	0.83	82.08	1.03	XGB	81.35	0.00	83.29	0.00
LR	80.17	0.00	82.43	0.00	LR	80.85	0.00	82.89	0.00
GNB	80.16	0.00	81.99	0.00	HT	80.80	0.80	82.20	0.79
INB	80.16	0.00	81.99	0.00	RandFC	80.51	0.78	82.45	0.78
RandFC	79.33	1.05	81.59	0.82	GNB	80.16	0.00	81.99	0.00
GradB	76.81	0.14	78.86	0.21	INB	80.16	0.00	81.99	0.00
NN	76.54	1.66	80.05	1.17	GradB	79.49	0.00	81.53	0.00
XGB	75.60	0.00	77.67	0.00	NN	78.25	1.19	80.59	0.43
DT	70.80	0.91	72.87	0.88	DT	73.64	0.00	75.07	0.00
KNN	62.39	0.00	66.95	0.00	KNN	62.39	0.00	66.95	0.00
SGD	57.72	3.65	62.18	3.47	SGD	57.72	3.65	62.18	3.47
SVC	51.00	0.00	62.23	0.00	SVC	51.00	0.00	62.23	0.00

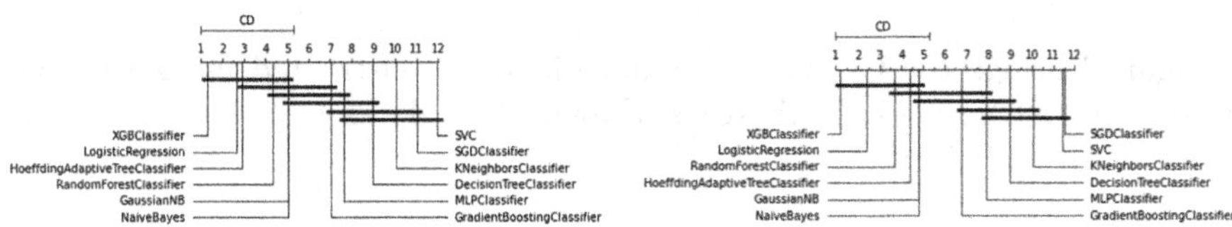

Fig. 6. Critical difference diagram over F1 and AUROC measures obtained by Optuna experiment in the Heart Disease dataset.

statistical difference is observed among them. Thus, the above methods can be used with similar efficiency for classifying who has heart disease.

The *Ionosphere* dataset consists of a phased array of 16 high-frequency antennas with a total transmitted power in the order of 6.4 kW. The targets are free electrons in the ionosphere. "Good" radar returns are those showing evidence of some type of structure in the ionosphere. "Bad" returns are those pass through the ionosphere. The purpose of using the dataset is to classify what is returned from the radar.

The dataset is multivariate with 351 instances (224 instances are "Good"), 34 attributes (32 real and 2 integer), and no missing values. The dataset is available at https://archive.ics.uci.edu/ml/datasets/ionosphere [11].

For the Ionosphere dataset, we just employed the predictive methods without preprocessing the data. The results are shown in Table 7.

As one can observe in Table 7, the lowest values were obtained by kNN. The best results were achieved by XGBoost, with respect 0% of standard deviation for both metrics, using Optuna hyperparameters.

Table 7. Ionosphere dataset experiments.

7a - Cross-validation					7b - Optimization with Optuna				
Method	F1		AUROC		Method	F1		AUROC	
	Mean	Std	Mean	Std		Mean	Std	Mean	Std
XGB	90.82	0.00	92.47	0.00	XGB	92.23	0.00	93.61	0.00
SVC	90.56	0.00	91.87	0.00	GradB	90.83	0.32	92.58	0.30
GradB	90.24	0.36	91.68	0.35	RandFC	90.62	0.20	92.20	0.11
RandFC	90.07	0.40	92.11	0.22	SVC	90.56	0.00	91.87	0.00
NN	87.89	0.84	89.70	0.60	NN	88.45	1.03	90.46	0.60
GNB	83.41	0.00	86.06	0.00	DT	86.92	0.34	90.59	0.25
DT	83.38	1.28	86.71	0.85	GNB	83.41	0.00	86.06	0.00
LR	78.57	0.00	82.65	0.00	LR	82.35	0.00	85.52	0.00
SGD	78.38	1.86	83.28	1.40	SGD	78.68	1.84	82.21	1.68
INB	77.10	0.00	82.75	0.00	HT	77.17	1.78	82.43	1.88
HT	74.91	1.88	80.03	3.27	INB	77.10	0.00	82.75	0.00
KNN	73.51	0.00	79.12	0.00	KNN	77.08	0.00	81.55	0.00

Figure 7 brings the Critical difference diagram constructed using Nemenyi test between F1 Scores and AUROCs of Ionosphere dataset.

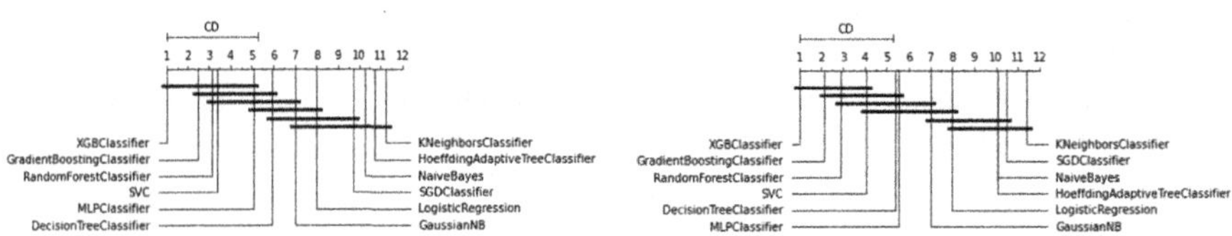

Fig. 7. Critical difference diagram over F1 and AUROC measures obtained by Optuna experiment in the Ionosphere dataset.

As one can note in Fig. 7, XGBoost, Gradient Boosting, Random Forest, SVC and Multilayer Perceptron methods are considered the best, but no significant statistical difference can be observed among them. Thus, the above methods can be used with similar efficiency to classify what is returned from the radar.

The *Blood Transfusion Service Center* dataset is intended to evaluate the blood donation marketing model. To build the model, 748 donors were selected from the donor database. The donor dataset includes the following information: months since last donation, total number of donations, volume of blood donated, months since first donation [39]. The purpose of using the dataset is to classify who can donate blood.

For the Transfusion dataset, we applied the SMOTE algorithm, and employed the methods. The results are shown in Table 8.

Table 8. Transfusion dataset experiments.

8a - Cross-validation					8b - Optimization with Optuna				
Method	F1		AUROC		Method	F1		AUROC	
	Mean	Std	Mean	Std		Mean	Std	Mean	Std
XGB	77.44	0.00	77.28	0.00	XGB	78.32	0.00	78.26	0.00
RandFC	76.57	0.52	76.54	0.50	RandFC	76.77	0.18	76.45	0.31
GradB	75.98	0.03	75.99	0.04	GradB	76.72	0.00	76.75	0.00
KNN	72.40	0.00	71.52	0.00	DT	73.13	0.00	73.30	0.00
LR	71.98	0.00	70.53	0.00	LR	72.03	0.00	70.53	0.00
DT	71.89	0.23	72.76	0.34	HT	70.40	0.50	66.40	0.63
HT	68.93	1.91	66.05	0.85	KNN	69.56	0.00	67.90	0.00
INB	68.58	0.00	68.02	0.00	INB	68.58	0.00	68.02	0.00
GNB	68.51	0.00	67.92	0.00	GNB	68.51	0.00	67.92	0.00
SVC	64.22	0.00	62.84	0.00	NN	64.57	0.89	64.76	0.59
NN	62.89	5.06	64.74	1.93	SVC	64.22	0.00	62.84	0.00
SGD	62.09	7.25	59.78	2.09	SGD	60.97	2.87	59.92	2.24

As one can observe in Table 8, the lowest values were obtained by Stochastic Gradient Descent, and the best values were achieved by XGBoost, with respect 0% to standard deviation for both metrics, using Optuna hyperparameters.

Figure 8 shows the Critical difference diagram constructed using Nemenyi test between F1 Scores and AUROCs of the Transfusion Blood dataset.

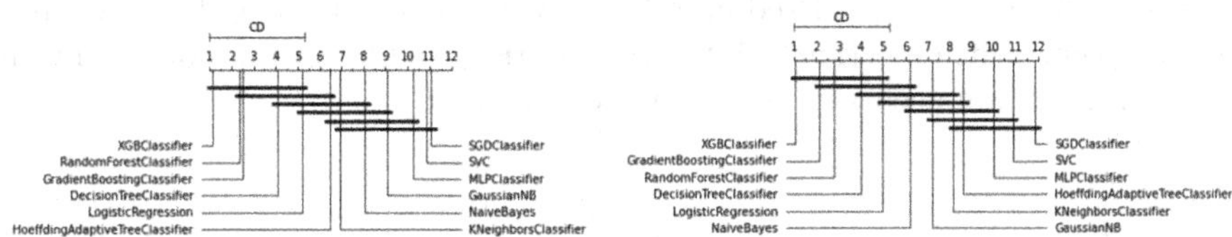

Fig. 8. Critical difference diagram over F1 and AUROC measures obtained by Optuna experiment in the Transfusion Blood dataset.

As one can note, XGBoost, Random Forest, Gradient Boosting, Decision Tree, and Logistic Regression obtained the best results, but no significant statistical difference could be observed among them. Therefore the aforementioned methods can be used with similar efficiency for classifying if a person can donate blood.

7 Evaluation of Concept Drift on Financial Domains

For the concept drift evaluation, it is necessary to keep the statistical streaming behavior of the dataset avoiding the use of the shuffle and balancing techniques. The Bank Marketing and Default of Credit Card Clients datasets have enough size and balance, allowing us to do the incremental learning adequately.

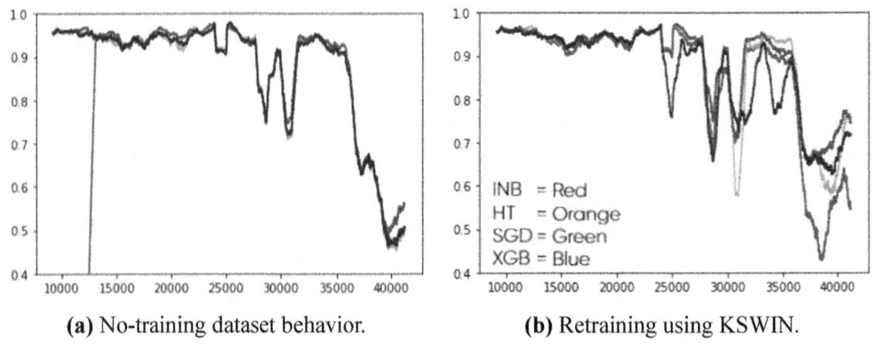

(a) No-training dataset behavior. **(b)** Retraining using KSWIN.

Fig. 9. Bank Marketing dataset concept drift detection evaluation experiment.

In this section, we present the results of application of seven concept drift detection methods with four incremental predictors (Naive Bayes, Hoeffding Tree, Stochastic Gradient Descent, and XGBoost). We tested using two steps. In the first step, we did an initial training (with 20% size of the dataset). In second step, we performed the experiments using the following schemes: (i) no retraining, (ii) retraining at fixed intervals (5% of the datasets), and (iii) retraining at concept drift detection points. The metrics used for evaluation are mean accuracy, minimum accuracy, and mean square error.

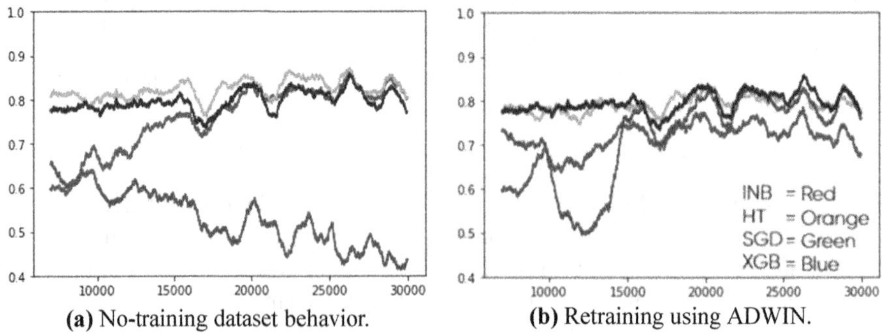

(a) No-training dataset behavior. **(b)** Retraining using ADWIN.

Fig. 10. Default of Credit Card Clients dataset concept drift detection evaluation experiment. (Color figure online)

Table 9. Incremental Learning scores for Bank Marketing dataset.

	NO_RETR	FIXED	ADWIN	DDM	EDDM	HDDMa	HDDMw	KSWIN	PAGE	
Method	Naive Bayes (INB)									Mean
mean acc	78.34	79.40	84.66	78.72	85.36	78.34	83.36	85.61	78.46	81.61
min acc	5.50	22.10	35.90	43.40	20.00	5.50	41.00	42.69	20.10	26.76
mse	11.17	7.21	4.53	7.94	4.59	11.17	5.13	4.10	8.57	7.15
instances	0/0	7/7	7/12	5/5	4/11	0/0	13/14	14/15	13/18	
Method	Hoeffding Tree (HT)									Mean
mean acc	87.11	87.05	87.34	87.59	87.79	87.15	87.17	88.19	86.57	87.36
min acc	45.30	47.50	53.70	47.59	46.80	45.30	51.20	57.59	47.50	49.37
mse	3.53	3.39	2.87	3.27	3.31	3.52	2.98	2.63	3.56	3.21
instances	0/0	7/7	12/16	6/8	1/1	0/0	9/13	9/17	10/12	
Method	Stochastic Gradient Descent (SGD)									Mean
mean acc	88.04	87.88	88.11	86.92	87.61	87.37	88.35	89.10	87.90	87.93
min acc	48.40	49.70	50.90	48.69	46.20	45.70	63.90	64.90	48.90	50.38
mse	3.08	3.07	2.96	3.23	3.41	3.46	2.21	2.03	3.10	2.94
instances	0/0	7/7	10/13	9/11	2/9	0/0	7/8	6/8	7/7	
Method	XGBOOST (XGB)									Mean
mean acc	87.16	87.71	88.61	87.42	87.38	87.16	87.15	86.33	87.23	87.31
min acc	46.00	49.70	60.80	48.19	48.10	46.00	60.19	62.50	48.40	51.10
mse	3.45	2.80	2.35	3.21	3.27	3.45	2.78	2.94	3.25	3.09
instances	0/0	7/7	11/14	6/8	2/3	0/0	11/12	9/11	6/7	
	Concept drift detection method score (means)									
mean acc	85.16	85.51	87.18	85.16	87.04	85.01	86.51	87.31	85.04	
min acc	36.30	42.25	50.33	46.97	40.28	35.63	54.07	50.14	41.23	
mse	5.31	4.12	3.18	4.41	3.65	5.40	3.28	2.93	4.62	

Figure 9 and Table 9 show the performance of the Naive Bayes (INB), Hoeffding Tree (HT), Stochastic Gradient Descent (SGD), and XGBoost (XGB) methods for the Bank Marketing dataset. The figure shows an abrupt concept drift behavior as one can see by the sharp forecast changes at 25,000 and 35,000 instances. Small retraining windows (1,800–2,000) and small delay size (100–500) were used to deal with this behavior. All the concept drift detection methods provided better results with this dataset when compared with the results obtained with the schemes no retraining or retraining with a fixed window size. The best accuracy and mean square error values were achieved by KSWIN, with 87.31% and 2.93% respectively, followed by ADWIN and HDDMw. The Stochastic Gradient Descent, XGBoost and Hoeffding Tree showed similar good results on these experiments, but the best values were achieved by combining SGD and KSWIN, with 89.10% of mean accuracy, 64.90% of minimum accuracy, and 2.03% of mean square error.

Figure 10 and Table 10 show the performance of the Naive Bayes (INB), Hoeffding Tree (HT), Stochastic Gradient Descent (SGD), and XGBoost (XGB) methods for the Default of Credit Card Clients dataset. It shows no concept drift with some oscillation without compromising the long term accuracy. Long retraining periods (5,000) and detection delays (3,000) are preferable for this dataset. Table 9 shows no significant statistical difference on all metrics for the

Table 10. Incremental Learning scores for Default of Credit Card Clients dataset.

	NO_RETR	FIXED	ADWIN	DDM	EDDM	HDDMa	HDDMw	KSWIN	PAGE	
Method	Naive Bayes (INB)									Mean
mean acc	75.75	68.85	70.82	54.27	60.75	75.75	70.92	68.19	62.83	67.69
min acc	58.19	56.49	58.19	40.69	53.20	58.19	58.19	57.49	53.70	56.15
mse	6.40	10.15	8.71	21.29	15.53	6.40	8.69	10.56	13.97	11.03
instances	0/0	7/7	3/6	1/1	1/10	0/0	6/12	4/14	2/4	
Method	Hoeffding Tree (HT)									Mean
mean acc	82.32	82.00	79.15	82.32	78.29	82.32	81.38	79.86	79.23	80.22
min acc	75.90	75.90	74.40	75.80	73.30	75.90	76.00	74.30	74.60	74.82
mse	3.16	3.28	4.37	3.20	4.74	3.16	3.51	4.12	4.35	3.98
instances	0/0	7/7	2/3	1/3	1/16	0/0	2/3	2/3	1/2	
Method	Stochastic Gradient Descent (SGD)									Mean
mean acc	52.96	68.47	72.46	69.41	68.54	63.30	60.67	61.34	77.48	66.27
min acc	41.19	48.40	49.40	55.40	56.20	54.00	32.40	19.60	69.00	46.24
mse	22.52	11.05	8.47	9.58	10.10	13.64	19.04	19.66	5.14	13.52
instances	0/0	7/7	3/11	0/0	0/0	0/0	4/10	3/11	1/1	
Method	XGBOOST (XGB)									Mean
mean acc	79.59	79.96	79.59	80.54	79.59	79.59	80.36	80.53	79.99	80.01
min acc	73.20	73.90	73.20	73.20	73.20	73.20	72.80	73.20	73.20	73.12
mse	4.21	4.06	4.21	3.87	4.21	4.21	3.93	3.87	4.06	4.06
instances	0/0	7/7	0/0	2/7	0/0	0/0	2/4	1/2	1/1	
	Concept drift detection method score (mean)									
mean acc	72.66	74.82	75.51	71.64	71.79	75.24	73.33	72.48	74.88	
min acc	62.12	63.67	63.80	61.27	63.98	65.32	59.85	56.15	67.63	
mse	9.07	7.14	6.44	9.49	8.65	6.85	8.79	9.55	6.88	

Hoeffding Tree and XGBoost methods. The exception is the Stochastic Gradient Descent method (green line on Fig. 10) that was able to recover the performance after initial retraining.

We also experienced the retraining and delay configuration used in Bank Marketing to train this experiment. Statistical similar results were obtained. We obtained best results for HDDMw with 77.05% of accuracy, 65.62% of minimum accuracy and 5.62% of mean square error, followed by KSWIN (75.37%, 64.40%, 6.48%) and ADWIN (75.53%, 62.48%, and 6.49%). This shows that even for conceptually stable datasets this configuration might be used.

The tables show the mean scores obtained with the concept drift detection methods tested with corresponding base learners. The footer part of the tables named Concept Drift Score present the average score result of each drift detector.

8 Concluding Remarks

This study investigated supervised classification methods for financial problems with focus on risk, fraud and credit analysis. Twelve supervised predictive methods were employed in five public financial datasets. The methods were evaluated using the classification performance metrics *F1 Score* and *AUROC*. The non-parametric Friedman Test was used to infer hypotheses and the Nemeyni test to validate it with Critical difference diagram.

We were able to confirm the results in [29]. In the financial domain, best results were obtained with the decision tree family of classification methods, XGBoost regularly showing good results. In other domains (health care/ionosphere), XGBoost also showed good results, although not systematically better than the other methods. Optuna optimization also showed better results in all evaluations, keeping XGBoost as the best method for the financial domain.

We also evaluated the concept drift phenomenon. For that, we investigated incremental learning for financial problems, applying seven concept drift detection methods in the Bank Marketing and Default of Credit Card Clients datasets. Four incremental supervised predictive methods were evaluated: Naive Bayes, Hoeffding Tree, Stochastic Gradient Descent, and XGBoost. The methods were evaluated using *mean accuracy*, *minimum accuracy*, and *mean square error* scores. We obtained the best results by combining the concept drift detection methods HDDMw, KSWIN and ADWIN with the classification methods Hoeffding Tree and XGBoost. Table 11 summarizes the experiments' results. The tables show the mean scores obtained with the concept drift detection methods tested with corresponding base learners, employed over both experiments.

Table 11. Average performance measured from concept drift experiments.

11a - Concept drift detection methods

	NO_RETR	FIXED	ADWIN	DDM	EDDM	HDDMa	HDDMw	KSWIN	PAGE
Mean acc (avg)	78.91	80.16	81.35	78.40	79.41	80.12	81.77	81.34	79.96
Min acc (avg)	49.21	52.96	56.40	54.12	52.12	50.47	59.80	57.27	54.42
Mse (avg)	7.19	5.63	4.83	6.95	6.15	6.13	4.45	4.70	5.75

11b - Predictive methods

	INB	HT	SGD	XGB
Mean acc (avg)	74.65	83.79	77.09	83.66
Min acc (avg)	41.46	62.10	48.31	62.11
Mse (avg)	9.09	3.59	8.22	3.57

Nielsen [26] explains that there are some reasons for the good performance of XGBoost. XGBoost can be seen as a Newton's method of numerical optimization, using a higher-order approximation at each iteration, being capable of learning "better" tree structures. Second, it provides clever penalization of individual trees, turning it to be more adaptive than other boosting methods, because it determines the appropriate number of terminal nodes, which might vary among trees. Finally, XGBoost is a highly adaptive method, which carefully takes the bias-variance trade-off into account in nearly every aspect of the learning process.

Other non-tree methods investigated, Naive Bayes, Support Vector Classifier and k-Nearest Neighbors, have shown performance worse than the decision tree based classification methods, indicating that the non-tree methods are not the recommended ones for the financial problems we investigated.

Confirming [29], we conclude that XGBoost is the recommended machine learning classification method to be overcome when proposing new methods for problems of analyzing risk, fraud, and credit.

In the financial domain, data statistics typically change over time diverging from the one used for the classification method training. As a result of this characteristic, the prediction of previously trained models becomes less accurate. The concept drift detection methods are used to identify when this behavior occurs. We investigated seven concept drift detection methods in order to identify which supervised methods perform best on problems of the financial domain. We verified that they help to provide better accuracy when compared to the no retraining or the fixed interval retraining approaches. We also identified HDDMw as the recommended concept drift detector and XGBoost as the preferred base learner for datasets with concept drift in financial domain.

References

1. Akiba, T., Sano, S., Yanase, T., Ohta, T., Koyama, M.: Optuna: a next-generation hyperparameter optimization framework. In: Proceedings of the 25th ACM SIGKDD International Conference on Knowledge Discovery and Data Mining, pp. 2623–2631. ACM (2019)
2. Bache, K., Lichman, M.: UCI machine learning repository, vol. 28. School of Information and Computer Science, University of California, Irvine, CA (2013). http://archive.ics.uci.edu/ml
3. Barros, R.S.M., Santos, S.G.T.C.: A large-scale comparison of concept drift detectors. Inf. Sci. **451**, 348–370 (2018)
4. Bifet, A., Gavaldà, R.: Adaptive learning from evolving data streams. In: Adams, N.M., Robardet, C., Siebes, A., Boulicaut, J.-F. (eds.) IDA 2009. LNCS, vol. 5772, pp. 249–260. Springer, Heidelberg (2009). https://doi.org/10.1007/978-3-642-03915-7_22
5. Bottou, L.: Large-scale machine learning with stochastic gradient descent. In: Lechevallier, Y., Saporta, G. (eds.) Proceedings of COMPSTAT 2010, pp. 177–186. Springer, Heidelberg (2010). https://doi.org/10.1007/978-3-7908-2604-3_16
6. Bouazza, I., Ameur, E.B., Ameur, F.: Datamining for fraud detecting, state of the art. In: Ezziyyani, M. (ed.) AI2SD 2018. AISC, vol. 915, pp. 205–219. Springer, Cham (2019). https://doi.org/10.1007/978-3-030-11928-7_17
7. Breiman, L.: Random forests. Mach. Learn. **45**(1), 5–32 (2001)
8. Chen, T., Guestrin, C.: XGBoost: a scalable tree boosting system. In: Proceedings of the 22nd ACM SIGKDD International Conference on Knowledge Discovery and Data Mining, pp. 785–794. ACM (2016)
9. Dal Pozzolo, A., Caelen, O., Johnson, R.A., Bontempi, G.: Calibrating probability with undersampling for unbalanced classification. In: 2015 IEEE Symposium Series on Computational Intelligence, pp. 159–166. IEEE (2015)
10. Damodaran, A.: Corporate Finance. Wiley, Hoboken (1996)
11. Dua, D., Graff, C.: UCI machine learning repository (2017). http://archive.ics.uci.edu/ml
12. Frías-Blanco, I., del Campo-Ávila, J., Ramos-Jimenez, G., Morales-Bueno, R., Ortiz-Díaz, A., Caballero-Mota, Y.: Online and non-parametric drift detection methods based on Hoeffding's bounds. IEEE Trans. Knowl. Data Eng. **27**(3), 810–823 (2014)

13. García, S., Fernández, A., Luengo, J., Herrera, F.: Advanced nonparametric tests for multiple comparisons in the design of experiments in computational intelligence and data mining: Experimental analysis of power. Inf. Sci. **180**(10), 2044–2064 (2010)
14. Géron, A.: Hands-On Machine Learning with Scikit-Learn, Keras, and TensorFlow: Concepts, Tools, and Techniques to Build Intelligent Systems. O'Reilly Media, Newton (2019)
15. Gonçalves, P.M., Jr., de Carvalho Santos, S.G., Barros, R.S., Vieira, D.C.: A comparative study on concept drift detectors. Expert Syst. Appl. **41**(18), 8144–8156 (2014)
16. Hofmann, H.: Statlog (German credit data) data set. UCI Repository of Machine Learning Databases (1994)
17. Hulten, G., Spencer, L., Domingos, P.: Mining time-changing data streams. In: ACM SIGKDD International Conference on Knowledge Discovery and Data Mining, pp. 97–106. ACM Press (2001)
18. Islam, M.J., Wu, Q.J., Ahmadi, M., Sid-Ahmed, M.A.: Investigating the performance of Naive-Bayes classifiers and k-Nearest Neighbor classifiers. In: 2007 International Conference on Convergence Information Technology (ICCIT 2007), pp. 1541–1546. IEEE (2007)
19. Lavanya, D., Rani, K.U.: Performance evaluation of decision tree classifiers on medical datasets. Int. J. Comput. Appl. **26**(4), 1–4 (2011)
20. Liaw, A., Wiener, M., et al.: Classification and regression by randomForest. R News **2**(3), 18–22 (2002)
21. Lin, W.Y., Hu, Y.H., Tsai, C.F.: Machine learning in financial crisis prediction: a survey. IEEE Trans. Syst. Man Cybern. Part C (Appl. Rev.) **42**(4), 421–436 (2011)
22. Montiel, J., Read, J., Bifet, A., Abdessalem, T.: Scikit-multiflow: a multi-output streaming framework. J. Mach. Learn. Res. **19**(72), 1–5 (2018). http://jmlr.org/papers/v19/18-251.html
23. Moro, S., Cortez, P., Rita, P.: A data-driven approach to predict the success of bank telemarketing. Decis. Support Syst. **62**, 22–31 (2014)
24. Natekin, A., Knoll, A.: Gradient boosting machines, a tutorial. Front. Neurorobot. **7**, 21 (2013)
25. Ng, A.Y., Jordan, M.I.: On discriminative vs. generative classifiers: a comparison of logistic regression and Naive Bayes. In: Advances in Neural Information Processing Systems, pp. 841–848 (2002)
26. Nielsen, D.: Tree boosting with XGBoost-why does XGBoost win "Every" machine learning competition? Master's thesis, NTNU (2016)
27. Pearl, J., Glymour, M., Jewell, N.P.: Causal Inference in Statistics: A Primer. Wiley, Hoboken (2016)
28. Peterson, L.E.: K-nearest neighbor. Scholarpedia **4**(2), 1883 (2009)
29. Pugliese, V.U., Hirata, C.M., Costa, R.D.: Comparing supervised classification methods for financial domain problems. In: ICEIS (1), pp. 440–451 (2020)
30. Quinlan, J.R.: Simplifying decision trees. Int. J. Man Mach. Stud. **27**(3), 221–234 (1987)
31. Raab, C., Heusinger, M., Schleif, F.M.: Reactive soft prototype computing for concept drift streams. Neurocomputing **416**, 340–351 (2020)
32. Ridgeway, G.: The state of boosting. In: Computing Science and Statistics, pp. 172–181 (1999)
33. Rish, I., et al.: An empirical study of the Naive Bayes classifier. In: IJCAI 2001 Workshop on Empirical Methods in Artificial Intelligence, vol. 3, pp. 41–46 (2001)

34. Roberts, S.: Control chart tests based on geometric moving averages. Technometrics **42**(1), 97–101 (2000)
35. Sinayobye, J.O., Kiwanuka, F., Kyanda, S.K.: A state-of-the-art review of machine learning techniques for fraud detection research. In: 2018 IEEE/ACM Symposium on Software Engineering in Africa (SEiA), pp. 11–19. IEEE (2018)
36. Suykens, J.A., Vandewalle, J.: Least squares support vector machine classifiers. Neural Process. Lett. **9**(3), 293–300 (1999)
37. Wang, G., Hao, J., Ma, J., Jiang, H.: A comparative assessment of ensemble learning for credit scoring. Expert Syst. Appl. **38**(1), 223–230 (2011)
38. Webb, G.I., Hyde, R., Cao, H., Nguyen, H.L., Petitjean, F.: Characterizing concept drift. Data Min. Knowl. Disc. **30**(4), 964–994 (2016). https://doi.org/10.1007/s10618-015-0448-4
39. Yeh, I.C., Yang, K.J., Ting, T.M.: Knowledge discovery on RFM model using Bernoulli sequence. Expert Syst. Appl. **36**(3), 5866–5871 (2009)
40. Yu, Q., Miche, Y., Séverin, E., Lendasse, A.: Bankruptcy prediction using extreme learning machine and financial expertise. Neurocomputing **128**, 296–302 (2014)
41. Zareapoor, M., Shamsolmoali, P.: Application of credit card fraud detection: based on bagging ensemble classifier. Proc. Comput. Sci. **48**(2015), 679–685 (2015)
42. Žliobaitė, I., Pechenizkiy, M., Gama, J.: An overview of concept drift applications. In: Japkowicz, N., Stefanowski, J. (eds.) Big Data Analysis: New Algorithms for a New Society. SBD, vol. 16, pp. 91–114. Springer, Cham (2016). https://doi.org/10.1007/978-3-319-26989-4_4

Sammon Mapping-Based Gradient Boosted Trees for Tax Crime Prediction in the City of São Paulo

André Ippolito(✉) and Augusto Cezar Garcia Lozano

Tax Intelligence Office, Under-secretariat of Municipal Revenue, Secretariat of Finance, São Paulo City Hall, São Paulo, Brazil
{aippolito,alozano}@prefeitura.sp.gov.br

Abstract. With the currently vast volume of data available, several institutions, including the public sector, benefit from information, aiming to improve decision-making. Machine Learning enhances data-driven decision-making with its predictive power. In this work, our principal motivation was to apply Machine Learning to ameliorate fiscal audit planning for São Paulo's municipality. In this study, we predicted crimes against the service tax system of São Paulo using Machine Learning. Our methodology embraced the following steps: data extraction; data preparation; dimensionality reduction; model training and testing; model evaluation; model selection. Our experimental findings revealed that Sammon Mapping (SM) combined with Gradient Boosted Trees (GBT) outranked other state-of-the-art works, classifiers and dimensionality reduction techniques as regards classification performance. Our belief is that the ensemble of classifiers of GBT, combined with SM's ability to identify relevant dimensions in data, contributed to produce higher prediction scores. These scores enable São Paulo's tax administration to rank fiscal audits according to the highest probabilities of tax crime occurrence, leveraging tax revenue.

Keywords: Data · Decision-making · Machine Learning · Fiscal · Audit · Tax · Crime · Sammon Mapping · Gradient Boosted Trees · Revenue

1 Introduction

With the huge amount of data contemporaneously available, various types of organizations try to profit from data with competitive purposes [1]. Recent advances in technologies for data processing and analysis allow companies to perform faster and more accurate analysis, providing valuable insights for decision-making. One way to enhance decision-making based on data analysis is to apply

Supported by São Paulo City Hall.

J. Filipe et al. (Eds.): ICEIS 2020, LNBIP 417, pp. 293–316, 2021.
https://doi.org/10.1007/978-3-030-75418-1_14

predictive analytics. This branch of analytics permits to unveil business opportunities, predict risks, prevent churn, segment customers, among other practical use cases, leading to an effective rise in market value.

The government sector is already taking advantage of data-driven decision-making. Public administrations ingest data from a multitude of areas, e.g. education, environment, traffic and healthcare. Analysis of these data and its insights enable governments to offer more effective and efficient public services [2]. In the public sector, tax administration is an area that can benefit from data analysis.

Tax revenue loss due to tax crimes represents a heavy burden to governments worldwide. Recent estimates surmount a US$500 billion global annual loss in tax revenue [3]. In Brazil, studies reckon that the annual loss with tax evasion in 2015 is about 7.7% of the Brazilian Gross Domestic Product (GDP) [4]. For the year 2020, statistics of the total number of active companies in Brazil amount to 21,371,725 [5]. Governments strive to identify tax crimes and they are usually immersed in an avalanche of fiscal data that turns the manual work of investigation and fiscal audit planning unfeasible.

Predictive techniques and algorithms originated from Machine Learning can help to accurately predict tax crimes, contributing to reduce government's monetary losses, by means of more effective planning of fiscal audits. With Machine Learning, it is possible to predict taxpayers' actions that are not compliant with tax laws, such as crimes against the tax system. This prediction allows tax administrations to plan their fiscal audits with more accuracy, prioritizing actions on taxpayers with higher probability of committing a crime, resulting in potentially higher levels of compliance and revenue.

The central motivation of our work is to use Machine Learning to predict tax crimes in the city of São Paulo. In our study, we use data from fiscal audits of São Paulo's tax administration, regarding service taxes. The municipality of São Paulo plays an important role in Brazil's economy. For instance, if we take the Brazilian GDP of 2016, we verify that the city of São Paulo participates with 33.71% [6]. As regards the total tax income of the Brazilian cities, São Paulo's tax revenue equals to 20% of the total amount [7]. In the year of 2018, São Paulo's tax income corresponded to 56.72% of the total revenue of the city [8].

Some tax administrations applied Machine Learning to predict tax crimes. The works of [9,10] used Neural Networks and Principal Component Analysis (PCA). The study of [11] is a cross feature-based solution with GBT that detected illicit trades in Nigeria. The use of cross features tends to cause noise and classification performance depreciation caused by the interplay between irrelevant attributes [12,13]. In our work, studies lead to the conclusion that using SM for dimensionality reduction, combined with GBT for classification, surpasses the classification performance scores of the related works. Aside from this prevalence, our work has the goal of predicting distinct kinds of crimes that are peculiar to the service tax system of São Paulo's tax administration.

In this work, we use Machine Learning and dimensionality reduction aiming to predict service tax crimes against the tax system of the municipality of São Paulo. In our methodology, we apply the following steps: data extraction; data preparation; dimensionality reduction; model training and testing; model evaluation; model selection. The results of our conducted experiments corroborate that the combination of SM and GBT provides the highest classification performance scores. To our knowledge, there are not works that predicted tax crimes against São Paulo's service tax system, based on SM and GBT. This work builds upon our previous work [14]. Comparatively, the present study has the following extensions and improvements:

- We train and test a broader range of classifiers (Deep Learning, Support Vector Machines, AdaBoost, LogitBoost, GBT and XGBoost);
- We apply dimensionality reduction techniques;
- We enhance our data preparation step with techniques such as oversampling, normalization and outliers' exclusion;
- We use more data features, including a feature that models the influence of time in tax crime prediction;
- We extract fiscal data of a wider interval of time (five years);
- We evaluate the classifiers with more performance measures;
- We achieve higher classification scores.

This paper is structured as follows: Sect. 2 gives the conceptual framework of Machine Learning, dimensionality reduction, crimes against the tax system, tax audit and tax compliance actions, Sect. 3 reviews the main state-of-the-art works, Sect. 4 explains our methodology, Sect. 5 discusses the results of our experiments, Sect. 6 encompasses the conclusion and future works.

2 Conceptual Foundations

In this section, we define the main concepts applied in our study. Initially, we explain the main theory that grounds Machine Learning, including its definition, different types, main performance measures and classifiers. In the sequence, we define dimensionality reduction, its motivations and two of its principal techniques, which we used and compared in this study. Subsequently, we briefly conceptualize crimes against the tax system, tax audit and tax compliance actions.

2.1 Machine Learning

Machine Learning is the science area in which researchers develop algorithms that learn from experience, regarding a task and performance measure [15]. An algorithm learns if its performance concerning a specific task augments with experience. Algorithms adjust its parameters based on the patterns they learn from data, with the goal of yielding the highest performance as possible.

Classification is one of the most usual tasks in Machine Learning. In this task, algorithms aim to categorize data, labelling data points (a.k.a. instances) with classes. The most common types of classification are supervised and unsupervised. In supervised classification, instances have class labels and algorithms can learn patterns associated with each of the classes. This learning enables them to classify unseen data into groups according to the learned patterns. In unsupervised classification, labels are not available and algorithms have to cluster data instances according to their similitude, which results from mathematical distances between features values of the instances.

Performance Measures for Supervised Classification. To evaluate the classification performance, there are some measures that are calculated based on the following metrics. We explain these metrics according to our case study scenario.

True negatives (TN): Total of not criminals correctly classified.

True positives (TP): Total of criminals correctly classified.

False negatives (FN): Total of criminals that are misclassified.

False positives (FP): Total of not criminals that are misclassified.

Based on these values, we can calculate performance measures that are usually applied in supervised classification, such as precision (P), recall (R), F-measure (F), accuracy (ACC) and specificity (S) [16]. If we consider a dataset with N instances, the calculation of these measures results from the following formulas:

$$P = \frac{TP}{TP + FP} \tag{1}$$

$$R = \frac{TP}{TP + FN} \tag{2}$$

$$F = \frac{2RP}{R + P} \tag{3}$$

$$ACC = \frac{TP + TN}{N} \tag{4}$$

$$S = \frac{TN}{TN + FP} \tag{5}$$

Besides the above-mentioned performance measures, the area under the receiver operating characteristic (AROC) curve and the Cohen's kappa are also traditional measures to evaluate supervised classifications.

AROC. It is a rank-based measure that orders instances according to predicted values, with the purpose of evaluating how well the ordering ranks positive instances above negative instances [17]. It provides an overview of the classification performance for different thresholds. Considering s_p as the sum of all positive instances ranked by their predicted values, n_p the number of positive instances and n_n the number of negative instances, AROC is equal to:

$$AROC = \frac{s_p - n_p(n_n + 1)/2}{n_p + n_n} \tag{6}$$

Cohen's kappa. Data collectors (a.k.a. raters) are individuals who collect data related to a specific industry (e.g. healthcare) [18]. As human observers, they can assign different scores to the same variable, leading to data inconsistency. On the other hand, when raters give the same score to the same feature, they achieve interrater reliability. Cohen's kappa functions as a test of interrater reliability, also pondering the chances of raters guessing some variables because of uncertainty. If we consider $P_r(a)$ the actual observed agreement and $P_r(e)$ the chance agreement, Cohen's kappa (k) equals:

$$k = \frac{P_r(a) - P_r(e)}{1 - P_r(e)} \tag{7}$$

In our case study, we applied supervised classification to predict tax crimes, comparing the aforementioned performance measures for the following classifiers: Neural Networks, Deep Learning, Logistic Regression, Naive Bayes, Support Vector Machines, Decision Trees, Random Forests, AdaBoost, LogitBoost, GBT and XGBoost. In the sequence, we define each of these algorithms.

Neural Networks. In a neural network, connections between nodes form a net that simulates human neurons' behaviour [19]. The net input data are the attribute values of a dataset, whereas the output is the classification of the data instances. The network assigns weights for each of the attributes, aiming to minimize the error between the predicted class and the real class for all the instances. First, weights are randomly assigned for each attribute. With the goal of minimizing the prediction error, an iterative procedure starts to calibrate these weights, based on an optimization criterion, e.g. gradient descent.

Deep Learning. It is composed of very large and dense neural networks that are able to achieve scalable performance [20]. The network becomes deep by adding more layers to it. It learns deep feature hierarchies, such that features at higher levels are compositions of lower level features. This structure enables the network to learn complex functions. One example of Deep Learning are large convolutional neural networks, which are usually applied in image recognition problems and use filters to reduce the number of input parameters without losing relevant feature information.

Logistic Regression. It is a statistical classifier that applies a logistic function based on the natural logarithm [21]. In its elementary form, it classifies instances in two groups (a.k.a. binary classification). Its logistic function calculates statistics that correspond to the probability of an instance being a member of a class. This classifier computes weights to attributes, aiming to minimize the error between class predictions and the real class labels.

Naive Bayes. It is a probabilistic classifier whose main premise is the statistical independence between variables [22]. This independence signifies that the existence of a specific attribute is not probabilistically related to other attributes. Naive Bayes uses conditional probabilities such that it computes the probability of classification in one of the data categories, given the value of an attribute, a.k.a. the posterior probability. The best classification hypothesis, given training data, corresponds to finding the most probable class for an instance, after observing examples.

Support Vector Machines. This algorithm models attributes and weights related to instances as vectors [23]. Then it imposes that the inner product of these vectors is zero to define a hyperplane. The purpose of this hyperplane is to separate instances into their respective classes. Instances that are the nearest to the hyperplane define the support vectors. The distance between these vectors and the hyperplane delimits the decision boundary for classification. One of the advantages of support vector machines is their power to map input attributes to higher dimensions, enabling to separate instances that are not linearly separable. Nevertheless, this transformation (a.k.a. kernel method) generates complex models, which may lead to overfitting and higher execution times.

Decision Trees. This classifier is a model comprised of hierarchies in the form of trees whose nodes represent data attributes [24]. Edges connect nodes such that each edge represents a distinct range of attribute values. Leaves in the bottom of the structure correspond to different classifications for data instances. The top node of the tree represents the attribute that best divides data instances into homogeneous categories. Entropy is a measure for this group homogeneity such that the less entropy a category has the more homogeneous it is. The node below the top node represents the attribute among the remaining ones that best segments data into homogeneous classes. This procedure of attribute selection based on entropy goes on until all attributes are represented in the tree. Pruning strategies exist in order to reduce possible overfitting.

Random Forests. They result from the combination of decision trees [25]. Random Forests belong to a group of classifiers based on ensemble learning [26]. This strategy applies a combination of various classifiers, forming a committee with the objective of performing better than a single classifier. Random Forests randomly draw a training set of instances and a subset of attributes for each tree. In the sequence, this classifier grows trees using these subsets without pruning. It is usually advantageous to adopt an ensemble of distinct trees, in order to benefit from the different characteristics of each tree. The final classification results from majority voting among the decision trees' classifications.

AdaBoost. It is an algorithm that starts by assigning weights to instances [27]. Then a sequence of iterations begins. In each iteration, AdaBoost computes a distribution based on the normalized weights and feeds a learner (a.k.a. weak learner, which is usually a short decision tree) that generates a hypothesis for the class labels. It calculates the error between the predicted labels and the real labels using an exponential loss function. Then it assigns new weights to the instances based on the minimization of these errors. These new weights are the input for a new weak learner, in the next iteration, and the process repeats. The main idea is that the next learner improves the classification performance of the previous learner. The final classification is a combination of the outputs of the learners in each iteration, corresponding to a weighted majority vote of the sequence of classifiers that are applied in each iteration. Thus, AdaBoost is an algorithm that adjusts adaptively to the errors in each iteration, adding new models that attempt to minimize the exponential loss function.

LogitBoost. It is an evolution of AdaBoost. LogitBoost is an iterative algorithm whose routine follows a structure that is similar to AdaBoost. It differs from AdaBoost because it applies a loss function based on logistic regression [28]. LogitBoost minimizes a logistic loss function in order to produce new guesses for a leaner, minimizing the error between the predicted class label and the real class label of the instances. The final classification corresponds to a combination of the results obtained by each learner in each iteration.

GBT. In this algorithm, instances in the trees' leaves will be attached with weights that correspond to prediction scores [29]. This classifier is also an evolution of Adaboost. In an iterative fashion, it adds decision trees to the final model, aiming to progressively strengthen the classification results. In each iteration, new guesses for the trees' parameters are based on the minimization of a loss function. To accomplish these estimates, GBT calculate the first and second order derivatives of the loss function, with respect to the predictions.

XGBoost. It is an extension of GBT. The main improvements regard optimizations in algorithms and systems. XGBoost [30], which stands for eXtreme Gradient Boosting, is capable of handling instance weights and sparse data. It also applies parallel and distributed computing, aiming to construct a scalable tree boosting system with parallel learning. Besides, XGBoost has a cache-aware access, which helps to fetch gradient statistics more rapidly.

2.2 Dimensionality Reduction

Dimensionality reduction techniques apply data transformations that reduce data to its most relevant dimensions, in such a way that it facilitates the separation of instances into homogeneous groups [31]. Dimensionality reduction helps to overcome the hindrances imposed by dimensionality, as it becomes increasingly difficult to select features, visualize data patterns and distinguish clusters when we deal with more dimensions. This obstacle is mainly due to the fact that data instances tend to be statistically uniform when we work with various attributes. Dimensionality reduction also helps to decrease the computational costs of data analysis. In the sequence, we define two main techniques used for dimensionality reduction: PCA and SM.

PCA. In PCA, data dimensions are reduced by means of linear transformations that convert a coordinate system, associated with the original data features, into a new system that corresponds to the directions of the highest data dispersion [32]. The transformed variables are the principal components that are linear combinations of the original features. These components form vectors, whose directions are given by eigenvectors and whose sizes are proportional to eigenvalues.

Data dispersion is assessed with the calculation of variance, such that the higher the variance the higher the dispersion is. Also in PCA, in order to reduce dimensionality, highly correlated variables are eliminated, since they tend to operate the same influence on the target attribute. To measure correlation, PCA calculates covariance between variables. PCA helps identifying latent data patterns, reducing data noise and eliminating variables' redundancy. It has been used in a wide range of areas, including risk management, neuroscience, image compression and anomaly detection.

SM. It is a nonlinear mapping algorithm that maps data instances from their original dimensional space to a space with lower dimensions, maintaining the inherent data structure [33]. SM accomplishes this transformation by imposing that distances between instances in the lower dimensional space are approximately equal to the distances in the original dimensional space.

Let us consider d_{ij} as the distance between an instance i and an instance j in the original space and d^*_{ij} as the distance between these instances in the lower dimensional space. SM calculates a mapping error E by means of the following formula, where n is the number of instances:

$$E = \frac{1}{\sum_{i<j}^{n} d^*_{ij}} \frac{\sum_{i<j}^{n} [d^*_{ij} - d_{ij}]^2}{d^*_{ij}} \tag{8}$$

SM minimizes E using the gradient descent. In an iterative process, SM computes estimates for the lower dimensional space based on the first and second order derivatives of E. Distances among instances are usually calculated using the Euclidean distance. SM has several practical applications, such as exploratory data analysis in genomics, text mining, localization of wireless sensors and multichannel speech separation.

2.3 Crimes Against the Tax System

Crimes against the tax system have the objective of reducing or suppressing taxes. In the context of the city of São Paulo, these taxes are related to services or real state. Examples of these crimes occur when taxpayers deny to supply invoices to fiscal authorities, create inaccurate or flawed documents, make declarations that are false or counterfeit fiscal documents.

2.4 Tax Audit

A tax audit is a procedure comprised of verification and analysis regarding compliance to tax laws. Fiscal authorities are exclusively responsible for performing tax audits. In a tax audit, these authorities must properly identify the taxpayer, compute possible tax debits and impose legal penalties. According to their competencies, fiscal authorities have the legal obligation to require taxes. Besides, their activities prevail over actions that emanate from other public sectors.

2.5 Tax Compliance Actions

Tax compliance actions usually orientate taxpayers about tax legislation or aim to cross-check the information obtained from taxpayers and their hired employees, clients or suppliers. These actions have educational purposes, with fiscal authorities guiding taxpayers through new and complex tax laws, but also envisage to highlight fiscal presence. Compliance actions redound to an increase in taxpayers' compliance to tax regulations, leveraging tax revenue. One of the paradigms that rises tax compliance is to provide taxpayers with assistance concerning the filing of tax returns and the due payment of taxes [34].

3 Literature Review

Some governments already applied Machine Learning in order to predict tax frauds. The work of [9] tested the capabilities of Neural Networks, Bayesian Networks and Decision Trees w.r.t. the task of predicting the use of false invoices, in the context of the tax administration of Chile. As regards dimensionality reduction, their work applied PCA, using a number of components that explained about 60% of the variance in data. Their study led to the conclusion that Neural Networks achieved the highest scores, with 92.6% of recall and 72.9% of specificity.

Another study that used Machine Learning to predict tax fraud emanates from the Spanish Institute of Fiscal Studies [10]. They applied Neural Networks to predict tax evasion and PCA to reduce dimensionality, using fiscal data from the Spanish Revenue Office. Their study resulted in 84% of correct predictions.

The work of [35] is a case study based on fiscal data of the Treasury Office of the State of Ceará, in Brazil. They proposed ALICIA, a feature selection method based on association rules and propositional logics, with centrality measures. In their study, they applied and compared Support Vector Machines, Neural Networks, Decision Trees, Probit Model and Markov Blanket Filter. The combination of ALICIA and Support Vector Machines was the most outstanding approach, yielding 0.7688 of F-measure and 0.79 of AROC. They also performed an experimental study, using a credit card dataset provided by Kaggle, in which ALICIA combined with XGBoost resulted in 0.8913 of F-measure and 0.87 of AROC.

In the study of [11], the authors developed the model DATE, which was tested with data related to import trade flows to Nigeria. The paper aimed to classify illicit trades and predict tax revenue. Their methodology encompassed three main steps: GBT are trained with the goal of generating cross features; DATE learns interactions among cross features and importers; DATE predicts illicit transactions and tax revenue. In their work, each decision path of the trees corresponds to cross features, which are combinations of multiple feature ranges. The use of cross features has some drawbacks already highlighted by the feature engineering literature [12,13]. To name a few, we can enumerate two of its main disadvantages: higher computational cost and noises due to interactions among irrelevant features, which downgrade classification performance. DATE achieved 92.66% of precision, 90.29% of recall, 96.79% of AROC and 75.32% of F-measure.

Comparatively, our best evaluated solution with SM and GBT surpassed the above-mentioned classification performance scores, as our experimental results attested. Higher performance scores for tax crime prediction enable more precise tax fraud detection and more assertive tax audit plans. Besides, our work is more embracing, since it compared a wider scope of classifiers and we evaluated these classifiers based on more performance measures. Additionally, if we consider the specific tax scenario of the city of São Paulo, we conclude that taxpayers have to comply with different obligations when they execute services in the city of São Paulo. This background results in distinct kinds of crimes, e.g. denial to provide documents to fiscal authorities, and taxpayers' behaviours. Hence, the type of crimes we aim to predict are peculiar and more specific to the service taxes' context of São Paulo's municipality. We are not aware of any previous work with the goal of predicting crimes against São Paulo's service tax system, using an SM-based approach combined with GBT for classification.

4 Methodology

In order to execute our proposed solution, we applied a traditional methodology for predictive analytics, inspired by the CRoss Industry Standard Process for Data Mining [36], using the open source tool KNIME [37]. Figure 1 gives an overview of the methodology we used, which is comprised of the following steps: data extraction; data preparation; dimensionality reduction; model training and testing; model evaluation; model selection.

4.1 Data Extraction

We extracted fiscal data related to tax fines values, fiscal audits and tax invoices. These data are available in our data stores, composed of relational databases modelled according to the entity-relationship paradigm.

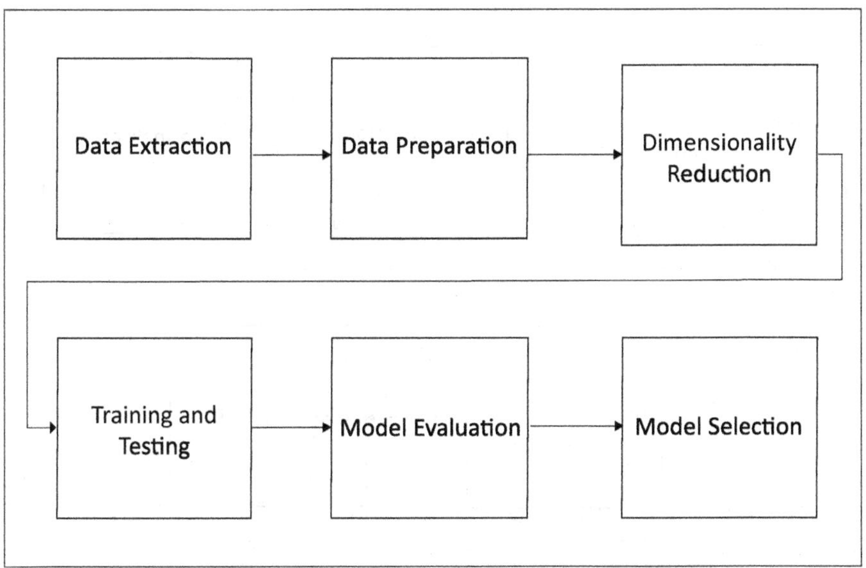

Fig. 1. Workflow of the methodology: we extracted data from our databases, prepared data, then reduced the dimensionality of our dataset. In the sequence, we trained, tested and evaluated predictive models based on Machine Learning classifiers and performance metrics. Finally, we selected the best evaluated model.

Based on our domain knowledge, we initially selected a collection of ten features w.r.t. the information contained in tax invoices, tax fines and fiscal audits. Our target attribute, a.k.a. explained variable, is a binary field that expresses whether or not the taxpayer committed a crime. The remaining features correspond to the preliminarily selected explanatory variables.

We constrained our data extraction to a five-year interval, ranging from 2015 to 2019. The extraction resulted in a dataset of 604 rows. Due to the fact that more detailed characteristics are rated as classified information, we cannot give further details about the features and its values.

4.2 Data Preparation

In this phase, we made five transformations on our original dataset, as illustrated in Fig. 2: conversion of categorical fields to number; exclusion of outliers; normalization; conversion of a numeric field to string; data balancing.

Category to Number Conversion. In order to adequate the data types of the explanatory variables to all the learning algorithms applied in this study, we needed to work with numeric attributes. Thus, we transformed categorical explanatory variables from string to number.

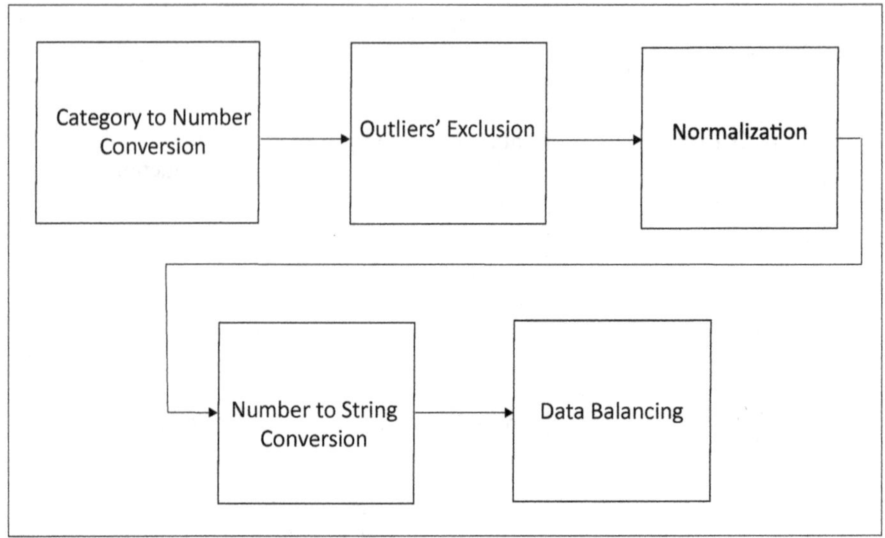

Fig. 2. Data preparation: we transformed categorical explanatory attributes to numeric data type, purged outliers, normalized data, converted the target field to string and balanced the class attribute with SMOTE.

Outliers' Exclusion. Aiming to exclude data instances that deviate from the main data patterns of our dataset, we excluded numerical outliers using the interquartile range [38]. This range is based on the numeric differences between the third and first statistical quartiles, also considering a multiplier. We used 1.5 for this multiplier, enabling us to work with values ranging from the lower to the upper limits of a boxplot's whisker.

Normalization. We normalized all the explanatory variables, with the goals of standardizing input data, reducing execution time and avoiding possible misleading influences from high attribute values in the final results. For this purpose, we worked with a range of values varying from zero to one.

Number to String Conversion. Aiming to enable the automatic identification of the target attribute by the tool we used in this study, we converted the target data type from numeric to string.

Data Balancing. After the previous transformations, our dataset became unbalanced for the class attribute. With the goal of balancing the categorical values' distribution of our target attribute, we applied the Synthetic Minority Over-sampling Technique (SMOTE) [39]. This procedure aims to improve the generalization performance of classifiers. SMOTE generates synthetic instances of the minority class until there is an equal number of occurrences for each class.

This equalization is operated based on the nearest neighbours of the minority class examples. SMOTE calculates the difference between attribute values of an example and its nearest neighbour and multiplies it by a random number between 0 and 1, creating new examples for the minority class.

4.3 Dimensionality Reduction

With the goal of discovering the most relevant dimensions of our previously prepared dataset, we applied dimensionality reduction to it. In order to have means to compare and evaluate the classification performance based on different dimensionality reduction techniques, we used two distinct approaches to reduce dimensions: one is PCA-based, whilst the other is SM-based. This process resulted in datasets with reduced dimensions, which were input for training and testing the classifiers.

PCA-Based Approach. Aiming to discover the number of components that explains the most relevant percentage of variance, we analysed the eigenvalues of the principal components. In Table 1, we discriminate the percentage of variance explained by each principal component and the accumulated percentage of variance achieved with these components. We calculated the percentage of explained variance as the ratio of each eigenvalue w.r.t. the sum of all eigenvalues.

Table 1. Eigenvalues and percentages of explained variance for the principal components.

Principal component	Eigenvalue	Percentage of explained variance	Accumulated percentage of explained variance
1st	0.1127	39.43%	39.43%
2nd	0.0865	30.27%	69.70%
3rd	0.0374	13.07%	82.77%
4th	0.0223	7.80%	90.57%
5th	0.0176	6.16%	96.73%
6th	0.0093	3.26%	99.99%
7th	$2.89X10^{-5}$	0.01%	100.00%
8th	$1.56X10^{-17}$	0.00%	100.00%

We verified that if we worked with a number of components ranging from four to six, we would be able to explain the most relevant percentage of variance (above 90%). Thus, we imposed these number of components as the number of output dimensions to be used in the PCA-based dimensionality reduction approach. We disregarded the seventh and eighth components because of their insignificant participation in the variance explanation.

SM-Based Approach. PCA-based initialisations are usually applied to initialise SM [33,40–42]. Inspired by this trend, for the SM-based approach, we imposed for SM the same number of dimensions to output as the ones we specified for PCA. Additionally, we parameterized SM using KNIME's default settings:

- Random seed: 1,000,000
- Epochs: 50
- Learning rate: 1.0
- Distance metric: Euclidean

Hence, for each dimensionality reduction approach, resulted three reduced datasets, having four, five and six dimensions respectively. A scheme of this procedure is shown in Fig. 3.

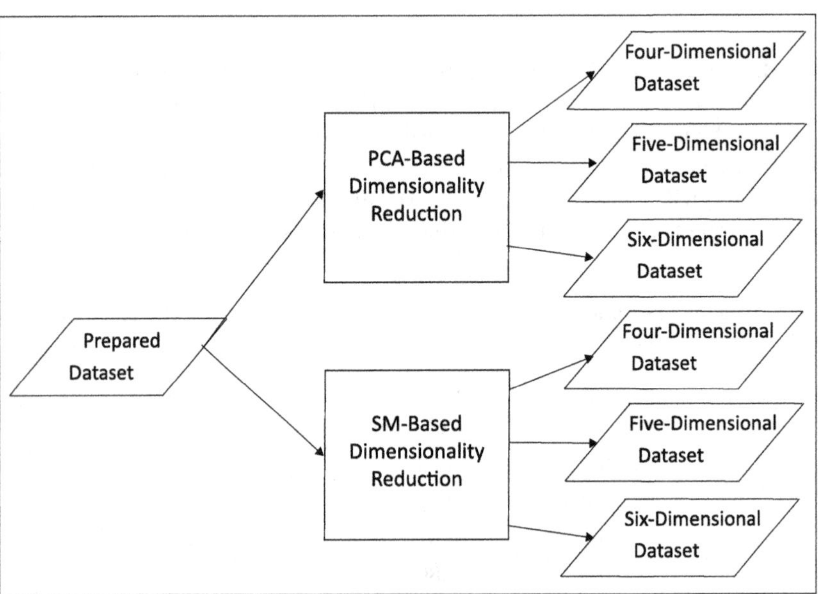

Fig. 3. In the dimensionality reduction step, we applied PCA- and SM-based approaches, both producing four-, five- and six-dimensional datasets as output.

4.4 Model Training and Testing

In this step, we used as input the datasets that resulted from each of the dimensionality reduction approaches. For each of these datasets, we applied a 10-fold cross validation method in which we trained and tested the following classifiers: Neural Networks, Deep Learning, Logistic Regression, Naive Bayes, Support Vector Machines, Decision Trees, Random Forests, AdaBoost, LogitBoost, GBT and XGBoost. We represent this process in Fig. 4.

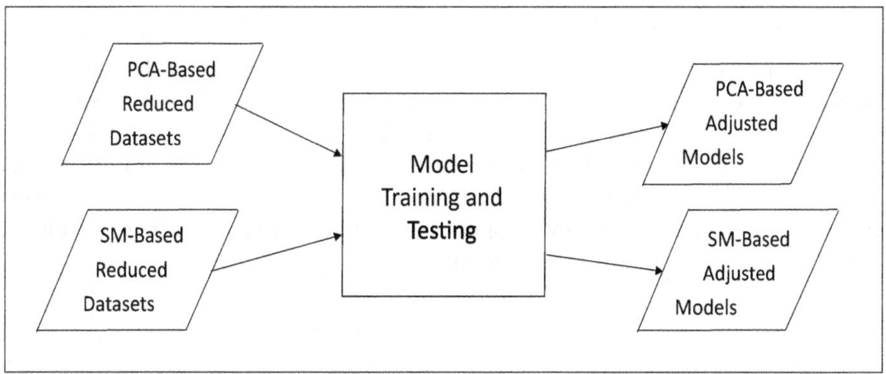

Fig. 4. Model training and testing: using the reduced datasets as input, emanating from each dimensionality reduction approach, we trained and tested the classifiers, providing calibrated models to be assessed in the evaluation phase.

4.5 Model Evaluation

In the evaluation phase, we evaluated the adjusted models from both dimensionality reduction approaches. To assess the classification performance of the learning algorithms and the different dimensionality reduction approaches, we compared the classifiers w.r.t. the following performance measures: precision, recall, F-measure, accuracy, specificity, AROC and Cohen's kappa. This process is illustrated in Fig. 5.

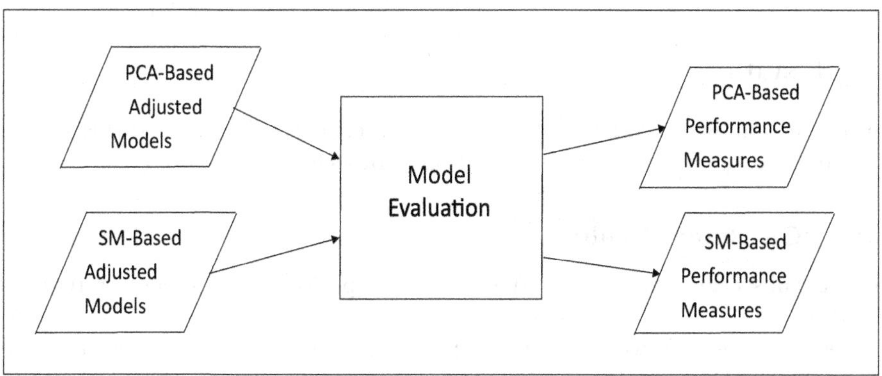

Fig. 5. Model evaluation: we evaluated the models that resulted from each dimensionality reduction approach and computed their performance measures.

4.6 Model Selection

In this step, we selected the overall best evaluated model and the associated dimensionality reduction technique, considering the classification scores. To accomplish this election, first we selected the best evaluated model based on PCA. Then, we selected the best one based on SM. Finally, we compared the performance measures of these selected models, to reach a consensus on the overall best evaluated classifier and respective dimensionality reduction technique. An overview of these steps is represented in Fig. 6.

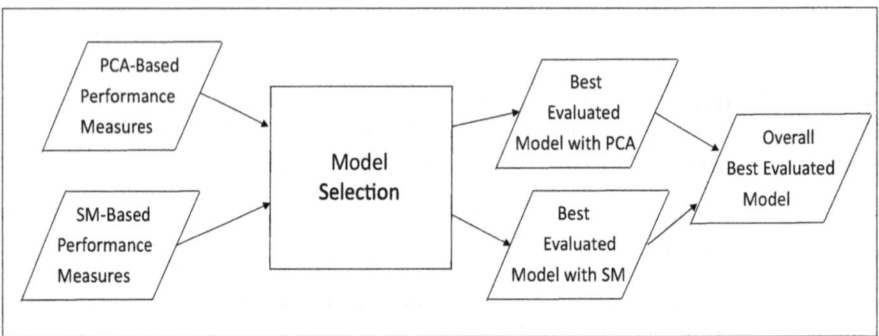

Fig. 6. Model selection: we compared the performance measures of the classifiers based on different dimensionality reductions, selected the best performing classifier of each reduction technique and then selected the classifier and respective reduction approach that yielded the overall highest scores.

5 Results

In this section, we discuss the results of our experiments. We divided this section according to the dimensionality reduction approaches we adopted.

5.1 PCA-Based Results

In this subsection, we analyse the classifiers' performance scores for reduced datasets outputted by PCA. We worked with four, five and six dimensions. For the four-dimensional reduced dataset, the outcome for the classifiers' scores is shown in Table 2. Bold numbers indicate the highest performance measures. Considering the number of times a classifier yielded the highest score as a ranking criterion, we verified that Random Forests outstripped the other classifiers, since it achieved the highest values in five out of seven scores calculated.

Analysing the five-dimensional reduced dataset, the resulting performance metrics for each classifier are presented in Table 3. Random Forests achieved the highest scores most of times, outranking the other classifiers.

Table 2. Performance measures of the classifiers for the PCA-based approach using a four-dimensional reduced dataset.

Classifier	Recall	Precision	Specificity	F-measure	Accuracy	Cohen's kappa	AROC
Decision Trees	0.843	0.898	0.904	0.869	0.873	0.747	0.865
Neural Networks	0.686	0.755	0.777	0.719	0.731	0.463	0.779
Random Forests	0.834	**0.965**	**0.969**	**0.895**	**0.902**	**0.803**	0.912
Naive Bayes	0.620	0.861	0.900	0.721	0.760	0.520	0.708
Logistic Regression	0.694	0.740	0.755	0.716	0.725	0.450	0.688
Deep Learning	0.144	0.541	0.878	0.228	0.511	0.022	0.387
AdaBoost	0.742	0.899	0.917	0.813	0.830	0.659	0.731
GBT	0.847	0.928	0.934	0.886	0.891	0.782	0.909
Support Vector Machines	**1.000**	0.500	0.000	0.667	0.500	0.000	0.213
XGBoost	0.843	0.923	0.930	0.881	0.886	0.773	**0.919**
LogitBoost	0.755	0.887	0.904	0.816	0.830	0.659	0.783

Table 3. Performance measures of the classifiers for the PCA-based approach using a five-dimensional reduced dataset.

Classifier	Recall	Precision	Specificity	F-measure	Accuracy	Cohen's kappa	AROC
Decision Trees	0.865	0.880	0.882	0.872	0.873	0.747	0.832
Neural Networks	0.690	0.806	0.834	0.744	0.762	0.524	0.816
Random Forests	0.834	**0.946**	**0.952**	0.886	**0.893**	**0.786**	0.938
Naive Bayes	0.607	0.832	0.878	0.702	0.742	0.485	0.830
Logistic Regression	0.690	0.778	0.803	0.731	0.747	0.493	0.830
Deep Learning	0.148	0.479	0.838	0.227	0.493	-0.001	0.397
AdaBoost	0.747	0.881	0.900	0.809	0.823	0.646	0.818
GBT	0.821	0.945	**0.952**	0.879	0.886	0.773	**0.972**
Support Vector Machines	**1.000**	0.502	0.009	0.669	0.504	0.009	0.166
XGBoost	0.852	0.929	0.934	**0.888**	**0.893**	**0.786**	0.895
LogitBoost	0.747	0.886	0.904	0.810	0.825	0.651	0.918

For the six-dimensional dataset that resulted from the application of PCA, we see in Table 4 that XGBoost performance metrics prevailed over the other algorithms.

Table 4. Performance measures of the classifiers for the PCA-based approach using a six-dimensional reduced dataset.

Classifier	Recall	Precision	Specificity	F-measure	Accuracy	Cohen's kappa	AROC
Decision Trees	0.852	0.907	0.913	0.878	0.882	0.764	0.908
Neural Networks	0.677	0.795	0.825	0.731	0.751	0.502	0.881
Random Forests	0.847	0.933	0.939	0.888	0.893	0.786	0.952
Naive Bayes	0.594	0.855	0.900	0.701	0.747	0.493	0.749
Logistic Regression	0.668	0.750	0.777	0.707	0.723	0.445	0.897
Deep Learning	0.170	0.600	0.886	0.265	0.528	0.057	0.555
AdaBoost	0.725	0.912	0.930	0.808	0.828	0.655	0.804
GBT	0.860	0.934	0.939	0.895	0.900	0.799	0.962
Support Vector Machines	**1.000**	0.501	0.004	0.668	0.502	0.004	0.336
XGBoost	0.882	**0.957**	**0.961**	**0.918**	**0.921**	**0.843**	**0.972**
LogitBoost	0.742	0.895	0.913	0.811	0.828	0.655	0.928

Table 5. Performance measures of the best-ranked classifiers for the PCA-based approach using four, five and six output dimensions.

Dimensions	Classifier	Recall	Precision	Specificity	F-measure	Accuracy	Cohen's kappa	AROC
4	Random Forests	0.834	**0.965**	**0.969**	0.895	0.902	0.803	0.912
5	Random Forests	0.834	0.946	0.952	0.886	0.893	0.786	0.938
6	XGBoost	**0.882**	0.957	0.961	**0.918**	**0.921**	**0.843**	**0.972**

To reach a consensus on the best performing classifier using the PCA-based approach, we isolated the best-ranked classifier for each of the reduced datasets. We concluded that the scoring metrics achieved with XGBoost were the highest in the prevalent number of times, as Table 5 shows.

5.2 SM-Based Results

As in PCA, we reduced the prepared dataset to four, five and six dimensions. Considering the four-dimensional reduced dataset, as Table 6 indicates, GBT yielded the highest performance measures in six out of seven scores.

For the five-dimensional reduced dataset, each classifier yielded the scores shown in Table 7. GBT results outscored the other classifiers in six out of seven performance measures.

Using the six-dimensional dataset that resulted from SM, we have the scores presented in Table 8. GBT prevailed over the other classifiers, achieving the highest scores in five metrics.

We compared the best-ranked classifier for each of the reduced datasets (see Table 9). Both the solutions with four and six dimensions yielded the highest scores in four out of seven metrics, outstripping the other algorithms. In order to untie these scores, we isolated the results for four and six dimensions and compared them. As a consequence, we concluded that the four-dimensional solution prevailed, since it outperformed the six-dimensional solution in recall, F-measure

Table 6. Performance measures of the classifiers for the SM-based approach using a four-dimensional reduced dataset.

Classifier	Recall	Precision	Specificity	F-measure	Accuracy	Cohen's kappa	AROC
Decision Trees	0.900	0.920	0.921	0.909	0.910	0.821	0.945
Neural Networks	0.686	0.773	0.799	0.727	0.742	0.485	0.824
Random Forests	0.904	0.976	0.978	0.939	0.941	0.882	**0.998**
Naive Bayes	0.454	0.920	0.961	0.608	0.707	0.415	0.702
Logistic Regression	0.686	0.751	0.773	0.717	0.729	0.459	0.777
Deep Learning	0.205	0.516	0.808	0.294	0.507	0.013	0.749
AdaBoost	0.847	0.975	0.978	0.907	0.913	0.825	0.976
GBT	0.948	**0.986**	**0.987**	**0.967**	**0.967**	**0.934**	**0.998**
Support Vector Machines	**1.000**	0.511	0.044	0.677	0.522	0.044	0.362
XGBoost	0.921	0.977	0.978	0.948	0.950	0.900	0.980
LogitBoost	0.821	0.879	0.886	0.849	0.854	0.707	0.910

Table 7. Performance measures of the classifiers for the SM-based approach using a five-dimensional reduced dataset.

Classifier	Recall	Precision	Specificity	F-measure	Accuracy	Cohen's kappa	AROC
Decision Trees	0.886	0.910	0.913	0.898	0.900	0.799	0.949
Neural Networks	0.681	0.768	0.795	0.722	0.738	0.476	0.814
Random Forests	0.891	0.976	0.978	0.932	0.934	0.869	0.977
Naive Bayes	0.445	0.927	0.965	0.602	0.705	0.410	0.834
Logistic Regression	0.681	0.768	0.795	0.722	0.738	0.476	0.856
Deep Learning	0.109	0.472	0.878	0.177	0.493	-0.013	0.528
AdaBoost	0.865	0.947	0.952	0.904	0.908	0.817	0.971
GBT	0.926	**0.991**	**0.991**	**0.957**	**0.959**	**0.917**	**1.000**
Support Vector Machines	**1.000**	0.512	0.048	0.678	0.524	0.048	0.225
XGBoost	0.921	0.977	0.978	0.948	0.950	0.900	0.996
LogitBoost	0.838	0.941	0.948	0.887	0.893	0.786	0.951

and AROC. On the other hand, the six-dimensional approach only prevailed over the four-dimensional solution in precision and specificity.

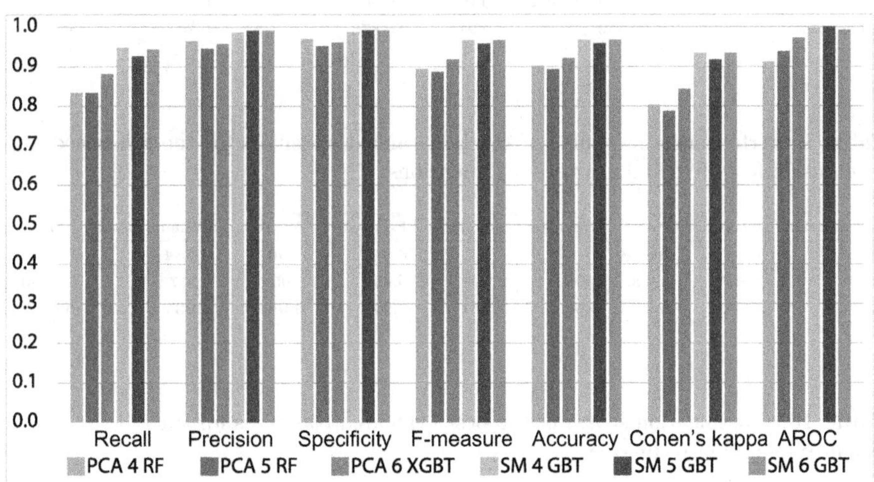

Fig. 7. Performance measures of the best evaluated classifiers for PCA and SM for each output dimensions.

5.3 Overall Results

In a broad view, comparing PCA's and SM's best results for each reduced dataset, we see that SM-based solutions contributed with the highest performance measures (see Fig. 7).

More specifically, with the goal of reaching a consensus on the overall best evaluated classifier and respective dimensionality reduction approach, we took the best-ranked solution with PCA and the best-ranked solution with SM.

The performance metrics in Table 10 indicate that the SM-based approach with four dimensions, using GBT, excelled in the classification task, outstripping the PCA-based approach with six dimensions and XGBoost. On the other hand, Support Vector Machines and Deep Learning achieved the lowest scores most of the times, both with SM and PCA, for all the output dimensions we considered.

Table 8. Performance measures of the classifiers for the SM-based approach using a six-dimensional reduced dataset.

Classifier	Recall	Precision	Specificity	F-measure	Accuracy	Cohen's kappa	AROC
Decision Trees	0.869	0.957	0.961	0.911	0.915	0.830	0.912
Neural Networks	0.664	0.768	0.799	0.712	0.731	0.463	0.874
Random Forests	0.917	0.977	0.978	0.946	0.948	0.895	0.994
Naive Bayes	0.467	0.915	0.956	0.618	0.712	0.424	0.727
Logistic Regression	0.681	0.768	0.795	0.722	0.738	0.476	0.757
Deep Learning	0.367	0.618	0.773	0.460	0.570	0.140	0.383
AdaBoost	0.873	0.976	0.978	0.922	0.926	0.852	0.962
GBT	0.943	**0.991**	**0.991**	**0.966**	**0.967**	**0.934**	0.992
Support Vector Machines	**1.000**	0.517	0.066	0.682	0.533	0.066	0.227
XGBoost	0.939	**0.991**	**0.991**	0.964	0.965	0.930	**1.000**
LogitBoost	0.852	0.907	0.913	0.878	0.882	0.764	0.899

Table 9. Performance measures of the best-ranked classifiers for the SM-based approach using four, five and six output dimensions.

Dimensions	Classifier	Recall	Precision	Specificity	F-measure	Accuracy	Cohen's kappa	AROC
4	GBT	**0.948**	0.986	0.987	**0.967**	**0.967**	**0.934**	0.998
5	GBT	0.926	**0.991**	**0.991**	0.957	0.959	0.917	**1.000**
6	GBT	0.943	**0.991**	**0.991**	0.966	**0.967**	**0.934**	0.992

Table 10. Classification scores of the best evaluated classifiers for each of the dimensionality reduction techniques.

Approach	Classifier	Recall	Precision	Specificity	F-measure	Accuracy	Cohen's kappa	AROC
SM with 4 dimensions	GBT	**0.948**	**0.986**	**0.987**	**0.967**	**0.967**	**0.934**	**0.998**
PCA with 6 dimensions	XGBoost	0.882	0.957	0.961	0.918	0.921	0.843	0.972

6 Conclusion

With recent advances in computer processing capabilities, together with more powerful predictive algorithms, analysts can perform more rapid and accurate data analysis. This scenario turns decision-making more feasible and competitive for institutions in several areas. Public administrations, for instance, are gradually exploiting data analysis, aiming to augment its productivity.

A category of data analysis that enhances decision-making is predictive analytics, with techniques mainly proliferating from the Machine Learning field. Predicting tax crimes with the aid of Machine Learning permits tax administrations to accurately program their fiscal audits, focusing on actions with higher probability of crime occurrence and, thus, rising tax compliance. As a consequence, tax administrations can lessen their revenue losses due to tax crimes.

In this work, we predicted crimes against the tax system of the city of São Paulo with the use of Machine Learning. In order to accomplish these predictions, we structured our methods with the following steps: data extraction; data preparation; dimensionality reduction; model training and testing; model evaluation; model selection. In our experiments, we applied two dimensionality reduction approaches: one using PCA and the other using SM, both tested with the same output dimensions. Using the reduced datasets as input, we applied the following classifiers: Neural Networks, Deep Learning, Logistic Regression, Naive Bayes, Support Vector Machines, Decision Trees, Random Forests, AdaBoost, LogitBoost, GBT and XGBoost. We compared and evaluated the classification scores of these classifiers w.r.t. tax crime prediction.

The results of our experiments revealed SM's supremacy over PCA w.r.t. classification performance. The strategy of reducing dimensionality, keeping the same distance between instances, was more efficient and adherent to fiscal data patterns than projecting data on the directions of the highest data dispersion. Additionally, our results indicated that a four-dimensional feature set based on SM, combined with GBT for tax crime classification, surpassed the other classifiers and the state-of-the-art works, as regards classification performance measures. Besides, this four-dimensional approach with SM outranked classification scores that resulted from other dimensions for the feature set, obtained with SM and PCA. We believe that the interplay between trees, inherent to GBT, aided to leverage the performance results, enhanced by the synergy between GBT and SM, hence providing more accurate crime predictions. We are unaware of any previous work whose goal was to predict crimes against the service tax system of São Paulo's municipality, using a combination of SM and GBT.

Aside from these contributions, we concluded that Machine Learning helped to intensify the success of our fiscal audit plans. With more precise tax crime predictions, São Paulo's tax administration orientates its audits' planning based on the ranking of tax audits according to the highest probability of tax crime occurrence. The adoption of this new criterion potentially leverages tax revenue.

Besides, we based our experiments on the use of a software tool. This procedure enabled us to automate and accelerate tax crime prediction. As a consequence, São Paulo's tax administration decision process becomes faster and more precise, increasing taxpayers' monitoring effectiveness and fiscal authorities' ubiquity. Also, our methodology constituted a framework in which other techniques can be added and tested, permitting to enrich its structure and results. One shortcoming of our work is the fact that, although we were able to predict whether or not a tax crime is probable to happen, we did not quantify the tax fines that outcome from each of the predicted tax crimes.

As a future work, we plan to calibrate regression models to predict service tax fines for São Paulo's municipality, which can generate fruitful results for fiscal audit planning and tax revenue. Another interesting study that we can consider in future analyses is applying Machine Learning to predict if taxpayers will change their behaviour after a tax compliance action. This prediction helps São Paulo's tax administration to focus their planning on more beneficial actions, with higher probability of effectiveness. We can base this future work on data about our previous compliance actions and the tax payment evolution in the elapsed time.

References

1. Provost, F., Fawcett, T.: Data science and its relationship to big data and data-driven decision making. Big Data **1**(1), 51–59 (2013). https://doi.org/10.1089/big.2013.1508
2. Matheus, R., Janssen, M., Maheshwari, D.: Data science empowering the public: data-driven dashboards for transparent and accountable decision-making in smart cities. Gov. Inf. Q. (2018)
3. Cobham, A.: Tax avoidance and evasion - the scale of the problem. Tax Justice Network (2017). https://www.taxjustice.net/wp-content/uploads/2017/11/Tax-dodging-the-scale-of-the-problem-TJN-Briefing.pdf. Accessed 03 Aug 2020
4. SINPROFAZ. Sonegação no Brasil – uma estimativa do desvio da arrecadação do exercício de (2015). http://www.quantocustaobrasil.com.br/artigos/sonegacao-no-brasil-uma-estimativa-do-desvio-da-arrecadacao-do-exercicio-de-2015. Accessed 6 Aug 2020
5. Empresômetro Homepage. https://www.empresometro.com.br. Accessed 6 Aug 2020
6. São Paulo State Government. Conheça São Paulo. Sistema Estadual de Análise de Dados (SEADE) (2019). https://www.seade.gov.br/wp-content/uploads/2019/01/Conheca_SP_2019_jan29.pdf. Accessed 15 Jan 2020
7. São Paulo Commercial Association. Impostômetro (2019). https://impostometro.com.br. Accessed 15 Jan 2020
8. São Paulo City Hall. Relatório Técnico do Balanço Geral de 2018. Accounting Department (2019)
9. González, P.C., Velásquez, J.D.: Characterization and detection of taxpayers with false invoices using data mining techniques. Expert Syst. Appl. **40**, 1427–1436 (2013)
10. López, C.P., Rodríguez, M.J.D., Santos, S.L.: Tax fraud detection through neural networks: an application using a sample of personal income taxpayers. Future Internet **11**, 86 (2019)
11. Kim, S., et al.: DATE: dual attentive tree-aware embedding for customs fraud detection. In: Proceedings of the 26th ACM SIGKDD Conference on Knowledge Discovery and Data Mining. ACM, USA (2020)
12. Weiyu, C., Yanyan, S., Linpeng, H.: Adaptive factorization network: learning adaptive-order feature interactions. In: Proceedings of the 34th Association for the Advancement of Artificial Intelligence (AAAI) Conference on Artificial Intelligence, New York, USA (2020). https://arxiv.org/pdf/1909.03276.pdf

13. Shan, Y., Hoens, T.R., Jiao, J., Wang, H., Yu, D., Mao, J.: Deep crossing: web-scale modeling without manually crafted combinatorial features. In: Proceedings of the 22nd ACM SIGKDD International Conference on Knowledge Discovery and Data Mining. ACM (2016). https://doi.org/10.1145/2939672.2939704
14. Ippolito, A., Lozano, A.C.G.: Tax crime prediction with machine learning: a case study in the municipality of São Paulo. In: Proceedings of the 22nd International Conference on Enterprise Information Systems - Volume 1: ICEIS, pp. 452–459. SciTePress (2020). https://doi.org/10.5220/0009564704520459
15. Mitchell, T.M.: Machine Learning. McGraw-Hill, New York (1997)
16. Hossin, M., Sulaiman, M.N.: A review on evaluation metrics for data classification evaluations. Int. J. Data Min. Knowl. Manag. Process **5**, 1 (2015)
17. Caruana, R., Niculescu-Mizil, A.: Data mining in metric space: an empirical analysis of supervised learning performance criteria. In: Proceedings of the 10th ACM SIGKDD International Conference on Knowledge Discovery and Data Mining, pp. 69–78. ACM, New York, USA (2004)
18. McHugh, M.L.: Interrater reliability: the kappa statistic. Biochemia Medica **22**(3), 276–82 (2012)
19. Hardesty, L.: Explained: Neural Networks. MIT News (2017). http://news.mit.edu/2017/explained-neural-networks-deep-learning-0414. Accessed 10 Jan 2020
20. Goodfellow, I., Bengio, Y., Courville, A.: Deep Learning, 2nd edn. MIT Press, Cambridge (2016)
21. Russell, S.J., Norvig, P.: Artificial Intelligence: A Modern Approach, 3rd edn. Pearson, London (2010)
22. Migon, S.H., Gamerman, D., Louzada, F.: Statistical Inference: An Integrated Approach. CRC Press, Boca Raton (2015)
23. Ben-Hur, A., Ong, C.S., Sonnenburg, S., Scholkopf, B., Ratsch, G.: Support vector machines and kernels for computational biology. PLoS Comput. Biol. **4**(10), e1000173 (2008)
24. Poole, D., Mackworth, A.: Artificial Intelligence: Foundations of Computational Agents, 2nd edn. Cambridge University Press, Cambridge (2017)
25. Breiman, L.: Random forests. Mach. Learn. **45**(1), 5–32 (2001)
26. Rokach, L.: Ensemble-based classifiers. Artif. Intell. Rev. **33**, 1–39 (2010)
27. Freund, Y., Schapire, R.E.: A decision-theoretic generalization of on-line learning and an application to boosting. J. Comput. Syst. Sci. **2**(55), 119–139 (1997)
28. Hastie, T., Tibshirani, R., Friedman, J.H.: Additive logistic regression: a statistical view of boosting. Ann. Stat. **2**(28), 337–407 (2000)
29. Hastie, T., Tibshirani, R., Friedman, J.: Boosting and additive trees. In: The Elements of Statistical Learning. SSS, vol. 2, pp. 337–387. Springer, New York (2009). https://doi.org/10.1007/978-0-387-84858-7_10
30. Chen, T., Guestrin, C.: XGBoost: a scalable tree boosting system. In: Proceedings of the 22nd ACM SIGKDD International Conference on Knowledge Discovery and Data Mining. ACM (2016). https://doi.org/10.1145/2939672.2939785
31. Steinbach, M., Ertöz, L., Kumar, V.: The challenges of clustering high-dimensional data. In: Wille, L.T. (ed.) New Vistas in Statistical Physics: Applications in Econophysics, Bioinformatics and Pattern Recognition, pp. 273–309. Springer, Heidelberg (2003). https://doi.org/10.1007/978-3-662-08968-2_16
32. Johnson, R.A., Wichern, D.W.: Applied Multivariate Statistical Analysis, 5th edn. Prentice Hall, Hoboken (2002)
33. Sammon, J.W., Jr.: A nonlinear mapping for data structure analysis. IEEE Trans. Comput. **18**(5), 401–409 (1969)

34. Alm, J.: What motivates tax compliance. Tulane Economics Working Paper Series, Working Paper 1903. Tulane University (2019)
35. Matos, T., et al.: Leveraging feature selection to detect potential tax fraudsters. Expert Syst. Appl. **145**, 113128 (2020). https://doi.org/10.1016/j.eswa.2019.113128
36. Wirth, R., Hipp, J.: CRISP-DM: towards a standard process model for data mining. In: Proceedings of the 4th International Conference on the Practical Applications of Knowledge Discovery and Data Mining, pp. 29–39 (2000)
37. Berthold, M.R., et al.: KNIME - the Konstanz information miner: version 2.0 and beyond. SIGKDD Explor. Newsl. **11**(1), 26–31 (2009)
38. Tukey, J.W.: Explanatory Data Analysis. Addison-Wesley, Boston (1977)
39. Chawla, N.V., Bowyer, K.W., Hall, L.O., Kegelmeyer, W.P.: SMOTE: synthetic minority over-sampling technique. J. Artif. Intell. Res. **16**, 321–357 (2002). https://doi.org/10.1613/jair.953
40. Lerner, B., Guterman, H., Aladjem, M., Dinstein, I.: On the initialisation of Sammon's nonlinear mapping. IEEE Trans. Comput. Pattern Anal. Appl. **3**(1), 61–68 (2000)
41. Lerner, B., Guterman, H., Aladjem, M., Dinstein, I., Romem, Y.: On pattern classification with Sammon's nonlinear mapping - an experimental study. Pattern Recogn. **31**, 371–381 (1998)
42. Mao, J., Jain, A.K.: Artificial neural networks for feature extraction and multivariate data projection. IEEE Trans. Neural Netw. **6**, 296–317 (1995)

Extraction of Speech Features and Alignment to Detect Early Dyslexia Evidences

Fernanda M. Ribeiro[1], Alvaro R. Pereira Jr.[1], Débora M. Barroso Paiva[2], Luciana M. Alves[3], and Andrea G. Campos Bianchi[1](✉)

[1] Computing Department, Federal University of Ouro Preto, Ouro Preto, Brazil
{alvaro,andrea}@ufop.edu.br

[2] School of Computing, Federal University of Mato do Grosso do Sul, Campo Grande, Brazil
debora.paiva@ufms.br

[3] Speech and Language Therapy Department, Federal University of Minas Gerais, Belo Horizonte, Brazil

Abstract. Specific reading disorders are conditions caused by neurological dysfunctions that affect the linguistic processing of printed text. Many people go untreated due to the lack of specific tools and the high cost of using proprietary software; however, new audio signal processing technologies can help identify genetic pathologies. The methodology developed by medical specialists extracts characteristics from the reading of a text aloud and returns evidence of dyslexia. This work proposes an improvement of the research presented in [25], extracting new features and improvements serving as a tool for dyslexia indication efficiently. The analysis is done in recordings of the reading of pre-defined texts with school-age children. Direct and indirect characteristics of the audio signal are extracted. The direct ones are obtained through the methodology of separation of pauses and syllables. Simultaneously, the indirect characteristics are extracted through the alignment of audio signals, the Hidden Markov Model, and some heuristics of improvement. The indication of the probability of dyslexia is performed using a machine learning algorithm. The tests were compared with the specialist's classification, obtaining high accuracy on the evidence of dyslexia. The difference between the values of the characteristics collected automatically and manually was below 20% for most features. Finally, the results show a promising research area for audio signal processing concerning the aid to specialists in the decision making related to language pathologies.

Keywords: Signal processing · Features · Mel cepstral frequencies · Supervised learning · Decision making system

1 Introduction

Dyslexia is one of the pathologies of the language rarely addressed in underdeveloped countries by professionals, mainly due to the high time required for

J. Filipe et al. (Eds.): ICEIS 2020, LNBIP 417, pp. 317–335, 2021.
https://doi.org/10.1007/978-3-030-75418-1_15

its evaluation. It requires a lot of research involving several professionals and, mostly, because the diagnosis is only a probability analysis, as described by Prates et al. [21] and Shaywitz [27].

Dyslexia is a disease caused by dysfunction or interruption of the brain connectors that connect the anterior and posterior areas of the brain [12,18]. In dyslexia, the person feels learning and reading difficulties, which is quite evident in the oral reading of a text, i.e., the person feels difficulty in understanding and emitting the various sounds of a word [27] (but has no physical anomaly), influencing strongly in the learning process of a child in training.

During the identification of the pathology, it is necessary to interact not only with the speech therapist but also with the psychology and the neurologist professionals. Each specialist has its analysis methodology, but the diagnosis is only concluded after confirmation by all specialists. The rapid identification of this pathology provides the child with a better quality of school life and, possibly, improvement in his evolution as a whole. Thus, many specialists are looking for methods that accelerate and/or facilitate the identification of this pathology.

There are several studies on the processing of audio signals for different applications but mainly for the indication of physical pathologies [13,19,26,32]. Regarding digital signal processing, most methods use methodologies related to signal winding and the extraction of characteristics in the domain of time and frequency, for example, the $HiddenMarkovModels$ (HMM) [18] and the Virtebi algorithm [11].

Our proposal uses the audio signal processing and machine learning techniques to delimit and model characteristics present in the reading audio of dyslexic individuals, proposing a solution to automate the identification and indication of dyslexia based on audios of readings aloud. The measures obtained from the audio processing make it possible to indicate this pathology so that the proposal aims to support and make the preliminary evidence of patients with dyslexia more reliable. Then, the patient can go to other professionals and be appropriately treated.

Although Zavaleta et al. [32] perform the identification of dyslexia using measures related to audio, our proposal is more generic and independent of questionnaires. Preliminary studies presented in [25] presented the audio signal processed by removing the pauses, which is one of the characteristics of analysis and counting syllables by varying a window in the audio signal. The results indicate a need for some improvements and extension of the characteristics and methods of classification.

In this paper, after the audio signal separation and segmentation, the subsequent analysis is the audio signal processing, where the alignment between two audio signals is performed. In a more recent work on alignment [10], a model of alignment of audio signals by phonemes is presented. It does not need a manually aligned base for training, so the models are directly trained on the audio signal. The method was applied to the French and English languages to extract characteristics and classification through HMM.

Thus, through these alignment processes, it is sought to return the sentence segments, the readings that are within the standard and the wrong words, which represent the problem in reading. Thus, it can serve as a basis for the analysis of the methodology applied in the indication of the risk of dyslexia.

This paper is organized as follows: in Sect. 2 we present the related works and our proposal; in Sect. 3 we present the background; in Sect. 4 we present our computational methodology; in Sect. 5 we describe the experiments; in Sect. 6 we discuss the experimental results and in Sect. 7 we present the concluding remarks.

2 Related Works

There are some works in detecting pathologies of language and voice. Marinus et al. [19] use methods of analysis of voice pathologies based on the Mel frequency Cepstral Coefficients to represent the signals of voice audios and Multilayer Neural Networks for the classification between normal voice, voice affected by edema and voice affected by other pathologies.

On the other hand, linguistic pathologies affect the reading and writing of a text and cause difficulties in interpreting and representing the syntactic and morphological part of a readable text. About dyslexia, we can cite the work of Zavaleta et al. [32] that proposes a technological tool to support the diagnosis of dyslexia. The authors collect response data on a specific questionnaire, related to factors indicative of the pathology, with questions about how are the reading, diseases, and pathological problems that exist in the family. In the audio responses, a neural network is applied and, through decision-making metrics, a classification is performed considering two groups, with or without dyslexia. The results were not completely accurate because four patients with dyslexia were diagnosed without dyslexia.

Considering the articulatory disorder, we can cite the research developed by Santos [26], which proposed a mobile application that analyzes the patient with Dyslalia and presents its evolution over time, helping the professional and offering measures of the pathology level.

In [28], the authors proposed a retinal processing based on the fixation time in parts of the text and the reading time. The additional value to an established pattern indicates dyslexia and or other detectable pathologies of reading. Besides, it can be done from anywhere with a simple computer camera. They extract patterns based on high levels of values but using patterns obtained from voice signals instead of the eye position to get these data. Both methods being two promising technologies with easy accessibility.

In the work of Rahman et al. [22] an application was created for the user to read aloud while the system evaluates images of the ocular variation. Being composed of several processing steps, and having its result presented to the doctor to give its approval from the data obtained. The system is very accessible as it runs on android phones. The system helps indicate dyslexia just looking for eye and voice patterns, such as time and articulation, which are also covered in this proposal, but with additional features, as presented in [3].

Another similar work is that of Rello et al. [24], where the identification of dyslexia is proposed using Support Vector Machine (*SVM*) in the score obtained from specific activities applied to the user by a virtual game. The system applies tests indicated by a specialist, where the system speaks the words or syllables and asks the person to identify them and some interpretations in text form. This information also takes into account the response time. Thus, after training, the system can predict a dyslexia classification pattern.

Ariste et al. [7] use a game on the web to extract fundamental characteristics for dyslexia identification. The system presents statistics using an interactive platform, which makes it more peaceful for children. The system displays the results, statistical data obtained from the responses, and the audios recorded during the game, but it does not perform the automatic dyslexia indication process. It is only to streamline the specialist's identification process, making identification less long and tiring.

There are other computational proposals for early detection of dyslexia [1, 13,20], including games [5] and mobile apps [15,31]; and there are also systems that support the treatment and evolution of the disease [2,23,29].

However, none of the alternatives presented has all the features that we propose in this article. In our work, in addition to providing information, the system automatically indicates the pathological condition of the patient, so that the specialist has complete data. We use feature extraction of the audio signal, including the number of syllables through HMM, alignment, and heuristics improvement to implement SVM classification. We also compared our results with the specialists.

3 Medical Approach to Dyslexia Identification

The voice is defined as the sound signal emitted by the vocal folds and the movement of the larynx [8]. Speech, in turn, is the articulatory sound produced by several vocal muscles. Language is the production of sound emitted based on the understanding of what was read, seeking to represent a thought or an idea, as stated by Prates et al. [21].

When there is a dysfunction in the emission of voice, language and/or speech, the patient has some pathology, which may have physical causes [16] or neurological causes such as dyslexia and stuttering. Linguistic pathologies affect the reading and writing of a text, leading to difficulties in interpreting and representing the syntactic and morphological part of a reading text.

Alves [3] believes the previous discovery of dyslexia through phonetic characteristics extracted from reading aloud. In his work, a collection of audios of readings aloud of a specific text is made with children from the clinical (with dyslexia) and non-clinical (without dyslexia) group, and such phonetic measures allowed the creation of a model for identifying the evidence for dyslexia. The methodology used is based on manual analysis of characteristics extracted from the audio, to classify individuals with or without the pathology.

It was also noticed that the group of young people who had speech therapy treatment presented better temporal and prosodic characteristics than the group without treatment, but still out of expectations when compared to the subjects in the control group (without language and learning changes). The prosodic characteristics refer to the intonation, the formants, and the frequencies of the audio signal, while the temporal characteristics refer to the audio reproduction time, such as the total audio duration and the articulated audio time directly without pauses. The acoustic characteristics extracted manually from the audio are the number of syllables (QS), the number of pauses (QP), and the total time of pauses (TTP).

After these measures, it was calculated the speech (TTE) and articulation (TTA) times, and the speech (TE) and articulation (TA) rates. The speaking time (TTE) is the total time spent (in seconds) by the reader to read the text aloud. The articulation time (TTA) is the total time of the spoken audio signal without pauses, also in seconds. The speech rates (TE) and articulation (TA) are related to the number of syllables emitted per second, according to the speech and articulation times, respectively.

In addition to the reading time measurements, made in the time domain, [3] also proposed measures of frequency of the audio as essential for the definition of models of dyslexia levels. The frequencies are related to the intonation and formants of a segment are expressed in Hertz [Hz]. Four measurements are extracted from the fundamental frequency of the audio signal: F_i = initial fundamental frequency, F_f = final fundamental frequency, F_{max} = maximum fundamental frequency and F_{min} = minimum fundamental frequency. The tessiture of the audio signal, $Tess$, was calculated from these measurements, which is the difference between the highest frequency produced and the lowest, that is, the fundamental frequency variation.

According to Alves [3] and Brezntiz and Leikin [9] these measures indicate the level of difficulty of prosodic interpretation in reading a text, and patients with a high probability of dyslexia, in general present higher values (QP, QS, TTE, TTA, TTP) or lower (TA, TE, $Tess$) than is expected according to the read text, observed from the non-clinical group. For example, in their work, the clinical group (with dyslexia) presented QP and TTP with high values, which demonstrate a longer time for interpretation and textual sequence. The higher value of QS is due to the tendency to keep repeating the previous syllable while trying to read the next syllable, demonstrating the difficulty of visualization and interpretation as a whole.

Alves [3] analyzes all the data and standardizes the values through the verification carried out concerning a control group, without a systematic methodology of the final analysis. It was checked which data is above or below the expected value, seeking to characterize the dyslexia pathology on an aspect not yet described in the literature.

4 Computational Methodology

Section 3 present a proposal for the identification of patients with dyslexia based on the audio analysis of texts read aloud. However, the most significant difficulty was the time to analyze each patient and the researcher's tiredness to collect and measure the characteristics, which were done manually. Figure 1 shows the flowchart of the automatic methodology proposed in this article for solving the problem using signal analysis for the extraction of characteristics and machine learning for the classification of evidence of dyslexia. There are an improvement and extension of the study presented in [25], adding the analysis by aligning the audios and the indication of dyslexia.

The audio of texts read aloud for each child is used as the input signal. We perform a pre-processing, and from this result, the characteristics (QP, QS, TTE, TTA, TTP, TA, TE) are extracted, obtained directly from the audio signal and indirect characteristics ($Tess$, QS e frequencies) obtained from aligned audio. The measures were extracted from audio (aligned or not) and after using classification returns evidence of dyslexia. The analyses used in each of the proposed methodology stages are described in detail in the next sub-sections.

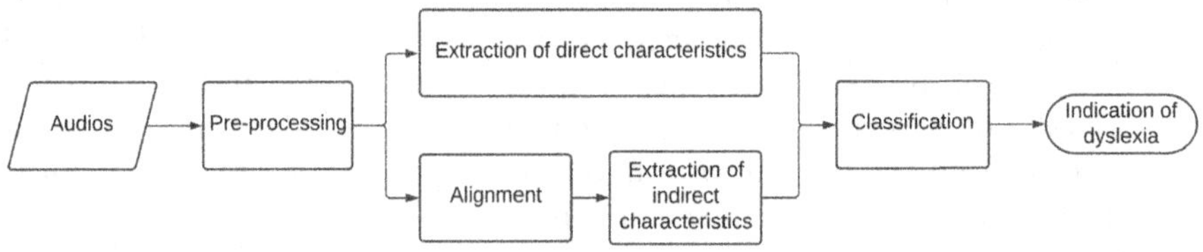

Fig. 1. Audio signal processing methodology.

4.1 Pre-processing of Audio Files

For the audios in the database, noise filtering is necessary due to the environment in which they were recorded. Two types of filters on the database were applied, high pass and low pass. The input signal is transformed into the frequency space using the FFT (Fast Fourier Transform). After this calculation, low-pass and high-pass filters are applied, filters that eliminate high and low-frequency noise.

The Inverse Fourier Transform (IFFT) is applied to this result, which returns the filtered signal to the time domain.

Pause Segmentation. As previous reported in [25] it was established to measure in the audio the number of pauses (QP) and the total time of these pauses (TTP), without considering the pause at the beginning and end of the audio signal. Figure 2 presents a flowchart of the algorithm used, which is based on the work of Barbedo et al. [6].

Once the audio signal was pre-processed, the signal winding is made using *Hamming windowing* of size w_s. Two characteristics are extracted from each window: Power Spectrum E and Spectral Centroid C, both measured from the frequency domain, which is obtained using the Discrete Fourier Transform (DFT) of the signal. The *Spectral Centroid* is a central average about the frequencies of the audio signal in each Hamming window, performing the location of the maximum and minimum frequencies peaks. The power spectrum of the signal is an average of the signal amplitude.

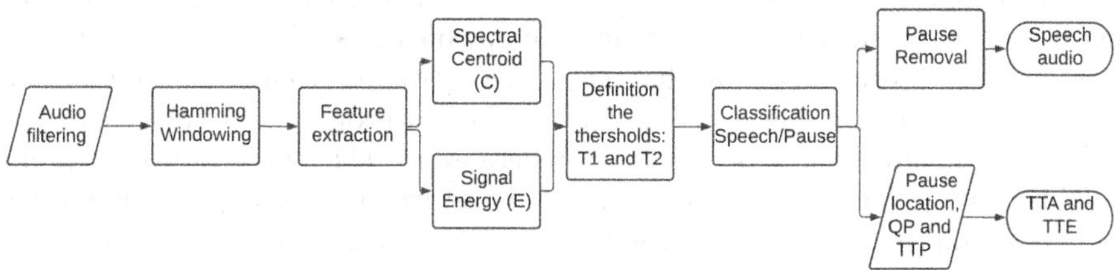

Fig. 2. Flowchart of the extraction process of direct characteristics, source: [25].

From the data obtained in each window, a histogram is generated for the two characteristics, Energy and Spectral centroid, extracting respectively their local maximum ($Tp1$ and $Tp2$) to be used as thresholds in the Hamming window. The cut-off thresholds for pause and voice are defined from these values, respectively.

After defining the thresholds considering the entire audio, an analysis is performed on signal segments to identify the pauses. From a fixed size window (w_s) with jumps (sp) the audio signal is traversed, and the characteristics mentioned above are extracted. Each value found per window is compared with the values of defined thresholds $Tp1$ and $Tp2$, signs with speech reach values above the cut threshold $Tp2$, being classified as 1 and pauses the values below the cut threshold $Tp1$, classified as 0.

The result of the pauses identification allows to return the number of pauses (QP) by counting the zeros obtained, the total time of the pauses (TTP) by the sum of the duration of each identified pause, the total articulation time (TTA) and the total speaking time (TTE). These values will be used later as input into the syllabic separation methodology. More details can be obtained at Barbedo et al. [6].

4.2 Direct Characteristics-Syllabic Segmentation

Once the audio signal has pre-processing to eliminate noise, the next step is to extract the characteristics of the signal by identifying and segmenting pauses and syllables, i.e., identifying voice and non-voice along with the audio.

The uninterrupted audio signal was obtained in Sect. 4.1, the segmentation into syllables is represented not only by grammatical separation but also by the emission of phonemes. Based on existing works [30] we can measure the number of syllables (QS). The original methodology was adapted so that it was possible to obtain a syntactic separator and the measure of (QS) on a text and not just on words, as suggested by Silva and Oliveira [30].

The audio without pauses is transformed through the half wave signal rectification function, which converts the audio signal into a positive signal, as exemplified in Silva and Oliveira [30], making the audio continuous and based on the positive frequencies of the signal.

In [25] we present the main structure of the code, where the audio signal is segmented into N windows of *Hamming* with size (w_s). In each window, it is extracted the Mel frequency cepstral coefficient (MFCC) - frequency and amplitude [10], and for each segment, the cut-off threshold is obtained by the average of the variation of the characteristics extracted from each window. The final count of the syllable is carried out through a grouping on values equal to 1, which symbolize a part of a syllable, where every 0 is considered the end of a syllable unit. From this grouping, a vector is returned with its position and the number of samples, which contains each syllable, thus counting the number of syllables (QS).

Considering these data, the metrics of TA and TE are calculated. The values of these variables are used to define the probability of dyslexia. These rates suggest domain over language and general diction, so minimal small values indicate a higher likelihood of dyslexia.

4.3 Indirect Characteristics: Alignment and Syllabic Segmentation

Even though the set of features contributed to the implementation of a classification modeling, we believe that the methodology can be improved if we carry out audio alignment training, i.e., the signal segments are aligned before the QS extraction, since it provides the correct words after certain syllables, improving the values found for QS. After filtering the audio signals and removing the silence, extraction of indirect characteristics began, that is, features that need the application of the alignment methodology to be obtained.

In alignment, the received signal is the incoming audio signal without pauses and the standard audio signal also without pauses, as described in Sect. 4.1. The standard sign is chosen as the basis for the analysis, as it is error-free and properly labeled as syllables. These audio signals are used at the end of the evaluation of results. According to the characteristics, a comparison is made with the segment of the incoming audio signal, checking whether it belongs to that part of the standard or not. Depending on the characteristics, a comparison is made with the incoming audio signal segment, checking whether it belongs to that part of the standard or not. According to Fellow [14], the method finds the ideal word sequence that best corresponds to a chain of unknown connected words.

First, the $s_{(i)}$ states that make up the Hidden Markov Models (HMM) were defined, the set of states was defined as the word unit, that is, each segment of the word is an internal state. The standard text allows HMM to train the probabilities of the audio signal to go from a state $s_{(i)}$ to a state $s_{(i+1)}$. From the analysis of the standard text, each word became a state to be reached, words that are not in this set of states are defined as words that were read incorrectly.

Figure 3 shows an outline of the alignment code at each stage of execution. The methodology was divided into two stages, the first step calculates the observation matrix (B) and the second step calculates the transition probabilities matrix between words (p).

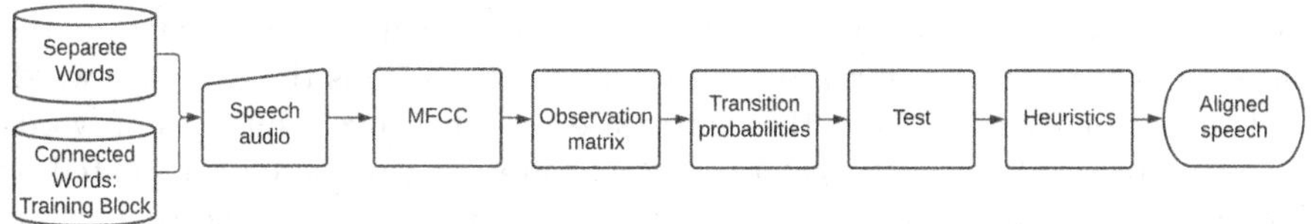

Fig. 3. Fluxogram representing the alignment.

The observations matrix (B) uses the mean and log gaussian of the MFCC coefficients extracted from speech audio without pauses, the audio signal is segmented into N windows of *Hamming* with size (w_s). In each window, it is extracted the Mel frequency cepstral coefficient (MFCC) - frequency and amplitude. These values were a vector of attributes representing the observations.

During the training the words are manually cut, each state is represented by a set of observations that vary depending on the size of the word Tam, which is the number of windows of size w_s, necessary to contain the word $DurP$, that is, $Tam = (DurP/w_s)$.

The probability transition matrix MT is established using the *Forward Backward* algorithm [14]. With fixed word sizes for each audio signal, the word most likely to happen at that moment is searched for, forming probability matrices based on the signals of trained audios. Word sizes are also defined, the maximum and minimum size that each word could contain is trained and adjusted. The chance of those transitions happening at that moment, $p = p_{01}, p_{02} ... p_{0n}$ considering that p_{01} are the transition probabilities from state 0 to 1 and n is the number of words.

From the probability matrices of the observations and transitions, the outputs most likely to be the answer of that segment are calculated. In the output, the *Viterb* model is applied, which analyzes the most likely response as a whole, returning the best path found for each segment, and finally, for the set of words they represent, based on in the sizes obtained in training.

Preliminary results using the traditional methodology proved to be insufficient for [4], so we made a proposal for improvement using two heuristics, considering the analysis of the nominal value of the probabilities and evaluating possible variations according to the expected word and words out of context, and the error heuristic, which proposes a new analysis in the part of the text that is very out of the standard and readjust its values.

4.4 Heuristic Analysis

The proposal to improve the nominal value of the probabilities is based on two distinct intervals defined as CE, error cut, and CD, cut of the difference. CE

represents the maximum and minimum values that the word reaches for the probability of being in a segment, i.e., if the probabilities are very different from the standard average for the word that was classified, becoming an error. The CD range represents the average value between the difference of two consecutive words. Not exceeding this threshold, the heuristic continues to analyze, even the four closest words that can represent that segment, which is the expected word.

The last analysis is about the heuristic error when five sequences of words do not correspond to the expected word and are not considered a repeated or skipped word; it undergoes a new analysis. The first word of the segment is considered to be a wrong word, and this goes on to a re-analysis within this segment, readjusting the value of the expected term. Their cut-off thresholds are reduced by the amount of their deviation, calculated in training, performing the reclassification, and adjusting the final values obtained.

Syllable Extraction Through Alignment. Through the aligned text, the number of syllables is counted for each word found in the obtained text. From these results, a certain distance from the comparison test value was also observed. Figure 4 shows the flowchart of the QS result improvement methodology. It has the audio signal without pauses as input, and it is obtained (QS_1) through the method of extraction of direct characteristics and (QS_2) through alignment. Of these values, the average is realized, obtaining QS.

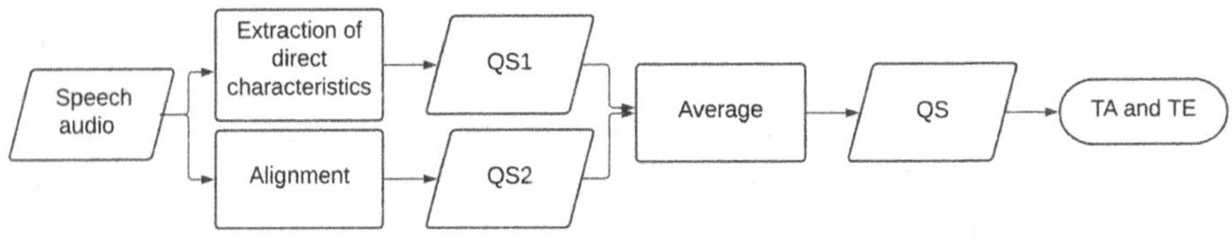

Fig. 4. Flowchart of number of syllables using alignment.

4.5 Extraction of Frequencies from the Audio Signal

In addition to the measurements extracted from the audio signal's amplitude, its frequencies are also important for the construction of the characteristic vector, as shown in Fig. 5. Instead of the total audio signal, just the segments of the audio signal in sentences are used.

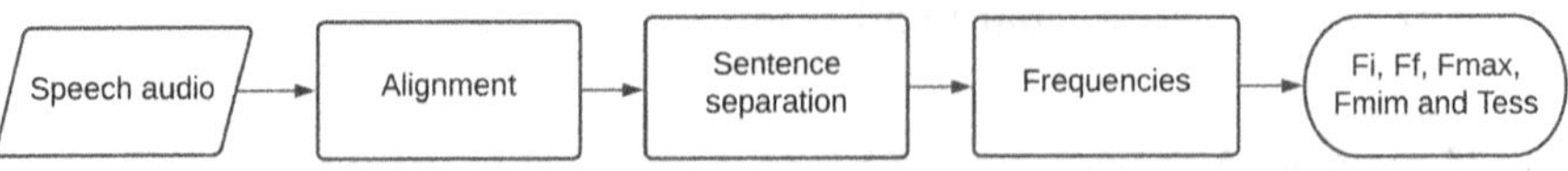

Fig. 5. Extraction flowchart on frequencies.

In the segmentation in phrases, the average size is analyzed and the beginning and ending words, dividing the text obtained from the alignment, in a set of more likely phrases. On each of these phrases, its fundamental frequency is measured F_0, calculated using the *Cepstrum* method, which determines the logarithm of the signal magnitude. The audio signal is analyzed on windows of 200 ms, and the cepstro is extracted from each audio signal sample. The maximum peak and location are calculated, converting at the end to Hertz and forming the vector corresponding to the audio signal's fundamental frequency.

In each segment of F_0, the first three frequencies are disregarded, at the beginning and at the end of the sentence, to remove micro melodic effects, which are involuntary fluctuations of F_0, as explained by Alves [3]. From this frequency curve, four more values are used to demonstrate the variation of intonation of the audio signal. They are: initial fundamental frequency (F_i), final fundamental frequency (F_f), maximum fundamental frequency (F_{max}) and minimum fundamental frequency (F_{min}). The last measure is tessiture ($Tess$), which, according to Alves [3], is one of the most important parameters for determining the likelihood of dyslexia.

5 Computational Experiments

The database used is the same as Alves [4] obtained through the base text **O Tatu Encabulado**, in Portuguese. The experiments were assembled from recordings of the reading audio aloud of school children. Of the total of 40 records, 10 (ten) are from children diagnosed with dyslexia, called clinical group (CG) and 30 (thirty) without dyslexia or language changes, called non-clinical group (NCG), varying between school grades, 3rd to 6th year, between 9 and 14 years old, male and female.

As the pause and syllable algorithm depends on some parameters, validation was performed using four audio signals, so that the database was divided between training and testing (90%) and validation (10%), ensuring GC and NCG in all samples. Thus, the values of the parameters should maximize the agreement between the results obtained automatically for the proposed methodology, and those derived from the manual annotation in the validation base, done by specialists.

The validation step consists of a set of experiments to obtain the optimal values of the $Tp1$, $Tp2$, and w_s parameters. They are varied, and the proposed methodology is used to calculate values for QP, TTP, and QS variables. The parameters are chosen when the variables are closest to the specialist results.

After validation, the testing phase uses the parameters to obtain results of QP, TTP, QS, TTE, TTA, TA, and TE using testing audios. The results were compared with the manual annotation made by specialists. The comparison metric is the absolute difference between the automatic value found by the proposed methodology and the manual annotation—the smaller the difference, the better the similarity between the data. Since the data are similar, the automatic values will be used as features for classification algorithms based on supervised learning. The classification allows identifying which category the data belongs based on a training set. There are a large number of algorithms for classification, and the one that returned higher predicted class was Support Vector Machine using a Gaussian kernel.

The entire methodology was developed using *Matlab*[1], a mathematical analysis and programming tool, using predefined signal processing functions of audio.

6 Results

One of the essential steps related to the results is the adjustment of the parameters, $Tp1$, and $Tp2$ for pause segmentation and w_s for syllable segmentation, called validation. As mentioned before, it was made empirically with the variation of the parameter values and the calculation of TTP, QP, and QS variables. For each one, the absolute difference between the values automatically calculated with the one provided manually by the specialist is calculated and used for comparison.

It is important to notice that we have a problem of minimizing multiple variables, which is challenging to solve and which solutions allow different heuristics. The proposed solution is to vary the set of parameters and find the values that minimize the difference between real and automatic values for the variables QP, TTP, and QS. Differences closer to zero represent more similar results.

6.1 Parameter Estimation

As mentioned, the identification of pauses is based on the Hamming window, which was fixed w_s equal to 0.13 ms and jump size *sp* of 0.04 ms. These values were obtained after a sequence of tests that resulted in less loss of information about each fragment. Table 1 presents the results of the absolute difference between the automatic values of TTP found by the proposed methodology and the manual annotation for variation of parameters $Tp1$ and $Tp2$. The best results for $Tp1$ are between 0.11 and 0.15 ms, and for $Tp2$ between 0.02 and 0.1 ms. The results were separated into clinical (CG) and non-clinical groups (NCG).

After delimiting the parameters of the segmentation of pauses, we continue to determine QP through the segmentation of syllables. Considering the variation of parameters delimited above, we obtained the values of QS for different values of w_s. In this experiment the size of the w_s window was varied from 16 ms to 30 ms. From the parameter variations, optimal values were chosen for $Tp1 = 0.13$, $Tp2 = 0.04$ and $w_s = 20$ ms. Tables 2 show the results obtained for QP, QS, TTP, TTE, TTA, TA and TE, on the audios that were used in the validation process. The table presents the manual value obtained by the specialist [3] and the result of the automatic proposed methodology.

[1] http://www.mathworks.com/products/matlab/.

Table 1. Absolute difference for manual and automatic TTP values, source: [25].

$Tp1(ms)$	0.11		0.12		0.13		0.14		0.15	
$Tp2(ms)$	GC	NCG	GC	NCG	GC	NCG	GC	NCG	GC	NCG
0.02	12.7	2.2	4.2	2.3	13.2	2.5	9.9	2.4	19.0	1.5
0.03	20.6	5.0	8.6	9.6	29.8	12.1	9.8	9.1	21.2	11.4
0.04	17.0	2.0	13.1	2.3	14.2	1.8	11.8	2.0	14.0	3.1
0.05	11.7	1.5	12.7	1.4	13.7	1.7	14.0	1.9	13.8	3.3
0.06	12.3	2.3	13.3	1.6	12.2	1.8	14.1	3.7	13.7	3.8
0.07	13.3	1.8	13.8	1.5	17.1	2.8	16.4	2.3	14.0	2.3
0.08	12.9	1.2	13.5	1.6	15.3	2.1	12.0	2.9	10.6	4.8
0.09	9.5	3.3	9.5	3.3	9.5	3.3	26.9	2.0	26.9	2.0
0.1	15.3	2.5	15.3	2.5	15.3	2.5	4.2	2.3	4.2	2.3

Table 2. Comparison between manual and automatic acoustic characteristics, source: [25].

Method	Audio	QP	TTP	QS	TTE	TTA	TE	TA
Manual	1	28	13.2	145	53.6	40.5	2.7	3.6
	2	40	17.9	173	73.6	55.8	2.3	3.1
Automatic	1	27	15.6	169	52.3	36.7	3.2	4.6
	2	39	19.5	165	72.8	53.3	2.3	3.1

6.2 Testing Phase

After setting parameters, tests were performed with the rest of the audios. Table 3 presents the mean values and standard deviation of the absolute differences over all variables. The variables results variables showed to be close to the values obtained in the research of Alves et al. in [3]. However, the results of QP and QS, Table 3 have higher mean values of absolute difference, indicating that the proposed methodology can still be improved.

Table 3. Mean and standard deviation for difference between manual and automatic values of variables using all database.

	TTP	TTE	TTA	TE	TA	QP	QS
Mean (μ)	2.22	2.16	3.65	0.69	1.06	2.78	24.05
Std (σ)	6.02	4.47	6.18	0.51	0.74	19.04	26.30

Among the classification methods tested using direct characteristics, the Support Vector Machine (SVM) using a Gaussian kernel obtained the best result, the experiments took place with k-fold cross-validation using the 36 audios, resulting 94.4% of accuracy, 80% of precision, 100% of recall and F1-score equal 0.88. Precision is related to the number of patients incorrectly classified as non-dyslexic, while the high recall value means high sensitivity; all dyslexic patients were correctly identified.

6.3 Indirect Characteristics

In this section, the results of the application of the alignment method are addressed, using HMM and improvement heuristics. We obtain the structure of the spoken text and the standard text, extracting the number of wrong words, and the beginning and end of sentences to extract frequencies and improve the number of syllables.

The accuracy of the tests is measured by the number of words that are correctly identified. The first test performed on 14 audio signals from the non-clinical group varied between 3 and 4 hidden states. Using 3 hidden states we obtained 51% of accuracy without heuristics and 89% of accuracy with four hidden states.

From these analyzes, the metric of three hidden states were established as alignment values. Thus, the tests were applied to the rest of the audio signals. Table 4 shows the accuracy of the alignment of the entire text on the 40 test audio signals, with a higher average of 43% in total. Error heuristics and window adjustment were applied to these data, thus not losing segments of audio signals from windows. After these processing, the accuracy increase for 60%, with a maximum value of 100% and a minimum of 14%. The distance between the maximum and minimum accuracy values is due to the accuracy of the GC audio signals, with lower values.

Table 4. Accuracy of QS for entire text.

Test	Heuristic	Mean	Max	Min
Fixed size	CD	43	96	4
	CE	43	90	5
Adjustable size	CD	51	100	3
	CE	60	100	14

It should be noted that the threshold obtained for the entire text is lower than that of a single sentence, since the database does not contain all the words of the vocabulary in Portuguese and in the misspellings of pronounced words that still interfere in the final accuracy.

6.4 Alignment Parameter Estimation

The alignment of the audio signal allowed a new calculation of QS for the audio signals. The result of QS obtained according Fig. 4, significantly improving the result. Table 5 shows the absolute difference between the values of QS and the standard value, so that values closer to zero represent more similar results.

It is noteworthy that the table shows the averages between all the audio signals and for each applied method, which are the extraction of QS directly, through the alignment and the average between the two. Table 5 shows that the average between the two methods generated data closer to that of manual data, with the smallest absolute difference for QS.

From the alignment, the fundamental frequency per phrase was obtained, from which the other frequency measures are extracted. Table 6 shows the average difference (in percentage) of the results found compared to that of the speech therapist [3].

Table 5. Table of the average difference of the audio signals over the parameter QS.

Audio	Automatic	Aligned	Mean
GNC	33.6	30.3	9.1
GC	23.3	47.4	27.5
Total	31.1	34.6	13.7

Table 6. Table of frequencies difference between automatic and measure values.

Audio	F_i	F_f	F_{min}	F_{max}	Tess
test	35%	29%	36%	21%	30%

The values were very close to the original data for some characteristics. From the data, it can be seen that the alignment with the base audio signal is worse for the GC, which counts as a metric for indicating the probability of dyslexia.

The data generated from the audio signal are used as decision parameters for the probability of dyslexia in each individual, being leveled in the high probability or low probability of being dyslexic. According to the data of the hypotheses suggested by [4], nine characteristics are essential for differentiation, which are: (QP), (TTP), (QS), (TTE) (TTA), (TE), (TA), $(Tess)$ and number of wrong words $(Error)$.

For the *KNN* and *SVM* methods, a selection of characteristics was performed to assess which has the greatest influence on the classification of data. This selection was carried out by comparing the linear difference of the results of [3], making the comparison between the data of the GC and GNC. The characteristics selected are those that have the greatest absolute difference from the data average, which are: QS, TTE, TTA, TE, TA, Fi and $Tess$.

The *KNN* was run with the raw data, which are the values obtained from the system without any analysis processing, obtaining 100 % accuracy for the feature set QS, TTA, TE, TA and Fi. In the SVM, the raw data result in an accuracy of 97.5 %.

The results show that for some characteristics, being an average reader interferes with the characterization and final classification. The wrong classifications are due to the use of all characteristics, where when selecting the most relevant attributes in terms of distances from the differences, the classification method classifies correctly, obtaining a better accuracy value, with most of the selected features dependent on the main ones.

From these data, there is a good agreement for the probability of dyslexia, raising promising results, for other improvements, including methodologies from different areas and expansion of research for other pathologies.

6.5 Statistical Evaluation

The last analysis is to measure and test the concordance between the data of the manual and automatic methods. Table 7 shows this analysis, where the *kappa* test is first presented, used in several studies to compare significance between two samples. According to Landis and Koch [17], the kappa test represents the degree of agreement of two groups of qualitative assessments on the same samples, where the results are evaluated according to the range of values found. In the kappa test, the higher the kappa value, the stronger the agreement. The kappa coefficient for the data was 0.87, where between the range of 0.81 and 1 there is a perfect agreement.

Table 7. Agreement values between the data obtained automatically and the manuals.

kappa	TTE	TTA	TA	TE	*Tess*
0.87	0.92	0.80	0.54	0.78	0.55

Individual concordances were greater than 0.78 in the case of TTE, TTA and TA and are below 0.55 for TE and $Tess$. The disagreement is that such measures did not have automatic values close to the standard's manual values.

7 Conclusion

This work proposed an automatic methodology for the extraction of audio characteristics from reading aloud and its use as variables in an intelligent model for the identification of young people with dyslexia. Among the main contributions can be highlighted the automation of the process of extracting features from the audio signal and its modeling using machine learning techniques.

The automatic TTP and TTP characteristics reached very close values when compared to manual measurements, while the number of QS syllables had a higher difference in the comparison. Even though the set of features contributed to the implementation of a classification model between indicative or not of dyslexia with an accuracy around 94%.

From the extraction of the indirect characteristics, TTP, DP and QP reached very close values when compared to manual measurements. In alignment for the GNC, an accuracy of 100% was achieved, which already shows the distinction and the importance of adding the amount of wrong words. The results proved to be promising since they already showed the distinction between the GC and GNC for each characteristic. Besides, the use of more comprehensive methods such as alignment proved to be effective in improving the prediction of values. After the alignment, the automatic method's results started to have a better agreement with the manual values.

Among the main contributions of this work, we can highlight the automation of extracting measurements from the audio signal, which are useful metrics for assessing the probability of dyslexia in audio signals. Finally, even though the results presented in the article relate to a text read in Portuguese, the proposed methodology also allows its use in other languages since the leading hypothesis for detecting dyslexia is related to the speed and the way words are spoken and not about the content itself.

We believe that the methodologies presented need a database with a more significant number of recording samples and a diversity of audience also may result in better training and testing of machine learning algorithms. Another alternative is to expand the metrics from the audio signal processing, aiming indication of dyslexia from other methodologies. For example, the intensity of the frequencies of each word could be a characteristic to be evaluated, in such a way that the range of characteristics extracted from the alignment would be expanded.

During the development of this research, ideas related to applications emerged, such as the construction of learning games for people with dyslexia. The proposal focuses on exercises that allow benefits in development and help to identify factors in which the patient has less control, related to improvements in quality of life.

Acknowledgements. This study was financed by *Fundação de Amparo à Pesquisa do Estado de Minas Gerais* (FAPEMIG) and *Universidade Federal de Ouro Preto* (UFOP).

References

1. Al-Barhamtoshy, H.M., Motaweh, D.M.: Diagnosis of dyslexia using computation analysis. In: 2017 International Conference on Informatics, Health & Technology (ICIHT), pp. 1–7. IEEE (2017)
2. Alghabban, W.G., Salama, R.M., Altalhi, A.H.: Mobile cloud computing: an effective multimodal interface tool for students with dyslexia. Comput. Hum. Behav. **75**, 160–166 (2017)
3. Alves, L.M.: A prosódia na leitura da criança disléxica. Ph.D. thesis, Universidade Federal de Minas Gerais - Faculdade de Letras, Belo Horizonte, May 2007. www.bibliotecadigital.ufmg.br/dspace/bitstream/

4. Alves, L.M., da Conceição Reis, C.A., Pinheiro, Â.M.V., Capellini, S.A.: Aspectos prosódicos temporais da leitura de escolares com dislexia do desenvolvimento. Revista da Sociedade Brasileira de Fonoaudiologia **14**(2), 197–204 (2009). http://www.scielo.br/pdf/rsbf/v14n2/10.pdf
5. Van den Audenaeren, L., et al.: DYSL-X: design of a tablet game for early risk detection of dyslexia in preschoolers. In: Schouten, B., Fedtke, S., Bekker, T., Schijven, M., Gekker, A. (eds.) Games for Health, pp. 257–266. Springer, Wiesbaden (2013). https://doi.org/10.1007/978-3-658-02897-8_20
6. Barbedo, J.G.A., Lopes, A.: Discriminador voz/música baseado na estimação de múltiplas frequências fundamentais. IEEE Lat. Am. Trans. **5**(5), 294–300 (2007)
7. Bartolomé, N.A., Zorrilla, A.M., Zapirain, B.G.: Dyslexia diagnosis in reading stage though the use of games at school. CGmaes 2012: The 17th International Conference on Computer Games, pp. 12–16 (2012)
8. Behlau, M.P.: Voz: o livro do especialista, vol. 1. Revinter (2001)
9. Breznitz, Z., Leikin, M.: Effects of accelerated reading rate on processing words' syntactic functions by normal and dyslexic readers: event related potentials evidence. J. Genet. Psychol. **162**, 276–296 (2001)
10. Brognaux, S., Drugman, T.: HMM-based speech segmentation: improvements of fully automatic approaches. IEEE/ACM Trans. Audio Speech Lang. Proces. **24**(1), 5–15 (2016)
11. Cano, P., Loscos, A., Bonada, J.: Score performance matching using HMMs. In: Proceedings of the International Computer Music Conference, San Francisco, pp. 441–444 (1999)
12. Deuschle, V.P., Cechella, C.: O déficit em consciência fonológica e sua relação com a dislexia: diagnóstico e intervenção. Revista CEFAC - Speech Lang. Hear. Sci. Educ. J. **11**(Supl 2), 194–200 (2009)
13. Drigas, A.S., Politi-Georgousi, S.: ICTs as a distinct detection approach for dyslexia screening: a contemporary view. IJOE: Int. J. Online Biomed. Eng. **15**(13), 46–60 (2019)
14. Fellow, L.R.R.: A tutorial on hidden Markov models and selected applications in speech recognition. IEEE **77**(2), 257–286 (1989)
15. Geurts, L., et al.: DIESEL-X: a game-based tool for early risk detection of dyslexia in preschoolers. In: Torbeyns, J., Lehtinen, E., Elen, J. (eds.) Describing and Studying Domain-Specific Serious Games. AGL, pp. 93–114. Springer, Cham (2015). https://doi.org/10.1007/978-3-319-20276-1_7
16. Gusso, G., Lopes, J.M.C.: Tratado de Medicina de Família e Comunidade: Princípios, Formação e Prática, vol. 2. Artmed (2012)
17. Landis, J.R., Koch, G.G.: The measurement of observer agreement for categorical data. Biometrics **33**(1), 159–174 (1977)
18. Leon, P., Pucher, M., Yamagishi, J., Hernaez, I., Saratxaga, I.: Evaluation of speaker verification security and detection of HMM-based synthetic speech. IEEE Trans. Audio Speech Lang. Process. **20**(8), 2280–2290 (2012)
19. Marinus, J.V.M.L., Araújo, J.M.F.R., Gomes, H.M., Costa, S.C.: On the use of cepstral coefficients and multilayer perceptron networks for vocal fold edema diagnosis. In: ITAB 2009–9th International Conference on Information Technology and Applications in Biomedicine, pp. 1–4 (2009)
20. Jothi Prabha, A., Bhargavi, R.: Prediction of dyslexia using machine learning—a research travelogue. In: Nath, V., Mandal, J.K. (eds.) Proceedings of the Third International Conference on Microelectronics, Computing and Communication Systems. LNEE, vol. 556, pp. 23–34. Springer, Singapore (2019). https://doi.org/10.1007/978-981-13-7091-5_3

21. Prates, L.P.C.S., Martins, V.O.: Distúrbios da fala e da linguagem na infância. Revista de Medicina de Minas Gerais **21**(4), 54–60 (2011)
22. Rahman, A., Hassanain, E., Rashid, M., Barnes, S.J., Hossain, M.S.: Spatial blockchain-based secure mass screening framework for children with dyslexia. IEE Access: Spec. Sect. Mob. Multimed. Healthc. **6**, 61876–61885 (2018)
23. Rahman, M.A., Hassanain, E., Rashid, M.M., Barnes, S.J., Hossain, M.S.: Spatial blockchain-based secure mass screening framework for children with dyslexia. IEEE Access: Multidiscip. Open Access J. **6**, 61876–61885 (2018)
24. Rello, L., Romero, E., Ali, A., Williams, K., Rauschenberger, M., Bigham, J.P., White, N.C.: Screening dyslexia for English using HCI measures and machine learning. In: DH 2018: 2018 International Digital Health Conference, pp. 23–26 (2018)
25. Ribeiro, F.M., Pereira Jr., A.R., Paiva, D.M.B., Alves, L.M., Bianchi, A.G.C.: Early dyslexia evidences using speech features. In: Proceedings of the 22nd International Conference on Enterprise Information Systems, ICEIS, vol. 1, pp. 640–647. INSTICC, SciTePress (2020). https://doi.org/10.5220/0009574906400647
26. Santos, M.C.S.: Disvoice: Aplicativo de apoio à Fonoaudiologia para dispositivos móveis. Mathesis, Fundação de Ensino Eurípides Soares da Rocha - UNIVEM (2013)
27. Shaywitz, S.: Entendendo a dislexia : um novo e completo programa para todos os níveis de problemas de leitura. Artmed, Porto Alegre, 1 edn. (2006). trad. sob a direção de Vinicius Figueira
28. Shrestha, S., Murano, P.: An algorithm for automatically detecting dyslexia on the fly. Intl. J. Comput. Sci. Inf. Technol. (IJCSIT) **10**(3), 1–18 (2018)
29. Sidhu, M.S., Manzura, E.: An effective conceptual multisensory multimedia model to support dyslexic children in learning. IJICTE - Int. J. Inf. Commun. Technol. Educ. **7**(3), 34–50 (2011)
30. Silva, E.L.F., Oliveira, H.M.: Implementação de um algoritmo de divisão silábica automática para arquivos de fala na língua portuguesa. Anais do XIX Congresso Brasileiro de Automática, CBA 2012, pp. 4161–4166 (2012). www2.ee.ufpe.br/codec/CBA2012_vf.pdf
31. Zarim, A., Azimah, N.: Android based dyslexia early screening test. Ph.D. thesis, UTeM (2016)
32. Zavaleta, J., Costa, R.J.M., da Cruz, S.M.S., Manhaes, M., Alfredo, L., Mousinho, R.: Dysdtool: Uma ferramenta inteligente para a avaliação e intervenção no apoio ao diagnóstico da dislexia. CSBC (2012) XXXII Congresso da Sociedade Brasileira de Computacao: XII WorKshop de Informatica Medica, WIM 2012 (2012)

Information Systems Analysis and Specification

An Extended Secondary Study to Characterize the Influence of Developers Sentiments on Practices and Artifacts in Open Source Software Projects

Rui Santos Carigé Júnior[1] and Glauco de Figueiredo Carneiro[2(✉)]

[1] Instituto Federal da Bahia (IFBA), Seabra-Bahia, Brazil
[2] Universidade Salvador (UNIFACS), Salvador-Bahia, Brazil
glauco.carneiro@unifacs.br

Abstract. *Context:* Sentiment Analysis applies computational techniques for both automated and semi-automated identification of human behavior. There is a trend to use such techniques in Sentiment Analysis tasks in the Software Engineering context. *Objective:* Characterize the influence of developers sentiments on software practices and artifacts in open source software projects. *Methods:* We conducted a Systematic Literature Review (SLR) to identify references in the literature related to the influence of developers sentiments on software practices and artifacts. *Results:* Evidence showed an increasing number of studies in this theme shedding light on issues related to the influence of developers sentiments on software practices. Practices focusing on developers productivity and collaboration, as well as source code, are the most vulnerable to sentiments variation. *Conclusions:* Based on the results provided in this SLR, we intend to present an updated and comprehensive overview regarding how the sentiments of developers can positively or negatively impact software practices and artifacts.

Keywords: Sentiment polarity · Secondary study · Open source software projects

1 Introduction

Emotions can be considered as behavioral motivators [7] and are associated with cognition [7], productivity [13], and decision making [18]. Studies on emotions are increasingly gaining focus in software engineering. Sentiment Analysis (SA) has been increasingly used in topics related to Computing, and Software Engineering (SE) is not an exception. Interactions among programmers through different ways and formats can unveil perceptions and behaviors that might influence software projects choices and results. The analysis of these perceptions requires resources not usually associated with traditional data analysis techniques.

The discussion of human aspects in Software Engineering has gained increasingly larger room in recent years [3, 6, 12]. Researchers have published studies discussing the effects of developers personality traits on software projects. Most of the results apply a holistic view of the subject, in this case, the developers.

J. Filipe et al. (Eds.): ICEIS 2020, LNBIP 417, pp. 339–359, 2021.
https://doi.org/10.1007/978-3-030-75418-1_16

Despite published studies reported that developers sentiments influence the way software is developed from different perspectives, there is still the need for a proper understanding of how the area as a whole deal with this issue. This is an extended version of a SLR published earlier. At the time of submission of the original secondary study, to the best of our knowledge, there was no secondary study that investigated how data related to developers interactions have been used by the research community. For this reason, we extended the original Systematic Literature Review study (SLR) to gather evidence provided by papers published in peer-reviewed conferences and journals from January 2019 to August 2020, considering that the previous work covered studies published from January 2000 to August 2019. In the original paper [17], we found 229 papers as a result of the applied search strings in specific electronic databases, from which we selected 11 studies to answer the stated research question. In this extended version, we included three additional papers from IEEE Xplore, and two additional papers from ACM Digital Library. Moreover, we had the opportunity to deepen the discussion of the impact of sentiments in adopted software practices and artifacts produced by programmers in open source software projects. Findings suggest gaps in current research agenda that unveil opportunities to further analysis of how developers sentiments data vary throughout the software life cycle. Special attention should be deserved for this variation on software practices and artifacts.

This SLR study is part of an ongoing larger project to discuss issues related to the identification, collection and analysis of the impact of developers sentiments on practices and artifacts in the open source software context. The project also aims at implementing an technological infrastructure to support the accomplishment of these goals. The Research Question (RQ) of this SLR is as follows: "What is the impact of sentiments in adopted software practices and artifacts produced by programmers in open source software projects?". This research question is in line with the goal of this secondary study. The motivation behind the research question is to characterize sentiments of developers as leading positive or negative effects on the quality of the software. We hope to strengthen the discussion to understand the possible roots and circumstances of positive and negative sentiments in open source software projects.

The structure of the rest of the paper is as follows. Section 1 presents Open Source Software (OSS) projects main concepts and characteristics. The Sect. 3 describes the design we followed in this secondary study. The Sect. 4 presents a panoramic view of evidence from selected papers. We discuss these results in Sect. 5, presenting the answer to the stated research question. Section 6 evaluates potential threats to the validity to the results. Finally, Sect. 7 presents conclusions and future work.

2 Open Source Software Projects

Open source project refers to any software made public and open for others to modify [2]. Stability and economic benefits are key issues for commercial software. However, for open source software projects, the satisfaction of users is on the foreground to prioritize filed issues and its corresponding implementation [26]. Effective practices and products from OSS projects are worldwide recognized [21]. Companies can draw lessons from open source software projects by studying their best practices and applying them internally [21,31]. It is a common practice in OSS projects for programmers to review code from their counterparts, to focus on quality attributes, among others activities [24], besides coding. These tasks rely on interaction and communication among programmers that take place through text messages registered in issue comments, wikis, forums, just to name a few. In fact, OSS communities are self-managed, and rely on consensus to reach decisions [2]. The structure of each OSS community is unique, reflecting its principles, values, beliefs, and norms [2]. This motivated us to investigate the way programmers interact through comments to shed lights on details, challenges and the rationale behind decisions of OSS projects.

As being part of the communication process, comments can reveal the way programmers feel during coding related activities. Researchers have already dedicated efforts to examine this in the context of software engineering. The term Behavioral Software Engineering (BSE) was coined to refer to the emphasis sentiments and therefore behavior have on software engineering activities [19]. We identified a SLR that selected studies reporting the influence of the emotions of software developers in their daily activities and indicators to assess them. The SLR provided a comprehensive view of the relationship of sentiments and software engineering [28]. The authors of the secondary study argued for the need of a better understanding of emotions in the context of software development activities, including issues related to performance, productivity, quality, and well-being. We tackled these issues in the SLR presented in the following sections.

We also found another SLR that investigated the effect of personality traits and team climate on software team performance [30]. The secondary study analyzed 35 studies that focused on this effect to conclude that team climate is to some extent related to several factors associated with the fields of management and behavioral sciences [30]. The influence of personality in Software Engineering was the goal of the SLR reported in [8]. The authors reported evidence of used methods, addressed topics, applied personality tests, among others. In an extended version of the mentioned paper, the authors found complemented that issues related to pair programming, education, team effectiveness, software process allocation, software engineer personality characteristics, and individual performance was discussed in 88% of 90 studies selected, while team process, behavior and preferences, and leadership performance were topics not focused by the majority of researchers [9].

Due to the absence of secondary studies focusing on the influence of sentiments in open source projects practices and artifacts, we decided to conduct this research as a contribution to the academic field. OSS projects allow their data to be scrutinized. We used it for research purposes. This is an effective practice of OSS projects, considering that sharing the results of the analysis of these data contributes to open source development. Another important reason to target on the analysis of sentiments in open source projects is its inherently collaborative environment. In these projects, there is the need for the contribution of programmers with a comprehensive range of characteristics, allowing the plurality of sentiments.

3 Research Design

This paper reports a Systematic Literature Review based on evidence related to the impact of sentiments on practices and artifacts in open source software projects. In the next subsections we present the research design we adopted.

3.1 Planning

We followed a protocol including objectives, criteria to select studies, research questions, target electronic databases, corresponding search strings to support the answer to the stated research questions [35]. The protocol and related artifacts we adopted are available in a public repository[1]. Table 1 presents the goal of this study according to the GQM approach [4].

Table 1. The SLR goal [17].

Analyze	Sentiments
for the purpose of	identification of its impact
with respect to	adopted practices as well as software artifacts produced
from the point of view of	programmers
in the context of	open source software projects

The stated Research Question (RQ) is as follows: *What is the impact of sentiments in adopted software practices and artifacts produced by programmers in open source software projects?* We derived RQ into two specific research questions: SRQ1 and SRQ2. The Specific Research Question 1 (SRQ1) is *What is the impact of sentiments of programmers in open source software projects in adopted software practices?* The Specific Research Question 2 (SRQ2) is *What is the impact of sentiments of programmers in open source software projects artifacts?*

Understanding to which extent the sentiment of developers can positively or negatively affect software practices and artifacts is a relevant requirement to improve software quality. The specific research questions are the support to obtain evidence to answer the stated research question. We adopted the PICO criteria [32] to define the search strings, as presented in Table 2. We used the search strings to select studies for this systematic literature review.

Table 2. PICO criteria for search strings [17].

(P)opulation & software engineering papers focusing on open source	Software projects
(I)ntervention & influence of sentiments in adopted software practices	And artifacts produced by programmers
(C)omparison & not applicable	
(O)utcomes & impacts (positive and negative) of the influence of	Sentiments on software practices and artifacts

We present in Tables 3 and 4 how we build the search string to search studies in the electronic databases. The Table 3 presents the major terms aligned to the research

[1] https://github.com/impactsentimentanalysis/iceisExploratoryStudy2020.

objectives based on the PICO criteria. To obtain a more effective search string, we included possible alternative terms and synonyms. For illustrative purposes, the term *sentiment* can be related to terms such as *feeling*, *emotion*, and *opinion mining*. The Table 4 presents the so called alternative terms. We also included these terms in the final search string that was built by joining the major terms with the boolean "AND" and joining the alternative terms to the main terms with the boolean "OR".

Table 3. Major terms for the research objectives [17].

Criteria & major terms	
(P)opulation & AND "software engineering"	
(I)ntervention AND "sentiment"	AND "open source software project"
(C)omparison & not applicable	
(O)utcomes & AND "software practice" AND "software artifact"	

Table 4. Alternative terms from majors terms [17].

Major Term & Alternative Terms
"open source software project" & "Free and open-source software" OR "free/libre and open-source software" OR "OSS" OR FOSS" OR "F/LOSS" OR "FLOSS"
"sentiment" & "feeling" OR "emotion" OR "opinion mining"
"software practice" & "software activity"
"software artifact" & "software asset"

Table 5 presents the digital libraries (ACM Digital Library, IEEE Xplore and ScienceDirect), we target to retrieve the papers along with the respective search strings used to retrieve the papers. All searches were performed on August 22, 2019.

Table 5. Electronic databases selected for this SLR [17].

Database and URL & Search Strings	
ACM Digital Library portal.acm.org & (+"software engineering" +("open source	software" "OSS" "open source projects" "OSP" "Free and open-source software" "FOSS" "free/libre and open-source software" "F/LOSS" "FLOSS") +("sentiment" "feeling" "emotion" "personality" "opinion mining"))
IEEE Xplore ieeexplore.ieee.org & ("software engineering" AND ("open source	software" OR "OSS" OR "open source projects" OR "OSP" OR "Free and open-source software" OR "FOSS" OR "free/libre and open-source software" OR "F/LOSS" OR "FLOSS") AND ("sentiment" OR "feeling" OR "emotion" OR "personality" OR "opinion mining")))
ScienceDirect sciencedirect.com & ("software engineering" ("open source	software" OR "OSS" OR "open source projects" OR "OSP" OR "Free and open-source software" OR "FOSS") ("sentiment" OR "feeling" OR "emotion" OR "personality" OR "opinion mining"))

We present the criteria for exclusion and inclusion of papers in this secondary study in Table 6. The term *OR* is used as exclusion criteria. In this case, the exclusion criteria are independent, i.e., meeting only one criterion is enough to exclude the paper. The *AND* connective is associated with the inclusion criteria, meaning that all inclusion criteria must met to select the paper under analysis. Table 6 lists the quality criteria adopted in this SLR as questions adjusted from Dyba and Dingsoyr [11]. We performed a detailed analysis based on the quality criteria on all remaining papers that passed the exclusion and inclusion criteria. Permanent selected papers needed to meet these criteria. In this case, the answer must be YES for each of the questions. Just one answer NO is enough to exclude a specific paper. The steps to apply the exclusion, inclusion and quality criteria in the selection process are presented in Table 7. The final classification of each retrieved paper is one of the three options as presented in Table 8: *Excluded*, *Not Selected* and *Selected*.

Table 6. Exclusion, inclusion and quality criteria [17].

Type	Id	Description	Connective or answer
Exclusion	E1	Published earlier than 2000	OR
Exclusion	E2	The paper was not published in a peer-reviewed journal or conference	OR
Exclusion	E3	The paper is not written in English	OR
Exclusion	E4	The paper has less than 4 pages	OR
Exclusion	E5	The paper does not present a primary study	OR
Inclusion	I1	The paper should present a study on the influence of sentiments on software practices or artifacts produced by programmers	AND
Inclusion	I2	The study should be conducted within the scope of open source software projects	AND
Inclusion	I3	Sentiments should be detected from records made by developers	AND
Quality	Q1	Are the aims of the study clearly specified?	YES/NO
Quality	Q2	Is the context of the study clearly stated?	YES/NO
Quality	Q3	Does the research design support the aims of the study?	YES/NO
Quality	Q4	Does the study have an adequate description of the impact of sentiments on open source software project practices and artifacts?	YES/NO
Quality	Q5	Is the data analysis of the study is rigorous and based on evidence theoretical of reasoning instead non-justified or ad hoc statements?	YES/NO

Table 7. Steps of the selection process [17].

Step	Description
1	Apply the search strings to obtain a list of candidate papers in specific electronic databases
2	Remove duplicated papers from the list
3	Apply the exclusion criteria in the listed papers
4	Apply the inclusion criteria after reading abstracts, introduction and conclusion in papers not excluded in step 3
5	Apply quality criteria in selected papers after step 4

Table 8. Classification options for each retrieved paper [17].

Classification	Description
Excluded	Papers met the exclusion criteria
Not selected	Papers not excluded due to the exclusion criteria, but did not meet the inclusion or quality criteria
Selected	Papers did not meet the exclusion criteria and met both the inclusion and quality criteria

3.2 Execution

Figure 1 presents the performed steps and the respective number of documents for each phase in the execution of this SLR. The template adopted in the figure is based on the PRISMA flow diagram [22].

Table 9. Effectiveness of the search [17].

Database	Papers retrieved by the search string	Selected papers	Search effectiveness
ACM Digital Library	66	7	10.6%
IEEE Xplore	23	4	17.4%
ScienceDirect	143	3	2.1%
Total	**232**	**14**	**6%**

We present the effectiveness of the search string in each digital library in Table 9. The table portrays the 232 retrieved papers that after removing the duplicates provided 223 papers. The ACM Digital Library contributed with seven papers, corresponding to a search effectiveness of 10.6%. The IEEE Xplore also stands out with 17.4% of effectiveness of the adopted search string. The 14 selected papers represented 6% of all 232 retrieved papers.

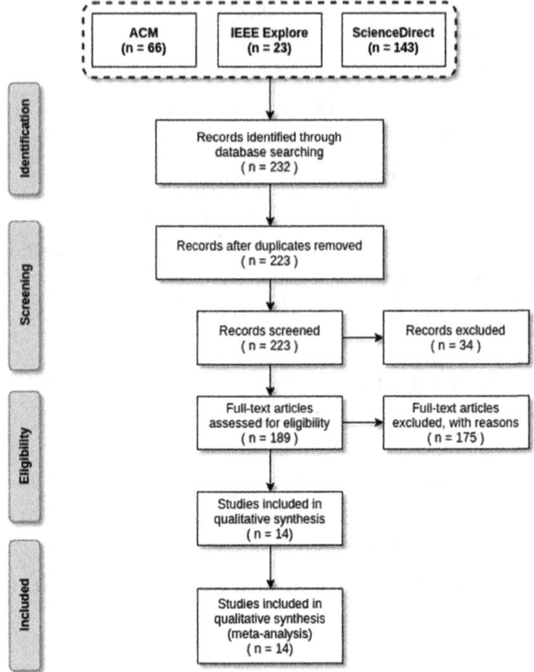

Fig. 1. Procedures and its results in the papers selection process. Adjusted from [17].

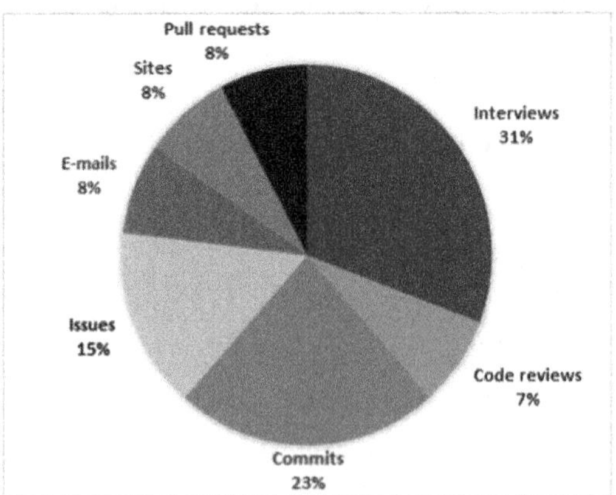

Fig. 2. Artifacts from which sentiments were identified [17].

4 Results

In the Appendix A, we present a list of the 14 selected papers of this secondary study. All studies are named in the format "SP" followed by the paper reference number. The complete list is available at the end of this document. The selected papers were published in conferences and journals.

From the 14 selected studies, we obtained the sources from which sentiments were analyzed as represented in Fig. 2. *Interviews* and *commits* were the main sources of evidence, representing 54% of the sources of sentiments in these studies. Figure 3 presents the distribution of sources of evidence we considered in this SLR to answer the Specific Research Questions SRQ1 and SRQ2. Each branch represents evidence and the corresponding studies from which they were collected to answer a Specific Research Question. The same figure highlights the positive and negative influence of sentiments on practices and artifacts.

In Table 10 it is possible to analyze the impact of sentiment on software practices based on collected evidence. The table presents the relationship among polarity, sentiment, software practice and corresponding impact according to the selected studies. Table 11 has the same goal, but now focusing on software artifacts instead of practices. We can conclude that majority of the articles discussed how practices have been affected, and only a few discussed the influence on software artifacts. In this case, **SP03**, **SP10**, **SP12** and **SP14** were the studies to explicitly discussed the influence on practices and software artifacts as well. We mention the term *explicitly* because the authors recognize in the text of the respective papers the impact of sentiments also on software artifacts. Despite being indicated only in these papers, the artifacts might also implicitly suffer the influence of sentiments as a consequence of performing the practices. For example, from data represented in Table 10, we can conclude that productivity can decrease, resulting in delay of the release expedition and corresponding not full

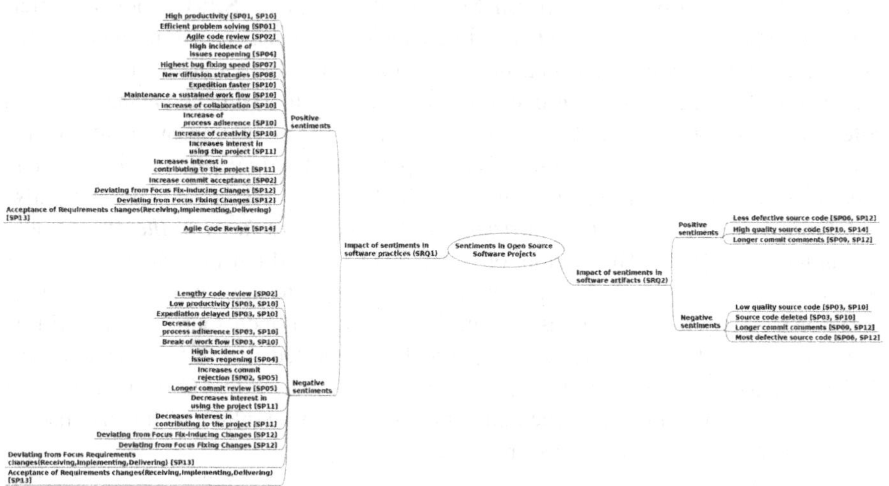

Fig. 3. Impact on software practices and artifacts. Adjusted from [17].

adherence to the process. This, in general, can lead to a possible source code of low quality. It is worth to mention that the sentiment of unhappiness can be a possible root cause of what was discussed before, as mentioned by the authors of **SP03** and **SP10**.

Tables 10 and 11 show a tendency of positive sentiments to positively impact software practices and/or artifacts. The results of **SP06** indicated that positive emotions often lead to few defects. Likewise, authors have argued that sentiments of negative polarity have the possibility to increase the number of defects. Moreover, according to **SP04**, issue reopening is not a positive result due to the indication that something has gone wrong in the issue handling. This means that additional effort must be made to fix it definitely. The results found by the authors showed that sentiments with negative and positive polarity can increased rate of issues reopening. In Tables 10 and 11, we mark sentiments from studies **SP02**, **SP04**, **SP06**, **SP07** and **SP09** as *Not specified*. The tools used in each study is described as follows: Senti4SD [5], SentiStrengthSE [16], LIWC [23], RNTN [29] and SentiStrength [33]. The tools supported researchers to automatically identify the polarity (positive or negative) of the sentiments expressed by the programmers.

We answer the specific research questions (SRQ1 and SRQ2) in the following subsections, respectively.

4.1 Impact of Positive Sentiments on Software Practices

We represented the influence of sentiments in software practices in the left side of Fig. 3. The upper left side of the figure represents the impact of positive sentiments. According to this figure and Table 10, 25% of this impact is associated with the Positive *Contentment* sentiment. The other registered sentiments *Calm*, *Trust* and *Happiness* account for approximately 12% of the collected evidence. Roughly 38% of the sentiments associated with positive polarity is marked as *not specified*.

Impacts of the Positive *Contentment* Sentiment. The paper **SP01** conducted interviews with open source software teams to characterize their perception regarding the collaborative virtual work in which they were engaged. The result of the interviews revealed that the contentment sentiment was associated with a suitable balance between professional and personal life. The consequence was productivity improvement and efficiency of the professionals. One interviewee reported that *"My impression is that I can be more productive working from home that working from the office. I probably also work for more hours than if I was working in the office. So when I in the office, I was interrupted very often"*. The authors of paper **SP11** pointed out that GitHub provides social coding features, including the assignment of stars to a given repository which indeed presumably express interest or satisfaction with an open source project. Based on the feedback of 791 programmers, the authors reported that 73% considered the number of stars as a relevant criterion for using or contributing to a GitHub project. Of these, 26.5% said that the larger the number of stars the more influence this fact had on the decision to use or contribute to a project.

Impacts of the Positive *Calm* Sentiment. In **SP01** the authors reported that when programmers did not felt rushed, they can solve problems much more efficiently. One of the programmers informed that *"I found that at the end I can really start to answer questions and solve problems a lot better than when I don't feel rushed we have an office and everybody needs to be here at the same time kind of rules"*.

Impacts of the Positive *Trust* Sentiment. The paper **SP08** investigated the influence of open source project developers' trust on global software engineering processes. The study found evidence of trust as a possible factor to increase developer confidence. The consequence is the diffusion of new strategies, including social and technical innovations within the distributed team. In fact, study revealed that the diffusion trajectory becomes diverse when considering individual variations on baseline trust.

Impacts of the Positive *Happiness* Sentiment. The paper **SP10** extracted data from interviews with software developers and analyzed the possible consequence of happiness and unhappiness in software processes and artifacts. The most reported consequence was the increase of productivity, as shown in these statements: *"When I have this [happy] feeling I can just code for hours and hours. (...) I felt that my productivity grew while I was happy. (...) The better my mood, the more productive I am"*. The authors also argued that another process influenced by the happiness of developers was software expedition, that became faster. The tasks were sped up without sacrificing quality, and one of the programmers commented that *"it seems more likely to reach my goals faster"*. Programmers from the same study claimed to be full of energy and with strong focus as a result of happiness. Thus they maintained a sustained flow and were *"unaware of the passing time. (...) I can continue to code without anymore errors for the rest of the day, and just knock out lines of code all day"*.

Based on the results, the authors stated that happy developers can also encourage more collaborative team members, leading to increased collaboration. The authors realized that happiness leads to more willingness to share knowledge and to bring colleagues together to solve a problem.

Adherence to the process was another consequence of the happiness that was reported by the interviewees in **SP10**, as explained by a programmer: *"when I am happy to work, I usually try new things and follow best practices and standards as much as possible"*. Finally, the authors of **SP10** stated that the creative process can also be a positive consequence of developers being happy, as stated in this response: *"if [...] I have a general good mood, the software process gets to be creative and very good"*.

Impacts of the Positive *Motivation* Sentiment. In **SP13**, the authors argued that tasks related to code review are mainly performed with motivation, supported by several motivation mechanisms, such as desire to learn, to grow reputation, or to improve one's positioning on the job market [20]. In fact, these are key motivators that attract programmers for OSS projects communities. They identified three stages of requirements changes (RC), where participants reported agile teams responding emotionally: receiving, implementing, and delivering an RC [20]. Positive and neutral emotional responses of agile teams varied in cycles from high to low and then high again as teams go through

Table 10. Impact of sentiments on software practices. Adjusted from [17].

Polarity	Sentiment	Software practice	Impact	Studies
Positive	Contentment	Productivity	Increase	SP01
		Use of project	Increased interest	SP11
		Contribution to the project	Increased interest	SP11
	Calm	Problem solving	Efficiency	SP01
	Trust	Diffusion of technology	New strategies	SP08
	Happiness	Productivity	Higher	SP10
		Expedition	Faster	SP10
		Work flow	Sustained	SP10
		Collaboration	Increased	SP10
		Process adherence	Increased	SP10
		Creativity	Increase	SP10
	Motivation	Requirements changes (receiving, implementing, delivering)	Acceptance	SP13
	Passion	Code review	Agility	SP14
	Not specified	Code review	Agility	SP02
		Commit acceptance	Increase	SP02
		Reopening of issues	Highest incidence	SP04
		Bug fixing speed	Highest	SP07
		Fix-inducing changes	Efficiency	SP12
		Fixing changes	Efficiency	SP12
Negative	Discontent	Use of project	Decreased interest	SP11
		Contribution to the project	Decreased interest	SP11
	Rudeness	Commit rejection	Increase	SP05
		Commit review	Longer	SP05
	Unhappiness	Productivity	Lower	SP03, SP10
		Expedition	Delayed	SP03, SP10
		Process adherence	Decrease	SP03, SP10
		Work flow	Broken	SP03, SP10
	Resistance	Code reviews	Lengthy	SP02
		Requirements changes (receiving, implementing, delivering)	Deviating from focus	SP13
	Demotivation	Requirements changes (receiving, implementing, delivering)	Acceptance	SP13
	Not specified	Commit rejection	Increase	SP02
		Reopening of issues	Highest incidence	SP04
		Fix-inducing changes	Deviating from focus	SP12
		Fixing changes	Deviating from focus	SP12

Table 11. Impacts of sentiments on software artifacts. Adjusted from [17].

Polarity	Sentiment	Software artifact	Impact	Studies
Positive	Happiness	Source code	High quality	SP10
	Passion	Source code	High quality	SP14
	Not specified	Source code	Less defective	SP06, SP12
		Commit comments	Longer	SP09, SP12
Negative	Unhappiness	Source code	Low quality	SP03, SP10
			Deleted	SP03, SP10
	Not specified	Source code	More defective	SP06, SP12
		Commit comments	Longer	SP09, SP12

the three RC stages. In contrast, negative emotional responses are high when receiving an RC and move towards positive as teams move to implementing and delivering, concluding that agile teams feel emotionally positive as they deliver an RC [20].

Impacts of the Positive *Passion* Sentiment. The well-known success of peer code review, frequently adopted in open source projects, is an emotionally intensive practice, public exposing the ones involved in the tasks, due to proving and giving the chance to reinforce their reputation and also ample opportunities for conflict [1]. The authors of **SP14** investigated the reasons for code review success in OSS projects communities, despite its inherent challenge. It is true that OSS project contributors experience rejection and negative feedback regularly [1]. However, they do not eliminate this negative experience, as this seems to be the core improvement mechanism of code review [1]. Developers in these communities are mainly passionate about all aspects of the project, including the reviews. The passion allows them to dedicate efforts into code review tasks, making them resilient to negative interactions [1].

Impacts of Other Positive Sentiments. In **SP02**, the authors studied the sentiments expressed by contributors during code review activities using the Senti4SD Sentiment Analysis tool [5]. The authors claimed that reviews with positive comments have a shorter duration on average. They argued that the presence of positive sentiments in comments related to source code reviews seems to contribute to reducing the review time by an average of 0.4 day. In the same paper, the authors stated that the presence of positive sentiments in code review comments seems to have positively affected the results of these reviews. They found that 91.81% of successful reviews were identified with positive sentiments, and 64.44% of aborted reviews contained negative sentiments.

The paper **SP04** analyzed data available in the JIRA issue tracking system from eight Apache Software Foundation open source projects. Sentiments were detected in developers comments through the use of the SentiStrengthSE tool [16]. The results indicated that sentiments with very high positive or high negative scores impact in the rate of issues reopening.

In the paper **SP07**, the authors analyzed comments from Github to investigate the correlation between emotional factors and the speed of bug fixes in open source software development. The authors realized that the Bug Fixing Speed (BFS), i.e., the average number of problems that have been resolved within a certain period of time, was lower when the developers' sentiment proved positive.

4.2 Impacts of Negative Sentiments on Software Practices

The bottom left side of the Fig. 3 shows evidence of the impact of negative sentiments on software practices. Among these sentiments, *Unhappiness* stood out in the percentage of appearances in the selected studies (33%). *Discontent* and *Rudeness* represented each 14% of the negative sentiments that influence software practices. We also identified that 33% of these negative sentiments were not specified.

Impacts of the Negative *Discontent* Sentiment. The results of studies by Borges et al. (**SP11**) show that three out of four developers consider the number of stars assigned to GitHub projects before using or contributing to those projects. The developers commented that the number of stars has a high influence on their decision of using or contributing to a project.

Impacts of the Negative *Rudeness* Sentiment. In the study **SP05**, the researchers used a tool proposed by Danescu-Niculescu-Mizil et al. [10] to identify a binary output of politeness (polite or uneducated) of particular text and found that developers who are less active in open source software projects, when committing with less polite comments, have a higher likelihood that your commits will be rejected in the main project repository. Ortu et al. (**SP05**) also found that developers who are less active with lower levels of politeness were more likely to be unmerged, with a longer reviewing process.

Impacts of the Negative *Unhappiness* Sentiment. In the studies **SP03** and **SP10**, Graziotin et al. extracted sentiments of developers from interviews. Low productivity was a consequence of the unhappiness most reported by interviews participants.

The result of the research by Graziotion et al. (**SP03** and **SP10**) show reports that unhappiness is delaying the execution of process activities: *"In both cases [negative experiences] the emotional toll on me caused delays to the project"*.

In **SP03** and **SP10**, developers declared that unhappiness caused them to have decreased process adherence, i.e., deviating from the agreed set of practices. It was reported that unhappiness led developers to compromise in terms of actions, in order to just get rid of the job.

Graziotion et al. (**SP03** and **SP10**) stated that unhappiness causes interruptions in developers' flow, resulting in adverse effects on the process. As put by a participant of interview, *"things like that [of unhappiness] often cause long delays, or cause one getting out of the flow, making it difficult to pick up the work again where one has left off"*.

Impacts of Other Negative Sentiments. The researchers of **SP02** found that reviews with negative comments take longer to complete. Negative reviews required a supplementary time of 1.32 day on average to be closed than positive ones. In other words, the average of durations for positive reviews is less than the average for negative reviews. For instance, in the Eclipse project, positive reviews last a maximum of 2.89 d, while reviews containing negative comments took approximately 5 d of review.

The results from **SP02** show that the sentiment expressed by developers also affects the code review result. For instance, in Eclipse project, over 93% of successfully merged reviews were tagged as positive, while 55% out of all abandoned reviews have negative sentiments into their comments.

Cheruvelil et al. (**SP04**) found evidence that suggests that negative sentiment correlates with issue reopening, although the effect size seems to be rather small.

Developer performance can be evaluated from software Commits associated with the type of code written in that Commit. For example, buggy code can be extracted from Commits by finding Fix-Inducing Changes (FIC) [15]. FICs are Commits that induce fixes and cause errors in the system [27]. Messages in FICs can contain sentiments expressed by programmers othan other Commits itself, to indicate complex assignments or imperfect work. According to **SP12**, the analysis of Fix-inducing Changes performed by the authors concluded that negative sentiment exceeds positive ones by at least 40% [15]. Commits that introduce bugs tend to be more negative and hence associated with less neutral message than regular Commits. The authors of **SP12** argue that this can be used for precautionary measures with sensitive changes. For example, Commits with negative messages can be emphasized during reviews. This gives room for more priority on both monitoring and maintaining the sentiment of developers, as that relates to problematic performance [15].

4.3 Impacts of Positive Sentiments on Software Artifacts

The impact of sentiments of programmers in open source software projects in software artifacts was registered in the right side of Fig. 3. In this case, the upper right side of the same figure portrays the impact of positive sentiments.

We observed that 33% of the selected articles indicate that the sentiment of happiness impacts the artifacts of open source software projects. Other positive sentiments affecting artifacts (67%) were not specified.

Impacts of the Positive *Happiness* Sentiment. The high quality of the source code was reported as a consequence of happiness in the study **SP10**. The authors stated that higher quality of code is generally realized when developers are happy, because they tend to make less mistakes, see solutions to problems more easily, and make new connections to improve the quality of the code. As a result, the code is cleaner, more readable, better commented and tested, and with less errors and bugs.

Impacts of Other Positive Sentiments. In the study **SP06**, Zhang et al. adopted a Linguistic Inquiry and Word Count (LIWC) tool [23] to recognize the developers' sentiment from email list. It can be observed that the emails whose emotion value is larger tend to be linked to less defective source code files.

Islam and Zimbra (**SP09**) extracted emotions from the developers' commit messages using SentiStrength tool [33]. They found that developers' emotions generate statistically significant impacts on the size of commit messages. Developers post longer comments when they are emotionally active.

4.4 Impacts of Negative Sentiments on Software Artifacts

We also present in Fig. 3 evidence of the impact of negative sentiments on software artifacts. Evidence from selected studies indicated signals of unhappiness (around 50%) among the sentiments of negative polarity prone to impacts software artifacts.

Impacts of the Negative *Unhappiness* Sentiment. The authors of **SP03** and **SP10** observed from their interviews that unhappiness negatively impacts the quality of the code. The interviewees declared that *"eventually [due to negative experiences], code quality cannot be assured. So this will make my code messy and more bug can be found in it"*.

According to **SP03** and **SP10**, programmers can discard the source code due to unhappiness. The following statement can illustrate this scenario: *"I deleted the code that I was writing because I was a bit angry". (...) "I deleted the entire project to start over with code that didn't seem to be going in a wrong direction"*.

Impacts of Other Negative Sentiments. Evidence from the study **SP09** suggest that when developers remain emotionally active, they usually tend to write longer commit comments. In this same direction, Zhang et al. (**SP06**) found evidence that revealed relationship of defective source code files with email messages emotionally loaded from developers.

5 Discussion

We now have enough evidence to answer the stated research question: *RQ: "What is the impact of sentiments in adopted software practices and artifacts produced by programmers in open source software projects?"* Evidence provided by the selected studies enable us to conclude that the sentiment of developers has the potential to affect productivity. It is worth to mention that higher productivity can also be the consequence of developers' sentiments when executing other practices. For instance, when reducing problem correction time, there is a tendency to increase productivity. Similarly, there is evidence that lengthy code reviews tend to decrease productivity.

We identified from the studies a relationship between the impact of sentiments on software practices with a corresponding impact on artifacts. There are also consequences of sentiments on artifacts that can somehow affect software practices. For instance, poor quality and source code disposal can decrease the productivity of the software development team.

There is an implicit vicious cycle between software practices and artifacts that can be positively or negatively fed by the sentiments of the programmers. These findings corroborate the understanding of the need for project managers to engage in promoting a healthy software development environment.

Studies **SP04** and **SP09** stand out from the others because they present a paradox: the impacts caused by the sentiments revealed by them are the same, regardless of the polarity of sentiment. The results of **SP04** indicated that sentiments with very high positive or high negative scores have a higher incidence of reopening of issues. **SP09** concluded that developers tend to write longer comments when they remain emotionally active (positively or negatively).

The fact that articles **SP04** and **SP09** have the same impacts for sentiments of inverse polarity indicates the need for studies on the balance of sentiments of software developers or on the neutrality of sentiments, revealing new possibilities for research in the area. We found both studies investigating the influence of sentiments on practices and artifacts as studies that investigated the influence of practices and artifacts on sentiments [14,25,34,36].

6 Threats to Validity

We identified possible threats to validity in our study. In the following paragraphs, we present the strategies we propose for their mitigation.

Research Question: The stated research questions may not encompass the impact of developer sentiments on open source software project practices and artifacts. For this reason, we used the GQM approach to effectively define the study objective and corresponding research questions.

Search Strings: There is the possibility that the search strings is not able to identify all studies in the area. We deal with this threat by expanding the number of electronic repositories searched to three. All repositories used are originally from the area of Computing.

Publication Bias: It is not possible to assure that the adopted search string identified all relevant primary studies available in electronic repositories. Some of these studies may not have been found to provide evidence for this secondary study. We manage to mitigate this potential threat through the use of alternative terms and corresponding synonyms.

Searching Process: Each searched electronic repository has its own searching process and this process can differ from one repository to another, We mitigate this by adapting the search string for each electronic repository and assume that equivalent logical expressions work consistently well across the selected electronic repositories.

Data Extraction: The selection of studies was not automatic. We manually applied inclusion, exclusion and quality criteria. Thus, not all relevant studies may have been selected. To deal with this threat, the selection was carried out by two researchers from different organizations.

7 Conclusions

In this study we investigated how the literature reported the impact of developers' sentiments on open source software projects. To accomplish this goal, we conducted a Systematic Literature Review of peer-reviewed papers available in three electronic repositories. The analysis of evidence gathered from the selected papers indicated that sentiments influence software practices and artifacts in issues such as productivity, collaboration, and also in the resulted source code. The analysis was able to map to which extent positive and negative sentiments tend to impact software practices and artifacts.

Not all sentiments were associated with positive or negative polarity. In these cases, they were marked as *not specified*. This brings the possibility for further investigation on theses sentiments and their respective impacts. We also plan to investigate sentiments in the context of open source software projects that adopt the so called frequent releases and to which extent they influence software productivity. Another opportunity of future research is the investigation of how programmers sentiments vary between releases.

Appendix

A. List of Selected Papers

SP01 & Johri, A. and Teo, H. J. (2018). Achieving equilibrium through coworking: Work-life balance in floss through multiple spaces and media use. In Proceedings of the 14th International Symposium on Open Collaboration, OpenSym '18, pages 7:1–7:11, New York, NY, USA. ACM.

SP02 & Asri, I. E., Kerzazi, N., Uddin, G., Khomh, F., and Idrissi, M. J. (2019). An empirical study of sentiments in code reviews. Information and Software Technology, 114:37–54.

SP03 & Graziotin, D., Fagerholm, F., Wang, X., and Abrahamsson, P. (2017). Consequences of unhappiness while developing software. In 2nd IEEE/ACM International Workshop on Emotion Awareness in Software Engineering, SEmotion@ICSE 2017, Buenos Aires, Argentina, May 21, 2017, pages 42–47.

SP04 & Cheruvelil, J. and C. da Silva, B. (2019). Developers' sentiment and issue reopening. In 2019 IEEE/ACM 4th International Workshop on Emotion Awareness in Software Engineering (SEmotion), pages 29–33.

SP05 & Ortu, M., Hall, T., Marchesi, M., Tonelli, R., Bowes, D., and Destefanis, G. (2018). Mining communication patterns in software development: A github analysis. In Proceedings of the 14th International Conference on Predictive Models and Data Analytics in Software Engineering, PROMISE'18, pages 70–79, New York, NY, USA. ACM.

SP06 & Zhang, Y., Shen, B., and Chen, Y. (2014). Mining developer mailing list to predict software defects. In 2014 21st Asia-Pacific Software Engineering Conference, volume 1, pages 383–390.

SP07 & Yang, B., Wei, X., and Liu, C. (2017). Sentiments analysis in github repositories: An empirical study. In 2017 24th Asia-Pacific Software Engineering Conference Workshops (APSECW), pages 84–89.

SP08 & Wang, Y. and Redmiles, D. (2016). The diffusion of trust and cooperation in teams with individuals' variations on baseline trust. In Proceedings of the 19th ACM Conference on Computer-Supported Cooperative Work & Social Computing, CSCW '16, pages 303–318, New York, NY, USA. ACM.

SP09 & Islam, M. R. and Zibran, M. F. (2016). Towards understanding and exploiting developers' emotional variations in software engineering. In 2016 IEEE 14th International Conference on Software Engineering Research, Management and Applications (SERA), pages 185–192.

SP10 & Graziotin, D., Fagerholm, F., Wang, X., and Abrahamsson, P. (2018). What happens when software developers are (un)happy. Journal of Systems and Software, 140:32–47.

SP11 & Borges, H. and Valente, M. T. (2018). What's in a github star? understanding repository starring practices in a social coding platform. Journal of Systems and Software, 146:112–129.

SP12 & Huq, Syed Fatiul, Ali Zafar Sadiq, and Kazi Sakib. (2020). Is Developer Sentiment Related to Software Bugs: An Exploratory Study on GitHub Commits. In 2020 IEEE 27th International Conference on Software Analysis, Evolution and Reengineering (SANER), pp. 527–531. IEEE, 2020.

SP13 & Madampe, K., Hoda, R. and Singh, P., 2020. Towards Understanding Emotional Response to Requirements Changes in Agile Teams. In New Ideas and Emerging Results track of the 42nd IEEE/ACM International Conference on Software Engineering, ICSE2020.

SP14 & Alami, A., Cohn, M.L. and Wąsowski, A., 2019, May. Why does code review work for open source software communities? In 2019 IEEE/ACM 41st International Conference on Software Engineering (ICSE) (pp. 1073–1083). IEEE.

References

1. Alami, A., Cohn, M.L., Wąsowski, A.: Why does code review work for open source software communities? In: 2019 IEEE/ACM 41st International Conference on Software Engineering (ICSE), pp. 1073–1083. IEEE (2019)
2. Alami, A., Nielsen, P.A., Wasowski, A.: A tailored participatory action research for FOSS communities. Empir. Softw. Eng. **25**(5), 3639–3670 (2020)
3. Asri, I.E., Kerzazi, N., Uddin, G., Khomh, F., Idrissi, M.J.: An empirical study of sentiments in code reviews. Inf. Softw. Technol. **114**, 37–54 (2019). https://doi.org/10.1016/j.infsof.2019.06.005. http://www.sciencedirect.com/science/article/pii/S0950584919301387
4. Basili, V.R., Rombach, H.D.: The tame project: towards improvement-oriented software environments. IEEE Trans. Softw. Eng. **14**(6), 758–773 (1988)
5. Calefato, F., Lanubile, F., Maiorano, F., Novielli, N.: Sentiment polarity detection for software development. Empirical Softw. Eng. **23**(3), 1352–1382 (2018). https://doi.org/10.1007/s10664-017-9546-9. https://doi.org/10.1007/s10664-017-9546-9
6. Cheruvelil, J., da Silva, B.C.: Developers' sentiment and issue reopening. In: 2019 IEEE/ACM 4th International Workshop on Emotion Awareness in Software Engineering (SEmotion), pp. 29–33 (May 2019). https://doi.org/10.1109/SEmotion.2019.00013
7. Colomo-Palacios, R., Hernández-López, A., García-Crespo, Á., Soto-Acosta, P.: A study of emotions in requirements engineering. In: Lytras, M.D., Ordonez de Pablos, P., Ziderman, A., Roulstone, A., Maurer, H., Imber, J.B. (eds.) WSKS 2010. CCIS, vol. 112, pp. 1–7. Springer, Heidelberg (2010). https://doi.org/10.1007/978-3-642-16324-1_1

8. Cruz, S.S.J.O., da Silva, F.Q.B., Monteiro, C.V.F., Santos, P., Rossilei, I.: Personality in software engineering: preliminary findings from a systematic literature review. In: 15th Annual Conference on Evaluation Assessment in Software Engineering, EASE 2011, pp. 1–10 (April 2011). https://doi.org/10.1049/ic.2011.0001
9. Cruz, S., da Silva, F.Q., Capretz, L.F.: Forty years of research on personality in software engineering: A mapping study. Comput. Hum. Behav. **46**, 94–113 (2015). https://doi.org/10.1016/j.chb.2014.12.008. http://www.sciencedirect.com/science/article/pii/S0747563214007237
10. Danescu-Niculescu-Mizil, C., Sudhof, M., Jurafsky, D., Leskovec, J., Potts, C.: A computational approach to politeness with application to social factors. In: Proceedings of the 51st Annual Meeting of the Association for Computational Linguistics (Volume 1: Long Papers), Sofia, Bulgaria, pp. 250–259. Association for Computational Linguistics (August 2013). https://www.aclweb.org/anthology/P13-1025
11. Dybå, T., Dingsøyr, T.: Empirical studies of agile software development: a systematic review. Inf. Softw. Technol. **50**(9), 833–859 (2008)
12. Graziotin, D., Fagerholm, F., Wang, X., Abrahamsson, P.: Consequences of unhappiness while developing software. In: 2nd IEEE/ACM International Workshop on Emotion Awareness in Software Engineering, SEmotion@ICSE 2017, Buenos Aires, Argentina, 21 May 2017, pp. 42–47 (2017). https://doi.org/10.1109/SEmotion.2017.5. https://doi.org/10.1109/SEmotion.2017.5
13. Graziotin, D., Fagerholm, F., Wang, X., Abrahamsson, P.: What happens when software developers are (un)happy. J. Syst. Softw.d **140**, 32–47 (2018). https://doi.org/10.1016/j.jss.2018.02.041. http://www.sciencedirect.com/science/article/pii/S0164121218300323
14. Guzman, E., Azócar, D., Li, Y.: Sentiment analysis of commit comments in GitHub: an empirical study. In: Proceedings of the 11th Working Conference on Mining Software Repositories, MSR 2014, pp. 352–355. ACM, New York, NY, USA (2014). https://doi.org/10.1145/2597073.2597118. http://doi.acm.org/10.1145/2597073.2597118
15. Huq, S.F., Sadiq, A.Z., Sakib, K.: Is developer sentiment related to software bugs: an exploratory study on GitHub commits. In: 2020 IEEE 27th International Conference on Software Analysis, Evolution and Reengineering (SANER), pp. 527–531. IEEE (2020)
16. Islam, M.R., Zibran, M.F.: Leveraging automated sentiment analysis in software engineering. In: Proceedings of the 14th International Conference on Mining Software Repositories, MSR 2017, Piscataway, NJ, USA, pp. 203–214. IEEE Press (2017). https://doi.org/10.1109/MSR.2017.9
17. Junior, R.S.C., de Figueiredo Carneiro, G.: Impact of developers sentiments on practices and artifacts in open source software projects: a systematic literature review. In: Filipe, J., Smialek, M., Brodsky, A., Hammoudi, S. (eds.) Proceedings of the 22nd International Conference on Enterprise Information Systems, ICEIS 2020, Prague, Czech Republic, 5–7 May 2020, vol. 2, pp. 31–42. SCITEPRESS (2020). https://doi.org/10.5220/0009313200310042
18. Kołakowska, A., Landowska, A., Szwoch, M., Szwoch, W., Wróbel, M.R.: Emotion recognition and its application in software engineering. In: 2013 6th International Conference on Human System Interactions (HSI), pp. 532–539. IEEE (2013)
19. Lenberg, P., Feldt, R., Wallgren, L.G.: Behavioral software engineering: a definition and systematic literature review. J. Syst. Softw. **107**, 15–37 (2015). https://doi.org/10.1016/j.jss.2015.04.084. http://www.sciencedirect.com/science/article/pii/S0164121215000989
20. Madampe, K., Hoda, R., Singh, P.: Towards understanding emotional response to requirements changes in agile teams. In: New Ideas and Emerging Results track of the 42nd IEEE/ACM International Conference on Software Engineering, ICSE2020 (2020)

21. Michlmayr, M., Fitzgerald, B., Stol, K.J.: Why and how should open source projects adopt time-based releases? IEEE Softw. **2**, 55–63 (2015)
22. Moher, D., Liberati, A., Tetzlaff, J., Altman, D.G., Group, P., et al.: Preferred reporting items for systematic reviews and meta-analyses: the PRISMA statement. PLOS Med. **6**(7), e1000097 (2009)
23. Pennebaker, J., Francis, M., Booth, R.: Linguistic inquiry and word count (LIWC) (Janurary 1999)
24. Santos, C., Kuk, G., Kon, F., Pearson, J.: The attraction of contributors in free and open source software projects. J. Strateg. Inf. Syst. **22**(1), 26–45 (2013)
25. Singh, N., Singh, P.: How do code refactoring activities impact software developers' sentiments? - An empirical investigation into GitHub commits. In: 2017 24th Asia-Pacific Software Engineering Conference (APSEC), pp. 648–653 (December 2017). https://doi.org/10.1109/APSEC.2017.79
26. Singh, V., Sharma, M., Pham, H.: Entropy based software reliability analysis of multi-version open source software. IEEE Trans. Softw. Eng. **44**, 1207–1223 (2017)
27. Śliwerski, J., Zimmermann, T., Zeller, A.: When do changes induce fixes? ACM Sigsoft Softw. Eng. Notes **30**(4), 1–5 (2005)
28. Sánchez-Gordón, M., Colomo-Palacios, R.: Taking the emotional pulse of software engineering - a systematic literature review of empirical studies. Inf. Softw. Technol **115**, 23–43 (2019). https://doi.org/10.1016/j.infsof.2019.08.002. http://www.sciencedirect.com/science/article/pii/S0950584919301661
29. Socher, R., et al.: Recursive deep models for semantic compositionality over a sentiment treebank. In: Proceedings of the 2013 Conference on Empirical Methods in Natural Language Processing, EMNLP 2013, Seattle, Washington, USA, pp. 1631–1642. Association for Computational Linguistics (October 2013). https://www.aclweb.org/anthology/D13-1170
30. Soomro, A.B., Salleh, N., Mendes, E., Grundy, J., Burch, G., Nordin, A.: The effect of software engineers' personality traits on team climate and performance: a systematic literature review. Inf. Softw. Technol. **73**, 52–65 (2016). https://doi.org/10.1016/j.infsof.2016.01.006. http://www.sciencedirect.com/science/article/pii/S0950584916000082
31. Stol, K.J., Fitzgerald, B.: Inner source-adopting open source development practices in organizations: a tutorial. IEEE Softw. **32**(4), 60–67 (2015)
32. Stone, P.W.: Popping the (PICO) question in research and evidence-based practice. Appl. Nurs. Res. **15**(3), 197–198 (2002). https://doi.org/10.1053/apnr.2002.34181. http://www.sciencedirect.com/science/article/pii/S0897189702000101
33. Thelwall, M., Buckley, K., Paltoglou, G.: Sentiment strength detection for the social web. J. Am. Soc. Inf. Sci. Technol. **63**(1), 163–173 (2012). https://doi.org/10.1002/asi.21662
34. Trainer, E.H., Kalyanasundaram, A., Herbsleb, J.D.: e-mentoring for software engineering: a socio-technical perspective. In: Proceedings of the 39th International Conference on Software Engineering: Software Engineering and Education Track, ICSE-SEET 2017, , Piscataway, NJ, USA, pp. 107–116. IEEE Press (2017). https://doi.org/10.1109/ICSE-SEET.2017.19
35. Wohlin, C., et al.: Experimentation in Software Engineering. Springer, Heidelberg (2012). https://doi.org/10.1007/978-3-642-29044-2
36. Zhao, M., Wang, Y., Redmiles, D.F.: Using playful drawing to support affective expressions and sharing in distributed teams. In: 2nd IEEE/ACM International Workshop on Emotion Awareness in Software Engineering, SEmotion@ICSE 2017, Buenos Aires, Argentina, 21 May 2017, pp. 38–41 (2017). https://doi.org/10.1109/SEmotion.2017.3

Improving Quality of Use-Case Models by Correlating Defects, Difficulties, and Modeling Strategies

Cristiana Pereira Bispo[1,3](✉), Ana Patrícia Magalhães[1,2], Sergio Fernandes[1], and Ivan Machado[3]

[1] Salvador University, Salvador, Brazil
cristiana.bispo@ufba.br, {ana.fontes, sergio.fernandes}@unifacs.br
[2] Department of Exact Sciences and Earth, State University of Bahia, Salvador, Brazil
[3] Computer Science Department, Federal University of Bahia, Salvador, Brazil
ivan.machado@ufba.br

Abstract. Use case (UC) models play an essential role in software specification since they describe system functional requirements. A UC model should be free of defects due to its relevance and impact throughout the software development life cycle. However, inspections in UC models frequently identify defects related to modelers' difficulties in different activities during the modeling process. The quality of a UC model is usually analyzed based on quality criteria such as ambiguity and inconsistency. Several strategies in the literature assist use case modeling in mitigating defects, but these strategies do not identify which potential defects they aim to prevent or eliminate. In this context, we proposed a correlation between UC modeling difficulties and strategies to mitigate these difficulties based on UC's quality criteria. In this paper, we describe each strategy contained in the correlation and present, in detail, the controlled experiment that assesses the correlation effectiveness, including the discrete data collected in analyzing the participants' models and statistical analysis performed in these data. Besides, we also propose a mechanism to guide in elaborating checklists to identify defects in UC models focusing on quality criteria. Through a controlled experiment, we evaluate the Antipattern strategy, and the results showed a clear indication that this strategy mitigates the difficulties in which it is related according to the correlation. Besides, the UC models developed in the experiment were evaluated using the checklist generated based on the proposed mechanism.

Keywords: Use case modeling · Use case inspection · Controlled experiment

1 Introduction

In Information System (IS) development, a technique commonly used to capture and describe systems' functional requirements is Use-Case Modeling (UCM), which produces a Use-Case (UC) model as output. Due to their relevance and impact throughout the software development life cycle, UC models should be free from defects.

J. Filipe et al. (Eds.): ICEIS 2020, LNBIP 417, pp. 360–385, 2021.
https://doi.org/10.1007/978-3-030-75418-1_17

The quality of a UC model can be analyzed based on quality attributes already defined in the literature, as summarized by Tiwari and Gupta [2], such as ambiguity, inconsistency, among others. In fact, several authors have reported that the inspection of use-case models concerning quality attributes frequently identifies the occurrence of defects [2–6]. The low quality of UC models has been related to modelers' difficulties in different activities, such as understanding the requirements and representing them in UC models [7]; and understanding the problem domain [5].

To mitigate defects in UC models, several authors propose strategies that help UCM. In a prior work, we carried out a systematic literature review (SLR) [26] and identified 39 studies that present UCM strategies. For example, some authors indicate the use of Antipattern [3, 9–13], a technique that brings evidence on common modeling mistakes so that developers do not make the same mistakes in their models. Business Process Notation, a standard notation to model business process, is also considered as a feasible strategy to identify requirements [14, 15]. The authors of these works agree that using a UCM strategy can lead to an improvement in the UC model concerning the quality attributes enhanced by the strategy. However, none of these works identifies which potential defects each of these strategies aims to prevent or eliminate. Nor do they identify which difficulties of the modelers the strategy can alleviate.

UCM strategies do not have a systematic method to inspect UC models built using the respective strategy, i.e., do not provide a formalism to assist in identifying defects in UC models. Additionally, they do not generate a list of identified defects that can be reused to inspect other use-case models in the same domain. In summary, it is challenging to assess a UC model's quality by the occurrence of defects because the works found in the literature do not show how the UC models can be corrected.

To assist UCM, we proposed a correlation between UCM difficulties and strategies to mitigate these difficulties based on quality attributes for UC [8]. With this correlation's support, it is possible to identify which strategies are most appropriate to mitigate the defects that affect quality specific attributes. This current investigation elaborates on such preceding work by detailing each literature strategy to support UCM. We also detail the controlled experiment that assesses the correlation effectiveness, including the discrete data collected in analyzing the participants' models and statistical analysis performed in these data. To assist in the inspection of the UCs created based on a strategy, in this article, we also propose a mechanism to guide in elaborating checklists to identify defects in UCMs with a focus on quality criteria. The mechanism comprises a set of steps for the generation of the checklist to inspect UC models. We validated the mechanism in the same experiment to assess the correlation.

The remainder of this chapter is organized as follows: Sect. 2 presents the concepts related to requirements specification and use-case modeling. Section 3 discusses related work. Section 4 details the strategy found in the literature to assist UCM. Section 5 presents the correlation between UCM difficulties and strategies to mitigate these difficulties, and Sect. 6 presents the proposed mechanism to generate a UC inspection checklist. Section 7 details the controlled experiment, and Sect. 8 draws concluding remarks and pinpoints opportunities for further research.

2 Requirements Specification

According to Sommerville [17], a system requirement represents the description of functionality or constraint that the system must fulfill. The process of identifying, analyzing, documenting, and verifying these requirements is called Requirement Engineering. It produces a Software Requirement Specification (SRS), which is an essential input for subsequent software development activities. Essential activity in the Requirement Engineering process are requirements identification, analysis, and validation. There are different techniques in the literature to specify requirements, among them the UC modeling.

UC modeling is the activity of designing a use-case model, which describes in detail the software functional requirements. They make use of graphic and textual notation to, respectively [18] (i) create the UC diagram that provides a visual summary of the system services and their interaction with the environment and users (called actors); and (ii) describe the interactions between the system and its actors.

The Unified Modeling Language (UML) [19] is widely adopted to represent use case diagrams. The UML UC diagram is mainly composed of the following elements: actors, use case, communication relationships, inclusion, extension, and generalization. The description of Use Cases is a textual notation, in natural language, that describes the behavior of each UC in the diagram. Cockburn [20] and Jacobson et al. [21] define the following elements to describe a UC: Name of the UC; Short description of the UC's objective; Actor (s) participating in the UC; Precondition to start the UC; Post-condition that must be met after the execution of the UC; Main Flow describing the main UC scenario; Alternative flows with alternative UC usage scenarios; Exception flows for unexpected occurrences; and Rules that must be considered when executing the UC.

2.1 Difficulties in UCM

Difficulties regarding the syntax and semantics of graphic and textual elements in elaborating the UC model compromise the quality attributes [22] of use case models, such as completeness, ambiguity, and inconsistency. In this paper, a difficulty is defined as any lack of knowledge of requirement modelers that prevent them from modeling UCs meeting specified quality requirements.

There are some works in the literature that present difficulties in UCM, such as difficulties to understand requirements and represent them in use cases [22]; difficulties in understanding the problem domain [5]; difficulties in specifying information unambiguously [23]; difficulties in representing information in a diagram [24]; among others. Nascimento et al. [5] summarize all of these in a model of difficulties in UCM.

Writing UC is an exploratory and visionary task, and a good modeler should have skills, such as the ability to write well; the ability to systematically address a problem; the ability to synthesize user needs; in addition to specific knowledge of the problem domain and understanding of software development, among other skills.

2.2 Rules and Guidelines for UCM Modeling

There are several guidelines and rules in the literature to writing effective UC [19, 21], and [25]. The work of Gregolin [45] presents a synthesis of these rules and guidelines. Some of these are:

- A UC should add value to the related actors, grouping atomic functions in a single functionality, avoiding functional decomposition;
- A UC should avoid individual features of CRUD (Create, Retrieve, Update, Delete). These functionalities should have a single generic UC, such as *Manage Registrations*, or *Keep Registrations*, and to use interface prototypes for each registration.
- The actor's name should reflect their role in the system and avoid titles of positions, organizations, or activities related to an organizational structure.
- The UC's name usually contains a verb followed by nouns, and the sub-nouns can have adjectives. The verb must be in the infinitive or present tense and must use the active voice instead of the passive voice, among others.

2.3 Quality Attributes in UCM

There are many recommendations in the literature on what constitutes quality in a use-case model. Through a Systematic Literature review (SLR), the authors of [2] bring these attributes together. However, different authors often give each attribute a particular understanding. Therefore, in this work, we perform a synthesis of each of them, as Table 1 shows.

Table 1. Synthesis of quality attributes.

Quality attribute	Description
Accuracy or completeness or integrity	There should be no missing information nor elements in the UC diagram and in the corresponding textual descriptions
Consistency	The UC model information should have the expected semantics. There should not be any conflicting elements in the diagrams and in their textual descriptions
Correctness	The UC diagram and its descriptions must correctly represent the requirements
Understandability	The information and rules contained in the UC diagrams and textual descriptions must be accurate and clearly defined
Ambiguity	There should be no information in the UC diagram and textual descriptions with more than one meaning
Redundancy	There should be no excessive, repetitive or superfluous information in the UC diagram and descriptions
Abstraction level	The UC diagram and descriptions should present only what the software should do at an appropriate granularity level. That is, the UC should not be broken down into parts that have no value in themselves

3 Related Work

The studies deemed as related to the purpose of this research focus on the same aspect: the difficulties of UCM requirements specifier that prevent them from building UC models meeting the defined quality requirements.

Nascimento et al. [5] sought to explore and understand difficulties in UCM by conducting four experimental studies. As a result, they presented a model of difficulties. The works [22, 23] and [24] also investigated and reported difficulties in UCM. These works do not present any strategy to mitigate these difficulties.

To mitigate the difficulties that requirements specifier faces when modeling UCs, several authors propose applying resources already used in other domains to verify their effectiveness in UCM. The work presented in [14] employed business process models to derive UCs because these models are often available in a company as work instructions or administrative manuals in a clear and structured manner. Conversely, the authors of [3] presented an Antipattern-based strategy for UCM, in which bad practices are identified to be replaced by recommended solutions. We identified other strategies in a previous investigation and their respective contributions to UCM [8]. However, these studies do not indicate which strategies could mitigate the UCM difficulties.

The difficulty-strategy correlation proposed in this paper guides the requirements specifier in selecting the most appropriate strategy to mitigate a given difficulty. It avoids adopting ineffective practices and presents various alternatives for applying tested and evaluated procedures to assist UCM.

4 Strategies for UCM

Several strategies in the literature support the use-case modeling, as identified in the SLR that we present in [26].

One of the first strategies proposed in the literature was that of **Ontology** [27–33]. It represents the domain concepts and their relationships, allowing automated reasoning. Consequently, it can minimize problems concerning requirements ambiguity, inconsistency, and incompleteness. Ontology strategy contributes to UCM, making it possible to specify UC requirements more completely and unambiguously.

Antipattern is one of the most common strategies found in the literature [3, 9–13]. It focuses on identifying deficiencies in UC, emphasizing the human cognitive abilities that allow identifying these deficiencies. According to it, a *bad* UC modeling structure does not necessarily indicate a defect, but it can lead to possible harmful consequences. This way, it shows dubious UCM structures and their harmful effects and is useful in detecting possible doubtful structures to perform corrective actions.

Another technique, called **Role Interpretation** [34–39], is used in teaching modeling. Students and instructors assume different roles in a modeling experiment. In this way, it aims to simulate the industrial environment so that students obtain a description very close to the system to be modeled.

The **Natural Language Processing** (NPL) strategy provides semi-automated assistance, through an algorithmic approach, for developers to generate UC models from standardized natural language requirements [46–50]. It inspects the requirements document to find nouns to add to the list of actors, and verbs to add to the list of UCs. The UC diagram is sketched from these lists. NPL protects the developer from the ambiguity, redundancy, and incompleteness inherent in the requirements specifications written in natural language. It uses advanced word processing techniques to rapid discover duplicate functionality, identify and extract actors and UCs, and to clarity specifications that are difficult to understand and communicate.

Scenario Pattern [51–54] uses standard UC specification scenarios that describe requirements and interactions between system actors in a standardized way. Pattern matching algorithms are automatically used to check if there is an omission of any necessary step in the UC specification. It provides automatic tool support to detect the missing part in the requirements specification and recommend appropriate instructions for including it to make the UC specification as complete as possible.

Business Process Notation is also used for modeling use cases by [14, 15]. The proposal uses business processes to derive UCs, through an algorithm that implements meta-models for use case diagrams (UCD) and for business process models (BPM). Thus, it creates UC diagrams more quickly because BPM is often available in a company in work instructions or administrative manuals in a clear and structured way. In addition, it tends to produce consistent and complete specifications.

Domain-Specific Languages (DSL) are used to describe UC [16], specifying the user and system actions clearly and precisely. This language uses a specific text syntax that allows a certain formalization in the UC model description. In this way, it can increase productivity by promoting understandable communication between engineers and domain experts, in addition to allowing the removal of ambiguities and inconsistencies observed in natural language texts.

Another strategy adopted is the **Fragment of Use Cases** [55, 56]. In this, the UC text is written using fragment composition in which each fragment represents a recurring set of interactions necessary to achieve a sub-objective. Each fragment can be customized and is coded using the best practices for writing the UC steps. Thus, the aim is to reduce the time required for the preparation of high-quality UC specifications. This strategy seeks to eliminate ambiguities, redundancies, inconsistencies, and conflicts with domain terminology (UCs contaminated by jargon). In this way, it standardizes and promotes the concise specifications of UCs, facilitating the maintenance and understanding of UCs by those involved with the system.

The **Communication Media** strategy [1] replaces face-to-face communication with Think-Pair-Square, a structured text-based chat, suitable for solving problems in the learning field. Its great application is in offering appropriate resources for distributed modeling, common in current development projects.

The **Mental Models** strategy [6] uses virtualization in UCM to propose a conceptual mental model representing the user's thinking of how it works. It refines the requirements phase by structuring the *imagination* process in a formal visualization stage, to be carried out before creating the UC diagram. In this way, helps inexperienced developers to overcome their difficulties in defining functional requirements at UCM and produces a tangible visual result of what the developer perceives as the user mental model of how something works.

The use of **Visual Languages** to describe UCs, replacing the UML language, is proposed in [43]. The authors consider that UML does not satisfactorily address the concerns (in terms of requirements) of human-computer interaction (HCI) professionals. Thus, they provide a common mechanism of communication and understanding between IHC professionals and software engineers (ES) so that the development and description of UC portray the interests of both together.

Reverse Engineering is a strategy that aims to systematically extract a large amount of information from UC descriptions [44]. The information can be read by machine and serves as a guideline for the "assembly of the UC diagram." Alternatively, the reverse process can be used: to decompose the diagram to assemble the UC specification. It seeks to systematically provide a minimal skeleton as a starting point for UCM.

Finally, [57] proposes the use of **Automatic Layout** with guidelines for defining a UC diagram. It focuses on the lack of appropriate layout mechanisms to provide an understanding of the system graphically and seeks to make up for the significant lack of resources for diagramming UCs as well as for displaying differences visually.

Table 2 summarizes the quality attributes enhanced by each strategy.

5 Correlation Between *Difficulties* and Strategies for UCM

This section presents an overview of the correlation proposed in [8] between modeling difficulties and strategies to mitigate these difficulties concerning the quality attributes for UC presented in Sect. 2.3.

The methodology used to establish the correlation between difficulties and strategies uses as input the information contained in the papers retrieved from the SLR [26]. We analyzed the experiments described in the SLR selected papers with the support of an inductive theory based on data analysis named Grounded Theory (GT) [41] to categorize the difficulties in UCM that we identified in the literature (Sect. 2.1). The result of this analysis made it possible to identify which quality attributes can be affected by each of the defined difficulty categories. Details of this process can be found in [8].

Table 3 shows the relationship established between difficulty and quality attributes. The first and second columns show the difficulty identifier and its description. Similarly, the third and fourth columns show the quality attribute's identifier affected by the difficulty and its description.

The combination of the results presented in Tables 2 and 3 allowed the definition of the correlation. Figure 1 shows part of the correlation, which is described in detail in [8]. On the top, there is the difficulty in identify/Extract/Discover UCs, actors and relationships. This difficulty affect the completeness-accuracy-integrity quality attribute, and can be mitigated by four different UCM strategies (on the botton): Role-playing, Natural Language Processing, Business Process Modeling, and Scenario pattern.

6 The Mechanism to Assist UCM Inspection

The correlation proposed in Sect. 5 helps modelers identify which strategies can mitigate specific difficulties in use-case modeling to enhance certain quality attributes.

Table 2. UC modeling strategies and quality attributes.

Strategies	Ambiguity	Inconsistency	Redundancy	Incompleteness	Incomprehensibility	Communicability
Ontology					X	
Antipattern	X	X				
Roleplaying				X	X	
Natural language processing	X	X	X	X	X	
Scenario patterns		X		X		
Business process notation	X		X	X	X	
DSLs	X	X				
Use case fragment	X	X	X		X	
Communication media						X
Mental model			X			
Visual language		X	X			X
Reverse engineering		X				
Automatic layout					X	

Table 3. Difficulties that affect quality attributes.

Id	Difficulty	Id At	Quality attribute affected
D1	Identify/extract/discover UC, actor and relationship	Q1	Completeness-accuracy-integrity
D2	Represent/express model elements	Q3	Correctness
D3	Write in detail the semantics of the UC model	Q2	Consistency
D4	Understand/interpret the problem domain.	Q1	Completeness - accuracy - integrity
D5	Perceive implicit requirements	Q4, Q5	Understandability; integrity
D6	Synthesize use cases	Q7	Abstraction level
D7	Condense the various information from the UC model	Q6	Redundancy

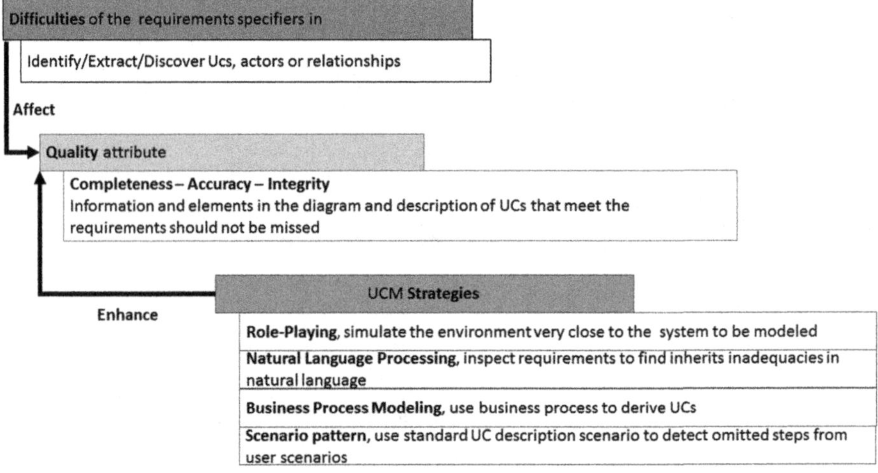

Fig. 1. Part of the proposed correlation.

According to [5], the modeler difficulties in UCM usually insert defects in UC models. Thus, in this paper, we consider the hypothesis that reducing the defects in UC model after using a strategy means that the difficulty that would potentially generate such defects is mitigated. However, to assess this hypothesis, we need a systematic mechanism to inspect a UC model and detect possible defects after applying a strategy.

Inspection is a static activity based on the visual examination of development products to detect defects, violations of development patterns, and other problems without trying to solve the identified problems [45]. In UCM, this activity consists of checking if the model expresses the requirements and can be understood by all involved. A checklist defines a list of questions that the inspectors must answer yes or no, must be prepared to detect the defects in UC model during the inspection [42].

In this section, we propose a mechanism for UCM developers/instructors to build a checklist to detect defects in the UC model based on the strategies presented in Sect. 4. The generated checklist will evaluate the quality attributes enhanced by the strategy adopted in its construction. We consider a mechanism a set of previous and necessary procedures to be performed for the construction of the checklist. The mechanism purpose is to guide the definition of a checklist from which to measure the effect of a specific strategy on difficulties of the students at UCM.

The checklist must contain questions to identify defects related to quality attributes. For example, if a difficulty affects the completeness of a UC model, it is because this difficulty inserts defects that make the model incomplete. The checklist must then contain questions that allow finding the defects related to the attribute of quality completeness. Thus, the mechanism's procedures are as follows (Fig. 2):

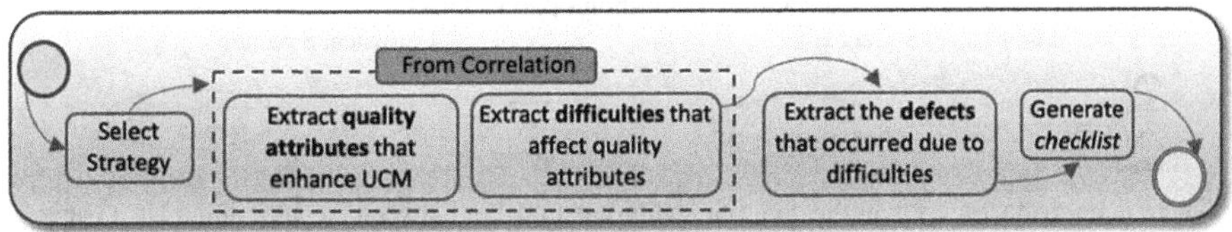

Fig. 2. Mechanism to generate the inspection checklist.

(a) Select Strategy - This procedure aims to choose the strategy to be used for UCM. Not enough criteria were found in the literature to recommend one strategy more than another. Therefore, the selection must be conditioned to previous knowledge of what difficulties affect the group of modelers at the moment of UCM and then, consulting the correlation (Sect. 5), select the corresponding strategy for such difficulties. It is essential to be aware of the strategy goal, specific procedures, and what contribution should be expected from it (Sect. 4).
Procedures (B) and (C) must be based on the correlation.
(b) Extract quality attributes enhanced by the strategy - Knowing what the attributes are, we aim to identify in the rules and guidelines for UCM (Sect. 2.2) conditions related to these attributes. It is recommended to elaborate a small synthesis of the understanding of each attribute extracted in this phase.

- Extract the difficulties that affect the quality attributes, after identifying the difficulty in the correlation, and

(c) Extract the defects inserted according to the difficulties. Add the knowledge obtained in (B) and (C), and draw up the list of defects for each quality attribute of the procedure (B).
(d) Generate checklist. In this procedure, questions are prepared to answer whether each defect generated in procedure (D) exists or not when the inspection of the UC diagram or description is performed.

6.1 Applying the Mechanism in a UCM Strategy

To assess the mechanism proposed initially, we defined the mechanism for a specific strategy. Then we apply the mechanism to evaluate the models created in the controlled experiment presented in [8] (Sect. 7).

The following steps illustrate the execution of the procedures defined in Fig. 2:

(a) Select Strategy - Antipattern.
(b) Extract Quality Attributes Enhanced - Consistency and Ambiguity.

The Consistency attribute refers to structure, elements, language, grammar, and any information in the diagram and description of a UC, which must have the semantics expected of them. They need to be coherent, logical, and consistent. To illustrate this attribute, consider Fig. 3.

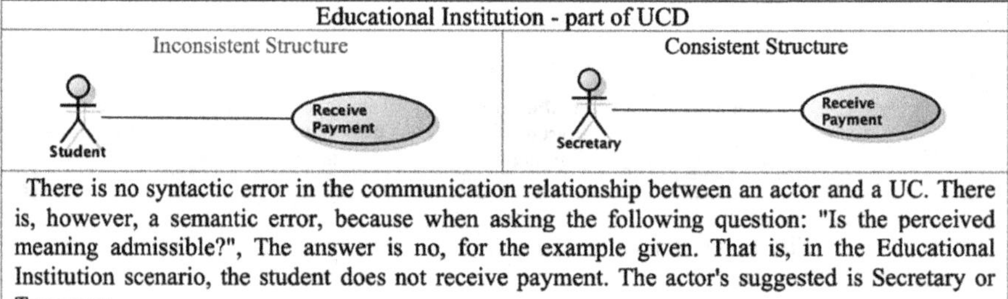

Fig. 3. Example of inconsistent and consistent diagrammatic structure.

The Ambiguity attribute refers to structure, elements, language, grammar, and any model information (present in the UC diagram or description) must be clear so that the interpretation is unique and the understanding is the same for both the client and for the developer. To illustrate this attribute, consider Fig. 4.

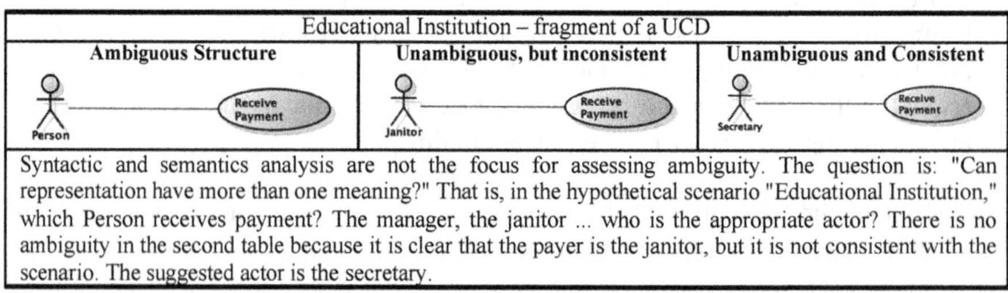

Fig. 4. Example of unambiguous and ambiguous diagrammatic structure.

(c) Extract Difficulties that Affect Quality Attributes.

The Consistency attribute is affected when the specifier has difficulty describing/specifying the semantics of the UC model. This difficulty presents itself as an inability to attribute value or highlight the meaning of the information. The specifier finds it challenging to convey what is necessary with logic and coherence not to compromise the modeled scenario with inconsistencies.

The Ambiguity attribute is affected when the specifier has difficulty perceiving implicit requirements. This difficulty presents itself as an inability or doubt to *accurately* visualize a hidden requirement. For example, in Fig. 4, it is possible that in the requirements document, it was not explicit that the secretary is responsible for the *receiving payment* task. Thus, the modeler can associate the functionality *receive payment* to a generic actor *person*, believing that any person can perform this use case. Ambiguity happens because it leads to different interpretations.

(d) Extract the Defects Inserted According to the Difficulties.

For illustration purposes, defects that appear in the UC diagram are listed in Tables 4 and 5 that make it inconsistent and ambiguous, respectively. The first column identifies the defect from an id (e.g. DfC1), and the second column shows the defect.

Table 4. Defects that compromise the quality attribute Consistency in the UC diagram.

Id	Description
DfC1	Relate actor and UC, when the actor is incompatible to interact with UC
DfC2	Name UC with a name inconsistent with the purpose of the UC
DfC3	Appoint an actor with job titles and not with his role in the system
DfC4	The UC diagram is not plausible with the list of requirements
DfC5	Decompose a UC when its parts alone do not represent value to an actor
DfC6	Use CRUD (Create, Retrieve, Update, Delete) functionality instead of a single generic UC (Manage … or Maintain …)
DfC7	Establish a communication relationship between two UCs instead of inclusion or extension
DfC8	Relate UC and actor through generalization
DfC9	Define an inclusion, extension or generalization relationship between UCs whose removal prevents understanding of the main UC's objective
DfC10	Have an inclusion UC that relates to only one UC
DfC11	Have an extension UC that does not add functionality to the base UC
DfC12	Define a UC as inclusion and extension at the same time

Table 5. Defects that compromise the quality attribute ambiguity in the UC diagram.

Id	Description
DfA1	Name actor or UC with long or inexpressive terms, which do not reflect their role or goal, with varying meaning or adverbs, synonyms, adjectives, pronouns, homonyms, and references
DfA2	Define a single UC as inclusion and extension simultaneously
DfA3	Define a generic UC and another UC that is part of the generic one without relating one with the other
DfA4	Specialize an actor who cannot establish inherited relationships
DfA5	Add brief description notes, without which it is not possible to understand the diagram
DfA6	Do not clearly show users and system functionality according to the list of requirements
DfA7	Have two or more UCs that perform the same functionality
DfA8	Modeling two or more actors who have the same role in the system with a different name
DfA9	Not making the relationships between the actors and the UCs clear

(e) Generate Checklist.

A question is formulated related to each defect to verify the defects presented in Tables 4 and 5. For example, for defect DfC1: Was a relationship between actor and UC found in the inspected diagram when the actor is incompatible with the UC? The answer must always be Y or N, yes or no, respectively.

7 Proposal Evaluation

This section details the controlled experiment briefly presented in [8]. The defect detection mechanism presented in Sect. 6 was also used to inspect the models generated by the experiment participants. The experiment was carried out following [40] guidelines and is structured in four stages: scope, planning, operation, and data analysis and interpretation, detailed in the following subtopics.

7.1 Experiment Scope

Each UMC strategy identified in Sect. 4 must be assessed individually. We started this validation with the Antipattern strategy because it is the strategy most found in the literature and the most detailed in terms of steps to guide the use of the strategy. Following the proposed correlation (Sect. 5), this strategy is related to the mitigation of difficulties that affect the consistency and ambiguity attributes of a UC models.

Experiment Goal. The experiment goal is defined according to the GQM (Goal Question Metric) template. It consists of *analyzing* the Antipattern-based strategy *for the* *purpose of* assess its effectiveness in mitigating the difficulties of describing/detailing the semantic of UC models, and implicit understanding requirements *concerning* consistency and ambiguity *from the point of view of* undergraduate students in the software engineering discipline.

Research Questions. Based on the objective of this experiment, the following research questions (RQ) were defined: *RQ1: Is the diagram produced with the support of the strategy free from defects that would make it inconsistent and ambiguous?* This question sought to assess whether the use of the strategy corrected defects or prevented the appearance of new ones. *RQ2: Does using the Antipattern strategy mitigate the difficulties that affect the consistency and ambiguity of use-case diagram?* With this question, we tried to verify if the difficulties of students that affect consistency and ambiguity disappeared or reduced.

Metric. The metric used to assess the correlation was the number of defects observed in the diagrams produced. Modeled diagrams were inspected using the defect detection mechanism defined in Sect. 6, based on the defect count according to Tables 4 and 5. These defects are inserted due to difficulties in describing/specifying the UC diagram's semantic and understanding implicit requirements. It was assessed whether the Antipattern strategy mitigates these difficulties by reducing these defects.

7.2 Experiment Planning

The experiment was planned in terms of context selection, type of experiment, formulation of hypotheses, dependent and independent variables, and instrumentation.

Context Selection. The experiment was conducted in an academic environment formed by undergraduate students with knowledge of the basic syntax and semantics of UML use-case diagram. To ensure that students had this knowledge, only those studying or who had completed the Software Engineering discipline in Computer Science courses were selected.

Experiment Type. The type of design used in this experiment was based on [40], comprising one factor, the strategy for UCM based on Antipattern, and two treatments: (1) the modeling of the UC diagram without the Antipattern strategy and (2) the modeling of the UC diagram with the Antipattern strategy. Each participating student used the two treatments to model the same scenario. Two sets of diagrams were generated: one without the Antipatterns strategy help (we called this set of UCD_Controlled); and another set based on Antipattern (called this set of UCD_Antipattern).

Hypotheses Formulation. Hypotheses were formulated to conduct the evaluation: null (H0) and alternative (HA), corresponding to the existence of defects in the UCD_Controlled and UCD_Antipattern, as showed in Fig. 5.

Variables Definition. The independent variable consisted of modeling the UC diagram that assumed two levels, modeling without the strategy and modeling using the Antipattern. The dependent variables were the attributes of UCM quality, consistency, and ambiguity, directly related to the measures used to test the hypotheses.

- Attribute ***Consistency***
  H1$_0$: The use of the antipattern strategy *does not influence* the consistency of the UC diagram.
  $$Consistency\_Defect_{UCD\_Antipattern} = Consistency\_Defect_{UCD\_Controlled}$$
  H1$_A$: The use of the Antipattern strategy *influences* the consistence of the UC diagram.
  $$Consistency\_Defect_{UCD\_Antipattern} <> Consistency\_Defect_{UCD\_Controlled}$$
  H1$_{A1}$: $Consistency\_Defect_{UCD\_Antipattern} > Consistency\_Defect_{UCD\_Controlled}$
  H1$_{A2}$: $Consistency\_Defect_{UCD\_Antipattern} < Consistency\_Defect_{UCD\_Controlled}$

- Attribute ***Ambiguity***
  H2$_0$: The use of the antipattern strategy *does not influence* the clarity of the UC diagram.
  $$Ambiguity\_Defect_{UCD\_Antipattern} = Ambiguity\_Defect_{UCD\_Controlled}$$
  H2$_A$: The use of the antipattern strategy *influences* the clarity of the UC diagram.
  $$Ambiguity\_Defect_{UCD\_Antipattern} <> Ambiguity\_Defect_{UCD\_Controlled}$$
  H2$_{A1}$: $Ambiguity\_Defect_{UCD\_Antipattern} > Ambiguity\_Defect_{UCD\_Controlled}$
  H2$_{A2}$: $Ambiguity\_Defect_{UCD\_Antipattern} < Ambiguity\_Defect_{UCD\_Controlled}$

Fig. 5. Data collected on defect inspection that makes the UC diagram inconsistent [8].

Instrumentation. The resources used to carry out this experiment involved: guidance for carrying out the experiment; a scenario for modeling the UC diagram; training to present and understand the strategy; selection of the material needed to execute the strategy; checklists to inspect the modeled diagrams. The experiment was conducted in two stages: stage 1, where all students modeled the UC diagram based on the same specification, and step 2, in which students were trained to understand and use the strategy for UCM and the artifacts to support its application. After completing the two stages, diagrams modeled with and without the strategy for inspection, comparison, and subsequent analysis were collected from each student.

7.3 Experiment Operation

The operation stage comprised the following steps: data preparation, execution, collection, and validation.

Data Preparation. The experiment participants were 16 students from two different software engineering classes from a Computer Science course.

Execution. The experiment was carried out in two replications, one for each class, in different days. Each student initially modeled the UC diagram without using the Antipattern strategy (in the UCD_Controlled group). Diagrams generated were collected for later evaluation. The strategy was then explained, and material supporting the strategy was provided to the students, e.g., a list of Antipattern with examples of its application. Using this material, the students improved the first modeled diagram, when necessary (in the UCD_Antipattern group). So as, each student produced a second UC diagram, all of which were also collected by the researchers for subsequent analysis.

Data Collection. The researchers inspect the diagrams collected in both groups using the checklist defined according to the mechanism showed in Sect. 6. Figures 6 and 7 show the data collected in this inspection.

Defect Consistency	P1		P2		P3		P4		P5		P6		P7		P8		P9		P10		P11		P12		P13		P14		P15		P16		Total DfC/Tipo	
	ID	FD	ID	FD	ID	FD	ID	FD	ID	FD	ID	FD	ID	FD	ID	FD	ID	FD	ID	FD	ID	FD	ID	FD	ID	FD	ID	FD	ID	FD	ID	FD	ID	FD
DfC1	Y	N			N	N	N	N	N	N	N	N	N	N	N	N	N	N	N	N	N	N	N	N	N	N	N	N	N	N	N	N	1	0
DfC2	Y	Y			Y	N	N	N	Y	N	Y	N	Y	N	N	N	Y	N	Y	N	Y	N	Y	N	Y	N	N	N	Y	N	Y	N	12	1
DfC3	Y	N			Y	N	Y	N	Y	N	Y	Y	Y	N	Y	N	Y	N	Y	Y	Y	N	Y	Y	Y	N	Y	N	Y	N	Y	Y	15	4
DfC4	N	N	Y	Y	N	N	N	N	N	N	N	N	N	N	N	N	N	N	N	N	N	N	N	N	N	N	N	N	N	N	N	N	1	1
DfC5	N	N			Y	N	Y	N	Y	N	N	N	Y	N	N	N	N	N	N	N	N	N	N	N	N	N	Y	N	N	N	N	N	5	0
DfC6	N	N			N	N	N	N	N	N	N	N	N	N	N	N	N	N	N	N	N	N	N	N	N	N	N	N	N	N	N	N	0	0
DfC7	N	N			Y	N	Y	N	Y	N	N	N	Y	N	Y	N	N	N	N	N	N	N	N	N	N	N	Y	N	N	N	N	N	6	0
DfC8	N	N			N	N	N	N	N	N	N	N	N	Y	N	Y	N	N	N	N	N	N	N	N	N	N	N	N	N	N	N	N	0	2
DfC9	N	N			N	N	N	Y	N	N	N	N	N	N	N	N	N	N	N	N	N	N	N	N	N	N	N	Y	N	N	N	N	0	2
DfC10	Y	N			Y	N	Y	N	N	N	Y	N	Y	N	Y	N	N	Y	Y	N	N	Y	Y	N	N	Y	Y	N	N	Y	Y	N	10	4
DfC11	Y	N			N	N	N	N	N	N	N	N	N	N	N	N	N	N	N	N	N	N	N	N	N	N	N	N	N	N	N	N	1	0
DfC12	Y	N			Y	N	Y	N	N	N	Y	Y	N	N	Y	N	Y	N	Y	Y	Y	N	Y	Y	Y	N	Y	N	Y	N	Y	Y	13	4
Total DfC/Part.	0	0	0	0	0	0	0	0	0	0	0	0	0	0	0	0	0	0	0	0	0	0	0	0	0	0	0	0	0	0	0	0	Tot. DfC/Group 64	18

Labels:

DfC --> Defect that makes the UC diagram inconsistent.
P --> Participant.
DI --> Initial diagram modeled without the UCM strategy.
DF --> Final diagram modeled with the UCM strategy.
Y, N --> Y - Yes, there was a defect. N - There was no defect.
Total DfC/Part. --> Total defects found in the initial and final diagram of each participant.
Total DfC/Type --> Total of each type of defect found in the total diagrams of all participants.
Total DfC/Grp --> Total defects found in the modeled diagrams without and with the strategy.
Red color represents the results of the diagram modeled without the strategy.
Black color represents the results of the diagram modeled **with** the strategy.

Fig. 6. Data collected on defect inspection that makes the UC diagram inconsistent.

Defect Ambiguity	P1		P2		P3		P4		P5		P6		P7		P8		P9		P10		P11		P12		P13		P14		P15		P16		Total DfA/Type	
	ID	FD	ID	FD	ID	FD	ID	FD	ID	FD	ID	FD	ID	FD	ID	FD	ID	FD	ID	FD	ID	FD	ID	FD	ID	FD	ID	FD	ID	FD	ID	FD	ID	FD
DfA1	Y	N			Y	N	N	N	Y	N	Y	N	Y	N	Y	N	Y	Y	N	N	Y	N	N	N	Y	Y	Y	N	Y	Y	N	N	11	3
DfA2	Y	N			Y	N	Y	N	N	N	Y	N	Y	N	Y	N	N	N	Y	N	Y	N	Y	N	N	N	Y	N	N	N	Y	N	11	0
DfA3	N	N			Y	N	Y	N	Y	N	Y	N	N	N	N	N	N	N	Y	N	Y	N	Y	N	N	N	Y	N	N	N	Y	N	9	0
DfA4	N	N			N	N	N	N	N	N	N	N	N	N	N	N	N	N	N	N	N	N	N	N	N	N	N	N	N	N	N	N	0	0
DfA5	N	N			N	N	N	N	Y	N	N	N	N	N	N	N	N	N	N	N	N	N	N	N	N	N	N	N	N	N	N	N	1	0
DfA6	Y	Y	Y	Y	N	N	N	N	Y	Y	N	N	N	Y	N	Y	N	N	N	N	N	N	N	N	N	N	N	N	N	N	N	N	3	5
DfA7	Y	N			Y	N	Y	N	N	N	Y	N	N	N	N	N	Y	N	Y	N	Y	N	Y	N	Y	N	Y	N	Y	N	Y	N	12	0
DfA8	N	N			N	N	N	N	N	Y	N	N	N	N	Y	Y	Y	Y	N	N	N	N	N	N	Y	Y	N	N	Y	Y	N	N	4	5
DfA9	Y	N			N	N	Y	N	Y	N	N	N	Y	Y	Y	N	Y	N	Y	N	N	N	Y	N	Y	N	N	N	Y	N	Y	N	11	1
Total DfA/Part.	0	0	0	0	0	0	0	0	0	0	0	0	0	0	0	0	0	0	0	0	0	0	0	0	0	0	0	0	0	0	0	0	Total 62	14

Labels:

DfA --> Defect that makes the UC diagram ambiguous.
P --> Participant.
ID --> Initial diagram modeled without the UCM strategy.
FD --> Final diagram modeled with the UCM strategy.
Y, N --> Y - Yes, there was a defect. N - There was no defect.
Total DfA/Part. --> Total de defeitos encontrados no diagrama inicial e final de cada participante.
Total DfA/Type --> Total de cada tipo de defeito encontrado no total de diagramas de todos os participantes.
Total DfA/Grp --> Total de defeitos encontrados nos diagrama modelados sem a estratégia e com a estratégia.
Red color represents the results of the diagram modeled without the strategy.
Black color represents the results of the diagrammed **with** the strategy.

Fig. 7. Data collected in the inspection of defects that make the UC diagram ambiguous.

In both Figs. 6 and 7, column 1 shows the code corresponding to each defect listed in Tables 4 and 5. For example, DfC1 indicates the defect *Relating actor and UC, when the actor is incompatible with interacting with UC*, which makes the diagram inconsistent. Columns 2 to 17 show the participants of the experiment. For example, P1 is participant 1. Each participant modeled a diagram without knowing the strategy, represented by DI (Initial Diagram) and the diagram using the strategy, represented by DF (Final Diagram). Each S and N indicates the researcher response when inspecting the presence or absence of defect in the initial and final diagram. The last line shows the total of defects found in the initial and final diagram of each participant. At the end of this line is the total of defects for the entire group. Finally, the last column shows the total of each type in the set of diagrams.

Data Validation. The validation was performed to assess any inconsistency in the data collected. The diagrams of participant P2 were discarded because he did not perform the modeling for the scenario presented. No other discrepancies were found. Therefore, all other answers were used in the data analysis.

7.4 Data Analysis and Interpretation

This step is responsible for the analysis of the data collected so that conclusions can be drawn. Figure 8 illustrates the result of the diagrams inspection. Eighty-two defects related to the inconsistency, of which 64 were identified in the UCD_Controlled group and 18 in the UCD_Antipattern group. Regarding ambiguity, 76 defects were found, of which 62 were identified in the UCD_Controlled group and 14 UCD_Antipattern group. The defects found in UCD_Antipattern group represented by the number 18 and 14, related to inconsistency and ambiguity, respectively, are defects found in UCD_Antipattern. They were not corrected with the strategy usage or are new defects that emerged after using the strategy.

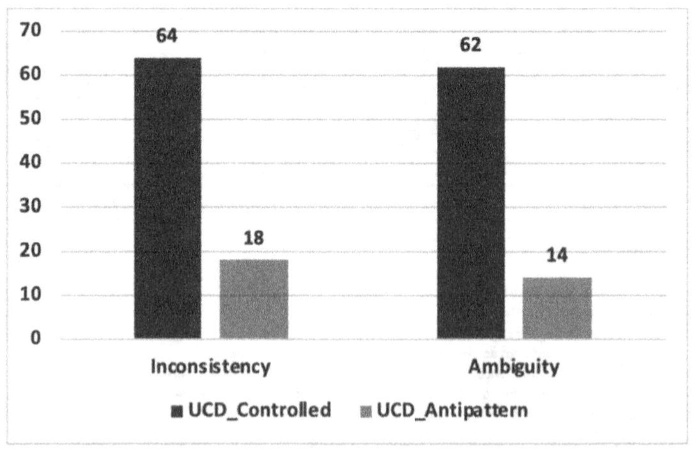

Fig. 8. Number of defects found in UC diagrams inspection.

In the UCD_Controlled group, the number of occurrences of each defect that made the diagrams inconsistent is illustrated in Fig. 9. The type of defect that was most evident was *DfC3 (Name actor with job titles and not with his role in the system)* with 15 occurrences, followed by *DfC12 (Define a UC as inclusion and extension at the same time)* with 13 occurrences, and *DfC2 (Name UC with a name inconsistent with the purpose of the UC)* with 12 occurrences. Three types of defects occurred only once, *DfC1 (Relate actor and UC, when the actor is incompatible with interacting with the UC)*, *DfC4 (The UC diagram is not plausible with the list of requirements)*, and *DfC11 (Have an Extension UC that does not add functionality to the base UC)*. The defect *DfC6 (Use CRUD, i.e. Create, Retrieve, Update, Delete, functionality instead of a single generic UC)* was not found in any diagram.

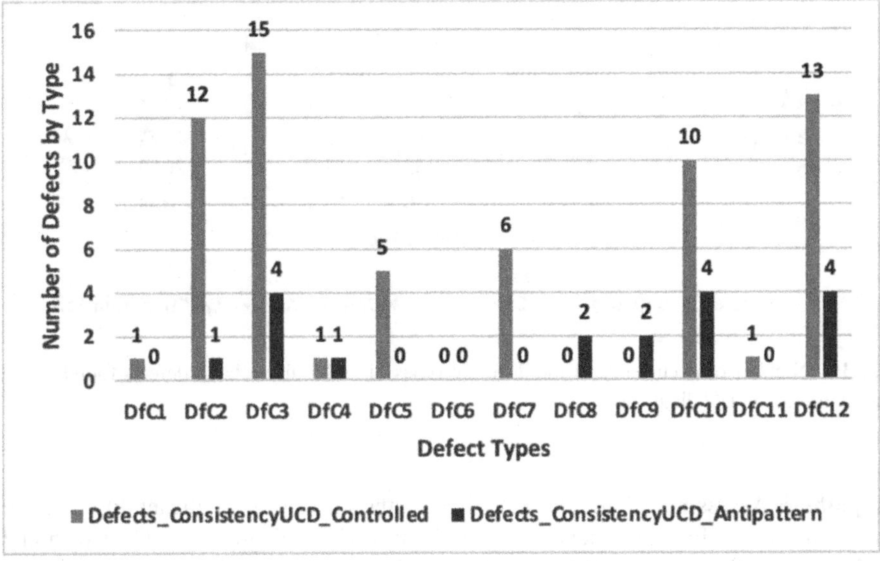

Fig. 9. Number of occurrence of the different types of defects related to consistency found in the inspection in the UC diagrams.

Figure 9 also shows the number of occurrences of each defect in the UCD_Antipattern. In this data set, in particular, two types of defects, *DfC8 (Relate UC and actor through generalization)*, and *DfC9 (Define inclusion, extension or generalization relationship between UCs whose removal prevents the understanding of the main UC objective)* occurred two times, however, not previously found in UCD_Controlled group.

Figure 10 illustrates the number of occurrences of each type of defect that made the diagrams ambiguous. In the UCD_Controlled, group the type of defect that was most evident was DfA7 (Having two or more UCs that perform the same functionality) with 12 occurrences, followed by DfA9 (Do not make the relationships between the actors and the UCs clear), DfA2 (Define a single UC as inclusion and extension simultaneously), and DfA1 (Name actor or UC with long or inexpressive terms), which occurred the same number of times, 11 occurrences. Various types of defects that appeared in the UCD_Controlled group disappeared in the UCD_Antipattern, such as DfA2 (Define a single UC as inclusion and extension simultaneously), DfA3 (Define a generic UC and another UC that is part of the generic without having a relationship between them), and DfA5 (Add brief description notes, without which it is not possible to understand the diagram). Two types of defects, DfA6 (Do not clearly show the users and the functionality of the software according to the list of requirements) and DfA8 (Have two or more actors with the same role in the system with a different name) increased in UCD_Antipattern.

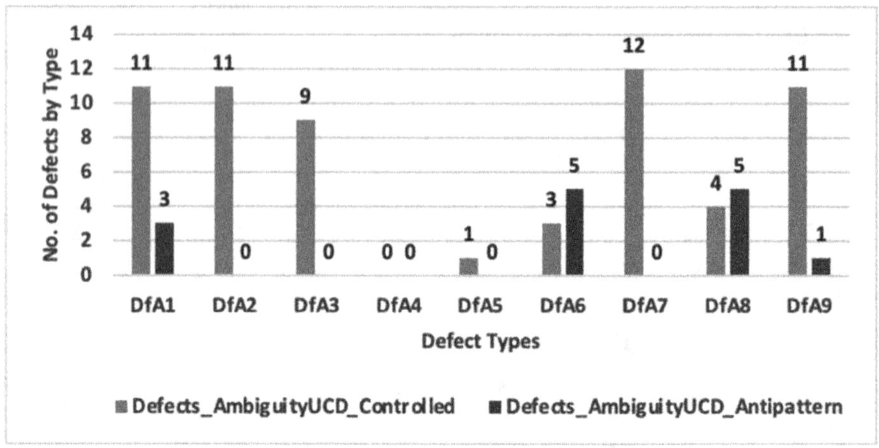

Fig. 10. Number of occurrence of the different types of ambiguity-related defects found in the inspection in the UC diagrams.

Hypothesis Assessment. In order to verify whether there is a significant difference in the quality (regarding consistency and ambiguity) of the diagrams modeled with and without the Antipattern based strategy, the hypothesis test was performed [40]. To select the test to be used, the normality of the data was verified. The tests showed that the distributions of the number of defects per diagram are expected in the UCD_Controlled group and in the UCD_Antipattern group. Figures 11 and 12 present the scatter plot showing that the number of defects generated by each participant is close to the average (linear across the points of the graph), concerning consistency and ambiguity.

In Fig. 11, for example, for participant P1, six defects were found in the UCD_Controlled group and one in the UCD_Antipattern group. On average, there are four defects in the UCD_Controlled group, and one in the UCD_Antipattern. The standard deviation is 1 and a tolerance of ± 1 for the two data sets. Hence, both six defects in the UCD_Controlled group and one defect in the UCD_Antipattern group are in the normality of the data.

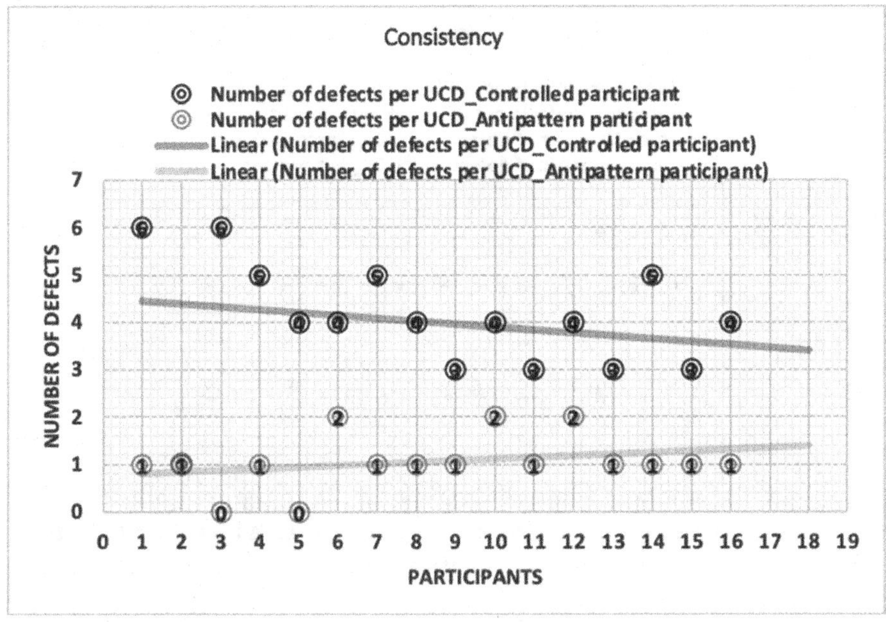

Fig. 11. Scatter plot of the number of defects related to consistency in the initial and final diagram of each participant.

Figure 12 presents the same analysis for ambiguity attribute. The average defect number for UCD_Controled group is 3.875, with a standard deviation 1 and tolerance of ±1. For UCD_Antipattern group, the average is 0.875, with a standard deviation 1 and tolerance of ±1. In both Figs. 11 and 12 x-axis show the participants P1, P2, up to P16, and the defects in the initial and final diagrams of each are the graph points.

As the data obey a normal distribution, the Shapiro-Wilk [58] test was used with the p-value (probability on the null hypothesis) equal to 0.05 to accept or reject the hypotheses. If the p-value < 0.05, the result is significant, and the null hypothesis can be rejected. After the test was carried out, the result showed that all null hypotheses (H10: Consistency_DefectsUCD_Antipatten = Consistency_ Defects UCD_Controlled) and H20: Ambiguity_ Defects UCD_Antipattern = Ambiguity_ Defects UCD_Controlled must be rejected according to the following: p-value (p-value) = 0.034 for the UCD_Controled group and p = 0.040 for the UCD_Antipttern; in relation to ambiguity, p = 0.035 for the UCD_Controlled group and p = 0.037 for the UCD_Antipattern group.

Having rejected the null hypotheses H10 and H20 that the use of the Antipattern-based strategy does not influence, respectively, the consistency and/or ambiguity of the UC diagram, the alternative hypotheses, Ambiguity_ DefectsUCD_Antipattern <> Ambiguity_ DefectsUCD_Controlled, were accepted. That is, the use of the Antipattern-based strategy influences the consistency and/or ambiguity of the UC diagram, which can be confirmed in the boxplot graph in Fig. 13.

Research Questions Answers. Having evaluated the hypotheses, we sought to answer the research questions formulated for this experiment. Related to RQ1, as showed in

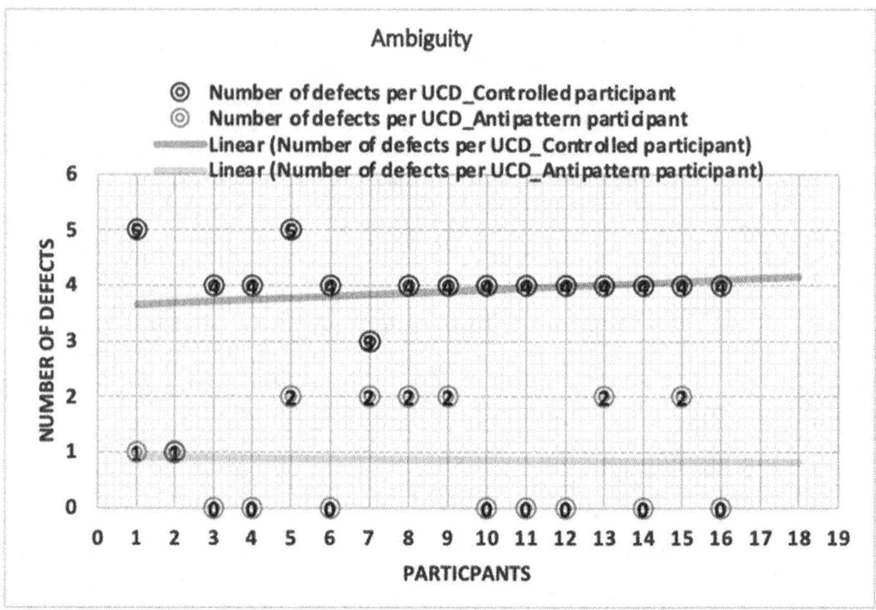

Fig. 12. Scatter plot of the number of defects related to ambiguity in the initial and final diagram of each participant.

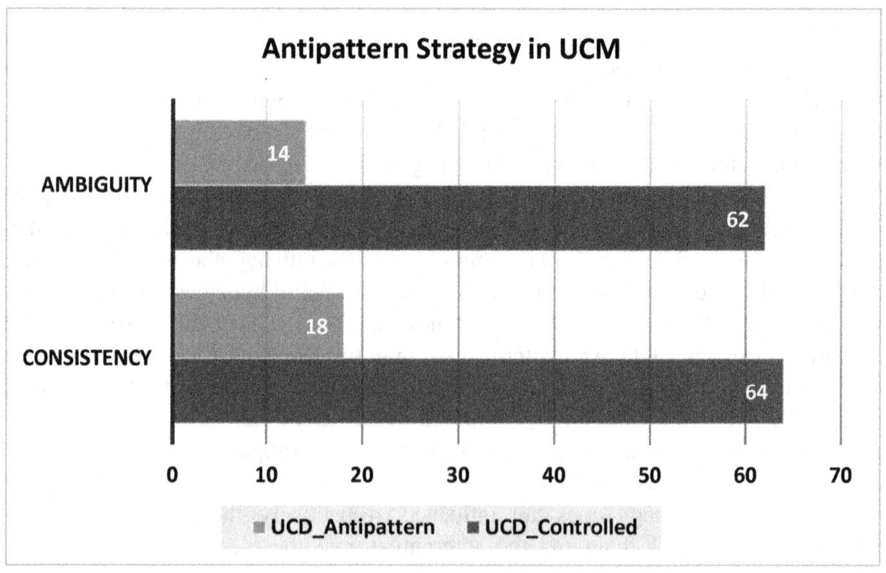

Fig. 13. Effect of Antipattern-based strategy on reducing ambiguity and inconsistency in the UC diagram [8].

Fig. 13, the use of the Antipattern based strategy in modeling the UC diagram considerably reduced the defects that make the diagram ambiguous and inconsistent. Concerning ambiguity, total defects reduced from 62 to 14 and, concerning consistency, total defects reduced from 64 to 18, when modeling with Antipattern. However, not all defects have been corrected, nor has it prevented the appearance of new ones.

Concerning RQ2, in this work, it was considered that the difficulties of the modelers generate the defects. Hence, if using Antipatterns reduces defects that affect consistency and ambiguity, then the difficulties that insert these defects were mitigated. However, there is a limitation in the results, which are considered as evidence and not conclusive.

7.5 Threats to Validity

To prevent bias in the validation, we now discuss some threats to this empirical study validity. Regarding internal validity, as the level of knowledge of the participant and experience in use case specification may influence the study results, we provided training for every participant in modelling use cases. Besides, we only selected participants who were enrolled in the software engineering discipline. Concerning external validity, to minimize the risk of sample representativeness, the diagrams produced by the participants were also inspected by people who did not participate in the experiment. The representativeness of the chosen domain and the size and complexity of the scenario are other threats to the experiment. As there was no possibility of using a real case, a scenario widely used in software modeling was chosen. However, experiments using real scenarios are necessary to validate our proposal better. Related to construction validity, we performed two pilot studies to validate the material used in the experiment. Distortions in understanding the antipattern strategy were minimized through a summary of examples of its use. Finally, concerning conclusion validity, the statistic method used may influence on the conclusion. Therefore, we consulted a specialist to define which method to adopt.

8 Conclusions and Future Works

Recent studies show the various difficulties of developers to model use cases. Therefore, the quality of the models is compromised due to the occurrence of defects. To mitigate these difficulties, we proposed a correlation between UCM defects and strategies to mitigate these defects based on UC' quality attributes.

This paper presented the controlled experiment performed to evaluate the correlation concerning the Antipattern strategy. In the experiment, participants developed a UC model in two stages, first without using the Antipattern strategy, and then assisted by this strategy. The produced UC models were analyzed using a checklist, which construction was guided by the proposed mechanism. A set of possible difficulties were identified in order to generate questions to be used in models inspection.

Shapiro-Wilk test was used with the p-value equal to 0.05 to accept or reject the hypotheses defined in the experiment. The results showed that the Antipattern-based strategy influences the consistency and ambiguity of quality criteria for the adopted scenario. We considered this a clear indication that Antipattern mitigates the difficulties related to it: difficulty to describe or to detail semantics in the UC model, and difficulty in understanding implicit requirements. Other difficulties may be mitigated by other strategies indicated in the correlation.

The strategy-difficulty correlation proposed in this paper organizes and guides the requirements specifier to select the most appropriate strategy to mitigate a given difficulty. This oriented indication that the correlation provides avoids adopting ineffective practices and makes the requirements specifier aware of several possibilities of applying tested and evaluated procedures to assist UCM. We are now working on new experiments to validate the other strategies contained in the correlation.

References

1. Erra, U., Portnova, A., Scanniello, G.: Comparing two communication media in use case modeling: results from a controlled experiment. In: ESEM 2010 Proceedings of the ACM-IEEE International Symposium on Empirical Software Engineering and Measurement (2010)
2. Tiwari, S., Gupta, A.: A systematic literature review of use case specifications research. In: Information and Software Technology, vol. 67, pp. 128–158 (2015)
3. El-Attar, M., Miller, J.: Constructing high quality use case models: a systematic review of current practices. Requir. Eng. **17**(3), 187–201 (2012)
4. Liu, S., Sun, J., Xiao, H., Wadhwa, B., Dong, J.S., Wang, X.: Improving quality of use case documents through learning and user interaction. In: 21st International Conference on Engineering of Complex Computer Systems (ICECCS), Dubai, pp. 101–110 (2016)
5. Nascimento, E.S., Silva, W., França, B.B.N., Gadelha, B., Conte, T.: Um Modelo sobre as Dificuldades para Especificar Casos de Uso. In: Conference Ibero-American on Software Engineering (CIBSE), Argentina (2017)
6. Beimel, D., Kedmi-Shahar, E.: Improving the identification of functional system requirements when novice analysts create use case diagrams: the benefits of applying conceptual mental models. Requir. Eng. **24**, 483–502 (2019). https://doi.org/10.1007/s00766-018-0296-z
7. Anda, B., Dreiem, H., Sjøberg, D.I.K., Jørgensen, M.: Estimating software development effort based on use cases—experiences from industry. In: Gogolla, M., Kobryn, C. (eds.) UML 2001. LNCS, vol. 2185, pp. 487–502. Springer, Heidelberg (2001). https://doi.org/10.1007/3-540-45441-1_35
8. Bispo, C., Magalhães, A., Fernandes, S., Machado, I.: Mitigating difficulties in use-case modeling. In: Proceedings of the 22nd ICEIS, 22nd International Conference on Enterprise Information Systems, 2020, Prague, vol. 2, pp. 43–52 (2020)
9. El-Attar, M., Miller, J.: Improving the quality of use case models using antipatterns. Soft. Syst. Model. **9**(2), 141–160 (2010)
10. Khan, Y.A., El-Attar, M.: A model transformation approach towards refactoring use case models based on antipatterns. In: 21st International Conference on Software Engineering and Data Engineering, Los Angeles, California, USA, pp. 49–54 (2012)
11. Khan, Y.A., El-Attar, M.: Using model transformation to refactor use case models based on antipatterns. Inf. Syst. Front. **18**(1), 171 (2016)
12. El-Attar, M.: Improving the quality of use case models and their utilization in software development. Department of Electrical and Computer Engineering, Alberta University (2009)
13. Fourati, R., Bouassida, N., Abdallah, H.B.: A metric-based approach for anti-pattern detection in UML designs. In: Lee, R. (ed.) Computer and Information Science 2011, vol. 364, pp. 17–33. Springer, Heidelberg (2011). https://doi.org/10.1007/978-3-642-21378-6_2

14. Bouzidi, A., Haddar, N., Abdallah, M.B., Haddar, K.: Deriving use case models from BPMN models. In: IEEE/ACS 14th International Conference on Computer Systems and Applications (AICCSA), Hammamet, 2017, pp. 238–243 (2017)
15. Cruz, E.F., Machado, R.J., Santos, M.Y.: From business process models to use case models: a systematic approach. In: Aveiro, D., Tribolet, J., Gouveia, D. (eds.) EEWC 2014. LNBIP, vol. 174, pp. 167–181. Springer, Cham (2014). https://doi.org/10.1007/978-3-319-06505-2_12
16. Saviä, D., Vlajiä, S., Lazareviä, S., Antoviä, I., et al.: Use case specification using the SilabReq domain specific language. Comput. Inf. **34**(4), 877–910 (2015)
17. Sommerville, I.: Software Engineering, 10th edn. University of St Andrews, Pearson, Scotland, London (2016)
18. Jacobson, I.: Use cases - yesterday, today, and tomorrow. Soft. Syst. Model. **3**(3), 210–220 (2004)
19. OMG Unified Modelling Language Superstructure - version 2.3. http://www.omg.org/spec/UML/2.3/ (2010)
20. Cockburn, A.: Writing Effective Use Cases. Addison Wesley, Reading (2000)
21. Jacobson, I., Christerson, M., Jonsson, P., Overgaard, G.: Object-Oriented Software Engineering: A Use-Case Driven Approach. Addison-Wesley, Reading (1992). Edition
22. Anda, B., Hansen, K., Sand, G.: An investigation of use case quality in a large safety-critical software development project. Inf. Soft. Technol. **51**(12), 1699–1711 (2009)
23. Bolloju, N.: Exploring quality dependencies among UML artifacts developed by novice systems analysts. In: 12th Americas Conference on Information Systems, p. 472 (2006)
24. Siau, K., Poi-Peng, L.: Identifying difficulties in learning UML. Inf. Syst. Manag. **23**(3), 43–51 (2006)
25. Spence, I., Bittner, K.: Use Case Modeling. Addison-Wesley, Reading (2003)
26. Bispo, C., Fernandes, S., Magalhães, A.P.: Strategies for use case modeling: a systematic literature review. In: Proceedings of the XXXIII Brazilian Symposium on Software Engineering (SBES 2019), pp. 254–263. ACM, New York (2019)
27. Ahmed, E.: Use of ontologies in software engineering. In: SEDE, pp. 145–150 (2008)
28. Dermeval, D., Vilela, J., Bittencourt, I.I., et al.: Applications of ontologies in requirements engineering: a systematic review of the literature. Requir. Eng. **21**, 405 (2016)
29. Gašević, D., Kaviani, N., Milanović, M.: Ontologies and software engineering. In: Staab, S., Studer, R. (eds.) Handbook on Ontologies. IHIS, pp. 593–615. Springer, Heidelberg (2009). https://doi.org/10.1007/978-3-540-92673-3_27
30. Couto, R., Ribeiro, A.N., Campos, J.C.: Validating an approach to formalize use cases with ontologies. In: Proceedings of the 13th International Workshop on Formal Engineering Approaches to Software Components and Architectures, vol. 205, pp. 1–15 (2016)
31. Yuan, X., Tripathi, S.: Combining ontologies for requirements elicitation. In: IEEE International Model-Driven Requirements Engineering Workshop, Ottawa, ON, pp. 1–5 (2015)
32. Bagiampou, M., Kameas, A.: A use case diagrams ontology that can be used as common reference for software engineering education. In: 6th IEEE International Conference Intelligent Systems, Sofia, pp. 035–040 (2012)
33. Dzung, D.V., Ohnishi, A.: Ontology-based reasoning in requirements elicitation. In: 2009 7th IEEE International Conference on Software Engineering and Formal Methods, pp 263–272 (2009)
34. Portugal, R.L.Q., Engiel, P., Pivatelli, J., do Prado Leite, J.C.S.: Facing the challenges of teaching requirements engineering. In: IEEE/ACM 38th International Conference on Software Engineering Companion (ICSE-C), Austin, TX, pp. 461–470 (2016)
35. Nkamaura, T., Tachikawa, Y.: Requirements engineering education using role-play training. In: IEEE International Conference on Teaching, Assessment, and Learning for Engineering (TALE), Bangkok, pp. 231–238 (2016)

36. Biddle, R., Noble, J., Tempero, E.: Role-play and use case cards for requirements review. In: Proceedings of the 12th Australasian Conference on Information Systems (2012)
37. Costain, G., Mckenna, B.: Experiencing the elicitation of user requirements and recording them in use case diagrams through role play. J. Inf. Syst. Educ. **22**(4), 367–380 (2011)
38. Kumar, B.S., Krishnamurthi, I.: Improving user participation in requirement elicitation and analysis by applying gamification using architect's use case diagram. In: Vijayakumar, V., Neelanarayanan, V. (eds.) Proceedings of the 3rd International Symposium on Big Data and Cloud Computing Challenges (ISBCC – 16'). SIST, vol. 49, pp. 471–482. Springer, Cham (2016). https://doi.org/10.1007/978-3-319-30348-2_39
39. Costain, G.: Cognitive support during object-oriented software development: the case of UML diagrams. Doctoral thesis. Auckland University, New Zealand (2008)
40. Wohlin, C., Runeson, P., Host, M., Ohlsson, M.C., Regnell, B., Wesslén, A.: Experimentation in Software Engineering. Springer, Heidelberg (2012). https://doi.org/10.1007/978-3-642-29044-2
41. Corbin, J.M., Strauss, A.: Basics of Qualitative Research: Techniques and Procedures for Developing Grounded Theory, 3rd edn. SAGE Publications, Thousand Oaks (2008)
42. Kalinowski, M., Card, D.N., Travassos, G.H.: Evidence-based guidelines to defect causal analysis. IEEE Softw. **29**, 16–18 (2012)
43. de Souza, A.J., Cavalcanti, A.L.O.: Visual language for use case description. Softw. - Pract. Exp. **46**(9) 1239–1261 (2016)
44. El-Attar, M., Miller, J.: Producing robust use case diagrams via reverse engineering of use case descriptions. Softw. Syst. Model. **7**(1), 67–83 (2008)
45. Gregolin, R.: Uma proposta de inspeção em modelos de caso de uso. São Paulo. Dissertação (Mestrado em Engenharia de Computação) – Instituto de Pesquisas Tecnológicas do Estado de São Paulo, 108 p. (2007)
46. Jebril, E.M., Imam, A.T., Al-Fayuomi, M.: An algorithmic approach to extract actions and actors (AAEAA). In: Proceedings of the International Conference on Geoinformatics and Data Analysis, Prague, Czech Republic, 20–22 April (2018)
47. Sawant, K.P., Roy, S., Parachuri, D., Plesse, F.: Enforcing structure on textual use cases via annotation models. In: ISEC 2014 Proceedings of the 7th India Software Engineering Conference, Chennai, India, 19–21 February (2014)
48. Rago, A., Marcos, C., Diaz-Pace, J.A.: Identifying duplicate functionality in textual use cases by aligning semantic actions. Softw. Syst. Model. **15**(2), 579–603 (2016)
49. Deeptimahanti, D.K., Sanyal, R.: Semi-automatic generation of UML models from natural language requirements. In: Proceedings ISEC 2011 the 4th India Software Engineering Conference, Thiruvananthapuram, Kerala, India, 24–27 February, pp. 165–174 (2011)
50. Liu, S., Sun, J., Xiao, H., Wadhwa, B., Dong, J.S., Wang, X.: Improving quality of use case documents through learning and user interaction. In: 21st International Conference on Engineering of Complex Computer Systems (ICECCS), Dubai, 2016, pp. 101–110 (2016)
51. Ko, D., Kim, S., Park, S.: Automatic recommendation to omitted steps in use case specification. Requir. Eng. (2018)
52. Ochodek, M., Koronowski, K., Matysiak, A., Miklosik, P., Kopczyńska, S.: Sketching use-case scenarios based on use-case goals and patterns. In: Madeyski, L., Śmiałek, M., Hnatkowska, B., Huzar, Z. (eds.) Software Engineering: Challenges and Solutions. AISC, vol. 504, pp. 17–30. Springer, Cham (2017). https://doi.org/10.1007/978-3-319-43606-7_2
53. Issa, A.A., Alali, A.I.: Automated requirements engineering: use case patterns-driven approach. IET Softw. **5**(3), 287–303 (2011)
54. Silva, A., et al.: Patterns for better use cases specification. In: Proceedings EuroPLOP' 2015. Hillside Europe (2015)

55. Dias, F., Schmitz, A., Campos, M., Correa, A., Alencar, A.: Elaboration of use case specifications: an approach based on use case fragments. In: ACM Symposium on Applied Computing (SAC), Fortaleza, Ceará, Brazil, pp. 614–618 (2008)
56. El Miloudi, K., Ettouhami, A.: A multiview formal model of use case diagrams using Z notation: towards improving functional requirements quality. J. Eng. **2018**, 9 (2018). Article ID 6854920
57. Holger, E.: Automatic layout of UML use case diagrams. In: SoftVis 2008 Proceedings of the 4th ACM symposium on Software visualization, Ammersee, Germany, 16–17 September, pp. 105–114 (2008)
58. Shapiro, S.S., Wilk, M.B.: An analysis of variance test for normality (complete samples). Biometrika **52**, 591–611 (1965)

iVOLUNTEER - A Platform for Digitization and Exploitation of Lifelong Volunteer Engagement

Elisabeth Kapsammer[1], Birgit Pröll[1], Werner Retschitzegger[1], Wieland Schwinger[1], Markus Weißenbek[1], Johannes Schönböck[2(✉)], Josef Altmann[2], and Marianne Pührerfellner[3]

[1] Johannes Kepler University, Linz, Austria
{elisabeth.kapsammer,birgit.proell,werner.retschitzegger, wieland.schwinger,markus.weissenbek}@jku.at

[2] Upper Austrian University of Applied Sciences, Hagenberg, Austria
{johannes.schoenboeck,josef.altmann}@fh-hagenberg.at

[3] Kunstuniversität, Linz, Austria
marianne.puehrerfellner@ufg.at

Abstract. Volunteering is an important cornerstone of our society, from social welfare to disaster relief, supported by a variety of volunteer management systems (VMS). These VMS focus primarily on centralized task management within non-profit organizations (NPOs) but generally do not provide mechanisms that allow volunteers to privately digitize and exploit their engagement assets in terms of digital badges for activities accomplished or competences acquired, in a trustful manner. This lack of sovereignty hampers volunteers in the exploitation of their engagement assets wrt. self-exploration, but also with regard to possible transfers of assets to other NPOs and beyond, e.g., to the education or labor market.

We put volunteers in the middle of concern by investigating "how engagement can be digitized and exploited in a lifelong way", thus adhering to the idea of human-centric personal data management. First we propose a conceptual architecture for a web-based volunteer platform based on a systematic identification of the requirements for trustworthy digitization and exploitation of engagement assets. Second, to address the massive heterogeneity that prevails in different areas of volunteering, a generic and extensible engagement asset model is proposed. Third a reification-based configuration mechanism is proposed so that each NPO can adapt the proposed model to its specific needs. Finally, a prototypical web application is presented which allows to »blockchainify« lifelong volunteer engagement in order to establish trust between all stakeholders.

Keywords: Human-centric personal data management · Digital badges in volunteering · Reification-based configuration mechanism · Blockchains

This work has been funded by FFG COIN 871494.

J. Filipe et al. (Eds.): ICEIS 2020, LNBIP 417, pp. 386–411, 2021.
https://doi.org/10.1007/978-3-030-75418-1_18

1 Introduction

Omnipresent and Manifold Volunteering. In times of refugee and health crises, volunteering is a vital cornerstone of our society and covers a broad part of our lives, from social welfare and disaster relief to cultural activities. More than 10% of the world's population are already engaged in voluntary work, in the EU it is 23%, in Austria even more than 46% [37]. In Austria, 2.25 million formal volunteers contribute more than 400 million voluntary hours per year, supplemented by more than 2.35 million "informal" (i.e., NPO-independent) volunteers who also contribute 351 million hours, with at least every second informal volunteer being formally engaged [21]. At present, new forms of volunteerism are emerging [21], which are carried out by *(i) patchwork volunteers*, who are involved in different NPO/life phases (e.g., from scouts to senior help), to *(ii) Engagement Hoppers*, who work informally, i.e., NPO-independent and ad-hoc (e.g., disaster relief), and *(iii) Crowd Volunteers* (e.g., open source projects).

Awareness Deficits About Engagements. Engagements lay the foundation for lifelong learning according to the *"Learning by Doing"* principle. The potential for experience-based acquisition of *informal competences* is a commonly agreed fact of all voluntary sectors [13]. However, volunteers are often *unaware* of their *accomplishments* and *achievements* they have earned during their engagement (e.g. tasks completed or competences acquired) [29]. This hinders volunteers from using their engagement assets for *self-exploration* and *personal development*, but also from transferring assets to other NPOs and beyond, e.g., to *education or labor markets*, although they are an invaluable substitute for certain formal qualifications [13]. However, these issues are not the focus of existing VMS's, even though *adequate IT support* is of primary importance from an economic point of view, and is in the very best interest of all those involved, i.e., the NPOs, the (informal) people seeking help, and the volunteers themselves [37]. Current *VMS* focus rather either on *central task management within NPOs* or on *coordination of informal voluntary work* [35].

Digitization and Exploitation of Life-Long Engagement. In contrast, the goal of our iVOLUNTEER project is to *put volunteers in the middle of concern* in order to increase the *appreciation of volunteerism*, as this is the only available reward in this area. This would not only be a step towards increasing the motivation of volunteers for their commitment, but would also strengthen the importance of *lifelong learning* for the benefit of all stakeholders involved. In context of this overall goal, the first important prerequisite is to facilitate the digitization of engagement assets by NPOs, being challenged by the massive heterogeneity of these engagement assets in different forms and areas of volunteer work (e.g. disaster relief vs. cultural or sports volunteering). The second complementary challenge is, according to the metaphor *»I am what I do«*, to enable volunteers themselves to make use of their deserved engagement assets in various forms, ranging from (i) self-exploration mechanisms that review past engagements from different perspectives, which facilitate the identification of new informal competences and the planning of future personal development objectives, to (ii) the transfer of engagement assets beyond the volunteer sector, which allow their further use, e.g. in the education or labor market.

Main Contributions and Paper Structure. In summary, the main contributions of this article, which go beyond our previous work [23–25,34,35] base on a systematic identification of requirements (Sect. 2) and an evaluation of related approaches (Sect. 3). They include *first*, a *conceptual architecture of a web-based volunteer platform* (cf. Sect. 4), which is *secondly*, complemented by a generic and extensible engagement asset model and an according *reification-based configuration mechanism* (cf. Sect. 5) to cope with the massive heterogeneity that prevails in different areas of volunteering, and *finally*, a prototypical web application which allows for *»blockchainification«, self-exploration, and transfer of engagement assets* to enhance exploitation possibilities and to establish trust between all stakeholders, so that each asset can be irrefutably traced, in terms of its *content* and its *existence* (cf. Sect. 6). A comparative evaluation together with a discussion of lessons learned conclude the article (cf. Sect. 7).

2 iVolunteer System Requirements

In order to focus on the intended end-user, i.e., mainly volunteers, NPOs and help-seekers, we followed the methodology of design science research [10,20], including a requirements engineering phase together with our demonstrators (Red Cross and Fire Brigade). This builds the basis for the evaluation of related approaches (Sect. 3) and for the development of concepts and the architecture (Sect. 4) of iVolunteer.

The requirements depicted in Fig. 1 arise primarily from our goal of *digitizing engagement assets*, including the need to *overcome scatteredness and diversity* and the demand to *maintain their lifelong evolution.* In addition, the goal of the *exploitation of engagement assets* by volunteers gives rise to the need for *sovereignty* and, at the same time, to the need to *establish trust* for each stakeholder who receives assets transferred by a volunteer.

Overcoming Scatteredness and Diversity. Not least due to the new forms of volunteering [21] as discussed above, the engagements of volunteers are manifold over time, resulting in the following requirements:

[REQ 1.1]. Engagement assets earned through various volunteering work are *scattered* across *data silos* of *proprietary VMS* in different NPOs, representing *partial views*, only. In order to obtain a comprehensive overview of all engagement assets earned in the life of a volunteer, a *global view* is required which forms the basis for further exploitation.

[REQ 1.2]. By their very nature, engagement assets of the engagement are *diverse* from each other under various aspects, as volunteering and NPOs are different. Their *level of detail* can range from simple confirmations of engagement, over detailed task accomplishments to comprehensive achievements. Furthermore, the *evidence* for asset earnings may range from simple textual justifications to formal, NPO-specific rules. To cope with these diversities when establishing a global view, generic representation mechanisms together with appropriate means for configuration are needed that allow NPO-specific extensions at run-time to incorporate the assets' peculiarities.

Maintaining Evolution. As (L)earning-by-doing is an evolutionary process, engagement assets need to co-evolve, leading to the following requirements:

[REQ 2.1]. Engagement assets should not only be *issued* once by NPOs, but must be *maintained* to reflect evolution history at the asset's instance level. Thus, *updates* of already existing assets (e.g., increasing a competences' proficiency level), their *withdrawal* (e.g., due to a missing refreshment training), or *deprecation* in case the assets' status is no longer maintained by the issuer (e.g., if a volunteer resigns engagement for the issuer) should be supported.

[REQ 2.2]. Engagement assets should be *traceable throughout their whole life-span* comprising not only their different states with respect to instance evolution (cf. REQ2.1), e.g., when they became *available*, but also indicating the assets' *validity* start and possible end time. In addition, traceability should also be ensured concerning potential evolution of the asset's structure, i.e., changes at the type level by e.g., adding additional properties to better describe a certain asset.

Achieving Sovereignty. To counteract the prevailing trend of centralized storage of personal data in volunteering or HRM [1,2], volunteers should be empowered to achieve *sovereignty* over their assets, which leads to the following requirements:

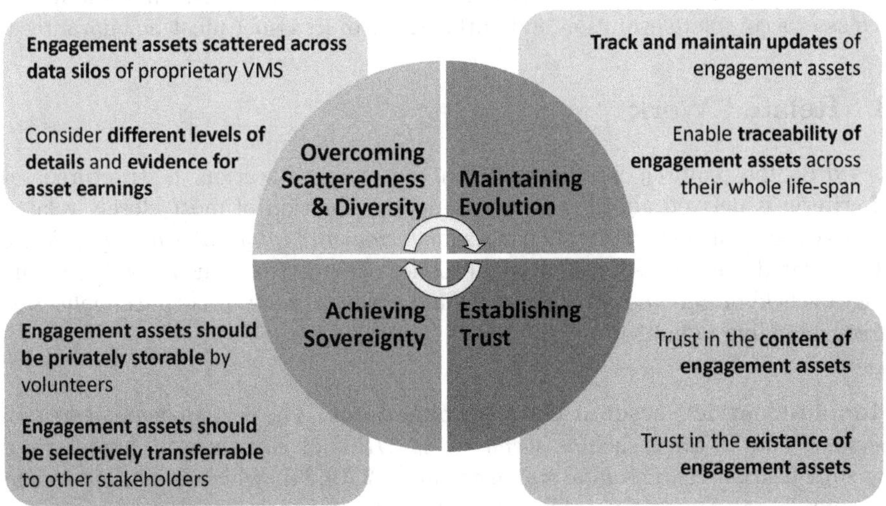

Fig. 1. iVOLUNTEER system requirements.

[REQ 3.1]. Engagement assets should be *privately storable* by volunteers at an arbitrary location (e.g., local NAS) and it should be definable *which assets to store* (e.g., tasks history, or certain awards).

[REQ 3.2]. Engagement assets that are already stored should be (i) *visually explorable* by volunteers in various ways and (ii) *selectively transferable to other stakeholders* in our digital ecosystem, like NPO's or job recruiters. This is a prerequisite not only to obtain personally satisfying volunteering tasks compatible with competences, but also to be able to claim the possession of a certain competence, e.g., when applying for jobs. If the volunteer revokes the transfer from an NPO, further maintenance of these assets by the NPO (e.g., updating the proficiency level of competences) must be prohibited.

Establishing Trust. In order to provide volunteers with sovereignty over their assets, it is crucial for their successful exploitation to build trust among all stakeholders. This is aggravated by the fact that sovereignty over the transfer of assets to certain receivers entails their uncertainty about the actual existence of assets, as these assets are claimed by the volunteers themselves, leading to the following requirements:

[REQ 4.1]. Engagement assets of a volunteer should be described in such a way that the receiver is given an adequate basis to gain *trust in their content.* For example, evidence about their justification, context of emergence, topicality, issuer, and evolution should be provided to further strengthen plausibility of their emergence. However, it naturally lays in the eyes of the beholders and their perspectives to trust in the content of an asset, not least since the level of trust strongly depends on the reputation of the issuer, i.e., a well-known NPO might be more trustful than an unknown informal help seeker.

[REQ 4.2]. Engagement assets should be transferable by the volunteer such that *trust in their actual existence*, despite their transfers by the volunteers themselves, can be irrefutably established for receivers. Thus, immutability of the asset's complete evolution and authenticity of its issuer must be guaranteed.

3 Related Work

Based on the requirements discussed in the previous section, a structured set of criteria is derived allowing for an in-dept evaluation of most closely related approaches in the field of *digital badges* for *formal and informal learning.* Prior to that, related work is discussed from a broader perspective, focusing on the highly relevant field of *human-centric personal data management* in *various application areas* including volunteering.

Human-Centric Personal Data Management. The current *management of personal data* in areas such as *social media*, *HRM* or *education* is characterized by *organisation-centric data management* [1,2,19,31], which is also prevalent in the field of volunteering. The available systems are designed as black-boxes, which means that the engagement assets stored in them (e.g., tasks carried out by a particular volunteer or competences acquired) cannot be exploited across the boundaries of the NPO in a trustworthy and confident way, as we found out during our in-depth evaluation of 18 VMS based on a reference model with more than 100 evaluation criteria in [35]. Not least since the emergence of the BC-paradigm [24], recent research efforts have focused on re-empowering users to manage their personal data, often referred to as *human-centric data management* [1], putting forward approaches for *decentralized social networks* [12,19], *personal information management* [36,39] or *digital badges* [4,14]. In particular, the field of *digital badges* is closely related to the intentions pursued by iVOLUNTEER, as digital badges are symbols defined and managed by an issuer and recognized within a community [4].

Table 1. Comparison of related approaches, based on [23].

Criteria			Approaches	UZHBC [Gresch et al., 2019]	SPROOF [Brunner et al., 2018]	BCE [Gräther et al., 2018]	Blockcerts blockcerts.org, 2019	EICS [Liu et al., 2018]
Overcoming Scatteredness & Diversity	[REQ1.1]	Global View		✗	✗	✗	✗	✗
	[REQ1.2]	Asset	Representation	schemaless	implicit schema	explicit schema	explicit schema	explicit schema
				PDF	n.a.	Open Badges based	Open Badges based	propriatary
			Taxonomy	✗	✗	✗	✗	✗
			Extensibility	✗	✗	~	~	✗
Maintaining Evolution	[REQ2.1]	Maintainability	Updateable	✗	✗	✗	✗	✓
			Withdrawable	✗	✓	✗	✓	✓
			Deprecatable	✗	✗	✗	✗	✗
	[REQ2.2]	Traceability	State Evolution	✗	✗	✗	✗	✓
			Schema Evolution	✗	✗	✗	✗	✗
			Availability Time	✓	✓	✓	✓	✓
			Validity Time	✗	✓	✓	✓	✓
Achieving Sovereignty	[REQ3.1]	Private Store	External	✓	✓	✓	✓	✗
			Technology	file system	IPFS (P2P filesystem)	BSCW (groupware)	mobile wallet app	n.a.
	[REQ3.2]	Transfer	Explorability	✗	✗	✗	✗	✗
			Selectivity	✓	✓	✓	✓	✓
			Revokability	✗	✗	✗	✗	✗
Establishing Trust	[REQ4.1]	Trust in AssetContent	Issuer	privileged users	everybody	privileged users	privileged users	privileged users
			Receipient	everybody	everybody	registered users	registered users	registered users
			Evidence	textual	textual	textual + URI	textual + URI	textual
	[REQ4.2]	Trust in Asset Existence	BC Platform	Ethereum	Ethereum	Ethereum	Bitcoin	HLF
				public	public	public	public	private
			Authenticity	✗	✓	✓	✓	✓
			Validity	✗	✓	✓	✓	✓
			Topicality	✗	✗	✗	✗	✓
			Completeness	✗	Asset Bundle	✗	✗	✗

~ *although Open Badges allows, per definition, for extensibility, it is not fully exploited by these approaches*

Digital Badges in Formal and Informal Learning. Digital badges gain increasing interest in the area of *formal and informal learning* for promoting learner engagement, participation, motivation and achievement [14]. In the following, related approaches for *BC-based personal data management* in terms of *digital badges* in the areas of *formal and informal learning* are investigated, namely *UZHBC (University of ZüricH BlockChain)* [18], SPROOF [11], *BCE (Blockchain for Education* [17], *Blockcerts* [32] and *EICS (Education-Industry Cooperative System)* [28]. These approaches are of particular interest because they (i) use different kinds of mechanisms to *represent their assets*, ranging from unstructured, schemeless to structured, schema-based, (ii) address *evolutionary* aspects to a certain extent, (iii) go at least partially *beyond the simple issuance* *and verification* of assets, and finally, (iv) provide for a mix of both, *permissionless and permissioned BC implementations*. The evaluation results are structured according to a set of criteria derived from our requirements (cf. Sect. 2). They are summarized in Table 1 and briefly discussed below on the basis of the individual requirements [REQ1-REQ4].

Overcoming Scatteredness and Diversity. Regarding our first category of requirements, scatteredness and diversity is not explicitly focused on by the evaluated approaches, i.e., establishing a global view on assets (i.e., digital badges) [REQ1.1] stemming from various sources is not an issue. The mechanisms for representing diverse assets [REQ1.2] are manifold, ranging from simply processing PDF-documents (UZHBC), to schema-less asset descriptions (SPROOF) and different schema-based ones, being partly based on Mozilla Open Badges (BCE and Blockcerts). Neither of them provides an engagement asset model, and thus a taxonomy of assets, nor do they fully support extensibility, allowing NPOs to do configurations of the asset taxonomy at run-time.

Maintaining Evolution. Maintenance of asset evolution [REQ2.1] is supported in different, often quite restricted forms. In particular, three systems support *withdrawal* of already issued assets, i.e., revising a previously stated claim (SPROOF, Blockcerts and ECIS), *updates* thereof are only supported by EICS. Regarding traceability of evolutions [REQ2.2], all approaches allow to capture an assets' *availability time*, and, with the exception of UZHBC, also its *validity time*. Tracing back the *evolution states* of assets is supported by ECIS, only. None of inspected systems address *different schema versions* during asset evolution.

Achieving Sovereignty. To achieve sovereignty, all systems except ECIS employ external private storage for assets [REQ3.1], using the BC for storing their hash values. External storage technologies used are manifold, ranging from distributed (P2P-)file systems to specific groupware. Finally, although all systems allow for a *selective transfer* of received assets, none of them provide means to *revoke this transfer* [REQ3.2] to prevent an assets' further maintainability.

Establishing Trust. Regarding *trust in an assets' content* [REQ4.1], support is restricted in almost all systems to the provision of *textual evidence and URIs to external justifications* only. BCE envisions to provide *algorithmic evidence* by means of smart contracts but misses a *structured representation of the emergence* of assets based on certain accomplishments and/or achievements. For establishing *trust in an assets' existence* [REQ4.2], the majority bases on the public BCs Bitcoin (Blockcerts) and Ethereum (UZHBC, SPROOF, BCE), only *EICS* employs the permissioned BC HLF. BC-entries can be *issued* in all systems, except SPROOF, by privileged users, only, whereas *receiving* assets is destined to registered users in UZHBC and SPROOF. *Authenticity* of the asset issuer can be verified by all systems applying asymmetric cryptography, except UZHBC - which, due to its schemaless representation, also lacks ability to verify validity. Beyond, only EICS provide means for establishing trust in the *topicality* of an asset throughout its evolution. With respect to *completeness of an asset*, SPROOF allows to verify asset bundles.

4 Conceptual Architecture of iVOLUNTEER

Based on the requirements identified in Sect. 2 and the deficiencies of existing systems identified in Sect. 3, this section introduces the conceptual architecture of the iVOLUNTEER platform (cf. Fig. 2).

Overall Usage Scenario. Following the metaphor *"I am what I do"*, volunteers are empowered to acquire certified *digital engagement assets* issued by NPOs by first downloading them into a *private repository* and then exploiting them through (i) *exploration mechanisms* that allow past engagements to be viewed from different perspectives and (ii) *transfer mechanisms* that allow assets to be *transferred* to third parties (e.g. job recruiters or social networks) in a self-determined manner. These can *verify* the credibility of assets *certified* by the issuing NPO on basis of a *blockchain*. Since iVOLUNTEER follows the principle of subsidiarity and data sovereignty, which puts the volunteer in the middle of concern, we rely on *decentralized system components*, as discussed in the following, comprising *VolunteeringHubs*, *PrivateAssetRepositories* and an *AssetBlockchain*.

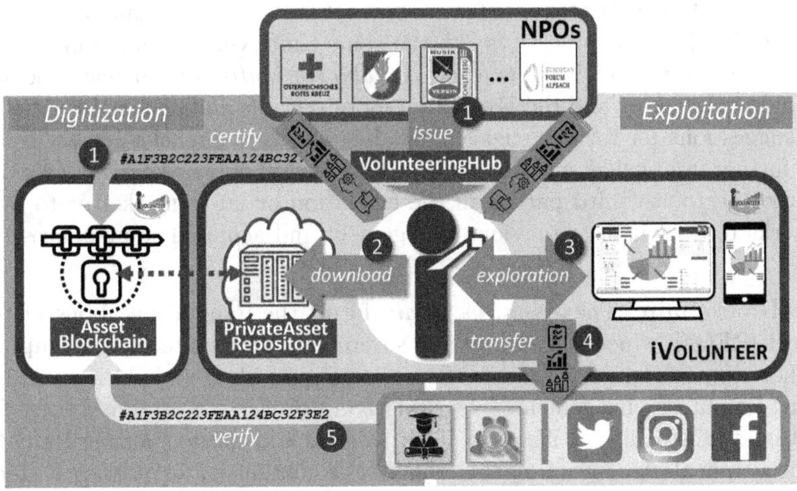

Fig. 2. Conceptual architecture of iVOLUNTEER.

VolunteeringHubs. *VolunteeringHubs* serve foremost as central component for *issuing* and *certifying* assets to appreciate the engagement of subscribed volunteers. This core function can be optionally complemented by three optional modules (not shown in Fig. 2) that provide typical VMS functionality for *offering volunteering opportunities, matchmaking with suitable volunteers and task management* (cf. [25,34] for details). The rationale behind this modularity is that NPOs, already employing some other VMS are able to use these modules selectively to complement the functionality of their own systems. For such usage scenarios, we also provide the possibility to import and certify assets which were earned outside of our platform and digitized in external VMSs. Finally, to further increase the value of volunteer work, *VolunteeringHubs* also allow to connect *donors* which are willing to provide assets in form of rewards for certain challenges, e.g., restaurants may provide vouchers if volunteers meet the challenge of more than 50 h of volunteer work per month.

PrivateAssetRepository. The *PrivateAssetRepository* constitutes the core storage module for the personal management of assets in the sole sovereignty of a volunteer to meet [REQ3]. Specifically, volunteers can (i) *download newly earned assets* issued and certified by an NPO from a *VolunteeringHub* into their *PrivateAssetRepository*, thereby providing a global view across all local views that might exist at different NPOs [REQ1.1], and (ii) *request certification of self-issued assets*, thus taking the burden from NPOs to issue every asset and supporting a simple negotiation process to facilitate consensus on the actual content of an asset. Once assets are stored in the *PrivateAssetRepository*, *self-exploration mechanisms* in the sense of interactive visualizations (cf. Sect. 6) can be employed to review past engagements from different perspectives, which facilitate the identification of new informal competences and the planning of future personal development objectives. At the same time, volunteers are empowered to (iii) selectively *publish assets* at other *VolunteeringHubs* to increase the accuracy of matchmaking and thereby obtain more suitable and personally satisfying task suggestions for further engagement, but also (iv) *unpublish assets* to prohibit further use by *VolunteeringHubs* and finally, (v) *transfer assets* beyond the volunteer sector to third parties in the education or labor market or to Social Networks to demonstrate previous engagement and acquired competences.

AssetBlockchain. The *AssetBlockchain* forms the integral backbone of trust to meet [REQ4], which even allows the sovereignty of volunteers, i.e., empowering them to maintain *PrivateAssetRepositories* providing global views on their lifelong volunteer engagement. The underlying process to establish trust foresees that the explicit certification of an asset at a *VolunteeringHub* leads to a »blockchainifcation«, meaning that the asset's content is, for privacy and storage space reasons, first hashed and then recorded within the *AssetBlockchain*, allowing to irrefutably verify the existence, immutability and authenticity of each single asset (cf. Sect. 6). Granting the permission for verification of certain assets is again under the sole control of volunteers and is, e.g., valuable when publishing assets to other NPOs or transferring them to third parties, thus allowing to establish trust.

5 Extensible Engagement Asset Model

After dealing with the *functional aspects* of iVOLUNTEER in the previous section, we now turn the focus on the *data aspect* by putting forward an *extensible model for engagement assets* that form the central subject of digitization and exploitation of volunteer work.

5.1 Core Engagement Asset Model and NPO Extensions

Rationale Behind. The primary intention of our engagement asset model is to benefit digitization and exploitation of engagement throughout a volunteer's life by (i) providing a common ground for diverse assets of various NPOs and (ii) allowing volunteers to explore all of their earned assets and making them exploitable across different NPOs and beyond, e.g., to education or labor markets. With this rationale in mind, we tried to formalize the core concepts of volunteering in terms of *engagement* and its *justified recognition by others* as well as the relationships in-between in terms of a UML class diagram (cf. the upper part of Fig. 3) and especially to put forward a configuration mechanism allowing for NPO-specific extensions at run-time (cf. the lower part of Fig. 3).

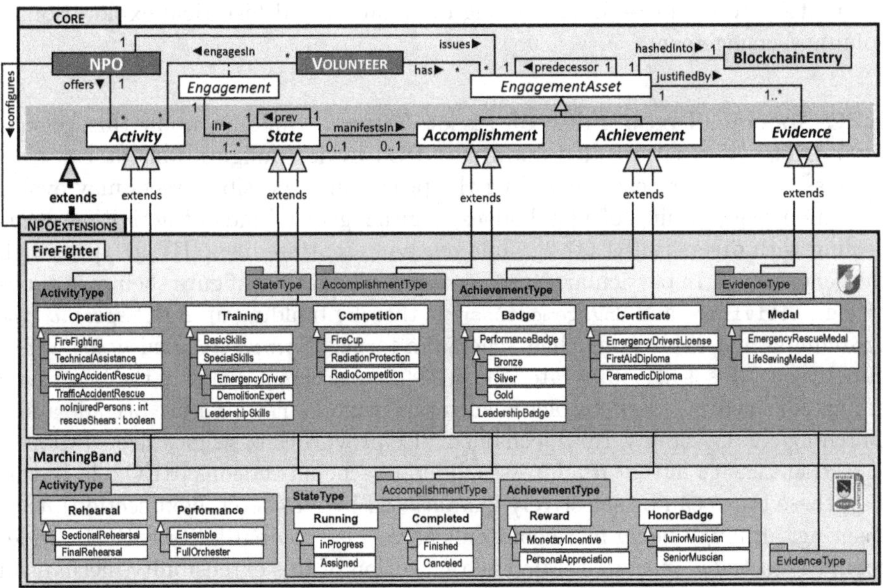

Fig. 3. Core engagement asset model and NPO extensions, based on [23].

Note that, for the sake of simplicity, asset properties are largely omitted. Following the metaphor *»I am what I do«*, a basis for the derivation of the common core concepts of engagement assets has been found in the area of *linguistic research*, especially in the prominent work of Vendler on the aspectual classification of verbs [38], as well as by considering well-known *upper ontologies* such as SUMO [33] or DOLCE [16].

The Generic Core Package. Adhering to [38], the generic concepts of our model depicted in the `Core` package in the upper part of Fig. 3, express the simple facts, that (i) `Engagement` of a `Volunteer` in an `Activity` offered by a `NPO`, which is running through certain `States` (e.g., offered, assigned, completed) can lead to the issuing of `EngagementAssets` through NPOs, comprising at least an (ii) `Accomplishment`, i.e., a manifestation of a certain activity state (e.g., engagement in an activity or its completion), and possibly several (iii) `Achievements` (e.g., award of a honoring medal). These assets are (iv) justified by some `Evidence` (e.g., textual or rule-based, eventually based on other assets) underpinning the level of confidence to achieve trust in their content [REQ4.1] and they are (v) manifested as `BlockchainEntry`, to ensure trust in their existence [REQ4.2]. In order to cope with the evolutionary nature of engagement assets [REQ2.1-2.2] (e.g., increase of a competence's proficiency level) the pre- and post-states of `EngagementAssets` are connected by a reflexive association. Consequently, `EngagementAssets` are considered to be immutable, meaning that every time an evolution takes place, a new `EngagementAsset` is created and connected to its predecessor. This design decision resembles the functionality of temporal databases [9], thereby establishing the pre-requisite for a life-long digitization and historical exploration of volunteer engagement.

NPO-Specific Taxonomies of Asset Types. Building upon this generic core, *extensibility* is provided allowing NPOs to do configurations at *run-time* not least since their needs regarding the provision of possible assets may evolve over time (e.g., issuing of new badges, digitizing new kinds of activities), thus dealing with diversity [REQ1.2] while resolving scatteredness [REQ1.1] through the generic core. In particular, NPOs are empowered to configure their own types of, e.g., `Activities` and `EngagementAssets`, thereby building up *NPO-specific type taxonomies* and explicating their schema in terms of *properties* and *associations* in-between. This is also illustrated in the `NPOExtensions` package in the lower part of Fig. 3, showing a small extract of such type taxonomies for a fire brigade and a marching band. Notably, this mechanism also provides a basis for the evolution of `EngagementAssets` in that it addresses different schema versions [REQ2.2]. In this way, these types can be selectively instantiated by NPOs (cf. next section), thus ensuring, although being more comprehensive, also compatibility with the widespread *open badge initiative* [14]. The realization of this extensibility mechanism is described in more detail in the next section.

5.2 Reification-Based Realization of Extensibility

The Dual Nature of Core Model Concepts and Extensions - "Clabjects". In order to allow the configuration of new types at *run-time*, e.g., activities or assets like introducing a new fire brigade-specific `LeadershipBadge`, types need to be represented as run-time objects, i.e., *reified*, since naturally, only objects can be "configured" at run-time. At the same time, although existing in form of objects, they need to act as *types* for respective *instantiations* thereof, i.e., in order to manifest the asset being finally digitized by iVOLUNTEER (e.g. a specific `LeadershipBadge` awarded to a certain volunteer, holding concrete property values). Thus, each of the NPO-specific types shown in the `NPOExtensions` package in Fig. 3 act, according to [6], as so-called *"clabject"*, having a dual nature in terms of both, instance and class (type) characteristics. Figure 4 shows at the left-hand side a small fraction of the core model concepts and their extensions illustrated in Fig. 3 and on the right hand side the realization of our extensibility mechanism as described in the following.

It has to be noted at this time that, we decided to realize also the concepts of the generic core on basis of this "clabject" mechanism, mainly for two reasons: (i) this allows for a homogeneous handling of all assets, no matter if they are NPO-specific or not, throughout the different digitization and exploration functionalities of iVOLUNTEER (e.g., blockchainification, visual exploration or transfer to third parties) and (ii) this would also provide the possibility to employ iVOLUNTEER for other domains, since also the generic core can be fully configured to the needs at hand, thus contributing also to research efforts of *integrating formal and informal learning engagement assets* [29], with particular emphasis on *transferable competencies* and the *open badge* initiative [14].

"Clabjects" Through Ontological and Linguistic Instantiations. To realize this "clabject" mechanism, we build upon a combination of design patterns proposed in the research field of Multi-Level Modeling (MLM) [6,7,27], cf. below. To put things in order, we distinguish, according to [6,7] between two orthogonal dimensions of instantiations (cf. Fig. 4): (i) *ontological instantiations* within core model concepts and its extensions (e.g., a concrete `LeadershipBadge` is an instance of a `LeadershipBadge` type) and (ii) *linguistic instantiations* using *reification* to represent both, the explication of core model concepts together with its extensions as well as instantiations thereof as run-time objects. According to this rationale, we put forward a simple *reification model* (cf. upper part of Fig. 4) providing appropriate linguistic classes which can be used to reify our ontological concepts, i.e., the core model and its extensions, at both, type and instance level, as explained in the next subsections.

Reifying the Type-Level of "Clabjects". In particular, a class `TypeDefinition` is provided (cf. Fig. 4) to allow for configuration of new types of activities or assets represented as linguistic instances thereof. The aggregated property explicates its `name`, e.g., an activity `TrafficAccidentRescue` and a reference to its respective ontological `superType` (`Operation` in our example in Fig. 4)

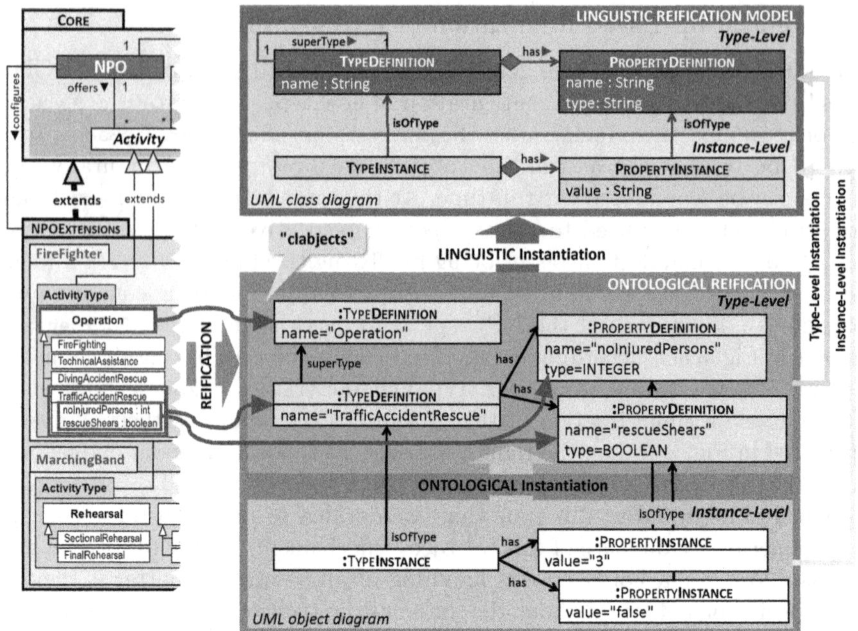

Fig. 4. Reification-based realization of extensibility.

provides for the *taxonomic aspect* of the NPO-specific configuration. To further configure the schema of this new type, it may posses properties (cf. reference `has` in Fig. 4) that themselves are represented on basis of linguistic instances of a `PropertyDefinition` class explicating the property's `name` and `type` (e.g., boolean, integer, date and string), in order to allow, e.g., the type `TrafficAccidentRescue` to manifest a `noInjuredPersons` integer property and a boolean one `rescueShears` allowing to indicate for the volunteering activity whether this special equipment had to be employed. Finally, note that, for the sake of simplicity, the realization of associations between asset types is not shown in Fig. 4. In order to allow NPOs to configure their type taxonomies, a web-based graphical configurator is provided, adhering to the syntax of UML class diagrams (cf. Fig. 8).

Reifying the Instance-Level of "Clabjects". To provide also the basis for the ontological instantiations of the configured types (e.g., certifying that a group of certain volunteers has participated in a concrete `TrafficAccidentRescue` operation), the linguistic `TypeInstance` class is introduced, explicitly referring to its respective ontological `TypeDefinition` (cf. `isOfType` in Fig. 4). This naturally goes along with the necessity to represent the types' properties in form of a `PropertyInstance` class containing the property's values along with a reference to their respective ontological `PropertyDefinition`. Overall, as one would expect from a non-reified object-oriented solution, creating, updating and deleting assets of a certain type are supported as a matter of course.

Ontological Instantiation Process. Since we have realized our own ontological type level in a reified way, we cannot rely on the instantiation mechanism which is naturally an inherent facility in any object-oriented language. Thus, it is necessary to realize a dedicated ontological instantiation process which is described further-on. Whenever an instantiation of a `TypeDefinition` takes place, two steps are performed.

Step 1: Type Schema Collection. First, since the schema information of types (i.e., properties and associations) which is necessary for the actual instantiation (cf. Step 2) is scattered across the inheritance hierarchy of the taxonomy, all aggregated properties and associated `TypeDefinitions` of its supertypes need to be collected first. The algorithm depicted in Algorithm 1 shows this process, i.e., from a given `TypeDefinition`, all supertypes are traversed in a recursive manner to collect all of their aggregated properties and find all associated `TypeDefinitions`. For handling associated `TypeDefinitions` two major object-oriented principles need to be considered, namely *inheritance* and *polymorphism.* Regarding inheritance, the taxonomy needs to be traversed upwards to collect the properties of all associated `TypeDefinitions` in a recursive manner as well. Simultaneously, to cope with polymorphism, the taxonomy of associated `TypeDefinitions` is traversed downwards in order to collect all possible `TypeDefinitions` that can be instantiated in its place.

Algorithm 1. Type Schema Collection.

```
function COLLECT(type, aggregation)
    aggregation.properties ← aggregation.properties ∪ type.properties
    if type.superType ≠ ∅ then
        aggregation ← collect(type.superType, aggregation)
    for association : type.associations do
        aggregation.associatedType ← collect(association.target, aggregation)
    return aggregation
```

Step 2: Actual Type Instantiation. Second, after collecting and assembling all relevant schema information needed to instantiate one specific `TypeDefinition`, this information is used to perform the actual instantiation through instantiating all of its aggregated properties (i.e., `PropertyInstances`) and associated types (i.e., `TypeInstances`) of the type itself and its supertypes (cf. Algorithm 2). Instantiation of associated `TypeDefinitions` can either be realized through referencing an already existing `TypeInstance` conforming to the referenced `TypeDefinition` or by newly instantiating that `TypeDefinition`, thus generating a new `TypeInstance` according to the Algorithms 1 and 2.

Algorithm 2. Type Instantiation.

```
function INSTANTIATE(aggregation)
    instance ← ∅
    for property : aggregation.properties do
        instance.properties ← instance.properties ∪ instantiate(property)
    for type : aggregation.associatedTypes do
        instance.assocInstances ← instance.assocInstances ∪ instantiate(type)
    return instance
```

Ontological Instantiation from a User's Point of View. During the creation of a new `TypeInstance`, either from instantiating a `TypeDefinition` or any of its associated `TypeDefinitions`, a user interface (UI) form is generated based on the collected properties and their data types wrt. whether allowed values for the property are pre-defined or not (e.g., weekdays as a text property). Figure 5 shows combinations of property's data types and their respective generated UI form controls. When instantiation takes place, a UI form control is built from each collected property, constructing a form from these controls. Instantiation of the `TypeDefinition's` properties naturally takes place when the user inputs data into each generated UI form control. For its associated *TypeDefinitions*, the user first has to choose which `TypeDefinition` should be instantiated (due to the polymorphism of `TypeDefinitions`, it can be either the directly associated `TypeDefinition` or any of its children). Following this user selection, for the chosen `TypeDefinition`, the instantiation process is reiterated, leading to a new subform, where the user can instantiate the associated `TypeDefinition`. Alternatively, the user can instead select an already existing `TypeInstance`, which is a valid instantiation (i.e., its type is either the associated *TypeDefinition* or one of its subtypes) of the associated `TypeDefinition`. In this case, no further instantiation of the associated `TypeDefinition` is necessary.

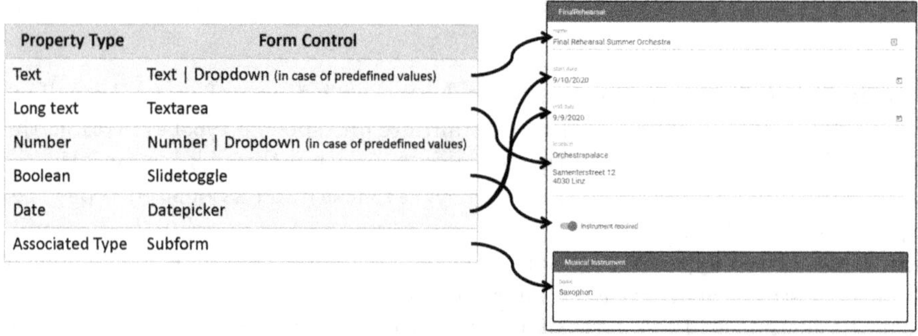

Property Type	Form Control
Text	Text | Dropdown (in case of predefined values)
Long text	Textarea
Number	Number | Dropdown (in case of predefined values)
Boolean	Slidetoggle
Date	Datepicker
Associated Type	Subform

Fig. 5. Concrete syntax mapping.

Rationale Behind Our Solution and Reflection. As already mentioned at the beginning of this subsection, our reification mechanism to realize "clabjects" along with linguistic and ontologial instantiations was inspired by the research field of MLM [6,7]. The rational behind was to follow, in principle, the *Type-Object pattern* and its solution as *Explicit Dynamic Types* [27]. The reification of the schema aspects of types in terms of `PropertyDefinitions` and `PropertyInstances` associated to the corresponding `TypeDefinitions` and `TypeInstances`, respectively, relates to the pattern of *Dynamic Features* and its realization as *Explicit Dynamic Features*. Finally, the reification of the taxonomic aspect through explicit representation of inheritance through ontologically referencing other "super" `TypeDefinitions` allows to realize the pattern of dynamic *Element Classification* [27].

Finally, reflecting on our solution, by sticking to the realization of just two ontological meta-levels and two orthogonal linguistic meta-levels (i.e., type-level and instance-level for each of them) we do not fully unleash MLM capabilities. This is mainly since, an arbitrary number of meta-levels as put forward by MLM, seems not to be necessary for our domain because reducing accidental complexity, which would be induced by arbitrary meta-levels, is not an issue. Instead, our rational behind was to employ MLM patterns to allow for type configurations at runtime. At the same time, we traded this meta-level flexibility of MLM for the benefit of a quite simple realization, which especially allowed us to put core concepts of MLM into a running web-based prototype allowing configuration of types and instantiation at runtime, which has been up to now, not the core focus of existing MLM approaches like *MetaDEPTH* [26], *Melanee* [5], *MultEcore* [30] and *DeepTelos* [22].

6 Proof-of-Concept Prototype

This section presents the iVOLUNTEER proof-of-concept prototype based on the requirements and the conceptual architecture outlined in Sect. 4 as well as the extensible engagement asset model presented in Sect. 5. The architecture of the iVOLUNTEER-prototype builds upon the inter-working of three layers (cf. Fig. 6), namely *Trust Layer*, *Service Layer* and *Client Layer*, ensuring a decoupled and decentralized architecture. This allows independent components whereby communication between them bases on the *REST paradigm* [15].

Trust Layer. The *Trust Layer* consists of the *AssetBlockchain (BC)*, in which immutable, obfuscated replicas of `EngagementAssets`, so-called `BlockchainEntries`, are stored. The modular, extensible and open source platform *Hyperledger Fabric (HLF)* has been used as the technological basis, enabling the operation of a permissioned BC [3]. HLF was chosen because (i) as a third generation BC, it provides a universal, general-purpose, distributed application development platform that is characterized by a high degree of configurability, especially in terms of consensus and storage mechanisms, (ii) is already used by major companies in real world settings, and (iii) support and maintenance is guaranteed by

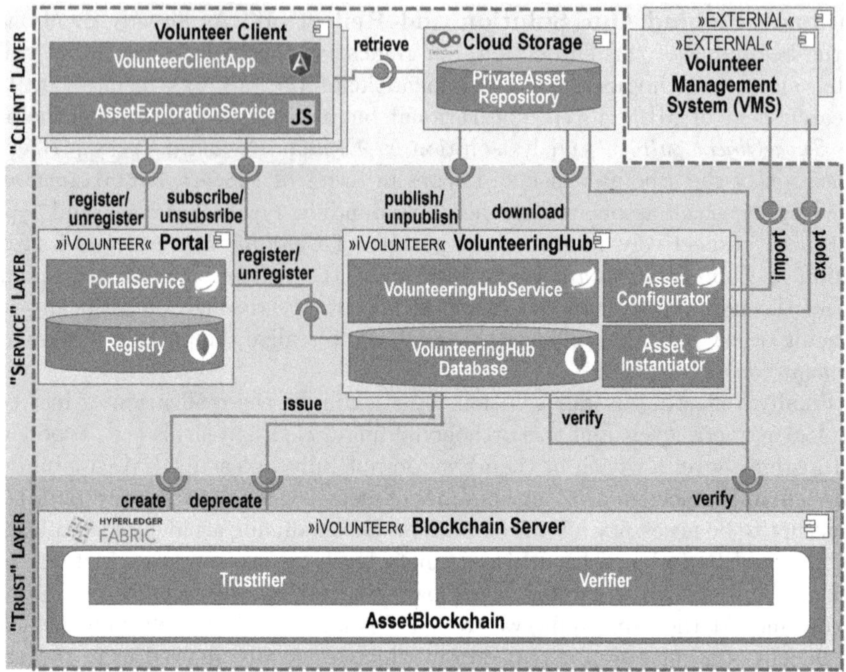

Fig. 6. iVOLUNTEER implementation architecture, based on [23].

IBM. The reason for using a private, permissioned BC is to minimize transaction costs and response time and to avoid possible limitations of transaction size that may occur in public BCs. The HLF implementation of the *AssetBlockchain* includes the *Verifier* and *Trustifier* components, which provide REST endpoints to read from and write to the BC (i.e., `create`, `deprecate`, and `verify`) and trigger their smart contract implementations, which are described in the following.

To protect the privacy of volunteers, their `EngagementAssets` are hashed (`EngagementAsset.generateHash()` function) and stored as `BlockchainEntry. hash` (cf. Fig. 7), which are reused throughout the BC as unique ID, allowing for fast retrieval by comparing them to the calculated hash of an asset, while at the same time *verifying the existence* of the asset (cf. `verify` method of *Verifier* in Fig. 6). To further guarantee and verify the *authenticity* of the issuer, asymmetric cryptography is used by signing each entry (cf. `BlockchainEntry.signature`), which has previously been calculated by the issuer of each asset by digitally signing the concatenation of its respective public key and a `BlockchainEntry.timestamp` with the issuer's private key using DSA. The inclusion of the timestamp results in different signatures each time an entry is created, thus precluding the signature's reuse.

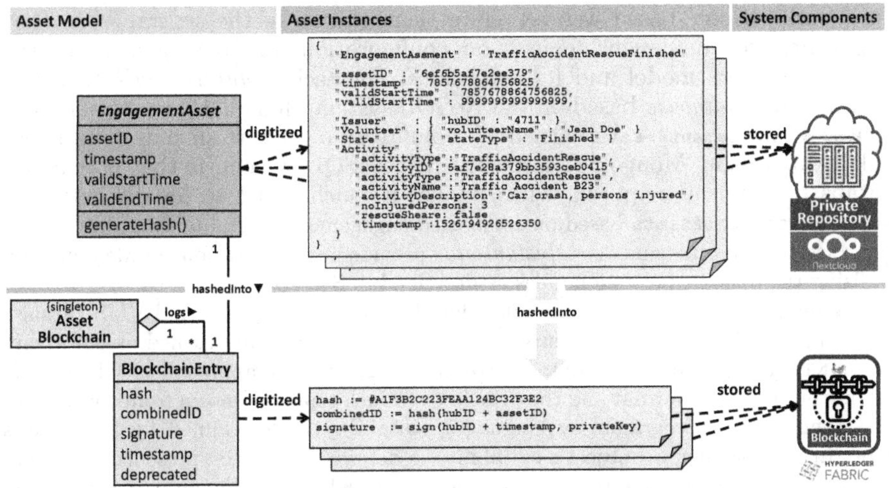

Fig. 7. Blockchainification of engagement assets.

In order to enable a *verification that goes beyond mere existence*, a verification of the history of an asset is necessary, which requires a linking of the individual evolution steps that belong together. This linkage is maintained by the `BlockchainEntry. combinedID`, which is calculated as a hash from the issuer's ID and the `Engagement Asset.assetID`, thus limiting the possibilities of retrieving information about the issuer (cf. Fig. 7). The combination of `BlockchainEntry-.hash` and `BlockchainEntry.combinedID` enables the verification of an asset's (i) *topicality*, (ii) *temporal dependencies*, by further exploiting the provided `Block-chainEntry.timestamp`, (iii) *completeness* of its past evolution, by being able to query all entries for an asset and finally, (iv) *future maintenance* by using the dedicated property `BlockchainEntry. deprecated`, which indicates the termination of an asset's evolution.

Figure 7 shows an exemplary asset instance in JSON format, typed to a customized asset model, representing the earning of an `Accomplishment` in terms of a `TrafficAccidentRescueFinished` asset being a result of participating in an `Activity` with the `ActivityType` of a `TrafficAccidentRescue` by a volunteer on the FireFighters' *VolunteeringHub*. This `EngagementAsset` may be `downloaded` to a volunteer's *PrivateAssetRepository* (see below) to provide a global view by employing a REST endpoint. Additionally, the evolution of the `EngagementAssets` is also stored in the blockchain (cf. `create` and `deprecate` REST endpoints in Fig. 6).

Service Layer. The *Service Layer* enables (i) volunteer engagement through *VolunteeringHubs*, (ii) the import of `EngagementAssets` from external VMS and (iii) export of `EngagementAssets` to third parties. For *VolunteeringHubs*, the

Java Spring Boot[1]-based web server implementation of the central *VolunteeringHubService* component focuses on configuration and management of the whole asset core model and its extensions. For each *VolunteeringHub*, a *VolunteeringHubDatabase* based on the NoSQL system MongoDB[2] stores the local views of `EngagementAssets` earned by volunteers on a certain hub. The rationale behind using MongoDB, respectively a NoSQL system, is that no explicit database schema is required, which makes it much easier to handle heterogeneous `EngagementAssets` based on configured engagement asset models. Additionally, the *AssetConfigurator* and *AssetInstantiator* provide implementations for defining and utilizing the reified asset taxonomies as described in Sect. 5.

VolunteeringHubs additionally allow for the integration of external VMS through their import REST endpoints, thus enabling usage of externally earned assets within iVOLUNTEER. For importing data, currently a light-weight approach is followed by relying on the JSON format and the principle of "duck typing"[3] known from JavaScript. In strongly typed programming languages, suitability is determined by an object's type. In duck typing, an object's suitability is determined by the presence of certain properties, rather than the type of the object itself, which is common in weakly typed programming languages like JavaScript. Thus, when exporting data from an external VMS, the JSON objects need to provide the properties defined in the NPO's customized engagement asset taxonomy (by using the means presented in Sect. 5.2).

Last, a registry of all *VolunteeringHubs* as well as the registration of users is maintained by the `iVolunteerPortal` (cf. Fig. 6). Every NPO that wants to make use of the iVOLUNTEER functionality needs to register its own `VolunteeringHub` at the `iVolunteerPortal`, e.g., Fire Fighters, Marching Bands or Red Cross. This component not only provides a lookup mechanism for all registered users, informing them, e.g., about new asset earning possibilities, but also to build-up cross-component functionality, like the provision of NPO-independent core competence models or asset generation rules for all `VolunteeringHubs`.

Client Layer. Finally, the *Client Layer* comprises access to functionalities provided by iVOLUNTEER for its users being it volunteers, NPOs, help seekers or other third parties. First, it provides volunteers the *VolunteerClientApp*, enabling them to communicate with the hubs, thereby providing means for engaging in volunteering activities as well as downloading their assets from the hubs to their *PrivateAssetRepository* and vice versa (i.e., publish assets to a hub). Second, the *AssetExplorationService* allows volunteers not only a simple filter- and search-based list-view of assets stored in their *PrivateAssetRepository* (cf. Fig. 7), but also to visually explore them. For the latter we employ JavaScript chart frameworks for displaying appropriate charts and diagramms, including *ngx-charts*[4], *ngx-beautiful-charts*[5] and *highcharts*[6]. These frameworks were chosen because of their extensive chart diversity and support for integration into an Angular web-application. The *PrivateAssetRepository* itself is deployed on a *NextCloud*[7] cloud storage, allowing users to either (i) select their very own hosted

[1] https://spring.io/projects/spring-boot.
[2] https://www.mongodb.com.
[3] "If it walks like a duck and it quacks like a duck, then it must be a duck".
[4] https://github.com/swimlane/ngx-charts.
[5] https://github.com/snowPu/ngx-beautiful-charts.
[6] https://github.com/highcharts/highcharts-angular.
[7] https://nextcloud.com/.

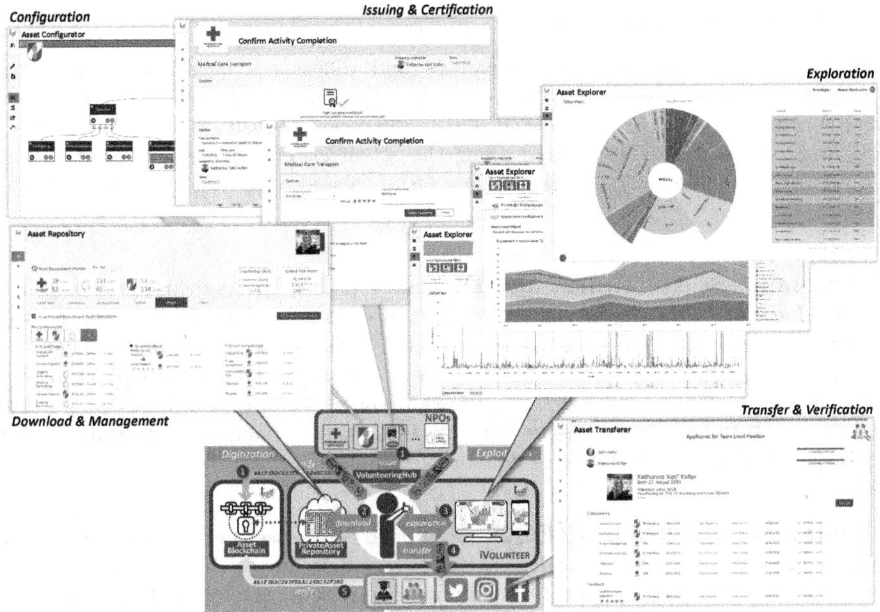

Fig. 8. Screenshots of iVOLUNTEER web application.

NextCloud and thus increase sovereignty and privacy of their downloaded engagement assets even more or (ii) any arbitrary *NextCloud* hosting provider. From a technical point of view, the engagement assets are stored as JSON to deal with the diversity of engagement assets [REQ1.2].

Figure 8 shows screenshots of the current prototype visualizing the volunteer's, NPO's and a recruiter's view, thus covering the main use cases already illustrated in Fig. 2, revolving around digitizing and exploiting engagement assets including: (i) engagement asset configuration, (ii) asset issuing and certification (iii) download into a *PrivateAssetRepository* and management, (iv) visual exploration, (v) transfer to other NPOs or third party stakeholder and (vi) verification. The app is implemented following the Progressive Web App (PWA) principle [8]. While still being a web app, PWAs allow for a more native feeling by providing means to start the app on a smartphone's home screen, offline availability, push notifications, and giving access to common hardware (e.g., sensors like GPS). However, by following the idea of responsive design, the app may still be used in a browser on a desktop PC or laptop, which might be more convenient to use in certain situations, e.g., for NPOs that might want to create tasks or that want to explore more complex visualizations (that would be too small on a smartphones' screen).

7 Comparative Evaluation and Lessons Learned

This section is dedicated to a brief comparative evaluation of iVOLUNTEER to related approaches considered in Sect. 3 and presents a discussion of lessons learned during conceptualizing, implementing and experimenting with our iVOLUNTEER prototype.

7.1 Comparative Evaluation

The following discussion of commonalities and differences of related approaches (*UZHBC (University of ZüricH BlockChain)* [18], SPROOF [11], *BCE (Blockchain for Education* [17], *Blockcerts* [32] and *EICS (Education-Industry Cooperative System)* [28]) to iVOLUNTEER is structured along the set of criteria being derived from our requirements (cf. Sect. 2) and the evaluation of related approaches (cf. Sect. 3 and Table 1)

Overcoming Scatteredness and Diversity [REQ1]. As mentioned in Sect. 3 scatteredness and diversity is not in the focus of existing system. In this respect, the existings systems still represent "walled gardens" in the sense that volunteers can not incorporate engagement assets earned outside a certain system. Neither of them provides an engagement asset model, and thus a taxonomy of assets, with a configurable and extensible type hierarchy as provided by iVOLUNTEER, being a unique selling point and urgently needed in the diverse domain of volunteering.

Maintaining Evolution [REQ2]. Maintenance of asset evolution is still in it's infancy in existing system, e.g., *deprecating* assets is unique to iVOLUNTEER. Tracing back the *evolution states* of assets is supported in its entirety by iVOLUNTEER, only. Specifically, iVOLUNTEER addresses *different schema versions* during asset evolution, whereas existing approaches focus, if at all, on instance evolution, only.

Achieving Souvereignity [REQ3]. While all systems except ECIS [28] employ external private storage for assets to achieve sovereignty, only iVOLUNTEER bases on the, in the meantime well established, NoSQL paradigm, allowing for convenient storage of *different asset versions* resulting from both, *structural-* and *state-based evolution* and the convenient import of external data. Finally, although all systems allow for a *selective transfer* of received assets, none of them provide means to *revoke this transfer* (in contrast to the *revokation of an asset*), as available in iVOLUNTEER, thus preventing an assets' further maintainability.

Establishing Trust [REQ4]. Regarding *trust in an asset's content*, iVOLUNTEER goes beyond existing approaches by providing *algorithmic evidence* by means of smart contracts allowing for a *structured representation of the emergence* of assets based on certain accomplishments and/or achievements. Similar to *EICS*, iVOLUNTEER bases on a permissioned blockchain, allowing for an independent operation (of several instances) and avoiding transactions costs, being crucial in the domain of volunteering. Only EICS and iVOLUNTEER provide means for establishing trust in the *topicality* of an asset throughout its evolution. With respect to *completeness of an asset*, SPROOF [11] allows to verify asset bundles, whereas iVOLUNTEER additionally allows to verify completeness of an assets' evolution states.

7.2 Lessons Learned

In the following, lessons learned tackle all of our four requirement categories, but focus especially on our *reification* approach and the employment of *BC technology*, thereby explicitly considering issues being also valuable to domains beyond volunteering.

Reification as Enabler Uniting Scattered Assets While Keeping Diversity. Reification of extensible types allows NPOs to configure their diverse engagement asset schema, adopt them to their own internal taxonomies and utilize them throughout volunteering processes, e.g., for task management or task recommendations through matchmaking. Volunteers that participate on several VolunteeringHubs are empowered to earn engagement assets from multiple NPOs, since these diverse engagement assets can be exploited on a united ground, allowing for an inspection and visualization of the global view across the volunteer's life-long engagements.

Reification as Method to Support Asset Evolution. The realization of our engagement asset model on basis of plain object-oriented technologies would allow for two levels (type and instance) only, which would not only hinder asset extensions at run-time, but especially also lead to an extensive migration effort of already issued assets, as soon as schema evolution of certain asset types take place (e.g., adding additional properties). Through schema reification, the migration effort of previous assets referring to outdated engagement asset schemas is kept to a minimum, since the VolunteeringHubs themselves are able to work with any, even former, schema transparently.

Reification as Possible Stumbling Block for Performance. Reification naturally exhibits an inherent trade-off between schema extensibility and system performance. Since every time new assets are digitized at the NPO's side (i.e., through ontological instantiation – cf. Sect. 5.2, Algorithm 1 and Algorithm 2) or existing assets are exploited at the volunteer's side (i.e., visually explore a volunteer's life-long engagement) the reified schema has to be assembled in a first step, before conducting the actual processing subsequently. To reduce this overhead at the NPO's side, iVOLUNTEER adheres to simple *load balancing* and *caching*, employing a decentralized system architecture as described in Sect. 4. First of all, only the affected *VolunteeringHub* is encumbered with schema assembly, compared to a centralized system, where the core application would be responsible for all registered NPO's reified schemas. Second, through in-memory data structures, assembled schemas are cached, reducing necessity to reassemble them every time these schemas or their instances are utilized.

Besides processing at the NPO's side, processing is also done locally at the client side during exploration of assets. This seams to be feasible also from a performance point of view, since the amount of data which has to be processed appears to be quite manageable. As estimation base: in Austria, being one of the leading volunteering countries world-wide as discussed in Sect. 1 [21], volunteers act in average about 5.3 h per week which would just lead to about 180 activities per year that have to be processed (precluding dependent assets that need be digitized, e.g., achievements).

Reification as Stepping Stone Enabling Asset Derivation at Run-Time. The acquirement of new achievements based on previous accomplishments lays the foundation for life-long volunteer engagement enhancing motivation and exploitation possibilities beyond mere finished activities. It occurred, however, that a mechanism would be beneficial, allowing NPOs to set up regulations in terms of pre- & post-conditions for awarding an achievement (e.g., competence earning), potentially also depending on already earned assets. In this respect, our reification mechanism provides a stepping stone towards enabling the realization of such *asset derivation rule* mechanism, which would allow the configuration of these derivation rules even at run-time, thus also enabling to reflect an NPOs "individual culture of appreciation".

BC as Backbone for Transparent Asset Derivation. To provide trust in the content of an asset and its evolution, a transparent asset derivation and evolution process needs to be provided by the BC. In particular, a library of asset derivation rules realized as smart contracts could be provided, allowing the provision of NPO-specific asset earning regulations in a transparent way through the BC. This could also lead to the emergence of cross-NPO asset derivation rules, e.g., the fire brigade considers qualifications earned at the red cross as pre-requisite for certain activities, and at the very end to some standardized procedures across different informal learning domains, resulting in a "homogenization" of assets and their evolution.

BC as Mechanism for Ensuring Assets' Existence. Although a BC can ensure the existence of assets, it can not entail *trust in the actual content* of an asset nor prevent *fake assets*. Despite of comprehensively representing an asset's context of emergence, there is no doubt that the issuer's reputation is *the* crucial impact factor. Due to openness of our platform where every registered user is allowed to act in the role of a volunteer or help seeker and since it is uncertain if consensus mechanisms of BCs may always prohibit fake assets, it always lays in the eye of the beholder to assess the plausibility of the provided assets. These are without a doubt some of the most crucial vulnerabilities of our system, although the special culture prevalent in the volunteering domain may reduce these risks to some extent.

BC as Battering Ram to Break Walled Silos. Today's application landscape in social media and HRM, which bases on data silos and walled gardens, could be turned upside down by empowering users to get back control over their data by employing BC as a means for establishing trust. Although we tried to take a first step in this direction with iVolunteer, further research is needed to investigate on possible use cases, technological requirements and borders of applicability for different application domains.

BC as Archive for Future Asset Certification Standards. A crucial hurdle on the way to achieve a widely accepted means for the digitization and exploitation of engagement assets in general and informal learning assets in particular is still the lack of standardized ways for their assessment and subsequent certification. Despite manifold efforts, no common agreement could be reached up to now. Since our BC-based digital volunteer platform allows at least to transparently and irrefutable document engagement assets and its context of emergence, the archived data could serve as input for future standardization and certification.

BC as Means for Joint Trust Establishment. One crucial issue when employing a private, permissioned BC like HLF is to decide about distribution and replication of system components to establish the required amount of trust. Therefore, the BC network's components, the committing peers, as well as the ordering service have to be hosted across different trustworthy organizations, preferably with different intentions like the national ministry of social affairs or trustworthy NGOs like the Red Cross.

BC as Enabler for Digitizing and Exploiting Further Assets. Leveraging our volunteer platform further beyond the volunteering domain, it would be valuable to allow e.g., educational institutions (providing, e.g., qualification certificates), companies (providing, e.g., job references) or governments (providing, e.g., a driving license) to issue assets. Besides these possibilities for digitization, exploitation could be further enhanced by providing functionality for recruiters to upload job profiles and carry out matchmaking.

References

1. Abiteboul, S., André, B., Kaplan, D.: Managing your digital life with a personal information management system. Commun. ACM **58**(5), 32–35 (2015)
2. Allard, T., Bouadi, T., Duguépéroux, J., Sans, V.: From self-data to self-preferences: towards preference elicitation in personal information management systems. In: Guidotti, R., Monreale, A., Pedreschi, D., Abiteboul, S. (eds.) PAP 2017. LNCS, vol. 10708, pp. 10–16. Springer, Cham (2017). https://doi.org/10.1007/978-3-319-71970-2_2
3. Androulaki, E., et al.: Hyperledger fabric: a distributed operating system for permissioned blockchains. In: Proceedings of EuroSys 2018, pp. 1–15 (2018)
4. Araújo, I., Santos, C., Pedro, L., Batista, J.: Digital badges on education: past, present and future. In: Proceedings of 4th European Conference on Social Media, pp. 27–35 (2017)
5. Atkinson, C., Gerbig, R.: Flexible deep modeling with Melanee. In: Betz, S., Reimer, U. (eds.) Modellierung 2016 - Workshopband, pp. 117–121. Gesellschaft für Informatik e.V., Bonn (2016)
6. Atkinson, C., Kühne, T.: The essence of multilevel metamodeling. In: Gogolla, M., Kobryn, C. (eds.) UML 2001. LNCS, vol. 2185, pp. 19–33. Springer, Heidelberg (2001). https://doi.org/10.1007/3-540-45441-1_3
7. Atkinson, C., Kühne, T.: Reducing accidental complexity in domain models. Softw. Syst. Model. **7**(3), 345–359 (2008)
8. Biørn-Hansen, A., Majchrzak, T.A., Grønli, T.M.: Progressive web apps: the possible web-native unifier for mobile development. In: Proceedings of International Conference on Web Information Systems and Technologies, vol. 2, pp. 344–351. SCITEPRESS (2017)
9. Böhlen, M.H., Dignös, A., Gamper, J., Jensen, C.S.: Temporal data management-an overview. In: Proceedings of eBISS 2017, pp. 51–83 (2017)
10. vom Brocke, J., Maedche, A.: The DSR grid: six core dimensions for effectively planning and communicating design science research projects. Electron. Markets **29**(3), 379–385 (2019). https://doi.org/10.1007/s12525-019-00358-7
11. Brunner, C., Knirsch, F., Engel, D.: SPROOF: a platform for issuing and verifying documents in a public blockchain. In: Proceedings of ICISSP 2019, pp. 15–25 (2019)
12. Chao, L., Palanisamy, B.: Incentivized blockchain-based social media platforms: a case study of Steemit. In: Proceedings of WebSci 2019, p. 10 (2019)
13. Deloitte: Building leadership skills through volunteerism (2016). www2.deloitte.com/content/dam/Deloitte/us/Documents/us-deloitte-impact-survey.pdf. Accessed 07 Sep 2020
14. Facey-Shaw, L., Specht, M., Van Rosmalen, P., Brner, D., Bartley-Bryan, J.: Educational functions and design of badge systems: a conceptual literature review. IEEE TLT **11**(4), 536–544 (2017)
15. Fielding, R.T.: Architectural styles and the design of network-based software architectures. Ph.D. thesis, University of California, Irvine (2000)

16. Gangemi, A., Guarino, N., Masolo, C., Oltramari, A., Schneider, L.: Sweetening ontologies with DOLCE. In: Proceedings of EKAW 2002, pp. 166–181 (2002)
17. Gräther, W., Kolvenbach, S., Ruland, R., Schütte, J., Torres, C., Wendland, F.: Blockchain for education: lifelong learning passport. In: Proceedings of ERCIM Blockchain Workshop 2018 (2018)
18. Gresch, J., Rodrigues, B., Scheid, E., Kanhere, S.S., Stiller, B.: The proposal of a blockchain-based architecture for transparent certificate handling. In: Abramowicz, W., Paschke, A. (eds.) BIS 2018. LNBIP, vol. 339, pp. 185–196. Springer, Cham (2019). https://doi.org/10.1007/978-3-030-04849-5_16
19. Guidi, B., Conti, M., Passarella, A., Ricci, L.: Managing social contents in decentralized online social networks: a survey. Online Soc. Netw. Media **7**, 12–29 (2018)
20. Hevner, A.R., March, S.T., Park, J., Ram, S.: Design science in information systems research. MIS Q. **28**, 75–105 (2004)
21. Holzer, M.: Volunteering IN Austria (2017). http://www.frei-willigenweb.at/sites/default/files/Volunteering%20in%20Austria_1.pdf. Accessed 07 Sep 2020
22. Jeusfeld, M.A., Neumayr, B.: DeepTelos: multi-level modeling with most general instances. In: Comyn-Wattiau, I., Tanaka, K., Song, I.-Y., Yamamoto, S., Saeki, M. (eds.) ER 2016. LNCS, vol. 9974, pp. 198–211. Springer, Cham (2016). https://doi.org/10.1007/978-3-319-46397-1_15
23. Kapsammer, E., et al.: (L) earning by doing - »Blockchainifying« life-long volunteer engagement. In: Proceedings of 22nd International Conference on Enterprise Information Systems, pp. 67–79 (2020)
24. Kapsammer, E., Pröll, B., Retschitzegger, W., Schwinger, W., Weißenbek, M., Schönböck, J.: The blockchain muddle: a bird's-eye view on blockchain surveys. In: In Proceedings of iiWAS 2018, pp. 370–374 (2018)
25. Kapsammer, E., et al.: iVOLUNTEER: a digital ecosystem for life-long volunteering. In: Proceedings of iiWAS 2017, pp. 366–372. ACM (2017)
26. de Lara, J., Guerra, E.: Deep meta-modelling with MetaDepth. In: Vitek, J. (ed.) TOOLS 2010. LNCS, vol. 6141, pp. 1–20. Springer, Heidelberg (2010). https://doi.org/10.1007/978-3-642-13953-6_1
27. de Lara, J., Guerra, E., Cuadrado, J.S.: When and how to use multilevel modelling. ACM Trans. Softw. Eng. Methodol. (TOSEM) **24**(2), 1–46 (2014)
28. Liu, Q., Guan, Q., Yang, X., Zhu, H., Green, G., Yin, S.: Education-industry cooperative system based on blockchain. In: Proceedings of HotICN 2018, pp. 207–211 (2018)
29. Livingstone, D.: Informal learning: conceptual distinctions and preleminary findings. Counterpoints **249**, 203–227 (2006). The Informal Education Reader
30. Macías, F., Rutle, A., Stolz, V.: MultEcore: combining the best of fixed-level and multilevel metamodelling. In: CEUR Workshop Proceedings of MULTI@MoDELS 2016 (2016)
31. Markoulli, M.P., Lee, C.I., Byington, E., Felps, W.A.: Mapping human resource management: reviewing the field and charting future directions. Hum. Resour. Manag. Rev. **27**(3), 367–396 (2017)
32. MIT Media Lab: Blockcerts: The open standard for blockchain credentials (2019). https://www.blockcerts.org. Accessed 07 Sep 2020
33. Niles, I., Pease, A.: Towards a standard upper ontology. In: Proceedings of FOIS 2001, pp. 2–9 (2001)
34. Schönböck, J., et al.: A semantic MatchMaking framework for volunteering MarketPlaces. In: Proceedings of WorldCIST 2018, pp. 701–711 (2018)
35. Schönböck, J., et al.: A survey on volunteer management systems. In: Proceedings of HICSS 2016, pp. 767–776 (2016)

36. Sjöberg, M., et al.: Digital me: controlling and making sense of my digital footprint. In: Proceedings of WS on Symbiotic Interaction, pp. 155–167 (2016)
37. UN Volunteers: State of the World's Volunteerism Report (2018). https://www.unv.org/sites/default/files/UNV_SWVR_2018_English_WEB.pdf. Accessed 07 Sep 2020
38. Vendler, Z.: Verbs and times. Philos. Rev. **66**(2), 143–160 (1957)
39. Zyskind, G., Oz, N., Pentland, A.: Decentralizing privacy: using blockchain to protect personal data. In: Proceedings of SPW 2015, pp. 180–184 (2015)

An Investigation of Currently Used Aspects in Model Transformation Development

Ana Patrícia Magalhães[1,2(✉)], Rita Suzana P. Maciel[3], and Aline Maria S. Andrade[3]

[1] Salvador University, Salvador, BA, Brazil
ana.fontes@unifacs.br
[2] Department of Exact Sciences and Earth, State University of Bahia, 2555 Silveira Martins St. Cabula, Bahia, Brazil
[3] Computer Science Department, Federal University of Bahia, Bahia, Brazil
{ritasuzana,aline}@ufba.br

Abstract. In Model driven development, a transformation chain is responsible for the conversion of high abstraction level models into other models until code generation. It comprises a set of transformation programs that automates a software development process. The development of a transformation chain is not a trivial task as it involves metamodeling, specific languages, knowledge of specific tools, and other issues that are not common in traditional software project development. Therefore, the adoption of software engineering facilities such as development processes, modeling languages, patterns, among others, have been proposed to assist in transformation development. This paper discusses currently facilities used to develop model transformations from a systematic literature review result. To better organize the currently aspects found, we structure the proposals according to a classification concerning software engineering development technologies in order to help developers and researchers in searching by solutions and challenges.

Keywords: Model transformation · Classification of model transformation development aspects · Transformation development strategies · Transformation modeling languages · Model transformation development approaches · Model transformation development methods

1 Introduction

Model Driven Development (MDD) is a software development approach that considers models as its principal artifacts. Models at high abstraction levels are (semi) automatically converted into other models until generating the application code. A key element of this approach is the transformation chain responsible for model conversions among different abstraction levels [2].

J. Filipe et al. (Eds.): ICEIS 2020, LNBIP 417, pp. 412–436, 2021.
https://doi.org/10.1007/978-3-030-75418-1_19

The MDD approach involves elements that are not usually part of traditional software development. For example, models might be specified using domain-specific modeling languages (DSMLs), with well-defined syntax and semantic. The abstract syntax and static semantic of a DSML is usually expressed by metamodels [1]. Thus, models are created as instances of metamodels and must be in conformance with them. Specific languages, e.g., ATL (Atlas Transformation Language) and QVT (Query View Transformation), appropriate for model manipulation, require the use of new tools and engines, among other elements.

In this work, we are interested in model transformation development, which is the first stage of the transformation chain in MDD process automation. Specially, we treat about investigating existent proposals to give support to model transformation development. In the concern of this, we performed a systematic literature review [18] covering the period from 2003, when MDD was standardized in MDA (Model Driven Architecture), to Dec 2019, including research databases of most publications related to MDD. We selected and analyzed 23 works in the final process so that we could outline the current situation in the state of the art in transformation design. Among some finds, we can highlight that works concentrate on different aspects of development without a consensus about which strategies to adopt.

Model transformation domain have been investigated through systematic literature reviews [26–29] in several points of view. Some works identify and classify concepts related to transformation foundation, such as models, metamodels, and rules. Others, like us, are concerned about transformation development and investigate the programming style adopted, e.g., relational and graph-based, the use of MDA in model transformation development, and the applicability of the proposals, i.e., how reliable it is, the development life cycle, among other aspects. These works usually propose a classification to compare the selected papers according to their specific goals. However, we did not find a classification based on software engineering foundations, such as development approaches, methods, and notations, to facilitate comparison among them.

Thus, in this work, we consider software engineering facilities to classify the current aspects related to transformation development, such as development approaches, strategies to guide development, phases of a transformation life cycle, methods to assist specific tasks, notations adopted in the specification, and automation resources. This classification can help better identifying which strategies are more suitable for certain types of transformation as well as features not yet covered that can arouse research questions.

This paper presents an updated release of the systematic literature in [18], also complementing the description of the selected papers. In addition, the selected works are presented through the classification created so that they can be better identified according to application requirements.

The rest of this text is organized as follows: Sect. 2 introduces some background about model driven development approach; Sect. 3 analyzes related works; Sect. 4 presents the proposed classification; Sect. 5 presents the research method used to guide the systematic review; Sect. 6 the execution and results of the systematic review; in Sect. 7 we discuss some threats to our review explaining how they were prevented; in Sect. 8, we discuss the results of the review identifying some current gaps in model transformation development; finally, Sect. 9 presents our conclusions.

2 MDD Basic Concepts

In MDD models are abstract representations of a system, which comprise its structure and behavior [54]. They are artifacts used as input of the transformations to generate other models or application code. Thus, they should be defined in modeling languages with a well-defined syntax and semantic.

MDD usually uses Domain-Specific Modeling Languages (DSML) for model specification, comprising four main elements [2]: an abstract syntax, in which the language constructors are defined; static semantics, in which well-formed rules and constraints are specified; a concrete syntax, in which concrete notation is defined; and behavioral semantics, to define how instantiated models are executed. In MDD, the abstract syntax and the static semantics of a modeling language are expressed in metamodels. Thus, models are designed conforming to metamodels [2]. UML can be customized to specific domains through the definition of UML profiles [54] to be used as a modeling language in MDD projects.

In MDD, transformations can be classified according to the artifact produced as output from model transformation (called model to model or M2M transformation) or program transformation (called model to text or M2T transformation). A model transformation processes input models in order to generate other models as output. A program transformation processes input models in order to generate code as output.

A model transformation (MT) can be seen from two different viewpoints, as a function that maps models between domains or as a terminating algorithm that applies structural and/or semantic changes to models [19]. This work focuses on the first viewpoint called relational transformation. Therefore, a MT comprises a set of rules that describes how models, instances of source metamodels, are converted into models, instances of target metamodels [20]. Thus, the definition of a MT is done at the metamodel level through rules that map how constructors from the source metamodels are converted into constructors of target metamodels. Model transformation can also be specified as a model, called model transformation model [21], enabling the adoption of the MDD approach itself in transformation development. Thus, model transformation models are transformed through a transformation chain until they generate transformation code. In this scenario, the transformation chain comprises a set of High Order Transformations (HOT), i.e., model transformations that analyze, produce, and manipulate other transformations [22].

3 Related Works

Different works report on literature review and classifications of mechanisms concerning MT. Some authors deal with aspects related to transformation concepts [20,26,53]. However, we are interested in works that address aspects of MT development, such as life cycle and methods.

In [28] the author carries out a systematic literature review aiming to identify proposals that use Model Driven Engineer (MDE) to develop MT. Seven features based on the Model Driven Architecture (MDA) [54] are defined and used to analyze each selected work. We consider our work encompass this research since we investigate different phases of a life cycle, not restricted to MDA abstraction levels. We cover aspects such as notations and methods, not mentioned in this work.

In the same direction of our proposal, [29] presents a literature review of MT development approaches, which, among other aspects, investigates features implemented in the approaches, e.g., phases of the life cycle, artifacts, and tools to aid development; and the applicability of approaches, e.g., evaluation and level of detail. The work defines a set of phases of the development life cycle to evaluate the proposals, e.g., specifying requirements, modeling source and target metamodels, and creating mappings between metamodels. In this, the concepts of phase and activity are placed at the same level of abstraction, making difficult comparisons with other works. In our review, we considered the development life cycle according to SPEM [55], i.e., a life cycle is composed of phases where a set of activities are performed producing artifacts. The use of a standard definition like SPEM makes it easier to understand and compare with other works. We also explore modeling language adopted that [29] do not.

A survey is presented in [27] which organizes the MT generation in three dimensions: what is used as input to generate the transformation, e.g., models and metamodels; what mechanisms can be used, i.e., ad-hoc algorithms, artificial intelligence techniques and formal techniques; and what to generate, i.e., executable transformation rules and not executable transformation rules. Our work complements this one, since details other aspects of development, e.g., notations, also identifying other methods adopted in development.

A systematic mapping to understand the trends and characteristics of MT in concrete problems, i.e., real-world problems, is presented in [52]. Among other aspects, the authors are interested in identifying languages used to implement MT, i.e., dedicated languages, like ATL, programming languages, like Java, or any other. Although this work presents a literature review and a structure organization of characteristics related to MT development, like our proposal, they explore implementation languages currently used while we investigate notations to transformation specification.

4 Model Transformation Development Classification

Model transformations development, as any software, should consider a development life cycle with the necessary elements to guide software engineers [3,12,16,23,24]. In [17] we defined a framework, which relates the relevant elements of model transformation development. This framework was considered to define features and assess the works selected in our systematic literature review.

Based on that framework, this section presents a classification (Fig. 1) that groups the main aspects related to model transformation development towards a taxonomy. The classification is divided into six categories: development approach, development strategy, phases of the life cycle, methods to assist the development, notation used in the specification, and level of automation provided.

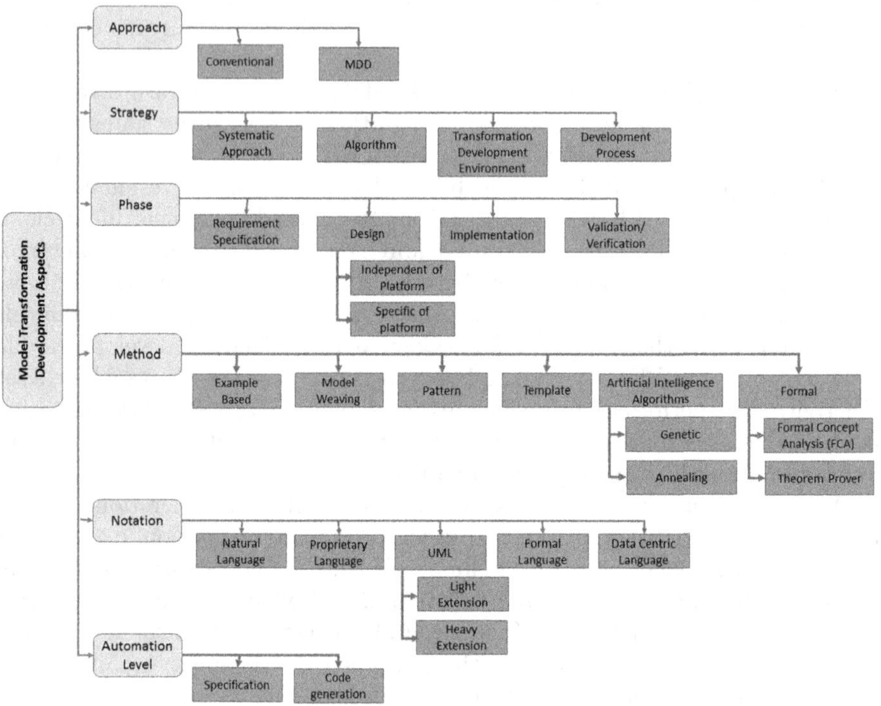

Fig. 1. Classification of aspects related to MT development.

Typical approaches for software development include a waterfall, evolutionary, incremental, and, more recently, the MDD approach, among others. They show, from a generic perspective, ways to organize tasks related to a software development process. Based on these, systematized strategies are specified according to aspects of a particular domain [5]. This is the case of model transformation development. Our classification organizes the proposals to develop model transformations in two approaches: *MDD*, when MDD principles are used, such as, converting of models into other models or code, and *Conventional* for the others that do not use MDD. It is also possible to adopt a hybrid approach mixing both types.

Another aspect of model transformation development concerns the strategy adopted. We consider a development strategy a set of elements, e.g., tasks and artifacts, well organized to be followed as a guide, such as development processes. In this direction, we identify four groups of strategies: *Systematic Approach*, which comprises a description of steps, written in natural language, to be followed by developers; *Algorithm*, with steps described in natural language and organized using control structures such as conditionals and loops; *Transformation Development Environment*, with an embedded methodology to drive development; and *Development Process* specified in a Process Modeling Language (PML) with the relevant elements of a process and their relationships.

Concerning the development life cycle, our classification identifies five phases for transformation development. The first one is *Requirement Specification*, which may involve tasks such as identifying the transformation requirements, selecting metamodels, among others. The other one is *Design phase*. Considering that a strategy may adopt MDD approach in transformation development, design is split into *Design Independent of Platform* and *Design Specific of Platform*. The *Implementation* phase concerns about transformation codification. Finally, there is the *Validation/Verification* phase related to tasks that evaluate if the transformation works as specified. Different strategies may comprise one or more phases, depending on their focus. For example, a systematic approach may comprise the Requirements Specification and Verification phases, if its focus is on properties verification. In contrast, another systematic approach may comprise Design and Implementation, if it intends to produce code.

Different methods have also been experimented to assist model transformation development. We consider a method, any technique to support specific tasks in development. Our classification organizes the most used methods in six categories: *Example Based*, when examples of models (also called concrete models) are used to guide development. These examples are potential candidates of input/output models of a transformation used to assist in identifying links between their source and target metamodel; *Weaving Model*, which uses a model to represent correspondences among elements of input and output models. The transformation is defined based on these correspondences; *Pattern* method, which uses transformation patterns, i.e. current solutions for common problems related to transformation context that can be reused in developing new transformations; *Template* method, which provides predefined formats to be used in specification. These templates are usually processed by an engine to generate the transformation; *Artificial Intelligence Algorithms*, methods from artificial intelligence domain to support metamodel analysis and transformation development, e.g. *Genetic Algorithm* and *Annealing Algorithm*; and *Formal* methods, such as *Formal Concept Analysis (FCA)*, a mathematical formalism for data analysis to discover conceptual abstraction in datasets, ordering them hierarchically in a net structure [8], and *Theorem Prover*, a software tool to assist with the development of formal proofs.

The transformation specification in different abstraction levels may require the adoption of specific notations. Different notations were identified in the literature, which we classify in five categories. Specification can be done in *Natural Language*, e.g., when focus on documentation; *Proprietary Language*, usually defined according to domain specificities; *UML* extensions, which can be a *Light*

Extension, also called UML Profile, or *Heavy Extension*; *Formal Language*, to formally specify the transformation enabling verification; and *Data-Centric* language, a language that works with management and manipulation of data.

Model transformation development can be complex, and automation is important to decrease the complexity of development. For example, engines to (semi) automatically generate code in a specific transformation language might reduce the need for expertise in transformation languages. Therefore, the proposed classification groups the development level of automation in two levels, the automation during specification, e.g., mapping design independent of platform to design in a specific platform, and the automation for code generation in specific programming languages.

5 Systematic Review Protocol

This section briefly describes the protocol definition used to perform our systematic literature review, following the guidelines of [4]. More details about this protocol are presented in [18].

The review aims to analyze proposals of model transformation development to identify strategies used in its development concerning phases of software development life cycle considered in each strategy, modeling languages adopted in the specification, and the automated resources to support the development. To achieve this goal, the following hypothesis and research questions (RQ) were formulated:

Hypothesis 1: Different strategies have been proposed in the literature to support model transformation development. However, there is still no well-defined strategy that can be widely used by the community. This hypothesis motivated us to construct the following research questions: RQ01: What are the current strategies used to guide the development of model transformations? RQ02: Which phases of the software development life cycle have been considered in model transformations development? RQ03: How automated is the proposed strategy? RQ04: Which validated methods have been used to evaluate the current proposals? RQ05: How many examples of model transformations are tested in each validation?

Hypothesis 2: In model transformation specification, there is no consensus on a notation to be adopted as a standard. This hypothesis led us to construct the following research question. *RQ06:* Which languages/notations have been used for model transformations specification?

The review considers four relevant research databases of Software Engineering, ACM Digital Library, IEEE Library, Science Direct, and Springer. The following search string is adopted and applied to the title, abstract, and keywords.

("MDE" or "MDD" or "MDA" or "Model Driven Development") and ("model transformation") and ("specification" or "development" or "approach" or "strategy" or "framework" or "systematic" or "process" or "methodology" or "method" or "life cycle")

Besides, this review analyzes the references of the selected papers and uses the snowballing method to find other relevant works.

The review includes works written in English published between 2003 and July 2020. The studies selected from the automatic search are refined manually according to the following inclusion/exclusion criteria: *Inclusion Criteria 1*: The study identifies and organizes the activities to develop model transformations; *Exclusion Criteria 1:* The study presents a strategy to develop model-to-text transformations. *Exclusion Criteria 2:* The study focuses on non-relational transformations because we focus on relational-transformation; *Exclusion Criteria 3:* The study is about transformations for a specific domain (e.g., transformations for web domain).

To help in the analysis of the selected works, we defined certain features based on the previously defined research questions, such as: What is the strategy used to guide development? Does the study consider the requirements specification phase? Does the study consider the design phase independent of platform? Does the study provide instructions to map requirements specification into design? Among others. The complete list of features can be seen in [18].

All the selected papers must fulfill the following quality criteria: have a proper introduction with a clear idea of the research objectives, present the results and contributions, and present a validated work.

6 Conducting and Reporting the Review

We performed this systematic review following the protocol presented in Sect. 5. In Fig. 2, we show the records returned from our search according to the stages of the study selection process.

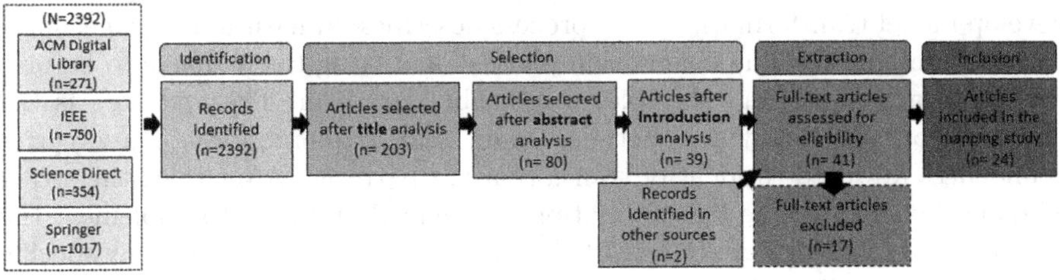

Fig. 2. Stages of the study selection process.

A total of 2392 papers were screened by analyzing the title, abstract, and introduction. Two works were also added, shown in box *Records identified in other sources*, chosen from the list of references of the previously selected papers. As a result, 41 papers were selected for full-text analysis. Eight papers were excluded as they were about previous proposals we had already selected. The others were excluded after applying the exclusion criteria. In the end, 24 works were considered relevant in the context of strategies to guide model transformation development.

6.1 Selected Works

Table 1 lists the selected papers in chronological order with an id (SP), publication year, title, and the search base. These works were analyzed based on the features defined in [18], briefly described in Sect. 5, in order to confirm/refuse the hypothesis previously formulated.

The first initiatives concerning the systematization of activities involved in model transformation development come from the work presented in [3] (SP1 in Table 1), a systematic approach covering phases such as requirements specification, design, implementation, and test. In this approach, development starts with the identification of requirements and metamodels involved in the transformation. Then, the design is performed at two levels of abstraction, high level, to document the rules that compose the transformation, and low level, where the rules are detailed in terms of mapping between the metamodel elements. The design phase finishes with a syntactic verification of the transformation using inspection technique, and semantic verification using test cases. Finally, the transformation is manually coded in the implementation phase.

In 2007, the authors of [30] (SP2 in Table 1) proposed the use of concrete examples of models to develop a model transformation instead of the direct manipulation of metamodels. In this proposal, the developer defines mappings between examples of input and out models, representing correspondences (links) among its elements. Together, these correspondences and the mappings between the abstract and concrete syntax of the modeling language are used to generate model transformation rules.

The idea of a process to guide development is introduced in [31] (SP3 in Table 1), where the authors propose an iterative and incremental process to develop model transformation. The process describes, in natural language, specification, design, and implementation phases identifying the artifacts to be produced in each phase and the roles responsible for them. The first step is the definition of correspondences between input and output models. These correspondences are used to identify transformation patterns. After that, the structure and behavior of the transformation are defined and used to implement the code in a specific language. The work does not recommend the use of any specific notation in modeling phases nor provides any automation.

Model Transformation by Demonstration (MTBD) is an approach proposed in [32] (SP4 in Table 1) to develop model transformations, The examples demonstrated by the developer are recorded by a tool and used to infer a model transformation automatically.

Table 1. Selected studies (updated from [18]).

Id	Year	Title	Search base
SP1	2005	A systematic approach to design model transformation [3]	IBM
SP2	2007	Towards Model Transformation Generation By-Example [30]	IEEE
SP3	2008	Transformation have to be developed ReST assured [31]	Springer
SP4	2009	Model Transformation by Demonstration [32]	Springer
SP5	2009	Towards the efficient development of model transformations using model weaving and matching transformation [13]	Springer
SP6	2010	Method of constructing model transformation rule based on reusable pattern [33]	IEEE
SP7	2011	Model transformation specification for automated formal verification [15]	IEEE
SP8	2012	A model based development approach for model transformation [34]	ACM
SP9	2013	Engineering model transformation with transML [23]	Springer
SP10	2013	Applying MDE to the (semi-) automatic development of model transformation [12]	Science direct
SP11	2013	The general algorithm for the MDA transformation models [14]	IEEE
SP12	2015	Specifying model transformation by direct manipulation using concrete visual notation and interactive recommendation [35]	Science direct
SP13	2016	Requirements engineering in model transformation development: a technique suitability framework for Model Transformation Applications [36]	ACM
SP14	2016	Multi-Step learning and adaptive search for learning complex model transformations from examples[37]	ACM
SP15	2016	Design pattern oriented development of model transformation [38]	Science direct
SP16	2016	Model-based M2M transformations based on drag-and-drop actions: Approach and implementation [39]	Science direct
SP17	2016	A Model-Driven approach for model transformation [40]	IEEE
SP18	2016	Designing and describing QVTo model transformation [43]	Scopus
SP19	2017	Formal concept analysis for specification of model transformation [44]	IEEE
SP20	2018	EVL-Strace: a novel bidirectional model transformation approach [50]	Science direct
SP21	2018	A generic approach to model generation operation [46]	Science direct
SP22	2019	Model driven transformation development (MDTD): An approach for developing model to model transformation [47]	Science direct
SP23	2019	Applying a Data-centric framework for Developing Model Transformations [48]	ACM
SP24	2020	CoqTL: a Coq DSL for rule-basedmodel transformation [51]	Springer

The proposal presented in [13] (SP5 in Table 1) uses the concept of a weaving model to develop model transformation. The development starts identifying the source and target metamodels, which are automatically analyzed through so-called matching transformations, i.e., transformations that identify mappings between metamodels using ontology or other methods. The result of this analysis is the weaving model. Developers analyze this weaving model, correct possible inconsistencies in these links, and use it to generate a model transformation model. The approach provides conversion to ATL or XSTL language.

Transformation patterns are also used in [33] (SP6 in Table 1), which propose a framework for model transformation development, based on patterns, that uses an algorithm to guide development. It defines seven transformation patterns and an algorithm to guide the extraction of rules based on these patterns.

Trans-Dv (Transformation Specification Development and Verification Framework) [15] (SP7 in Table 1) is a systematic approach to specify model transformations focusing on the application of formal methods. The specification starts with the definition of the Model Transformation Requirement Model (MTRM), which specifies transformation requirements. Then, in the design phase, these requirements are refined to define the structural and behavioral aspects of the transformation, represented as a Model Transformation Specification Model (MTSM). The notation used in these models is TransML [11], a proprietary language composed by diagrams. The approach contains a set of templates, which represent patterns of model transformation and verification properties. Templates are instantiated in a Model Transformation Formal Specification Model (MTFM), and Trans-Dv uses it to generate the specification in Alloy notation.

With the same focus on formal specification, the authors of [34] (SP8 in Table 1) propose the use of an MDD approach, called UML-RSDS (Reactive System Development Support) for model transformation development. UML-RSDS comprises a UML-based notation for model specification in different abstraction levels and support for verification. Notation includes diagrams, such as state machines, activity and sequence diagrams, and a subset of OCL language. The process, informally described, starts with requirements specification phase where functional and non-functional requirements, source and target metamodel, and constraints are specified. Then, there is the abstract specification phase to map relationships between metamodels. It is modeling as an abstract state machine through the definition of predicates, pre and post conditions. In the third phase, transformation rules are defined as specifying operations and their execution sequence in a state machine. Finally, in implementation phase, the code is generated in languages such as Java and ATL.

Concerning notation for transformation specification, the authors of TransML [23] (SP9 in Table 1) propose a family of modeling language to support model transformation specification through requirements, analysis, design, implementation, and test phases. It comprises metamodels and diagrams, e.g., requirements diagram to specify functional and non-functional requirements, mapping diagram to specify mappings among metamodels, and others. TransML is not a methodology, it provides a notation for model transformation specification in any methodology. It does not include an implementation language but supports conversion to ETL and QVT-R languages.

MeTAGen [12] (SP10 in Table 1) is a methodological approach that uses MDD itself in model transformation development. In this work, model transformation models are specified using a textual notation at different abstraction levels: Platform Independent Transformation Model (PIT), to map elements of source and target metamodels; Platform Specific Transformation Model (PST), which represents the transformation in a particular approach of development, e.g., declarative, imperative or hybrid; and Platform Dependent Model (PDM), which represents a transformation in a specific platform. At the end of the process, code is generated in ATL or RubTL language. The proposed methodology is encapsulated in a tool.

A generic algorithm to design MDA model transformations is presented in [14] (SP11 in Table 1). It organizes the steps towards the specification of mappings among different metamodels involved in a transformation. An algorithm sequentially organizes the steps to produce the design of a transformation. The algorithm is composed of eight steps and produces a metamodel as output with stereotypes and target values of the transformation, and a model transformation model, with the mapping rules.

Concrete examples of models are used in CONVErT tool [35] (SP12 in Table 1) to develop model transformation. The development process is encapsulated in the tool and is based on examples of input and output models. These examples of models are specified using a concrete notation provided by the tool. The notation enables the visual specification of mappings between models, according to a set of correspondences automatically identified by the tool. Correspondences are suggestions of mappings between metamodel elements based on heuristics, e.g. similarities in names, values and structure of the metamodels elements. In the end, the tool generates a script code of the transformation in XSTL language.

The use of requirements engineering (RE) techniques in model transformation development is investigated in [36] (SP13 in Table 1). Based on the 4-phase RE process proposed by [5] the authors define a RE framework for the model transformation domain. In this direction, techniques of RE are adapted, aiming to better capture the requirements of a transformation. For example, considering the structured interview technique, they suggest a catalog concerning relevant aspects of MT development, such as refactoring and refinement, to aid interviews.

Model Transformation by Examples (MTBE) is used by [37] (SP14 in Table 1). This work discusses fundamental issues related to the efficiency of MTBE for complex transformation and proposes a process to alleviate this problem. The process starts analyzing and refining mappings between input and output models to generate pools of model examples. Then, these pools are processed by a genetic program [7] to derive a set of transformation rules that best transform these examples. The rules are merged into a transformation program that is refined using a simulated annealing algorithm [6].

Transformation patterns is also considered in [49] (SP15 in Table 1). The authors first propose a notation for the specification of transformation patterns independent of a specific transformation language. Then they define a catalog with 15 patterns and propose a method to develop model transformations using these patterns. Finally, the authors demonstrate, through an example, how to automatically generate code in different languages using patterns.

The work presented in [39] (SP16 in Table 1) consists in a tool to customize and develop model transformations. It is based on drag-and-drop actions, i.e., the developer selects and drags elements of an input model and drops them in elements of the output model. These actions invoke an engine, which triggers a transformation rule appropriated for that action. A relationship between the concepts involved is created in order to develop the new transformation. Like other works, the approach makes use of transformation patterns to assist model transformation specification.

The use of MDD in model transformation development is also explored in [40] (SP17 in Table 1). Model transformations elements are analyzed, concerning modeling language, development process, and support facilities. The paper first describes some characteristics of a model transformation, such as understandability, composition, and reusability. Then, the diversity of technologies concerning programming languages, notation style (e.g., graphic and textual), specification style (e.g., imperative and declarative) and directionality (e.g., unidirectional and bidirectional) are discussed. After that, it presents an approach to develop model transformation using MDD, which covers Platform-independent transformation (PIT), Platform-specific transformation (PST), and code generation, following MDA principals. The authors suggest using a view model, named 4+1 [41], for the description of the transformation architecture and proposes a modeling language named VisTML [42]. There is also a modeling tool and a transformation engine to generate script in specific languages, such as QVT and ATL.

Mathematical notation is used in [43] (SP18 in Table 1) to design model transformations. The authors argue that such mathematical concepts are familiar to most engineers. The proposal guides the reader on how to align the notation to QVTo concepts in order to document a transformation, e.g., how to specify metamodels and mappings between metamodel elements. It presents a set of steps to guide the design of the transformation. It was validated in a proof of concept, where two examples of model transformation are developed through interviews with QVTo practitioners.

The work proposed in [44] (SP19 in Table 1) uses the Formal Concept Analysis (FCA) technique to generate transformation rules semi-automatically from a set of models. The approach uses pairs of source-target models and techniques of FCA to classify model elements according to their attributes to facilitate the interpretation of mappings between them and optimize the generation of rules. Two algorithms are proposed to automate the process, and the work was validated in a single example.

EVL+Strace [45] (SP20 in Table 1) is an approach to develop bidirectional transformations that use Epsilon Validation Language (EVL) [9], to express constraints, and a domain-specific trace metamodel, called Strace, which defines links between source and target elements of the involved metamodels. The approach is supported by an Eclipse-based toolkit called MoDEBiTE, where developers can define a model connecting source and target meta-elements based on a general weaving metamodel. The toolkit has a validator that checks the model and generates both a trace metamodel and a bidirectional transformation. A process with nine steps on how to develop using MoDEBiTE is also illustrated in a workflow. Additionally, the proposal was validated in six case studies.

The work presented in [46] (SP21 in Table 1) proposes an approach and a framework to specify and execute model operations. This is a kind of operation that has the ability to semi-automatically generate or complete a model, in MDD context defined as a model transformation. The authors argue that model operations may have different uses, such as model verification and system definition. As a result, a combination of techniques might be adopted in its development, such as the use of constraint programming solver. However, different problems might require different solvers. The authors propose a generic approach to specify and execute operations that can be represented as a model independent of the solver. The approach is organized into layers. A layer, named Model Search, provides the extraction/injection of the input and output models into/from the solver format. On top of this layer, there is the transformation layer, which converts the specification into a model search layer, independent of any solver. The approach defines the concepts involved in both the model search layer and in the transformation layer and describes in steps how to use it.

Model Driven Transformation Development (MDTD) process, [47] (SP22 in Table 1), is an iterative and incremental development process for model transformation construction based on MDD approach. MDTD is specified in SPEM [55] and covers different phases of the development life cycle, from requirements specification until code. The process is supported by a UML profile, to enable the definition of model transformation models, and by an Eclipse-based tool. Therefore, model transformation models are specified at a high abstraction level and semi-automatically converted in models at a lower abstraction level until they generate code in ATL or QVT-R language.

The authors of [48] (SP23 in Table 1) presents a study on a data-centric language, named Bloom, to develop model transformations. The first step of the process is to extract the input model in the Bloom data format. To support it, the work provides a data translator to Bloom format. After that, they use this language to implement and validate the transformation rules. The authors argue that Bloom has a simple semantic and syntax that motivates its use.

Finally, the authors of [51] (SP24 in Table 1) propose a domain-specific language, named CoqTL, for the specification of declarative model transformation using Coq, an interactive theorem prover, and its Gallina specification language [10]. Their main goal is to use the same language for both model transformation specifications and properties to prove, avoiding errors in model transformation translation. To support it, they provide a transformation engine to interpret programs written in CoqTL.

6.2 Results Summarization

This subsection presents the results of our review according to the classification proposed in Sect. 4.

Initially, model transformations used to be developed through a conventional approach. MDD was introduced in 2009, and its adoption has been intensified since 2011. Nowadays, both approaches are used relatively in the same proportion (Fig. 4).

Concerning strategies adopted (respected to RQ01), the first initiatives found in the literature organize the tasks to develop model transformation in systematic approaches (15 works, 63%) and algorithms (2 works). More recently, environments have been proposed (6 works, 25%) embedding the activities in specific tools. There is also one work that proposes a development process. A strategy may assist one or more phases of the transformation development life cycle. The proposed classification (Sect. 4) identifies five different phases, which we use to analyze each strategy and to answer RQ02. Figure 4 shows the relationship among strategies and phases. Most of the strategies emphasize *Design* and *Implementation* phases. For example, analyzing the phase *Design (independent of platform)*, we observe that among the 24 works selected, 23 of them (represented by the circles labeled with numbers 14, 1, 6, and 2), consider this phase. Similarly, *Implementation* phase is considered in 19 of the 24 works (represented by the circles labeled with numbers 11, 1, 6 and 1). An important observation is that despite the relevance of requirements specification in software quality, the phase *Requirements* is not considered in any tools or algorithms, probably because the high level of abstraction makes it difficult its automation (Fig. 3).

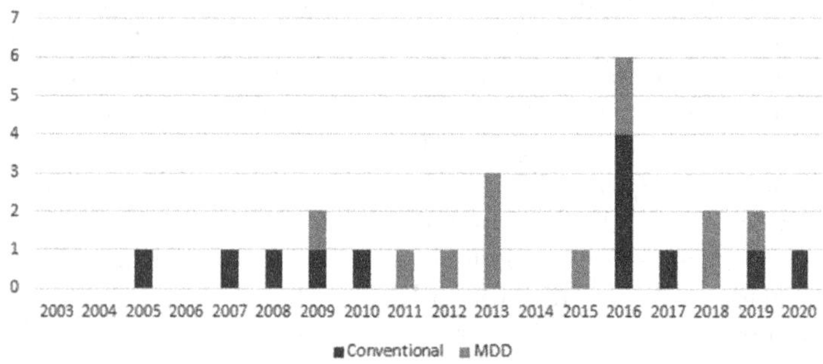

Fig. 3. Adoption of Conventional and MDD approach in model transformation development along the years.

We found different proposals among the selected papers related to methods adopted to support specific tasks in transformation development. For example,

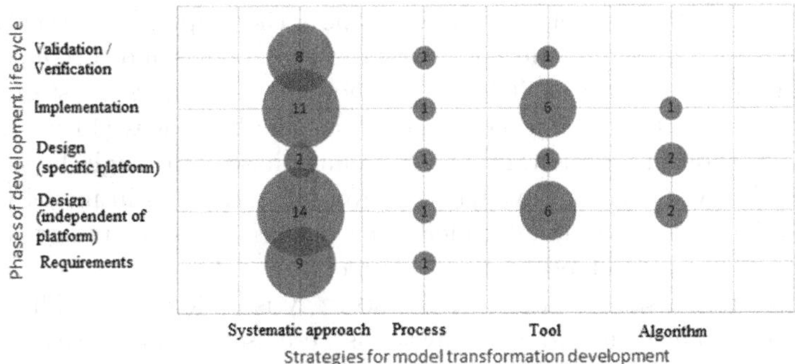

Fig. 4. Development phases covered by the strategies.

six works use the example based method to assist requirement and design specifications, and two adopt weaving models to define correspondences between source and target metamodels. A detailed view of the methods adopted in each selected paper can be seen in Fig. 7.

To specify transformations in different abstraction levels, languages and notations have been proposed (investigated in RQ6). As shown in Fig. 5, among the analyzed strategies, two of them (8%) adopt UML or a light extension of this (called UML profiles) and the other two works (8%) use a heavy extension of UML. Formal languages are still used in seven works (29%). Most works (nine, which represent 37%) propose new languages specific for the transformation domain (shown as a Proprietary language in the graph). There is also one work that uses natural language. The other works do not mention which languages are adopted. In summary, we observe many different languages and notations used.

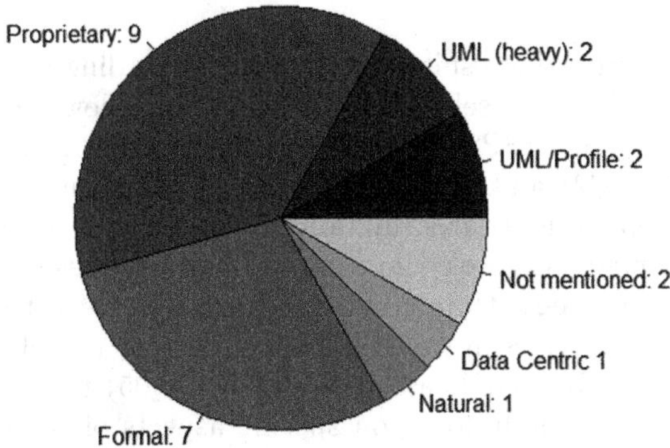

Fig. 5. Notations adopted in model transformation specification.

Concerning the level of automation provided by each strategy (investigated in RQ3), according to the organization proposed in (Sect. 4), two levels were considered: *Specification* for those that support the mapping between requirements models and design models or between design models and specific platform models; and *Code generation*. The results show that nowadays, most of the proposals offer support in code generation for specific languages (14 works), only five works provide automation through modeling phases. Although the significant difference observed, we can say that the level of abstraction in model transformation development is gradually increasing because some years ago, transformations used to be developed manually directly in code.

An important issue when adopting a strategy is its feasibility. Therefore, in this RSL we also investigated which methods were used to validate each strategy (related to RQ04). As showed in Fig. 6, examples are used to validate 38% of the works (9 works) while (25%, six works) adopts case studies and (38%, nine works) controlled experiments. Despite these methods, important information that took our attention was the few transformations developed in each validation, which seemed to indicate a low level of confidence. According to our review, only two works use more than 10 model transformations in their validation (RQ05).

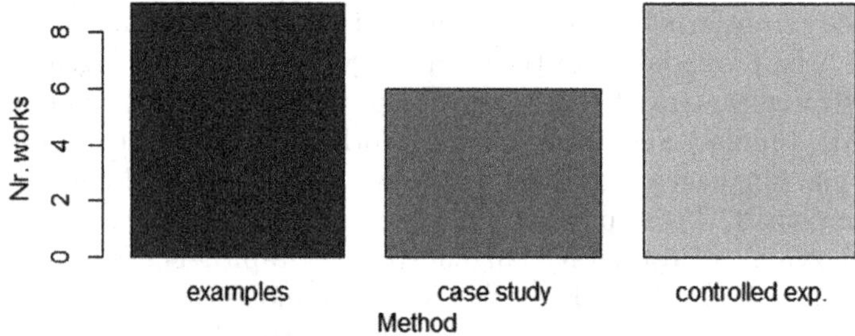

Fig. 6. Validation methods of the selected works (updated from [18]).

Figure 7 summarizes the results of our review according to the classification presented in Sect. 4. For each selected paper, the figure shows how it is classified. For example, papers SP1, SP2, SP3, SP4, SP7, SP8, SP9, SP13, SP14, SP15, SP17, SP18, SP19, SP21 and SP23 (first column of the figure) use a systematic approach to develop model transformation (fourth column of the figure).

From the presented data, we could confirm the **hypothesis 1 - there is still no well-defined strategy that can be widely used by the community**. We found different strategies in the literature to develop model transformations, but they are not widely tested (as evidenced in RQ05) to be used on a large scale. Moreover, they usually focus on specific aspects of the development (as observed in RQ02), like Design and Implementation, not covering the entire development life cycle. Finally, the **hypothesis 2 - In model transformation specification, there is no consensus on a notation to be adopted as a standard** was confirmed as most of the works adopted a proprietary language.

	Approach		Strategy				Phases of Life Cycle					Method							Notation						Autom.	
Id	Conventional	MDD	Systematic aproach	Algorithm	Transformation Development Environment	Development Process	Requirement Specification	Design Independent of Platform	Design Specific of Platform	Implementation	Validation / Verification	Example Based	Model weaving	Pattern	Template	Artificial Intelligence Algorithms	Formal Concept Analisys (FCA)	Theorem Prover	Natural Language	Proprietary Language	UML Light Extension	UML Heavy Extension	Formal Language	Data Centric Language	Specification	Code generation
SP1	X		X				X	X		X	X															
SP2	X		X					X		X	X	X								X						
SP3	X		X				X	X		X		X		X												
SP4	X		X					X			X	X								X						
SP5		X			X			X		X			X							X						X
SP6	X			X				X	X					X						X						
SP7		X	X				X	X			X				X							X				
SP8		X	X				X	X		X	X												X		X	X
SP9		X	X				X	X		X	X											X				X
SP10		X			X			X	X	X										X					X	X
SP11		X		X				X	X	X													X			
SP12		X			X			X		X		X								X						X
SP13	X		X				X												X							
SP14	X		X				X	X		X		X				X							X		X	
SP15	X		X				X	X		X				X									X			X
SP16		X			X			X		X		X									X					X
SP17		X	X				X	X	X	X	X									X					X	X
SP18	X		X					X												X						
SP19	X		X					X		X							X						X			X
SP20		X			X			X		X			X							X						X
SP21		X	X					X	X	X	X												X			X
SP22		X				X	X	X	X	X	X										X				X	X
SP23	X		X					X		X														X		X
SP24	X				X			X		X	X							X					X			X

Fig. 7. Selected works organized according to the proposed classification.

7 Limitations

This section discusses some threats to the validity of our review, namely construction, internal and external validity, and reliability.

Considering construction validity, to decrease bias of the reviewers, we established a protocol for how the reviews should be conducted. This defines each parameter of the review, such as search string, based on the main goal. As there are essential differences in transformation development concerning transformation viewpoints, the protocol restricts the review to relational model-to-model transformation. Therefore, there might be different development strategies for other kinds of transformations, e.g., graph transformation or model-to-text transformation, which were not considered.

Related to internal validity, to reduce the chance of relevant papers being excluded, extraction stages were analyzed by at least two researchers. To prevent distortions in data collection, we used a conceptual framework that specifies relevant elements related to the model transformation development domain. Based on this framework, we defined features to be observed in each study analyzed in our review.

Concerning external validity, we conducted a systematic literature review considering studies in four major research databases however, we may have missed some relevant studies. Therefore, there is a threat to validity related to the generalization of the conclusions of the study. However, this threat is minimized if we consider that the research databases used contain the main publications in MDD.

Finally, in order to enable future replications of our review, we used the Start tool. So all the definitions, i.e., goal, research questions, and the protocol information, as well as the metadata of the analyzed works in each stage, e.g., identification, selection, and extraction, are stored and can be used by other researchers.

8 Discussions

In this section we make a discussion about transformation development state of art based on our systematic review, considering some software engineering foundations. We group the argument in subsections according to similar subjects.

8.1 Development Approach, Strategy and Life Cycle

It is well known that the quality of a product is influenced by the process used to produce it [5]. Through our review we observed an increasing number of approaches and strategies used to support model transformation development. They differ from each other in the level of specification granularity and in the coverage of phases concerning to development life cycle.

According to [5], a software development process must define what should be done, how to perform it, when and by whom. Most proposals focus on the definition of what should be performed, i.e. identify and organize the necessary tasks, but they do not detail how to perform them. For example, in [3] the requirements and analysis phase comprises the definition of source and target metamodels involved in the transformation, but it does not explain how to define these metamodels, nor does it clarify what the necessary steps are to help in performing this task. We perceived that this scenario is directly related to the level of formalism used to specify the proposed method. The first initiatives, e.g. [3,30–32], provide a description of the method in natural language, what we call in our review a systematic approach. Over the years the proposals have become more formal e.g. algorithms have been used to structure the method, as in [33] and in [14], giving better support. Recently, some tools have been proposed as in [12] and in [35] as well as a development process [16] specified in PMLs such as SPEM, providing resources for automation and enactment that facilitate the adoption by others. In this direction, improving the strategies towards a well-defined process is important to enable its replication.

Concerning the development life cycle, we perceive that works usually focus on specific phases but do not cover an entire life cycle. For example, [13] focus on design and implementation, but do not support the requirements specification phase, and [15] focus on formal specification and verification. This scenario leads to the need to adopt more than one method to cover all the development. Consequently, problems such as selection of tools and document interchange, may occur. Leaving the responsibility to solve this to the Software Engineer may lead them to adopt ad-hoc methods. Therefore, strategies should be specified to cover the entire development cycle of model transformations.

8.2 Notation for Specification

The specification of transformations, as in other software, requires the adoption of modeling languages to produce the necessary documentation. This is also essential for model refinement and code generation. We found no consensus on the adoption of a modeling language suitable for model transformation development. As a result, different proprietary modeling languages (and also UML profiles and heavy extensions of UML) have been used which may hamper communication and interchange of documents between tools. QVT (Query View Transformation), proposed by OMG as a standard is in fact not widely adopted. As a result, the definition of a modeling language which can be used for both industry and academia is still required.

8.3 Methods to Assist Specific Development Tasks

In order to assist specific development tasks, methods originated from other areas of knowledge have been adapted for transformation development. In the proposed classification we could identify six categories of methods that is currently applied and observed that new ones have been experimented each day. There are methods related to manipulation of big amount of data and process models, methods related to artificial intelligence to infer relationships among metamodels elements, patterns to reuse solutions of common problems, among others. Each one of these has its strengths and weaknesses, and we can say this is still an open research area.

8.4 Validation

Model transformation also has special issues related to verification and validation. For example, model transformations must guarantee some properties (e.g. syntax and semantic correctness and completeness [25]), which require the use of special techniques in order to assure quality. Model transformation specification using formal languages have been considered in some studies, however, they usually require special expertise in these languages. Strategies to reduce the need for such expertise, as for example the work in [15] which proposes templates to generate formal specifications, are required.

In addition, more effort to validate the current proposals is required. We observed that despite the wide range of initiatives in experimenting with different approaches to model transformation development, there have been few results that assure the feasibility of them.

8.5 Automation

Finally, the development of model transformation involves different elements, e.g. development processes, modeling languages, modeling environments, formal languages and automation, and the current strategies usually do not cover all of these aspects. Thus, developers have a hard job choosing and integrating them. Therefore, integrated approaches are also required in order to reduce the complexity of the development.

9 Conclusions

Over the last decade, many proposals have emerged to reduce transformation development complexity as well as improve transformation quality. Motivated by the lack of consensus about which techniques and methods are more appropriate to use, we performed a systematic literature review in an attempt to investigate aspects related to model transformation development. This systematic review allowed the identification of approaches and strategies most frequently used in transformation development, as it raises relevant issues to support transformation development such as development phases, methods, modeling languages, and the level of automation provided.

The review results showed that proposals for model transformation development usually consider different concerns. In order to better organize these proposals, we made a classification based on software engineering common development features, grouping the works according to specific development aspects. This classification can assist developers in model transformations to analyze which aspects are relevant to their needs. Moreover, it might be useful for researchers to identify gaps and opportunities for new researches.

For future work, we intend to improve the classification towards a taxonomy considering other aspects of model transformation and giving more details about the current aspects considered in our classification, as for example, to investigate methods for validation and verification, and add other phases of the software life cycle, e.g., evolution and maintenance.

References

1. Stahl, T., Volter, M.: Model-Driven Software Development. Technology, Engineering, Management. Wiley, Hoboken (2010)
2. Brambilla, M., Cabot, J., Wimmer, M.: Model-Driven Software Engineering in Practice, 1st edn. Morgan & Claypool Publishers, San Rafael (2012)
3. Kuster, J., Rynduna, K., Hauser, R.: A systematic approach to designing model transformations. Computer Science (2005)
4. Kitchenham, B.: Procedures for performing systematic reviews. Keele University Technical report TR/SE-0401/NICTA Technical Report 0400011T.1 (2004)
5. Sommerville, I.: Software Engineering, 2nd edn. Addison-Wesley, Boston (1999)
6. Aarts, E., Korst, J.: Simulated Annealing and Boltzmann Machines: A Stochastic Approach to Combinatorial Optimization and Neural Computing. Wiley, New York (1998)
7. Mitchel, M.: An Introduction to Genetic Algorithms (Complex Adaptive Systems). MIT Press, Cambridge (1998)
8. Valtchev, P., Missaoui, R., Godin, R.: Formal concept analysis for knowledge discovery and data mining: the new challenges. In: Eklund, P. (ed.) ICFCA 2004. LNCS (LNAI), vol. 2961, pp. 352–371. Springer, Heidelberg (2004). https://doi.org/10.1007/978-3-540-24651-0_30
9. Kolovos, D., Paige, R., Polack, F.: TransML: on the evolution of OCL for capturing structural constraints in modelling languages, rigorous methods for software construction and analysis. LNCS, vol. 5115, pp. 204–216 (2009)
10. Avigad, J., Mahboubi, A.: Interactive theorem proving, rigorous methods for software construction and analysis. In: 9th International Conference (2018)
11. Guerra, E., de Lara, J., Kolovos, D.S., Paige, R.F., dos Santos, O.M.: *trans*ML: a family of languages to model model transformations. In: Petriu, D.C., Rouquette, N., Haugen, Ø. (eds.) MODELS 2010. LNCS, vol. 6394, pp. 106–120. Springer, Heidelberg (2010). https://doi.org/10.1007/978-3-642-16145-2_8
12. Bollati, V., Vara, J., Jiménez, A., Marcos, E.: Applying MDE to the (semi-)automatic development of model transformations. Inf. Softw. Technol. **55**, 699–718 (2013)
13. Del Fabro, M., Valduriez, P.: Towards the efficient development of model transformations using model weaving and matching transformations. Softw. Syst. Model **8**, 305–324 (2009). https://doi.org/10.1007/s10270-008-0094-z
14. Tavac, M., Tavac, V.: The general algorithm for the design of the MDA transformation models. In: Fifth International Conference on Computational Intelligence, Communication Systems and Networks, pp. 171–176 (2013)
15. Sani, A., Polack, F., Paige, R.: Model transformation specification for automated formal verification. In: 5th Malaysian Conference in Software Engineering, pp. 76–81 (2011)

16. Magalhaes, A.P., Andrade, A., Maciel, R.S.P.: A model driven transformation development process for model to model transformation. In: 30th Brazilian Symposium of Software Engineering, pp. 3–12 (2016)
17. Magalhaes, A.P., Andrade, A., Maciel, R.S.P.: Model driven transformation development (MDTD): an approach for developing model to model transformation. Inf. Softw. Technol. **114**, 55–76 (2019)
18. Magalhaes, A.P., Maciel, R.S.P., Andrade, A.: Developing model transformations: a systematic literature review. In: Proceedings of the 22nd International Conference on Enterprise Information Systems (ICEIS), pp. 80–89 (2020)
19. Ma, K., Chen, Z., Yang, B., Abraham, A.: A relational approach to model transformation with QVT relations supporting model synchronization. J. Univ. Comput. Sci. **17**(13), 1863–1883 (2011)
20. Mens, T., Czarnecki, K., Gorp, P.: A taxonomy of model transformation. In: Proceedings of the International Workshop on Graph and Model Transformation (GraMoT 2005) Graph and Model Transformation, pp. 125–142 (2006)
21. Bézivin, J., Büttner, F., Gogolla, M., Jouault, F., Kurtev, I., Lindow, A.: Model transformations? Transformation models!. In: Nierstrasz, O., Whittle, J., Harel, D., Reggio, G. (eds.) MODELS 2006. LNCS, vol. 4199, pp. 440–453. Springer, Heidelberg (2006). https://doi.org/10.1007/11880240_31
22. Tisi, M., Jouault, F., Fraternali, P., Ceri, S., Bézivin, J.: On the use of higher-order model transformations. In: Paige, R.F., Hartman, A., Rensink, A. (eds.) ECMDA-FA 2009. LNCS, vol. 5562, pp. 18–33. Springer, Heidelberg (2009). https://doi.org/10.1007/978-3-642-02674-4_3
23. Guerra, E., de Lara, J., Kolovos, D.S., Paige, R.F., dos Santos, O.M.: Engineering model transformations with transML. Softw. Syst. Model. **12**(3), 555–577 (2013). https://doi.org/10.1007/s10270-011-0211-2
24. Vignaga, A.: A methodological approach to developing model transformations. Model Driven Engineering Languages and Systems, pp. 1–5 (2007)
25. Braga, C., Santos, C., Da Silva, V.T.: Consistency of model transformation contracts. Sci. Comput. Program. (Print) **92**, 86–104 (2014)
26. Silva, A.R.: Model-driven engineering: a survey supported by the unified conceptual model. Comput. Lang. Syst. Struct. **43**, 139–155 (2015)
27. Berranla, K., Deba, E.A., Benhamamouch, D.: Model transformation generation. A survey of the state-of-the-art. In: International Conference on Information Technology for Organizations Development (IT4OD), pp. 1–6 (2016)
28. Bollati, V.: MeTAGeM: Entorno de Desarrolo de Transformaciones de Modelos Dirigido por Modelos. Ph.D. thesis, Universidad Rey Ruan Carlos (2011)
29. Silva, G.C., Rose, L.M., Calinescu, R.: A qualitative study of model transformation development approaches: supporting novice developers. In: Proceedings of Model-Driven Development Processes and Practices, pp. 1–10 (2014)
30. Wimmer, M., Strommer, M., Kargt, H., Kramler, G.: Towards model transformation generation by-example. In: Proceedings of the 40th Hawaii International Conference on System Sciences - HICSS, pp. 289–294 (2007)
31. Siikarla, M., Laitkorpi, M., Selonen, P., Systä, T.: Transformations have to be developed ReST assured. In: Vallecillo, A., Gray, J., Pierantonio, A. (eds.) ICMT 2008. LNCS, vol. 5063, pp. 1–15. Springer, Heidelberg (2008). https://doi.org/10.1007/978-3-540-69927-9_1
32. Sun, Yu., White, J., Gray, J.: Model transformation by demonstration. In: Schürr, A., Selic, B. (eds.) MODELS 2009. LNCS, vol. 5795, pp. 712–726. Springer, Heidelberg (2009). https://doi.org/10.1007/978-3-642-04425-0_58

33. Li, J., Yin, G.: Method of constructing model transformation rule based on reusable pattern. In: International Conference on Computer Application and System Modeling, no. 8, pp. 519–524 (2010)
34. Kolahdouz-Rahimi, S., Lano, K.: A model-based development approach for model transformations. In: Arbab, F., Sirjani, M. (eds.) FSEN 2011. LNCS, vol. 7141, pp. 48–63. Springer, Heidelberg (2012). https://doi.org/10.1007/978-3-642-29320-7_4
35. Avazpour, I., Grundy, J., Grunske, L.: Specifying model transformation by direct manipulation using concrete visual notation and interactive recommendations. J. Visual. Lang. Comput. **28**, 195–211 (2015)
36. Tehrani, S.Y., Zschaler, S., Lano, K.: Requirements engineering in model-transformation development: an interview-based study. In: Van Van Gorp, P., Engels, G. (eds.) ICMT 2016. LNCS, vol. 9765, pp. 123–137. Springer, Cham (2016). https://doi.org/10.1007/978-3-319-42064-6_9
37. Baki, I., Sahraoui, H.: Multi-step learning and adaptive search for learning complex model transformations from examples. ACM Trans. Softw. Eng. Methodol. **25**(3), 1–37 (2016)
38. Ergin, H., Sysiani, E., Gray, J.: Design pattern oriented development of model transformation. Comput. Lang. Syst. Struct. **46**, 106–139 (2016)
39. Skersys, T., Danenas, P., Rimantas, B.: Model-based M2M transformations based on drag-and-drop actions: approach and implementation. J. Syst. Softw. **122**, 327–341 (2016)
40. Ma, Z., He, X.: A model-driven approach for model transformations. In: SAI Computing Conference, pp. 1199–1205 (2016)
41. Kruchten, P.: Architectural blueprints: the "4+1" view model of software architecture. IEEE Softw. **12**(6), 42–50 (1995)
42. He, X., Ma, Z., Liu, Y., Chen, H., Shao, W.: VisTML: a visual modeling language for model transformation. In: 18th Asia-Pacific Software Engineering Conference, pp. 130–137 (2011)
43. Tikhonova, U., Willemse, T.: Documenting and designing QVTo model transformations through mathematics. In: Lorenz, P., Cardoso, J., Maciaszek, L.A., van Sinderen, M. (eds.) ICSOFT 2015. CCIS, vol. 586, pp. 349–364. Springer, Cham (2016). https://doi.org/10.1007/978-3-319-30142-6_19
44. Berranla, K., Deba, E.A., Benhamamouch, D., Touam, R., Brahimi, Y., Benhamamouch, D.: Formal concept analysis for specification of model transformations. In: First International Conference on Embedded & Distributed Systems (EDiS), pp. 1–6 (2017)
45. Semimi-Dehkordi, L., Zamani, B., Kolahdouz-Rahimi, S.: EVL+Strace: a novel bidirectional model transformation approach. Inf. Softw. Technol. **100**, 47–72 (2018)
46. Kleiner, M., Del Fabro, M.D.: A generic approach to model generation operations. J. Syst. Softw. **142**, 136–155 (2018)
47. Magalhaes, A.P., Andrade, A., Maciel, R.S.P.: Model driven transformation development (MDTD): an approach for developing model to model transformation. Inf. Softw. Technol. **114**, 55–76 (2019)
48. Camargo, L.C., Del Fabro, M.C.: Applying a data-centric framework for developing model transformations. In: ACM SAC 2019, Programming Languages Track, pp. 2–5 (2019)
49. Ergin, H., Syriani, E., Gray, J.: Design pattern oriented development of model transformations. Comput. Lang. Syst. Struct. **46**, 106–139 (2016)
50. Dehkordi, L., Zamani, B., Kolahdouz-Rahimi, S.: EVL-Strace: a novel bidirecional model transformation approach. Inf. Softw. Technol. **100**, 47–72 (2018)

51. Cheng, Z., Tisi, M., Douence, R.: CoqTL: a Coq DSL for rule-based model transformation. Softw. Syst. Model. **19**(2), 425–439 (2019). https://doi.org/10.1007/s10270-019-00765-6
52. Batot, E., Sahraoui, H., Syriani, E., Molins, P., Sboui, W.: Systematic mapping study of model transformations for concrete problems. In: International Conference on Model-Driven Engineering and Software Development (2016)
53. Czarnecki, K., Helsen, S.: Classification of model transformation approaches. In: Workshop on Generative Techniques in the Context of Model-Driven Architecture, pp. 1–17 (2002)
54. Model Driven Architecture. http://www.omg.org/mda/specs.htm. Accessed 23 Aug 2016
55. Software Process Engineering Metamodel Specification. http://www.omg.org/spec/SPEM/2.0/. Accessed 17 July 2017

Software Evolution and Maintenance Using an Agile and MDD Hybrid Processes

Elton Figueiredo da Silva[1(✉)], Ana Patrícia F. Magalhães[2,3], and Rita Suzana Pitangueira Maciel[1]

[1] PGCOMP Computação, UFBA Universidade Federal da Bahia, Salvador, Bahia, Brazil

[2] Exact and Earth Science Department, UNEB Universidade do Estado da Bahia, Salvador, Bahia, Brazil

[3] Post Graduated Program in Computing and Systems, Salvador University, Salvador, Brazil

Abstract. The growing software usage in modern society motivated several new development processes proposals aiming to increase productivity, reduce software delivery time, and improve the final product quality. In this context, while some software development processes emphasize source code production, such as the agile processes, others focus on modeling activities, such as the Model-Driven Development (MDD). Hybrid processes, which integrate different approaches into another one, aim to mitigate weaknesses and at the same time take advantage of the involved approaches strengths. This paper discusses the Agile and MDD approaches hybridization benefits concerning evolution and maintenance aspects through a controlled experiment. The influence of modeling tasks in development agility, as well as the consequences of model specification in product quality, are some of the issues discussed. The experiment results showed that the hybrid agile MDD process adopted conducted participants to produce a more correct and complete software release than the participants that developed the same release directly in code. So, we important indications that the hybrid process can assist developers during software evolution. We hope by sharing our results with the community, new replications could be performed towards news finds and generalizations.

Keywords: Hybrid process · MDD and Scrum integration · Software evolution · Software maintenance

1 Introduction

The growth in demand for new software worldwide is notorious. In order to attend this demand, software development processes have been proposed to serve this market quickly and improve software quality. Among these processes are agile and model-driven development (MDD) appraoches [4].

J. Filipe et al. (Eds.): ICEIS 2020, LNBIP 417, pp. 437–457, 2021.
https://doi.org/10.1007/978-3-030-75418-1_20

Agile approaches focus on the implementation phase, incrementally delivering software releases, reducing the time to market, in detriment of software documentation. Therefore, not assuring enough documentation to assist in the evolution and requirements tracking [4]. On the other hand, MDD approaches focus on the modeling phases, i.e., models are (semi) automatic transformed into other models until it generates application code. As a result, MDD provides enough documentation to assist in software evolution and requirements tracking tasks [5].

Software Process hybrid approaches have been proposed as a way to integrate different strategies strengths. Among them is the integration between Agile and MDD methods, i.e., integrating modeling tasks to the agile life cycle. The development process automation provided by MDD can contribute to avoid overloading the process with the manual construction of models, improve in productivity and quality, faster development rate, and better customer satisfaction [23]. However, there are a few works in the literature that demonstrate these advantages. Most of the works which proposes integrated approaches present an example of the use or the proposal feasibility. Nevertheless, these works do not concern evolution issues, such as the quality of the developed requirements and the time spent in development. Therefore, they do not demonstrate how the integration between MDD and the agile approaches can mitigate the problems related to the little documentation existing in the agile approach [23].

The work presented in [6] is an example of a hybrid approach. It proposes a meta-process, names ScrumDDM, to integrate Model-Driven Development practices with the agile Scrum framework [13]. This metaprocess is instantiated in the ArqProjProcess [6], a hybrid process for the service oriented architecture domain. By using ArqProjProcess, ScrumDDM effectiveness in developing a new software was evaluated [6]. Likewise the other works, the authors do not provide an evaluation of software maintenance and evolution aspects. Evidence that the models, produced during the development process, proper supports software's evolution and maintenance are still needed.

In [2] we presented the controlled experiment using ScrumDDM and ArqProjProcess, in which a new version of an already existing software was developed. The experiment maindgoal was to evaluate ScrumDDM effectiveness in supporting software maintenance and evolution. Besides, we instantiated another hybrid process already existing in the literature, named Qualitas [3], to asses the coverage of ScrumDDM concepts in another domain. This paper extends the work presented in [2] to: provide a more in-depth discussion about hybrid Agile and MDD processes and its strengths and weaknesses in software development; discuss the relevance of agile process and MDD approach hybridization in software evolution and maintenance, based on our controlled experiment results; and moreover presenting in details the Qualitas instantiation process.

The sections in this paper are organized as follows: Sect. 2 introduces the background concepts related to this paper and Sect. 3 presents the related works; in Sect. 4, hybridization of the software process is presented as a trend; Sect. 5 presents the used metaprocess ScrumDDM; then, Sect. 6, shows the ScrumDDM metaprocess evaluation; and Sect. 7, the discussion about hybridizations based on the result of the performed experiment; finally, Sect. 8, presents the conclusions and future works.

2 Background

Changes during software development are inevitable, as new user demands may arise during this development [22]. The updated documentation usually produced during the development process can be a viable alternative to support software maintenance.

To deal with requirement changes, different process models may have different strategies. In the so-called traditional methods, the execution of each step of a process is carried out strictly. So, changes along software development demands update the documentation, representing non-meeting of deadlines and rising costs. Agile methods, on the other hand, are lighter processes with reduced documentation, where changes are well come. However, over time, this reduced documentation can make maintenance difficult, as there is no traceability between the code and the requests of the users, impacting software evolution [1].

In this work, we discuss the integration between Agile and MDD methods. In the next subsection, the main concepts related to these topics are presented.

2.1 Agile Methodology

Differently from the software development methodologies existing at the time, a group of developers presented, in 2001, the **Manifesto for Agile Software Development** containing some ways to reduce the time between planning and delivery software to the user [1]. In the presented manifesto, the developers established as the most relevant values the **communication between people and their interactions**, the **software working**, the **client present and collaborating** and the **possibility of modifying the project**. The idea behind these values is to reduce development time and address the necessities of the clients focusing the software development on the implementation phase, i.e. code production [16].

After presenting the manifesto for agile software development, several methods were implemented, such as: Feature Driven Development (FDD) [17], Extreme Programming (XP) [14], Dynamic Systems Development Method (DSDM) [8], Adaptive Software Development (ASD) [14], Crystal Clear [14], the Scrum framework [13], and others. Among them, some methods enable the generation of software documentation (XP and Crystal Clear), methods that emphasize code quality (XP, FDD, Crystal Clear and ASD), in addition to a framework of agile practices (Scrum) that aggregates practices of different agile methods and other development approaches. Scrum stands out for being a framework focused on project management. Each agile method share principles and values defined by the Agile Manifesto [1].

In Scrum, a system is developed in versions called *Sprints*, with an average duration of one month. In general, the process starts defining the necessities of the clients, which in Scrum is represented by user stories in the *product backlog* artifact. These stories are prioritized and organized in parts, called *sprint backlog*, to be developed during a sprint. At the end of the sprint, the system code is delivered to the client. This cycle continues to implement all the product backlog stories (Fig. 1).

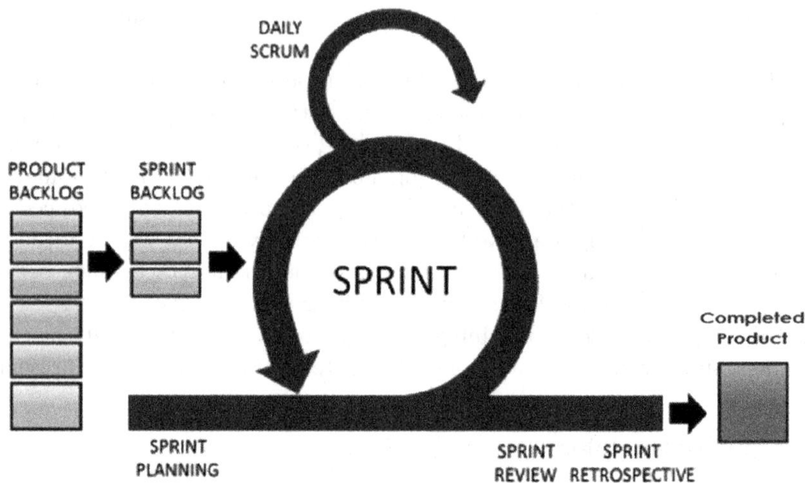

Fig. 1. Scrum agile framework [13].

2.2 Model-Driven Development (MDD)

MDD is a development approach that emphasizes the use of models as the primary artifact in software construction. In the development process with MDD, models are built at a high abstraction level, usually as platform-independent models, and (semi)-automatically transformed into less abstract models until the source code of the application is generated [5]. Thus, the activity of code development becomes something secondary. In addition, models can be reused to generate code on different platforms [9].

The best-known achievement of MDD is the standard proposed by the OMG called Model-Driven Architecture (MDA) [20]. This approach is designed to increase software portability, interoperability, and productivity, assuming that concepts are more stable than technology. In this way, the domain exists independently of the technology, making it possible to perform the modeling and then define which technology will be used in the software development.

In MDD, models are not just documentation. They are the inputs for generating other models or system code. Therefore, models must be written in a programming language with well-defined syntax and semantics. In MDD, it is common to use the metamodeling technique to define these languages. In this way, models are seen as instances of metamodels. Metamodels are also used in the construction of transformation programs, responsible for generating other models or the code [5].

Transformation programs map elements of one or more source metamodels to elements of one or more target metamodels. Thus, any model, instance of the source metamodel can be transformed into a model/code instance of the target metamodel [18]. Transformations are classified according to the artifact that is produced as an output. Transformations that take models as input and generate models as output are called model transformations or model to model transformation. Those that receive models as input and produce code as output are called program transformations or model to text transformations [5].

3 Related Works

This section presents the works found in the literature that concerns about hybridization between MDD approaches and Agile methods, and works that propose strategies to support software maintenance and evolution.

According to [36], the hybridization strategies of MDD and Agil approaches usually focus on modeling the problem rather than prioritizing the construction of the code. In this context, the authors of [15] guide software engineers in the development of efficient processes that prioritize the problem. However, the result of this research evidenced that requirements are usually specified in textual documents and used to promote traceability in the code, but these requirements specifications are rarely changed [15]. Therefore, over time, it becomes obsolete and useless.

The works proposed by [18,19,30] suggest the integration of Agile methods and MDD, in which they use MDD as the primary process in addition to agile practices. The goal is to integrate agile processes to specify models iteratively, i.e., convert, test, and modify them in short, interactive, and incremental cycles.

In another direction, the proposal of [39] and [38] integrate MDD practices during the Agile process life cycle. So the practices of MDD as metamodeling, modeling, and use of transformations are added to an Agile process. Besides, transformations are used to automate tasks.

There are also some works that concerns software evolution. The authors of [21] propose a systematic approach to evaluate software evolution. It uses code quality attributes to retrieve, document, and apply knowledge about how and why software systems evolve. In the same direction, the authors of [27], present a technique to evaluate the evolution of the software based on architectural metrics. In order to make this evaluation, the architecture evolution process is divided into stages. Then, an impact analysis is performed at each modified stage. The goal is to analyze how the architectural changes influence the relevant attributes of software quality to keep good quality software, enabling a healthy evolution. This proposal analyzes techniques to improve software quality concerning maintenance, stability, and reliability quality attributes.

Despite their interested in software evolution, none of these proposals provide controlled experiments in the context of agil MDD hybridization.

A Systematic Literature Review (SLR) is presented in [23] aiming to identify the main characteristics of the current agile MDD process and the benefits, problems, and challenges of this integration. This study included a total of 15 papers that were published from the year 2001 until 2016. The authors conclude that most of the works use an MDD based approach integrated into some agile practices. Some contributions provided by the integration include customer collaboration in the development process, code generation, and the accelerating development process. The most used agile method is Scrum, followed by extreme programming (XP). Selected papers highlighted benefits such as the increase of productivity in development and quality in the final product. The main problem reported was the steep learning curve, although many publications reported success projects using this hybridization.

The metaprocess ScrumDDM [6], used in this work, adopts the agile process Scrum as the primary approach, adding practices MDD to software development. Its main difference from the other works analyzed in this section is that ScrumDDM, as metaprocess, can be used for multiple domains (e.g., SOA, MDD, TDD, Informations Systems, Web Application). Concerning effectiveness in software evolution and maintenance, neither of the hybrid proposals covers software evolution aspects as an objective to be achieved in their development, nor demonstrates it in a controlled experiment.

4 Agile and MDD Software Processes Hybridization Aspects

Hybrid processes integrate characteristics from different development processes to mitigate their weaknesses and highlight their strengths. This section discusses some characteristics related to Agile and MDD hybridizations.

The interest in the hybridization of Agile and MDD has grown because it may offer advantages for both approaches. According to [23], MDD has been considered an inflexible process by the industry, and adding agility to it can facilitate its adoption. On the other hand, the benefits of using MDD in agile projects can improve development, for example, providing process automation resources, e.g., code generation, and supporting software evolution, e.g., using models for both documentation and code generation.

In general, hybridizations that integrate Agile and MDD processes are characterized by the following features: the involvement of the customer, who becomes part of the team, contributing in the system definition; the development carried out in short cycles with continuous product deliveries; the use of models as part of the development process; and the automation of the development process through model-to-model and model-to-text transformations; among other aspects that contribute to faster and more assertive project development.

In the next subsection, we discuss some benefits and problems addressed in Agile and MDD hybridization literature.

4.1 Agile and MDD Hybridization Benefits

Keeping the customer involved and contributing to the project development is one of the reasons to integrate agile and MDD [31,32,35], as well as the opportunity to accelerate the development process [33,34]. These reasons may influence many other issues as productivity, customer satisfaction, and quality of the final product.

In order to achive productivity, agile and MDD hybrid approaches usually adopt a shorter development life cycle and improve it with the automation resources provided by MDD. For example, code generation has been used to decrease the number of code lines manually produced. In addition, code can be generated using design patterns and applying implementation standards, improving modularization [35]. This may also contribute to better software documentation and understandability.

Concerning quality, MDD approach emphasizes the model as the central artifact in software development. This strategy contributes to a better understanding of the problem domain, which can lead to better solutions. The reuse of these models in code generation may reduce the modeling effort and increase productivity. In addition, the automatic code generation usually improves quality by decreasing the number of defects during the test phase. Customer involvement in the development team may also anticipate feedback and enables quick design, contributing to improving the final product quality [23].

Validation and verification activities may also take advantage of models produced along with development [37]. For example, test cases can be generated through the produced models improving test accuracy and coverage. Model simulation is another technique that can be adopted. It simulates the system execution allowing us to analyze how it behaves in specific scenarios. As consequence, anticipates errors that would only be found in test execution after code generation.

Related to system evolution and maintenance, models in MDD indeed reflect the current code. They are not obsolete documentation. So, the development of new requirements or changes in existing ones uses the models for both regenerate code and conduct manual implementations. This documentation also facilitates the introduction of new programmers to the development team.

Agile practices also may shorter the start curve of MDD adoption. For example, during the development process the involvement of users may facilitate the requirement specification, and pair development practice can make team more commited [23].

In summary, it is possible to highlight advantages in aspects such as productivity, quality in code, faster development rate, and customer satisfaction [23].

4.2 Agile and MDD Hybridization Problems

Despite the benefits presented before, the integration of different approaches may also cause some disadvantages.

First of all, new processes may have a steep learning curve due to the lack of experience and culture of the development team. When the approach integration is relatively new, it is less likely to produce benefits in the short-term [32].

Another problem concerns the adoption of specific tools for model development, management, and transformation [24]. The adoption of MDD requires specific modeling languages, i.e., models should be specified using a language with a well-defined syntax and semantic in order to be processed by transformations. So as it is necessary to use automated tools that support the modeling tasks. In addition, transformation programs should be implemented according to the adopted modeling language. Moreover, transformation engines are also necessary to execute these transformation programs.

In order to allow incremental development, as in the Agile approach, tools must provide code synchronization. In a new software release, code must be complemented, so as not to miss the manual implementations carried out in the previous version. Therefore, the absence of a synchronization mechanism that guarantees the tracking between models and code can impede responding to changes incrementally [31].

The lack of configuration management of different teams working in parallel on several models at the same time can still be a problem that needs to be addressed.

In summary, there are also some problems in Agile MDD integration, such as the learning curve, the necessity of a well-defined modeling language with customized transformation programs, specific tools, a synchronization mechanism between models and code, among others.

5 ScrumDDM

ScrumDDM is an agile meta-process proposed by [6], intending to hybridize practices of model-driven development (MDD) to the Scrum framework.

The hybridization proposed in ScrumDDM aims to provide interoperability, increase the portability and productivity of software systems, increase communication between those involved in the development of the project, enable modification at any stage of the development, and prepare only the necessary project documentation. Thus, the use of the meta-process guides those involved to focus on developing models, making the activity of building the source code something secondary in the development process. ScrumDDM is specified in *SPEM - Software Process Engineering Metamodel* [10], and developed in EPF[1]. It was sought to develop a meta-process for the development and simple management, which could be adapted and meet a specific or proposed domain.

ScrumDDM Life Cycle

The ScrumDDM meta-process follows the Scrum life cycle with the phases (*PreGame, Game, PostGame*) and adds, in the first two phases, the practices *Modeling, Transformation (M2M)* and *Transformation (M2T)* from MDD. The

[1] *Eclipse Process Framework Composer* (EPF, 2014).

set of phases, activities, tasks, and artifacts lead to the instantiation of software development processes.

PreGame Phase: This phase begins with the sprint planning meeting and aims to develop the list of requirements used to define the System architecture and create the *backlog*. The latter is used to estimate development time, prioritize what will be developed, and plan software versions.

The adding of MDD practices begins in the *PreGame* phase with the objective of meeting the CIM (*Computation Independent Model*) and supporting the PSM (*Platform Specific Model*), corresponding to the DDM process.

Game Phase: In this phase, the software is developed through the artifacts generated in the *PreGame* phase. In the *Game* phase, this development occurs through the execution of activities and tasks in one or multiple Sprints, until the delivery of the software project and its documentation to the user. This phase consists of three activities (*Development, Test and Integration*) and the tasks related to MDD, such as: *Modeling, M2M Transformation and M2T Transformation*, as well as tasks related to the agile methodology *Planning meeting*, Daily meeting, *Sprint Review Meeting* and the *Sprint Retrospective.*

PostGame Phase: MDD practices are not used in this phase because their objective is to deliver the software and its documentation to the user through the execution of the *Closing* activity.

The *Closing* activity consists of the following tasks: *Approval of Deliverables* and *Completion of the project.* At this stage, the project is closed.

As a meta-process, ScrumDDM needs to be instantiated in order to be used. Instances may have different objectives to address specific domains of software development, such as WEB, SOA, mobile applications, among others. This way, the processes, ArqProjProcess [6] and Qualitas [3] were instantiated from ScrumDDM.

6 ScrumDDM Metaprocess Evaluation

Software processes are essentially difficult to evaluate, as they involve: (i) people with different experiences, (ii) methodologies, and (iii) tools. In this way, the ScrumDDM evaluation was divided into two stages: the definition of instances of processes using the ScrumDDM metaprocess, to evaluate its ability in instantiating new processes; and a controlled experiment to evaluate its ability to evolve software and the impact of this evolution in development agility. Subsection 6.1 and 6.2 present the instances, especially in this paper, we detail the Qualitas instance (Sect. 6.1). Subsection 6.3 presents the controlled experiment.

6.1 ScrumDDMQualitas Instance

The Qualitas process [3] is an iterative and incremental software development process to develop model-driven projects of small and medium-size, through approaches to Model-Driven Development and Test-driven Development (TDD).

Table 1. Summary of the phases, activities with their objectives, in the Qualitas process.

Phases	Activities	Objective
Planning	Project Planning	Construct the project management plan
	Execution Planning	Create the execution plan
Execution	Requirements Engineering	Create requirements document
	CIM	Create independent computing model
	CITM	Generate independent computing test model Test the independent computing model Analyze test results
	PIM	Create platform independent model
	PITM	Generate platform independent test model Test the platform independent model Analyze test results
	PSM	Create specific platform model
	PSTM	Generate specific platform test model Test specific platform test model Analyze test results
	Code	Generate source code from PSM Complete source code
	Test	Generate test from souce code
	Version Validation	Validate that the version conforms to the specified Evaluate the results of test runs
Delivery	Version Release or End Product Liberation	Make the end software product available to the user

This process was developed, considering that it could be integrated into the agile methodology.

In Table 1, we can see the Qualitas process execution scheme through its phases, steps, and activities, as well as a summary of the process.

The Planning phase includes the *"Project Planning"* and *"Execution Planning"* activities, as well as the *"Project Management Plan"* and *"Execution of Plan"* artifacts. The Execution phase includes the activities *Requirements Engineering, Independent Computing Model (CIM), Independent Computing Test Model (CITM), Platform Independent Model (PIM), Platform Independent Test Model (PITM), Specific Platform Model (PSM), Specific Platform Test Model (PSTM), Code, Test Generation (GTs)* and *Version Validation*, in addition to the *CIM, CITM, PIM, PITM, PSM, PSTM Source Code* and *Test Code*, activities in this phase can occur incrementally, depending on the number of versions defined in the project scope. Finally, the Delivery phase includes the *Release Version* and *Final Product* activities, with the artifact of this phase being the developed, tested, and validated software.

The Qualitas process [3], needed to be modified to include the concept of Agile methodologies and thus be instantiated from the metaprocess ScrumDDM. The following modifications were done: the phases and activities of Qualitas became activities and subactivities respectively in ScrumDDM; phases were deleted, and some subactivities were renamed. The details on the modification of Qualitas represented in the instance defined as ScrumDDMQualitas are explained below.

The *Pregame, Game* and *PostGame* phases of ScrumDDM were maintained. The *Planning Execution* and *Delivery* of Qualitas phases became activities in ScrumDDMQualitas. The activity *Project Planning* was deleted and substituted by the task *Planning Meeting* existed in ScrumDDM, because in the Qualitas process it was optional and in ScrumDDM the process starts with the task *Planning Meeting*. The subactivities *Requirements Engineering*, *CIM* and *CITM* became subactivities of activity *Planning*, with the goal of grouping activities related to code design. The task *Retrospective of Execution* replaced the subactivity *Version Validation* in the phase *Game*, because they were conceptually equivalent. Finally, the subactivity *Version Release* was preserved in the activity *Delivery*. In Fig. 2, it is shown the life cycle of the instance ScrumDDMQualitas.

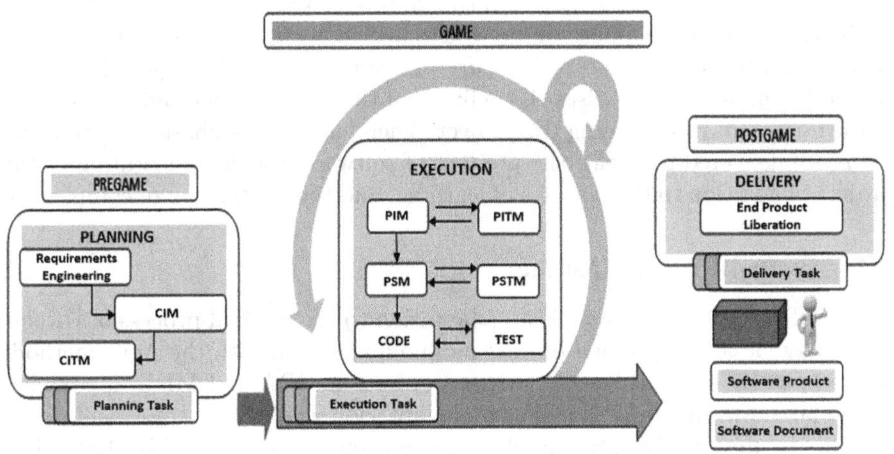

Fig. 2. Life cycle of the instance ScrumDDMQualitas.

The ***PreGame*** phase starts with the execution of the task *Execution Planning Meeting*, for that the subactivity *Requirement Engineering* can generate the artifacts *Product Backlog* and *Development Execution Plan.* With the *Development Execution Plan* generated, the artifacts "*Independent Computing Models* (CIM)" and "*Independent Computing Test Models* (CITM)" are developed.

***Requirements Engineering Subactivity*:** this Subactivity develops the artifacts *Development Execution Plan* and *Product Backlog*; the *Development Execution Plan* is an artifact of entrance for task *Generate CIM* in the subactivity *(CIM)*.

***Subactivity (CIM)*:** this Subactivity generates the artifacts Domain models or business models will be built. These models can be represented through

Business Process Modeling Notation (BPMN), Data Flow Diagrams or UML in Activity Diagrams.

***Subactivity (CITM)*:** this Subactivity develops the artifact *Independent Computing Test Models* (CITM). The objective here is to test, analyze and validate the results of the application of (CITM) in (CIM). The artifact of input is the (CIM); through the obtained results of this Subactivity, some modification on (CIM), or on (CITM) can be done. After the conclusion of the modifications, the *Game* phase is started with the execution of task *Generate PIM* by transforming (CIM) to (PIM).

In the ***Game*** phase, it starts with the execution of the task *Planning Meeting* for developing the *PIM, PITM, PSM*, and *PSTM*. After the source code and the test code are generated through a series of transformations among the developed models. In this phase there are tasks related to MDD such as *Generate PIM, Generate PSM, Generate Source Code* and *Transformation M2M and M2T*, tasks related to TDD such as *Generate PITM, Generate PSTM*, and *Generate Code Test*; and finally, tasks related to Agile methodologies, such as, for example: *Plan, Review Sprint* and *Perform Sprint Retrospective.*

In the ***Postgame*** phase there is a mark indicating the activity *Version Release* and two tasks: *Release Software Version* and *Provide Project Documentation.* Its milestone comprises the delivery of the final product and the delivery of the documentation, closing the project. Therefore, in this phase, the following items are delivered to client: the product of software, the documentation of the project in addition to the transfer of learning, and the celebration of completion.

6.2 ArqProjProcess Instance

The ArqProjProcess instance is an adaptation to the original process of [7] performed by [6], as the original process had no integration with the Agile methodology. In order to hybridize the original process in SOA and MDD approaches to the Scrum framework, the following adaptation was made.

The instance in Fig. 3, Scrum phases, and tasks, as well as the MDD practices of ScrumDDM have been preserved. In the *PreGame* phase, the *Architecture* activity was not preserved since the ArqProjProcess [6] instance is designed to develop architectural software design. As in the previous phase, the *Game* phase, the *Test* and *Integration* activities were not preserved, however, in this phase, the subactivities *Specify Business Model, Analyze Services* and *Design Services* were incorporated to *Development.*

The **PreGame** phase of this instance is initialized in *Vision* activity, with the *Sprint Planning Meeting* task. Its goal is to build the *Product Backlog* through the tasks, *Create, Estimate, and Prioritize Product Backlog.* The user stories and the use cases developed are the input artifacts for model building, such as the functionality model and the business information model, even though you can use the *Update and Write Use Case* tasks.

The **Game** phase of the instance starts with the task *Planning Meeting*, as seen in Fig. 4. It comprises subtasks related to software development, such as *Specify Business Model, Analyze Service* and *Design Service* (not shown in the

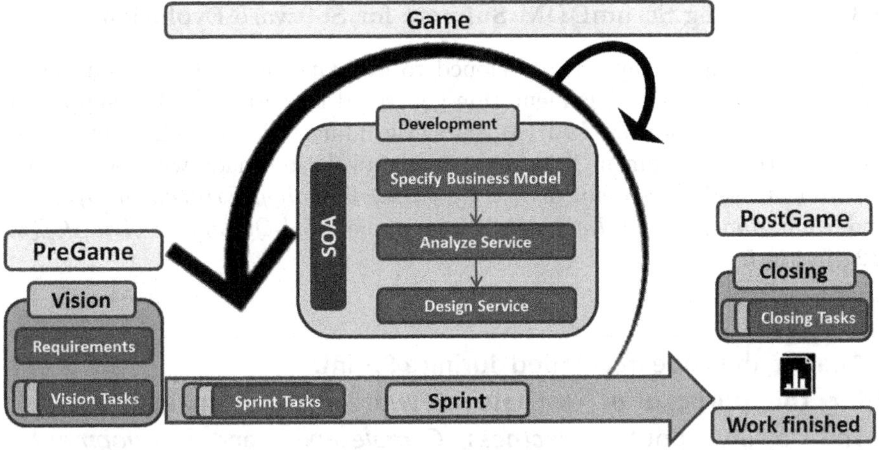

Fig. 3. Adapted from ArqProjProcess Instance [6].

figure). After these, the *Daily Meeting* task is triggered, and the *Sprint Review Meeting* will later be executed. Finally, the Sprint *Retrospective* task will be executed.

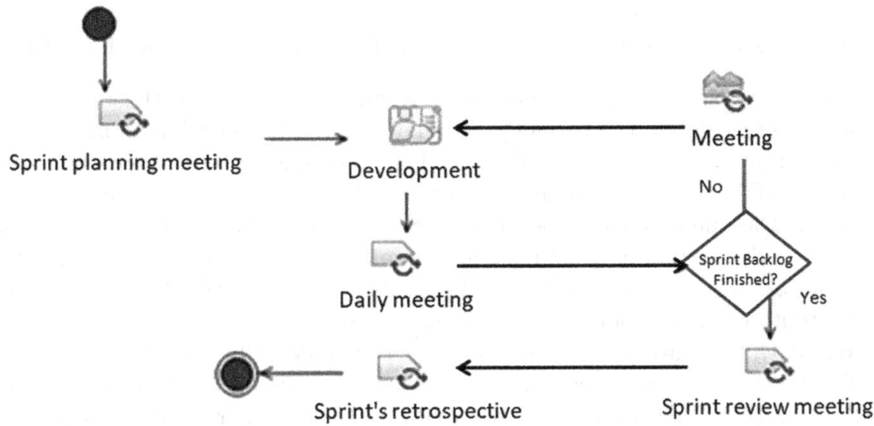

Fig. 4. Adapted from activity flow *Sprint* of the *Game* phase [6].

The **PostGame** phase of ScrumDDM instance [6] aims to deliver the software to the client. This project and the generated documentation must comply with the needs specified by the client in the *PreGame* phase and developed during the *Game* phase. This phase consists of the *Closing* activity and comprises the tasks *Deliverables Approval* and *Project Conclusion*.

6.3 Evaluating ScrumDDM Support for Software Evolution

A controlled experiment was developed to evaluate ScrumDDM regarding (i) the effectiveness of the documentation generated by ScrumDDM in supporting the evolution of software and (ii) if its use influenced the agility (effort and speed) of the development in relation to the evolution made with Scrum. The experiment used steps defined by [11]: *Scope, Planning, Experiment Operation* and *Data Analysis* and *Interpretation* and the *Goal-Question-Metric* (GQM) [12], in Fig. 5.

Analyze the code developed during a Sprint.
for the purpose of evaluate software process maintainability and agility about *Correctness, Completeness* and *Development Time* attributes of defined user story in a specific Sprint.
from the point of view of the used Development method.
in context of students graduating and post-graduating and professionals from technology area.

Fig. 5. Goal of the ScrumDDM metaprocess validation [2].

To guide the experiment, we also defined the following research questions (RQ):

- **RQ1:** Did the artifacts generated by ScrumDDM influence the number of user stories developed correctly in relation to the same development performed directly in the code (Scrum)? The objective is to investigate whether ScrumDDM influenced the number of user stories developed correctly;
- **RQ2:** Did the artifacts generated by ScrumDDM influence the integrity of user stories compared to the same development carried out directly in the code (Scrum)? The goal is to investigate whether ScrumDDM influenced the amount of fully developed user stories;
- **RQ3:** Did the artifacts generated by ScrumDDM influence the time spent developing user stories compared to the same development done directly in the code (Scrum)? The objective is to investigate whether ScrumDDM influenced development agility.

Three metrics were used to measure the objective of the experiment: for *Correctness*, the development of the evolution is considered correct if the number of stories implemented, divided by the total number of stories evolved, is equal to 1. For *Completeness*, the development is considered complete if all user stories are implemented. In this case, we consider the completeness value equal to 1, otherwise 0. Finally, *Timing* is measured in minutes.

The experiment was carried out in an academic environment, with undergraduate and graduate students related to computer science and professionals in

Table 2. Measures for the attributes *Correctness, Completeness* and *Timing* of the participant Groups (*ScrumDDM* and *Controlled*) [2].

ATTRIBUTE	Group	Descriptive Measures Calculated Group							
		Min	Q1	Median	Average	Q3	Max	SD	CV%
CORRECTNESS	Controlled	0,37	48%	64%	66%	86%	0,94	21%	32%
	ScrumDDM	0,69	79%	86%	86%	97%	1,00	11%	13%
COMPLETENESS	Controlled	0,00	0%	36%	36%	57%	0,86	32%	89%
	ScrumDDM	0,29	57%	79%	72%	86%	1,00	24%	33%
TIMING (In minutes) Partial Documentation Update	Controlled	35	43	52	55	68	87	17	31%
	ScrumDDM	12	27	42	38	48	64	17	45%
TIMING (In minutes) Full Documentation Update	Controlled	35	43	52	55	68	87	17	31%
	ScrumDDM	20	43	59	59	77	87	21	36%

the software development area. Knowledge in Unified Modeling Language (UML) and Java was a requirement, and a questionnaire was applied to verify it.

The design used in the experiment covered *one factor*, the development method, and *two treatments*, the ScrumDDM approach, and the SCRUM approach. Thus, divided into two groups, *ScrumDDM Group*, which used ScrumDDM in development; and *Controlled Group*, which used Scrum in development. The development scenario was the same for both groups.

A group of 20 people was selected to participate in the controlled experiment that was carried out between October and December 2018 and lasted, on average, 55 min of execution time in the Controlled Group and 38 min of execution time in the ScrumDDM Group.

The data collected during the experiment were used as input for the analysis of the experiment and analyzed through a grading scale with the expected results for each participant.

The data analysis considered the attributes in Table 2: *minimum observed value* (Min), *first quartile* (Q1), *median, average, third quartile* (Q3), *maximum observed value* (Max), *standard deviation* (SD) and the *coefficient of variation* (CV).

For the **Correctness attribute**, the average of the stories developed correctly indicates that the ScrumDDM Group (median = average = 86%) presented the development more correctly than the Controlled Group (median = 64%; average = 66%). When analyzing the coefficient of variation (CV), we observed that the stories developed by the ScrumDDM Group (13%) showed less variation than those developed by the Controlled Group (32%), thus confirming better conciseness in the evolution of user stories in results of the ScrumDDM Group.

For the **Completeness attribute**, the mean of the stories developed completely indicates that the ScrumDDM Group (median = 79%, mean = 72%)

obtained more complete stories (start-end) than the Controlled Group (median = mean = 36%). When analyzing the coefficient of variation (CV), we observed that the stories developed by the ScrumDDM Group (33%) showed less variation than those produced by the Controlled Group (89%). This lower variability in the ScrumDDM group may show a more standardized behavior in relation to the controlled group.

The **development time attribute** was analyzed considering two aspects: (i) when the partial documentation (*Class Diagram*) was updated and (ii) when all the documentation (*System Architecture Diagram, Component Diagram, Sequence Diagram, Diagram Architecture, Use Case Diagram*) were updated, in addition to the documentation that generates the project's source code.

Regarding the partial documentation update (i), the development time attribute indicates that the ScrumDDM Group (median = 42 min; mean = 38 min) performed the evolution faster than the Controlled Group (median = 52; mean = 55 min). When analyzing the coefficient of variation (CV), the ScrumDDM Group (43%) shows greater variability than the Controlled Group (31%). This can be explained due to the fact that half of the participants in this group were more agile than (75%) of the participants in the Controlled Group.

Regarding all the documentation update (ii), the development time attribute indicates that the Controlled Group (median = 52; mean = 55 min) made the evolution in less time than the ScrumDDM Group (median = mean = 59 min). The Controlled Group showed less variability than ScrumDDM group. According to a variation coefficient of 31%, while the ScrumDDM Group presented a variation coefficient equal to 36%. Even though the ScrumDDM Group consumed more time and varied more, it did not mean that this increase in time (4 min) influences the speed in using the meta-process. We can consider that all project documentation, not just the source code, has been updated.

To verify the existence of a significant difference in the developed code, the hypothesis test was performed [11] using *Student's t-test.* The mean differences whose p-value in the test was less than or equal to 0.05 were considered statistically significant.

The *Students t-test* observed for the *Correctness attribute* (0,200), meaning that the ScrumDDM Group obtained 20% more success than the Controlled Group, showed that this success is statistically significant (t (18) = −2,72; p-value = 0.014; d = 0.54).

The *Students t-test* observed for the *Completeness attribute* (0.360), thus means that the ScrumDDM Group obtained 36% more success than the Controlled Group, showed that this success is statistically significant (t (18) = −2.83; p-value = 0.011; d = 0.55).

The *Students t-test* observed for the *time attribute* (i), when the partial documentation was developed, and the differences observed between the two groups of average times indicated that this difference is statistically significant (t (18) = −2.18; the value of p = 0.042; d = 0.45), showed that the ScrumDDM Group carried out the development in less time than the Controlled Group.

The *Students t-test* observed for the *time attribute* (ii), when all the documentation was developed, indicated that the difference (4 min) observed between the two groups is not statistically significant t (18) = 0.43; p-value = 0.682 d = 0.10), showing that for the ScrumDDM Group, despite having carried out the development in a longer period of time than that of the Controlled Group, this time was not significant.

To minimize possible threats about *Internal validity* a survey was applied for identifying the level of knowledge of the participants about UML, and Java (Eclipse). The participants only had known about the experiment goal at the end of the development. In addition, the researcher accompanied both groups, and interaction between participants was not allowed. To minimize possible threats about *External validity*, 20 (twenty) people were selected and randomly divided into two groups (10 people each). The inability to experiment in the industry was minimized by selecting technology area professionals in both groups. To minimize possible threats about *Construction validity*, several documents and transformations were made available to the participants.

7 Effects of Agile and MDD Hybridization in Software Maintenance and Evolution

Evolution is natural in the entire software life cycle since the domain for which it was developed undergoes modifications [26]. According to Lehman [40], evolving software is necessary so that it remains useful. Inevitably, software must undergo evolutions motivated by changes in the domain environment and user requests, bug fixes, among other issues. The main goal in the software evolution process is to make it useful for longer, i.e., extend its life cycle [27].

Maintenance activities represent almost more than half of the total effort invested in any software system during its lifespan. So as a well-discussed issue within the community of researchers and engineers is how to use Agile methodologies for maintaining and evolution the developed software. The Agile software development life cycle does not have the specifically planned mechanism for this [25].

In software development that uses some Agile method, the source code is the main artifact of development and interacts with several phases of the process, e.g., planning, implementation, verification, and validation [29]. However, other artifacts in software development could be more efficient than code for planning or testing activities. This factor motivates to introduce of some documentation for to support maintenance phase.

The integration of Agile and MDD introduced the idea of saving the sprint state through models version control. So the model assists maintenance planning, registers which requirement was developed, and keeps documentation updated after each iteration. Moreover, the existence of the documentation produced along the development process and the well-structured and well-developed code makes possible traceability between what was developed and the user requests [27].

In this sense, we performed a controlled experiment to assess the ability of a hybrid process, Agile and MDD, in promoting changes in software requirements concerning corrective and evolutionary way. In this section, we discuss some finds of this experiment according to the issues discussed in Sect. 4.

Related to the infrastructure used in the experiment, i.e., modeling language, transformation programs, and tools, as we developed a new release of existing software, the structure was already available. However, we could perceive that much work was spent developing this structure, representing an initial effort necessary. It is important to highlight, however, that this structure is developed once and reused in different projects in the same domain.

In our experiment, software evolution was performed by a new development team. So, the models specified in the first release were fundamental not only to regenerate code but also for system comprehension.

Concerning quality, of the experiment data, the user stories developed by the ScrumDDM group were more complete and correct than the stories developed directly in code. We attribute this result to the available documentation, i.e., models that were reused from the previous version. ScrumDDM group did not insert any manual code. They developed the new requirements modifying the existing models. On the other hand, the controlled group had to understand code produced by other developers to perform the necessary modifications.

Related to the learning courve, it was necessary to provide training for the group that used ScrumDDM. We showed the tools, the development process, and an example of how to use it. It was not necessary to teach the modeling language since we adopted a UML profile. We only introduced profile stereotypes. Indeed, the training represents an extra effort in project development. However, once the team is familiar with the environment, development flows smoothly.

8 Conclusion and Future Works

The software development process does not end when delivered to the client but continues throughout its whole useful life. After delivery, maintenance is inevitable to keep its usefulness. Changes in requirements, for example, may be necessary to the adequate system to business dynamism and to correct problems that occurred during its operation.

This paper discussed the effects of hybrid processes Agile and DDM in software maintenance and evolution. In this direction, we used an Agile MDD process already specified in our research group and performed a controlled experiment. After that, we discussed the results concerning maintenance and evolution issues.

The hybrid process used in this paper takes advantage of models to provide updated software documentation without compromising productivity. Indeed, as models are input artifacts in code generation, they contributed to improving productivity, i.e., manually code was not necessary for the experiment project. Besides, the high abstraction level of the models facilitated project comprehension for all participants, including the users that in agile context are part of the team. In the experiment, we had a real user requesting for new user stories.

The performed controlled experiment evidenced improvements in quality and productivity. The conducted experiment pointed to the capacity of developing software and maintaining the agility, as evidenced by the presented data. The group that used the hybrid process, for example, developed the system more completely and produced the software artifacts more correctly than the other group. Besides that, the evolutions occurred quickly, indicating that the hybrid process maintained the Agile methodology aspect. The hybrid process usage made it possible to keep the updated documentation artifacts synchronized with software product source code.

Despite the advantages, it is essential to highlight that hybridizations usually demands changes in culture and a learning curve. Moreover, when using MDD, specific infrastructure must be prepared, which demands specific expertize, takes some time, and involves costs. However, once created, the infrastructure can be reused in many projects.

As future works, we intend to expand our experiment to industry context to amplify the usage possibility of hybrid processes in real projects. Moreover, new experiment replications are necessary to allow an in-depth analysis of the results in addition to generalizations. Besides, we want to analyze other aspects that could influence software maintenance and evolution, such as using design patterns in code generation and the adoption of test-driven methods.

References

1. BeckK, K., et al.: Manifesto para desenvolvimento ágil de software. Retirado em **20** (2001)
2. da Silva, E.F., Maciel, R.S.P., Magalhães, A.P.F.: Integrating Model-Driven Development Practices into Agile Process: Analyzing and Evaluating Software Evolution Aspects. ICEIS (2) (2020)
3. de Jesus Almeida, C.C.: Qualitas: uma modelo de processo de desenvolvimento de software orientado a modelos (2014)
4. Tomás, R.M.: Métodos ágeis: características, pontos fortes e fracos e possibilidades de aplicação (2009)
5. Brambilla, M., Cabot, J., Wimmer, M.: Model-driven software engineering in practice. Synth. Lect. Softw. Eng. **3**(1), 1–207 (2017)
6. Sales, P.M., Maciel, R.S.P.: Integrando práticas do Desenvolvimento Dirigido a Modelos ao Scrum (2017)
7. Braga, V.: Um processo para projeto arquitetural de software dirigido a modelos e orientado a serviços. Universidade Federal de Pernambuco, Brasil, Diss. Dissertação (2011)
8. Pressman, R., Maxim, B.: Engenharia de Software-8ª Edição. McGraw Hill Brasil (2016)
9. Beydeda, S., Book, M., Gruhn, V. (eds.): Model-driven software development, vol. 15. Springer, Heidelberg (2005). https://doi.org/10.1007/3-540-28554-7
10. OMG, SPEM, Notation, O.: Software and systems process engineering meta-model specification. OMG Std. Rev. **2**, 18–71 (2008)
11. Wohlin, C., et al.: Experimentation in software engineering. Springer Science & Business Media (2012). https://doi.org/10.1007/978-3-642-29044-2

12. Caldiera, V.R.B.G., Rombach, H.D: The goal question metric approach. Encycl. Softw. Eng. 528–532 (1994)
13. Schwaber, K., Beedle, M.: Agile Software Development with Scrum. Prentice-Hall, Hoboken (2001)
14. Wazlawick, R.: Engenharia de software: conceitos e práticas. Elsevier Editora Ltda. (2019)
15. Wagner, S., et al.: Status quo in requirements engineering: a theory and a global family of surveys. ACM Trans. Softw. Eng. Methodol. (TOSEM) **28**(2), 1–48 (2019)
16. Vilain, P., Fagundes, P.B., Machado, T.L.: A Framework for Selecting Agile Practices and Defining Agile Software Processes, p. 25–28 SEKE (2007)
17. De Lucia, A., Qusef, A.: Requirements engineering in agile software development. J. Emerg. Technol. Web Intell **2**(3), 212–220 (2010)
18. Mellor, S.J., et al.: MDA distilled: principles of model-driven architecture. Addison-Wesley Professional, Boston (2004)
19. Ambler, S.W., Agile Model Driven Development (AMDD), Symposium XOOTIC, vol. 12, No. 1, pp. 13–21, Eindhoven, Netherlands (2007)
20. Mellor, S.J., Kendall, S.: MDA Journal BPTrends, Agile MDA - A White Paper (2004)
21. Svetinovic, D., Godfrey, M.: Attribute Based Software Evolution: Patterns and Product Line Forecasting. ICSE 2002 Buenos Aires, Argentina (2001)
22. Sommerville, I.: Engenharia de Software. ed. (2011)
23. Alfraihi, H., Kevin, L.: The integration of agile development and model driven development. In: Proceedings of the 5th International Conference on Model-Driven Engineering and Software Development. SCITEPRESS-Science and Technology Publications, Lda (2017)
24. France, R., Rumpe, B.: Model-driven development of complex software: a research roadmap. In: Future of Software Engineering (FOSE 2007). IEEE (2007)
25. ur Rehman, F., et al.: Scrum software maintenance model: efficient software maintenance in agile methodology. In: 2018 21st Saudi Computer Society National Computer Conference (NCC). IEEE (2018)
26. Bennett, K.H., Václav, T.R.: Software maintenance and evolution: a roadmap. In: Proceedings of the Conference on the Future of Software Engineering (2000)
27. Li, B., Liao, L., Si, J.: A technique to evaluate software evolution based on architecture metric. In: 2016 IEEE 14th International Conference on Software Engineering Research, Management and Applications (SERA). IEEE (2016)
28. Grigera, J., Rivero, J.M., Robles Luna, E., Giacosa, F., Rossi, G.: From requirements to web applications in an agile model-driven approach. In: Brambilla, M., Tokuda, T., Tolksdorf, R. (eds.) ICWE 2012. LNCS, vol. 7387, pp. 200–214. Springer, Heidelberg (2012). https://doi.org/10.1007/978-3-642-31753-8_15
29. Svensson, H., Host, M.: Introducing an agile process in a software maintenance and evolution organization. In: Ninth European Conference on Software Maintenance and Reengineering. IEEE (2005)
30. Santos, N., Pereira, J., Ferreira, N., Machado, R.J.: Modeling in agile software development: decomposing use cases towards logical architecture design. In: Kuhrmann, M., et al. (eds.) PROFES 2018. LNCS, vol. 11271, pp. 396–408. Springer, Cham (2018). https://doi.org/10.1007/978-3-030-03673-7_31
31. Grigera, J., et al.: From requirements to web applications in an agile model-driven approach. In: International Conference on Web Engineering. Springer, Heidelberg (2012). https://doi.org/10.1007/978-3-642-31753-8_15

32. Basso, F.P., Pillat, R.M., Roos-Frantz, F., Frantz, R.Z.: Combining MDE and scrum on the rapid prototyping of web information systems. Int. J. Web Eng. Technol. **10**(3), 214–244 (2015)
33. Lano, K., Alfraihi, H., Tehrani, S.Y., Haughton, H.: Improving the application of agile modelbased development: experiences from case studies. In: Tenth International Conference on Software Engineering Advances. S International Academy, Research, and Industry Association (IARIA) (2012)
34. Eliasson, U., Heldal, R., Lantz, J., Berger, C.: Agile model-driven engineering in mechatronic systems - an industrial case study. In: Dingel, J., Schulte, W., Ramos, I., Abrahão, S., Insfran, E. (eds.) MODELS 2014. LNCS, vol. 8767, pp. 433–449. Springer, Cham (2014). https://doi.org/10.1007/978-3-319-11653-2_27
35. Rivero, J.M., Grigera, J., Rossi, G., Luna, E.R., Montero, F., Gaedke, M.: AMockup-driven development: providing agile support for model-driven web engineering. Inf. Softw. Technol. **56**(6), 670–687 (2014)
36. Ambler, S.W.: Agile model driven development is good enough. IEEE Softw. **20**(5), 71–73 (2003)
37. Zhang, Y., Patel, S.: Agile model-driven development in practice. IEEE Softw. **28**(2), 84 (2011)
38. Aguiar, G.: Uso em um framework para seleção de práticas ágeis (2012)
39. El-Sheikh, A., Omran, A.: Suggested framework for agile MDA and agile methodologies. The Research Bulletin of Jordan ACM, ISSN, pp. 2078–7952 (2004)
40. Lehman, M.M.: Laws of software evolution revisited. In: Montangero, C. (ed.) EWSPT 1996. LNCS, vol. 1149, pp. 108–124. Springer, Heidelberg (1996). https://doi.org/10.1007/BFb0017737

DSL Based Approach for Building Model-Driven Questionnaires

Luciane Calixto de Araujo[1], Marco A. Casanova[2(✉)], Luiz André P. P. Leme[3], and Antônio L. Furtado[2]

[1] Brazilian Institute of Geography and Statistics, Rio de Janeiro, Brazil
luciane.araujo@ibge.gov.br

[2] Department of Informatics, PUC-Rio, Rua Marquês de São Vicente, 225, Rio de Janeiro, Brazil
{casanova,furtado}@inf.puc-rio.br

[3] Institute of Computing, Federal Fluminense University, Av. Gal. Milton Tavares de Souza, s/n, Niterói, Brazil
lapaesleme@ic.uff.br

Abstract. Surveys are pervasive in the modern world, with its usage ranging from the field of customer satisfaction measurement to global economic trends tracking. Data collection is at the core of survey processes and, usually, is computer-aided. The development of data collection software involves the codification of questionnaires, which vary from simple, straightforward questions to complex questionnaires in which validations, derived data calculus, triggers used to guarantee consistency, and dynamically created objects of interest are the rule. Questionnaire specification is part of what is called survey metadata and is a key factor for collected data and survey quality. Survey metadata establishes most of the requirements for survey support systems, including data collection software. This article proposes a Domain Specific Language (DSL) for modeling questionnaires, presents a prototype, and evaluates DSL use as a strategy to reduce the gap between survey domain experts and software developers, improve reuse, eliminate redundancy, and minimize rework.

Keywords: Survey questionnaires · Domain-specific languages · Model-driven software engineering

1 Introduction

A survey is a systematic method for collecting data about (a sample of) entities to construct quantitative descriptors of the attributes of a larger population of which the entities are members. The usage of a questionnaire is, by far, the most common data collection strategy [1].

Highly influenced by recent information technology advances, developing software to support a questionnaire-based survey seems to be an ordinary software engineering task. After all, questionnaires are forms, for which a large number of different solutions and development strategies exist. However, as survey scales in size or complexity, this

J. Filipe et al. (Eds.): ICEIS 2020, LNBIP 417, pp. 458–480, 2021.
https://doi.org/10.1007/978-3-030-75418-1_21

ordinary task becomes daunting. A questionnaire can comprise hundreds of variables intertwined in a complex web of data quality controls implemented to guarantee that each question is fully understood and adequately answered. Hence, underestimating questionnaire design complexity is a common flaw that directly impacts survey quality [1].

In engineering, complexity is frequently handled by raising the level of abstraction. In particular, model-driven software engineering (MDSE) aims at raising computer language abstraction further by making models first-class citizens in the software development process. This article proposes the usage of a model-driven approach for designing complex questionnaires. Specifically, it proposes a Domain Specific Language (DSL) for modeling questionnaires, presents a prototype, and evaluates the use of the DSL as a strategy to reduce the gap between survey domain experts and software developers, improve reuse, eliminate redundancy and minimize rework.

In more detail, the article first describes a domain analysis that resulted in a model for the structure of questionnaires, including elements that allow the modeling of questionnaire data consistency and integrity rules, as well as the specification of behavioral aspects of the questionnaire required to capture navigation flow.

Next, based on the proposed model, the article describes the design of a prototype domain-specific language (DSL) for modeling complex questionnaires, called SLang, and its implementation using the MPS projectional language workbench. The SLang design process and the main decisions provide insights on how model-driven DSL approaches can be applied to real-world problems.

Finally, the article covers the evaluation of SLang, providing a good picture of the proposed solution in practice. In this evaluation, real-world questionnaires modeled with SLang were deployed in a complete setting, including both the SInterviewer app and a backend responsible for centralizing collected data, with real end-users executing mock interviews.

SLang prototyping has been previously described in [2]. This article contains additional insights on the survey questionnaires domain and how the domain analysis was performed, additional information on the SLang design, with improved examples, and details of the language evaluation process, including an IT environment that is closer to reality.

The rest of this paper is structured as follows. Section 2 discusses related work. Section 3 covers the domain analysis and the questionnaire model. Section 4 describes the SLang design decisions and its main features. Section 5 details the implementation and evaluation of SLang. Section 6 presents conclusions and further developments.

2 Related Work

2.1 Survey Questionnaires

Survey questionnaires are part of a broader context in which a survey is planned and executed and produces results based on the analysis of collected data. As such, the survey process comprises a series of activities, executed in three stages: survey planning, survey execution, and survey data publishing (Fig. 1).

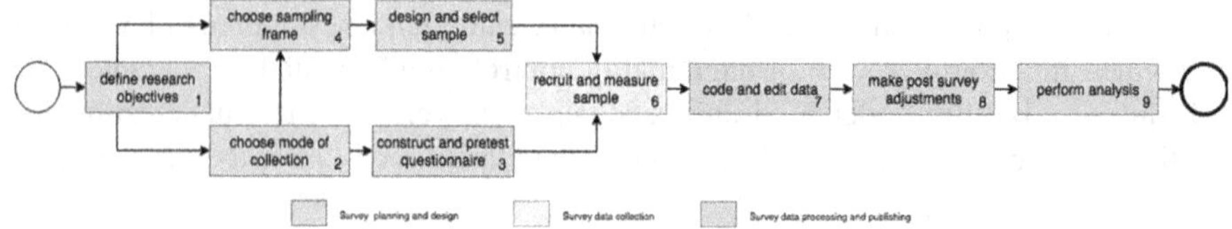

Fig. 1. Survey process [3].

From an IT point of view, the questionnaire has a role in each of the survey process phases. First, it must be created during the survey planning and design phase. During the questionnaire creation, metadata (data about the collected data), data (actual collected data), and paradata (data about the data collection process) are defined. Once the questionnaire has been created, the issue becomes how to use that questionnaire to support and control interviews during the data collection phase. Considering a reality where software is mandatory, supporting the interview means transforming the defined questionnaire into a data collection application that conforms to questionnaire specification. Finally, once data are collected, the questionnaire is a reference that gives meaning to the collected data, helping to understand and produce the answers to the questions raised for the survey.

Table 1 presents a list of survey software requirements related to questionnaires and points to the complexity of creating IT solutions to support survey questionnaires.

2.2 Survey Models

A few existing survey models exist in the literature, either in the form of standards, developed and maintained by entities related to the statistical survey community, or proposed as part of research endeavors.

Survey standards include GSIM, DDI, SDMX. GSIM is the first internationally endorsed reference framework of information objects, enabling generic descriptions of the definition, management, and use of data and metadata throughout the statistical production process [4]. DDI is an international standard for describing survey metadata, including questionnaires, statistical data files, and social sciences study-level information [5]. SDMX aims at standardizing the mechanisms and processes for the exchange of statistical data and metadata among international organizations and their member countries [6].

Since the goal of this paper is to provide a DSL-based approach for questionnaire design, GISM and DDI are relevant inputs for the domain analysis and parallel the development of the model proposed in this work.

Borodin and Zavyalova [7] described an ontology for survey questionnaires. The proposed ontology models questionnaires with simple navigation rules. However, it lacks support for grouping questions by theme, validating answers, and creating objects of interest, among other requirements for complex questionnaires. Still, the ontology is a useful reference for modeling questionnaires.

Table 1. Questionnaire related requirements for survey IT solutions.

Survey process phase	Requirement
Planning and design phase	1. Specify survey attributes
	2. Specify survey constants
	3. Specify survey tabulation
	4. Specify sampling and coverage parameters
	5. Specify themes, questions, and answers considering the diversity of questioning and answering strategies
	6. Specify measurements format
	7. Specify conditional question and answer options visualization
	8. Specify data imputation according to answer and survey metadata-based rules
	9. Specify conditional questionnaire navigation (questions might be skipped depending on previous answers)
	10. Specify conditional visualization of question and answer options
	11. Specify complex data validation
	12. Specify triggers for data adjustments as questions are answered
	13. Specify interview instructions
	14. Specify the creation of data derived from measurements
	15. Specify survey object of interest creation
Data collection phase	16. Support multiplatform data collection strategies
	17. Support integration with third-party software (sampling, object identification, data input software)
	18. Support question and answers customization according to context
	19. Specify paradata
	20. Support themes, questions, and answers presentation customization
Data analysis and publishing phase	21. Specify data imputation
	22. Specify coding of open text measurements

2.3 Questionnaire Design Tools

Questionnaire design and data collection tools can be divided into Web-based survey tools and frameworks. Web-based survey tools work by allowing the user to create a questionnaire that will be distributed and answered through the Web. Google Forms

(GF), SurveyMonkey (SM), Zoho (ZH), and Qualtrics (QT) are examples of Web-based survey tools. CSPro (CS), developed by the US Census Bureau, Blaise (BL), developed by Netherlands Statistics, and Open Data Kit (ODK) are examples of frameworks. They usually include at least a questionnaire design tool and a data collection tool [8].

2.4 Domain Specific Language

DSLs are languages tailored to a specific application domain that offer substantial gains in expressiveness and ease of use in their application domain, when compared with general-purpose programming languages [9]. DSLs are almost as old general-purpose languages (GPLs) and have been applied to domains ranging from bioinformatics through robotics, including Web applications, embedded systems, low-level software, control systems, parallel computing, simulation, data-intensive apps, real-time systems, security, education, and networks [10].

In the domain of complex survey questionnaires, the use of DSLs for questionnaire design is not new. Kim et al. [11] developed the Survey Design Language (SDL) and its supporting tool, SDLTool. SDL consists of a set of domain-specific visual languages. Each language in the set is designed to model a specific aspect of statistical surveys, providing high-level and low-level modeling facilities capable of matching expert cognitive models for statistical surveys. The SDLTool was the environment that tied together the different DSLs aspects through visual modeling of survey resources, the design of statistical survey elements, running modeled surveys on target population datasets, and providing visualization support features [2].

The 2013 Language Workbench Challenge (LWC) assignment consisted of developing a DSL for questionnaires, which had to be rendered in an interactive GUI that reacted to user input and had to store the answers of each question. The questionnaire definition was expected to be validated, detecting errors such as unresolved names and type errors. In addition to basic editor support, participants were expected to modularly develop a styling DSL that could be used to configure the rendering of a questionnaire. The proposed languages offered basic questionnaire functionality but lacked primordial features, such as the possibility of specifying questionnaire navigation, complex validation rules, and triggers, among others [12].

Zhou, Goto, and Cheng [13] presented QSL, a language to specify questionnaire systems that includes some aspects of questionnaire design. Still, it is not clear whether QSL supports a questionnaire complex logic for navigation flow and question presentation. Also, the dynamic creation of objects of interest is not mentioned. Finally, QSL presents a tight coupling between questionnaire presentation and modeling [2].

None of the aforementioned DSLs were designed having questionnaire specification as its main goal. As such, each of them has shortcomings. SDL avoids completely questionnaire specification having its focus on survey methodology and data analysis. 2013 LWC languages were focused on demonstrating language workbenches potential and provided limited support for questionnaire specification. Finally, QSL had its focus on e-questionnaire systems with limited support for questionnaire specification [2].

3 Complex Survey Questionnaire Domain Analysis and Model

3.1 Questionnaire Domain Analysis

Domain analysis is the activity of identifying the objects and operations of a class of similar systems in a problem domain [14]. It involves the identification, acquisition, and analysis of domain knowledge to be reused in software specification and construction [15]. Through domain analysis, information about the domain is identified, collected, organized, and represented. The domain analysis may be based upon: the study of existing systems and their development histories; knowledge captured from domain expert; or the characteristics of emerging technology within the domain [16].

Researchers and practitioners proposed multiple ways of doing a domain analysis, each of them focused on specific goals. Although domain analysis for DSL design and implementation is in its beginnings, there is enough clarity on the general activities to be performed: domain scoping, data collection, data analysis, classification, and domain model evaluation [17].

In this work, the analysis of the domain of complex questionnaires was performed in 4 stages: scope definition, data collection, data analysis and classification, domain modeling. The domain model was evaluated through its usage and suitability in SLang design.

The scope of the domain of complex questionnaires was defined as the universe of statistical surveys whose data collection step has the following characteristics: information is gathered primarily by posing questions to people; information is collected either by having interviewers ask questions and record answers using forms or by having people read the questions and record their answers using forms; information is collected from (a sample of) the population to be described [3].

Considering these characteristics of the universe of statistical surveys, data collection included: survey planning and document specification; statistical survey data collection software documents; data collection software source code; and statistical survey standards experts' interviews [8]. Collected data was processed through an iterative process that included analyzing a piece of domain data and evolving the domain model.

3.2 Questionnaire Model

As collected data was analyzed, domain concepts were extracted, and their attributes, relationships, and pertaining information were documented. The domain model was organized into two main components: the structural model and the behavioral model. The structural model comprises metadata, data, and paradata showing the concepts and their relationships of each area. The behavioral model details the questionnaire navigation flow specification and state evolution.

Metadata Structural Model. Metadata concepts define the questionnaire structure, including consistency and integrity rules, to be applied during data collection. Figure 2 presents the metadata structural model. This model evolved continuously together with the domain understanding and knowledge consolidation.

Survey and *Dictionary* are the model root concepts and tie together the metadata structure. *Survey* is directly related to the *Theme*, *QuestionSet*, *Question*, and *QuestionItem* hierarchy, which hold the questionnaire actual content in the form of question stating, answer options, etc. *Theme*, *QuestionSet*, *Question*, and *QuestionItem* are also *NavigationItems* and are responsible for providing information for navigation flow and questionnaire state controls.

Measurements define the pieces of data that a questionnaire aims at gathering. *Measurements* can be related to a *QuestionItem* in a one to one relationship or stand by themselves (that is the case of *CalculatedMesurements*). A survey *Dictionary* is basically a measurement list. A detailed account of the aforementioned domain concepts can be found in [2, 8].

It is important to single out the modeling of *ObjectOfInterest* relationships. Each *Survey* is related to one or more "objects of interest". An *ObjectOfInterest* is a concrete unity. A survey gathers information about one or more objects of interest. For example, in a hospital care medical drug usage survey, there can be three objects of interest: the hospital, the patient, and the drug. Objects of interest have two important roles in the survey domain. First, they segment questionnaire data by pointing to which real-world entity the data pertains. This segmentation is achieved through the relationship between *ObjectsOfInterest* and *Themes*. Second, objects of interest enable the ability to dynamically create entities about which questions are answered (Table 1). Considering the dynamic entity creation, the domain analysis pointed at two paths for entity creation: *themes* and *question items*. Theme-based entity creation happens when it is necessary to register a full account of the entities related to that object of interest, such as, for example, when it is necessary to build a patients list, with information such as name and age, before asking questions related to each patient drug usage while in hospital care. Question item-based entity creation happens when the answers to a question determine an entity creation. That would be the case when, in a question, a list of drugs (each a *QuestionItem*) is selected and used as input for making a set of questions about each drug and its application in patient care. This second scenario is the reason for the relationship between *QuestionItem* and *ObjectOfInterest*.

As previously mentioned, *NavigationItems*, whose concrete representations are *Theme*, *QuestionSet*, *Question*, and *QuestionItem*, hold the information necessary to model questionnaire navigation flow and state. As such, they must store expressions that define if each item should be visible or not, as well as a list of validations and triggers that allow to perform some level of answer quality control and data consistency guarantees. Model navigation items, validations, and triggers rely heavily on Boolean and arithmetic expressions to determine their behaviors.

Questionnaire metadata connects to the data model through 3 concepts: *Survey*, *ObjectOfInterest*, and *Measurement*, as shown in Fig. 3. Data and paradata structural models are closely related to a survey data collection operation (Fig. 1). As such, the relationship between concepts from metadata and concepts from data and paradata models is usually realized through conventions adopted in the data collection software. The usage of a dotted line indicates this lack of a concrete relationship between those concepts. A survey is basically a collection of observations made during an interview, which explains the relationship between *Survey* and *Interview*. Objects of interest are closely

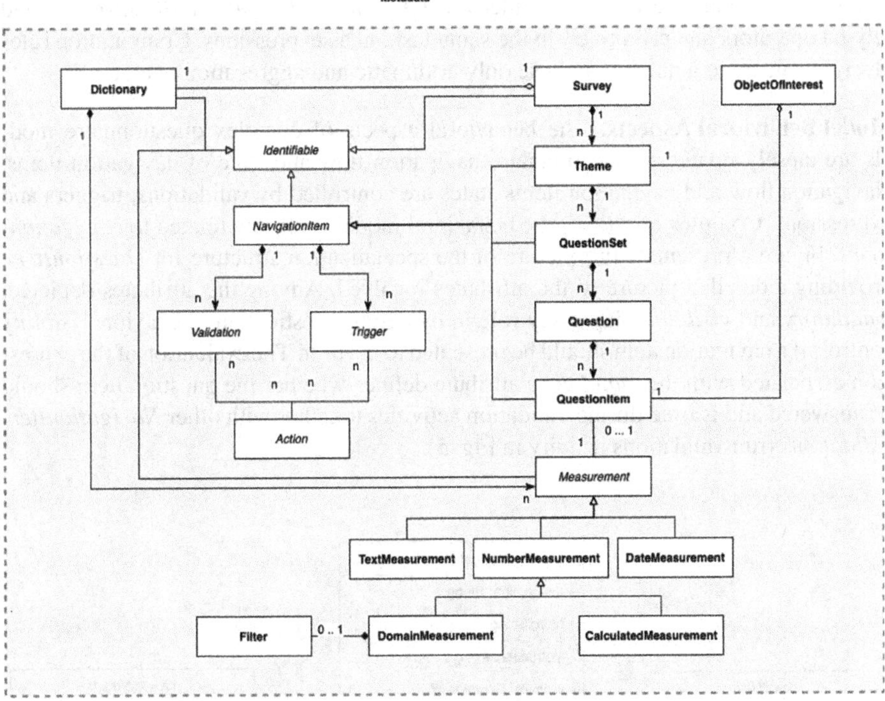

Fig. 2. Domain questionnaire metadata structural model.

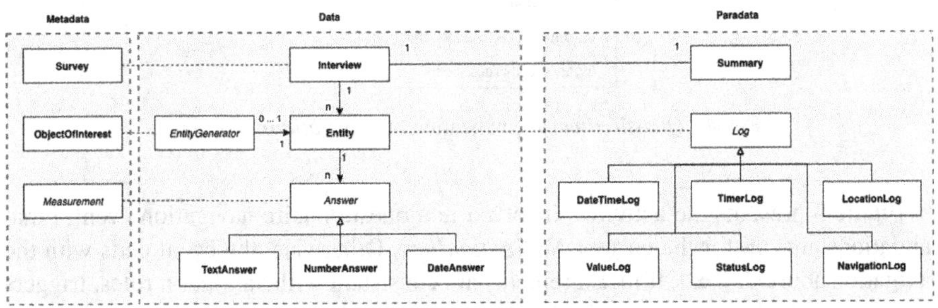

Fig. 3. Data and paradata structural models.

related to entity generators since an *EntityGenerator* defines rules for *Entity* creation. *Measurement* is the specification of an *Answer* format, type, meaning, and calculus rules.

Transversal Model Concepts. Expressions and rules are transversal concepts in the complex questionnaires domain model, frequently appearing as concept attributes. *Expressions* depend on primitives and operators. Operators mapped during domain analysis included boolean, arithmetic, comparison, set operators, and user-defined ones. Primitive Booleans, strings, and numbers coming from *Measurements* values or *Survey* constants are used in combination with operators in an *Expression* definition. *Rules* can be used to compute values in a rule-based *Action* or a *CalculatedMeasurement* and rely on operators and primitives, in the same fashion as expressions. Computation rules always evaluate to a number and use only arithmetic and aggregation operators.

Model Behavioral Aspects. The behavioral aspects of complex questionnaire models are closely related to questionnaire navigation flow and state of navigation items. Navigation flow and navigation items states are controlled by validations, triggers and expressions. Complex questionnaire behavioral model is closely related to *NavigationItems*. Figure 4 presents a full picture of the specialization structure for *QuestionItem*, providing a detailed picture of the attributes involved. Among the attributes depicted, *mandatory* and *visibility* play a key role in modeling questionnaire behavior. *Visibility* controls if a navigation item should be presented to the user. The evaluation of the expression associated with the *mandatory* attribute defines whether the question item should be answered and is used during validation activities together with other *NavigationItem validations* (run validations activity in Fig. 5).

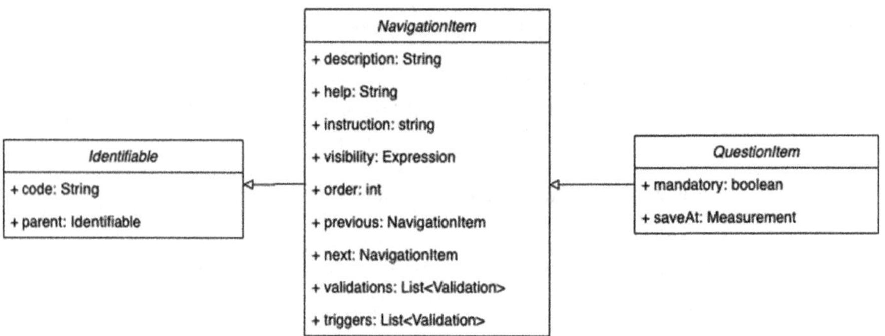

Fig. 4. *QuestionItem* specialization of *NavigationItem* [2].

Figure 5 presents the activities involved in a questionnaire navigation event. First, validations are run for the current *NavigationItem.* Otherwise, the event ends with the navigation aborted. Next, if measurements are consistent with validation rules, triggers are executed. If there is *NavigationItem* to be displayed after the current one, this item is retrieved. If not, the navigation is aborted. It is important to notice that navigation can happen up or down the questionnaire content hierarchy [2]. Finally, if a *NavigationItem* has been retrieved, its visibility is checked. If the item is visible, navigation is complete; if not, it is necessary to check again if there is a next *NavigationItem* to be presented.

Two additional aspects are important in the behavioral model. First, the questionnaire *NavigationItems* are organized in a hierarchy that affects navigation events resulting in horizontal and vertical navigation events. Horizontal navigation events happen when navigation occurs at the same hierarchy level. Vertical navigation events happen when there is a change in the hierarchy level. Second, the activities listed in Fig. 5 trigger transitions that affect the state of a *NavigationItem.* Mapped *NavigationItem* states are NOT ANSWERED, VIEWED, SKIPPED, and ANSWERED [2].

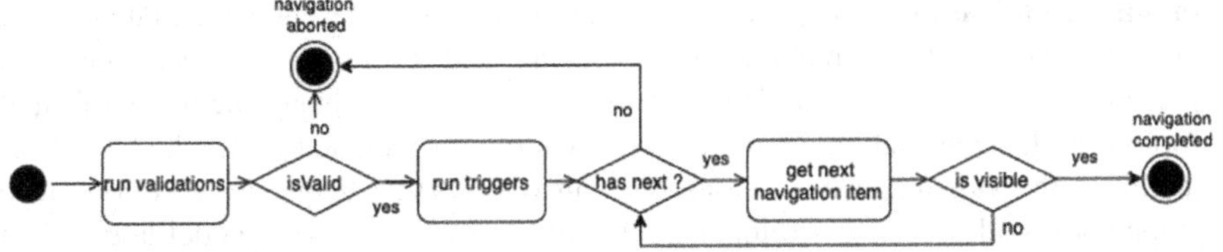

Fig. 5. Questionnaire navigation activities.

4 SLang Design

Research and industry practice bring little guidance on how to map domain models to DSLs concepts and syntax [18, 19]. This section describes the process of transforming the model described in Sect. 3 into a DSL for survey questionnaire specification, called SLang.

SLang was constructed to facilitate survey design and execution, focusing on questionnaire specification, and allowing to experiment with a model-driven approach and to probe its potential benefits. The focus on questionnaire specification was chosen due to its relevance to survey processes. SLang was designed by analyzing the model proposed in Sect. 3 and mapping its elements into potential language concepts. Next, potential language concepts were selected to be included and represented in SLang. This process is described in detail, starting with SLang's main design decisions in Sect. 4.2, followed by an overview of its abstract syntax and its concrete syntax in Sect. 4.3. Section 4.4 presents a survey questionnaire example, highlighting SLang's main features.

4.1 SLang Design Principles

Design guidelines for programming languages have been extensively discussed. Principles such as simplicity, security, fast translation, efficient object code, and readability are well-established programming language design guidelines. For DSLs, the general principles are simplicity, uniqueness, consistency, and scalability. Still, it is necessary to translate those general principles into guidelines for DSL design. Based on DSL design guidelines [20], the following principles guided SLang design.

Language Purpose. During domain analysis, questionnaire specification presented itself as the main challenge in the survey process, considering both the need to keep track of questionnaire design and the challenges in data collection systems development. When looking at the requirements involved in a survey process (Table 1), it becomes clear that there is a need for a formal communication tool that can tie different stages of the survey process helping avoid metadata scattering through the survey process, While improving communication among a multidisciplinary team that includes survey domain area experts, statistical and survey methodology experts, software engineers, IT infrastructure engineers, database administrators and business analytics. As such, SLang and its models were designed with the main purpose of survey questionnaire specification.

Since its purpose is complex questionnaire specification, SLang does not contemplate the questionnaire presentation aspects that directly affect the user interface, and general aspects related to data collection systems, such as security and data analysis.

Language Realization. Complex survey questionnaires present the challenge of expressing non-linear information flows, be it interactions, relations, or state changes that are better tackled by visual languages [21]. On the other hand, the textual format is more useful, scales better, and the necessary tools for a textual DSL take less effort to build. In the vast majority of cases, starting with textual languages is a good idea – graphical visualizations or editors can be built on top of the metamodel later, when and if a real need is established [22]. Besides, as described in Sect. 2, no DSLs were found with the specific purpose of questionnaire specification, limiting the possibilities for language reuse. As such, a decision was made to create a textual language. For simplicity reasons, it was decided that SLang would initially be a non-executable language with pre-processor characteristics [8]. This decision does not preclude the possibility of making SLang executable in the future.

The usage of more general-purpose languages, such as UML or XML, were considered, but language workbenches had a series of advantages. First, they offer DSL users better editing experience and allow DSL designers to create custom editors with functionality similar to those of modern IDEs [10, 23]. Second, language workbenches can make the development of new languages affordable and, therefore, support a new quality of language engineering, where sets of syntactically and semantically integrated languages can be built with comparably little effort [12]. Third, language workbenches bring the possibility of illustrative executions [24]. Finally, the available language workbenches offered the possibility of bringing SLang closer to IBGE social survey data collection infrastructure endpoints which is currently Android-based.

Language Content. Language content should reflect only the necessary domain concepts, keeping it simple, while avoiding unnecessary generality. The number of language elements should be limited by eliminating, whenever possible, conceptual redundancy and inefficient language elements [20]. With the stated language purpose in mind, the first decision made was to focus on metadata specification leaving data and paradata portions of the model presented in Sect. 3 untouched. The process of domain analysis and modeling eliminated most of the conceptual redundancy and unnecessary generalizations. Careful thought was also given to expressions. Questionnaire navigation is controlled by visibility attributes, usually expressed as Boolean expressions with *Measurement* values as operands. The level of complexity for those expressions is limitless, considering data collected during domain analysis. As such, the option was made to limit the operators available to those described in Sect. 4.3.

4.2 Abstract Syntax

An *abstract syntax tree* (AST) is a data structure that represents the abstract syntax used by compilers to represent and manipulate programs. Aspects of SLang AST are presented Fig. 6 and Fig. 7, in which rectangles represent nodes or concepts, arrows link parent and child concepts, forming the AST branches, and dotted arrows specify non-mandatory parent/child relationships.

The questionnaire hierarchy is presented in Fig. 6. The diagram was enriched to highlight two important aspects of SLang. *NavigationItems* generalization, represented as grayed out elements, and *ObjectOfInterest* generators, which are possibly either *Theme* or *QuestionItem.*

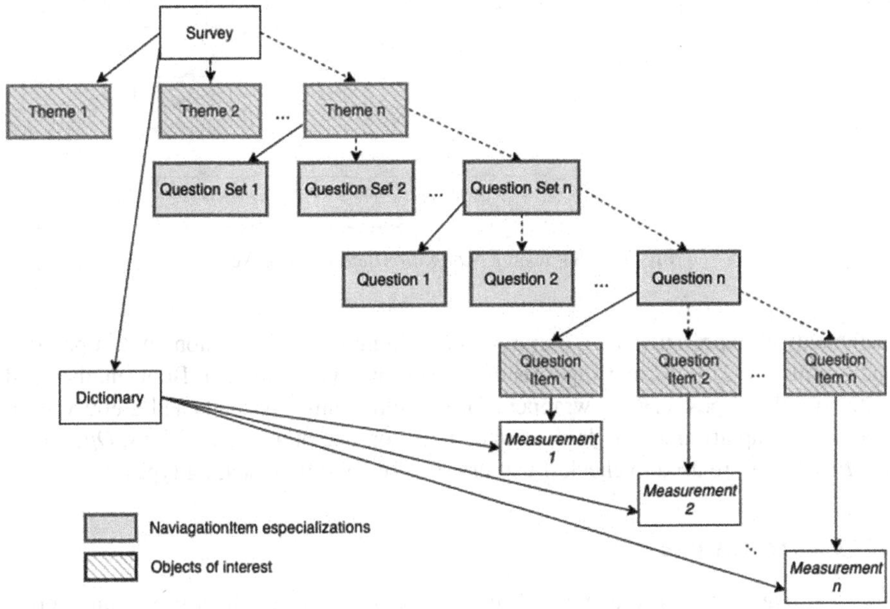

Fig. 6. AST from a questionnaire concept structure perspective.

The SLang root concept is *Survey. Dictionary* and *Survey* are related in a one-to-one composition, which is slightly different from the questionnaire model. Initially, SLang was designed to have *Dictionary* as a second root concept. The decision to model it as a child of *Survey* was taken after interviewing domain experts, who pointed out that dictionaries are survey-specific. There is limited opportunity for *Dictionary* model reuse and deriving it from *Survey* makes questionnaire modeling easier.

Figure 7 represents the navigation item ASTs. *NavigationItem* replicates the questionnaire model by providing the generic type that *Theme*, *QuestionSet*, *Question*, and *QuestionItem* specialize. Each *NavigationItem* can have multiple validations and triggers, which form branches of the AST. Each *Trigger* is connected to an *Action.* An *Action* affects one or more *Measurements.* The semantics of each one of those elements reflect the domain model. *NavigationItem* helps define the flow of answering a questionnaire. *Validation* is responsible for guiding how to provide adequate and valid answers. *Triggers* are responsible for guaranteeing data consistency among *Measurements.*

As presented in the model, *Expressions* are a transversal concept of SLang, providing means to represent intended questionnaire presentation flow and rule informed data consistency checks and actions. As such, most of the concepts are connected to one or more *Expression* concepts. SLang expressions follow the usual representation for

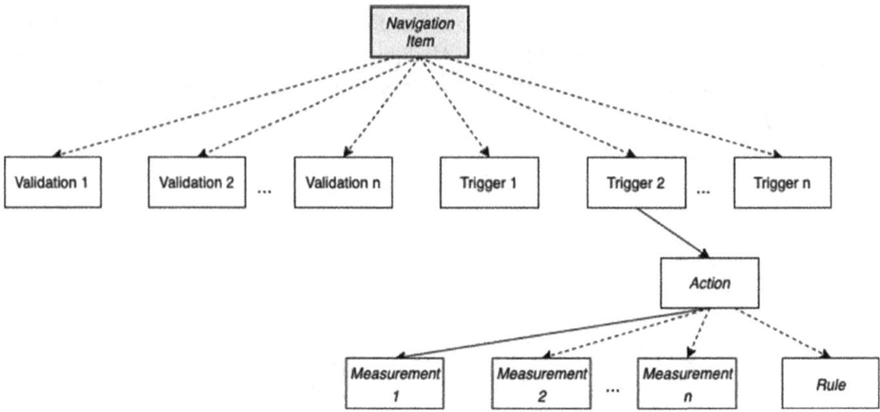

Fig. 7. AST from a *NavigationItem* perspective.

mathematical expressions as a hierarchy combining other expressions with operators and primitives [2, 8]. SLang supports three primitive types: number, Boolean, date and text. Primitive types serve as wrappers for measurements, constants and literal values. Operators wrap arithmetical, logic, and comparison operators. *Expressions*, *Operators*, and *Primitives* are used to check if expressions are valid by matching types.

4.3 Concrete Syntax

The *concrete syntax* of a language defines the notation with which users can express programs. In the case of SLang, a program is, actually, a questionnaire specification. The way questionnaires are expressed vary considerably, depending on the user background in terms of IT tools: some use spreadsheets, while others adopt text documents. Such a scenario makes SLang concrete syntax design decisions harder. In Slang, the concrete syntax is close to the textual writing of a questionnaire and uses sensible defaults, consistent choice of style, indentation, and conventions.

SLang concrete syntax is based on five main concepts: **Survey**, **Dictionary**, **Theme**, **QuestionSet**, **Question**, and **Item**. **Survey** is the root concept and, whenever created, a **Dictionary** is also created. **Theme**, **QuestionSet**, **Question**, and **Item** are nested elements from Survey and are progressively indented to represent their hierarchy. Each of those main elements has a list of attributes.

Attributes are the next level for SLang concrete syntax. For example, Table 2 shows **Theme** attributes. Only four attributes are mandatory: code, description, order, and object. The order attribute is grayed out because it is not defined by the questionnaire designer and is derived from the questionnaire hierarchical organization.

The *Identifiable* concept plays a key role in SLang concrete syntax. Its attributes, code and alias, are unique and play a key role in allowing the user to navigate the created model. *Identifiable* codes are automatically filled for all concept instances, except for *Survey*. An alias can be included whenever the questionnaire designer feels it is necessary for model clarity. *Survey*, *Theme*, and *Measurement* concepts require easy identification, as the user creates a questionnaire, and needs to review what was done. As such, either

Table 2. **Theme** attributes.

Theme Attribute	Mandatory	Description
code	x	string that holds entity unique identification
alias		String that holds a unique alias for the created theme
description	x	string for navigation item description
help		string explaining the navigation item
instruction		string for information regarding answering the navigation item
order	x	integer indicating the order of the navigation item within its level
visibility		*Expression* indicating whether the *NavigationItem* should be visible or not
triggers		list of *Triggers* for the navigation item
validations		list of *Validations* for the navigation item
questionSets		list of *QuestionSets* for the navigation item
creator		*Expression* that indicates if this theme should generate entities about which data will be collected
object	x	*ObjectOfInterest* to which the theme is associated.

code or alias values are presented always in green for those concepts, as can be seen in the Sect. 4.4 listings.

Another important aspect of the concrete syntax are the *Measurement* types. Given the importance of those types for questionnaire specification, the choice was made to give them special styling by using the orange color. SLang uses three types for those concepts: text, number, and domain. During domain analysis, modeling measurements was a challenge. For IT-related domain data pointed at multiple type possibilities many times in line with GPL primitive type system or relational database data types. Still, the complexity of those type systems was confusing for most survey domain experts. As such, when defining the abstract and the concrete syntaxes, the types remaining were those most easily absorbed by survey domain experts. To achieve the level of detailing additional attributes were included, with sensible default rules, to allow their mapping to the more complex software-related type systems. Table 3 presents the attributes of a *NumberMeasurement* with the associated default values.

Table 3. *NumberMeasurement* attributes.

NumberMeasurement Attributes	Default value	Description
code	-	*Measurement* unique identification
alias	-	*Measurement* unique alias for the created theme
type	-	number
size	3	integer representing the number of digits
decimal	0	Integer representing the number of decimal places

Expressions and *Rules* are determinant for specifying navigation flow [2, 8]. *Expressions* combine primitives and operators. Primitives might be literals or *Measurements*. As such, a Measurement code or alias can be used in expressions. *Rules* use the same paradigm. Table 4 presents a list of currently supported operators with concrete syntax examples.

4.4 SLang Code Snippets

When designing SLang the main idea was to allow the questionnaire designer to write its specification as text. Since this work started [2, 8], SLang syntax has already been adjusted to include some minor improvements to make it more friendly. Listing 1 presents how main concepts are expressed, including an example of the usage of **Survey**, **Theme**, **Question**, and **Item**. The language has been updated to allow suppressing the usage of **QuestionSet** for *Themes* with only one *QuestionSet*. The example also presents the usage of comments.

Table 4. SLang supported operators.

literal type	operator	example
Boolean	and	V0001 == 5 and V0002 != 0
	or	V0001 == 5 or V0001 == 6
	not	**not**(V0001 == 5 or V0001 == 6)
comparison	empty	**empty** V0001
	equal	V0001 **equal** 6
	not equal	V0001 **not equal** 6
	greater than	V0001 **greater then** V0002
	less than	V0001 **less than** 0
	greater than or equal to	V0001 **greater than or equal to** 6
	smaller than or equal to	V0001 **smaller than or equal to** 1
arithmetic	*	V0001 * 6
	/	V0001 / 5
	-	V0001 - 5
	+	V0001 + 6
set	min	**min** (V0001, V0002)
	max	**max** (V0001, V0002)
	in	V0001 in (1,3,5)
	out	V0001 out of (1,3,5)
aggregation	mean	**mean**(V0001, V0002)
	sum	**sum**(V0001, V0002)

```
Survey: A demographic survey (S0001)
  description: A demonstration survey
  version: 1 !! allows controlling changes in the survey specification
  constants:
    min_wage: number 7.0
    reference_date: date 08/01/2019
  object: household ( an address that serves as a dwelling for one family )
  Theme: Inhabitants information (T01)
    object: household
    instructions: Here go instructions for a potential interviewer
    help: Here goes an explanation of this theme
      Question: 1.1.1 - How many people lived in this household on {reference_date}?
        Item:
          saveAt: number qty_people_household
      Question: 1.1.2 - How many children with ages between zero and nine (include
      newborns) lived in this household on {reference_date}?
        Item:
          saveAt: number qty_children_household
```

Listing 1. Questionnaire content survey concepts.

Listing 2 presents an example of how a **Dictionary** is defined, including an example of a *CalculatedMeasurement*, which uses the *Rule* concept as an attribute of *Measurement*. Listing 3 presents the usage of *Validations* and *Triggers* in the context of specifying a question validation and measurement consistency through triggers.

```
Dictionary: A demographic survey dictionary (D0001)
  alias: household_proprietor
  code: V0003
  type: text
  size: 100
  alias: resident_income
  code:V0060
  type: number
  precision: 11
  scale: 2
  alias: household_income
  code: V0060t
  type: number
  precision: 14
  scale: 2
  rule: sum(resident_income)
```

Listing 2. SLang Dictionary example.

```
Question: 8.1.1 - How many sons and daughters born alive until {reference_date}?
   visibility: sex == 2 and age >= 10
   mandatory: false
   Item:
     saveAt: domain had_children_born_alive
        1: Had Children
        2: Didn't have children
     Triggers:
        action: clear
        measurements: V0802, V0803
        expression: had_children_born_alive == 2
   Item: How many man?
     visibility: V0801 == 1
     saveAt: number V0802
     Triggers:
        action: input(V0802 + V0803)
        measurements: qty_children_born_alive
     Validations:
        type: ERROR
        expression: qty_children_born_alive > 1 and qty_children_born_alive 30
        message: "The number of children born alive is invalid."
   Item: How many women?
     visibility: V0801 == 1
     saveAt: number V0803
     Triggers:
        action: input(V0802 + V0803)
        measurements: qty_children_born_alive
```

Listing 3. *Validations* and *Triggers* expressed in Slang.

5 SLang Implementation

SLang was prototyped using the MPS language workbench. This prototype provided the environment for experimenting and validating SLang as a model-driven approach, applied to the complex survey questionnaire domain. The validation process included two stages: real-world questionnaire specification using SLang and the usage of questionnaire model transformations applied in a software called SInterviewer used by real stakeholders to perform mocked interviews.

In the first stage, two real-world questionnaire specifications were encoded in SLang, using this prototype implementation. This part of the experiments validated the expressiveness of SLang. Then, mock-up surveys were run, using the encoded questionnaires, on top of a survey environment, called SInterviewer. This section describes some aspects of SLang prototype implementation, important aspects of SInterviewer, and the results of the prototype validation process.

5.1 SLang Implementations Highlights

The SLang prototype was built on top of JetBrains Meta Programming System (MPS). Five workbenches were first considered for the implementation of SLang: Spoofax, XText, Rascal, MetaEdit + and MPS. The decision to adopt MPS was influenced by the fact that only MPS had a projectional editor, that is, an editor that makes it possible to create, edit and interact with one or more ASTs, avoiding the need to use parser tools [8, 23]. Another aspect that influenced this decision was a broad Android-based data collection infrastructure used at IBGE and the fact that MPS provides Java compatibility, hence providing means for model transformations outputting Java code.

The process of prototyping SLang started with concepts mapping, which defined the AST creation rules and the SLang base structure. Then, the concrete syntax was enforced using the MPS editors. Finally, behavior and static semantics were added using the MPS Behavior, Constraint, and Type System aspects.

Behavior aspects made possible, for example, to attribute default values to questionnaire model properties, and to create and manipulate child nodes and references using MPS concept constructors and MPS concept methods. Static semantics was established through MPS Constraint Aspects and Type System Aspects. Constraint aspects provided, among other things, control of where concepts are allowed, validation for properties values, answer options control. Type system aspects were used for semantic aspects that could not be modeled using MPS base concepts, behavior aspects, and constraint aspects. For example, preventing nodes with the same name to exist in a specific scope could not be done using concept structure or constraint aspects. Constraint and type systems aspects provided hooks used by MPS to implement context assistance and error reporting, in the final language IDE generated using MPS [2].

5.2 SInterviewer

SInterviewer is a data-intensive mobile application, built on top of the Android platform, to collect questionnaire data and paradata. Its architecture is an evolution of the data collection software developed for IBGE Housing and Social Surveys, which included the 2017 Agricultural Census and the 2020 Demographic Census.

The SInterviewer engine is based on questionnaire metadata specified in Json format. As such, the application is configured during packaging for one survey at a time. That means it is impossible to run multiple surveys in the same Android application instance, that is, and each survey has its own app.

When the app is started, questionnaire Json metadata deserialization is triggered. The deserialized model is used through the services layer to control questionnaire navigation, persistence of answers and general application functionality, such as closing questionnaires and registering interview observations. Although being possible to configure the app differently, the SInterviewer presentation unit is a question, which means that the app user answers one question at a time. SInterviewer processes a questionnaire navigation request in a similar fashion to the behavioral model proposed in Fig. 5, but with additional activities for language persistence and the necessary checks to fully allow navigation item state control.

Two aspects are relevant for the usage of SInterviewer paired with SLang. First, SInterviewer mixes presentation aspects with questionnaire metadata. This poses a challenge that was mitigated by the usage of sensible defaults when implementing SLang to SInterview notation transformation. A second factor is that SInterviewer parses expressions for deciding if an answer is valid and if a question should be visible. The issue here is related to expression complexity. Not all aspects of SInterviewer are supported by SLang. As such, for certain behaviors possible in SInterviewer to be achieved from a SLang coded questionnaire, manual intervention is necessary for the outputs of the model transformation process. The option was made not to make this intervention. Still, this specificity should be addressed by a language evolution plan.

Data collected with SInterviewer can be extracted in two ways: local data extraction or synchronization with a backend. Local data extraction allows to export data in two formats: CSV or Json. If backend synchronization is enabled, data will periodically be sent to a backend and stored in relational databases.

5.3 SLang Usage in a Real-World Context

Two real-world questionnaire specifications were modeled using SLang and used in mock interviews using SInterviewer [2]. Table 5 presents some statistics about the modeled questionnaires providing some insights on their complexity. The number of *Themes* and *QuestionSets* were the same because both surveys had only one block of questions in each Theme.

Table 5. Modeled real-world survey questionnaire statistics [2].

Characteristic	Survey 1	Survey 2
Objects of interest	1	8
Themes	18	41
Question sets	18	41
Questions	102	185
Question items	102	893
Measurements	107	957
Validations	34	134
Triggers	28	102

Survey 1 was about K-12 students' health aspects and included themes such as students' socio-economical aspects, family context, eating habits, and physical activity, among others [25]. It included around 102 questions composed of single choice multiple-choice answers questions, which implies one *QuestionItem* per *Question*. Each *QuestionItem* was connected to a DomainMeasurement that defined the answer domain. Another aspect was the usage of derived measurements, which accounts for the fact that we have more measurements then question items. This survey navigation logic was simple and included mostly visibility expressions for skipping themes or questions and cross-validation checks among questions (questions with related answers).

Survey 2 was a preliminary version of the 2017 Brazilian Agricultural Census [26]. This survey had a more complex navigation structure with a larger number of question items due to the extensive usage of many items in one question (that happens when you have a multiple-choice question where the respondent can select more than one item). Survey 2 made extensive usage of the objects of interest creation support of SLang and SInterviewer. This allowed exercising theme-based object of interest dynamic creation and question item-based object of interest creation. In practice, the number presented in the table corresponds to the number of *QuestionItems* and *Measurements* from a questionnaire in SInterviewer. The number of coded *QuestionItems* corresponds to 305, and the number of coded *Measurements* amounts to 369.

5.4 Model Transformations and Mock Interviews

While working on SLang prototype model, transformations were experimented with the implementation of two model to text (M2T) transformations: one to generate questionnaire to SQL schema; and another to generate SInterviewer questionnaire to json metadata. Both transformations were implemented using the MPS TextGen aspect.

The questionnaire to SQL schema transformation aimed at creating a relational scheme to store collected data in backend databases. The mapping between SLang and SQL compliant code starts with a relational scheme for each survey and a table for each object of interest. In each table, one column was created to store each measurement value. Column creation code was derived from *Measurement* attributes according to each measurement type, with *Identifiable* code attribute being used to define column names. Each table has a column "id" marked as primary key. Currently, it is not possible to use compound primary keys or customize the primary key column name. Validations and triggers presented a challenge considering the complexity of expressions. Still, they are respectively mapped to "check" and "create trigger" SQL statements on a best effort fashion. SQL constraints were also included in the output. For example, whenever the *Measurement* associated *QuestionItem* was mandatory, the SQL "not null" constraint was added.

The SInterviewer questionnaire Json metadata transformation generates questionnaire specification in Json format, which can be submitted to SInterviewer (see Sect. 5.3), where it is parsed and used to feed SInterviewer questionnaire engine. Here, there was extensive usage of sensible defaults. As mentioned in Sect. 4, SLang does not contemplate presentation aspects, which are, to a small extent, covered in SInterviewer questionnaire metadata Json notation.

5.5 SLang and SInterviewer Validation

Starting from the surveys mentioned in Sect. 5.2, backend databases and SInterviewer apps were created and set up for performing mock interviews for each survey. Figure 8 presents the validation scenario.

Mock interviews focused on performing the main navigation flows of questionnaires to check how well the data collection system would work. Interviewers had previous

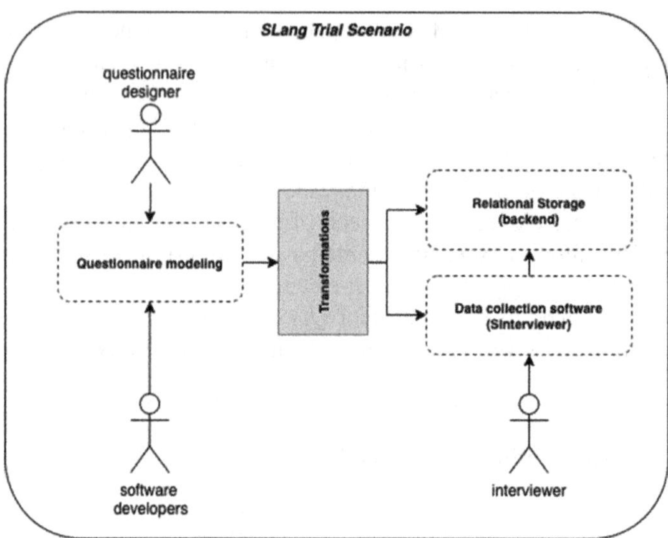

Fig. 8. SLang evaluation trial scenario.

experience with the original survey 1 and 2 data collection applications. Adjustments to questionnaire specification (required by interviewers) were simulated, to test the ability to quickly change questionnaire specifications and propagate them to the data collection system. In this scenario, the usage of the implemented transformations presented minor challenges specifically on *Expressions*, since excessively customized expressions do not have support both on SLang and SInterviewer. Also, the current version of SLang still lacks support for some operators and functions available on SInterviewer questionnaire metadata notation.

Overall, SInterviewer usage, with SLang generated input, performed well and received positive feedback from interviewers, when compared with previous data collection systems. Also, the time to deliver questionnaire changes was short, requiring only model adjustments, transformation execution and the publication of new versions of survey metadata. Some negative points were the impossibility of controlling presentation aspects on SInterviewer and the lack of support for highly customized expression rules.

The SQL generated schema also proved useful and avoided the need for domain expert data dictionaries that have to be translated into SQL schemas. It also standardized data constraints and integrity checks in schemas, which can potentially ease database schema administration for collected data relational databases. It also received positive feedback, since it eliminated the need to create and maintain data about the questionnaire design end, easing the integration between a mobile data collection and survey data collection management infrastructure.

The overall impression of stakeholders involved in the SLang mock trial was positive in the sense that having a questionnaire specification tool, and the possibility of developers working with model transformations generated a multitude of ideas that can be further explored in the future, such as: creating data tabulation systems input using model transformation, survey elements reuse through a survey models repository, survey comparisons, generation of inputs for legacy survey restoration, code generation to improve and ease systems integration, questionnaire version control support, among others.

6 Conclusions

This article presented the current status of SLang development, with improvements done to the domain model, and a broader scope evaluation of SLang in a mocked practical scenario. The effort to prototype and probe SLang as a model-driven approach to improve survey processes started with the goal of improving survey data collection systems through better communication between IT personnel and survey domain experts and through the improvement of software artifacts reuse, as well as elimination of unnecessary rework.

Questionnaire specification is the central point to achieving these goals. The usage of a model-driven approach proved a good strategy to maintain questionnaire specification centralized and to help generate artifacts that improved systems integration. Using the strategy of simulating a real survey scenario created an environment where stakeholders' feedback was rich in ideas for improvements, showing that there is considerable potential in the adoption of a model-driven approach.

Further work has been planned to further investigate and evaluate some points, before adopting SLang at an industrial scale. First, although SLang proved itself flexible enough for modeling complex questionnaires, usability evaluation is necessary to validate choices made in the concrete syntax and IDE approaches. Second, the evaluation stressed the potential benefits of combining a questionnaire presentation language with SLang. Third, survey IT infrastructure is diversified, and it is important to understand how far the usage of SLang can go. Further testing with code generation and transformations to integrate survey models with existing systems, such as survey metadata repository, survey data distribution, and publication systems, and data collection management systems, are necessary to better understand and evaluate SLang's full potential.

References

1. Saris, W.E., Galhofer, I.N.: Design, Evaluation, and Analysis of Questionaires for Survey Research, 2nd edn. Wiley, Hoboken (2014)
2. Araujo, L. Casanova, M.A., Leme, L., Furtado, A.: SLang: a domain-specific language for survey questionnaires. In: Proceedings of the 22nd International Conference on Enterprise Information Systems – Volume 2: ICEIS, pp. 133–144. Scitepress, Prague (2020)
3. Groves, R.M., et al.: Survey Methodology, 2nd edn. Wiley , Hoboken (2009)
4. GSIM and standards. https://statswiki.unece.org/display/gsim/GSIM+and+standards. Accessed 01 Aug 2020
5. DDI Alliance. https://www.ddialliance.org. Accessed 01 Aug 2020
6. SDMX. https://sdmx.org/. Accessed 01 Aug 2020
7. Borodin, A.V., Zavyalova, V.: Ontology-based semantic design of survey questionnaires. In: Proceeding of the 19th Conference of open Innovations Association (FRUCT), Jyvaskyla, pp. 10–15. IEEE (2016)

8. Araújo, L.C.: Model-driven questionnaires based on a domain specific language. Master dissertation presented to the Graduate Program in Informatics, Pontifical Catholic University of Rio de Janeiro, Rio de Janeiro (2019)
9. Mernik, M., Heering, J., Sloane, A.M.: When and how to develop domain-specific languages. ACM Comput. Surv. **37**(4), 316–344 (2005)
10. Nascimento, L.M.D., et al.: A systematic mapping study on domain-specific languages. In: The Seventh International Conference on Software Engineering Advances (ICSEA 2012), pp. 179–187. IARIA XPS Press, Lisboa (2012)
11. Kim, C.H., Grundy, J., Hosking, J.: A suit of visual languages for model-driven development of statistical surveys and services. J. Vis. Lang. Comput. **26**(99), 99–125 (2015)
12. Erdweg, S., et al.: Evaluating and comparing language workbenches: existing results and benchmarks for the future. Comput. Lang. Syst. Struct. **44**(A), 24–47 (2015)
13. Zhou, Y., Goto, Y., Cheng, J.: QSL: a specification language for e-questionnaire systems. In: IEEE 5th International Conference on Software Engineering and Service Science, Beijing, pp. 224–230. IEEE (2014)
14. Neighbors, J.M.: Software construction using components. Department of Information and Computer Science University of California, Irvine (1980)
15. Falbo, R.D.A., Guizzardi, G., Duarte, K.C.: An ontological approach to domain engineering. In: Proceedings of the 14th International Conference on Software Engineering and Knowledge Engineering (SEKE 2002), Ischia, pp. 351–358. ACM (2002)
16. Jatain, A., Goel, S.: Comparison of domain analysis methods in software reuse. Int. J. Inf. Technol. Knowl. Manag. **2**(2), 347–352 (2009)
17. Arango, G.: A brief introduction to domain analysis. In: SAC 1994 Proceedings of the 1994 ACM Symposium on Applied Computing, Phoenix, pp. 42–46. ACM (1994)
18. Czech, G., Moser, M., Pichler, J.: Best practices for domain-specific modeling. A systematic mapping study. In: 44th Euromicro Conference on Software Engineering and Advanced Applications, Prague, pp. 137–145. IEEE (2018)
19. Frank, U.: Some guidelines for the conception of domain-specific modeling languages. In: Proceedings of the 4th International Workshop on Enterprise Modelling and Information Systems Architectures, Hamburg, pp. 93–106 (2011)
20. Karsai, G., et al.: Design guidelines for domain specific languages. In: Proceedings of the 9th OOPSLA Workshop on Domain-Specific Modeling, Florida (2009)
21. Wegeler, T., et al.: Evaluating the benefits of using domain-specific modeling languages - an experience report. In: Proceedings of the 2013 ACM Workshop on Domain-Specific Modeling, Indianapolis, pp. 7–12. ACM (2013)
22. Voelter, M., et al.: DSL Engineering. CreateSpace Independent Publishing Platform (2013)
23. Campagne, F.: The MPS Language Workbench, 3rd edn. Campagnelab (2016). http://books.campagnelab.org
24. Fowler, M.: A pedagogical framework for domain-specific languages. IEEE Softw. **26**(4), 13–14 (2009)
25. PeNSE. https://www.ibge.gov.br/en/statistics/social/justice-and-security/16837-national-survey-of-school-health-editions.html?=&t=o-que-e. Accessed 15 Jan 2020
26. IBGE Census of Agriculture 2017. https://www.ibge.gov.br/en/statistics/economic/agriculture-forestry-and-fishing/21929-2017-2017-censo-agropecuario-en.html?=&t=o-que-e. Accessed 18 Dec 2019

Modeling of Robot Interaction in Coalition Through Smart Space and Blockchain: Precision Agriculture Scenario

Alexander Smirnov and Nikolay Teslya(✉)

SPC RAS, 14th line 39, Saint Petersburg, Russia
{teslya,smir}@iias.spb.su

Abstract. Modelling of interaction of intelligent agents is one of the urgent topics for verifying interaction models of joint task solving. It includes platform selection or new platform development that will provide functions for coalition formation, task decomposition and distribution, winnings sharing, and implementation of proposed techniques and models. This work focuses on modeling and visualizing the interaction of intelligent robots using open software Gazebo, and Robotic Operation System with information exchange between coalition members through distributed ledger technology and smart contracts using the Hyperledger Fabric platform. To adjust robot actions the ontologies of robot and context are used. Context ontology combines environmental characteristics with robots and task descriptions to provide a full situation context. The ontology of robot provides description of main robot functions and characteristics. The architecture of the modeling environment is described as the result of existing solutions overview and task required, as well as an example of modeling and visualization based on precision agriculture scenario.

Keywords: Coalition · Visualization · Modelling · Robot · Intelligent agent · Blockchain · Smart space

1 Introduction

One of the main directions in the field of collective work of robots is the study of interaction models of intelligent agents [1, 2]. Interaction of agents is most often required when solving a problem that cannot be solved by the efforts of one agent due to the lack of its capabilities. In this case, the task is divided into several independent subtasks. Each of them is then assigned to a separate agent [2, 3]. The importance and relevance of these studies are due to the development of robotic systems in which each robot can be considered as an independent intelligent agent. A robot can make decisions to achieve its own goals and a common goal. This interaction model is also called a coalition. A wide range of tasks in the development of coalition interaction models lies in the field of models that provide the most optimal choice of participants in a collective solution to the problem and ensure the interaction of coalition members [4, 5].

J. Filipe et al. (Eds.): ICEIS 2020, LNBIP 417, pp. 481–497, 2021.
https://doi.org/10.1007/978-3-030-75418-1_22

The formation of a coalition and the joint solution of tasks by robots can be demonstrated in most detail based on the following subject areas: precision agriculture, remote planet exploration, emergency medicine. In all these areas to date, many highly specialized robots have been developed. They effectively solve a limited set of tasks within the framework of the conditions for which they were developed. For example, to solve the problem of precision farming, some robots can conduct soil quality exploration, automated seeders, cultivators, tractors, etc. However, to solve a complex problem that goes beyond the conditions of each robot, it is required to ensure their joint work. So, to ensure the full cycle of growing crops, it is necessary to solve the problems of field exploration, selection of crops grown, sowing, watering, and harvesting. To do this, robots should form a coalition in such a way to solve a complex task, while ensuring the maximum overall effectiveness of the whole coalition and each robot within it. Efficiency for different subject areas is calculated depending on the gain that can be obtained when solving the problem: the maximum harvest in the problem of precision farming, the number of people saved (the number of survivors a month after the disaster) for disaster medicine, and the total amount and time spent on obtaining, processing and transmitting to the customer information about the state of the monitoring objects of interest to him for remote sensing tasks.

The solution to the problem of forming a coalition based on the mathematical theory of fuzzy cooperative games is presented in a previous work of the authors [6]. This work focuses on ensuring the interaction of coalition members through distributed ledger technology and smart contracts using the Hyperledger Fabric platform, as well as modeling and visualizing the interaction of intelligent robots using open software Gazebo, and Robotic Operation System.

Compared to the earlier work by the authors [7], this work presents a detailed description of the environment used to model coalition formation and interaction of coalition members during the complex task solving. The case study is based on the coalition formation scenario for precision agriculture presented earlier in work [6]. This work also provides an overview of existing solutions for interaction modeling and implementation of the basic interaction model using a cyberphysical space environment and Gazebo/ROS for result visualization.

The paper is structured as follows. Section 2 discusses the existing approaches to coalition formation and methods for modeling and visualizing robot interaction. Section 3 presents the development of a coalition interaction model based on the concept of smart spaces, including a description of the ontology-based context model used to form the coalition. Section 4 presents the architecture for the coalition members' interaction through a distributed ledger. Section 5 shows modeling of the interaction of coalition members for precision agriculture scenario using robotic models and simulated physical world in the Gazebo environment. Section 6 concludes the article and describes future work directions.

2 Related Work

Two major approaches can be distinguished in the process of coalition organizing: centralized organization and decentralized. In the centralized organization of the coalition, a command center is used where decisions are made on the composition of the coalition, the distribution of tasks, and a plan for solving the problem is formed. The command center can be hosted on a separate high-powered computing device that performs only the functions of the center, and a robot that performs tasks along with the rest [8]. In this case, the structure of the coalition can be multilevel hierarchical, in which robots at each underlying level obey only one center of their superior level [9]. The decentralized organization of the coalition usually implies the absence of a decision center, often focusing on bio-inspired methods of organizing collaboration, such as swarms and flocks [10, 11]. At the same time, robots are considered equal in a hierarchy and all of them are guided by the same algorithms when making a decision.

Human operators can also be involved in the process of coalition formation. Three levels of automatization in multi-robotic systems are identified in work [12] based on human involvement:

- 1st level: coalition formation, task solving, and decomposition are carried out by a human, and task solving is carried out by a system under human control.
- 2nd level: this level divided into two categories: (1) task decomposition and coalition formation and carried out by a human, but task distribution and task solving are carried out autonomously by the system; (2) task decomposition and distribution are carried out by a human, but coalition formation and task solving are carried out autonomously by a system. Paper describes a lot of existing systems, that achieved this level.
- 3rd level: coalition formation, task distribution between robots, and task solving passes without any intervention of a human. Paper describes existing systems, that achieved this level but all of them verified only in environment simulation.

Work [12] also provides a review, which concludes that in existing systems, a human plays a decisive role in the management, decomposition, and distribution of tasks. To reduce human intervention, it was proposed to use methods of autonomous machine learning, the dynamic formation of coalitions and the distribution of tasks, the creation of a unified framework using big data and Internet of Things technologies.

Hierarchical clustering can also be used in the management and coordination of several robots [13]. Hierarchical clustering is a formalism for identifying and representing cohesive groups of robots in the form of trees. The authors of the article formalize and use this relation, developing algorithms for navigation in a combinatorial space and a specific choice of hierarchy. In this paper, a mathematical model for disk-shaped robots was proposed that describes a specific hierarchical navigation environment with centralized control. By the moment of preparing this work, the system of [13] had been not physically implemented and existed in the form of a simulation of the environment.

Finding the most optimal coalition formation and the distribution of tasks between robots is also an urgent issue [14]. An important factor in optimizing task solving is to determine the number of robots needed to solve the task on time, which is achieved using the ANT dynamic coalition formation technology. The article describes the more efficient use of local search in the memetic algorithm, without using a global search.

The paper [15] describes the formation of a team based on swarm intelligence to solve RoboCup Rescue League search tasks. As a result, the eXtreme-Ants algorithm was developed based on the tokens distribution function that contains a list of tasks. The agent who received the token decides on the possibility of completing the task and has the opportunity to refuse the token by sending it to another random agent.

Then, an experimental comparison of the developed algorithm (eXtreme-Ants) was carried out, based on the analysis of a swarm and two task distribution algorithms based on a generalized assignment task. The experiment showed that using the distribution model of tasks in eXtreme-Ants allows you to perform more efficient coordinated actions. The decision-making system is probabilistic and requires high communication and computing power to increase efficiency. The implemented algorithm does not provide for the automatic decomposition of tasks.

The paper [16] presents a task distribution algorithm for a system with several robots, in which tasks are divided into disjoint groups, and there are also priority restrictions between task groups. The existing auction methods for distributing tasks assume the independence of tasks and the independence of robots from the tasks themselves. In the paper, each robot can perform a fixed set of tasks and receive benefits/costs for each task.

Tasks are divided into groups, and each robot can perform only one task from each group. These restrictions arise when robots have to complete a set of tasks with priority restrictions, and each task takes the same time.

As a result, an extended auction algorithm using the shared memory model was presented to provide an optimal solution to the task assignment task with established priority restrictions. The developed algorithm ends with a finite number of iterations.

The coalition formation model to increase the efficiency of joint search and pursuit could be created by minimizing the delay in the persecution and minimizing the size of the coalition to pursue [17]. An algorithm was developed that takes into account the limitations and delays of communication when searching for potential coalition members. To reduce computational complexity, a multistage non-optimal coalition formation algorithm with low computational complexity was used. For autonomous work of the coalition in solving search and prosecution problems, a strategy for managing a coalition of robots based on a finite state machine was adopted, that is, the functions of decomposition and distribution of tasks lie on the person.

A similar question was also discussed in work [18] in the scope of using UAVs for search and pursuit of a target. It analyzes two algorithms: (1) An algorithm for forming a coalition in polynomial time; (2) An algorithm for forming an optimal coalition. Both algorithms determine the optimal number of agents that have enough resources to pursue a goal.

As a result of the work, a method for optimizing a swarm of objects was developed, which estimates the global efficiency of algorithms. Through simulation, the effectiveness of the algorithms is evaluated in terms of the complexity of the algorithms and the time required to form a coalition. As a result of the simulation, it was accepted that the efficiency of these algorithms is close to the same computing resources.

The information exchange between coalition members is an important component of the joint problem solving by a coalition, since it requires notification of coalition members about the current state of the problem solution for organizing coordinated actions, or monitoring the implementation of the plan [19]. The information exchange can be organized through a common centralized repository of information on a separate device or by the distribution of information between coalition members [20]. It also considers the combination of two approaches with the formation of so-called smart spaces - the creation of a common repository of information that provides links to resources that are coalition members, which makes information distributed among all participants [21]. There are also solutions based on peer2peer networks and a distributed registry that provides a quick distribution of information between all participants while duplicating all the information on the device of each participant [22, 23]. The distributed ledger technology can be considered as an example of peer2peer based networks. In addition to the benefits of the peer2peer network, the distributed ledger solutions also provide information protection against rewriting, which can be useful when organizing a coalition with the requirement to ensure trust between the participants without a single certification center [23].

Table 1. Coalition formation systems.

Article, Year	Task	Coalition formation	Task distribution	Verification
[28], 2017	Overcoming obstacles	State machine	Human	Modeling in smart-M3
[29], 2016	Search	State machine	Human	Modeling in smart-M3
[30], 2016	Navigation	Swarm intelligence	–	Simulation modeling
[13], 2016	Navigation	Hierarchical clustering	Hierarchical clustering	Mathematical modeling
[14], 2015	Watching	Dynamic ANT coalition formation	State machine	Simulation modeling
[15], 2011	RoboCup rescue	Swarm intelligence	–	Simulation modeling
[16], 2011	Limited priority tasks	–	Auction	Mathematical modeling
[31], 2016	Search and pursuit	Multistage non-optimal coalition formation algorithm	–	Simulation modeling
[18], 2011	Search and pursuit	Object swarm optimization	–	Mathematical modeling

The type of coalition organization also influences how tasks and resources are distributed among coalition members. Centralized hierarchical coalitions usually organize work through centralized planning when receiving a task. At higher nodes, a work plan is built taking into account the capabilities of lower nodes and coalition resources, in which the performers and the procedure for solving problems are fixed [24]. This ensures that the stages of the plan and the entire plan are completed by a certain date with an accurate forecast of the expenditure of resources and the payment of remuneration if this is provided for by the conditions of the task. However, this solution is not flexible, because when an emergency occurs, it leads to a deviation from the plan, with the need for its correction or complete reorganization. Decentralized coalitions are based on the adaptation of participants to current conditions, with the absence of a single plan for solving the problem [25, 26]. This provides the flexibility to solve the problem under conditions of frequent changes in the composition of the coalition or available resources but limits the ability to predict the time of solving the problem [27].

Below is a Table 1 that provides overview of existing approaches to coalition formation and ways of verification of them. Based on the data in the table, it can be seen that existing multi-robotic systems, used for various tasks, don't have full automatization of core processes – coalition formation and task distribution between robots. Most of the existing systems are simulation or mathematical models and can't be applied to physical usage and solving physical tasks. Also, existing systems don't have a unified approach to the forms of a coalition, dynamic assessment of necessity to create a particular form of the coalition following its usefulness for solving a joint task.

Special attention should also be paid to simulation and visual modeling of the interaction of robots. This approach simplifies hypothesis testing by reducing development costs and allows to visually present the results of the proposed approaches. Some researchers develop their visualizations, displaying robots with conventional signs since the tasks under consideration do not require the detailed design of the robot [10]. For detailed visualization, the Gazebo visualization package is most used in combination with the robot operating system (ROS) for controlling a virtual robot [32–35].

3 Interaction of Coalition Participants in Cyberphysical Space

Robots are interacting through the cyberphysical framework presented previously in work [6]. The framework is based on the smart cyberphysical space created on the top of a smart space concept (based on the "blackboard") and blockchain. It provides the ability to organize the basic interaction of robots in the physical and cyber (virtual) spaces. The interaction includes solo and joint manipulations with physical objects, information exchange about the current state of robots, and objects for planning further joint actions during the coalition formation.

Coalition members are intelligent agents (robots in the scope of this work) with different equipment, environmental sensors, software modules. For their interaction, it is necessary to ensure semantic interoperability between them. This allows to specify a description of the properties of the coalition members and context in a format that is understandable to everyone, from machines to the system's operators, as well as automate the search for coalition members following the requirements for the task at hand. An ontology is used to ensure semantic interoperability. The model of the context, the coalition participant, and tasks are described using the ontological modeling apparatus, which allows the context-driven dynamic formation of the coalition and the distribution (redistribution) of tasks, roles, and system resources taking into account not only competencies but also the current situation in the coalition.

To build an ontological model of the context, the main scenarios of the interaction of robots, robot designs, and typical tasks that each type of robot solves for disaster medicine, precision farming, and remote sensing of the Earth had been analyzed. As a result, the basic concepts that should be present in the ontology were identified, among which three groups stand out: i) concepts describing the current situation; ii) concepts describing the design and functions of the robot; iii) concepts describing the requirements and conditions for solving specific applied problems posed to a given group of robots.

The ontology graph for context is presented on Fig. 1. For the current situation, the concepts describe the physical parameters of the environment (for example, temperature, wind speed and direction, humidity, current time, atmospheric pressure for precision farming and disaster medicine, for remote sensing - the tactical and technical characteristics of the onboard target and supporting equipment, potential areas of interaction with objects observation and ground-based points for receiving Earth remote sensing data, light level, solar activity, radiation level, quality indicators of function observation of spacecraft, etc.), the position of all the coalition robots, the position and properties of objects in the field of action of the coalition.

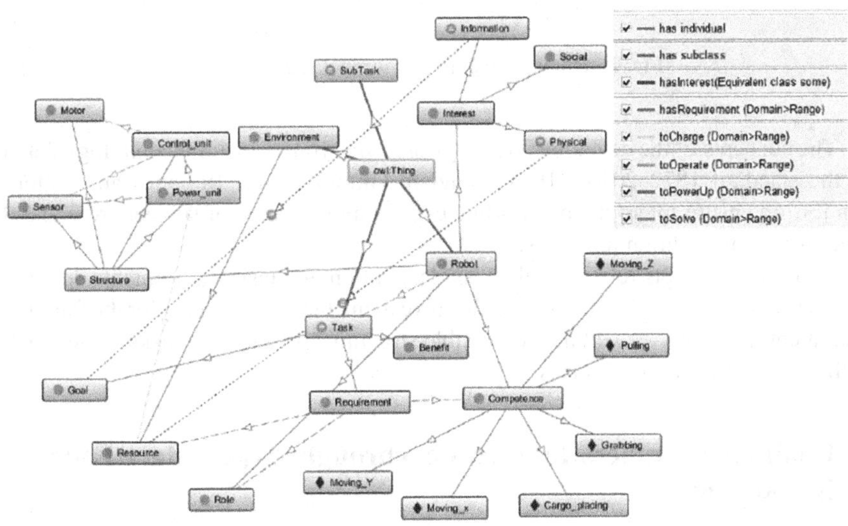

Fig. 1. Context ontology [7].

Ontology concepts for describing the design and functions of the robot provide characteristics of the hardware and software components of the robot (number, types, measuring ranges and current sensor readings, number, types, the current position of motors, state and current battery charge, fuel level (if there is a fuel tank) computing power (frequency of the processor, the amount of RAM, data storage, connection to data networks, information about the characteristics of the selected geolocation systems)) and many components (installed libraries, software modules, platforms, control code for hardware, action templates), as well as a description of the functions and possible technologies that the robot can perform using its hardware and software (see Fig. 2).

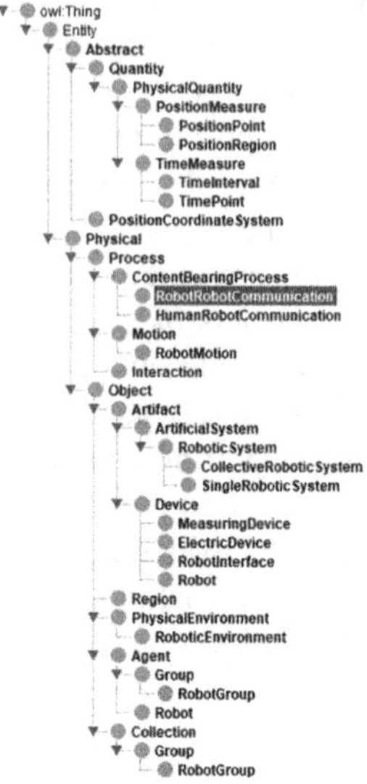

Fig. 2. Ontology hierarchy for robot representing.

The concepts of the developed ontology are synchronized with the ontology defined by the standard 1872–2015 - IEEE Standard Ontologies for Robotics and Automation [36] to ensure interoperability while expanding the range of subject areas and the composition of coalition members.

Ontology concepts related to the task description and task requirements determine the types of tasks, the resources needed to solve them, the requirements for the functional equipment of robots, as well as the possible structures for dividing tasks solved by the coalition of robots into sub-tasks.

4 Coalition Members Interaction Through Hyperledger Fabric Blockchain

In this work, the Hyperledger Fabric platform has been chosen for the blockchain network and smart contacts implementation. The choice of the platform is justified by its architecture, which makes it easy to adapt the coalition structure into the platform structure (see Fig. 3). The main elements of the architecture are nodes, divided into three levels: "Client", "Peer", "Orderer". Client level corresponds to robots whose main task is to send data from sensors or to perform operations and report on their performance. For example in precision farming, such robots can be tools of a combined harvester, scouts, and transport robots. On a higher level, there are devices that collect information and execute the smart contracts - "Peer". The example of a peer node is a control block of a combine harvester. Their main task is to collect information from the lower level, process it using smart contracts, and transfer it to the upper level, in which information will be disseminated and stored. The highest level is "Orderer". Its task is to store information in the appropriate chain of blocks, to ensure the coordination and distribution of the new block between other Orderers and corresponding Peers.

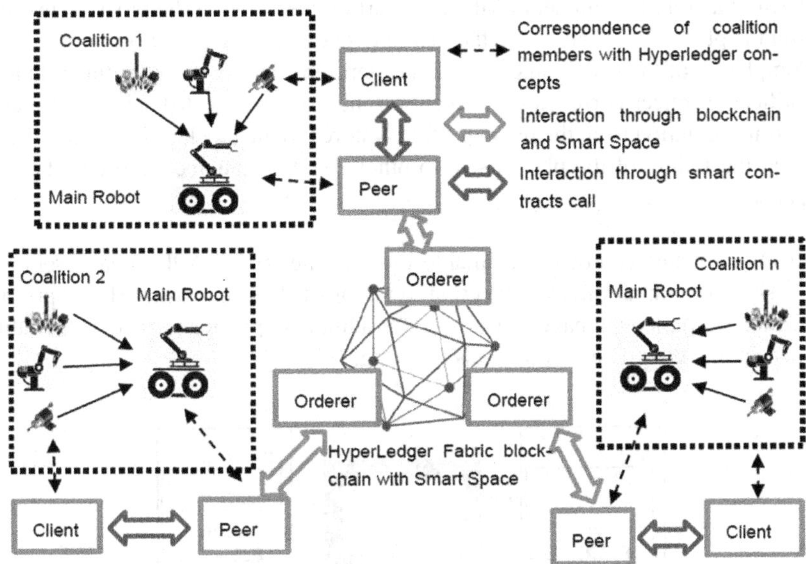

Fig. 3. Coalition members interaction through HyperLedger fabric with smart space support [7].

For the implementation of additional protocols over an ordered list of published statements, for example, ownership and transfer of consumables, it is necessary to keep a certain state and track its change according to generally accepted rules following published statements. All changes of coalition state are reflected in the smart space using adding/deleting relevant information according to the in the ontology. During the interaction between coalition participants through the IoT platform with the blockchain support, all smart contacts can be called either directly through the provided transaction initiation interface in the blockchain or using other smart contacts methods. To simplify the use of custom protocols in blockchain, the method of any contract placed in the contract chains should be called using the basic smart contract.

For this purpose, a basic smart contract is used, which provides the following functions of interaction between coalition participants:

1. Schedule upload using XML format. It provides:

   a. Receiving an execution plan in XML.
   b. Parsing the plan, extracting the robots, tasks associated with robots, the order of the task execution, and the timing of each task.

c. Generating entries in the blockchain from the extracted items. Each entry must contain one task, which is associated by the coalition member responsible for its execution, and timestamps of execution start and end for each task.
d. Sending notification to the blackboard through blockchain knowledge processor to start the plan execution.

2. Start the task execution. Accept a message from the robot about the start of the task execution, store the moment of the real start of execution in the blockchain, verify with the planned one, and store the fact in case of a strong deviation.
3. Completion of the task. Same as with the start the contract provides function to get notification, check correctness of execution and store this fact to the ledger. In case of a strong deviation from the plan, this fact is stored to the blockchain and notification about the failure of the plan is sent to other coalition members through the smart space.

The basic smart contract is available through the REST API so each robot can access the plan without direct connection to the blockchain just using HTTP protocol. In addition, a simple web page is available for a human operator to check the coalition state and follow the process of joint task solving.

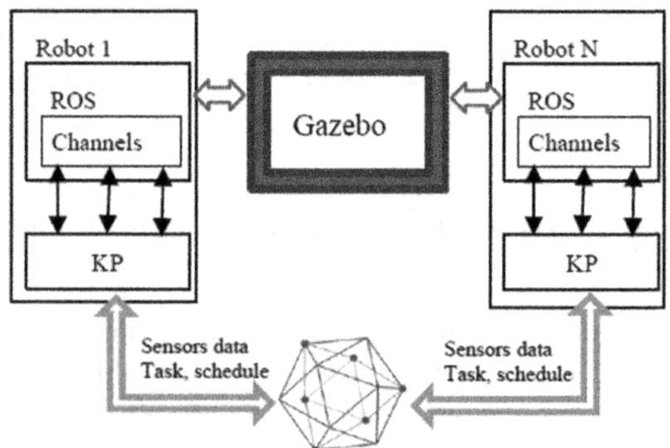

Fig. 4. Architecture for robot interaction modellings [7].

5 Visualization of Coalition Interaction in Precision Agriculture Scenario

For the experiments, a precision agriculture scenario of coalitional interaction of robots was implemented in a Gazebo modeling environment [37, 38] with ROS Melodic framework [39] (see Figs. 4 and 5). The use of the above software is currently widespread in the field of robotic modeling tools and includes a diverse library of elements and off-the-shelf devices, which can significantly reduce model development time. Due to the possibility of connecting third-party models, it is also possible to quickly create a three-dimensional physical world that simulates the real world with automatic object collision calculation, environmental physical properties and parameters. This feature allows to faithfully display the physical world and take into account a large number of parameters during modeling.

Using the ROS operating system also has several advantages. The first is that this operating system supports the basic functions of real-life robots related to controlling servo drives, receiving data from sensors and exchanging data with the external environment through an accessible communication channel. Interaction between robots is carried out through specially organized channels in the operating system, to which a program can be subscribed and receive sensor readings from them or send control commands. In the presented architecture, the robot knowledge processor (KP) subscribes to channels and transmits messages from the robot to the smart space and blockchain using the ontology, as well as receives back the task that needs to be completed, and the schedule and controls the actions of the robot.

The second important advantage is the ability to integrate ROS both in existing robots and in robot models created in the Gazebo simulation environment. This allows to develop a model of robots interaction in the virtual world and then, with minimal changes, transfer it to physical robots, and due to high-quality simulation of the physical world conditions, many features of the physical environment will be taken into account during virtual implementation and estimation.

The following robot models were developed:

1. Combined harvester robot with the following characteristics:
a. The front and back parts have lift mechanisms that allows to rise or down parts of robots to process field (plowing, seeding, harvesting).
b. On the front and back parts of the robot there are laser distance sensors to measure the distance to objects, directed in parallel to the surface on which the robot moves.
c. The robot is equipped with a coordinate sensor (GPS/GLONASS in real robot) for tracking location.
2. Quadcopter Robot.
a. The robot has the functionality of a standard quadcopter for moving in three-dimensional space.
b. The robot is equipped with a laser distance sensor to scan for obstacles and is directed downward perpendicular to the plane of rotation of the rotors.
c. GPS/GLONASS location sensor.
3. Scout Robot.
a. Four-wheel mobile robot with the function of moving on a two-dimensional surface without huge obstacles.
b. The robot is equipped with laser distance sensors located around the perimeter of the body to scan the space around the robot.
c. GPS/GLONASS location sensor.

The transmission of control signals to the robot and the receipt of data from the sensors is carried out through subscription to the channels in ROS associated with the equipment of the robots (Fig. 5). For example, for the all-terrain robot, channels have

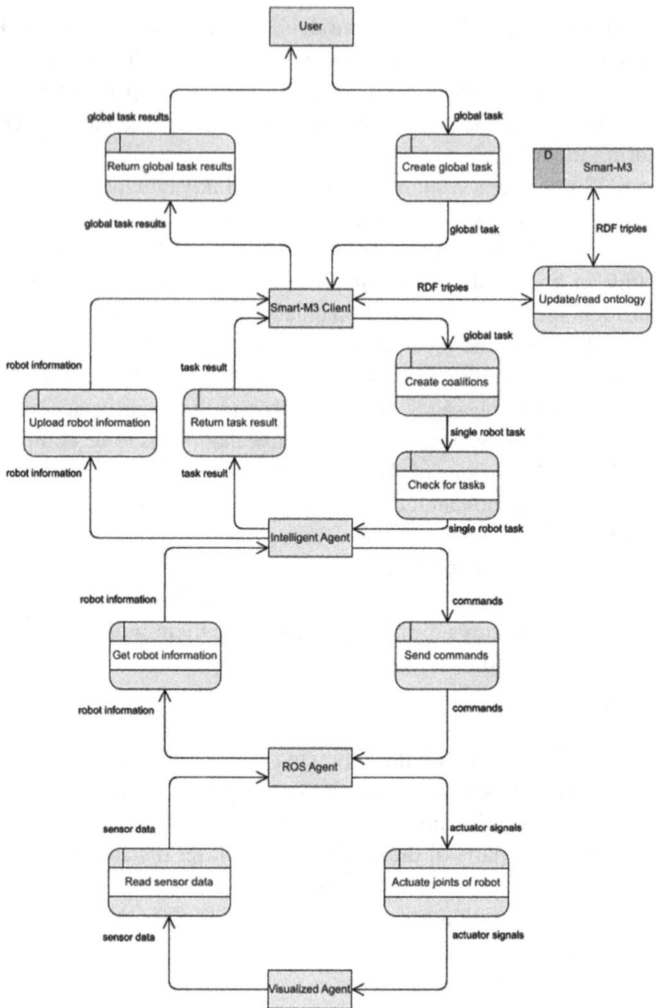

Fig. 5. Data flow of a single simulated robot.

been created through which data from the distance sensor and coordinates can be received and independent control actions to each of the robot servos can be sent. For each of the servos, it is also possible to obtain its status by accessing the corresponding channel.

Smart-M3 client is used as a centralized intelligent data storage. It receives information from intelligent agents such as their location, battery charge and so on. Information in Smart-M3 platform is fully mapped to ROS topics by using robot ontology. When a global task is given to the Smart-M3 client, intelligent agents coalitions are formed based on information gathered from intelligent agents. Then each intelligent agent receives an individual task and starts its execution.

Robot models that implement the functionality of quadrocopter and scout robots were taken from open libraries for ROS and Gazebo and are the most popular today for robotic modeling in these environments. The quadcopter model, which implements the quadcopter robot functionality, was implemented based of the hector_quadrotor free model package [40, 41]. This package was developed at the University of Darmstadt and contains a description of the robot in the URDF format for Gazebo simulation, parameters necessary for simulation, a remote-control package, as well as a large number of packages for simulating quadcopter movement physics.

Scout robot model was implemented based on the husky_gazebo package [42]. This package was developed by Clearpath Robotics to simulate its mass-produced Husky robot and, like the hector package, contains a description of the robot in URDF format and various examples and tutorials of working with the robot.

The first stage of the scenario is to explore an unknown territory and build a map of it. As the area is quite small and grouped in one place it is more efficient for one scout robot to explore it to provide a more accurate map. When scout receives an exploration task with the coordinates of the target territory it begins the exploration. After it explored the whole area the map is saved and sent to the Smart-M3 client. The generated map of the scene after exploration is presented on Fig. 6.

After exploration sowers are spawned together with some unknown obstacles (boxes) which also need to be avoided. This is the initial state for seeding stage. Each robot is registered in the Smart-M3 client, received a map of the area and its initial position on the map. They have their navigation configured and could go to any place on the map avoiding unknown obstacles on their way and planning their path using the known map.

When robots receive seeding tasks, they plan the path to the target location and start the execution. Robot distribution between the areas is shown on Fig. 7.

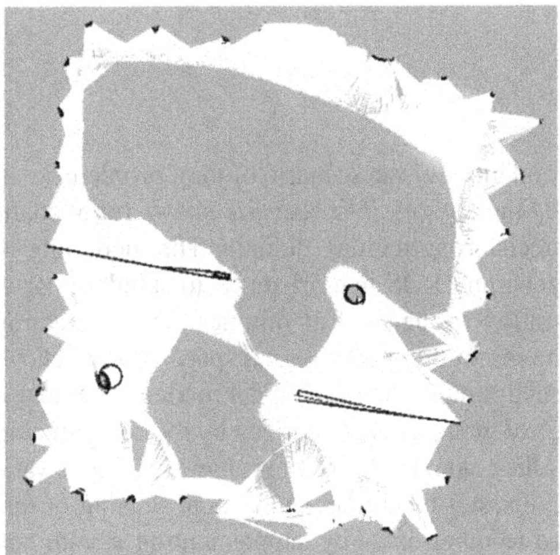

Fig. 6. Map created after exploration stage.

No seeding manipulators for robots have been modeled because of high complexity so robots imitate seeding by driving around the field. There are different sowing strategies, for example spread sowing or line sowing. During the spread sowing robots throw seeds on the ground. Line sowing is putting seeds in the ground with a certain depth and in lines. Implemented robots could imitate both seeding strategies depending on the path planner configuration.

During planning the path to the target location, the global planner of robots analyzes obstacles on the map and creates the shortest path to the target location. It could be seen on Fig. 7.

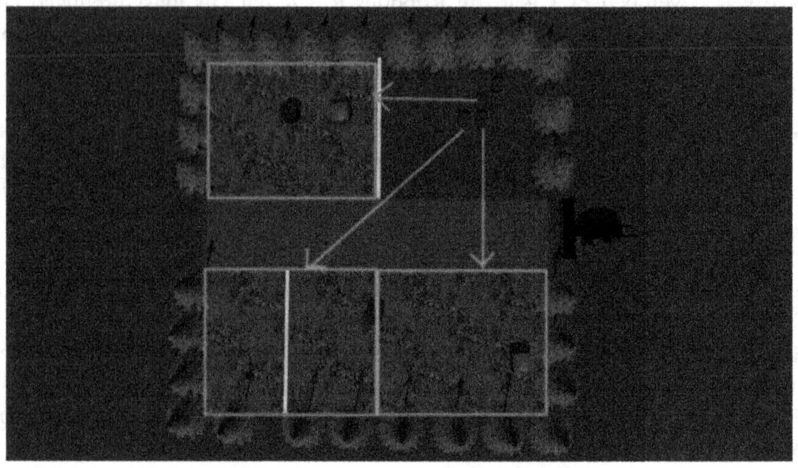

Fig. 7. Robot distribution between seeding areas.

6 Conclusions

The paper provides modelling of the scenario of joint problem solving by robots' coalition proposed early by the authors. The scenario shows robot coalition formation for a field processing in precision agriculture domain. The modelling is based on the open source packages Gazebo and ROS that are de-facto standards of virtual environments modelling in visualization in the field of robotics. The interaction between robots is implemented based on the smart cyberphysical space with blockchain support.

To describe the environment the ontological model of context has been developed. The obtained ontological model is characterized by the concretization of the parameters applicable to describe the context of the joint problem solving in various fields, including precision agriculture, disaster medicine, and remote sensing of the Earth. The context model can be adapted to other areas by supplementing it with concepts that describe the current situation, specific to these areas, while maintaining the context in terms of describing the structure and functions of robots and tasks. Also, the ontology for robot description has been developed. It provides main robot functions and characteristics that are used for choosing robots into coalition.

The visualization had been created that shows scenario of robot cooperation for field processing. The result shows specific of map generation with virtual robot sensors and task distribution between robots in coalition. The modeling shows that the selected approach can be further used to implement more detailed precision agriculture scenarios with higher number of robots and longer time interval of simulation.

Acknowledgements. This research was funded by RFBR Grant No. 17-29-07073 and by Russian State Research No. 0073-2019-0005.

References

1. Vig, L., Adams, J.: Issues in multi-robot coalition formation. Multi-Rob. Syst. Swarms Intell. Autom. **3**, 15–26 (2005). https://doi.org/10.1007/1-4020-3389-3_2
2. Bayram, H., Bozma, H.I.: Coalition formation games for dynamic multirobot tasks. Int. J. Rob. Res. **35**, 514–527 (2015). https://doi.org/10.1177/0278364915595707
3. Cui, R., Guo, J., Gao, B.: Game theory-based negotiation for multiple robots task allocation. Robotica **31**, 923–934 (2013). https://doi.org/10.1017/S0263574713000192
4. Klusch, M., Gerber, A.: Dynamic coalition formation among rational agents. IEEE Intell. Syst. **17**, 42–47 (2002). https://doi.org/10.1109/MIS.2002.1005630
5. Dukeman, A., Adams, J.: Hybrid mission planning with coalition formation. Auton. Agents Multi-Agent Syst. **31**(6), 1424–1466 (2017). https://doi.org/10.1007/s10458-017-9367-7
6. Smirnov, A., Sheremetov, L., Teslya, N.: Fuzzy cooperative games usage in smart contracts for dynamic robot coalition formation: approach and use case description. In: ICEIS 2019 - Proceedings of the 21st International Conference on Enterprise Information Systems. SCITEPRESS - Science and Technology Publications, pp. 349–358 (2019)
7. Smirnov, A., Teslya, N.: Modelling and visualization of robot coalition interaction through smart space and blockchain. In: Proceedings of the 22nd International Conference on Enterprise Information Systems. SCITEPRESS - Science and Technology Publications, pp. 278–286 (2020)
8. Smirnov, A., Kashevnik, A., Ponomarev, A.: Multi-level self-organization in cyber-physical-social systems: smart home cleaning scenario. In: Procedia CIRP. Elsevier B.V., pp. 329–334 (2015)
9. Guerrero, J., Oliver, G., Valero, O.: Multi-robot coalitions formation with deadlines: complexity analysis and solutions. PLoS One **12**, 1–26 (2017). https://doi.org/10.1371/journal.pone.0170659
10. Koes, M., Nourbakhsh, I., Sycara, K.: Heterogeneous multirobot coordination with spatial and temporal constraints. AAAI Work - Technical Report WS-05–06:9–16 (2005)
11. Yu, L., Cai, Z.: Robot exploration mission planning based on heterogeneous interactive cultural hybrid algorithm. In: 5th International Conference on Natural Computation, ICNC 2009, pp. 583–587 (2009)
12. Rizk, Y., Awad, M., Tunstel, E.W.: Cooperative heterogeneous multi-robot systems: a survey. ACM Comput. Surv. **522**, 1–31 (2019). https://doi.org/10.1145/3303848
13. Arslan, O., Guralnik, D.P., Koditschek, D.E.: Coordinated robot navigation via hierarchical clustering. IEEE Trans. Robot **32**, 352–371 (2016). https://doi.org/10.1109/TRO.2016.2524018
14. Padmanabhan, M., Suresh, G.R.: Coalition formation and Task Allocation of multiple autonomous robots. In: 2015 3rd International Conference on Signal Processing, Communication and Networking, ICSCN 2015. Institute of Electrical and Electronics Engineers Inc. (2015)

15. dos Santos, F., Bazzan, A.L.C.: Towards efficient multiagent task allocation in the RoboCup rescue: a biologically-inspired approach. Auton. Agent Multi Agent Syst. **22**, 465–486 (2011). https://doi.org/10.1007/s10458-010-9136-3
16. Luo, L., Chakraborty, N., Sycara, K.: Multi-robot assignment algorithm for tasks with set precedence constraints. In: Proceedings - IEEE International Conference on Robotics and Automation, pp. 2526–2533 (2011)
17. Liu, Z., Gao, X.G., Fu, X.W.: Coalition formation for multiple heterogeneous UAVs in unknown environment. In: Proceedings - 5th International Conference on Instrumentation and Measurement, Computer, Communication, and Control, IMCCC 2015. Institute of Electrical and Electronics Engineers Inc., pp. 1222–1227 (2016)
18. Manathara, J.G., Sujit, P.B., Beard, R.W.: Multiple UAV coalitions for a search and prosecute mission. J. Intell. Robot Syst. Theory Appl. **62**, 125–158 (2011). https://doi.org/10.1007/s10846-010-9439-2
19. Verma, D., Desai, N., Preece, A., Taylor, I.: A block chain based architecture for asset management in coalition operations. In: Pham, T., Kolodny, M.A. (eds.) Proceedings of SPIE 10190, Ground/Air Multisensor Interoperability, Integration, and Networking for Persistent ISR VIII, p 101900Y (2017)
20. Shabanov, V., Ivanov, D.: Organization of information exchange in coalitions of intelligent mobile robots. In: 2019 International Conference on Industrial Engineering, Applications and Manufacturing (ICIEAM-2019), pp. 1–5 (2019).https://doi.org/10.1109/ICIEAM.2019.8743043
21. Castelló Ferrer, E.: The blockchain: a new framework for robotic swarm systems. In: Arai, K., Bhatia, R., Kapoor, S. (eds.) FTC 2018. AISC, vol. 881, pp. 1037–1058. Springer, Cham (2019). https://doi.org/10.1007/978-3-030-02683-7_77
22. Qian, B., Cheng, H.H.: Bio-inspired coalition formation algorithms for multirobot systems. J. Comput. Inf. Sci. Eng. **18**, 1–8 (2018). https://doi.org/10.1115/1.4039638
23. Liang, X., Xiao, Y.: Studying bio-Inspired coalition formation of robots for detecting intrusions using game theory. IEEE Trans. Syst. Man, Cybern. Part B Cybern. **40**, 683–693 (2010). https://doi.org/10.1109/TSMCB.2009.2034976
24. Smirnov, A., Kashevnik, A., Teslya, N., Mikhailov, S., Shabaev, A.: Smart-M3-based robots self-organization in pick-and-place system. In: 2015 17th Conference of Open Innovations Association (FRUCT). IEEE, pp. 210–215 (2015)
25. Hartanto, R., Eich, M.: Reliable, cloud-based communication for multi-robot systems. In: 2014 IEEE International Conference on Technologies for Practical Robot Applications (TePRA). IEEE, pp. 1–8 (2014)
26. Tosello, E., Fan, Z., Castro, A., Pagello, E.: Cloud-based task planning for smart robots. In: Chen, W., Hosoda, K., Menegatti, E., Shimizu, M., Wang, H. (eds.) IAS 2016. AISC, vol. 531, pp. 285–300. Springer, Cham (2017). https://doi.org/10.1007/978-3-319-48036-7_21
27. Ivanov, D.: Decentralized planning of intelligent mobile robot's behavior in a group with limited communications. Springer International Publishing (2019)
28. Smirnov, A., Kashevnik, A., Petrov, M., Parfenov, V.: Context-based coalition creation in human-robot systems: approach and case study. In: Interactive Collaborative Robotics. ICR 2017. Lecture Notes in Computer Science. Springer, Cham, pp. 229–238 (2017). https://doi.org/10.1007/978-3-319-66471-2_25
29. Teslya, N., Savosin, S.: Smart-M3-based robot interaction in cyber-physical systems. In: Proceedings of 16th Conference of Open Innovations Association FRUCT. IEEE, pp. 108–114 (2014)
30. De Oliveira, L.F.F., De Lima, F.B., Oliveira, S.C., Bastos-Filho, C.J.A.: A fuzzy-swarm based approach for the coordination of unmanned aerial vehicles. J. Intell. Fuzzy Syst. **31**, 1513–1520 (2016). https://doi.org/10.3233/JIFS-151214

31. Liu, Z., Gao, X.G., Fu, X.W.: Coalition formation for multiple heterogeneous UAVs cooperative search and prosecute with communication constraints. In: Proceedings of the 28th Chinese Control and Decision Conference, CCDC 2016. Institute of Electrical and Electronics Engineers Inc., pp. 1727–1734 (2016)
32. Xue, F., Tang, H., Su, Q., Li, T.: Task allocation of intelligent warehouse picking system based on multi-robot coalition. KSII Trans. Internet Inf. Syst. **13**, 3566–3582 (2019). https://doi.org/10.3837/tiis.2019.07.013
33. Alkilabi, M., Narayan, A., Tuci, E.: Cooperative object transport with a swarm of e-puck robots: robustness and scalability of evolved collective strategies. Swarm Intell. **11**(3–4), 185–209 (2017). https://doi.org/10.1007/s11721-017-0135-8
34. Li, B., Moridian, B., Kamal, A., Patankar, S., Mahmoudian, N.: Multi-robot mission planning with static energy replenishment. J. Intell. Rob. Syst. **95**(2), 745–759 (2018). https://doi.org/10.1007/s10846-018-0897-2
35. Barbosa, F.S., Duberg, D., Jensfelt, P., Tumova, J.: Guiding autonomous exploration with signal temporal logic. IEEE Robot Autom. Lett. **4**, 3332–3339 (2019). https://doi.org/10.1109/lra.2019.2926669
36. IEEE Robotics and Automation Society: IEEE Standard Ontologies for Robotics and Automation (2015)
37. Koenig, N., Howard, A.: Design and use paradigms for Gazebo, an open-source multi-robot simulator. In: 2004 IEEE/RSJ International Conference on Intelligent Robots and Systems (IROS), pp. 2149–2154 (2004)
38. Agüero, C.E., et al.: Inside the virtual robotics challenge: simulating real-time robotic disaster response. IEEE Trans. Autom. Sci. Eng. **12**, 494–506 (2015). https://doi.org/10.1109/TASE.2014.2368997
39. Stanford Artificial Intelligence Laboratory et al. (2018) ROS.org|Powering the world's robots. https://www.ros.org/. Accessed 5 Jan 2020
40. Meyer, J., Sendobry, A., Kohlbrecher, S., Klingauf, U., Von Stryk, O.: Comprehensive simulation of quadrotor UAVs using ROS and Gazebo. In: Lecture Notes in Computer Science (including subseries Lecture Notes in Artificial Intelligence and Lecture Notes in Bioinformatics), pp. 400–411 (2012)
41. Meyer, J., Sendobry, A., Kohlbrecher, S., Klingauf, U., Von Stryk, O.: GitHub - tu-darmstadt-ros-pkg/hector_quadrotor: hector_quadrotor contains packages related to modeling, control and simulation of quadrotor UAV systems (2018). https://github.com/tu-darmstadt-ros-pkg/hector_quadrotor. Accessed 5 Jan 2020
42. Gariepy, R., Mukherjee, P., Bovbel, P., Ash, D.: GitHub - husky/husky: Common packages for the Clearpath Husky (2019). https://github.com/husky/husky. Accessed 6 Jan 2020

A Capability Based Method for Development of Resilient Digital Services

Jānis Grabis[1](✉), Janis Stirna[2], and Jelena Zdravkovic[2]

[1] Institute of Information Technology, Riga Technical University, Kalku 1, Riga 1658, Latvia
grabis@rtu.lv

[2] Department of Computer and Systems Sciences, Stockholm University, Postbox 7003, 16407 Kista, Sweden
{js,jelenaz}@dsv.su.se

Abstract. Capability Driven Development (CDD) is a capability-based method for developing context-aware and adaptive systems. This paper proposes to extend CDD to address security and resilience concerns in organizational networks. A method extension defining modeling concepts and development procedure is elaborated. It includes development of a data-driven digital twin, which represents the security and resilience concerns of the network and is used to diagnose security incidents and to formulate a resilient response to these incidents. Application of the proposed method extension is illustrated using examples of secure computer network governance and secure supplier onboarding.

Keywords: Capability · Networks · Security · Resilience · Digital twin

1 Introduction

Digital service security and resilience is an active research area [1]. Despite the fact that digital services are delivered jointly by networks of organizations, security and resilience are often addressed within organizational boundaries, or even boundaries of a specific service or information system. However, the magnitude and sophistication of the challenges faced by modern organizations can only be addressed if the organizational ecosystem is considered [2]. Every node of the network contributes to both strengths and weaknesses of the network and its services. Hence, it is a common practice for malicious actors to target the nodes having lax security practices or limited resilience, including organization's business partners, customer, contractors, and suppliers.

There are many examples of vulnerabilities in networked organizations [3]. For example, 40 million of Target's credit card data were stolen by exploring security in shortcoming in HVAC devices used at the stores and provided by an external vendor. Similarly, 40,000 U.K. users were affected by a data leak in a third party customer chat-bot. Experts predict that these problems will be amplified by a rush to create new devices and services relying on the 5G mobile network [4], which would open doors for malicious actors to exploit vulnerabilities.

J. Filipe et al. (Eds.): ICEIS 2020, LNBIP 417, pp. 498–516, 2021.
https://doi.org/10.1007/978-3-030-75418-1_23

This paper argues that the digital services provisioning network as a whole as well as individual nodes should possess security and resilience capabilities to ensure safe and sustainable delivery of the services. In this regard, capability is seen as *an ability and capacity to meet desired goals (including security and privacy requirements) in dynamic context* [5]. The security and resilience capabilities jointly allow for a quick recovery in the case of security incidents.

The capabilities are developed following the Capability Driven Development (CDD) methodology [6]. CDD defines capability design, delivery, and knowledge accumulation processes. These three activities are integrated in such a way that allows the company to capture contextual information in relation to capability objectives and to adapt capability delivery according to changes in the context and performance. CDD is suitable for design of secure and resilient systems because the concerns of context adherence and performance improvement are addressed at the organizational and inter-organizational level rather than just the technical level. CDD is a component-oriented methodology [7] and extensions can be developed for specific purposes.

The objective of the paper is *to elaborate a method extension of the CDD methodology for development of security and resilience capabilities*. The method extension is specifically targeted towards networked organizations. It includes inspection of the capability design to identify security and resilience concerns using both expert judgment and capability patterns. The method also considers using a data-driven digital twin for continuous exploration and prediction of newly arising threats. The digital twin uses live data to evaluate the current situation and to appraise suitable response mechanisms. The paper extends the previous work [8] on resilience of ICT product supply chains by generalizing the capability-based resilience management to different types of networked organizations.

The rest of the paper is structured as follows. Section 2 discusses foundations of the proposed method extension. Section 3 describes the component of the digital twin, focal for enabling security analysis in the network. The method extension is elaborated in Sect. 4. Sections 5 and 6 present two application examples. The related work is reviewed in Sect. 7. Section 8 discusses the contribution as well as provides concluding remarks and outlines issues for future work.

2 Foundations

Capabilities are designed on the basis of the capability meta-model [6] following guidelines of the CDD methodology. The CDD methodology also defines activities performed during capability delivery.

2.1 CDD Methodology

The capability model defines service provider's ability and capacity to deliver an appropriate solution in specific circumstances. Figure 1 shows a simplified overview of the key elements used in capability modeling. Goals are business objectives the capability allows to achieve. They are measured by key performance indicators (KPIs). The capability is designed for delivery in a specific context as defined using context elements. The context elements name factors affecting the capability delivery. The context elements take value from a context range and jointly define a context set (not shown in Fig. 1) describing the area of capability suitability. The process element specifies a capability delivery solution. The capability model does not model the whole delivery solution but only focuses on features that are context dependent and adaptive. The delivery solution is designed and implemented following the organization and platform specific engineering process.

In order to ensure that a capability is delivered as expected in different contextual situations, adjustments are used to adapt its delivery [9]. The adjustments take context data and KPIs as input and evaluate potential changes in the capability delivery. They are also used to implement complex context-dependent decision-making logics. The capability designs are aimed to be reusable across organizations. The reusable components are represented by patterns. The patterns provide solutions to capability design and delivery problems observed in similar contexts.

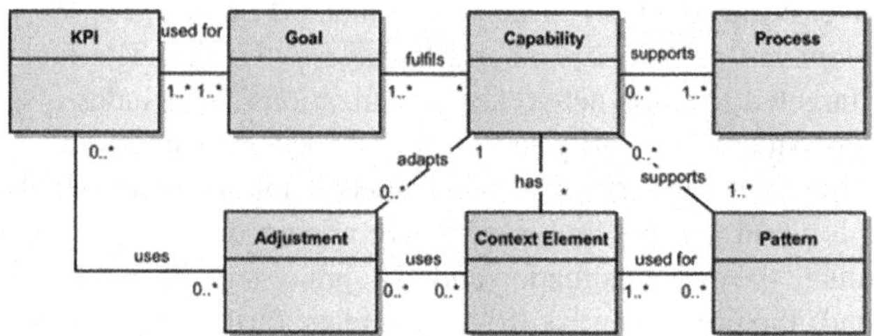

Fig. 1. Key elements of the CDD meta-model.

The process of capability development is performed according to the CDD methodology. Its main features are:

- Enterprise Modelling phase, which defines information about the digital enterprise necessary to specify requirements for development of capable information systems. It allows involving business people in information system development;
- Design phase, where the capability-enabled information system is designed on the basis of the existing knowledge in a model-driven manner;
- Delivery phase, where the information system is executed, monitored, and adjusted to changes in the operating environment if necessary;
- Feedback phase, where the information system delivery experiences are accumulated and changes in the design are requested.

The capability model is developed during the design phases. The model is used to configure the capability delivery solution. The adjustments are also implemented during the design phase. The CDD methodology focuses on development of adaptable components of the overall capability deliverer solution. Development of other components needs to be done with other engineering methods and tools specific for the problem domain and business case. The delivery phase concerns run-time aspects. The main aspect of capability delivery is the execution of capability delivery adjustments. The adjustments are executed in separate container and interact with the other parts of the capability delivery solution through well-defined interfaces.

2.2 Method Extensions

CDD consists of a number of method components each focusing on a specific task of the capability cycle, such as Capability Design, Context Modeling, Patterns and Variability Modeling, and Capability Adjustment Algorithm Specification. These method components are considered to comprise the regular CDD method. It can be extended with method extensions for dealing with certain business challenges such as supporting business process outsourcing, context-aware configuration of e-government services, industrial symbiosis, designing of new entrepreneurial ventures [10], as well as managing service configuration with the support of open data [11]. Method components and method extensions are described following the structure of a method component initially outlined in [7].

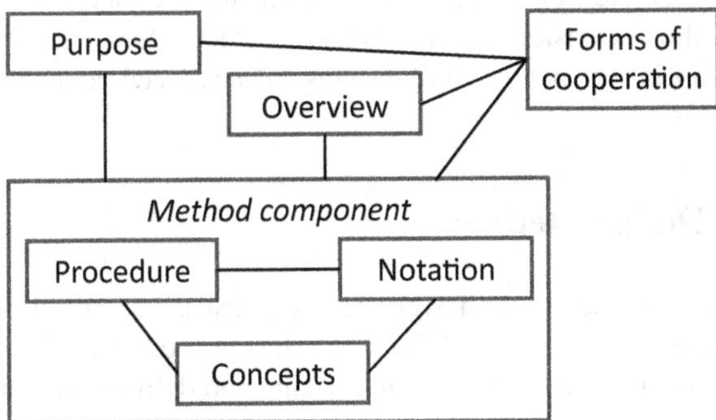

Fig. 2. Overview of the structure of a method component, adapted from [7].

Each method component and extension is to be described according to the structure shown in Fig. 2. More specifically:

- *concepts* specify what aspects of reality are regarded as relevant in the modelling process, what is important and what should be captured in a model.
- *procedure* provides guidance how to identify the concepts in practice, prerequisites, and resources. It should be described in terms of steps to be performed with input, output, and tool support.
- *notation* specifies how the result of the procedure should be documented, i.e. the graphical representation of concepts including relationships.
- *overview* describes the relationships between the individual method components, i.e. which components are to be used and under what conditions, as well as the sequence of the method components (if any).
- *forms of cooperation* describe the necessary skills, ways of cooperation, and organizational roles that should be involved in using the method component.

- *purpose* states what the purpose of the method is, what modelling or problem-solving task can be addressed by the method component.

The benefits of structuring a method in method components are: (i) the resulting method is not a monolith, which means that its parts can be combined and applied depending on the needs of the modeling domain; (ii) all components are based on a common meta-model, which allows efficient elaboration of new components and extensions, (iii) the method components or their parts can be easily updated and replaced, e.g. the goal modeling language of CDD is based on 4EM and can be replaced with another; (iv) the method can be extended with other methods offering modelling perspectives that are not currently addressed by it. With respect to the latter, this paper presents parts of the results of research project ARTSS among the objectives of which is to elaborate an extension of the CDD methodology for modeling secure and resilient business service ecosystems in crisis and post-crisis situations. To this end, the existing CDD method components will be interfaced with additional method extensions addressing capability design and management of (a) secure foundational IT services, (b) secure telemedicine services, (c) secure remote workplaces, (d) secure business services. In order to elaborate the envisioned method extensions additional three services should also be elaborated (e) specification of capability-enabled digital twins, (f) advanced context data processing for live data streams.

3 Resilience Design Method

The secure and resilient capability design method consists of a set of concepts supplementing the existing CDD meta-model and the design procedure. The purpose of the method extension is to ensure that capability design and delivery includes concerns for security and resilience.

3.1 Concepts

The core building blocks of the secure and resilient capability design method are (1) the principle of joint responsibility of members of service provision network to deliver secure and resilient services, (2) the notion of capability for configuring the supply chains with goal and context awareness, (3) the need to develop and manage digital twins to simulate and predict the behavior of the supply chain from the data of executing capabilities, and (4) a collaboration-based approach to the reconfiguration and adjustment of supply chains to ensure the needed security and optimal trust.

In this regard, capability defines an ability and capacity of the network and its nodes to provide secure and resilient services. The network has common goals, exchange contextual information as well as shares knowledge to a degree of trust required in the network. Figure 3 illustrates additional concepts introduced in the CDD meta-model to design secure and resilient service delivery capabilities. The network is required to possess security and resilience capabilities. Delivery of these capabilities is supported by services (used as an alternative implementation mechanism to processes in the original CDD meta-model). These services are provided by a network consisting of multiple nodes. The concept of security concern is introduced to identify goals dealing with security and resilience. In order to explore network behavior depending on contextual changes and adjustments used, its virtual representation or digital twin is used [12]. It uses live data to diagnose potential problems and to plan capability delivery. It plans the application of adjustments to optimize capability delivery according to the specified goals. The digital twin uses the actual network services in its simulations. It is not aimed to represent the whole network and its services. The digital twin focuses on the security concerns identified during the capability design.

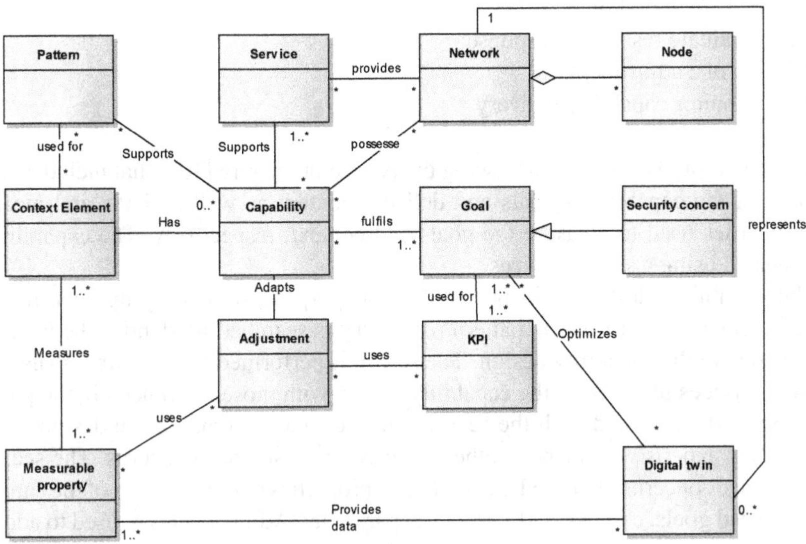

Fig. 3. Conceptual model of the approach.

3.2 Procedure

The capability design is an iterative and interactive processes. There are several pathways to design the capability and the modeling procedure defines the suggested modeling steps but not necessarily a fixed sequence of their execution. The following steps are performed to design a resilient capability:

1. Specify capability;
    1.1. Define capability;
    1.2. Identify goals and context;
    1.3. Identify services.
2. Identify security and resilience concerns;
    2.1. Search relevant patterns;
    2.2. Perform expert analysis.

3. Specify resilient response;

   3.1. Define context ranges and KPI thresholds.
   3.2. Select adjustments
   3.3. Specify twinning concerns.

4. Manage resilience;

   4.1. Set-up digital twin;
   4.2. Evaluate resilience response;
   4.3. Invoke adjustments;
   4.4. Monitor capability delivery.

The capability is designed following the typical procedure [10] what includes identification of the capability's goals and delivery context as well as KPIs and measurable properties used to measure the goals and context, respectively. The capability is implemented using a set of services.

The capability design is inspected for the purpose if identifying security and resilience concerns. At first the pattern repository is searched to identify the concerns associated with the capability design. The search is performed by matching goals, context, and services included in the capability design with those referenced in the pattern repository and associated with the security and resilience domain. The design is also inspected by experts who identify other security and resilience concerns. The security and resilience concerns identified are marked as properties of the elements of the capability design and goals, context, and services in particular. Adjustments are used to address the security and resilience concerns. They are selected from the pattern repository or specified by the domain experts. Context values or their ranges as well as KPI thresholds triggering the adjustments are also specified.

The adjustments change the capability delivery in response to changes in the context or performance. In order to appraise these changes in a virtual environment, a digital twin is developed to represent certain aspects of the capability delivery. The capability design is inspected to mark aspects of capability delivery requiring representation using a digital twin; these aspects are referred as to twinning concerns.

The digital twin is developed to represent the twinning concerns identified. It used live data and contextual information to suggest a suitable resilience response (i.e., triggering of appropriate adjustments). The capability delivery dashboard is used to continuously monitor the capability delivery performance.

4 Data-Driven Digital Twin

A data-driven digital twin is a digital representation of a service provisioning network on the basis of live data streams for the purpose of analysis of security concerns and experimentation (Fig. 4). The network is only partially observable and there is a degree of uncertainty and missing data in its representation. Therefore, it is referred as to data-driven because of relaxed requirements in terms of knowing the internal structure of the network. The digital twin receives data about a live network as specified in the capability model. The data streams are analyzed to identify irregularities and to benchmark them against the normal behavior captured in the patterns. The digital twin is then used to predict the behavior of the network and to propose adjustments to cope with security concerns and to adjust the network. The digital twin is based on a model representing the network. Various types of models including simulation, statistical, and machine learning could be used. The Design-of-experiments (DOE) module is used to specify scenarios to be explored to identify weaknesses and potential threats. The Simulator module is responsible for exploring the scenarios. The evaluation scenarios are visualized and the prediction results are transferred to the portal and the dashboard.

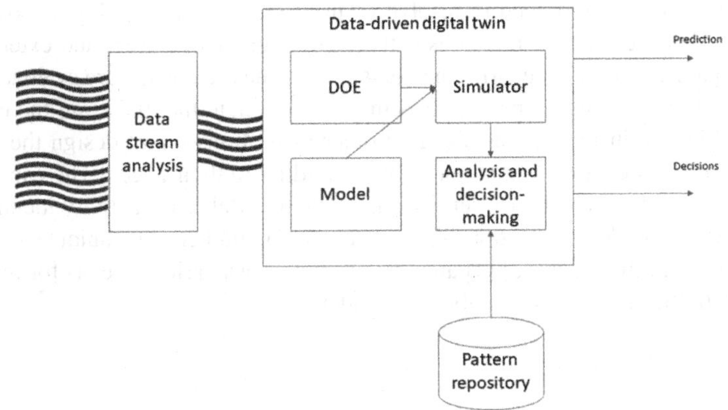

Fig. 4. Components of the data driven digital twin [8].

5 Secure Campus Area Network Governance Example

A campus area network (CAN) consists of multiple interconnected local area networks (LAN) in a limited geographical area. It is often characterized by a combination of different modes of management including access and security control. The example explores the case of CAN governance at a large higher education institution [13] that operates a CAN. The network is highly heterogeneous consisting of multiple LANs with different security settings and governance modes. One part of the network is governed by a centralized network management system while other parts are not. There is a variety of devices connecting to the network including private computers and specialized devices. Moreover, there are locations and situations permitting access without authentication. As a consequence of the COVID-19 crisis, studies are taking place on-line and employees work remotely, which puts strains on network performance and creates additional security risk exposures.

5.1 Capability Model

In order to ensure continuous operations of the higher education institution, the Secure campus network governance capability is deployed. Following the CDD methodology, a capability model is developed (Fig. 5). The capability fulfills goal 3.1 To provide secure IT environment supported by the goals 3.4 To provide high connectivity and 3.2 To prevent security incidents. Goal 3.4 specifically targets the ability of all users to access network, especially, during live events. Goal 3.2 concerns security. The top goal 3.1 is also supported by the goal 3.3 To minimize warnings by CERT, an external network security monitoring organization.

The capability delivery is affected by relevant context elements. The Device threat level indicates whether a network connected device is potentially infected; it is evaluated using multiple measurable properties. This evaluation might not always be 100% accurate because of obfuscation. The user threat level depends on availability of the user identity information and user's previously recorded behavior. The Urgency context element characterizes the need to resolve security incidents as quickly as possible. It is assumed that the urgency increases if there are many active users and external data indicates intensification of security incidents. The context elements are evaluated using the specified measurable properties. It is important to note that these are the elements considered by the institution of this case; other institutions might design the capability differently and choose context elements according to their needs. The capability is implemented using two primary services, namely, the Malicious activity identification service and the Incident resolution. The former uses a number of techniques to identify infected nodes in the network [14] and the latter provides various means for informing users on activities required to resolve the incident.

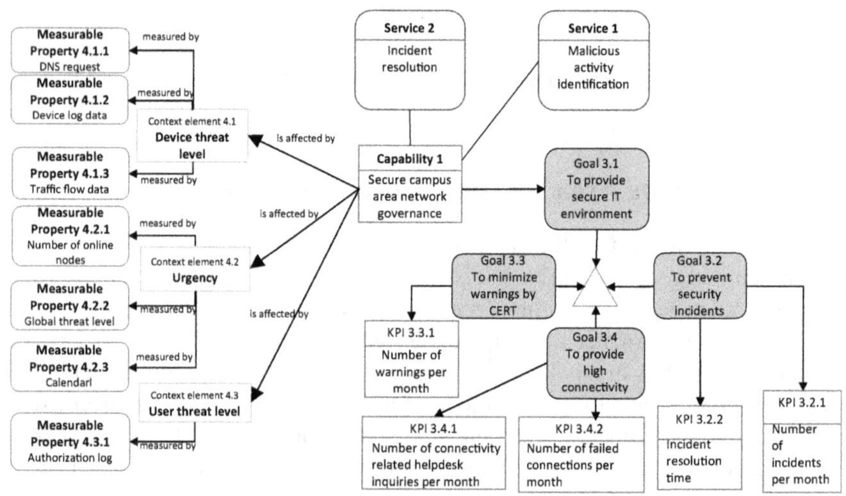

Fig. 5. The initial secure campus area network governance capability design.

In order to identify means for ensuring security and resilience, the capability model is further refined (Fig. 6). Potential values of the context elements are identified and context ranges are specified. These values guide the capability delivery in response to changes in the context. They are defined as categorical variables and the categories are defined by the modelers and using data analysis. The Device threat level assumes values {None,

Possible Low, Possible High, Definite low, Definite high}. The low values indicate that a threat identified does not create an immediate harm while the high values suggest that the node could affect the whole network. The latter case requires an immediate action to resolve the incident. The possible values suggest detection of suspicious activities though they cannot be classified as malicious. The definite values indicate that the node is certainly infected. The Urgency has categories {Low, Medium, High}. The high level is assigned if there are many active nodes during crucial operation hours. The User threat level is determined according to the previously observed behavior of the respective user and it is set to Unknown if the user cannot be identified. Adjustments to adapt the capability are identified. The Select response adjustment determines, which options provided by the Incident resolution service should be invoked depending on the context and KPI. In this case two types of responses could be used: (1) notify user; and (2) disconnect device. These responses are treated as sub-capabilities because the institution should possess abilities to carry out the response mechanisms and they are potentially reusable for other organizations.

The Select response adjustment invokes either the Notify user capability or the Disconnect device capability (1.1 and 1.2 respectively). The decision is made according to a decision-table specifying the selection outcome according to the context values. The Disconnect device capability is deployed if User threat level is Unknown and Urgency is Low or Medium because disconnecting one node could affect other nodes urgently requiring the network. The Notify user capability uses the Select notification type adjustment, which select among posting a message into user profile, sending an e-mail message or sending and SMS message. The latter has been observed as the most efficient notification type (i.e., user response time is the shortest) though it is also the most intrusive and its over-usage could decrease it efficiency. Therefore, it is preserved for cases when either of the context elements have value High. Selection of suitable response and notification means directly affects resilience of the system in terms of its ability to respond to security incidents and to ensure that ongoing network operations are uninterrupted.

The Urgency update adjustment is also introduced (Fig. 7). This adjustment dynamically changes classification of the incidents according to their urgency. If there are too many incidents and their resolution time is too long then more incidents are classified as urgent requiring efficient resolution mechanisms.

In order to identify the need for urgent update and to determine classification changes required, a digital twin of the CAN is created specifically to explore the Incident resolution service (Fig. 8). The digital twin uses live data from the CAN. That includes the evaluated device threat level and urgency as well as the observed number of incidents. The DOE specifies several alternative ways to define the urgency level. In order to evaluate these alternatives, a model representing the CAN and its Incident resolution service is constructed. The live data are fed into the model and simulated for a specific planning horizon. The model uses actual network governance services to discover incidents and to select response mechanisms. As the results, it presents the expected number of incidents depending on the way the urgency level is defined. It is important to note that the digital twin is constructed to represent specific aspects of the network.

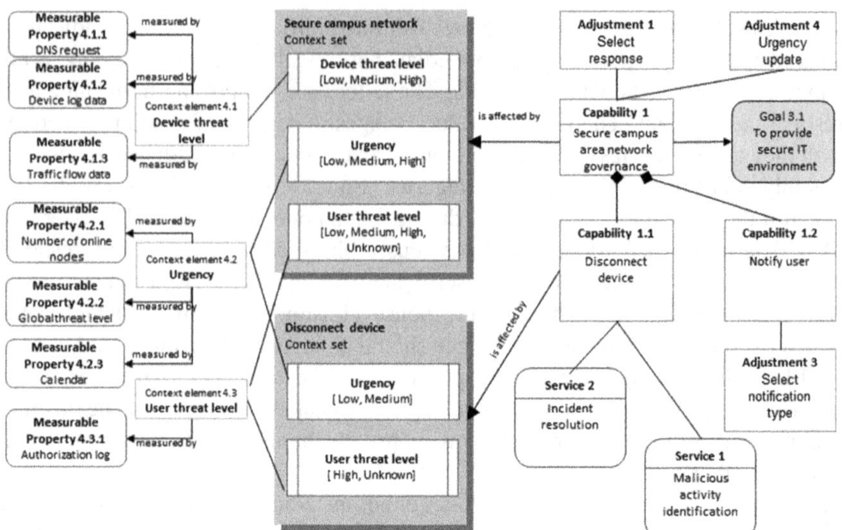

Fig. 6. The secure campus area network governance capability and its sub-capabilities.

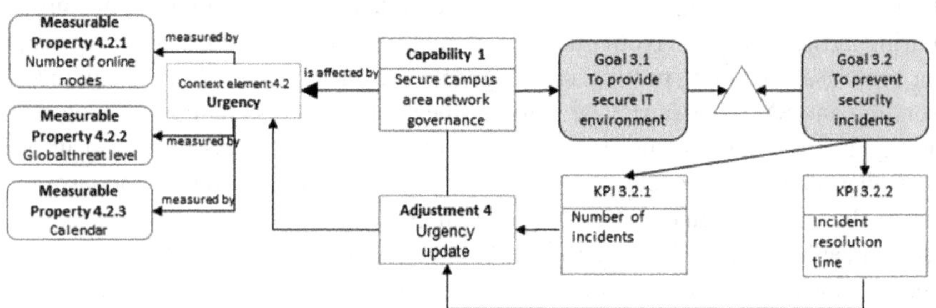

Fig. 7. The urgency update adjustment.

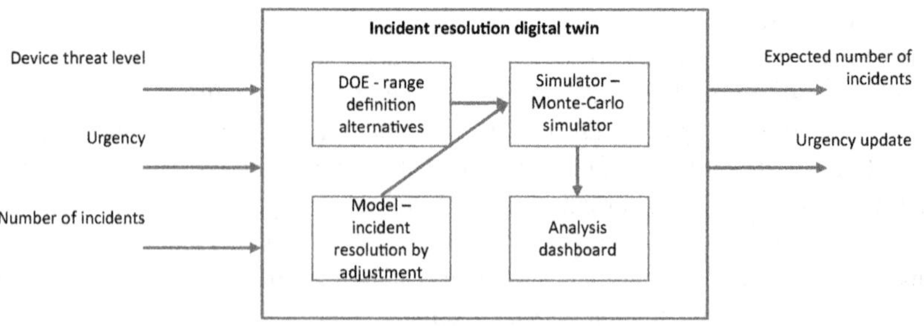

Fig. 8. Incident resolution digital twin.

5.2 Incident Resolution

To demonstrate the Urgency update adjustment, an example is elaborated. It evaluates alternative ways to evaluate the Urgency context element as follows:

1. The Urgency is set to High for most of the incidents (U1);
2. The Urgency levels are uniformly distributed (U2);
3. The Urgency is set to Low for most of the incidents (U3);

A network fragment of 1000 nodes is simulated over a fixed time period:

1. Start with the first period;
2. Infection of a fraction of the nodes is simulated;
3. The Malicious activity identification service is used to discover the infection. It can be discovered immediately or in one of the forthcoming time periods.
4. The infected node can infect a random number of other nodes as long as the incident is not resolved.
5. If the incident is discovered, the Select response adjustment is invoked to determine the response type;
6. Application of the response mechanism is represented as simulated time needed to resolve the incident;
7. The expected value of the Number of incidents KPI is evaluated;
8. The process is repeated from Step 2 till the end of the planning horizon.

The simulated evaluation is performed for all three alternatives. Figure 9 shows the incident resolution duration depending on the urgency evaluation policy. Consequently, if many incidents are classified as urgent and the SMS notification is used, the resolution speed increases. These results can be compared with the desired value of KPI. If KPI value is satisfactory then the capability can be adapted to use a more conservative classification (i.e., prefer the U3 policy). However, if the KPI value is not satisfactory, the U1 policy should be invoked.

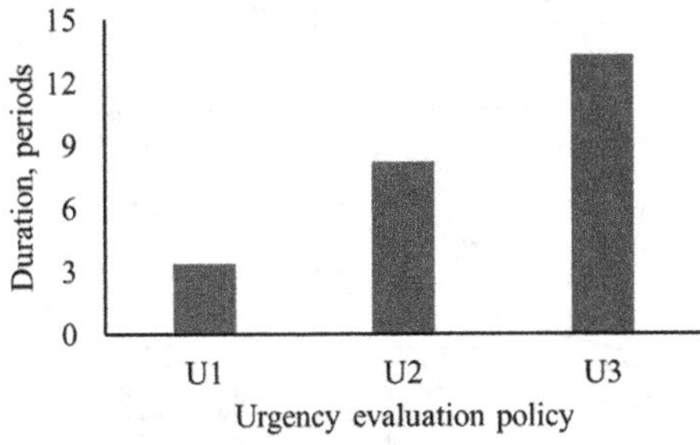

Fig. 9. Estimated average time of incidents resolution depending on urgency evaluation policy.

6 Secure Supplier Onboarding Example

The example case considers an ICT product called IoTool (https://iotool.io/), which is a lightweight IoT gateway. The IoTool uses various third-party components such as sensors and actuators supplied by different vendors. The tool vendor needs to make sure that the components are trusted and do not compromise IoTool end-users. Currently, the IoTool design implements fair-enough security, but needs to face the interoperation with additional untrustworthy IoT devices and protocols which usually are black-boxes in the sense that they offer little information with regard to what security and privacy protections they implement.

6.1 Capability Model

The IoTool solution uses out-of-the-box devices (e.g. smartphones) which are not 100% secure and users want to use them for different purposes. It is necessary to make sure that data transport protocol is secured to keep data integrity and confidentiality. By adopting the capability-based approach, the IoTool requires that vendors of the connected devices provide the secure sensing capability (Fig. 10). The goals of secure sensing capability are goal 2 To preserve data privacy and goal 3 to prevent using sensing devices for DOS or similar malicious activities, as well as goals 5 to provide the desired features requested by customers. Additionally, the overall supply chain risk level should not exceed a pre-specified threshold (goals 4). To ensure that the ICT product supply chain is able to achieve these goals, context is defined in terms of to how the vendors and devices are characterized by their trustworthiness and vulnerabilities.

Measurable properties are used to evaluate the context. Both internal and external data sources are used. It is assumed that the trustworthiness score is evaluated using data (i.e., measurable properties) provided from a credit scoring agency (e.g., https://www.companyhouse.de/) and the product vulnerability score is evaluated using data from the Common Vulnerabilities and Exposures database (https://cve.mitre.org/). The Product

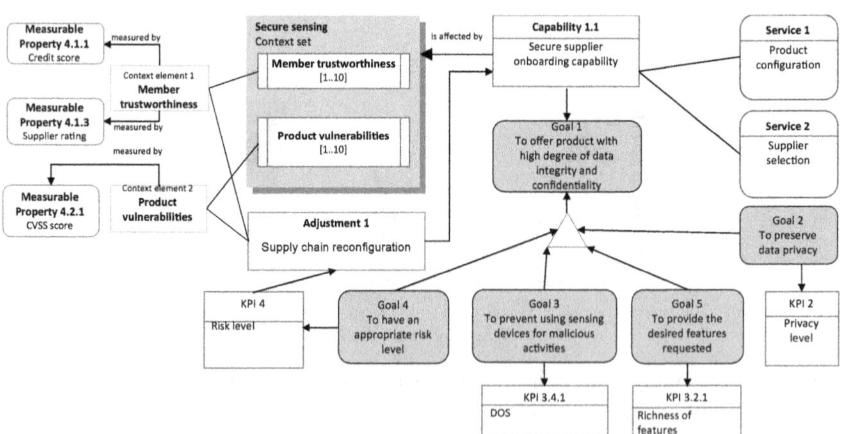

Fig. 10. Secure sensing capability model.

vulnerabilities assume values from the range 1...10 with categorical ratings from None to Critical.

The ICT product supply chain topology including its product design and supply chain configuration is developed (Fig. 11). There is a choice between various providers, for example, the IoTool can use different cloud storage vendors, which could be selected according to their ability to support the Secure sensing capability. Once the IT environment supporting the digital twin for the supply chain is put in place the context is continuously evaluated on the basis of the measurable property data and if irregularities are observed the supply chain is reconfigured. For example, in case external data sources provide a report that one of the vendors has supplied insecure sensors and the Vendor trustworthiness falls below a certain threshold set by the IoTool vendor, the sensors provided by this supplier are excluded from the offering.

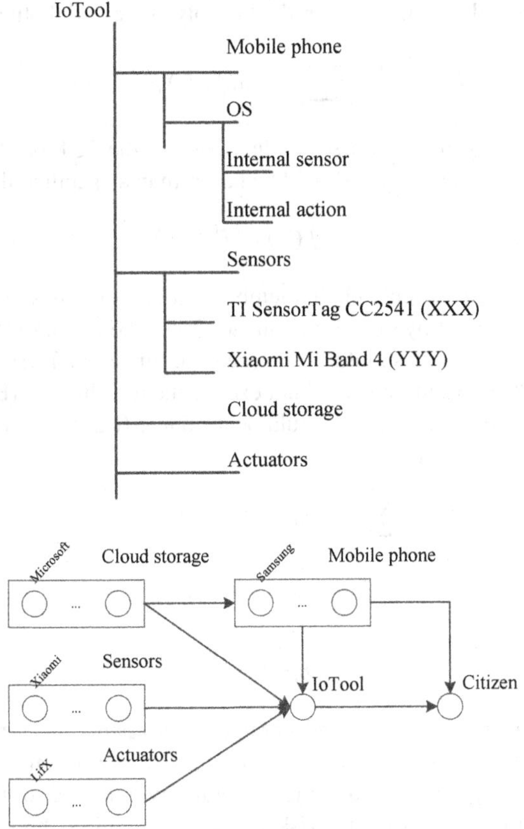

Fig. 11. Fragments of IoTool product design and supply chain. The rectangles represent products and circles represent vendors or supply chain members [8].

6.2 Suppliers Selection

The ICT supply chain is continuously monitored for security concerns. The digital twin employs various models to represent and to analyze the supply chain. A data-driven supply chain configuration model [15] is one of these models. It uses the current data to evaluate the agreement of the established supply chain typology with the goals defined for the capability. In the sample model provided, the supply chain risk minimization is considered. The model selects supply chain members (referenced using index j) and products (references using index i) to minimize the overall supply chain level of risk R (Eq. 1). The risk is calculated by taking into account the context factors, namely, the product vulnerability score a_i and the supply chain member trustworthiness score b_j. The parameters w_i indicate the relative importance of the product item in the ICT product and γ_{ij} is 1 if the ith product is provided by the jth supply chain member and 0 otherwise. The decision variable X_{ij} is 1 if the ith product provided by the jth supply chain member is included in the ICT product supply chain and 0 otherwise.

$$R = \sum\nolimits_{i=1}^{M} \sum\nolimits_{j=1}^{N} w_i a_i b_j \gamma_{ij} X_{ij} \rightarrow \min \tag{1}$$

Equation 2 requires the minimum set of features for the ICT product stating that a function f of the selected products should be larger than a specified threshold F.

$$f(\mathbf{X}) \geq F \tag{2}$$

Furthermore, any selected supply chain member is required to possess the secure sensing capability, represented by constraints in Eq. 3–5. The overall trustworthiness risk contribution for any supplier should not exceed the threshold T (Eq. 3). The product item individual risk contribution should not exceed the threshold U (Eq. 4). The supply chain member trustworthiness score should be available (Eq. 5), i.e. context data about the member should be available.

$$\sum\nolimits_{i=1}^{M} w_i b_j \gamma_{ij} X_{ij} \leq T, \forall j \tag{3}$$

$$a_i b_j \gamma_{ij} X_{ij} \leq U, \forall i, j \tag{4}$$

$$b_j \neg \emptyset, \forall j \tag{5}$$

The ICT supply chain monitoring and configuration results are represented using a dashboard (Fig. 12) The dashboard shows the current overall risk, the trustworthiness of individual supply chain members, and the vulnerability dynamics of the product items.

The example shows that the overall risk is below the required threshold, however, the trustworthiness of one of the supply chain members has declined. Therefore, a round of supply chain configuration model execution should be triggered and recommendations for supply chain reconfiguration should be generated.

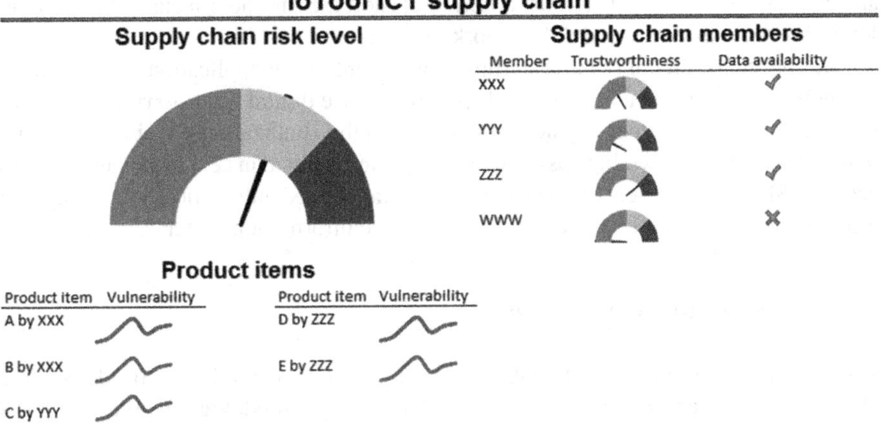

Fig. 12. Sample ICT supply chain dashboard [8].

7 Related Work

Design of networks providing digital services is a challenging research area [16]. Security concerns are particularly prevalent in such networks often due to their obscure nature [17]. Specialized modeling methods help to identify and highlight the security concerns [18]. Security Incident Response Modelling Language is proposed to model security and recovery issues [19]. Various aspects of designing secure systems can be captured using a unified modeling technique representing attack trees, vulnerability cause graphs, security activity graphs, and security goal indicator trees [20]. Insufficient transparency in networks also can be addressed by means of data analysis [21].

Similarly, resilience should be a measurable feature of network design [22]. Considering design of resilient systems, [23] suggested starting from a goal-framework for achieving: diversity (variety of actors), efficiency (productivity and resource utilization), adaptability (transparency and flexibility), and cohesion (alignment of actors and their capabilities), which then need to be realized by the goals and objectives specific to a business domain. ISO 22316:2017 provides a guideline for any size or type of organization. It is not specific to any industry or sector and it can be applied throughout the life of an organization. MITRE [24] has outlined a set of "design principles" for cyber resilience elicited from various domains including evolvability, survivability and security. In [25], the National Institute of Standards and Technology (NIST) provides a framework for improving the cybersecurity and resilience of critical infrastructures. Some research studies have proposed guidelines for organizational cyber resilience and measuring its maturity by extending the design of the information systems with, for example, additional interfaces to the systems' environment for increased situational awareness, design of alternative, modular software capabilities, development of security-related parameters and related metrics, and rich testing to different conditions [22]. Haque et al. [26] present a comprehensive cyber resilience framework for Industrial Control Systems by decomposing "resilience" into a hierarchy of several sub-metrics. Their resilience framework can serve as a platform for a multi-criteria decision aid and help technical experts in identifying the gap in the study of network resilience.

The concept of digital twins is starting to get attention for application in the security management domain. For instance, [12] propose to use digital twins to rise cyber situational awareness for cyber-physical systems through visualizations. Viability of using data streams in digital twining has been recently demonstrated in [27]. In advanced cases such as [28], digital models of service networks are created automatically by analyzing large sets of data available in the existing enterprise information systems.

8 Discussion and Conclusions

This paper has presented a method extension of the CDD methodology for design and delivery of secure and resilient capabilities. The CDD methodology is used as a basis for this development because it allows to consider security and resilience concerns from both organizational and technical perspectives. The capability delivery is supported by digital services, which are provided by a network of organizations. It is shown that the network level resilience can be achieved if individual nodes are able to meet security and resilience goals and to share contextual and performance information. The data-driven digital twin is a primary tool to ensure resilience. It is used to evaluate the current capability delivery situation as well as to suggest the necessary adjustments in response to contextual changes and performance goals.

The digital twin component is introduced in the paper. However, the digital twin development process is still highly case specific. Formalization of the digital twin development and exploration processes in an area for future research. It should also be considered in relation to knowledge reuse in the form of patterns as suggested by the CDD methodology. The CDD methodology is also supported by a toolset supporting capability design, delivery, and knowledge management. However, support for digital twin development of operation is currently not supported by this toolset and digital twins are developed for a specific purpose using they own technology stack. Integration between tools used in the CDD methodology and digital twin development technologies is also subject to further investigation.

Acknowledgments. This research is partially funded by the Ministry of Education and Science, Republic of Latvia, project ARTSS, project No. VPP-COVID-2020/1–0009.

References

1. Uday, P., Marais, K.: Designing resilient systems-of-systems: a survey of metrics, methods, and challenges. Syst. Eng. **18**(5), 491–510 (2015)
2. World Economic Forum. Advancing Cyber Resilience Principles and Tools for Boards (2017)
3. Kshetri, N., Voas, J.M.: Supply chain trust. IT Prof. **21**(2), 6–10 (2019)
4. Madnick, S.: 5G security concerns persist with new research pointing to critical flaw (2019). https://www.itpro.co.uk/mobile/32893/

5. Sandkuhl, K., Stirna, J.: Capability thinking. In: Sandkuhl, K., Stirna, J. (eds.) Capability Management in Digital Enterprises, pp. 1–24. Springer, Cham (2018). https://doi.org/10.1007/978-3-319-90424-5_1
6. Berziša, S., et al.: Capability driven development: an approach to designing digital enterprises. Bus. Inf. Syst. Eng. **57**(1), 15–25 (2015)
7. Goldkuhl, G., Lind, M., Seigerroth, U.: Method integration: the need for a learning perspective. IEE Proc. Softw. **145**(4), 113–118 (1998)
8. Grabis, J., Stirna, J., Zdravkovic, J.: Capability management in resilient ICT supply chain ecosystems. In: Proceedings of the 22nd International Conference on Enterprise Information Systems ICEIS, no. 2, pp. 393–400 (2020)
9. Grabis, J., Kampars, J.: Adjustment of capabilities: how to add dynamics. In: Sandkuhl, K., Stirna, J. (eds.) Capability Management in Digital Enterprises, pp. 139–158. Springer, Cham (2018). https://doi.org/10.1007/978-3-319-90424-5_8
10. Sandkuhl, K., Stirna, J.: Capability Management in Digital Enterprises. Springer, Cham (2018). https://doi.org/10.1007/978-3-319-90424-5
11. Kampars, J., Zdravkovic, J., Stirna, J., Grabis, J.: Extending organizational capabilities with open data to support sustainable and dynamic business ecosystems. Softw. Syst. Model. **19**(2), 371–398 (2019). https://doi.org/10.1007/s10270-019-00756-7
12. Eckhart, M., Ekelhart, A., Weippl, E.: Enhancing cyber situational awareness for cyber-physical systems through digital twins. In: IEEE International Conference on Emerging Technologies and Factory Automation, 1222 (2019)
13. Minkevics, V., Kampars, J.: IS security governance capability design for higher education organization. In: Proceedings of 59th International Scientific Conference on Information Technology and Management Science of Riga Technical University, ITMS 2018, pp. 1–5 (2018)
14. Minkevics, V., Kampars, J.: Methods, models and techniques to improve information system's security in large organizations. In: Proceedings of the 22nd International Conference on Enterprise Information Systems ICEIS, no. 1, pp. 632–639 (2020)
15. Grabis, J., Chandra, C., Kampars, J.: Use of distributed data sources in facility location. Comput. Ind. Eng. **63**(4), 855–863 (2012)
16. De Reuver, M., Sørensen, C., Basole, R.C.: The Digital platform: a research agenda. J. Inf. Technol. **33**(2), 124–135 (2018)
17. Mouratidis, H., Argyropoulos, N., Shei, S.: Security requirements engineering for cloud computing: the secure Tropos approach. In: Karagiannis, D., Mayr, H., Mylopoulos, J. (eds.) Domain-Specific Conceptual Modeling: Concepts, Methods and Tools, pp. 357–380. Springer, Cham (2016). https://doi.org/10.1007/978-3-319-39417-6_16
18. Elahi, G., Yu, E.: Modeling and analysis of security trade-offs - a goal oriented approach. Data Knowl. Eng. **68**(7), 579–598 (2009)
19. Athinaiou, M., Mouratidis, H., Fotis, T., Pavlidis, M., Panaousis, E.: Towards the definition of a security incident response modelling language. In: Furnell, Steven, Mouratidis, Haralambos, Pernul, Günther. (eds.) TrustBus 2018. LNCS, vol. 11033, pp. 198–212. Springer, Cham (2018). https://doi.org/10.1007/978-3-319-98385-1_14
20. Byers, D., Shahmehri, N.: Unified modeling of attacks, vulnerabilities and security activities. In: Proceedings - International Conference on Software Engineering, p. 36 (2010)
21. Lu, T., Guo, X., Xu, B., Zhao, L., Peng, Y., Yang, H.: Next big thing in big data: the security of the ICT supply chain. In: Proceedings - SocialCom/PASSAT/BigData/EconCom/BioMedCom 2013, p. 1066 (2013)
22. Fiksel, J.: Designing resilient, sustainable systems. Environ. Sci. Technol. **37**(23), 5330–5339 (2003)
23. Bodeau, D., Graubart, R.: Cyber resiliency design principles, pp. 1–90. Technical report, Report No: 17-0103. The MITRE Corporation, United States, January 2017

24. Korpela, K., Kuusiholma, U., Taipale, O., Hallikas, J.: A framework for exploring digital business ecosystems. In: 46th Annual Hawaii International Conference on System Sciences HICSS 2013, pp. 3838–3847. Institute of Electrical and Electronics Engineers Inc. (2013)
25. Ross, R., Pillitteri, V., Graubart, R., Bodeau, B., McQuaid, R.: Developing. Cyber Resilient Systems: A Systems Security Engineering Approach. SP 800-160, vol. 2 (2019). https://nvlpubs.nist.gov/nistpubs/SpecialPublications/NIST.SP.800-160v2.pdf. Accessed 28 May 2020
26. Haque, Md.A., Kamdem De Teyou, G., Shetty, S., Krishnappa, B.: Cyber resilience framework for industrial control systems: concepts, metrics, and insights. In: Proceedings of IEEE International Conference on Intelligence and Security Informatics Conference, ISI. IEEE (2018). https://doi.org/10.1109/ISI.2018.8587398
27. Kritzinger, W., Karner, M., Traar, G., Henjes, J., Sihn, W.: Digital Twin in manufacturing: a categorical literature review and classification. IFAC-PapersOnLine **51**(11), 1016–1022 (2018)
28. Murphy, A., et al.: Representing financial data streams in digital simulations to support data flow design for a future Digital Twin. Robot. Comput.-Integr. Manuf. **61**, 101853 (2020)

Adopting Agile Software Development Combined with User-Centered Design and Lean Startup: A Systematic Literature Review on Maturity Models

Maximilian Zorzetti[1(✉)], Cassiano Moralles[1], Larissa Salerno[1], Eliana Pereira[2], Sabrina Marczak[1], and Ricardo Bastos[1]

[1] MunDDoS Research Group – PPGCC, School of Technology, Pontifical Catholic University of Rio Grande Do Sul, Porto Alegre, RS, Brazil
{maximilian.zorzetti,cassiano.mora,larissa.salerno}@acad.pucrs.br, {sabrina.marczak,bastos}@pucrs.br

[2] Instituto Federal do Rio Grande do Sul, Porto Alegre, RS, Brazil
eliana.pereira@restinga.ifrs.edu.br

Abstract. The use of Agile in the software development industry in the past two decades revealed that it is lackluster in some aspects, such as in guaranteeing user involvement and assuring that the right software is being built. There are reports that combining Agile with Lean Startup and User-Centered Design (UCD) helps in overcoming these shortcomings while also yielding several other benefits. However, there is not much documentation on how to use this "combined approach" and adapting existing organizations to use it is a challenge in of itself, in which the use of an instrument to guide or assess such transformations is typically pivotal to their success. As such, in this paper we seek to identify maturity models that assess the use of Agile, Lean Startup, and UCD. We conducted a systematic literature review of maturity models for these three methods published between 2001 and 2020. We characterized the maturity models and determined how they see maturity, how they are applied, and how they were evaluated. As an extended version of a previous paper, we augmented our analysis criteria and further classified the models in how they interpret maturity and what strategy they suggest when undergoing an improvement process, in addition to providing new insight on various aspects of the models. We found 35 maturity models, of which 23 were for Agile, 5 for Lean thinking, 5 for UCD, and 2 for Agile and UCD combined. No models for the combination of the three methods were found (nor for Lean Startup), as expected due to the novelty of the approach. Existing models mostly focus on practice adoption and acquiring continuous improvement capabilities, and are typically developed with a specific context in mind. We also note a lack of proper evaluation procedures being conducted on the majority of models, which could be due to the lack of well-established maturity model development methods and guidelines.

J. Filipe et al. (Eds.): ICEIS 2020, LNBIP 417, pp. 517–541, 2021.
https://doi.org/10.1007/978-3-030-75418-1_24

Keywords: Systematic literature review · Maturity model · Agile software development · Lean startup · User-Centered Design

1 Introduction

Agile is extensively used by organizations today [26] as it serves as a powerful and adaptive alternative to the rigid and wasteful software development approaches that were previously used, e.g., waterfall. However, there are some issues with Agile that indicate that it might not be enough by itself—such as lack of user involvement [62] and clear identification of added value [35]. Recent industry cases [22,66] show that a combined use of Agile, Lean Startup, and User-Centered Design (UCD) can be a way to overcome the aforementioned issues: Lean Startup [57] focuses on adding value to business stakeholders through strategic experimentation, while UCD [43] puts the user at the center of the discussion to foster empathy.

Adopting such a combined approach can lead to several organizational challenges of different nature, such as cultural (e.g., trust), structural (e.g., roles), and technical (e.g., techniques), which are aggravated when dealing with large enterprises as new large-scale issues arise (e.g., inter-team coordination) [48], making instruments to guide and assess the transformation essential in these cases. An example of such instruments are maturity models, which can gauge the transformation in a not overly expensive and time-consuming manner [39] and provide guidance towards improving software engineering processes [18]. Maturity models are widely used in several domains as a means to improve something (e.g., processes or products) and are typically suggest "levels" of maturity to be achieved, as is the case for the well-known Capability Maturity Model Integration (CMMI) [12].

We aim to show what is the current state of the art in maturity models for a software development approach composed of Agile, Lean Startup, and UCD pillars through a systematic literature review following existing best practices for systematic reviews on the software engineering domain. We report on several aspects of maturity models found both in academic and gray literature. As an extended version of a previous paper [78], our study provides new analysis criteria and novel insight on currently available maturity models.

The remainder of this paper is organized as follows: Sect. 2 discusses the use of a combined approach of Agile, Lean Startup, and UCD in software development; Sect. 3 discusses related work; Sect. 4 explains how the systematic literature review was conducted and outlines the research questions; Sect. 5 presents our findings; Sect. 6 deliberates on our analysis; and Sect. 7 deliberates on this study and considers future work.

2 Agile, Lean Startup, and User-Centered Design

The Agile movement dates back to 2001 with the introduction of the Agile Manifesto [5], a result of the then-current wasteful and rigid software development culture and work processes. The extensive use of agile in the past two decades has brought to light some of its weaknesses, such as difficulty in increasing user involvement [62]. A development method composed of Agile, Lean Startup, and UCD is a novel approach that has been argued as a way to overcome such weaknesses [75] and that is drawing the attention of academics [13,75] and industry practitioners [22,66].

This combined approach tackles business-level issues with Lean Startup, an entrepreneurship method that focuses on developing a business plan iteratively through the use of a "build-measure-learn" loop, where business hypotheses are evaluated through carefully planned and efficient experiments that gather useful customer feedback, enabling organizations to pivot away from ideas that data suggests to be unfruitful and persevere on the ones most likely to succeed [57]. This idea pulls heavily from traditional Lean values, by way of reducing development waste on new products or features that do not earn back enough to warrant being successful. Although not specifically a software development method, studies have reported on it being a great driving force for software development [15,76], although embracing its continuous experimentation practices requires proper technological capabilities (e.g., continuous deployment) and organizational support (e.g., culture) [37].

To ensure that the software not only meets business demands but also the users', the combined approach enlists the use of UCD to enable developers to understand the user's real needs and create improved software with better usability and user satisfaction [59]. UCD consists of a set of procedures, processes, and techniques that focus on setting the user as the center of the design space or development process [43] at varying degrees of intensity, from consultation of their needs to having them actively participate in the design process [1]. Integrating UCD and software development can help developers with the difficult practice of involving customers, and the wider concern of how to integrate human-computer-interaction concerns with software engineering [9,61]. As it stands, UCD has evolved into an umbrella term for similar approaches, thus encompassing terms like Design Thinking and Human-Centered Design.

One successful example of the combined approach is fashion retailer Nordstrom's Discovery by Design. Grossman-Kahn and Rosensweig [22] report on the evolution of the Nordstrom Innovation Lab, a Nordstrom initiative to rapidly and cheaply test novel concepts internally. Each iteration of the lab improved upon the shortcomings of the former, turning what started as an isolated agile development team into an acclaimed innovation team with its own development methodology which encapsulates Design Thinking, Lean Startup, and Agile. The team with its "iterative mindset, relentless focus on the needs of the customer, and bias towards rapid experimentation, prototyping and testing" [22] emerged as a powerful and dynamic asset for Nordstrom.

In academia, Moralles et al. [41] conducted an empirical study to compare Extreme Programming (XP), Lean, and UCD concepts identified through literature reviews with what was being used in practice by two software development teams that use a development methodology that encompasses the three methods.

Their findings suggest that both teams use a complementary subset of concepts from each pillar, in addition to techniques and roles not found in the literature. Their study motivated us to seek maturity models that propose the combination of the three aforementioned pillars. Maturity models, which can be prescriptive or descriptive, aim to offer guidance on practices that are relevant to master. The Agile Compass [16], backed by an agile maturing framework [17], is an example of a checklist-based agile maturity model which introduces the category of outcomes an agile team should seek as it matures with regards to the use of practices. Such models can be of help to bring awareness to newcomers to the combined use of Agile, Lean Startup, and UCD.

3 Related Work

There has been several studies contemplating the integration of Agile Software Development with UCD [61]. Adding Lean Startup to this "method combo" is rather of a novelty given the time frame of the three approaches. Current literature encompasses studies contemplating several aspects of the approach itself (e.g., benefits, challenges, or use of experimentation) [66,67,74] and also studies that propose models to using the combined approach with varying degrees of abstraction [13,22,75]. The combined approach can be very different from typical agile development as it requires a certain degree of developer empowerment that larger organizations might not be used to, making adaptation efforts difficult and the use of maturity models and enticing choice.

Maturity models for agile development has been an interesting research subject ever since the rise of Agile: studies on such maturity models can be traced back to the early 2000s. Leppänen [36] reports that maturity models for Agile have varying levels of maturity, and Nurdiani et al. [44] compares the practice adoption order proposed in existing maturity models with that of industry experts. Ozcan-Top and Demirörs [46] evaluated the strengths and weaknesses of Agile maturity models and frameworks from a process assessment and improvement perspective. Fontana et al. [19] conducted a systematic literature review on Agile maturity models and delineated a classification criteria for how maturity can be defined based on the analysis of the identified models. Henriques and Tanner [25] bring to light that maturity models lack research providing them validation. Pereira and Serrano [52] analyze the main development and evaluation methods for IT maturity models.

4 Systematic Review Protocol

This study was conducted as a systematic literature review based on guidelines for conducting systematic literature reviews in software engineering [33]. Our first effort on mapping maturity models for a combined approach of the three aforementioned pillars found zero results, so we expanded our effort into 7 systematic literature reviews (SR) about maturity models for Agile, Lean Startup,

UCD, and their intersections: Agile combined with Lean Startup; Agile combined with UCD; Lean Startup combined with UCD; and Agile, Lean Startup, and UCD combined (each is hereinafter referred to as a *search context*). The goal of these SRs is to identify and assess primary and secondary studies regarding the use, structure, and evaluation of maturity models for the three pillars. The protocol for the systematic literature review is documented next.

4.1 Research Questions

All SRs address the same research questions, each related to their respective search context.

RQ1. What maturity models are available?
RQ2. How are these maturity models characterized?
RQ3. How do these maturity models envision maturity?
RQ4. How are these maturity models applied?
RQ5. How are these maturity models evaluated?

4.2 Search

As suggested by Kitchenham [33], we used the PICO criteria to guide the formulation of our search string.

Population: Primary and secondary studies related to their respective search context.
Intervention: Maturity models related to their respective SR context.
Comparison: This criterion does not apply to our RQs because the goal of this study is not to compare the identified maturity models.
Outcomes: Understanding of use, structure, and evaluation of identified maturity models.

All SRs followed the same search process. We retrieved studies from electronic databases that met the following source selection criteria:

- Databases that include journal articles, conference, and workshop papers related to their respective SR context;
- Databases with an advanced search mechanism that allows filtering of the results by keywords that address the research questions; and
- Databases that provide access to full papers written in English.

Based on these criteria, we selected the following databases: ACM Digital Library, IEEExplore, Science Direct, Scopus, and Springer Database. We adapted the search string (Eq. 1) for each database based on the search functionality offered by each. Each search string consisted of two parts—S1 and S2—defined as follows:

- S1 is a string composed of keywords related to maturity models, namely: maturity model, capability model, self assessment, health check, and team assessment; and

Table 1. Keywords used in the search string of each SR [78].

ID	SR	Keywords
K1	Agile	"Agile" OR "Agile Method*" OR "Agile Development" OR "Agile Software Development" OR "Agile Practice" OR "Test Driven Development" OR "Test-driven Development" OR "Behavior-driven Development" OR "Behavior Driven Development" OR "Behaviour-driven Development" OR "Behaviour Driven Development" OR "Extreme Programming" OR "Scrum" OR "Kanban"
K2	Lean	"Lean Startup" OR "Lean Start-Up" OR "Lean UX" OR "Lean User Experience" OR "Lean Software" OR "Lean Development"
K3	UCD	"Design Thinking" OR "*Centered Design" OR "*Centred Design" OR "User Experience" OR "Usability" OR "Human Computer Interaction" OR "Computer-Human Interaction" OR "Human Factor" OR "User Interface"
K4	Agile and Lean	K1 AND K2
K5	Agile and UCD	K1 AND K3
K6	Lean and UCD	K2 AND K3
K7	Agile, Lean, and UCD	K1 AND K2 AND K3

- S2 is a string composed of keywords related to the search context of each SR. Table 1 presents the keywords used.

As Lean Startup is the newest of the three pillars, we chose to broaden its search context by including other Lean thinking schools, such as Lean UX.

Equation 1. Search criteria boolean expression.

$$S1 \text{ AND } S2 \tag{1}$$

Afterwards, inclusion and exclusion criteria were applied by a varying number of researchers for each SR on the retrieved studies in two distinct rounds, as explained in Sect. 4.3. The first round consisted of title and abstract inspection to triage the candidate studies based on the inclusion and exclusion criteria. The second round consisted of a thorough inspection with full text reading to further filter the studies and to perform the data extraction procedure (Sect. 4.5).

4.3 Study Selection

To determine whether a study should be selected, all SRs applied the following selection criteria.

Inclusion Criteria: (I1) the study presents a maturity model for its SR context; (I2) the study is written in English; (I3) the study is fully written in electronic format; (I4) the study was retrieved from a conference, workshop, or journal.
Exclusion Criteria: (E1) the study does not present a maturity model for its SR context; (E2) the study is an extended abstract or editorial paper; (E3) the study is duplicated.

We only searched for studies published between 2001 and 2020. We chose 2001 as the lower bound as it is the publication date of the Agile Manifesto [5]. Additionally, we performed a manual, informal search on the internet and considered gray literature studies, as these concern very current issues which might have not yet been covered in academic literature [34].

4.4 Quality Assessment

We used a set of quality criteria proposed by Guyatt et al. [23]—later used by Dybå and Dingsøyr [14] in software engineering—to assess the methodological quality of the studies selected for review, as they cover thoroughness, trustworthiness, and significance of the studies [28]. The criteria are based on four quality assessment questions:

C1 – Is the research objective clearly defined?
C2 – Is the research context well addressed?
C3 – Are the findings clearly stated?
C4 – Based on the findings, how valuable is the research?

We scored the selected studies on each criterion using an ordinal scale instead of a dichotomous scale to obtain a more accurate assessment [28]. Table 2 shows the scoring scale for each criterion. When there was not an agreement on a study's score, we had meetings to discuss the issue until we agreed upon the same score.

Table 2. Quality criteria scoring scheme [78].

Criteria	Response scoring
C1	{1, 0.5, 0} (Yes, Moderately, No)
C2	{1, 0.5, 0} (Yes, Moderately, No)
C3	{1, 0.5, 0} (Yes, Moderately, No)
C4	(>80% = 1), (<20% = 0), (in-between = 0.5)

4.5 Data Extraction and Analysis

We performed a full text reading of each study to identify, categorize, and analyze the following items:

- **D1** – Study identification (RQ1);
- **D2** – Publication source and year (RQ1);
- **D3** – Audience: the expected users of the model (RQ2);
- **D4** – Aim: if the model determines necessary improvements for its use case (analysis) or if it presents best practices for comparison (benchmarking) (RQ2);
- **D5** – Scope: if the model is generic or limited to a specific method (RQ2);
- **D6** – Strategy: whether the model suggests a "big bang" or "gradual" approach to its adoption process (RQ2);
- **D7** – Maturity levels: if the model has specified quantifiable standards (levels) of maturity and has definitions for each (RQ3);
- **D8** – Maturity class: whether the model's characterization of maturity fits into "practices adoption", "continuous improvement", "sustaining approach", "project performance", or "highly productive teams"; (RQ3)
- **D9** – Administration mechanism: if the model has defined a mechanism to apply the model (RQ4).
- **D10** – Evaluation: if the model was evaluated, such as by having it applied in a real context (RQ5); and
- **D11** – Evaluation type: whether the model's evaluation process (if any) can be regarded as an "author evaluation", "domain expert evaluation", or "practical setting evaluation" (RQ5);

Most of these items were adapted from the guidelines for developing maturity grids by Maier, Moultrie, and Clarkson (2012). Although the guidelines concern maturity grids, we found them adequate to fulfill the needs of our study. We chose guideline elements that facilitate the categorization of maturity models.

New to this study are items D2, D3, D6, D8, and D11. We enhance our existing evaluation analysis (D10) by categorizing existing evaluation methods as defined by Helgesson, Höst, and Weyns [24] and using Salah and Cairn's [58] nomenclature: "author evaluation", an evaluation performed by the model's authors to assess the model's processes regarding its intended use or to compare it to similar models; "domain expert evaluation", an evaluation performed by domain experts external to the model's development; and "practical setting evaluation", an evaluation that involves applying the model in a practical setting.

We also make use of the maturity definitions outlined by Fontana et al. [19] and analyze each model's characterization of maturity, which can resolve into the following categories: "practices adoption", maturity is increased when new practices are adopted; "continuous improvement", maturity is similar to CMMI-DEV's, i.e., mature organizations/teams focus on process improvement; "sustaining approach", maturity implies organizations/teams with lean processes and adherence to agile values; "project performance", maturity is a way to obtain results; and "highly productive teams", maturity is related directly to team productivity.

Additionally, we draw on the work of Julian, Noble, and Anslow [30], who argue that having a strategy to adopting a new approach is crucial for its success, and categorize the suggested strategies of existing models as either "big bang", in which a set of practices are adopted all at once by-the-book for teams then to learn and modify, or "gradual", in which practices are gradually adopted and adapted alongside in-place non-agile practices in what is a typically longer transitioning period.

Each researcher received an equal amount of studies to extract data from and apply the study selection criteria again. We made use of the data found in a similar literature review study [19] that focused on Agile maturity models as our search resolved into a superset of the models it identified.

5 Results

This section summarizes the results of each SR. Table 3 presents the results of the search process in the electronic databases selected in Sect. 4.2. We analyze the studies in light of our research questions based on the data extracted using the procedure in Sect. 4.5 next. Entries marked as "—" on the following tables stand for "unspecified".

Table 3. Number of identified studies during the distinct rounds of our systematic search for maturity models [78].

Search context	Retrieved	Round 1		Round 2	
		Excl.	Incl.	Excl.	Incl.
Agile, Lean, and UCD	82	77	5	5	0
Agile and Lean	152	144	8	8	0
Agile and UCD	77	72	5	3	2
Lean and UCD	78	73	5	5	0
Agile	2188	2095	93	76	17
Lean	231	207	24	19	5
UCD	3194	3142	52	47	5
Total	**5920**	**5810**	**192**	**163**	**29**

Table 4. Selected maturity model studies and their sources.

Context	Study	Year	Source	Source type
Agile	[42]	2001	EUROMICRO	Conference
	[38]	2005	International Software Process Workshop	Conference
	[65]	2007	Innovations in Systems and Software Engineering	Journal
	[49]	2007	Agile Development Conference	Conference
	[56]	2008	Journal of Systems and Software	Journal
	[50]	2009	Int'l. Journal of Software Engineering	Journal
	[51]	2009	Journal of Software	Journal
	[27]	2009	Web publishing (ThoughtWorks)	Grey Lit
	[6]	2010	Hawaii Int'l. Conference on System Sciences	Conference
	[54]	2010	Web publishing (personal blog)	Grey Lit
	[77]	2011	Int'l. Conference on Software Engineering Advances	Conference
	[8]	2011	Int'l. Conference on Product Focused Software Development and Process Improvement	Conference
	[40]	2012	Hawaii Int'l. Conference on System Sciences	Conference
	[70]	2013	Agile Development Conference	Conference
	[20]	2014	Int'l. Conference on Agile Processes in Software Engineering and Extreme Programming	Conference
	[68]	2014	Int'l. Conference on the Quality of Information and Communications Technology	Conference
	[47]	2014	Int'l. Conference on Process Improvement and Capability Determination	Conference
	[69]	2015	Iberian Conference on Information Systems and Technologies	Conference
	[16]	2015	Journal of Systems and Software	Journal
	[72]	2015	EUROMICRO	Conference
	[3]	2016	Web publishing (CrossTalk)	Grey Lit
	[71]	2018	Telecommunications Forum	Conference
	[64]	2020	Int'l. Journal of Information Management	Journal
Lean	[29]	2007	Int'l. Federation for Information Processing	Book Chap.
	[31]	2012	Euromicro	Conference
	[11]	2012	International Journal of Advanced Manufacturing Technology	Journal
	[63]	2015	Int'l Conference on Industrial Engineering, Management Science and Applications	Conference
	[2]	2018	IEEE Transactions on Software Engineering	Journal
UCD	[73]	2009	Int'l Conference on Human Centered Design	Conference
	[10]	2014	Int'l. Conference of Design, User Experience, and Usability	Conference
	[32]	2016	Symposium on Applied Computing	Conference
	[45]	2017	Int'l. Conference on Software Process Improvement and Capability Determination	Conference
	[55]	2018	Int'l Conference on Human Computer Interaction	Conference
Agile and UCD	[53]	2014	Agile Development Conference	Conference
	[60]	2016	Int'l. Conference on Software Process Improvement and Capability Determination	Conference

Table 5. Selected maturity model studies and their respective quality scores [78].

Context	Study	C1	C2	C3	C4	Avg.
Agile	[42]	1.0	1.0	0.5	0.5	0.75
	[38]	1.0	0.5	0.5	0.5	0.625
	[65]	1.0	0.5	0.5	1.0	0.75
	[49]	1.0	0.5	1.0	1.0	0.875
	[56]	0.5	0.5	1.0	0.5	0.625
	[50]	1.0	1.0	1.0	1.0	1.0
	[51]	0.5	1.0	0.5	1.0	0.75
	[27]	0.5	0.5	0.5	0.5	0.5
	[6]	0.5	0.5	0.5	1.0	0.625
	[54]	0.5	0.5	0.5	0.5	0.5
	[77]	1.0	0.5	1.0	1.0	0.875
	[8]	0.5	1.0	0.5	1.0	0.75
	[40]	1.0	1.0	1.0	0.5	0.875
	[70]	1.0	1.0	0.5	1.0	0.875
	[20]	1.0	1.0	1.0	1.0	1.0
	[68]	1.0	1.0	0.5	0.5	0.75
	[47]	1.0	1.0	1.0	1.0	1.0
	[69]	1.0	1.0	1.0	0.5	0.875
	[16]	1.0	1.0	1.0	1.0	1.0
	[72]	1.0	0.5	1.0	1.0	0.875
	[3]	0.5	0.5	0.5	0.0	0.375
	[71]	1.0	1.0	1.0	1.0	1.0
	[64]	1.0	1.0	1.0	1.0	1.0
Lean	[29]	1.0	1.0	0.5	0.5	0.75
	[31]	1.0	1.0	1.0	1.0	1.0
	[11]	1.0	1.0	0.5	0.5	0.75
	[63]	1.0	1.0	0.0	0.5	0.875
	[2]	1.0	1.0	0.5	0.0	0.625
UCD	[73]	0.5	0.5	0.5	0.5	0.5
	[10]	1.0	1.0	1.0	1.0	1.0
	[32]	1.0	1.0	0.5	0.5	0.75
	[45]	1.0	1.0	0.5	1.0	0.875
	[55]	1.0	1.0	1.0	1.0	1.0
Agile and UCD	[53]	1.0	0.5	0.5	0.5	0.625
	[60]	1.0	1.0	1.0	1.0	1.0

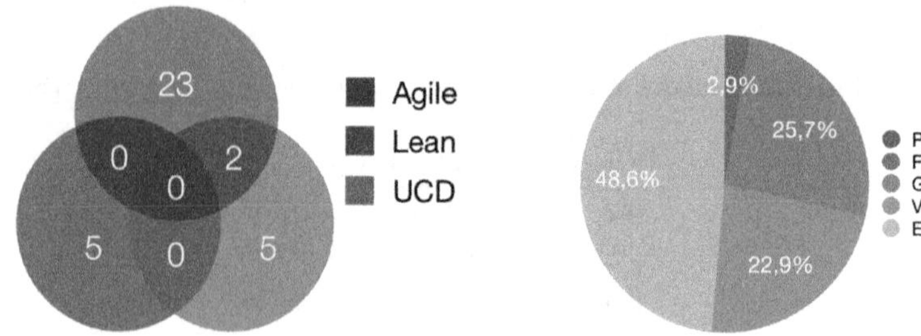

Fig. 1. Venn diagram of maturity models for Agile, Lean, and UCD (adapted [78]).

Fig. 2. Normalized quality score distribution for maturity model studies.

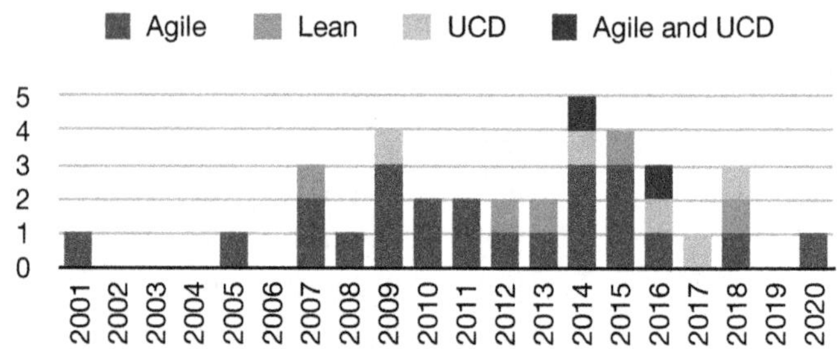

Fig. 3. Publication frequency of maturity models [78].

5.1 RQ1. What Maturity Models Are Available?

As mentioned in Table 3, our systematic literature review identified a total of 29 studies establishing maturity models for Agile, Lean Startup, UCD, and their intersections. From our manual search, we selected an additional 4 academic studies [50, 56, 65, 77] and 2 gray literature studies [3, 54] for a total of 35 studies. Table 4 shows the selected maturity models studies and their publication year and venue/source. Our initial objective was to identify maturity models for a combined approach of Agile, UCD, and Lean Startup. This search, however, proved fruitless. There are few maturity models for intersections of the pillars—only 2 for a combined use of Agile and UCD—with a notable absence of models for all three pillars combined. Figure 1 shows the number of maturity models for each category using a Venn diagram. The higher number of maturity models for Agile is expected, as it is the most dominant approach to software engineering worldwide. Of the existing Lean studies, we point out that none concern the use of Lean Startup.

We assessed the quality of the papers as per systematic literature review guidelines [33] (see Table 5). The papers scored approximately 0.78 on average, with at most a 0.03 score difference between identified categories (Agile, UCD, Lean Startup, and Agile with UCD). Studies with a low score (0.45 and below) tended to be short studies with a low page count. Figure 2 shows the studies' scores in a normalized fashion.

Figure 3 shows the publication frequency of the maturity model studies on a stacked bar chart. Agile maturity models see a fairly consistent publication rate throughout the years. Most Lean maturity models and all UCD ones were published in the past ten years, likely due to the rising popularity of Design Thinking and Lean Startup in software engineering.

5.2 RQ2. How Are These Maturity Models Characterized?

Table 6 shows an overview of the maturity models. While earlier agile maturity models focused on XP [6,38,42] due to it being one of the first agile methods to become popular [4] (and also harboring a strong influence from CMMI), the majority of the maturity models are generic, focusing on the general idea and values of the method they adhere to. Although most of them come from a more general need for maturity models, some originate from very specific demands, such as Lui and Chan's model [38], which was specifically developed to help Chinese companies dealing with commercial-off-the-shelf software; or Peres' [53], which focuses on integrating Agile and UCD on CMMI level two compliant organizations.

We identified some benchmark-based maturity models (two for Agile [6,49] and two for Lean [29,31]) that were based on CMMI-DEV [12] and adapted from industry models to be used in a generic fashion. The remaining models have analysis as their focus, evaluating the situation of teams and organizations and making use of the notion that each organization has its own unique context and characteristics, where comparisons can be inadequate and faced with resistance [17].

The Lean maturity models were mostly influenced by the manufacturing industry and propose gradually evolving circumstances, focusing on a sustainable adoption of their method of choice [2,11,31,63], with the exception of Cil and Turkan [11] with its analytic network process approach. Whilst Julian et al. [30] identifies two paths to minimize the effects of adopting "big bang" and "gradual" strategies, Lui and Chan [38] interestingly considers both approaches.

5.3 RQ3. How Do These Maturity Models Envision Maturity?

Maturity can be seen as a state to be achieved by teams or organizations, as such, maturity models mostly quantify this by defining maturity levels. Models usually define four to six levels, most commonly defaulting to five (see Table 7). Some authors do not use levels to quantify maturity, however, such as Fontana et al. [17], who argue that "maturity is obtained when results are accomplished in various aspects of software development" and propose an outcome-based model; Cil and Turkan [11], who propose a model based on analytic network process;

Table 6. Overview of data extracted from selected maturity model studies (adapted [78]).

Context	Study	Adoption process	Aim	Scope	Evaluation	Maturity level	Audience	Admin. Mech.
Agile	[42]	Big bang	Analysis	XP	Yes	Yes	Organizations adopting XP	Partial
	[38]	Big bang/gradual	Analysis	XP	No	Yes	Inexperienced XP Teams	No
	[65]	Gradual	Analysis	Generic	Yes	Yes	Agile Teams	Yes
	[49]	Gradual	Benchmarking	Generic	Yes	Yes	Agile Teams	Yes
	[56]	Gradual	Analysis	Generic	Yes	Yes	Agile Teams	Yes
	[50]	Gradual	Analysis	Generic	Yes	Yes	Agile Teams	Yes
	[51]	Gradual	Analysis	Generic	Yes	Yes	Agile Teams	Yes
	[27]	Gradual	Analysis	Generic	Yes	Yes	Agile Teams	Yes
	[6]	Big bang	Benchmarking	XP	Yes	Yes	XP Teams	Partial
	[54]	Gradual	Analysis	Generic	No	Yes	Any organization	No
	[77]	Gradual	Analysis	Scrum	Yes	Yes	Scrum Teams	Partial
	[8]	Gradual	Analysis	Generic	No	Yes	Agile Teams	No
	[40]	—	Analysis	Generic	No	Yes	Agile Teams	Yes
	[70]	Gradual	Analysis	Generic	No	Yes	Agile Teams	Yes
	[20]	—	Analysis	Generic	No	Yes	Agile Teams	No
	[68]	Gradual	Analysis	Quality Assurance	Yes	Yes	Agile and CMMI	No
	[47]	Gradual	Analysis	Generic	Yes	Yes	Implementing Agile	Yes
	[69]	Gradual	Analysis	Generic	No	Yes	Agile and CMMI	No
	[16]	—	Analysis	Generic	No	No	Agile Teams	Yes
	[72]	Big bang	Analysis	Generic	Yes	Yes	Scaling Agile	Yes
	[3]	Gradual	Analysis	Generic	No	Yes	Agile Teams	No
	[71]	Big bang	Analysis	Generic	No	Yes	Distributed teams	Partial
	[64]	Gradual	Analysis	Generic	No	Yes	No evidence	No
Lean	[29]	—	Benchmarking	Generic	No	Yes	Lean Sustainability	No
	[31]	Gradual	Benchmarking	Generic	Yes	Yes	Soft. Dev. Teams	Yes
	[11]	—	Analysis	Generic	Yes	No	Lean Transformation	Yes
	[63]	Gradual	Analysis	Generic	No	No	Lean Implementation	No
	[2]	Gradual	Analysis	Kaizen	Yes	No	Soft. Dev. Teams	Yes
UCD	[73]	Gradual	Analysis	Generic	No	Yes	All Organization	No
	[10]	Gradual	Analysis	Generic	No	Yes	Soft. Dev. Teams	No
	[32]	Gradual	Analysis	Generic	Yes	Yes	Soft. Dev. Teams	Yes
	[45]	—	Analysis	Generic	No	No	Soft. Dev. Teams	Yes
	[55]	Gradual	Analysis	Generic	Yes	Yes	Soft. Dev. Teams	Yes
Agile and UCD	[53]	Big bang	Analysis	Generic	Yes	Yes	Scrum Teams	Yes
	[60]	Gradual	Analysis	Generic	Yes	Yes	Agile Teams	Yes

and Schröders and Cruz-Machados [63], whose model individually assesses and quantitatively rates several criteria pertaining to leadership, culture, knowledge, and process.

5.4 RQ4. How Are These Maturity Models Applied?

Table 8 shows the administration mechanisms of the selected maturity models. Administration mechanisms tend to be simple, not deviating much from instruments similar to questionnaires or checklists, attesting to the inexpensive and somewhat swift quality maturity models are known for. Of note is the work of Patel and Ramachandran [51], which has the support of a web-based tool; and Cil and Turkan [11], which employs the analytical network process to resolve assessments. Almost half of the maturity models, however, provide no mechanism at all, perhaps overstepping on simplicity.

We notice a necessity for more studies and/or discussions on how administration mechanisms must designed or used in a way that is adequate for the specific needs for which their corresponding maturity model was designed for. As a mechanism sensitive to the context its applied on, it most likely needs to be highly dynamic features to properly assess and identify improvement needs and to perform follow-up measurements. Only a few maturity models present in-depth metrics, typically analysis-focused ones [11], but we speculate that using both a quantitative and qualitative approach to analysis might be a better path to proper assessment.

5.5 RQ5. How Are These Maturity Models Evaluated?

Table 9 shows how each maturity model was evaluated. Of the models that were in fact evaluated, most of them underwent either a domain-expert or practical setting evaluation. Given that author evaluations seem to be the cheapest alternative and that the majority of the models were not evaluated at all, this perhaps suggests that author evaluations are seen as not worthwhile by academics, or that there is a lack of industry involvement during the development of these artifacts. Indeed, only a few models seem to be actively used in the software industry and none had follow-up studies, revealing a concerning detachment between academia and industry with regard to maturity models.

Table 7. Maturity class and levels of maturity models.

Context	Study	Maturity class	Maturity levels
Agile	[42]	Practices adoption, project performance	4 levels, based on existence an adherence to a structured process
	[38]	Practices adoption	4 levels, based on learning
	[65]	Practices adoption, sustaining approach	5 levels, based on existence and adherence to a structured process
	[49]	Highly productive teams	5 levels, based on people
	[56]	Sustaining approach	6 levels, based on existence and adherence to a structured process
	[50]	Project performance	5 levels, based on existence and adherence to a structured process
	[51]	Practices adoption	4 levels, based on existence and adherence to a structured process
	[27]	Continuous improvement	5 levels, based on existence and adherence to a structured process
	[6]	Highly productive teams	6 levels, based on existence and adherence to a structured process
	[54]	Project performance	5 levels, based on existence and adherence to a structured process (practices)
	[77]	Continuous improvement	5 levels, based on existence and adherence to a structured process (Scrum practices)
	[8]	Continuous improvement	—
	[40]	Sustaining approach	—
	[70]	Practices adoption, continuous improvement	—
	[20]	Practices adoption	3 levels, but they mention adherence to a process should not be measured
	[68]	Practices adoption	6 levels, based on existence and adherence to a structured process
	[47]	Practices adoption, continuous improvement	4 levels, based on existence and adherence to a structured process (practices)
	[69]	Continuous improvement	5 levels, based on existence and adherence to a structured process
	[16]	—	—
	[72]	Sustaining approach	5 levels, based on existence and adherence to a structured process
	[3]	Practices adoption	—
	[71]	Continuous improvement	—
	[64]	Practices adoption	6 levels, based on existence and adherence to a structured process
Lean	[29]	Practices adoption, highly productive teams	5 levels, based on existence and adherence to a structured lean culture and process
	[31]	Practices adoption	5 levels, based on existence and adherence to a structured lean culture and process (focus on products)
	[11]	—	—
	[63]	Highly productive teams,sustaining approach	—
	[2]	Practices adoption,project performance, highly productive teams	3 levels, based on team productivity and team experience
UCD	[73]	Practices adoption	5 levels, based on existence and adherence to a structured user experience process (practices)
	[10]	Continuous improvement	5 levels, based on existence and adherence to a structured user experience process
	[32]	Continuous improvement	3 levels, based on existence and adherence to a structured usability process
	[45]	—	—
	[55]	Continuous improvement	5 levels, based on integration to a structured UCD process
Agile and UCD	[53]	Practices adoption,continuous improvement	—
	[60]	Practices adoption, continuous improvement, highly productive teams	—

Table 8. Administration mechanisms of maturity models [78].

Context	Study	Administration mechanism
Agile	[42]	Partial, provides a list of items to be observed by an XP tracker but leaves the definition of an assessment method open-ended
	[38]	–
	[65]	Uses a Goal-Question-Indicator-Metric approach to measure readiness for practice adoption; provides a 4-step process for organizations to adopt agile
	[49]	Reports a case that used user stories based on agile goals
	[56]	Provides a custom analytical tool that evaluates agile methods through four distinct perspectives
	[50]	Uses questionnaires that are distributed to a project's development team and any other associated personnel; and a roadmap for software process improvement
	[51]	Proposes the use of user stories for each desired maturity level; has a web-based tool to assess organizational suitability to use story card-based requirements engineering and agile practices
	[27]	Provides a somewhat superficial plan-do-check-act cycle to roll out improvements throughout an organization
	[6]	Partial, uses an undisclosed list of required measures and evidence to determine maturity levels that is to be used by a third party and as a self-assessment tool to allow for complementary views
	[54]	–
	[77]	Partial, uses an undisclosed checklist of Scrum practices for each maturity level
	[8]	–
	[40]	Reports a case that used a survey
	[70]	Collects data on indicators and through a series of computations resolves into numeric scores for strategies, principles, and objectives
	[20]	–
	[68]	–
	[47]	Uses a questionnaire about specific practices and generic agile practices
	[69]	–
	[16]	Provides a checklist that helps teams to identify which outcomes they have attained
	[72]	Provides indicators to be assessed in assessment meetings
	[3]	—
	[71]	Partial, describes two methods to interpret maturity parameters that are evaluated using discrete scoring, but does not specify how to obtain the latter
	[64]	—
Lean	[29]	—
	[31]	Provides assessment items for the lean practices in each of its process areas
	[11]	Uses the analytical network process
	[63]	—
	[2]	Specifies a process area for evaluation methods
UCD	[73]	–
	[10]	–
	[32]	Uses a questionnaire to assess organizational strategic usability
	[45]	Uses a questionnaire for peer and self-assessment
	[55]	Uses U+A SPICE, a mechanism adapted from ISO/IEC 15504
Agile and UCD	[53]	Synchronizes the iterative development cycle with its process improvement suggestions; provides a list of relevant measurements
	[60]	Provides a performance scale to rate organizational performance and an assessment procedure that contains a sheet template for information recording, maturity scores for comparison, guidelines, and other benchmarks

Table 9. Evaluations performed on maturity models (adapted [78]).

Context	Study	Evaluation type	Evaluation
Agile	[42]	Practical	Evaluated by 5 project teams composed of 6 students each in a university. Teams were asked to organize their work according to the model and apply as many XP practices as possible
	[38]	–	–
	[65]	Expert	Evaluated through questionnaires answered by 28 members of the agile community
	[49]	Practical	Evaluated through 20 teams using it for over six months
	[56]	Practical	Evaluated through 2 industry case studies
	[50]	Expert	Evaluated through a discussion with 3 different organizations
	[51]	Expert	Evaluated through case study in three companies
	[27]	–	–
	[6]	Practical	Evaluated through a case study in a multinational communication company
	[54]	–	–
	[77]	Expert/practical	Evaluated through action research, which incorporated interviews with Scrum, Agile, and CMMI experts
	[8]	–	–
	[40]	–	–
	[70]	–	–
	[20]	–	–
	[68]	Expert	Evaluated through a survey based on the opinion of experts
	[47]	Practical	Evaluated through an exploratory case study in a government organization
	[69]	–	–
	[16]	–	–
	[72]	Practical	Evaluated through a case study in a large organization
	[3]	–	–
	[71]	–	–
	[64]	–	–
Lean	[29]	–	–
	[31]	Author	Evaluated by comparison
	[11]	Practical	Evaluated through a case study in a company
	[63]	–	–
	[2]	Expert	Evaluated through a survey
UCD	[73]	–	–
	[10]	–	–
	[32]	Expert/practical	Evaluated survey between novices and experts
	[45]	Expert/practical	Evaluated through a survey with experts and a case study
	[55]	–	–
Agile and UCD	[53]	Expert	Evaluated by a panel of experts
	[60]	Expert	Evaluated by domain experts

6 Discussion

A large sum of the maturity models resort to the simple course of action of practice adoption as their main focus to improving maturity, though many also see the competence of being able to continuously improve as the key takeaway of maturity. We highlight that Lean maturity models generally consider the culture and behavior domains, while UCD models focus on development teams trying to adhere to UCD during their work process; outlining the disparity of what maturity even is among different studies. This "confusion" of maturity is not without reason (after all, different models try to solve different things), but makes for a difficult time in trying to establish guidelines on proper maturity model development [7]. Several maturity models are developed to solve the problems and fulfill the specific needs of certain contexts and/or problem domains identified by academia or requested by the industry, even though most of them make use of generic methods. Furthermore, many go without a suitable evaluation procedure and none were demonstrated being applied on real teams or organizations in one or more follow-up studies, casting some doubt on the actual capabilities of some models, despite the fact that our quality assessment resolved into generally positive scores. This could be due to the lack of established guides on maturity model development [52], which would reflect on the quality of the models themselves [21], the inaccessibility of domain-experts in some academia circles, or the higher costs of conducting a proper evaluation procedure with industry partners.

The combined approach of Agile, UCD, and Lean Startup seems most promising still, and it and its variants are being used by organizations worldwide. As larger institutions move to use it, the hardships of large-scale adoption will become more apparent, highlighting the need of maturity models to support them and likely instigating research on the topic. The maturity models reported in this paper garner a lot of knowledge that could help in developing a proper model for the combined approach, even if Lean Startup was not directly addressed by them.

6.1 Threats to Validity

As with any systematic literature review, most threats to validity concern study selection bias and inaccuracy during data extraction. We carried out procedures to reduce such threats, but our protocol is prone to faults: the first round of inclusion and exclusion criteria was applied only once by multiple researchers (no study was evaluated more than once); the studies that participated in the second round of inclusion and exclusion criteria were assessed by two researchers, but no metric to rate inter-rater agreement among the researchers was calculated; data extraction results obtained from a researcher were not checked by another; and no snowball search of any kind was executed.

The systematic literature review was conducted by one PhD, one PhD candidate, three graduate students, and two undergraduate students (all from IT-related education); with guidance from two senior researchers.

7 Conclusion

This paper reports on a systematic literature review of maturity models for Agile, Lean Startup, UCD, and their intersections in a software engineering context following existing systematic review guidelines [33]. We found a total of 35 maturity models, but none were of a combined approach of the three pillars. The methodological quality of the maturity model studies was evaluated using previously established criteria [14,23,28]. Then, we analyzed and categorized the maturity models using criteria adapted from maturity grid guidelines [39] and other studies [24,58].

The absence of maturity models for the combined approach of the three pillars is likely due to its infancy. Research on the use of Lean Startup in software development is not as extensive as the other two pillars, which also already have a subject area specific to researching integration efforts between the two, i.e., Agile User-Centered Design Integration (AUCDI). Additionally, all three pillars lack a widely accepted theoretical basis that properly defines each pillar, leading back to issues like "what is a mature agile team?" and what issues should each pillar tackle individually or together; making integration efforts difficult. The yet unexplained inner workings of the combined approach make the development of a maturity model for it a daunting task.

Although we found some maturity models for Lean thinking, none were specifically for Lean Startup, which seems to be a major driving force behind the combined approach of Agile, Lean Startup, and UCD [22]. Lean Startup deals heavily with continuous experimentation, a practice that is very much intertwined with the method's somewhat risk-tolerant mindset, perhaps making future maturity models for it focus on cultural concerns, much like the reported Lean models of this study.

The identified maturity models show a worrisome trend in evaluation procedures: about half of the studies did not report on evaluating their maturity model. Although many studies lack a sound theoretical basis and methodology [21], Pereira and Serrano [52] report that several maturity model development guidelines have been created, but that authors choose to follow their own method instead, which could be the cause of this trend. Nevertheless, the lack of evaluation shows an alarming disconnect from the industry, which is where the maturity models are to be applied in the first place.

For future work, the development of a maturity model for the combined approach of the three pillars is evident, although a better understanding of how the pillars interact should be attained first, even if it has been suggested that cultural and mindset factors should be one of the focus points of such a model [22]. The combined approach is a promising take on software development [22,66], albeit an understudied one. As practices from Lean Startup continue to be adopted by the software industry, we hope to see improved interest in this subject.

Acknowledgments. We acknowledge that this research is sponsored by Dell Brazil using incentives of the Brazilian Informatics Law (Law no. 8.2.48, year 1991).

References

1. Abras, C., Maloney-krichmar, D., Preece, J.: User-centered design. In: Encyclopedia of Human-Computer Interaction. SAGE Publications Inc, Bainbridge Island (2004)
2. Al-Baik, O., Miller, J.: Integrative double kaizen loop (IDKL): towards a culture of continuous learning and sustainable improvements for software organizations. IEEE Trans. Softw. Eng. **45**(12), 1189–1210 (2019)
3. Ambler, S., Lines, M.: The disciplined agile framework: a pragmatic approach to agile maturity. J. Defense Softw. Eng. - CrossTalk **29**, 25–31 (2016)
4. Beck, K.: Embracing change with extreme programming. Computer **32**(10), 70–77 (1999)
5. Beck, K., et al.: Manifesto for agile software development (2001)
6. Benefield, R.: Seven dimensions of agile maturity in the global enterprise: a case study. In: Hawaii International Conference on System Sciences, Koloa, Kauai, HI, USA, pp. 1–7. IEEE Computer Society, December 2010
7. de Bruin, T., Freeze, R., Kulkarni, U., Rosemann, M.: Understanding the main phases of developing a maturity assessment model. In: Australasian Conference on Information Systems, January 2005
8. Buglione, L.: Light maturity models (LMM): an agile application. In: Proceedings of the International Conference on Product Focused Software Development and Process Improvement, Torre Canne, Brindisi, Italy, pp. 57–61. ACM (2011)
9. Chamberlain, S., Sharp, H., Maiden, N.: Towards a framework for integrating agile development and user-centred design. In: Abrahamsson, P., Marchesi, M., Succi, G. (eds.) XP 2006. LNCS, vol. 4044, pp. 143–153. Springer, Heidelberg (2006). https://doi.org/10.1007/11774129_15
10. Chapman, L., Plewes, S.: A UX maturity model: effective introduction of UX into organizations. In: Marcus, A. (ed.) DUXU 2014. LNCS, vol. 8520, pp. 12–22. Springer, Cham (2014). https://doi.org/10.1007/978-3-319-07638-6_2
11. Cil, I., Turkan, Y.S.: An ANP-based assessment model for lean enterprise transformation. Int. J. Adv. Manuf. Technol. **64**(5–8), 1113–1130 (2013)
12. CMMI Product Team: CMMI for Development, 1.3 edn. Software Engineering Institute, Carnegie Mellon University (2010)
13. Dobrigkeit, F., de Paula, D., Uflacker, M.: InnoDev: a software development methodology integrating design thinking, scrum and lean startup. In: Meinel, C., Leifer, L. (eds.) Design Thinking Research. UI, pp. 199–227. Springer, Cham (2019). https://doi.org/10.1007/978-3-319-97082-0_11
14. Dybå, T., Dingsøyr, T.: Empirical studies of agile software development: a systematic review. Inf. Softw. Technol. **50**(9), 833–859 (2008)
15. Fagerholm, F., Guinea, A.S., Mäenpää, H., Münch, J.: The RIGHT model for continuous experimentation. J. Syst. Softw. **123**, 292–305 (2017)
16. Fontana, R.M., Reinehr, S., Malucelli, A.: Agile compass: a tool for identifying maturity in agile software-development teams. IEEE Softw. **32**(6), 20–23 (2015)
17. Fontana, R., Meyer Jr., V., Reinehr, S., Malucelli, A.: Progressive outcomes: a framework for maturing in agile software development. J. Syst. Softw. **102**, 88–108 (2015)
18. Fontana, R.M.: Maturity in agile software development. Ph.D. thesis, Pontifical Catholic University of Paraná, Curitiba, Brazil (2015)

19. Fontana, R.M., Albuquerque, R., Luz, R., Moises, A.C., Malucelli, A., Reinehr, S.: Maturity models for agile software development: what are they? In: Larrucea, X., Santamaria, I., O'Connor, R.V., Messnarz, R. (eds.) EuroSPI 2018. CCIS, vol. 896, pp. 3–14. Springer, Cham (2018). https://doi.org/10.1007/978-3-319-97925-0_1
20. Fontana, R.M., Reinehr, S., Malucelli, A.: Maturing in agile: what is it about? In: Cantone, G., Marchesi, M. (eds.) XP 2014. LNBIP, vol. 179, pp. 94–109. Springer, Cham (2014). https://doi.org/10.1007/978-3-319-06862-6_7
21. García-Mireles, G., Moraga, M., Garcia, F.: Development of maturity models: a systematic literature review. In: International Conference on Evaluation & Assessment in Software Engineering, Ciudad Real, Spain, vol. 2012, pp. 279–283, January 2012
22. Grossman-Kahn, B., Rosensweig, R.: Skip the silver bullet: driving innovation through small bets and diverse practices. Leading Through Design, p. 815 (2012)
23. Guyatt, G., Rennie, D., Meade, M., Cook, D.: Users' Guide to the Medical Literature Essentials of Evidence-Based Clinical Practice, 2nd edn., vol. 270, no. 21. McGraw-Hill, New York (2008)
24. Helgesson, Y.Y.L., Höst, M., Weyns, K.: A review of methods for evaluation of maturity models for process improvement. J. Softw. Evol. Process **24**(4), 436–454 (2012)
25. Henriques, V., Tanner, M.: A systematic literature review of agile and maturity model research. Interdiscip. J. Inf. Knowl. Manag. **12**, 53–73 (2017)
26. Hoda, R., Salleh, N., Grundy, J., Tee, H.M.: Systematic literature reviews in agile software development: a tertiary study. Inf. Softw. Technol. **85**, 60–70 (2017)
27. Humble, J., Russell, R.: The agile maturity model applied to building and releasing software (2009)
28. Inayat, I., Salim, S.S., Marczak, S., Daneva, M., Shamshirband, S.: A systematic literature review on agile requirements engineering practices and challenges. Comput. Hum. Behav. **51**, 915–929 (2015). Computing for Human Learning, Behaviour and Collaboration in the Social and Mobile Networks Era
29. Jørgensen, F., Matthiesen, R., Nielsen, J., Johansen, J.: Lean maturity, lean sustainability. In: Olhager, J., Persson, F. (eds.) APMS 2007. ITIFIP, vol. 246, pp. 371–378. Springer, Boston, MA (2007). https://doi.org/10.1007/978-0-387-74157-4_44
30. Julian, B., Noble, J., Anslow, C.: Agile practices in practice: towards a theory of agile adoption and process evolution. In: Kruchten, P., Fraser, S., Coallier, F. (eds.) XP 2019. LNBIP, vol. 355, pp. 3–18. Springer, Cham (2019). https://doi.org/10.1007/978-3-030-19034-7_1
31. Karvonen, T., Rodriguez, P., Kuvaja, P., Mikkonen, K., Oivo, M.: Adapting the lean enterprise self-assessment tool for the software development domain. In: Euromicro Conference on Software Engineering and Advanced Applications, Cesme, Izmir, Turkey, pp. 266–273, September 2012
32. Kieffer, S., Vanderdonckt, J.: STRATUS: a questionnaire for strategic usability assessment. In: Proceedings of the 31st Annual ACM Symposium on Applied Computing, Pisa, Italy, pp. 205–212. ACM (2016)
33. Kitchenham, B., Charters, S.: Guidelines for performing systematic literature reviews in software engineering (2007)
34. Kiteley, R., Stogdon, C.: Literature Reviews in Social Work. Sage, Thousand Oaks (2013)

35. Kuusinen, K., et al.: Knowledge sharing in a large agile organisation: a survey study. In: International Conference on Agile Software Development, Cologne, Germany, pp. 135–150 (2017)
36. Leppänen, M.: A comparative analysis of agile maturity models. In: Pooley, R., Coady, J., Schneider, C., Linger, H., Barry, C., Lang, M. (eds.) Information Systems Development, pp. 329–343. Springer, New York (2013). https://doi.org/10.1007/978-1-4614-4951-5_27
37. Lindgren, E., Münch, J.: Raising the odds of success: the current state of experimentation in product development. Inf. Softw. Technol. **77**, 80–91 (2016)
38. Lui, K.M., Chan, K.C.C.: A road map for implementing eXtreme programming. In: Li, M., Boehm, B., Osterweil, L.J. (eds.) SPW 2005. LNCS, vol. 3840, pp. 474–481. Springer, Heidelberg (2006). https://doi.org/10.1007/11608035_38
39. Maier, A.M., Moultrie, J., Clarkson, P.J.: Assessing organizational capabilities: reviewing and guiding the development of maturity grids. IEEE Trans. Eng. Manag. **59**(1), 138–159 (2012)
40. Medappa, P., Bhattacharya, S.: Towards a framework for assessing agility. In: Hawaii International Conference on System Sciences, pp. 5329–5338, January 2012
41. Moralles, C., et al.: On the mapping of underlying concepts of a combined use of lean and user-centered design with agile development: the case study of the transformation process of an IT company. In: Meirelles, P., Nelson, M.A., Rocha, C. (eds.) WBMA 2019. CCIS, vol. 1106, pp. 25–40. Springer, Cham (2019). https://doi.org/10.1007/978-3-030-36701-5_3
42. Nawrocki, J., Walter, B., Wojciechowski, A.: Toward maturity model for extreme programming. In: Proceedings EUROMICRO Conference 2001: A Net Odyssey, Warsaw, Poland, pp. 233–239, September 2001
43. Norman, D.A., Draper, S.W.: User Centered System Design: New Perspectives on Human-Computer Interaction. Lawrence Erlbaum Associates Inc., New Jersey (1986)
44. Nurdiani, I., Börstler, J., Fricker, S., Petersen, K., Chatzipetrou, P.: Understanding the order of agile practice introduction: comparing agile maturity models and practitioners' experience. J. Syst. Softw. (2019). https://doi.org/10.1016/j.jss.2019.05.035
45. Ogunyemi, A., Lamas, D., Stage, J., Lárusdóttir, M.: Assessment model for HCI practice maturity in small and medium sized software development companies. In: Mas, A., Mesquida, A., O'Connor, R.V., Rout, T., Dorling, A. (eds.) SPICE 2017. CCIS, vol. 770, pp. 55–69. Springer, Cham (2017). https://doi.org/10.1007/978-3-319-67383-7_5
46. Ozcan-Top, O., Demirörs, O.: Assessment of agile maturity models: a multiple case study. In: Woronowicz, T., Rout, T., O'Connor, R.V., Dorling, A. (eds.) SPICE 2013. CCIS, vol. 349, pp. 130–141. Springer, Heidelberg (2013). https://doi.org/10.1007/978-3-642-38833-0_12
47. Top, Ö.Ö., Demirörs, O.: Assessing software agility: an exploratory case study. In: Mitasiunas, A., Rout, T., O'Connor, R.V., Dorling, A. (eds.) SPICE 2014. CCIS, vol. 477, pp. 202–213. Springer, Cham (2014). https://doi.org/10.1007/978-3-319-13036-1_18
48. Paasivaara, M., et al.: Large-scale agile transformation at Ericsson: a case study. Empirical Softw. Eng. **23**, 2550–2596 (2018). https://doi.org/10.1007/s10664-017-9555-8
49. Packlick, J.: The agile maturity map a goal oriented approach to agile improvement. In: AGILE 2007, Washington, DC, USA, pp. 266–271. IEEE Computer Society (2007)

50. Patel, C., Ramachandran, M.: Agile maturity model (AMM): a software process improvement framework for agile software development practices. Int. J. Softw. Eng. **2**, 3–28 (2009)
51. Patel, C., Ramachandran, M.: Story card maturity model (SMM): a process improvement framework for agile requirements engineering practices. J. Softw. **4**(5), 422–435 (2009)
52. Pereira, R., Serrano, J.: A review of methods used on it maturity models development: a systematic literature review and a critical analysis. J. Inf. Technol. **35**, 161–178 (2020)
53. Peres, A.L., Silva, T.S.D., Silva, F.S., Soares, F.F., Carvalho, C.R.M.D., Meira, S.R.D.L.: AGILEUX model: towards a reference model on integrating UX in developing software using agile methodologies. In: Agile Conference, Orlando, FL, USA, pp. 61–63, July 2014
54. Proulx, M.: Yet another agile maturity model (AMM) - the 5 levels of maturity (2010)
55. Quintal, C., Macías, J.A.: A capability maturity proposal for usability and accessibility centered processes. In: Proceedings of the International Conference on Human Computer Interaction, Palma, Spain, pp. 15:1–15:8. ACM (2018)
56. Qumer, A., Henderson-Sellers, B.: A framework to support the evaluation, adoption and improvement of agile methods in practice. J. Syst. Softw. **81**(11), 1899–1919 (2008)
57. Ries, E.: The Lean Startup: How Today's Entrepreneurs Use Continuous Innovation to Create Radically Successful Businesses. Crown Business, New York (2011)
58. Salah, D., Paige, R., Cairns, P.: An evaluation template for expert review of maturity models. In: Jedlitschka, A., Kuvaja, P., Kuhrmann, M., Männistö, T., Münch, J., Raatikainen, M. (eds.) PROFES 2014. LNCS, vol. 8892, pp. 318–321. Springer, Cham (2014). https://doi.org/10.1007/978-3-319-13835-0_31
59. Salah, D., Paige, R., Cairns, P.: Patterns for integrating agile development processes and user centered design. In: Proceedings of the 20th European Conference on Pattern Languages of Programs, Kaufbeuren, Germany, p. 19. ACM (2015)
60. Salah, D., Paige, R., Cairns, P.: A maturity model for integrating agile processes and user centered design. In: Clarke, P.M., O'Connor, R.V., Rout, T., Dorling, A. (eds.) SPICE 2016. CCIS, vol. 609, pp. 109–122. Springer, Cham (2016). https://doi.org/10.1007/978-3-319-38980-6_9
61. Salah, D., Paige, R.F., Cairns, P.: A systematic literature review for agile development processes and user centred design integration. In: Proceedings of the International Conference on Evaluation and Assessment in Software Engineering. Association for Computing Machinery, New York (2014)
62. Schön, E.-M., Winter, D., Escalona, M.J., Thomaschewski, J.: Key challenges in agile requirements engineering. In: Baumeister, H., Lichter, H., Riebisch, M. (eds.) XP 2017. LNBIP, vol. 283, pp. 37–51. Springer, Cham (2017). https://doi.org/10.1007/978-3-319-57633-6_3
63. Schröders, T., Cruz-Machado, V.: Assessing lean implementation. In: Gen, M., Kim, K.J., Huang, X., Hiroshi, Y. (eds.) Industrial Engineering, Management Science and Applications 2015. LNEE, vol. 349, pp. 803–811. Springer, Heidelberg (2015). https://doi.org/10.1007/978-3-662-47200-2_84
64. Shukla, S.K., Sushil: Evaluating the practices of flexibility maturity for the software product and service organizations. Int. J. Inf. Manag. **50**, 71–89 (2020)
65. Sidky, A.S., Arthur, J.D., Bohner, S.A.: A disciplined approach to adopting agile practices: the agile adoption framework. Innov. Syst. Softw. Eng. **3**, 203–216 (2007)

66. Signoretti, I., Marczak, S., Salerno, L., d. Lara, A., Bastos, R.: Boosting agile by using user-centered design and lean startup: a case study of the adoption of the combined approach in software development. In: 2019 ACM/IEEE International Symposium on Empirical Software Engineering and Measurement (ESEM), Porto de Galinhas, Recife, Brazil, pp. 1–6, September 2019
67. Signoretti, I., Salerno, L., Marczak, S., Bastos, R.: Combining user-centered design and lean startup with agile software development: a case study of two agile teams. In: Stray, V., Hoda, R., Paasivaara, M., Kruchten, P. (eds.) XP 2020. LNBIP, vol. 383, pp. 39–55. Springer, Cham (2020). https://doi.org/10.1007/978-3-030-49392-9_3
68. Silva, F.S., Soares, F.S.F., Peres, A.L., de Azevedo, I.M., Pinto, P.P., de Lemos Meira, S.R.: A reference model for agile quality assurance: combining agile methodologies and maturity models. In: International Conference on the Quality of Information and Communications Technology, Guimaraes, Portugal, pp. 139–144. IEEE Computer Society (2014)
69. Soares, F., Meira, S.: An agile strategy for implementing CMMI project management practices in software organizations. In: Iberian Conference on Information Systems and Technologies, Águeda, Portugal, pp. 1–4, June 2015
70. Soundararajan, S., Balci, O., Arthur, J.D.: Assessing an organization's capability to effectively implement its selected agile method(s): an objectives, principles, strategies approach. In: 2013 Agile Conference, Leuven, Belgium, pp. 22–31, August 2013
71. Stanisavljevic, Z., Walter, B., Vukasovic, M., Todosijevic, A., Labedzki, M., Wolski, M.: GEANT software maturity model. In: Telecommunications Forum, Belgrade, Serbia, pp. 420–425, November 2018
72. Stojanov, I., Turetken, O., Trienekens, J.J.M.: A maturity model for scaling agile development. In: Euromicro Conference on Software Engineering and Advanced Applications, Madeira, Portugal, pp. 446–453, August 2015
73. Tyne, S.: Corporate user-experience maturity model. In: Kurosu, M. (ed.) HCD 2009. LNCS, vol. 5619, pp. 635–639. Springer, Heidelberg (2009). https://doi.org/10.1007/978-3-642-02806-9_74
74. Vargas, B.P., Signoretti, I., Zorzetti, M., Marczak, S., Bastos, R.: On the understanding of experimentation usage in light of lean startup in software development context. In: Proceedings of the Evaluation and Assessment in Software Engineering, Trondheim, Norway, pp. 330–335. ACM (2020)
75. Ximenes, B.H., Alves, I.N., Araújo, C.C.: Software project management combining agile, lean startup and design thinking. In: Marcus, A. (ed.) DUXU 2015. LNCS, vol. 9186, pp. 356–367. Springer, Cham (2015). https://doi.org/10.1007/978-3-319-20886-2_34
76. Yaman, S.G., et al.: Introducing continuous experimentation in large software-intensive product and service organisations. J. Syst. Softw. **133**, 195–211 (2017)
77. Yin, A., de Lemos Figueiredo, S., Mira da Silva, M.: Scrum maturity model. In: International Conference on Software Engineering Advances. IARIA, Barcelona, Spain, January 2011
78. Zorzetti, M., et al.: Maturity models for agile, lean startup, and user-centered design in software engineering: a combined systematic literature mapping. In: Proceedings of the 22nd International Conference on Enterprise Information Systems, vol. 2, pp. 145–156. SciTePress, Prague (2020)

Evaluating a LSTM Neural Network and a Word2vec Model in the Classification of Self-admitted Technical Debts and Their Types in Code Comments

Rafael Meneses Santos[1(✉)], Israel Meneses Santos[2], Methanias Colaço Júnior[2], and Manoel Mendonça[1]

[1] Graduate Program in Computer Science, Federal University of Bahia, Salvador, Brazil
manoel.mendonca@ufba.br

[2] Department of Information Systems, Federal University of Sergipe, Itabaiana, Brazil

Abstract. Context: Software development teams constantly opt for faster, lower quality solutions to solve current problems without planning for the future. This situation will have a negative long-term impact and is called technical debt. Similar to a financial debt, technical debts require interest payments and must be managed and detected so that the team can evaluate the best way to deal with them. One way to detect technical debts is through classification of source code comments. Developers often insert comments warning of the need to improve their own code in the future. This is known as Self-Admitted Technical Debt (SATD). **Objective:** Combine Word2vec for word embedding with a Long short-term memory (LSTM) neural network model to identify SATDs from comments in source code and compare with other studies and LSTM without word embedding. **Method:** We plan and execute an experimental process with model's effectiveness data validation. **Results:** In general, the classification improves when all SATD types were grouped in a single label. In relation to other studies, the LSTM model with Word2vec achieved better recall and f-measure. The LSTM model without word embedding achieves greater recall, but perform worse in precision and f-measure. **Conclusion:** We found evidence that LSTM models combined with word embedding are promising for the development a more effective SATD classifier.

Keywords: Mining software repositories · Self-admitted technical debt · Long short-term memory · Neural networks · Deep learning · Word embedding

1 Introduction

The evolution or maintenance of software is a fundamental step to guarantee the quality of the software. Sometimes, quick fixes and workarounds must be applied

J. Filipe et al. (Eds.): ICEIS 2020, LNBIP 417, pp. 542–559, 2021.
https://doi.org/10.1007/978-3-030-75418-1_25

in order to meet close deadlines. During these decisions, developers should consider a trade-off between meeting deadlines and the quality of the software in the long-term. This type of choice is called technical debt.

The term "technical debt", defined by Ward Cunningham, is a decision that the development team takes when it chooses to do something that is easy to implement to meet a short-term goal but can have a negative impact in the future [5]. When technical debts are not managed and corrected, they can have serious long-term consequences, increasing costs during the maintenance [25]. Although there are cases of technical debt occurring unintentionally, there are situations in which developers admit that they have found a technical debt. Developers can, at some point, find problems already present in the code and, mainly through comments, record details of the problem in the source code. This type of technical debt is called self-admitted technical debt (SATD) [21].

Given the need of better ways to deal with technical debt, some work have been done on how to detect and manage technical debt [4,8,19]. The studies have proposed ways of detecting technical debts through manual or automatic analysis of project artifacts. Source code is the most analyzed artifact when you want to find traces of technical debt. Some studies extracted metrics from the code to obtain indications of possible irregularities that may point to technical debt [15]. Another important contribution was the definition of various types of technical debt. Some of them are: design debt, requirement debt, defect debt, documentation debt and test debt [1].

Some works [7,21] have shown that technical debt can be found by analyzing comments in source code. Also, In this case, the technical debts found in the comments are SATDs because the developers explicitly indicate that parts of the code needs changes. Comments may indicate that the code is not complete, does not meet the requirements, needs refactoring or even needs to be completely redone. It has also been found that the most common types of SATDs are design and requirement ones [14].

Detecting technical debts in comments has some advantages over the source code approach [26]. Extracting comments is a simple task, which can be done using even a regular expression. When we use source code it is often necessary to assemble complex structures with high computational cost. Also, in cases of detection from code smells, it is necessary to set thresholds for the metrics been used, a problem that is still being researched.

The manual process of detecting technical debt in comments is problematic. In projects with thousands of comments, it becomes virtually impossible for developers to look into comments and classify whether that comment refers to a type of technical debt or not. Also, a manual analysis can be affected by error and bias of the person performing the task. In this way, an automatic process for detecting SATDs in comments is necessary.

Some approaches to automatic detect technical debt in comments have been proposed recently. Maldonado, Shihab, and Tsantalis [26] proposed an approach that uses natural language processing (NLP) and a maximum entropy classifier. Wattanakriengkrai et al. [27] proposed a combination of N-gram IDF and auto-sklearn classifier for detection of SATDs. Both studies obtained good results, and the approach from Wattanakriengkrai et al. presented an improvement over the maximum entropy classifier. In both cases, these work mainly used design and requirement SATDs for training and testing their models, in addition they made available their dataset so that other researchers can evaluate other classifiers.

Currently, deep learning neural networks have shown impressive results in classification tasks such as image recognition, speech, and text classification [12]. Long Short-Term Memory (LSTM) neural networks presented better results than traditional techniques in text classification and sentiment analysis [29,30]. The ability to capture temporal and sequential information makes Recurrent Neural Networks (RNN) and LSTMs ideal for text classification tasks [29].

For text classification, one of the best ways to process text data is through word embedding [16,18]. Mikolov et al. [17] propose a method called Word2vec that can transform a great amount of text in numerical representation. A neural network can use these numerical vector as input instead of other word representation with no contextual information.

Therefore, based on the results of [26,27], in this paper we evaluate a LSTM neural network model and Word2vec to identify SATDs from comments in source code. This work is an extension of previous works [23,24] with a new experiment that assesses the SATD classification while ignoring the type of technical debt. To make this comparison, all types of technical debt were grouped into a single type, dividing the dataset into comments that may or may not have technical debt. Thus, it was also possible to evaluate the efficiency of the model in the classification of SATD regardless of type.

First we train a LSTM neural network with and without Word2vec using the dataset made available by Maldonado and Shihab [14]. Then we apply the training model to classify a test set. The validation process was carried out using 10 projects from the dataset through a leave-one-out cross-project validation process. Finally, the results were compared to Maldonado, Shihab, and Tsantalis, (2017) and Wattanakriengkrai et al. (2018), to evaluate the performance of the LSTM network.

The results showed that the LSTM model with Word2vec have improved in recall and f-measure in design SATD classification. The LSTM model without word embedding achieve better recall, but perform worse in precision and f-measure. In requirement classification, Wattanakriengkrai et al. model [27] precision was greater than any LSTM model, however, the f-measure was similar to that of the LSTM with Word2vec, from the statistical significance point of view. This may have occurred because the database is imbalanced, having more comments without SATDs and little amount of training data in the requirement SATDs.

We also evaluate the SATD classification by grouping all different types of SATD into a single type. When we combine all types of SATDs, the model achieves a better prediction rate. The combination of LSTM and Word2vec achieves a f-measure of 70.8% and the LSTM model alone, a f-measure 62.5%. In both cases, there were improvements in the effectiveness of the classification when compared to the results of the classification of design and requirement SATDs.

Recall may be more important than precision depending on the problem being discussed [9]. Someone may accept a slightly higher rate of false positives to get more true positives, if the trade-off is considered interesting.

The rest of this paper is organized as follows. In Sect. 2, we discuss works related to LSTM and detection of SATDs. Section 3 presents our methodology, the dataset, and a introduction to LSTM and word embedding. Then, in Sect. 4, we discussed the planning and execution of the experiment. The results of the experiment are presented and discussed in Sect. 5 as well as the threats to validity. Finally, in Sect. 6, we present the conclusions of the paper and some possible extensions that can be researched in future works.

2 Related Works

In our work we use an LSTM neural network to classify SATDs in source code comments. There are related papers that talk about both LSTM in text classification tasks and technical debt detection in source code comments, especially self-admitted technical debts.

Zhou et al. [29] propose a combination of LSTM neural networks and Convolutional neural networks (CNN) to perform text classification and sentence representation. CNN is able to extract high-level information from sentences, forming a sequence of phrases representations, and then feeds an LSTM to obtain the representation of the complete sentence. This approach is particularly suitable for text classification because the model can learn local and global representation of the features in the convolutional layer and temporal representation in the LSTM layer.

The detection of SATDs has been the subject of some research using mainly natural language processing. It is possible to analyze comments automatically to identify technical debt, but the amount of false positives is a common problem [6,7]. Farias et al. [6] defined a contextualized vocabulary to identify technical debts in code comments. With the help of 23 participants, they were able to improve a contextualized vocabulary used to detect SATD. A contextualized vocabulary can offer a quick alternative for detecting SATD and, when added to an automatic classifier, it can positively contribute to its classification efficiency.

Potdar and Shihab [21] extracted 62 comment patterns from projects that can indicate SATDs. They found that technical debt exists in 2.4% to 31% of the files. In most cases, the more experienced developers tend to introduce comments in the code that self-admit a technical debt. Finally, their work presented that only 26.3% to 63.5% of SATDs are resolved in the project. Other works have show

Maldonado, Shihab, and Tsantalis [26] proposed an approach to detect SATDs using natural language processing (NLP) and a maximum entropy classifier. In this work, only design and requirement SATDs were analyzed because they are the most common and all the researched projects contains this type of

SATDs. They build a dataset of manually labelled comments from 10 projects: Ant, ArgoUML, Columba, EMF, Hibernate, JEdit, JFreeChart, JMeter, JRuby and SQuirrel SQL. The results show that the approach presented better results compared to the model that uses comments pattern. Words related to sloppy or mediocre code tend to indicate SATD design, whereas comments with words about something incomplete show indications of requirement SATD.

Wattanakriengkrai et al. [27] also worked with design and requirement SATD. They proposed a model that combines N-gram IDF and the auto-sklearn machine learning library and compared the results with Maldonado, Shihab, and Tsantalis [26]. The results show that they outperformed the previous model, improving the performance over to 20% in the detection of requirement SATD and 64% in design SATD.

Two studies used mining techniques to classify SATDs [11,13]. The first work proposed a model that uses feature selection to find the best features for training and uses these features in a model that combines several classifiers. The second one introduces a plugin for Eclipse to detect SATDs in Java source code comments. From this tool, the developer can use the model integrated to the plugin or another model for the detection of SATDs. The plugin can find, list, highlight and manage technical debts within the project.

Based on the studies of Maldonado, Shihab, and Tsantalis, and Wattanakriengkrai et al. [26,27], in our work we propose to evaluate an LSTM model with and without Word2vec and compare with the results obtained by these approaches. We think that an LSTM neural network can achieve better results in this type of classification task, based on previous work reports on text classification and LSTM [29,30].

This work is an extension of previous works [23,24] with a new experiment that assesses the SATD classification while ignoring the type of technical debt.

3 Methodology

The main objective of this work is to evaluate an LSTM model with Word2vec for detection of design, requirement SATDs and general SATDs in source code comments. The first step of the work is to load and clean a dataset of SATDs so that they can be properly used by the LSTM model. After cleaning the dataset, we trained the Word2vec and the LSTM network. We classify the test set using the trained model. To perform this procedure, a controlled experiment was defined and executed. This experiment is detailed in Sect. 4.

Experimentation is a task that requires rigorous planning with well-defined steps [28]. We have to elaborate planning, execution and analysis of the data. From this, it is possible to apply a statistical treatment of the data, with hypothesis tests, so that it can be replicated by others and produce reliable information.

3.1 Self-admitted Technical Debt Dataset

One of the main contributions of Maldonado and Shihab [14] was to build and make available a dataset of SATDs so that other researchers analyze and test their models. The dataset was created by extracting comments from 10 open source software projects. The selected projects were: Ant, ArgoUML, Columba, EMF, Hibernate, JEdit, JFreeChart, JMeter, JRuby and SQuirrel SQL. The criterion used for this selection was that the projects should be from different application domains and had a large amount of comments that can be used for classification and analysis of technical debt.

After extracting comments from the projects, the researchers labelled each comment manually with some type of technical debt. The classification was made based on the work of Alves et al. [1], who presented an ontology of terms that can be applied to define types of technical debt. The types defined were: architecture, build, code, defect, design, documentation, infrastructure, people, process, requirement, service, test automation and test debt. Not all types were used during the labelling process because some of them were not found in code comments. They found technical debt comments of the following types: design debt, defect debt, documentation debt, requirement debt and test debt. Table 1 shows the number of SATDs by type and project. This process resulted in the classification of 62275 comments, being 4071 comments of technical debt of different types and 58204 comments that indicate no technical deb.

Table 1. Total number of comments by type of project and technical debt [24].

	Defect	Test	Documentation	Design	Requirement	No. technical debt	**Total**
Ant	13	10	0	95	13	3967	**4098**
ArgoUML	127	44	30	801	411	8039	**9452**
Columba	13	6	16	126	43	6264	**6468**
EMF	8	2	0	78	16	4286	**4390**
Hibernate	52	0	1	355	64	2496	**2968**
JEdit	43	3	0	196	14	10066	**10322**
JFreeChart	9	1	0	184	15	4199	**4408**
JMeter	22	12	3	316	21	7683	**8057**
JRuby	161	6	2	343	21	4275	**4897**
Squirel	24	1	2	209	50	6929	**7215**
Total	**472**	**85**	**54**	**2703**	**757**	**58204**	**62275**

The most common types of SATDs found were design (2703 comments) and requirement SATDs (472 comments). The two types have some distinctions which are explained and exemplified below. Self-admitted design debts are characterized by comments that talk about issues in the way that the code was implemented. The comments usually indicate that the code is poorly constructed and that they need modifications to improve its quality. Modifications can be made through refactoring process or even redoing the code from the beginning. Some examples of comments that contain design SATDs are:

1. check first that it is not already loaded otherwise consecutive runs seems to end into an OutOfMemoryError or it fails when there is a native library to load several times this is far from being perfect but should work in most cases (from Ant project);

2. this is wrongly called with a null handle, as a workaround we return an empty collection (from ArgoUML project).

In the first comment, the developer explains some troublesome code, but that works most of the time. In addition to explaining the code problem, it is obvious to the reader that this must be solved to prevent the program from throwing exceptions. The second example shows a comment stating that code has a workaround to solve a problem. Words like workaround, stupid and needed? are good candidates to identify comments with SATDs design [14].

Self-admitted requirement debts are found in comments that talk about incomplete code that do not fully meet some requirement. Some examples are:

1. TODO: Why aren't we throwing an exception here? Returning null results in NPE and no explanation why. (from ArgoUML project);
2. Have we reached the reporting boundary?Need to allow for a margin of error, otherwise can miss the slot. Also need to check we've not hit the window already (from jMeter project)

In both cases, the developers explain in the comments that the code needs to be completed and question specific points that are missing in the code.

3.2 LSTM Neural Network

LSTM was initially proposed by Hochreiter and Schmidhuber [10] to solve the problem of long sequences by a recurring neural network. In the original approach, updating the weights, which is done by a variation of the Backpropagation algorithm, called Backpropagation Through Time, is affected when the level of recurrence is high, which makes it impossible for the network to memorize long sequences.

Its architecture consists of a set of subnets connected repeatedly, called the memory block, located in the hidden layer. This block contains memories cells with auto connections capable of storing the temporal state of the network, in addition to special units, called gates, which are responsible for controlling the flow of information, as shown in Fig. 1. Each block consists of one or more memory cells connected with three gates: forget (Eq. 1), input (Eq. 2) e output (Eq. 3). Each gate has a function that allows you to reset, write and read the operations inside the memory block. Each gate uses a sigmoid logistic function (σ) to flatten the values of these vectors between 0 (closed gate) and 1 (open gate).

$$f_t = \sigma(W_{xf} * x_t + W_{hf} * h_{(t-1)} + W_{cf} * c_{(t-1)} + b_f) \tag{1}$$

$$i_t = \sigma(W_{xi} * x_t + W_{hi} * h_{(t-1)} + W_{ci} * c_{(t-1)} + b_i) \tag{2}$$

$$o_t = \sigma(W_{xo} * x_t + W_{ho} * h_{(t-1)} + W_{co} * c_t + b_o) \tag{3}$$

$$\tilde{C}_t = tanh(W_{xc} * x_t + W_{hc} * h_{(t-1)} + b_c) \tag{4}$$

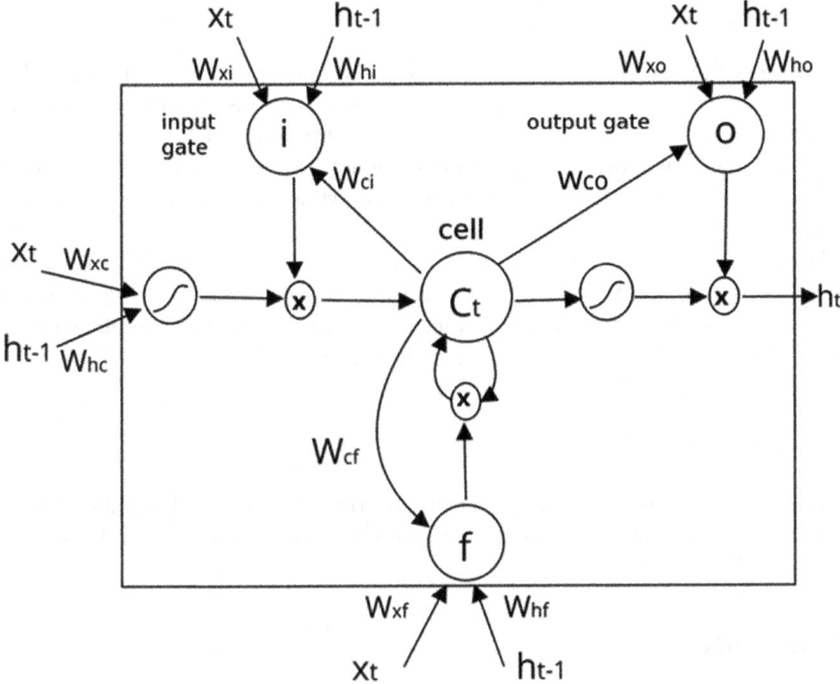

Fig. 1. LSTM memory block [24].

$$C_t = f_t * C_{(t-1)} + i_{t*}\tilde{C_t} \tag{5}$$

The forget gate defines whether previous state activation will be used in memory. The input gate defines how much of the new state calculated for the current input will be used, and finally, the output gate defines whether the internal state will be exposed to the rest of the network (external network). Then a hyperbolic tangent layer creates a vector of new candidate value $\tilde{C_t}$ (Eq. 4) that can be added in the cell.

C_t is the internal memory unit, which is a combination of the previous memory $C_{(t-1)}$ multiplied by the forget gate and the candidate values $\tilde{C_t}$t multiplied by the input gate (Eq. 5). Intuitively, one realizes that memory is a combination of memory in the previous time with the new one in the current time.

Given this memory C_t, it is finally possible to calculate the output of the hidden state h_t by multiplying the memory activations with the output gate (Eq. 6).

$$h_t = o_t * tanh(C_t) \tag{6}$$

The variables i_t, f_t, o_t, c_t, and h_t are vectors representing values in time t. W_* are matrices of weight connected to different gates, and b_* are the vectors that correspond to bias.

3.3 Word2vec

Word Embeddings are methods that transform a sequence of words into a low-dimensional numerical representation. In this way, it is possible to model a language or extract features from it to use as input in machine learning algorithms and other natural language processing activities. One of the most famous methods of word embedding was proposed by Mikolov et al. [17] and is called Word2vec.

Word2vec is a model based on neural networks that can process a large amount of text to and store context information of words. Finally, Word2vec returns a vector space with information for each word. For text classification, we can use Word2vec in the text pre-processing step.

4 Experiment Steps

We follow a experimental process to evaluate our LSTM model results based on Wohlin's guidelines [28]. In this section, we will discuss planning and execution of the experiment.

4.1 Objective

The objective of this study is to evaluate, through a controlled experiment, the efficiency of the LSTM neural network with Word2vec in SATDs classification in source code comments. The experiment was done by using a dataset build by Maldonado and Shihab [14] to train the LSTM model and we compare the results to those from the studies of Maldonado, Shihab, and Tsantalis, and Wattanakriengkrai et al. [26,27]. Wattanakriengkrai et al. combined N-gram IDF and auto-sklearn, and Maldonado, Shihab, and Tsantalis used maximum entropy classifier. This work is an extension of previous works [23,24] with a new experiment that assesses the SATD classification while ignoring the type of technical debt.

The objective was formalized using the GQM model proposed by Basili and Weiss [2]: Analyze the LSTM neural network with Word2vec, with the purpose of evaluating it (against results of algorithms evaluated in previous works), with respect to recall, precision and f-measure, from the viewpoint of developers and researchers, in the context of detecting self-admitted technical debts in open source projects.

4.2 Planning

Context Selection: We selected the dataset discussed in Sect. 3.1 for the classification of SATDs. We use SATDs to train the Word2vec model, and train and test the LSTM model. The model was validated using a leave one-out cross-project validation approach to two dataset groups. We trained the models using 9 of the 10 projects and tested on the remaining one. This procedure is repeated 10 times so that each project can be tested with the trained model.

Hypothesis Formulation: To reach the proposed goal, we define the following research question: Is the LSTM neural network with Word2vec better than previous works in terms of recall, precision, f-measure?

The following hypothesis was defined for each proposed metric, h_0: the algorithms have the same metric mean (Eq. 7) and h_1: the algorithms have distinct metric means (Eq. 8). Note that h_0 is the hypothesis that we want to refute.

$$\mu_1(metric) = \mu_2(metric) \tag{7}$$

$$\mu_1(metric) \neq \mu_2(metric) \tag{8}$$

Selection of Participants: We divided the dataset into two groups, the first (60,907 code comments) having comments with design SATDs and comments without any SATDs, and the second (58,961 code comments) with comments with requirement SATDs and comments without SATDs. evaluation of the models when we group all types of SATDs in a single label.

Experiment Project: The experiment project refers to the following stages: Preparation of the development environment, it means downloading and installation of all the items described in instrumentation. Subsequently, we implement and trained the model with the dataset. Finally, we run the experiments and perform statistical tests for the assessment of the defined hypotheses.

Independent Variables: The LSTM neural network, Word2vec model and the dataset.

Dependent Variables: Predictions made by the model, represented by: precision (Eq. 9), recall (Eq. 10) and f-measure (Eq. 11). True positives (TP) are cases in which the classifier correctly identifies a SATD comment and true negative (TN) corresponds to the correct classification of a comment without SATD. If the model classifies a SATD comment as without SATD, it is a case of false negative (FN), and the case of false positive (FP) is when the model classifies a comment without SATD as a SATD comment.

$$precision = TP/(TP + FP) \tag{9}$$

$$recall = TP/(TP + FN) \tag{10}$$

$$f - measure = (2 * precision * recall)/(precision + recall) \tag{11}$$

Instrumentation: The instrumentation process started with the environment configuration for the achievement of the controlled experiment; data collection planning; and the development and execution of the assessed algorithms. The used materials/resources were: Keras [3], Scikit-learn [20], Gensim [22], and a computer with Intel(R) Core(TM) i5-7400 CPU @ 3.00GHz, 16 GB RAM - 64 bits. The preparation of the test environment was done by downloading and installing all the mentioned libraries.

4.3 Execution

After all preparation, the experiment was performed. First, the dataset was loaded and a cleanup process was performed. Some special characters and numbers were eliminated so as not to confuse the training process and to improve the quality of the features. Then the data was submitted to the training model. At the end, leave-one-out cross-project validation was carried out on the 10 projects.

4.4 Data Validation

We used two statistical tests to validate our results: Student's t-test and Shapiro-Wilk test. Student's t-test is used to determine if the difference between two means is statistically significant. For this, it is necessary that the distribution of samples is normal. Normality is tested using Shapiro-Wilk.

5 Experiment Results

After performing the training and testing of the LSTM model, the results of the classification were obtained through the leave-one-out cross-project validation process. Table 2 and 3 present the results achieved for each project and metric in the classification of design and requirement SATDs respectively. We use the abbreviations for precision (Pr), recall (Rc), and f-measure (F1) because of the little space available in the table. The best result for each metric is highlighted in bold.

The results show that without pre-processing with Word2vec, LSTM has higher recall than Auto-sklearn (AS) and Maximum Entropy (ME), but loses in precision and f-measure. When pre-processing with Word2vec is applied, we have a significant improvement in precision, and consequently in the f-measure. The model without word embedding showed a 56% improvement in recall compared to Auto-sklearn and Maximum Entropy in the design SATD classification. With the application of Word2vec, the improvement in recall was approximately 36% in both cases. Despite having a lower recall, when we apply a pre-processing with

Table 2. Comparison of the metrics obtained in the design SATD classifications [24].

	LSTM + word2vec			LSTM			Auto-sklearn			Maximum entropy		
	Pr	Rc	F1	Pr	Rc	F1	Pr	Rc	F1	Pr	Rc	F1
Ant	0.621	0.608	**0.614**	0.228	**0.821**	0.357	**0.676**	0.301	0.360	0.554	0.484	0.517
ArgoUML	**0.940**	0.798	**0.863**	0.443	**0.963**	0.607	0.784	0.703	0.741	0.788	0.843	0.814
Columba	0.658	0.813	0.728	0.148	0.952	0.256	0.765	**0.940**	**0.842**	**0.792**	0.484	0.601
EMF	0.346	0.613	0.442	0.069	**0.910**	0.129	**0.802**	0.501	**0.604**	0.574	0.397	0.470
Hibernate	0.735	**0.912**	**0.814**	0.467	0.873	0.609	**0.833**	0.450	0.583	0.877	0.645	0.744
JEdit	0.316	**0.837**	0.459	0.214	0.744	0.333	**0.943**	0.701	**0.810**	0.779	0.378	0.509
JFreeChart	0.418	0.733	**0.532**	0.277	**0.885**	0.422	**0.872**	0.250	0.390	0.646	0.397	0.492
JMeter	0.734	**0.859**	**0.791**	0.233	0.854	0.367	0.706	0.420	0.530	**0.808**	0.668	0.731
JRuby	0.845	0.838	**0.841**	0.362	**0.932**	0.522	**0.856**	0.750	0.801	0.798	0.770	0.784
Squirrel	0.545	0.708	**0.616**	0.192	**0.894**	0.317	**0.903**	0.630	0.740	0.544	0.536	0.540
Average	0.616	0.772	**0.670**	0.263	**0.882**	0.391	**0.814**	0.564	0.640	0.716	0.560	0.620

word embedding, the precision of the LSTM network increases by approximately 135%. For design SATD, the Auto-sklearn classifier from Wattanakriengkrai et al. [27] obtained the best results in precision.

Table 3. Comparison of the metrics obtained in the requirement SATD classifications [24].

	LSTM + word2vec			LSTM			Auto-sklearn			Maximum entropy		
	Pr	Rc	F1	Pr	Rc	F1	Pr	Rc	F1	Pr	Rc	F1
Ant	0.230	0.500	**0.315**	0.013	**0.692**	0.026	**0.650**	0.136	0.226	0.154	0.154	0.154
ArgoUML	**0.839**	0.683	0.753	0.388	**0.854**	0.533	0.779	0.762	**0.771**	0.663	0.540	0.595
Columba	**0.860**	0.755	0.804	0.107	**1.000**	0.194	0.781	0.935	**0.851**	0.755	0.860	0.804
EMF	0.375	**0.750**	0.500	0.015	0.562	0.030	**0.826**	0.682	**0.747**	0.800	0.250	0.381
Hibernate	0.765	0.765	**0.765**	0.165	**0.921**	0.281	**0.809**	0.435	0.566	0.610	0.391	0.476
JEdit	0.571	0.666	0.615	0.014	**0.785**	0.028	**0.937**	0.715	**0.811**	0.125	0.071	0.091
JFreeChart	0.800	0.266	0.400	0.064	**0.800**	0.118	**0.846**	0.280	**0.421**	0.220	0.600	0.321
JMeter	0.619	0.565	**0.590**	0.029	**0.952**	0.057	**0.693**	0.418	0.522	0.153	0.524	0.237
JRuby	0.572	**0.900**	0.700	0.296	0.763	0.427	**0.859**	0.749	**0.800**	0.686	0.318	0.435
Squirrel	0.780	0.513	0.619	0.060	**0.760**	0.112	**0.848**	0.535	**0.656**	0.657	0.460	0.541
Average	0.414	0.636	0.606	0.115	**0.809**	0.180	**0.803**	0.565	**0.637**	0.482	0.416	0.403

Figures 2 and 3 show the results of the f-measure in the classification of design and requirement SATDs respectively. With the bar graph visualization, it is possible to see more clearly the impact of preprocessing with a word embedding algorithm. The LSTM model benefits from the learning done by the Word2vec model in addition to reducing the effects of the limited amount of positive data for network training.

The good results achieved in the SATD design classification were not maintained in the SATD requirement classification. The SATD requirement dataset has a lower number of positive cases, which makes it difficult to train the LSTM network. In this case, the f-measure results were statistically compatible between the LSTM and Auto-sklearn models. The Auto-sklearn classifier was superior in precision, but produced a lower recall than LSTM models with and without word embedding.

Although the LSTM model has a higher average recall and f-measure, it was necessary to follow a statistical validation to verify if the improvement was significant. The next step was to perform the Shapiro-Wilk test with the set of metrics. The Shapiro-Wilk test showed that the distribution of the metrics is normal. In this way, we applied the Student's t-test for paired samples to verify if the difference was statistically significant.

Table 4, 5, and 6 presents the p-values calculated from the average precision, recall and f-measure respectively obtained with the classification models for design and requirement SATDs detection. As can be seen, the improvements in recall and f-measure in the design SATD classification and improvement in recall in requirement SATD have p-values lower than the significance level of 0.05. This indicates that only for these cases the improvement was statistically

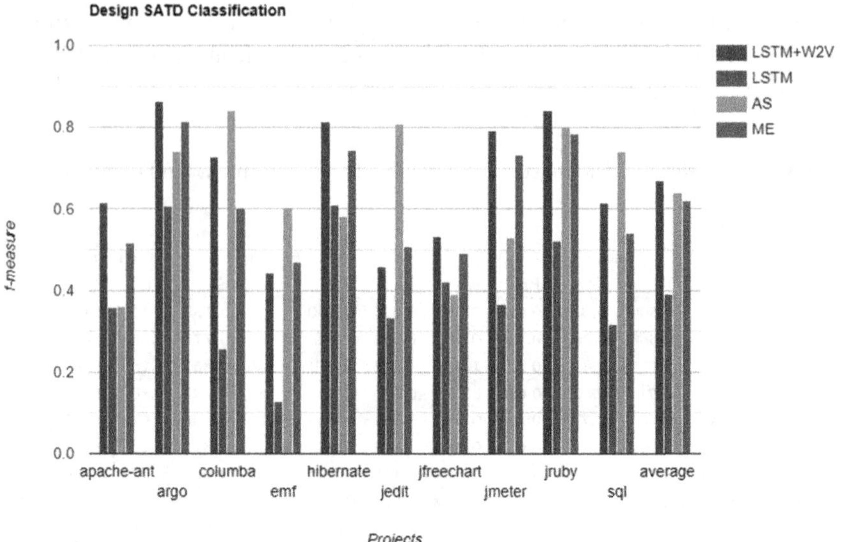

Fig. 2. F-measure results from design SATD classification.

significant. The reason for this may be related to the lower amount of requirement SATDs for training the LSTM network, which shows that the LSTM network is dependent on a larger dataset for better results.

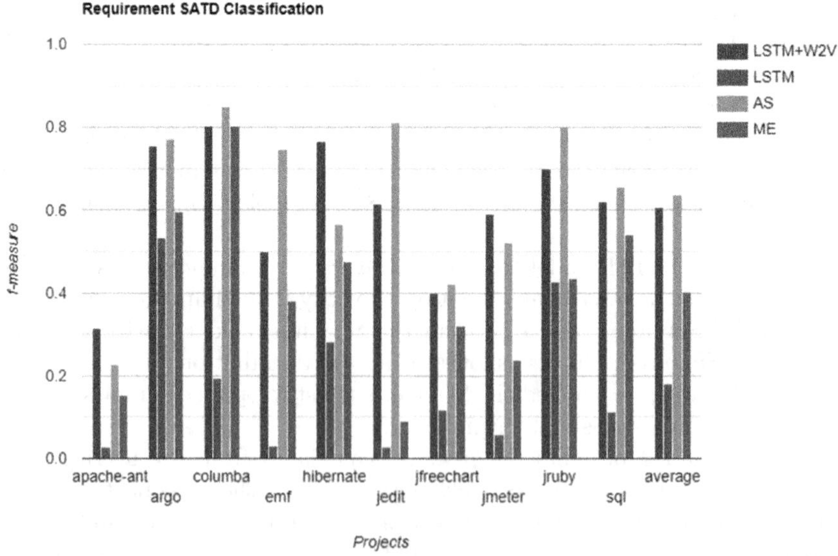

Fig. 3. F-measure results from requirement SATD classification.

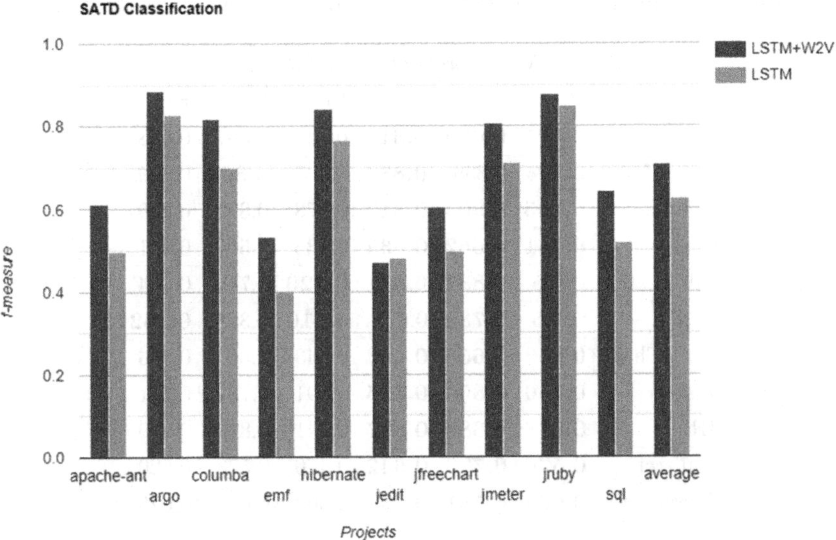

Fig. 4. F-measure results from SATD classification.

Table 4. Results from t-test for average precision [24].

LSTM + word2vec vs Classifiers	Design		Requirement	
	p-value	Result	p-value	Result
Maximum entropy	0.11	Retain H_0	0.11	Retain H_0
Auto-sklearn	0.03	Refute H_0	0.03	Refute H_0
Only LSTM	0.00	Refute H_0	0.00	Refute H_0

Table 5. Results from t-test for average recall [24].

LSTM + word2vec vs Classifiers	Design		Requirement	
	p-value	Result	p-value	Result
Maximum entropy	0.00	Refute H_0	0.05	Refute H_0
Auto-sklearn	0.01	Refute H_0	0.23	Retain H_0
Only LSTM	0.01	Refute H_0	0.03	Refute H_0

Table 6. Results from t-test for average f-measure [24].

LSTM + word2vec vs Classifiers	Design		Requirement	
	p-value	Result	p-value	Result
Maximum entropy	0.01	Refute H_0	0.00	Refute H_0
Auto-sklearn	0.66	Retain H_0	0.47	Retain H_0
Only LSTM	0.00	Refute H_0	0.00	Refute H_0

Table 7. Comparison of the metrics obtained in the SATD classification.

	LSTM + word2vec			LSTM		
	Pr	Rc	F1	Pr	Rc	F1
Ant	**0.595**	**0.629**	**0.611**	0.444	0.567	0.498
ArgoUML	**0.934**	0.839	**0.884**	0.795	**0.866**	0.829
Columba	**0.745**	**0.904**	**0.817**	0.578	0.882	0.699
EMF	**0.451**	**0.652**	**0.534**	0.335	0.500	0.401
Hibernate	0.805	**0.881**	**0.841**	**0.820**	0.716	0.764
JEdit	0.335	**0.789**	0.471	**0.610**	0.398	**0.482**
JFreeChart	**0.559**	**0.653**	**0.603**	0.443	0.564	0.496
JMeter	**0.756**	**0.860**	**0.805**	0.691	0.732	0.711
JRuby	**0.873**	**0.880**	**0.877**	0.841	0.860	0.850
Squirrel	**0,576**	**0,723**	**0.642**	0.476	0.573	0.520
Average	**0.663**	**0.781**	**0.708**	0.603	0.666	0.625

Table 8. Results from t-test for average precision, recall, and f-measure in SATD classification.

	Design		Requirement		Requirement	
	p-value	Result	p-value	Result	p-value	Result
LSTM + word2vec vs LSTM	0.09	Retain H_0	0.00	Refute H_0	0.00	Refute H_0

Another point evaluated was the classification of SATD without division by SATD type. To train the model, all types of SATDs were grouped into a single label that represents any type of SATD. In this way, we carry out the training and testing of the LSTM model with and without preprocessing with Word2vec. It was not possible to compare these results with previous works because they did not perform this experiment setup. Therefore, the comparison was only with the two variations of the LSTM model.

The results can be seen in Table 7 and Fig. 4. Table 7 shows all the metrics extracted when we applied the trained classifiers to the test projects. Figure 4 shows a bar graph with the results of the f-measure.

The results show that when we combine all types of SATDs, the model achieves a better prediction rate. The combination of LSTM and Word2vec achieves a f-measure of 70.8% and the LSTM model alone, a f-measure 62.5%. In both cases, there were improvements in the effectiveness of the classification when compared to the results of the classification of design and requirement SATDs. The combination of LSTM and Word2vec increased precision in 9.95%, recall in 17.26% and f-measure in 13.28% compared with LSTM alone. As shown in Table 8, only the improvement in recall and f-measure were statistically significant (p-value < 0.05).

5.1 Threats to Validity

There are some aspects of an experiment that define the validity of the results achieved during its execution. It is ideal that all threats to the validity of the experiment are known and that measures are taken to have them reduced or eliminated. The following are threats found during the planning and execution of the experiment:

1. Construct validity

a. The implementation of an LSTM neural network algorithm must meet the theoretical requirements and any changes may compromise its results. To ensure that a correct implementation of the LSTM neural network was evaluated, we used the Keras (Fraçois et al., 2015) library that has thousands of citations in study publications;

b. A manually annotated dataset may contain errors caused by human failure, such as incorrect labelling and labelling bias. This may compromise classifier performance. In this case, we compared the LSTM model with other classifiers that used the same dataset and followed the same process of validation of the experiment.

6 Conclusion

In order to increase the effectiveness of the self-admitted technical debt classification and create better alternatives to support their management, in this research, we combined Word2vec for word embedding with a Long short-term memory (LSTM) neural network and performed an experimental process to evaluate model predictive performance against two other natural language processing approaches using auto-sklearn and maximum entropy classifiers.

The results showed that the LSTM model achieved better recall and f-measure in design SATDs classification and recall in requirement SATD. The average f-measure was statistically the same as the Auto-sklearn classifier, which was the model with better precision.

We also noticed that the classification improves when all SATD types were grouped in a single label. With this grouping, the model achieves better effectiveness and the combination of LSTM and Word2vec obtained a f-measure of 70.8% and the LSTM model, alone, a f-measure of 62.5%. In both cases, there were improvements in the prediction when compared to the results of the project and requirement SATDs classification.

As in some medical diagnosis contexts, when healthcare professionals prefer to deal with some false positives than to lose some sick patients, recall can be more important than precision. The LSTM model without Word2vec has a better recall rate, but lower precision. The combination of Word2vec and LSTM has a great advantage for overall performance.

In future publications, we will evaluate and combine other deep learning neural networks and other word embedding methods. In addition, we will development a classifier that will use the best evaluated models for each specific context. The main objective will be to reduce the number of false negatives produced by the LSTM model in the SATDs case. Briefly, we found evidence that LSTM models combined with word embedding are promising for the development a more effective SATD classifier.

References

1. Alves, N.S., Ribeiro, L.F., Caires, V., Mendes, T.S., Spínola, R.O.: Towards an ontology of terms on technical debt. In: 2014 6th International Workshop on Managing Technical Debt, pp. 1–7. IEEE (2014)
2. Basili, V.R., Weiss, D.M.: A methodology for collecting valid software engineering data. IEEE Trans. Softw. Eng. **6**, 728–738 (1984)
3. Chollet, F., et al.: Keras. https://keras.io (2015)
4. Codabux, Z., Williams, B.: Managing technical debt: an industrial case study. In: 2013 4th International Workshop on Managing Technical Debt (MTD), pp. 8–15. IEEE (2013)
5. Cunningham, W.: The WyCash portfolio management system. ACM SIGPLAN OOPS Messenger **4**(2), 29–30 (1992)
6. de Freitas Farias, M.A., de Mendonça Neto, M.G., Kalinowski, M., Spínola, R.O.: Identifying self-admitted technical debt through code comment analysis with a contextualized vocabulary. Inf. Softw. Technol. **121**, 106270 (2020)
7. de Freitas Farias, M.A., de Mendonça Neto, M.G., da Silva, A.B., Spínola, R.O.: A contextualized vocabulary model for identifying technical debt on code comments. In: 2015 IEEE 7th International Workshop on Managing Technical Debt (MTD), pp. 25–32. IEEE (2015)
8. Guo, Y., Seaman, C.: A portfolio approach to technical debt management. In: Proceedings of the 2nd Workshop on Managing Technical Debt, pp. 31–34 (2011)
9. Hand, D., Christen, P.: A note on using the f-measure for evaluating record linkage algorithms. Stat. Comput. **28**(3), 539–547 (2018)
10. Hochreiter, S., Schmidhuber, J.: Long short-term memory. Neural Comput. **9**(8), 1735–1780 (1997)
11. Huang, Q., Shihab, E., Xia, X., Lo, D., Li, S.: Identifying self-admitted technical debt in open source projects using text mining. Empirical Softw. Eng. **23**(1), 418–451 (2017). https://doi.org/10.1007/s10664-017-9522-4
12. LeCun, Y., Bengio, Y., Hinton, G.: Deep learning. Nature **521**(7553), 436–444 (2015)
13. Liu, Z., Huang, Q., Xia, X., Shihab, E., Lo, D., Li, S.: Satd detector: a text-mining-based self-admitted technical debt detection tool. In: Proceedings of the 40th International Conference on Software Engineering: Companion Proceedings, pp. 9–12 (2018)
14. Maldonado, E.D.S., Shihab, E.: Detecting and quantifying different types of self-admitted technical debt. In: 2015 IEEE 7th International Workshop on Managing Technical Debt (MTD), pp. 9–15. IEEE (2015)
15. Marinescu, R.: Assessing technical debt by identifying design flaws in software systems. IBM J. Res. Dev. **56**(5), 1–9 (2012)
16. Mikolov, T., Chen, K., Corrado, G., Dean, J.: Efficient estimation of word representations in vector space. arXiv preprint arXiv:1301.3781 (2013)
17. Mikolov, T., Chen, K., Corrado, G.S., Dean, J.A.: Computing numeric representations of words in a high-dimensional space. US Patent 9,037,464, 19 May 2015

18. Mikolov, T., Sutskever, I., Chen, K., Corrado, G.S., Dean, J.: Distributed representations of words and phrases and their compositionality. In: Advances in Neural Information Processing Systems, pp. 3111–3119 (2013)
19. Nord, R.L., Ozkaya, I., Kruchten, P., Gonzalez-Rojas, M.: In search of a metric for managing architectural technical debt. In: 2012 Joint Working IEEE/IFIP Conference on Software Architecture and European Conference on Software Architecture, pp. 91–100. IEEE (2012)
20. Pedregosa, F., et al.: Scikit-learn: machine learning in python. J. Mach. Learn. Res. **12**, 2825–2830 (2011)
21. Potdar, A., Shihab, E.: An exploratory study on self-admitted technical debt. In: 2014 IEEE International Conference on Software Maintenance and Evolution, pp. 91–100. IEEE (2014)
22. Rehurek, R., Sojka, P.: Software framework for topic modelling with large corpora. In: In Proceedings of the LREC 2010 Workshop on New Challenges for NLP Frameworks. Citeseer (2010)
23. Santos, R.M., Junior, M.C.R., de Mendonça Neto, M.G.: Self-admitted technical debt classification using LSTM neural network. In: Latifi, S. (ed.) 17th International Conference on Information Technology–New Generations (ITNG 2020). AISC, vol. 1134, pp. 679–685. Springer, Cham (2020). https://doi.org/10.1007/978-3-030-43020-7_93
24. Santos, R.M., Santos, I.M., Júnior, M.C.R., de Mendonça Neto, M.G.: Long term-short memory neural networks and word2vec for self-admitted technical debt detection. In: ICEIS (2), pp. 157–165 (2020)
25. Seaman, C., Guo, Y.: Measuring and monitoring technical debt. In: Advances in Computers, vol. 82, pp. 25–46. Elsevier (2011)
26. da Silva Maldonado, E., Shihab, E., Tsantalis, N.: Using natural language processing to automatically detect self-admitted technical debt. IEEE Trans. Softw. Eng. **43**(11), 1044–1062 (2017)
27. Wattanakriengkrai, S., Maipradit, R., Hata, H., Choetkiertikul, M., Sunetnanta, T., Matsumoto, K.: Identifying design and requirement self-admitted technical debt using n-gram IDF. In: 2018 9th International Workshop on Empirical Software Engineering in Practice (IWESEP), pp. 7–12. IEEE (2018)
28. Wohlin, C., Runeson, P., Höst, M., Ohlsson, M.C., Regnell, B., Wesslén, A.: Experimentation in Software Engineering. Springer Science & Business Media, Heidelberg (2012). https://doi.org/10.1007/978-3-642-29044-2
29. Zhou, C., Sun, C., Liu, Z., Lau, F.: A C-LSTM neural network for text classification. arXiv preprint arXiv:1511.08630 (2015)
30. Zhou, P., Qi, Z., Zheng, S., Xu, J., Bao, H., Xu, B.: Text classification improved by integrating bidirectional lstm with two-dimensional max pooling. arXiv preprint arXiv:1611.06639 (2016)

Software Agents and Internet Computing

From Simple to Structural Clones: Tapping the Benefits of Non-redundancy

Stan Jarzabek(✉)

Faculty of Computer Science, Bialystok University of Technology, Bialystok, Poland
s.jarzabek@pb.edu.pl

Abstract. Similarities are common in software. We observe them at all levels, from software requirements, to design, to program code and to other software artifacts such as models or test cases. Similar program parts are termed *code clones.* Detection of clones in code, as well as methods to achieve non-redundancy in programs have been an active area of research for last decades. In this chapter, I discuss sources of redundancies, variety of forms of their manifestation, software productivity benefits that can be gained by avoiding clones, and difficulties to realize these benefits with conventional programming languages and design techniques. I point to generative techniques as a promising approach to tap the benefits of non-redundancy. Research so far has mainly focused on detection of similar code fragments, so-called *simple clones.* The knowledge and possible unification of simple clones can help in maintenance. Still, further gains can be obtained by elevating the level of software similarity analysis to the design-level, larger granularity similar program structures, such as recurring architecture-level patterns of collaborating components. Recurring patterns of simple clones often indicate the presence of interesting higher-level similarities that we call *structural clones.* Detection of structural clones can help in understanding the design of a system for quality assessment, better maintenance and re-engineering, opening new options for design recovery (reverse engineering) which has been an active area of research for last 25 years, with only limited impact on software practice. In this paper, I broadly discuss the landscape of clone research, with particular emphasis on structural clones (This study was supported by the grant of WZ/WI-IIT/3/2020 from Bialystok University of Technology and founded from the resources for research by Ministry of Science and Higher Education, Poland.)

Keywords: Software clones · Generic design · Software maintainability and reusability · Software complexity

1 Introduction

Much similarity within and across programs creates potential for program simplification and reuse. The extent to which similar program structures deliberately spread through programs indicates that this potential has not been fully exploited. In this paper, we discuss software similarity phenomenon and its manifestation in programs as software clones.

J. Filipe et al. (Eds.): ICEIS 2020, LNBIP 417, pp. 563–590, 2021.
https://doi.org/10.1007/978-3-030-75418-1_26

If not tackled with some form of generics, similarities show as program structures repeated many times within a program or across programs. We observe similar program structures of various types and granularity such as architectural patterns of components, patterns of collaborating classes, similar classes, source files, or code fragments. Recurring program structures are termed *software clones*. Clone detection, analysis and visualization has been an active area of research in last two decades, e.g., see survey 30.

In a number of studies, my team at the National University of Singapore (NUS) observed that extensive cloning sometimes occurred due to the lack of strong enough parameterization mechanisms to avoid repetitions without compromising other engineering goals that mattered to them. Much redundancy is common in old, heavily maintained programs. However, we found much cloning also in newly developed programs that, in our judgment, were well designed in view of their design goals and technology used.

As many of these recurring program structures represented important concepts from software requirements or design spaces, these observations seemed to point to some interesting and may be fundamental issue, worth investigation. In follow up research, my team developed a clone detection tool called Clone Miner 6 that allowed us to find and study large-granular clones 45 in addition to similar code fragments. We also developed an Adaptive Reuse Technology, ART[1] for representing groups of cloned code structures with parameterized, generic, adaptable, therefore, reusable meta-components. We applied these tools in projects across a wide range of application domains and programming platforms, observing consistently 50%–90% levels of redundancies. In this paper, I summarize our findings, with references to relevant publications. A detailed discussion of our earlier projects can be also found in a monograph 32.

Generic design can help avoid redundancies, reducing conceptual complexity, as well as the physical size of programs. STL 50 is a premier example of engineering benefits of generic design in the domain of data structures. However, in many other domains the potential of similarity patterns for program simplification and reuse remains often untapped. Software Product Line SPL approach 16 attempts to address the problem at the architecture-component level. As it is often the case, "the devil is hidden in detail", and we need much finer level variability management mechanism to tackle redundancies and fully reap their potential for program simplification and reuse.

In this chapter, I share experiences from my research on clones. In the remaining sections, I discuss the multi-faceted nature of software redundancy, software clone definition, the reasons why clones occur in programs, their impact on software development and maintenance, productivity benefits that can be gained by avoiding clones, and difficulties to realize these benefits with conventional programming languages and design techniques.

2 What Are Software Clones?

Similar program parts are termed *software clones*. Considering their form and size, we distinguish two types of clones: *Simple clones* are formed by textually similar code fragments. More precisely, simple clones are similar segments of contiguous code such

[1] Adaptive Reuse Technology, https://art-processor.org.

as program functions or any code fragments 2335. Larger program structures, formed by configurations of simple clones, are called *structural clones* 56. Examples of clones are similar class methods, classes, source files, directories, any similar software components or recurring patterns (configurations) of similar components.

We introduce clones informally as follows: Two program structures of considerable size are clones of each other if they meet a certain user-defined threshold of similarity measure. The required size and similarity threshold are subjective, depend on the context, and therefore must be set by a programmer to meet goals of a specific clone analysis exercise.

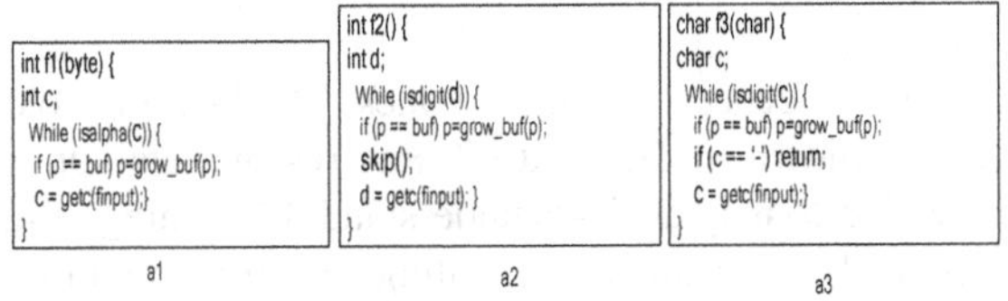

Fig. 1. Simple clones formed by similar code fragments 31.

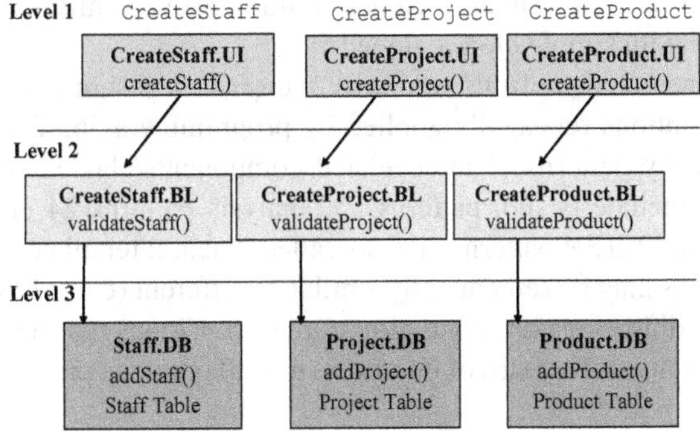

Fig. 2. Structural clones 31.

Figure 1 and Fig. 2 show intuitive examples of simple and structural clones. Three code fragments (a1, a2, a3) in Fig. 1 differ in code details highlighted in **bold**. We can consider them as simple clones of each other, provided they meet a user-defined similarity threshold. Figure 2 shows three structures implementing features `CreateStaff`, `CreateProject`, `CreateProduct` in a web portal for project management. Boxes are PHP files implementing user interface (at the top), business logic (in the middle), and database aspects of respective features. Each of the files consists of PHP functions. Functions across files at each level (UI, BL and DB) are similar to each other forming simple clones such as shown in Fig. 1. Suppose PHP files are densely covered by clones. Then, we could consider files at each level as similar (abstraction step). As there is a calling relation among functions in respective PHP files, the three structures form a collaborative structural clone class. We conclude that features `CreateStaff`, `CreateProject`, `and CreateProduct` are similar to each other (abstraction step).

Most of the interesting clones are similar, but not identical. Differences among clones result from changes in their intended behaviour, and from dependencies on the specific program context in which clones are embedded (such as different names of referenced variables, methods called, or platform dependencies).

Clone detection techniques 6102335 can automate finding clones in programs, and refactorings 24 can help us to free programs from some clones. At times, clone elimination may be hindered by risks involved in changing programs 18 or by other design goals that conflict with refactorings 3339.

3 Structural Clones

Much clone research has focused on simple clones. The lack of looking at the bigger picture of similarity is known in the field 62, and researchers applied classification, filtering, and visualization to help analysts understand the cloning of information 3661. The concept of s*tructural clones* aims at revealing the bigger picture of similarities. It is based on the observation that certain recurring configurations of simple clones form interesting higher-level similarities. This observation also forms the basis for detecting structural clones by finding recurring configurations of simple clones. Empirical studies confirmed that detection of structural clones could improve similarity analysis beyond what was possible with simple clones alone [6].

Structural clones emerge from similarity patterns in design and analysis spaces, or from design solutions repeatedly applied by programmers 56. Typically, structural clones are formed by patterns of inter-related components/classes, such as shown in Fig. 2. Examples include design patterns 25, analysis patterns 24 enterprise patterns using .NET™, core J2EE™ patterns, and so-called "mental templates" 10.

Structural clones may be semantically similar or different (e.g., clones diverge when many adaptive modifications are done after copying). Clones that do not perform any well-defined function at all are still of interest as they play role in program understanding and reuse.

Suppose code fragments (a1, a2, a3), (b1, b2, b3), (c1, c2, c3) … (e1, e2, e3) of Fig. 3 form simple clone classes, and they appear in files X1, X2 and X3, as shown in Fig. 3(a). Grey code is unique to X1 and X2, and black code is unique to X3. Then, we could consider the three groups [a1, b1, c1, d1, e1], [a2, b2, c2, d2, e2] and [a3, b3, c3, d3, e3] as a structural clone class.

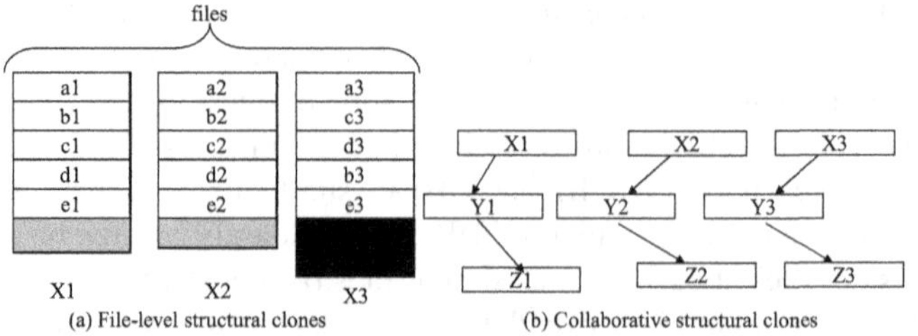

(a) File-level structural clones (b) Collaborative structural clones

Fig. 3. Structural clones.

Suppose a substantial part, say at least 80%, of each of the three files is covered by structural clones (meaning that less than 20% of code is contained in grey or black areas). Then, we could consider files X1, X2 and X3 a file-level structural clone class. This abstraction step allows us to build structural clones in a hierarchical way, whereby higher-level, larger-granular structural clones are formed in terms of smaller-granularity ones. The "80% coverage by clones" is an example of a user-defined similarity threshold value. We elaborate on clone similarity measures later in this paper.

Suppose further that (Y1, Y2, Y3) and (Z1, Z2, Z3) are also file-level structural clone classes and they form collaborative (via message passing) structures such as shown in Fig. 3(b). Then, we consider a group of such patterns a higher-level collaborative structural clone class.

4 Software Clone Definition

Structural clones can be modelled as graphs. We say that a *clone relation* exists between two *program entities* (*entities* for short) e1 and e2, if e1 and e2 satisfy certain thresholds of pre-defined similarity metrics. Human judgment is also an important factor in deciding whether two entities are clones of each other or not. We introduce preliminary terms first.

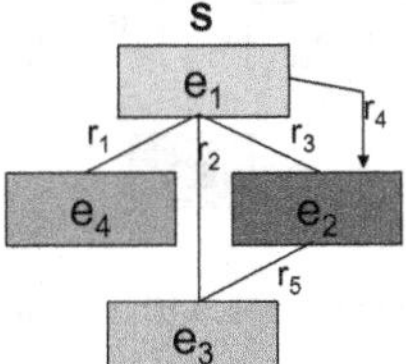

Fig. 4. A structure S.

In Fig. 4, we see structure S consisting of four entities (e_1, e_2, e_3, e_4), and five relationships (r_1, r_2, r_3, r_4, r_5) among them. Entities e_1 and e_2 are related by means of two relationships, namely r_3 and r_4, where r_4 is a directed relationship. An example of a relationship is the co-location of the interrelated entities. For example, each group of code fragments in Fig. 3(a) is co-located in the same file. A directed relationship among entities indicates the order in which code fragments appear in the file.

An entity can be any program element that we can clearly identify such as program statement, code fragment, file, directory, class method, interface, package, component, a group of collaborating components, sub-system or the whole system.

Following terms in 35, a *clone relation* is an equivalence relation (i.e., reflexive, transitive, and symmetric relation). For a given clone relation, a pair of entities is called a *clone pair* if a clone relation holds between the two entities. An equivalence class of a clone relation is called a *clone class*. A clone class is a maximal set of entities in which a clone relation holds between any pair of entities.

Code fragments (i.e., segments of contiguous code) are the simplest type of entities that can participate in a clone relation. Similar code fragments are called simple clones.

A *program structure* (*structure* for short) is a connected mixed multigraph where nodes are entities, and (directed or undirected) edges are relationships between entities. A *relationship* is any meaningful physical or logical connection between two entities in a structure. In a mixed multigraph, the same pair of nodes can be connected by multiple edges. This property is useful in characterizing certain types of structures.

We may be also interested in entities found in the "same function", "same class", "same module", "same sub-system" and so on. Other examples of relationships include "message passing", "inheritance", "association", or "hyperlink" among web pages. Any other physical, semantic or syntactic relationship among entities may be included in a definition of a structure depending upon the context of similarity analysis.

To explain the structure hierarchies, we introduce the terms *atomic entities* and *abstract entities.* An *atomic entity* is one which has no relevant internal structure. For example, a code fragment is often considered an atomic entity in clone analysis. An *abstract entity* is one whose internal structure is abstracted away, to form a building block for higher level structures. Abstract entities allow us to define higher-level structures in terms of lower-level structures at as many levels as is useful.

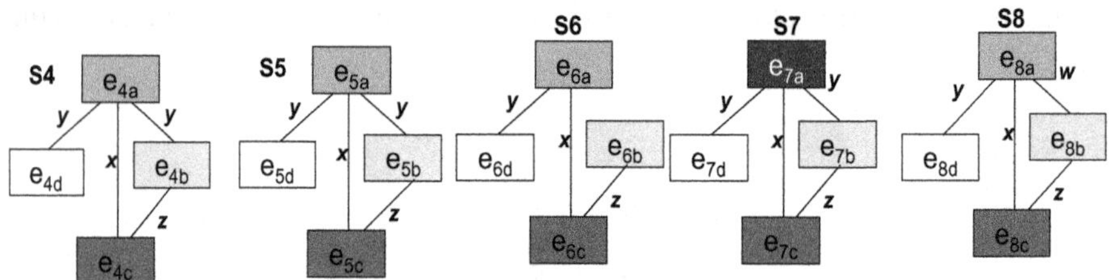

Fig. 5. Five structures with relational variants.

We can now define more precisely a structural clone relation as a clone relation between two structures as follows:

Definition: A *structural clone relation* holds between two structures S1 and S2 if (and only if):

(a) S1 and S2 have the same graph structure
(b) a clone relation has already been established between corresponding entities in S1 and S2, and
(c) corresponding relationships in S1 and S2 are of the same type.

In Fig. 5, we see five structures S4–S8, each consisting of four entities, where *x*, *y*, *z*, and *w* are different relationship types. Any two entities identified by the same subscript letter (i.e., a, b, c or d) are corresponding entities. For example, e_{4a} and e_{5a} are corresponding entities in S4 and S5.

Suppose entities of same shade are in clone relation with each other (e.g., e_{4a}, e_{5a}, e_{6a}, and e_{8a} are clones of each other, but e_{7a} is not in clone relation with the rest). Then, only S4 and S5 are in structural clone relation. By relaxing condition (a), we can include structures s6, s7 and s8 into a structural clone class together with s4 and s5. The above definition can benefit from more systematic treatment of clone similarity criteria, metrics and their threshold values.

5 Comments on Clone Similarity Criteria and Metrics

Clone detection has been an active area of research in software engineering for last 30 years. Still, we lack a standard definition for clones and well-defined similarity measures 5657. Software similarity is a multi-faceted phenomenon that escapes precise definition. The notion of similarity changes depending on the context: Whether or not we consider two code structures as similar depends on what we want to do with them.

We can characterize clones that are likely to meet our goals by metrics such as the minimum size of clones, the percentage of common code among clones 35, or the editing distance 46 among clones, measured in terms of editing operations required to convert one text fragment to another.

The definitions, types of clones detected by various techniques, and clone similarity metrics depend on the underlying program representations and algorithms used for clone detection. Code representations range from text 12, to token streams 735, to syntax trees 10, to program control/dependency graphs41 42, to neural networks 27. In all cases, multiple criteria and their respective metrics are used to characterize clones.

For structural clones, we can consider two structures characterized by similar, but not identical multigraphs as members of one structural clone class. Similarity measures have been proposed in research on semantic clones 42. Semantic similarity is used to analyse the matching between two domain concepts from differently conceptualized ontologies. Such similarity analysis is mostly done by employing statistical or probabilistic methods. Checking semantic relations between two ontologies involves comparison of graphs. Structural clones are also represented as graphs. Ontology graphs can be large, while clone graphs typically are small. Also, currently, all members of a structural clone class are represented by identical graphs. We expect that it may useful to relax this condition and to allow for variation in graph structure. Similarity measures developed in ontology research may become useful and we plan to explore this option.

6 Software Similarity Phenomenon

6.1 Reasons for Cloning

Similar design solutions are repeatedly applied to solve similar problems. These solutions are usually copied from the existing code, and modified to fit in the new context. Clones that result from poor design can be removed by refactoring the design. However, many clones cannot be removed from programs because they play some useful role 37, because of the limitations of a programming language 63339, risks involved in refactoring 18, or because of trade-offs involved in clone removal 53. Clones that stem from beneficial standardization of software solutions and from pattern-driven development on modern component platforms also belong to this category of clones. Still, even some of such "essential" clones can be tackled with macros 10 or using unconventional meta-level techniques such as Aspect-Oriented Programming 38 or ART 3345.

Similarities stem from different sources, depending on the nature of an application domain and design techniques used. Therefore, the form of clones, as well as reasons why they are there, vary across different program situations. Poor design and ad hoc maintenance are the two often-mentioned reasons for cloning. Such repetitions can often be avoided with good design or refactored.

Much cloning is found in system variants that originate from a common base of code during evolution. Often created by massive copying and modifying of program files, clones – small and large - are pervasive in such system variants. The Product Line approach 16 aims at transforming this primitive and ineffective form of reuse into systematic reuse that better exploits software similarities, and turns them into software productivity gains. Today's state-of-the-art in software reuse is based on adaptive reuse of architecture and code components shared by a family of similar systems.

While the knowledge of clones is usually evident at the time of their creation, we lack formal means to make the presence of clones visible in software, other than using external documentation or naming conventions. The knowledge of clones can be easily lost during subsequent software development and evolution.

Independently of the reasons why clones occur, it is useful to know about clones, their exact location and differences among their occurrences. In software maintenance, clones often signify inter-dependent program parts that must be consistently maintained. Knowing clones reduces the risk of update anomalies that occur due to errors in change propagation across remote and seemingly unrelated program parts. In maintenance of system families arising from evolution, the gains of clone knowledge is even bigger as changes must often be consistently propagated across many system variants.

Whether clones are good or bad it all depends on the motivation for cloning, the role clones play in a program, and the perspective from which we judge the impact of clones.

Despite the benefits of non-redundancy, at times, cloning is done in a good cause. Many clones stem from replicating code fragments (simple clones), or bigger program parts such as files or directories (structural clones). With *copy-paste-modify* practice, we can speed up development, achieving quick productivity gains. Developers also duplicate code to improve program performance or reliability. Such repetitions are intentional and should not be eliminated from a program even if a suitable refactoring could do the job. In maintenance of legacy software, changes involved in refactoring clones may create risks that are unacceptable for business reasons 18 – it is safer to maintain own piece of code rather than a generic solution shared with other developers who may also be changing the same functionality. Kapser and Godfrey 37 discuss a number of situations that justify cloning. Developers may choose to live with repetitions, as the lesser of the two evils, for variety of such reasons.

Modern component platforms (such as JEE™ or .NET™) encourage architecture-centric, pattern-driven design that naturally induces much redundancy to programs. Patterns lead to beneficial standardization of program solutions and are basic means to achieve reuse of common service components. Standardization of program solutions has many benefits, and creates an interesting case for our discussion of clones. At times IDEs support application of major patterns, or programmers use manual *copy-paste-modify* to apply yet other patterns. Representing patterns in generic form and enhancing their visibility can be beneficial, as it reveals a simpler view of a program. With generic pattern representation, we can provide better support for pattern instantiation and injection of pattern instances into a software system under construction. Generic design can help us avoid explosion of look-alike program structures, pattern instances. The knowledge of the location of pattern instances and the exact differences among them is helpful in understanding, maintenance and reuse.

Yet other repetitions occur because avoiding them with conventional approaches is either impossible or would require developers to compromise other important design goals. Kim estimates that 34% of clones cannot be refactored 40. This type of unavoidable repetitions is of our primary interest in this paper. We pay special attention to large-granularity program structures 45, signifying important design concepts, recurring many times in variant forms, whose noticing may bring significant engineering benefits.

Summarizing the above discussion, we classify clones into the following categories:

1. *Desirable.* Such clones are useful at runtime (e.g., for performance or reliability) and cannot be eliminated from programs. Intentional clones induced by an implementation technique (e.g., by J2EE or .NET architecture and patterns) also belong to this category.
2. *Avoidable.* These clones are caused by the programmer's carelessness. For example, similar code fragments introduced by poor design or ad hoc *copy-paste-modify* practice during maintenance often fall into this category.
3. *Problematic.* These are all the clones that are not *desirable* but are difficult to avoid using conventional design techniques, without compromising important design goals. As the name suggests, nothing definite can be said about problematic clones. They are relative to design techniques and design goals. Despite their enigmatic nature, we find the concept useful in discussing cloning problems. Most of the clones discussed in the Buffer library case study belong to this category.

Katsuro Inoue, one of the precursors of software clone research, is an author of CCFinder 35, a clone detection tool used by thousands of software companies in Japan and world-wide for software quality assessment 66. They consider the extent of cloning as one of the important indicators to estimate the expected maintenance cost in outsourcing software maintenance. The relation between cloning and the cost of changing programs can be explained as follows: Even if clones are created with good intentions, most of the clones increase the risk of update anomalies, and hinder program understanding during the maintenance in at least, two ways: (1) a programmer must maintain more code than he/she would have to maintain should the clones be removed, and (2) when one logical source of change affects many instances of a replicated program structure scattered throughout a program, to implement a change, a programmer must find and update all the instances of the replicated structure. The situation is further complicated if instances of an affected program structure must be changed in slightly different ways, depending on the context.

6.2 How Much Cloning?

Several studies suggest that as much as 20–50% of large software systems consist of cloned code 10235754. Other studies revealed lower, but still substantial rates of repetitions, 20%–30% 39, and indicated that 49%–64% of clones "were not easily refactorable due to programming language limitations". It is important to note that these other studies focused only on cloned code fragments, while our notion of similarity and studies cover large-granularity program structures, that may involve, for example, patterns of collaborating components recurring in variant forms.

In controlled lab studies and industrial projects, we typically observed 50%–90% rates of repetitions in newly developed, well-designed programs. Our studies covered a range of application domains (business systems, Web Portals, command and control, mobile device applications, class libraries), programming languages (Java, C++, C#, JSP, PHP) and platforms (J2EE, .NET, Unix, Windows). For example, the extent of similarities in Java Buffer library was 68% 3334, in parts of STL (C++) - over 50% 8, in Web Portal (J2EE) – 68% 67, and in certain .NET Web Portal modules – up to 90% 51. A survey of 17 Web Applications revealed 17–60% of code contained in clones 54. We measured the percentage of redundancies by comparing the subject program against a non-redundant representation for the subject program built with ART outlined later in this paper. Not always does size reduction lead to program simplification. However, we focused only on repetitions that created reuse opportunities, induced extra conceptual complexity into a program, and/or were counter-productive for maintenance.

7 How Do We Find Clones?

Many different clone detection algorithms have been proposed in the literature. For illustration, we briefly outline clone detection in Clone Miner 567. Clone Miner tokenizes source programs and uses suffix array to detect simple clones [7]. Then, Clone Miner detects the recurring configurations of simple clones using frequent item set mining [22] and clustering.

7.1 Simple Clone Detection

Clone Miner uses a token-based simple clone detector as the front-end. The front-end first tokenizes the input source code into a token string. Then, a suffix array based string matching algorithm is used to directly detect the simple clone classes, without computing them through the clone pairs. Clone Miner can tokenize source code written in Java, C++, Perl, and VB.net. The tokenizer also performs some simple parsing to detect method and function boundaries.

The usual output from simple clone detectors is in the form of clone pairs. Groups of similar code fragments form so-called *simple clone classes* (SCC). We construct SCCs by grouping clone pairs in such a way that every member in a group is a clone of every other member.

7.2 Detecting Structural Clones

Recurring configurations of simple clones often form interesting higher-level similarities. The structural clones embody this observation. Structural clone detection is also based on that observation. Clone Minder focuses on structural clones formed by configurations of simple clones located in the same physical source code unit such as a file or directory.

Therefore, Clone Miner finds simple clones first, and then uses data mining techniques to find candidates for structural clones [5]. Clone Miner finds location-based structural clones, such as shown in Fig. 3(a). Such structural clones are recursively defined as patterns of lower-level clones residing in the same location (such as file or directory). Clone Miner uses abstraction mechanism to detect higher-level structural clones based on the presence of similar patterns of lower-level clones. SCCs are the lowest-level clones that we consider in this analysis.

Higher level structural clones are detected in a similar way, with highly similar containers found from the previous lower level clone analysis, being treated as cloned abstract entities. Using the above mechanism, we detect structural clones composed of patterns of simple clones, method clones, file clones or directory clones.

The process of detecting structural clones consists of iteratively performing the following four steps:

Step 1: Find recurring patterns of lower-level clones in different containers.
Step 2: Calculate coverage of the containers by these patterns.
Step 3: Cluster similar containers based on the patterns with significant coverage.
Step 4: Abstract similar containers into clones for the next higher level of similarity analysis.

Using the above mechanism, we detect structural clones composed of patterns of simple clones, method clones, file clones or directory clones. Method clones, file clones and directory clones are clones formed by the abstraction mechanism.

7.3 Detecting Collaborative Clones

Other interesting types of structural clones include clones formed by recurring patterns of collaborating components (e.g., classes) such as illustrated in Fig. 2. We use the following technique to detect collaborative structural clones based on analysis of program execution traces: To detect such clones, we must know components' runtime relationships (e.g., method/function invocation). This can be inferred runtime relationships based on analysis of program execution traces. We insert monitors to a program to record any message passing among objects: A::f() -> B::g(), meaning that object of class A sends a message g() to an object of class B while executing method f(). Aspect-Oriented Programming (AOP) can be used to generate traces of message passing among objects such as A::f() - > B::g(), meaning that object of class A sends a message g() to an object of class B while executing method f(). AOP allows us to produce such traces without invading the subject code with instrumentation. Once we generate sufficient traces for the system, we remove duplicate traces. After that, we transform traces by substituting method names with their method clone class ID's, as generated by Clone Miner. The idea is that if two traces A()-> B() and A'()-> B'() exists, where A() and A'() belong to the same method clone class and B() and B'() belong to the same clone class, then both traces should have identical representation. Traces containing methods that are not part of any method clone class are also removed, as there can be no clone of such traces. After that, all transformed traces are concatenated together, with unique sentinels preserving the trace boundaries. Now the problem of finding cloned traces is again reduced to the problem of detecting matching substrings and we apply the same suffix array based algorithm that is used to detect simple clones from the token representation of source code.

Program trace analysis has been applied in testing (to check test coverage), in improving software reliability (to find anomalies during program execution, possibly leading to errors), and in program specification mining.

Clone detection is usually performed on source files. Evolving system versions are often stored in version control tool repositories. Much interesting research is being done on mining useful information from software repositories 49. It is interesting to explore the possibility of clone detection from systems stored in software repositories, without recreating source files. One would expect that the knowledge of deltas among file versions can provide extra inputs for clone detection.

8 Post-detection Clone Analysis

Clone detectors produce much information, including many false positives, or clones just not relevant to a program analysis task at hand. Therefore, the output of clone detectors must be further analysed to filter out false-positives and select useful clones for a given task (e.g., maintenance or re-engineering) 3335. This post-detection analysis must involve a human expert. Visualization, abstraction and filtering techniques are used for that purpose (see recently published comprehensive survey of clone visualization techniques 30).

Application domain knowledge is essential to filter clones that represent meaningful domain/design concepts from less useful ones. Visualization, abstraction and filtering methods assist human experts in design recovery for software reuse and re-engineering.

The number of potentially useful abstract views of cloning situation is endless. A query system of clone analysis tool 65 can allow the user to define a wide range of views in a descriptive way. Then, a tool displays views in a graphical form. The key concept here is a data model for cloning information produced by a clone detection tool. Based on that model, we can create relational database schema and store cloning information in a database. Standard SQL can be used to define abstract views to be produced from clone detector output. Abstract views (that is, query results) are then displayed in graphical form by public domain graph generation tools.

9 Generic Design

Generic design aims at achieving non-redundancy by unifying differences among similar program structures. The importance of generic design in managing software complexity have been recognized for long. Macros were one of the early attempts to parameterize programs and make them more generic. Goguen 28 popularized ideas of parameterized programming. Among programming language constructs, type parameterization (called generics in Ada, Eiffel, Java and C#, and templates in C++), higher-order functions, iterators, and inheritance can help avoid repetitions in certain situations 26. Design techniques such as design patterns 25, table-driven design, and information hiding are supportive to building generic programs.

There are three engineering benefits of generic design (and three reasons to avoid unnecessary repetitions): Firstly, genericity is an important theme of software reuse where the goal is to recognize similarities to avoid repetitions across projects, processes and products. Indeed, many repetitions merely indicate unexploited reuse opportunities. Secondly, repetitions hinder program understanding. Repeated similar program structures cause update anomalies complicating maintenance. Thirdly, by revealing design-level similarities, we reduce the number of distinct conceptual elements a programmer must deal with. Not only do we reduce an overall software complexity, but also enhance conceptual integrity of a program which Brooks calls "the most important consideration in system design" 14. Common sense suggests that developers should be able to express their design and code without unwanted repetitions, whenever they wish to do so.

10 Redundancies in STL

STL 50 is a classical and powerful example of what skilful generic design can do for complexity reduction. STL implements commonly used algorithms, such as sort or search, for a variety of container data structures. Without generic containers and algorithms, the STL's size and complexity would be enormous, hindering its evolution. Such simple-minded solution would unwisely ignore similarity among containers, and among algorithms applied to different containers, which offers endless reuse opportunities. Generic design with templates and iterators helped STL designers to avoid these complications, without compromising efficiency.

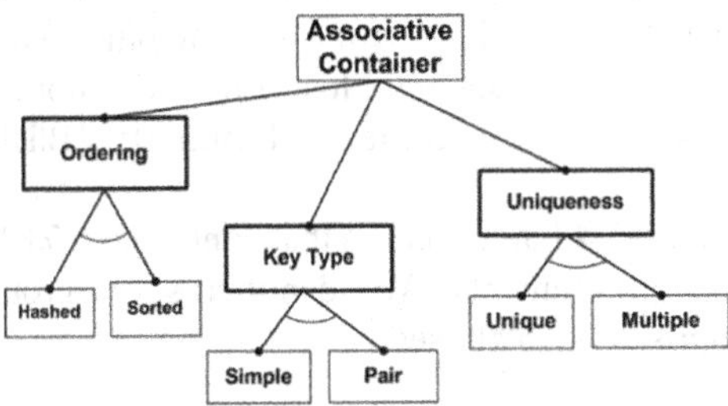

Fig. 6. Associative container features (STL) 31.

In STL, generic solutions are mainly facilitated by templates and iterators. We analysed associative containers - variable-sized containers that support efficient retrieval of elements based on keys. Figure 6 shows variant features of associative containers. There are eight STL templates, one for each of the eight legal combinations of features.

We focused on the STL regions that showed high cloning rates. We ran a clone detector to identify these regions. We found that container classes displayed a remarkable amount of similarity and code repetition. Four 'sorted' associative containers and four 'hashed' associative containers could be unified into two generic ART containers, achieving 57% reduction in the related code. Stack and queue classes contained 37% of cloned code. Algorithms set union, set intersection, set difference, and set symmetric difference (along with their overloaded versions) formed a clone class with eight instances. On overall, non-redundant representation of these parts of STL in ART contained 48% of code found in the original STL 8.

There were many non-type-parametric differences among associative container templates. For example, certain otherwise similar methods, differed in operators or algorithmic details. While it is possible to treat many types of non-parametric differences using sophisticated forms of C++ template meta-programming, often the resulting code becomes "cluttered and messy" 19. We did not spot such solutions in STL, and believe their practical value needs to be further investigated.

The reader may find full details of the STL case study in 8.

11 Redundancies in the Java Buffer Library

A buffer contains data in a linear sequence for reading and writing. Buffer classes differ in features such as a memory scheme: Heap or Direct; element type: byte, char, int, double, float, long, or short; access mode: writable or read-only; byte ordering: S – non-native or U – native; B – BigEndian or L – LittleEndian.

Each legal combination of features yields a unique buffer class, with much similarity among classes. As we combine features, buffer classes grow in number, as observed in 9. Some of the buffer classes are shown in Fig. 7. A class name, such as DirectIntBufferRS, reflects combination of features implemented into a given class. Class names are derived from a template: [MS][T]Buffer[AM][BO], where MS – memory scheme: Heap or Direct; T – type: int, short, float, long double, char, or byte; AM – access mode: W – writable (default) or R - read-only; BO – byte ordering: S – non-native or U – native; B – BigEndian or L – LittleEndian. All the classes whose names do not include 'R', by default are 'W' – writable. VB – View Buffer is yet another feature that allows us to interpret byte buffer as Char, Int, Double, Float, Long, or Short. Combining VB with other features, yields 24 classes ByteBufferAs[T]Buffer[R][B|L]. The last parameter [B|L] means "B or L".

The experiment covered 74 buffer classes that contained 6,719 LOC (physical lines of code, without blanks or comments). We identified seven groups of similar classes where each group comprised 7–13 classes:

1. [**T**]Buffer: 7 classes at Level 1 that differ in buffer element type, **T**: int, short, float, long double, char, or byte
2. Heap[**T**]Buffer: 7 classes at Level 2, that differ in buffer element type, **T**
3. Heap[**T**]BufferR: 7 read-only classes at Level 3
4. Direct[**T**]Buffer[**S**|**U**]: 13 classes at Level 2 for combinations of buffer element type, **T,** with byte orderings**:** **S** – non-native or **U** – native byte ordering (notice that byte ordering is not relevant to buffer element type 'byte')

5. Direct[**T**]BufferR[**S**|**U**]: 13 read-only classes at Level 3 for combinations of parameters **T**, **S** and **U**, as above
6. ByteBufferAs[**T**]Buffer[**B**|**L**]: 12 classes at Level 2 for combinations of buffer element type, **T,** with byte orderings: **B** – Big_Endian or **L** – Little_Endian
7. ByteBufferAs[**T**]BufferR[**B**|**L**]: 12 read-only classes at Level 3 for combinations of parameters **T**, **B** and **L**, as above.

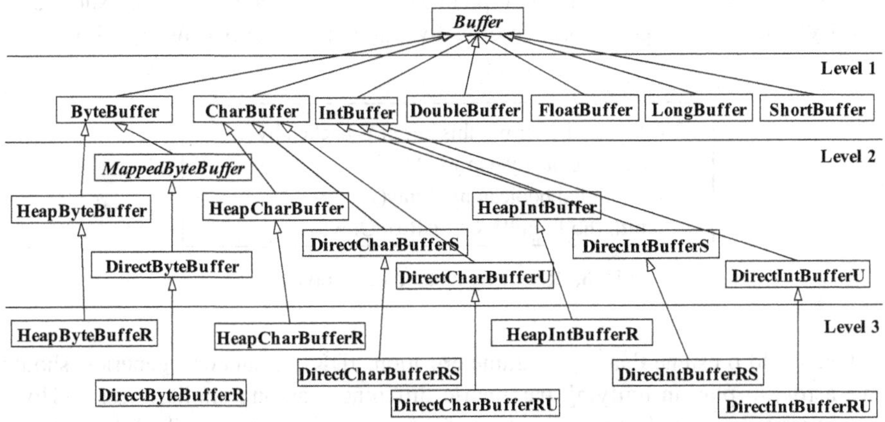

Fig. 7. A fragment of the buffer library 31.

Classes in each of the above seven groups differed in details of method signatures, data types, keywords, operators, and editing changes. We paid attention only to similarities whose noticing could simplify class understanding and help in maintenance. Some of the classes had extra methods and/or attributes as compared to other classes in the same group. Many similar classes or methods occurred due to the inability to unify small variations in otherwise the same classes or methods. Generics could unify 15 among 74 classes under study, reducing the code size by 27%. The solution with generics was subject to certain restrictions that we discussed in 34.

So why did Buffer library designers chose to keep redundancies?

Any solutions to unifying similarities must be considered in the context of other design goals developers must meet. Usability, conceptual clarity and good performance are important design goals for the Buffer library. To simplify the use of the Buffer library, the designers decided to reveal to programmers only the top eight classes (Fig. 7). For conceptual clarity, designers of the Buffer library decided not to multiply classes beyond what was absolutely needed. We see almost one-to-one mapping between legal feature combinations and buffer classes.

In many situations, designers could introduce a new abstract class or a suitable design pattern to avoid repetitions. However, such a solution would compromise the above design goals, and therefore was not implemented. Many similar classes or methods were replicated because of that many similarities in buffer classes sparked from feature combinations. As buffer features (such as element type, memory scheme, etc.)

could not be implemented independently of each other in separate implementation units (e.g., class methods), code fragments related to specific features appeared with many variants in different classes, depending on the context. Whenever such code could not be parameterized to unify the variant forms, and placed in some upper-level class for reuse via inheritance, similar code structures spread through classes.

Method **hasArray()** shown in Fig. 8 illustrates a simple yet interesting case. This method is repeated in each of the seven classes at Level 1. Although method **hasArray()** recurs in all seven classes, it cannot be implemented in the parent class Buffer, as variable **hb** must be declared with a different type in each of the seven classes. For example, in class **ByteBuffer** the type of variable **hb** is **byte** and in class **IntBuffer**, it is **int**.

```
/* Tells whether or not this buffer is backed by
    an accessible byte array. */
    public final boolean hasArray() {
    return (hb != null) && !isReadOnly; }
```

Fig. 8. Recurring method hasArray()31.

One could presume that type parameterization, JDK 1.5 supports generics, should have a role to play in unifying parametric differences among similar classes. However, generics have not been applied to unify similarity patterns described in our study. Groups of classes that differ only in data type are obvious candidates for generics. There are three such groups comprising 21 classes, namely **[T]Buffer**, **Heap[T]Buffer** and **Heap[T]BufferR**. In each of these groups, classes corresponding to Byte and Char types differ in non-type parameters and are not generics-friendly. This leaves us with 15 generics-friendly classes whose unification with three generics eliminates 27% of code. There is, however, one problem with this solution. In Java, generic types cannot be primitive types such as int or char. This is a serious limitation, as one has to create corresponding wrapper classes just for the purpose of parameterization. Wrapper classes introduce extra complexity and hamper performance. Application of generics to 15 buffer classes is subject to this limitation.

```
/*Creates a new byte buffer containing a shared
  subsequence of this buffer's content. */
public ByteBuffer slice() {
      int pos = this.position();
      int lim = this.limit();
      assert (pos <= lim);
      int rem = (pos <= lim ? lim - pos : 0);
      int off = (pos << 0);
      return new DirectByteBuffer(this, -1, 0, rem,
```

Fig. 9. Method slice() 31.

Repetitions often arise due to the inability to specify small variations in otherwise identical code fragments. Many similar classes and methods differ in parameters representing constants, keywords or algorithmic elements rather than data types. This happens when the impact of various features affects the same class or method. For example, method **slice()** (Fig. 9) recurs 13 times in all the Direct[T]Buffer[S|U] classes with small changes highlighted in bold in. Generics are not meant to unify this kind of differences in classes.

In summary, generics are rather limited in unifying similarity patterns that we find in practical situations, e.g., such as we observed in the Buffer library. It is interesting to note that repetitions occur across classes at the same level of inheritance hierarchy, as well as in classes at different levels of inheritance hierarchy. Programming languages do not have a proper mechanism to handle such variations at an adequate (that is a sufficiently small) granularity level. Therefore, the impact of a small variation on a program may not be proportional to the size of the variation.

Developers of the Buffer library used macros, scripts and makefiles in order to exploit similarities and write/maintain buffer classes with less effort (these macros and scripts can be found in the Community Source Release for the Buffer library). While the reasons why Sun developers escaped to non-OO solution and the solution itself are not explained or documented, its existence hints at difficulties to treat similarity patterns with conventional OO techniques, given the overall design goals the Buffer library had to meet.

12 Towards Non-redundancy

While practitioners are aware of much repetitions in software, they also know how difficult it is to avoid them. Problems with implementing effective reuse strategies 21 evidence these difficulties, as well.

Structural clones can be represented with generic, adaptable meta-components for meta-level reuse with ART 2 (in earlier papers ART was called XVCL, XML-based Variant Configuration Language). Applications of ART demonstrated that such unconventional reuse considerably raises reuse levels (and productivity levels) beyond what is possible with conventional architecture-centric and component-based approaches to Product Line reuse. Work on meta-level with ART inspired research into structural clones.

It is not clear if and how we could implement buffer classes without redundancies in any of the conventional programming languages. A possible solution calls for flexible parametrization unconstrained by the rules of a programming language. It is as if our need to express program behaviour was in conflict with our need to achieve non-redundancy. To resolve this conflict, generative approaches propose to think about programs at two levels: a meta-level that provides a platform for program construction, and a level of actual program that is compiled and executed. Program generation technologies offer solutions for specific application domains, with abstract notations to specify required program behaviour (a meta-level), and a generator that encodes the semantics of a given application domain, and generates a program ready for execution. Quite often much redundancy can be avoided in abstract program specifications. We comment further on generation approaches in the following section, and here we outline a general-purpose solution to non-redundancy, based on flexible parameterization at the meta-level, and code manipulation in pre-processing fashion. We explain the solution in a way that ART (Adaptive Reuse Technology) implements these concepts.

On the left-hand-side of Fig. 10, we see a non-redundant meta-level representation of buffer classes. Boxes are ART templates that represent building blocks for Buffer classes. As such, they contain relevant Java code instrumented (parameterized) with ART commands. The purpose of parameterization is to enable reuse of ART templates in multiple contexts of the situations when a given functionality is need for building buffer classes. ART Processor interprets ART commands embedded in templates and generates buffer classes on the right-hand-side of Fig. 10.

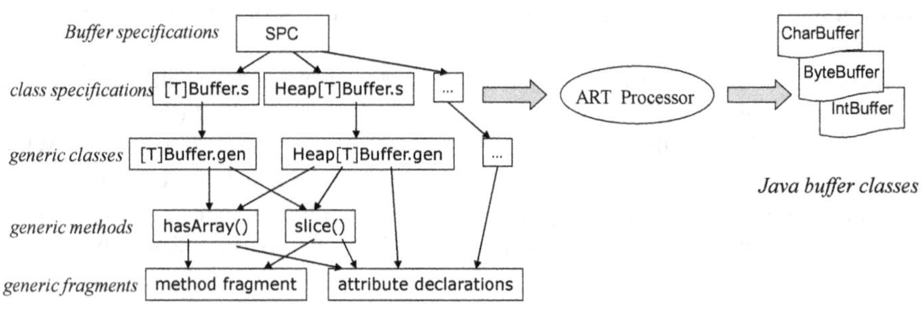

Fig. 10. Non-redundant representation of buffer classes in ART/Java 31.

An arrow between two templates: X → Y is read as "X adapts Y", meaning that X controls adaptation of Y. We have seven *generic class* templates, one for each of the seven groups of similar classes described in Sect. 11 (we show only two of them in Fig. 10). Each class template defines common part of classes in the respective group. The essence of a generic component (generic class, in our case) is that it can be adapted to produce its instances (specific classes in a group, in our case). Smaller granularity generic building blocks for classes are defined below, namely class methods and fragments of method implementation or attribute declaration sections. Therefore, lower-level templates are composed, after possible adaptations, to construct required instances of higher-level generic components. At the top, we have specification elements – they tell the ART Processor how to generate specific buffer classes, from templates. Top-most SPC, sets up global parameters and exercises the overall control over the generation process.

ART Processor interprets the template framework starting from the SPC, traverses templates below, adapting visited templates and emitting buffer class code. By varying specifications, we can instantiate the same template framework in different ways, deriving different, but similar, program components from it.

We now explain the parameterization and adaptation mechanism, which is the "heart and soul" of how ART achieves goals of non-redundancy:

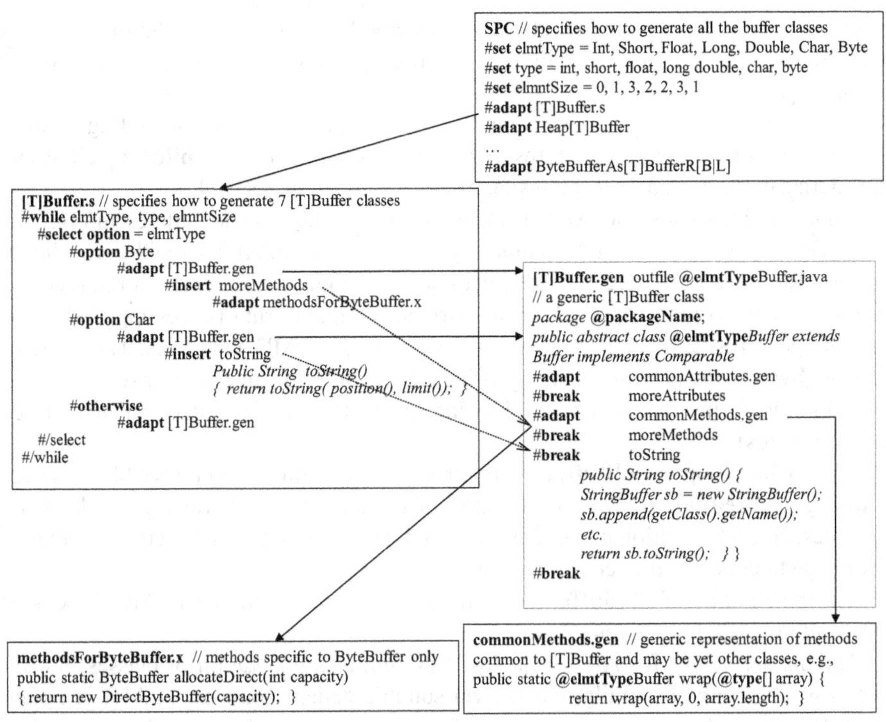

Fig. 11. Non-redundant representation for seven [T]Buffer classes in Java/ART (partial) 31.

ART variables and expressions provide a basic parameterization mechanism to make templates generic. **#set** command assigns a value to a variable. Typically, names of program elements manipulated by ART, such as components, source files, classes, methods, data types, operators or algorithmic fragments, are represented by ART expressions. Such expressions are then instantiated by the ART Processor, according to the context. For example, names and other parameters of the seven similar classes **[T]Buffer** are represented by ART expressions in the a template **[T]Buffer.gen**.

ART variables have global scope, so that they can coordinate chains of all the customizations related to the same source of variation or change that spans across multiple templates. During processing of templates, values of variables propagate from an template where the value of a variable is set, down to the lower-level templates. While each template usually sets default values for its variables, values assigned to variables in higher-level templates take precedence over the locally assigned default values. Thanks to this overriding rule, templates become generic and adaptable, with potential for reuse in many contexts.

Other ART commands that help us design generic and adaptable templates include **#select**, #insert into **#break** and **#while**. We use **#select** command to direct processing into one of the many pre-defined branches (called options), based on the value of a variable. With **#insert** command, we can modify templates at designated **#break** points in arbitrary ways. ART expressions, **#select** and **#insert** into **#brea**k are analogous to

AOP's mechanism for weaving advices at specified join points 38. The difference is that ART allows us to modify templates in arbitrary ways, at any explicitly designated variation points.

#while command iterates over template(s), with each iteration generating similar, but also different, program structures. A **#select** command in the **#while** loop allows us to generate classes in each of the seven groups discussed in Sect. 11.

Figure 11 illustrates how ART mechanisms realize the scheme outlined in Fig. 10.

ART template names, ART commands and references to ART variables are shown in bold. References to ART variables parameterize code. For example, a reference to variable **@elmtType** is replaced by the variable's value during processing. Values of variables set in **SPC** reach all their references in adapted ART templates. The value of variable **byteOrder** is set to an empty string, "S" or "U", in a respective **#set** command placed in one of the ART templates that **#adapt**'s ART template **slice.gen** (not shown in our pictures).

The **#while** loop in **[T]Buffer.s** is controlled by two multi-value variables, namely **elmtType** and **elmtSize**. The i'th iteration of the loop uses i'th value of each of the variables. In each iteration of the loop, the **#select** command uses the current value of **elmtType** to choose a proper #option for processing.

Attribute *outfile* of **[T]Buffer.gen** defines the name of a file where ART Processor will emit the code for a given class.

Having set values for ART variables, **SPC** initiates generation of classes in each of the seven groups of similar classes via suitable #adapt commands. ART template **[T]Buffer.gen** defines common elements found in all seven classes in the group. Five of those classes, namely **DoubleBuffer**, **IntBuffer, FloatBuffer**, **IntBuffer,** and **LongBuffer** differ only in type parameters (as in the sample method **wrap()** shown in ART template **commonMethods.gen**). These differences are unified by ART variables, and no further customizations are required to generate these five classes from ART template **[T]Buffer.gen**. These five classes are catered for in **#otherwise** clause under **#select**. However, classes **ByteBuffer** and **CharBuffer** have some extra methods and/or attribute declarations. In addition, method **toString()** has different implementation in **CharBuffer** than in the remaining six classes. Customizations specific to classes **ByteBuffer** and **CharBuffer** are listed in the **#adapt** commands, under **#option** s **Byte** and **Char**, respectively.

We refer the reader to 3334 to further the details of this study.

A shorter program without redundancies does not automatically mean that such a program is easier to understand and maintain than a longer program with redundant code. For example, compressed code is short but impossible to read and understand. To further support claims of easier maintainability of the ART solution, we extended the Buffer library with a new type of buffer element – Complex. Then, we compared the effort involved in changing each of the two solutions, Java classes and Java/ART representation. Many classes must be implemented to address the Complex element type, but in this experiment we concentrated only on three of them, namely **ComplexBuffer**, **HeapComplexBuffer** and **HeapComplexBufferR**. In Java, class **ComplexBuffer** could be implemented based on the class **IntBuffer**, with 25 modifications that could be automated by an editing tool, and 17 modifications that had to be done manually. On the other hand, in the ART representation, all the changes had to be done manually, but only 5 modifications were required. To implement class **HeapComplexBuffer**, we needed 21 "automatic" and 10 manual modifications in Java, versus 3 manual modifications in ART. To implement class **HeapComplexBufferR**, we needed 16 "automatic" and 5 manual modifications in Java, versus 5 manual modifications in ART.

13 Clones in Web Portals

13.1 ASP.NET Portal

In the ASP Web Portal (WP) Product Line project, our industry partner ST Electronics Pte. Ltd.,

Singapore, applied state-of-the-art conventional methods to maximize reusability of a Team Collaboration Portal (TCP). Still, a number of problem areas were observed that could be improved by applying ART to reduce redundancies. The benefits of ASP/ART TCP were the following:

- Short time (less than 2 weeks) and small effort (2 persons) to transform the ASP TCP into the first version of a mixed-strategy ASP/ART Product Line architecture.
- High productivity in building new portals from the ASP/ART solution. Based on the ASP/ART solution, ST Electronics could build new portal modules by writing as little as 10% of unique custom code, while the rest of code could be reused. This code reduction translated into an estimated eight-fold reduction of effort required to build new portals.
- Significant reduction of maintenance effort when enhancing individual portals. The overall managed code lines for nine portals under the ASP/ART were 22% less than the original single portal.
- Wide range of portals differing in a large number of inter-dependent features supported by the ASP/ART solution.
- The reader may find full details of this project in 51.

13.2 JEE Portal

In the follow up project, we evaluated J2EE™ as a platform for Product Line development. Unlike ASP, J2EE supports inheritance, generics and other OO features via Java.

Component platforms such as J2EE or .NET encourage organizing software around standard architectures. Patterns help programmers solve routine tasks in pre-defined ways in conformance to architectures. Application of patterns further standardizes software at macro and micro levels. Not surprisingly, we find much similarity in software developed in that way. Such uniformity of software structure is beneficial, as similar problems are always solved in a similar way across a system. It also facilitates easy reuse of common services/components provided by a platform. However, not always are pattern instances clearly visible in code. Pattern-driven development could be even more beneficial if we knew the exact location of pattern instances and how instances are similar and different one from each other. This would help in the future maintenance: When the pattern-related code is to be changed, it would be clear which of the pattern's instances should be changed and how. Currently, pattern-driven development is mainly limited to the middleware areas such as database communication, coordination between requests, application model and views (e.g., implied by the MVC organization) or reuse of common services. In application domain-specific areas, the benefits of patterns are less. At times, application of patterns may even scatter domain-specific functionality across many components (or classes), which complicates reuse of domain-specific code, and magnifies problems of tracing requirements to code.

In J2EE project, we applied ART to enhance the visibility of patterns and to achieve reuse in application domain-specific areas. We worked with a portal developed by ST Electronics, a variant of TCP. The portal supported collaborative work and included 14 modules such as Staff, Project and Task. We studied similarity patterns in presentation and business logic layers.

Within modules, we found 75% of code contained in exact clones, and 20% of code contained in similar clones (leaving only 5% of code unique). Analysis across modules, revealed design-level similarities, with 40% of code contained in structural clones. Both intra- and inter-module similarities were important for clarity of the design, however they could not be unified with generic design solutions expressed by J2EE mechanisms.

In the second part of the experiment, we applied ART to unify similarity patterns. Unification reduced the solution size by 61%, and enhanced the clarity of portal's conceptual structure as perceived by developers. In a controlled experiment, we found that to implement the same enhancement, J2EE/ART portal representation required 64% less modifications that the original J2EE portal.

The reader may find full details of this project in 67.

14 Clone Detection Versus Plagiarism Detection

Plagiarism detection 11117524858 is closely related to clone detection. In cloning, code is copied and then modified to fit the new context. In plagiarism, copied code is changed to hide similarity. Therefore, plagiarism detection is adversarial in nature, as opposed to clone detection which is not. The plagiarized code must behave exactly as the original one. In clone detection, on the other hand, we are primarily interested in structural similarities of semantically diverse program parts. Plagiarism detection can determine code similarity among program files, but does not tell us about the internal structure of clones inside files, which is the objective of clone detection. A comparative study 12 involving three clone detection tools and two plagiarism detection tools, *JPlag* 52 and Moss 1 revealed that involved tools are equally effective in finding cloned files.

Despite the differences between goals of cloning and plagiarism, some clone detection techniques, e.g., token-based, have been also used for plagiarism detection 5852, after applying extensive normalization to the source code. Applying plagiarism detection technique at the level of code fragments (simple clones) is possible in theory, but impractical for performance reasons. Clone detection is also related to the problems of genetic sequencing, spam and malware detection, and data compression.

15 Generators

Powerful domain-specific solutions can be built by formalizing the domain knowledge, and using generation techniques to produce custom programs in a domain. Advancements in modelling and generation techniques led to Model-Driven Engineering (MDE) 59, where multiple, inter-related models are used to express domain-specific abstractions. Models are used for analysis, validation (via model checking), and code generation. Platforms such as Microsoft Visual Studio™ and Eclipse™ support generation of source code using domain-specific diagrammatic notations.

By constraining ourselves to a specific application domain, we can make assumptions about its semantics. A domain engineer encodes domain-specific knowledge into a generic, parameterized program solution. A developer, rather than working out all the details of the program solution, writes a concise, declarative problem description in a Domain-Specific Language (DSL). A generator uses DSL specifications to instantiate parameters of a generic solution to produce a custom program. Problem specifications in DSL are much smaller and simpler than the instantiated, complete and executable program solution. While we do not reduce the overall program complexity, generation-based solutions shield a programmer from complexities of the domain-specific code that is now manipulated by a generator. DSL may take many different forms, depending on a domain, from a formal text (e.g., BNF for parser generator), to visual interface (e.g., GUI) and to models (in Model-Driven Engineering approaches).

This is in contrast with ART which is an application domain- and programming language-independent technique. There is no concept of DSL in ART. Generators can be built in well-understood and fairly stable application domains. On the other hand ART, performs best in domains where frequent changes occur at both large and small granularity levels.

Generators must overcome a number of challenges to have a greater impact on practice. A common pitfall of generators is that abstract program specifications in DSL can get easily disconnected from the generated code. This happens when the generated code is modified by hand to accommodate changes not catered for by the DSL. As any re-generation of code would override such modifications, future maintenance must be done by hand and developers can't benefit from the generator anymore. Round-trip engineering could overcome this problem, but is difficult to achieve. This problem is particularly acute in the situation when we need to evolve multiple generated programs differing in certain features, as it is often the case of a Product Line. Implementing variant features in the generator will propagate all the variant features to all the programs, which may not be desirable. On the other hand, implementing variant features directly into generated programs that need them, automatically disconnects those programs from the generator.

Another problem faced by generators is that a problem domain served by a generator is often only a part of an overall programming problem developers need to solve. Strategies for integrating multiple domain-specific generators and embedding them into systems implemented using yet other techniques have yet to be developed. One of the reason for success of compiler generators is that compilation on its own is a self-contained domain.

Rich abstractions lead to powerful generators. Without sufficient abstractions, there is not much we can automate. We believe not enough of general-purpose abstractions is the main reason why, despite much research, we have not achieved success in domain-independent, generation-based automatic programming. This also reminds us Brooks' doubts about reducing essential program complexity by means of abstraction 14.

16 Non-redundancy and Software Reuse

Software reuse and maintenance are two important themes in software development. Avoiding repetitive work - a basic goal of reuse - is a common sense, appealing way to improve software productivity. That's why reuse ideas have attracted much research. Modern component-based and architecture-centric technologies aim at reuse. Still, reuse of software is hard. Despite much research in academia and industry (e.g., Product Line initiative at Software Engineering Institute, Carnegie-Mellon), so far, reuse has not penetrated the industrial software development deep enough to become a standard practice. Many problems with realizing reuse strategies have been reported 19. On the other hand, software maintenance (or: evolution) has become a single most expensive software process, averaging at 50–60%, and in some companies mounting to 80% of total software budgets. We can't avoid maintenance, and as long as maintenance is done as it is today - mostly manually, at the level of code, with little support from tools – its cost is likely to further grow in the future.

Reuse is based on the premise that much similarity exists in software systems. This is particularly true for systems in a given application domain. The Product Line approach 16 is based on adaptive reuse of architecture and code components shared by a family of similar systems. Similarity analysis of legacy systems often is a prelude to re-engineering them into a Product Line that makes their components easier to reuse. Similarity analysis also helps maintain a single system, as many similar parts are inter-dependent, and must remain consistent after changes.

Finding structural clones is also the first necessary step in re-engineering legacy software into reusable ART software representations. Structural clones offer new, attractive options for reuse reaching beyond software components. Unconventional forms of reuse facilitated by ART can bring substantial productivity gains 5164. Suppose we have a family of similar software systems that are maintained as separate products, ignoring similarities. This often happens in software systems arising from evolution, whereby different variants of the same system are released to various customers overtime. Structural clones found across a family or similar systems hint at reuse opportunities. Generic ART representations unifying such structural clones can be reused across similar systems. Gradually, we start managing a software system family from generic ART representations rather than in an ad hoc way. Such process is the essence of re-engineering for reuse with ART (that is, migrating existing software system from ad hoc to reuse-based development/maintenance). It leads to a software Product Line 16 realized with ART.

A family of similar systems is often a starting point for re-engineering into a software Product Line for reuse. Similarity analysis of such systems is then a prelude to re-engineering. While it is impossible to fully automate this process, we can try to automate construction of first-cut meta-component design of a generic representation, with identified differences among its instances. Fine-tuning would then be done by a human expert.

17 Conclusions

I discussed software similarity phenomenon and its manifestation in programs as software clones. Much clone research has focused on similar code fragments, called simple clones. In many situations, programmers are interested in a bigger picture of software similarity that simple clones alone do not reveal. This is especially true when we modernize old programs, or re-engineer families of similar program variants for reuse, into a Software Product Line, whereby similar systems are built and maintained from a common base of highly parametrized, reusable software artefacts. To this end, I defined and discussed structural clones representing higher level similar program structures. In the paper, I also discussed common reasons for cloning, and techniques to eliminate redundancies in programs.

Acknowledgements. Author thanks PhD students and research assistants at the National University of Singapore who developed clone detection tools, implemented ART Processor, and participated in studies on software redundancies.

References

1. Aiken, A.: A system for detecting software plagiarism. Moss homepage: https://theory.stanford.edu/~aiken/moss/
2. ART Adaptive Reuse Technology. https://art-processor.org
3. Baker, B.S.: On finding duplication and near-duplication in large software systems. In: Proceedings of 2nd Working Conference on Reverse Engineering, pp. 86–95 (1995)
4. Bassett, P.: Framing Software Reuse – Lessons from Real World. Yourdon Press, Prentice Hall (1997)
5. Basit, H.A., Jarzabek, S.: Detecting higher-level similarity patterns in programs. In: Proceedings of European Software Engineering Conference and ACM SIGSOFT Symposium on the Foundations of Software Engineering, pp. 156–165 (2005)
6. Basit, H.A., Jarzabek, S.: Data mining approach for detecting higher-level clones in software. IEEE Trans. on Soft. Eng. **35**(4), 497–514 (2009)
7. Basit, H., Puglisi, S., Smyth, W., Turpin, A., Jarzabek, S.: Efficient token based clone detection with flexible tokenization. In: ESEC-FSE'07, European Software Engineering Conference and ACM SIGSOFT Symposium on the Foundations of Software Engineering, Dubrovnik, pp. 513–516, September 2007
8. Basit, H.A., Rajapakse, D.C., Jarzabek, S.: Beyond templates: a study of clones in the STL and some general implications. In: Proceedings of the 28th International Conference on Software Engineering (ICSE 2005), pp. 451–459 (2005)
9. Batory, D., Singhai, V., Sirkin, M., Thomas, J.: Scalable software libraries. In ACM SIGSOFT'93: Symposium on the Foundations of Software Engineering, Los Angeles, California, pp. 191–199 (1993)
10. Baxter, I., Yahin, A., Moura, L., Anna, M.S.: Clone detection using abstract syntax trees. In: Proceedings of International Conference on Software Maintenance (ICSM 1998), pp. 368–377 (1998)

11. Belkhouche, B., Nix, A., Hassell, J.: Plagiarism detection in software designs. In: ACM Southeast 42nd Regional Conference, Huntsville, Alabama, USA, pp. 207–211, April 2004
12. Baker, B.: A program for identifying duplicated code. In: Proceedings of Computing Science and Statistics: 24th Symposium on the Interface, vol. 24, p. 4957, March 1992
13. Brooks, F.: The Mythical Man-Month. Addison Wesley, Boston (1995)
14. Brooks, F.P.: No silver bullet 1986. In: Kugler, H.K. (ed.) Proceedings of IFIP 10th World Computing Conference. Elsevier Science, pp. 1069–1076 (1986)
15. Burd, E., Bailey, J.: Evaluating clone detection tools for use during preventative maintenance. In: Proceedings of the 2nd IEEE International Workshop on Source Code Analysis and Manipulation (SCAM 2002), Montreal, Canada, pp. 36–43, October 2002
16. Clements, P., Northrop, L.: Software Product Lines: Practices and Patterns. Addison-Wesley, Boston (2002)
17. Clough, P.: Old and new challenges in automatic plagiarism detection. National Plagiarism Advisory Service (2003). https://ir.shef.ac.uk/cloughie/index.html
18. Cordy, J.R.: Comprehending reality: practical challenges to software maintenance automation. In: Proceedings of the 11th International Workshop on Program Comprehension, pp. 196–206 (2003)
19. Czarnecki, K., Eisenecker, U.: Generative Programming: Methods, Tools, and Applications. Addison-Wesley, Boston (2000)
20. Dagenais, M., Merlo, E., LaguÄe, B., Proulx, D.: Clones occurrence in large object oriented software packages. In: Proceedings of the 8th IBM Centre for Advanced Studies Conference (CASCON 1998), Toronto, Ontario, Canada, p. 192200, October 1998
21. Deelstra, S., Sinnema, M., Bosch, J.: experiences in software product families: problems and issues during product derivation. In: Nord, R.L. (ed.) SPLC 2004. LNCS, vol. 3154, pp. 165–182. Springer, Heidelberg (2004). https://doi.org/10.1007/978-3-540-28630-1_10
22. De Lucia, A., Scanniello, G., Tortora, G: Identifying clones in dynamic web sites using similarity thresholds. In: Proceedings of the International Conference on Enterprise Information Systems (ICEIS 2004), pp. 391–396 (2004)
23. Ducasse, S., Rieger, M., Demeyer, S.: A language independent approach for detecting duplicated code. In: Proceedings of the 15th International Conference on Software Maintenance (ICSM 1999), Oxford, England, pp. 109–118, September 1999
24. Fowler, M.: Analysis Patterns: Reusable Object Models. Addison-Wesley Fowler M. 1999. Refactoring - improving the design of existing code. Addison-Wesley, Boston (1997)
25. Gamma, E., Helm, R., Johnson, R., Vlissides, J.: Design Patterns – Elements of Reusable Object-Oriented Software, Addison-Wesley, Boston (1995)
26. Garcia, R., et al.: A comparative study of language support for generic programming. In: Proceedings of the 18th ACM SIGPLAN Conference on Object-Oriented Program, Systems, Languages, and Applications, pp. 115–134 (2003)
27. Gitchell, D., Tran, N.: Sim: a utility for detecting similarity in computer programs. ACM SIGCSE Bull. **31**(1), 266–270 (1999)
28. Goguen, J.A.: Parameterized programming. IEEE Trans. Softw. Eng. **SE-10**(5), 528–543 (1984)
29. Gray, J. (ed.): Data Mining: Concepts and Techniques. Series Editor Morgan Kaufmann Publishers, Burlington, March 2006
30. Hammad, M., Basit, H., Jarzabek, S., Koschke, R.: A mapping study of clone visualization. Com. Sci. Rev. **37**, pp. 1–55 (2020). https://doi.org/10.1016/j.cosrev.2020.100266
31. Jarząbek, S.: Software similarity patterns and clones: a curse or blessing? In: Joaquim, F., et al. (ed.) Proceedings of the 22nd International Conference on Enterprise Information Systems, vol. 1, pp. 5–17 (2020). ISBN 978-989-758-423-7. https://doi.org/10.5220/000982000005 0017

32. Jarzabek, S.: Effective Software Maintenance and Evolution: Reused-based Approach. CRC Press, Taylor and Francis (2007)
33. Jarzabek, S., Shubiao, L.: Eliminating redundancies with a composition with adaptation meta-programming technique. In: Proceedings European Software Engineering Conference and ACM SIGSOFT Symposium on the Foundations of Software Engineering (ESEC-FSE 2003), pp. 237–246 (2003)
34. Jarzabek, S., Li, S.: Unifying clones with a generative programming technique: a case study. J. Softw. Maintenance Evol. Res. Pract. **18**(4), 267–292. John Wiley & Sons (2006)
35. Kamiya, T., Kusumoto, S., Inoue, K.: CCFinder: a multi-linguistic token-based code clone detection system for large scale source code. IEEE Trans. Softw. Eng. **28**(7), 654–670 (2002)
36. Kapser, C., Godfrey, M.W.: Toward a taxonomy of clones in source code: a case study. In: Evolution of Large Scale Industrial Software Architectures (2003)
37. Kapser, C., Godfrey, M.W.: Cloning considered harmful' considered harmful. In: Proceedings of the 13th Working Conference on Reverse Engineering, pp. 19–28 (2006)
38. Kiczales, G., et al.: Aspect-oriented programming. In: Akşit, M., Matsuoka, S. (eds.) ECOOP 1997. LNCS, vol. 1241, pp. 220–242. Springer, Heidelberg (1997). https://doi.org/10.1007/BFb0053381
39. Kim, M., Sazawai, V., Notkin, D., Murphy, G.: An empirical study of code clone genealogies. ACM SIGSOFT Softw. Eng. Notes **30**(5), 187–196 (2005)
40. Kim, M., Bergman, L., Lau, T., Notkin, D.: An ethnographic study of copy and paste programming practices in OOPL. In: Proceedings of International Symposium on Empirical Software Engineering. ISESE 2004, Redondo Beach, California, pp. 83–92 (2004)
41. Komondoor, R., Horwitz, S.: Using slicing to identify duplication in source code. In: Cousot, P. (ed.) SAS 2001. LNCS, vol. 2126, pp. 40–56. Springer, Heidelberg (2001). https://doi.org/10.1007/3-540-47764-0_3
42. Kontogiannis, K., DeMori, E., Galler, M.M., Bernstein, M.: Pattern matching for clone and concept detection. In: Automated Software Engineering, vol. 3(1–2), 77–108, June 1996
43. Koschke, K., Falke, R., Frenzel, P.: Clone detection using abstract syntax suffix trees. In: Proceedings of the 13th Working Conference on Reverse Engineering (WCRE 2006), pp. 253–262, Benevento, Italy, October 2006
44. Krinke, J.: Identifying similar code with program dependence graphs. In: Proceedings 8th Working Conference on Reverse Engineering, pp. 301–309 (2001)
45. Kumar, K., Jarzabek, S., Dan, D.: Managing big clones to ease evolution: Linux kernel example. In: Federated Conference on Computer Science and Information Systems, FedCSIS, 36th IEEE Software Eng. Workshop, pp. 1767–1776 (2016)
46. Levenshtein, I.: Binary codescapable of correcting deletions, insertions and reversals. Cybern. Control Theory **10**(8), 707–710 (1966)
47. Liu, C., Chen, C., Han, J., Yu, P.S.: GPLAG: detection of software plagiarism by program dependence graph analysis. In: Proceedings of the 12th ACM SIGKDD International Conference on Knowledge Discovery and Data Mining (KDD 2006), Philadelphia, USA, pp. 872–881, August 2006
48. Madhavji, N.: Compare: a collusion detector for Pascal. Techniques et Sciences Informatiques **4**(6), 489497 (1985)
49. Mining Software Repositories MSR, 5th Working Conference, 10–11 May, Leipzig, Germany (2008)
50. Musser, D., Saini, A.: STL Tutorial and Reference Guide: C++ Programming with Standard Template Library. Addison-Wesley, Boston (1996)
51. Pettersson, U., Jarzabek, S.: An industrial application of a reuse technique to a web portal product line. In: ESEC-FSE 2005, European Software Engineering Conference and ACM SIGSOFT Symposium on the Foundations of Software Engineering, ACM Press, Lisbon, pp. 326–335, September 2005

52. Prechelt, L., Malpohl, G., Philippsen, M.: Finding plagiarisms among a set of programs with JPlag. J. Univ. Comput. Sci. **8**(11), 10161038 (2002)
53. Rajapakse, D.C., Jarzabek, S.: Using server pages to unify clones in web applications: a trade-off analysis. In: International Conference on Software Engineering, ICSE 2007, Minneapolis, USA, May 2007
54. Rajapakse, D., Jarzabek, S.: An investigation of cloning in web portals. In: International Conference on Web Eng, ICWE 2005, Sydney, pp. 252–262 (2005)
55. Rajapakse, D., Jarzabek, S.: Towards generic representation of web applications: solutions and trade-offs. Softw. Pract. Experience **39**(5), 501–530 (2009). Published Online: 27 November 2008
56. Roy, C., Cordy, J.R.: Scenario-based comparison of clone detection techniques. In: Proceedings ICPC 2008, IEEE International Conference on Program Comprehension, Amsterdam, pp. 153–162, June 2008
57. Roy, C., Cordy, J.: A survey on software clone detection research. Technical Report 2007-541, School of Computing, Queen's University, November 2007
58. Schleimer, S., Wilkerson, S., Aiken, A.: Winnowing: local algorithms for document fingerprinting. In: Proceedings of the 2003 ACM SIGMOD International Conference on Management of Data (SIGMOD 2003), San Diego, California, p. 7685, June 2003
59. Schmidt, D.: Model-driven engineering. IEEE Comput. **39**, 25–31 (2006)
60. Smyth, B.: Computing Patterns in Strings. Addison-Wesley, Boston (2003)
61. Ueda, Y., Kamiya, T., Kusumoto, S., Inoue, K.: Gemini: maintenance support environment based on code clone analysis. In: Proceedings of the 8th IEEE Symposium on Software Metrics, pp. 67–76 (2002)
62. Walenstein, A., Lakhotia, A., Koschke, R.: The second international workshop on detection of software clones: workshop report. SIGSOFT Software. Eng. Notes **29**(2), 1–5 (2004)
63. Zhang, H., Jarzabek, S.: A mechanism for handling variants in software product lines. Spec. Issue Softw. Variability Manage. Sci. Comput. Program. **53**(3), 255–436 (2004)
64. Zhang, W., Jarzabek, S.: Reuse without compromising performance: experience from RPG software product line for mobile devices. In: 9th International Software Product Line Conference, SPLC 2005, Rennes, France, pp. 57–69, September 2005
65. Zhang, Y., Basit, H., Jarzabek, S., Anh, D., Low, M.: Query-based filtering and graphical view generation for clone analysis. In: Proceedings 24th IEEE International Conference on Software Maintenance, ICSM 2008, Beijing, September 2008
66. Yamanaka, Y., Choi, E., Yoshida, N., Inoue, K., Sano, T.: Industrial application of clone change management system. In: Proceedings of the 6th International Workshop on Software Clones, IWSC, pp. 67–71 (2012)
67. Yang, J., Jarzabek, S.: Applying a generative technique for enhanced reuse on J2EE platform. In: 4th International Conference on Generative Programming and Component Engineering, GPCE 2005, 29 September–1 October 2005, Tallinn, pp. 237–255 (2005)

SACIP: An Agent-Based Constructionist Adaptive System for Programming Beginners

Adson M. da S. Esteves[1(✉)], Aluizio Haendchen Filho[1,2], André Raabe[1], and Rudimar L. S. Dazzi[2]

[1] Laboratory of Innovation on Technology in Education, Itajaí Valley University (Univali), Rua Uruguay, 458, Itajaí, Brazil
raabe@univali.br

[2] Laboratory of Applied Intelligence, Itajaí Valley University (Univali), Rua Uruguay, 458, Itajaí, Brazil
rudimar@univali.br

Abstract. Brazilian universities have a high dropout rate in Computing courses. We believe E-learning personalized solutions can help to reduce this problem. This paper presents an architectural model for an adaptive system called SACIP that uses learning paths to deliver personalized assistance for students learning to program. Constructivist and constructionist theories were used as guidelines for the system modeling, and a collaborative multiagent system was developed to assist students in their choice of paths. Details of the SACIP implementation on different platforms are described, as well as its benefits and advantages over similar adaptive systems that use learning paths in a distinct manner. The application of SACIP with beginners in programming is aimed to facilitate learning, allow curricular flexibility and help to reduce dropout rates in Computing courses.

Keywords: E-Learning · Intelligent tutoring systems · Adaptive system · Constructivism · Constructionism

1 Introduction

In Brazilian universities, dropout is an existing negative phenomenon that creates problems for students, educational institutions and the market. Students are unable to grow professionally, and institutions fail to educate and prepare students for the labor market. However, recently there has been a significant increase in interest for professionals in the field of computing and information technology shown in several job search websites [1–5]. To cover this market demand, young people seek to learn more about these areas through courses related to computing.

The dropout rates in computer science courses remain very high in Brazil. They vary from 22% to 32% and are the second highest rate among Brazilian universities courses [6]. One of the main reasons for this, is the difficulty of the initial subjects that are the pillars of the course: Algorithms and Programming. For new students starting computer courses, these subjects are considered hard due to the need for abstraction and logical thinking skills that not common in student's daily life [7].

J. Filipe et al. (Eds.): ICEIS 2020, LNBIP 417, pp. 591–611, 2021.
https://doi.org/10.1007/978-3-030-75418-1_27

In order to better understand the causes of this problem, Giraffa and Mora [8] conducted a survey with students who stopped attending a Computer Science graduation course in the period 2012–2013. They found that the main reasons that contributed to the dropout were: (i) lack of time to study; (ii) difficulty in understanding the activities; and (iii) teachers underqualified to teach students with different profiles.

The difficulty of classroom questions may be related to the different types of students who enter computer courses. Heterogeneous profiles such as gender, age, level of education, form of learning and problem-solving skills make it difficult to create unique content that meets all of them [10]. Specific teachers for each student would be ideal educational scenario, but the cost makes this inviable [9].

Two alternatives can be applied to reduce the problem: changing teaching methodologies and e-Learning. Changes in teaching methodology are not the focus of this work. E-Learning is a web-based learning ecosystem that integrates various technologies, processes and techniques. It provides a flexible and personalized way of learning, allowing learning on demand and reducing its cost [11].

Intelligent Tutorial Systems (ITS), also called Adaptive Systems, have been applied in e-Learning. Artificial Intelligence (AI) techniques are used to develop tutors who know what, for whom and how they teach [12], considering the student's particularities.

The properties and characteristics of multiagent systems, such as autonomy, collaboration and learning seem appropriate for the development of ITS. Many authors [13–18] have successfully used Multi-Agent Systems (MAS) for the development of such systems. The use of MAS can help solve complex tasks, such as monitoring student activity, capturing information about their dynamic contexts, recommending content based on their profiles and much more [19].

Another way to create more adaptive solutions can be carried out by changing teaching methodologies. Constructivism [20] and constructionism [21] theories suggests a general change in the form of educating to facilitate learning. In the constructivist approach, during the learning process, the student should be able to decide how to learn and act proactively in the construction of knowledge [22]. In addition, constructionism points out that building a product related to students' interests helps learning to occur in a more efficient way.

Constructionism concepts can be applied for build learning environments [23]. These concepts have been used in several domains, such as microworld building, hypermedia text, and programming environments. In this, it can help the student facilitating the learning of logic.

One way to enable students to conduct their learning is to provide choices on learning paths. Paths are different types of skills or knowledge that follow a user-definable sequence. Thus, led by the adaptive system, the students can define their own learning path, facilitating the skills acquisition.

This paper presents an architecture model of a constructionist adaptive system using learning paths called SACIP. The initials stand for the Brazilian Portuguese words for Constructionist Adaptive System for Programming Beginners.

The IDE Portugol Studio [24] is used for instantiation of the SACIP architecture. It is a tool developed for facilitating the learning of programming logic in Brazil and has architectural features that facilitates building adaptable learning paths for the students.

SACIP was designed with features of an adaptive STI. It offers content on various topics with different themes. This enables the student to choose the content they want to learn and set up their own learning path. Meanwhile the system helps the student to choose the content that better fits his/her profile. In addition, if the students experience difficulties the system provides solving tips and support materials as well as exercises tailored to their profile. The following subsections present the learning path structure, the generic architecture, and the proposed role model for the agents.

2 Background

Concepts related to constructivism, constructionism, e-Learning, learning paths, and adaptive systems, are following presented. At the end of the section, some important related works are described.

2.1 Constructivism and Constructionism in Learning Paths

Piaget's constructivism [20] points out that the individual is already born with the capacity to learn reality. He learns by experiencing reality by means of actions called schemes, which makes him reflect on what happens and formalize his model of reality. Piaget declares that the existing knowledge in reality can be obtained by carrying out these actions. When analyzing Piaget's work with constructivism, Philips [25] realizes that the idea is to deliver the learning protagonism for the student. This allows him to face the results, errors and contradictions of his ideas about how the world works and to remodel them.

Papert [21] was heavily influenced by constructivist ideas. It corroborates the idea that knowledge is not transmitted from one individual person to others but is formed by a mental model of the student. This model is built by the student himself when learning. In addition, he points out one of the concepts not widely recognized in Piaget's research, that there are several forms of learning. According to Papert, the correct way of teaching should not be through "pieces of facts", which are passed on and repeated to the student until he learns (or records). The student must look for relationships among those facts, the real world and what he intuitively understands about the subject. The connection among all of these must be made by the student, and not given ready to him. For this reason, an adaptable system must allow the student to learn while making his own connections and experiences.

Considering the ideas about constructivist learning, the following characteristics can be established:

- Contents must be related to the student's history
- Content must be relevant to the student
- Students should be encouraged to make their own decisions
- The Tutor as an aid, the student as a driver
- Multiple forms of content representation

These characteristics can be explored through models of learning paths. The paths are sequences of content on different topics that the student chose to perform. This will allow the student to learn from the content that interests him most, facilitating and making his learning more natural. A path is a sequence of contents connected together in order. Details on learning trails can be seen in Sect. 3.3.

2.2 e-Learning and Adaptive Systems

E-learning is an alternative form of learning that does not require real-life meetings. The concepts vary between online classes with multimedia content or simply any learning where technology has an interference [26]. Today E-learning also seeks new forms of learning, in order not to become inflexible, closed and reducing teaching [27].

With the advancement of technology and the world seeking to ensure education for the entire population with safety and quality, investments in E-learning tools have been made. Virtual learning environments allow greater accessibility, communication and flexibility through the union of advanced teaching techniques and connected devices.

The E-learning format is already used in several universities, being an extra teaching alternative for students. The teaching via internet allows students to select different forms of content and learning objects. E-Learning formats can vary as standalone courses, simulations and games, mobile, social learning and virtual classrooms [28].

Intelligent Tutoring Systems (ITS) [29] are among the most accepted systems for the application of these technologies. These systems can help with solutions to reduce students' difficulties with excessive information in activities [30]. Several STI models have been modeled by different authors in the literature. Some works with similar ideas that helped to develop the system are described below.

2.3 Related Work

Using MAS techniques, De Meo *et al.* [31] developed the X-Learn system. Comprised of an XML learning structure, the system consists of a knowledge repository and three main agents: (i) User Device Agent (UDA); (ii) Skills Management Agent (SMA); and (iii) Learning Program Agent (LPA). The UDA is responsible for communication with the user. It collaborates with SMA to discover the skills that the student can learn. When the user selects a skill he wants to learn, SMA will collaborate with LPA to define the best learning program for the student and will present a list of materials that he will need to learn the desired skill.

Another system developed was the APLe (Agents for Personalized Learning) by Panagiotis *et al.* [32]. The system is divided into: (i) Learner Model and (ii) Learning Object and Outcomes. They deal with the characteristics and development of the students, and the knowledge of the system. This knowledge uses units of digital educational content called Learning Objects and the knowledge that the student is expected to learn called Learning Outcomes. For each student, the tutor agent assigns an LSM (Learning Space Management) and an LTC (Learning Tactic Control). Using learning graphs with Learning Objects and Outcomes, they allow the tutor to find out what content the student needs to learn.

Cabada *et al.* [33] developed an adaptive system called Java Sensei for teaching Java programming language. Based on the characteristics of the students and the similarities between them and other students in the system, the system recommends exercises that best suit the student in need. In addition, it uses image processing to check students' feelings when solving exercises and being able to send personalized tips. The system's agent layer consists of three modules: (i) domain module, (ii) tutor module, (iii) student module and (iv) exercise recommendation system.

The systems here presented use Multi-Agent Systems allowing greater flexibility, scalability, error tolerance, robustness and security. The characteristics of these systems influenced the structures and architectures that will be presented in Sect. 4.

3 Specification and Design

This section describes the design of the system, that was planned in several activities, performed in three steps: specification and design, implementation, and integration of the solution as a plugin for Portugol Studio tool. In the first step, following described, the definition of the learning paths structure, the design of the system architecture, and specifications of the tutor and agent workflow.

3.1 Definition of the Learning Paths Structure

Inside the system knowledge domain, there is a structure that allows the creation of learning paths. It has several contents about programming, which can be of two types: exercises and support materials. Exercises are related to the objectives that the students must fulfill, and support materials are texts or videos that it should only read or watch. The structure is shown in Fig. 1.

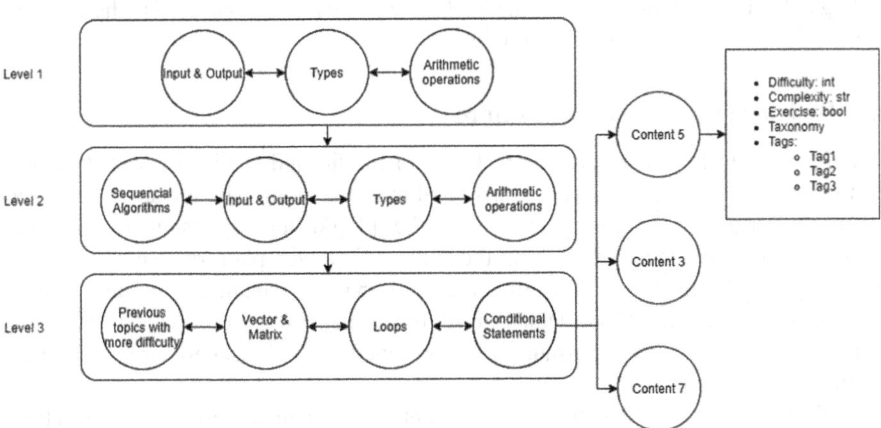

Fig. 1. Structure of knowledge domain.

The contents are organized in a hierarchy of three levels: Level, Topic and Content. At the content level it has five attributes: (i) Difficulty, (ii) Complexity, (iii) Content

Type, (vi) Taxonomy, and (v) Tags. The first four use the exercise taxonomy of Santos *et al.* [43] to create learning paths. The last attribute Tags refers to themes not related to the topic present in the content. It is used to identify whether a certain content uses didactics that have topics of interest to students.

Every time a student consumes content in the system, that content is registered in his account and connected to the previous content he used. The learning trails that students create are a sequence of connected content that the student has consumed. This can be seen in Fig. 2.

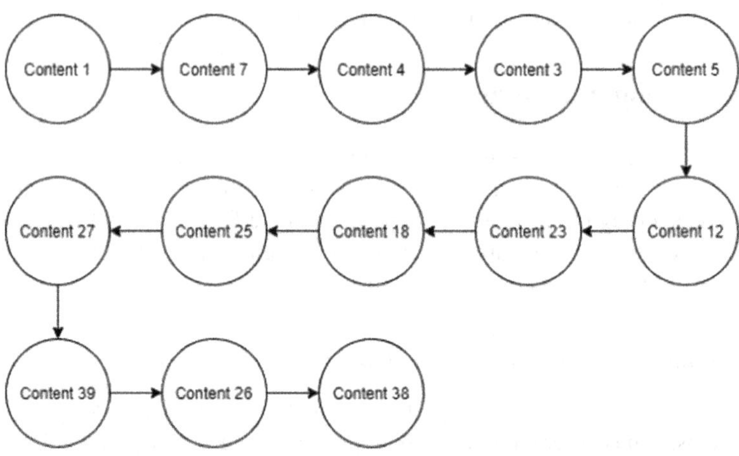

Fig. 2. Generic example of a learning path [42].

The system also allows the student to revisit topics that he has seen before, thus adding content in different orders in the path. Each path that is registered in the student's account is publicly available to everyone.

3.2 Designing the System Architecture

SACIP works in two instances: (i) in Portugol Studio and (ii) in the Web Platform. Portugol Studio is installed on the user's local machine along with the adaptive system plugin. In Portugol Studio, the system uses other plugins already installed, such as the correction and error messages to help the student. The web platform uses the MIDAS framework [44] which provides infrastructure services to facilitate the development of multi-agent systems. More details on its functionalities are described in Sect. 4.1. Considering the two referred instances, a representation of the generic architecture can be drawn as shown Fig. 3.

On the web platform, the domain and student models contain, respectively, structured content and data for all students. To use the system, the student must create a profile account, which will register its characteristics. This registration is done by the plugin and sent to agents in the cloud who will organize this data for use in future recommendations.

Five agents work on the web platform: (i) Pedagogical Tutor, (ii) Interface, (iii) Tracker, (iv) Classifier and (v) Recommender. The agents collaborate by making services

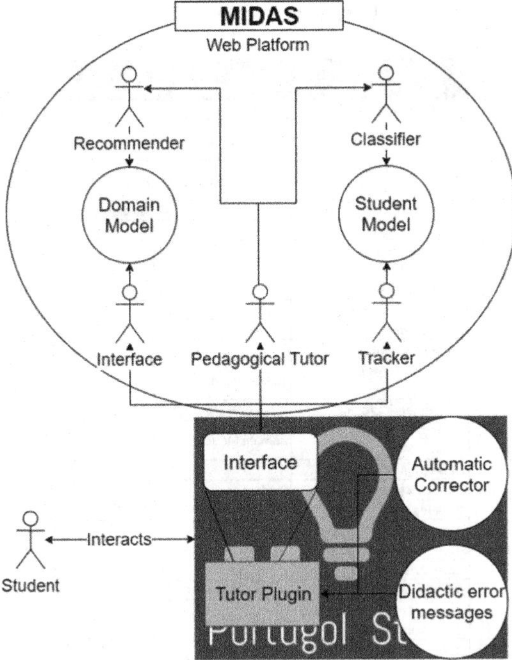

Fig. 3. System generic architecture [42].

available to each other, allowing the contents to be recommended according to the interests and characteristics of the students. The permissions and responsibilities of each agent can be seen in the next section.

3.3 Specification of the Tutor Role Model

The role model was selected to define the specifications of the agents in the tutor module. This model has been used [45, 46] as a way of defining a summary of the responsibilities that the agent needs to fulfill. To define a role, two attributes are used: (i) Responsibilities and (ii) Permissions. Responsibilities define the agent's obligations regarding a feature. The permissions establish the rights of the agents, indicating which characteristics and components they have access to. Table 1 shows the role model of the system agents. Yet another optional attribute that can complement the model are the collaborations, that is, the agents withal which the Tutor needs to collaborate in order to achieve its purposes.

Tracking Agent. Placed on the web platform, it is responsible for capturing and organizing the interactions of each student with Portugol Studio. this responsibility gives it access to the student model that contains user data. The following data is captured by this agent:

1. Clicks on the interface;
2. Length of stay in the system;

Table 1. Agents role model [42].

Organization: Adaptive System		
RESPONSABILITIES	PERMISSIONS	COLABORATION
Pedagogical Tutor Agent		
Suggest next contents to the student	Domain Model	Classifier
Suggest exercise resolution tips	Student Model	Recommender
Identify exercises errors		Tracker
Identify preferences		
Identify current knowledge		
Calculate Average Exercise Resolution Time		
Sort resolution time by subject type		
Classifier Agent		
Cluster students by their characteristics	Student Model	
Recommender Agent		
Recommend content by a group	Domain Model	
Tracker Agent		
Store clicks on interface	Student Model	
Store log time		
Store exercise solving time		
Store solved exercises		
Store chosen tracks		
Interface Agent		
Create Interaction Interface to the Admin	Domain Model	
Store user questions	Student Model	
Answer user questions		

3. Exercise response time;
4. Solved exercises;
5. Chosen paths.

The plugin receives these interactions through the communication interface, which makes a direct call to the agent. The tracking agent can also collaborate with the pedagogical agent to verify difficulties that the student may have with the exercises.

Interface Agent. It communicates with the administrator, delivering an interface that makes it possible to manage the system without the need to change the database directly. This agent also makes a connection between the system administrators and the students. It allows to receive questions from students and send a message to the teacher or administrator to answer. If student questions have been answered previously, the agent will skip the step of sending to the administrator and will respond to the student immediately. Therefore, their responsibilities are:

1. Record student's questions;
2. Notify the administrator about unresolved student's questions;
3. Send to the student's questions the respective answers of the administrators;
4. Answer student's questions without the need for an administrator if a similar question has already been asked by another student and answered.

For the students' questions to be stored, this agent must have access to the domain and student models. Each question will be stored in the domain model but will be referenced by attributes in the student model.

Tutor Agent. This agent collaborates with two other agents, mainly: Classifier and Recommender. The collaboration between these agents makes it possible to recommend content according to the interests and characteristics of the students. With the collaboration of Classifier, similarities between students are discovered, and with the collaboration of Recommender, the most used content by similar students is found. Finally, these contents are used by the tutor agent as recommendations for the student. Thus, this agent's responsibilities are:

1. Suggest learning paths based on student preferences and profile (collaboration with Classifier and Recommender Agent);
2. Suggest real-time exercise solving tips with the aid of the automatic correction module to help the learning process become easier;
3. Identify errors in exercises (collaboration with Tracking Agent);
4. Identify student preferences in content (collaboration with Tracking Agent);
5. Identify the average exercise resolution time (collaboration with Tracking Agent).

This is the main agent of the system. Collaborates with other agents to discover knowledge about students and improve learning strategies. Therefore, in addition to collaborations, it has access to domain and student models.

Classifier Agent. This agent is responsible for grouping students by similarities in their characteristics. It collaborates with the pedagogical agent, which will use these groups to discover knowledge about the student. For this, the Classifier uses clustering and machine learning techniques, with the data obtained on the students' profile, and their stored paths. The data that allow classification are:

1. Age,
2. Gender,
3. Educational level,
4. Personal preferences,
5. Chosen paths,
6. Resolved exercises,
7. Average time exercise resolution and
8. Most frequent types of errors.

The first four attributes are obtained from users when they create their accounts in the system. Everyone should make their data available to allow the system to make individual recommendations. The following four attributes are obtained in collaboration with the tutor agent, which analyzes this data when captured by the tracking agent.

Recommender Agent. The Recommender agent collaborates with the pedagogical agent to find out what are the most recommended content to be delivered to the student. To this end, it analyzes the learning paths that resemble the student who requires the content. In this analysis it discovers the contents that were most used in these paths, and which contents were not used by the student, therefore, finding which are the most recommended for him.

3.4 Specification of Agent Services Workflow

The agents described in the previous section work collaboratively to recommend content to students. The pedagogical agent organizes the calls with the other agents of the system in order to obtain the best content from the particularities of the student's use of the system.

Communication between agents is performed by calls to microservices that the MIDAS platform provides. Blackboard provides a powerful mechanism for asynchronous communication between agents (more details can be seen in Sect. 4.1). Communication through microservices can then be described using a BPMN (Business Process Modeling Notation) model [47].

Process modeling using the BPMN tool was developed so that all business users could easily read the processes of a system [48]. Due to its wealth of representation elements, the tool makes it possible to have several models in a single representation. It is especially effective for representing the exchange of messages, which in the context of our research is very important.

Figure 4 presents a representation of the user request content in BPMN model process. SACIP, and the agents are represented in the four swim-lanes. The first refers to the SACIP plugin, which is located on the user's machine and has a communication interface with the agents. The second, third and fourth swim-lanes represent the Pedagogical, Classifier and Recommender agents, respectively.

The process starts at the machine plugin with the PS, with the user requesting new content. Upon receiving the message, the interface agent sends a request for content to the pedagogical agent in MIDAS. After receiving this request, the pedagogical agent needs to discover groups of students who have made similar requests. For this, the pedagogical agent requests the classifying agent to cluster this student with the others. The classifying agent will analyze in the student domain, the characteristics of each other student in the system, and look for similarities between them and the requested student. These similar students will be grouped and sent to the pedagogical agent.

After finishing the task, the pedagogical agent sends a message to the recommender with the student groups and the requesting student. The recommender will then analyze the learning paths of each student in the group, looking for similar content. These similar contents are filtered to consider only contents that the student has not yet used. After receiving the recommended content, the agent will organize it according to the difficulty and send it to the plugin interface as the content that best suits the profile, and the degree of evolution of the student. After this process, the system will show the contents on the user interface so that the student can choose the one it wants to learn.

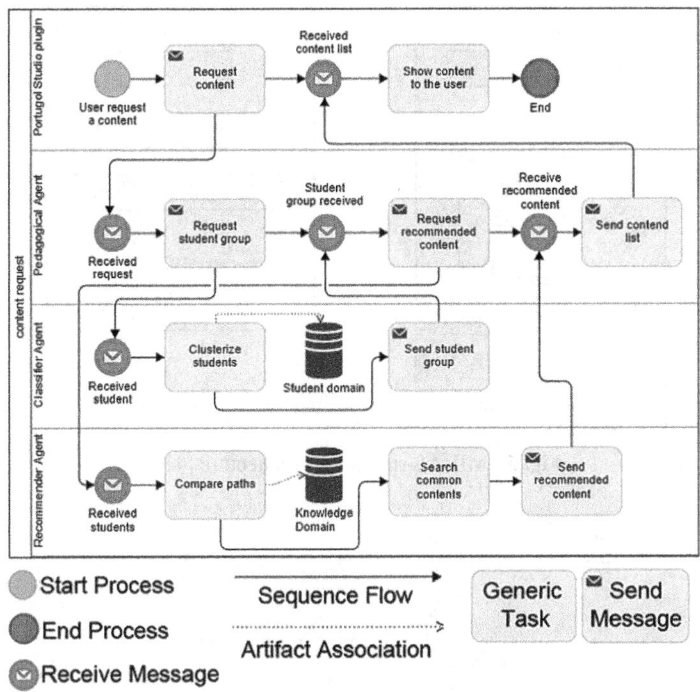

Fig. 4. The process *user request content* in BPMN model [42].

4 Implementation

This section explains how agents were instantiated on the platform and implemented using the facilities provides by the Midas. The graphical interface developed for the user to navigate using learning paths is also explained.

4.1 Implementation of the Agents

MIDAS platform allows the development of multiagent systems using microservices as a base. The platform is composed of a main server and one or more containers of agents. This structure can be seen in Fig. 5. MIDAS server is synchronized with the agent containers, and its responsibility is to apply system rules, synchronize agents and communicate with external applications. It reduces the complexity of communication among agents by automatically integrating containers.

Internal communication is performed through an HTTP interface that allows containers to communicate transparently with each other using the server. In addition to this interface, the system has: (i) a REST interface for communication with external systems and (ii) a web interface for system management, exemplified in Fig. 6.

Server communications depend on five middleware agents: Broker, Catalog, Proxy, Manager and Blackboard. The Broker is responsible for the transmission of messages, translating the messages exchanged by agents in HTTP. The Proxy instantiates the classes

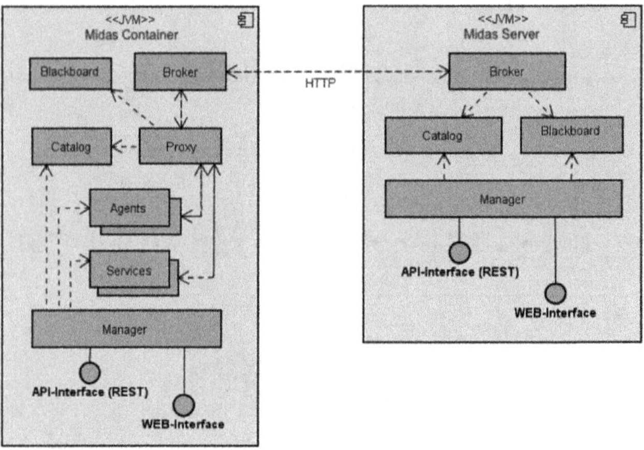

Fig. 5. MIDAS simplified architecture [41].

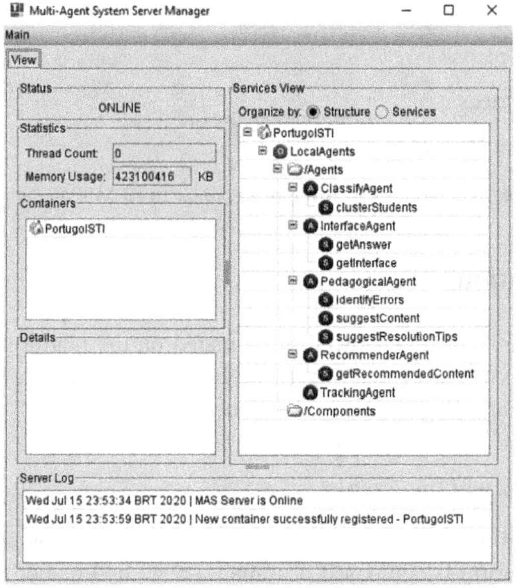

Fig. 6. MIDAS server web interface.

and has a dynamic configuration that allows you to redirect messages if necessary. The Catalog works with the data representation model allowing humans and software to locate the services and agents of the system. The Manager is responsible for maintaining the continuous activities of the system, such as Lifecycle, QoS (Quality of Service), statistics and graphical interfaces. Finally, Blackboard is the agent that works with the form of communication between agents. It allows each agent to report messages to all other agents in the system either synchronously or asynchronously.

MIDAS containers carry software agents that may have microservices. The container catalogs all the services that its agents have and makes available on the MIDAS server. With that, all agents of other containers or external agents can also use the services that those agents in the container have. The agents are instantiated by extending an abstract class that has the implementable *provide* and *lifecycle* functions. These functions are the hotspots for implementing the lifecycle of agents.

The *provide* function will have the implementation of the agent's services. Each time an agent in the system requests a service from another agent, the MIDAS server will search for which agent has that service and call the *provide* function for that agent. Figure 7 shows the partial code of the *provide* function of the pedagogical agent.

```
@Override
public void provide(String service, Map in, List out) throws ServiceException {
    // TODO Auto-generated method stub

    if(service.equals("suggestExercise"))
    {
        try
        {
            System.out.println("verificando informacões do aluno");
            ServiceWrapper serviceWrapper = require("LocalAgents", "getAluno");
            List alunoInfo = serviceWrapper.run();

            System.out.println("requisitando exercicio");
            ServiceWrapper serviceWrapper2 = require("PublicAgents", "getRecommendedExercises");
            serviceWrapper2.addParameter("aluno", alunoInfo.get(0));
            List exercicio = serviceWrapper2.run();

            System.out.println("apresentando exercício");
        }
        catch(Exception e)
        {
            throw new ServiceException("Não foi possível pegar o exercício recomendado - PAgent",e);
        }
```

Fig. 7. Partial code of the *provide* function of the pedagogical agent.

One of the pedagogical agent's responsibilities is to suggest exercises to the student. The code for this service shows that it will first make a synchronous call to an agent service that will deliver student data. Then it will request another service from another agent in the system, which will inform what are the recommended exercises. These synchronous service requests go through the system's Blackboard.

In the *lifecycle* function, the agent's workflow is implemented. In this function, Blackboard is normally invoked to check the system status. So, the rules define what actions the agent should take depending on the state of the system. The *lifecycle* function enables the autonomy of the software agent in the application.

In the Fig. 8 example, the *lifecycle* function checks the state of Blackboard and performs the task of checking student errors. The pedagogical agent, when discovering the difficulties, will look for the best tips for the student, and will present them to him/her at the moment. This action of analyzing the system context, without the need for another agent, is characteristic of autonomous agents.

```
@Override
protected void lifeCycle() throws LifeCycleException, InterruptedException {

    Board.addMessageListener("PortugolSTI", this);
    for (String aluno : alunosOnline) {
        try
        {
            while(Board.getContextAttribute(aluno+"eventState").equals("Programando"))
            {
                ServiceWrapper serviceWrapper = require("LocalAgents", "getAlunoData");
                serviceWrapper.addParameter("aluno", aluno);
                List data = serviceWrapper.run();
                List dicas = checkDicas(data);
                if(dicas.size()>0)
                {
                    sendDicas(dicas);
                }
                Thread.sleep(2000);
            }

        }
        catch(Exception e)
```

Fig. 8. Partial view of the *lifecycle* function on the pedagogical agent.

4.2 Development of the Graphical User Interface

The SACIP plugin located in Portugol Studio provides communication of students with the system. A graphical interface that meets the main needs of students is necessary to have a good experience. The following screens were developed in the Portugol Studio plugin. They allow students to select the content he wants to see. As the system was made for Brazilian students, the screens will be in Portuguese.

Figure 9 presents the login screen. This is the first screen the student will see when accessing the system. As explained in Sect. 3.1, the user will have his/her usage data saved in the system. Therefore, it is necessary that the student has the credentials to use the system.

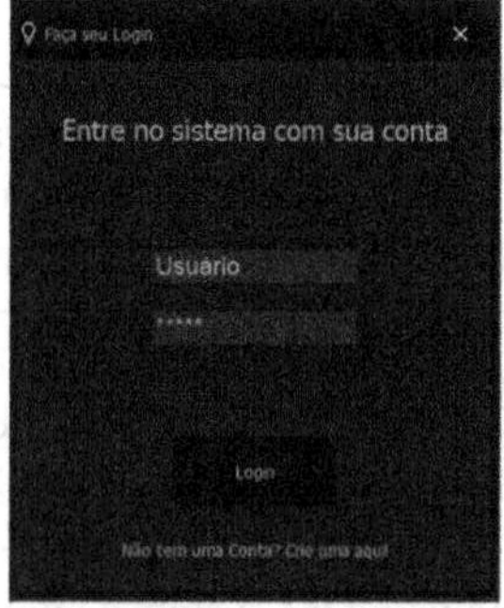

Fig. 9. Adaptive system login screen.

When the user logs in to the system, it is directed to the main menu screen. This screen shows the actions the user can perform on the system (Fig. 9). In the center, the opening screen presents the student path so far. It can navigate through its own path from the main menu, and view all the used content so far, being able to review any case he wants. The options in the left panel take the student to other menus. The four options with darker colors in Fig. 10 refer to the actions: (i) *See Path*, (ii) *Select Content*, (iii) *Ask the System* and (iv) *Help* respectively.

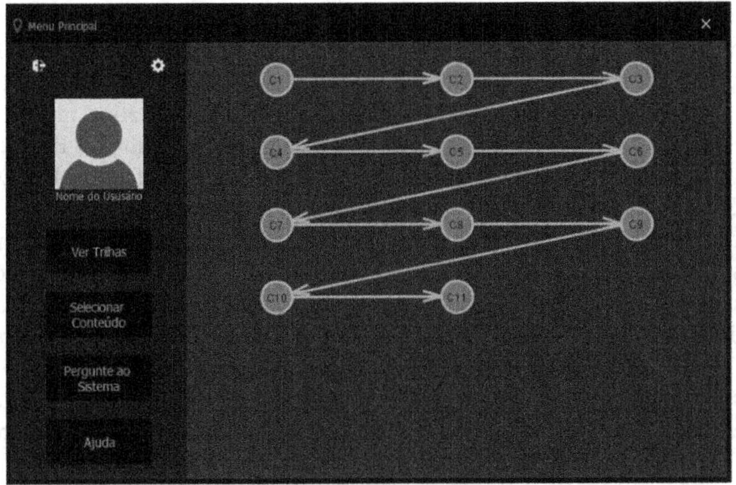

Fig. 10. Adaptive system main menu screen.

The *See Path* action allows the student to see tracks from other students who have already used the system. The system will also recommend similar trails for him/her. Thus, if the student wishes, he or she can follow the path that another student has taken with a view to improving their skills.

The *Select Content* action will send the student to the content selection screen, where the system will make recommendations. When the content is displayed, by clicking on it, a short description of what it is about will be displayed. Thus, the student can select what he wants to learn, adding knowledge to his/her learning path.

The *Ask the System* action will allow the student to ask questions to the system administrators. These questions will not be answered in real time, as it will need to wait until administrators respond. However, the system will search for questions already asked by other students with similar content answered. Thus, repeated questions are answered instantly without the need of the administrator.

The *Help* action has system documentation. It will explain how each part of the system works and how the student should use it. The Help structure is divided into tree-shaped topics to make it easier to find specific content.

Beyond of these four main actions, there are three more actions that the student can perform. These actions are performed on the user's image and on the two icons above the user's image. When clicking on the user image, the student is taken to a screen where he can change the following settings (i) username; (ii) password; (iii) image; (iv) privacy details; (v) preferences; or (vi) delete the account. The icons above the user's image enable logout actions and system settings. In the first, the student will only log out of his account and, in the second, he will be able to change some system settings such as screen size and themes.

5 Plugging the Architecture on the Portugol Studio Tool

Several authors have already created tools to facilitate the learning of programming [34–39]. Among them, Portugol Studio (PS) [24], emerges as a programming IDE focused on facilitating programming learning for Brazilian students. PS was designed to lower barriers at the beginning of learning, such as language. Its documentation, menus, screens and the language itself are in Portuguese. Noschang [24] states that in Brazil, one of the major barriers in learning programming is the language. Professional IDEs are in English, and many Brazilians still have difficulties with foreign languages.

In addition to the language, PS has a syntax like JavaScript and PHP, several installed libraries and complete syntax documentation. It has didactic errors messages, which point out the errors to the student and indicate possibilities of how to solve them. It also has several initial examples of how to program in Portuguese, from simple examples like writing "Hello World" to more complex ones like animations and games.

PS supports the addition of plugins to it. The plugins are programmed in Java and have access to several PS components. This allowed different authors to do research with PS [40, 41]. Some of these works have turned into open plugins and are already available in the tool itself as a usable plugin.

IDE is open source and has been used by several universities throughout Brazil, reaching more than 400,000 downloads with constant updates. It has been presented in scientific conferences and its core is used for different projects, some not linked to the main IDE.

SACIP will use the PS plugin system, which has three main points for installation: libraries, actions and license [49]. The license core is mandatory for any plugin on PS, identifies the details of the plugin. Actions and libraries are optional, as they are inside the plugin development scope is and not every need both. SACIP does not need to add libraries. It only has the action buttons to confirm or not that the system boots and communicates with agents in MIDAS.

It is also necessary that some functions of the PS source code are available open for use, that is, that they are programmed as public. The system to be plugged in needs access to specific PS components, such as the editor and the output panel.

The main part programmed in the plugin is the communication interface, as shown in Subsect. 3.1. The PS is a local tool installed on the user's machine that does not need servers or the internet to work. Therefore, it is necessary that the plugin to be added on PS be able to communicate with agents in the cloud for the system to work.

The interface uses HTTP connections to bridge the gap between PS and agents in the cloud. The main communication functions with agents are presented in Table 2.

With this type of communication interface, it may be possible for other systems in the future to use the SACIP, as long as they adhere to the interface API. There are other

Table 2. Communication interface functions in the plugin.

+ createAccount()	Create a user account
+ loginUser()	Log the user in using their username and password
+ getUserPath()	Request the path made by the user so far
+ getPopularPaths()	Request paths from other users that are more popular and like the one logged in
+ getRecommendedContent()	Request to the tutor agent recommended content according to the user's preferences and path
+ getContent()	Request to tutor agent the content selected by the user
+ setNewContentOnPath()	Add user-selected content to the last path node
+ getAnsweredQuestions()	Search the database for similar questions already answered by the user
+ sendQuestion()	Send the new question asked by the user to the system
+ changeUserInformation()	Edit user information according to what it wants
+ checkUserTips()	Check whether the tutor agent has user tips and displays them

secondary functions of communication with the web system, however they are not as relevant to the system as the ones presented.

6 Discussion

Since its creation, Portugol Studio has been used to research better ways to teach programming to Brazilian students. Thinking about the possibility of improvements in the tool (not coming from the research group that developed the project), the PS added the possibility of adding plugins. Thus, the possibility of the adaptive system described here being added to the PS became viable.

Constructivist and constructionist theories were used to structure the learning paths. Theories were developed by Piaget and Papert, defining how the student can learn when facing his knowledge in the real world and having the power to choose what he wants to learn. Learning paths in the literature are not cited as concepts like described, but their characteristics can be seen in works such as De Meo [27] and Panagiotis [28].

In De Meo's work, the learning paths are trails ready for the student. Each path leads to a finalizable project, with each step making it possible to learn new knowledge. Panagiotis, on the other hand, allows the student the option of following a recommended system trail, that is, the student has freedom of choice.

Both works have concepts of a learning path that can be followed by the student. With De Meo, the student needs to follow a predefined path, even being able to choose a project of interest. On the other hand, Panagiotis allows the student a greater option of choosing its learning, with the possibility of the student not following the system paths. In these works, the possibility of the student having his its path is not envisaged.

To define the learning paths, the system uses a mix of the student's personal choices with recommendations. The exercises are selected according to the characteristics of the students, and the contents are chosen by the student himself. Thus, learning paths are formed. Each content recommended by the system has several features, and these are used to define which are best for the student. Part of these features were based on the work of Santos *et al.* [39] and expanded to the context of this work.

The approach in this paper follows the standard of multiagent systems. In Sect. 1, several authors were presented who follow this same pattern to develop adaptive systems. As presented in the paper, the MIDAS service-oriented platform was used for the development of agents. The agent's development, responsibilities and actions were influenced by the related work.

Analyzing the study of the Cabada [29] it was possible to see how the use of agents that recommend content can facilitate the work. The proposal distributes the recommendation and assistance work among the recommending and tutoring agents. Assistance became a task of the tutor only in the moments of activity, while the recommender analyzes the characteristics of the students and delivers the exercises.

The recommender agent used in the approach presented in this article, has only the function of recommending exercises. In addition, it has part of its work carried out in collaboration with the classifier agent. Unlike Cabada's work, here the classifier agent is responsible for analyzing the characteristics of the students. This facilitates the "agent recommender" job of finding suitable exercises for students, as it limits the search for similar students.

In addition to the recommender and classifier agents, two other agents were used in the approach: the tracking agent and the interface agent. A tracking agent is needed to update the database with all the student's actions in the system. On the other hand, the interface agent has more functions related to that of administrators. It is responsible for managing both the screens that the system administrator will use as well as organizing the questions that users will ask the system that the administrator must answer.

The agents and content structure used in conjunction with a learning tool such as Portugol Studio can therefore help facilitate the learning of Brazilian students' programming. The possibility of using successful learning trails and content adapted to students can make students more interested and facilitate the development of each other's mental learning model, better fixing the content. In the same vein, Portugol Studio inherently already eliminates some barriers in the beginning of programming, making the learning process easier and more productive.

With the success of Portugol Studio, the plugin can be easily accessed in Brazil, and can influence the development of new research related to student learning based on adaptive paths.

7 Conclusion and Future Works

Research on programming learning is a traditional field in the area of Computer Science. Many students have difficulties with abstraction and personal barriers that hinder their learning. Tools to assist this process are desired, and adaptive systems are a possibility. Alternative learning methods such as constructivism and constructionism can produce new tools for learning programming in the context of adaptive systems. As the PS has characteristics of a constructionist system, the integration of SACIP is facilitated.

This work presents some contributions to the area of teaching logic and programing languages. One is the proposal of an architectural model with low-level diagrams representing the functionalities of the agents for adaptive system using learnings paths. This can generate gains in preparing a generation in a high demand market for this knowledge type. Another contribution is the modeling of the tutor, which adapts the contents to the students' interests. It recommends content according to the characteristics of the student.

After the implementation is completed, the project goes on trial. Groups of students from the local university will be used in the experiments. These students will be in the initial classes of computer science courses because they have classes in initial programming. The system will be used by them in order to help them understand their programming difficulties.

With the launch of the public open plugin, all students with updated versions of the PS will already be able to use the adaptive system. As the PS is used in several universities in Brazil and by independent users, a lot of data about learning to program can be acquired.

References

1. Guia da Carreira: Saiba quais as 8 profissões que mais crescem no Brasil (2018). https://www.guiadacarreira.com.br/profissao/profissoes-que-mais-crescem/. Accessed 30 Dec 2019
2. MichaelPage: THE WORLD'S MOST IN DEMAND PROFESSIONS (2019). https://www.michaelpage.co.uk/minisite/most-in-demand-professions/. Accessed 30 Dec 2019
3. Pattabiraman, K.: LinkedIn's Most Promising Jobs of 2019 (2019). https://blog.linkedin.com/2019/january/10/linkedins-most-promising-jobs-of-2019. Accessed 30 Dec 2019
4. CareerCast: The Toughest Jobs to Fill in 2019 (2019). https://www.careercast.com/jobs-rated/2019-most-difficult-jobs-to-fill. Accessed 30 Dec 2019
5. Trade Schools: 31 High-Demand Jobs in 2019 for Almost Every Type of Person (2019). https://www.trade-schools.net/articles/high-demand-jobs.asp. Accessed 30 Dec 2019
6. Lobo, R.: 'A Evasão No Ensino Superior Brasileiro – Novos Dados', Estadão, 07 October (2017)
7. Raabe, A.L.A., Silva, J.D.: Um ambiente para atendimento as dificuldades de aprendizagem de algoritmos. In XIII Workshop de Educação em Computação (WEI'2005). São Leopoldo, RS, Brasil, vol. 3, p. 5. Sn (2005)
8. Giraffa, L. M., & da costa Mora, M. (2013). Evasão na disciplina de algoritmo e programação: um estudo a partir dos fatores intervenientes na perspectiva do estudante. In Congresos CLABES.
9. Weragama, D.S.: Intelligent tutoring system for learning PHP (Doctoral dissertation, Queensland University of Technology) (2013)
10. Oliveira, C.M., Pimentel, A., Krynski, E.M.: Estudo sobre o sequenciamento inteligente e adaptativo de enunciados em programaçao de computadores. In: Anais dos Workshops do Congresso Brasileiro de Informática na Educação, vol. 4, no. 1, p. 1320,October 2015
11. Cidral, W.A., Oliveira, T., Di Felice, M., Aparicio, M.: E-learning success determinants: Brazilian empirical study. Comput. Educ. **122**, 273–290 (2018). https://doi.org/10.1016/j.compedu.2017.12.001
12. Nwana, H.S.: Intelligent tutoring systems: An overview. Artif. Intell. Rev. **4**(4), 251–277 (1990)

13. Giraffa, L.M.: Uma arquitetura de tutor utilizando estados mentais. 1999 (Doctoral dissertation, Tese (Doutorado em Ciências da Computação) – Instituto de Informática, UFRGS, Porto Alegre) (1999)
14. Yaghmaie, M., Bahreininejad, A.: A context-aware adaptive learning system using agents. Expert Syst. Appl. **38**(4), 3280–3286 (2011)
15. Dolenc, K., Aberšek, B.: TECH8 intelligent and adaptive e-learning system: Integration into technology and science classrooms in lower secondary schools. Comput. Educ. **82**, 354–365 (2015). https://doi.org/10.1016/j.compedu.2014.12.010
16. Hooshyar, D., Ahmad, R.B., Yousefi, M., Yusop, F.D., Horng, S.-J.: A flowchart-based intelligent tutoring system for improving problem-solving skills of novice programmers. J. Comput. Assisted Learn. **31**(4), 345–361 (2015)
17. Harley, J.M., Bouchet, F., Hussain, M.S., Azevedo, R., Calvo, R.: A multi-componential analysis of emotions during complex learning with an intelligent multi-agent system. Comput. Hum. Behav. **48**, 615–625 (2015). https://doi.org/10.1016/j.chb.2015.02.013
18. Vaidya, N.M., Sajja, P.S.: Agent based system for collaborative learning environment in an educational habitat. In: 2016 International Conference on ICT in Business Industry & Government (ICTBIG) (2016). https://doi.org/10.1109/ictbig.2016.7892644
19. Frade, R.V.C.: "UNIVIRTUAL – Ambiente Virtual 3D Multiagente com Recomendação Personalizada de Objetos de Aprendizagem", Dissertação (Mestrado em Ciência da Computação) - Universidade Estadual do Rio Grande do Norte. Universidade Federal Rural do Semi-Árido, Mossoró (2015)
20. Piaget, J.: The psychogenesis of knowledge and its epistemological significance. In: Piatelli-Palmarini, M. (ed.) Language and Learning, pp. 23–34. Harvard University Press, Cambridge (1980)
21. Papert, S.: A máquina das crianças. Porto Alegre: Artmed (1994)
22. Bada, S.O., Olusegun, S.: Constructivism learning theory: a paradigm for teaching and learning. J. Res. Method Educ. **5**(6), 66–70 (2015)
23. Baranauskas, M.C.C., Vieira, H., Martins, R.M.C., D'ABREU, J.V.: Uma taxonomia para ambientes de aprendizado baseados no computador. O computador na sociedade do conhecimento, 45 (1999)
24. Noschang, L.F., Pelz, F., Raabe, A.: Portugol studio: Uma ide para iniciantes em programaçao. Anais do CSBC/WEI, pp. 535–545 (2014)
25. Philips, D.C.: The Good, the Bad, and the Ugly. The many Faces of Constructivism. I (1995)
26. Al-, D., Joy, M., Masa'deh: Evaluating E-learning systems success: an empirical study. Comput. Hum. Behav. (2019). https://doi.org/10.1016/j.chb.2019.08.004
27. Mota, J.: Da web 2.0 ao e-learning 2.0: aprender na rede (Doctoral dissertation) (2009)
28. Horton, W.: E-learning by design. John Wiley & Sons (2011)
29. Phobun, P., Vicheanpanya, J.: Adaptive intelligent tutoring systems for e-learning systems. Procedia Soc. Behav. Sci. **2**(2), 4064–4069 (2010). https://doi.org/10.1016/j.sbspro.2010.03.641
30. Tarus, J., Niu, Z., Mustafa, G.: Knowledge-based recommendation: a review of ontology-based recommender systems for e-learning. Artif. Intell. Rev. **50**(1), 21–48 (2017). https://doi.org/10.1007/s10462-017-9539-5
31. De Meo, P., Garro, A., Terracina, G., Ursino, D.: Personalizing learning programs with X-Learn, an XML-based, "user-device" adaptive multi-agent system. Inform. Sci. **177**(8), 1729–1770 (2007). https://doi.org/10.1016/j.ins.2006.10.005
32. Panagiotis, S., Ioannis, P., Christos, G., Achilles, K.: APLe: agents for personalized learning in distance learning. In: Zvacek, Susan, Restivo, Maria Teresa, Uhomoibhi, James, Helfert, Markus (eds.) CSEDU. CCIS, vol. 583, pp. 37–56. Springer, Cham (2016). https://doi.org/10.1007/978-3-319-29585-5_3

33. Cabada, R.Z., Estrada, M.L.B., Hernández, F.G., Bustillos, R.O., Reyes-García, C.A.: An affective and Web 3.0-based learning environment for a programming language. Telematics Inf. **35**(3), 611–628 (2018). https://doi.org/10.1016/j.tele.2017.03.005
34. Cooper, S., Dann, W., Pausch, R.: Alice: a 3-D tool for introductory programming concepts. J. Comput. Sci. Coll. **15**(5), 107–116 (2000)
35. Ng, S.C., Choy, S.O., Kwan, R., Chan, S.F.: A web-based environment to improve teaching and learning of computer programming in distance education. In: Lau, R., Li, Q., Cheung, R., Liu, W.. (eds.) ICWL. LNCS, vol. 3583, pp. 279–290. Springer, Heidelberg (2005). https://doi.org/10.1007/11528043_28
36. Resnick, M., et al.: Scratch: Programming for all. Commun. ACM **52**(11), 60-67 (2009)
37. Wolber, D.: App inventor and real-world motivation. Proceedings of the 42nd ACM Technical Symposium on Computer Science Education - SIGCSE 2011 (2011). https://doi.org/10.1145/1953163.1953329
38. Paiva, J.C., Leal, J.P., Queirós, R.A.: Enki. In: Proceedings of the 2016 ACM Conference on Innovation and Technology in Computer Science Education - ITiCSE 2016 (2016). https://doi.org/10.1145/2899415.2899441
39. Romagosa i Carrasquer, B.: The Snap! Programming System. In: Tatnall, A. (eds.) Encyclopedia of Education and Information Technologies. Springer, Cham (2019)
40. Pelz, F.D.: Correção automática de algoritmos no ensino introdutório de programação. 2011. TCC (graduação em Ciência da Computação) - Universidade do Vale do Itajaí, Itajaí (2011)
41. Hodecker, A.: Aprimoramento e avaliação do corretor de questões do Portugol Studio. 2014. TCC (graduação em Ciência da Computação) - Universidade do Vale do Itajaí, Itajaí (2014)
42. Esteves, A.S., Haendchen Filho, A., Raabe, A., Dazzi, R.: An adaptive system architecture model for the study of logic and programming with learning paths. In: Proceedings of the 22nd International Conference on Enterprise Information Systems - Volume 1: ICEIS, pp. 679–690 (2020). https://doi.org/10.5220/0009412406790690.ISBN 978-989-758-423-7
43. Santos, A., Gomes, A., Mendes, A.: A taxonomy of exercises to support individual learning paths in initial programming learning. 2013 IEEE Frontiers in Education Conference (FIE) (2013). https://doi.org/10.1109/fie.2013.6684794
44. Haendchen Filho, A.: Um Framework do tipo Middleware para Sistemas Multi-Agentes na Internet (Doctoral dissertation, PUC-Rio) (2005)
45. Gonçalves, E.J.T.: Modelagem de arquiteturas internas de agentes de software utilizando a linguagem MAS-ML 2.0 (Doctoral dissertation, Dissertação de Mestrado. Universidade Estadual do Ceará. Centro de Ciência e Tecnologia. Fortaleza) (2009)
46. Filho, A., Thalheimer, J., Dazzi, R., Santos, V., Koehntopp, P.: Improving decision-making in virtual learning environments using a tracing tutor agent. In: Proceedings of the 21st International Conference on Enterprise Information Systems - Volume 1: ICEIS, pp. 600–607 (2019). https://doi.org/10.5220/0007744006000607.ISBN 978-989-758-372-8
47. Küster, T., Lützenberger, M., Heßler, A., Hirsch, B.: Integrating process modelling into multi-agent system engineering. Multiagent Grid Syst. **8**(1), 105–124 (2012). https://doi.org/10.3233/mgs-2012-0182
48. OMG Business Process Modeling Notation. "Version 1.0." OMG Final Adopted Specification. OMG (2006)
49. Esteves, A., Noschang, L., Raabe, A., Filho, A.: Portugol Studio: Em direção a uma comunidade aberta para pesq40uisa sobre o aprendizado de programação. In: Anais do XXVII Workshop sobre Educação em Computação, pp. 513–522. Porto Alegre: SBC (2019). https://doi.org/10.5753/wei.2019.6656

An Adaptive and Proactive Interface Agent for Interactivity and Decision-Making Improvement in a Collaborative Virtual Learning Environment

Aluizio Haendchen Filho[1](✉), Karize Viecelli[1], Hercules Antonio do Prado[2], Edilson Ferneda[2], Jeferson Thalheimer[1], and Anita Maria da Rocha Fernandes[1]

[1] Laboratory of Applied Intelligence, University of the Itajaí Valley (UNIVALI), Rua Uruguay, 458, Itajaí, SC 88302-901, Brazil

[2] Catholic University of Brasilia (UCB),, QS 07, Lote 01, EPCT, Taguatinga, Brasília, DF 71966-700, Brazil

Abstract. According to a report published by the Brazilian Association of Maintainers of Higher Education, the number of places offered in distance learning courses (aka distance education) in 2018 exceeded the places offered in classroom courses. At the same year, there was an increase of 20.6% in enrollments in distance education, with a forecast that exceeds classroom attendance by 2023. The same report showed concerns about the high dropout rates and low graduation rates in distance learning. The research indicates that the possible reasons for this are the students' feelings of loneliness, isolation, and demotivation. This paper presents an Interface Agent in the context of collaborative agents which main responsibility is to make the Virtual Learning Environment (VLE) more interactive, proactive, and able to adapt to the student's profile. This research is expected to contribute for making VLE more attractive, improve the teaching process, motivate students and, consequently, reduce dropout rates.

Keywords: Interface agent · Adaptive e-learning · Virtual learning environments · E-Learning

1 Introduction

According the National Institute of Educational Studies and Research (INEP), the remarkable expansion of Distance Learning (DL), in Brazil, has led this modality of higher education to surpass the presential courses in 2018 [20]. In 2019, the Brazilian Association of University Education Maintainers (ABMES) [3] has found that 30% of all vacancies in higher education were fulfilled by DL. It has found an increasing of 20.6% (from 7.8 million to 9.3 million) of enrollments across education modalities, and 78.7% (from 1.3 million to 2.3) in DL. Additionally, the INEP found an increasing of 51% in the number of vacancies offered in DL courses, from 2017 to 2018 [20]. In keeping this growth rate, ABMES [3] predicts that distance education will surpass presential university education by 2023.

J. Filipe et al. (Eds.): ICEIS 2020, LNBIP 417, pp. 612–634, 2021.
https://doi.org/10.1007/978-3-030-75418-1_28

However, there are major challenges. In the 6th Distance Learning Census, the Brazilian Association of Distance Learning (ABED) [2], pointed out a high dropout rate and a low graduation completion rate. According to ABMES [3], the indicators of the completion rate of distance learning students in 2016 was 35% and the dropout rate reached 62%, tending to increase. For Open University (UK), this is a globalized scenario, as internationally graduation rates would be close to 10% and tending to decline [27].

It was observed that the main reasons for high dropout rates are the feeling of demotivation and isolation of students [2]. From this perspective, the suggested actions to reduce these rates are similar for both Open University researchers and ABED and ABMES. They are unanimous in recommending the development of proactive motivational support from institutions for student retention. In contrast to this idea, it is clear that the vast majority of institutions are reactive, that is, they wait for students to contact them for help [27].

Some works found in literature focused on improvements on interaction and/or interactivity. Simbine et al. [23] implemented a model for visualization of student interactions based on their learning trajectory. By monitoring the way the student interacts in the VLE, the model generates graphical information with their interaction characteristics. Vaidya and Sajja [25] proposed an adaptive learning agent platform that produces and assesses learning contents. It applies analytical reasoning about content before and after presenting it to the student in VLE. Data related to student interaction monitoring is used for planning and organizing the content of the VLE.

The presented solution seeks for improving the interaction and interactivity actions in VLE using collaborative software agents, and proposes an agent that encompasses a set of characteristics recommended by ABED, Open University, and ABMES. These characteristics refer to more interactive environments, mainly adaptive and proactive. More adaptive and proactive environments can greatly contribute to mitigate the problems of demotivation, lack of involvement and feeling of isolation and abandonment by stimulating the student engagement. So, it was developed an intelligent software agent that operates in a collaborative organization of agents, playing the responsibilities defined in the role model and enabling the student to be protagonist on his/her own learning. Additionally, the agent can improve the decision-making process by providing visual information related to the student, tutor, and teacher activities.

Next, Sect. 2 presents the background, Sect. 3 discusses some related works, Sect. 4 explains the methodological approach, Sect. 5 discusses some preliminary results, and Sect. 6 concludes with some remarks and future work.

2 Background

In this section a description of MIDAS, the microservice-oriented platform, applied as the basis for the solution development is presented. Next, the data model, that relies on the Data Webhouse concepts (a kind of web Data Warehouse), is described. After that, the Case-Based Reasoning (CBR) technique, adopted for problem solving and knowledge acquisition, is described. Finally, the Felder and Silverman's profile model [9] is presented. This model is fundamental for dealing with the adaptability characteristics, once is based on the different ways' students learn.

2.1 Microservice-Oriented Multi-agent System

Despite the drawbacks reported about Service-Oriented Architecture (SOA), it remains. the best option available for system integration and leverage of legacy systems [14] due to its inherent ability to compose applications, processes, and assemble new functionalities from existing services. In the industry segment, the SOA approach has evolved as microservices, an architectural paradigm based on fine-grained and independent software components that interact to build highly scalable distributed systems [8].

Although not so spread in the industry realm [4], several principles for MAS development have been observed in the microservices model [8]. Several authors have studied the applications of microservices paradigm as a framework for building modern MAS, in an attempt to shorten this gap between industry and academic efforts [10, 14] To assure that microservices can meet these expectations, multiple specifications and standards have been proposed and created, and middleware products are becoming more robust [8]. MIDAS [13] is a platform that relies on microservices as the basis for development of MAS. Its architecture is composed by a front-end server called MIDAS Server (MS) and one or more agent containers here denominated MIDAS Container (MC), as shown in Fig. 1.

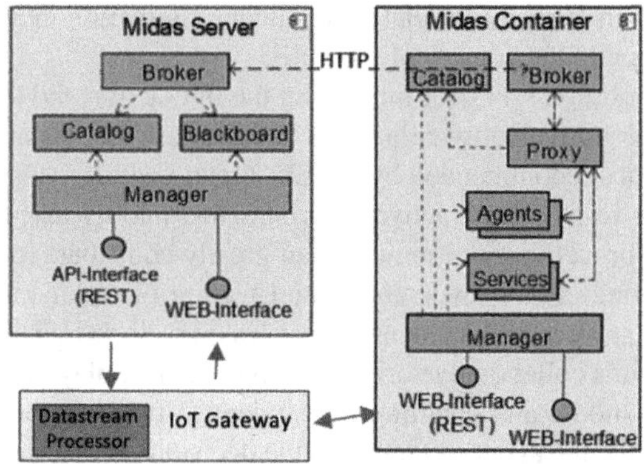

Fig. 1. MIDAS Generic Architecture [26].

MS is responsible for the platform integration rules, synchronizing the containers and interoperating with external applications. It contains four main interfaces: (i) an HTTP interface for intra-platform communication among MS and the MCs; (ii) a REST interface that allows communication with external applications; (iii) an interface for IoT interactions; and (iv) Web interface for human management and configuration.

MC is a lightweight container that houses software agents and/or microservices. It is capable of cataloguing its own interfaces services. The containers may register themselves on MS, exposing their services and agents, allowing for distributed collaboration with other containers within the same server domain.

IoT Gateway receives protocols such as MQTT, Bluetooth low energy, NFC, Wi-Fi, among others. The DataStream Processor performs data compression, storage, and provides data and real-time communication to the other layers.

Also, MS performs the integration on its synchronized containers, eliminating the complexity of service lookup and remote requests between containers. Application agents are instantiated in containers and developed by extending the abstract class, from which specific business rules for application can be implemented.

The middleware agents are located both in the MS and in the MC. They provide infrastructure services, playing the rules defined by the reference architecture [13], described next.

a) *Broker*: focuses on the architectural aspects related to the message transport - send/receive, pack/unpack, and managing exceptions - and translates agents and services request in HTTP streams.
b) *Proxy*: plays the role defined by the service-oriented model, which focuses on the architectural aspects related to the messages processing. It acts as a service provider representative, being responsible for the dynamic configuration and creation of instances. Dynamic configuration focuses on the capacity of redirecting messages to different providers during runtime, whenever the Catalog agent updates the resource-oriented model.
c) *Catalog*: responsible for aspects related to the resources concept of a resource-oriented model [13]. A resource description is a machine runnable metadata representation that makes possible for a human or software program to locate services and agents within the ecosystem.
d) *Manager*: play the roles defined at the management and policy levels. It involves a set of tasks that enable the control over the platform, such as the life cycle management, communication control checking activities, statistics, and GUI wizards.
e) *Blackboard*: takes responsibility for information exchange in symbolic cognitive MAS, and mainly for asynchronous communication among agents. Its structure follows the basic blackboard pattern: the knowledge sources represent the agents, the data structure is visible to all agents, and the controller is responsible for notifying the agents about the changes in the environment. When a MC is running in stand-alone mode, it has a local Blackboard agent that deals with intra-container communication. When a MC connects to a MS, each local Blackboard of each MC synchronizes with the MS Blackboard in order to provide transparent communication within the whole ecosystem.

The introduction of the agent concept to play these roles complies with the current tendency and non-functional requirements for microservices-oriented architectures: flexibility, dynamic behavior, pro-activity, and adaptability. They abstract the standard code and the complexity required to implement characteristics, such as communication, concurrency control, lifecycle management, services discovery and interoperability, enabling the developer to focus only in specific business rules of the application.

2.2 Data Webhouse

By the end of nineties, when the wide spreading of Internet started, researchers from the field Data Warehouse (DW) focused on studying data transmission and handling in the Internet, eventually adopting a DW. This way, a new field of study, known as

Data Webhouse (DWH), emerged sharing the same principles of DW but rather with a different architecture that includes Internet. The Web allows recording practically all behavioral actions of the user in a single click [21]. It means that one can capture not only the page accessed but also navigability information. The recording of all interactions made by anyone via application or web site, is called clickstream. Activities carried out by the user such as clicks, form filling, and others, create conditions for analysis, profile identifications, preferences and trends of each particular user.

Dimensional modelling is a discipline that seeks to model data for the purposes of understandability and performance. All dimensional models rely on the concept of measured facts. The Facts table, represented by the entity Clickstreams, stores users clicks on the VLE. The dimensions relate to the entities that serve as perspectives of analysis in any subject of the model.

Snowflaking is a method of normalizing the dimension tables in a star schema [16]. When the dimensional model is normalized along the dimension tables, the resulting structure resembles a snowflake with the fact table in the center. The principle behind snowflaking is the normalization of the dimension tables by removing low cardinality attributes and forming separate tables [21]. Figure 2 shows a very simple example of a dimensional snowflake model for a DWH.

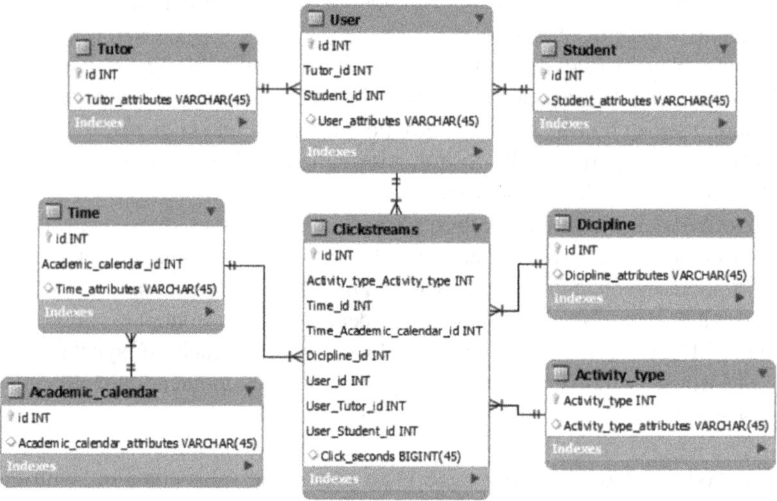

Fig. 2. Simple dimensional snowflake DWH example for a VLE.

User, Discipline, Time, and Activity are the dimensions connected to the Clickstreams fact table. *User* dimension has two snowflake tables (Student and Tutor) connected as example, among the different representations it can assume. VLE administrators, teachers, and multidisciplinary EAD team, among others, are examples of entities that can interact with the platform. The *Time* dimension represents the exact click time that will be recorded in the fact table and has the academic calendar connected. The *Discipline* table represents the content involved in each discipline. Actually, there are other tables connected in the Discipline snowflake, such as course, module, class, and so on. The *Activity_type* dimension represents the different types of activities that the user can perform, depending on their permission level.

Besides manipulating information and discovering knowledge, a VLE needs to be prepared to react immediately to students' actions in the environment, reducing the time between the occurrence of an event and the execution of an action [22]. This is called Zero Latency Enterprise (ZLE) strategy and uses a DWH integrated with other Business Intelligence tools to deliver real-time, zero-latency information for much faster decision making.

2.3 Case-Based Reasoning

Case-Based Reasoning (CBR) is an Artificial Intelligence technique for problem solving and knowledge acquisition based on the principle that "similar problems have similar solutions" [1]. The use of CBR methodology and its application in VLE is based on a broad cognitive theory that involves the process of remembering, as a problem-solving phenomenon, and the process of reusing past episodes to solve new problems, that corresponds to a frequent and powerful way of human reasoning.

The basic elements of a CBR system are: (i) knowledge representation, carried out by means of concrete experiences; (ii) similarity measure, which looks for similar situations for the current problem in a knowledge base; (iii) adaptation, where past situations not identical to the current problem can be adapted to find a suitable solution for the new one; and (iv) learning, which occurs every time a case is resolved and a new experience is retained and integrated into the knowledge base. A conceptual model for the CBR cycle (Fig. 3) was proposed by Aamodt and Plaza [1].

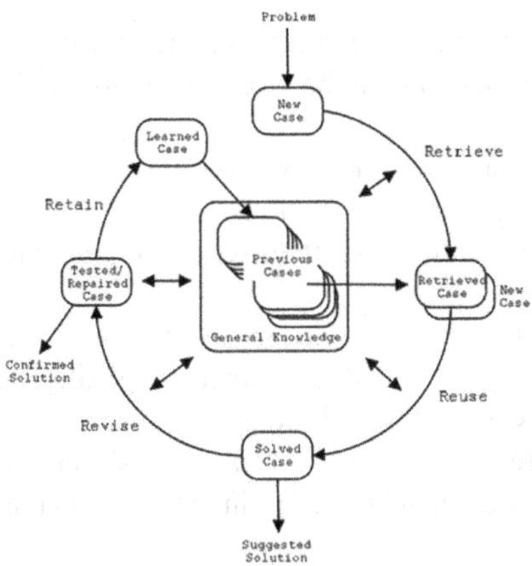

Fig. 3. Cycle of case-based reasoning [1].

A CBR system encompasses a continuous cycle of reasoning, consisting of four main tasks: (i) recovering the most similar cases from the case base, in which the goal is to find a case or a small set of cases in the base that contains a problem description near to the current problem or situation; (ii) reusing these cases to solve the problem; (iii) review the proposed solution in order to transfer it to the present situation; if necessary, the recovered solution can be adapted to fully meet the requirements of the present situation; and (iv) retaining the experience represented by the current case (or parts of that experience) for future reuse.

In CBR approach, knowledge maintenance is simplified by the ability to learn new information in the form of cases. Other advantage is the fast response time and the ability to work in domains that are not completely known. These features enable its application in many types of tasks such as diagnostic systems, help desk systems, evaluation systems, decision support systems, and project systems [18].

2.4 Felder-Silverman Model

Felder and Silverman [9] developed a theory that states there is difference in the way students learn: seeing or hearing; reflecting or acting; reasoning logically or intuitively; memorizing, visualizing, drawing analogies or building mathematical models; steadily or not. The authors mapped learning styles and created a questionnaire entitled Index of Learning Styles (ILS) based on them. The structure of ILS is shown in Table 1, in where two dimensions - Learning Style and Teaching Style – are depicted. Each dimension includes four types of learning / teaching styles, and each type has two features. This model has been widely used to classify profiles [10].

The features identified by ILS for learning style are:

a) *Active*: tend to understand and retain information if they can turn that knowledge into action.
b) *Reflective*: prefer to think about information before acting and tend to enjoy working alone.
c) *Intuitive*: they like innovations, but not repetitions; they may be better at understanding new concepts and tend to be more innovative and work faster than sensory students.
d) *Sensory*: they like to learn facts and solve problems by established hands-on methods and don't like surprises and complications.
e) *Visual*: they easily remember what they see such as movies, photos, diagrams and demonstrations.
f) *Verbal*: acquire and assimilate knowledge based on written and spoken explanations.
g) *Sequential*: tend to gain knowledge in linear, logically interconnected steps, and follow step-by-step ways to find solutions.
h) *Global*: use to acquire knowledge in a systemic fashion, considering the singularity of each component under an integrated point of view and their inherent relationships.

For the ILS teaching style, there are the following features:

a) *Concrete*: explore resources of sight, sound, physical sensations.
b) *Abstract*: use possibilities, hunches, perceptions
c) *Visual*: use figures, diagrams, films, demonstrations.
d) *Verbal*: lectures, readings, discussions.

Table 1. Learning dimensions [10].

Learning Style preference		Corresponding Teaching Style	
Sensory Intuitive	Perception	Concrete Abstract	Content
Visual Verbal	Fundraising	Visual Verbal	Presentation
Active Reflective	Processing	Active Passive	Student Participation
Sequential Global	Understanding	Sequential Global	Perspective

e) *Active*: phenomena that lead to principles.
f) *Passive*: principles that lead to phenomena.
g) *Sequential*: students speak, move, reflect.
h) *Global*: students watch and listen.

These features are used to characterize the students. Based on this assessment, according to the answers provided, an index is calculated that establishes the predominant dimension present in each profile. The index ranges from 1 to 11 and represent the intensity of the categories.

3 Literature Review

A systematic review of literature was carried out to identify the main techniques applied in VLE (particularly by means of interface agents or adaptive e-Learning) from 2012 to 2019. The search strings used (Interface AND agent* AND adaptive AND ("Virtual learning" OR e-Learning)). Results are as follow:

Repository	Number of Papers
Science Direct	399
IEEE Explore Digital Library	397
Brazilian Symposium on Informatics in Education	14
Brazilian Journal of Informatics in Education	17

Next, the following exclusion criteria were applied:

- Short papers (expanded abstracts);
- Papers that do not address the performance of intelligent agents;
- Lack of adherence of Title, Abstract, Introduction or Conclusions to the paper focus;

After applying the exclusion criteria, seven studies remained and are summarized below.

3.1 Reported Works

Simbine [23] proposed a model for student interaction visualization on the basis of their learning trajectory, along with an interactive visualization system of learning trajectories in VLE. Thus, it generated a model of data collection, visualization and analysis in the form of graphs, according to the characteristics of student interaction. By analyzing these data, it was possible to verify the order of access of students' interaction with existing content, which can be used to improve the organization of educational content in the VLE. The model presents the interaction in two ways: guided and open to the user. Also, analytical reports provide the sequence of student interaction. They can collaborate in planning educational activities.

Referring to the issue of interaction with students in VLE, Maciel et al. [19] propose a virtual assistant integrated with the Moodle environment in order to offer daily support to the academic activities of distance learning students. This wizard exposes the content orally through a visual avatar, making it more interesting for students. Besides, it allows the human tutor to contact the student through this avatar, sending messages in BackOffice. The author points out some limitations in the solution that has impacted the results, mainly the absence of a responsive GUI, which allows the student to access from a device of his choice (e.g., a mobile device).

Regarding the adaptability of a VLE, Vaidya and Sajja [25] proposed an agent-based system for collaborative learning environment in an educational habitat. The approach provides an agent that not only offers the student learning facilities, but also calibrates content and learning outcomes. The solution evaluates the knowledge, behavior, pedagogy, and activities carried out by the student in VLE. Thus, based on the selected preferences, learning style, and pedagogy adopted in the environment, the system is capable of producing dynamic content for students.

Dorça [7] presents a probabilistic approach using reinforcement learning, in which a dynamic, interactive and gradually updated student model is implemented through a stochastic process. Model updating occurs based on information about student performance within the learning environment. This approach adopts the ILS, presenting a probabilistic model based on the combination of ILS Learning Styles (LS). Reinforced machine learning is applied to create a dynamic and interactive model for the student's style that is gradually updated by means of a stochastic process. This model takes into account the uncertainties in relation to the student's LS during the course and seeks to correct dynamically the inconsistencies in the model.

Zapparolli et al. [28] developed a tool called FAG that uses Business Intelligence and Learning Analytics techniques to assist in knowledge management and decision support in a VLE. The tool provides analytical and consolidated reports with cross-sectional and systemic views, considering all virtual classrooms and contexts of a specific teacher. It enables corrective actions to be taken, ensuring quality work and preventing dropout risk. FAG also offers the manager decision support in the VLE, with reports on the actions taken by the teachers. For example, the actions related to a procedure for clarifying doubts, and interactions between teacher and student in the VLE can be evaluated.

Dantas [5] presents a system to assist the teacher for recommending pedagogical strategies in virtual learning environments. The system considers characteristics of Multiple Intelligences for each student detected during the cognitive process. It also includes

CBR techniques in order to develop new cases, using the information stored in the student model, with respect to these characteristics.

In this sense, do Nascimento et al. [6] apply CBR in order to suggest a pedagogical action for a student-learning problem. The work emphasizes problem solving based on analogy, being focused on code. This system also uses learning objects to support pedagogical actions aimed at facilitating the understanding of complex concepts of the Introduction to Programming discipline. It uses the RBC technique to suggest a pedagogical action for a given student's learning problem. The research seeks to fill the deficiency in the learning of logical mathematical reasoning. To make easy the understanding of complex concepts of the Introduction to Programming the author uses ILS to support pedagogical actions.

Although the many efforts to improve interaction, interactivity, and decision-making processes in VLE, none of them provide an integrated platform with the most relevant functionalities, as proposed in the present work. It is also important to note that neither approach adopts a proactive procedure, they are all reactive.

4 Specification and Design

In the specification and design level, the research is carried out by designing the system architecture, the data model, and specification of the IAg role model. The section is closed with an example of the specification of the IAg services workflow.

4.1 Designing the System Architecture

The architecture comprises the following modules: (i) a microservice-oriented platform for MAS development and management [15]; (ii) a DWH structure for data storage and retrieving; (iii) an organization of collaborative agents, instantiated on the platform; and (iv) the Moodle virtual environment, used as case study. The platform facilitates the development of agents, providing communication, management, and database access, as described in Sect. 2. Figure 4 shows the generic architecture, in which the IAg is included with the other collaborative agents.

The following collaborative agents are instantiated in the organization VLE MAS:

a) *Interface Agent (IAg)*: accounts for the communication among platform and the environment, performing interactions with human actors and interactivity with other agents, login procedures, application of the ILS questionnaire, among other responsibilities.
b) *Tracing Agent (TAg)*: responsible for storing and managing the data structure of the Dimensional Model. This involves populating the dimensional tables and registering the clickstreams. The Clickstreams fact table contains logs extracted from the Moodle, and are stored in the DWH.
c) *Pedagogical Agent (PAg)*: responsible for content management, creation of adaptive learning trails and objects, and for providing resources for motivational and proactive procedures.

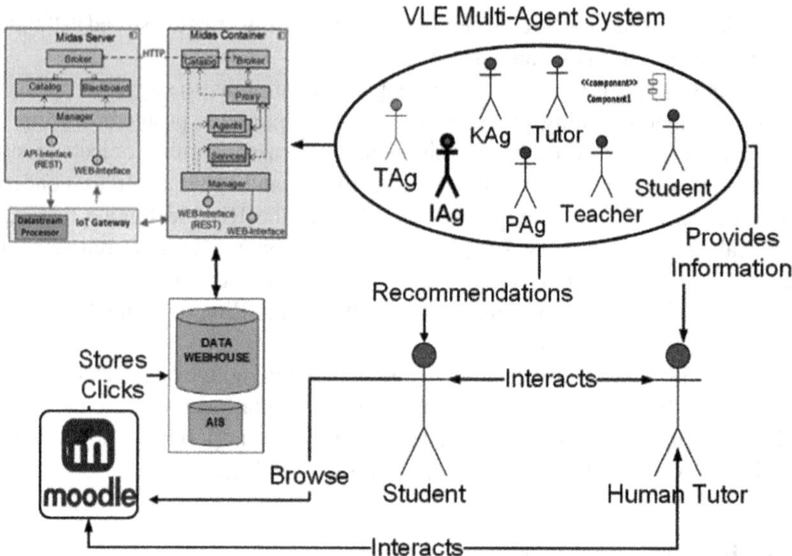

Fig. 4. Generic architecture [26].

d) *Knowledge Agent (KAg)*: manages AI techniques to perform predictions and prescriptions using the data storage in the Dimensional Model.
e) *Student*, *Tutor* and *Teacher*: represent virtual instances of these human actors.

The databases are represented by the Academic Information System (AIS) and the DWH. The AIS contains academic data, such as student profile and history, data from tutors, teachers, discipline/courses, and so on. The DWH is a dimensional model described in previous section.

Listeners are placed in relevant spots in the interface, waiting for the clicks triggering the script to store the information. Locations do not necessarily have to be on the links since data can be stored with simple interactions.

4.2 Designing the Data Model

The information collection, transformation, integration, analysis and distribution, and the knowledge extraction are not easy tasks. So, for reaching more efficiency, the DW can be combined with Business Intelligence (BI), Online Analytical Processing (OLAP), and Data Mining.

Figure 5 shows the data structure stored in the dimensional model. A central fact table is connected with the dimensions relevant to the context of the VLE. The model has a central fact table called Clickstream connected to five dimensional tables: User, Session, Discipline, Time and Activity. Each dimension has several aggregated tables, in the snowflake style. The model consists of set of dimension tables linked to a central fact table, named *ClickStream*. The model tables are described below:

(i) *ClickStream*: stores the users' clicks with information extracted from the Moodle log file associated with the five primary keys of the dimensions.
(ii) *User*: can be a student, a teacher, a tutor or an administrator, among others.
(iii) *Time*: aggregate dimensions, *Calendar*, includes day, quarter, semester, etc., and *Academic_calendar*, identifies holidays, seasons, academic actions, events, etc.
(iv) *Discipline*: contains all information related to a discipline and may include class hours, credits, opening and closing dates, and so on; it aggregates tables *Course* and *Module.*
(v) *Session*: is the set of actions taken by a user while it is continuously connected to the VLE.
(vi) *Activity*: related to the activities the user performs in the VLE; it was modeled with several tables in the style of snowflakes, such as video lesson, participating in a forum, chat, and quiz.

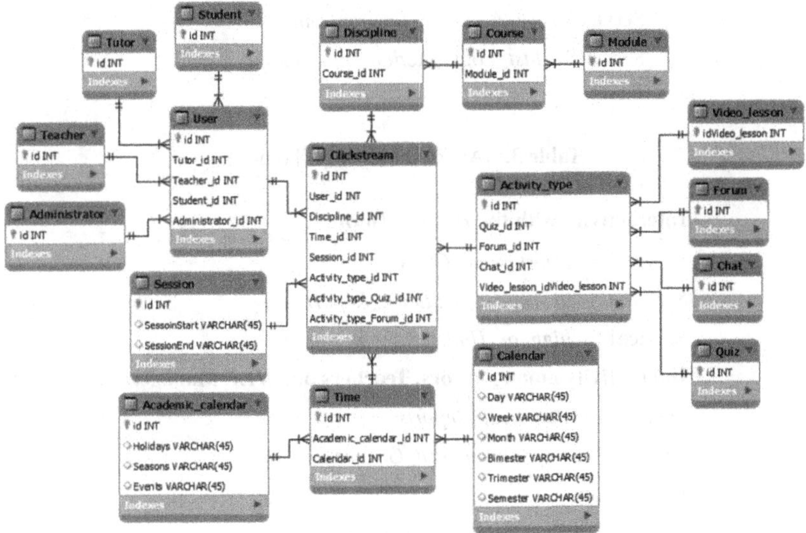

Fig. 5. VLE dimensional model.

4.3 Specification of IAg Role Model

The role model [12, 13] provides a summary of software agents behavior. A role can be described by means of responsibilities, that indicate functionality or service. Interaction and interactivity are key concepts in the IAg role model (Table 2). Interactivity can be defined as the communication human-machine and refers to a mediated environment in where participants can communicate synchronous or asynchronously and participate in reciprocal message exchanges [17]. Interaction occurs among same nature entities: human-human or machine-machine. Below, the agent´s responsibilities shown in Fig. 2 are described.

Table 2. IAg role model [26].

ORGANIZATION: VLE MULTAGENT SYSTEM
AGENT: INTERFACE AGENT
Login Procedures
Service1 – *system_Login*
Service2 – *apply_ILS_Questioner*
Service3 – *create_Interactive_Message*
Knowledge Representation
Service4 – *doubts_Registration*
Service5 – *lookup_Cases*
Interactivity with the Student
Service6 – *track_Progress_Notification*
Service7 – *set_Adaptive_Environment*
Service8 – *last_Login_Action_Information*

Table 3. IAg Role Model [24] (cont.).

Interactivity with the Human Tutor
Service9 – *busiest_Time_Class*
Service10 – *absence_Attendance_Report*
Service11 – *high_or_Underperformer_Info*
Interactivity among Tutors, Teachers and VLE Managers
Service12 – *statistical_Reports*
Service13 – *olap_Dashboard_Generation*
Interaction
Service14 – *send_Student_ILS_Profile*
Service15 – *missing_Student_Notification*
Service16 – *chat_Forums_Invitations*
Service17 – *help_Low-frequency_Students*

I. Login Procedures.
IAg has three services under this responsibility:

(i) *system_Login*: check login procedures and the permissions of each user.
(ii) *apply_ILS_Questionnaire*: occurs when the C student's first login occurs. In this procedure, he works in collaboration with the PAg. According to the information

collected in the questionnaire, the student is inserted in one of the profiles provided for in the Felder-Silverman model.

(iii) *create_Interactive_Message*: each time a student logs in, this service is executed and recovers and exhibits the set of information about their last access.

II. Knowledge Representation.

In representing knowledge, its responsibilities are:

(i) *doubts_Registration*: collects and records new cases and questions.
(ii) *lookup_Cases*: is the ability to consult via (KAg) the most similar cases that can be reused for a specific problem, allowing the student, the tutor or the teacher to consult on cases that have already been resolved.

III. Interactivity with the Student.

The interaction with students occurs by means of:

(i) *track_Progress_Notification*: generates notification on the learning track position after the last access.
(ii) *set_Adaptive_Environment*: for fresh students, the IAg applies the ILS questionnaire, sends a welcoming message, and forwards him/her to an adapted environment.
(iii) *last_Login_Action_Information*: maintains a proactive instance, informing the student on their latest actions in the discipline.

IV. Interactivity with the Human Tutor.

The responsibilities for enabling the tutor to interact with students by sending notifications with relevant information include:

(i) *busiest_Time_Class*: allows the tutor and teacher to interact in real time with students by means of chats, forums and any other activity that can motivate interaction with groups.
(ii) *absence_Attendance_Report*: generates notification of absence list with frequency rates (low/high) in the discipline.
(iii) *high_or_Underperformer_Info*: generates notifications on students with high or low performance in the discipline.

V. Interactivity among Tutors, Teachers and VLE Managers.

The services related to this responsibility are:

(i) *statistical_Reports*: provides analytical and consolidated reports with cross-sectional and systemic views, considering all dimensions and contexts of a specific classes and students.
(ii) *olap_Dashboard_Generation*: provides information to VLE tutors, teachers, and managers decision-making.

VI. Interaction.

This responsibility involves interaction among IAg and other collaborative agents and includes the services:

(i) *send_Student_ILS_Profile*: means that IAg collects information from the ILS questionnaire, scores the answers, and identifies the student profile in one of the seven possible Felder-Silverman Model categories. With this information, PAg can define which learning paths best fit this profile.
(ii) *missing_Student_Notification*: sends notification to students that has not logged in for a while in order to stablish a dialog to help on possible difficulties. Agent interactions can occur synchronously or asynchronously. In asynchronous mode, Blackboard is used as a mediator, as will be shown in the following section.
(iii) *chat_Forums_Invitations*: invites the student when logging in to participate in an ongoing chat or forum, groups or chat with tutors.
(iv) *help_Low-frequency_Students*: sends a message inviting students to return to the environment, schedules a conversation with a tutor or registers a request for help.

4.4 Agent Services Workflow Specification

The Business Process Model and Notation (BPMN) for process diagram (orchestration) (applying the HEFLO tool) was used to specify IAg's responsibilities. BPMN diagramming is intuitive and allows the representation of complex process details as a standard language. As presented, it is easy to understand the message flow, and the asynchronous communication system between agents via Blackboard. Synchronous communications are carried out via Proxy, ensuring the requirement of low coupling between services.

Figure 6 presents a workflow for the Service1 (*system_Login*). Links and messages describe how the interaction and collaboration among agents and other actors occurs. In the horizontal swimlanes the actors that participate in the task are represented: Tracking Agent, Blackboard, Interface Agent, Pedagogical Agent and the Moodle Environment in the right corner of the first swimlane. The workflow starts after the student logs in to the Moodle environment.

After logon, TAg runs the *profile_Identify* service, checking if there exists a link profile. In this case, TAg runs a *last_Actions* service, in order to retrieve data from its last login. After that, the service *create_Interactive_Message* sends the data to IAg, that begins an interactive task. It performs the *create_Welcome_Message* service that informs the student its last login and the executed activities. It also interacts with the student to take it directly to the point where he previously stopped.

However, if the student does not have a profile linked to his/her account, TAg writes on Blackboard (BB) that it is necessary to apply the questionnaire for this student, and BB generates a notification for IAg to apply the questionnaire. Then, IAg begins applies ILS questionnaire to rate the student profile. When finished, it posts a message on the BB, notifying TAg for storing the questionnaire data. After reading the message on BB and storing the profile data, TAg writes on the BB a message for the PAg informing that there is a new student with a defined profile. Thus, PAg will be able to offer this new student the layout and the most appropriate track for his/her learning. The IAg now

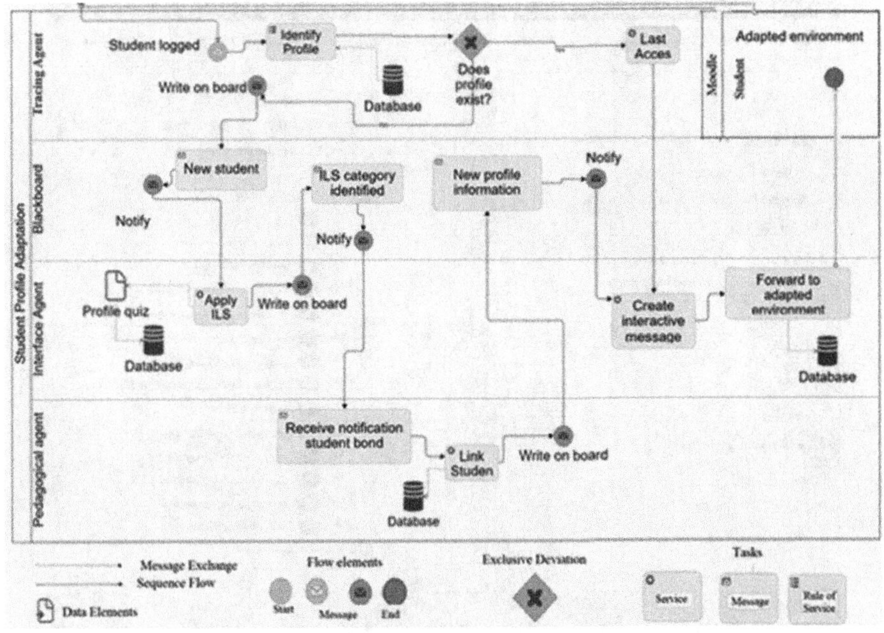

Fig. 6. Workflow for the Service1 - *system_Login* [24].

performs an interactive welcoming message for the student and forwards his/her to its adapted environment.

5 Implementation

The implementation initially shows how agents were instantiated on the platform. Following, it is described how the dimensional model is populated with data obtained from the Moodle log tables and the implementation of BI interfaces. Finally, the CBR techniques applied to recover suggestions for doubts clearing are depicted.

5.1 Implementation of the Agents

The VLE system was instantiated and implemented on the MIDAS platform. It allows the development of multi-agent systems using microservices as a base. The platform structure is composed of a main server and one or more containers of agents, as described in Subsect. 2.1. The platform provides a GUI wizard to assist in global management tasks, as shown in Fig. 7. All agents that make up the VLE can be seen in the panel on the right side.

The panel shows a navigable resource hierarchy organized by agents. When an agent is selected, details are displayed in the Details panel on the left side of the window. At the bottom of the window, the Server Log panel shows details of all transactions being executed. In the Containers panel on the top left, you can see the two containers registered on the platform and used in this domain: the instantiated VLE and the Academic

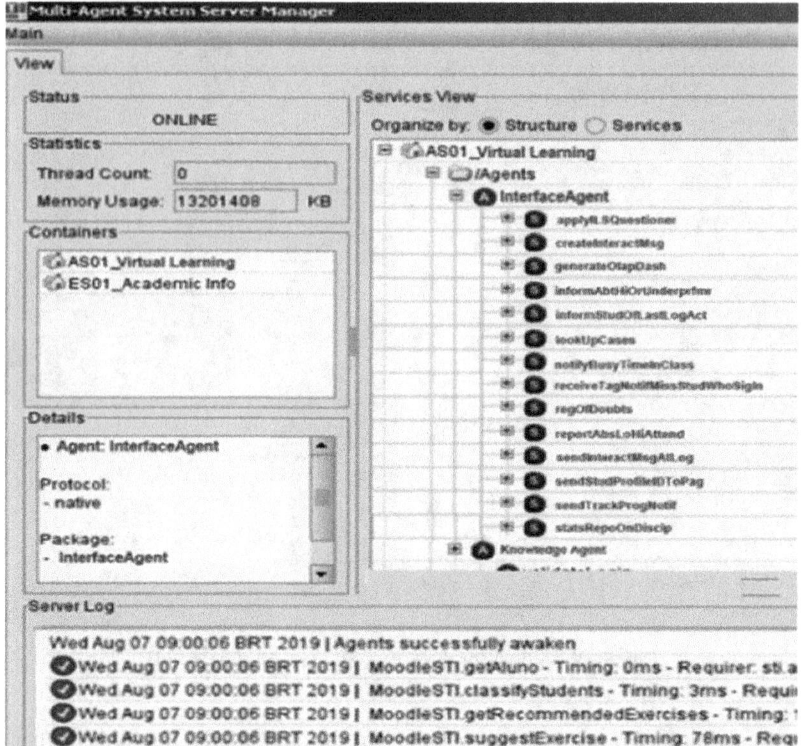

Fig. 7. GUI wizard to view instantiated agents and its services [24].

Information System, which works integrated with the VLE, from which the data for the DW is extracted.

The details of the Interface Agent in the right-hand panel show the services it provides. The + sign in front of the services indicates that these services are broken down into microservices at a lower level of detail, as previously shown.

5.2 Populating Database and Development of BI Interfaces

Moodle stores the activities performed by the users and their interactions in a relational database. This data structure is composed by approximately 422 interrelated tables. The fact table in the dimensional model is populate with data obtained of these tables. Specific data, such as the profile of students, teachers, tutors, as well as the student's academic record, are updated periodically from DWH to Academic Information System (AIS) and conversely. It is necessary to adapt the integration module to the type of AIS used by the institution.

Populating the dimensional model involves three stages: (i) to select the tables of the relational model to compose each dimension; (ii) when appliable, to normalize the data on the selected tables to populate the dimensions; and (iii) after populating all dimensions, the facts table is populated with the clickstreams (see Subsect. 4.2). After completion of these processes, the tables are structured in an integrated and concise model. This enables the obtention of strategic information from the data analysis under different views, and the ability to drill-up and drill-down on the different levels of granularities. It is important to notice that this model is able to cope with the data heterogeneity obtained from disparate sources.

BI allows looking at the organization as a whole, looking for points that can be used to obtain knowledge and competitive advantages in DL. This is how administrators, tutors, and teachers obtain knowledge to make decisions in order to anticipate changes and actions. Many alternatives for visualizing are available and the related fields can be selected in the side menu, as shown in Fig. 8.

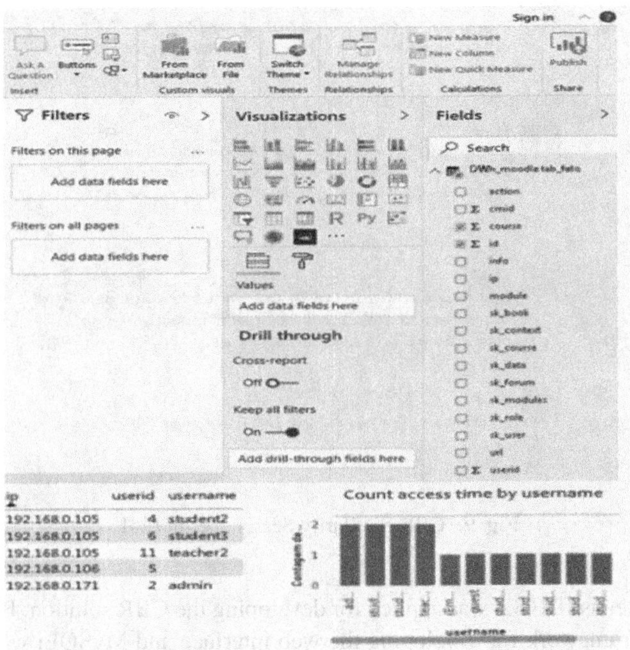

Fig. 8. Power BI filters and the partial view of a query [24].

The Power BI JavaScript API provides bidirectional communication between Power BI reports and the application. The JavaScript API enables easier embedding of reports into applications and to programmatically interact with those reports so that the applications and the reports are more integrated.

5.3 Clearing Doubts with CBR

Knowledge representation was implemented with the development of a base of questions, previously fed, that allows the student to carry out research. This part of the solution was developed using CBR techniques and rules, applying similarity-based recover procedure for answers retrieval. Figure 9 shows the results of a search, presenting several questions organized by degree of similarity to the searched subject.

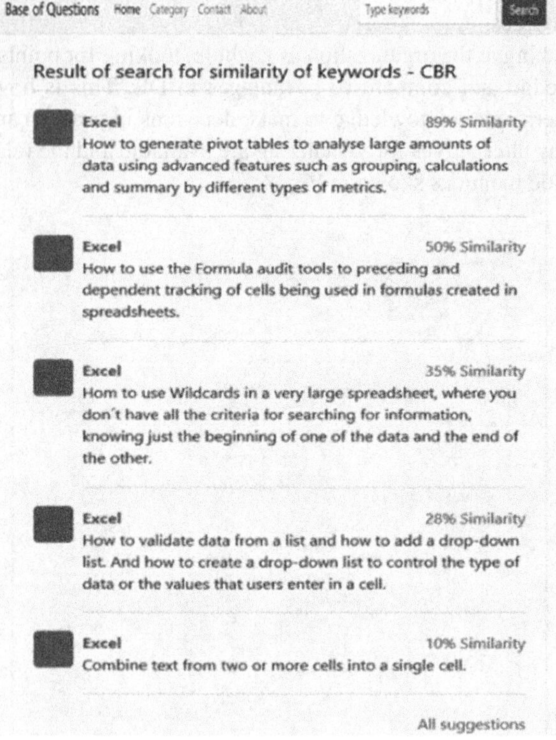

Fig. 9. CBR Similarity Search Result [24].

The NetBeans IDE 8.1 was applied for developing the CBR solution. Bootstrap was chosen as a framework for developing the web interface and MySQLi was adopted as the database management system.

The results achieved so far have shown to be adequate to the proposal of this work. Beyond its proactive nature, the solution offers yet another tool for student learning, by providing a knowledge base that will assist the course with several classes of students. The recovery, adaptation, and learning of registered cases are still under development.

6 Discussion

This work is motivated by suggestions from Open University (UK), Brazilian Association of Distance Learning Maintainers [3], Brazilian Association of Distance Learning [2], and the previously cited authors, beyond other researchers that unanimously recommend the development of proactive solutions for VLE. The literature shows that the vast majority of existing solutions do not meet this requirement, or only partially.

The solution characteristics were compared to the most known VLE approaches for interface agents. The following characteristics were considered to classify the related works:

(i) *Artificial Intelligence* (AI): does the system use AI techniques?
(ii) *Data Model* (DM): is there a data model defined?
(iii) *Interaction Model* (IN): is there interaction among agents?
(iv) *Interactivity* (IT): is there interactivity among stakeholders?
(v) *Proactivity* (PA): does the system use proactive resources to interact with stakeholders?
(vi) *Adaptability* (AD): is the system adaptable to the user's profile?
(vii) *Software Agents* (SA): does the system use software agents technology?
(viii) *BI reports* (BI): does the system offer BI capabilities?
(ix) *ILS questionnaire* (ILS): does it use the ILS questionnaire to classify the student's teaching profile?
(x) *Learning trails* (LT): does it use learning trails to adapt the interface paths to the student's profile?

Table 3 presents the results comparison

Table 3. Comparing the IAg characteristics with related works.

	AI	DM	IN	IT	PA	AD	SA	BI	ILS	LT
Simbine [22]	**yes**	**yes**	no	no	no	**yes**	no	no	no	no
Maciel et al. [19]	no	no	**yes**	**yes**	no	no	no	no	no	no
Vaidya & Sajja [23]	**yes**	no	no	**yes**	**yes**	**yes**	**yes**	no	no	**yes**
Dorça [7]	**yes**	**yes**	**yes**	no	**yes**	**yes**	no	no	**yes**	no
Zapparolli [26]	no	no	**yes**	no	no	no	no	**yes**	no	no
Dantas [5]	yes	yes	no	no	no	no	no	no	no	no
do Nascimento [6]	**yes**	**yes**	no	**yes**	no	no	no	no	no	no
Our approach	**yes**	**yes**	**yes**	**yes**	**yes**	**yes**	**yes**	**yes**	**yes**	**yes**

The most common characteristics are AI, DM, IN, and IT. The remaining characteristic are shared by a small set of solutions. It is important to notice that Proactivity, the most relevant missing characteristic according the literature, is satisfied by only two of the previous approaches. Moreover, AD, AS, BI, ILS, LT, that are barely present in the previous approaches are also in the present proposal.

It was not found in the literature a solution that includes all functions incorporated in the proposed solution. The approaches that have the greatest number of functions are Vaidya, Sajja [23] and Dorça [7].

The present approach encompasses a broader set of functions in a single approach, offering a complete and comprehensive proposal. In addition, it offers an environment composed by a set of collaborative agents which aims to create a virtual environment for proactive learning, in contrast to all approaches that act in a reactive way.

7 Conclusions and Future Works

As reported by the Brazilian Association of Maintainers of Higher Education [3], current virtual learning environments are predominantly reactive. This characteristic is identified as one of the reasons for students' demotivation and feeling of abandonment, and possible cause for the high dropout rate in higher distance education modality. In view of the great progress in distance education modality, the main challenges involved are increasing students' permanency and motivation in order to complete their graduation. Such benefits can be addressed by more adaptable environments to student profile, that can bring large-scale interactivity and proactivity. Dissonance and dissatisfaction among stakeholders, especially students, are well-known consequences of high unrealistically technology expectations [11].

The solution proposed in this work, based on the needs of the educational market, aims to meet students' expectation and generation of adequate levels of motivation and satisfaction. This solution includes proactive characteristics as adaptation, interactivity, and interaction in order to promote a strong sense of satisfaction among students. This adaptative interface enables the creation of learning objects based on the student's profile, that was characterized by means the ILS questionnaire. It was developed in the Applied Intelligence Laboratory at University of Vale do Itajaí over a platform for building multi-agent systems, that has been successfully applied to implement collaborative agents systems [15].

Considering that this solution was already applied in a controlled environment, the next step in this research is to deploy it a real-world situation. By such way, it will be possible to observe its results, aiming to improve the system.

References

1. Aamodt, A., Plaza, E.: Case-Based reasoning: Foundational issues, methodological variations, and system approaches. AI Commun. **7**(1), 39–59 (1994). https://doi.org/10.3233/AIC-1994-7104
2. Brazilian Association of Distance Learning – ABED: Census EAD.BR 2018 Analytical report on distance learning in Brazil (2019). https://abed.org.br/arquivos/CENSO_DIGITAL_EAD_2018_INGLES.pdf
3. Brazilian Association of Higher Education Maintainers – ABMES: Distance learning should surpass the face-to-face in a few years Estimate predicts that the recent method exceeds the traditional in 2023. Brazilian Higher Education Census (2018). https://abmes.org.br/noticias/detalhe/2991
4. Collier, R., Russell, S., Lillis, D.: Reflecting on Agent Programming with AgentSpeak(L). In: Chen, Qingliang, Torroni, Paolo, Villata, Serena, Hsu, Jane, Omicini, Andrea (eds.) PRIMA 2015. LNCS (LNAI), vol. 9387, pp. 351–366. Springer, Cham (2015). https://doi.org/10.1007/978-3-319-25524-8_22

5. Dantas, A.C., de Melo, S., Fernandes, M., Lima, L., do Nascimento, M.Z.: Recommendation of pedagogical strategies by means of emotions, personality profiles, and multiple intelligences using case-based reasoning. In: Brazilian Symposium on Informatics in Education (SBIE 2018). pp. 1213–1222 (2018). https://doi.org/https://doi.org/10.5753/cbie.sbie.2018.1213. (In Portuguese)
6. do Nascimento PB, de Oliveira EHT, da Silva FS, Ramos IMM, Ramos DB, : System for Recommendation of Pedagogical Action using Case-Based Reasoning in Teaching Introduction to Programming. Nuevas Ideas en Informática Educativa **12**, 624–629 (2016)
7. Dorça, F.A.: A stochastic approach based on reinforcement learning for automatic and dynamic modeling of student learning styles in adaptive and intelligent systems for distance education. PhD Thesis, Universidade Federal de Uberlândia, Brazil. https://repositorio.ufu.br/handle/123456789/14314(2012). (in Portuguese)
8. Dragoni, N., et al.: Microservices: Yesterday, today, and tomorrow. In: Meyer, B. (ed.) Mazzara M, pp. 195–216. Present and Ulterior Software Engineering, Springer International Publishing (2017)
9. Felder, R., Silverman, L.: Learning and teaching styles in engineering education. Eng. Educ. **78**(7), 674–681 (1988)
10. Freitas, R.V., Dornellas, A.A., Belhot, D.V.: Production engineering student professional requirements: a view through learning styles. Revista GEPROS **1**(2), 125–135 (2006). (In Portuguese)
11. García-Álvarez, M.T., Novo-Corti, I., Varela-Candamio, L.: The effects of social networks on the assessment of virtual learning environments: a study for social sciences degrees. Telemat. Inf. **35**(4), 1005–1017 (2018). https://doi.org/10.1016/j.tele.2017.09.013
12. Gonçalves, E.J.T.: Modeling internal architectures of software agents using the language MAS-ML 2.0. MSc Thesis, State University of Ceará, Brazil (2009). https://siduece.uece.br/siduece/trabalhoAcademicoPublico.jsf?id=57200. (in Portuguese)
13. Haendchen Filho A (2017) Sistemas Multiagentes na Internet: Uma arquitetura orientada a serviços para o desenvolvimento de agentes inteligentes de software na Internet. Novas Edições Acadêmicas.
14. Haendchen Filho A, do Prado HA, Ferneda, E (2015) A Resource-Oriented Model for Services Management and Discovering in SOA. In: IEEE International Conference on Services Computing, pp. 782–787.
15. Haendchen Filho, A., Thalheimer, J.M., Dazzi, R.L.S., Santos, V., Koehntopp, P.I.: Improving decision-making in virtual learning environments using a tracing tutor agent, In: Proceedings of the 21st International Conference on Enterprise Information Systems (ICEIS 2019), vol. 1, pp. 588–595 (2019). https://doi.org/10.5220/0007744006000607.
16. Kimball, R., Mertz, R.: The Data Webhouse Lifecycle Toolkit. Wiley , New York (2000)
17. Kiousis, S.: Interactivity: a concept explication. New Media Soc. **4**(3), 355–383 (2002). https://doi.org/10.1177/146144480200400303
18. Kolodner, J.L., Leake, D.B.: A tutorial introduction to Case-Based Reasoning. In: Leake, D.B. (ed.) Case Based Reasoning: Experiences, Lessons, and Future Directions, AAAI Press/MIT Press, pp. 31–65 (1996)
19. Maciel, A.M.A., Rodrigues, R.L., Carvalho, E.C.B.: Development of a virtual assistant integrated with moodle to support online learning. In: XXV Brazilian Symposium on Informatics in Education (SBIE 2014) (2014). https://doi.org/10.5753/cbie.sbie.2014.382. (in Portuguese)
20. National Institute of Educational Studies and Research "Anísio Teixeira" – INEP / Brazilian Ministry of Education – MEC (2019) Censo da Educação Superior 2018: notas estatísticas. https://download.inep.gov.br/educacao_superior/censo_superior/documentos/2019/censo_da_educacao_superior_2018-notas_estatisticas.pdf
21. Ponniah, P.: Data Warehousing Fundamentals for IT Professionals. Wiley (2010)

22. Sassi, R.J.: Data webhouse and business intelligence operational: revisiting the technology and analyzing data warehouse trends. In: National Meeting on Production Engineering (2010). https://www.abepro.org.br/biblioteca/enegep2010_tn_sto_120_781_14912pdf. (In Portuguese).
23. Simbine, F.B., de Lima, J.V., Torres, M.A.R., Chiguv, S.J.S.: Analysis of learning paths in virtual learning environments through information visualization. Rev. Bras. Des. Information **15**(2), 183–196 (2018). (In Portuguese)
24. Trevelin, A.T.C., Pereira, M.A.A., Oliveira Neto. J.D.: The use of 'inverted classroom' in technology courses: a comparison between the traditional model and the flipped classroom inverted model adapted to learning styles. Revista de Estilos de Aprendizaje **6**(12) (2013). (in Portuguese)
25. Vaidya, N.M., Sajja, P.S.: Agent based system for collaborative learning environment in an educational habitat. In: Proceedings of 2016 International Conference on ICT in Business, Industry, and Government (ICTBIG 2016), pp. 1–5 (2017). https://doi.org/10.1109/ICTBIG.2016.7892644.
26. Viecelli, K., Haendchen Filho, A., Antonio do Prado, H., Ferneda, E., Thalheimer, J., Fernandes, A.: Collaborative agents in Adaptative VLEs: towards an interface agent for interactivity and decision-making improvement. In: Proceedings of the 22nd International Conference on Enterprise Information Systems - Volume 1: ICEIS, pp. 691–702 (2020). https://doi.org/10.5220/0009415006910702.ISBN 978-989-758-423-7
27. Woodley, A., Simpson, O.: Student Dropout: The elephant in the room. In: Distance, Online (ed.) Zawacki-Richter O, Anderson T, pp. 459–485. Towards a Research Agenda, AU Press, Education (2014)
28. Zapparolli, L., Stiubiener, I., Pimentel, E., Braga, J.C.: Applying business intelligence and learning analytics technics in distance learning environments. In: Brazilian Symposium on Informatics in Education (SBIE 2017) (2017). https://doi.org/10.5753/cbie.sbie.2017.536.

Human-Computer Interaction

Classification and Synthesis of Emotion in Sign Languages Using Neutral Expression Deviation Factor and 4D Trajectories

Diego Addan Gonçalves[1](✉), Maria Cecília Calani Baranauskas[2,3], and Eduardo Todt[1,3]

[1] Department of Informatics, Universidade Federal do Paraná, Curitiba, Brazil
dagoncalves@inf.ufpr.br

[2] Institute of Computing, University of Campinas, São Paulo, Brazil

[3] Post-graduation Program in Informatics, Federal University of Paraná (UFPR-PPGInf), Curitiba, Brazil

Abstract. 3D Avatars are an efficient solution to complement the representation of sign languages in computational environments. A challenge, however, is to generate facial expressions realistically, without high computational cost in the synthesis process. This work synthesizes facial expressions with precise control through spatio-temporal parameters automatically. With parameters compatible with the gesture synthesis models for 3D avatars, it is possible to build complex expressions and interpolations of emotions through the model presented. The built method uses independent regions that allow the optimization of the animation synthesis process, reducing the computational cost and allowing independent control of the main facial regions. This work contributes to the definition of non-manual markers for 3D Avatar facia expression and its synthesis process. Also, a dataset with the base expressions was built where 4D information of the geometric control points of the avatar built for the experiments presented is found. The results of the generated outputs are validated in comparison with other expression classification approaches using Spatio-temporal data and machine learning, presenting superior accuracy for the base expressions. The rating is reinforced by evaluations conducted with the deaf community showing a positive acceptance of the facial expressions and synthesized emotions.

Keywords: 3D Avatar · Sign language · Facial expression · Spatio-temporal · 4D

1 Introduction

Sign languages use facial expressions as part of the information conveyed, adding context and meaning to the message. Facial details are a constant challenge in animating avatars in virtual environments, since a balance between realism and computational cost is necessary.

J. Filipe et al. (Eds.): ICEIS 2020, LNBIP 417, pp. 637–659, 2021.
https://doi.org/10.1007/978-3-030-75418-1_29

Studies on gesture synthesis in 3D space have an immediate impact on the development of new essential technologies and deserve attention and investment [1,3,4]. Several computational systems have been developed for Sign Languages based on signal synthesis through the user interaction [3,5–7]. In general, those systems use configuration parameters for hand-based gestures, body and arms positioning, aiming at the fidelity between the virtual and real representations.

According to Neidle *et al.* [10], it is remarkable that an addressee in a sign language dialogue tends to look more to the eyes of the partner than to the hands, reinforcing the importance of facial expressions in the communication. Although facial expressions provide message context and define part of the information transmitted [8], *e.g.*, irony or sarcasm, there is a serious lack of such features in most of the existing software tools. In this direction, systems that use virtual interpreters need a greater focus on the parameterized representation of facial expressions, which is an indispensable element in the effective transmission of a message [9].

Even with the advances for Sign Language 3D Avatar, facial parameters are scarce and lack details that correspond with the models applied to the manual elements of the language. Facial expressions, that also provides the emotion, should consider cultural elements, morpho-syntactic elements and facial expression interpolations. Although some systems implement facial expressions in their avatars, they often do not follow any models integrated with the other parameters and do not associate the emotional expression with the message. To this end, it is necessary a control model that understands the relationship of each facial region with facial expressions for the generation of a virtual environment. A model with these characteristics facilitates the integration of complex features in the system.

The manual sign language parameters has well-defined models, such as the CORE-SL [11], that describes the hand parameters as shape, location, movement and orientation. Each parameter have values and hierarchy relationship that compose a signal. These models do not incorporate detailed facial parameters, or Non-Manual Markers (NMM), a fact that unfortunately holds for models and systems for other sign languages as well.

In this work, we propose and validate a novel parametric model of facial expression synthesis in a virtual environment, based on the Neutral Expression Deviation Factor (NEDeF) calculation using Centroids and facial regions to extract 4D meshes. This work is an extension of the one presented in [63], where details of the classification methods are included, comparisons with the results of works with similar approaches and techniques, in addition to the inclusion of new references that update and contextualize the concepts used.

In particular, we propose a functional facial model for Sign Languages 3D animation that considers parameters of the face using Spatio-temporal information. The Spatio-temporal information enables to control the behavior of the geometric landmarks and to generate an automatic synthesis of emotions through a 3D avatar.

We aimed to enhance the animation process supporting a more accurate representation of sign language messages and integrating into the Core-SL hierarchic model.

The conducted methodology is synthesized as follows:

- Proposed a parameterized computational model for facial representation in sign language 3D avatar.
- Identified the main components of deformation landmarks in the avatar face to define the relationship between facial regions and base expressions. Base expressions are the main emotions used to generate all the secondary interpolations. This enables to generate expressions that follow the descriptive sign language model by associating the expression with the signal message. This allows full control of the geometric mesh of the face.
- Applied concepts of temporal data related to a geometric mesh to control and optimize the 4D actions of a 3D avatar.
- Extraction of Spatio-temporal meshes referring to trajectories of main facial regions associated with geometric controllers. These meshes can be used in the classification and interpolation of 3D deformations using machine learning algorithms.
- Evaluated the proposed system through a data set based on interpolation of parameters and generation of simplified and adapted expressions using machine learning to classify avatar generated expressions.
- Refinement of the calculated interpolations based on artificial intelligence classification using base emotions 2D Data-sets and an evaluation of the avatar outputs based on the Deaf community feedback.

The main contributions are the definition of a Non-Manual Model (NMM) for sign languages avatar and a process to synthesize emotions and expressions in a 3D avatar for synthesis systems, supported by the extraction of the spatio-temporal data of facial animation that defines facial behavior.

This work is organized as follows: Sect. 2 presents a synthesis of related work. Section 3 presents the built model for facial parameters in the 3D avatar. Section 4 reports on our technique for the automatic generation of complex facial expressions based on behaviors of facial landmarks. Section 5 describes the evaluations conducted to assess our proposal. Section 6 refers to the final remarks.

2 Related Work

The use of virtual agents for educational or entertainment systems has increased as well as the interest by target users [7,13,14]. Basawapatna *et al.* [15] reinforce the importance of 3D avatars in educational environments, comparing the impact of more dynamic virtual environments with traditional concepts in programming teaching.

The avatar motions may be generated by controllers associated with the 3D geometric mesh using techniques such as Blend Shapes or Morph Targets as well as the tracking of geometric controllers based on a set of features and classifier cascade [16–18].

The base classes of emotional states used in interactive scenarios are the *Positive* (joy, surprise and excited emotions), *Neutral* (calm and relaxed expressions) and *Negative* (afraid, anger and sadness expressions) classes [21]. Ekman's model [22] identifies base expressions such as *Anger*, *Fear*, *Sadness*, *Surprise*, and *Joy*. Other representations of emotions are identified as a combination and interpolation of basic expressions also called Plutchick's Wheel of Emotions.

The face can be decomposed into specific and independent regions handling the classification of key points in groups [32]. These key points are frequently used in Facial Expressions Recognition (FER) [23], where mathematical spatial models can define, for example, a distance for eyes and nose, that can be applied to the synthesis process as well. MPEG-4 Facial Points is a broadly used standard set of points of interest in the face [24]. It provides 84 points mapped on a model with a neutral expression, including areas such as tongue, lips, teeth, nose, and eyes, with points distributed along the perimeter of these regions, particularly at the corners.

It is also possible to use specific points of each facial region where the displacements can indicate which emotion is being represented [47]. Kakarla and Reddy, 2014, used the corners of their mouths and eye regions as a facial point for tracking and classifying emotions using a depth camera. These displacements were reproduced in a 3D avatar for the inclusion of emotions in a virtual environment using a simplified coordinate conversion system [47]. This technique is similar to those used in cinema as Motion Capture where points tracked by an interpreter are converted and used as a controller in a virtual environment.

There are studies that observe the effect of avatar appearance and posture and how the user interprets these signs [49]. Artificial intelligence techniques can identify emotions through the context of textual messages [48], and it can affect both the user's attention and reaction, guiding a feeling through visual outputs (e.g. expressions of joy and sadness in an avatar) and voice modulations in synthesized outputs, where context-based emotions can also be included.

This type of technique can be integrated into a process of synthesis of sign languages through 3D avatars by the parser of textual inputs, enabling the generation of emotions based on information and not arbitrarily or randomly. This integration can be done by parser of textual inputs, analyzing the morphology of phrases and assigning meaning to them, by audio analyzing distortions and patterns in the signals or by user inputs, such as tracked expressions or interactions in a system interface [48, 51].

There are models that define the main facial landmarks for tracking and classifying a human face, defining main regions [34] or Points of Interest (POI) [52]. Silva *et al.* 2016, define ten main POI on the face that can be used as controllers for identifying and tracking the main emotions [52]. The authors used these points to extract coordinates that aided in the synthesis of emotions in a humanoid Robot, applying the restrained displacements of users in the same POIs.

Shang et al. 2017, represent facial expressions as a vector in a 3D continuous space and different sets of static visual features, which the authors validate for facial expression recognition. This work proposes a probabilistic framework in order to simulate the interaction between the user and the virtual avatar. This framework uses feature extraction that has its dimensions reduced with the use of the Principal Component Analysis (PCA) algorithm and manages to continuously identify the user's emotion, defining the most appropriate reaction of the 3D avatar [50]. Other similar methods use pattern recognition and artificial intelligence to classify the user's emotion, such as Support Vector Machine (SVM) [53]. The points of interest used in the classification serve as a controller applied to a 3D avatar in a process that begins with geometric transformations based on similarity and linear deformation.

Models such as MPEG-4 Facial Points integrated with a feature extraction and emotion classification process with machine learning are current models that provide the generation of realistic animations of facial expressions in avatars [54]. The great challenge in this case, and a gap in the works cited, is the proposal of models that define interpolations of emotions and understand the behavior of the face in the synthesis of these emotions. For this, it is important to understand the process of generating emotion temporally, from the neutral state to the complete synthesis.

There are many models for looking at a geometric point in a time window. These objects can be called 4D objects, or spatio-temporal points, which can easily be viewed on a 2D plane as projections of hypercubes in incremental dimensions. Since this type of representation doubles the number of points of interest by $t_(n)$, some authors propose the use of size reduction calculations [26] or Convolutional Neural Networks [55] so that It is possible to use spatio-temporal data, or 4D, without the computational cost making the solution unfeasible.

Spatio-temporal data are widely used in the identification of human actions or environment changes, using image processing techniques such as Dense Optical Flow applied in conjunction with 3D mesh visualization algorithms temporally, or geometric reconstruction works using RGB-D channels [56,57]. An example of this technique is the 4-dimensional Color-Depth (CoDe4D) method for recognizing Human Activity, where the authors propose a descriptor that classifies actions based on tracking points of interest using SVM algorithm, followed by a projection of 4D meshes where the action is mapped. These meshes are 3D regions observed temporally as an object during the sequence of displacement of the activity.

Most approaches use a similar spatio-temporal information to classify facial data [25,26]. Different mathematical models for temporal data representation are found in the literature [27–30]. The general definition commonly used for 4D data is for an object observed temporarily in a defined space.

The spatial data type, as well as the temporal environment, define the specificity of the models developed and can help to understand the behavior of emotion animation.

Algorithms such as Factor Analysis (FA) and Principal Components Analysis (PCA) can be used to understand the correlation between landmarks and the expressions through their behavior [29,31].

These techniques can be applied together with conceptual models of facial landmarks in order to define controllers of the geometrical mesh in a sectorized way, allowing interpolations and syntheses of well-defined animations that allow the inclusion of complex emotions in avatars using independent controllers as with Manual Models for sign languages avatars.

3 Model for Automatic Synthesis of Facial Expressions in Sign Languages Through 3D Avatar

In the following sections, a novel model for the automatic synthesis of facial expressions in sign languages is presented using the proposed non-manual markers based on regions and Spatio-temporal facial behavior. The NMM model follows the manual parameters format defined for sign language computational models around the world. Using this model, methods for facial behavior analysis was applied to extract 4D meshes that detail the relationship between facial landmarks and base expressions. With this model, an automatic facial expression generation and parser system have been developed that can be integrated into automatic sign language synthesis systems.

3.1 Proposed Sign Language Non-manual Markers Based on Region Centroid

The model proposed in this section relies on hierarchical modeling, compatible with concepts widely used in the computational representation of the manual parameters of sign languages, improving systems that do not use expressions or emotions on message transmission.

A humanoid model was built with 598 facial polygons and a rigging supporting the aforementioned base expressions. The facial model MPEG-4 FP [33] was used as a points of interest reference in avatar modeling. According to Obaid *et al.* [34], the main regions for facial expression recognition are: forehead, eyes, cheeks, nose, and mouth.

Other studies using facial regions for the extraction or classification of characteristics argue for the local division as a fundamental resource to better understand the behavior and relation between a geometric area or related spatial points [32,35]. A process for generating the base expression animations was performed using the morph target method [36], where geometric deformations are processed using tracking of points of interest from an input sequence. In the process of tracking and measuring the point of interest, distances were used as a reference two 2D data-sets with images of the base expressions states and representations.

Japanese Female Facial Expression (JAFFE) data-set and the Averaged Karolinska Directed Emotional Faces (AKDEF) data-set provide samples of the base expressions interpreted by more than 20 subjects providing approximately 120 samples [37,38]. These data-sets classify the images using averaged semantic ratings, which define expressions by points of interest values, statistically identifying which emotions each image is and its general intensity(the displacement of the facial muscles and facial points of interest define the expression represents and the value of intensity).

For the input and output parser the target descriptive model used was the CORE-SL, which already implements manual and non-manual sign language parameters and elements, being the most complete model today.

3.2 Neutral Expression Deviation Factor (NEDeF)

The next methods were built aiming to relate each facial region with each base expression. This process is fundamental to identify values for the parameters defined in the previous section and can be used in automatic synthesis in the same way as the manual elements. A metric was proposed to measure the global normalized region deformation, representing the influence on the mesh for each expression, defined by the Neutral Expression Deviation Factor (NEDeF) as follows:

$$\frac{\sum_{v_{ir} \in R_r} \frac{|d_n v_{ir} - d_e v_{ir}|}{d_n v_{ir}}}{NV_r} \quad (1)$$

For each expression (e) a measure of the distortion relative to the neutral expression (n) is computed. This is done, for each facial region (R_r), by the normalized sum of differences of the Euclidean distances in the 3D space from the centroid region to each respective landmark. A second normalization is computed considering the number of landmarks (reference points) defined for each region (NV_r). The distances were taken in absolute values because the distortion of the regions is assumed to be additive.

The spatial location of the landmarks in the same region (vertexes placed as MPEG-4 Point of Interest) is defined respective to the centroid. The absolute values of the displacements were used to calculate the NEDeF considering that one distortion in the mesh should be considered in any direction in the virtual environment in order to evaluate deviation from the neutral expression.

Table 1 shows the NEDeFs of the regions and their distortion points compared to the same landmarks with the synthesized expressions, together with the normalized values of intensity of influence in the 3D mesh that was extracted in each region.

Table 1. NEDeF of the base expressions for main facial regions, indicating the relation between the facial regions and each expression. The displacements are calculated using centroid position for facial region point of interest [63].

Geometrical comparison of facial regions					
	Foreh	Eyes	Cheeks	Mouth	Nose
Joy	0.18	1.00	0.80	0.79	0.37
Anger	0.62	0.90	0.95	0.70	0.08
Surp.	0.36	0.46	0.00	0.97	0.00
Fear	0.25	0.48	0.67	1.00	0.09
Sadn.	1.00	0.98	0.30	0.40	0.09
Disg.	0.73	0.05	0.66	1.00	0.00

The results presented in Table 1 allow the identification of the more affected regions for each emotion. The forehead region is a highlight in the expressions of anger and sadness and the mouth region is of great importance in joy, surprise, and fear expressions. For the joy emotion, the regions with more distortion in the mesh were the cheeks; on the negative expressions (anger, fear, sadness) the nose region tends to have a more noticeable change in comparison with the positive emotions (joy and surprise).

This is important to identify the general weight of each region, which will allow prioritizing regions of greater importance in the synthesis of facial animations. For this process the Principal Components Analysis (PCA) and Factor Analysis (FA) algorithms were used, reducing the data dimensions, making it easier to observe the regions with the most relevant influence in the facial expression animations.

The facial landmarks were used as the PCA variables with their Euclidean values ranging from a neutral expression to the extracted synthesized expressions. The covariance matrix was calculated as the eigenvalues and eigenvectors of the average vector of the samples, relating the landmarks and the euclidean displacement.

The Explained Variance Ratio (EVR) was obtained based on the eigenvectors and eigenvalues; it was observed that 58% of the variance of the data is in the direction of the main components of the Forehead region, being the Mouth region with the second most expressive value with 26%, followed by Cheek's with 11% of the variance of the data directed to its components. The Eye region had less expression in the tests, followed by the nose region which had the EVR value lower than 1%.

The normalized values for the analyzed factors eigenvalues are 0.79 for Forehead, 0.0 for the Eyes, 0.85 for Cheeks, 1.0 for Mouth, 0.13 for Nose region, which show that using the five main facial regions of the proposed sign language NMM and the six base expressions of the Ekman's model, it's correct to consider three main factors. The reduced factors based on the eigenvalues reinforces that the absolute values of mouth and nose regions can be reduced in a single factor, as well as values for cheeks and forehead. This means that, in the synthesis of base emotion expressions, using the points of interest landmarks and regions of the proposed 3D avatar facial model, defined in the previous sub-section, we can point out that the Forehead and Cheeks regions had a similar displacement expressiveness in the geometric mesh.

The Mouth region was the most expressive with the most noticeable details of geometric change and Eyes and Nose can be considered as a factor with less perceptive displacement. The PCA results reinforce the proposed NEDeF function and point to the more expressive facial regions and centroids in base expression animation.

4 A Model for Facial Landmarks Trajectory Behavior

The following methods were built aiming at the extraction of the displacement geometric meshes of each controller regarding the centroids of the main facial regions. These results allow to define the behavior of each controller and generate trajectories that, observed as curves, allow the complete analysis of the synthesis process of emotions and the generation of complex facial expression interpolations.

For this process, the parameters of the NMM for sign language 3D avatar defined in the previous sections were transformed into function curve structures. In order to represent a function curve, the values of the points represented by t are a part of the sequence called Knot Vector and determine the base function that influences the shape of the B-Spline trajectory.

Knot vector is represented by $t = (t_0, t_1, ...t_n)$ in range $t \in [t_0, t_n]$ [39]. Each centroid is a spatial point in a trajectory in synthesis process. In this way, for a trajectory referring to the synthesis of an expression E of degree n and Controller Points represented by the centroid of the region α observed as a time point $t(\alpha)$ as follows:

$$t = p_0, p_1, ...p_n$$

$$t = \forall p(\alpha) \in [0, 1] = (p_0, p_1, ...p_n) p_i \geq p_{i-1}$$

$$t(C) = \frac{\sum_{v_{ir} \in R_r} \frac{|d_n v_{ir} - d_e v_{ir}|}{d_n v_{ir}}}{NV_r}$$

$$B(t) = \sum_{i=0}^{n} N_{i,k}(t) C(t) + W_{e(t)}$$

For all control points p_i as centroid $t(C)$ and the knots in B-Spline consider the parameter W based on the value of influence extracted from PCA and FA analysis. The influence parameter W is considered in the animation generation process, where a lower value of W corresponds to less important regions that can be ignored in the expression synthesis, reducing the computational load.

4.1 Spatio-Temporal Centroid Extraction

The Spatio-temporal concept used in this work follows the presented by Erwig and Güting [40] where the trajectory of a spatial point, based on temporal readings can be observed as a region if its spatial displacement is considered. When the shape of the curve changes, the region displacements is expanded or retracted.

The centroids trajectories along the generated facial expression animation can be represented by dynamic curves, controlling the edges that connect the vertexes landmarks, producing the movement or deformation of facial regions. The relevant coordinates occur in the transition between the neutral expression to the synthesis of one of the six base emotions of the Ekman model, since with these landmarks it is possible to observe the specific impact of these expressions on the 3D mesh. The trajectories of the Facial Expression Landmarks (FEL) of the facial NMM model defined in the previous section were extracted using intervals of 60 frames for each Expression by:

$$Traj = \sum_{i=1}^{n} Cent_{R_1}, [E_0, E_1]$$

When in a range between the neutral expression and the base expression $[E_0, E_1]$, the coordinates of the FEL are extracted by the region centroid $Cent_{R_1}$, and their displacement. The centroids displacements of the region define their behavior and influence for each expression.

The splines shown in Fig. 1 represent the geometric displacement calculated by the Three-dimensional Euclidean Distance (3DED) of the Centroid coordinates $C_{E_1 1}$ observed at 50 ts (t relating to an element of the Keyframes vector for the emotion synthesis). Each spline in images defines the Centroid trajectory by the expression, the displacement projection can be considered the 4D data as a trajectory mesh [41], once represent a geometric controller in a time slice.

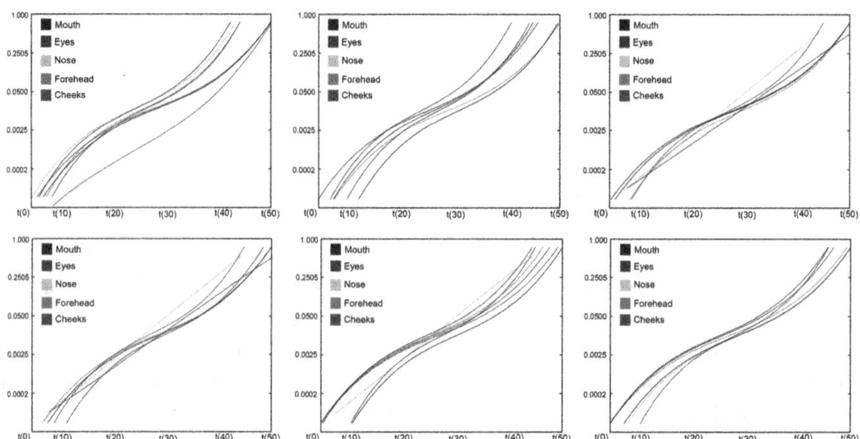

Fig. 1. Spatio-temporal Centroid trajectories, from left-top to right bottom: Joy, Anger, Surprise, Sadness, Fear, Disgust expression synthesis. The Spatio-temporal trajectories for each Centroid follow the displacement of 3D Euclidean Distance Variance as shown in Table 2 [63].

Figure 1 shows the Spatio-temporal trajectories of the five main facial regions for the base expressions. Graphically it is possible to observe which region has more spatial variation considering its geometric coordinates and the shape of the curve.

The 3D Euclidean Distance values of each Centroid reading are assigned to the matrix sequenced by the Keyframe of the observed animation, where Euclidean Distance Variance values of a mouth facial region Centroid are displayed and their overall variance value. Based on 3D Euclidean Distances, Table 2 shows the Variance Analysis Matrix for each Centroid in the time slice of the synthesis of each base expression.

Most of the values in the EDVA table are reinforced by PCA tests, such as the fact that the Nose region is more significant in negative expressions (such as Anger and Sadness) with a variance of 0.00236 and 0.004911 respectively, significant among the matrix average.

Table 2. Centroid trajectory EDVA values. For each base expression, as Joy, Anger, Surprise, Fear, Sadness and Disgusting, the variance value of each facial region is shown (M. for mouth region, C. for cheeks region, N. for nose region, F. for forehead region and E. for eyes region) [63].

	EDVA values					
C_n	Joy	Anger	Surp.	Fear	Sadn.	Disg.
M.	.123	.001	.002	.010	.197	.182
C.	.163	.001	.002	.002	.001	.003
N.	.002	.000	.000	.000	.000	.002
F.	.052	.001	.076	.014	.004	.011
E.	.006	.002	.001	.001	.004	.001

The forehead region has a greater variance in the positive expressions (Joy and Surprise) as well as the mouth region, reinforcing the variance obtained with the PCA Algorithm. For negative expressions, the mouth region has more significant variance values. The expression Anger, according to the analyzed matrix (Table 2) indicates that the regions of the eyes with 0.002369 and cheeks with 0.001781 are those that have a greater variance in the displacements also reinforcing the values tabulated by the PCA test.

The avatar's mouth presented high values of geometric displacement variance in practically all base expressions, highlighting the disgust expression, with 0.1820 of variance and sadness expression, with 0.1972, reinforcing their relationship between these expressions and facial regions. The cheek region has a greater variance in the Joy expression synthesis with 0.16371, and its lower values are in the expressions of surprise and fear, expressions that depend more on the forehead region that presented high values of 0.076342 and 0.014019 respectively.

Table 3 presents the normalized variance data for the PCA and EDVA test relating the centroids of the facial regions and the base expressions. Although extracted values cannot be compared directly, due to the difference between the algorithms, the data presented for the influence of facial region on the synthesis of expressions reinforce the main regions that characterize the negative and positive emotions.

There are many approaches to interpolate curves or Spatio-temporal data where intermediate points are calculated in order to obtain a predicted trajectory that extends the original shape [42,43]. From a weight parameter, simplifies curves by interpolating duplicated or less relevant edges shortening the number of controller points in the spline [44], a useful resource for interpolated expressions generation.

Table 3. Centroid displacement variance for Facial Regions in expression synthesis. The shown values were normalized and separated by the facial region and animated expression. For regions with more than one Centroid, the mean-variance values were calculated ((M. for mouth region, C. for cheeks region, N. for nose region, F. for forehead region and E. for eyes region) [63].

PCA normalized variance values						
	Joy	Anger	Surp.	Fear	Sadn.	Disg.
M.	.626	.005	.011	.053	1.0	.923
C.	.829	.009	.003	.003	.005	.020
N.	.001	.003	.002	.003	.003	.002
F.	.267	.006	.387	.071	.020	.057
E.	.003	.012	.009	.009	.024	.007
3D EDVA normalized values						
M.	.791	.702	.970	1.00	.400	1.00
C.	.800	.953	.000	.673	.307	.667
N.	.376	.080	.000	.095	.091	.000
F.	.118	.628	.363	.250	1.00	.730
E.	1.00	.900	.460	.484	.980	.050

The proposed process was applied to create a data-set with different outputs representing interpolated emotions or simplification of base expressions parameters. The Newton Polynomial Interpolation method was used in the process. Through a list containing the values of the coordinates separated by sequential readings of the expressions, the simplification method of the 4D regions was implemented.

In this process, the readings must contain values in the three geometric axes for each key-frame, defined by 50 frames for each expression, plus 10 frames to return to the neutral state, exporting a different list for each landmark. The landmarks are further grouped by facial region, in order to parameterize the simplification process by data-sets.

With a list of intermediate values u, spatial vertexes displacements are simplified generating a new Spatio-temporal trajectory. With the exported coordinates it is possible to intensify the animations by increasing the variance in the displacements or optimize the generation of the expression by excluding 4D regions with a lower w parameter. The value of w represents the weight extracted from the EDVA, PCA and FA results, which define the weight relationship between the expression and the facial region.

Once the value of an intermediate point u is calculated, its coordinates can be inserted in Centroid values list of the morph target by the distance between Centroid and the data matrix using the function that calculates the distance between C_u and C_j in $[C_0, C_{n-1}]$.

The whole process of outputting and calculating base expression animation or expression interpolation has been extracted from a parser process that follows the CORE-SL framework structure. This structure follows an XML notation model where parameters, relationships, and values are defined in order to generate a sign. For the facial model proposed by this work, the parameters are the points of interest and Centroid of each facial region, and their values are the geometric displacement relations defined by the 4D meshes. In summary, the presented generation process is complete and can generate any interpolation variation proposed by the Plutchik Wheel of Emotions, or intensify and alter the base expressions to suit local and regional representations.

5 Emotion Synthesis Classification and Interpolation

A training and classification process was built to validate the outputs generated by the proposed method. This process was carried out using the pattern recognition algorithms K-nearest neighbors algorithm (KNN) and Support Vector Machine (SVM) from the refined controllers from the space-time trajectories.

5.1 Using Machine Learning and 4D Trajectories for Spatio-Temporal Facial Expression Recognition and Interpolation Synthesis

For the training procedure for Facial Expression Recognition, using Spatio-Temporal points of interest, Zhao *et al.* strategy was adapted [58], which presents high accuracy when recognizing base expressions. Algorithm 1 presents the applied strategy, which can be parallelized considering the facial region Centroid defined in the previous section.

The shown algorithm uses facial landmarks coordinates for each Centroid of the five facial regions defined in Sect. 3. The classification process repeats for each geometric independent area, being able to evoke the function whenever necessary to label an animation. As the coordinates of each point of interest set an example in the machine learning process, interpolations based on these patterns can be applied and the training process can be refined with a high number of examples of emotions on a 2D data-sets. The algorithm calculates the training error of each feature as follows:

$$s(x) = \begin{cases} 1 & \text{if} p.f_{p,i}(x) > p.\theta_{p,i} \\ 0 & \text{otherwise} \end{cases} \qquad [58]$$

Algorithm 1. Algorithm for Spatio-Temporal Facial Expression Recognition adapted from that proposed by Zhao *et al*, [58].

Input: Image samples, labels $(x_1, y_1), (x_2, y_2), ..., (x_n, y_n), x_i$ is a expression sequence, $y_i \in \{-3, -2, -1, 0, 1, 2, 3\}$
Output: Classifier $P(y|x, \Theta)$
foreach $t(C) = \frac{\sum_{v_{ir} \in R_r} \frac{|d_n v_{ir} - d_e v_{ir}|}{d_n v_{ir}}}{NV_r}$ **do**
  for $i = 1$ *to* n **do**
    Build candidate set $\{F\}$ and encode it as a binary vector;
    foreach $f_i \in F$ **do**
      if $\varepsilon[f_1, ..., f_m, f_i,] < \varepsilon[f_1, ..., f_m]$ **then**
        $f_1 \rightarrow [f_1, ..., f_m]$;
        Update training error process;;
      end
    end
    According to last two sub-blocks conditions, calculate $P(y_i|x_i, \Theta)$;
  end
  Return Θ
end
$t(C)_i = t(C)_{i+1}$;

Where θ minimizes the weighted error, and the feature f_i is to be added to the combined feature if the training error is reduced after condition:

$$f_1 \rightarrow [f_1, ..., f_m]$$

$$If(\varepsilon[f_1, ..., f_m, f_i,] < \varepsilon[f_1, ..., f_m])$$

This modeling considers the temporal elements described in the previous section, and can be applied for tracking 2D images of facial expressions, defining coordinates of controllers that can be applied to the avatar in a virtual environment. The entire classification process does not consider elements of a specific data-set and the generated controllers can be adapted to different avatars with small adjustments in the neutral position of the facial regions.

Although there are no models that define detailed NMM parameters for sign languages 3D avatar, some models and landmarks are widely used for facial expression recognition (Fig. 2), and can be used in order to point out facial behaviors that define emotions and expressions for the virtual environment generation.

KNN and SVM are pattern recognition algorithms that use machine learning to classify an input based on a previously trained space of hypothesis H_d to recognize patterns that define classes in a database.

According to Phillipp and Kaliouby [61], SVMs perform an implicit embedding of data into a high dimensional feature space, where linear algebra and geometry may be used to separate data that is only separable with nonlinear rules in input space. The authors use SVM for the classification of the six base

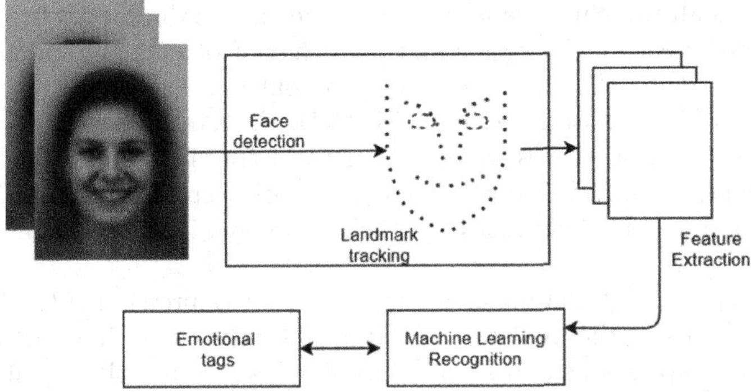

Fig. 2. Facial expression recognition using Machine Learning techniques such as Support Vector Machine.

emotions, using 2D sequential images, where the method separates the training data in feature space by a hyper-plane defined by the type of kernel function used finding the hyper-plane of maximal margin, defined as the sum of the distances of the hyper-plane from the nearest data point of each of the two classes as in the equation:

$$f(x) = sgn(\sum_{l}^{i=1} \alpha_i y_i K(x_i.x) + b)[61]$$

Using Lagrange multipliers (α_i) it is possible to show that only some of the αi are non-zero in the optimal solution, namely those arising from training points nearest to the hyper-plane, called support vectors [61]. The authors used the standard SVM classification algorithm together with a linear kernel, achieving good classification results that reach 87.5% accuracy in the classification of base emotions.

Other machine learning techniques can also be used to classify facial expressions like Convolution Neural Network (CNN) or KNN. Kin *et al.* 2016, use tracked spatio-temporal elements to classify facial expressions using strategies such as Long Short-Term Memory (LSTM) + CNN [60]. This technique uses the micro-expression spatial feature learned using CNN observed in a sequence of frames that represent a base expression. There are also strategies that use CNN and LSTM in conjunction with a second classification that can use, for example, audio signals to classify emotions and reinforce the labels defined by facial landmarks [59].

The KNN algorithm does not learn any hypothesis model being a distance-based strategy of elements by comparing their attributes and defining their dimensional position based on their values. In this sense, the advantage of using KNN in these scenarios is that their classification is more agile since the training stage is eliminated. Applied in conjunction with SVM, the classification process is strong enough to define the labels for the geometric displacement patterns of the centroids of each facial region and the behavior of each landmark.

In order to train the classifier that points out the displacements that define emotions, two facial bases, JAFEE and AKDEF, were used defining the facial model in Sect. 3. Both bases were used in SVM training, where characteristics such as eye position, mouth corners, nose, and other spatial characteristics were defined as parameters of variables and the base expressions were defined as labels [46] and [45].

After training with 2D image bases, 3D Facial Expressions (3DFE) dataset was generated using the model proposed in this work, with 21 examples of synthesized base expressions and some interpolations between them. Although the model generates virtual environment animations as outputs, the final output of each rendered expression can be used as an example and applied to SVM and KNN for labeling, once the coordinates of the landmarks can be identified. In this way, based on H_d, which is adaptable and learns from new classifications, it is possible to label each output of the dataset identifying if the intended expression was generated correctly.

Figure 3 shows the accuracy of Machine Learning algorithms in the labeling of facial expressions generated.

It is possible to identify that the KNN algorithm had less accuracy in classifying the output images, perhaps because its hypothesis space is totally defined by distance calculations. The SVM results were supported by the training stage based on AKDEF and JAFEE databases, and have shown high positive classifications.

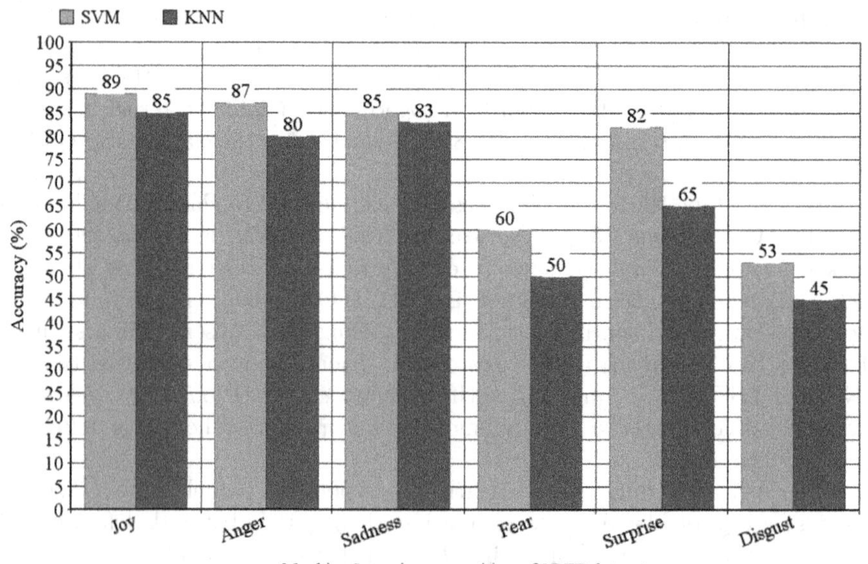

Fig. 3. Average accuracy's of 3DFE classification [63].

Table 4 presents the results of accuracy compared to the classification of emotions considering approaches that use similar models for the classification of facial expressions using image processing methods combined with artificial intelligence. In the second column, the results are presented using the approach of Phillip *et al.* [61] where the SVM algorithm is applied with its own model for tracking facial landmarks. In the second case, Lu *et al.* [61] proposed a Multiple Spatio-Temporal Feature (MSTF) learning for base emotion recognition method and Yang *et al.*, [62] used a method of Facial Expressions classification with Compositional Features (CF) that tracked from a displacement model of facial reference points and Action Units (AU).

Table 4. Accuracy results in the classification of base emotions considering the proposed method that extracts the Spatio-Temporal Trajectories using the Centroids of the Facial Regions. The compared methods are not exactly compatible and the comparison is only considering the classification results.

Emotion	Accuracy using SVM [61]	Accuracy using MSTF [59]	Accuracy using CF [62]	Accuracy using our method
Anger	66,7%	74,4%	81,1%	84,2%
Disgust	64,3%	0,0%	55,2%	48,5%
Fear	66,7%	27,2%	81,3%	55,5%
Joy	91,7%	79,2%	94,9%	87,2%
Sorrow	62,5%	45,0%	45,0%	84,2%
Surprise	83,3%	70,2%	74,3%	73,2%
Average	71,8%	49,3%	71,8%	72,2%

Although the extraction of the 4D trajectory may suggest an increase in the computational cost, the process is optimized by calculating independent facial regions that can be parallelized, greatly reducing the computational cost. In addition, this technique allows to observe the behavior of the base expressions and to define, in future works, controllers for interpolations of secondary emotions. The average accuracy of the proposed technique had results above expectations with 72.2% in the classification using machine learning. The process presented in this work considers an optimized calculation compared to the use of MSTF since it comprises independent controllers although the use of MSTF can generate more stable results. Regarding the use of Compositional Features and Action Units, the 4D trajectories allow the analysis of behavior in the synthesis of emotions, being a way to define interpolations between the basic emotions, for example.

The results demonstrate that the expressions generated by the methods proposed in this work are correct and highlight an important first step in the integration of complex expressions in avatars for sign language automatic synthesis systems, and can also be integrated into other sign language systems around the world. A dataset with the base expressions and fifteen interpolations was generated aiming to test the automatic generation process and to present to the deaf community through sign language synthesis systems through 3D avatar.

5.2 3DFE Dataset and User Evaluation

A form containing a collection of 21 outputs synthesized expressions generated using the proposed technique was made available to Sign Language users from the Federal Technological University of Paraná (UTFPR) in order to evaluate the system and its outputs. The form was made available online, and did not require users identification or personal information once it was the system being analyzed and not the people. Users should identify the facial expressions presented, including some interpolations with intensity variations in facial regions, and simplifications by excluding regions with the less relevant w parameter.

The application of the form has received 30 anonymous responses. It was reported that 46.7% of the respondents already used software with 3D avatars as sign language interpreters, 10% had already used software that uses 3D avatars but not in the context of sign languages and 43.3% had never used software with these characteristics.

Following, 86.7% of respondents consider facial expressions a very important feature in a sign language conversation, 10% consider it important and 3% consider it less important. Table 5 presents Spearman's Rank Correlation coefficient and significance applied to the data extracted from the answers feedback. The data below were organized in order to relate the importance of the parameters used to define the outputs in the user's responses.

The positive correlation between the basic expressions and the hits of 0.9 justify the use of the proposed and implemented system, being a viable model in the first moment for the parametrized representation of facial expressions in signal language systems.

Other variables in Table 5 highlight the strong correlation of interpolated expressions with N.I. responses and the strong correlation between N.I. and Errors. Therefore, the system would be improved if there were image bases with representations of secondary expressions to serve as hypothesis space helping to classify these variations geometrically.

Table 5. Spearman's rank correlation coefficient applied to validated parameters in the applied formulas. The presented values vary between 1 for a direct relation between the parameters and −1 for inverse relation. 0 represents no relation. The parameters presented are: "B.E." for base expressions, Interpolations (I.), 'N.I.' for unidentified expressions, 'E.' and 'H.' for the number of errors and hits by output image [63].

	B.E.	I.	N.I.	E.	H.
B.E	1.00	−0.75	−0.33	−0.09	0.91
I.	−0.75	1.00	−0.09	0.13	−0.06
N.I.	−0.33	−0.09	1.00	0.15	−0.51
E.	−0.09	0.13	0.15	1.00	−0.90
H.	0.91	0.06	−0.51	−0.90	1.00

Some relationships can be made between classifications using artificial intelligence and user responses. Negative expressions had more false positives, especially expressions of fear and disgust. The most successful expressions in both tests were the expressions of joy and anger, as well as sadness. One hypothesis is that these emotions have less cultural variation and are more recognizable regardless of facial and local characteristics. The Deaf community response with the generated expressions was positive as well as the SVM and KNN results; this represents an important step in integrating complex facial expressions into avatars for automatic signal language synthesis, using computational parameters, as it is already done with manual elements.

6 Conclusion

This work presented a model for 3D Avatar automatic facial expression synthesis based on spatio-temporal parameters. A técnica apresentada utiliza controllers baseado em regiões faciais e dados temporais que facilitam a integração com modelos de síntese de animaçoes em ambiente virtual e permite otimizar animações complexas processando as regiões faciais independentemente. A principal contribuição deste trabalho é o modelo de sintese de animações faciais em avatares 3D que utiliza dados spatio-temporal e permite um cotrole preciso de animação.

An evaluation was performed using machine learning algorithms with the classification that presented an accuracy superior to approaches that use sectorized facial control techniques for the classification of emotions. Also, the outputs generated were presented to the deaf community with positive acceptance of the facial expressions and emotions synthesized. Future works can define relations between base expressions and interpolations and apply the parameters presented in other 3D meshes reinforcing the generality of the model. Besides, it is essential that morphosyntactic elements, defined in each gestural language, be integrated into the model.

Acknowledgments. This work was financially supported by the São Paulo Research Foundation (FAPESP) (grants #2015/16528-0, #2015/24300-9 and Number 2019/12225-3), and CNPq (grant #306272/2017-2). We thank the University of Campinas (UNICAMP) and Universidade Federal do Paraná (UFPR) for making this research possible.

References

1. Lombardo, V., Battaglino, C., Damiano, R., Nunnari, F.: An avatar-based interface for the Italian sign language. In: 2011 International Conference on Complex, Intelligent and Software Intensive Systems (CISIS), pp. 589–594 (2011)
2. Punchimudiyanse, M., Meegama, R.: 3D signing avatar for Sinhala Sign language. In: 2015 IEEE 10th International Conference on Industrial and Information Systems (ICIIS), pp. 290–295 (2015)
3. Kacorri, H., Huenerfauth, M., Ebling, S., Patel, K., Willard, M.: Demographic and experiential factors influencing acceptance of sign language animation by deaf users. In: Proceedings of the 17th International ACM SIGACCESS Conference on Computers and Accessibility, pp. 147–154. ACM (2015)

4. Sofiato, L.: Brazilian sign language dictionaries: comparative iconographical and lexical study. Educação e Pesquisa **40**(1), 109–126 (2014). https://doi.org/10.1590/S1517-97022014000100008
5. Bento, J., Claudio, A., Urbano, P.: Avatars on Portuguese sign language. In: 2014 9th Iberian Conference on Information Systems and Technologies (CISTI), pp. 1–7 (2014)
6. Adhan, S., Pintavirooj, C.: Thai sign language recognition by using geometric invariant feature and ANN classification. In: 2016 9th Biomedical Engineering International Conference (BMEiCON), pp. 1–4 (2016)
7. Ratan, R., Hasler, B.: Playing well with virtual classmates: relating avatar design to group satisfaction. In: Proceedings of the 17th ACM Conference on Computer Supported Cooperative Work; Social Computing, pp. 564–573. ACM (2014)
8. Elons, A., Ahmed, M., Shedid, H.: Facial expressions recognition for Arabic sign language translation. In: 2014 9th International Conference on Computer Engineering Systems (ICCES), pp. 330–335 (2014)
9. Hyde, J., Carter, E., Kiesler, S., Hodgins, J.: Evaluating animated characters: facial motion magnitude influences personality perceptions. ACM Trans. Appl. Percept. 13(2), 8:1–8:17 (2016)
10. Neidle, C., Bahan, B., MacLaughlin, D., Lee, R., Kegl, J.: Realizations of syntactic agreement in American sign language: Similarities between the clause and the noun phrase. Studia Linguistica **52**(3), 191–226 (1998)
11. Iatskiu, C., Garcia, L., Antunes, D.: Automatic SignWriting generation of libras signs from CORE-SL. In: Proceedings of the XVI Brazilian Symposium on Human Factors in Computing Systems, pp. 55:1–55:4. ACM (2017)
12. Griffin, H., Aung, M., Romera-Paredes, B., McLoughlin, C., McKeown, G., Curran, W., Bianchi-Berthouze, N.: Perception and automatic recognition of laughter from whole-body motion: continuous and categorical perspectives. IEEE Trans. Affective Comput. **6**(2), 165–178 (2015)
13. Grif, M., Manueva, Y.: Semantic analyses of text to translate to Russian sign language. In: 2016 11th International Forum on Strategic Technology (IFOST), pp. 286–289 (2016)
14. Wiegand, K.: Intelligent assistive communication and the web as a social medium. In: Proceedings of the 11th Web for All Conference, pp. 27:1–27:2. ACM (2014)
15. Basawapatna, A., Repenning, A., Savignano, M., Manera, J., Escherle, N., Repenning, L.: Is drawing video game characters in an hour of code activity a waste of time? In: Proceedings of the 23rd Annual ACM Conference on Innovation and Technology in Computer Science Education, pp. 93–98. ACM (2018)
16. Feng, R., Prabhakaran, B.: On the "Face of Things". In: Proceedings of the 2016 ACM on International Conference on Multimedia Retrieval, pp. 3–4. ACM (2016)
17. Ahire, A., Evans, A., and Blat, J.: Animation on the web: a survey. In: Proceedings of the 20th International Conference on 3D Web Technology, pp. 249–257. ACM (2015)
18. Kacorri, H.: TR-2015001: A Survey and Critique of Facial Expression Synthesis in Sign Language Animation. CUNY Academic Works (2015)
19. Kaur, S., Singh, M.: Indian Sign Language animation generation system. In: 2015 1st International Conference on Next Generation Computing Technologies (NGCT), pp. 909–914 (2015)
20. Huenerfauth, M., Lu, P., Rosenberg, A.: Evaluating importance of facial expression in American sign language and pidgin signed english animations. In: The Proceedings of the 13th International ACM SIGACCESS Conference on Computers and Accessibility, pp. 99–106. ACM (2011)

21. Alkawaz, M., Basori, A.: The effect of emotional colour on creating realistic expression of avatar. In: Proceedings of the 11th ACM SIGGRAPH International Conference on Virtual-Reality Continuum and Its Applications in Industry, pp. 143–152. ACM (2012)
22. Szwoch, W.: Model of emotions for game players. In: 2015 8th International Conference on Human System Interactions (HSI), pp. 285–290 (2015)
23. Happy, S., Routray, A.: Automatic Facial Expression Recognition Using Features of Salient Facial PatchesAffective Computing. IEEE Transactions on **6**(1), 1–12 (2015)
24. Bouzid, Y., El Ghoul, O., Jemni, M.: Synthesizing facial expressions for signing avatars using MPEG4 feature points. In: 2013 Fourth International Conference on Information and Communication Technology and Accessibility (ICTA), pp. 1–6 (2013)
25. Sikdar, B.: Spatio-temporal correlations in cyber-physical systems: a defense against data availability attacks. In: Proceedings of the 3rd ACM Workshop on Cyber-Physical System Security, pp. 103–110. ACM (2017)
26. Yu, S., Poger, S.: Using a temporal weighted data model to maximize influence in mobile messaging apps for computer science education. J. Comput. Sci. Coll. **32**(6), 210–211 (2017)
27. Erwig, M., Güting, R., Schneider, M., Vazirgiannis, M.: Abstract and discrete modeling of spatio-temporal data types. In: Proceedings of the 6th ACM International Symposium on Advances in Geographic Information Systems, pp. 131–136. ACM (1998)
28. Lee, J., Han, B., Choi, S.: Interactive motion effects design for a moving object in 4D films. In: Proceedings of the 22Nd ACM Conference on Virtual Reality Software and Technology, pp. 219–228. ACM (2016)
29. Mahmoud, M., Baltrusaitis, T., Robinson, P.: Automatic detection of naturalistic hand-over-face gesture descriptors. In: Proceedings of the 16th International Conference on Multimodal Interaction, pp. 319–326. ACM (2014)
30. Suheryadi, A., Nugroho, H.: Spatio-temporal analysis for moving object detection under complex environment. In: 2016 International Conference on Advanced Computer Science and Information Systems (ICACSIS), pp. 498–505 (2016)
31. Oliveira, M., Chatbri, H., Little, S., O'Connor, N.E., Sutherland, A.: A comparison between end-to-end approaches and feature extraction based approaches for Sign Language recognition. In: 2017 International Conference on Image and Vision Computing New Zealand (IVCNZ), pp. 1–6 (2017)
32. Lemaire, P., Ben Amor, B., Ardabilian, M., Chen, L., Daoudi, M.: Fully Automatic 3D Facial Expression Recognition Using a Region-based Approach. In Proceedings of the 2011 Joint ACM Workshop on Human Gesture and Behavior Understanding, pp. 53–58. ACM (2011)
33. Kacorri, H., Huenerfauth, M.: Implementation and evaluation of animation controls sufficient for conveying ASL facial expressions. In: Proceedings of the 16th International ACM SIGACCESS Conference on Computers and Accessibility, pp. 261–262. ACM (2014)
34. Obaid, M., Mukundan, R., Billinghurst, M., Pelachaud, C.: Expressive MPEG-4 facial animation using quadratic deformation models. In: 2010 Seventh International Conference on Computer Graphics, Imaging and Visualization (CGIV), pp. 9–14 (2010)
35. Lv, S., Da, F., Deng, X.: A 3D face recognition method using region-based extended local binary pattern. In: 2015 IEEE International Conference on Image Processing (ICIP), pp. 3635–3639 (2015)

36. Dailey, M.N., et al.: Evidence and a computational explanation of cultural differences in facial expression recognition. Emotion, vol. 10(6) (2010)
37. Lyons, M.J., Akemastu, S., Kamachi, M., Gyoba, J.: Coding facial expressions with gabor wavelets. In: 3rd IEEE International Conference on Automatic Face and Gesture Recognition (1998)
38. Lundqvist, J.: The averaged karolinska directed emotional faces - AKDEF. In: CD ROM from Department of Clinical Neuroscience, Psychology section (1998)
39. Aldrich, J.: Doing Least Squares: Perspectives from Gauss and Yule (1998)
40. Erwig, M., Güting, R.H.: Temporal objects for spatio-temporal data models and a comparison of their representations. In: International Workshop on Advances in Database Technologies, pp. 454–465 (1998)
41. Le, V., Tang, H., Huang, T.: Expression recognition from 3D dynamic faces using robust spatio-temporal shape features. In: 2011 IEEE International Conference on Automatic Face Gesture Recognition and Workshops (FG 2011), pp. 414–421 (2011)
42. Gloderer, M., Hertle, A.: Spline-based Trajectory Optimization for Autonomous Vehicles with Ackerman drive (2010)
43. T. Jusko, E. Scalable Trajectory Optimization Based on Bézier Curves (2016)
44. Li, H., Kulik, L., Ramamohanarao, K.: Spatio-temporal trajectory simplification for inferring travel paths. In: Proceedings of the 22Nd ACM SIGSPATIAL International Conference on Advances in Geographic Information Systems, pp. 63–72. ACM (2014)
45. Song, N., Yang, H., and P. Wu (2018). A Gesture-to-Emotional Speech Conversion by Combining Gesture Recognition and Facial Expression Recognition
46. Adil, B., Nadjib, K.M., Yacine, L.: A novel approach for facial expression recognition propagation for core extraction (2019)
47. Kakarla, M., Reddy, G.R.M.: A real time facial emotion recognition using depth sensor and interfacing with Second Life based Virtual 3D avatar. In: International Conference on Recent Advances and Innovations in Engineering (ICRAIE-2014), Jaipur, 2014, pp. 1–7 (2014). https://doi.org/10.1109/ICRAIE.2014.6909153
48. Wan, Y., Chiu, C., Liang, K., Chang, P.: Midoriko chatbot: LSTM-based emotional 3D avatar. In: 2019 IEEE 8th Global Conference on Consumer Electronics (GCCE), Osaka, Japan, 2019, pp. 937–940. https://doi.org/10.1109/GCCE46687.2019.9015303
49. Pandita, S.: Affective embodiment: the effect of avatar appearance and posture representation on emotions in VR. In: 2020 IEEE Conference on Virtual Reality and 3D User Interfaces Abstracts and Workshops (VRW), Atlanta, GA, USA, 2020, pp. 539–540 (2020). https://doi.org/10.1109/VRW50115.2020.00121
50. Shang, Z., Joshi, J., Hoey, J.: Continuous facial expression recognition for affective interaction with virtual avatar. In: 2017 IEEE International Conference on Image Processing (ICIP), Beijing, pp. 1995–1999 (2017). https://doi.org/10.1109/ICIP.2017.8296631
51. Angga, P.A., Fachri, W.E., Elevanita, A., Suryadi, Agushinta, R.D.: Design of chatbot with 3D avatar, voice interface, and facial expression. In:2015 International Conference on Science in Information Technology (ICSITech), Yogyakarta, pp. 326–330 (2015). https://doi.org/10.1109/ICSITech.2015.7407826
52. Silva, V., Soares, F., Esteves, J.S.: Mirroring emotion system - on-line synthesizing facial expressions on a robot face. In: 8th International Congress on Ultra Modern Telecommunications and Control Systems and Workshops (ICUMT), Lisbon 2016, pp. 213–218 (2016). https://doi.org/10.1109/ICUMT.2016.7765359

53. Cho, T., Choi, J.-H., Kim, H.-J., Choi, S.-M.: Vision-based animation of 3D facial avatars. In: 2014 International Conference on Big Data and Smart Computing (BIGCOMP), Bangkok, 2014, pp. 128–132 (2014). https://doi.org/10.1109/BIGCOMP.2014.6741422
54. Zhao, Y., Jiang, D., Sahli, H.: 3D emotional facial animation synthesis with factored conditional Restricted Boltzmann Machines. In: 2015 International Conference on Affective Computing and Intelligent Interaction (ACII), Xi'an, 2015, pp. 797–803 (2015). https://doi.org/10.1109/ACII.2015.7344664
55. Choy, C., Gwak, J., Savarese, S.: 4D spatio-temporal ConvNets: Minkowski convolutional neural networks. In: 2019 IEEE/CVF Conference on Computer Vision and Pattern Recognition (CVPR), Long Beach, CA, USA, 2019, pp. 3070–3079 (2019). https://doi.org/10.1109/CVPR.2019.00319
56. Ivson, P., Nascimento, D., Celes, W., Barbosa, S.D.: CasCADe: a novel 4D visualization system for virtual construction planning. IEEE Trans. Visualization Comput. Graph. **24**(1), 687–697 (2018). https://doi.org/10.1109/TVCG.2017.2745105
57. Zhang, H., Parker, L.E.: CoDe4D: color-depth local spatio-temporal features for human activity recognition from RGB-D videos. IEEE Trans. Circuits Syst. Video Technol. **26**(3), 541–555 (2016). https://doi.org/10.1109/TCSVT.2014.2376139
58. Zhao, S., Yao, H., Sun, X., Xu, P., Liu, X., Ji, R.: Video indexing and recommendation based on affective analysis of viewers. In: Proceedings of the 19th ACM International Conference on Multimedia (MM 2011), pp. 1473–1476. Association for Computing Machinery, New York (2011)
59. Lu, C., et al.: Multiple spatio-temporal feature learning for video-based emotion recognition in the wild. In: Proceedings of the 20th ACM International Conference on Multimodal Interaction (ICMI 2018), pp. 646–652. Association for Computing Machinery, New York (2018)
60. Kim, D.H., Baddar, W.J., Ro, Y.M.: Micro-expression recognition with expression-state constrained spatio-temporal feature representations. In: Proceedings of the 24th ACM International Conference on Multimedia (MM 2016), pp. 382–386. Association for Computing Machinery, New York (2016)
61. Michel, P., El Kaliouby, R.: Real time facial expression recognition in video using support vector machines. In: Proceedings of the 5th International Conference on Multimodal Interfaces (ICMI '03), pp. 258–264. Association for Computing Machinery, New York (2003)
62. Yang, P., Liu, Q., Metaxas, D.N.: Exploring facial expressions with compositional features. In: CVPR (2010)
63. Gonçalves, D., Baranauskas, M., Reis, J., Todt, E.: Facial expressions animation in sign language based on spatio-temporal centroid. In: Proceedings of the 22nd International Conference on Enterprise Information Systems - Volume 2: ICEIS, pp. 463–475 (2020). https://doi.org/10.5220/0009344404630475.ISBN 978-989-758-423-7
64. Liu, Z., Zhang, C.: Spatio-temporal analysis for infrared facial expression recognition from videos. In: Proceedings of the International Conference on Video and Image Processing (ICVIP 2017), pp. 63–67. Association for Computing Machinery, New York (2017). https://doi.org/10.1145/3177404.3177408

How Communication Breakdowns Affect Emotional Responses of End-Users on Progressive Web Apps

Giulia de Andrade Cardieri(✉) and Luciana A. M. Zaina

Federal University of São Carlos, Sorocaba, São Paulo, Brazil
lzaina@ufscar.br

Abstract. Progressive Web App (PWA) is a recent approach made of a set of techniques from both web and native apps. End-User Development (EUD) is an approach from which end-users are allowed to express themselves. The impacts of associating EUD and PWAs have been little exploited. With this in mind, we proposed the PWA-EU approach in previous work. In this paper, we compare the communication breakdowns and the users' emotional responses with the aim of finding if both aspects are correlated. We conducted a study with 18 participants that interacted with Calendar, a mobile app based on the PWA-EU approach. We carried out a qualitative analysis based on the communication breakdowns and emotional responses of the participants' interaction. Our findings point out that common issues occurred and affected both the emotional response and breakdowns of the participants. Still, how these common issues affected the participants were different between individuals.

Keywords: Progressive web apps · End-user development · Experimental study

1 Introduction

Mobile devices are used daily by most individuals in the world to access different information systems [8]. Even so, there still browser and devices limitations that affect end-user experience. One way to overcome these issues is by adapting user interfaces (UIs) for different screen sizes, resources and dimensions [5,16].

Progressive Web App (PWA) is a recent approach for the development of web mobile apps that is made of a set of techniques from both web and native apps. Based on the progressive enhancement technique, PWAs gain more features to become a more complete application [25]. A PWA may be accessed on any device with a browser, which makes it easily connectable via an uniform resource locator (URL) [15]. This definition brings an adaptive nature to PWAs, as the growth of user interactions affects the number of available features on the app.

End-user Development (EUD) can be defined by methods, situations, and socio-technical environments that allow and empower end-users to express themselves. Moreover, users can be independent of developers while performing EUD

J. Filipe et al. (Eds.): ICEIS 2020, LNBIP 417, pp. 660–682, 2021.
https://doi.org/10.1007/978-3-030-75418-1_30

activities [14]. Meta-design is an EUD approach in which the end-user can participate actively in app development and is not restricted to its use. Users are in charge of solving their own problems, instead of relying on technology experts [13].

Although PWAs have an adaptive nature, its traditional approach does not take EUD concepts into account. For instance, end-users cannot select their preferences on the app, due to the progressive enhancement features being the same to all users. Hence, we proposed the PWA-EU approach [2,6]. Our proposal extends the traditional PWA architecture by including meta-design in a way that users can include their preferences to adapt interfaces on run-time. Another of the motivation for the development of PWA-EU is the small number of studies on PWAs, especially when we consider the user perspective. In this paper we take a different approach on the same sets of data explored on a previous study [2]. Previously, we analyzed how communication breakdowns affected the experience of end-users acting as co-designers. Unlike this paper, in which we compare the communication breakdowns and the user experience together with the aim of finding if both aspects are correlated. We defined one research question (RQ) to guide our study: *How do communication breakdowns affect the end-users emotional responses?*.

The main contribution of our work is providing a discussion of how the communication breakdowns experienced by end-users on their interactions with PWAs affect their emotional response. Moreover, we analyze the difference between the interaction of participants who had previous experience with technology and the ones who did not, and also between participants who actively participated on the design choices as co-designers. Furthermore, our findings can aid developers in the creation of apps more self-guided from which end-users with no previous experience with technology can perform the role of co-designers.

The rest of the paper is organized as follows: Sect. 2 presents the related work; PWA-EU, the approach we developed is presented in Sect. 3; the details of our study are discussed in Sect. 4 and its results are in follow Sect. 5; in Sect. 6 we return to our research questions to point out the important results and make a comparison with the literature; In Sect. 7 we presents the limitations of our study; and finally Sect. 8 discusses the conclusion and future work.

2 Background

2.1 Related Work

Even though PWAs are a recent technology, some authors explored its technical properties and compared them to other mobile and web apps development methods. Biørn-Hansen et al. presented a technical comparison between PWA, Native, Hybrid, and Interpreted apps [3]. Adetunjia et al. provided a systematic literature review of the technical achievements and limitations of PWAs. They compare mobile and web development approaches in other to provide an overview of the benefits and disadvantages on each development process [1]. Sharma et al. discusses the definition of PWAs and compares their performance to native and hybrid apps [23]. Overall, all authors mention the potential of PWAs to unify web and native development. On a UX perspective, our previous paper [7] brought a comparison between PWA, Native and Web apps, in which we found out that all three approaches provided a satisfying experience to end-users. As of today, we could not find works in the literature that joins the PWA area with both interface adaption or EUD fields. In contrast, there are many works on interface adaption for mobile apps, and on EUD.

On the UI adaptation area, we can mention authors who worked with mobile and web apps. Yigitbas et al. proposes AdaptUI, a Domain Specific Language that adapts the UI at run-time based on the device's context. The adaptation changes the navigation and layout based on rules pre-defined by domain experts. When we compare AdaptUI and PWA-EU, there is no user interference on AdaptUI since the changes relate to the device environment [26]. Peissner et al. developed MyUI, a framework that adapts UIs based on end-users' needs, devices and environmental conditions during run-time. The adaptation takes place in a smooth way to the navigation, layout, and interaction methods. Correlating MyUI and PWA-EU, we observe MyUI's user profile is mostly obtained from pre-defined data on the user's mobile device and that the proposed solution was not tested with users or mobile apps [20]. HyMobWeb is an approach that adapts UIs from web apps on mobile devices based on context sensibility and multimodality. The proposal extends front-end frameworks enabling developers to include adaptation points on the code. A hybrid strategy was adopted to combine both adaptions on run-time and on design time [5]. PWA-EU, besides providing a solution for both web and mobile dimensions, provides a more customized adaptation as the selection of the visual choices are fully defined by the end-user.

On the EUD field, we can highlight the work of some authors. Danado & Paternó proposes Puzzle, a framework in which end-users with no programming experience can develop or customize complex mobile apps and connect their apps to web services and smart devices [12]. Costabile et al. propose a EUD desktop app based on the software shaping workshop (SSW) method in which a meta-design participatory approach is adopted during the entire software life cycle [11]. Namoun et al. (2016) propose a model linking the features of performing EUD in mobile devices to end-users' attitudes towards and intent of doing this. In the studies, the authors considered participants with experience or not on EUD in mobile. The results show that rather than creating apps the end-users have more interest in customizing apps to improve their experience. Nonetheless, these works do not take PWA and interface adaptation into account.

3 PWA-EU Approach

We adopted PWA-EU to conduct our study. We have proposed it [6] and discussed it [2] in previous work. To provide a better understanding, we will explain the approach briefly. PWA-EU approach is an extension of the PWA architecture in which users' preferences are taken into account on the interface adaptation at run-time. Developers and designers should follow the architectural proposal to construct apps from which end-users can modify the interface in accordance with their preferences. The conception initial steps of design is not a part of our proposal. Rather than use methodologies such as user-centered design or design thinking [24]. PWA-EU provides a software architecture that allows including features in the apps from which the end-user can act as a co-designer.

Looking at Fig. 1, we see on the left side the traditional PWA is shown while our extension is on the right side.

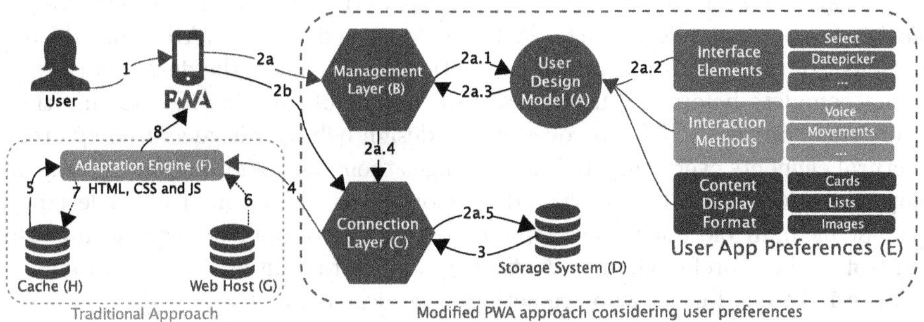

Fig. 1. PWA-EU approach [2].

The **User Design Model (UDM)** (A) establishes and combines different *User App Preferences* (E), some examples are UI elements appearance, interaction methods, and content display formats such as data and images. The *User App Preferences* are connected to the app domain. For instance, on a PWA in which end-users create lists, their preferences could be linked to how each list element is displayed. Moreover, the relation between the app and *User App Preferences* should be previously defined by designers and developers. These preferences have default settings and follow meta-design principles, in which users can change their preferences. Developers must select a technology to store these preferences.

The **Management Layer** (B) is responsible for including, editing and deleting the *User App Preferences* on the *UDM*. It can receive two different requests from the PWA:(i) receiving and sending modifications to the *UDM*, and (ii) for sending an updated *UDM* to the *Connection Layer*. The **Connection Layer** (C) has three functions: (i) sending the *UDM* to be stored, (ii) retrieving the *UDM* from the storage, and (iii) sending the latest *UDM* to the *Adaptation Engine*. The first scenario only occurs when the *UDM* was updated by the user on the *Management Layer*. The other scenarios happen when is necessary to load a new UI on the PWA, which can occur when a new *UDM* preference is set or when the user is browsing through the app. The **Adaptation Engine** is the link between the PWA-EU and the traditional PWA architecture. It takes into account the *UDM User App Preferences* to alter the app UI. Moreover, this engine should store the app UI on cache memory and recover it when necessary.

We defined the PWA-EU architecture into these three elements and the connection with the *Adaptation Engine* following an adaptation of the event-driven and the layered architectures for software development [21]. The event-driven part is relevant since the PWA-EU architecture depends on events triggered by users, such as modifying the *UDM* or loading a new UI.

With regards to EUD, PWA-EU is based on the guidelines presented in the meta-design framework proposed by Fisher, Nakakoji & Ye (2009) as described in the following: *Support Human-problem Interaction* - Designers and developers should analyze the app domain and its end-users. After this analysis, the *User App Preferences* are defined and included as part of PWA-EU's *UDM*. *Underdesign for Emergent Behaviour* - A PWA developed with PWA-EU has default options before the end-user initial interaction. Due to that, the design of the app is not complete before end-users modified the default options by selecting their preferences on the *UDM*. Moreover, a new design behavior is generated by architectural elements every time the user changes their *User App Preferences* on the app. *Share Control* - Designers and developers are the original meta-designers who define multiple *User App Preferences* during development. At run time, the control of these preferences, that affect app's functionalities and appearance, is transferred to end-users who are acting as co-designers.

4 Exploring Users Emotional Responses and Communication Breakdowns

In next section we describe the planning, conduction and analysis our study guided by [17]. The goal of this section is to take a different approach exploring data we obtained in a previous study [2]. On this new perspective, we compare the end-users emotional responses with communication breakdowns in order to find out they are correlated.

4.1 Calendar - A PWA-EU App

A Calendar app following the PWA-EU approach was developed to support our study (App available at https://h1eneas.dlvr.cloud/). Calendar is a PWA that allows users to manage events and their time (i.e. event name, date, location, category, start, and end time). Calendar was chosen because it represents a common-sense domain avoiding the need for users to learn about the app domain. To build the Calendar, we considered the most popular frameworks in web mobile app design, such as Bootstrap[1], Foundation[2], Materialize[3]. As PWA is a Google proposal based on Material Design, we chose Materialize as the default design for Calendar's UI elements. Material Design consists of guidelines, components, and tools that support the best practices of UI design available as open-source code

[1] https://getbootstrap.com/.
[2] https://get.foundation/.
[3] http://materializecss.com.

[9]. We chose Vue.js[4] to develop the front-end because it has a fast learning curve, its applications are smaller to store when compared to the other frameworks, it is easy to use in projects and availability of an official Vue.js PWA template[5] [22,25].

At the start we set up the *User App Preferences* (see Fig. 1). After conducting an investigation on the Material Design guide and considering the nature of the app (i.e. Calendar), we decided to cover five *UI elements*: calendar, checkbox, input, select and timepicker. We also introduced a second option in which users acting as co-designers could select. For each element, users acting as co-designers could choose between a Material Design element and a browser-default appearance as part of the *User App Preferences* of the UDM. A browser-default look changes according to the device and browser the user is accessing. For instance, the input from Safari Mobile for iOS shows a vertical scroll for the day, month and year, while on Chrome Mobile for Android it displays a pop-up calendar.

Gestures and voice *Interaction Methods* (see Fig. 1) were available. The swipe gesture was selected due to some Material Design components recommending its use, such as lists and cards. The voice synthesis and speech recognition were included considering they became a standard feature on smartphones [10]. Complementing the *User App Preferences*, *Content Display Formats* were selected to provide users with distinct ways to read the information on the app. Following Material Design guidelines, we defined that users can choose between list and card formats to display data, besides showing or hiding images on the cards.

Each architectural element from PWA-EU and the *Adaptation Engine* are represented on the Calendar app. All three PWA-EU architectural elements (UDM, Management, and Connection Layers) are defined in the My Design section. In this section, participants could select and edit their *user app preferences*. While the *UDM's* values were visually represented by My Design's UI, the *Connection* and *Management Layers* were built as sets of JS functions. *Connection Layer's* functions store and retrieve data from LocalStorage and the *Management Layer's* code changes the values of *user app preferences* on the *UDM*. On the other hand, the *Adaptation Engine* is formed by multiple CSS and JS files that modify *user app preferences* on Home and Include New Event sections.

4.2 Planning

Participants were invited to take part in the study voluntarily via social networks like Twitter and Facebook. The participants were selected by convenience, according to their availability to take part in the study [17]. We obtained two distinct groups of participants who were separated based on their academic and professional experience [19]. One group had participants from the technology field (i.e. developers and designers), we named *tech*, and the other had participants from other professional and academic fields, we named *non-tech*. This

[4] https://vuejs.org/.
[5] https://github.com/vuejs-templates/pwa.

separation avoids that only individuals with previous knowledge on UI design or development acted as co-designers and indicates whether or not a difference stands between end-users *co-designers* who have a technical profile and those who do not have.

We selected the Communicability Evaluation Method (CEM) [18] to conduct our study and guide our analysis of communication breakdowns. We used the method to separate into codes the recurring issues and interactions regarding users' communication breakdown situations. CEM is a semiotic engineering-based method that aims to explore communication breakdowns between the designed system and the user through the observation of how a group of users interacts with a particular system. Considering that in our study part of the users are also the designers (i.e. co-designers), these breakdowns can point out whether these individuals' roles as co-designers lead to more satisfying interactions, and understand if even acting as co-designers users still have communication breakdowns.

Moreover, we took the Self Assessment Manikin (SAM) [4] to collect the participants' experience after their interactions with the application. SAM is a pictograph evaluation method to measure emotional responses from some sort of stimulus. Three dimensions are considered by this technique: pleasure (if the participant had a positive or negative reaction), arousal (body stimulation level from an event or object) and dominance (feeling in control of the situation or controlled by it). The user chooses a value on a scale of one to nine on each dimension, using images, to represent their emotions after interactions. Participants answered questions regarding SAM after each task. Furthermore, when all tasks were accomplished we conducted an interview with four questions to have a broader comprehension of the end-users' perspective.

A pilot test was carried out with two participants. We concluded that no changes were necessary and the study could be run.

4.3 Conduction

The study was conducted over four days in Sorocaba, São Paulo, Brazil. We had a total of 20 participants who were between 18 and 59 years old, median 23 years old. Most participants were undergraduate students or had a degree. All participants accepted the term of consent about the use of data and images for academic ends. Besides the technical profile separation (see Planning section), half of the participants were randomly selected and named *co-designers* due to having an active participation in the setting of the UI design, by selecting their preferences and customizing the app's design. The other ten participants interacted only with features that did not make changes on the UI and were named *non-designers*.

The participants' observation took place individually and the researcher made notes regarding each communication breakdown. Initially, each participant received instructions about Calendar app and the conduction of the study. For the *co-designers*, three tasks were proposed. The tasks were specified as indicated in Table 1 and Fig. 2 displays examples of the Calendar app screen on each task.

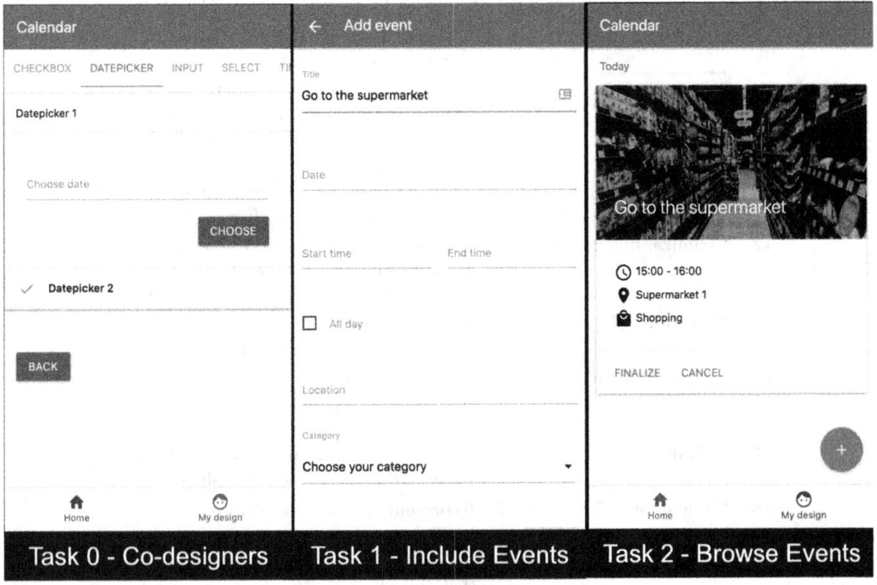

Fig. 2. Calendar app screens on each task.

Table 1. Tasks conducted by the participants.

Task	Description	Group
Task 0 - Co-designers	Set the UDM by selecting preferences regarding UI elements, interaction methods, and display format options	Only Co-designers
Task 1 - Include events	Include two new events on the app	Both
Task 2 - Browse Events	Cancel and mark as finished the previous included events	Both

Only the description of the current task was displayed in text format on a laptop screen. We did not introduce a time limit to the participants accomplished the tasks. All the participants used the same mobile device, a Motorola Moto G4 Play running Android 8.1. This avoided that differences in Android operating system versions could introduce bias on interaction data. The participants' interactions with the PWA were recorded by using DU Recorder[6], installed on the mobile device. In addition, the participants' facial expressions and voice were

[6] http://www.duapps.com/product/du-recorder.html.

Table 2. Common observed causes of Communication Breakdowns.

Tag	Description
T1 - "I give up."	The user admits s/he is unable to achieve his/her goal. S/he interrupts his/her activity without accomplishing what was proposed
T2 - "Looks fine to me."	The user is convinced the goal was accomplished, when, in fact, it has not
T3 - "Thanks, but no, thanks."	The user is aware of what should be accomplished. Still, s/he decides to do something different than is expected
T5 - "Where is it?"	The user expects to see a specific sign or element, but cannot find it
T6 - "What happened?"	Users repeat an operation because they could not see or understand the effects of their actions
T7 - "What now?"	The user is temporarily does not know what to do next to accomplish the task
T8 - "Where am I?"	Represents the communication breakdown when a user did not find a particular feature by pausing and searching it
T9 - "Oops"	A user momentarily makes a mistake and immediately corrects it. S/he sees that s/he has made a wrong step and usually activates the "undo" function immediately.
T12 - "Help!"	The user explicitly call for help in order to restore a productive interaction
T13 - "Why doesn't it?"	The user is trying to make sense of the designer message by repeating the steps of previous unsuccessful communication in order to find out what went wrong

captured by a laptop camera placed in front of the participant. The think-aloud protocol [24] was adopted to provide a better understanding of user interaction and to assist the application of the CEM method. This protocol requires users to speak their thoughts during their interactions. Following the CEM method, we prepared a semi-structured interview that asked a question about the difficulties the participants had in using the app. The set of structured questions was: *(i) Do you think that the choices you made regarding UI elements, interactions methods and content display format contributed for a better UX during your interactions with the app? (ii) Did you have difficulties while selecting design options? (iii) Did you have difficulties while adding events on the app? (iv) Did you have difficulties while browsing through events on the app?*.

4.4 Analysis

The analysis was conducted aiming to observe and correlate aspects covered both by SAM and CEM methods. We decided to keep part of the analysis we did in our previous work [2] and complement it with an extra round of analysis. Overall, our findings were obtained by crossing-over four data sources (i.e. SAM questionnaire, video recordings, screen recordings, and the main researcher's notes). Approximately, we collected five hours of video recordings from user interactions on the mobile device and recordings of user's faces, and around two hours of audio from the interviews. During a pre-analysis of all video and audio recordings, two samples were discarded. These participants were *co-designers*, from both *tech* and *non-tech* groups, who did not complete Task 0. In the end, we had 18 participants, from which eight acted as *co-designers* and the others as *non-designers*. Due to the sample size of 18 participants, we decided not to conduct an inferential statistical analysis. We carried out four rounds of analysis with at least two cycles of data exploration for each one. In the first cycle, the first author of this paper explores the data, and after, in the next cycle, the first and second authors together revised, refined, and consolidated the outcomes.

In the first round of analysis, the videos containing the users' interactions on apps and facial expressions were put together on the same video aiming to enable easier and more precise identification of each evidence of a communication breakdown. For this, we used the iMovie Software[7]. Regarding the interview, we transcribed to text all conversations recorded on audio between the users and the researcher. After all the videos were combined, we started CEM's tagging step. The first author of this paper watched all video recordings twice, took notes of when, why and where it occurred, and assigned a tag when a communication breakdown was identified. A spreadsheet was used to support all the analysis. We used the 13 tags proposed by the CEM method [18]. These tags are natural language expressions commonly found in human communications, which the participants might utter during the thinking aloud process. On Table 2 we describe each identified tag.

In the second round, the chunks that were tagged were revisited for the identification of the main communication breakdowns. Each tag was listed and analyzed in two perspectives: (i) the frequency and context of occurrence which was identified by listing when, where and the probable reason each tag occurred, and (ii) the existence of pattern in sequences of tag types, such as grouping tags connected to the similar communication breakdowns on multiple or single participants. In each step of tagging, the first author conducted a double-checking of the results.

In the third round, we did a crossing-over among the issues that have caused the observed breakdowns, with the participants' profile, SAM answers, recordings and interview data.

In the fourth round, we grouped together the issues of each communication breakdown. First, we observed the similarities between the issues that occurred within different participants but were categorized in the same tag from the CEM method. Secondly, we looked for causes or issues that were similar but could be categorized into different tags. With this round of analysis, we could build a list of common observed causes for the communication breakdowns. This list is available at Table 3.

[7] https://www.apple.com/imovie/.

Table 3. Common observed causes of Communication Breakdowns.

Issue	Participants	Tags	Task
C1 - Difference between voice recognition and voice synthesis	P2, P5, P7	T1, T2	Task 0
C2 - Location is a mandatory field	P3, P6, P15	T2, T7	Task 1
C3 - Did not follow the instructions to include an event	P2, P3	T2	Task 1
C4 - Did not open all of the options in one of the design selections	P1, P6, P8	T3	Task 0
C5 - Repeated the selection of one of the design choices	P1, P2, P3, P4, P5, P6, P7, P8	T6	Task 0
C6 - Repetition during category selection	P6, P9, P13, P14, P15, P17, P18	T6	Task 1
C7 - Issues with marking an event as finished	P4, P7, P8	T6, T8, T9, T13	Task 2
C8 - Opened the timepicker before selecting the 'All day' checkbox	P1, P2, P3, P7, P8, P10, P15, P16, P17, P18	T9	Task 1
C9 - Issues with the form UI elements	P1, P7, P8, P9, P15, P16, P17, P18	T9	Task 1
C10 - Issues selecting the design choices	P2, P4, P5, P8	T5, T6, T12	Task 0
C11 - Thought today's date would be automatically selected by the datepicker	P1, P4, P10, P11, P13, P15, P17	T13	Task 0 and Task 1

5 How Do Communication Breakdowns Affect the Emotional Responses?

We decided to combine the analysis of the communication breakdowns and the participants' UX perception in order to understand if the both dimensions were related. Table 3 presents the common causes of some of the communication breakdowns. Table 4 displays the list of participants divided by group (i.e. Co-designers or Non-designers), their technical profile (TP) (i.e. Tech or Non-tech), the selected values for each SAM dimension and the identified CEM tags on each Task. Following, we discuss the findings obtained by crossing over data from both tables.

5.1 Common Causes During Task 0

Four common causes, C1, C4, C5, and C10 (See Table 3) were identified exclusively on Task 0. On this subsection, we will discuss each issue in details and correlate it with the participants' emotional responses and communication breakdowns.

C1 - Difference between voice recognition and voice synthesis an issue that occurred with three participants from the *co-designers* group (P2, P5, and P7) during Task 0. P2's interaction was different from the other participants who suffered from *C1.* His/hers experience can be classified as *T1 - "I give up."* because the participant gave up on trying to understand what Voice Recognition means. The recordings showed us that s/he was trying to interact with the Voice Recognition but s/he did not understand it was necessary to click on a microphone icon in order to activate it. In contrast, P5 and P7 breakdowns were categorized as *T2 - "Looks fine to me."*. The participants did not play with both interaction methods options, they just read the name of each method and skipped this part. Since the names "Voice Recognition" and "Voice Synthesis" sound similar, we believe the participants understood both relate to a voice or audio option, but they did not seem interested in finding out the difference between them.

Relating C1 and SAM, we observe the participants selected values between 6–9 for SAM dimensions, as showed in Table 4. P2 and P5 were non-tech participants who had multiple breakdowns during Task 0. Besides *C1*, both experienced *C5* and *C10*, and P5 also suffered from *C10.* We believe their lower SAM values on this task were a combination of all three common issues. We assumed that because all are somewhat related to asking for help to understand the goal of the Task, and to the UI elements behavior. In the other hand, P7 is a tech participant who experienced only *C5* in addition to *C1* during Task 0. This participant selected 6 for the pleasure index, but 8 for the other two indexes. His/her breakdowns could have affected how much s/he enjoyed performing the task. However it seemed to affect less how in control s/he felt during the task. This remark could be connected to his/her tech experience, which could have led him/her to not feel the need to ask for help as the other two participants did.

C4 - Did not open all of the options in one of the design selections issue happened on Task 0 with three *co-designers* participants, P1, P6, and P8. In every situation, the participants did not interact with the all of the options for Interaction Methods. Still, they did interact with the other design options available at the My Design section of the Calendar app. That's why we classified their breakdowns as *T3 - "Thanks, but no, thanks."*. The selection of Voice Recognition and Voice Synthesis did not bring interest in any of the three participants. P6 did not interact with the Swipe option either. Meanwhile, P8 did not feel the need to interact with any of the Interaction Methods. Despite having similarities with *C1*, the interactions reported as *C4* did not happen because the participants were confused about the meaning of each method. Instead, our assumption is that these participants seem to understand exactly what the Interaction Meth-

Table 4. SAM and Communication Breakdowns by participants.

	Group	TP	Task 0				Task 1				Task 2			
			PI	AI	DI	Tags	PI	AI	DI	Tags	PI	AI	DI	Tags
P1	Co-designer	Non-tech	8	7	9	T3, T6	9	9	9	T9, T13	9	9	9	
P2	Co-designer	Non-tech	7	6	8	T1, T6, T12	9	8	8,5	T2, T9, T12	9	9	9	
P3	Co-designer	Tech	8	9	9	T6, T9	7,5	7,5	8,5	T2, T7, T9	9	9	9	T9
P4	Co-designer	Non-tech	6	6	3	T5, T6, T9, T12, T13	8	8,5	8,5		8	8	7,5	T6
P5	Co-designer	Non-tech	8	9	7	T2, T6, T9, T12, T13	9	9	9		9	9	9	
P6	Co-designer	Tech	7	9	8	T3, T6	8	9	8	T2, T6	8	9	7	
P7	Co-designer	Tech	6	8	8	T2, T3, T6	5,5	5,5	6,5	T9, T13	5,5	4	5,5	T9, T13
P8	Co-designer	Tech	7	5	9	T3, T6, T12	9	9	7,5	T9	9	9	9	T8
P9	Non-designer	Non-tech					9	9	9	T3, T5, T6, T9, T13	9	9	9	
P10	Non-designer	Non-tech					8,5	9	8,5	T3, T9, T12	9	9	9	
P11	Non-designer	Non-tech					7,5	7,5	7,5	T8, T13	9	8	9	
P12	Non-designer	Tech					7,5	8,5	9		9	9	9	T5, T9
P13	Non-designer	Non-tech					7	7,5	6,5	T5, T6, T13	7	9	6	T5
P14	Non-designer	Tech					7,5	9	9	T6	8	9	9	
P15	Non-designer	Tech					7	8,5	7,5	T2, T6, T9, T13	7,5	8,5	7,5	
P16	Non-designer	Non-tech					7,5	7,5	8	T6, T9	9	9	9	
P17	Non-designer	Tech					7	8	5	T6, T9, T13	5	7,5	3,5	
P18	Non-designer	Tech					8	9	8	T6, T9, T13	8,5	8	9	

ods meant and decided, on purpose, not to interact with it because the options did not seem interesting enough.

In reference to SAM and C4 (See Table 4), P1 and P6 selected values between 7-9 on each dimension and experienced the situation defined as *C5* (See Table 3) as well. P1 selected 7 for the arousal index on Task 0. With this in mind, we believe the pre-defined Interaction Methods could have influenced in his/her decision. Similarly, P6 choose 7 as the value for the pleasure index, a fact that indicates s/he could have felt less satisfied when realized some of the design choices were options s/he is familiar but do not enjoy using on his/her mobile apps. P8 also selected lower values for both pleasure (7) and arousal (5) indexes. Still, this participant suffered from *C5* and *C10* on Task 0. Her/his experience with *C10*, that will be discussed in the end of this section, could have had a larger influence on his/her UX perception related to Task 0.

For *C5 - Repeated the selection of one of the design choices*, participants from the *co-designers* group suffered from a communication breakdown categorized as *T6 - "What happened?"* during Task 0. In every case, the participants repeated their interactions, at least once, with one of the design choices. We consider this a breakdown as they needed to repeat their actions to decide their preferences, however, it did not seem to affect negatively the participants. This breakdown was not the only one identified on participants on Task 0. All *co-designers* suffered from at least another communication breakdown. We believe the repetition to select an element was not the main reason to select a lower value on SAM, as it was not mentioned as a problem in all participants videos or during the interview. For this reason, we will not discuss in details the participants' SAM values (See Table 4) on this subsection.

The communication breakdowns categorized into *C10 - Issues selecting the design choices* happened with four *co-designers* participants during Task 0. While they interacted with the design choices, they had to ask for help. For this reason, we classified the breakdowns with *T12 - "Help!"*. Some of the breakdowns also received other tags such as *T5 - "Where is it?"* and *T6 - "What happened?"*

P2, P4, and P5 had issues connected to C10 while trying to understand how to select the design options. Even though they had access to the instructions and the researcher gave a brief explanation of the study, they could not understand how to start or why they should select design choices. With every participant, the researcher repeated the goal of the study and mentioned the design choices would be visible on the following tasks. Unlike P2 and P5, P4's breakdown was also categorized into T5 because the participant was looking for a visual instruction on the app in order to accomplish the task. However, the P2 and P5 had more questions during the conduction of Task 0. P2 asked for help other four times. His/her issues included specific questions about UI elements (i.e. Datepicker), and also about interaction methods (i.e. Swipe and Voice Recognition). Likewise, P5 also had questions about Voice Recognition. On *C1* we discussed about these and other issues participants had with Voice Recognition. Conversely, P8 had doubts about the selection of Content Display Formats. S/he questioned the choice of cards and lists because s/he assumed that only one event would be displayed if s/he selected cards. The researched clarified both cards and list would always display all events s/he included. His/her breakdown can be categorized into *T6* as s/he repeated the interaction with the card and list example.

Taking into account the participants technical profile, we observe P8 was the only individual with technical experience who experienced this C10. Nonetheless, s/he asked for help with a very specific question, unlike the non-tech participants who had issues understanding the main goal of Task 0. Hence, we believe the technical experience could have helped participants, since they might be familiar with some of the UI elements, Interaction Methods, and Content Display Formats.

Regarding SAM, all participants who experienced *C10* chose values 7 and below for, at least, one SAM dimension (See Table 4). Based on the selected values, we believe this issue did affect participants experience with the app on Task 0. In particular, P4 selected 6 for both pleasure and arousal indexes, and 3 for the dominance index. Even though s/he suffered from other issues, his/her initial questioning could be one of the reasons for the lower values in comparison with other participants. Concerning the interview, the participants did not mention any good or bad situation that could be associated with this issue during Task 0.

5.2 Common Causes During Task 1

The common causes C2, C3, C6, C8, and C9 were observed only during Task 1. On the next paragraphs, we will discuss each issue taking into account the participants' results from SAM and CEM methods.

C2 - Location is a mandatory field occurred with three participants during Task 1. Whereas participants P6 and P15's breakdown was categorized as *T2 - "Looks fine to me."*, P3's breakdown could be defined as a *T7 - "What now?"*. This issue occurred when participants had to include their first event on Task 1. The instructions did not indicate an specific name for the location, instead, the task requested participants included an event to take their pets to the veterinarian. For this reason, all three participants concluded the location field was not mandatory. Still, we classified P3's action as T7 instead of T2 because s/he had issues with the form button. This means s/he included all data related to the event and then realized the "Save" button was disabled. After a few seconds, the participant realized the location field was mandatory. As opposed to the other two participants, who had different experiences. Both said out loud they thought the location field was not mandatory. In addition, we observed they interacted with the field and left it blank, an action that triggered a visible validation error message before they could click on the "Save" button. By considering the participants' profile, we note all three participants have technical experience. With this in mind, we can discuss if the participants' experience interfered with their expected behavior of this UI element, as they might be used to a specific method to indicate the field is mandatory.

Regarding participants' SAM values, we notice all three participants who suffered C2 selected values between 7.5 and 9 for all three dimensions (see Table 4). Besides *C2*, the participants experienced other issues during Task 1. For instance, P3 suffered from *C3* and *C8*, yet P6 experienced *C6*. In contrast, *C6*, *C8*, *C9*, and *C11* were identified in P15. Moreover, P15 had a higher number of communication breakdowns while s/he performed Task 1, hence, those affected his/hers SAM values as s/he selected 7 for the pleasure index and 7.5 for the dominance index. Similarly, we observed one evidence of a disruption on P3's experience. During the interview, s/he mentioned that having to fill all form fields to include an event was not enjoyable. Based on this fact, we assume the breakdowns that occurred in Task 1 influenced on the SAM values as s/he selected 7.5 for both pleasure and arousal indexes. Even though we observed *C2* and *C6* during P6's interaction on Task 1, we believe these issues did not affect this participant's experience as much as the others as all SAM dimensions had values between 8–9.

Participants P2 and P3, both from the *co-designers* group, suffered from *C3 - Did not follow the instructions to include an event* during Task 1. Even though both situations were classified as *T2 - "Looks fine to me."*, the participants' actions were different. P2 did not read carefully the instructions to include an specific event related to taking pets to the veterinary. At first, s/he included an event not related to the request and, after a couple of minutes, s/he said *"Did I have to include the event mentioned on the instructions?"*. Consequently, P2 repeated Task 1 and included the requested event. Conversely, P3 included the event theme according to the instructions. However, s/he did not include the correct start time. We requested the event started at 2 PM as a way to observe the interaction of the participants with the timepicker element. Yet, s/he included the event starting at 2:25 PM. We believe this action is connected to the fact that P3 was from the *co-designers* group. Being part of this group, s/he could select his/hers preferred timepicker on Task 0. A fact that could have influenced s/he wanting to interact more with the UI element and not including the precise event time.

On top of that, every participant who experienced *C3* could be categorized in CEM's T2. This tag usually represents a breakdown that the participant do not realize s/he did something wrong during the task. As both participants suffered from other breakdowns during Task 1, we believe this cause did not have a big influence on the participants emotional response.

We observed *C6 - Repetition during category selection* with seven participants from both *co-designers* and *non-designers* group on Task 1. Every participant who experience *C6* interacted with the first option of the select element, which was the one provided by Materialize. Regarding CEM, we defined all interactions as *T6 - "What happened?"*. The participants needed to select a category for each event on the Calendar app. The category was not pre-defined on the instructions and should be chosen by the participant. We believe the lack of a category in the instructions made the participants confused about which was the best choice given the events' domain. For instance, P18 mentioned during the interview: *"I felt lost browsing through the categories .. after interacting with the select for some time, it was easier to choose a category."*. Moreover, P13 mentioned s/he had difficulties trying to select a Category that fit what s/he wanted.

Considering the emotional response values during C6, we notice most participants selected values between 6.5 - 9 (See Table 4). The only participant who did not have other communication breakdowns on Task 1 was P14, who selected 7.5 as the pleasure index. At the same time, s/he stated during interview that s/he wish s/he could type the date instead of using the Datepicker. Based on this remark, we assume the lower value is not connected to *C6*. The other participants had at least one other situation described in Table 3. Taking into account the participants' recordings and interviews, we consider only participants P13 and P18 had their experiences affected by this breakdown. Still, issues related on *C9* and *C11* seemed to have a bigger impact on the overall experience of these participants.

An interesting remark about *C6* is that only one participant from the *co-designers* group experienced this situation. After analyzing P6's recordings and interviews, we did not find any piece of evidence of why s/he selected Materialize's select element instead of the browser-default option. Moreover, s/he was not the only *co-designer* participant who made this selection. With this in mind, we wonder whether some of the *non-designer* individuals would have chosen the browser-default option if they could choose their design preferences.

C8 - Opened the timepicker before selecting the 'All day' checkbox was identified with ten participants from both *co-designers* and *non-designers* group on Task 1. On the second step of Task 1, participants should include an event that would occur during an entire day. The include an event form allowed users to include a date with a start and finish time using the timepicker, or mark a checkbox for an event that would take the entire day. We were expecting users to go directly to the checkbox as the instructions made it clear it was an "all day" event. However, all ten participants first selected the timepicker, and only after a few seconds realized the checkbox was the ideal option for this event. On every occurrence, this communication breakdown was categorized as *T9 - "Oops!"*. Moreover, some of the participants even reacted loudly to this breakdown, for instance, P8 said *"Oops!"* right after s/he located the checkbox. In addition, one participant who did not experienced C8 mentioned on the interview s/he expected some kind of message or action when the checkbox was selected. P6 said s/he wish the timepickers were hidden when the checkbox was selected.

Every participant who suffered C8 had other breakdowns on Task 1 as well. When we associate the participants' emotional response, recordings and interviews, we cannot identify evidence that this breakdown directly affected most participants' experience. For this reason, we will not discuss the details of each participant emotional response on this subsection. Conversely, other common causes such as *C9* and *C11* were mentioned by the participants.

Eight participants experienced the *C9 - Issues with the form UI elements* during Task 1. This issue is connected to multiple issues related to the form UI elements. P1, P7, P9, P15, and P17 had problems while interacting with the timepicker. On the other hand, P8 had difficulties interacting with the select element. On top of that, P16 and P18 had issues with the datepicker. Even though the participants' had distinct experiences, all breakdowns were classified as *T9 - "Oops!"*.

The participants who had problems with the timepicker can be divided into two groups. P1, P7, and P17 initially selected the wrong time for their event, however, they realized the mistake and included the correct time. At the end of the conduction, P7 stated s/he thought the timepicker was "strange", since s/he is used to the timepicker used in iOS. On the other hand, P17 mentioned s/he felt the timepicker was too precise. The second group had a different issue. While P9 and P15 tried to include the start and end time, they accidentally closed the timepicker element before finishing their interactions. When asked about the timepicker during interview, P15 mentioned *"The timepicker movement made me confused as I'm not used to it. But it is precise"*. P9 also described his/her issues with this element on the interview, *"There's too much sensibility on the timepicker."*. Regarding the emotional response, the values were distinct between the participants. While some participants selected values below 7 for all dimensions, others selected 9 for all SAM dimensions. With this in mind, we believe the distinct causes and experiences affected differently each participant. One example of this contrast is from the participants from the first group. P1 selected 9 for all dimensions, unlike P7, who selected 5.5 for both pleasure and arousal indexes and 6.5 for the dominance index.

Concerning the datepicker group, P16 and P18 had distinct issues while interacting with the datepicker. P16 accidentally selected the date of the following day, action that s/he corrected after a few seconds by selecting the current day date. In the other hand, P18 misclicked on the title of the event when s/he wanted to click on the datepicker. The emotional response from these participants might have being slightly affected by this situation, as both selected values 8 and below for one of the SAM dimensions.

P8 was the only participant who suffered *C9* related with the select element. S/he had just interacted with the timepicker and the operating system keyboard was still visible on the UI. When s/he clicked on the select element the keyboard closed, a fact that surprised him/her. The participant said *"Oops"*, and continued his/her interaction with the select. Even though the participant did not mention this issue during the interview, s/he selected 7.5 as the dominance index on SAM.

5.3 Common Causes During Task 2

C7 - Issues with marking an event as finished was the only common cause that occurred exclusively while participants performed Task 2. It happened with three participants from the *co-designers* group. During this task, participants were asked to mark one event as completed, and then mark the other one as finished. All three participants selected the list format during Task 0, a format that changes how the buttons were displayed. On Task 2, P4 and P7 repeated the action that opened the modal confirming if the event should be marked as finished. Meanwhile, P7 and P8 had doubts about the location of the finalize event button. Based on the participants' recordings and the answers during interview, we concluded distinct causes were responsible for each of the breakdowns. P4 mentioned s/he just wanted to test the mark as finished action before actually finishing the event, which led us to understand s/he experienced *T6 - "What happened?"*. In addition, P7 stated s/he was confused with the difference between the "cancel" and "mark as finished" actions in an event. With this in mind, we categorized P7's experience as both *T9 - "Oops!"* and *T13 - "Why doesn't it?"*. Similarly, P8 was initially confused about what "mark an event as finished" meant, but as soon as s/he looked at the app s/he identified the finish button. For this reason, we classified P8 as *T8 - "Where am I?"*. These breakdowns were the only ones that occurred with these participants during Task 2.

When we consider the SAM values (See Table 4), we observe a difference in the range of values selected by each participant. This remark can be associated with the fact that each breakdown experienced by the participants was related to a different CEM tag. Despite, P4 selecting values between 7.5 and 8 for all dimensions, P7 choose values between 4 and 5.5 for all dimensions. During his/her experience with the app, P7 had two breakdowns that were mentioned during the interview. The participant stated *"Even though the action buttons were not very clear, I could understand one was positive (finish) and one was negative (cancel). However, the messages were very similar and confused me."*. Based on this statement, we assume P7's experience was affected negatively with the breakdown. In contrast, P8 selected 9 for all SAM dimensions, a remark that indicates his/her brief moment of doubt indicated by T8 did not affect his/her experience with the app.

5.4 Common Causes on Multiple Tasks

The situation experienced by seven participants and represented by *C11 - Thought today's date would be automatically selected by the datepicker* occurred on both Task 0 and Task 1. It was the only common cause that was observed on multiple tasks. In every case, the participants were expecting the current date would be automatically selected by closing the datepicker without choosing an specific date. After a review of the recordings of the participants' interactions, we noticed Materialize's datepicker default behavior is to display the current day with a distinct font color. This fact led participants to believe a date was pre-selected, while, in fact, the selected date had a different background color. We classified this issue with *T13 - "Why doesn't it?"*.

The only participant who experienced this issue on Task 0 was P4. S/he repeated the action of selecting a date on Materialize's datepicker and realized s/he needed to touch over a date on the calendar in order to select it. Based on this experience, s/he opted to use the browser-default option for the datepicker.

The other *co-designer* participant who experienced this situation was P1. On the recordings, s/he did not realize Materialize's datepicker behavior on Task 0. Instead, s/he encountered the issue on Task 1. S/he stated: *"When I open it (datepicker) doesn't allow me to select the current date by just clicking on 'ok' like the Timepicker does."*. In contrast, P1 selected 9 for all SAM dimensions on Task 1 and mentioned this situation did not affect his/her interaction, as it was just a detail during his/hers interaction. When we associate this piece of evidence with his/her interview, we assume this issue did not affect his/her overall experience on Task 1.

Five participants from the *non-designers* group encountered the issue with the
datepicker. This entire group suffered from other issues on Task 1, but the problems with the datepicker were mentioned by most of them. For instance, P13 complained about this issue while s/he interacted with the app. S/he said: *"UGH! .. Ok .. I found something that stressed me out!"*. Likewise, P15 mentioned the issues as well: *"When I clicked on the datepicker the current date was pink, I thought it was selected, but it wasn't."*. Besides P10, all participants selected values 7 and below for, at least, one index from SAM. We believe the default behavior of Materialize's datepicker was unexpected for many participants and ended up affecting their overall experience while including new events.

6 Discussion

Considering the discussion from the previous section, we answer our RQ *How do communication breakdowns affect the end-users emotional responses?*. By looking at the different situations and communication breakdowns the participants experienced, we can point out that even participants who had comparable issues did not have a similar emotional response. Moreover, we believe the participants' previous experience, both as a user or a developer/designer in case of the *tech* group, could have affected how they understood each task and their perception of each UI element, interaction method and content display formats.

In addition, our findings indicate all of the common issues could have affected the participants' UX. However, some had a bigger impact than others. For instance, *C9 - Issues with the form UI elements* was mentioned by many participants during interview, while *C5 - Repeated the selection of one of the design choices* did not get attention by any participant. With this in mind, we can discuss that some of the observed issues are connected to specific UI elements behaviour, such as Materialize's date picker and Select. As reported in our previous work [7], despite bringing satisfaction, datepickers can cause confusion and frustration on PWA users. Overall, these issues seemed to impact most participants' UX and may be connected to the participants' previous experience interaction or/and working with specific UI elements or interaction methods.

Similarly to our previous work [2], the *tech* participants had fewer or less impacting breakdowns (i.e. the ones that were not mentioned during interview and seemed not to affect the UX). Furthermore, when both *tech* and *non-tech* participants had a similar breakdown, there were cases in which the emotional response of the *tech* individuals was more negative when compared to the other group. We believe this remark is linked to how they had more expertise in the behavior and functionality of different UI elements than *non-tech* ones. In contrast, some of the *non-tech* participants from the *co-designers* group seemed to have a rough start. At least three of them asked for help to understand how to start Task 0. In addition, the *non-tech* group may not be as motivated as the *tech* group because they might not see how it will improve their experience. This remark is important as one of the main motivation for end-users engaging in EUD actions is the improvements that it can bring to their experience [14]. With this in mind, the lack of technical skills on the *non-tech* group makes them not engage as *co-designers* [19].

7 Limitations

As the sample of our work can be considered a limitation (i.e. 18 participants) we took different actions to mitigate this as discussed below. First, we adopted four distinct data sources (an online questionnaire application, researcher' notes, and two recording apps) in order to triangulate data that provides reliability to the analysis. Yet, about the data gathered we assured all participants used the same mobile device, a Motorola Moto G4, avoiding bias which can be caused by different Android versions. By adopting different methods to conduct the data analysis (i.e. SAM and CEM), we could explore the data on different lenses and perspectives and consequently getting rich discussions of the results. Besides, we divided participants into two different conduction groups. The first completed all three tasks and acted as *co-designers* on the app. The second group was the *non-designers* and they did not actively participate in the application's design choices. All selected participants were familiar with mobile devices and apps. We ensured they were frequent users in order to avoid problems related to the interaction with the mobile platform. Still, our work is limited to the use of certain UI elements. These elements were part of the user app preferences and are connected to the user's experience.

8 Final Considerations

This paper presented a comparison of how participants' emotional response is connected to the communication breakdowns they experienced during interactions with a PWA. The findings revealed that even when individuals had similar issues, their emotional response was not the same. Further, participants who had previous technical experience seemed to have fewer or less impacting breakdowns.

Our work contributes to the discussion of how end-users perceive their experience with PWAs. The results show that individuals who have no technical experience had more difficulties in playing the role of co-designers and interacting with the PWA. Those issues can have a direct impact on the overall UX. Working on more guided mobile apps could be a start point for achieving an inclusive environment for end-users. We also present an important discussion regarding the potential of combining PWAs and EUD. PWA-EU could be seen as a contribution for developers of apps in the sense it arranges the responsibilities from the architectural perspective of the application.

As future work, first, we intend to refine the PWA-EU approach by inviting end-users to join developers and designers while they define which UI elements, interaction methods and content display formats will be part of the User App Preferences. Moreover, we would like to include recommendations on how developers can provide more guidance during interaction and consequently support end-users in their role of co-designers. For instance, only after the user interacts with a user app preference (i.e. a UI element) the different preference options are displayed. Further, the app section in which the user can select their preferences is progressively built according to user interactions.

Acknowledgment. This study was financed in part by the Coordenação de Aperfeiçoamento de Pessoal de Nível Superior - Brasil (CAPES) - Finance Code 001.

References

1. Adetunjia, O., Ajaegbub, C., Otunemec, N., Omotosho, O.J.: Dawning of Progressive Web Applications (PWA): edging Out the Pitfalls of Traditional Mobile Development. Am. Sci. Res. J. Eng. Technol. Sci. (ASRJETS) **68**(1), 85–99 (2020)
2. de Andrade Cardieri., G., Zaina., L.A.M.: Playing the role of co-designers on mobile pwas: an investigation of end-users interaction. In: Proceedings of the 22nd International Conference on Enterprise Information Systems - Volume 2: ICEIS, pp. 476–486. INSTICC, SciTePress (2020). https://doi.org/10.5220/0009350804760486
3. Biørn-Hansen, A., Majchrzak, T.A., Grønli, T.M.: Progressive web apps: the possible web-native unifier for mobile development. In: Proceedings of the 13th International Conference on Web Information System and Technology (Webist), pp. 344–351 (2017). https://doi.org/10.5220/0006353703440351
4. Bradley, M., Lang, P.J.: Measuring emotion: the self-assessment manikin and the semantic differential. J. Behav. Therapy Exp. Psychiatry **25**(1), 49–59 (1994). https://doi.org/10.1016/0005-7916(94)90063-9
5. Bueno, D.C., Zaina, L.M.: HyMobWeb: a hybrid adaptation of context-sensitive Web interfaces with multimodality support in mobile devices. SBC Journal on Interactive Systems (2017)
6. Cardieri, G.A., Zaina, L.A.M.: PWA-EU: Extending PWA approach for promoting customization based on user preferences. In: Proc. ACM Hum.-Comput. Interact. 3(EICS), June 2019. https://doi.org/10.1145/3331152
7. Cardieri, G.A., Zaina, L.M.: Analyzing user experience in mobile web, native and progressive web applications: a user and HCI specialist perspectives. In: Proceedings of the 17th Brazilian Symposium on Human Factors in Comp. Sys. IHC 2018, pp. 9:1–9:11 (2018)
8. Casadei, V., Granollers, T., Zaina, L.: Investigating accessibility issues of UI mobile design patterns in online communities: a virtual ethnographic study. In: Proceedings of the XVI Brazilian Symp. on Human Factors in Comp. Sys., IHC 2017, pp. 33:1–33:10 (2017)
9. Clifton, I.G.: Android User Interface Design: Implementing Material Design for Developers. Addison-Wesley Professional (2015)
10. Corbett, E., Weber, A.: What can I say?: Addressing user experience challenges of a mobile voice user interface for accessibility. In: Proceedings of the 18th International Conferenc on HCI Interaction with Mobile Devices and Services, MobileHCI 2016, pp. 72–82 (2016)
11. Costabile, M.F., Mussio, P., Parasiliti Provenza, L., Piccinno, A.: End users as unwitting software developers. In: Proceedings of the 4th International Workshop on End-user Software Engineering - WEUSE 2008, pp. 6–10 (2008). https://doi.org/10.1145/1370847.1370849
12. Danado, J., Paternò, F.: Puzzle: A mobile application development environment using a jigsaw metaphor. J. Vis. Lang. Comp. **25**(4), 297–315 (2014). https://doi.org/10.1016/j.jvlc.2014.03.005
13. Fischer, G.: End-user development and meta-design: foundations for cultures of participation. In: Pipek, V., Rosson, M.B., de Ruyter, B., Wulf, V. (eds.) IS-EUD 2009. LNCS, vol. 5435, pp. 3–14. Springer, Heidelberg (2009). https://doi.org/10.1007/978-3-642-00427-8_1
14. Fischer, G., Fogli, D., Piccinno, A.: Revisiting and broadening the meta-design framework for end-user development. In: Paternò, F., Wulf, V. (eds.) New Perspectives in End-User Development, pp. 61–97. Springer, Cham (2017). https://doi.org/10.1007/978-3-319-60291-2_4

15. Frankston, B.: Bits Versus Electrons. IEE Consumer Electronics Magazine **106–108**, March 2018. https://doi.org/10.1109/MCE.2017.2776463, https://rmf.vc/iEEEHTML5
16. Gullà, F., Ceccacci, S., Germani, M., Cavalieri, L.: Design adaptable and adaptive user interfaces: a method to manage the information. In: Biosystems & Biorobotics, vol. 11, pp. 47–58. Springer Internatinal Publishing (2015). https://doi.org/10.1007/978-3-319-18374-9
17. Lazar, J., Feng, J.H., Hochheiser, H.: Research Methods in Human-Computer Interaction. 2 edn. (2017)
18. Leitão, C.F., Souza, C.S.: Semiotic Engineering Methods for Scientific Research in HCI. Morgan & Claypool (2009)
19. Namoun, A., Daskalopoulou, A., Mehandjiev, N., Xun, Z.: Exploring mobile end user development: Existing use and design factors. IEEE Trans. Softw. Eng. **42**(10), 960–976 (2016). https://doi.org/10.1109/TSE.2016.2532873
20. Peissner, M., Janssen, D., Peissner, M., Häbe, D., Janssen, D., Sellner, T.: MyUI : Generating accessible user interfaces from multimodal design patterns MyUI: generating accessible user interfaces from multimodal design patterns. In: Proceedings of the 4th ACM SIGCHI Symposium on Engineering Interactive Computing Systems, pp. 81–90. No, June 2012 (2012). https://doi.org/10.1145/2305484.2305500
21. Richards, M.: Software Architecture Patterns. O'Reilly Media (2015)
22. Rojas, C.: Making your first progressive web app. In: Building Progressive Web Applications with Vue. js, pp. 1–46. Springer (2020)
23. Sharma, V., Verma, R., Pathak, V., Paliwal, M., Jain, P.: Progressive Web App (PWA) - One Stop Solution for All Application Development Across All Platforms. Int. J. Sci. Res. Comp. Sci. Eng. Inform. Tech. **5**(2), 1120–1122 (2019). https://doi.org/10.32628/cseit1952290
24. Sharp, H., Preece, J., Rogers, Y.: Interaction Design-beyond human-computer interaction. John Wiley & Sons (2019)
25. Sheppard, D.: Beginning progressive web app development creating a native app experience on the web. Apress (2017). https://doi.org/10.1007/978-1-4842-3090-9
26. Yigitbas, E., Stahl, H., Sauer, S., Engels, G.: Self-adaptive UIs: integrated model-driven development of UIs and their adaptations. In: Anjorin, A., Espinoza, H. (eds.) ECMFA 2017. LNCS, vol. 10376, pp. 126–141. Springer, Cham (2017). https://doi.org/10.1007/978-3-319-61482-3_8

Visualizing Deliberation and Design Rationale: A Case Study in the OpenDesign Platform

Fabrício Matheus Gonçalves[1(✉)], Alysson Prado[1], and Maria Cecília Calani Baranauskas[1,2]

[1] Institute of Computing, University of Campinas, Campinas, Brazil
{fabricio.goncalves,aprado,cecilia}@ic.unicamp.br
[2] PPGInf, Federal University of Paraná, Curitiba, Brazil

Abstract. The open phenomenon coming from the free-software movement has gained several fields, including services, digital and physical products. Nevertheless, some authors point out the limited availability of supporting methods and online tools to face the challenges of the distributed collaboration of volunteers with diverse backgrounds and motivations. In this paper, we present the OpenDesign Platform and its potential to support distributed co-creation. A case study conducted with 22 participants attending a Conference in Organizational Semiotics illustrates their use of the platform to clarify the tensions and ideas towards the conception of a community-driven solution to a given design challenge. Results of their participation through the platform analyzed through graphical representations based on concepts of the Actor-Network Theory provided a visual representation of the network constituted by both the participants and the artifacts (boundary objects). These analyses, corroborated by the perception of the participants on their use of the platform, have shown the effectiveness of the OpenDesign Platform to afford online deliberation and communicate elements of the design rationale between participants. The QUID tool, used for the network visualization, revealed its representational power as an instrument for visualization and analysis. Further studies include investigating how the integration of the visualization tool into the OpenDesign platform may increase awareness of other's contributions during the (open) design process.

Keywords: Open phenomena · Open design · Open source · Online deliberation · Design rationale · Organizational semiotics · Actor-network theory · Participatory design

1 Introduction

The open phenomenon to systems design comes from the free-software movement and got spread over several fields including open data, open science, open governance, to name a few. Its origin goes back to the 1970 decade, with the political movement that occurred in reaction to the proprietarization of software source code, chaired by Stallman through the 'GNU Project' [27]. The open-source software, as defined by Warger [32], p.18 "an approach to software development and intellectual property in which program code is available to all participants and can be modified by any of them", focuses

J. Filipe et al. (Eds.): ICEIS 2020, LNBIP 417, pp. 683–707, 2021.
https://doi.org/10.1007/978-3-030-75418-1_31

on the process of software code development and intellectual property. Since then, the open concept has been widened to reach other domains, including the broad cycle of product design [5].

Open design, in its broader sense, has been pointed out as promising and disruptive, although acknowledged as a phenomenon that has been yet little studied by the scientific community [5]. One of the main reasons for this effort seems to be the democratization of design; design here referring to physical as well as digital products, and services. The motivations of stakeholders to get involved in open initiatives range from ideological to the direct and indirect benefits perceived by participants. In this sense, the open phenomenon shares some principles and can learn from Participatory Design (PD) practices and related issues [26]. An example of this match is the user's role bringing his/her expertise to the design process, going beyond the object-for-money trade relation to other forms of contributions in the process and product of design. The tradition of participatory design is to ensure that end users are involved in the design process bringing the tacit and contextual knowledge to help shape design toward the most meaningful solution [26]. However, changes in information and communication technology, consumer culture, communities of interest, manufacturing processes, economies, and global markets have brought new opportunities to extend this tradition [16]. As stated by Frauenberger et al. [10], PD practices must be extended to increase the democratization of technology design, allowing a broader range of stakeholders to participate in the design process. Frauenberger et al. [10] propose to understand this movement through strategies such as scale and dialectics. Scale asks for ways to extend PD reach without giving up on its core qualities. Dialectics is about creating and maintaining spaces for constructive conflict by networking and linking with other stakeholders, organizations and domains. These demands pose difficulties and challenges for PD practitioners and researchers to conduct practices synchronously and in the same geographical place.

Achieving larger scale and improved dialectic requires tools that, on the one hand, afford creative and collaborative behavior, fostering the participation of anyone who feels affected by the proposed construct. On the other hand, such tools should avoid excessive or inadequate actions, such as having the voice monopolized by someone. Moreover, access for anyone to join the discussion at any point in time must be assured. Online Deliberation - OD - tools address the issues related to the online discussion process, helping participants to clarify a subject, by separating pros and cons arguments and opinions from each participant [17], equalizing biases and promoting awareness of points of view. However, used in isolation they lack support for more focused actions such as those needed in contexts of design, in which a group of people is creating something (an artifact, a concept, a system) together.

The Socially Aware Design – SAwD – design model [2] is a theoretical and methodological framework rooted on Participatory Design [26], Universal Design [30] and Organizational Semiotics [20] approaches. Its goal is to allow a collective construction of meaning, encompassing the diverse point of views from people involved and affected by the design of an information system or a digital artifact. The convergence of opinions and objectives occurs through a series of the so-called Semio-Participatory workshops, in-place activities conducted in face-to-face meetings where discussion and expression of all interested parties are promoted and mediated. Buchdid et al. [7] illustrates the

SAwD in a work involving the situational context of designing iDTV applications as a new object in the production chain of a Brazilian broadcasting TV organization. The work, which lasted seven months, aimed to develop an iDTV application for the TdG (*Terra da Gente*) TV show. A group of 10 interested parties from both within and outside the organization took part in the Semio-Participatory workshops and co-design activities. While that work illustrates the situated aspects of SAwD inside the organization, we wonder how it could gain scale, maintaining discussion around the main artifacts used in the synchronous and face-to-face meetings.

The OpenDesign project[1] was proposed to scale the SAwD, maintaining its structure and boundary objects, while also inspired by the Open Source (OS) philosophy. The OS phenomenon has provided a number of high quality software products, gathering and coordinating efforts from people with different skills and from different places. The OpenDesign main idea is to bring the same type of collaborative phenomenon to the activities that precede coding. The project's objective is to formalize a community-driven design process for interactive system design and to provide tools for its accomplishment. In the context of this project, a web platform was developed, intended to enable the Semio-Participatory workshops from SAwD to be carried out in asynchronous and distributed scenarios. One of the products of the OpenDesign Project is the OpenDesign platform, a web-based system enabling a community of participants the experience in open design.

In this paper, we investigate the importance of deliberation and design rationale in the (open) design process, through the use of the OpenDesign Platform in a case study conducted with participants of a Conference Summer School [25]. The platform we are addressing in this work can be seen as a purposeful social technology, where participants voluntarily interact towards a shared design goal, starting from the early stages of clarifying a design problem, by identifying the interested parties, and anticipating their potential issues regarding a prospective design solution. By 'open' we mean a design process that allows contribution of the volunteers to the product design since its conception, not only in a prototyping stage. The process is supported by well-established methods and tools, materialised in an actual online system: the OpenDesign platform and its artifacts (boundary objects).

The contributions of the work can be summarized as: 1. the presentation of a platform for open design, its architecture and main artifacts, including its deliberation and rationale aspects; 2. A case study revealing a preliminary use of the OpenDesign Platform; 3. A graphical instrument of visualization based on the Actor-Network Theory, for analysis of the platform use. This investigation advances our previous work [11] by further discussing the association network of participants and their contributions (boundary objects), highlighting the connectedness of their participation and the intersubjective relation of participants while contributing through the platform. Also, we further describe the platform architecture that enabled such co-creation to take place. Thus, this extended version reflects the new content also with a new structure for the text. The paper is organized as follows: In Sect. 2 we present the background and related work context. Then, in Sect. 3 we present an overview of the OpenDesign Platform, including its architecture and technological aspects, and its boundary objects. A case

[1] https://opendesign.ic.unicamp.br.

study on the Platform Usage follows in Sect. 4, with discussion on the main results. The final section concludes pointing out further work.

2 Background and Related Work

Bonvoisin et al. [6] argue that the spread of ICT and cheap low-size production tools like 3D-printers led to the community-based and open source development of physical products. This innovative organization of product development (open design) offers a great opportunity for continuous improvement of products as well as a potential for product innovation and, in consequence, incubation of new businesses. The authors also point out the limited availability of supporting methods and online tools for helping to face the organizational challenges raised by distributed collaboration of non-experts, non-professional and non-contractually engaged volunteers. They claim that online collaborative platforms are still needed with special features to build and keep the community active, providing mechanisms for the convergence of the design process, for knowledge management, and for supporting co-creation. Those features would be essential to the rise of open design.

In the tradition of PD, some efforts have been conducted with social technologies to increase people's participation regarding information production, publication, and sharing [13]. Hargreaves and Robertson [14] propose the use of social technologies (Skype video calls, screen sharing and email) to allow discussions between researchers and participants who are remotely located, and prototyping activities to occur at a distance. The interaction among participants is structured in regular cycles of reflective discussion and prototype modification. While social technologies are participatory by their nature as they require and depend on people's involvement to take shape, they also have drawbacks. For instance, regarding power, it is not clear who exactly benefits from people's participation, how to value participation without exploitation. Problems with privacy, ownership, deletion and sharing of personal information might be some issues raised when developing participatory systems. Bringing a participatory approach to the design of such systems is critical to ensure that people have the ability to negotiate, control and understand the implications of participation as they evolve [12].

In a study on the open design state of the art review, Boisseau et al. [5] illustrate that the subject of open design of products started in the early 2000s and is still a growing phenomenon. They have shown that the limited number of published papers suggests the concept has not spread over traditional design communities yet, still being restricted to a few research groups. Drawing on the design science approach, they argue that the subject that designers have to address is to provide a *plan* based on a *gap* (a design problem), through the development of a solution. Three elements are proposed to describe the product design process: a) the phases and activities that constitute the process, b) the boundary objects that constitute the information formalized and carried from one phase to the next one, and c) the participants (or stakeholders) taking part in activities of the design process. We should notice that the boundary objects are used for sharing a common understanding of the solution being constructed among the participants and that the plan is the final boundary object.

Fischer [9] studied design communities and identified types of common barriers they must cope with in order to work together: spatial (across distance), temporal (across time), conceptual (across different communities of practice), and technological (between persons and artifacts). For spatial barriers, it is straightforward to propose the use of computer-mediated communication as a solution. On the temporal domain, he stresses that "long-term collaboration requires that present-day designers be aware of the rationale behind decisions that shaped the artifact, and aware of information about possible alternatives that were considered but not implemented" [15, p. 155]. Conceptual barriers must be overcome by humans serving as knowledge brokers and by integrating diversity, making all voices to be heard. Overcoming the technological barriers depends on approximating people and technological artifacts and this approximation can be facilitated through their involvement in the design process itself.

Several attempts have already been made to provide a distributed platform for online collaborative design. Heintz et al. [15] searched for tools to support such tasks, and analyzed six applications:

- GABBEH [23] mimics paper prototyping by enabling users to comment on the current design by drawing with a software tool, but its technical requirements made it too restricted.
- DisCo [31] supports distributed PD sessions, but was not publicly available for use at the time of this study.
- Appotate (appotate.com) brings together different stakeholders, allowing them to give feedback on a prototype.
- MarkUp (markup.io) allows the user to draw and write on a website; however, it does not offer a structured way to store and retrieve this feedback.
- MyBalsamiq (mybalsamiq.com) mock-up software offers a wide range of common interface elements to create feedback on prototypes.
- Webklipper (webklipper.com) is an online application that enables the user to annotate websites and share the results.

Most of the mentioned tools mimic paper prototyping and enable users to comment on the current design. Most of them focus on the support for sharing the artifact being built - a prototype or a mock-up, for instance. They also propose a tool that, beyond the already found features, provides a like/dislike heatmap and enhanced interactivity.

More recently, tools such as UXPin (www.uxpin.com), Figma (www.figma.com) and Proto.io (www.proto.io) became available online as solutions for collaborative design. As a common feature, they provide real time editing of prototypes for all members of a team of designers, allowing the addition of comments and keeping the change history. Similarly to the tools studied by Heintz et al. [15], their focus is mostly on how the final product will look like or behave, and participation and deliberation occurs over an already materialised mockup. However, a collaborative, geographically sparse, culturally diverse design process asks for tools that allow the convergence of opinions and concepts also about what is being designed, for what purposes, uses and contexts, and who will be involved in its lifecycle.

Bjögvinsson et al. [4] discussed that a fundamental challenge for designers and the design community is a change of focus from designing objects to designing socio-material assemblies, which encompass heterogeneity of perspectives among actors who engage in attempts to align their conflicting objects of design. In their proposal, the role of non-human participants in the design process, such as prototypes, mock-ups, models, and diagrams is to act as "presenters" of the evolving object of design, supporting communication and participation in the design process, potentially binding different participants together.

We agree with Bonvoisin et al. [6] in acknowledging the concept of open design as a significant phenomenon, supported by trends in contemporary digital technology and organization, which faces significant challenges of interest for several scientific disciplines. For example, they cite (p. 3): "*understanding the dynamics of online communities, developing motivation models for contributors, identifying business models that allow to create sustainable economic value with open source products, understanding the decision processes in horizontal work organizations, clarifying legal issues of intellectual property, identifying ways to ensure and validate product quality, liability and safety*", among others.

As for the open design of products proposed by Boisseau et al. [5] in the three elements that constitute it (the gap, the process itself, and the plan), the authors state that the gap is contingent, and the actors of the design process have no influence on it. In our approach, the addressed design problem (the gap for Boisseau and colleagues), is open too, as it is open to the interpretation and clarification by the interested parties, as part of the design process. Although our concept of design may reach the plan (i.e. for example, the drawings of a design product), differently from these authors, we are not limiting the object of design to a (material) 'product' of industrial design. In our work, we associate to 'design' the activities that precede the code production in a digital information system design, encompassing problem discussion, deliberation and clarification, ideation of solutions, requirements elicitation, design rationale. The next section provides an overview of our proposal for the OpenDesign Platform and its boundary objects.

3 Deliberation/Rationale in the OpenDesign Platform

In this section we present an overview of the OpenDesign Platform based on the phases and activities that constitute the (SAwD) process; the boundary objects that constitute the information formalized (the artifacts); and the dynamics of activities with participants (or stakeholders) taking part in the design process. Processes of Deliberation and Rationale underlying the use of the artifacts are indicated with illustration of an specific artifact (Deliberation Frame). Technical details of the Platform development are also provided showing how it was made possible.

3.1 The Platform Overview

The Socially Aware Design model has inspired the process which we have incorporated into the OpenDesign Platform. Traditionally, this model is inspired by some Organizational Semiotics [19,29] artifacts and consists of three phases, each guided by one specific artifact. The first phase is the elicitation of stakeholders, which is made using the artifact known as Stakeholders Identification Diagram (SID) adapted from Xiaojia [33]. This artifact allows us to indicate all those who will affect or will be affected by the product of the design, which can be categories of individuals (like developers, designers, etc.), or entities (such as universities or corporations). SID's graphical representation usually has five sequential layers that contain each other, like an onion, as we can see in Fig. 1.

The idea is that the innermost layers contain stakeholders that are more directly involved with the design product. From the center to the border, we have the Operation Layer, where stakeholders who operate the envisioned artifact are placed, followed by the Contribution Layer, where are represented the actors directly involved to the current situation or future solution; next, the Source and Market Layer represent, respectively, those who provide information and products to the discussed situations, and the related to the market; finally, the Community Layer is used to represent the broader social environment in which the problem and its solution are placed, being direct or indirectly affected by them.

During this phase of the SAwD process, participants discuss who they believe affects or is affected by the designed solution, and to which layer each stakeholder belongs. After ideas of several possible stakeholders are placed into the artifact, a deliberation can be conducted on whether or not the identified stakeholders are correctly named, categorized or even if they actually should be there. This provides a broader understanding for all participants of who are the person and entities interested or being affected by the subject, and the diagram materialises this shared and negotiated knowledge, as well as some of the rationale behind the decisions taken by the participants to materialize it in such a specific way.

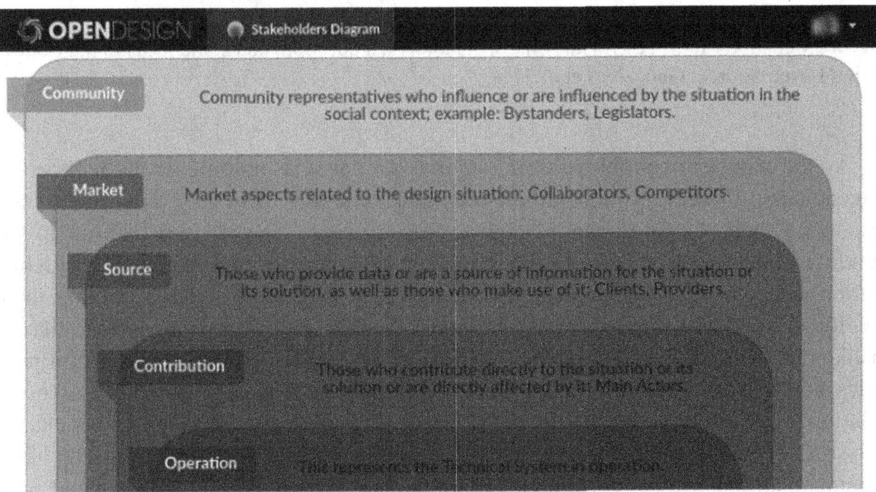

Fig. 1. Stakeholder Identification Diagram (SID) with five nested layers.

The next phase is guided by the Evaluation Frame (EF), a table-like artifact that supports the reasoning of problems and solutions associated with each stakeholder identified in the SID adapted from Baranauskas et al. [3]. Usually, it contains one column for raising issues, and another column for solutions or ideas associated with the issues. Then, each row represents one layer of the SID. Therefore, the contribution of the proposed artifact to the solution of problems of each stakeholder, and conversely, the contributions of the stakeholders to the existence and operation of the artifact are clarified, from the point of view of each participant. On the OpenDesign platform, such a format was slightly adapted to allow users to navigate through the layers from a left-side menu, allowing to tag a specific stakeholder to an issue and creating links between problems and proposals as first steps towards revealing and reconstructing the elements of the rationale the participants took to make their contributions (Fig. 2).

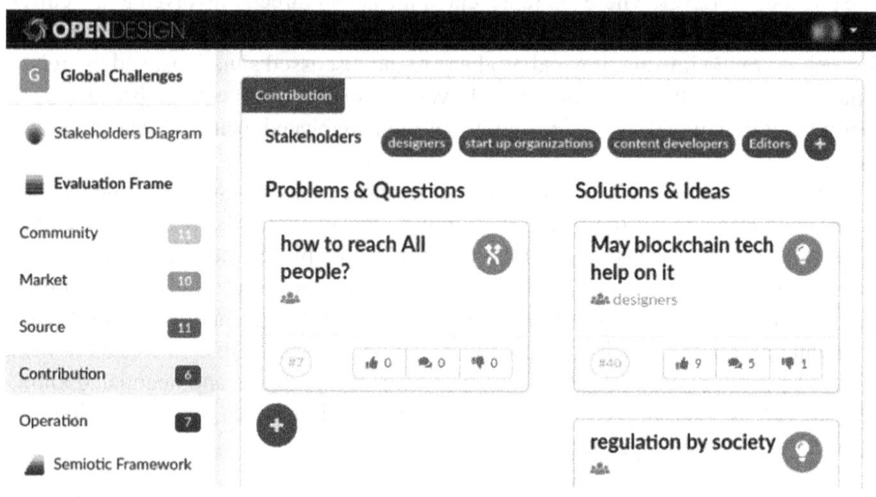

Fig. 2. Evaluation Frame (EF) with some stakeholders, a question, and some ideas in the Contribution Layer. Source: Gonçalves et al. [11].

As the number of problems and solutions grows as a multiple of the number of stakeholders, it becomes more difficult to manage a face-to-face or videoconference discussion about the participant's opinions for each issue. Therefore, in addition to the building of the EF itself, the OpenDesign Platform proposes and implements a deliberation mechanism, in which each user, interacting with the platform, provides their opinion. The deliberation aspect of this stage is reinforced by allowing users to like or dislike entries, and to add comments justifying their votes. Hence, the platform provides a **deliberation frame** (Fig. 3), where participants can raise pros and cons about a solution that was placed on the EF, or they can add neutral comments to further clarify the issue. They may also endorse each other's arguments by hitting the thumbs up of an argument, all this activity is summarized by the numbers of thumbs up/down and conversations in the card. The main idea, then, is to provide a way to document not only the deliberation process but also elements of the rationale behind decisions, either by registering participants' votes and arguments but also inviting the registering of decisions taken without enforcing a specific strategy to reach the final decision. The focus was to listen to all interested participants, avoiding speech monopoly and other face-to-face deliberation problems with the added benefit of keeping some of the trails left by the participants without increasing much the effort taken.

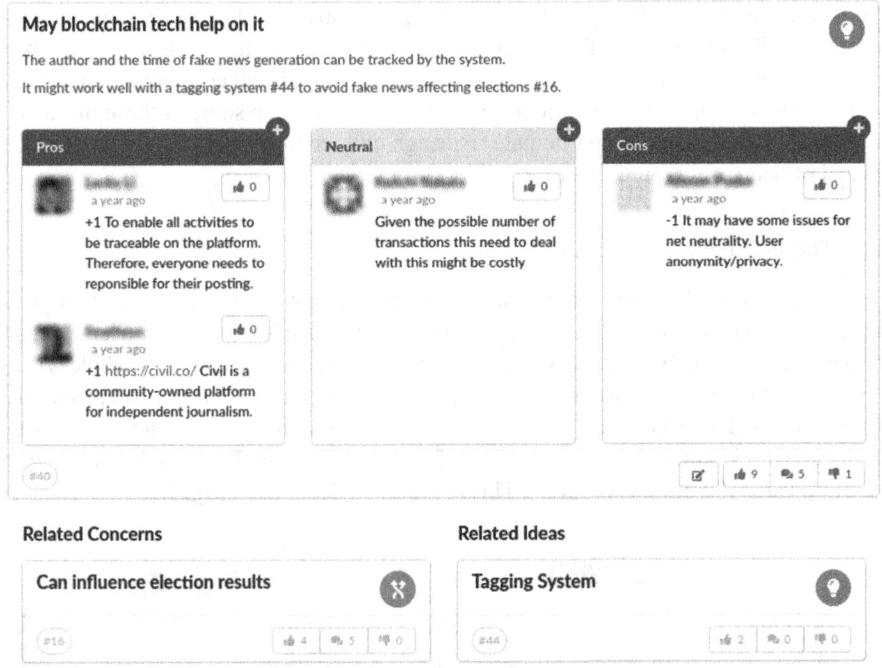

Fig. 3. Deliberation Frame (DF) with pros/cons arguments received by a proposal, followed by related concerns and alternative ideas.

Finally, the third stage is guided by the Semiotic Framework (SF) artifact adapted from Stamper [28], which provides six levels of knowledge, stacked on top of each other in a progressive manner, similar to a ladder [11]. The bottom three levels are related to the structure of signs, how they are organized and transmitted: physical, regarding their material support, density, hardware; empirics, referring to statistical properties of signals and codes, noise, medium, entropy; and syntactics, related to the combination of signs and their formal structure, regardless of their meaning. In turn, the upper three levels are related to how signs are used, in terms of meanings, intentions, and social impact they have: semantics, concerning the meanings of signs, propositions, their validity, truth; pragmatics, regarding the intentional use of signs and its practical consequences, as well as intentions, conversations and negotiation; and the toppermost social world, where consequences of the solution system in human activities are anticipated. Therefore, the SF can be a useful instrument for identifying and organizing the requirements of the design product.

Together these three artifacts provide criteria and structure to promote the process of deliberation and rationale, where participants are engaged in discussing particular aspects of the design, which they might not think of without the artifacts. Furthermore, the platform instantiates these artifacts in a way that encourages and documents deliberation and rationale. For instance, besides designers, apprentices and domain knowledge's holders taking part in a given OpenDesign session, the participants are prompted to remember and represent the concerns and hopes of absent stakeholders. This remembering is facilitated by the SID boundary object that frames 5 distinct levels of involvement with the system, from daily operationalization to the distant community with spectators and legislators. The platform aims to give direct voice to some of these more distant stakeholders to directly participate in design and feedback about a design product, instead of being only represented.

3.2 The Platform Dynamics of Use

The OpenDesign platform can be used following certain steps. First, a challenge is proposed by one or more key participants, representing each local group of users. Not all local groups participants are necessarily operating the platform online, nevertheless they might participate in local discussions and may contribute to the solutions and proposals. Moreover, a video conference can create another communication channel, supporting all the activities, but mainly the selection of an issue from the EF to be worked out in the SF towards the design solutions. This dynamic is depicted in Fig. 4.

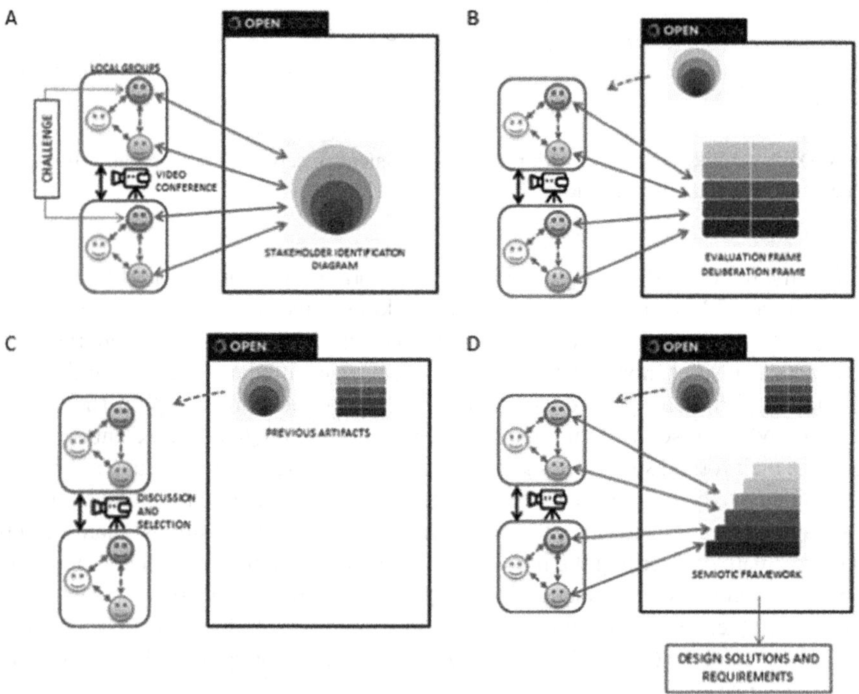

Fig. 4. Open Design Semiotic Workshop dynamics synthesized: A) a challenge is proposed by key participants in each local group and then the SID is populated with related Stakeholders; B) evaluation frame and deliberation are carried out; C) based on the rationale captured by the artifacts available at the platform, issues are discussed and a focal idea of solution is selected for the next phase; D) semiotic framework is generated, giving rise to the proposal of solutions and requirements. Source: Gonçalves et al. [11].

3.3 Platform Architecture and Technical Details

Since OpenDesign aims to support projects that need a continuous design development experimentation flow, but our main focus is on the design aspects, we choose to sum efforts with a well-established platform for online and distributed development, deployment, and operation of software products as Gitlab (gitlab.com). Since it is a loose coupling, we envisage it would be possible, in the future, to integrate OpenDesign with other platforms like Github (github.com), an alternative in the same niche. Also, even other platforms not only focused on software but in other human endeavors, e.g. Trello (trello.com), could benefit from participatory and distributed design articulated with deliberation mechanisms and rationale support provided by the OpenDesign.

Gitlab has a hybrid open source and commercial license that leverages both open contribution and financial sustainability providing software as service with enterprise premium features. Besides the source code that could allow more coupled customization, they also provide an extensive application program interface (API) allowing loose coupling with almost any functionality provided in the platform, taking advantage of the fast pace of their development; e.g. since we started this project they have even introduced a new feature with initial support for design they call "Design Management" to facilitate communication flow between designers and engineers allowing to upload design assets, e.g. wireframes, mockups, to GitLab issues. This feature will be useful for a new OpenDesign module than allows online braindraw [21,22], where distributed participants edit each other basic drawing mockups in fast-paced turns.

OpenDesign Platform has three main artifacts or boards where the unit of information is manipulated as 'cards' in its user interface, e.g. problems, proposals, and requirements. Each card is mapped into a Gitlab 'issue' and shares elements as numeric identification (id), title, description, author, creation date, and up/down votes. This allows searching, filtering, or even informing Gitlab boards layout with design labels. This integration is achieved using the Gitlab Representational State Transfer (REST) API (Fig. 5), which also allows us a more independent choice of frontend technology. OpenDesign uses the Meteor Javascript framework (1.10.2), React (16.8.6) and Semantic UI

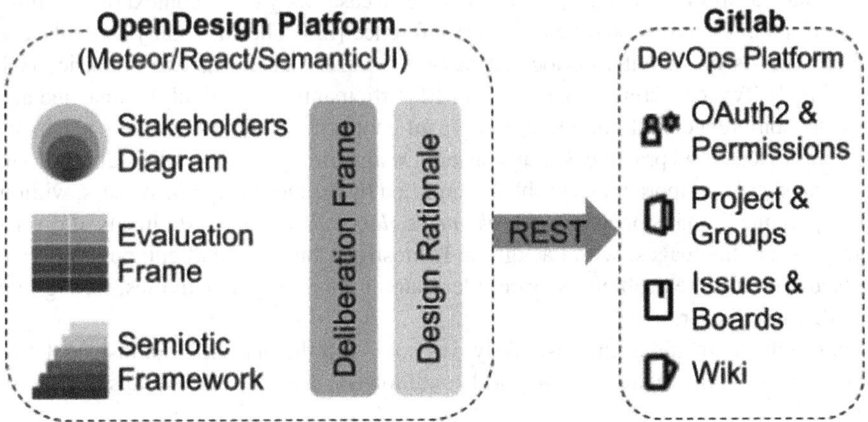

Fig. 5. OpenDesign Platform architecture and Gitlab integration.

(2.4.1) than combined provides the base for a modern-looking, fast and responsive user interface and still easy to maintain source code, once it is an academic and open source collaborative project[2]. We also take advantage of project and group organization from Gitlab, and Oauth2 authentication and access level permissions for a smooth transition between the two platforms using the same user credentials.

Other elements of OpenDesign artifacts that have no direct mapping are transformed into a Gitlab issue label to allow their usage after design, e.g. Gitlab Kanban board where cards showing the project development flow (to do, doing, done) are enriched by meaningful design labels coming from OpenDesign. Thus, each OpenDesign card becomes a design issue in Gitlab, stamped with labels of the OpenDesign artifact it comes from: 'StakeholderDiagram', 'EvaluationFrame', or 'SemioticFramework'. A Gitlab issue that comes from the OpenDesign SID or EF artifacts, receives a label stating the layer it occupies in the diagram ('Operation', 'Contribution', 'Source', 'Market', or 'Community'). Finally, in the case of an EF issue, it is also labeled with its category ('Problem' or 'Proposal') and if attributed to a Stakeholder, it also gets a label with the stakeholder title, preceded with a '@' to indicate it refers to a stakeholder. SF issues are labeled with the level it occupies in the semiotic ladder ('PhysicalWord', 'Empirics', 'Syntactics', 'Semantics', 'Pragmatics', or 'SocialWorld').

4 A Case Study on the Platform Usage

In this section we present the context and participants of our case study and introduce the representation used to analyze the participants' usage of the platform as a network of associations constituted among the participants and the artifacts they collectively constructed; then we present a qualitative synthesis of the participants' perceptions about the platform, raised through their answers to an evaluation questionnaire; a discussion of these results follows.

4.1 Context and Participants

To evaluate the proposed platform we conducted a case study in the context of a Summer School in the Organizational Semiotics Conference [25]. The case study took place as a 3h15' hands on activity, co-located in two research centers in two countries (UK and Brazil). Two conference rooms (9 and 12 participants, respectively) connected also by videoconference, and one more participant connected from home. The participants goal was to learn and practice Socially-aware Design through the OpenDesign Platform, discussing a contemporaneous problem amplified by technology pervasiveness, without corresponding human concerns - *the fake news global challenge*. Participants of diverse countries and languages, with backgrounds mostly in business and computer science, worked on the same problem, sharing ideas along the hands on activities, through the OpenDesign Platform.

Since the main aim of the case study was not to test the platform, but instead, to use it in a scenario more related to real world conditions, to drive the participants to focus on

[2] https://gitlab.ic.unicamp.br/opendesign/opendesign.

the proposed 'problem' (the fake news global challenge), the platform was previously fed in a so-called "warmup session" involving only some of the participants and other volunteers (5 co-located and 4 digitally-located, in a 1 h activity, one week before the event), providing 24 stakeholders, 13 issues and 7 proposals of solutions. The purpose of this warmup was just to seed some elements for an initial discussion, hence the SF was intentionally left out of the warmup to keep it open until further deliberation and participation of all volunteers.

The use of the collaborative platform was organized into phases: clarifying the problem, raising the main interested parties, raising issues the interested parties might have, proposing ideas of design solutions for facing the issues, discussing and selecting one potential solution to carry on, and organizing requirements for the selected solution idea. At the end, participants were invited to express their opinion on the experience with the Platform. Table 1 illustrates the Agenda of activities carried out in the 3h15' hands on meeting.

Table 1. Agenda for the hands on meeting. Source: Gonçalves et al. [11].

Boundary objects	Time spent	Activity description
Videoconference	20	Greetings; platform overview and challenge presentation
Stakeholder Identification Diagram	20	Making sense of previews entries and raising new Stakeholders
Evaluation Frame 1	35	Raising most issues and arguments
Evaluation Frame 2	30	Creating solution proposals and arguments
Semiotic Framework 1	10	Discussing and selecting ideas from the EF to SF
	30	Coffee-Break
Semiotic Framework 1	15	Discussing and selecting idea of solution from EF to SF
Semiotic Framework 2	25	Filling the SF with requirements
Videoconference	10	Wrap-up and invitation for the Evaluation Questionnaire

Table 2 synthesizes the participants' main contributions using the platform before (warmup) and during the workshop. Some participated (6) in both moments, while in the workshop some participants did not interact through the platform (4), although they exchanged ideas presentially with collocated colleagues. The deliberation (arguments) and SF filling (requirements) was performed only during the workshop.

Table 2. Participants and information they formalized into the platform. Source: Gonçalves et al. [11].

	Warmup	Workshop	Total
Participants	9	22	25 (6 in both)
Active in Platform	9	18	24 (3 in both)
Duration	1 h	2 h 45 min	3 h 45 min
Stakeholders	24	26	50
Issues	13	12	25
Solution proposal	7	13	20
Arguments	–	61	61
Requirements	–	23	23

After the hands on activity, the participants were invited to give feedback about their experience with the platform through an online questionnaire; our aim was to raise the platform's boundary objects capacity to promote deliberation, rationale and awareness, from the point of view of participants.

4.2 Data Representation for Analysis

Social networks are not just made up of persons, they consist of people who are connected by the so-called social objects [8], which we are understanding as content in boundary objects. From this perspective, Engeström [8] argues that what causes the failure of many social networking sites is the lack of shared objects acting as hubs for people's interaction [1].

In order to understand the interplay between people and non-human entities in a social scenario, the Actor-Network Theory (ANT) proposes to study social phenomena as heterogeneous networks where both human and non-human can contribute [18]. This approach allows one to acknowledge the mediation role of objects that propagate human intentions. Such heterogeneous social networks can provide a visual representation of both the participants and the artifacts they produce and interpret. For instance, scientific social networks are mediated by publications, and by analyzing both entities together we can highlight structures of scientific communities [24].

In our study, this approach is employed to illustrate the interactions between participants and the diverse artifacts (boundary objects) used in the Semiotic Workshops. Whenever a participant creates or edits a stakeholder in the Stakeholder Identification Diagram, an issue in the Evaluation Frame, or a requirement in the Semiotic Framework, a bond is established between them. Moreover, issues of the EF related to each stakeholder are also linked together. Social relations between participants arise also when a participant makes an argument on the issues created by another person. These structures allow us to evaluate the interactions afforded by the platform. For instance, the presence of cliques (short loops with a single participant) or many disconnected vertices may indicate a poor discussion, leading to less representative design proposals. Conversely, a richer discussion can emerge from associations with no single hubs, nor disconnected sub-groups.

The visual representation of these social interactions and the mediating artifacts are represented in Fig. 6 through the QUID[3] tool: participants are depicted as red circles, stakeholders from SID as blue circles, issues and solutions pointed out in EF as blue squares and Semiotic Framework items as dark blue squares. For the arguments, drawn in green, triangles mean positive, diamonds mean neutral, and crosses mean negative. When any of the elements was created during the warmup phase, it will be depicted as a dashed line linking it to the creator. Node sizes are proportional to the number of other vertices attached to it.

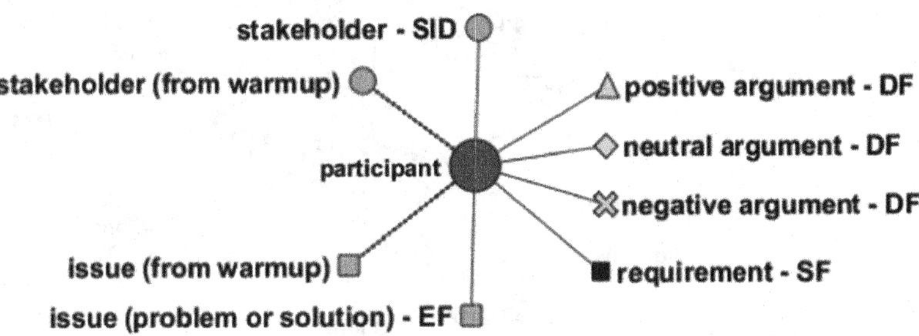

Fig. 6. Visual representation for the network of interactions between people and types of artifacts created on the platform.

Figure 7 shows an example of what can be represented using this graph notation, based on actual data from the case study to illustrate. The participant P15 pointed out the "Citizens" stakeholder in the Stakeholder Identification Diagram; afterwards, through the Evaluation Frame, attached the issue "use common sense". This user also created a positive argument to the issue. On the Semiotic Framework, user P15 added a "secured unchangeable information" requirement to the issue "may blockchain tech help on it" suggested by P17.

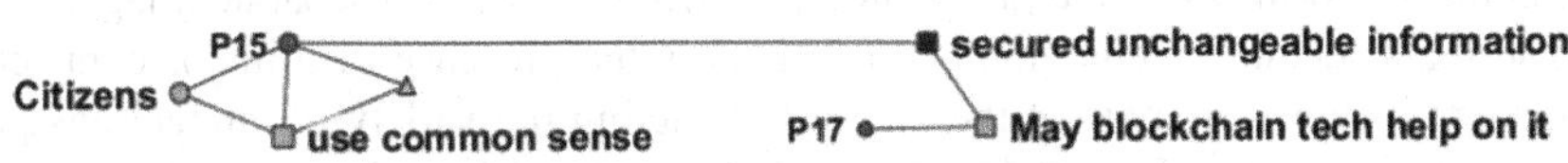

Fig. 7. Example of the elements used in the graph of relations.

4.3 Results

In this section we first analyze the use of the platform based on the network of associations constituted among the participants and their contributions through the artifacts; then we synthesize the perceptions of the participants about the platform, raised through their answers to the questionnaire.

[3] http://www.quid.net.br.

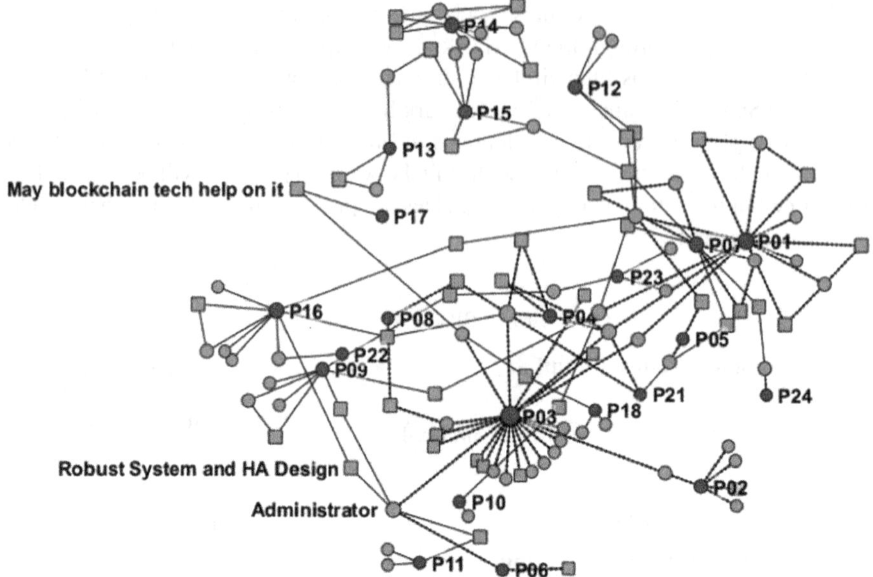

Fig. 8. Complete graph of relations between participants (red circles, randomly numbered and labeled as P*nn* to anonymize participants), stakeholders they raised or edited (blue circles) and problems and solutions pointed out (blue squares). (Color figure online)

Analyzing Interactions Through Graphs. From the interaction logs collected by the platform and processed by the Quid tool, it was possible to visually represent who contributed and, from the social objects perspective presented earlier, how their work builds up to the workshop outcomes. Figure 8 shows the main artifacts with which the participants interacted - namely the stakeholders, issues from the Evaluation Frame, and requirements from the Semiotic Frame - and the paths of interactions they create. This picture does not contain the positive, negative and neutral arguments from the deliberation phase; despite the fact that comments and issues could be created at any time in the platform, with no *a priori* order, we will depict it in a separate image.

The main feature of this graph of relations is the presence of a major connected component comprising most of the vertices, depicting the variety of interactions provided by the platform. Except for P14 and their related boundary objects, all other participants engaged in interactions with the content created by others, creating paths of associations linking most of the participants. This suggests a successful sharing of ideas and concepts among participants.

Although many contributors interacted with their own content - for instance, adding a problem to a stakeholder proposed by themselves, creating the "triangles" on the graph - they also contributed to others' contents. Also noticeable is the lack of correlation between the physical location of users during the workshops and their placement on the graph; for instance, although P03 and P16 were on different places, there is a short path between them on the graph, as they constructed collaboratively some content involving the "Administrator" stakeholder and the "Robust System and HA Design" issue.

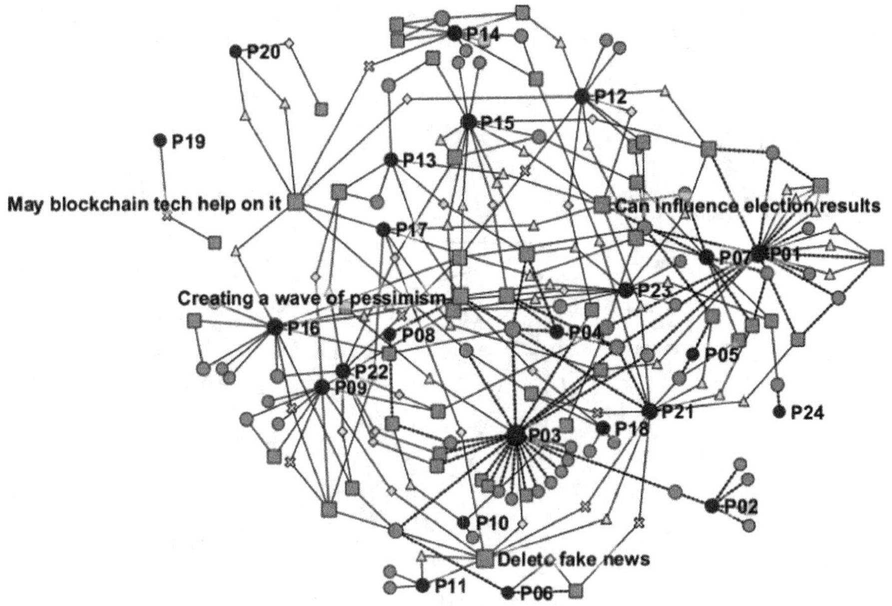

Fig. 9. Relationship between participants (red circles), problems and solutions (blue squares) and arguments (green shapes). Positive arguments are depicted as triangles, neutral ones as diamonds, and negative arguments as crosses. The issues with a greater number of associations are labeled. (Color figure online)

It is also noticeable that more eloquent participants do not necessarily attract more attention. For instance, the sole contribution of P24 was the stakeholder "social groups", which was linked by P07 to the issue "popularity of fake news". Conversely, participant P03 provided several stakeholders that did not receive further interaction by other users.

Regarding the deliberation phase, Fig. 9 brings a view on how each argument is related to the issues and solutions raised in the Evaluation Frame, and to their authors as well. As the node sizes grew proportionally to the number of other vertices now attached to them, the graph reveals the most active contributors and, according to their interaction, the most interesting subjects raised.

Issues that received most arguments are: "Delete fake news", suggested by P11, "May blockchain tech help on it", created by P17, "Can influence election results", provided by P07 during warmup phase, and "Creating a wave of pessimism", created by P08 also during the warmup. It is noticeable that most commented problems and solutions were not proposed by the most eloquent participants suggesting different profiles of participation, and showing that the tool provided a balanced discussion environment. Authors also did not try to overemphasize solely the positive aspects of their proposals. It is also noticeable that "old" and "new" issues, that is, created in different moments of the system usage, received similar attention, suggesting a successful support for asynchronous discussion and preserving its rationale.

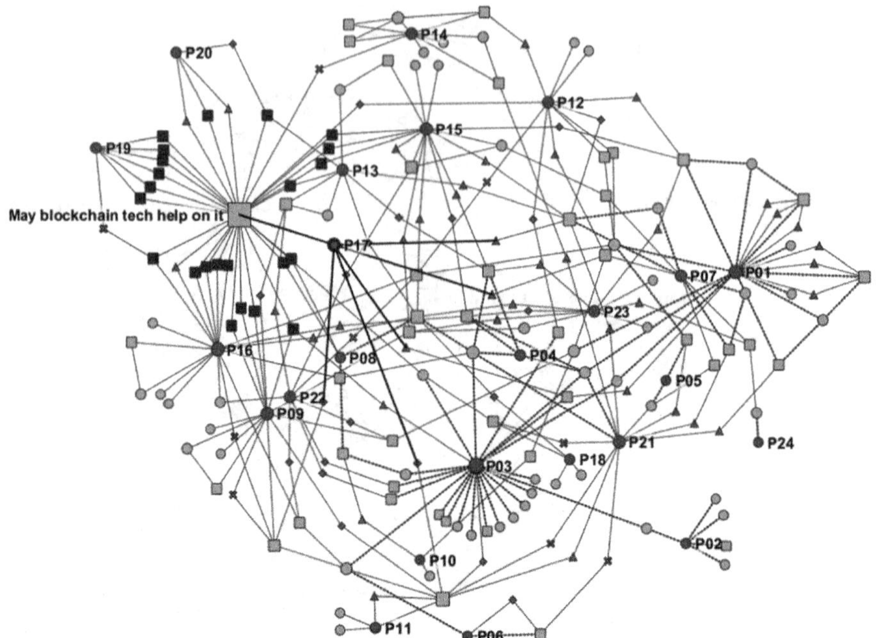

Fig. 10. Complete graph of interactions after the Semiotic Framework and Deliberation phases were carried out. Requirements of the SF are shown as dark blue squares. In bold, highlight of the interaction of the participant who contributed to the idea selected as the focal point in the SF. As the size of vertices is proportional to the number of connections, the "May blockchain tech help on it" node grew even bigger. (Color figure online)

The graph including the requirements in the Semiotic Framework (white circles) highlights the interaction of the participant P17 who contributed with the main idea ("May blockchain tech help") for the final phase. This participant registered only this proposal, after engaging in deliberation with other 5 issues and ideas proposed by 4 other participants with 3 positive and 2 neutral arguments Fig. 10. This same behavior is shared by other participants (P16, P09, P15, P13) while other participants were more engaged in deliberation (P22, P21, P23, P12) and others contributed more to SF but not in the deliberation (P19, P8).

Figure 11 brings the same content of Fig. 10, but highlighting the content created during the warmup phase in distinction to the workshop carried out afterwards. It is clear that the final solution grew over the warmup content, but most of the content was created during the workshop phase, including the final solution: in workshop phase, P17 provided the solution "May blockchain tech help on it" based on the stakeholder "designers" proposed by P03 in warmup phase. The "designers" stakeholder was also linked to a "peer review system" solution by P18, but the discussion did not go further.

Zooming in on the graph we can pick examples of distinct behavior during the activities Fig. 12. While P17 spread their contributions among several other users' contents, P01 advocated on its own side, giving positive feedback to issues he/she created

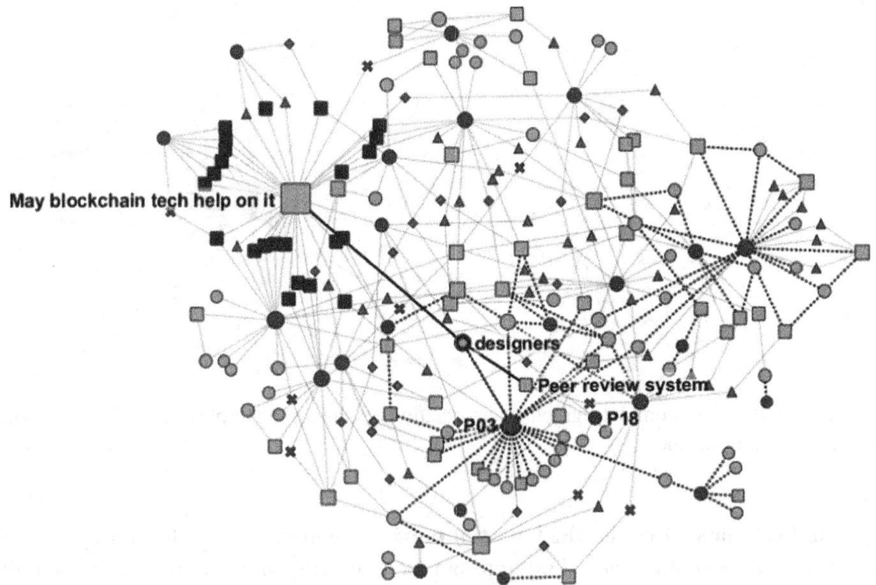

Fig. 11. Complete graph of associations, highlighting the distinction between the warmup phase (dashed edges) and the workshop phase (light gray edges). Although warmup content is concentrated on the lower right side of the graph, there are several links between it and the content created afterwards, including the final solution. (Color figure online)

Table 3. Mode of the answers for the Part 1 closed questions. Source: Gonçalves et al. [11].

Question	Mode
The format of the argumentation (Pros/Neutral/Cons) is useful for collective decision making	7
The format of the argumentation (Pros/Neutral/Cons) facilitates collective decision making	8
Voting (Like/Dislike) is useful for collective decision-making	9
Voting (Like/Dislike) facilitates collective decision-making	7
I discussed with another workshop participant regardless of the platform	9
I recorded the result of a face-to-face discussion on the platform	8
I considered arguments recorded on the platform to build my own opinion	7
I can easily relate a requirement to a stakeholder	6
A solution proposal is always related to the problem that it seeks to solve	7
It was easy to relate a requirement on the Ladder to the proposed solution that gave rise to it	8
Arguments of other participants influenced my opinion about the importance of a problem	8
Arguments of other participants influenced my opinion about the importance of a stakeholder I had not considered	9
Arguments of other participants influenced my opinion about the value of a solution	6

him/herself ("Remember to tell everyone" and "tell me more about that" - which looks more like a menta-communication). The diversity of interactions may be influencing the final choice of the group towards P17's proposal.

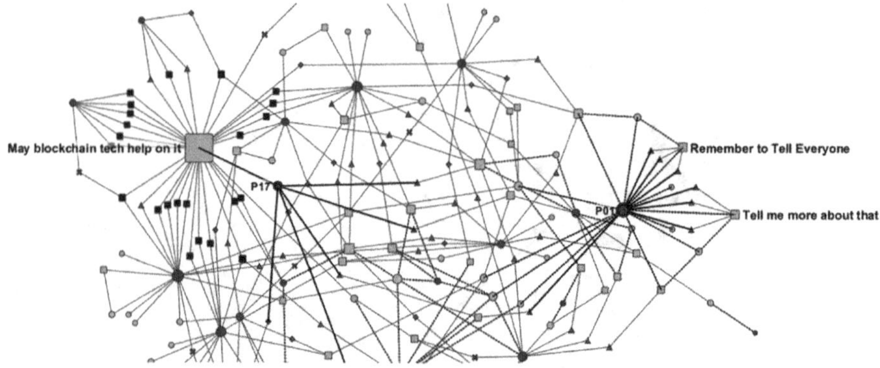

Fig. 12. Detail of the complete graph of associations, showing examples of participants with distinct behaviour profile.

Getting Feedback Through the Questionnaire. Regarding the platform usage evaluation, after the activities carried out through the platform, an invitation was made only for those in one of the two groups, not involved with the OpenDesign Project or its platform design, development or test. Eight volunteer participants answered the evaluation questionnaire. The objective of this questionnaire was to understand the ability of the platform artifacts to promote deliberation, rationale and awareness, from the point of view of activists and the hands-on course participants (platform users).

The questionnaire encompasses 28 questions organized into two different parts. Part 1 with 19 questions regarding Deliberation, Rationale, and Awareness, and Part 2 with 9 questions regarding Feelings and Usability. The questions in Part 1 used a 9 point Likert scale ranging from 'completely disagree' to 'completely agree' for 13 questions, and included 6 open questions. The questions in Part 2 used a 5 point Likert scale ranging from 'completely unhappy' to 'completely happy', using a manikin with different facial expressions for expressing the feeling. Table 3 shows the closed questions of Part 1 and the mode of answers.

For the Part 2, in the 9 questions related with feelings and usability, in a 5 points Likert scale of satisfaction with different aspects of interaction (e.g. collaboration, self presence in the solution, facility of use, feeling able to contribute, etc.), five (of eight respondents) gave the highest value (5) to them. The lowest value attributed to an item was 3 (in 5). Table 4 shows the mode of the responses for the questions in Part 2.

As for the open questions, we asked if the platform: allowed them to deliberate on the proposed ideas, made the rationale visible, and promoted the awareness of others and their ideas. Inviting them to justify their impression highlighting the platform elements that afforded or prevented each of these goals. We also asked what other mechanisms could be more appropriate, in their opinion, to achieve each of these three aspects (deliberation, rationale, and awareness). All participants agreed that the platform contributed to all these aspects for different reasons and also suggested potential improvement we present in the following section.

Table 4. Mode of the answers for the Part 2 questions. Source: Gonçalves et al. [11].

Question	Mode
Did you feel like collaborating with others?	4
Did you feel represented in the discussion?	4
Did you feel the presence of others in the discussion?	4
Did you feel you achieved something collaborative?	5
Did you feel you were able to contribute?	5
Did you feel you were free to express your ideas?	4
Was the collaboration spontaneous?	4
Would you use the system to solve another problem?	4
How easy-difficult was it to use the platform?	4

4.4 Discussion of Results

In this paper, we investigate the importance of deliberation and design rationale in the (open) design process presenting a case study conducted among participants of a Conference Summer School [25] using the OpenDesign Platform tools. Our findings point out towards the platform being able to provide mechanisms for the convergence of the design process, for knowledge management, and for supporting co-creation. Regarding the content generated through the platform and represented as associations networks, some aspects of the effectiveness of the platform tools can be highlighted:

- *Connectedness.* The connectedness of the graphs presented reflect the successful sharing of ideas and concepts regarding the problem discussed among the participants, independently of the physical location of the different working groups, their culture, and the specific professional profiles (academy, industry, practitioners, etc.) considering the participants were part of audience attending the Conference. In addition to the different physical location, different time of creation of the ideas were not barriers for the interconnectedness of the contributions, showing synchronous and asynchronous use of the platform equally useful.
- *Intersubjectiveness.* The graphs reveal the network of associations of the participants (and their contributions) who, being in different groups and despite not having previous acquaintances of each other, collaborated in the ideas of each other and reached a consensual solution. Moreover, they reached a common ground discussion independently of the prolific (eloquent) voices of some.
- *Scale.* On the scale aspect, once it is not effective or even feasible to participate in all discussions, the platform provided frames to gather organic interest of subgroups formed around an issue or idea proposal and their interconnections. To handle dialectics the platform provided a frame structure that affords capturing benefits and drawbacks of each proposed idea. By mediating discussion with a semi-structured argumentation, it was possible to collaboratively elicit which pros and cons resonate more around an idea or issue.

As for the perception of the participants for the work through the OpenDesign Platform, (raised by the open questions), the main findings regarding deliberation, rationale and awareness can be summarized as:

Deliberation Aspects

- All respondents somewhat agree that those in the same physical location have discussed regardless of the platform, but some agree their discussion was moved by the artifacts and they have recorded the results of the face-to-face discussion on the platform. The recording is necessary and hopefully sufficient to make what happened out of the platform to become a fact capable of impacting those who have not participated in that particular face-to-face discussion.
- All respondents agree that the artifacts in the platform allowed them to deliberate on the proposed ideas. They emphasized the structure of the platform as a key aspect to orient, provide sequence and overview of contributed information, keeping good flexibility and interaction among people. Besides structure, they mention aspects of flexibility of use, visibility of ideas and of thinking processes, e.g.: *"They facilitate the visualization of the ideas", "ideas and thinking stimulation"*.
- When asked how deliberation and decision-making could be improved, they pointed out that consolidation still happens to depend on video conference support and that could be improved with mechanisms such as: digital mediator, discussion turns, focus groups.

Rationale Aspects

- Most respondents agree to be easy to relate a requirement to a stakeholder, a solution proposal to the issue it seeks to solve, and to a lesser degree to link a requirement on SF to the proposed solution that gives rise to it. Regarding the artifacts, "*They contribute to build a more complete perspective of what is being discussed, in different dimensions that may not have been considered.*"
- When asked what other mechanisms could be more appropriate for making the rationale visible, they suggest "*schemas, maps, visual mind maps", "It will be helpful if there is a graph network that can show the relationships between different solutions*". This last aspect is something an integration of the QUID tool to the Platform could offer, visually facilitating a global view of the human and non-human actors in the solution tracking, as Figures 7 to 10 show us.

Awareness Aspects

- According to the respondents, the platform collaborates to understand the others' point of view, making visible their arguments in a non-biased way. Visibility of all ideas and their influence on the others is mentioned by the majority of respondents, e.g. "*The ideas are all shown on the platform, which are easy to check out.*" Nevertheless, they acknowledge there is still room for improvement: "*Yes, in some way because we can see productions (stakeholders, problems.) of other people and their comments. We can see how problems, ideas and other productions are related as well. But the artifacts can improve to make awareness even better.*

These results, added to the content generated through the platform during the activity, has shown the use of the platform was smooth, and the boundary objects were valuable as mediators along the process that initiated with a challenge and ended up with requirement specifications of an elected idea of solution.

5 Conclusion

The concept of open design, with roots in the free-software movement, has been recognized as a meaningful phenomenon, supported by trends in contemporary digital technology and organizations. The nature of open design presupposes the interaction of diverse people towards a co-creation of the design product (e.g. a plan, a drawing, a requirements list for the intended product or service). Moreover, these people with diverse backgrounds need to align different motivations (e.g. to influence the final result, to voluntarily contribute to the process, to get a benefit of it) to collaborate. The research community has pointed out the lack of online platform tools to support meaningful interaction and co-construction in open design, especially in the activities related to the early clarification on a design problem towards the conception of community-driven solutions.

In this work, we presented the OpenDesign platform, characterizing it by its boundary objects (artifacts), participants, and design process. OpenDesign draws on artifacts and practices of the Organisational Semiotics and Participatory Design to conduct a Socially-aware Design enhanced with online deliberation and cues to capture and reveal elements of the design rationale while participants use it. We carried out a case study with participants in two geographically distant sites attending a summer school in a Conference. They interacted through the platform artifacts to discuss a proposed design open challenge and together have evolved the conception and requirements of a design solution that emerged from collaboration.

We adopted the lens of the Actor-Network Theory to analyze the data generated from the design process through the platform artifacts. These data were processed by the QUID tool to build a graphical representation in which both the participants as well as the boundary objects are part of the same network. This visual representation allowed us to reveal the different paths the participants weave along the interaction through the platform, the tracks they leave while discussing, proposing ideas, deliberating, interacting with the others mediated by different boundary objects. Our further analysis revealed several aspects of the association between participants and the boundary objects: the connectedness of the contributions, the intersubjective relation of participants while sharing information through the platform, and the tracing back of the solution and of specific discussions along the way.

Furthermore, we presented a qualitative synthesis of the participants' feedback about the platform, raised through their answers to an evaluation questionnaire. Results reveal the participants' perception of the platform's value related to deliberation, rationale, and awareness, while they also gave valuable opinions on how to improve these aspects. Their answers also reveal very positive feedback on their feelings on the experience and usability of the platform.

We expect that this work contributes to a more systemic vision for the design of interactive systems that considers the technical solution as part of a more complex social system and welcomes open participation. Further studies involve: a) integrating the visualization tool into the OpenDesign platform and investigating whether it can increase awareness of others' contributions; b) investigating the visual map of the network of participants and boundary objects as a mediator along the (open) design process.

Acknowledgements. This work is financially supported by the São Paulo Research Foundation (FAPESP) (grants #2015/24300-9, #2018/25972-9, and #2015/16528-0) and by the National Council for Scientific and Technological Development (CNPq) (grant #306272/2017-2).

References

1. Avram, G.: At the crossroads of knowledge management and social software. Academic Conferences Ltd (2005)
2. Baranauskas, M.C.C.: Social awareness in HCI. Interactions **21**(4), 66–69 (2014). https://doi.org/10.1145/2621933
3. Baranauskas, M.C.C., Schimiguel, J., Simoni, C.A., Medeiros, C.: Guiding the process of requirements elicitation with a semiotic approach. In: 11th International Conference on Human-Computer Interaction, pp. 100–111 (2005)
4. Bjögvinsson, E., Ehn, P., Hillgren, P.A.: Design things and design thinking: contemporary participatory design challenges. Des. Issues **28**(3), 101–116 (2012). https://doi.org/10.1162/desi_a_00165
5. Boisseau, É., Omhover, J.F., Bouchard, C.: Open-design: A state of the art review. Des. Sci. **4**, (2018). https://doi.org/10.1017/dsj.2017.25
6. Bonvoisin, J., Boujut, J.F., et al.: Open design platforms for open source product development: current state and requirements. In: DS 80–8 Proceedings of the 20th International Conference on Engineering Design (ICED 15), vol 8: Innovation and Creativity, Milan, Italy, 27–30.07. 15, pp. 011–020 (2015)
7. Buchdid, S.B., Pereira, R., Baranauskas, M.C.C.: Pro-IDTV: a sociotechnical process model for designing IDTV applications. J. Syst. Softw. **154**, 234–254 (2019). https://doi.org/10.1016/j.jss.2019.04.078
8. Engeström, J.: Why some social network services work and others don't–Or: The case for object-centered sociality. Blog posting **13** (2005) http://www.zengestrom.com/blog/2005/04/why-some-social-network-services-work-and-others-dont
9. Fischer, G.: Social creativity. In: Proceedings of the Eighth Conference on Participatory Design Artful Integration: Interweaving Media, Materials and Practices - PDC 04, pp. 152–161. ACM, ACM Press (2004). https://doi.org/10.1145/1011870.1011889
10. Frauenberger, C., Foth, M., Fitzpatrick, G.: On scale, dialectics, and affect. In: Proceedings of the 15th Participatory Design Conference on Full Papers - PDC 2018, vol. 1, p. 12. ACM, ACM Press (2018) https://doi.org/10.1145/3210586.3210591
11. Gonçalves, F., Prado, A., Baranauskas, M.: OpenDesign: analyzing deliberation and rationale in an exploratory case study. In: Proceedings of the 22nd International Conference on Enterprise Information Systems, pp. 511–522. INSTICC, SCITEPRESS - Science and Technology Publications (2020). https://doi.org/10.5220/0009385305110522
12. Greenbaum, J., Halskov, K.: PD a personal statement. Commun. ACM **36**(6), 47 (1993). https://doi.org/10.1145/153571.214816
13. Hagen, P., Robertson, T.: Social technologies. In: Proceedings of the 11th Biennial Participatory Design Conference on - PDC 2010, pp. 31–40. ACM, ACM Press (2010). https://doi.org/10.1145/1900441.1900447

14. Hargreaves, D.M., Robertson, T.: Remote participatory prototyping enabled by emerging social technologies. In: Proceedings of the 12th Participatory Design Conference on Exploratory Papers Workshop Descriptions Industry Cases - Volume 2 - PDC 2012, pp. 25–28. ACM, ACM Press (2012). https://doi.org/10.1145/2348144.2348153
15. Heintz, M., Law, E.L.C., Govaerts, S., Holzer, A., Gillet, D.: Pdot. In: Proceedings of the Extended Abstracts of the 32nd Annual ACM Conference on Human Factors in Computing Systems - CHI EA 2014, pp. 2581–2586. ACM, ACM Press (2014). https://doi.org/10.1145/2559206.2581266
16. Hulkko, S., Mattelmäki, T., Virtanen, K., Keinonen, T.: Mobile probes. In: Proceedings of the Third Nordic Conference on Human-Computer Interaction - NordiCHI 2004, pp. 299–300. Indiana University, ACM Press (2004). https://doi.org/10.1145/1028014.1028020
17. Kriplean, T., Morgan, J., Freelon, D., Borning, A., Bennett, L.: Supporting reflective public thought with considerit. In: Proceedings of the ACM 2012 conference on Computer Supported Cooperative Work - CSCW 2012, pp. 265–274. ACM, ACM Press (2012). https://doi.org/10.1145/2145204.2145249
18. Latour, B.: Reassembling the Social. An Introduction to Actor-Network-Theory. Oxford University Press, New York (2005)
19. Liu, K.: Semiotics in Information Systems Engineering. Cambridge University Press (2000). https://doi.org/10.1017/cbo9780511543364
20. Liu, K.: Organisational Semiotics for Business Informatics. Routledge (2014). https://doi.org/10.4324/9780203550977
21. Magalhães, L.A.F., dos Reis, J.C.: Design participativo na plataforma opendesign. Technical report, PFG-19-35, Institute of Computing, University of Campinas (2019)
22. Muller, M.J., Haslwanter, J.H., Dayton, T.: Participatory practices in the software lifecycle. Handbook Hum.-Comput. Interaction **2**, 255–297 (1997)
23. Naghsh, A.M., Andy, D.: GABBEH: a tool to support collaboration in electronic paper prototyping. In: CSCW 2004 the ACM Conference on Computer Supported Cooperative Work, Chicago, USA (2004)
24. Prado, A.B., Baranauskas, M.C.C.: Representing scientific associations through the lens of actor-network theory. In: 2012 Fourth International Conference on Computational Aspects of Social Networks (CASoN), pp. 3–12. Springer, IEEE, November 2012. https://doi.org/10.1109/cason.2012.6412383
25. of Reading UK, T.I.R.C.H.B.S.U.: Henley symposium and summer school on organisational semiotics 2019 (2019). https://www.henley.ac.uk/school/organisational-semiotics-summer-school
26. Schuler, D., Namioka, A.: Participatory Design: Principles and Practices. CRC Press (1993)
27. Stallman, R.: The GNU manifesto. Dr. Dobb's J. Softw. Tools **10**(3), 30–35 (1985)
28. Stamper, R.: Information in Business and Administrative Systems. Wiley, New York (1973)
29. Stamper, R.: A semiotic theory of information and information systems. In: Applied Semiotics. In Invited Papers for the ICL/University of Newcastle Seminar on" Information (1993)
30. Stephanidis, C., Akoumianakis, D., Savidis, A.: Universal Design in Human-Computer Interaction, pp. 741–745, January 2001
31. Walsh, G., et al.: DisCo. In: Proceedings of the 11th International Conference on Interaction Design and Children - IDC '12, pp. 11–19. ACM, ACM Press (2012). https://doi.org/10.1145/2307096.2307099
32. Warger, T.: The open-source movement. Educ. Q. **25**(3), 18–21 (2002)
33. Xiaojia, L.: Employing MEASUR methods for process reengineering in china, June 2001

Tabletop Interface Design: Developing and Using Usability Heuristics

Vinícius Diehl de Franceschi[1], Lisandra Manzoni Fontoura[1(✉)], and Marcos Alexandre Rose Silva[2]

[1] Programa de Pós-Graduação em Ciência da Computação, Federal University of Santa Maria (UFSM), Santa Maria, Brazil
v.f.diehl@gmail.br

[2] Colégio Politécnico, Federal University of Santa Maria (UFSM), Santa Maria, Brazil
marcos.silva@ufsm.br

Abstract. Tabletops enable the development of applications with high interactivity and simultaneous collaboration between users. Due to the size of these devices, they differ considerably from other touchscreen devices such as tablets and smartphones. Therefore, understanding these differences to define appropriate heuristics for the evaluation of tabletop interface designs is necessary. This paper presents a set of heuristics to be considered in tabletop interface design and a checklist composed of a set of questions for each proposed heuristic that aims to assist in the evaluation of interfaces for tabletops. Based on the literature, we analyze different proposals for heuristics for specific applications to identify the main concepts involved in defining a heuristic and how to write heuristics for specific applications. Also, we researched the particular features of tabletop applications that need to be considered in interaction design. From these characteristics, Nielsen's ten heuristics were specifically adapted for use with tabletop devices and two new heuristics were defined. A case study was carried out in the SIS-ASTROS project, the proposed heuristics were considered to design a virtual tactical simulator, and an evaluation of the interface by the military was performed. Observing the militaries using them, we have gathered evidence that these heuristics can help designers to think about essential interface characteristics to support users to realize and understand the interface goal and how to interact with it.

Keywords: Tabletop · Heuristics · Usability

1 Introduction

Different technological devices as laptops, smartphones, and tablets are changing people's behaviors and activities [20]. Then, systems for these devices must be designed to support people's use. On the other hand, this design can be a challenge because each device may have different characteristics and specificities to be considered.

Tabletops, large dimension devices, have been used in different contexts, and their use shows satisfactory results, e.g., with systems related to maps, because of the birds-eye view. In health, due to the number of elements that can be displayed on the screen,

J. Filipe et al. (Eds.): ICEIS 2020, LNBIP 417, pp. 708–729, 2021.
https://doi.org/10.1007/978-3-030-75418-1_32

allowing to see more details about the medical image and supporting better analysis, among other contexts [11, 22].

Tabletops size and weight represent different characteristics from other devices as smartphones, and this paper describes research that allowed noticing these characteristics to be considered, and they were useful to adapt Nielsen's heuristics for tabletop context.

According to Prates and Barbosa [17], heuristic evaluation is one of the evaluation methods more widespread and better known by researchers and professionals from human-computer interaction. This evaluation examines the interface and judges its compliance with recognized usability principles, which are the heuristics [15].

It is important to say that these heuristics support the evaluation but also since the beginning of the design, because these principles can be used to create and/or chose interface elements, as well as their organization, among others, saving considerable amounts of time during early usability testing.

Dourado [4] says heuristics are easy to understand and useful to identify usability problems with low costs. In contrast, they can be very general, causing the recognition of specific problems a difficult task. The evaluators need much experience with these heuristics and device to judge the specific characteristics and needs of a device. Therefore, it is possible to improve the effectiveness of the method significantly using adapt heuristics with problems and examples related to a specific context [4, 19].

2 Methodology

The potential and use of Nielsen's heuristics, heuristics adapted from Nielsen or others formalized to different context as smartphone and tablet are described in papers as Chuan et al. [2], Tehrani et al. [21], Joyce et al. [10], Masip et al. [12], Neto et al. [13], Humayoun et al. [8], Shneiderman et al. [20], Durães et al. [5]; however they do not describe the heuristics adaption or developing process. In this context, Rusu et al. [19] report a methodology in six steps considering the whole process since initial research to write the heuristics.

First step, called Exploratory, to collect and organize bibliography related to topics of the research, such as concepts; specific systems and/or applications, i.e., different kind of software and devices; their characteristics; examples of use; and general or related usability heuristics.

Second step, called Descriptive, to highlight the most important information collected from the First step to formalize the main concepts associated with the research.

Third step, called Correlational, to identify the characteristics that the heuristics for specific applications should have, based on other heuristics and observation analysis.

Fourth step, called Explicative, to specify the set of the proposed heuristics, i.e., writing the adapted or new heuristics.

Fifth step, called Validation, to observe the heuristics use, for example, considering them in interface design, and then, analyzing interface-user interaction through user tests.

Sixth step, called Refinement, to improve or change heuristic writing based on the feedback from the previous stage.

Rusu et al. [19] did not present an image to explain the methodology, but Fig. 1 was made in order to show the sequence of the steps. The arrows reinforce the possibility of changing the order, if necessary, because of its iterative processes. For example, the results of the users' observation (Step 5) can show a different experience or situation that needs to be better investigated, so it is necessary to explore that situation (Step 1). After refinement (Step 6), another validation can be done (Step 5) to analyze and compare the results.

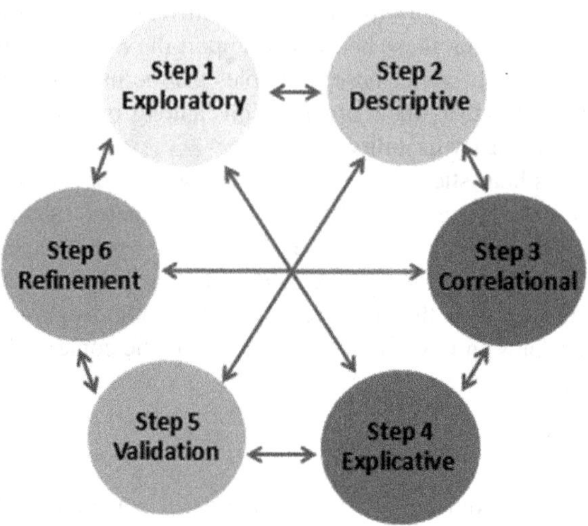

Fig. 1. Methodology.

Each one of these six steps, considered to develop and adapt heuristics to tabletop context, is also described below.

2.1 First Step: Exploratory

It was intended to investigate researches related to use, developing, and/or adaptation of interface heuristics for multi-touch tabletop, but few researches were found. Because of that, researches related to any touchscreen devices were considered in this investigation.

Shneiderman [20] formalized eight heuristics to be considered in any interactive technology. For example, reduce short-term memory load, reinforcing the limitation of human attention, and the capacity of maintaining around five items in short-term memory. Neto et al. [13] propose eleven heuristics to mobile devices like smartphones and tablets describing also characteristics related to their small size as good use of screen space to describe about quantity of elements and how to organize them.

D'Carlo et al. [3] present fourteen heuristics to evaluate the usability of mobile educational devices, e.g., learning considering a goal to emphasize how to plan interfaces to allow users to notice and understand application goals and what can be done thought it. Humayoun et al. [8] describe fifteen heuristics to validate mobile device multi-touch gestures as fatigue to avoid defining a difficult gesture to understand what needs to be done, as well as how to do it considering the screen size, among others.

Inostroza et al. [9] defined twelve heuristics for touchscreen-based mobile devices, e.g., physical interaction and ergonomics to provide a physical button or similar for main functionalities. Chuan et al. [2] proposed four usability heuristics for testing gestural interaction considering phases of use as before using, during using, after using, and prolonged using as Gesture Learnability to emphasize the learning of gestures by users.

Rusu et al. [19] defined twelve heuristics to evaluate grid computing systems, e.g., metaphors, indicating the application must use appropriate metaphors, making actions easy to understand, through familiar images and objects. Joyce et al. [10] formalized eleven heuristics for native smartphone mobile applications as facilitate effortlessness input by displaying large keyboard buttons.

Heuristics for tabletop context were not found, and then works about tabletop systems design were collected to observe important information and experience. Madni et al. [11] reported the use of tabletop for supporting medical diagnostics based on images where a group of doctors could see and interact at the same time. Yang et al. [22] describe a tabletop system design about GIS (Geographical Information System). Bortolaso et al. [1] present a multi-display tabletop simulator to support military training.

2.2 Second Step: Descriptive

Table 1 contains a comparison considering investigated researches related to Nielsen's heuristics and others heuristics to identify the most essential characteristics, interface elements discussed by these heuristics, among others.

This comparison showed that Nielsen's heuristics are widely used in them. Some authors usually adapt their heuristics with detailed descriptions and examples considering a specific context or create new heuristics, in this case, explaining why these new ones are different from Nielsen.

Nielsen developed 10 heuristics based on work together with Rolf Molich. According to him, these heuristics represent general principles for interaction design and they are called heuristics because they are rough rules of thumb and not specific guidelines, and these broad rules allow contextualize them taking into consideration users, devices, among others [16].

For example, a heuristic called Match between system and the real world describes that the system should speak user's language, i.e., it does not describe which word, options, images, among others, should be used, but reinforces the need to identify users' characteristics and necessity to define the name of the options, instructions, as well as their organization.

This kind of writing allows using these heuristics in different contexts of systems and users. On the other hand, depends on a lot of evaluator' experience and knowledge about users, devices, among others, to use them. Because of that, adapted and new heuristics were developed, in order to support evaluators to identify and know some specific characteristics of users and/or devices in different contexts.

According Table 1, the first heuristic proposed by Nielsen is Visibility of system status which describes the system should always keep users informed about what is going on. Joyce et al. [10], [R6] in Table 1, also describe that the system should show new users a welcome message, at their first heuristic (H1), as well as their second heuristic (H2) is related to show the status of system as soon as users interaction happens.

Table 1. Heuristics comparison [7].

Nielsen's heuristics	[R1]	[R2]	[R3]	[R4]	[R5]	[R6]	[R7]	[R8]
Visibility of system status	H1	H4	H2	H4	H1	H1, H2	H9, H3	H1
Match between system and the real world	H2	H1, H2	–	H5	H2	H3	H5	H2
User control and freedom	H3	H8, H9	–	–	H3	H4	H8	H3, H6
Consistency and standards	H4	H5	H3	H1	H4	H5	H2	H4
Error prevention	H5	H10	–	H6	H5	H6	H6	H5
Recognition rather than recall	H6	H7	H1	H8	H6	–	H11	H11, H13
Flexibility and efficiency of use	H7	H6	H4	H3, H9	H7	H7, H8	H7, H1	H9, H10
Aesthetic and minimalist design	H8	H3	–	–	–	H9	–	H7
Help users recognize, diagnose, and recover from errors	H9	H11		H10	H8	–	H4	H13
Help and documentation	H10, H11	H12	–	H14	–	–	H10	H8
Cooperative/collaborative Usability	–	–	–	H11	–	–	–	–
View and functionality adaptation	H12	–	–	–	–	–	–	H15

Authors - [R1] Inostroza et al. [9]; [R2] Rusu et al. [19]; [R3] Chuan et al. [2]; [R4] D'Carlo et al. [3]; [R5] Shneiderman et al. [20]; [R6] Joyce et al. [10]; [R7] Neto et al. [13]; [R8] Humayoun et al. [8].

It is important to highlight that this reasoning was considered with all investigated heuristics, i.e., each heuristic was analyzed and compared to each other, and then, similar heuristics were inserted in the same table line. Table 1 shows all ten Nielsen's heuristic in the first column, and other columns represent others investigated researches.

Some heuristics describe specific contexts and examples not related to Nielsen's heuristics as information related to Cooperative/Collaborative Usability and View and functionality adaptation. In these cases, two lines were created at first Table 1 column with them, e.g., Humayoun et al. [8], [R8] in Table 1, described as the fifteenth heuristic Screen orientation that gestures, images, and information should work on vertical and horizontal screen orientations as well as on all screen sizes. Because of that, this heuristic is on View and functionality adaptation line.

Franceschi et al. [6] describe the systematic review to identify these works, describing their names and more details about their content. Taking into account, this paper aims to present the use of this information to develop and use these heuristics.

Works about tabletop were also analyzed and compared among them, as shown in Table 2. Each work designed system to be used in a different context, but they share some common characteristics, as simultaneous and collaborative interaction, drawing and manipulation of objects, and big sizes. These tabletops sizes are smaller than tabletop, 84-in., used by the project described in this paper. On the other hand, there are many common characteristics among their works and the one presented in this paper as radial menu, geographic information at different angles as described in Yang et al. [22]; route planning, geographical charts as described by Bortolaso et al. [1]; and manipulating digital content and images as in the work of Mandi et al. [11].

Table 2. Tabletop characteristics comparison [7].

Yang et al. [22]	Bortolaso et al. [1]	Mandi et al. [11]
• Simultaneous and collaborative interaction • Interactive table 40-in. • Blocks can be dynamically oriented • Radial menu • Can be used on multiple devices simultaneously • Works with geographic information at different angles and dimensions	• Shared interaction for multiple users • 55-in. 2D tabletop • Multi-touch inputs • Bifocal lenses (Zoom) • Route planning by drawing waypoints • Can be used on multiple devices simultaneously • Uses geographical charts • Allows future and past locations, tracking line, visibility, and range of a military unit	• Collaborative interaction • Samsung SUR40 (40-in. interactive tabletop) • Manipulation of digital content and images • 2D/3D information display • Multi-touch interaction, tangible objects, pens, or mouse • Zoom in/on objects

2.3 Third Step: Correlational

This step is to identify the useful characteristics for tabletop usability heuristics based on other heuristics and observation analysis. According to Preece et al. [18], the observation supports collecting data from real situations in order to perceive user computer interaction problems or success.

The observation occurred in a military simulator project [6]. This project means to develop a virtual tactical simulator for teaching military doctrines[1] related to recognition,

[1] Military doctrine is the expression of how military forces contribute to campaigns, battles and engagements.

choice, and position occupation of a missile and rocket battery for commanders from these areas.

The idea was to teach these doctrines conceptually, and then militaries could practice them on the tabletop before going to the military training camp. Figure 2 shows the tabletop where the simulator is working. It is TV 84-in. with a capacitive sensor to recognize users' interaction by touch on the screen.

Fig. 2. Tabletop virtual tactical simulator [7].

There were maps, cards, spatial/geographical information in this simulator where a group of five users interacted with during observation. According to Nielsen [14], the best results come from testing no more than 5 users.

The users are military instructors who use this simulator to support military training in technical and tactical operations related to the military use of an ASTROS battery. One interesting characteristic to be designed in this simulator is related to its functionalities, the sequence of instructions to be executed must not appear in the interface, these depend on the tactical training and are the responsibility of the instructing.

On the other hand, the functionalities, images, options, among others should be clear, e.g., it is not right to show a functionality as the first step, because militaries must know it, but after clicking on it, militaries can see clearly the options related to it, and according to their knowledge and experience to choose one or more options, i.e., interface do not show which option must be chosen either.

It is important to say that this simulator was a prototype because a few functionalities were working, and it was developed by other professionals with no influence by the authors of this paper.

Figure 3 shows one of the perceptions where users did not find where to click because after pressing a button with an icon, its color changed and its name shows up in the middle of the radio button, and users tried to click on it, but the name is just an instruction, not a button, so users need to click on icon button again. The five users wrongly clicked on the name.

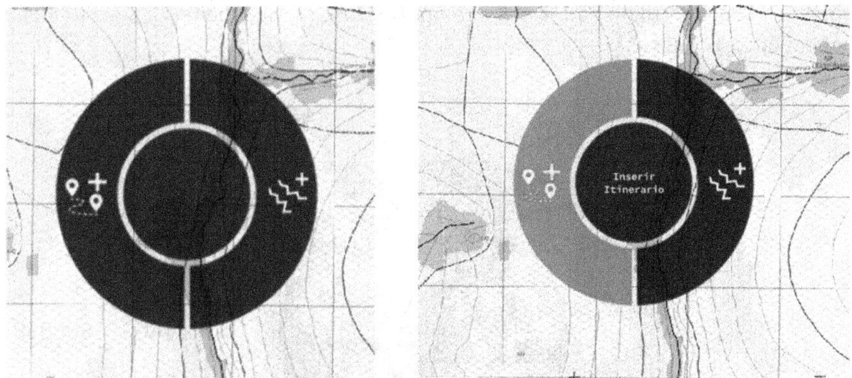

Fig. 3. Left radio button before and after pressed.

Another noticed problem is related to identify sub buttons, i.e. buttons which appear when one button is pressed because there is no sign or visual element to identify when there are more options. Figure 4 shows what happens when the right button is pressed, options like LP, LC, among others show up. In this case, users did not recognize sub buttons before clicking on buttons.

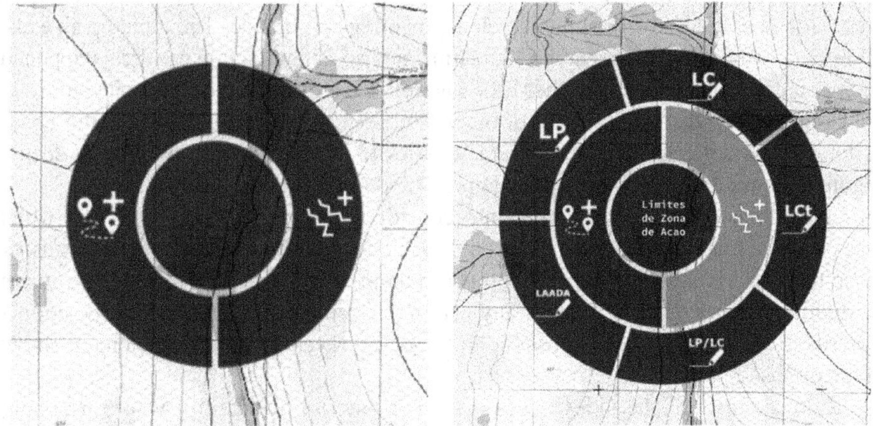

Fig. 4. Right radio button before and after pressed.

Interface elements also overlay to each other, e.g., it is necessary to click on an interface element to see a radial button with options related to it, but this radio appears overlay to the element, then identify what element was clicked is not an easy task, as shown at Fig. 5, and there is no sign to identify where to click to close this radio, in this case, it is necessary to click off the radio.

On the other hand, there are many interface elements nearby, and there is no sign indicating which element is a clickable option or just a visual element. So, sometimes users clicked off the radio to close it, but with no intention clicked on another clickable element and another radio button showed up. Because of that, there is a need to identify when a button is clickable.

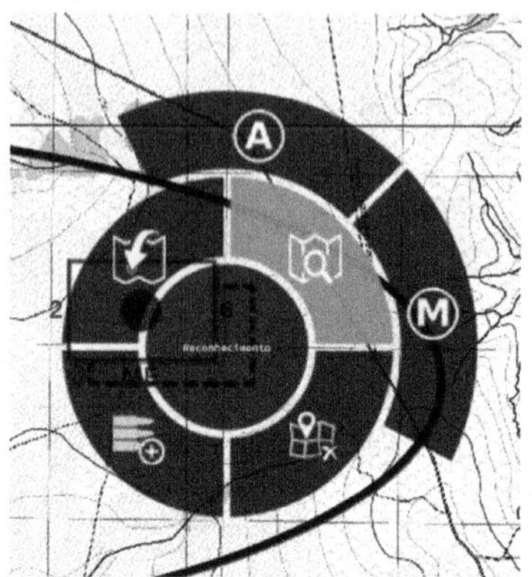

Fig. 5. Interface elements overlay.

Interface element can also be showed upside down considering user' view in some times because the interface of the simulator was programmed considering the proximity of the sides, that is, if the user is on one side but extends his hand to be closer to the other side, the tabletop will show the elements in this other side. The tabletop size also allows many users to interact at the same time, and some interaction conflicts happened, such as a user could cancel another user's action, even they were on opposite sides.

Considering characteristics from other heuristics, tabletop studies, and the observation, it was collected essential characteristics to be considered on tabletop design. Figure 6 shows them into two parts, elements and quantity.

Elements can be shown, taking into consideration where the user is, i.e., which side occurred touch on the screen. Through this information, tabletop recognizes users' place to show visual elements (proximity) to them and oriented (orientation) to them, facilitating the interaction. That is, any action performed by users must be positioned towards them, and as close as possible to them, to prevent an interface element shown in the opposite position.

The proximity of the sides is not a perfect criterion to identify where users are. Because of that, there is a necessity to allow rotating the visual elements 360°. Zoom out and zoom in represent other useful strategies to consider on design, allowing wider view, for example, to see a country, or detailed view, to see a street in a city. Moving elements is also another required to be thought because if an element appears over what users want to see, they can move it, as well as bring the element closer to where they are (zoom/move).

Tabletop sizes allow displaying a lot of information at the same time, so it is important to highlight which information can be clicked so that interactivity is highlighted, and the user does not have to look for options. Whenever items and sub-items exist, it is necessary to visually inform the user (highlighted interactivity).

The (quantity) is a tabletop characteristic that stands out from other types of smaller touchscreen devices, such as smartphones and tablets, which usually only one person uses at a time. In tabletop, it is important to identify how many users can use at the same time, and how many features can be performed at the same time, and the action of one user should not interfere with the action of another.

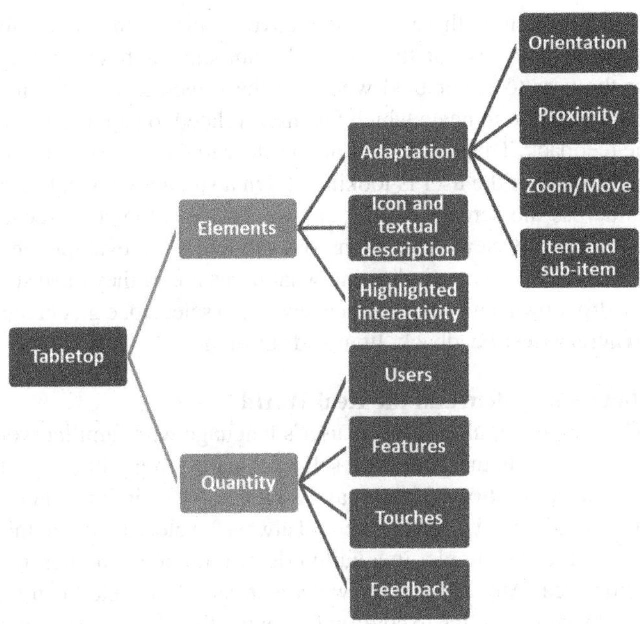

Fig. 6. Characteristics of tabletop design [7].

The quantity of (touches) should also be considered, because users, with no intention to enter in an option, can unintentionally click on it, while they are explaining and just pointing with their fingers what they want to show. It is necessary to identify how to show the feedback for each system feature and possible click (highlighted interactivity) for each user who is interacting with, as well as always indicate when an item was selected and if there are other sub-items available (feedback).

It is needed to say that these characteristics are not for defining specific images, elements, among others, to be inserted in the tabletop interface. These elements assist in the process of thinking about design, allowing designers to know and consider important features of the tabletop, and they can think about how to do it according to system and user context.

2.4 Fourth Step: Explicative

This paper proposes adaptations to the ten heuristics aimed by Nielsen and two new heuristics were created for the tabletop context [7]. This set was elaborated considering related works, studies and observation as mentioned before.

H1 - Visibility of System Status
Definition: System status visibility refers to how well system status is transmitted to users.
Adaptation: In a tabletop, there are interactive elements that blend in with non-interactive elements. Because of their size, the amount of these elements can make it difficult for the user to understand where on the screen to look for feedback, so it should be clear enough to indicate which feature is related to and be concerned with the location of the feedback. The location should be close to the clicked component because it is possibly the location the user is looking at. On a smaller device, the user can view what happens across the screen. On the other hand, on a tabletop, feedback may be outside the user's field of view. Also, items and sub-items, for example, in menus, need to be visually informed to users, indicating what items are, if they have sub-items, e.g., with arrow like dropdown, and whether they have been selected, e.g., changing the color.
Related Characteristics: Feedback, Item and sub-item.

H2 - Match between System and the Real World
Definition: The system should speak the user's language with familiar words.
Adaptation: Evaluate whether icons are self-explanatory regarding their function and application domain conventions. There may be a textual description to help users understanding the meaning of the icon, but it should always be close to and in the same visual element as the icon. For example, in a button design, the icon and description may be within the button area. Any distance between icon and description can make the user confused as to which one to click to confirm the interaction. Because tabletops are touch, some gestural conventions of these devices should be used for sliding, dragging, and so on.
Related Characteristics: Icons and textual description.

H3 - User Control and Freedom
Definition: Users can choose system functions by mistake and need to undo and redo options without a lengthy process.
Adaptation: Enable the user to undo their actions even when using multiple users at the same time, ensuring that undoing one user's actions will not impact another user's actions. Undo and redo options should be easily understood by all users interacting at the table.
Related Characteristics: Quantity of Users, Quantity of Features.

H4 - Consistency and standards
Definition: The device must follow established conventions, allowing the user to do things in a standardized and consistent manner.
Adaptation: Conventions established for other touchscreen devices should be followed, such as gestures to select, execute, zoom in/out, slide and drag used on smartphones, for

example. It is emphasized that the proportion of the movement will not always be equal to the desired result. On the smartphone, the user can pinch the element and open their fingers to the desired size. In the case of the tabletop, the user may be interested in greatly enlarging an element, so with a little opening between fingers could allow enlarging the image to a much larger size than is between your fingers. Interactions throughout the system must be consistent.
Related Characteristics: Proximity Adaptation, Zoom/Move Adaptation.

H5 - Error Prevention
Definition: Verify that the interface contains only essential elements and eliminate error-prone conditions.
Adaptation: Context-associated menus can be used to reduce user choice options, disabling options not currently available. Depending on the outcome of the interaction, confirmations must be requested for the user to have the option to rethink an action. For example, the user may tap the screen only to point out an element during the explanation and not to activate a feature, so it is essential to question if the user wants to do the action. In case of just opening a feature or a menu option, no problem, since realizing that opened the option, the user can click the close or cancel icon. Besides, because of potential touch accuracy issues, it may be desirable to identify the user's intent.
Related Characteristics: Quantity of Touches.

H6 - Recognition Rather than Recall
Definition: The user should not have to remember information for system use.
Adaptation: The icons should be accompanied by textual descriptions making it easy for the user to understand the action that will be performed. When selecting an element, before executing, it is essential to make it clear which action the element will perform. It is also necessary to highlight the next actions to be performed. Interactive elements should be distinct from other elements, not requiring the user to remember. Because of their size, which allows many more elements to be displayed than other touch devices, the difficulty in identifying what can and cannot be clicked impairs interaction. User-tabletop interaction may occur similarly to smartphones. On these devices, users do not analyze the entire system to understand it before starting the interaction, but find clickable items and compare it with their intention of finding out where to click. On a smaller device, when the user does not identify what can be clicked, the user can click on what is being displayed to identify resources by trial and error. This possibility can also occur on a tabletop, but more clicks will be done. Therefore, the interactive components must be distinct from the other components. Icons and descriptions should be defined to clearly illustrate their goals.
Related Characteristics: Icon and textual description, Highlighted interactivity.

H7 - Flexibility and Efficiency of Use
Definition: The device must be able to load and display information within a reasonable period and minimize the steps required to perform a task.
Adaptation: The interaction elements (menus) must be distributed considering each context and close to the respective feature. This enables fewer options in each menu and fewer interactions to perform the task because each feature will be close to its interface

element. It is also necessary to ensure that the information is displayed in sufficient detail for the correct operation of the application, and zoom operations will perform satisfactorily.
Related Characteristics: Quantity of Features.

H8 - Aesthetic and Minimalist Design
Definition: The device should avoid displaying unwanted information by overloading the screen.
Adaptation: Tabletops allow users to view a large amount of content at the same time, but too much information can compromise the viewing of information. Whenever possible, enable the user to access information without overloading the interface. Swipe menus can be hidden whenever not in use, making it easy to access when needed. Also, the menus should be aesthetically simple and describe the features clearly. The buttons, text, and colors should be a contrast to the background and, when necessary, be transparent enough to see what is in the background, allowing the information to be viewed.
Related Characteristics: Icon and textual description, Highlighted interactivity.

H9 - Help Users Recognize, Diagnose, and Recover from Errors
Definition: Error messages should be expressed in plain language (no codes), accurately indicate the problem, and constructively suggest a solution.
Adaptation: Messages indicating errors should be clearly defined to assist the user in identifying and correcting them. The error may have been caused by another user's interaction with the tabletop so that it may be distant from the user's current interaction location. In this case, it is necessary to indicate the error information so that all users can view it.
Related Characteristics: Quantity of Feedback.

H10 - Help and Documentation
Definition: Provide user-friendly, task-focused help, and documentation mechanisms.
Adaptation: Assist the user in interacting with the system, providing relevant information. It is noteworthy that the use of instruction manual or any other option with a lot of content/text is not always the best way, because just like smartphones, users tend not to read all the help information to start the interaction. Therefore, help may be indicated according to the user's need and interaction. For example, by designing a symbol that represents the action that should be taken, such as a hand sign with the finger-pointing at an item may indicate that it is clickable.
Related Characteristics: Quantity of Feedback, Highlighted interactivity.

H11 – View Adaptation
Definition: Whereas several people use the table at the same time, the system must identify each user's position and display interactive visuals oriented to it. If the guidance is not efficient or the user wishes to orient the visual element to another person, the system must be able to change this orientation simply. Because tabletops can be large, interactive elements should be displayed close to and oriented to the interacting user. For devices of this type, it is relevant to be able to zoom in on the viewer allowing users to display

more detail or show more interface elements with less detail (multiresolution). These operations must be associated with conventions established by touchscreen devices.
Related Characteristics: Orientation Adaptation, Proximity Adaptation, Zoom/Move Adaptation.

H12 - Cooperative/Collaborative Usability
Definition: Collaborative devices must allow multiple users to interact at the same time. It is essential to evaluate whether the tabletop allows more than one user to interact with the same or different elements at the same time. The execution of a feature by one user may not affect the execution of another user. Actions taken by one user must be visible to other users. The use of elements to explicitly express the user's intent is required. For example, a close button to close a feature or option. In this case, an explicit touch on close is required to close the feature. Another user interaction with other elements cannot close it.

It is crucial to verify that an action that a user is performing does not cancel another action that is being performed by another user. Therefore, if a user is choosing or has already chosen an option, the other options, which may cancel or directly affect it, must be blocked. If they become available, when chosen, there should be a warning that this option will influence another user's action. In this case, explicitly report which option will be influenced.

Related Characteristics: Quantity of Touches, Quantity of Users, Quantity of Features, Quantity of Feedback.

2.5 Fifth Step: Validation

The validation of the proposed heuristic set was performed through another observation analysis, in which a prototype was elaborated by applying adapted heuristics and tabletops characteristics.

Prototyping allows the creation, evaluation, and refinement of design options until usability is achieved at the desired level. There is low fidelity as well as high fidelity prototypes [18]. The high fidelity prototyping utilizes materials that are expected to be in the final product. For this case study, a high fidelity prototype was developed, allowing direct interaction with the user based on a final version of the product.
Prototype Design: A prototype was developed to observe user interaction and behavior during interaction and validate the set of heuristics. An instance of the simulator was created to perform validation tests with users.

This prototype was developed based on tabletop heuristics, as well as the characteristics are shown in Fig. 6. The following is a brief description of how each characteristic was contemplated.
Orientation Adaptation: Icons are always displayed considering the proximity of the touch and oriented to the user's position (compared to center and edges). If the click occurs between the center and the left edge, the elements are initially facing left; however, there is an arrow to rotate the view 360° so that the user can adjust to the view, as shown in Fig. 7 with an arrow above the Rescue vehicle option.
Proximity Adaptation: Menus are displayed where the user clicks, but he can drag them on the screen with one tap. Feature-specific icons appear next to that feature, without

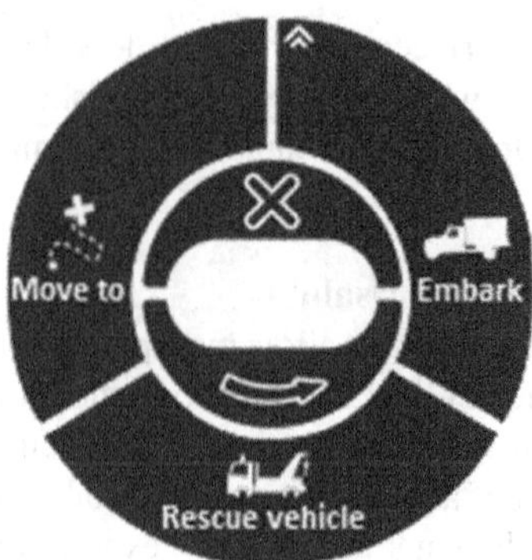

Fig. 7. Radial interaction menus [7].

the option to move it because, without the feature, the option could lose context, and the user might have difficulty recognizing its purpose later.

Zoom/Move Adaptation: There is a possibility to move the elements of the screen by dragging them with one touch. You can zoom in by zooming in on a specific area and showing more detail about the displayed scenario, as well as zooming out, showing less detail and more elements with multi-resolution rendering that performs well.

Icon and Textual Description: Each menu item contains the icon associated with its description, making it easy for the user to understand.

Highlighted Interactivity: When a radial menu item contains interaction sub-items, there is an indicator (arrow, as illustrated in Fig. 7 - Embark option). Besides, there is a hand sign for all clickable options. We chose to have a visible hand, but not so prominent in color, so as not to detract from the background view.

Figure 8 shows the previous interface with no sign to identify clickable options, and Fig. 9 shows the similar interface, used in this observation, with hands to highlight clickable options. A hand with the pointed finger sign was chosen because express the gesture that needs to be done, but other signs can be used depends on the user and system context.

Quantity of Users and Features: The simulator is intended for use by more than two people, so menus are closed only when explicit close commands are executed (click the X icon) and not the next click, which could interrupt or cancel another user's interaction. Therefore, more than one menu may be visible at a time, and commands may be entered into each menu by different users.

Quantity of Touches: The click hand sign (Fig. 9) indicates that only one tap is required to open the options.

Quantity of Feedback: Feedbacks are always displayed next to the clicked items, and when an option is active, it turns a different color to stand out from the default background color.

Observation.

The task to be performed was informed, but without explaining how to do it. The goal was to observe if users would recognize on interface what they needed to do to perform the task. As mentioned before, militaries must know the sequence to do the actions and

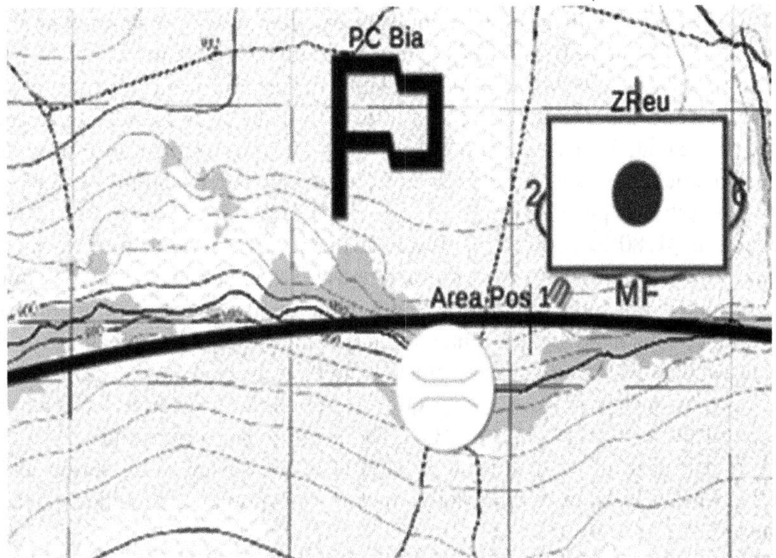

Fig. 8. Interface elements with no visible hand.

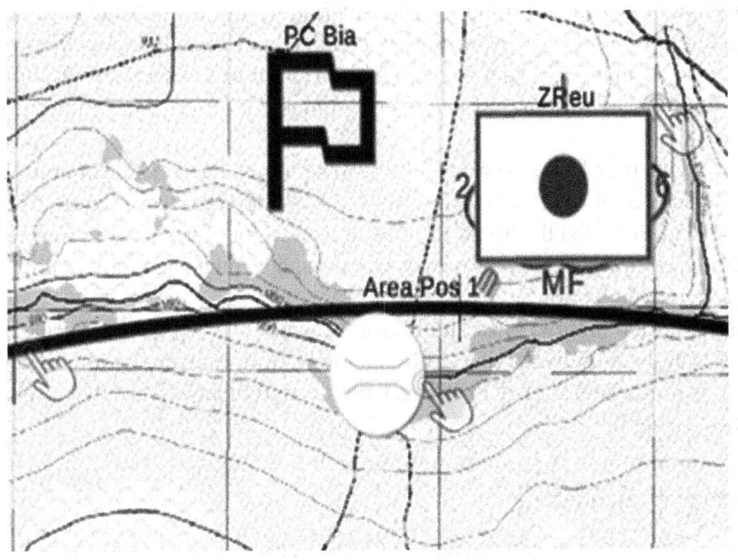

Fig. 9. Interface elements with visible hand.

no tips about what to do and how to do should display on tabletop. On the other hand, all functionalities, images, etc., should be clear.

All users have a military background and they frequently use touchscreen applications: three captains, two from Artillery (user 1 and 2), and one from Computer Engineer (user 3) and two Artillery sergeants (users 4 and 5). User 1 uses the simulator very easily and often; User 2 is familiar but does not use very often; User 3 is unfamiliar and uses infrequently. Users 4 and User 5 never interacted with the simulator.

During the observation, the ease of completion of the tasks was assessed by observing the interaction and understanding of the user to perform the tasks, as well as the feedback provided. In the end, the users were asked about: Q1) It is easy to identify which interface elements are clicked, i.e., with interaction; Q2: You can understand the content of each interface and what its purpose is.

Regarding Q1, 80% of users strongly agreed that it was easy to identify the interactive elements, while 20% partially agreed i.e., one user who had never used the simulator. He commented about some difficulty initially in finding the elements of interaction, but when he noticed the hand sign, the interaction became easier. All users agree that it was possible to understand the interface elements and their purposes (Q2).

Later, other questions were asked: Q3: It is possible to identify the features and objectives of the menus and options; Q4: The menu icons correspond to the purpose itself; Q5: It is possible to understand the goal of the system, i.e., during and after interacting what can be done through it is noticed; Q6: Report the difficulties and facilities encountered.

In Q3 and Q4, all users strongly agreed that the features, their names, objectives, and icons presented are consistent, e.g., all names were consistent with the icons, as well as users noticed their purposes after seeing them. In Q5, all users strongly agreed to say they could compare and see similarities considering what is learned and what is showing at the tabletop, on other words, militaries could see all need functionalities to do the work, because of that, the goal was got. Through Q6, it was possible to obtain feedback pointing out the facilities and difficulties encountered by users during interaction with the simulator.

For example, user 1 liked to see hand sign to highlight the clickable option, according to him, there are a lot of interface elements, so it is important to easily see what can be touched. User 3 reported the arrow to rotate the view 360°, as shown in Fig. 7, left him a little confused in the beginning because he thought it means going to a previous step and not to rotate the option.

User 5 reported that the most significant difficulty was not having frequent contact with the simulator, initially making it difficult to use. Still, later, he was able to understand the functionalities and started to interact without problems. The other users had no difficulties during the test.

2.6 Sixth Step: Refinement

By observing the use of the prototype, it was necessary to refine some information of the adapted heuristics. It is noteworthy that all refinements have already been included in the description of heuristics in Sect. 2.4 so that each description already represents the final version for ease of understanding and use by other researchers, designers, among others.

A refinement occurred in H10, as users reported the hand sign as a decisive factor to indicate clickable options, as the icon somewhat illustrates the action they should take. Therefore, in this heuristic, it was explicitly described the possibility of using a sign that represents the action that must be done.

The H12 was changed to reinforce that multiple users can interact at the same time, but one user's action cannot interfere with another user's action. The use of icon and textual description was rewritten on H2 because icons like arrows "menu rotate - arrow" can allow many interpretations such as returning an action or rotating, then a text can facilitate this interpretation.

As Refinement step, we also decided to define some questions for each tabletop heuristic in order to support its understand and use, because designers and evaluators can read the heuristic, as well as the questions to know more details about each one. A similar strategy was done by Tehrani et al. [21] to define some questions to support Nielsen's heuristics, i.e., for general context.

Table 3 shows the heuristics and questions considering the tabletop context. There are 3 columns for each heuristic and question to designers or evaluators mention if it is been considered on interface design (Yes), it is not been considered on interface design (No), or if it is not applicable (Not applicable), i.e., there is no necessity to be considered on design.

Table 3. Checklist for tabletop context.

Heuristics	Yes	No	Not applicable
Visibility of system status			
a) Is there a visual indicator that an item has sub-items?			
b) Is there a distinction between interactive and non-interactive components?			
c) Is there feedback for each user action showing whether the action was successful or not?			
d) Is it possible to identify which functionality, action, or interface element is associated with feedback?			
e) Does the feedback appear closer to the position where the action was taken?			
f) Is it possible to identify the necessary steps to carry out a task?			
Match between system and the real world			
a) Do system status or feedback messages have clear, concise language that makes them easier to understand?			
b) Are the icons self-explanatory?			
c) Is the textual description close to or in the same element of the interface that has a symbol/icon?			
d) Are the commands and actions self-explanatory in relation to their function?			
e) Do the gestures necessary to manipulate the objects match the user's mental model?			

(*continued*)

Table 3. (*continued*)

Heuristics	Yes	No	Not applicable
User control and freedom			
a) Can the user cancel an action that is being performed?			
b) Is there an icon or description that shows the possibility to cancel the action?			
c) When a user cancels an action, does that action not impact another user's actions?			
Consistency and standards			
a) Are the conventions established for other touchscreen devices followed?			
b) Are the movements necessary to perform an action consistent with the size of the device?			
c) Does the same gesture/action to interact with an interface element remain to interact with all other similar elements?			
Error prevention			
a) Are context-associated menus been used to reduce user choice options?			
b) Are disabled options not currently available?			
c) Are confirmations required for the user to have the option to rethink an action?			
Recognition rather than recall			
a) Is there a textual description to help identify the purpose or meaning of the icons?			
b) Are all elements of the interface understandable by users?			
c) When the user selects an interface element, is it clear which action will be performed?			
d) Are interactive components distinct from non-interactive?			
Flexibility and efficiency of use			
a) Are the interaction elements (menus) distributed considering the context and close to the feature?			
b) Is the information displayed in sufficient detail for the correct operation of the application?			
c) Are zoom operations performed satisfactorily??			
Aesthetic and minimalist design			
a) Is the information available to users without overloading the interface?			

(*continued*)

Table 3. (*continued*)

Heuristics	Yes	No	Not applicable
b) Can swipe menus be hidden whenever not in use, making it easy to access when needed?			
c) Are the menus aesthetically simple and describe the features clearly?			
d) Do the buttons, text and colors contrast with the background and, when necessary, are they transparent enough to allow background information to be viewed?			
Help users recognize, diagnose, and recover from errors			
a) Are messages clearly defined to assist the user in identifying and correcting them?			
b) Can all users see the error information?			
Help and documentation			
a) Is the system providing just relevant information?			
b) Is help information being showed according to the user's need and interaction?			
View Adaptation			
a) Is the system identifying each user's position and display interactive visuals oriented to him/her?			
b) Is the system able or allow users to change this interface elements orientation in a simple way?			
c) Is the system able to zoom in on the viewer allowing users to display more detail or show more interface elements with less detail (multiresolution)?			
d) Are these operations associated with conventions established by touchscreen devices?			
Cooperative/Collaborative Usability			
a) Cannot execution of a feature by one user affect the execution of another user?			
b) Does an action that a user is performing no cancel another action that is being performed by another user?			
c) If execution/action affects the execution/action of another user, is there a warning that it will influence another user's action?			
d) Are actions taken by one user visible to other users?			
e) Are the interface elements to explicitly express the user's intent?			

3 Conclusions

In short, adapting usability heuristics for tabletop applications is relatively new compared to the proposed web and mobile heuristics. There is a need for new usability evaluation heuristics to tabletop devices, as their characteristics differ from others, and designers must be aware of these differences to define better forms of interaction.

In this context, this research adapted usability heuristics and defined some characteristics of tabletop design that can be useful during the design of interactive interfaces for the tabletop context. This work was evaluated using observations, in which the interface of a simulator was changed based on the proposed heuristics and characteristics. Users using the system evaluated that the simulator was easy going.

As future work, it is proposed to use these heuristics for other systems contexts and to invite other researchers/developers to use these heuristics to design and evaluate other systems.

Acknowledgements. We thank the Brazilian Army for the financial support through the SIS-ASTROS Project (813782/2014), developed in the context of the PEE-ASTROS 2020.

References

1. Bortolaso, C., Oskamp, M., Graham, T.C.N., Brown, D.: OrMiS: a tabletop interface for simulation-based training. In: ACM Interactive Tabletops and Surfaces (2013)
2. Chuan, N.K., Sivaji, A., Ahmad, W.F.W.: Proposed usability heuristics for testing gestural interaction. In: Artificial Intelligence with Applications in Engineering and Technology (2014)
3. D'Carlo, D., Barbosa, G.A.R., Oliveira, E.R.: Proposta de Um Conjunto de Heurísticas para Avaliação da Usabilidade de Aplicativos Móveis Educacionais, 2nd edn. Abakós, Belo Horizonte (2017)
4. Dourado, M., Canedo, E.: Usability heuristics for mobile applications - a systematic review. In: Proceedings of the 20th International Conference on Enterprise Information Systems (2018)
5. Durães, D.M., Dias, C.E.: Usability heuristics for mobile applications. In: Proceedings of the 20th International Conference on Enterprise Information Systems (2018)
6. Franceschi, V.D., Fontoura, L.M., Silva, M.A.R.: Heurísticas para o Design de Sistemas Educacionais em Tabletop. In: Nuevas Ideas en Informática Educativa (2018)
7. Franceschi, V.D., Fontoura, L.M., Silva, M.A.R.: Usability heuristics for tabletop systems design. In: International Conference on Enterprise Information Systems (2020)
8. Humayoun, S.R., Chotala, P.H., Bashir, M.S., Ebert, A.: Heuristics for evaluating multi-touch gestures in mobile applications. In: British Computer Society Human Computer Interaction (2017)
9. Inostroza, R., Rusu, C., Roncagliolo, S., Jimenez, C., Rusu, V.: Usability heuristics for touchscreen-based mobile devices. In: Proceedings - 9th International Conference on Information Technology: New Generations (2012)
10. Joyce, G., Lilley, M.: Towards the development of usability heuristics for native smartphone mobile applications. In: Marcus, Aaron (ed.) DUXU 2014. LNCS, vol. 8517, pp. 465–474. Springer, Cham (2014). https://doi.org/10.1007/978-3-319-07668-3_45
11. Madni, T.M., Nayan, Y.B., Sulaiman, S., Abro, A., Tahir, M.: Usability evaluation of orientation techniques for medical image analysis using a tabletop system. In: 3rd International Conference on Computer and Information Sciences (2016)

12. Masip, L., Granollers, T., Oliva, M.: A heuristic evaluation experiment to validate the new set of usability heuristics. In: Information Technology: New Generations (2011)
13. Neto, M. José, O.: Usabilidade da interface de dispositivos móveis: heurísticas e diretrizes para o design. In: Phd thesis – São Paulo University (2013)
14. Nielsen, J.: Why You Only Need to Test with 5 Users (2000). https://www.nngroup.com/articles/why-you-only-need-to-test-with-5-users/
15. Nilsen, J.: How to conduct a heuristic evaluation. In: Nielsen Norman Group (1994). https://www.nngroup.com/articles/how-to-conduct-a-heuristic-evaluation/
16. Nilsen, J.: 10 Usability heuristics for user interface design. In: Nielsen Norman Group (1994). https://www.nngroup.com/articles/ten-usability-heuristics/
17. Prates, R.O., Barbosa, S.D.J.: Avaliação de interfaces de usuário–conceitos e métodos. In: Jornada de Atualização em Informática do Congresso da Sociedade Brasileira de Computação (2003)
18. Preece, J., Sharp, H., Rogers, Y.: Interaction Design: Beyond Human-Computer Interaction, 4th edn. Wiley, Hoboken (2015)
19. Rusu, C., Roncagliolo, S., Rusu, V., Collazos, C.: A methodology to establish usability heuristics. In: 4th International Conferences on Advances in Computer-Human Interactions (2011)
20. Shneiderman, B., Plaisant, C., Cohen, M., Jacobs, S., Elmqvist, N., Diakopoulos, N.: Designing the User Interface: Strategies for Effective Human-Computer Interaction, 6th edn. Pearson, London (2016)
21. Tehrani, S.E.M., Zainuddin, N.M.M., Takavar, T.: Heuristic evaluation for virtual museum on smartphone. In: User Science and Engineering (2014)
22. Yang, Q., Liu, J., Qin, Y., Yu, C., Yuan, Q., Shi, Y.: Studying accessible states of user interfaces on tabletops. In: IEEE 11th International Conference on Ubiquitous Intelligence and Computing and IEEE 11th International Conference on Autonomic and Trusted Computing and IEEE 14th International Conference on Scalable Computing and Communications and Its Associated Workshops (2014)

User Stories and the Socially-Aware Design Towards the OpenDesign Platform

Julio Cesar dos Reis[1,3](✉), Andressa Cristina dos Santos[1], Emanuel Felipe Duarte[1], Fabrício Matheus Gonçalves[1], Breno Bernard Nicolau de França[1], Rodrigo Bonacin[2], and M. Cecilia C. Baranauskas[1,4]

[1] Institute of Computing, University of Campinas, Campinas, SP, Brazil
{jreis,andressa.santos,emanuel.duarte,fabricio.goncalves, breno,cecilia}@ic.unicamp.br

[2] UNIFACCAMP and Center for Information Technology Renato Archer, Campinas, SP, Brazil
rodrigo.bonacin@cti.gov.br

[3] Nucleus of Informatics Applied to Education, University of Campinas, Campinas, SP, Brazil

[4] PPGInf, Federal University of Paraná UFPR, Curitiba, PR, Brazil

Abstract. Successful design projects require the involvement of stakeholders to adequately capture their needs, desires and objectives. However, literature lacks studies and open software platforms to support in this task to help solve wide-ranging design problems. In this paper, we explore user stories articulated with Socially-aware Design artifacts for the ideation and construction of the OpenDesign platform. We present how the platform was conceptualized and its key features obtained. Our study explored participatory practices in the elaboration of user stories and involved stakeholders for achieving problem clarification in a Socially-aware Design perspective. Our results illustrate the developed platform to support open design processes.

Keywords: Socially-aware design · Participatory design · User stories · Requirements engineering · OpenDesign

1 Introduction

In the 1970s, Papanek [33] highlighted the impacts caused by a culture driven by economic and technical issues that neglects the social context of design, the intended audience and society in general. Even though Papanek's focus at the time was on industrial design, his reasoning remains not only relevant in the present, but can also be extended to other areas, such as the design of computational systems. To illustrate, in recent years, the unwanted impacts that computer systems can have on economic, ethical, political, and societal issues have become more evident. Several cases reported in international and national media illustrate these

J. Filipe et al. (Eds.): ICEIS 2020, LNBIP 417, pp. 730–752, 2021.
https://doi.org/10.1007/978-3-030-75418-1_33

impacts, such as the Volkswagen Scandal, one of the largest auto companies, which admitted to have developed a software that alters pollutant emissions results to fabricate false results [13]. In the Brazilian context, the eSocial platform was also criticized by renowned Brazilian researchers, highlighting what were called "gross errors", such as not considering some kinds of zip codes, various formats of valid email addresses, and overall not taking into account the different levels of computer proficiency on the part of potential users [14]. Such problems seem to stem from a software development view that does not privilege the social world in which solutions are used and are supposed to make sense; that is, a view that does not consider software development as a socio-technical activity.

In this sense, open-source communities sometimes excel at creating quality software products that emerge from a combination of a philosophy, some kind of governance and means of participation. For this reason, the phenomenon of open-source distributed software development has drawn the attention of the scientific community to ways in which open-source communities articulate themselves to create quality software. However, besides code development activities, scientific literature does not report similar community efforts in the design of interactive systems. It is important to emphasize that what we mean by design here is much more than user interfaces and aesthetics, actually it includes the activities that precede code development itself, such as clarifying the design problem and proposing possible solutions.

The OpenDesign Project [2] is situated in this context, which stems from the absence of open environments for software system design that involves stakeholders with a socio-technical perspective, and integrates various artifacts, methods, guidelines, and tools to support different stages of a design project. OpenDesign is based on principles associated with agile [5] and lean [35] methods of software development, as well as open-source movements. The aim is to obtain open collaboration, egalitarianism, meritocracy and self-organization. Additionally, the project is based on theories, practices and methods of Organizational Semiotics [23], Universal Design [9], and Participatory Design [30]. In general, OpenDesign aims to provide an approach and software platform to support the design of computational solutions in a systemic and socially aware view, starting from the initial activities of understanding the problem and proposing solution ideas. Existing literature lacks studies in the direction of our investigation. Using similar terminology, Boisseau, Omhover and Bouchard [6] presented a literature review related to the concept of open-design. However, although it contributes to disseminate the open approach, their work emphasized product design. Other investigations have described the benefits of participatory practices and user stories in software development: Kautz [20] presented an exploratory study to assess user participation in Agile Software Development (ASD) projects; and Lucassen *et al.*. [26] studied the perceived effectiveness of user stories by showing benefits in employing user stories in everyday work environments. Our work, in turn, presents further concerns related to technology design, and our approach to design includes addressing early problem clarification activities and proposal of possible solutions.

In this article, we advance the state of the art by proposing an ideation process that supports OpenDesign practices. The concept of OpenDesign was clarified in a collaborative dynamic based on the Socially-aware Design. We conducted the clarification, creation and prioritization of user stories for the OpenDesign platform. The process featured a practice inspired by Participatory Design and resulted in a set of user stories from various stakeholders involved in the process. Based on user stories, a concept map was collaboratively built by organizing feature groups for the platform. This investigation advances our previous work [36] by further discussing the way the OpenDesign concept is placed in the existing literature. Also, we further describe the implemented technological platform and how it is connected to the organized concept map.

This work provides the following contributions: (1) a literature review situating the proposal of our open design project in the state of art; (2) an instance of a Socially-aware Design process for ideation to be used in situations involving uncertainties about requirements, as we experienced in the OpenDesign platform proposal, promoting solutions that make sense to stakeholders; (3) the clarification and organization of the requirements for building our platform that can openly promote system design; (4) the description of features in the implemented open design platform.

The remaining of this paper is organized as follows: Sect. 2 provides a detailed description on related work regarding the open design concept. Section 3 presents the theoretical and methodological foundations of Participatory Design, Socially-aware Design and User Stories; Sect. 4 presents the dynamics of activities conducted, which reflects the methodological path taken in the ideation of the OpenDesign Platform; Sect. 5 presents the results of this process with respect to the OpenDesign concept, the feature concept map and key features illustrating the OpenDesign platform; Sect. 6 discusses our findings and Sect. 7 presents the conclusion remarks.

2 Related Work

The nature of the systems design, by the large, and its sub-areas including engineering (requirement engineering, software engineering) and the role of its methods and practices are in continuous transformation, showing important niches formed by its practitioners, who move towards User-Centered Design (UCD), ASD, Design Thinking (DT), to name a few. The practical contributions of the OpenDesign Project are revealed in their fundamentals and practices, when we contrast our approach to system design and development, with more recent initiatives where the human has a special role, such as ASD, UCD and DT, briefly highlighted below.

Several authors have recognized the challenges of requirements engineering in the specification of requirements for the system (*e.g.*, [1,11,15,24,25,27,31,34, 43,45,47]) and in the fluidity of these requirements throughout a development process (*e.g.*, [12,16,18,19,38,40,48,50]; agile methods deal with the volatility of requirements, rejecting the idea of precisely defined a priori specifications. Agile methods focus on "customer" engagement, synthetic documentation efforts and iterative and incremental work on code production and software development [12,18,27,38,48,50].

One key issue in requirement engineering is how to select and prioritize functionalities to be developed. Oulasvirta *et al.* [32], for instance, argue that UCD techniques and concepts should be used for selecting systems functionalities. The authors proposed a method and computational tools to this end. The literature also points out key aspects to be taken into account to introduce UCD in a software engineering process, such as: to create suitable environments for encouraging the engagement of users and designers in a co-creation process [7], and to provide techniques for stimulating creativity during the requirement engineering process [34].

According to Teixeira *et al.* [48] ASD processes and UCD have different approaches, but they also have similarities, for instance, both are based on an iterative process with small steps, which provides opportunities for validation. The integration of ASD and UCD may bring benefits such as to increase the product user experience, to make the development process more human-centric as well as to speed up product development [40]. Such benefits, similarities and differences lead to several studies on how to integrate UCD and ASD [50]. Salah, Paige and Cairns [38], for instance, present patterns for integrating ASD and UCD processes based on a systematic literature review and interviews with experts. According to the authors there are common aspects in both approaches, such as they are both iterative, promote the end user involvement, have empirical measurement, and focus on team coherence and people. However, they also highlight significant differences, such as agile development is based on software engineering culture while the UCD considers psychology and social science culture. Losada [25] argues that hybrid methods can take advantage of both approaches, however the differences between the approaches bring difficulties to develop an agile-UCD project. The author emphasizes the need for flexible integrating strategies, which depends, for instance, on whether the requirements are clear as well as on the development stage. Marques *et al.* [27] emphasize the need to include usability aspects from the initial projects ASD sprints before the development activities. Kropp and Koishwitz [21] propose that a separate team member of an ASD project should be trained in UCD methods and practices, and then (s)he should act as consultant, with a focus on the user's needs and expectations during the entire development process.

User stories are a largely adopted technique in ASD. In fact, user stories play a central role in the development process as communication artifacts that should be understandable by both users and software developers [28]. In a Human-Computer Interaction (HCI) perspective, as emphasized by Lopes *et al.* [24], user stories should be enriched by usability and user experience aspects to produce usable software systems. Harbers, Derweiler e Neerincx [15] investigate how users stories created using techniques from UCD (during workshops) can represent values, as well as how they can be used by developers in the requirement engineering process. Melegati and Wang [29] argue that experiment-driven software development results on features most suited to users' wishes/needs. However new guidelines and templates should be used to write users stories from the UCD perspective.

DT is a concept that proposes a more user-centered approach to requirements engineering, offering a methodology for understanding what the user "really needs", and producing a series of quick and simple prototypes that eventually converge into innovative solutions [8,16,19,22,25]. DT can be understood and described at three different levels: as a way of thinking (mindset), as a process to be integrated into development processes (*e.g.* agile) or as a "toolbox", useful at different times in a process [16,22]. Carell, Lauenroth and Platz [8], for instance, present how DT can be used as a framework for eliciting requirements in digitalization and digital transformation projects. Recently, several researchers have been discussing how to facilitate the integration of DT with software development, to bring more diverse knowledge to the challenges of the process (*e.g.*, [10,16,17,19]). According to Hehn and Uebernickel [17] DT can be integrated into the requirements engineering process to deal with fuzzy stakeholders' needs. To this end, the authors argue that the human-oriented working mode of DT must be integrated with the more formal and technology-driven world of requirement engineering. Hehn *et al.* [16] argue that DT can bring a UCD approach to requirements engineering, but different integration strategies should be considered according to the project' objective and context. While Levy and Huli [22], instead of providing a large integration strategy, focus on how specific concepts and activities of DT (in workshops) can be integrated to facilitate requirement elicitation in a business process.

Some authors have proposed combined process models that integrate DT with ASD models (*e.g.*, [10,31]). Possible models for the integration of DT in systems development processes involve: its use as a separate process before development itself; its use as an initial phase of the process with the participation of one or more members of the development team; its use as an initial phase of the process and a great overlap between the DT team and the development team; as methods that developers can use to overcome problems they encounter throughout the process [10]. Murugesan, Hoda and Salcic [31] compare three requirement elicitation techniques (systematic review, crowdsourcing and DT) in the identification of design features in a UCD view. According to the authors the combination of those techniques effectively supported identifying the system requirements, in a study of system development in the context of domestic electricity consumption. DT, as a "user-centric" practice, is consistent with the early stages of requirements engineering, rapid prototyping and "customer" engagement of ASD methods.

While UCD includes a set of methods, techniques and processes that puts an (epistemic) user at the center of the systems development process, in the "open design" proposed in this work, the "user" is not just a figure imagined, but has its real multifaceted role in different "stakeholders" who participate in the process as "codesigners"; this goes beyond participation only as an informant of needs. Unlike the aforementioned approaches, the focus is on the open (co)design process, and the code (or coding phase) integrates with this larger design process (and not vice versa).

3 Theoretical and Methodological Background

In the OpenDesign perspective, the design of interactive systems is understood as a social process that involves both the characterization of the design problem and its solution. It involves a dialogue not only of the designer with the design materials as proposed by Schön [39], but also mainly among the individuals participating in the process (*e.g.*, designers, developers, users and other stakeholders). This dialogue contrasts various design views and various ways of framing design situations. Several artifacts (informal, formal and technical) are used as communication and mediation tools among participants during the process of creating the interactive system [3,4]. Thus, system functionalities should be defined based on the understanding of the stakeholders' needs, and their aspirations and objectives. To this end, it is necessary to enable the effective participation of stakeholders in the process, from joint clarification of the problem to the definition of system functionalities in design. In this paper, this process is supported by techniques and practices based on Participatory Design, Socially-aware Design, and collaboratively constructed User Stories. These concepts are described in the following subsections.

3.1 Participatory Design

According to Schuler and Namioka, pioneered in Scandinavia, Participatory Design represents an approach towards computer systems design in which *"the people destined to use the system play a critical role in designing it"* [41]. Participatory Design is, therefore, characterized by the active participation of system end users throughout the software design and development cycle [49]. More than sources of information, end-users (and other stakeholders) perform effective contributions at every stage of the design and development cycle, reflecting their perspectives and needs. The quality in the design process and in the resulting system is a consequence of the mutual learning and combination of backgrounds from the various participants.

Participatory Design methods are characterized by the use of simple techniques and affordable materials or digital resources; for instance, techniques such as brainstorming, storyboarding, and workshop, all widely used in Participatory Design, require mostly simple materials such as pen and paper, whiteboard and post-it notes. Literature has shown that Participatory Design practices can be used throughout the (conventional) software life cycle. As an example, the work of Muller, Haslwanter and Dayton [30] presents a collection of 61 participatory practices used at different stages of a system life cycle (pre-design, design, evaluation, post-design). Our work, however, is not focused on applying specific Participatory Design practices, but on using it as a theoretical and methodological background to promote a more systemic view of the design of computational system, using ideas coming from the effective participation of stakeholders in the understanding of the design problem and proposal of the system functionalities.

3.2 Socially-Aware Design

The Socially-aware Design [4] recognizes the complexity of the social context in which a system is created and subscribes to a systemic view for technology design, which involves the informal, formal and technical levels of a design situation or design problem. Inspired in Stamper's semiotic perspective [46], which introduced the design of information systems seen by its semiotic aspects, taking the sign as a basic unit of information, in the Socially-aware Design, it is assumed that a design solution depends on knowledge of the informal and formal aspects of organizations and society, and that solution impacts back in the formal and informal layers of information of a social group.

From a practical point of view, the Socially-aware Design consists of three basic components: first, the interested parties: users, stakeholders, co-authors. Second, artifacts, systems and methods that enable communication and expression of everyone involved with different skills in a common language. Third, face-to-face workshops that we call (semi)participatory in reference to Semiotics and Participatory Design. In these workshops, stakeholders work in the design process while there is mutual learning and co-construction of the design product by the parties. Some of its artifacts and practices for clarifying the problem, some of which originated in Organizational Semiotics [23,46]. These artifacts are used to support the design problem clarification, the anticipation of problems related to the interested parties, and the proposition of ideas and solutions for the system. The conceptual proposition for design, the proposed artifacts and practices support the understanding of social and technical issues involved in the problem, without losing their relation to the situated context of the design, which gives meaning to the problem and stakeholders' demands.

The Socially-aware Design, its methodological approach, artifacts and practices laid the foundation for the OpenDesign Platform design.

3.3 User Stories

User stories are one of the key development practices for teams using ASD and project management methods (*e.g.*, Scrum or eXtreme Programming). A User Story is a representation that aims to capture a description of a software resource from an end-user perspective. The user story spells out the user's role, what they want, and why. A user story helps create a simplified description of a requirement and can fit into agile process structures. Therefore, a "user story" represents a definition of a high-level requirement, containing enough information for developers to produce a reasonable estimate of the effort to implement it.

User stories can be considered a starting point for a conversation that establishes the actual requirements of the product. In this paper, they were written collaboratively by product stakeholders, also helping to prioritize prospective functionalities during the system design period. Identifying and describing user stories thus serve as reminders of what is still missing and why of those needs to be developed. Coupled with user stories, product backlog, and sprint practices are accomplished [42]. A backlog acts as an ordered list of stories to be developed based on priorities set by their business value and technical feasibility [42]. In our work, user stories were aligned with Participatory Design practices because user stories specify involved roles, the notion of purpose and business value.

4 OpenDesign Ideation Process

The activity dynamics adopted in this research reflect the path conducted by the OpenDesign project team aiming to articulate the Socially-aware Design model with open-source practices and lean agile methods. This dynamic considers the insertion of collaborative and participatory practices, used to assess their relevance to the design process of the OpenDesign platform. The elicitation and understanding of features in the OpenDesign followed a process in which the project team and other researchers, as stakeholders, were invited to collaborate on participatory practices for the creation, prioritization, and organization of features. Participants included 7 researchers of HCI and Software Engineering fields, 13 graduate students, and 1 undergraduate student. Stakeholders contributed to include their vision and needs in the context of the project.

The process described in Fig. 1 does not repeat conventional software development. In our work, there was an initial effort in its conception, that is, in understanding the concept to be implemented through design activities (rounded blue boxes with dotted border in Fig. 1). Subsequently, these were associated with the development of the software itself (red rounded boxes with a continuous border in Fig. 1). The activities represented in gray and dashed boxes have both design and software development nature. The SD, EF, and SF are artifacts of Socially Aware Design. User stories were also used in the process in a participatory and online manner.

Each of the defined activities (rectangles in Fig. 1) was carried out collaboratively (*i.e.*, involving multiple stakeholders) and at least in two moments: (1) in activities, we call *warm-ups*, in which participants contributed individually and previously to the elaboration of artifacts (documents), usually using online tools; and (2) in face-to-face meetings, in which participants interacted to generate a consolidated artifact. At both times, collaborative tools were used to provide a shared view of the project to all participants. Among the collaborative face-to-face meetings and online activities, we highlight the following: (1) anticipation of problems and solutions; (2) initial definition of requirements; (3) creation of user stories; (4) prioritization of stories; (5) refinement of functionality; and (6) summary of features.

1. **Anticipating Problems and Solutions:** From the identification of stakeholders, it becomes possible to anticipate prospective problems and their associated solution ideas. To this end, the Evaluation Frame was used for describing, for each layer of the Stakeholder Diagram, a set of problems or questions. Also, it associates possible solutions or ideas on how to solve these problems succinctly. This artifact was completed online and collaboratively with the support of the SAwD tool [44].

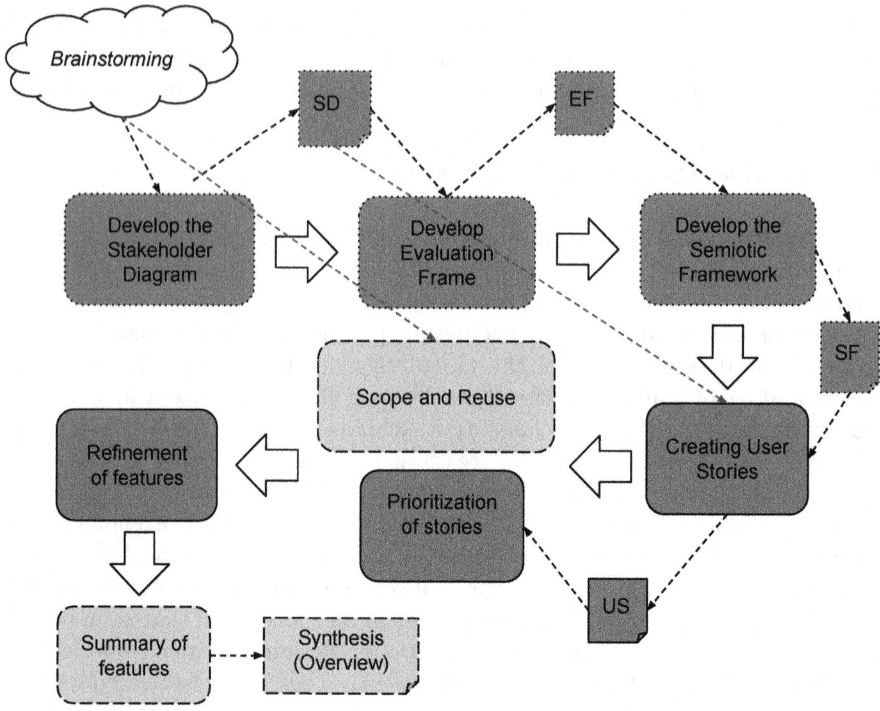

Fig. 1. Process conducted in the OpenDesign Project [37]. (Color figure online)

2. **Initial Requirements Definition:** Possible solutions and ideas described in the Evaluation Framework were used to support the initial identification of requirements at a high level of abstraction. For this purpose, the requirements were described based on the different steps of the Semiotic Framework. The framework is typically populated from the highest step (Social World) to the lowest step (Physical World), in a refinement approach in which a requirement on a higher step may derive requirements on the immediately lower step. However, the up and down movements of the ladder can be performed iteratively, as the understanding of the initial requirements matures as they are listed. This was the third and final artifact to be completed online and collaboratively by the project participants.
3. **Creating User Stories:** For this first step, Participatory Design contributed with action dynamics that made it possible to bring stakeholders into the design process. We proposed a participatory practice that combined the concept of user stories with a collective brainstorming session to create stories for OpenDesign. The session was facilitated and used a whiteboard on the wall, visible among the participants, on which user stories written with post-its were glued and organized by all participants. User stories were designed and crafted for potential OpenDesign stakeholders, previously raised using

the Stakeholder Diagram. All user stories were later inserted into the Trello[1] tool, which allowed us to visually organize the project in the form of charts and tasks in the form of cards. Such an organization helped to categorize user stories and to obtain an overview.

4. **Prioritization of Stories:** We used the ConsiderIt[2] tool to prioritize user stories, inserted into the tool for discussion and voting by project members. ConsiderIt provides a slider where users can vote for how much they agree or disagree with an idea (in this case, a user story), while listing positive or negative points for that idea. ConsiderIt enables participants to justify their rating, or comments on points raised by others. Thus, user stories were put into ConsiderIt. Those involved in the project were asked to rate them according to their priority for developing the OpenDesign platform.
5. **Refinement of Features:** To better define the features raised and prioritized, we performed a collective and face-to-face refinement of user stories. To this end, involved researchers in the project looked at the list of user stories prioritized in ConsiderIt and deliberated on what were, in fact, the key stories for the OpenDesign platform to be implemented. Following this deliberation, an online form was created that allowed participants to associate features with more detailed specifications to each user story; so we come to a lower granularity and therefore closer specification to implementation.
6. **Summary of Features:** Finally, in the last step, the participants jointly and in person created a synthesis of backlog in a concept map. The creation of a functional synthesis map involved the collaborative analysis of the user stories and the suggestion of categories to organize them. This synthesis map plays a key role in the elaboration of an overview of the platform's potential modules. The categories emerged from the analysis of the stories and they were refined through face-to-face discussions among the participants. Researchers iteratively suggested new categories and changes of elements from one category and to another until the conclusion of the map.

5 OpenDesign Concept and Platform

This section presents the results of the process definition and instantiation in the OpenDesign project. Among our results, we highlight the clarification of the OpenDesign concept (*cf.* Subsect. 5.1) and the prioritized set of user stories (*cf.* Subsect. 5.2). In addition, Subsect. 5.3 presents key features of the implemented open design platform.

5.1 OpenDesign Concept Clarification

The collaborative use of the Socially-aware Design provided clarification of the OpenDesign concept among the project stakeholders. Several categories of professionals and organizations were indicated as stakeholders. For example, on the

[1] https://trello.com/.
[2] https://consider.it/.

one hand, even distant from the implementation of the OpenDesign platform itself, educational institutions such as the University of Campinas (UNICAMP), and funding such as São Paulo State Research Support Foundation (FAPESP), affect and are affected by the OpenDesign to some degree, so they should not be overlooked. On the other hand, programmers and HCI professionals involved in the project have a direct relationship with the project, and consequently, with the OpenDesign platform.

Several questions and problems were raised for the identified stakeholders. The participants proposed ideas and solutions. For example, for the stakeholder "computer science student" the problem arose that *"they (the students) might find that design is just the "make-up" of a computer system, can be done later, on top"* (Translation from Portuguese language of the original phrase by the researchers). For this problem, the following solution/ideas were proposed: *"Providing teaching materials, tutorials and concrete examples; Providing simplified versions and grounded and detailed versions of the materials, with practical tips and examples; Suggest 'next' steps for the work done"* (Translation from Portuguese language of the original phrases by the researchers). The anticipation of potential problems and issues provided supported to raise initial requirements for a solution within a design project, in this particular case, initial requirements for the OpenDesign platform.

This "on top" means the frontend is constructed on top (and after) of the backend of the system.

Definition of requirements encompassed requirements from the social world (*e.g.*, contributing to distributed design and providing universal access to people) to the physical world requirements (*e.g.*, server needs and platform compatibility with assistive technologies). Obtained requirements inspired a participatory approach, described in the next subsection. Based on the identification of user stories, we seek a level of abstraction closer to the prospective use of the OpenDesign platform by the identified stakeholders.

5.2 User Stories

In the user story creation stage, a total of 95 user stories were created after the participatory practice, already considering the removal of redundant/duplicate stories. In the next step, we chose the user stories that would be essential for the development of the OpenDesign platform and then associated system functionalities with each one. Table 1 lists those user stories considered essential. With this practice, participants collectively produced a map that synthesizes the key concepts involved in the OpenDesign concept, as well as groups desired functionalities for the OpenDesign platform. Figure 2 graphically illustrates the concept map and requirement groups from user stories. In the following, we describe the conceptual groups, including functional and non-functional requirements associated to the number of the user stories:

- **OpenDesign:** The platform should present a list of featured projects, as well as allow its users to search existing projects; The aesthetics of the platform must be pleasant and customizable; The philosophy behind the platform

Table 1. User stories considered essential [37].

#	User stories
#31	**As** designer **I want** design process documentation **so** I can understand how to replicate that proposal
#33	**As** amateur **I want** filter projects and backlogs **to** choose how and what to contribute
#35	**As** amateur/enthusiast **I want** to see images and videos of projects on the platform **so** I can get to know the enabled jobs
#38	**As** OpenDesigner **I want** to look for problems/projects **so** I can contribute to it
#41	**As** OpenDesigner **I want** to know the artifacts available on the platform **so** that I can use in my project
#47	**As** an OpenDesigner **I want** to share my project **so** that I can engage the community
#54	**As** the project owner **I want** set the license **to** secure and set limits for contributors
#67	**As** an IHC teacher **I want** the platform supports multiple design paradigms **so** I can work multiple with my students
#92	**As** a user, **I want** an FAQ **to** resolve recurring questions by collaboratively building solutions

should be communicated, as well as tutorials and demonstrations on how to use the platform; the platform must have a support forum, terms of use and a mechanism for reporting terms violations; Finally, the platform must follow accessibility criteria and recommendations. This description relates to some of the essential stories (#33, #38, #41 and #92) that appear in this group.

- **OpenDesigners:** Platform users can have a portfolio that highlights projects they have contributed to; users can bookmark projects for follow-up and later reference; users can communicate with each other privately via chat; the platform should provide a form of reputation among its users, according to their contributions and their peer recognition. This description relates to some of the essential stories (#23, #84 and #87) that appear in this group.
- **Project:** The platform should allow the creation and maintenance of projects, by considering the choice of a desired software license; the platform must allow the control of project visibility (public or private); a project overview as well as a dashboard with metrics; project access control; Possibility to export, version, compare, diverge and combine projects; ways to publicize (share) an

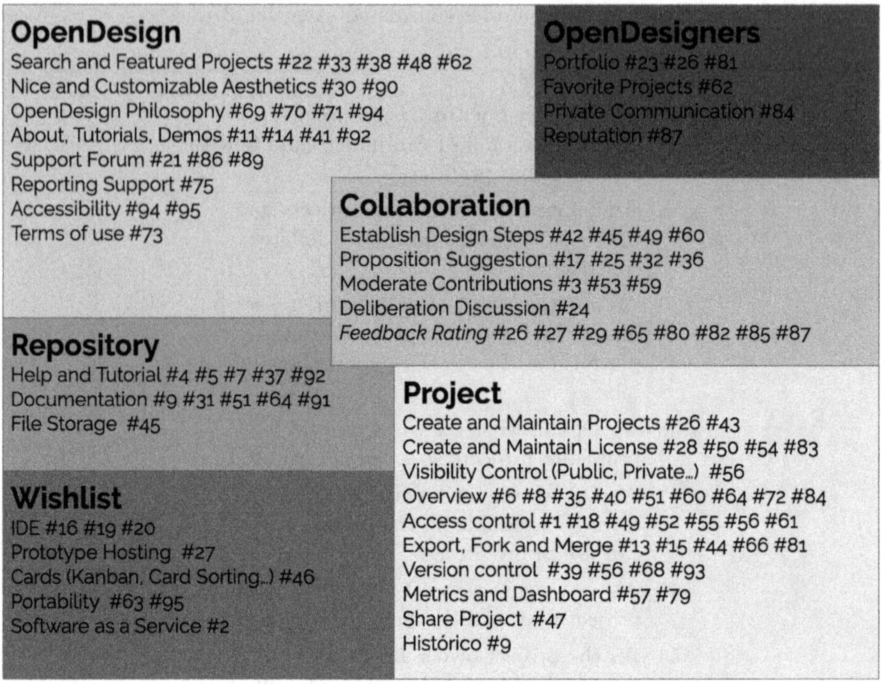

OpenDesign
Search and Featured Projects #22 #33 #38 #48 #62
Nice and Customizable Aesthetics #30 #90
OpenDesign Philosophy #69 #70 #71 #94
About, Tutorials, Demos #11 #14 #41 #92
Support Forum #21 #86 #89
Reporting Support #75
Accessibility #94 #95
Terms of use #73

OpenDesigners
Portfolio #23 #26 #81
Favorite Projects #62
Private Communication #84
Reputation #87

Collaboration
Establish Design Steps #42 #45 #49 #60
Proposition Suggestion #17 #25 #32 #36
Moderate Contributions #3 #53 #59
Deliberation Discussion #24
Feedback Rating #26 #27 #29 #65 #80 #82 #85 #87

Repository
Help and Tutorial #4 #5 #7 #37 #92
Documentation #9 #31 #51 #64 #91
File Storage #45

Project
Create and Maintain Projects #26 #43
Create and Maintain License #28 #50 #54 #83
Visibility Control (Public, Private...) #56
Overview #6 #8 #35 #40 #51 #60 #64 #72 #84
Access control #1 #18 #49 #52 #55 #56 #61
Export, Fork and Merge #13 #15 #44 #66 #81
Version control #39 #56 #68 #93
Metrics and Dashboard #57 #79
Share Project #47
Histórico #9

Wishlist
IDE #16 #19 #20
Prototype Hosting #27
Cards (Kanban, Card Sorting...) #46
Portability #63 #95
Software as a Service #2

Fig. 2. Map of features defined for the OpenDesign platform (with indication of number of the User Stories) [37].

existing project; and a view of the history of existing projects on the platform. This description relates to some of the stories considered essential (#35, #47 and #54) that are in this group.

- **Collaboration:** Within a project, the platform should enable design steps to be established; it should be possible to suggest or propose ideas to which users can contribute by voting, rating or feedback; There should also be a way to moderate contributions as well as deliberate ideas and make decisions. It should still be possible to create evaluations within the project and request feedback from any input, product, and/or artifact. This description relates to some of the stories considered essential (#49 and #3) that appear in this group.
- **Repository:** Within a project, the platform should enable the creation of a repository of knowledge and design artifacts, where users can store digital files, manage project documentation, and write help documentation and tutorials. The story considered essential #31 appears in this group.

Other desired aspects were elicited, but not considered priorities for the first versions of the platform, such as integrated code creation functionality, prototype hosting, and *kanban* style card functionality for various uses. Also portability so that the platform can be used on different devices, the capability for communicating with different devices, and finally the desire for the platform to be made available as a service.

5.3 OpenDesign Platform

After accessing the platform's homepage and logging in, the participant can view all the projects in which (s)he participates (*cf.* Fig. 3). In this project interface, the participant can also create a new project or join an existing project.

Fig. 3. Interface to create and join a new project.

4 PARTICIPANT(S)
50 STAKEHOLDER(S)
26 PROBLEM(S)
20 IDEA(S)
25 REQUIREMENT(S)
OpenDesign for Global Challenges
Henley Symposium and Summer School on Organisational Semiotics - July 11th, 2019 Hands-on Short Course
Facilitators: Lara Piccolo and Angela Nobre (UK), Cecilia Baranauskas and Fabricio Gonçalves (Brazil), with support of LInterHAD, IC Unicamp
Goal: learning and practice of the socially aware design instantiated in a common problem nowadays - the fake news global challenge. Through the use of this collaborative platform, we will clarify the problem, raising the main interested parties, issues they might have and ideas of design solutions for facing the problem. The short course participants are co-located part in Reading (UK) and part in Campinas (Brazil); both groups will work in the same problem, sharing ideas along the short course.
References
1. Stamper, R.K. (1973) Information in Business and Administrative Systems. John Wiley and Sons, NY Liu, K. (2000) Semiotics in Information Systems Engineering. Cambridge University Press, Cambridge
2. Baranauskas, M.C.C. (2014) Social Awareness in HCI. In ACM Interactions 21(4), 66-69
3. da Silva J.V., Pereira R., Hayashi E.C.S., Baranauskas M.C.C. (2018) Design Practices and the SAwD Tool: Towards the Opendesign Concept. In: Liu K., Nakata K., Li W., Baranauskas C. (eds) Digitalisation, Innovation, and Transformation. ICISO 2018. IFIP Advances in Information and Communication Technology, vol 527. Springer, Cham, p. 208-217
#72
0
0
0
Pros
Neutral
Cons

Fig. 4. Interface presenting a project overview.

Upon selecting/joining a specific project, the participant accesses the project overview interface (*cf.* Fig. 4). It presents the project description and presents ongoing indicators, such as the number of participants involved, the number of registered stakeholders, identified problems and candidate solutions, as well as elicited requirements. In this interface, the participant can access the Socially-aware Design artifacts through the menu on the left side.

Figure 5 shows the diagram of stakeholders. With it, project members (participants) can register stakeholders in different layers, from the operational (more technical) layer to the community layer, with no ordering constraints. They are able to browse in the different layers of stakeholder categories. The purpose of this diagram is to support the identification of diversified stakeholders who may affect or be affected by the project under development.

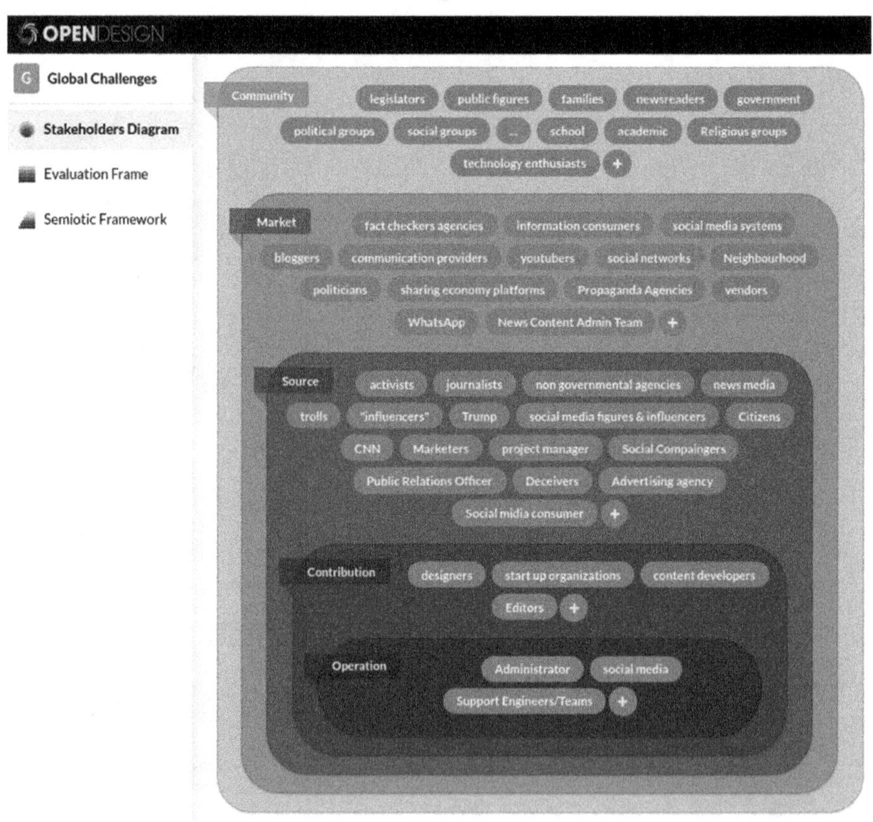

Fig. 5. Stakeholders diagram (with examples of Stakeholders).

Figure 6 presents the interface of the Evaluation Frame, in which participants can register design issues and solutions associated with the stakeholders identified in the layers. This interface provides functionalities to allow participants to vote on issues and solutions, as a way to prioritize them.

Figure 7 shows the interface for the Semiotic Framework for the organization of requirements in the project in various categories associated with the ladder, considering from the physical (physicalinfrastructure part of the system) until the social step, which represents requirements with direct implications in the social world of the system under design. The OpenDesign platform allows associating explicitly the requirements created with issues and solutions raised in the Evaluation Framework, characterizing part of the design rationale of a solution.

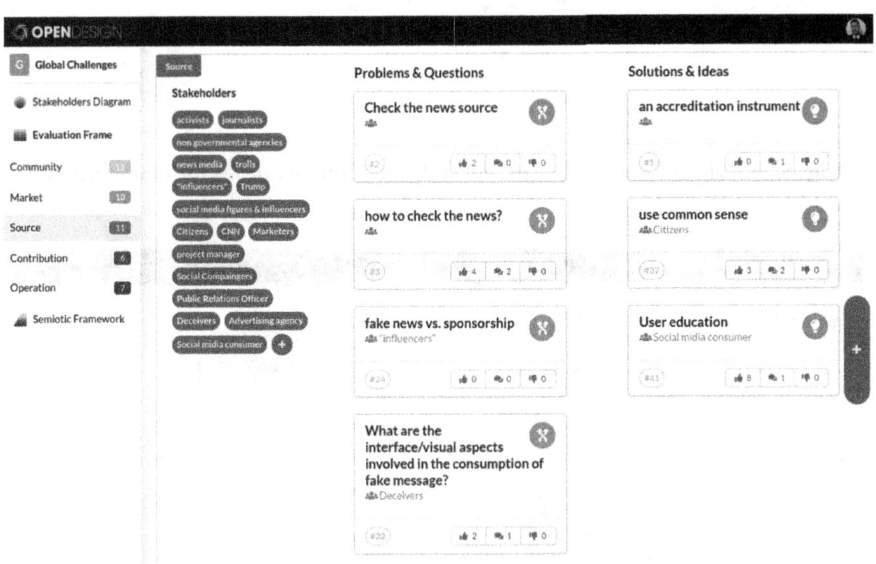

Fig. 6. Evaluation frame with example of cards.

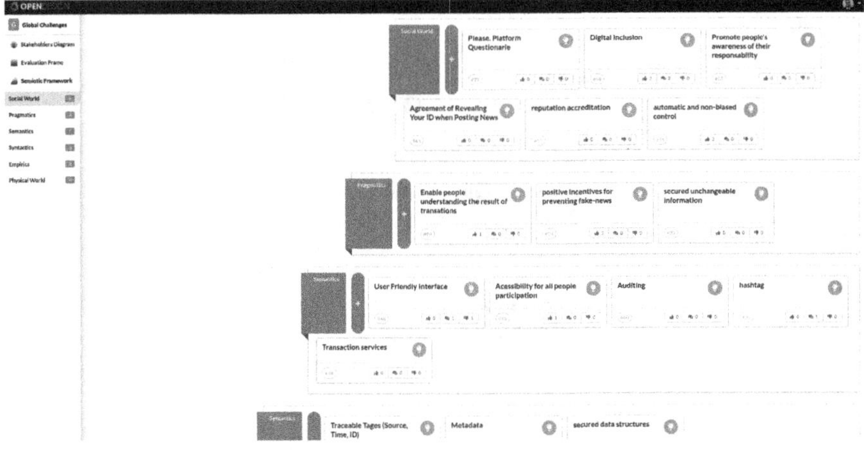

Fig. 7. Interface of the Semiotic Framework (with examples) ework, characterizing part of the design rationale of a solution.

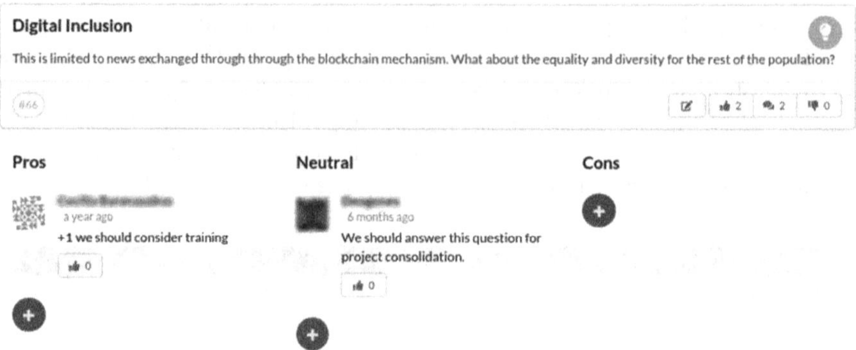

Fig. 8. Interface for adding pros/neutral/cons commentaries for deliberation.

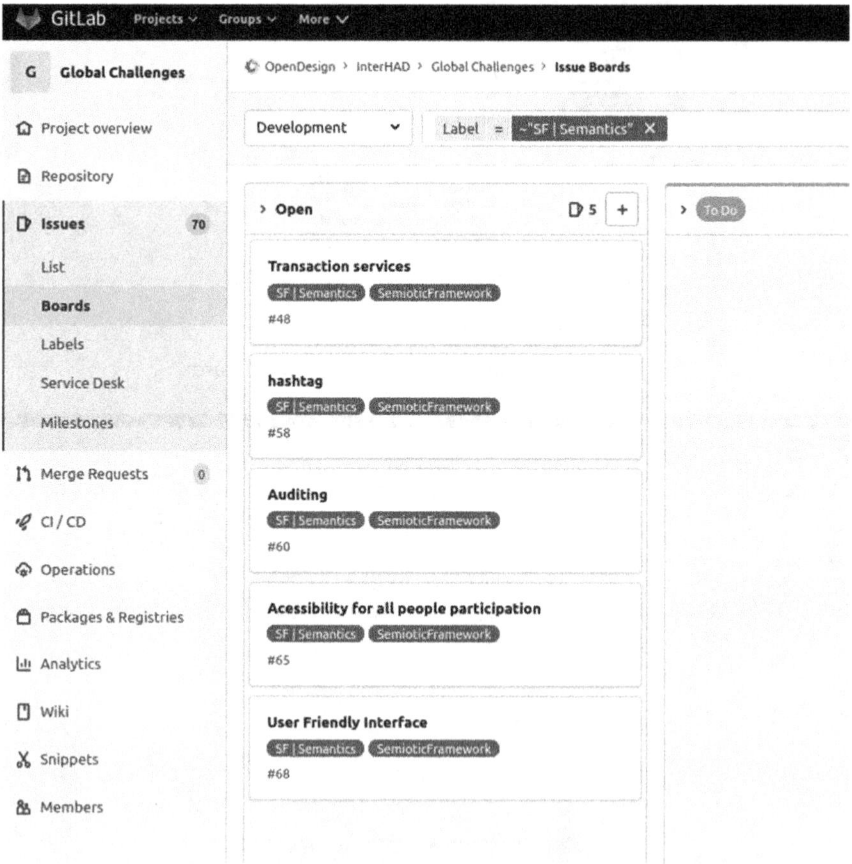

Fig. 9. Requirements from the Semiotic Framework viewed in Gitlab as issues.

The platform aims to foster discussions among participants. To this end, after the creation of a problem, solution, or requirement, users can add pros/neutral/cons commentaries on the elements (*cf.* Fig. 8). This supports a deliberation regarding that requirement because it allows people to express themselves on the proposed requirement. This feature was also implemented in the evaluation frame, in which participants can further discuss and deliberate on ideas/solutions and problems.

Additionally, elements in the evaluation frame and in the Semiotic Framework can be managed in the Gitlab issue tracker (*cf.* Fig. 9). This way, requirements defined in the design phase of an open project move towards development and other downstreams phases.

6 Discussion

We proposed the concept and a supporting software platform for OpenDesign. We understand that our proposal extends the state of the art because our design approach occurs as open and participatory (co)design process, highlighting the real participation and involvement of stakeholders.

From the methodological point of view, the main feature of the proposed process is the integration between design and requirements engineering. The motivation for this has its roots in the light of agile methods and open source software development, without losing sight of the practices of Socially-aware Design. This integration occurs primarily in the planning and at the end of each iteration with ongoing reviews and stakeholder involvement. In this sense, the different levels of abstraction used to describe functionalities (Semiotic Framework, User Stories, and Functionality Map) were necessary to think both design and development, in an integrated way. Whereas one abstraction suits better to software development, the others favor design and prototyping of interaction with the platform and user interfaces.

We understand that the identification of stakeholders in several community layers was a valuable source for reflecting on the necessary roles in defining user stories (epics, at the beginning) and, hence, the platform. Our design approach allowed participants to anticipate problems that permeate the platform and potential solution referrals, both from a divergent perspective, where alternative paths are identified in an ideation process. Afterwards, they need to converge in a deliberation process to a solution. In this stage, it was possible to understand the value or type of contribution of each functionality to the OpenDesign platform.

Practices conducted to prospect and refine stakeholder requirements on the OpenDesign platform culminated in two complementary views: 1) the feature map being an overview of the platform concept; 2) the user stories set in a backlog, expected to continue with software development supported by design practices. By adopting user stories, identified in a participatory approach with different stakeholders, it was possible to identify candidate features for the platform that were refined in participatory analysis and detailing activities. Participatory practices fostered the emergence and discussion of new ideas that took shape and evolved during the discussions. This favored the maturation of user stories and served as an object for the construction of shared meaning among participants.

Refined user stories during participatory practices resulted in functionalities related to several aspects of the platform, which included: project management; people's engagement and participation; collaboration and sharing of knowledge; and technical aspects involved in building the platform. These user stories proved adequate to form a basic set of functionalities and indicated possible extensions.

We understand this process can be an alternative for the current demand of the software industry regarding design approaches (*e.g.*, DT and Design Sprint), considering its features on Socially-aware Design and further development. Additionally, the result of this process delivered a more concrete concept (and a platform) of what an open design approach may represent, targeting other "open" initiatives, such as open source software development. Therefore, it advances the idea of working design as an open and social-aware process.

We obtained the OpenDesign software platform[3] (still in beta version). We showed and exemplified key features implemented in the platform. We plan keeping its development to reach and offer further features to the community. For instance, we aim to improve the experience of newcomers via onboarding features to help participants in understanding the artifacts and the design process. Also, we aim to implement evaluation tools in the platform to designers report field studies based on prototypes.

7 Conclusion

The development of systems that do not privilege the social world in which solutions are used might trigger unwanted impacts on economic, ethical, political and society in general. Our work investigated the concept of OpenDesign as an approach to design systems by involving stakeholders' perspectives in the ideation of systems. Our solution adopted participatory practices to understand the problem and identify requirements involving different stakeholders. Our research showed how the OpenDesign concept is situated in literature, and illustrated a methodological process and artifacts used in the design of a platform to support the ideation and collaborative creation of systems aligned with this concept. We presented a beta version of the platform, which is currently being tested for evaluation in relevant cases. The synergy between the employed practices favored the identification of different interests and perspectives, which resulted in the specification of requirements and the construction of a software platform that implements the concept. Future work will involve the investigation of several use cases with the platform.

Acknowledgments. This work is financially supported by the São Paulo Research Foundation (FAPESP) through grants #2015/24300-9, #2017/06762-0, #2018/25972-9, and #2020/04242-2, by the National Council for Scientific and Technological Deve-

[3] http://opendesign.ic.unicamp.br.

lopment (CNPq) through grant #306272/2017-2, and by the Coordenação de Aperfeiçoamento de Pessoal de Nível Superior (CAPES) - Finance Code 001.

References

1. Bano, M., Ikram, N.: Addressing the challenges of alignment of requirements and services: a vision for user-centered method. In: Zowghi, D., Jin, Z. (eds.) Requirements Engineering. CCIS, vol. 432, pp. 83–89. Springer, Heidelberg (2014). https://doi.org/10.1007/978-3-662-43610-3_7
2. Baranauskas, M.C.C.: Opendesign: Técnicas e artefatos para o design socialmente consciente de sistemas computacionais. https://bv.fapesp.br/pt/auxilios/98824/opendesign-tecnicas-e-artefatos-para-o-design-socialmente-consciente-de-sistemas-computacionais/ 2017, more info like: Regular Fapesp research Grant: #2015/24300-9. Accessed Oct 2019
3. Baranauskas, M.C.C.: Socially aware computing. In: da Rocha Brito, C., Ciampi, M.M. (eds.) Proceedings of the ICECE'2009 - VI International Conference on Engineering and Computer Education, vol. 6, pp. 84–88. Council of Researches in Education and Sciences (COPEC), Buenos Aires, Argentina (2009)
4. Baranauskas, M.C.C.: Social awareness in HCI. Interactions **21**(4), 66–69 (2014)
5. Beck, K., et al.: Manifesto for agile software development (2001). http://agilemanifesto.org/. Accessed Oct 2019
6. Boisseau, E., Omhover, J.F., Bouchard, C.: Open-design: a state of the art review. Design Sci. **4** (2018). https://doi.org/10.1017/dsj.2017.25
7. Budweg, S., Draxler, S., Lohmann, S., Rashid, A., Stevens, G.: Open design spaces: socially crafting interactive experiences. In: Proceedings of the 8th ACM Conference on Designing Interactive Systems, pp. 442–443 (2010)
8. Carell, A., Lauenroth, K., Platz, D.: Using design thinking for requirements engineering in the context of digitalization and digital transformation: a motivation and an experience report. In: The Essence of Software Engineering, pp. 107–120. Springer, Cham (2018). https://doi.org/10.1007/978-3-319-73897-0_7
9. Connell, B.R., et al.: The principles of universal design (1997)
10. Dobrigkeit, F., de Paula, D.: Design thinking in practice: Understanding manifestations of design thinking in software engineering. In: Proceedings of the 2019 27th ACM Joint Meeting on European Software Engineering Conference and Symposium on the Foundations of Software Engineering, pp. 1059–1069 (2019)
11. Doerr, J.: Towards acceptance of socio-technical systems-an emphasis on the requirements phase. In: Socioinformatics-The Social Impact of Interactions between Humans and IT, pp. 145–152. Springer (2014)
12. Espinoza, J., Loarte, P., Espinoza, C., Paz, F., Arenas, J.: A new software development model: innovation through mobile application with UCD. In: Marcus, A., Wang, W. (eds.) DUXU 2018. LNCS, vol. 10918, pp. 673–692. Springer, Cham (2018). https://doi.org/10.1007/978-3-319-91797-9_47
13. G1: Erros grosseiros: esocial rejeitou e-mail com hífen: posted in em 05/11/2015. (2015). Accessed Oct 2019
14. G1: 'dieselgate': veja como escândalo da volkswagen começou e as consequências: posted in 23/09/2015 15h48 and updated 05/02/2019 20h51 (2019). Accessed Oct 2019

15. Harbers, M., Detweiler, C., Neerincx, M.A.: Embedding stakeholder values in the requirements engineering process. In: Fricker, S.A., Schneider, K. (eds.) REFSQ 2015. LNCS, vol. 9013, pp. 318–332. Springer, Cham (2015). https://doi.org/10.1007/978-3-319-16101-3_23
16. Hehn, J., Mendez, D., Uebernickel, F., Brenner, W., Broy, M.: On integrating design thinking for human-centered requirements engineering. IEEE Softw. **37**(2), 25–31 (2019)
17. Hehn, J., Uebernickel, F.: The use of design thinking for requirements engineering: an ongoing case study in the field of innovative software-intensive systems. In: 2018 IEEE 26th International Requirements Engineering Conference (RE), pp. 400–405. IEEE (2018)
18. de la Hidalga, A.N., Hardisty, A., Jones, A.: Scram-ck: applying a collaborative requirements engineering process for designing a web based e-science toolkit. Requirements Eng. **21**(1), 107–129 (2016)
19. Kahan, E., Genero, M., Oliveros, A.: Challenges in requirement engineering: could design thinking help? In: Piattini, M., Rupino da Cunha, P., García Rodríguez de Guzmán, I., Pérez-Castillo, R. (eds.) QUATIC 2019. CCIS, vol. 1010, pp. 79–86. Springer, Cham (2019). https://doi.org/10.1007/978-3-030-29238-6_6
20. Kautz, K.: Investigating the design process: participatory design in agile software development. Inf. Technol. People **24**(3), 217–235 (2011)
21. Kropp, E., Koischwitz, K.: Experiences with user-centered design and agile requirements engineering in fixed-price projects. In: Ebert, A., Humayoun, S.R., Seyff, N., Perini, A., Barbosa, S.D.J. (eds.) UsARE 2012/2014. LNCS, vol. 9312, pp. 47–61. Springer, Cham (2016). https://doi.org/10.1007/978-3-319-45916-5_4
22. Levy, M., Huli, C.: Design thinking in a nutshell for eliciting requirements of a business process: A case study of a design thinking workshop. In: 2019 IEEE 27th International Requirements Engineering Conference (RE), pp. 351–356. IEEE (2019)
23. Liu, K.: Semiotics in Information Systems Engineering. Cambridge University Press, Cambridge (2000)
24. Lopes, L.A., Pinheiro, E.G., Silva da Silva, T., Zaina, L.A.M.: Adding human interaction aspects in the writing of user stories: a perspective of software developers. In: Proceedings of the 31st Brazilian Symposium on Software Engineering pp. 194–203 (2017)
25. Losada, B.: Flexible requirement development through user objectives in an agile-ucd hybrid approach. In: Proceedings of the XIX International Conference on Human Computer Interaction, pp. 1–8 (2018)
26. Lucassen, G., Dalpiaz, F., Werf, J.M.E.M., Brinkkemper, S.: The use and effectiveness of user stories in practice. In: Daneva, M., Pastor, O. (eds.) REFSQ 2016. LNCS, vol. 9619, pp. 205–222. Springer, Cham (2016). https://doi.org/10.1007/978-3-319-30282-9_14
27. Marques, A.B., Figueiredo, R., Amorin, W., Rabelo, J., Barbosa, S.D., Conte, T.: Do usability and agility combine? investigating the adoption of usability modeling in an agile software project in the industry. In: Proceedings of the 17th Brazilian Symposium on Human Factors in Computing Systems, pp. 1–11 (2018)
28. Medeiros, J., Vasconcelos, A., Goulão, M., Silva, C., Araújo, J.: An approach based on design practices to specify requirements in agile projects. In: Proceedings of the Symposium on Applied Computing pp. 1114–1121 (2017)
29. Melegati, J., Wang, X.: Quest: New practices to represent hypotheses in experiment-driven software development. In: Proceedings of the 2nd ACM SIGSOFT International Workshop on Software-Intensive Business: Start-ups, Platforms, and Ecosystems, pp. 13–18 (2019)

30. Muller, M.J., Haslwanter, J.H., Dayton, T.: Participatory practices in the software lifecycle. In: Helander, M.G., Landauer, T.K., Prabhu, P.V. (eds.) Handbook of Human-Computer Interaction (Second Edition), 2nd edn., pp. 255–297. North-Holland, Amsterdam (1997). https://doi.org/10.1016/B978-044481862-1.50077-7
31. Murugesan, L.K., Hoda, R., Salcic, Z.: Identifying design features using combination of requirements elicitation techniques. In: 2017 IEEE/ACM 1st International Workshop on Design and Innovation in Software Engineering (DISE), pp. 6–12. IEEE (2017)
32. Oulasvirta, A., Feit, A., Lähteenlahti, P., Karrenbauer, A.: Computational support for functionality selection in interaction design. ACM Trans. Comput. Hum. Interaction (TOCHI) **24**(5), 1–30 (2017)
33. Papanek, V.: Design for the real world, including less developed countries (1971), Accessed October 2019
34. Pinto, R., Silva, L., Valentim, R.: Managing sessions of creative requirements elicitation and assessment. In: Proceedings of the 35th Annual ACM Symposium on Applied Computing, pp. 1355–1362 (2020)
35. Poppendieck, M., Poppendieck, T.: Lean Software Development: An Agile Toolkit: An Agile Toolkit. Addison-Wesley, Boston (2003)
36. Reis, J.C.D., Santos, A.C.D., Duarte, E.F., GonÇalves, F.M., de FranÇa, B.B.N., Bonacin, R., Baranauskas, M.C.C.: Articulating socially aware design artifacts and user stories in the conception of the opendesign platform. In: Proceedings of the 22nd International Conference on Enterprise Information Systems - Volume 2: ICEIS, pp. 523–532. INSTICC, SciTePress (2020). https://doi.org/10.5220/0009418205230532
37. dos Reis, J.C., dos Santos, A.C., Duarte, E.F., Gonçalves, F.M., de França, B.B.N., Bonacin, R., Baranauskas, M.C.C.: Articulating socially aware design artifacts and user stories in the conception of the opendesign platform. In: ICEIS (2), pp. 523–532 (2020)
38. Salah, D., Paige, R., Cairns, P.: Patterns for integrating agile development processes and user centred design. In: Proceedings of the 20th European Conference on Pattern Languages of Programs (EuroPLoP '15). pp. 1–10. Association for Computing Machinery, New York, NY, USA (2015). https://doi.org/10.1145/2855321.2855341
39. Schön, D.A.: The design process. In: Howard, V.A. (ed.) Varieties of Thinking: Essays From Harvard's Philosophy of Education Research Center. Routledge, Abingdon, United Kingdom (1990)
40. Schön, E.M., Winter, D., Uhlenbrok, J., Escalona Cuaresma, M.J., Thomaschewski, J.: Enterprise experience into the integration of human-centered design and kanban. In: ICSOFT-EA 2016: 11th International Joint Conference on Software Technologies (2016), pp. 133–140. ScitePress Digital Library (2016)
41. Schuler, D., Namioka, A.: Participatory Design: Principles and Practices. CRC Press (1993)
42. Schwaber, K., Beedle, M.: Agile Software Development with Scrum, vol. 1. Prentice Hall, Upper Saddle River (2002)
43. da Silva, A.P.C., Hirata, C.M.: Requirements elicitation method for designing virtual collaborative systems with collaborative sensemaking. In: WEBIST pp. 24–35 (2012)

44. da Silva, J.V., Pereira, R., Buchdid, S.B., Duarte, E.F., Baranauskas, M.C.C.: SAwD - socially aware design: an organizational semiotics-based CASE tool to support early design activities. In: Baranauskas, M.C.C., Liu, K., Sun, L., Neris, V.P.A., Bonacin, R., Nakata, K. (eds.) ICISO 2016. IAICT, vol. 477, pp. 59–69. Springer, Cham (2016). https://doi.org/10.1007/978-3-319-42102-5_7
45. da Silva, L.M., Bezerra, C.I., Andrade, R.M., Monteiro, J.M.S.: Requirements engineering and variability management in dspls domain engineering: A systematic literature review. In: ICEIS (1), pp. 544–551 (2016)
46. Stamper, R.: Information in Business and Administrative Systems. A Halsted Press Book, Wiley (1973). https://books.google.com.br/books?id=qQQmAQAAIAAJ
47. Sun, L., Ousmanou, K., Cross, M.: An ontological modelling of user requirements for personalised information provision. Inf. Syst. Front. **12**(3), 337–356 (2010)
48. Teixeira, L., Saavedra, V., Santos, B.S., Ferreira, C.: Integrating human factors in information systems development: user centred and agile development approaches. In: Duffy, V.G.G. (ed.) DHM 2016. LNCS, vol. 9745, pp. 345–356. Springer, Cham (2016). https://doi.org/10.1007/978-3-319-40247-5_35
49. Vieira, H., Baranauskas, M.C.C.: Design e Avaliação de Interfaces Humano-Computador. Unicamp, Campinas, SP, Brazil (2003)
50. Zapata, C.: Integration of usability and agile methodologies: a systematic review. In: Marcus, A. (ed.) Integration of usability and agile methodologies: a systematic review. LNCS, vol. 9186, pp. 368–378. Springer, Cham (2015). https://doi.org/10.1007/978-3-319-20886-2_35

Electronic Coupon Management System for Regional Vitalization

Hiroki Satoh[1(✉)], Toshiomi Moriki[1(✉)], Yuuichi Kurosawa[1(✉)], Tasuku Soga[1(✉)], and Norihisa Komoda[2(✉)]

[1] Research and Development Group, Hitachi, Ltd., 1-280, Higashi-Koigakubo, Kokubunji, Tokyo 185-8601, Japan
{hiroki.satoh.yj,toshiomi.moriki.jn,yuichi.kurosawa.fv, tasuku.soga.ts}@hitachi.com

[2] Code Solutions, Co., Ltd, Daidoseimei South Building 9th Floor, 1-2-11 Edobori Nishi-ku, Osaka, Osaka 550-0022, Japan
komoda@codesol.jp

Abstract. We propose an electronic coupon system to support regional vitalization. This system is designed for a number of events held in Japan, where local shopping districts offer special menus to rediscover the shops and their at-tractions. The system has a web application for customers that incorporates a mechanism for interaction between customers and shop clerks, and a function for event organizers to visualize the success of the event. The proposed system was used in an actual event and its effectiveness was confirmed using a user survey.

Keywords: Electronic currency · Electronic coupon · User interface · Regional vitalization

1 Introduction

The number of regional vitalization events is increasing in Japan. Japan's declining birthrate and aging society have reduced the number of young people and regional vitality [1]. The decline in regional vitality will lead to a decline in the regional economy. As a solution to this problem, the introduction of a local currency and regional vitalization events are being considered [2–4].

In recent years, a series of events called "Town Bar" has become very popular, as part of the efforts to prevent the decline of the local economy. The aim of these events is for residents and citizens to rediscover the charm of the area by buying tickets and walking around the restaurants in the target shopping district [5–7].

Electronic money is becoming more and more popular in Japan. With the spread of smartphones in Japan, anyone can easily use e-money. In recent years, the government has been aiming for the spread of e-money and has implemented measures to reduce the consumption tax from 10% to 5% for payments made with e-money.

This study proposes a system for utilizing electronic money for "Town Bar," a regional vitalization event. Conventionally, paper coupons have been used for "Town

J. Filipe et al. (Eds.): ICEIS 2020, LNBIP 417, pp. 753–768, 2021.
https://doi.org/10.1007/978-3-030-75418-1_34

Bar" events, but this has been changed to electronic coupons. This system provides customers and shops with an automated payment system and event organizers with a visualization of the entire event.

In addition, we propose a new payment user interface to support regional vitalization. A smartphone-based user interface will be implemented to embed collaboration between customers and shop clerk at the time of payment, and to facilitate some kind of conversation between customers and shop clerk. Our research aims to pro-mote interaction between customers and shop clerk, which will activate the community and strengthen the human network.

2 Outline of Regional Electronic Coupon

2.1 Regional Vitalization Event "Town Bar"

There is an event called "Town Bar" as part of preventing the decline of the local economy. Figure 1 shows an overview of the "Town Bar" event.

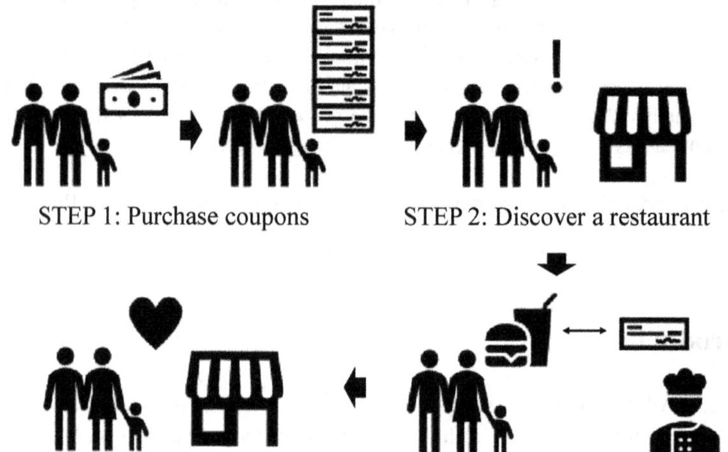

Fig. 1. Conceptual diagram of a "Town Bar" [7].

The aim of this project is to encourage residents to rediscover the charms of the area by buying tickets and walking around the restaurants in the target shopping district. In the conventional "Town Bar" event, the event organizer sells paper coupons to customers, who then pay the stores with the paper tickets. By paying with the paper coupon, customers are offered a special menu at a discount. If customers enjoy this special menu, they will continue to be fans of the restaurant and will be encouraged to keep coming back.

2.2 Validation and Implementation Issues of Regional Electronic Coupons

We examined the effectiveness of electronic coupons to solve the challenges of operating events with paper coupons. Next, we identified the implementation challenges of electronic coupons.

The number of paper coupons must be counted manually, and it is difficult to ascertain the status of the event. In the past, stores had to take the paper coupons received from customers to the event organizer, count the number of coupons, confirm sales, and exchange the coupons with the organizer for money. Sometimes the coupons were found to be lost and counting the number of coupons was a huge burden. Event organizers were also unable to monitor customer activities and real-time sales as they could only check sales results.

The regional electronic coupons are effective in solving this problem. Electronic coupons are a means of payment in the same way as digital currency [9]. Instead of paying with a physical paper coupon, the customer pays with an electronic coupon on a smart phone. Because it is electronic, the system can record who paid how much and when at the same time as the payment. This makes it possible to automatically calculate sales and monitor the status of events based on sales data. Stores will have less manpower and event organizers will be able to monitor the status of the event in detail.

However, there were three challenges to implementing the regional electronic coupons. The first is to examine how to introduce digital coupons without installing new digital devices in the stores. Most of the participating shops in the "Town Bar" are individual shops, so it is difficult to bear the additional cost of introducing local electronic coupons. The second point is the creation of new payment methods to sup-port regional vitalization. Conventional electronic payment methods are designed for the purpose of quick payment. On the other hand, the main purpose of the "Town Bar" method is to vitalize the local community through interaction between customers and shop clerk, and it is important to promote interaction between customers and shop clerk [8]. The third point is to provide a means of understanding the entirety and details of the event. It is not possible for an event organizer who is not an expert in data analysis to output payment data to an electronic file and analyse it in a spreadsheet.

Table 1. Functional requirements of regional electronic coupon management system [7].

System requirements	Description
Payment with a single smartphone	The ability to make payments using only the smartphones owned by customers
Interaction between customers and stores	The system has a function that allows customers and stores to interact with each other during settlement
Event management and analysis	The ability to manage and analyze the status of the entire event

2.3 Proposal of Regional Electronic Coupon Management System

We propose a regional electronic coupon management system to solve the implementation challenges of regional electronic coupons presented in Sect. 2.2. Table 2 shows the functions that the proposed system needs to perform in order to solve the challenges in Sect. 2.2 (Table 1).

The ability to complete payment with a single smartphone solves the problem of implementing local electronic coupons without installing new digital devices in stores. With the implementation of this feature, local electronic coupons can be introduced without placing an economic burden on local stores.

The payment function, which allows customers and stores to interact with each other, solves the problem of establishing a new payment method that supports local community vitalization. The realization of this function will help rediscover local attractions and strengthen the human network through local community revitalization.

The ability to visually identify an event solves the challenge of providing a means of understanding the entirety and details of an event. The implementation of this feature allows event organizers to analyze the overall event and use it for the next event and to promote the success of the event.

3 System Configuration

3.1 System Overview

Figure 2 shows a conceptual diagram of the entire electronic coupon system. Customers pay cash to the event organizer to receive an electronic coupon [10]. These electronic coupons can only be used at the local stores participating in the event.

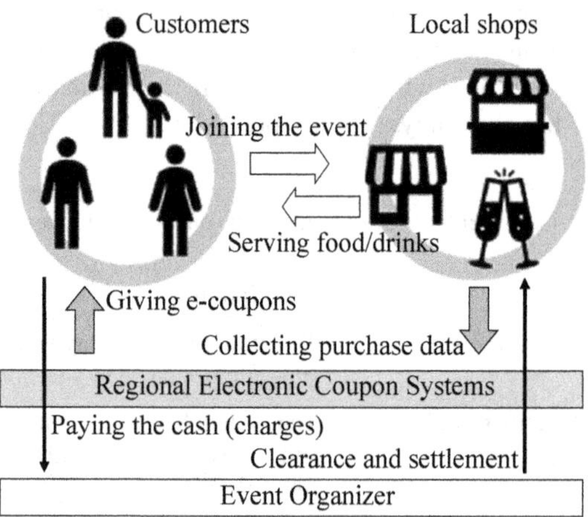

Fig. 2. Regional electronic coupon management system overview [7].

The electronic coupon system consists of a website for mobile devices, which customers can use on the same day without downloading an application if they have a smartphone and an e-mail address. In addition, the event organizer can monitor the usage of the "Town Bar" events at any time via the electronic coupon system. Table 2 shows a list of data obtained by the electronic coupon system.

Table 2. List of the regional electric coupon management system acquisition data [7].

Name	Data description
E-mail address	"Town Bar" customer's email-address (which used as user-ID)
Charges	Times of electric coupon charges
Information of food/drink menus	Coupons used for food/drink orders – Times of food/drink orders – Number of shops visited – Number of coupons used each order
Comments	Word-of-mouth information by visitors. They can leave a comment each time the coupon used and show thanks to the shop owners and advertisements to other users

3.2 Functional Features

The system consists of three main functions: the first is to provide users with enjoyment, the second is to manage money transfers between customers and stores for stores and event organizers, the third is to provide event organizers with information on customer movement, customer interaction, and money transfers between customers and stores. It is a function to analyse information and display the results.

There are three functions to provide users with fun: the "information function," the "comment posting function," and the "exchange payment function. The "information function" is a function to provide information about the menu and recommendations from the store to customers, which encourages customers to visit the store easily. The "comment-posting function" allows customers to post their impressions of the store or event, which can be seen by all participants of the event and the store. The "Exchange Payment Function" allows customers and store staff to simultaneously press and hold their smartphone screens to complete their payments, which is intended to promote conversation and exchange. This function is described in Sect. 3.4.

There are three functions to manage money transfer between customers and stores: the QR code settlement function, the settlement data management function, and the money transfer display function. In the QR Code Settlement Function, customers can use QR codes to read special menu items provided by stores to complete payment. This function eliminates the need for stores to prepare digital devices. It can show customers their own coupon usage history and event organizers their entire coupon usage history.

The event information analysis function includes the "customer analysis function," "coupon usage analysis function," and "analysis result display function. The "customer analysis function" is a function that displays and compares the number of times a digital coupon has been purchased and used by each customer. The 'coupon usage analysis function' is a function that displays and compares the usage history of digital coupons by customer, store and time. The "Analysis result display function" is a function that displays the results of event analysis on a map. It allows the user to check how the entire event has changed by time.

3.3 Functional Configuration and Database Design

Figure 3 shows the proposed system configuration. The proposed system consists of a server and a client-type IT system with a smartphone as the client. The system is provided as a web application to the smartphone used by the customer for screen operations and data input/output. The server is executed in the cloud to input/output the database and display the results of data analysis. Among the functions described in Sect. 2.3, the functions executed by the smartphone are the "information provision function," "comment posting function," "payment settlement function," "QR code settlement function," and "money transfer display function. The functions executed on the server are "Financial Data Management Function", "Money Transfer Display Function", "Customer Analysis Function", "Coupon Usage Analysis Function" and "Analysis Result Display Function".

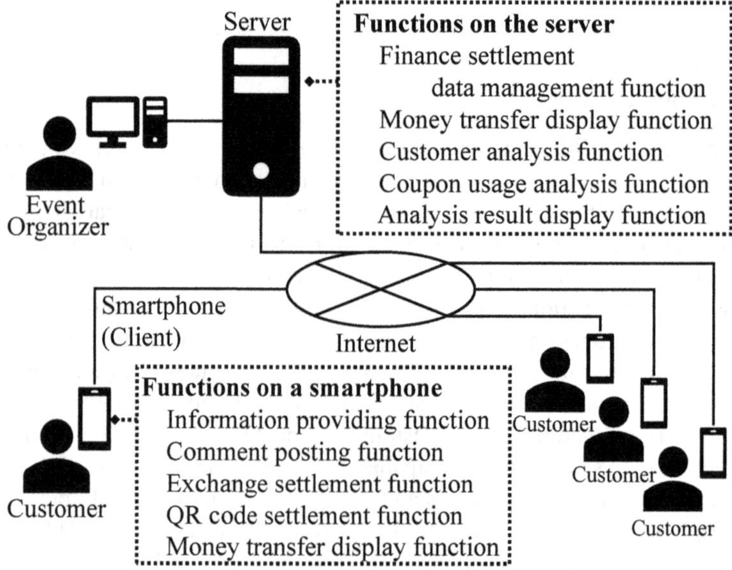

Fig. 3. Functional configuration diagram [7].

Tables 3, 4, 5 and 6 show the database design. In this example, the subjects of the data are shown as "Businesses," "Consumers," "Transactions," and "Stores".

3.4 User Experience Design

Conventional e-money payment user interfaces are designed for the purpose of quick payment in terms of processing efficiency. On the other hand, since the main purpose of

Table 3. Database definition of the "Businesses" table.

No.	Name	Description	SQL Data Type
1	business_id	Identification	bigint
2	business_uid	Unique identification	varchar
3	business_name	Name	varchar
4	business_email	Email address	varchar
5	business_password	Password	varchar
6	created	Registration date	datetime
7	modified	Update date	datetime

Table 4. Database definition of the "Consumers" table.

No.	Name	Description	SQL Data Type
1	consumers_id	Identification	bigint
2	consumers_uid	Unique identification	varchar
3	consumers_name	Name	varchar
4	consumers_email	Email address	varchar
5	consumers_password	Password	varchar
6	created	Registration date	datetime
7	modified	Update date	datetime

"Town Bar" is to vitalize a community through interaction between customers and shop clerk, we think it is important to promote interaction between customers and shop clerk. In this paper, we propose a new user interface that incorporates the collaboration between users and shop clerks at the time of payment to facilitate some kind of conversation between customers and shop clerks.

Figures 4 and 5 show the proposed user interface. We propose an interaction system in which the customer and the shop clerk simultaneously press and hold a screen of a smartphone to stimulate conversation between the customer and the shop clerk. The design of the screen shows a vertical bar on the screen, and the bar advances by pressing and holding it, and the payment is completed when the bar fills up. This screen design is intended to provide an opportunity for two people to have a conversation by continuously pressing a button. To make it easier to understand the payment process on a smartphone, we designed the screen so that the customer using the smartphone to pay the electronic

Table 5. Database definition of the "Transactions" table.

No.	Name	Description	SQL Data Type
1	transaction_id	Identification	bigint
2	transaction_uid	Unique identification	varchar
3	transaction_datetime	Transaction date	timestamp
4	transaction_status	Transaction status	enum
5	transaction_point_quantity	Trading points	int
6	transaction_point_price	Transaction point amount	int
7	transaction_summary	Outline of the transaction	varchar
8	user_uid	Unique user identification	varchar
9	created	Registration date	datetime
10	modified	Update date	datetime

Table 6. Database definition of the "Stores" table.

No.	Name	Description	SQL Data Type
1	transaction_id	Identification	bigint
2	transaction_uid	Unique identification	varchar
3	transaction_datetime	Transaction date	timestamp
4	transaction_status	Transaction status	enum
5	transaction_point_quantity	Trading points	int
6	transaction_point_price	Transaction point amount	int
7	transaction_summary	Outline of the transaction	varchar
8	user_uid	Unique user identification	varchar
9	created	Registration date	datetime
10	modified	Update date	datetime
11	store_url	URL	varchar
12	store_map_no	Store map number	int
13	store_lat	Latitude	int
14	store_lng	Longitude	int
15	created	Registration date	datetime
16	modified	Update Date	datetime

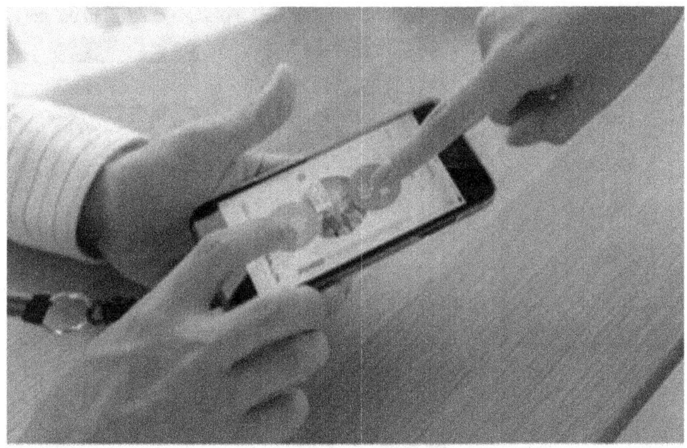

Fig. 4. Touch user interface [7].

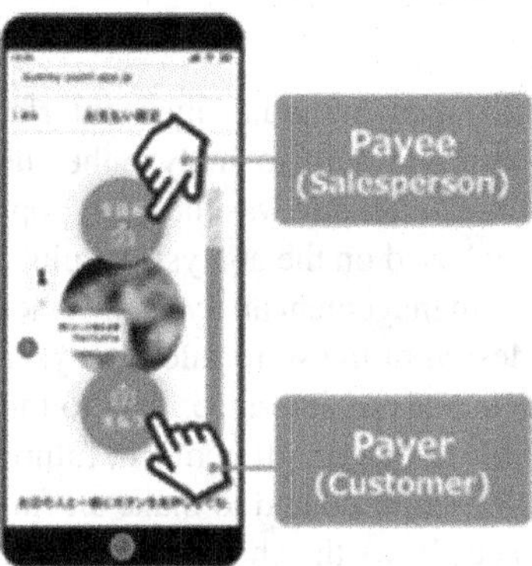

Fig. 5. Payment user interface [7].

coupon is on the front side and the shop clerk receiving the coupon is on the opposite side.

Figure 6 shows an example of several smartphone screens. The home screen displays the currently held coupons. In the payment confirmation screen, you select how many tickets and how many people you want to pay for. After this, the screen switches to the screen shown in Fig. 5, where the customer and the shop clerk press and hold simultaneously. And customers can select a store and post their impressions of the store.

4 Interface for Management and Analysis

4.1 Management Analysis Interface

The management and analysis interface are used by the organizer to understand the status of the event and to analyze the results of the event. The management and analysis interface allow us to visually check the behavior of customers, sales of the store, and the

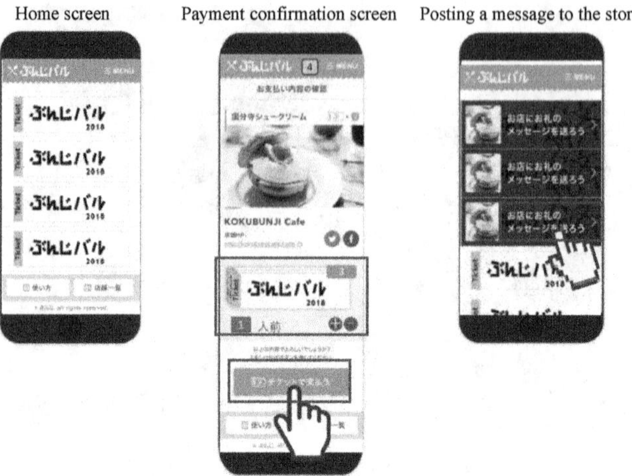

Fig. 6. Examples of smartphone application screens.

degree of interaction. The management and analysis interface can dynamically display changes over time. This screen allows us to analyze the customers' movement at each time of the day, and which time of the day was the most popular, so that we can manage the next event more smoothly based on the analysis results.

Table 7 shows the types of management analysis interfaces and their purposes of use. Figure 7 shows the screen design of the store sales analysis interface and Fig. 8 shows the screen design of the sales analysis interface. Due to the small number of acquired data items, it is not yet possible to display all items. In future events, we plan to increase the number of data items to be acquired and to make all functions available. Currently, the user data analysis interface allows the changes in people and store sales throughout the entire event.

Table 7. Types of management analysis interface [7].

Type	Description
Store sales analysis	Displays sales and customer attribute information for each store
Sales and marketing analysis	Displays the number of visitors, attributes, and store sales rankings for the entire event
User Data Analysis	Displays the movement of each user or all users on the map

4.2 User Data Analysis Viewer

Figures 9 and 10 show the screen for analyzing and displaying user data. On the left side of the screen, the date and the customer's activity information for each store are displayed. On the right side of the screen, there is a map, and on the map, the shops participating in

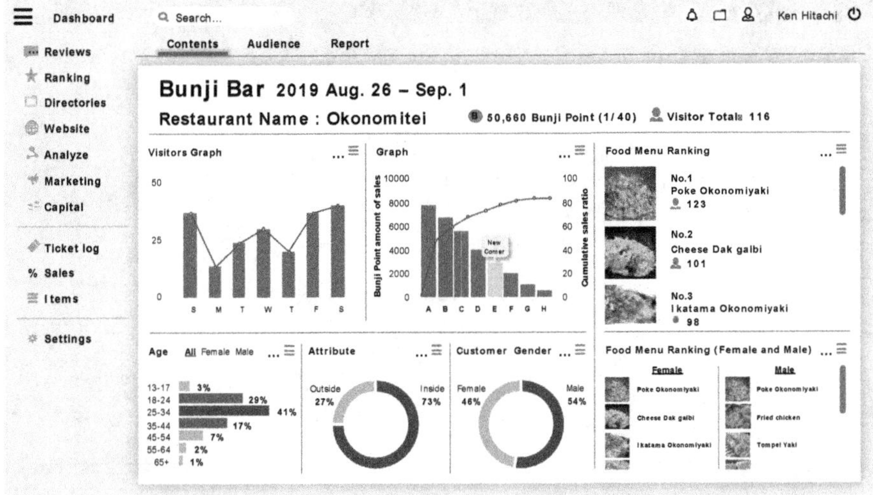

Fig. 7. Store sales analysis interface screen.

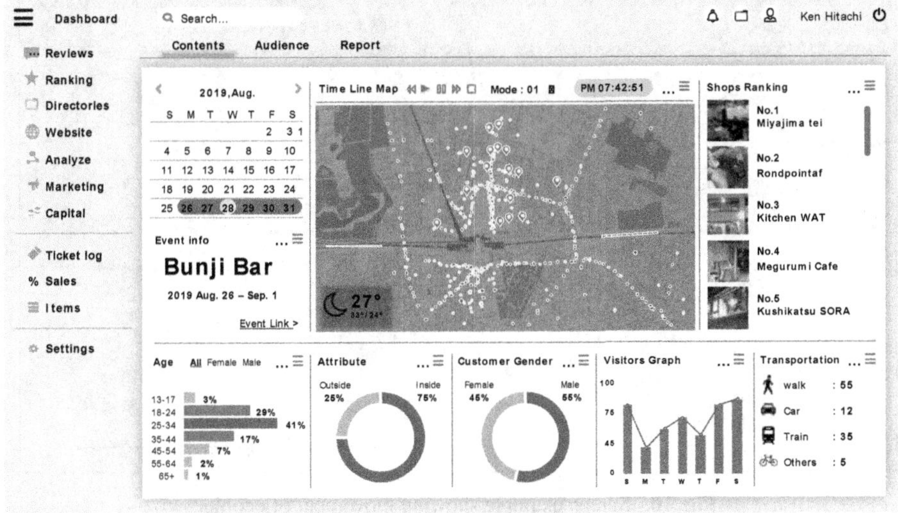

Fig. 8. Sales and marketing analysis interface screen.

the event, user comments, payment and purchase of electronic coupons are shown. The lower part of the screen shows the time and date information of the selected date, and at what time the customer paid with the electronic coupon or purchased the coupon.

The screen changes dynamically as the time transitions. This allows us to see on the map where customers are paying, purchasing coupons, and sending out comments. Using Fig. 9 as an example, we can show that after 3:30 p.m., an electronic coupon has

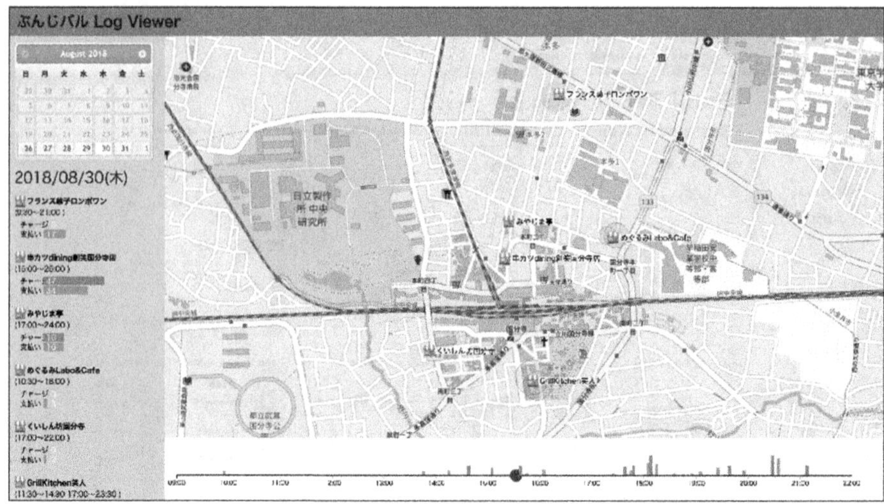

Fig. 9. Example of user data analysis viewer screen.

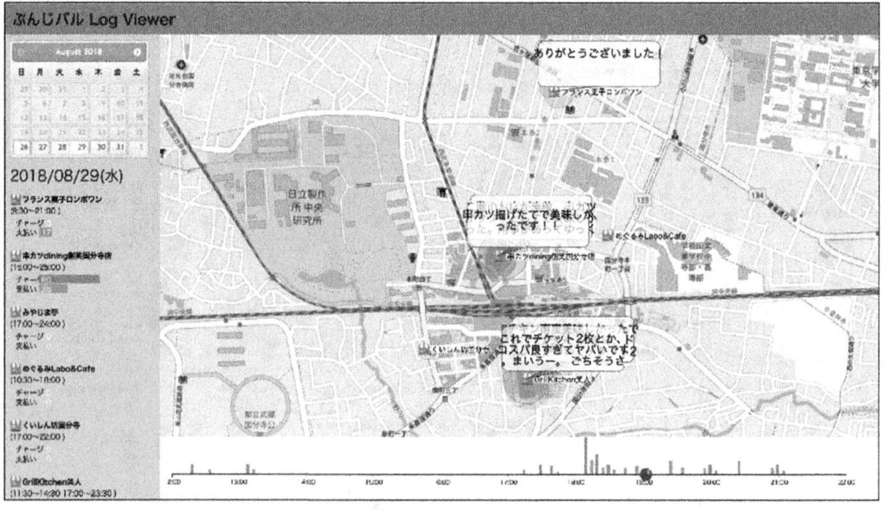

Fig. 10. Screen example showing movement of people.

been purchased and multiple coupons have been paid for at the store at the bottom of the screen.

If the same user is paying at different stores at different times, the system determines that the user has moved and displays the move on the screen. Using Fig. 10 as an example, because a customer paid at the upper store with an electronic coupon and then at the lower store with an electronic coupon, a small circle is displayed on the screen. This small circle moves dynamically from the upper store to the lower store. This is a simple display because it does not use GPS information, but it allows the user to quickly see that he or she is using more than one store.

4.3 Usage Scenario

The management and analysis capabilities of the proposed electronic coupon system will be used by event organizers to analyze the improvement and effectiveness of their events and by event organizers to promote the success of their events to event participants and supporters.

When analyzing the improvement and effectiveness of an event, you can display payment by customer, customer comments and customer journeys by time. Checking the payment allows you to see which customers visited the store. By checking the comments, we can check whether the interaction between the customer and the store was active. By checking the movement of customers, you can anticipate the movement of people in future events and devise a response.

When advertising the success of an event, the event analytics screen will show the customer's payment, the distinctive comments posted by the customer, and the customer's movement through the event. Unlike in the case of analysis, what has been done is clearly displayed. Multiple payment information should be dis-played in batches, the number of comments should be carefully selected, movements of people should be highlighted, and the content of the event should be clearly represented to give the impression of success.

5 Demonstration and Evaluation of the Proposed System

5.1 Demonstrating at a "Town Bar" Event

The proposed region-specific electronic coupon system was introduced on a trial basis at "Bunji-Bar", a "Town Bar" event held in Kokubunji City, Tokyo, Japan. The "Bunji -Bar" has been held in the shopping district around Kokubunji Station in Tokyo since 2013 and will be held for the 9th time in 2019. The proposed system was introduced on a trial basis in 2018 and 2019. Table 8 and Table 9 show an over-view of "Bunji-Bar 2018" and "Bunji-Bar 2019", respectively.

The proposed system was first tested in 2018. Although there were six stores that supported electronic coupons, electronic coupon-enabled stores ranked first and second

Table 8. Summary of event, "Bunji-Bar" in 2018 [7].

Name	Description
Event period	During 26th Aug to 1st Sep., 2018
Place	Around Kokubunji Station, Tokyo
Format of food/drink coupons	Paper-based or electronic coupons provided. (400 JPY x 7/set)
Num. of shops	42 shops. (6 shops: e-coupon ready)
Number of visitors	Approx. 450 visitors (42: e-coupon users)

(continued)

Table 8. (*continued*)

Name	Description
Types of shops	Bar and Izakaya: 18 shops (43%) Restaurant: 12 shops (29%) Café and coffee: 6 shops (14%) Others: 6 shops (14%)

Table 9. Summary of event, "Bunji-Bar" in 2019 [7].

Name	Description
Event period	During 25th to 31st Aug., 2019
Place	Around Kokubunji Station, Tokyo
Format of food/drink coupons	Paper-based or electronic coupons provided. (400 JPY x 7/set)
Num. of shops	49 shops. (22 shops: e-coupon ready)
Number of visitors	Approx. 512 visitors (52: e-coupon users)
Types of shops	Bar and Izakaya: 22 shops (45%) Restaurant: 16 shops (33%) Café and coffee: 6 shops (12%) Others: 5 shops (10%)

in sales. In 2019, the number of stores participating in electronic coupons increased to 22, and again, stores that support electronic coupons ranked first and second in sales.

5.2 User's Evaluation Results

In order to investigate whether the proposed system meets the functional requirements presented in Sect. 2.3, we conducted a survey of comments from customers, shop clerk and event organizers.

Table 10 shows comments on payment with a single smartphone. Table 11 shows comments on customer/store interaction. Table 12 shows comments on event management and analytics.

5.3 Discussion of Evaluation Results

The results of the questionnaire show that the restaurant has received high marks from customers and staff, as expected. In addition, we believe that we can apply the results of the event management and analysis to future events and improve the menu. We will further improve the table and the results of the analysis for stores and event organizers.

Table 10. Comments on payment methods on a single smartphone.

No.	Comment
1	It was nice and simple to finish the process with just my smartphone
2	Downloading the application is a hassle, so it's nice to have it available in the browser alone
3	I was able to easily support digitalization by simply sticking a QR code on the existing menu
4	Not having to install an electronic money reader device was very helpful in terms of management and cost
5	Less time spent counting paper coupons after the event
6	A conversion process that used to take three days has been reduced to one day

Table 11. Comments on customer and store interactions.

No.	Comment
1	It's fun to work with the shop clerk and it's also interesting to see other people paying with their smartphones
2	If there was some sense of embarrassment or agreement on the part of the payer/receiver and some sense of confusion about the operation, I felt that it was not just a mechanical settlement
3	The event was a lot of fun because doing touch together creates a conversation with the restaurant staff every time
4	Maybe it was because we happened to be a fun group of customers, but every time I pushed a button, the conversation got better
5	It was interesting to see how the concentration gathered around choosing a menu and paying for it, so it was an interesting conversation starter between ourselves and the waitress
6	The customer seemed embarrassed to press the button, and that's when the conversation started

Table 12. Comments on event management and analysis.

No.	Comment
1	It's great to get an overview of the event
2	It was useful to clarify that customers use different coupons at different dates and times of day
3	I've been able to figure out which customers use coupons at multiple stores on the same day and which customers use coupons at different stores on different days. I'd like to think about providing a different benefit for these customers
4	I'd like to see the display linked to the weather and temperature. I would like to change the menu offered on weekdays and weekends, or if you are participating in more than one store, I would like to see you offer special menu items
5	It's even better if you know the customers that come through after the event. I would like to use the local electronic coupons for everyday use

6 Conclusion

In this study, we proposed an electronic coupon management system for regional vitalization events. The proposed system was successfully demonstrated at an actual event, and was well received by participants and organizers alike, as it contributed to the excitement of the event. The data accumulated by this system has the potential to analyze social capital [11], which is the basis of regional vitalization, and is expected to have various applications other than the continuous improvement of events.

References

1. Muramatsu, N., Akiyama, H.: Japan: super-aging society preparing for the future. Gerontologist **51**(4), 425–432 (2011)
2. Kido, K., Hasegawa, S., Komoda, N.: Enhanced Community Currency Trading through Transaction Group Formation. CIMCA **1**, 229–234 (2005)
3. Matsui, N., Yukio Ikemoto,Y.: Solidarity Economy and Social Business New Models for a New Society. 1st edn. Springer, Japan (2015)
4. Onitsuka, K., Hoshino, S.: Inter-community networks of rural leaders and key people: Case study on a rural revitalization program in Kyoto Prefecture. Japan. Journal of Rural Studies **61**, 123–136 (2018)
5. Hall, C., Stefan G.: Food Tourism and Regional Development: Networks, products and trajectories, 1st edn. Routledge (2016)
6. Civelek, M., Ključnikov, A., Krajčík, V., Žufán, J.: The Importance of discount rate and trustfulness of a local currency for the development of local tourism. J. Tourism Serv. **10**(19), 78–93 (2019)
7. Satoh, H., Moriki, T., Kurosawa, Y., Soga, T., Kobayashi, M., Komoda, N.: Management system for regional electronic coupon. In: Proceedings of the 22nd International Conference on Enterprise Information Systems - Volume 2, pp. 571–578 (2020)
8. Nakazato, H., Lim, S.: Community rebuilding processes in a disaster-damaged area through community currency: the pilot project of Domo in Kamaishi. Japan. Disaster Prevention Manage. **26**(1), 79–93 (2017)
9. Kim, S.M., Lough, B., Wu, C.-F.: The conditions and strategies for success of local currency movements. Local Econ. **31**(3), 344–358 (2016)
10. Kurita, K., Miyazaki, Y., Nishibe, M.: CC coupon circulation and Shopkeepers' behaviour: a case study of the city of Musashino, Tokyo, Japan. Int. J. Commun. Currency Res. **14**(1), 136–145 (2012)
11. Moriki, T., Satoh, H., Maki, H., Komoda, N.: Estimation of social capital based on user behavior information on regional electronic coupon In: Proceedings of ICT, Society and Human Beings (ICT-conf 2020), pp. 11–18 (2020)

A Modern Approach to Personalize Exergames for the Physical Rehabilitation of Children Suffering from Lumbar Spine

Cristian Gómez-Portes(✉), Carmen Lacave, Ana I. Molina, David Vallejo, and Santiago Sánchez-Sobrino

School of Computer Science, University of Castilla-La Mancha, Paseo de la Universidad 4, 13071 Ciudad Real, Spain
{Cristian.Gomez,Carmen.Lacave,AnaIsabel.Molina,David.Vallejo, Santiago.Sanchez}@uclm.es

Abstract. Physical rehabilitation of people with injuries or illnesses related to the lumbar spine involves an intensive treatment to reduce pain or improve mobility. Research studies have evidenced the benefits of complementing the patient's regular treatment with exercise routines at home. However, in the case of children and adolescents, there is a risk of abandoning the exercise routine if it is not motivating enough. Currently, there is a trend which consists in using games for rehabilitation exercises, called exergames, as a possible solution for motivating patients while they perform physical rehabilitation. However, both customizing and creating them is still a task that requires considerable investment both in time and effort. Thus, this paper presents a language along with a system based on the physical rehabilitation of children suffering from some sort of lower back pain, which enables the customization and the automatic generation of exergames. We have conducted an experiment with children for evaluating the capabilities of our approach. The obtained results show that the tool is fun, interesting and easy to use.

Keywords: Virtual rehabilitation · Personalization · Exergames · Lumbar spine pain · e-Health

1 Introduction

Lumbar spine injury is considered as one of the most frequent diseases worldwide, according to the research [25]. This type of condition results in a reduction of movement of certain parts of the human body. In this case, its treatment consists in performing rehabilitation therapies that try to recover or improve the patient's motor function. However, there are several important threats beyond the disease that hamper the therapies not to be successful. Actually, one of biggest problems is the economic burden, which hinders to sustain therapies in the long-term because therapists need to visually assess how a patient is performing an exercise [5]. This drawback is more than obvious in countries with low-and middle income [32], where access to clinicians is commonly non-existent.

The original version of this chapter was revised: Special characters in the names of Cristian Gómez-Portes and Santiago Sánchez-Sobrino have been corrected. The correction to this chapter is available at https://doi.org/10.1007/978-3-030-75418-1_42

J. Filipe et al. (Eds.): ICEIS 2020, LNBIP 417, pp. 769–790, 2021.
https://doi.org/10.1007/978-3-030-75418-1_35

In this line, the fact that patients need to be moved to the rehab center is considered another threat, being most exacerbated in the case of dependents, which has become even more evident because of physical restriction caused by COVID-19. On the other hand, motivation is a fundamental key so that patients are engaged in their therapy. Generally, rehabilitation sessions consist in repeating exercises several times, which can be monotonous and boring. In fact, this can lead to a loss of interest in therapy over time and may result in the abandonment of the treatment.

These previous problems are especially aggravated when patients are children or adolescents, since it is frequent they suffer from some sort of pain related to lumbar spine [14]. However, there has been put a great effort in this field over the last decade to solve some of the threats mentioned above. Firstly, the improvements made in sensors and computing technology for recognizing motion capture have advanced drastically in the past few years.

This has led to the marketing of devices, such as Kinect, which has been massively used in physical rehabilitation [23]. Consequently, this result has produced some advantages in this field, for example, the reduction in costs or transportation [17,26]. Secondly, a relatively new trend has been adopted for promoting entertainment while patients perform exercises, which consists in using gamification techniques and serious games in combination with the use of motion capture devices [12,28]. Indeed, this approach has been successfully demonstrated over the years, providing results that state it improves the motivation and adherence of patients in rehabilitation treatments [11,19].

However, games introduce some problems when they are used for rehabilitation. Generally, they are designed to be fun and challenging, being the main objective engaging people to play as much as they can. Taking this consideration into account, games can be detrimental for rehabilitation because patients can forget they are in a rehabilitation session. Consequently, it can cause a loss of fear and produce a new injury or accentuate the damage. In this line, some authors have shown the consequences of misusing games for rehabilitation. By way of example, [34] reported four game-related injuries by using comercial Nintendo Wii games for rehabilitation, and [24] stated that Wii gaming can lead to severe injuries if they are not used correctly.

One of the reasons of this problem is led to the design of the games, fundamentally because they are not usually developed based on a clear target of population. Nevertheless, this problematic can be solved following precision rehabilitation principles [16], that is, a philosophy which consists in creating rehabilitation activities that are customized according to the patient needs and their type of illnesses. Therefore, the main target of this idea is to create therapies both safe and effective [29].

In this particular context, the present work proposes the design, developing and validating a system focused on rehabilitating children and adolescents suffering from lumbar spine, making use of a device which allows to track accurately the movements of a person. In our work, we use Azure Kinect DK™ device, a non-intrusive one which avoids attaching any type of sensor to the patient's body for detecting easily his/her movements. In this way, a patient naturally interacts with the system, carrying out exercises understood as an immersive game that recreates, with virtual elements, a physical exercise linked to a therapeutic treatment. The rehabilitation tasks are customized with a language we have defined, called Personalized Exergames Language (PEL), on the basis of the GL Transmission Format specification (glTF) [30] to store and transmit 3D models information. Essentially, PEL contains the customized structure of an exercise, while glTF stores the multimedia resources of the game.

The structure of PEL is clearly defined in three cores: (1) scene, (2) actors and (3) gameplay. The first element is the responsible for defining the information of each view in the game. The second is used to store the elements with which the patient must interact. The third contains the sentences for specifying the behavior and the dynamic of the previous elements in the game. It must be noted that, even though the current version of PEL is simple enough to understand how to specify exergames, all this information has to be defined with the support of developers. Therefore, a co-creative approach must be established between therapists, developers and patients to accelerate the process for generating exergames.

In order to validate the proposal, and with the aim of evaluating the current capabilities of PEL for generating personalized exergames, a study has been conducted with 23 potential patients from a rhythmic gymnastics club.

This article is based on the work [9], but clearly differentiated by adding these contributions:

- Firstly, a schema of the PEL language has been added, giving details of each sentence defined for describing the components used to create games for rehabilitation.
- Secondly, a section has been added to explain why we use the new Kinect and the advantages and disadvantages with respect to use other wearables available in the market.
- Thirdly, we extend the result section adding the specifications in PEL to create the exergame for lumbar spine used for the experiment conducted.
- Fourthly, we add another section to explain the improvements made after the analysis of the results obtained from the experiment.

The rest of the article is structured as follows. Section 2 positions our research within the context of other works in the field. Then, Sect. 3 sets out the system architecture. Subsequently, Sect. 4 describes the conducted experiment and the obtained results. Finally, in Sect. 5, the conclusions drawn are described as well as future lines of research.

2 Related Work

Swayer and Smith presents in [31] a taxonomy for serious games, where the most salient type of games are those related to health. Among them the authors describe games which promotes entertainment to patients while they perform physical exercises.

Since then, there are several tools for rehabilitation whose exercises are automatically analyzed by machines which use precise skeleton tracking or computer vision techniques. The latter have traditionally been based on using the Kinect device [2,36], a low cost hardware solution whose effectiveness has been seen in the field of physical rehabilitation [1,22,23]. Similarly, there are solutions on the market which are orientated at training and fitness, whose aim is to attract specific targets, apart from providing remote rehabilitation from home [4,7]. However, some authors shows scepticism towards this approach, claiming that the lack of customization of these games may make injuries more likely [34].

In light of this issue, various authors have set out a series of considerations to ensure that exergames are both efficient and safe. In [20], the author discusses some proposals to successfully create games. One of the most outstanding ones is to control the game experience. That is, the game needs to be focused on an specific target of user, such as child or adult. Otherwise, the original concept of the game may be misinterpreted, being the game noneffective. Furthermore, the author states that the patient cannot be felt engaged and motivated if the quality of the game is poor. A slightly different approach is presented in [37], which provides recommendations for the optimal design of exergames. One of them is to design games following the user-centered design approach. The authors also recommend adapting the game design to the patient needs, which may result in having an effective therapy. Similarly, [29] presents a methodology for creating and designing efficient and safe therapeutic exergames. It is split into four phases: i) set the therapeutic objective, ii) implement the objectives into virtual exercises, iii) transform virtual exercises into exergames adding graphics elements, and iv) add game dynamics and provide feedback to the patient. Likewise, [27] presents a set of guidelines and best practices for the design of exergames. The authors suggest designing games based on sport, hobbies or daily activities. Besides, they recommend creating game narrative which provides playful experiences or games that are challenging and promote competitiveness.

The use of exergames and gamification techniques has also been studied in the literature to assess their effectiveness in the rehabilitation process. A systematic review, shown in [19], analyzes different types of systems and their context, the game elements used and the results yielded from them, providing positive outcomes both in patient conduct and their well-being. In a similar vein, the research presented in [11] explores studies on games and gamification applied to physical exercise, which finds positive effects in a reduction in body weight and in encouraging physical exercise. Similarly, the paper [15] shows a set of exergames evaluated by a group of thirty participants, which concludes that video games are useful in the physical training of elderly people. There are even other works, such as [6,28], that use gamification techniques to motivate children and adolescents to perform rehabilitation exercises and which provide great benefits.

There is also a state of the art angle in which technological solutions orientated at facilitating or automating the generation of exergames is envisaged. This research line consists in providing tools to clinicians so that they can configure game parameters and thus reducing the cost and time involved in their production. In this line, the TANGO:H (Tangible Goals:Health) platform [10] has a range of functions and contains a graphics editor to help experts create exercises adapted to patient needs. In this case, it refers to hospitalized children. Similarly, in [13], the StoryTec tool has been envisaged. This was designed to support experts in the field, such as doctors or therapists, so that they can adapt and customize game-based training programmes for elderly and disabled people. There is a slightly different approach called VirtualGym [8], which presents a system that helps therapist to design exercises routines. This approach analyzes a video from a rehabilitation exercise, generating a grammar used to produce animations applied to a 3D avatar which can guide patients in the exercise. In a similar vein, [18] describes a cooperative environment, where clinical professionals record postures later reproduced by a 3D avatar, making patients to understand easily the exercises they have to perform.

3 Architecture

Figure 1 graphically shows the architecture designed for the rehabilitation process for bone-marrow injuries. This approach mainly requires an expert in the field to participate, i.e. a therapist who is responsible for assigning an exercises routine to a patient. The exercises routine is defined by means of data modeling in *.gltf* format extended with our PEL language, which stores all information relating to the exergame. In other words, the game dynamics, graphics, gamification elements, and the different metrics by which patient participation can be evaluated. In the Subsect. 3.1 there is an in-depth explanation on the structure of the language we have defined, so as to provide details in the Subsect. 3.2 on the translation process to automatically generate exergames.

The capture module is that which interacts with the hardware device in order to obtain the information associated with the position and orientation of the bones that make up the patient skeleton. This module is responsible for filtering the data provided by the device, excluding any information that is not related to the bones in the rehabilitation exercise. For example, if patients have to lie down to exercise their backs, and raise the upper part of their torsos, data related to the lower part of the hip is omitted. Apart from that, the capture module also manages voice recognition, as it enables the voice of the patient to be analyzed in order to perform certain tasks, namely, starting a rehabilitation routine or indicating that the activity has finished.

The processing module carries out two specific tasks: i) it evaluates the exercise performed by the patient and ii) it monitors motivation according to the activity. A games scenario by definition in our system is made up of an avatar and a sequence of nodes, called actors, which make up the path a specific part of the patient's body must take. To evaluate the exercise, the module analyses the set of bones that have interacted with the sequence of actors and the order in which this interaction has taken place. In this way, patients are aware of whether they are performing the exercise correctly, as the module shows scores as the activity progresses. Furthermore, it helps to keep them motivated, whether the exercise is being performed correctly or not, since a high score may help maintain the rhythm, or even a low score may help incentive them to improve.

Finally, the displaying module is responsible for showing on the screen how the patient is progressing in respect to the task performed. This module is interconnected with the processing module, which provides information in numerical format (score the patient has, percentage of progress for the activity or achievements, among other items) in a visually attractive format which can easily be interpreted by the patient in order to capture and maintain their motivation and attention.

3.1 PEL Language

PEL is supported by a higher level specification based on 3D graphics, popularly known as glTF[1], which is a relatively new, open format, based on the JSON standard, for distributing 3D scenes in an efficient and interoperable way. Essentially, PEL contains the

[1] https://www.khronos.org/gltf/.

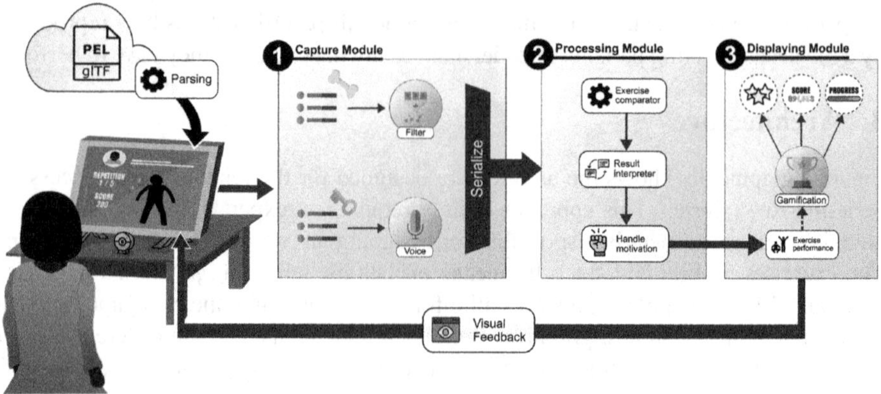

Fig. 1. General overview of the proposed architecture [9].

customized structure of an exercise, while glTF stores the multimedia resources of the game. Fundamentally, glTF was chosen because i) it is an open source, which means that the project can be freely changed to be customized; ii) its format is efficient and interoperable; and iii) the extensibility of its specification allows to add new properties which provide new opportunities for improving the schema.

Basically, our language is made up of a series of components that potentially enable tools to be developed, which through parsing, automatically generate games-based exercises focused on rehabilitating patients. The following points briefly summarize the elements which define the nature of an exergame with PEL (see Fig. 2):

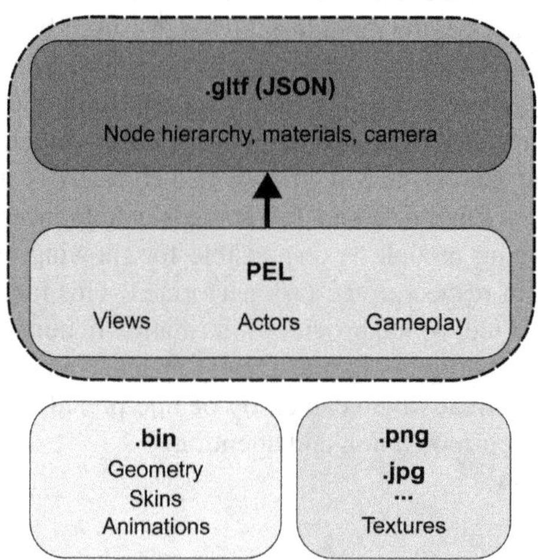

Fig. 2. Exergame structure: glTF specification extended with PEL [9].

- **Scene.** This component contains three different views which the game is split into. That is, the language envisages a tutorial view, where an avatar, using animation, shows the patient the activity to be carried out; a scene in which the patient performs the activity that has been visualized previously, with movements replicated by the avatar; and finally a view where the results are shown as a consequence of the actions carried out by the patient.
- **Actors.** They represent the elements of the game that the avatar must interact with. These objects essentially make up a sequence of the activity the patient must perform, with behavior that may be either static or dynamic. That is, elements that move in a 3D space through animation, or which are just fixed elements suspended over certain points in the 3D scene. Moreover, these elements provide visual feedback, which in itself, indicates that the node must be or has interacted with the target joint.
- **Gameplay.** This component defines a set of actions the user must take to complete a repetition of the exergame. In other words, it specifies the sequence of the actors and the user interaction mode with them. Correct execution of the game mechanics deploys gamification elements to capture, interest and motivate the user, which results in increasing the score the patient achieves for each repetition, the progress of the activity or on unblocking achievements, among others.

In effect, PEL schema is divided into 3 cores, as described above. PEL is based on JSON syntax, as it is supported by glTF. Therefore, each component is defined through a set of objects encapsulated between brackets which store key-value pairs. In the following description, if a field is not explicitly REQUIRED or described with a MUST or SHOULD, it can be considered OPTIONAL (Table 1).

Table 1. Fields of the scene component.

Field name	Description
Description	**SHOULD**. Provides the description of the exergame
Setup	**REQUIRED**. An element to hold the configuration of the exergame
Constraints	Includes the joints that the patient does not need to use to compensate his o her lack of movement
Metrics	**MUST**. Contains the different indicators to monitor the progress of a patient

While the object *description* is a simple object which stores a text field to describe the exergame, the sub-component *setup* defines a more complex structure. Therefore, Table 2 shows the sentences that can be used to initially configure an exergame.

Table 2. Fields of the setup sub-component.

Field name	Description
Avatar	**REQUIRED**. Contains information about the avatar, such as an identifier, initial posture, position and rotation
Camera	**REQUIRED**. Defines an identifier, position and rotation of a given camera in the 3D space
Repetition	Allows to specify a maximum number of repetitions
Score	If it is defined, it means that score feature is enabled

Thirdly, the field *constraints* is an enum list which specifies those joints that the patient does not need to use. Basically, the main objective of this component is to prevent patients from helping to perform an exercise with the help of some part of their body. The joints that can be added are those available for Azure Kinect DK[2].

Fourthly, the field *metrics* is the last sub-component of scene component. This element defines a set of Key Performance Indicators (KPIs) which allow to monitor patients and their level of improvement as rehabilitation progresses. This element, at least, must define one indicator. Currently, PEL defines the following KPIs, described in Table 3.

Table 3. Fields of the metric sub-component.

Field name	Description
Performance	**REQUIRED.** Defines the patient's performance taking into account the score obtained
Rehabilitation_time	**REQUIRED.** Allows to record the time it takes for a patient to complete an exercise
Mobility	**REQUIRED.** Allows to obtain the degree of mobility of upper and lower extremities
Displacement	**REQUIRED.** Measures aspects related to gestures of movements, that is, gestures of static walking or lateral steps
Balance_control	**REQUIRED.** Evaluates aspects related to balance

The second core of the PEL structure is related to *actors*. They represent virtual elements basically composed of spheres suspended in 3D space. The objective consists in the patient moving one or more limbs, depending on the exergame, to collide with the virtual spheres. The language contemplates two types of actors, which cannot be added at once in an exergame. Table 4 describes each type.

Table 4. Fields of the actors component.

Field name	Description
Virtual_nodes	**REQUIRED**. Contains information to define an identifier, the type of the actor, the position and the radius. This field is represented as an array, where each element is a single virtual sphere
Trajectories	**REQUIRED**. Contains information to define an identifier, a list, which specifies the position of each sphere belonged to the trajectory, and their radius

Finally, gameplay element is the last core of the PEL structure, where the information related to game dynamics is found, that is, the events that are triggered when the patient interacts with the actors in the 3D space. It must be noted that the behavior associated with each 3D object is thanks to the extension mechanism that glTF offers. At a higher level, gameplay element allows to define three fundamental aspects, which are described in Table 5.

[2] https://docs.microsoft.com/es-es/azure/kinect-dk/body-joints.

Table 5. Fields of the gameplay component.

Field name	Description
Actors	**MUST**. Contains the identifiers associated with virtual nodes or trajectories
Interaction	**MUST**. Defines the interaction mechanisms the patient needs to use in the exergame
Feedback	**MUST**. Allows to define the feedback for the patient when an action is completed

3.2 Automatic Generation of Exergames

An exergame is automatically generated from a hierarchical analysis of a file with a gltf extension, whose structure is defined by means of the JSON syntax. As noted above, the language is intuitive and simple enough to specify rehabilitation task. However, it is no intended to be used by therapists, but for developers who translate the description made by an expert into PEL sentences.

This system has been implemented as an application developed with Unity 3D[3] game engine, which allows to easily deploy solutions both for Windows and for Linux. This application receives the multimedia contents, and their behavior, of a 3D scenario in the way of a URI (*Uniform Resource Identifier*), which is subsequently analyzed by means of parsing technique in order to build the game.

The mechanism for adding glTF extensions is made by means of the "extra" field supported by each 3D graphic element (scene, node, camera, material, animation, etc.), which has conducted to create virtual rehabilitation scenarios. This customized behavior can be added through a plug-in developed for Blender[4]. This tool supports multimedia content specified in glTF format. The plug-in allows us to extend the information of a 3D model, adding the values of properties that PEL supports. Then, a scene can be exported, making the plug-in to generate a file which extends glTF content with PEL sentences.

The process for analyzing the "extras" field is integrated into the parsing module, which takes the form of a syntax analyzer, based on the glTF implementation, in order to effectively de-serialize the elements included in the JSON file. In this way, 3D scenes can be built for particular therapies depending on the needs of patients.

The idea of PEL is to provide an intermediate language that can be manipulated by advanced systems. The specification of PEL enables the development of visual tools which allow therapists to design and create new exergames without the rol of developers. Therefore, a visual tool is thought to be created as a future step to help therapists to easily define the dynamics of an exergame.

3.3 Azure Kinect DK

As mentioned previously, this architecture is based on a hardware component for accurate skeleton tracking: Azure Kinect DK[5], a device recently released by Microsoft

[3] https://unity.com/es.

[4] https://www.blender.org/.

[5] https://azure.microsoft.com/es-es/services/kinect-dk/.

(see Fig. 3). This device includes a RGB (Red, Green and Blue) camera, a depth sensor and several microphones, apart from the software necessary for capturing movements and recognizing voice commands.

Fig. 3. Azure Kinect DK.

Azure Kinect DK offers many possibilities that make this device suitable to complement traditional rehabilitation programs. Firstly, the device is enough versatile because of its size and price, since it can be used in hospitals, clinics, and private homes, making tele-rehabilitation services more affordable, accessible, and effective.

Secondly, the motion capture system and the advanced AI behind Azure Kinect DK allow to track accurately simple and hard movements of a person. It enables to create natural user interfaces which let patients use systems easily, even in the case when they have difficulties. In addition, this device offers to track multiple skeletons in real time, which can be useful for some therapies.

Thirdly, Azure Kinect DK provides software developer kits to develop solutions to fully take advantage of the hardware device. Apart from this, this device can be exploited in combination with the cloud resources and services that Microsoft provides hosted on Azure. Among them, computer vision and speech services can be used.

However, Azure Kinect DK shows some disadvantages from its predecessor. For example, this device must be run on a computer with a powerful graphic card to use the Azure Body Tracking SDK[6]. Unlike its predecessor, Azure Kinect DK does not support the execution of applications which uses body tracking feature on the CPU. In physical rehabilitation, common users are elderly people who do not usually have powerful computers. In this case, this requirement may prevent real users from using applications which they need in their daily lives.

4 Experiment Conducted with Potential Patients

In order to evaluate the designed system, a quasi-experiment with potential patients has been carried out. The purpose of this was to validate the system in terms of understanding and suitability. The results of such experiment are described and discussed in this section.

[6] https://docs.microsoft.com/es-es/azure/kinect-dk/system-requirements.

4.1 Exergame for Lumbar Spine Pain

The exercise designed for the experiment consists in facing up from a lying down position, raising the upper part of the torso by about 25 cm with knees bent and feet resting on the ground. Patients must maintain this position for 2 s and then return to their original position. In essence, the exercise is repeated 5 times, where the satisfactory realization of each one increases their score by 200 points.

Figure 4 shows the different scenes which make up the exergame designed for the physical rehabilitation of bone-marrow injuries. In (1) there is a screenshot of the tutorial view, where the avatar is seen carrying out the movement, which the patient must replicate later on. The participant view (2) reflects how the patients repeats the movement shown in the previous view, increasing his or her score each time the repetition is performed correctly. Note the spherical blue object (actor), which is used both as a guide for performing the exercise and as a mechanism for determining the stopping conditions in a repetition. In this case, once the head of the avatar collides and remains on the sphere for 2 s, the repetition finishes and the score increases. The results view (3) reflects the indicators provided by the therapist. Finally, the screenshot (5) shows a girl using the system. It must be noted that the equipment used is a tripod that holds the Azure Kinect DK, which is placed facing the floor to fully detect the movements of a patient lying down.

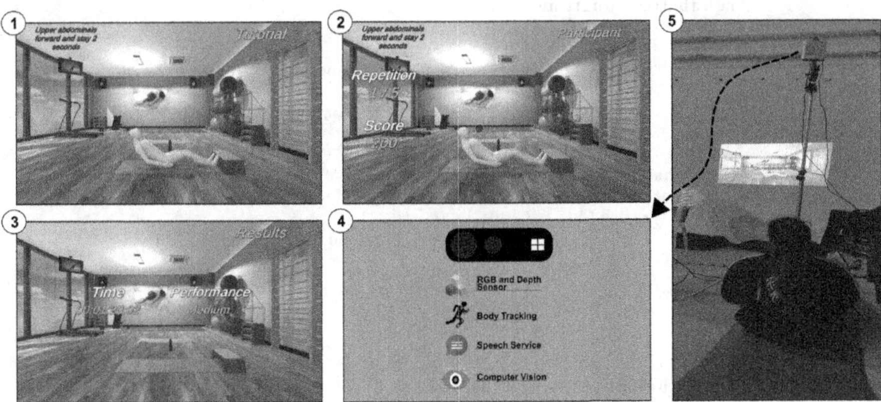

Fig. 4. Rehabilitation system in action. (**1**) tutorial view. (**2**) Participation view. (**3**) Result view. (**4**) Features of the Azure Kinect DK™ device. (**5**) User using the system [9].

This exergame was developed to motivate potential patients to perform exercises aimed at the physical rehabilitation of the lumbar spine. The exergame was customized by a therapist, who defined the therapeutic objective, the trajectory of the rehabilitation movement and the metrics to measure the progress of the patient, being this information translated into PEL sentences to automatically generate the game. The full PEL specification of the exercise is shown in Listing 1.

Listing 1. Full PEL definition for the exergame upper abdominals.

```
{
  ''description'': {
      ''text''   : ''Upper abdominals forward and stay 2 seconds''
  },
  ''setup'' : {
      ''avatar''  :  { ''id_avatar'' :  ''avatar-1'',
                       ''posture''   :  ''lying'',
                       ''position''  :  [7.298, 0.14, 5.762],
                       ''rotation''  :  [-90, 0.00, 0.00] },
      ''camera'':   { ''id_camera'' :  ''camera-1'',
                       ''position''  :  [9.782, 0.845, 4.96],
                       ''rotation''  :  [0.00, -90, 0.00] },
      ''repetitions'' : { ''activated'' : true, ''max_number_repetitions'' 5 },
      ''score''       : { ''activated'' : true }
  },
  ''virtual_nodes'': [
      {
          ''id_virtual_node'' : ''touch-head'',
          ''position''        : [7.35, 0.596, 4.678],
          ''radius''          : 1.00
      }
  ],
  ''constraints'': {
      ''joints'': [''shoulder-left'', ''shoulder-right'', ''elbow-left'',
      ''elbow-right'',''knee-left'', ''ankle-left'', ''knee-right'', ''ankle-
           right'']
  },
  ''metrics'' : [
      {
          ''rehabilitation_time'' : {
              ''time_activity'' : {
                  ''time''   : [180, 120, 90],
                  ''labels'' : [''not-bad'', ''good'', ''perfect'']
              }
          }
      },
      {
          ''performance'' : {
              ''score_performance'' : {
                  ''score''  : [200, 600, 1000],
                  ''labels'' : [''medium'', ''good'', ''perfect'']
              }
          }
      }
  ]
  ''gameplay'' : [
      {
          ''id_virtual_node'' : ''touch-head'',
          ''involved_joint'' : ''head'',
          ''interaction'' : {
              ''type'' : ''stay'',
              ''interaction_time'' : 2.0
          },
          ''on_repetition_completed'' : {
              ''repetition_increment'' : 1,
              ''score_increment''      : 100
          },
          ''on_repetition_failed'' :{
              ''action'' : ''retry''
          }
      }
  ],
}
```

4.2 Participants and Method

The experiment was carried out by two instructors and some girls from a rhythmic gymnastics club. Twenty-three girls were randomly selected, of whom 20, aged between 11 and 19, agreed to participate in the experience. Of these, 8 suffered from some type of injury, mainly related to the spine, which requires rehabilitation exercises at home. In order to avoid biasing the results [21] and to motivate their participation [33], we explicitly stated at the beginning of the experience the information collected would be treated confidentially and used exclusively for this research. After being informed, girls and their parents gave their consent to use their data.

The quasi-experiment was divided into two phases, conducted in one session of 120 min:

- **Phase 1 (Preparation Phase).** One instructor presented the system to all participants for ten minutes. An example of using the system was projected onto the wall so that the girls may understand the explanation.
- **Phase 2 (Development Phase).** Each girl participating individually engaged in two activities: First, they completed the prepared exergame, which required about 2 min; then, they filled in a questionnaire, which was provided to them through *One Drive Forms*. These concerned their perception of the activity developed, their understanding and how suitable they thought the tool was.

The exergame, used as an exercise for rehabilitating the spine, was projected onto the wall so that the girls may better visualize their avatar, as well as the gamification elements provided by the system.

The questionnaire consisted in 21 items, shown in Table 6, rated on a five-point Likert scale (1: totally disagree; 5: totally agree) grouped into five blocks or dimensions: *performance subjective ratings*, *cognitive load and effort*, *system utility*, *usefulness of the user interface components* and *TAM-based questionnaire* (abbreviated by TAM). The first dimension (performance subjective ratings), composed of four items, allowed us to measure the subjective perception of the users regarding their performance during the activity, assessing aspects such as their interest during its execution, their commitment to doing it correctly, as well as the user-friendliness of the system. The next block (cognitive load) was formed by four items, inspired by the Cognitive Load Theory (CLT) [35], which allowed us to measure two of the types of cognitive load about the use of a software system: the complexity imposed by the task to be performed (intrinsic load) and the complexity imposed by the use of the software and the interaction devices used during the performance of the task (extraneous load). In addition, two questions were included related to the effort the users had to make to complete the task.

The third block consisted in four questions related to certain users' views: preference for the use of this type of systems over face-to-face assistance in rehabilitation centers, the system's game format and finally, whether they considered that its use might improve their motivation and constancy in rehabilitation tasks. In the following block, five items were included in which users had to assess the degree of usefulness of each of the main elements of the application's user interface (virtual representation of the user, number of repetitions, score, etc.). Finally, users completed a questionnaire based on the Technology Acceptance Method (TAM) framework [3], which included four items to measure the perceived user-friendliness, usefulness and intended uses for the system being evaluated.

To speed up the process of filling in the questionnaire, we used three laptops so that girls may do it in parallel once they finished the exergame. There was no time limit for this but, on average, each girl used about 6 min to complete the questionnaire. In addition, another instructor was in charge of explaining the meaning of some questions to the youngest girls who requested them.

Table 6. Descriptive statistics of the dimensions evaluated. *Mean and standard deviation are shown (in parentheses).

Dimension	Item	Mean		Mode	
		Injured	Not injured	Injured	Not injured
Activity perception	1. INT1	**5.00** (0.00)*	4.33 (0.65)*	5	4
	2. INT2	**4.88** (0.35)*	4.75 (0.62)*	5	5
	3. EFF	4.88 (0.35)*	4.92 (0.29)*	5	5
	4. LEA	4.88 (0.35)*	4.83 (0.39)*	5	5
Cognitive load	5. TD	3.00 (0.76)*	3.00 (1.48)*	3	3
	6. E1	3.75 (0.89)*	**4.17** (1.03)*	3	3
	7. E2	3.13 (1.36)*	**2.17** (1.27)*	4	1
	8. DD	**1.25** (0.53)*	1.50 (0.45)*	2	1
Utility	9. PREF	**4.13** (1.13)*	4.00 (1.54)*	5	5
	10. CONS	**4.75** (0.46)*	4.25 (1.06)*	5	5
	11. PSA	**5.00** (0.00)*	4.67 (0.49)*	5	5
	12. GAM	4.50 (1.41)*	**5.00** (0.00)*	5	5
Interface elements	13. IE1	**4.88** (0.35)*	4.42 (1.24)*	5	5
	14. IE2	4.50 (1.41)*	4.58 (1.00)*	5	5
	15. IE3	**5.00** (0.00)*	4.58 (0.79)*	5	5
	16. IE4	**4.88** (0.35)*	4.58 (1.16)*	5	5
	17. IE5	**4.88** (0.35)*	4.75 (0.45)*	5	5
TAM	18. PEU	4.88 (0.35)*	4.92 (0.29)*	5	5
	19. PU	**5.00** (0.00)*	4.83 (0.39)*	5	5
	20. ITU1	5.00 (0.00)*	5.00 (0.00)*	5	5
	21. ITU2	**5.00** (0.00)*	4.83 (0.39)*	5	5

4.3 Results and Discussion

The results obtained from the data collected, illustrated in Table 6, show that the tool has been very well received in all dimensions. It should be stressed that the activity was more fun (item 1) and interesting (item 2) for those girls who had some injury and, therefore, needed rehabilitation, which seems logical. Moreover, all participants thought they had tried to do the activity well (item 3), they had found the system useful and it had been user-friendly (item 18).

All participants considered that the cognitive load of the activity was not very high (items 5, 8). Furthermore, most of them tried to concentrate on the activity (item 6). However, item 7 received a very high score. This may be because the question was posed in the opposite way to the others. Therefore, the results show inconsistencies; probably because some girls were confused when interpreting the values of the answers.

On the other hand, the injured girls valued the system more useful (items 9, 10) because they were aware of what it meant for them to perform exercises at home (without going to rehab centre). One highly important point to consider is that the system can be motivating (item 11), since one of the disadvantages to rehabilitation for young people is the lack of motivation to do the exercises at home. Perhaps motivation is key to understanding the greater enthusiasm of the injured girls to using the system at home (item 20), as well as to recommending to friends (item 21).

As for, the usefulness of the interface elements (items 13–17), they received positive appraisal with the "score" being the best-rated one (item 15). Apart from this, it is needed to bear in mind that gamification has a crucial role in motivating a patient, especially children.

We also have studied the existence of correlations among items using the Kendall tau-b correlation coefficient, a non-parametric measure association as the variables do not follow a normal distribution. We have obtained some interesting findings, shown in Table 7.

Table 7. Significant correlations among items (*:$p<0.05$; **:$p<0.01$).

	TD	CONS	PSA	IE2	IE3	IE4	IE5
INT1			.457*				
INT2	.468*	.509*					.498*
E1		.584**					.427*
PREF		.520*					
CONS							.634**
PSA				.442*	.824**	.498*	.688**

The positive correlations between item 1 and item 11 reflect that fun and motivation are closely related factors. The correlation between item 2 and item 5 shows that the greater the interest in the activity, the greater the concentration of girls on their performance. Correlations between item 10 and items 2, 6 and 9 indicate that the idea of having this system at home makes them be more motivated to carry out rehabilitation activities. On the other hand, the positive correlations among item 11 and items 14, 15,

16 and 17, indicate that interface elements contribute to motivate the girls to do their rehabilitation exercises. The absence of correlation with the item related to the avatar (item 13) can give us some extra information in the sense that its appearance should be adapted to the user. Finally, the positive correlations among item 17 and items 2, 6, 10 and 11 seem to indicate that the interface contributes to capture the user's interest. All of these correlations can be observed graphically in Fig. 5, grouped into three categories: activity perception, usefulness, interface.

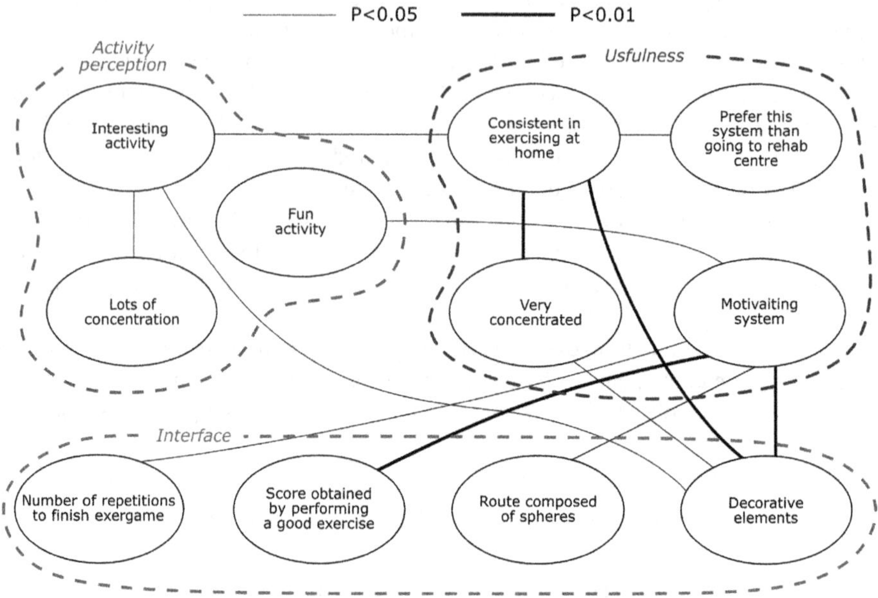

Fig. 5. Correlation between variables grouped into activity perception, usefulness and interface [9].

4.4 Improvements of the Rehabilitation System

In view of the results obtained from the experiment conducted, we have identified the drawbacks the system presents, implementing the most important features and leaving the less ones as future lines of research. Figure 6 shows some of the improvements which are clearly visible in the interface.

The main improvements that have been incorporated are summarized below:

- **New Interaction Mechanisms.** We have extended the voice commands feature so that patients can navigate easily through the system. In the previous version, this feature was only supported during the rehabilitation task, providing few voice commands, such as start or finish the game. At this point, a wide number of commands have been incorporated so that patients can select an exergame, move through the

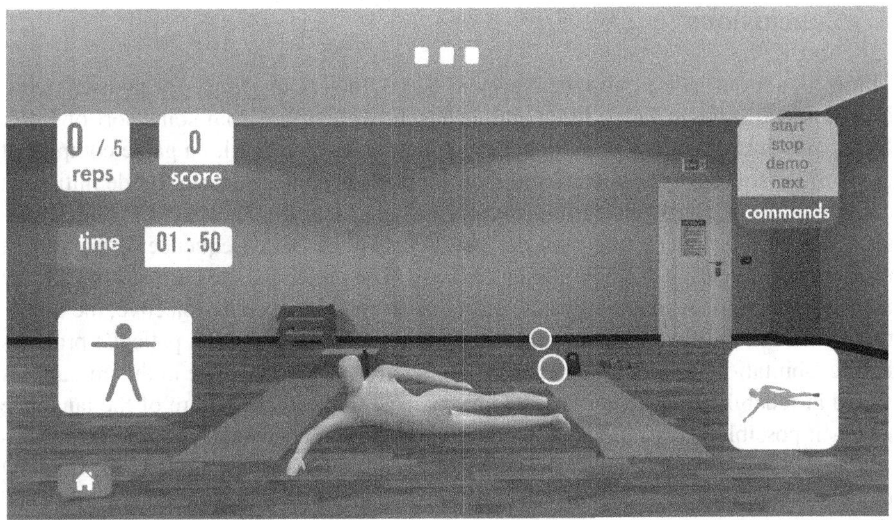

Fig. 6. New interface of the rehabilitation system.

interface, play animations and much more. This functionality is essentially important for patients who present motor disabilities. At the top of the interface, shown in Fig. 6, a boometer is displayed, which serves to provide visual feedback to patients when they speak. This improvement has been made possible thanks to the speech cognitive service[7] provided by Microsoft on Azure.

- **New Interface.** The interface has been re-designed based on the comments from the experiment. In some cases, some of the participants were not able to see the score or the repetitions labels, which clearly denoted the interface was not correctly designed. At this moment, all visual elements are perfectly visible, providing more information about the exercise adding some extra widgets. For example, on the left side of the screenshot there is a visual element which contains information about the execution of the exergame and a timer with a countdown. On the right side, there is an element which shows the possible voice commands the user can use, and a video element, which makes possible the user to see an animation of the exercise.
- **New Visual Components.** A few visual components have been added to allow patient to understand better what they need to do in each exergame. For example, the integration of colored joints allows users to know which part of the body they have to use to perform the exercise. Moreover, the design of the actors has been changed, using their size to represent the sequence the user must follow to complete a repetition. As before, each time user's joint collides with an actor its color changes to white, which means that actor has been touched.

[7] https://azure.microsoft.com/es-es/services/cognitive-services/speech-services/.

5 Conclusions

The work set out herein presents a system based on serious games for physical rehabilitation and focused on children and teenagers who suffer from some sort of lower back pain. Such games are traditional rehabilitation exercises with a game component to motivate and engage patients into their therapy. A game scenario, by definition, is composed of an avatar and virtual spheres suspended in the 3D space. The idea consists in the avatar interacting with the virtual spheres as a consequence of the patient movements recreating a rehabilitation exercise. The created games are customized by the use of a language, named PEL, to easily define the therapeutic objective, the trajectory of the rehabilitation movement, and the metrics to measure the patient's progress in a rehabilitation task. Although the main objective of the language in this research is aimed at rehabilitating patients with bone-marrow injuries, the nature of the language makes it possible be extrapolated to other rehabilitation contexts. A note to be considered is that PEL is supported by the glTF specification for encapsulating scenes and 3D models. Therefore, PEL contains the customized structure of an exercise, while glTF stores the multimedia resources of the game. As a result, glTF injects the PEL statements through an extension mechanism to allow the complete and automatic generation of a rehabilitation game.

Our software makes use of the recent Azure Kinect DK device released by Microsoft, a low-cost hardware solution to track accurately the patient's skeleton. It does not demand to use wearable devices, which means that patients prevent from wearing sensors attached to their body. Thanks to it, our system can be used in hospitals, clinics, and private homes, promoting rehabilitation at home and increasing flexibility when patients perform exergames.

Moreover, this article shows a preliminary experiment with girls from a rhythmic gymnastics club; 8 of them suffering from some type of injury mainly related to the spine. Its objective was to evaluate the system in terms of understanding and suitability, providing an exergame designed with PEL to motivate them to perform a common rehabilitation exercise for lumbar spine. As a result of the conducted experiment, the main finding obtained was that most of the participants identified the system as useful to facilitate and motivate physical rehabilitation. Besides that, they considered the tool was fun, interesting, and easy to use. However, the experiment presents several threats that may affect the results. For example, girls may be influenced, unconsciously, by the level of perfection that demands rhythmic gymnastics sport. They may thought that the only value which reflects a positive evaluation was the highest. Moreover, the sample is not representative enough for the general population. Therefore, data collected may not be generalized to other educational scenarios.

Regarding future research lines, we consider three major aspects. First, we aim to develop a module, to be integrated in the current version of the proposed system, which is able to inform the therapist about the progress level of each teenager. The interaction with the therapist can be implemented through notifications, while a personalized supervision can be carried out. Second, a visual tool must be designed and developed to graphically allow the therapists to establish the exergame parameters and the game dynamics. This is an important issue to facilitate the definition of exergames and avoid the direct use of the proposed language. Then, this visual representation can be automatically translated into PEL sentences, which requires the design and development of a custom parser. Third, a complete experiment which can be as representative as possible for the general population must be conducted, taking into account the aforementioned threats to avoid erroneous results.

Acknowledgments. This research was funded by Instituto de Salud Carlos III, grant number DTS18/00122, co-funded by the European Regional Development Fund/European Social Fund "Investing in your future".

Appendix

A Description of the Evaluated Items

See Table 8.

Table 8. Description of the evaluated items.

Dimension	Item	Description
Activity perception	1. INT1	This activity has been fun for me
	2. INT2	I found this activity interesting
	3. EFF	I have worked to do it well
	4. LEA	It has been easy for me to learn how to use this system
Cognitive load	5. TD	The activity required a lot of concentration
	6. E1	I have been very concentrated during the activity
	7. E2	I have had to work pretty hard to get the activity done
	8. DD	I have found difficult to perform rehabilitation exercises using this system
Utility	9. PREF	I would rather use this system at home than have to go to a rehab center
	10. CONS	This system would make me more consistent in performing the exercises at home
	11. PSA	I believe that using this system to do rehabilitation exercises can be motivating
	12. GAM	I like the application has the format of a game
Interface elements	13. IE1	The design of the avatar is appropriate
	14. IE2	The information about the repetition is useful
	15. IE3	I like to get score every time I perform a good exercise
	16. IE4	The route composed of spheres helps to perform the exercises
	17. IE5	The gym has immersed me in a rehabilitation environment
TAM	19. PEU	This system is easy to use
	20. PU	Using this system could help me in performing the rehabilitation exercises
	21. ITU1	If I could borrow this system, I would use it at home
	22. ITU2	I would recommend my friends to use this system to do the exercises at home

References

1. Clark, R.A., et al.: Validity of the microsoft kinect for assessment of postural control. Gait Posture **36**(3), 372–377 (2012). https://doi.org/10.1016/j.gaitpost.2012.03.033
2. Da Gama, A., Fallavollita, P., Teichrieb, V., Navab, N.: Motor rehabilitation using kinect: a systematic review. Games Health J. **4**(2), 123–135 (2015). https://doi.org/10.1089/g4h.2014.0047

3. Davis, F.D.: User acceptance of information technology: system characteristics, user perceptions and behavioral impacts. Int. J. Man-Mach. Stud. **38**(3), 475–487 (1993)
4. Deutsch, J.E., Robbins, D., Morrison, J., Bowlby, P.G.: Wii-based compared to standard of care balance and mobility rehabilitation for two individuals post-stroke. In: 2009 Virtual Rehabilitation International Conference, pp. 117–120. IEEE (2009). https://doi.org/10.1109/ICVR.2009.5174216
5. Duthey, B.: Background paper 6.24 low back pain. In: Priority medicines for Europe and the world. Global Burden of Disease (2010),(March), pp. 1–29 (2013)
6. Eckert, M., Gómez-Martinho, I., Meneses, J., Martínez, J.F.: New approaches to exciting exergame-experiences for people with motor function impairments. Sensors **17**(2), 354 (2017). https://doi.org/10.3390/s17020354
7. Esculier, J.F., Vaudrin, J., Beriault, P., Gagnon, K., Tremblay, L.E.: Home-based balance training programme using wii fit with balance board for Parkinson's disease: a pilot study. J. Rehabil. Med. **44**(2), 144–150 (2012). https://doi.org/10.2340/16501977-0922
8. Fernandez-Cervantes, V., Neubauer, N., Hunter, B., Stroulia, E., Liu, L.: Virtualgym: a kinect-based system for seniors exercising at home. Entertainment Comput. **27**, 60–72 (2018). https://doi.org/10.1007/978-3-658-07141-7_11
9. Gómez-Portes, C., Lacave, C., Molina, A., Vallejo, D., Schez-Sobrino, S.: Personalising exergames for the physical rehabilitation of children affected by spine pain. In: Proceedings 22nd International Conference on Enterprise Information Systems(ICEIS 2020), vol. 2, pp. 533–543 (2020)
10. González, C.S., Toledo, P., Padrón, M., Santos, E., Cairos, M.: TANGO:H: creating active educational games for hospitalized children. In: Casillas, J., Martínez-López, F., Vicari, R., De la Prieta, F. (eds.) Management Intelligent Systems, pp. 135–142. Springer, Heidelberg (2013). https://doi.org/10.1007/978-3-319-00569-0_17
11. González, C.S.G., del Río, N.G., Adelantado, V.N.: Exploring the benefits of using gamification and videogames for physical exercise: a review of state of art. IJIMAI **5**(2), 46–52 (2018). https://doi.org/10.9781/ijimai.2018.03.005
12. González-González, C.S., Toledo-Delgado, P.A., Muñoz-Cruz, V., Torres-Carrion, P.V.: Serious games for rehabilitation: gestural interaction in personalized gamified exercises through a recommender system. J. Biomed. Inf. **97**, 103266 (2019). https://doi.org/10.1016/j.jbi.2019.103266
13. Hardy, S., Dutz, T., Wiemeyer, J., Göbel, S., Steinmetz, R.: Framework for personalized and adaptive game-based training programs in health sport. Multimedia Tools Appl. **74**(14), 5289–5311 (2014). https://doi.org/10.1007/s11042-014-2009-z
14. Jones, M., Stratton, G., Reilly, T., Unnithan, V.: A school-based survey of recurrent non-specific low-back pain prevalence and consequences in children. Health Educ. Res. **19**(3), 284–289 (2004). https://doi.org/10.1093/her/cyg025
15. Katajapuu, N., et al.: Benefits of exergame exercise on physical functioning of elderly people. In: 2017 8th IEEE International Conference on Cognitive Infocommunications (CogInfoCom), pp. 000085–000090. IEEE (2017). https://doi.org/10.1109/CogInfoCom.2017.8268221
16. König, I.R., Fuchs, O., Hansen, G., von Mutius, E., Kopp, M.V.: What is precision medicine? Eur. Respir. J. **50**(4), 1700391 (2017). https://doi.org/10.1183/13993003.00391-2017
17. Lai, C.L., Huang, Y.L., Liao, T.K., Tseng, C.M., Chen, Y.F., Erdenetsogt, D.: A microsoft kinect-based virtual rehabilitation system to train balance ability for stroke patients. In: 2015 International Conference on Cyberworlds (CW), pp. 54–60. IEEE (2015). https://doi.org/10.1109/CW.2015.44

18. Li, B., Maxwell, M., Leightley, D., Lindsay, A., Johnson, W., Ruck, A.: Development of exergame-based virtual trainer for physical therapy using kinect. In: Schouten, B., Fedtke, S., Schijven, M., Vosmeer, M., Gekker, A. (eds.) Games for Health 2014, pp. 79–88. Springer, Wiesbaden (2014). https://doi.org/10.1007/978-3-658-07141-7_11
19. Matallaoui, A., Koivisto, J., Hamari, J., Zarnekow, R.: How effective is "exergamification"? a systematic review on the effectiveness of gamification features in exergames. In: Proceedings of the 50th Hawaii International Conference on System Sciences (2017). https://doi.org/10.24251/HICSS.2017.402
20. McCallum, S.: Gamification and serious games for personalized health. In: pHealth, pp. 85–96. IOS Press (2012). https://doi.org/10.3233/978-1-61499-069-7-85
21. McCambridge, J., Witton, J., Elbourne, D.R.: Systematic review of the hawthorne effect: new concepts are needed to study research participation effects. J. Clin. Epidemiol **67**(3), 267–277 (2014). https://doi.org/10.1016/j.jclinepi.2013.08.015
22. Mobini, A., Behzadipour, S., Saadat Foumani, M.: Accuracy of kinect's skeleton tracking for upper body rehabilitation applications. Disabil. Rehabil. Assist Technol. **9**(4), 344–352 (2014). https://doi.org/10.3109/17483107.2013.805825
23. Mousavi Hondori, H., Khademi, M.: A review on technical and clinical impact of microsoft kinect on physical therapy and rehabilitation. J. Med. Eng. **2014**, 1–16 (2014)
24. Müller, S.A., Vavken, P., Pagenstert, G.: Simulated activity but real trauma: a systematic review on nintendo wii injuries based on a case report of an acute anterior cruciate ligament rupture. Medicine **94**(12), e648 (2015). https://doi.org/10.1097/MD.0000000000000648
25. Murray, C.J., et al.: Disability-adjusted life years (DALYs) for 291 diseases and injuries in 21 regions, 1990–2010: a systematic analysis for the global burden of disease study 2010. Lancet **380**(9859), 2197–2223 (2012). https://doi.org/10.1016/S0140-6736(12)61689-4
26. Palacios-Navarro, G., García-Magariño, I., Ramos-Lorente, P.: A kinect-based system for lower limb rehabilitation in Parkinson's disease patients: a pilot study. J. Med. Syst. **39**(9), 1–10 (2015). https://doi.org/10.1007/s10916-015-0289-0
27. Paraskevopoulos, I.T., Tsekleves, E., Craig, C., Whyatt, C., Cosmas, J.: Design guidelines for developing customised serious games for Parkinson's disease rehabilitation using bespoke game sensors. Entertainment Comput. **5**(4), 413–424 (2014). https://doi.org/10.1016/j.entcom.2014.10.006
28. Pérez-Munoz, A., Ingavélez-Guerra, P., Robles-Bykbaev, Y.: New approach of serious games in ludic complements created for rehabilitation therapies in children with disabilities using kinect. In: 2018 IEEE XXV International Conference on Electronics, Electrical Engineering and Computing (INTERCON), pp. 1–4. IEEE (2018). https://doi.org/10.1109/INTERCON.2018.8526464
29. Pirovano, M., Surer, E., Mainetti, R., Lanzi, P.L., Borghese, N.A.: Exergaming and rehabilitation: a methodology for the design of effective and safe therapeutic exergames. Entertainment Comput. **14**, 55–65 (2016). https://doi.org/10.1016/j.entcom.2015.10.002
30. Robinet, F., Arnaud, R., Parisi, T., Cozzi, P.: gltf: designing an open-standard runtime asset format. GPU Pro **5**, 375–392 (2014)
31. Sawyer, B., Smith, P.: Serious games taxonomy. In: Slides from the Serious Games Summit at the Game Developers Conference, vol. 5 (2008)
32. Semrau, M., et al.: Strengthening mental health systems in low-and middle-income countries: the emerald programme. BMC Med. **13**(1), 79 (2015). https://doi.org/10.1186/s12916-015-0309-4
33. Shull, F., Singer, J., Sjøberg, D.I.: Guide to Advanced Empirical Software Engineering. Springer, Heidelberg (2007). https://doi.org/10.1007/978-1-84800-044-5

34. Sparks, D.A., Coughlin, L.M., Chase, D.M.: Did too much wii cause your patient's injury? motion-controlled game consoles like wii may be used to play virtual sports, but the injuries associated with them are real. Here's what to watch for-and a handy table linking specific games to particular injuries. J. Fam. Pract. **60**(7), 404–410 (2011)
35. Sweller, J., Van Merrienboer, J.J., Paas, F.G.: Cognitive architecture and instructional design. Educ. Psychol. Rev. **10**(3), 251–296 (1998). https://doi.org/10.1023/A:1022193728205
36. Webster, D., Celik, O.: Systematic review of kinect applications in elderly care and stroke rehabilitation. J. Neuroeng. Rehabil **11**(1), 108 (2014). https://doi.org/10.1186/1743-0003-11-108
37. Wiemeyer, J., et al.: Recommendations for the optimal design of exergame interventions for persons with disabilities: challenges, best practices, and future research. Games Health J. **4**(1), 58–62 (2015). https://doi.org/10.1089/g4h.2014.0078

Enterprise Architecture

Digital Innovation and Transformation to Business Ecosystems

Kecheng Liu(✉) and Hua Guo(✉)

Informatics Research Centre, University of Reading, Reading, U.K.
k.liu@reading.ac.uk, h.guo@pgr.reading.ac.uk

Abstract. Digital technologies have been penetrating every aspect of business. Such pervasive deployment of digital technologies enables organisations to reinvent themselves in defining and conducting business. Leveraging the value of information as the key resource, through digital innovation such as digitisation and servitisation of products and services, becomes necessary for a business to survive and to remain competitive. Business ecosystems of the organisations and their partners have been formed through digital connectivity and digital platform which offer clearly strategic advantages and competitiveness. Interconnectivities between entities in the ecosystem are realised through multiple flows such as goods, finance, and information. For a business organisation to gain competitive advantages, a successful transformation of the organisation in many dimensions is essential to allow value co-creation with other members in the ecosystem. These dimensions include mindset, culture, values, leadership, structure, process and IT systems. This keynote will discuss the notions of digital innovation and transformation, and the prerequisites and readiness for the transformation towards business ecosystems. By examining the current practice of successful examples, key findings will lead to the principles and models of organisational transformation for value co-creation and optimising benefits in the business ecosystems. This keynote will hopefully inspire practitioners and researchers to benefit from existing theoretical lenses and methods and to derive their own guidelines and models to support organisations in digital transformation.

Keywords: Digital Business Ecosystems · Digital innovation · Digitisation · Servitisation · Value co-creation · DBE typology · Organisational onion

1 Introduction

ICT has become strategic weapons for business organisations to enhance their business competitiveness in forming an alliance through effective communication and collaboration (Bensaou and Earl 1998). More importantly, ICT has also become essential tools for the survival of the business which has deployed and embedded in organisations and business processes to perform necessary functions and maintain expected efficiency and effectiveness for the set goals and objectives (Venturini 2009, Lyver and Lu 2018, Gërguri-Rashiti et al. 2017). With the help of digital technologies, business organisations

J. Filipe et al. (Eds.): ICEIS 2020, LNBIP 417, pp. 793–803, 2021.
https://doi.org/10.1007/978-3-030-75418-1_36

gain additional capabilities of innovation in research, development, production, distribution and delivery of products and services. Moreover, these also lead to the innovations in products, services and even new business sectors which were not possible without the underpinning of digital technologies, e.g. electronic commerce, online education, digital entertainment and many other digital economic activities (Bloom et al. 2016).

Meanwhile, challenges still remain in how digital technologies can best fit into organisations in all dimensions (Camarinha-Matos et al. 2009, Shachaf 2008). As new technologies emerge from time to time, every adoption of a new technology may require changes in organisations and business processes which generate impact on the company and sometimes affect the way of working with partners. Fraction and resistance within the company at leadership or other levels of workforce may occur. All have to be handled properly if the benefits of ICT are to be optimal.

To a business organisation, the deployment of digital technology gives it more capabilities to extend or to enhance their performance into digital arena or virtual world where signs such as data and information can be utilised to create or add value to business. Such extended capabilities can be seen as new affordances that can be studied by the discipline of informatics. From the technical perspective, informatics focuses on the development of technologies and enhances the functions to aid organisations and end-users; and from the business perspective, informatics places emphasis on the value added by ICT and relationship between business and technology. Digital technology enables the creation of the virtual world that corresponds to its physical counterpart in the physical world. With the disposal of signs in the virtual world representing the physical world, experiments and inventions become less expensive and much quicker, for example, typically through digital design, prototyping, simulation and validation (Liu and Li 2015).

2 Characteristic of Digital Innovation

Digital innovation often refers to the innovation carried out with the help of digital technology, and its characteristics are as follows.

A digital innovation may require one to re-envision familiar products and services empowered by digital technology. Unlike the technological innovation in a traditional sense, some digital innovations enable one to conceive completely different features and even change the nature of familiar products and services, resulting in disruptive changes (Christensen et al. 2015). For example, with the help of digital connectivity, entertainment (e.g. gaming), finance (e.g. online banking), consultancy and training can deliver their products and services all online. Traditional shopping can be become online shopping. The e-commerce, online-payment, unmanned store and chatbot have brought in new features to the familiar business which are only possible by enjoying the unique capabilities of digital technology.

A digital innovation involves a combination of digital and physical components by wrapping the physical product with a service layer, which is also referred to as servitisation. Such an innovation requires not only the understanding of the products and users' needs, but also the capabilities of an organisation to create mutual value through leveraging digital innovations to undergird their competitive advantages.

A digital innovation aims at creating new types or forms of business previously not existed. For example, the emergent digital platforms have made possible for a new type of business and economic activities which is called the platform economy. Two sides of the economic players, i.e. the suppliers and customers are facilitated to identify their ideal counterparts and conduct their business transactions through the support of the technology platforms.

A digital innovation can lead to radically different business models. Business operations and business models have been transformed with a more focus on the added value to the end users as well as business partners. A successful digital innovation can breed open processes of value co-creation which generates win-win outcomes for all participants in collaboration within a business ecosystem. The transformations for business models and the new user experiences are underpinned and driven by digital innovations, although digital technology is a necessary, not sufficient condition (Yoo et al. 2010).

3 Digitisation and Servirisation

Digitisation and servitisation are the two critical phases for an enterprise to join the digital business ecosystem cooperation and competition. Digitisation emphasises using ICT to extend the affordance of a company by performing business functions (e.g. marketing, sales, design, production and delivery); while servitisation, focusing much on the value and experience for customers, treats all transactions as services by wrapping a service layer to each product the company offers to the customer. With servitisation, a one-off sale of a product, which may be a physical device or a digital product, will be transformed into an offering of user experience over a lasting period of time which encamps for the user to enjoy the utility of the core commodity as well as a repertoire of collateral activities (e.g. services).

Digitisation makes it more possible for servitisation of products and goods by adding service wrappers enabled by digital means. The digitization-enabled servitisation is a journey of transformation with the purpose of creating the mutual values. The ideology of servitisation goes hand in hand with the shift of emphasis from product to service, and requires the producers and their partners in the business ecosystem to focus on the value and experience for the end-users. The value co-creation activities between the partners in the business ecosystem expand from the physical space as well as much in the virtual space. Such combination of virtual and physical dimensions is coined as servgoods by Tien (2015, 2017). With the tangible goods transformed to servgoods, the variability of products has been shifted from identical to assorted in order to better meet the diversified demands. The servgood aligns the interests of clients and providers.

Digitisation and servitisation open up new value propositions for the enterprises and a whole range of new possibilities for them to position and pursuit (Table 1).

Table 1. Tangible goods, intangible services, servgoods and connected servgoods (Tien 2015).

Focus	Tangible goods	Intangible services	Servgoods	Connected servgoods
Production	Pre-produced	Co-produced	Demand-produced	Internet-connected
Variability	Identical	Heterogeneous	Assorted	Assorted
Physicality	Tangible	Intangible	Mixed	Mixed
Product	"Inventoryable"	Perishable	Identifiable	Identifiable
Objective	Reliable	Personalizable	Adaptable	Connectable
Satisfaction	Utility-related	Expectation-related	Satisfaction-related	Status-related
Life cycle	Recyclable	Reusable	Flexible	Agile
Example	Car	Electronic-assists	Internet of Things	Connected autonomous cars

4 Digital Business Ecosystems

A digital ecosystem often refers to a virtual environment populated by digital entities such as software applications, hardware and processes which serves as an interconnected environment for creating and disseminating digital services over the Internet (Nachira et al. 2007). Meanwhile a business ecosystem is a community of individuals and organisations that operate outside their traditional industry boundaries (Moore 1993). A digital business ecosystem (DBE) is a socio-technical environment supported by digital technologies within which business organisations operate. The members of the DBE collaborate or compete and form dependent and interrelationships with their customers (Senyo et al. 2019, Liu et al. 2018).

4.1 Main Drives of Digital Business Ecosystems

Three types of flows can be identified in business ecosystems: information, finance and physical goods. These three flows commonly appear in most types of enterprises with different proportions. For instance, the Internet companies may not have physical goods flow because the products and the services they provide are all digital or virtualized. While, when analysing the types of Digital Business Ecosystems, these three flows will be seen as the principal driving forces behind the formation and growth of each digital ecosystem.

As shown in Table 2, the drivers of the business ecosystem can vary. Even the Internet companies, with similar business models, may have different motivations. The principal drivers of forming individual business ecosystems are highly associated with business operating.

Table 2. The types of typical DBE.

Principal driver	Typical enterprise	Typical DBE typology	Operating philosophy
Financial flow	Amazon	Mesh topology	Discover profitable businesses
Information flow	Tencent	Bus topology	Traffic monetization
Physical goods flow	Mecedes	Star topology	Increase sales
Financial flow	Huawei	Ring topology	Raise barriers, secure leadership

4.2 Topologies of DBE and Examples

Amazon's (https://www.aboutamazon.com/) business diversity is no less than the ecological diversity of the Amazon Basin, where is the world's largest rainforest and a web of ecosystems. Amazon builds its various business ecosystems around customers in a wide range of perspectives. These business sub-ecosystems symbiotically interact with each other while maintaining their separated business objectives at the same time. Each of them provides users with its unique business value and enriches the user experiences by the support of other business sub-ecosystems, and then eventually, these multiple sub-ecosystems involved in value integration forms a symbiotic and prosperous ecosystem. For example, the Kindle-centric e-book sub-ecosystem, on the one hand, is an integration platform for content innovation, and on the other hand, is the e-book distribution channel for the Amazon online shopping platform. Therefore, Kindle and Amazon shopping website forms a typical symbiotic ecology. In addition, the Amazon Prime sub-ecosystem, derived from the Amazon online shopping platform, provides value-added services to enhance customer loyalty and promotes website sales.

Relying on the large customer base, Amazon generates substantial intermediary profits by providing the platform services, which contributes the majority of the margin in the whole ecosystem. The sub-ecosystems of home assistant and Amazon Web Services are all the extensions from its core businesses to enrich the user experiences of their customer. From this case, it can be reckoned that the mutual synergy of the symbiosis provides a perfect basis for companies to maintain their competitiveness through entering new fields, figuring out the new profit growth points and keeping the vitality of the company.

The topology of Amazon ecosystem mainly displays the characteristics of a mesh type (Fig. 1). The sub-business entities within the ecosystem are connected with each other in symbiotic relationships. Although the products or services offered by each sub-business system come in different forms, the purpose of the links among them is the same: to support each other and grow together. The sub-business system can be generated based on business envisioning or the users' needs, but the main criterion for whether it can grow into a symbiotic sub-business entity is whether it can bring cash profit now or in the future.

Tencent (https://www.tencent.com/en-us/about.html) is one of the Internet giants and builds its mega internet ecosystems to interconnect its social networking, fintech, games,

Fig. 1. The Mesh topology of Amazon ecosystem.

videos and cloud services all altogether. Tencent's business philosophy is to provide free services to the vast majority in order to accumulate huge customer base and monetize the enormous volume of internet traffic bringing by the users.

WeChat and QQ are Tencent's star products which link more than 1 billion Internet users in China. Both of them are free to use and provide a variety of services covering almost all dimensions of the users' daily life, including social, online shopping, mobile payment, investment and travel services. The users contribute their time, attention and their data to the service provider, so that Tencent made considerable profits from advertisement. Also, these large numbers of participants provide a better experience for high-end users. For example, the free gamers increase the popularity of game products and create a promising gaming atmosphere for the high-end fee-paying game players.

Based on the huge customer base, Tencent builds its profit model cautiously to maximize the value of directed traffic between sub-business systems and also to avoid ruining the user experiences. The business sub-ecosystems in Tencent are orchestrated and interdependent.

As shown in Fig. 2, the topology of Tencent ecosystem shows the key features of a bus type. The core platform in bus topology is the huge customer base acquired by the free apps. The business sub-ecosystems that grow on this nutrient block are the various methods of monetizing the vast internet traffic by providing different services. Therefore, the types of business sub-ecosystems are diversified, as long as the user base can be leveraged to support its business growth.

Daimler (https://www.la.mercedes-benz.com/en/passengercars/mercedes-benz-cars/models/e-mobility/project-eq/models/electric-intelligence.module.html), has been evolving around its new strategy. The automotive industry giant is repositioning its competitiveness from four aspects: the fields of networking (Connected), autonomous driving (Autonomous), flexible use (Shared and Services) and electric drive systems (Electric). The abbreviation of CASE depicts the Daimler's understanding of the future of the automotive industry. Comparing with the dumb vehicle, the smart car largely extends the scope of the ecosystem and grows to diversify. In this case, the characteristics of the digital business ecosystem are fully expressed. The first is digital enabler (Selander et al. 2013), which is the foundation of servitisation. The digital enabler enables the physical vehicle to turn into a carrier of services. From the angle of connectivity, users' driving experiences are redefined by personalised services via the app, website, or straight from the assistance software in automotive. The second and the third characteristics of DBE is reflected by the participants of ecosystems. For instance, the feature of autonomous driving introduces software platform, data analytics, image identification, artificial intelligence and connective technologies into the new automotive ecosystem. The participants are highly interdependent with each other, while sharing the benefits of the ecosystems (Senyo et al. 2019, Stanley and Briscoe 2010). The last characteristic is self-organising (Senyo et al. 2019) which elaborates the spontaneous behaviours of the participants to maintain the balance of ecosystems in term of service providing.

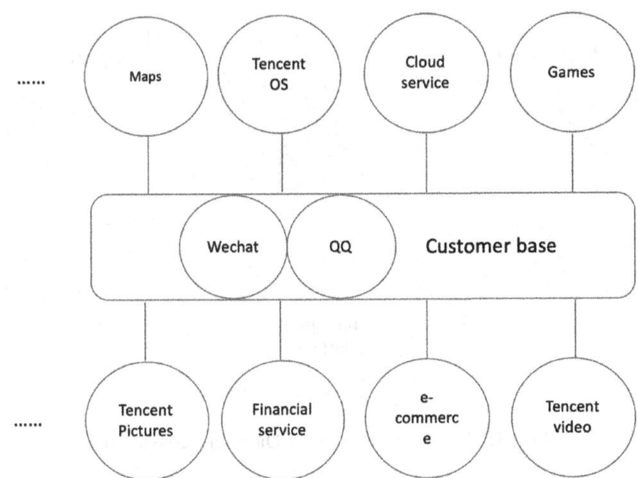

Fig. 2. The bus topology of Tencent ecosystem.

The topology of Daimler ecosystem can be best represented as a typical star type (Fig. 3). All the sub-business systems are tightly integrated around the main product deliverable: the Mercedes automobile. In this topology, the Mercedes automobile are both the product for external users and the platform for its symbiotic partners.

Huawei (https://www.huawei.com/en/fully-connected-intelligent-world), has become a telecom giant and occupies an important position in fields of telecommunication and computing. In order to maintain the industry leadership status, Huawei develops its own ecosystem tightly surrounding the core business. As Huawei dedicated in the telecom industry for decades and gradually becomes a global leader, its first core business is the telecom equipment which bridges telecom operators and users, and the positioning of this core business lasts for a few decades. After they gains notable advantages in telecom equipment, encircling tis core business, the smart phone business start to extend Huawei's business model from 2B to 2C, and then the cloud service which can further enhance the end-to-end user experience is established immediately. These two new fileds of business not only open up the market space and bring new points of profit growth, but also will gradually build up the core competitiveness on the sub-ecosystems of smart phone and cloud services respectively. With the computing power becoming the core competitiveness of internet companies, AI has been introduced into its various products and services as another core competitiveness. Therefore, as shown in Fig. 4, the topology of Huawei ecosystem can be typified by a ring type. The four business fields are developing independently and at the same time relying on each other to form a symbiotic relationship.

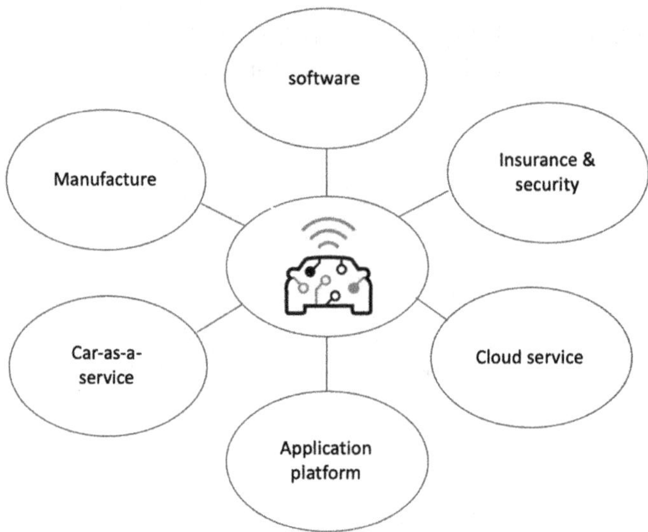

Fig. 3. The star topology of Daimler ecosystem.

5 Impact of Digital Transformation on Enterprises

Digital transformation is a journey of an enterprise to gain a competitive advantage and to maximise its business value. However, this journey seems never to end, because of the continued development of ICT and also due to ever-changing, complicating and challenging business world. Organisational semiotics, as the discipline to study the nature, characteristics and effect of information in organised activities, offers us an effective approach to study the interplay between physical and virtual world. To fully understand the impact of digital transformation, an organisation can be seen as three interdependent layers (Fig. 5). The outer layer encompasses the whole organisation where organisational values, trust, culture and other informal components exist and

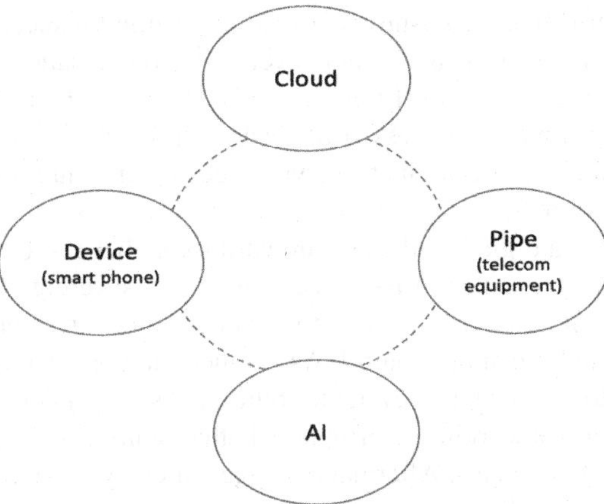

Fig. 4. The ring topology of Huawei.

determine the mission, vision and value propositions. On the premise of the informal layer, the middle layer comprises of formal components such organisational structure, policies, procedures and work processes. In the inner layer of an enterprise, some formal elements as defined in the middle layer are digitised for automatic execution. Changes in any one part will generate impact on others and induce necessary alignment adjustment, possibly sometimes radical and disruptive (Stamper 1991, Liu and Li 2015).

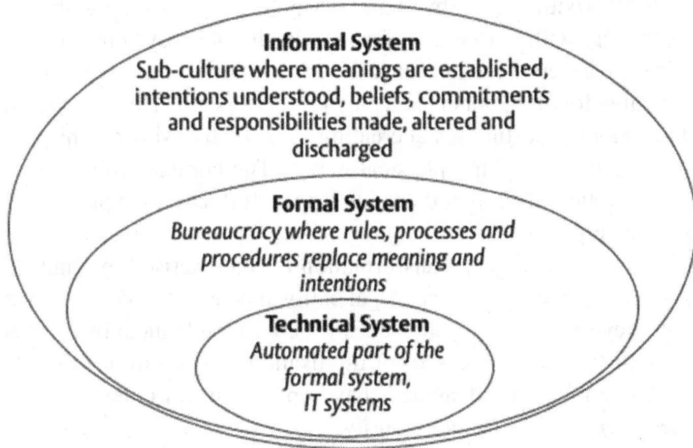

Fig. 5. The organisation onion (Stamper 1991, cited from Liu and Li 2015).

The process of digital transformation is complex. The success rate of digital transformation based on industry statistics is relatively low, because the process of transformation requires participation from actors in all levels of the organisation and brings in profound impact on entire organisation, including all dimensions such as parts from technical, formal to informal parts.

Therefore, an effective leadership is critical in ensuring a successful digital transformation. Such leadership at the senior level in an organisation, as what is called digital leadership or e-leadership (Li et al. 2017), should be able to plan, orchestrate and direct the transformation process. This leadership involves understanding of digital innovations, technologies, products, services, customer requirements, and even the whole business ecosystem.

The challenges to a digital leadership are hard to predict and often multifaced. For example, due to the emergence of issues such as digital sovereignty, data privacy and network security, digital leaders may have to rethink whether an enterprise has the capabilities to respond and remain in control. As another example, external factors such as Covid-19 are hard to foreseen; however, the ability to respond quickly with digital capabilities to support remote working determines whether or not the company could survive when facing such emergencies. Aftermath changes in every state in the value network including production, marketing, sales, working with partners and services delivered to the end-customers are expected which may require further transformation of business in many aspects that are facilitated by digital technologies. The digital leadership will play a key role in facing and responding to the challenges.

6 Discussion and Conclusion

Digital technologies can offer organisations capabilities of expanding their access and powers in the virtual world. With the acquired repertoire of capabilities, organisations can enjoy digital innovation to increase the productivity and competitiveness. Sertivisation of physical goods and products aided by digital technologies makes it possible for the enterprise to work closely with partners by forming business ecosystems. The competition between business partners become win-win, collaborative competition (or copetition) that maximise value for all members in a business ecosystem. Through servitisation, the enterprises have more opportunities of creating value by focusing on enhancing the user experience instead of just selling physical goods. The combination of goods and services leads to the notion of servgoods which inspire further emergence of new business possibilities for enterprises.

An advanced stage of digital transformation is characterised by digitally interconnected enterprises. Each enterprise is a legal entity and acts as a member in the business ecosystem; yet they share visions and values which is enable them in value co-creation. By going through the three stages from digitisation, servitisation to digital business ecosystem, a whole range of advantages have opened up for enterprises to gain more competitiveness, returned value and security.

Challenges for digital transformation of an enterprise come from many dimensions. A successful transformation will continue for aligned changes in all aspects which can be represented as the three interdependent layers of the enterprise. The informal part of the organisation encompasses the leadership's vision and mindset, values and organisational culture. The formal part consists of organisational structures, policies, procedures, work processes and formally defined functions and responsibilities. The technical part is constituted by automated processes and functions as defined in the formal part of the enterprise. The automation and digitisation of the processes and functions are realised through the adoption of digital technologies and digital platforms which offer pervasive connectivity for working collaboratively within the company and in the business ecosystems. The understanding of the interrelationship between these parts guides us in preparing for and coping with the complexity and profound impact of digital transformation from the perspective of individual companies, and more widely, digitally connected business ecosystems.

References

Bensaou, M., Earl, M.: The right mind-set for managing information technology. Harvard Bus. Rev. **76**, 119–130 (1998)

Bloom, N., Draca, M., Van Reenen, J.: Trade induced technical change? the impact of Chinese imports on innovation, IT and productivity. Rev. Econ. Stud. **83**, 87–117 (2016)

Camarinha-Matos, L.M., Afsarmanesh, H., Galeano, N., Molina, A.: Collaborative networked organizations–concepts and practice in manufacturing enterprises. Comput. Ind. Eng. **57**, 46–60 (2009)

Christensen, C.M., Raynor, M.E., Mcdonald, R.: What is disruptive innovation. Harvard Bus. Rev. **93**, 44–53 (2015)

Gërguri-Rashiti, S., Ramadani, V., Abazi-Alili, H., Dana, L.P., Ratten, V.: ICT, innovation and firm performance: the transition economies context. Thunderbird Int. Bus. Rev. **59**, 93–102 (2017)

Li, W., Liu, K., Ghobadian, A., Belitski, M., O'regan, N. : e-Leadership through strategic alignment: an empirical study of small and medium sized enterprises in the digital age. J. Inf. Technol. **31**, 185–206 (2017)

Liu, K., Li, W.: Organisational semiotics for business informatics. Routledge, Abingdon (2015)

Liu, K., Nakata, K., Li, W., Baranauskas, C.: Digitasation, Innovation and Transformation. In: 18th IFIP WG8.1 International Conference on Informatics and Semiotics in Organisations, ICISO 2018, Reading, UK, 16–18 July 2018, Proceedings (IFIP AICT 527). Springer, Heidelberg (2018)

Lyver, M.J., Lu, T.-J.: Sustaining innovation performance in SMEs: exploring the roles of strategic entrepreneurship and IT capabilities. Sustainability **10**, 442 (2018)

Moore, J.F.: Predators and prey: a new ecology of competition. Harvard Bus. Rev. **71**, 75–86 (1993)

Nachira, F., Nicolai, A., Dini, P., Le Louarn, M., León, L.R.: Digital business ecosystems. European Commission (2007)

Senyo, P.K., Liu, K., Effah, J.: Digital business ecosystem: literature review and a framework for future research. Int. J. Inf. Manag. **47**, 52–64 (2019)

Shachaf, P.: Cultural diversity and information and communication technology impacts on global virtual teams: an exploratory study. Inf. Manag. **45**, 131–142 (2008)

Stamper, R.: The semiotic framework for information systems research. In: Information Systems Research: Contemporary Approaches and Emergent Traditions, pp. 515–528 (1991)

Stanley, J., Briscoe, G.: The ABC of digital business ecosystems (2010). arXiv preprint arXiv: 1005.1899.

Tien, J.: Internet of connected ServGoods: considerations, consequences and concerns. J. Syst. Sci. Syst. Eng. **24**(2), 130–167 (2015)

Tien, J.M.: Internet of things, real-time decision making, and artificial intelligence. Ann. Data Sci. **4**, 149–178 (2017)

Venturini, F.: The long-run impact of ICT. Empirical Econ. **37**, 497–515 (2009)

Yoo, Y., Henfridsson, O., Lyytinen, K.: The new organizing logic of digital innovation: an agenda for information systems research. J. Inf. Syst. Res. **21**(4), 724–735 (2010)

Reference Architecture for Project, Program and Portfolio Governance

Gonçalo Cordeiro, André Vasconcelos(✉), and Bruno Fragoso(✉)

INESC-ID, Instituto Superior Técnico, Lisbon University, Avenida Rovisco Pais 1, Lisbon, Portugal
{goncalo.cordeiro,andre.vasconcelos, bruno.fragoso}@tecnico.ulisboa.pt

Abstract. This paper presents a reference architecture on projects, programs and portfolios (PPP) governance models. Usually, organizations have projects organized into programs that in turn are organized into portfolios. Within organizations it is necessary to establish project, program and portfolio management policies, so that all the expected benefits of projects are achieved and the previously imposed constraints (time, costs, etc.) are met. It is therefore also important for organizations to have a governance model for projects, programs and portfolios. A governance model entails how decisions are made and how actions are carried out, considering organizational values. The required roles, responsibilities, and performance criteria should be an integral part of the governance model for projects, programs, and portfolios. However, given the permeability of projects, programs, and portfolios to several factors (e.g. regulatory factors and the project's project management maturity), several roles can be considered about governance models, thus promoting the existence of different project, program and portfolio governance models. This research proposes a reference architecture that allows the comparison of governance models to verify deviations and detect which model best suits the context of an organization.

Keywords: Projects · Programs · Portfolios · Project management · Governance · Reference architecture · Enterprise Architecture · Competences · Layers · Roles

1 Introduction

Nowadays, organizations live immersed in a constantly changing and extremely competitive market. This change and competitiveness influence the definition of the organizations' strategy. An organization's strategy, in turn, contributes to the continuous improvement of the business [20], supporting the development of organizations [15]. Projects must bring value to organizations as well as the expected return on investment. For this to happen, it is necessary to carry out a correct management of projects, programs and portfolios (PPP) and to follow a correct PPP governance model. Portfolio consists of structured knowledge about the programs and projects underway in the organization, views of costs, terms, scope, among others, are characteristics to which the portfolio has

J. Filipe et al. (Eds.): ICEIS 2020, LNBIP 417, pp. 804–832, 2021.
https://doi.org/10.1007/978-3-030-75418-1_37

access [17]. Programs aim to organize projects according to their dominant scope [16]. A project is a temporary effort made to create a product, service or result [15].

For the success of projects, programs and portfolios it is necessary a proper project, program and portfolio (PPP) management. "Project management performance explained 44,9% of the variance in project success", [21] and follow a correct PPP governance model, "approximately 6.3% of the variation of project success correlates with the stakeholder-orientation of the governance structure" [19].

The term governance in this investigation is considered as "a formal or informal model that determines how decisions are made and how actions are carried out, in the perspective of maintaining organizational values in face of changes, whether these are caused by problems, changes in actors or changes in environments", OECD definition [26].

"An effective PPP governance model ensures that the project portfolio is aligned with the organization's objectives (…) and aligns the interests of the directors, project team and other stakeholders" [2], "the governance of a project must influence the selection of the project management methodology and the way it evolves" [31].

However, for each company and its context, defining the best PPP governance model is really a problem.

This research is a extension of the paper "*Project, Program, Portfolio Governance Model Reference Architecture in the Classic Approach to Project Management*" [6] that leads with a governance model related to PPP management systems. We take a PPP governance model as regulations of PPP management, it can include policies, procedures, roles, responsibilities and other things [18]. Considering the probably of the environment affect projects, programs and portfolios, different roles can be considered when referring to PPP governance models.

To allow PPP governance help PPP management is essential that the roles associated with the PPP governance model has the competencies associated with them.

This research proposes a reference architecture that can be used to find comparisons between PPP governance models and help organizations find the best PPP governance model that best suits their needs.

Next section presents this research related work. In Sect. 3 the governance model reference architecture is proposed. The demonstration and evaluation are performed in Sects. 4 and 5, respectively. Section 6 presents the conclusions and the future work.

2 Related Work

This section introduces the main areas relevant for the development of this research.

2.1 PPP Governance

The concept of governance is understood in this research as "a formal or informal model that determines how decisions are made and how actions are taken, with a view to maintaining organizational values in face of change, whether caused by problems, changes in actors or changing environments"- OECD [13].

An effective PPP governance model "ensures that the project portfolio is aligned with the organization's objectives (...) and will align the interests of directors, project team and other stakeholders" [2]. The governance model gains importance regarding the alignment between projects and organizational objectives and the delivery of project results in an efficient and sustainable manner.

"Roles, responsibilities and performance criteria should be clearly defined in the PPP governance model" [2]. "Role" in an organization is defined within this research as a "function or something someone has" [28].

It is therefore essential that PPP governance is aligned with organizational governance, respecting its principles, decisions and processes [1].

Organizational Governance is a model that includes, "rules, processes, norms, relationships, systems and procedures" [1]. This model allows to steer an organization [18] by influencing how the organization's objectives are defined and achieved, how risk is monitored and evaluated, and hoe performance is optimized [1].

In the governance model present in ISO 21505 [18], four key roles and their responsibilities related to PPP governance and two related Organizational Governance are presented. The roles are identified in Table 1.

Table 1. Roles related to PPP governance and organizational governance [6].

PPP governance	Organizational governance
PPP governing body	Governing body
Project governing body	Organizational governing body
Program governing body	
Portfolio governing body	

2.2 PPP Management

Project management gains importance in organizations, enabling strategy implementation, business transformation and the development and continuous improvement of new products [21].

The "Rethinking Project Management (RPM) [34] research line contributed to "complement the classic approach to project management represented in PMBOK [1]" [29] and similarly in ISO 21500 [15], 21503 [16], 21504 [17] in PM2 [8] and APMBoK [2].

The line of research proposed by Sauer and Horner [35], aim to "highlight the criticisms of the project management knowledge base, (...) provide a set of possible research directions that can be followed in order to enrich and increase the project management knowledge base" [35]. The knowledge base that was criticized at the time is still used in many organizations and is considered in this research as a classical approach to project management, which is characterized by:

- Project and Project Management Lifecycles;
- Projects as Instrumental Processes;
- Focus on Product Creation;
- Restricted Project Conceptualization;
- Practitioners as Trained Technicians;

Knowing that there are several PPP management models, each with their own defined set of roles, it was decided to use as reference in this research the model presented in ISO's, namely ISO 21500 [15], for project management related roles, ISO 21503 [16], for program management related roles and ISO 21504 [17] for portfolio management roles, because the set of roles presented is referenced in different methodologies such as [1, 2], and is summarized in Table 2.

To help organizations compare their practices with the best ones, project management maturity models, "enable an organization to evaluate and compare its own practices with best practices and purpose improvements. A maturity model is a framework that describes the optimal progression using various levels until the desired improvement is achieved" [23]. Examples of project management maturity models are: i) the Organizational Project Management Maturity Model (OPM3) [9]; ii) the Capability Maturity Model Integration (CMMI) [30]; and iii) the IT Score for Program & Portfolio Management [12, 24].

Table 2. Roles related to PPP management [6].

ISO 21500 (Projects)	ISO 21503 (Programs)	ISO 21504 (Portfolio)
Project sponsor	Program sponsor	Portfolio manager
Project manager	Program manager	Portfolio management team
Project management team	Program management team	
Project team		

2.3 Organizational Layers

"Structuring an organization is important because it defines how tasks are divided and how organizational resources will be implemented" [7]. The organizational structure is defined as, (1) "the set of tasks assigned to individuals and departments"; (2) "dependency relationships within organizations, including lines of authority, decision-making responsibility, number of hierarchical levels, and the number of people that managers are responsible for controlling"; (3) and finally the "systems used by the organization to ensure coordination among departmental officials" [3].

For PPP governance and management, the PM2 methodology [8] creates a layered structure where it is possible to relate each role to its layer, taking into account the characteristics of both.

These layers are:

- Business Governing Layer – "Determines the vision and strategy for the entire organization. It consists of one or more management committees that operate at the level of the directors. This is where priorities are set, investment decisions are made and resources are allocated".
- Steering Layer – "Provides guidance and general project guidance, keeping the project focused on its goals".

- Directing Layer – "Sponsors the project and has its Business Case. Mobilize necessary resources and monitor project performance to achieve project objectives".
- Managing Layer – "It focuses on the daily operations of the project, organizing, monitoring and controlling the work to produce the desired deliverables and implement them in the organization. Managing Layer members report to Directing Layer".
- Performing Layer – "Performs the project work, producing deliveries and implementing them in the organization. Members belonging to the Performing Layer report to Managing Layer".

Other methodologies indirectly also present a layered structure. For example, the ISO suggests for governance a layered structure where there is a main Organizational Governance layer, which encompasses PPP Governance which in turn encompasses portfolios, programs and projects. For PPP management the ISO suggests a similar structure for projects, programs and portfolios, with a sponsorship layer, managing layer and developing layer (Fig. 1).

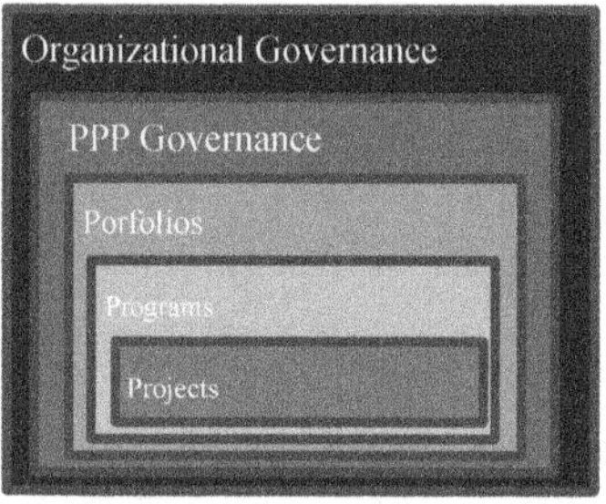

Fig. 1. Layered structure related to the PPP governance model present in ISO 21505 [5].

2.4 PPP Management Competences

The training of people involved in projects must be a concern of project managers as well as other aspects of human resource management practices related to projects that until then had been neglected, including the delegation of roles and responsibilities, the identification of informal communication networks and their risks and constant monitoring of the requirements required for the project and the capabilities of human resources [27].

IPMA ICB is a "global standard that defines the competences required of individuals working in the field of project, program and portfolio management" [14].

Regarding projects, programs and portfolios, the competencies in IPMA ICB are divided into three domains: i) "People competences", the personal or interpersonal skills required to successfully participate or lead a project, program or portfolio [14]; ii) "Practice competences", specific methods, tools and techniques used in projects, programs or portfolios to make them successful [14]; and iii) "Perspective Competences", specific methods, tools and techniques by which individuals interact with their surroundings and lead organizations to initiate and support projects, programs or portfolios [14] (Fig. 2).

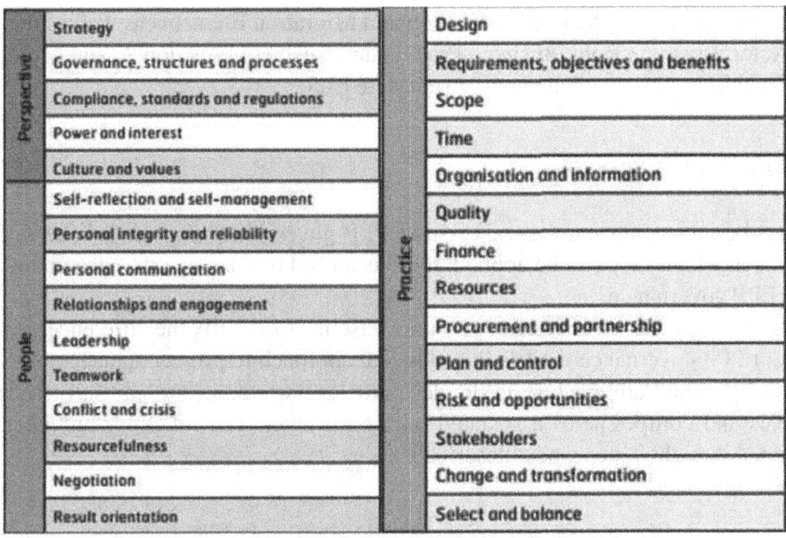

Fig. 2. The 3 areas of competence associated with project, program and portfolio management.

2.5 Reference Architecture of the PPP Governance Model

"A reference architecture captures the essence of other existing architectures, as well as their vision and evolving needs to provide guidance in developing new architectures" [4]. Reference architectures "allow to be effective, (…) discuss future modifications and extensions" [4], however this type of architecture must "be based on proven concepts" [4] and is considered a "standard method (…)" in line with best practice in the area" [4].

There are already investigations aimed at building a reference architecture of the PPP governance model, one of which is [33] that propose four key elements for improving project performance and therefore create value for organizations. These four elements are: (1) Portfolio Management; (2) Project Sponsor; (3) Project Management Office (PMO); (4) Project and Program Support. Also the research [25], with the aim of creating a framework that is useful for project governance, clarifies the various dimensions that the governance concept can acquire: 1) sovereignty, a popular dimension of governance often found in the general management literature, which describes the levels of external autonomy and internal control granted to projects by the governance system; 2) governance mechanisms, predominantly control oriented; and 3) the number of governance models as an indicator of the complexity of project governance implementation.

2.6 Enterprise Architecture

"An organization-level architecture is called Enterprise Architecture" [11], which by definition, according to Lankhorst, consists of "a coherent set of principles, methods and models used in the design and realization of the organizational structure of a company, business processes, information systems and infrastructure" [22]. ArchiMate [32], is a language that provides a uniform graphical representation of Enterprise Architecture [22].

This research uses the Implementation and Migration Elements as well as the Business Actor, Business Role, Business Process and Business Function, business elements and Capability, strategy element, of ArchiMate [32].

3 Solution Proposal

Due to the increasing number of existing PPP governance models and the different contexts in which they can be applied, there is a need to create a reference architecture to the PPP governance.

This architecture is expected to provide a baseline to identify the differences between different PPP governance models in a classical approach to project management.

The reference architecture is divided into layers, where one can have access to the roles and competencies associated with each layer. The reference architecture is developed thought 7 steps, next described.

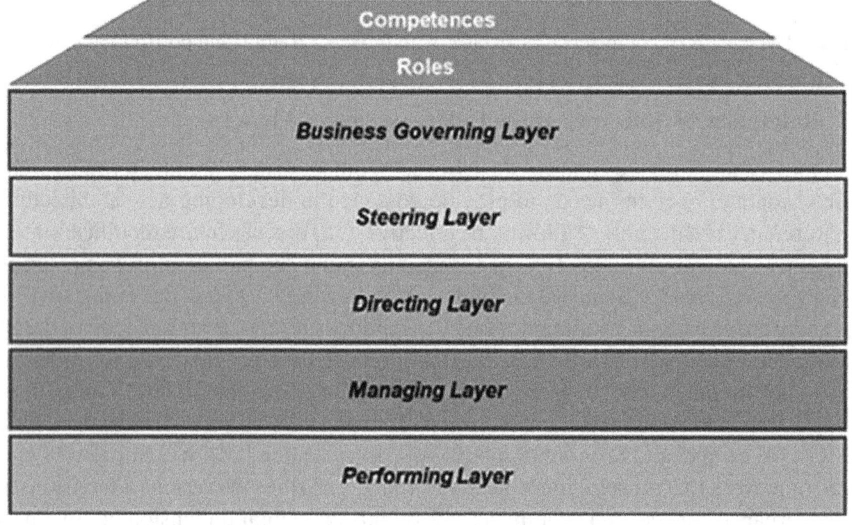

Fig. 3. Reference architecture layers [6].

3.1 Define Architecture Layers

The decision to divide the architecture into layers arises due to the fact that a structured architecture in layers "defines how tasks are divided and how the organization's resources will be implemented" [27], thus allowing a better understanding of the organization's command line as well as the specialization of the work.

In order to choose layers for the reference architecture, the different existing layers are analyzed. The PM2 methodology [8] introduces the division of the project management organization in layers that allow a clear relationship between the definition of functions and the roles associated with them and the level that they integrate within organizational structures and processes. More specifically the PM2 layers, allow a more organizational-related approach that integrates the Business Governing Layer while the other layers are more directed to the project domain, programs and portfolios – see Fig. 3.

3.2 Define Architecture Roles

Organizational governance, as the system by which organizations are directed and controlled [13], has in the roles associated with PPP governance and management processes the core activities carried out under the organizational governance system. These organizational roles, defined as "the roles played by individuals or groups in a project" [2], program or portfolio, are considered fundamental and essential in creating a reference architecture of the PPP governance model. An analysis of existing governance and management models is performed (Fig. 4).

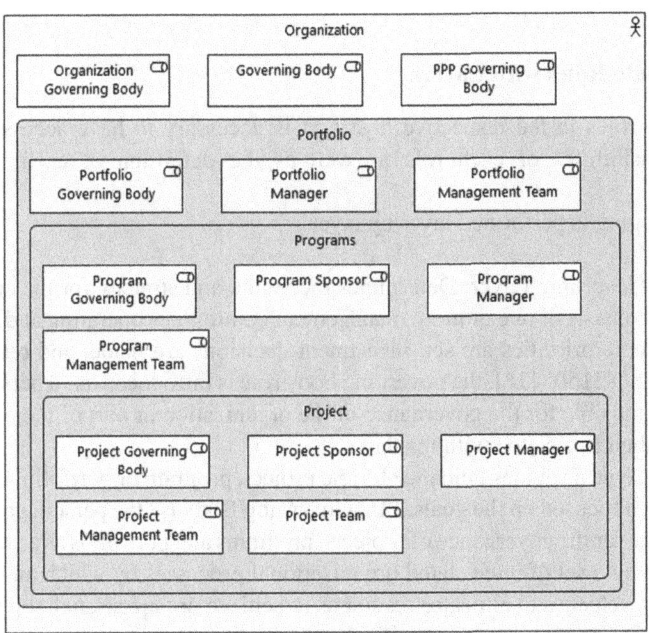

Fig. 4. ArchiMate representation of the framework of the roles associated with PPP governance within an organization [5].

As mentioned in [10], the roles present in ISO's 21500 [15], 21503 [16], 21504 [17] and 21505 [17], are used as a reference in other methodologies. These roles are presented in Sect. 2.2 for PPP and organizational governance and Sect. 2.3 for project, program and portfolio management, and are used in the reference architecture.

3.3 Identify Role Concerns

To identify the concerns of each role, the organization documentation is analyzed (where the roles are identified) and a table is filled, where it is possible to check which concerns and responsibilities associated with each role.

This approach provides an easy access to the concerns of each role, which is useful for the next steps (Fig. 5).

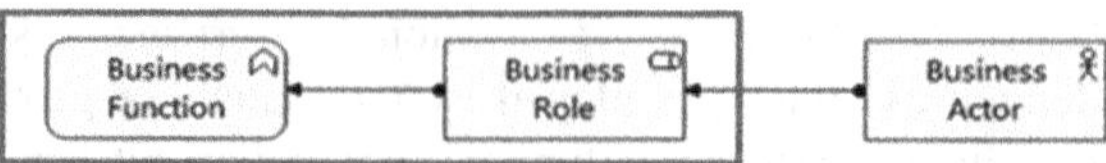

Fig. 5. View of role matching with their concerns identified in the ArchiMate Business Function viewpoint [6].

3.4 Associate Roles with Layers

To organize roles in the respective layers, it is necessary to have access to the concerns/responsibilities of each role as well as the definition of each layer of the architecture.

The mapping is performed layer by layer:

- Business Governing Layer: Determines the vision and strategy for the entire organization. It consists of one or more management committees operating at director level. This is where priorities are set, investment decisions are made, and resources allocated. In ISO 21505 [18], the governing body role is introduced as "a person, group or entity responsible for the governance of the organization or part of the organization" which is then associated with this layer.
- Steering Layer: Provides guidance for the project, program or portfolio, keeping each one of them focused on the goals. PPP Governing Body is "the person, group or entity responsible for the governance of projects, programs and portfolios as an organization, supported by a set of interrelated organizational processes by which an organization prioritizes, selects and allocates resources to achieve organizational objectives" [18]. An organization may have a specific governing body for the project ecosystem, the program ecosystem and the portfolio ecosystem.
- Directing Layer: Sponsors the project, program or portfolio. It mobilizes the necessary resources and monitors the performance of the project, program or portfolio to achieve its objectives. Project Sponsor is defined as "the person who authorizes the project, makes executive decisions and solves problems above the authority of the project manager" [15]. The Program Sponsor is defined as the person "responsible for the overall program strategy and its implementation defence" [16].
- Managing Layer: is focus on the daily operations of the project, program or portfolio, organizing, monitoring and controlling the work to produce the intended deliverables and implement them in the organization. Managing Layer members report to

Directing Layer. In this layer, the project manager is the person who "leads and manages the project activities and is responsible for its completion" [15], supported by the project management team. The program manager is the person "responsible for project performance. program and coordinating its components" [16], supported by the program management team. Finally the portfolio manager is the person "responsible for applying the portfolio management supported by the portfolio management team" [17].

- Performing Layer: responsible for the work of the project, program or portfolio, producing the deliverables and implementing them in the organization. Performing Layer members report to the Managing Layer. Associated with this layer is the Project Team role that "carries out the project activities" [15] (Fig. 6).

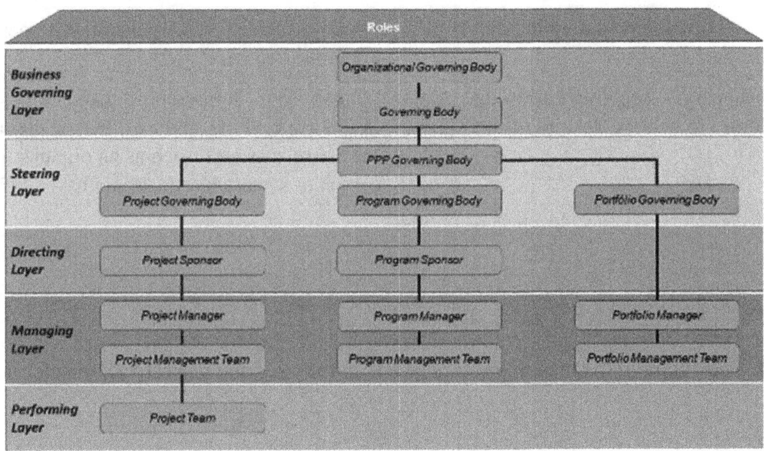

Fig. 6. Mapping the roles associated with PPP Governance in the respective layers [6].

3.5 Associate Competences with Roles

To associate which competencies are required for each role, it is essential to consider the competency indicators present in IPMA ICB, as well as the concerns of each of them.

Through a comparative analysis between the concerns of each role as well as the indicators of each competence, a mapping between the roles and the competences is performed (Appendix A).

3.6 Associate Competences with Layers

Through the mapping of roles in layers and the association of competencies with roles, a mapping of layered competencies.

For each layer of the architecture, the roles that belong are identified, after that the competencies of these same roles are also identified and finally the competencies are associated with these same layers.

3.7 Model in ArchiMate by Layer the Roles and Their Competencies

Through the mapping of roles in layers and the association of competencies with roles, a mapping of layered competencies. For each layer of the architecture, the roles that belong are identified, after that the competencies of these same roles are also identified and finally the competencies are associated with these same layers.

To correctly model in ArchiMate, it is necessary relate the terms "role" and "competence" with the elements present in ArchiMate, namely the elements "business role" and "capability" – see Table 3 (Fig. 7).

Table 3. Mapeamento entre termos da arquitetura de referência e elementos do ArchiMate [6].

Reference architecture concept	Element in ArchiMate
"Role" – "Function or something that someone has" [28])	"Business Role" – "Responsibility to perform a specific behavior" [32]
"Competence" – "applying knowledge, skills and abilities to achieve desired results" [14]	"Capability" - "represents a resource (knowledge, skills, abilities) that an element of the active structure, such as an organization, person or system has or should have" [32]

Fig. 7. Viewpoint of the mapping in ArchiMate of the roles and respective competences [6].

4 Demonstration

The demonstration of the use of the PPP governance methodology is performed in a Government owned company which for confidentiality reasons is named DemoCorp. DemoCorp is an organization with 740 employees, that has in the PMO (Project Management Office) the governance model that controls an ecosystem with portfolios, programs and projects, promoting organizational alignment and increasing the levels of efficiency and effectiveness in project development. DemoCorp is rated at maturity level 4 according to Gartner's IT Score for Program & Portfolio Management maturity model.

The demonstration is carried out step by step, through a process similar to the one described in the "Solution Proposal" section. The DemoCorp methodology is represented in the layers of the reference architecture supporting a comparative analysis between the methodology of DemoCorp and the PPP governance model used in the reference architecture.

With the demonstration one may conclude that the number of roles is less in the DemoCorp model compared to the reference architecture governance model. Upon further analysis the following findings are identified:

- In the business governing layer, there is one more role in the DemoCorp model than in the reference architecture's PPP governance model.
- There are also fewer roles in the project ecosystem than in the DemoCorp model. In the Managing Layer there is no project management team that assists the project manager which can cause work overload on the project manager. Steering Layer also doesn't have a Project Governing Body, with Project Management Officer's role being responsible for its functions.
- However, a project sponsor is added to realize the benefits of the project in the organization linking the project and its implementation in the organization.
- In both the program and portfolio ecosystems, a role in the Managing Layer is removed and there is no program management team or portfolio management team, which means that both program and portfolio managers may be overloaded. In these ecosystems there is no program governing body or portfolio governing body. Similar to the project ecosystem, its functions are performed by the project management officer (Fig. 8).

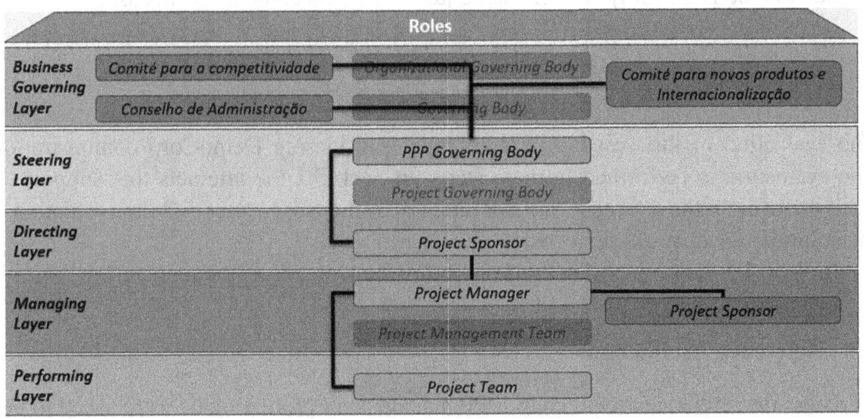

Fig. 8. Analysis of role deviations [5].

At the level of missing competencies or competencies added through the demonstration one can conclude that the number of competences is less in DemoCorp model compared to the reference architecture governance model (respectively 91 and 122 competences), see (Appendix B).

Performing a further analysis, one found that in the set of both governance models, 156 competences were identified, of which 64 are in accordance with both models, 27 are introduced by DemoCorp governance model and 65 only appear in the governance model of the reference architecture (Fig. 9).

Through the demonstration it is possible to verify that DemoCorp PPP governance methodology has some gaps, both in terms of roles and competencies associated with it. In terms of roles, there is an absence of teams that help portfolio managers, programs and projects, as well as a specific governing body for both projects, programs and portfolios.

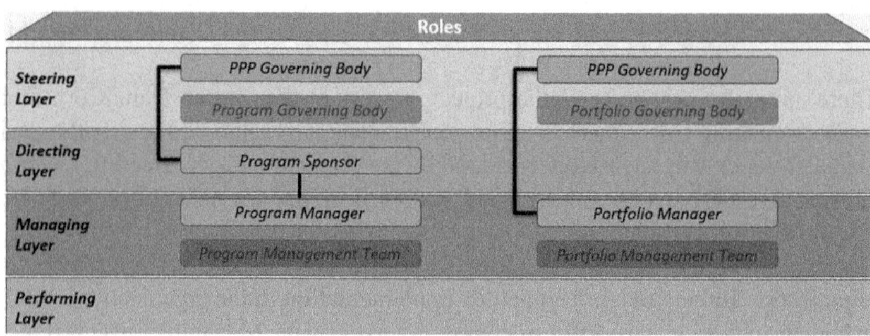

Fig. 9. Analysis of role deviations (program and portfolio ecosystem) [5].

Regarding competencies, it is noted that there are 65 competencies not considered in DemoCorp methodology, suggesting a revision of the methodology processes in order to decrease this figure.

It is important to notice that the addition of roles and the 27 skills that are only included in the DemoCorp model can be justified by its context and needs.

5 Evaluation

The evaluation of this research is performed considering DemoCorp demonstration. The evaluation is performed in three steps. In Sect. 5.1 the artefacts that support the solution, namely the reference architecture, are evaluated. In Sect. 5.2 the evaluation of the solution objectives is performed.

In Sect. 5.3 the proposed solution is compared with existing ones.

5.1 Reference Architecture

To assess the reference architecture, PPP management professionals were asked to perform a similar process to steps 3.4 - Associate Roles with Layers and 3.5 - Associate Competencies with Roles.

These two steps were chosen because they are the structural steps in the creation of the reference architecture.

The result of step 3.4 coincided 100% with the reference architecture and with the step 3.5 we concluded that of the total of 317 competences addressed by the reference architecture and by the respondent or both, 57.4% comply, i.e. according to both (reference and respondent architecture), 6.6% are only addressed by the reference architecture and 36% are only addressed by the respondent.

The fact that only a low percentage (6.6%) of competences is not addressed by the respondents is a good indicator for validating the association of skills with roles in the reference architecture (because it shows that most of the competences present in the reference architecture were identified by the respondents).

5.2 Solutions Benefits

The proposed solution provides a reference architecture that can serve as a basis for comparing PPP governance models. This reference architecture allows to verify the failing elements that may be influencing the functioning of the PPP governance model and also to find deviations that may be related to the context of the organization that holds the PPP governance model to be compared.

It facilitates the identification/understanding of failing elements in PPP governance models in a classical approach to project management. Through the demonstration in Sect. 4 we find that it is possible to perform an analysis of both competencies and roles associated with the PPP governance model.

We can see which roles and competencies have been added or are flawed in DemoCorp methodology based on the reference architecture presented in Sect. 3.

It highlights additional roles and relationships by comparison, allowing to explore role patterns or redundancy, better understanding the value offer of one methodology over another. By adding and removing roles (that the method allows to identify), patterns may be created considering the different contexts of project execution. An example of this is the addition of the project sponsor, and associated skills, working with the project manager to ensure that all project benefits are absorbed by the organization. This addition can be considered a necessary standard when dealing with a certain type of projects.

Another benefit of the proposed solution is on the verification on how the skills required for proper PPP governance are addressed when roles are added or removed. ArchiMate modelling makes it easier to analyse competency handling when roles are added or removed.

Finally, the proposed solution allows access to a reality that comprises only one project, or one project ecosystem and its associated methodologies, including programs and portfolios. The method allows comparing PPP governance models that encompass both a project-only ecosystem, an ecosystem with programs and projects, and an ecosystem with portfolios, programs and projects.

5.3 Related Work Comparison

5.3.1 Roles

In contrast to the studied PPP governance models, namely the APMBoK and PM2 models, which address an ecosystem where there are only projects, the PPP governance model used in the reference architecture, similar to the PMBOK model, besides supporting a project ecosystem, it also supports a program ecosystem and a portfolio ecosystem.

5.3.2 Competences

IPMA ICB, which is the main source for the competencies used in the reference architecture, presents the cross-referencing of the competences presented with ISO 21500 (from which the roles and their concerns/responsibilities associated with project management were taken) and with ISO 21504 (from which the roles and responsibilities/concerns associated with portfolio management were taken).

After analysing the result of this intersection with what is presented in the reference architecture, we find that with respect to the ISO 21500 the identified roles don't perform two competences ("Selfreflection and Self-management" and "negotiation") which are addressed in the intersection of ISO with IPMA ICB. With respect to ISO 21504, the roles identified therein perform all the competences addressed at the intersection of this ISO with the IPMA ICB. The reference architecture by also mapping the competencies of IPMA ICB with the roles taken from ISO 21503 and ISO 21505 also makes a possible intersection of these ISOs with IPMA ICB.

5.3.3 Architecture Levels

None of the studied researches present simultaneously the two levels present in the reference architecture or present only the level with the essential roles for the correct governance of PPP or present only the level with the competences.

Therefore, this architecture is innovative because it allows the roles to be associated with the skills they must perform.

5.3.4 Other Comparing

The reference architecture proposed may also be compared with maturity models since they the evaluation of PPP governance methodologies by assigning them a level of maturity.

All maturity models are concerned with PPP governance processes and with the creation the creation of a structure, that allows for this same governance. None of them deepens this same structure, namely the roles it should contain and the skills they must perform.

Therefore this research allows to compare the PPP governance structures considered essential in the maturity models, at a level of roles and competencies that it should contain.

6 Conclusions

Projects are important for organizations to reach their strategy and keep up with the competitive and changing market.

For projects to be successfully delivered, complying with estimated time, cost and resources, it is necessary that correct project management is followed, in an ecosystem with projects, programs and portfolios.

All the roles and responsibilities associated with correct PPP management must be established by the PPP governance model, which must follow the principles and values of the organizational governance model.

This research presents a reference architecture, this architecture makes possible to analyze two or more PPP governance models and find differences between them. For that, this research allows to check differences between models and choose the model that best suit with each organization scope.

The reference architecture consists in a layered architecture, when in each layer we can find the respective roles identified in ISO's 21500, 21503, 21504 and 21505. These documents are quite referred in so much other project management methodologies. In the reference architecture we can find too the competences related to project management divided in the architecture layers. This research adopts an Enterprise Architecture approach for modeling roles and the respective competences in the different ecosystems (portfolios, programs or projects), using ArchiMate modeling language.

For prove the useless of the reference architecture created, we use a DemoCorp and its governance model. In fact, when analyze the different governance models (Reference Architecture Governance Model VS DemoCorp Governance Model) we find differences between roles and competences.

6.1 Contributions

This research contributes to the scientific community in the different ways. The major contribution is the reference architecture created. This architecture is the first architecture that through the relation between the roles presented in ISO's and the competences put the competences organized in the project management layers. The layered architecture structure is another innovation that allows a better comprehension of the differences between PPP governance models and can help to choose the best model for each company, due its context.

While we work on the main objective that was a PPP reference architecture we contribute with other things, including:

- Highlighting new roles, new competences and exploring patterns or redundancy, better understanding the advantages among governance models.
- Crossing between IPMA ICB with ISO 21503 and ISO 21505. When analyzing the IPMA ICB, it was found that the Appendix resents the crossing of the competences presented with ISO 21500, from where the roles and their concerns/responsibilities associated with project management were taken and with ISO 21504 from where the roles and their concerns/responsibilities in relation to portfolio management. The fact that the reference architecture presents the mapping of the competences of the IPMA ICB with the roles taken from ISO 21503 and ISO 21505 makes a possible crossing of these ISOs with the IPMA ICB that had not yet been carried out.
- The capacity of verifying how the competences for proper PPP governance is addressed when roles are added or removed.

6.2 Limitations

During our research several limitations were identified, presenting new research opportunities to improve the reference architecture.

The limitations found are as follows:

I. the fact that one role may have much competences and responsibilities and may be associated with different layers;
II. when we map competencies to roles, through the IPMA ICB, we can find a set of indicators for the same competence. Considering it, the same role may be associated with different competences taking into account differ-ent indicators;

III. the attribution of competencies to roles, in most cases, is not something that is straightforward after analyzing concerns and responsibilities, bringing some degree of uncertainty to the approach proposed.

6.3 Future Work

This research and its limitations can allow some future work opportunities. This reference architecture can be used in different contexts with the goal of find and establish different patters of roles and competencies needed for each context. Due to the limitation of the mapping between roles and competences, a deeper investigation must be taken to link roles with competences in a stronger way.

Acknowledgments. This work was supported by national funds through Fundação para a Ciência e a Tecnologia (FCT) with reference UIDB/50021/2020 and by the European Commission program H2020 under the grant agreement 822404 (project QualiChain).

Appendix A

See Table 4.

Table 4. Mapping the competences associated with each role taking into account the indicators.

Role	Competence	Indicator
Global		
Organization Governing Body	Result Orientation	1, 2, 3
	Organisation and information	2, 4, 5
Governing Body	Strategy	2, 3
	Governance, structures and processes	1, 2, 3, 4
	Compliance, standards and regulations	1, 2, 3, 4, 5, 6
	Culture and values	1, 2, 3
	Personal integrity and reliability	3
	Leadership	1, 3, 5
	Teamwork	4
	Conflict and crisis	3
	Relationships and engagement	1
	Result Orientation	3
	Requirements, Objectives and Benefits	5, 6
	Scope	6
	Organisation and information	3, 4, 5
	Plan and Control	1, 5, 7
	Stakeholders	1, 2, 3, 4, 5
PPP Governing Body	Strategy	1
	Culture and values	1, 2, 3
	Compliance, standards and regulations	1, 2, 3, 4
	Governance, Structures ans Processes	1, 2, 3, 4
	Result Orientation	2
	Leadership	4
	Resourcefulness	5
	Personal integrity and reliability	1, 2, 3, 4, 5
	Personal Communication	1, 2, 3
	Finance	7
	Resources	2, 3, 4, 6
	Requirements, Objectives and Benefits	3, 7
	Scope	2, 4
	Plan and Control	7
	Organisation and information	1, 2, 3, 4, 5
	Quality	1
	Procurement	1, 2
	Risk and opportunities	1
	Stakeholders	1, 2, 3, 4, 5

(continued)

Table 4. (*continued*)

Role	Competence	Indicator
Projects		
Project Manager	Resul orientation	1, 4, 5
	Personal integrity and reliability	3, 4, 5, 1
	Personal Communication	1, 2, 3
	Leadership	1, 2, 3, 4, 5
	Teamwork	1, 2, 3, 4
	Conflict and crisis	1, 4
	Resourcefulness	4, 5
	Relationships and engagement	1, 2
	Governance, structures and processes	1
	Compliance, standards and regulations	1, 2, 3, 4, 5
	Power and interest	1, 2, 3
	Culture and values	1, 2, 3
	Requirements, Objectives and Benefits	1, 3, 5, 7
	Design	2, 3, 6, 7
	Scope	1, 3, 4, 5, 6
	Time	2, 3, 4, 5, 6
	Quality	1, 2, 5
	Finance	4
	Organisation and information	2
	Plan and Control	2, 7, 9, 10, 11, 12,13
Project Manager	Resources	1, 2, 3, 4, 5, 6
	Stakeholders	2
	Procurement	5
	Risk and opportunities	1, 2, 3, 4, 5
Project Management Team	Resul orientation	1, 4, 5
	Personal integrity and reliability	1, 3, 4, 5
	Personal Communication	1, 2, 3
	Leadership	1, 2, 3, 4, 5
	Teamwork	1, 2, 3, 4
	Conflict and crisis	1, 4
	Resourcefulness	4, 5
	Relationships and engagement	1, 2
	Governance, structures and processes	1, 4
	Compliance, standards and regulations	5
	Power and interest	1, 2, 3
	Culture and values	1, 2, 3
Project Management Team	Requirements, Objectives and Benefits	1, 3, 5, 7
	Design	2, 3, 6, 7
	Scope	1, 3, 4, 5, 6
	Time	2, 3, 4, 5, 6

(*continued*)

Table 4. (*continued*)

Role	Competence	Indicator
Projects		
Project Management Team	Quality	1, 2, 5
	Finance	[4]
	Plan and Control	2, 7, 9, 10, 11, 12,13
	Resources	1, 2, 3, 4, 5, 6
	Stakeholders	2
	Procurement	5
	Risk and opportunities	1, 2, 3, 4, 5
Project Team	Personal integrity and reliability	5
	Risk and opportunities	1, 2, 3, 4, 5
	Plan and Control	11, 12, 13
Project Sponsor	Result Orientation	1, 5
	Personal integrity and reliability	1, 3, 4
	Personal Communication	1, 3
	Relationships and engagement	1
	Leadership	5
	Conflict and crisis	1, 2, 3, 4
	Governance, structures and processes	4
	Strategy	2, 3
	Scope	3
	Requirements, Objectives and Benefits	2, 4, 6
Project Sponsor	Plan and Control	11, 12, 13
	Risk and opportunities	1, 2, 3, 4, 5
	Organisation and information	2
Project Governing Body	Strategy	1, 3
	Governance, structures and processes	4, 5
	Compliance, standards and regulations	1, 2, 3, 4, 5
	Relationships and engagement	2
	Leadership	1, 2, 3, 4, 5
	Requirements, Objectives and Benefits	5
	Organisation and information	2, 4, 5
	Quality	1, 2, 5, 6
	Resources	4
	Procurement	2
	Plan and Control	9
	Risk and opportunities	1, 2, 3, 4, 5
	Stakeholders	2
	Change and transformation	5

(*continued*)

Table 4. (*continued*)

Role	Competence	Indicator
Programs		
Program Sponsor	Strategy	1, 3
	Personal integrity and reliability	4
	Personal Communication	1, 2
	Relationships and engagement	5
	Leadership	5
	Teamwork	4
	Conflict and crisis	1, 2, 3, 4
	Result Orientation	1
	Design	5
	Requirements, Objectives and Benefits	5
	Organisation and information	2, 3
	Plan and Control	2, 11
	Stakeholders	1, 2, 3, 4, 5
Program Manager	Governance, structures and processes	2
	Power and interest	1, 2, 3
	Personal integrity and reliability	1, 3, 4
	Personal Communication	1, 2
	Relationships and engagement	3, 4
	Leadership	1, 3, 4, 5
	Teamwork	3, 4
	Conflict and crisis	1, 2, 3, 4
	Negotiation	5
Program Manager	Result Orientation	1
	Design	6, 7, 8
	Requirements, Objectives and Benefits	1, 2, 4, 5
	Organisation and information	1, 4
	Resources	2, 4, 5, 6
	Plan and Control	2, 5, 7, 8
	Finance	2
	Stakeholders	1, 2, 3, 4, 5
	Select and balance	1, 5, 7
Program Management Team	Governance, structures and processes	1, 2, 4
	Design	6, 7
	Requirements, Objectives and Benefits	3
Program Governing Body	Strategy	1, 3
	Governance, structures and processes	4, 5, 6, 7
	Compliance, standards and regulations	1, 2, 3, 4, 5, 6
	Culture and values	2, 3
	Personal integrity and reliability	1, 3, 4
	Personal Communication	1

(*continued*)

Table 4. (*continued*)

Role	Competence	Indicator
Programs		
Program Governing Body	Leadership	1, 2, 3, 4, 5
	Teamwork	4
	Result Orientation	1
	Design	4, 5
	Metas Objetivos e Benefícios	1, 5
	Scope	4
	Time	7
	Organisation and information	2
	Quality	3
	Finance	3, 4
	Resources	1, 5
	Plan and Control	1, 2, 5, 11
	Risk and opportunities	1, 4
	Change and transformation	1, 2, 4, 5, 6
	Select and balance	1, 2
Portfolios		
Portfolio Manager	Strategy	1, 2, 4, 5
	Governance, structures and processes	3, 5, 6, 7
	Compliance, standards and regulations	1, 2, 3, 4, 5
	Culture and values	2
	Leadership	5
	Conflict and crisis	2
Portfolio Manager	Negotiation	1, 4, 5
	Result Orientation	1
	Design	1, 7
	Requirements, Objectives and Benefits	1, 2, 3, 5
	Time	1, 4, 5, 6
	Organisation and information	1
	Resources	1, 3, 5
	Plan and Control	1, 3, 4, 5, 6, 8, 11
	Risk and opportunities	1, 4, 5
	Stakeholders	1, 3
	Change and transformation	1, 4, 5, 6
	Select and balance	1, 3, 5, 6, 7
Portfolio Management Team	Governance, structures and processes	4
Portfolio Governing Body	Strategy	1, 3
	Governance, structures and processes	3, 5, 6, 7
	Compliance, standards and regulations	1, 2, 3, 4, 5, 6
	Culture and values	2

(*continued*)

Table 4. (*continued*)

Role	Competence	Indicator
Portfolios		
Portfolio Governing Body	Personal integrity and reliability	1, 3, 4
	Personal Communication	1
	Leadership	1, 2, 3, 4, 5
	Teamwork	5
	Negotiation	4
	Result Orientation	1
	Design	4
	Requirements, Objectives and Benefits	1
	Organisation and information	2
	Quality	3
	Finance	8
	Resources	4, 5, 6
	Procurement	4
	Plan and Control	7
	Risk and opportunities	1, 4
	Select and balance	1, 2, 6, 7

Appendix B

See Table 5.

Table 5. Differences between the reference architecture and the model to be compared at the level of competencies by layer.

Competences/Layer		
Reference Archtecture	**Competence**	**Democorp Architecture**
	Global	
	Business Governing Layer	
X	Strategy	X
X	Governance, structures and processes	X
X	Compliance, standards and regulations	
X	Culture and values	
X	Personal integrity and reliability	
	Personal Communication	X
X	Relationships and engogement	
X	Leadership	X
X	Teamowork	
X	Conflict and Crisis	
X	Result orientation	
X	Requirements, objectives and benefits	X
X	Scope	
X	Organisation and information	X
	Resources	X
X	Plan and control	X
X	*Stakeholders*	
	Select and balance	X

(*continued*)

Table 5. (*continued*)

Competences/Layer		
Reference Archtecture	**Competence**	**Democorp Architecture**
	Global	
	Steering Layer	
X	Strategy	X
X	Governance, structures and processes	
X	Compliance, standards and regulations	
X	Culture and values	X
X	Personal integrity and reliability	X
X	Personal Communication	X
	Relationships and engogement	X
X	Negotiation	
X	Leadership	X
X	Teamowork	
X	Resourcefulness	
X	Result orientation	X
X	Requirements, objectives and benefits	
X	Scope	X
X	Resources	
X	Time	X
X	Organisation and information	X
X	Quality	
X	Finance	
X	Design	
X	Resources	X
X	*Procurement*	
X	Plan and control	
X	Risk and opportunitiess	
X	*Stakeholders*	
X	Change and transformation	
X	Select and balance	
	Projetos	
	Directing Layer	
X	Strategy	
X	Governance, structures and processes	X
	Compliance, standards and regulations	X
	Power and interest	X
	Culture and values	X
X	Personal integrity and reliability	X
X	Personal Communication	X
X	Relationships and engogement	X
X	Leadership	X
	Teamowork	X
X	Conflict and Crisis	X
X	Result orientation	
	Design	X
X	Requirements, objectives and benefits	X
X	Scope	
X	Organisation and information	X
	Finance	X
	Resources	X
	Procurement	X

(*continued*)

Table 5. (*continued*)

Competences/Layer		
Reference Archtecture	Competence	Democorp Architecture
	Projetos	
	Directing Layer	
X	Plan and control	X
X	Risk and opportunitiess	
	Stakeholders	X
	Change and transformation	X
	Managing Layer	
	Strategy	X
X	Governance, structures and processes	X
X	Compliance, standards and regulations	X
X	Power and interest	X
X	Culture and values	
X	Personal integrity and reliability	X
X	Personal Communication	X
X	Relationships and engogement	X
X	Leadership	X
X	Teamowork	X
X	Conflict and Crisis	X
X	Resourcefulness	
X	Result orientation	
X	Design	X
X	Requirements, objectives and benefits	X
X	Scope	X
X	Time	
	Organisation and information	X
X	Quality	X
X	Finance	X
X	Resources	X
X	*Procurement*	
X	Plan and control	X
X	Risk and opportunitiess	X
	Managing Layer	
X	*Stakeholders*	
	Change and transformation	X
	Performing Layer	
X	Personal integrity and reliability	X
X	Plan and control	X
X	Risk and opportunitiess	
	Programas	
	Directing Layer	
X	Strategy	
	Governance, structures and processes	X
X	Personal integrity and reliability	X
X	Personal Communication	X
X	Relationships and engogement	X
X	Leadership	
X	Teamowork	
X	Conflict and Crisis	
X	Result orientation	
X	Design	X
X	Requirements, objectives and benefits	

(*continued*)

Table 5. (*continued*)

Competences/Layer		
Reference Archtecture	Competence	Democorp Architecture
Programas		
Directing Layer		
	Time	X
X	Organisation and information	X
X	Plan and control	X
X	*Stakeholders*	
Managing Layer		
X	Governance, structures and processes	X
X	Power and interest	
X	Personal integrity and reliability	X
X	Personal Communication	X
X	Relationships and engogement	X
X	Leadership	
X	Teamowork	X
X	Conflict and Crisis	
X	Negotiation	
X	Result orientation	
X	Design	
X	Requirements, objectives and benefits	X
	Scope	X
	Time	X
X	Organisation and information	X
X	Finance	
X	Resources	
X	Plan and control	X
X	*Stakeholders*	
X	Select and balance	
X	Strategy	
X	Governance, structures and processes	X
X	Compliance, standards and regulations	
X	Culture and values	
	Personal integrity and reliability	X
	Personal Communication	X
	Relationships and engogement	X
X	Leadership	
	Teamowork	X
X	Conflict and Crisis	
X	Negotiation	
X	Result orientation	
X	Design	
X	Requirements, objectives and benefits	X
	Scope	X
X	Time	X
X	Organisation and information	X
	Finance	X
X	Resources	X
X	Plan and control	X
X	Risk and opportunitiess	
X	*Stakeholders*	
X	Change and transformation	
X	Select and balance	

References

1. (PMI), P. M. I.: Project Management Institute (PMI): The Standard for Portfolio Management - Fourth Edition, vol. 4, no. 3 (2017)
2. Association for Project Management. APM Body of Knowelge, 5th ed. (2006)
3. Child, J.: Organization: A Guide to Problems and Practice, 2nd edn. Sage, Thousand Oaks (1984)
4. Cloutier, R., Muller, G., Verma, D., Nilchiani, R., Hole, E., Bone, M.: The concept of reference architectures. Syst. Eng. **13**(1), 14–27 (2010). https://doi.org/10.1002/sys.20129
5. Cordeiro, G.: Arquitetura de Referência do Modelo de Governance de Projetos, Programas e Portfolios na Abordagem Clássica à Gestão de Projetos (2019)
6. Cordeiro, G., Vasconcelos, A., Fragoso, B.: Project, Program, Portfolio Governance Model Reference Architecture in the Classic Approach to Project Management (2020). https://doi.org/10.5220/0009155706190630
7. Daft, R.L.: The new era of management (2008)
8. Edition, O.: Project Management Methodology Guide (2016)
9. Foundation, K. (n.d.). Organizational Project Management Maturity Model (OPM3)
10. Fragoso, B., Vasconcelos, A., Borbinha, J.: On the roles of project, program and portfolio governance. In: Shishkov, B. (ed.) BMSD 2019. LNBIP, vol. 356, pp. 221–228. Springer, Cham (2019). https://doi.org/10.1007/978-3-030-24854-3_15
11. Garcia, I.: Alignment between Organization Projects and Strategic Objectives (Issue October). Instituto Superior Técnico (2017)
12. Gartner. IT Score for Program & Portfolio Management (2019)
13. Guria, A.: G20/OECD Principles of Corporate Governance. In: OECD Secretary-General (2015). www.oecd.org/daf/ca
14. Individual Competence Baseline for Project, Programme and Portfolio Management, (2018).
15. ISO 21500. Project, programme and portfolio management - Guidance on project management, 1 (2012)
16. ISO 21503. Project, programme and portfolio management - Guidance on programme management (2017)
17. ISO 21504. Project, programme and portfolio management - Guidance on portfolio management (2015)
18. ISO 21505 - Project, programme and portfolio management—Guidance on governance (2017)
19. Joslin, R., Müller, R.: The relationship between project governance and project success. Int. J. Proj. Manag. **34**(4), 613–626 (2016). https://doi.org/10.1016/j.ijproman.2016.01.008
20. Kaplan, R., Norton, D.: The Balanced Scorecard. HBS Press, Boston (1996)
21. KPMG. Programme Management Survey (2003)
22. Lankhorst, M.: Enterprise Architecture at Work. Springer, Heidelberg (2009). https://doi.org/10.1007/978-3-642-01310-2
23. Man, T.: A framework for the comparison of Maturity Models for Project-based Management, vol. 114 (2007)
24. Mieritz, L.: ITScore Overview for Program and Portfolio Management (2017). https://www.gartner.com/en/documents/2837917
25. Müller, R., Zhai, L., Wang, A.: ScienceDirect governance and governmentality in projects: profiles and relationships with success. Int. J. Proj. Manag. **35**(3), 378–392 (2017). https://doi.org/10.1016/j.ijproman.2017.01.007
26. OECD. Modernising government, the way forward (2005). www.oecd.org
27. Pournader, M., Akhavan, A., Baloh, P.: ScienceDirect a three-step design science approach to develop a novel human resource-planning framework in projects : the cases of construction projects in USA, Europe, and Iran. JPMA **33**(2), 419–434 (2015). https://doi.org/10.1016/j.ijproman.2014.06.009

28. Priberam. (n.d.). Significado de Papel. https://dicionario.priberam.org/papéis
29. Sauer, C., Horner, B.: Rethinking IT project management: evidence of a new mindset and its implications. Int. J. Proj. Manag. **27**(2), 182–193 (2009). https://doi.org/10.1016/j.ijproman.2008.08.003
30. SEI. CMMI ® for Development, Version 1.3 Improving processes for developing better products and services Software Engineering Process Management Program. November, vol. 468 (2010). https://www.sei.cmu.edu
31. Sirisomboonsuk, P., Ching, V., Qing, R., Burns, J.R.: ScienceDirect relationships between project governance and information technology governance and their impact on project performance. Int. J. Proj. Manag. **36**(2), 287–300 (2018). https://doi.org/10.1016/j.ijproman.2017.10.003
32. The Open Group. ArchiMate® 3.0.1 Specification (2017)
33. Too, E., Weaver, P.: The management of project management: a conceptual framework for project governance. Int. J. Proj. Manag. **32**(8), 1382–1394 (2014). https://doi.org/10.1016/j.ijproman.2013.07.006
34. Winter, M., Smith, C.: Final Report (2006)
35. Winter, M., Smith, C., Morris, P., Cicmil, S.: Directions for future research in project management: the main findings of a UK government-funded research network. Int. J. Proj. Manag. **24**(8), 638–649 (2006). https://doi.org/10.1016/j.ijproman.2006.08.009

Pre-modelled Flexibility for the Control-Flow of Business Processes: Requirements and Interaction with Users

Thomas Bauer(✉)

Hochschule Neu-Ulm, Wileystr. 1, 89231 Neu-Ulm, Germany
thomas.bauer@hnu.de

Abstract. At process-aware information systems (PAIS), it is sometimes necessary to deviate from the predefined process. Otherwise the users are restricted too much. This paper presents an approach that allows to pre-model predictable flexibility already at build-time. An advantage, compared to completely dynamic changes at run-time, is that the effort for the end users necessary to trigger a deviation is reduced significantly. Furthermore, process safety is increased since, for instance, it can be predefined which users are allowed to perform which modifications. The corresponding requirements for the control-flow perspective are presented in this paper, with a special focus on the kind of information that shall be predefined at build-time. Examples from practice are presented in order to illustrate the necessity of the requirements. Furthermore, the interaction with the users is explained in order to show that triggering a flexible deviation causes only little effort at run-time.

Keywords: Process modelling · Process execution · Process engine · Pre-modelled flexibility · Flexibility by design · User interaction

1 Introduction

For enterprises, business processes (BP) are an important topic. They are also covered in scientific literature, but often only the aspects modelling, optimization, and simulation of BP are respected. In addition to these aspects, the automatic control of BP by process management systems (PMS) offers many advantages. Such an active control results in process-aware information systems (PAIS) [1] that guarantee that a process is executed exactly as defined at build-time (i.e. process reliability). Additionally, end users are unburdened from non-productive tasks as searching for the right function of the application or the data required in the current process step. PAIS perform such actions automatically. But they also have disadvantages: Some users dislike the reduced freedom that results from the active process control by the PMS. Additionally, in exceptional cases, restricting the possible execution orders of the process activities may result in situations where orders are not possible which would be advantageous for the business. In order to avoid such disadvantages, there must exist the flexibility to vary from the rigidly designed BP [2–4].

J. Filipe et al. (Eds.): ICEIS 2020, LNBIP 417, pp. 833–857, 2021.
https://doi.org/10.1007/978-3-030-75418-1_38

A special case of flexibility are predictable deviations, which are pre-modelled already at build-time in order to apply them later on at run-time of the process instances (Pre-Designed Flexibility [5], Flexibility by Design [3]). In scientific literature, however, this categorization is only mentioned. Details of the corresponding requirements and approaches for their realization are hardly content of existing research.

The project CoPMoF (Controllable Pre-Modelled Flexibility) addresses this aspect. The flexibility of PMS shall be increased, but deviations shall not be defined arbitrary (i.e. completely dynamic) by the end users. Instead, already at build-time, it is pre-modelled which predictable flexibility is required at run-time. This allows the BP-designer and the BP-owner (the responsible person) to evaluate the consequences of these possible deviations. In addition, process reliability is guaranteed since only intended deviations are possible and only users with the required rights are allowed to trigger deviations.

The main advantage, however, is that performing a deviation causes less effort for the end user compared to a dynamic change. Eventually, it would be even too complicated to define a change dynamically. Assume for a telephone enquiry that it fails for a specific customer. Then, one possible solution is to insert an activity "enquiry by mail" dynamically into the BP. At this solution, the end user of the PAIS has to define all the specifications described in the following: The location of the new activity within the control-flow has to be defined (i.e. the preceding and successive activities). Furthermore, the data-flow has to be specified; i.e., the mapping of the input and output parameters of the activity to BP-variables. As an example, the field Street of the input parameter Address shall get its content from the attribute CustomerStreet of the BP-variable CustomerAddress. In addition, it is necessary to define an appropriate actor assignment for this activity; e.g. "role = credit-approver and department = x", where x is read from the attribute ExecutingUnit of the BP-variable CreditApplication.

Such dynamic changes [1] allow inserting new activities into a process instance. Furthermore, they can be deleted and moved. This functionality is indispensable for not predictable modifications. As explained, they are not appropriate for predictable exceptional situations, since dynamic changes cause much effort for the users at run-time. For a predictable deviation, it is better to pre-model the required flexibility only once at build-time.

This paper is an extended version of the conference paper [6] published at ICEIS 2020. Both publications describe BP of different domains in order to explain scenarios of predictable flexibility. Thereby, many requirements and their variants are explained, in order to present the scenarios in an exhaustive and understandable manner. As an extension compared to [6], this paper describes the interaction with end users that becomes necessary when he or she[1] uses a pre-modelled deviation at run-time. This is a very important topic since main advantages of CoPMoF are that changes cause only little effort and are easy to handle for the end users. In the following, solely the control-flow perspective is respected (for other process perspectives see [7]). The development of detailed realization concepts (e.g. execution semantics for a process engine), prototypical realizations, and case studies based on such prototypes are future work.

The following section discusses related work and explains the research gap and research question. In Sect. 3, several terms are introduced, principles of PAIS are

[1] In order to improve readability, only the male form is used in the following.

ex-plained, and challenges are demonstrated at an example process from practice. Section 4 presents the requirements, corresponding practical examples, and describes the interaction with the users. The paper concludes with a summary and an outlook.

2 State of the Art and Resulting Research Gap

In this section, scientific literature is analyzed that concerns pre-modelling flexibility for BP. This analysis is used to demonstrate the research gap. Afterwards, a research question is presented that has the goal to reduce this research gap.

2.1 Related Work

[5] distinguishes different types of flexibility for BP. The category that corresponds to CoPMoF is called "Pre-Designed Flexibility". The categories are refined in [3] with the resulting categories "Flexibility by Design" and "Flexibility by Underspecification". [8] defines categories of dynamic changes and schema evolution. [4] distinguishes between flexibility at build-time and flexibility at run-time. The approach for the first category, however, is to use appropriate verifications and tests in order to achieve that changed process templates become executable as soon as possible. The papers do not discuss requirements for pre-modelled flexibility or appropriate interactions with users, as presented for CoPMoF in this paper.

A literature review[2] has shown that, until now, it was hardly examined what shall be pre-modelled at build-time in order to reach much flexibility and low effort for the end users at run-time. That means, no scientific work explicitly handles the topic of pre-modeling flexibility. Instead, flexibility papers in the BP domain [1] concern topics as dynamic changes (e.g. ADEPT$_{flex}$ [9], Breeze [10], Wasa [11], Spade [12]), schema evolution and its propagation to running process instances (e.g. ADEPT2 [13], Breeze [10], MOKASSIN [14], TRAM [15], WASA2 [16], WIDE [17]), or managing BP variants (e.g. [18, 19], ADOM [20], C-YAWL [21], Provop [22]). Therefore, the following describes approaches that handle topics similar to pre-modeled flexibility in a wide sense, and approaches that can be used to realize some of the presented requirements.

An approach, that may be used to pre-model special cases, is exception handling based on events and exception handlers [1, 23]. Thereby, for a single activity or a whole process region, an event is modelled. If it occurs at run-time (throw), an exception handler is executed (catch). This is the same technique as for try-catch-blocks in programming languages. It is well suited to handle technical errors; e.g., the crash of an activity program. The same technique can be used to handle business exceptions during process execution; e.g., alternative activities (cf. CF-2 presented in Sect. 4) may be modelled this way [1, 23]. Embedding events and exception handlers into the process model, however, results in a more complex process graph. Such a graph may be too complicated for many BP-designers and business users since they typically do not have the required IT

[2] The search was performed with the following terms, all in combination with business process: flexibility by design, pre-designed flexibility, pre-modelled flexibility, flexibility build-time, flexibility control flow. Furthermore, [1] as an "overview book for flexibility in BP" was examined with respect to hints to relevant approaches.

background. But it is possible to use events and exception handlers as workaround: The requirements presented in Sect. 4 may be automatically mapped to such building blocks of an existing BP language (e.g. BPMN) in order to enable a corresponding process engine to execute such processes.

In [24] it is described how pre-modelled jumps may be mapped to the regular building blocks of the ADEPT meta model. This covers the requirements CF-3a and CF-3b (forward and backward jumps), but without parallelism.

Complex control-flow patterns offer a special type of flexibility. [1] describes several patterns that enable many execution orders and, therefore, offer a type of pre-modelled flexibility. In addition to modelling of BP, the aspects execution semantics and verification (i.e. checking correctness of a process model) are respected in [1]. But the requirements that concern flexibility by design are not presented comprehensively. [25] describes further control-flow patterns. This work, however, does not have the focus on requirements for pre-modeled flexibility as well. "Pattern of predefined change" are presented in [26]. They allow to pre-model that specific decisions shall only be made at run-time. This includes Multi-Instance-Parallelism (cf. CF-4).

[27] defines additional quality of service goals for BP (e.g. process execution time, costs). At this approach, the PMS can automatically deviate from the standard process, if necessary in exceptional cases, in order to reach these goals. Three types of "flexible elements" are offered as basis for this flexibility: alternative activities (cf. CF-2), non-vital (i.e. optional) activities (CF-1), and optional execution orders (they shall be respected but also parallel execution is allowed; this is similar to CF-6e).

[2] offers a special kind of pre-modelled flexibility. The execution order of a BP is defined by the processing order of its business objects (data). Different types of business objects interact with each other and their signals define the resulting execution order. Optional signals, that are triggered by a user, enable flexibility. These "dynamic signal types" enable pre-modelled flexibility and, therefore, cover some CoPMoF requirements: An activity may be "delegated" to a different (alternative) activity type (cf. CF-2). Furthermore, additional (optional) activity instances and sub-processes (CF-1) can be created. Their type is pre-modelled and this action is only allowed in predefined process states (cf. process regions).

A goal of CoPMoF is to reach flexibility while retaining a process structure; i.e., a process graph is modelled. The following approaches have a different goal: a much higher degree of flexibility. At [28] only process fragments are (pre-)modelled, not the whole BP. Constraints (i.e. rules, conditions) define which fragments shall be used, which dependencies exist between these fragments, and when a process instance is finished. Using this information, at run-time, the user may create process instances (manually) that fulfil his needs. Case Handling [29] is an approach for knowledge intensive BP, with the focus on data. The state of a process instance results from the content of its data objects. They determine the activities that are currently executable; i.e., the control-flow is not modelled explicitly. The users decide (autonomously) to execute, skip, or repeat activities (cf. CF-7).

In literature, there are several approaches that are constraint-based (e.g. [30–32], Freeflow [33], Tucupi [34]) or rule-/goal-based [35]. They have in common that the control-flow is not modelled as a graph. Instead, constraints (rules) are defined which restrict the set of allowed execution orders. That means, all executions orders are allowed that do not violate a constraint. If only a few constraints are defined, a large number of execution orders are possible. This way, it is possible to reach high flexibility with little effort. An overview of such approaches and their principles can be found in [1]. If a constraint-based approach is used, alternative execution paths are not modelled explicitly as special cases. Therefore, they cannot be distinguished from normal execution paths. Furthermore, there does not exist a graphical representation of the process structure. Thus, constraint-based approaches may be not suited well for many domains and BP-designers since they do not possess the required IT skills. Additionally, even business users (who may not have any IT skills at all) shall be able to understand, discuss, and improve BP models. For instance, [36] present a case study where this is very important, with the conclusion that corresponding building blocks (e.g. optional activities, cf. CF-1) shall be offered by a graphical process modelling language.

As explained, there does not exist literature that handles pre-modelled flexibility explicitly. Therefore, the topic of user interaction in such cases was not respected in research as well.

2.2 Research Question

Scientific literature only defines the category pre-designed flexibility, but there exists no work concerning details. That means, the required functionality is not examined in a detailed and comprehensive way until now. As a consequence, there exists no literature concerning an appropriate interaction with the users when pre-modelling or using such functions. Therefore, no answer to the following research question exists: At which scenarios is it advantageous to pre-model flexibility for the control-flow perspective of a BP at build-time, which requirements exist, which information shall be provided for this purpose, and which is the best way to interact with the users? In CoPMoF, an approach with the following properties is developed:

- The requirements shall cover as many scenarios as possible. However, because of the research design, completeness cannot be reached. In order to identify a large number of requirements, several BP are analyzed with respect to their flexibility requirements. Some of these BP are known by the author because of his long-term work in industry and research. In addition, generally known BP and processes described in scientific literature (e.g. credit application) were respected.
- Following the presented approach, BP templates are enriched with pre-modelled flexibility. Thereby, they shall stay well-understandable for BP-designers and "normal users". This is especially necessary for semantic process models (the business view), but also for technical models (the process implementation); e.g., to enable users to detect errors in the process models.
- Despite the desired simplicity, the building blocks for pre-modelled flexibility must have a clear execution semantics, since a vague modelling technique does not allow automatic execution of process instances by a process engine.
- Finally, triggering a flexible deviation during process execution shall cause only very little effort for the end users.

This paper presents scenarios and requirements for pre-modelled flexibility. A large part was already published in [6], but in the following additional requirements and the completely new aspect of unplanned user actions are presented as well. Furthermore, in each subsection of Sect. 4, a completely new topic is handled: the appropriate interaction with end users when using pre-modelled flexibility at run-time and with BP-designers when defining such flexibility.

3 Basics and Challenges

Section 3.1 describes the basics of PMS. In Sect. 3.2, several problem statements for pre-modelled flexibility are explained at an example scenario from practice.

3.1 Business Process Management

PMS consist of a build-time and a run-time component. At build-time, a process template is designed that describes the BP. For this purpose, a process graph is modelled that contains activities (nodes). Their execution order is determined by edges and conditions. This process template is used at run-time to create process instances. A process engine controls the execution of these process instances. For each currently executable activity instance (often named short: activity) it inserts a work-item into the worklists of the potential actors. One of these end users selects the work-item and performs this activity (instance). Such an activity execution is often performed by filling a form.

The process perspective control-flow defines the execution order with a process graph (cf. Figure 1). Its nodes represent activities (human tasks performed by users) or automatically executed program code (e.g. service calls). In addition, the process graph contains gateways (Split- and Join-Nodes). In literature [25] many control-flow patterns are described. Commercial PMS typically offer Split- and Join-Nodes with XOR-(one branch is chosen based on a rule), OR-(several branches), and AND-Semantics (all branches are executed). Loops are typically supported as well. Branches and loops represent a simple form of pre-modelled flexibility, since the set of executed activities and their execution order may differ at each process instance. Additionally, some PMS allow to define a variable number of identical parallel branches. This number must be determined at least when starting this "Multi-Instance-Parallelism". This corresponds to the control-flow pattern "Multiple Instances with a priori Run-Time Knowledge" [25].

3.2 Pre-modelled Flexibility

This subsection demonstrates the need for pre-modelled flexibility at an example process from practice. As already mentioned in Sect. 1, completely dynamic changes are not in the focus of this paper. Instead, predictable exceptional cases are inspected; i.e., an appropriate behavior of the BP can be pre-modelled already at build-time.

Figure 1 shows a simplified Change Management Process (CMP) as used to request product changes in the automotive domain. The notation is similar to BPMN 2.0 (but extended). With Act. A, an arbitrary employee of the automobile manufacturer can request a change of a vehicle part (e.g. the shape of the engine bonnet). Since an execution

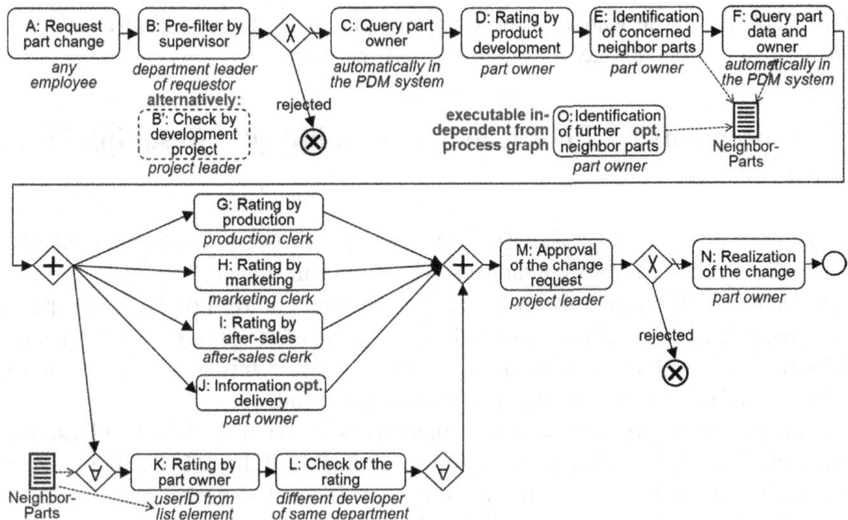

Fig. 1. Change management process (CMP) for product modifications, similar in [6].

of the CMP causes much effort, it can be stopped with Act. B resp. B' by a manager. Act. C automatically determines the owner of the concerned part with a query to the product data management (PDM) system. In Act. D this owner rates the effort and the benefits of the change from the viewpoint of the development domain. In Act. E he identifies neighbor parts (e.g. car wing, radiator) that must be changed additionally because of the modified shape of the engine bonnet. Act. F queries the corresponding part details and part owners and stores these data in the list NeighborParts.

At the activities G to I, clerks of several domains rate (in parallel) whether the change can be realized and they estimate the resulting costs. Act. J enables the part owner to provide additional information to these clerks (for instance, because of a request by phone). Act. K is the rating from the viewpoint of the neighbor parts by the respective part owners. This activity is instantiated multiple times (one time for each neighbor part). The same applies to the check of the rating by another developer in Act. L. Act. M decides on the approval of the change request. The parts may be changed in Act. N.

At the execution of a CMP, flexibility is required at several points: Act. B' is an alternative to Act. B and it is used if Act. B is not appropriate in this case. This may happen, for instance, if the department leader does not have sufficient technical competences for this decision. Act. J is optional (flag opt.), therefore, it appears in the worklist of the part owner with a corresponding label. He decides whether additional information is required, or whether he wants to omit this activity. The activities K and L are part of a Multi-Instance-Parallelism. Thus, the ∀-Split creates a number of branches that corresponds to the length of its input list NeighborParts. This list was filled by Act. E and Act. F. It is also possible to extend it later with Act. O. This activity is independent of the process graph. Therefore, it can be executed at an arbitrary point in time. Its execution, however, does only make sense before the Multi-Instance-Parallelism is finished (i.e.

before the ∀-Join). Afterwards, an additional neighbor part cannot result in an additional branch (of Act. K and L); i.e. it is not respected any more.

4 Requirements for Pre-modelled Flexibility and Resulting User Interaction

This section explains the approach of CoPMoF; i.e., which aspects can be pre-modelled at build-time to achieve flexibility at run-time without causing much effort for the end users. As mentioned, in this paper only the process perspective control-flow (CF) is respected. Other process perspectives are described in [7]. Some of the presented scenarios have similarities with well-known control-flow patterns [25]. In this section, however, they are discussed from the viewpoint of pre-modelling flexibility.

The requirements were explained and their necessity is demonstrated with examples from practice for all identified aspects of pre-modelled flexibility. Most of these requirements were already described in [6], but for jumps (Sect. 4.3) further requirements are presented and user actions (Sect. 4.7) are a completely new aspect. Compared to [6], however, the main extension is that the interaction with the users is described for all aspects. In many cases, this concerns the interaction with the end user when he uses the pre-modelled flexibility. For some aspects, the interaction with the BP-designer when he pre-models flexibility is relevant in addition. Especially the first topic is very important because one of the main advantages of CoPMoF is that using pre-modelled flexibility at run-time is easy, compared to completely dynamic changes.

4.1 Optional Activities

An optional activity can be executed, but it is not necessary to execute it at all process instances. Normally the end users decide, whether it shall be executed, or whether it is not relevant for this process instance. An optional activity is displayed in the worklists of its potential actors. Then, these end users may start this activity. The worklist-item contains an additional label that indicates that this activity is optional. Furthermore, there exists a possibility to omit this activity; e.g. by using a button. An optional activity may also be a composed activity; i.e., a whole sub-process can be omitted this way.

Requirements. It shall be possible to predefine at build-time that a specific activity is optional (cf. Act. J in Fig. 1); e.g. by setting a flag. It is even possible that an optional activity shall be skipped automatically (see below: CF-1b); i.e., without a user omitting it explicitly. It must be possible to define such a behavior as well.

CF-1a: The normal case for optional activities is that they are executed in fact. In order to omit an optional activity, an end user has to perform an action actively. For instance, at the CMP depicted in Fig. 1, Act. J is offered to the part owner until he executes it or omits it explicitly. This is meaningful since the provided information may be used by each of the actors of the activities G to I. For other scenarios, a different behavior is more appropriate:

CF-1b: At Fig. 2, the original part description is created in Act. A. It is possible to modify (typically slightly) this description with the optional Act. C, if this is necessary because of the design decisions made in Act. B. Act. E transmits this part description to the BoM system. But Act. E does not wait for the execution or omission of Act. C since the new BoM data are urgently required by other BP and delays are not acceptable. After completion of the lower branch (i.e. Act. E is finished) the execution of Act. C is no longer meaningful since its output data are only used by Act. E. Then, Act. C shall be omitted by the process engine automatically; i.e., Act. E has the role of a "milestone" for Act. C. It must be possible to pre-model such a behavior at build-time as well.

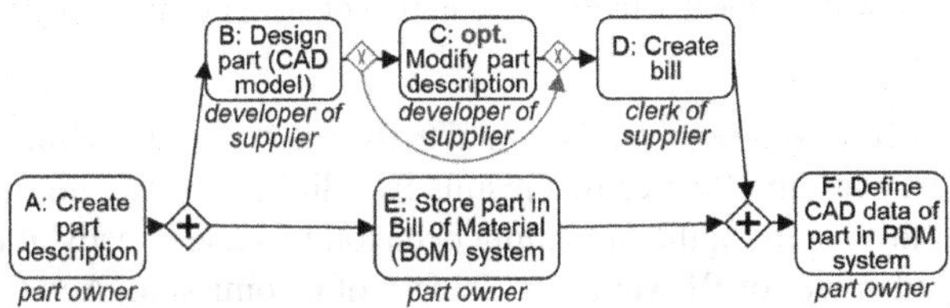

Fig. 2. Development of a part by a supplier, similar in [6].

User Interaction. One main advantage of pre-modelled flexibility is that it is easy to use for the end users (compared to dynamic changes). In order to explain this fact, in the following, the interaction of the PMS with the end user is explained. At run-time, an optional activity is offered to an end user in his worklist similar to a normal activity (cf. the first worklist-item in Fig. 3a). Since it is an optional activity, it is especially marked; e.g. with a red label "opt.". Additionally, a pre-modelled hint may be displayed, that explains the cases when the activity shall be executed and the cases when it shall be omitted. To keep the worklist clear, this hint may only appear at "mouse-over" (cf. Figure 3a). The end user has the possibility to perform this activity by using the button "Start". In addition, he has the possibility to skip this activity with the button "Skip". That means, skipping the activity (i.e. using the flexibility) causes only minimal effort for end users.

For CF-1a, the BP-designer only has to mark the activity with a flag in order to define it as optional. Then, the process engine possesses all information required at run-time. For CF-1b, additional information is required: A milestone has to be defined to describe the point in the process until which using the optional activity is allowed. To keep modelling simple and clear, the process modelling tool shall allow to define such milestones graphically. Figure 3b shows a possible realization: The optional activity is marked with a red X. The milestone activities (here only Act. E) are marked with a blue arrow pointing to the left side (only during configuring the optional activity). This arrow indicates that the optional Act. C may be only executed before the milestone Act. E is finished. To keep this graph visualization clear, only the relevant part of the graph may be visualized [37]. For instance, the activities proceeding the optional activity cannot be used as milestones and are not part of the depicted graph.

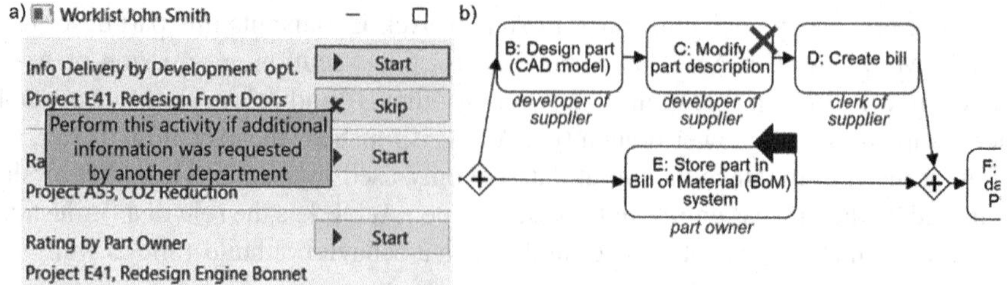

Fig. 3. a) Worklist with optional activity b) Definition of an optional activity with milestone.

Realization with XOR-Gateways. Optional activities cannot be realized with normal XOR-Nodes[3] (grey in Fig. 2) since this results in a different (resp. no) user interaction for CF-1a: With this solution, the branching decision is made already at the XOR-Split based on rules and values of BP-variables. In case of its omission, Act. C never appears in the worklists of the potential actors. In the other case, it is not possible to skip it at all. The reason for this problem is that the branching decision is made too early. For the same reason, CF-1b cannot be realized this way as well: The decision whether Act. C shall be offered to the users is made already at the XOR-Split. This is too early because the completion of Act. E occurs later. That means, a workaround with XOR-Nodes does not result in the desired behavior; i.e., optional activities must be offered as separate building block for the control-flow of BP.

4.2 Alternative Activities

A scenario with alternative activities was already presented for the CMP in Sect. 3.2: A department leader may not be able to execute Act. B. Therefore, an alternative Act. B' is executed (that may use a different form and is performed by a different person). In such a case, the PMS inserts the standard activity into the worklists of its potential actors. The resulting worklist-items contain a label that indicates that an alternative activity exists. The user has the possibility to switch to the alternative by performing an active action (e.g. using a button, cf. Figure 4).

Requirements. At build-time, the BP-designer has to define for each alternative activity, which of the following types shall be used; i.e., when and how it is decided to switch to the alternative activity.

CF-2a: The user decides to switch the alternative Act. Y before he reserves (resp. starts) the regular activity X. This is the standard and simplest case.

CF-2b: The user can decide to switch to the alternative Act. Y even after he started Act. X (i.e. while he executes Act. X). Then, Act. X is aborted automatically and the alternative Act. Y is inserted into the worklists of its potential actors.

[3] A realization would be possible with a XOR-Split with "Deferred Choice" semantics [25]. This type of XOR, however, is typically not supported by commercial PMS. As an alternative, in BPMN, a two-way event-based XOR-Split may be modelled in combination with intermediate throw and catch events. Defining such a sophisticated BP graph, however, may overwhelm "normal" BP-designers.

CF-2c: It is possible that any user detects after completion of Act. X (and maybe its successors) that the alternative activity would have been the better choice. For such cases it may be allowed that the alternative Act. Y is executed in addition to the regular Act. X. This is meaningful, for instance, if an Act. X captures data and an alternative Act. Y captures more or different data. Now assume that someone realizes, at a successive activity, that the output data of the alternative Act. Y is required. Then it is executed in addition. That means the originally captured process data is complemented.

CF-2d: It may be even allowed that the process engine switches to an alternative activity automatically. This is meaningful if the execution of the regular activity fails. This situation may be caused by a failed service call performed by an automatically executed activity. At manually executed activities (human tasks), post-conditions may be used to identify failures; e.g., missing or inconsistent output data may be detected.

For each activity with alternatives, it must be possible to pre-model at build-time, who has the right to switch to the alternative activity. In many cases, a (potential) actor of the regular activity is allowed to make this decision. However, this set of persons may be restricted; e.g., to (a few especially competent) persons who possess a special role. Furthermore, this set may be extended; e.g., if actors of succeeding activities may detect that the alternative is the better choice (cf. the example for CF-2c).

An alternative activity may be a composed activity as well. Then, a switch to an alternative sub-process (instead to a single activity. is performed. In such cases, this decision does not only concern one actor, but all actors performing activities of this sub-process. It may be necessary, therefore, that this must be decided by an especially responsible person (e.g. project leader). This person may not even be an actor of the regular activity.

For one regular activity, several alternative activities may be pre-modelled as well. Then, the user has to select the most appropriate alternative activity.

User Interaction. Figure 4a shows the worklist of a department leader. For the upper worklist-item (Pre-filter by Supervisor) an alternative activity was pre-modelled. Thus, it additionally contains a button that allows switching to this alternative. Furthermore, a predefined hint may be used to explain the scenarios where this alternative activity shall be used. It is almost no effort (only one click) to switch to this alternative in the case CF-2a. For CF-2b, as depicted in Fig. 4b, such a button may be used to switch to the alternative activity as well. This requires that the worklist also contains already started activities. As an alternative user interaction method for CF-2b, a similar button may be part of the application that is used to execute the regular activity; e.g. a form may contain this additional button.

For CF-2c, the worklist cannot be used to switch to the alternative activity since the regular activity is already completed and normally no longer part of any worklist. Therefore, it is necessary for the user to search the process instance, as described for jumps in the next section, in order to start the alternative activity (cf. Figure 6). For CF-2d, no user actions are required at all since switching to the alternative activity is performed automatically.

Alternative activities, again, cannot be realized with (normal) XOR-Nodes. Similar as for CF-1, the decision would be made too early; i.e., before the regular activity

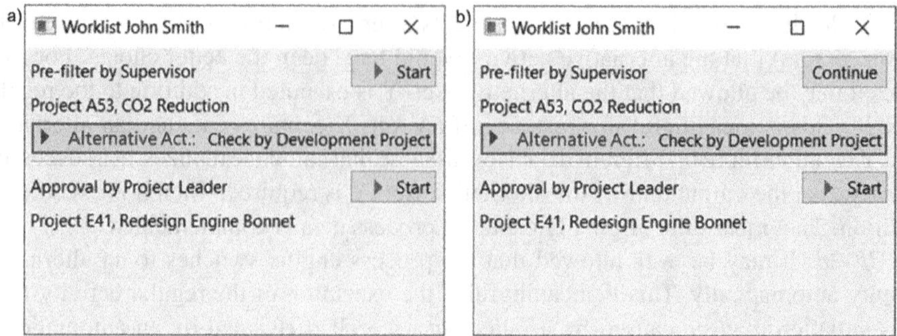

Fig. 4. Worklist with an alternative activity a) before, b) after starting of the regular activity.

was inserted into the worklists. That means, the end users would never see the regular work-items as well as the buttons depicted in Fig. 4.

4.3 Jumps Within the Process Graph

In exceptional cases, forward jumps may be used to save time at process execution by skipping activities. Backward jumps allow to repeat activities in order to correct errors that were made at their original execution.

Requirements. Expectable jumps are pre-modelled by the BP-designer at build-time. In the following, requirements for different types of jumps are presented.

CF-3a (Forward Jumps): A travel application process may contain several evaluations and cost ratings with a long duration. After their completion, the approval of the journey is performed. As an exceptional case, assume a travel application for a near-term appointment. Then it is necessary to jump directly to the approval and omit the remaining evaluation and cost rating activities. Otherwise, an important appointment is missed what results in a large economic loss.

A jump may be triggered before its source activity (i.e. the starting point of the jump) is started. Otherwise, if the source activity is already running, it may be aborted automatically when performing the jump. At build-time, it must be possible to pre-model whether a jump is allowed with this source activity, and whether the source activity shall be aborted automatically. Furthermore, the set of possible target activities as well as rights have to be defined; i.e., who is allowed to trigger a jump. This may be the actor of the source activity, the starter of the process instance, or the owner of the process template.

In addition, for the bypassed activities (between the source and the target node), it must be possible to define whether they shall be caught up. In many scenarios they shall be omitted. But it may be also necessary to perform a bypassed activity later on. Assume, for instance, that one of the cost ratings is necessary to calculate the travel expenses (an activity performed after the approval). Therefore, it must be possible to pre-model that this activity must be caught up and, in addition, that this is not allowed at any time, but must occur before a predefined successive activity can be started.

To improve clarity, the process modelling tool shall be able to hide jump edges. If the user decides to display such edges, they shall be visualized distinguishable from regular control-flow edges. In addition, it must be possible to pre-model whole regions as possible sources resp. targets of a jump (cf. the grey edges and blocks in Fig. 5).

A realization of jumps with normal XOR-Splits is a hardly meaningful workaround. As already explained at CF-1, the conditions of XOR-Splits are evaluated too early. Furthermore, this requires many XOR-Gateways and edges that result in a confusing process graph. At the travel application process, an additional XOR-Split in front of all evaluations and cost ratings would be necessary to enable the required forward jumps.

CF-3b (Backward Jumps): During the execution of Act. E to L of the CMP (Fig. 1), a user may detect that incorrect data were captured in Act. D. Thus, the request cannot be approved (Act. M) and the process must jump back to Act. D in order to repeat it. After the jump, the activities of this process part are executed again (with correct data).

As necessary in this scenario for Act. E to L, it must be possible to pre-model several potential source activities for a backward jump. Again, it is no good idea to insert a XOR-Split after each of these activities. Instead, a source and a target region shall be pre-modelled. In addition, it must be definable, who has the right to trigger this jump.

After a backward jump, the process graph is traversed forward again. In order to realize variants for this forward execution, for each activity, the way to handle the original results (output data) of this activity is pre-modelled. For this purpose, three variants are offered. It depends on the nature of the activity which one is suited best.

1. Discard: The original results of this activity shall be discarded at the jump. Later, at forward execution, it is executed "normally" (i.e. repeated) as at its first execution.
2. Control: The activity shall be executed again, but the original output data are kept. The user can inspect these data; e.g., a pre-filled form may contain these values. If necessary, the user is able to modify each single value.
3. Keep: The activity shall not be executed again; i.e., its output data stay unchanged.

The necessity of all these variants is explained at the CMP: Assume that the activities G to I are already completed, when the backward jump from Act. L to Act. D is triggered. For Act. I (after-sales), the variant Discard is used since its original output data are no longer valid after the jump because a changed part causes different purchase and installation costs at a repair. Control is selected for Act. G (production) since the results of this activity may be influenced seldom by changed development data. Therefore, the output data of Act. G have to be controlled and modified sometimes. Such a modification becomes necessary, for instance, if changing the part results in a more difficult assembly procedure. Keep is specified for Act. H (marketing). It is not necessary to repeat it after the backward jump since changed development details are never relevant for marketing. The advantage of the variants 2 and 3 is that time and effort can be saved for the concerned activities at their later forward execution.

At a backward jump, the compensation of an activity execution may be necessary as well. Assume an order process with a backward jump to a target activity in front of the activity "Place order at supplier". Because of this jump, the order shall be suspended since it will be changed in future. In order to realize this action, a compensation activity is pre-modelled. It has the same behavior as a normal activity, but it is connected with the backward jump; i.e., it is not part of the regular control-flow.

CF-3c (Jumps and Parallelism): Several additional aspects have to be respected at jumps into and jumps out of regions with parallel branches. The source and target regions (grey background in Fig. 5) of such a jump are defined at build-time again, as well as all other aspects already described (e.g. user rights). The activities of the source resp. the target region belong to several parallel branches. At the example of Fig. 5a, potential target activities are Act. C and D in the upper branch, Act. F and G in the middle branch, and Act. I in the lower branch. The user has to select one activity of each branch as target when he triggers a jump.

Forward Jump: For the forward jump of Fig. 5a, target nodes must be selected for three branches; e.g., the activities C, G, and I. To reduce the effort for the end user when triggering a jump, a default target node may be pre-modelled for each branch.

In addition, as at CF-3a, it can be pre-modelled whether the bypassed activities between the source and the target nodes of the jump shall be caught up or skipped.

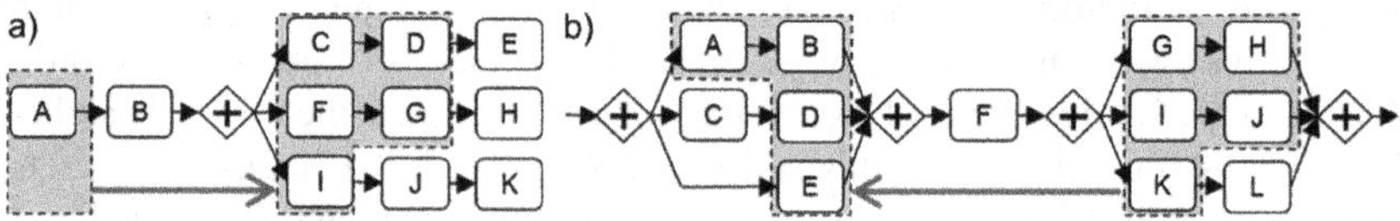

Fig. 5. Parallelism with a) Forward b) Backward jumps, similar in [6].

Backward Jump: Assume for Fig. 5b that Act. G still has to be started and that other users currently execute the activities I and K. Furthermore, assume that a potential actor of Act. G triggers the depicted backward jump. For each activity that is concerned by the backward jump, its behavior can be pre-modelled at build-time. For this purpose, one of the following variants is selected:

1. Abort (no Start, no Complete): An activity of a parallel branch, that was already started, shall be aborted automatically. For instance, Act. I shall be aborted since its results will be discarded (cf. CF-3b: Discard) at the forward execution (after the backward jump) anyway. No further activities are started in this branch (i.e. Act. J).
2. Complete (no Start): An already started activity may be completed, but it is not allowed to start it. Assume for the already started Act. K that it has the type Control (CF-3b). To avoid the loss of the already performed work, its completion is allowed. If Complete was also selected for the next Act. L, it cannot be started. This makes sense since Act. L is currently not executed by a user; i.e., no effort can be lost.
3. Start&Complete: Assume this type for the activities K and L of the lower branch. Then, it is even possible to start Act. L, when Act. K finishes. This is meaningful if their output data will be used later on, for instance, at the type Keep (CF-3b). The advantage is that much time is available to execute these activities, till the other branches reach the AND-Join after the backward jump. The execution of a branch can be continued by the users until an activity of type Abort or Complete is reached.

This parameter is specified for the potential source activities of a jump and for their successors. These variants can be used for backward jumps without parallelism as well.

In combination with Control (CF-3b) all variants are meaningful. The BP-designer can select between fast process execution at Start&Complete (i.e. no delays are caused by waiting) and reduced effort for the end users at Abort (i.e. no work that will be performed in future will be discarded). Complete is a compromise where already spent effort and time may be saved, but no effort is spent for still not started activities.

Jumps in sequences are the basic case and jumps into resp. out of parallel branches are especially relevant, since even activities of parallel branches can be continued, repeated, etc. In addition, jumps shall be possible within conditional branches and loops as well (these topics were omitted in [6] due to lack of space):

CF-3d (Jumps between Conditional Branches): A user may detect that the wrong conditional branch is executed currently. Therefore, he decides to abort this branch by jumping into another branch. Internally, the process engine can realize such a jump by a backward jump to the XOR- resp. OR-Split node, followed by a jump to the target activity. Therefore, all requirements presented for forward and backward jumps are relevant for conditional branches as well.

CF-3e (Jumps to other Loop Iterations): A forward jump to the next iteration of a loop may be necessary to save time. Thereby, bypassed activities of the original iteration may be caught up later, as described for CF-3a. A backward jump to an activity of a previous iteration of the loop may be necessary if errors were made at the execution of this activity. It shall be possible, for instance, to continue the source activity of the jump and its successors as described above (CF-3b).

User Interaction. Compared to the topics presented in the previous sections, jumps occur rather seldom and the source region may contain several activities (cf. Figure 5). Therefore, it is not meaningful to extend the work-items with Jump-Buttons for all these activities, since these buttons would be used very seldom. In addition, jumps often cannot be triggered by normal process participants but only by process administrators with special rights, because a jump concerns many activities and many users. Since such a process administrator normally does not execute activities, he has no such work-item.

In order to select the concerned process instance, without having a corresponding entry in a worklist, the administrator must be able to search for it. As depicted in the upper left part of Fig. 6, search attributes as for instance the name of the process type or the user who started the process instance may be used for this search. In addition to such "formal process attributes", it shall be possible to use application data of the process instance for a search as well; e.g., the name of the part that shall be changed, the name of the supplier that develops a part, or the destination at a travel application process. Such a search may result in several process instances (lower part of Fig. 6). The user selects one process instance and, as depicted in the right part of Fig. 6, possible actions are offered, including jumps if available for the current execution state of this process instance. The user starts one of these actions; e.g. with the context menu.

Especially at parallelisms, many activities may be allowed as jump targets (cf. Figure 5) and it may be necessary to select several of these activities. In some scenarios, it may be difficult for the end user to select these activities from a list with activity names, because he may not know which activities belong to which parallel

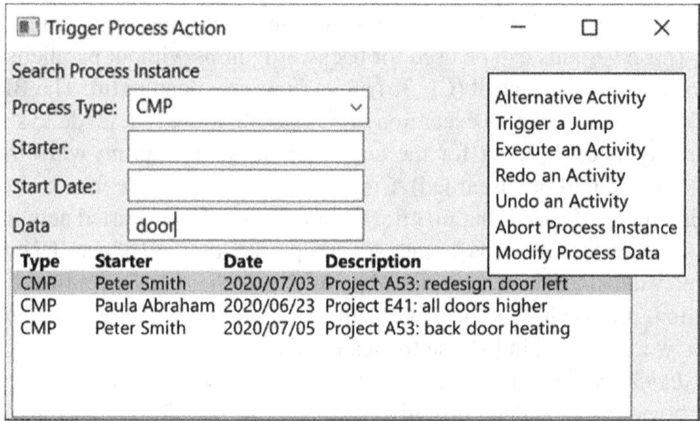

Fig. 6. Searching a process instance in order to trigger a process action.

branch. Then, a graphical process visualization can be used, as depicted in Fig. 7 for the jump of Fig. 5a. This allows easily, for instance, to identify activities of the same branch, the default target activities, and whether one activity is already selected for each parallel branch. Bypassed activities may be visualized in grey color. In addition, the current execution state may be visualized to give a better understanding of the consequences of the intended jump. As already mentioned, such a process visualization can be generated automatically [37] based on the process template, the information pre-modelled for the jump, and the current execution state of the process instance.

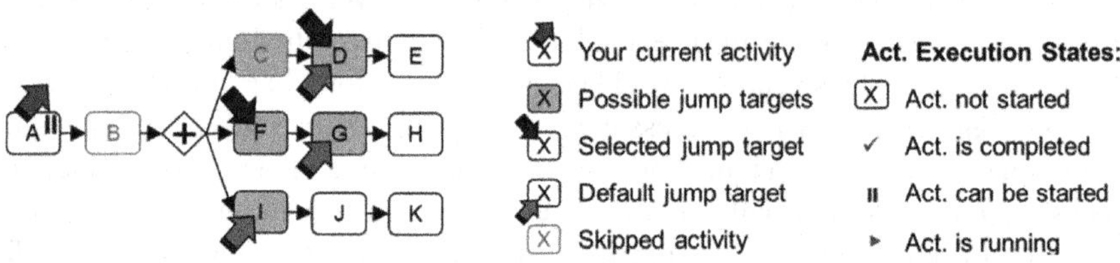

Fig. 7. Selection of the target activities for a jump based on a graphical process visualization.

4.4 Multi-instance-Parallelism

The CMP of Fig. 1 contains an example for this aspect: The modification of a part may affect its neighbor parts since its shape may change. Act. E determines the list of these neighbor parts. The number of instances required for Act. K and L corresponds to this list length. All these activity instances have different input data and actors.

Requirements. *CF-4a:* As basic case, it is known in advance how many parallel branches (instances of Act. K and L) are required. That means, this number is determined when the ∀-Split node is executed and not changed later on. This case corresponds to the control-flow pattern "Multiple Instances with a priori Run-Time Knowledge" [25].

CF-4b: It is possible to append additional parts to the list NeighborParts by executing Act. O. A dynamic data modification operation [7] may be used to extend this list as well. In both cases, additional parallel branches are required. They result in further instances of Act. K and L. It may be meaningful to create such instances until all branches of the Multi-Instance-Parallelism have finished (i.e. reached the ∀-Join). This corresponds to the pattern "Multiple Instances without a priori Run-Time Knowledge" [25].

CF-4c: As an extension, a user-defined rule may be used to specify whether it is possible to create additional branches anymore. This may be allowed only until a milestone in one multi-instance branch is reached: For instance, new branches can only be created as long as no rating of a part owner (Act. K) has finished. This rating was not yet checked by a colleague (Act. L); i.e., the branch has not finished (cf. CF-4b). Act. K uses the (original) list NeighborParts as input data, therefore, an extended list may result in a different rating. Thus, it is prohibited to change this list subsequently. The milestone may belong to a parallel branch outside the Multi-Instance-Parallelism as well: If the after-sales clerk has completed his rating in Act. I (based on the original list NeighborParts), it is not allowed to change this list anymore; i.e., no new branches can be created later on.

It shall be possible to pre-model at build-time, which type of Multi-Instance-Parallelism shall be used, and in addition the milestone for CF-4c. Furthermore, it has to be defined which users have the right to create additional branches at run-time. In the given scenario this may be realized by the actor assignment of Act. O.

User Interaction. At run-time, the process engine automatically creates the correct number of parallel branches; i.e., no user interaction is necessary for this purpose. For CF-4b, Sect. 4.5 describes how an activity can be started that extends the list NeighborParts. The other possibility, the modification of process data (i.e. this list), may be performed by searching the process instance (cf. Figure 6), triggering the operation, and selecting the concerned process variable. At build-time, for CF-4c, it may be necessary to specify milestones that define how long it is allowed to create additional branches. These milestones may be selected graphically as sketched for optional activities in Fig. 3b.

4.5 Activities Independent from the Process Graph

Some activities may be offered to the users additionally to the regular activities (which are part of the process graph). The users can decide whether such an activity is necessary in the current situation. At the execution of the CMP (cf. Figure 1), a user may recognize at a later point in time, that a neighbor part was forgotten in Act. E. In this case, he executed the process graph independent (i.e. additional) Act. O. It complements the list NeighborParts, with the consequence that an additional branch of the Multi-Instance-Parallelism will be created (cf. CF-4b in Sect. 4.4).

Requirements. *CF5a:* As regular activities, process graph independent activities are pre-modelled at build-time. Thereby, all necessary aspects are specified; e.g., an actor assignment, the mapping of input/output parameters to BP-variables (e.g. the list NeighborParts). Process graph independent activities can be executed at run-time as normal activities, but they are started in a different way (see below). Since all relevant information is pre-modelled, this causes less effort for the end users (cf. Section 1) than a dynamic insert [9] of an additional activity.

CF5b: At build-time, it shall be possible to specify for a process graph independent activity, whether it can be executed multiple times. Furthermore, a process region may be pre-modelled, where it is allowed to start this activity. For instance, Act. O of the CMP may be started several times, until completion of the Multi-Instance-Parallelism.

In principle, it is possible to realize such an activity with a parallel branch that contains an optional activity; i.e., Act. O in the example of the CMP. This optional activity has to be embedded in a loop if it may be executed several times. The parallel branch surrounds the process region where Act. O may be started, at the CMP from Act. F to the ∀-Join. This workaround results in a more complex process graph, that becomes even more confusing if several process graph independent activities are required, especially if a different process region is used for each of these activities.

User Interaction. A user can start a process graph independent activity using an entry of his program menu or with a special button of his user interface. For this purpose, he has to search the concerned process instance as described for jumps (cf. Figure 6). Then, he can start this activity with the context menu ("Execute an Activity" in Fig. 6). Another possibility is to offer this function in the worklist of the user (similar to an optional activity, cf. Figure 3a) during the whole execution of the corresponding process region. At build-time, again, such a process region may be pre-modelled as sketched in Fig. 3b.

4.6 Start-End-Dependencies Between Activities

Currently published control-flow patterns [25] allow many orders at process execution, but these patterns always concern whole activities: At a sequence, for instance, an activity must be completely finished before the next one can be started. In the following, this is extended by respecting the start and the end events of activities separately. Since this results in additional execution orders, flexibility at process execution is increased.

Requirements. With respect to sequences, there exist four possibilities to define execution orders (the first one corresponds to a normal sequence; i.e. is the standard case):

- *CF-6a (EndBeforeStart):* End of Act. A must happen before start of the next Act. B
- *CF-6b (StartBeforeStart):* Start of Act. A must happen before start of Act. B
- *CF-6c (EndBeforeEnd):* End of Act. A must happen before end of Act. B
- *CF-6d (StartBeforeEnd):* Start of Act. A must happen before end of Act. B

In the example of Fig. 8a, the type EndBeforeEnd (CF-6c) is used between Act. B and Act. C. Act. B (e.g. Clean vehicle) must be completed before Act. C (Deliver vehicle to customer) finishes. Cleaning the vehicle afterwards is not possible. With this type of "sequence", however, the activities may be executed partially concurrently: It becomes possible to clean the vehicle during a transportation break. Figure 8b shows all allowed execution orders of these activities B and C.

The type StartBeforeStart (CF-6b) is used in Fig. 8a as well: the vehicle delivery (Act. C) must be started before Act. D (Inform customer about upcoming delivery) can be started. Informing the customer earlier bears a high risk of misinformation since transportations are often cancelled; e.g., because the truck is not available or broken.

Optional activities (cf. CF-1) can be used in combination with these new types of dependencies as well. Act. B (Clean vehicle) may be omitted if the vehicle is already clean. Figure 8c shows that one additional execution order is allowed in this case.

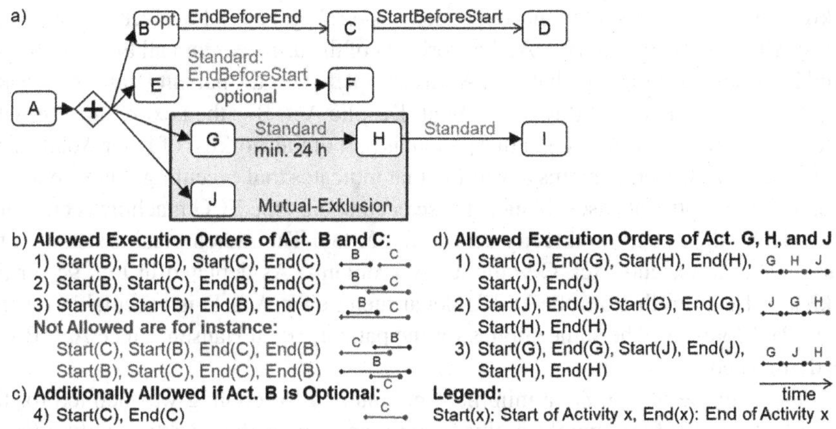

Fig. 8. Process with start-end-dependencies, similar in [6].

CF-6e (Optional Dependencies): In addition, also dependencies can be optional (cf. the dotted edge to Act. F in Fig. 8a). This means that this order is desired, but not absolutely necessary. Then, the end users have the possibility to start Act. F before Act. E if this is necessary in exceptional cases.

The hospital process of Fig. 9a is used to explain the necessity of optional dependencies: After a specific diagnosis (Act. A), normally first an electrocardiogram (ECG) is made in Act. B, then an X-ray in Act. C, and finally a magnetic resonance tomography imaging (MRTI) in Act. D. In the exceptional case that one of these examination facilities is not available or overburdened, however, it is allowed to deviate from the standard order, with the result that Act. C and/or Act. D are performed earlier.

CF-6f (Time Intervals): It shall be possible to define minimal and maximal time intervals between activities. Assume for Fig. 8a that a part is hardened in Act. G. Then, it must cool for at least 24 h before it is possible to paint it in Act. H. That means, Act. H can be only started after these 24 h have elapsed. Such time intervals are pre-modelled at build-time and guaranteed by the process engine at run-time. They may refer to the start and the end event of the activities.

CF-6g (Mutual-Exclusion): Figure 8a shows a mutual-exclusion for the activities G, H, and J (visualized as green rectangle here). It is only allowed to execute one of these activities at any point in time. All other activities must be executed completely before or completely after it (cf. Critical Section and Interleaved Routing [25]). A part is hardened in Act. G and painted in Act. H. Then, a bill is created in Act. I. In parallel, the customer controls the part in Act. J. Since the activities G, H, and J are performed at different locations, but the part must be located there (physically), the execution of these activities cannot overlap in time. A mutual-exclusion specifies this restriction; i.e., that no activity that is part of this building block can be started while another one is running. At the process example of Fig. 8a, the execution orders sketched in Fig. 8d are allowed. Thereby, Act. I is not part of the mutual-exclusion, thus, its execution can overlap with Act. J.

User Interaction. For CF6a to 6d and CF-6g, the PMS inserts "normal entries" into the worklists of the end users. Optional dependencies (cf. Figure 9a), however, cause special entries: After completion of Act. A, the worklists of the users contain all the activities B, C, and D. Assume for Fig. 9b, that Act. A was completed for the patient Paul Johnson and Peter Adam works in the X-Ray department. Despite Act. B is the next regular activity, the X-Ray examination (Act. C) of this patient is part of the worklist of Peter Adam. This worklist-item, however, contains a warning that indicates that executing this activity is a measure for exceptional cases. If such a case occurs; e.g., the ECG machine is currently not available, the patient is directly sent to the X-ray. The radiological assistant Peter Adam is able to execute Act. C without any problems. As depicted in Fig. 9c for the worklist of Lisa Smith from the NMRI department, even Act. D is executable for this patient Paul Johnson. The same applies for the patient Adam Hansen. Since Act. B was already finished for.

In the example of Fig. 8a, a minimal time interval for cooling was defined for the start of Act. H (CF-6f). Therefore, this activity appears in the worklists only after 24 h have elapsed; i.e., no special interaction with the users is required. At minimal time intervals that concern the end of an activity, the Complete-Button of the corresponding activity program may be locked (e.g. visualized in grey) until the required point in time is reached. Maximal time intervals can be enforced with escalation mechanisms. For instance, a message may be sent to the actor of the activity or his supervisor in case of a delayed start or completion, or better some time before the deadline will be missed.

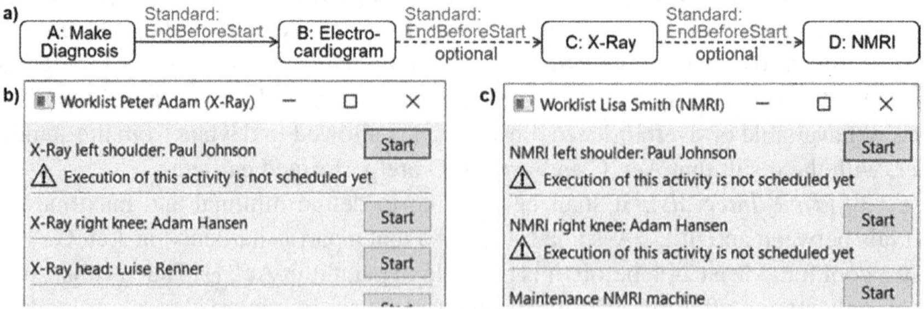

Fig. 9. a) Process with optional dependencies with b) and c) the resulting worklists.

4.7 User Actions

A user can perform actions (cf. [38]) spontaneously, that change the execution state of a single process instance. Such an action is not really predictable and not explicitly pre-modelled; i.e., it is not contained in the process model and the user does not have any special item in his worklist. Instead, he may trigger such an action from the menu of his application (due to lack of space, this aspect was completely omitted in [6]).

Requirements. Some definitions have to be made for user actions already at build-time despite they are unplanned; e.g., whether, when, and by whom it may be performed.

CF-7a (Abort a Process Instance): A user can abort a whole process instance, i.e., stop it immediately. All currently running activity instances (performed by different users) shall be aborted as well. At least, these users shall be informed that activity execution is not required anymore. At the CMP depicted in Fig. 1, it may be necessary (e.g. because of a management decision) to abort a running change request since this change will not be realized anyway.

At build-time, for each process template it has to be specified, when such an abort is allowed. One of the following variants may be selected:

1. An abort is not allowed for this process type at all.
2. An abort is possible at each execution state.
3. An abort is only allowed if the process execution is currently in a specific state. Therefore, it may be specified for each activity whether an abort is allowed during its execution.

For the variants 2 and 3, it has to be defined in addition, who is allowed to trigger the abort; i.e., rights for users have to be defined.

CF-7b (Abort an Activity): The actor of a running activity instance may want to abort it; e.g., since he is not able to complete this activity because of his workload. Afterwards, it can be executed by a different user. For each activity, it has to be defined whether an abort is allowed. This is necessary since there may exist activities that must be completed (when started) because of their irreversible effects (e.g. cooling an incandescent part during hardening).

CF-7c (Skip an Activity): An activity may be skipped, since it is not necessary in the current business transaction, or since it is not possible to execute it. This user action concerns non-optional activities; i.e., activities that must be executed (usually). Therefore, skipping such an activity is a much more extreme intervention than omitting an optional activity (CF-1). Assume for the CMP, that Act. L cannot be executed since there is no developer of the same department available (e.g. because of illness). To avoid delays, this check shall be omitted for this change request; i.e., this instance of Act. L is skipped. At build-time, it has to be defined for each activity whether a skip is allowed, and additionally, who has the right to perform this skip action.

CF-7d (Undo an Activity): An already finished activity shall be made undone. Assume for the CMP that, before starting Act. E, the part owner detects that he made an error at the already completed Act. D. Then, he triggers an undo action for Act. D in order to become able to execute Act. D again. At build-time, it has to be defined for each activity, whether an undo is allowed and who has the right to perform it; e.g. the original actor of the activity or the process owner.

Often, an undo action is only allowed as long as no successive activities are started. However, even a later undo may be possible. Then, it must be defined additionally, how long this undo action is allowed. For instance, at the CMP, a faulty rating of the part owner (Act. D) can only be undone as long as the request is not approved (Act. M), since the approval is based on this rating.

Similarly as for bypassed activities at forward jumps (CF-3a), for such a late undo, it shall be defined, whether the activity has to be caught up (i.e. executed again) or omitted. If it is caught up, the improved rating is used for the approval decision (Act. M). Otherwise, the decision is made without this information. As also already mentioned for jumps, for each activity a compensation activity may be modelled. Its purpose is to reverse the effects of the original activity execution. It may be used for an undo action as well. For instance, the Act. X "Fetch part from stock" can be compensated by the Act. Y "Bring part back to stock". Later, when catching up Act. X, the right part is fetched.

CF-7e (Redo an Activity): An already executed activity shall be executed again; e.g., to correct its output data. For instance, at the CMP, a part owner may modify his rating by a repeated execution of Act. K. Furthermore, it can be necessary to fix real world problems with a redo action: Assume the Act. X that fetches a part from the stock. If it is detected later that the part has a malfunction, Act. X must be repeated. This shall not only happen in the real world but the process activity (controlled by the PMS) has to be repeated as well, since the execution of Act. X decreases the number of available parts stored in a database. Furthermore, the resulting process costs can be captured this way. At build-time, it has to be defined for an activity, whether a redo action is allowed and who may perform it (e.g. the original actor of the activity, each potential actor, or a process owner). Additionally, the process region for a redo may be restricted. For instance, at the CMP, a redo of Act. K (rating) is not meaningful any more, if Act. M (approval decision) was already performed, since the new rating will not be respected anymore.

User Interaction. As already mentioned, these user actions are intended for very seldom exceptional cases. Therefore, it is not meaningful to offer buttons for all these actions at all worklist-items. Instead, an abort (CF-7b) and a skip (CF-7c) action for the current activity may be triggered with a respective menu entry. For a redo (CF-7d) and undo (CF-7e) of an already completed activity as well as for an abort of a process instance (CF-7a), it is necessary to search the concerned process instance as depicted in Fig. 6. Then, the intended action can be triggered with the context menu.

5 Summary and Outlook

In PMS, it shall be possible to deviate from the modelled process. Otherwise, such a system is not usable in practice. For this purpose, dynamic changes can be used. For predictable deviations, however, this results in much effort for the end users and may cause mistakes. To avoid such disadvantages, predictable exceptions and special cases should be pre-modelled already at build-time. The corresponding requirements and examples from practice are presented in this paper. In addition, the resulting interaction with the end users is explained for each aspect. For this purpose, for instance, additional buttons may be included into the work-items of the end users. For other functions it is more appropriate to use entries of the menu. For the interaction with BP-designers, graphical processes visualizations may be used; e.g., to select several activities of the process graph (i.e. to pre-model a process region for a deviation).

Hopefully, this work will motivate tool manufacturers to support the described scenarios in commercial PMS. A direct support is even necessary for requirements that are currently realizable with workarounds, since they typically use complex constructs; e.g., event-based gateways and catching events of BPMN for the realization of optional activities as described in Sect. 4.1. Such a complicated modelling method overwhelms "normal" BP-designers. Furthermore, business users are no longer able to understand and check correctness of a process graph.[4]

Generalizability and relevance of the presented scenarios has to be verified with further practical examples, ideally from other business domains. Thereby, the presented requirements for pre-modelled flexibility can be complemented. This, however, is impeded by the fact that several of the presented concepts are not available in current BP modelling languages; e.g., edges for start-end-dependencies (CF-6b to d) are not offered by EPC and BPMN. That means, using a documentation-oriented analysis method (i.e. to analyze existing BP documentations) is not appropriate for this purpose, since corresponding scenarios are probably not captured in these documents even if they exist in reality. This may be solved, for instance, by using expert interviews.

One of the next steps of CoPMoF is to examine some of the presented requirements in more detail. For instance, this is necessary for jumps (CF-3) where the desired execution semantics has to be defined formally since it is not obvious (especially at parallelisms). Additionally, this is necessary for start-end-dependencies (CF-6).

References

1. Reichert, M., Weber, B.: Enabling Flexibility in Process-Aware Information Systems. Challenges, Methods, Technologies. Springer, Berlin (2012)
2. Redding, G., Dumas M., Ter Hofstede, A.H.M., Iordachescu, A.: Modelling flexible processes with business objects. In: Proceedings IEEE Conference on Commerce and Enterprise Computing, pp. 41–48 (2009)
3. Schonenberg, M.H., Mans, R.S., Russell, N.C., Mulyar, N.A., van der Aalst, W.M.P.: Towards a Taxonomy of Process Flexibility (Extended Version). Eindhoven Univ. Technol. 10, 41–56 (2007)
4. Dadam, P., Reichert, M., Rinderle-Ma, S.: Process Management Systems. Only a bit Flexibility will not be enough (in German). Informatik-Spektrum **34**, 364–376 (2011)
5. Kumar, K., Narasipuram, M.M.: Defining requirements for business process flexibility. In: Workshop on Business Process Modeling, Design and Support, Proc. of CAiSE06 Workshops, pp. 137–148 (2006)

[4] In order to realize process execution by a process engine, however, it is an option to map the "easy to understand" modelling constructs (presented in this paper) automatically to such already existing constructs of the BP execution engine, even in a sophisticated way.

6. Bauer, T.: Business processes with pre-designed flexibility for the control-flow. In: Proceedings 22nd International Conference on Enterprise Information Systems, pp. 631–642 (2020)
7. Bauer, T.: Pre-modelled flexibility for business processes. In: Proceedings of 21th International Conference on Enterprise Information Systems, pp. 547–555 (2019)
8. Regev, G., Soffer, P., Schmidt, R.: Taxonomy of flexibility in business processes. Workshop on business process modeling, design and support. In: Proceedings of CAiSE06 Workshops, pp. 90–93 (2006)
9. Reichert, M., Dadam, P.: ADEPTflex - supporting dynamic changes of workflows without losing control. J. Intell. Inf. Syst. Spec. Issue Workflow Manage. Syst. **10**, 93–129 (1998)
10. Sadiq, S., Marjanovic, O., Orlowska, M.: Managing change and time in dynamic workflow processes. Int. J. Coop. Inf. Syst. **9**, 93–116 (2000)
11. Weske, M.: Formal foundation and conceptual design of dynamic adaptations in a workflow management system. In: Proceedings of 34th Hawaii International Conference on System Sciences (2001)
12. Bandinelli, S., Fugetta, A., Ghezzi, C.: Software process model evolution in the SPADE environment. IEEE Trans. Soft. Eng. **19**, 1128–1144 (1993)
13. Rinderle, S.: Schema evolution in process management systems. Ph.D. thesis, Universität Ulm (2004)
14. Joeris, G., Herzog, O.: Managing evolving workflow specifications. In: Proceedings of International Conference on Cooperative Information Systems, pp. 310–321 (1998)
15. Kradolfer, M., Geppert, A.: Dynamic workflow schema evolution based on workflow type versioning and workflow migration. In: Proceedings of International Conference in Cooperative Information Systems, pp. 104–114 (1999)
16. Weske, M.: Flexible modeling and execution of workflow activities. In: Proceedings of 31th Hawaii International Conference on System Sciences, pp. 713–722 (1998)
17. Casati, F., Ceri, S., Pernici, B., Pozzi, G.: Workflow evolution. Data Knowl. Eng. **24**, 211–238 (1998)
18. La Rosa, M., van der Aalst, W.M.P., Dumas, M., Ter Hofstede, A.H.M.: Questionnaire-based variability modeling for system configuration. Soft. Syst. Model. **8**, 251–274 (2009)
19. Schobbens, P.Y., Heymans, P., Trigaux, J.C.: Feature diagrams: a survey and a formal semantics. In: 14th IEEE International Requirements Engineering Conference Minneapolis/St. Paul, pp. 136–145 (2006)
20. Reinhartz-Berger, I., Soffer, P., Sturm, A.: Extending the adaptability of reference models. IEEE Trans. Syst. Man Cybern. **40**, 1045–1056 (2010)
21. Gottschalk, F.: Configurable process models. Ph.D. thesis, Eindhoven University of Technology (2009)
22. Reichert, M., Hallerbach, A., Bauer, T.: Lifecycle management of business process variants. In: vom Brocke, J., Rosemann, M. (eds.) Handbook on Business Process Management 1. IHIS, pp. 251–278. Springer, Heidelberg (2015). https://doi.org/10.1007/978-3-642-45100-3_11
23. Lerner, B.S., Christov, S., Osterweil, L.J., Bendraou, R., Kannengiesser, U., Wise, A.E.: Exception handling patterns for process modeling. IEEE Trans. Softw. Eng. **36**, 162–183 (2010)
24. Reichert, M., Dadam, P., Bauer, T.: Dealing with forward and backward jumps in workflow management systems. Softw. Syst. Model. **2**, 37–58 (2003)
25. Russell, N., Ter Hofstede, A.H.M.: Workflow control-flow patterns. A revised view. BPM Center Report BPM-06–22 (2006)
26. Weber, B., Reichert, M., Rinderle-Ma, S.: Change patterns and change support features - enhancing flexibility in process-aware information systems. Data Knowl. Eng. **66**, 438–466 (2008)

27. Klingemann, J.: Controlled Flexibility in Workflow Management. In: Wangler, B., Bergman, L. (eds.) CAiSE 2000. LNCS, vol. 1789, pp. 126–141. Springer, Heidelberg (2000). https://doi.org/10.1007/3-540-45140-4_10
28. Mangan, P., Sadiq, S.: On building workflow models for flexible processes. Aust. Comput. Sci. Commun. **24**, 103–109 (2002)
29. van der Aalst, W.M.P., Weske, M., Grünbauer, D.: Case handling: a new paradigm for business process support. Data Knowl. Eng. **53**, 129–162 (2005)
30. Montali, M.: Specification and verification of declarative open interaction models. Springer, Berlin (2010)
31. Pesic, M., Schonenberg, M.H., Sidorova, N., van der Aalst, W.M.P.: Constraint-based workflow models: change made easy. In: Proceedings of 15th International Conference on Cooperative Information Systems, pp. 77–94 (2007)
32. Sadiq, S., Sadiq, W., Orlowska, M.: Pockets of flexibility in workflow specification. In: S.Kunii, H., Jajodia, S., Sølvberg, A. (eds.) ER 2001. LNCS, vol. 2224, pp. 513–526. Springer, Heidelberg (2001). https://doi.org/10.1007/3-540-45581-7_38
33. Dourish, P., Holmes, J., MacLean, A., Marqvardsen, P., Zbyslaw, A.: Freeflow: mediating between representation and action in workflow systems. In: Proceedings of ACM Conference on Computer Supported Cooperative Work, pp. 190–198 (1996)
34. Wainer, J., Bezerra, F., Barthelmess, P.: Tucupi: a flexible workflow system based on over ridable constraints. In: Proceedings of ACM Symposium on Applied Computing, pp. 498–502 (2004)
35. Burmeister, B., Steiert, H.-P., Bauer, T., Baumgärtel, H.: Agile processes through goal- and context-oriented business process modeling. In: Eder, J., Dustdar, S. (eds.) BPM 2006. LNCS, vol. 4103, pp. 217–228. Springer, Heidelberg (2006). https://doi.org/10.1007/11837862_22
36. Laue, R., Kirchner, K.: Using patterns for communicating about flexible processes. In; Proceedings of 18th International Conference on Business Process Modeling, Development and Support, pp. 12–19 (2017)
37. Reichert, M., Kolb, J., Bobik, R., Bauer, T.: Enabling personalized visualization of large business processes through parameterizable views. In: Proceedings of 27th Symposium On Applied Computing, pp. 1653–1660 (2012)
38. Russell, N., van der Aalst, W., ter Hofstede, A., Edmond, D.: workflow resource patterns: identification, representation and tool support. In: Pastor, O., Falcão e Cunha, J. (eds.) CAiSE 2005. LNCS, vol. 3520, pp. 216–232. Springer, Heidelberg (2005). https://doi.org/10.1007/11431855_16

On Enterprise Architecture Patterns: A Tool for Sustainable Transformation

Roberto García-Escallón(✉), Adina Aldea, and Marten van Sinderen

University of Twente, Enschede, The Netherlands

Abstract. Organizations across the world today face similar problems, thus, solutions to these problems could also be re-used across many organizations. Climate change is one such common problem to organizations, where any solution implemented by one could be re-used by many others. These re-usable solutions could be Enterprise Architecture Patterns, but there is a lack of guidance on how to use them in practice. This study proposes an extension to the commonly used TOGAF, to better leverage these reusable solutions. Thus, this enables organizations using this framework to enhance it with Enterprise Architecture Patterns. As part of the proposed methodology, re-usable patterns supporting sustainable characteristics are built. The resulting methodology is validated with an expert panel, gathering positive comments.

Keywords: Enterprise Architecture · Enterprise Architecture Patterns · Sustainability

1 Introduction

Today, climate change calls for a widespread transformation. A transformation for a sustainable society that is being demanded by millions around the globe [20]. Many projects are underway to make said change a reality, but time is running out, as the looming deadline of 2030 for reducing carbon emissions by 45% worldwide, to avoid warming higher than 1.5 °C, nears [2]. Although, the kind of changes required is similar for all organizations. For example, all organizations can reduce their CO2 footprint by avoiding using paper in their internal processes, something that can be solved in a reusable manner with technology. Because some of the changes needed to diminish emissions are common throughout organizations, solutions developed for one organization may apply to others. For these common challenges and repeatable solutions, the concept of patterns is of great value [1].

As Enterprise Architecture (EA) practitioners aim to steer changes in organizations, they are pushed to do so at an increasing pace. Besides climate change, the pace of technological change and the threat of being disrupted is driving organizations to change faster. To enable EA practitioners to bring about these transformations in organizations, this study proposes a methodology for the specification of EA for sustainable organizations using patterns. This study is a

J. Filipe et al. (Eds.): ICEIS 2020, LNBIP 417, pp. 858–882, 2021.
https://doi.org/10.1007/978-3-030-75418-1_39

followup of [7], and [6], and is organized as follows: First, a description of the background is provided; second, the research methodology is described; third, the results of a systematic literature review are reported; fourth, the methodology is proposed; fifth, a validation of the methodology is shown; finally, conclusions are drawn.

2 Background

Enterprise Architecture (EA) is a relatively new field of research that aims to steer the change in an organization [13]. For example, by aligning the goals of different layers of an organization. As a field, it has developed multiple tools to help practitioners in their activities, and as a result, there are many frameworks and methods. Such methods include the Zachman framework and The Open Group Framework (TOGAF), among others [21,27]. Most methods and frameworks depend on descriptions of the organization, both of how it is working at the moment and how it should be working in the future [12]. The methods and frameworks define the tools to design the desired future version of the organization, as a result, EA has been applied to drive change in organizations.

2.1 Patterns

The organizations' EA is applied in, as well as all other organizations, have similarities in their structure and behaviour with other organizations, similarities that could be also called patterns. In the words of Alexander [1], the authors of the book A Pattern Language, "Each pattern describes a problem which occurs over and over again in our environment, and then describes the core of the solution to that problem, in such a way that you can use this solution a million times over, without ever doing it the same way twice" [1]. From this definition we can extract four key attributes, first, patterns are a solution to a recurring problem. Second, the pattern is the core of the solution, meaning the solution has a scope and does not aim to describe anything that is not needed to solve the problem. Third, the pattern should be usable as many times as needed. And fourth, each specific use of the pattern might look different than the last. Thus, patterns provide reusable solutions to problems that occur repeatedly.

Patterns have been used in other fields, most famously in Computer Science, where patterns describe solutions to common problems when developing software [5]. These patterns helped to hasten the development of software, as they offered ready to use solutions to some common problems. Thus, similar effects are expected in EA by applying the concept of patterns, helping drive change faster.

2.2 Sustainability

In the global context, Sustainability has become the goal to many, ranging from individuals to the United Nations as a response to climate change and other obstacles [23]. The effects of climate change are clear: climate change, impacts to human health, mass extinction of species, among others [3,4,19].

Climate change calls for a transformation, one that is widespread and towards sustainability, which is being demanded by millions around the globe [20]. The rate at which needs to increase, as the looming deadline of 2030 for reducing carbon emissions by 45% worldwide, to avoid a warming higher than 1.5 °C, draws closer [2]. Although, the kind of changes required are similar for all organizations. For example, all organizations can reduce their CO2 footprint by avoiding using paper in their internal processes, something that can be solved in a reusable manner with technology.

3 Research Methodology

The approach followed throughout this study is Design Science Methodology [25]. As a methodology, Design Science describes the designing and investigating of an artifact and its interaction with its context. These activities correspond with the research problems in design science, they are either Design Problems or Knowledge Questions. The former focuses on designing a solution that will act within a problem context, the latter focuses on knowledge without the influence of the context [25].

The implementation of this methodology is based on a cycle, the engineering cycle. It is composed of four steps, the investigation of the problem, the design of the solution, its validation and finally its implementation. Within the scope of this study, the solution in question was not implemented [25].

To achieve the study's goal, these steps are followed, starting with the definition of the design problem: the need for an increased pace of EA change projects, and the difficulty practitioners face when treating sustainability concerns in organizations. To solve the problem, an artifact designed to enable practitioners to prepare an EA specification for sustainable organizations using patterns is prepared.

In order to design this artifact, first, an investigation on the problem is performed, to which two knowledge questions were identified, *what EAPs are there in literature?* and *what characterizes a sustainable organization?*. These two questions are answered using the Systematic Literature Review (SLR) methodology shown below, and their results are shown in Sect. 4.

After gathering the required knowledge during the problem investigation step the artifact is designed. This design is described in Sect. 5. Finally, The artifact is then validated, using a case study and interviews with experts as shown in Sect. 6. A more detailed description of this step can be found in [7].

3.1 Systematic Literature Review

In order to answer the knowledge questions, two separate SLR are performed. The SLRs follow the methodology of Rouhani et al. [18], as it is a very thorough work in the EA field. This method outlines three stages to a SLR: a planning phase, an execution phase and a result analysis phase. In order to enhance this method and arrive at a richer set of literature, the backwards and forwards citation techniques will be included, to include additional articles that might be relevant but were not present in the results of the initial query [26]. In order to execute this last step, Google Scholar was used, which shows both, backward and forward citations, easily. For all other steps leverage Scopus, ACM Digital Library, IEEE Xplore, Science Direct - Elsevier, Springer Link, Taylor and Francis and Web of Science. The process is shown in Fig. 1. For the first knowledge question, the keywords used *"enterprise architecture" AND "Pattern"*. While for the second knowledge question the steps were more relaxed as the time constraints on the study drew nearer. As such, the backward and forward citation techniques were not performed. Finally, the keywords used for the second review were *"characteristics of sustainability" or "characteristics of circular economy" or "characteristics of sustainable business" or "sustainability characteristics" or "circular economy characteristics" or "sustainable business characteristics"*.

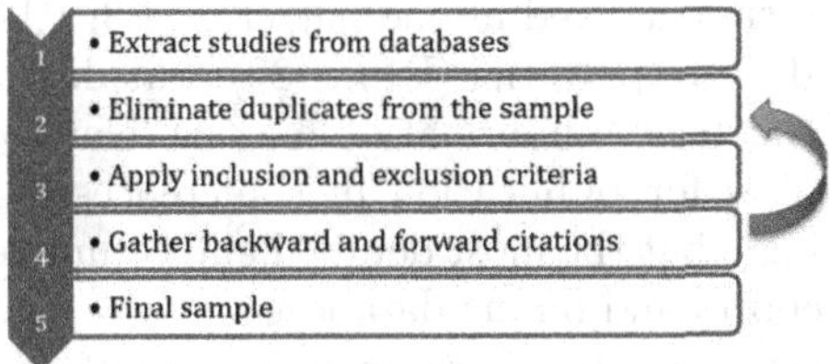

Fig. 1. Overall process of the SLR - adapted from [6].

4 Systematic Literature Reviews

In this section, the two SLR executed are described, first on EAPs in literature, second on Characteristics of sustainable organizations. The results of the former SLR was first published in [6], but the main findings are included here. The results of the latter SLR can be found in [7], but the results are included below.

4.1 EAPs in Literature

During the overall execution of the SLR, it was noticed that the concept of patterns has been used by all fields in the sample with different periods. The latest one being the studies on BMI, particularly in the Sustainable Business Development field. The recent surge of the BMI field's effort on documenting the patterns has been welcomed with open arms by the academic community, as seen by the citation count mentioned in earlier sections. Such interest extends to the sub-field of SBMs, that expands the Business Model Patterns. P2 which expands upon the Repository and adds more Business Model Patterns. Contrary to all other studies in this sample, all the studies related to BMI, SBM and Circular Economy are based around the framework proposed by [16]. Although the representation of the patterns may differ, their basic constructs are the same, which would make it possible to translate these patterns into an EA representation of them, based on the work of [10]. This method would pave the way to take these Business Model Patterns repositories into EAPs.

Being able to relate the patterns extracted from SBMs, and translating them to EA, is aligned with the overall interest of society to a more sustainable world. With the call for sustainability, as seen by the Sustainable Development Goals of the UN [23], it means that organizations will need to develop new functions or transform their current ones. This change could be supported by EA, and, being a generalized need, would benefit from having a repository of patterns to draw from.

State of the Literature on Enterprise Architecture Patterns. Each of the 24 studies reviewed in this SLR have described patterns, however, not all of them seem to be written in a way that can be used by future works. For example in the conference proceedings and journal articles, where the authors report mostly on how they arrived at the patterns (P3, P17), or describes how one could extract and write patterns (P16, P22), but do very little in actually documenting them. This lack of information may be related to space limitations when submitting studies for publication in conferences or journals. Which is aligned with our findings that the most detailed and complete patterns are found in books, technical reports and online databases.

With space being such a valuable resource in journal articles and conference papers it raises the question of what is the best way to gather patterns in a way that is usable for future research as well as practitioners. Within this SLR, books and technical reports focused more on the patterns themselves, while journal articles and conference proceedings focused more on methods or presented sample patterns. Although P24 fused the two, by publishing each new kind of pattern in journal articles while at the same time keeping the online repository updated they were able to present a high amount of information on their work while avoiding the space limitations scientific publishing implies.

Based on the initial definition, patterns solve a repeating problem. In the case of EAPs then, the problem is a deficit in the organization as perceived by the stakeholders. Thus, EA practitioners would be in the best position to detect both the problems that repeat themselves, as well as the solutions that could be reused to meet them. This line of thinking means that researchers must be in contact with practitioners far and wide to expand patterns, or that researchers must be practitioners as well. This poses a limitation, or it could be taken as an opportunity to include practitioners in future works on patterns.

Fields Researching Patterns. On Fig. 2 we have classified the four main types of patterns found through the SLR in terms of the four main layers of the Archimate Language [13]. Based on [10] and the Archimate constructs they use to describe a Business Model [16] it can be concluded that Business Model Patterns are confined to the Strategy and Business layers. While the Business

Process Patterns was mapped to the Business Layer due to its' scope, Business Processes, which is enclosed in this layer.

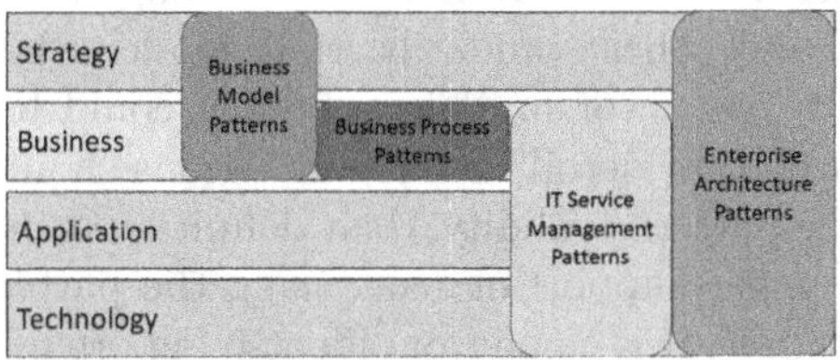

Fig. 2. Classification of fields - adapted from [6].

Enterprise Architecture Management Patterns are omitted from this graph. This is due to their focus on the practice of EA itself, the methods enterprise architects use to gather information, as well as how they present it to stakeholders. As such these patterns are more akin to an EA framework and methodology. Thus, although the focus of the EAM patterns is the EA practice itself, this is different to the EAPs, which is the enterprise.

Methodologies for Pattern Extraction. A Challenge faced by all the fields in this SLR was the gathering of the patterns. While the Business Model Patterns can be extracted from an organization through literature reviews and researching real-world organizations. On the other hand, the patterns presented by the EA field are based on the authors' experience. For example P14 mentioning that the source of these patterns is the day to day experience of the practitioner and detecting a repeating problem. This mention of a repeating problem is also present in other works on patterns [1]. Another avenue seen is presenting a framework that is built with discreet choices of concepts and then building patterns exhausting all possible combinations. This method is used by P2, P11, P12, and in a more limited way P24 which strives for an exhaustive work but do not explicitly show all possible combinations. Finally, P22 took current standards and made them into patterns.

The 50% of studies in the sample-based the patterns from the author's experience. When reviewing the definitions by [1] the source of patterns is an experienced professional experiencing the same problem again and again, so this would explain this method's commonality. What is missing, however, is the argumentation on the existence of the problem, one that fosters the need for a pattern in the first place. With most of the studies in the sample were missing a framing of the problem they're set to solve.

The second most common method of extracting patterns is the literature review (25%), which extracts patterns from current literature. Determining what method is used in the sources of these literature reviews is outside the scope of this SLR. These sources apply their own methods.

The studies that deviate from the literature review and author's experience are P5 and P17 which base their patterns on standard practices. In the case of P5, it's the ITIL library, which dictates practices on how to operate the IT function of an enterprise. With P17 it's a framework of their own which extends upon ITIL, Cobit, CMMI and other standards, that models the entire IT function as an enterprise by its' own worth. This approach could be expanded upon to include other standards that detail how organizations should act.

In order to give the patterns validity, their definition must come accompanied by some kind of argument supporting that using the pattern indeed solves the problem [1]. In the sample, this came from identifying organizations that worked under patterns in question (P20, P1, P9, P11, P14, P16, P21), from personally applying the patterns (P18), from having practitioners apply the pattern (P8), from building business cases (P5), or from having other researchers validate the patterns found (P1, P24). However, the majority of the works in the sample describe no manner of validation (P2, P3, P4, P6, P7, P8, P10, P12, P13, P15, P17, P19, P22, P23), a majority of the sample.

4.2 Characteristics of Sustainable Organizations

In this section, the results of an SLR performed to find the characteristics of sustainability is described. First, a classification of the characteristics found is offered. The resulting classes are then compared to concepts of the Archimate language, in order to identify how feasible it is for an architect to express the characteristics of sustainability. From this comparison, the final set of classes, as well as their descriptions and constraints, is produced.

Classification of Characteristics. To better analyze, and apply, the set of characteristics shown above, they are classified. To do so, the classification used by S2 is taken as a basis, which uses an extended version of the BMC. The changes introduced in S2 are two new components: Take-back systems and Adoption factors. Where the former describes the mechanisms needed for some sustainable organizations that need to recall products, e.g. for maintenance or replacement. The latter describes factors that can not be attributed to any of the other components, mostly related to the capabilities of the organization. These, and all other classes, are explained in further detail below. In Tables 1 and 2 this classification is shown.

The new building block, *take-back systems*, is in it is core a combination of Customer Channels and Customer Relations but in a reverse direction. It describes, for example, how an organization manages the end of life of their products. In order to maintain the BMC succinct, the take-back building block will be fused together with Customer Channel and Customer Relations.

4.3 Description

In Fig. 3 each class and the total amount of characteristics found to belong to it are shown. Also shown is whether the classification comes from the source or the author. It must be mentioned that the ratio of characteristics classified under the classes mentioned above by the source, when compared to the ones classified by the author, are 58:151. Where the classification of adoption factors

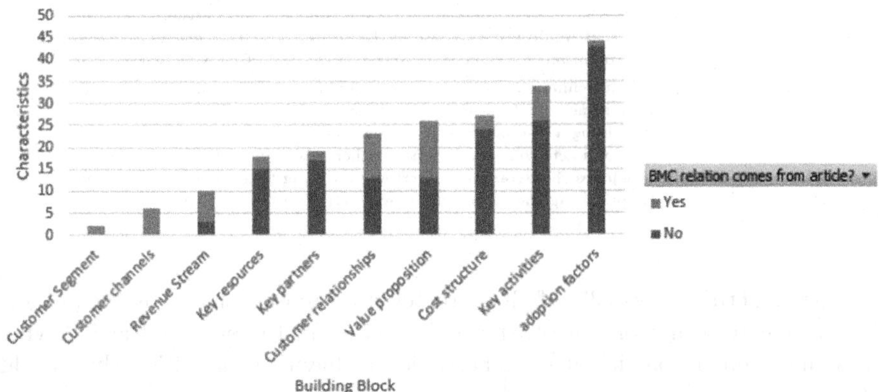

Fig. 3. Characteristics classified.

Table 1. Classification of characteristics.

Building block	Description of characteristics classified to this block
Customer segments	In this SLR there are few characteristics found that are attributed to this class, one is environmentally aware customers for example. It could be argued that environmentally aware customers is a segment made of innovators and early adopter, the first two segments to adopt new services [17]
Customer relationships	Co-creation of the value proposition, community relationships
Channels	In this SLR the characteristics found to belong to this class all focus on the relationship between the organization and their customer. Among them some are focusing on the co-creation of the value proposition, this could be including customers in the product design phases for example (C31, C32, C77). Some others focus on broadening the concept to include relationships with the community (C8, C110, C111, C112), which reflects one of the pillars of sustainability, the society the organization embeds itself in (P21) [9]. Also in this block are the characteristics related to take-back systems (C27,C49)
Value proposition	Those that focus mainly on digital communications (C3, C28, C29, C30). And, as mentioned before the building block focusing on how the organization manages the reverse channels (C47, C48)
Key activities	Varied, those focusing on the implementation of agile practices, process re-engineering (C194, C41, C65, C124, C147, among others). Another example is the automation of tasks using Information Systems (C117, C104, C105, C42). Many others describe activities related directly to sustainability, e.g. harvesting rainwater, conserving natural resources, and avoiding using slave labour (C126, C128, C131, C204, C152, C151)
Key Resources	Varied, some refer to the IT systems for communication, Business Process Management, multimedia, among other uses (C120, C121, C122, C199...). Others refer to the sustainability of the inputs of the organization, using bio-materials and waste (C200, C15). In a similar vein, some characteristics refer to the use of better-insulated spaces, using natural light, using water-efficient appliances, using solar panels, among others; all to diminish energy consumption (C55, C127, C206, C207, C205, C208, C209)

(continued)

Table 1. (*continued*)

Building block	Description of characteristics classified to this block
Key partnerships	Characteristics that focus on coordination along the supply chain and adopting Supply Chain Management practices (C72, C114, C116, C119, C130, C192, C138, C85, C16). Others focus on the selection of these partners, describing that organizations must choose suppliers and partners that are efficient, that would reduce waste in the entire chain, that ensure dignified working conditions for their employees (C133, C134, C54, C58, C66, C72)
Revenue stream	Those that define new ways of generating revenue stream, or that affect the stream in some way. For the former, examples include the revenue generated from providing a product as if it were a service, engaging in the circular economy, and achieving funding from other actors (C33, C34, C35, C67). The latter includes characteristics describing reducing time to market, avoiding waste, optimizing the supply chain, etc. (C18, C20, C17, C19)

and cost structures have 33% of all characteristics, most assigned by the author. This is due to many of the other studies referring to some of the following ideas: management participating in activities, or having capabilities, that would catalyze the adoption of sustainable practices; management including the needs of stakeholders beyond the customer; business flexibility, or the ability to change. On the cost class, many sources describe the organizations wastewater, physical waste, emissions, and end of life of products of organizations.

Table 2. Classification of characteristics cont'd.

Building block	Description of characteristics classified to this block
Cost structure	Divided into two. The first are those that focus on the costs of production and sourcing green materials (C21, C22, C46, C86, C87, C88). The second are those that focus on the wastes of the organization, e.g. the emissions it generates and physical wastes, like paper or (C95, C155, C156, C53, C68, C69). The latter also includes characteristics describing the use of triple bottom line practices, where the accounting of the organization also includes environmental and social costs (C196, C175, C178)
Adoption factors	An example are the characteristics focusing on learning at an organizational level, as well as educating the community it is embedded in (C52, C74, C75, C107, C136, C189). Another example are the characteristics describing the awareness of the organization, the knowledge the organization already has on sustainability (C144, C76, C71, C50). Lastly, there are characteristics describing the organization's management and how they conduct business, e.g. the flexibility of the organization to change how they operate and implement any of the characteristics; and explicitly including sustainability in business models (C193, C188, C189, C190, C146, C149, C153, C154, C123, C125, C84, C106, C70, C62, C63, C64)

4.4 Applicable Characteristics

The following section relates the characteristics with Archimate concepts, mainly through the use of their assigned classes. Previous research is identified and expanded.

Previous research has analyzed the BMC meta-model and mapped it into Archimate concepts [10]. They found that, as the BMC represents an entire organization in the most abstract terms, that the Archimate concepts each building block relates to are concentrated in the Strategy and Business layers. Although, thanks to the flexibility of the language, one can specify the Strategy layer's resource into the lower layer's concepts, e.g. an application component. In Table 3 the BMC building blocks, including the two extensions mentioned above, are shown. With an X the relationship studied by [10] is shown and with a C the relationship proposed after reviewing the Sustainability Characteristics. The reasoning behind each assignment is explored below.

Table 3. Relation between extended BMC and Archimate. X means relationship is defined by [10], C means it is proposed by this study.

Class/Layer	Motivation	Strategy	Business	Application
Adoption factors	C			
Cost structure	X			
Customer channels		X	X	C
Customer relations	C	X	X	
Customer segment			X	
Key activities	C	X	C	
Key partners	C		X	C
Key resources	C	X	X	X
Revenue stream		X		
Value proposition	C	X	X	C

The adoption factors building block, as it's a new concept being introduced in S2, is not present in the previous studies [10]. As such these characteristics have not been related to Archimate concepts before and had to be analyzed to find which layer's concepts relate. As most of these relate to intangible things, e.g. leadership creating strategies for reducing waste (C62, C63, C64), that describe how the organization should act then it is closely to with the Motivation layer concepts. In this layer, the architect can represent the drivers of multiple stakeholders, as well as define constraints, goals, and value. With these concepts the characteristics found in this class can be built. However, it calls for a wide implementation of the motivation concepts to the more traditional layers (Business, Application and Technology).

The cost structure is found to be problematic within the concepts of Archimate, as no concept ties directly to it. The language specification mentions that object attributes could be used for costs [22]. [10] also faced difficulties when defining cost, they decided on using *negative* value. These obstacles are maintained in this study, as the concept is expanded to include environmental and social costs as part of the triple bottom line approach described by some characteristics (C175, C177, C178). A solution

As the Key Partnerships building block refers to the actors, organizations, and other parties the business has to partner with; then it had to be extended beyond previous studies [10]. For this study, in order to avoid overloading the other building blocks, the characteristics assigned to Key partnerships goes beyond external parties. This extension is seen in the addition of the motivation and application layers.

Along the verticals, it can be seen that two layers are added to multiple building blocks: Motivation and Application. The former reflects the characteristics that call for activities, or resources, to be included with a sustainability goal or constraint in mind, e.g. avoid using slave labour in the business processes (C152). The latter reflects the characteristics that call for virtualization (C28, C42, C37, C25), digitalization or the implementation of IT in some way (C29, C3, C30, C6, C7, C117, C104, C105, C120, C121, C122, C199, C90).

The need to include motivation concepts is interesting in its own right, as it shows the limitation of the Archimate language to show the impacts or consequences of the objects. This may be simple to assume under normal circumstances, the basic goals and constraints of what an organization does is to generate revenue and will generate a cost. While in sustainability the costs go beyond monetary and have to include every impact the organization produces on the environment and society. This change may call for an extension of the language, or for EAPs to be written in a way that they include sustainability characteristics explicitly.

5 Methodology for the Specification of EA Using Patterns

Although the first two knowledge questions posed above were answered, there were still a few building blocks missing before a methodology could be designed. First, from the sample of EAPs gathered, a small subset of them were described using a common language. Second, besides the BMI patterns, no EAPs were focusing on sustainability. Third, any methodology would depend on the combination of multiple EAPs, alas no studies in the sample mentioned ways to do so. Thus, the following subsection focuses on these design problems and the solutions proposed.

5.1 Designing the Building Blocks

Translating the Subset of Patterns to Archimate. The EAPs found in the SLR were described in multiple ways, written and with diagrams, some were described using Archimate, while some others were already described using other modeling languages. A subset of the latter were translated by the author concentrating on being as aligned with the source as possible. All other patterns translated had to be drawn based on their description, this was done by the author at the best of his abilities. The resulting diagrams can be accessed in the Appendix. There, both the source file and extracted images are present.

Proposing New Patterns. After performing the translation, some gaps were identified and in some of the EAPs, i.e. the applications are deployed to nodes in the internal facilities of the organization. However, as is found in the Characteristics found in the second knowledge question, sharing and virtualization is a characteristic of sustainable businesses, and currently, a popular representation of this are Cloud Services.

With this gap in mind, three new EAPs are proposed: NewPattern001, NewPattern002, and NewPattern003. Representing the use of a SaaS, PaaS, and IaaS service respectively. These were built using the guidelines established by [15]. In Fig. 4 the NewPattern001 is shown, the other two follow the same structure. Under the Appendix, the diagrams and source files of these EAPs can be found.

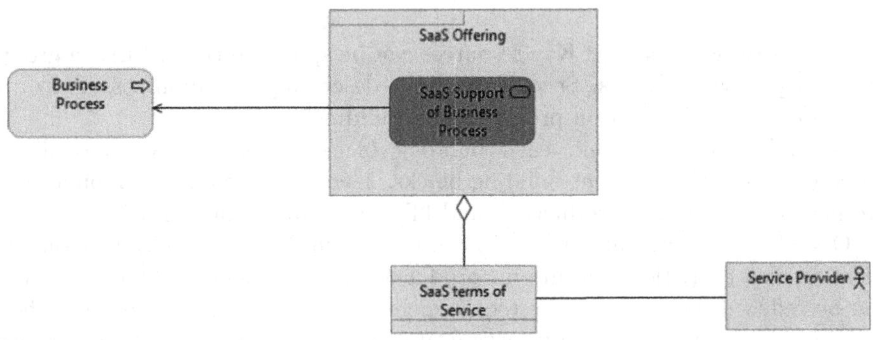

Fig. 4. NewPattern001.

Proposing Sustainable Patterns. Following this gap, the need arises for another set of patterns. Patterns that describe the characteristics of sustainability that an organization wants to implement.

In order to be applicable in a large range of settings, an approach similar to P12 is taken. Here the different characteristics, as they were compared to Archimate concepts in Table 3 are used as a basis. In terms of Archimate, the scope of these patterns is limited to the Business and Application layers of the organization for this exercise. In terms of sustainability, the scope is defined by the characteristics that can apply to the layers mentioned: Customer channels, Customer relations, Key activities, Key resources, Key partners, and value proposition. From these building blocks, and within the scope, some sustainable patterns are proposed. These were cataloged in the Appendix.

The resulting patterns use Archimate concepts, except for Key partners which does not include application layer concepts. Instead, the characteristics describe business concepts with certain requirements, however, it's arguably impossible to fulfil the requirements and constraints of the characteristics without application support.

After developing the patterns it was clear that within the scope of some of the building blocks there were separate groups of objects. The building blocks where this is seen are Customer channels, Key activities, and Key resources. All other blocks were able to be described as one coherent pattern. These groupings are explained as follows.

The patterns describing customer channels may be divided into two: one focusing on the customer as the buyer of the value proposition; and another focusing on the relationship with other stakeholders like government.

The patterns describing Key activities may be split into six groups. First, a group detailing the requirements the waste management process must follow. Second, the requirements related to the implementation of new capabilities. Third, activities and requirements related to product design and production activities. Fourth, a group that applies to any activity. Fifth, the goal of being a sustainable organization and the principles that go with it. Sixth, requirements related to IT systems.

The patterns describing Key resources can be split into three. First, a group describing service centres. Second, a group describing IT resources. Third, a group describing production processes and buildings.

Beyond divisions within each building block, there were commonalities observed across the different building blocks. There were mainly two found, the product, its design and production; and IT aspects of the organization.

One of these sustainable patterns can be seen in Fig. 5, where the characteristics for key activities focusing on products are implemented. These patterns can be reduced, i.e. some of the requirements and goals can be removed, when needed by architects. They serve organizations as they need to explicitly present their sustainability requirements for the future.

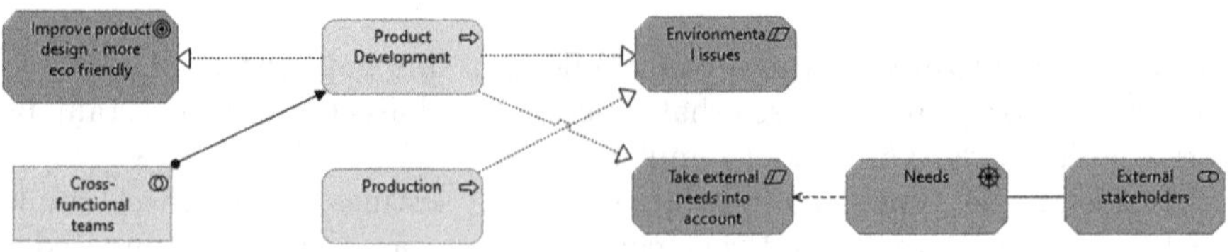

Fig. 5. Key activities - product.

Requirements. As part of the preliminary phases the requirements of the guiding pattern and the patterns added to the subset have to be found. This can be achieved using Archimate diagrams, as shown in Fig. 6. By describing the patterns that are derived from pattern X and the patterns Y that fulfil it.

It is possible to document all the requirements that each pattern introduces, which would make the creation of the subset a matter of choosing the guiding pattern. The guiding pattern would have its' requirements previously defined, as well as those of all other patterns. Thus, creating the subset based on the relationships between patterns documented previously would be automated.

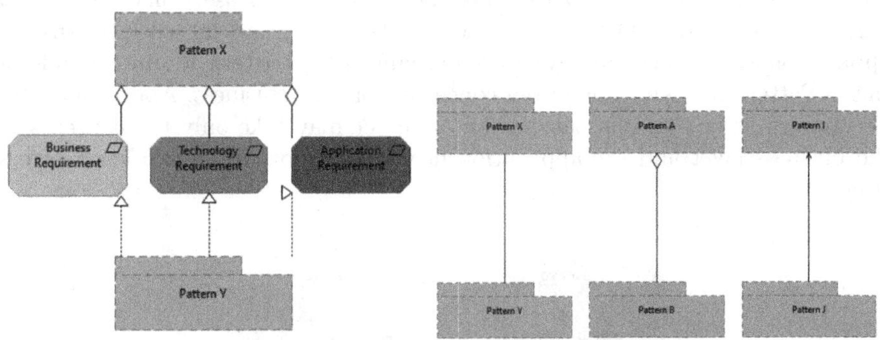

Fig. 6. Requirements and patterns & patterns combined.

Combining Patterns. As previously stated in the specification phases, patterns have to be combined to form a final EA specification. During this study, five ways of combining patterns were found. The first is a parallel combination, in which there is no relationship between the patterns.

The second way of combining patterns is shown between pattern X and Y in Fig. 6. Here, a description of two patterns that are related to an association is shown. This is the most complex form of combination as it revolves around the contents in common between the patterns. These commonalities may be present in one or many objects, where the higher the overlap the more complex it is to combine them. An example would be combining patterns of supply chain and production, as they are closely related concepts.

The third way of combining patterns is shown in Fig. 6 between pattern I and J. This combination happens mainly when pattern J is focused on a lower layer than pattern I. For example, a pattern showing how a business process is supported by applications. In this form of combination, P12 defines exhaustively the ways concepts of the Business and Application layers can be combined with each other. However, such patterns do not exist between the application and technology layers. Between the business and the motivation and strategy layers, there is also a noticeable lack of patterns describing how one layer may serve the next.

The fourth way of combining patterns is through the aggregation relationship. This case is shown in Fig. 6, between pattern A and pattern B. Here pattern B provides a narrower focus and higher level of detail to the ideas presented in pattern A. An example would be adding individual Business Functions to a more abstract pattern like a Business Model Pattern.

A final aspect of the combination of patterns is their decomposition. As shown in Fig. 6. In some cases, EAPs describe multiple layers of EA, Business, Application, and even Technology. However, due to the ways patterns can be combined with each other, and especially the way this happens between layers, an architect may decompose a pattern in its' layers. This would then give the architect freedom of building the lower layers as they please. This is shown in Fig. 6, where Pattern Q is decomposed in its layers, and its sub-patterns for application and technology layers are exchanged for Pattern R. An example of this is Pattern173, which describes concepts from the Business, Application and Technology layers. In this pattern, an architect may take only the concepts of the business layer and the application needs, and replace the two lower layers (Fig. 7).

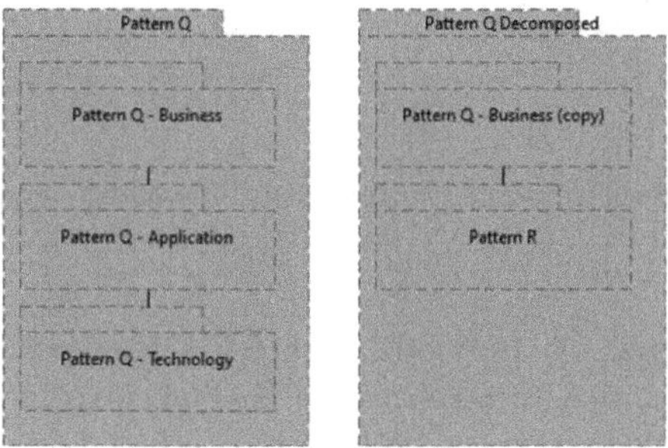

Fig. 7. Patterns decomposed.

Following the ways of representing patterns, and their relationships, one may model the specification in terms of the patterns and their relationships. This would offer a high abstraction view of the organization as defined by the solutions being re-used.

To avoid falling into the Yet Another Model trap, the proposed methodology is based on the ADM. Focusing on the first phases: Preliminary phase, Phase A Architecture Vision, Phase B Business Architecture, Phase C Application Architecture, and Phase B Technology Architecture. During an EA specification project, each of these phases will be executed in sequence by a team of architects, where phases B, C and D focus on the description of the organization's future (or old) state. It is during these phases, that the use of the Archimate language is concentrated.

5.2 Preparatory Phases

The proposed methodology assumes that the architect wishes to describe a future state of the organization that is sustainable. As such, the method starts with the preliminary Phase and Architecture Vision phases. During these phases the focus will be on defining the main constraints of the EA. With the most important one being that the resulting organization must be sustainable. This constraint will frame all other tasks in the methodology.

In the same preliminary phase, a second constraint will be added, a guiding Pattern. In this method, the guiding pattern will serve as a general abstraction of the target's organization. Within the findings of this thesis, the authors find that an acceptable guiding pattern is a Business Model pattern in combination with an operational model pattern (as defined by P11).

Next is Phase A Architecture Vision, in which a subset of EAPs is selected. The constraints resulting from the previous phase introduce problems and questions as is natural from their level of abstraction. E.g. a Business Model Pattern that calls for interactions with customers will introduce questions like How will the customers be contacted? Who will do it? And so on. These questions and problems will feed a search for EAPs that may solve them. At this phase, the search is performed only to form a subset of EAPs with all possible solutions to problems, the constraint, and guiding pattern pose. This is similar to the Palette artists have at hand while they're developing their works.

The detailed process of finding the problems and questions is similar to that of requirement elicitation. The precise nature of how these should be is outside the scope of this study and is left to the experience of each architect when following the proposed method. However, it is an iterative process in two aspects. First, as the selection of EAPs introduce problems and questions of their own solutions for these, in turn, have to be included. Second, by nature of the layered approach of ADM, the architect must add patterns that solve the problems in the lower layers of each solution, i.e. in a manner of supporting pillars, Business Layer patterns will need Application Layer patterns to stand upon.

In Fig. 8 a view into such a phase is given. This was taken from a case study performed using the methodology. In the figure, the guiding pattern can be found at the top left, from which a series of business layer requirements are detected and listed using the before mentioned notation. On the right side of the figure, the patterns found to fulfil the requirements are gathered. These patterns are linked using the Realization relationship as proposed in the notation above. Similar diagrams are prepared for each layer, Application and Technology.

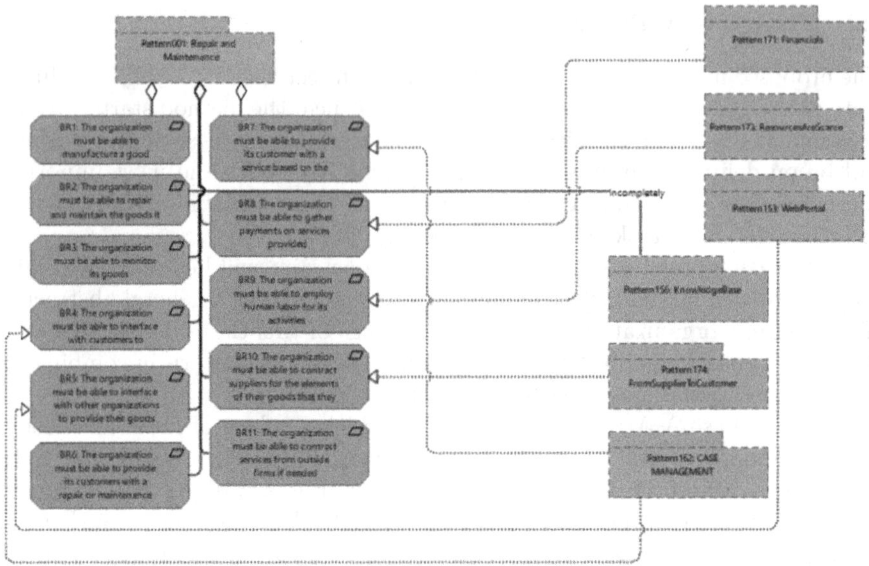

Fig. 8. Preliminary phase example.

5.3 Specification Phases

Following on comes the specification of the EA, through phases B, C, and D. These phases each focus on a different layer, on the Business Layer, the Application Layer, and finally, the Technology Layer. These layers will be built using the patterns in the selected subset, by selecting and combining them, thus producing the EA specification. The selection of each pattern will be based on its' compliance to the constraints defined in the previous Phase and each other, e.g. when presented with different EAPs in the subset that can solve a specific requirement, the one that can arguably support a more sustainable organization will be chosen. Multiple patterns can be selected to solve the same problem, in the same way that organizations may have their own retail stores and sell their goods in other stores at the same time.

In Fig. 9 the result of these phases can be seen. Using the proposed notation and methodology in a case study, a specification of EA was built [7]. With this single view, practitioners could have a simple way to communicate how an entire organization functions, by leveraging a shared understanding of the solutions being reused.

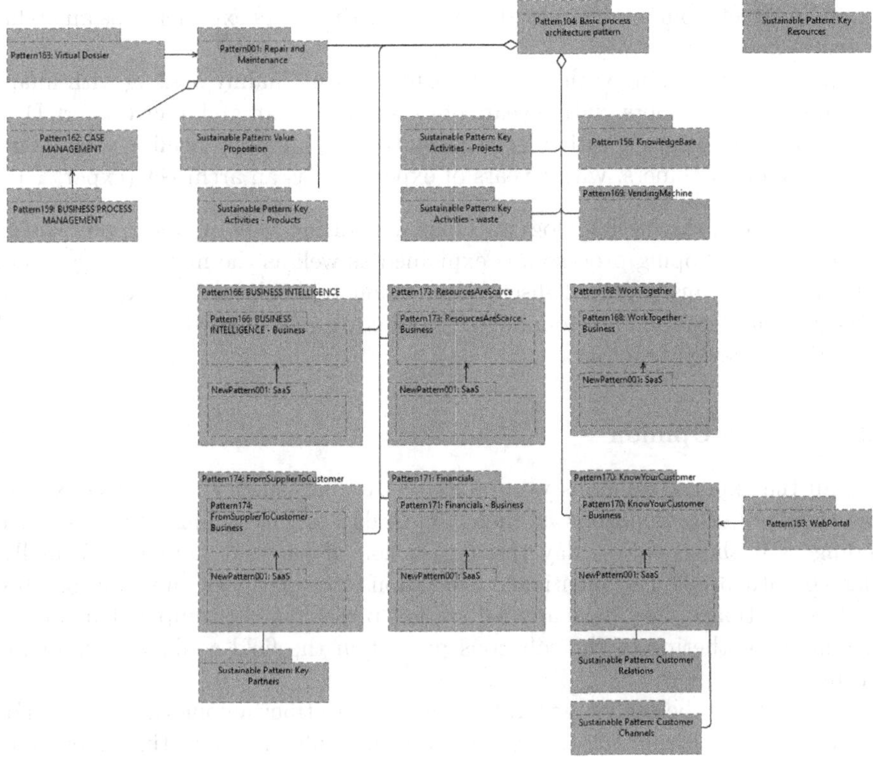

Fig. 9. EA specification using patterns.

6 Validation of the Methodology

Following the Design Science methodology, after developing a solution it must then be validated, prior to its implementation [25]. In order to do so, a panel of experts is assembled and shown the method. These same experts are then asked to partake in a questionnaire based on the Unified Theory of Acceptance and Use of Technology (UTAUT). This theory is selected as it has been proven to be effective in predicting the adoption of Information Systems, with an r-squared of 0.5 [24]. However, the sample size of this study means that the results are valid for qualitative analysis only. This questionnaire is present in the Appendix.

6.1 Expert Panel

The five experts that agreed to take part in the validation process are five:

1. A professor for EA in a dutch technical university, with more than 10 years experience (expert A)
2. An architect from an important bank in the UK, with 8 years experience as an architect (expert B)

3. An architect from a major cloud provider, with 2 years experience as an architect (expert C)
4. An architect working with a large consulting firm, mainly dealing with financial sector customers, with 8 years of experience as an architect (expert D)
5. An architect working with a large consulting firm, mainly dealing with financial sector customers, with 6 years of experience as an architect (expert E)

For each expert, the methodology was explained using a set of powerpoint slides. The overall developing process was explained as well as the methodology itself. Afterwards, an unstructured discussion session took place. Finally, they were asked to complete the questionnaire. It is important to note that expert D and E were interviewed together.

6.2 Experts Opinion

Overall the experts' opinions were positive. A commonality across the experts was their re-use of solutions in their day-to-day practice. This was a positive finding as it shows that today the application of patterns happen informally. Aligned with previously mentioned limitations, the patterns and solutions the experts mentioned they have applied are not present in the sample. This means that in the gathering of the solutions present in the field could be a valuable future project.

The patterns themselves generated different reactions among the experts. For expert B, standardizing all solutions to problems could arguably stifle innovation. As all organizations would be simply a combination of publicly known solutions. However, they also mention that what would make them different is how they implement them. For expert C, patterns should easily communicate their value and applicability, otherwise any architect would rather use either a different pattern or design their own solution. For expert A, translating patterns from different fields was of interest, mainly the method chosen to do so.

As for the sustainability focus of the methodology, experts C, D, and E made mention of it. For experts D and E, this could be due to their own experiences, where they have seen sustainability being used as a criterion in one of their customers. As for expert C, this could be their relation to commercial teams made them think of a pattern-based approach to present a highly customized sustainable transformation project to an organization. By using publicly available reports of a specific organization as a starting point and then applying the pattern-based methodology for a hastened EA specification of a future sustainable state.

The focus on implementation was something common for experts B and C. Both thought of it as a central differentiating aspect of organizations, as assuming that all organizations use patterns then what makes them different is how they implement them. Expert B even went as far as proposing that a possible next step to this study is the identification of implementation patterns that organizations could use depending on their own capabilities.

However, exceptions were experts D and E that mentioned their worry on the sustainability concepts were introduced to EA. This is reflected in earlier chapters of this study, where the concepts of sustainability are hard to relate but are finally described using Motivation layer concepts. Experts D and E also mentioned that they have not seen such concepts used in their practice. They also commented on their confusion regarding the relationship between ADM and the proposed methodology. This specific question was also asked by expert C, which means that further improvements to the methodology's description are necessary to make this relationship more clear.

6.3 Questionnaire Results

Table 4. Questionnaire results - constructs.

Construct	Average	St. dev.	Min.	Max.
Performance expectancy (PE)	3.9375	2.205107707	1	7
Effort expectancy (EE)	5	1.673320053	1	7
Attitude toward using technology (ATUT)	4.125	1.821171784	1	6
Social influence (SI)	2.875	1.962141687	1	5
Facilitating conditions (FC)	3.5	2.033060091	1	7
Self efficacy (SE)	4.4375	1.412739655	2	7
Anxiety (ANX)	1.375	0.6191391874	1	3
Behavioral intention (BI)	3.75	2.050498831	1	7

Overall the results of the questionnaire are positive. With the highest score being **EE**, which is aligned with the contribution goal of this study, helping architects drive change faster. There is an exception to be made for Experts D and E, as their evaluation of the method was negative (Table 4).

As can be seen in the main results, the **EE** and **SE** constructs are the highest ones at and also have a low standard deviation. With this result for **EE**, the experts evaluated that the methodology would be clear or easy to use and that they perceive that mastery at using the methodology is easy to obtain. This corresponds with their comments on the methodology, that they re-use previous solutions in their day-to-day activities. And, with this result in **SE**, the experts evaluated that they were capable of using the methodology. Inversely, the **ANX** construct has the lowest score and the lowest deviation. This means that experts evaluated that using the methodology produces no fear or intimidation. This reflects the high degree of experience all the experts approached have, they have confidence in their capabilities.

The results in the **PE** construct are positive, except for the evaluation of expert D and E. This means that experts evaluated the methodology as being useful and increasing their productivity. The **ATUT** construct had a similarly positive result meaning that experts perceive the methodology interesting, fun, or a good idea.

The **BI** and **SI** constructs had a split score, half of the experts evaluated them positively and half as negative. For the former, representing that they are not certain they will use the methodology in the future. For the latter, it may be because this method was not shown to any of their superiors, as all items relate to external social influences.

7 Conclusions

The result of this study is a methodology for the specification of EA for sustainable organizations using patterns. In preparation for its design, there was a need for the characteristics of sustainability to be described explicitly in the Archimate language. This was due to the lack of sustainability concerns in the EAPs found, beyond the sustainability business model patterns. The methodology was designed extending the ADM, building upon its same phases to make its adoption simpler to organizations already using the TOGAF. As part of the methodology, a notation is proposed for describing the relationships between patterns implemented in an organization. This specific artifact could also be of use in organizations that are implementing recommendations from multiple standards, e.g. ISO certification-related aspects.

First, an SLR was performed to gather the state of the literature for EAPs. The patterns found were heterogeneous in multiple ways, like the field of research they originated from, their scope, the way they're represented, and how they're extracted. However, the sources used were scientific literature, which meant practitioner sources were excluded. For example, industry standards and best practices were not included in the set of patterns. This meant that for some of the studies in the sample the space limitations of journals and conferences were an obstacle. As these are usually limited to a relatively short number of pages, it is difficult to present all the information that may be necessary when describing patterns. This obstacle is missing in the books in the sample, where the authors could elaborate on the details of the patterns, making them more valuable.

Second, an SLR was performed to gather a set of characteristics of sustainable organizations present in literature, classified based on the BMC building blocks. All of the characteristics were relevant, but by using the Archimate language components a way was found to include them into EA artifacts for each building block. This set is likely incomplete, as the literature on sustainability is increasingly wide and complex. For example, there are more than 100 different definitions on Circular Economy alone [8,11]. This makes the selection of characteristics difficult, as just the choice of the word "characteristic" already filters concepts and requirements from this study. A more comprehensive and exhaustive study may result in a more valuable set of characteristics.

Among the characteristics of sustainability found there were commonalities in the concepts. E.g. a common concept across many characteristics was the use of digital tools. Applying text mining techniques might produce clusters of characteristics, which could then be used to propose more abstract characteristics.

In the field of EA, the use of sustainability concepts is not common, and including them in the practice was not straightforward. This difficulty to express sustainability concepts using EA artifacts may be because of, or causes, the absence of sustainability aspects in EA practice. In this study, the obstacle of expressing these concepts was surmounted by using the motivation layer concepts of Archimate. With them, it was possible to describe the requirements sustainability imposes on the operation of an organization, e.g. select suppliers that are not using slave labour. However, a step further would be to design structural and behavioural components in an organization that implicitly implements the requirements used in this study. These newly designed components could then become patterns to be adopted widely.

7.1 Contributions

The contributions to academia are multiple. First, a systematic literature review producing an extensive set of EAPs. By applying an exhaustive approach, the resulting review has expanded the existing definitions of EAP beyond just the field of EA, including fields like BMI and BPM. Second, a systematic literature review producing a set of characteristics of sustainability and circular economy. Extending the knowledge of the aspects organizations need to be sustainable, as today there are many definitions of sustainability, having an aggregated list of such characteristics is needed. Third, extending the EA field, a novel notation based on the Archimate language is proposed to describe the ways architects may combine patterns, as well as a categorization of the ways they may be combined with each other.

As for practitioners, the methodology proposed can be applied together with the patterns found in scientific literature and industry standards, simplifying EA projects by using it as a common language describing solutions proven to work in other organizations. Second, the set of sustainability characteristics classified in terms of the building blocks of the BMC offer organizations today a classified list, based on scientific literature, of characteristics they can implement to be more sustainable. Third, a notation was proposed to describe the relationships of patterns used in an organization. This allows for organizations to describe the patterns (best practices, standards, and other re-usable solutions) they are implementing in their current and future states, resulting in a clear abstraction useful.

7.2 Limitations

The main limitation is the language used, where the concept of the *pattern* may not be used commonly, especially in other fields of research. An example is the Human Resource (HR) Architectures proposed by [14], which show patterns in HR management. This is present also with the sustainability keywords, where a more in-depth exploration of the terms used in the field may have gleaned more characteristics. However, the main limitation of the methodology, which is confirmed by the panel of experts, is the lack of a real-world case study. Until this methodology is applied, in a real organization looking into changing towards a more sustainable future version, then its' true weaknesses will not be perceived.

7.3 Future Work

For the academic community possible future work continuing with this study could be its' application in a real-world organization that is looking for a sustainable transformation. This would reveal the weaknesses of the methodology if any.

Another avenue of research is to define a way in which patterns can be extracted from organizations, either through literature or with the help of its members. This is similarly to the approach taken by P20, which compiled a classification of Business Models, and offered a base by which both academics and practitioners can study and compare organizations. Another way to extract them would be to focus on architects. How they re-use their previous solutions, including those found in literature or those their peers propose. This would offer more understanding of how to better serve these architects with more patterns than those they can already access.

Finally, the implementation of this methodology into a traditional EA project using TOGAF is needed. Through its real-world usage, a more thorough validation would be achieved, allowing its improvement into a more valuable tool for the field.

Appendix

Under the git repository https://github.com/robertorgarcia/EAPatterns the appendices to this study can be found. Specifically the following:

- The study samples used in the SLRs
- The results of the SLRs (EAPs and Characteristics)
- The intermediary classifications of the results
- A subset of EAPs translated to Archimate
- A set of proposed EAPs
- A case study built using EAPs

References

1. Alexander, C., Ishikawa, S., Silverstein, M.: A Pattern Language: Towns, Buildings, Construction. Oxford University Press, New York (1977)
2. Allen, M., et al.: Technical summary: global warming of 1.5 c. An IPCC Special Report on the impacts of global warming of 1 (2019)
3. Braungart, M., McDonough, W., Bollinger, A.: Cradle-to-cradle design: creating healthy emissions - a strategy for eco-effective product and system design. J. Cleaner Prod. **15**(13–14), 1337–1348 (2007). https://doi.org/10.1016/j.jclepro.2006.08.003

4. Ceballos, G., Ehrlich, P.R., Dirzo, R.: Biological annihilation via the ongoing sixth mass extinction signaled by vertebrate population losses and declines. Proc. Natl. Acad. Sci. **114**(30), E6089–E6096 (2017). https://doi.org/10.1073/pnas.1704949114
5. Gamma, E., et al.: Design Patterns - Elements of Object Oriented Software Architecture (1995)
6. García-Escallón, R., Aldea, A.: On enterprise architecture patterns: a systematic literature review. In: Proceedings of the 22nd International Conference on Enterprise Information Systems, pp. 666–678. SCITEPRESS - Science and Technology Publications (2020). https://doi.org/10.5220/0009392306660678
7. Garcia-Escallon, R.: Enterprise architecture patterns: Supporting sustainable development. Ph.D. thesis, University of Twente, Enschede, The Netherlands (2020)
8. Geissdoerfer, M., Savaget, P., Bocken, N.M., Hultink, E.J.: The circular economy - a new sustainability paradigm? J. Cleaner Prod. **143**, 757–768 (2017). https://doi.org/10.1016/j.jclepro.2016.12.048
9. Geissdoerfer, M., Vladimirova, D., Evans, S.: Sustainable business model innovation: a review. J. Cleaner Prod. **198**, 401–416 (2018). https://doi.org/10.1016/j.jclepro.2018.06.240
10. Iacob, M.E., Meertens, L.O., Jonkers, H., Quartel, D.A.C., Nieuwenhuis, L.J.M., van Sinderen, M.J.: From enterprise architecture to business models and back. Softw. Syst. Model. **13**(3), 1059–1083 (2012). https://doi.org/10.1007/s10270-012-0304-6
11. Kirchherr, J., Reike, D., Hekkert, M.: Conceptualizing the circular economy: an analysis of 114 definitions. Resour. Conserv. Recycl. **127**(September), 221–232 (2017). https://doi.org/10.1016/j.resconrec.2017.09.005
12. Lankhorst, M.M., Proper, H.A., Jonkers, H.: The architecture of the archimate language. In: Halpin, T., Krogstie, J., Nurcan, S., Proper, E., Schmidt, R., Soffer, P., Ukor, R. (eds.) BPMDS/EMMSAD -2009. LNBIP, vol. 29, pp. 367–380. Springer, Heidelberg (2009). https://doi.org/10.1007/978-3-642-01862-6_30
13. Lankhorst, M.: Enterprise Architecture at Work. The Enterprise Engineering Series, 3rd edn. Springer, Heidelberg (2017). https://doi.org/10.1007/978-3-662-53933-0
14. Lepak, D.P., Snell, S.A.: Examining the human resource architecture: the relationships among human capital, employment, and human resource configurations. J. Manag. **28**(4), 517–543 (2002). https://doi.org/10.1177/014920630202800403
15. Nardi, J.C., et al.: Service commitments and capabilities across the archimate architectural layers. In: 2016 IEEE 20th International Enterprise Distributed Object Computing Workshop (EDOCW), vol. 2016-Septe, pp. 1–10. IEEE (2016). https://doi.org/10.1109/EDOCW.2016.7584386
16. Osterwalder, A., Pigneur, Y.: Business Model Generation: A Handbook for Visionaries, Game Changers, and Challengers. John Wiley & Sons, Inc., Hoboken (2010)
17. Rogers, E.M.: Diffusion of Innovations, 5th edn. Free Press, New York (2003)
18. Rouhani, B.D., Mahrin, M.N., Nikpay, F., Ahmad, R.B., Nikfard, P.: A systematic literature review on enterprise architecture implementation methodologies. Inf. Softw. Technol **62**, 1–20 (2015). https://doi.org/10.1016/j.infsof.2015.01.012
19. Stahel, W.R.: Circular economy. Nature **531**, 435–438 (2016)
20. Taylor, M., Watts, J., Bartlett, J.: Climate crisis: 6 million people join latest wave of global protests (2019). https://www.theguardian.com/environment/2019/sep/27/climate-crisis-6-million-people-join-latest-wave-of-worldwide-protests

21. The Open Group: The TOGAF® Standard, Version 9.2 (2018). https://www.vanharen.net/the-togaf-standard-version-9-2/
22. The Open Group: ArchiMate® 3.1 Specification: Chapter 3 Language Structure (2019). https://pubs.opengroup.org/architecture/archimate3-doc/chap03.html
23. United Nations: Transforming Our World: The 2030 Agenda for Sustainable Development. Technical report, United Nations Publishing, New York, New York, USA (2015). https://sustainabledevelopment.un.org/?menu=1300
24. Venkatesh, V., Morris, M.G., Davis, G.B., Davis, F.D.: User acceptance of information technology: toward a unified view. MIS Q. Manag. Inf. Syst. (2003). https://doi.org/10.2307/30036540
25. Wieringa, R.J.: Design Science Methodology: For Information Systems and Software engineering. Springer, Heidelberg (2014). https://doi.org/10.1007/978-3-662-43839-8
26. Wolfswinkel, J.F., Furtmueller, E., Wilderom, C.P.M.: Using grounded theory as a method for rigorously reviewing literature. Eur. J. Inf. Syst. **22**(1), 45–55 (2013). https://doi.org/10.1057/ejis.2011.51
27. Zachman, J.A.: A framework for information systems architecture. IBM Syst. J. **26**(3), 276–292 (1987)

Security Architecture Framework for Enterprises

Michelle Graham(✉), Katrina Falkner(✉), Claudia Szabo(✉), and Yuval Yarom(✉)

School of Computer Science, University of Adelaide, North Terrace, Adelaide, Australia
{michelle.graham,katrina.falkner,claudia.szabo}@adelaide.edu.au, yval@cs.adelaide.edu.au

Abstract. Security is a complex issue for organisations, with its management now a fiduciary responsibility as well as a moral one. Without a holistic robust security structure that considers human, organisational and technical aspects to manage security, the assets of an organisation are at critical risk. Enterprise architecture (EA) is a strong and reliable structure that has been tested and used effectively for at least 30 years in organisations globally. It relies on a holistic classification structure for organisational assets. Grouping security with EA promises to leverage the benefits of EA in the security domain. We conduct a review of existing security frameworks to evaluate the extent to which they employ EA. We find that while the idea of grouping security with EA is not new, there is a need for developing a comprehensive solution. We design, develop, and demonstrate a security EA framework for organisations regardless of their industry, budgetary constraints or size; and survey professionals to analyse the framework and provide feedback. The survey results support the need for a holistic security structure and indicate benefits including reduction of security gaps, improved security investment decisions, clear functional responsibilities and a complete security nomenclature and international security standard compliance among others.

Keywords: Design science research · Information systems security policy · Enterprise architecture

1 Introduction

The Australian Cyber Security Centre had more than 13,672 reports of cybercrime from July to September 2019 and of those 11,461 were of sufficient merit to be referred to Australian law enforcement agencies [1]. High profile American security breaches such as the Verizon breach releasing more than 14 million customer records[1], the WannaCry ransomware computer hack giving access to NSA files[2] and the iCloud accounts extortion[3] highlight the global need for increased security resilience. These startling statistics

[1] http://www.zdnet.com/article/millions-verizon-customer-records-israeli-data/.
[2] http://www.wired.co.uk/article/wannacry-ransomware-virus-patch.
[3] https://www.theregister.co.uk/2017/04/07/icloud_wipe_threat/.

J. Filipe et al. (Eds.): ICEIS 2020, LNBIP 417, pp. 883–904, 2021.
https://doi.org/10.1007/978-3-030-75418-1_40

highlight that effective security has never been more important to businesses and therefore individuals [2], however very few companies have adopted a cohesive security strategy that encompasses the protection of all assets whether they be physical, digital or cognitive [3]. The benefits of a holistic approach could be used to mitigate these risks and requires all aspects of security to be considered and risk managed based on the budget, size and mechanisms of the organisation, and provides a reduction in responsibility confusion and appropriate resourcing [4, 5]. Enterprise Architecture (EA) is a holistic method to guide the enterprise's people, information, processes and technologies, to achieve the most effective execution of the corporate vision and strategy [6]. The development of the concept of a holistic security structure using EA would demonstrate that security is not just technical but requires a focusing on all the organisational assets of people, information, technology and processes and will provide enterprise security management guidance to contemporary digitalised organisations of the 21st Century.

The resulting research question for this work therefore is:

Will a holistic security model, using Enterprise Architecture, provide security benefits to an organisation more effectively than a piecemeal approach?

This paper builds on and extends our past work [7]. The main new contribution is an expanding the original four design principles to five, adding a requirement for ontological phrases and providing a broader explanation of the principles. This paper further expands the analysis of the evaluation of the artifact and revises the related works section identifying and analysing five analogous security frameworks surveys. In addition, this paper includes a more detailed explanation of the theoretical foundations including Design Science Research, philosophy approach and qualitative analysis method.

The opportunity for a reduction of security breaches, increased economic security and cyber resilience in organisations through a holistic approach to an organisational security framework with methodological supporting documentation, the importance and benefits of which have been mentioned in research, needs to be tested [7, 8]. We develop a novel, fully researched enterprise security architecture (ESA) framework for organisations. The framework is analysed by industry professionals to determine if a holistic security model can address the much needed solution to the identified organisational security gaps and provide security benefits. The framework, the Security Architecture Framework for Enterprises (SAFE), is a comprehensive security solution based on the enterprise architecture methodology. Our analysis, backed by feedback from industry professionals, supports our hypothesis that a holistic security design using EA will provide security benefits to an organisation more effectively than a piecemeal approach. This research is a complete security solution and provides organisational defense-in-depth and in the current world climate, what could be more necessary to business [5].

The paper is organised as follows. The background discusses the methods used and the sources consulted to meet the research goals. To describe the ESA artifact, a description of the design search (development) process and procedures that led to the artifact design principles is provided as well as a detailed description of the artifact itself. The assessment includes a description of the evaluation tool – an Oppenheim Survey including how the survey questions were developed, written and mapped to the research motivation; the qualitative analysis process which used Grounded Theory Methodology coding and how this provided cyclical results through each coding phase, iterating to a rich data set for analysis. The explanation also demonstrates the chosen evaluations' utility, validity, quality and efficacy [9] and discusses the significance and real world applications that have been identified from the design evaluation outcomes. The discussion links together the research question and design goals to the artifact and show how the novelty of the artifact design has bridged the research gap. Through research, five security framework surveys were identified and analysed, a discussion of each will be provided. Finally, we conclude the research with the key findings, noting the artifact demonstrates the success of the design and describes future work options to expand and develop the research further.

2 Background

Notwithstanding the popularity and recent adoption of EA, the majority of EA frameworks do not have a security component [8]. The Zachman ontological framework [9] is one of the most widely accepted and implemented EA frameworks, however despite Zachman's success, it does not include security in any form [10]. This lack of security has been identified by others [5] who have used Zachman to create an enterprise security architecture (ESA) [11]. However the results have been limited and none of the ESA's to date have utilised the Zachman concept of an ontology or ensured a strict adherence to the original definitions of Zachman [9]. Zachman is the ontological language of EA and building on this concept, a security implementation of Zachman would be the first security ontology available – a defined organisational security language. Furthermore, most existing ESAs are from business white papers, and thus lack in-depth case study analysis, experimental replicability and research exploration [12]. The use of EA in security will also provide a single capture of all the organisation's security – a holistic security structure that is not yet available in a mature form. A Design Science Research (DSR) study [10, 11] suited the research due to the emphasis on the design and creation of an artifact to test a research question [12] and the research rigor the DSR methodology provided [13, 14]. The philosophy for this work is constructivist, the approach is inductive and the choice of data analysis is qualitative using the grounded theory methodology to analyse an Oppenheim [15] qualitative questionnaire. We will explain the history of EA, the rationale behind the choice of DSR and how it is used in this research as well as the selection for the philosophy, approach and method.

2.1 Enterprise Architecture

The enterprise architecture (EA) domain began with Zachman's seminal work in 1987 [16]. The paper notates the construction process done by all industries that design, engineer, and build large scale objects, e.g., airplanes and buildings. The notation, or architecture, is then applied to the engineering of organisations, specifically focusing on the advent of computing. The theory states that an organisation is at least as complex as a large construction project and should be engineered using the same process; the context, the concept, the design, the build, the implementation and the use. EA provides a link between organisational goals and mission statements, through the organisational layers, down to the project level, just as an initial engineering concept document is traceable to a final built product. The organisation's assets are defined in EA as people, information, process and technology, and these are used to implement the vision of the organisation.

EA frameworks fall into two categories, ontologies and prescriptive methodologies. An ontology or classification structure is a recognized vocabulary used to describe objects in a particular domain [17]. A prescriptive methodology describes how to create the artifacts and with what tools or describes which artifacts are required to be in compliance with the framework. The Zachman framework is an example of an ontology and is now the adopted vocabulary for EA. The Zachman is also a structure independent of the tools and methods used in any particular business. This is useful because it can be adopted by any organisation without the need for specific, proprietary tools.

The implementation of the Zachman 6×6 framework grid would require an enterprise architect to use all 36 cells of the framework as a guide to describe a complex item like an organisation. The cells are called primitive models. Primitive models are the classification name of a required element in an EA framework. For example a primitive model for an organisation's security could be "access control", and an organisation might decide on specific artifacts to fulfill it, e.g., security guard, firewall, door locks etc., depending on organisational needs and budget constraints. The rows of the framework are the views of an organisation, for example the executive view would be the management of the organisation. The columns are English interrogatives which describe the details of each view e.g., the what, how, where, who, when and why of the management perspective. The result is a complete explanation of the particular view of the organisation. The ontology is used to organise and categorise an organisation's artifacts which are notated in the framework's grid.

2.2 Security

The need for organisational security initially began with the protection of information stored on computers and the physical security of organisations however this has broadened to include almost all departments within an organisation. The difficulty is that due to their evolution most departments have retained the individual control of the security measures they have put in place and this has meant that each security solution is managed separately. The overall effect is a lack of a cohesive strategy for organisational security [18].

Looking at frameworks that address the need for enterprise-wide security, holistic frameworks for organisational security are limited. One example is governance frameworks which are defined by the IT Governance Institute [19] as the "set of responsibilities and practices exercised by the board and executive management". However governance frameworks focus on management fulfilling their legal requirements, which does include security; however they do not address security any lower in the organisation than management.

The other most common response to organisational security has a technical focus such as computer and information security [20]. Unfortunately it is still very common for a company to believe that organisational security is solely about virus defense and firewalls. When asked, most do not include broader security mechanisms in their definitions of security, other than computer security and the effect is a lack of awareness for the need of a broader security strategy until a security incident occurs [21]. As Anderson [7] states,

"Security engineering requires cross-disciplinary expertise, ranging from cryptography and computer security through hardware tamper-resistance and formal methods to a knowledge of economics, applied psychology, organizations and the law". The solutions are not just technical and require a broader response.

2.3 Design Science Research

Vaishnavi and Kuechler [22] describe the body of DSR knowledge as man-made objects – artifacts – that are designed to meet specific goals. It creates novel contributions through the design of new artifacts including the analysis of their operation using evaluation and abstraction. DSR uses design as a research method that maps functional requirements on to a fulfilling artifact. As indicated in Fig. 1 there are five steps.

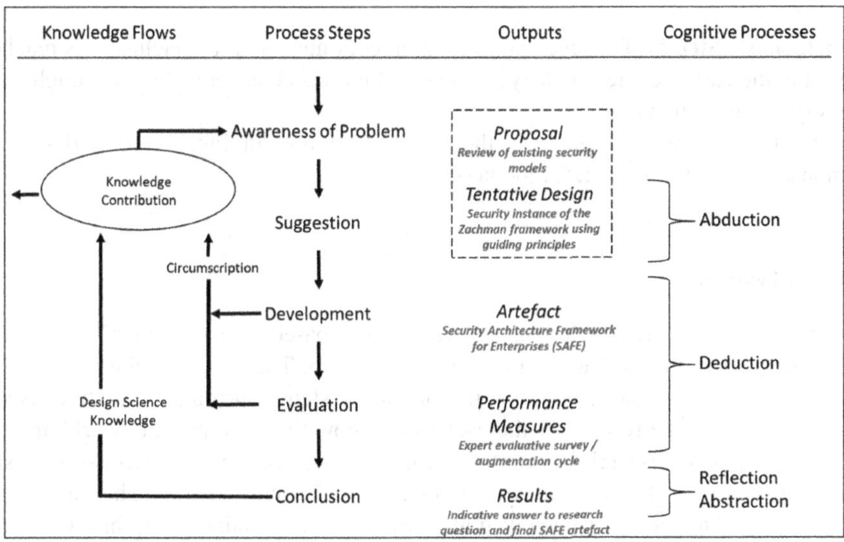

Fig. 1. SAFE outputs in design science research cycle [22, 23].

Awareness of the Problem – Step 1. An individual becomes aware of a problem that doesn't appear to have an existing solution and therefore a research proposal is written. For this research the problem was identified whilst the researcher was working in security and wanted to use a holistic security model for the organisational security approach. The problem was then confirmed through a literature review of holistic security models, that there is a lack of fully researched security models looking at security from a complete organisational perspective and not just a category of problem e.g. computer security or human resource security.

Suggestion – Step 2. Suggestion is the second stage and indications of the first sample of the design idea including performance needs of a prototype are developed. Through the literature review, recommended principles for a security model were identified and developed to provide both the design and the performance needs of the model. The principles included purpose, type, assurance, kernel theory reference and coverage.

Development – Step 3. At step three the artifact is created using the design from step two however it is important to note that the emphasis is on the novelty of the design not the creation of the artifact. Using the principles to guide the design, a security architecture framework artifact was created – a thirty-six cell security instantiation or ontology.

Evaluation – Step 4. Performance measures are placed on the artifact from the initial proposal to evaluate and at this stage, any changes are fed into the design towards a new design process. Using the design principles and security domain guidelines, the security framework was given to managers, security professionals, IT professionals and researchers, along with a questionnaire about the utility of the framework, to evaluate and provide the cyclical feedback to inform the artifact change process.

Conclusion – Step 5. This cyclical evaluation continues until a conclusion is reached – usually the end of the research cycle or the solution is considered "good enough" and the results are written up.

Our research is overlaid onto the Outputs column in Fig. 1 (colored red) to demonstrate our use of the DSR methodology.

2.4 Philosophy

The philosophy of the development of research is based on a system of beliefs and assumptions that are used at all stages of the research. The shaping belief system can be drawn from the research field e.g. information systems, surrounding realities or the human aspects of the researcher themselves and how they interpret the world and the findings [24]. This research is based in the underlying assumptions of constructivism, which are often used in concert with DSR and qualitative research. This provides a basis for how knowledge is perceived and how it can be obtained [25] throughout the DSR activity. Constructivism describes truths not as discovered but as reliant on human awareness and the struggle of the conflict between personal models and discrepant new insights which create new representations of reality and therefore new models using cultural tools and symbols bringing meaning and finding authentication through discussion in communities of practice [26]. DSR and our research is representative of this world-view due to the nature of the conception of a problem statement idea (the current world-view model is challenged), development of a design to address the challenge (incorporating the conflict between what we knew and what we now know), the artifact to test the design and the cyclical analysis of the artifact until the design is satisfied (new knowledge and models are created) [27, 28].

2.5 Approach

Inductive reasoning is a logical thought progression in which various propositions, all believed true or found true the majority of the time, are combined to develop a definitive assumption or likely conclusion [29]. Inductive reasoning was chosen for this research because of the nature of DSR and the research itself began with specific observations – the design principles for the security artifact, and used those principles to develop a recommended way forward for the development of a likely artifact that could provide an assumptive solution.

2.6 Qualitative Research Method

Grounded Theory was identified as the best suited for this research and for the qualitative data set that would be developed from the Oppenheim [15] questionnaire evaluation of the security artifact.

The Oppenheim approach was chosen because it provides clear advice on the length, clarity, grammar and specificity of the questions and attempts to avoid such bias in questions as social desirability, double barreled questions and negatively worded questions [15, 30]. The type of responses gathered from an Oppenheim also lend itself to provide effective inputs to a grounded theory qualitative analysis [31].

Grounded theory is a methodology by which qualitative analysis is iterative – the data (meaningful concepts from the texts) are collected and separated from the conversation and each data unit is assigned codes [32]. The codes are inspected for patterns and then reintegrated to form dominant thematic subjects and connections [33, 34]. The code inspection or coding, is done iteratively to a level of detail that provides a thematic essence of the original data set or texts. According to Martin and Turner [35] grounded theory is "an inductive, theory of discovery methodology that allows the researcher to develop a theoretical account of the general features of a topic while simultaneously grounding the account in empirical observations or data". This method was chosen to analyse the results from our Oppenheim questionnaire about our security artifact because it provided a detailed methodology which, through each coding phase, produced a synthesis of the themes that became richer and more meaningful about the artifact's utility.

3 Enterprise Security Architecture Design

As described previously, this is a DSR study therefore the development of the artifact to test a theory which was based on principles was critical. The question being tested is "will a holistic security model using EA provide security benefits in an organisation more effectively than a piecemeal approach". An analysis of 25 security frameworks, detailed in our past work [23], provided a set of five guiding design principles that would help guide an organisation towards a more secure corporate posture and concurrently support the achievement of corporate vision and strategy. The five design principles were identified to inform the design, development and evaluation of the artifact to test this question. Those principles will be used to frame the following discussion, describing the artifact, how it was developed and the three layers of abstraction achieved.

3.1 Guiding Recommendations

The following discussion explains the five design principles which have guided the development of the enterprise security architecture artifact.

Framework Purpose (Principle 1). The purpose of an effective framework should be to support the organisation's vision. To do this all assets of a company should be employed e.g., people, technology, process and information. The recommendation derived from this criterion is therefore a holistic framework will include security mechanisms for all of the assets. Providing a separate security strategy for each department or asset; or choosing a select few assets to secure would not provide full security coverage and therefore lacks defense-in-depth.

Framework Type (Principle 2). EA ontologies are a classification system that provides a structured way to articulate the required organisational assets for the purpose of alignment to the corporate vision. Whilst allowing the company to choose the implementation based on its specific needs. In contrast artifact-based framework types require a company to purchase or produce specific artifacts or methodologies to be in compliance. Artifact-based frameworks are restrictive and difficult to comply with particularly if the company is small and has budget constraints and doesn't take into account the nuances and individuality of each company. The principle from this criterion is to provide the organisation a framework type that is secure but also works with the individual organisational needs and uses ontological phrases and constructs.

Security Compliance/Assurance (Principle 3). The concept of security is not new and there are very effective security standards available. Security standards are used as a benchmark by organisations to provide a level of assurance for their security programs [36]. For compliance and assurance purposes, a framework should be in compliance with at least one security standard. From the framework reviews, two standards are used more than any others and either or both would provide an effective compliance tool. The two recommendations are ISO/IEC 27000 and NIST. The recommendation from this criterion is the use of an internationally recognized standard to provide security assurance.

Enterprise Architecture Reference (Principle 4). EA is a proven structure for organisations to use to effectively complete their mission [37]. From the analysis of the 25 frameworks, seven of the 25 reference EA in some form, with two frameworks specifically named – the Zachman and the FEAF. Moreover some of the most implemented EA structures have used the Zachman as a basis for the development of significant frameworks. These are the TOGAF [38], the GEAF [37], the FEAF [39] and the DoDAF [40]. The recommendation from this criterion is therefore the use of an existing and well referenced EA for the basis of the development of a framework.

Framework Coverage (Principle 5). The need for security initially began with the protection of information stored on computers however this has broadened to include all departments within an organisation. The difficulty is that most departments have retained the individual control of the security measures they have put in place and this has meant that each security solution is managed separately (see Table 1). The overall effect is a lack of a cohesive strategy for organisational security [18]. To provide effective security for any entity, the whole entity needs to be considered. The same is true of an organisation. If we choose to only secure a department, the rest of the organisation remains insecure. The recommendation from the criterion is for a framework to regard the whole of the organisation, not just singular departments or assets. A structure that provides an integrated view of all security instances will give a credibility and confidence to business security responsibility [41].

Table 1. Artifact survey questions and design principle mapping.

Artifact survey questions	Design principle mapping/Question purpose
Q1. What is the biggest security challenge facing organisations today?	Background security question to help participants begin thinking about security in preparation for completing the survey
Q2. Referencing the background information and the framework, please indicate if you believe any security categories or elements are missing and should be included?	Principle 1, Principle 2, Principle 3
Q3. Do you believe a holistic approach to security is likely to provide a more secure organisation? Why or why not?	Principle1, Principle 5
Q4. Do you believe a holistic approach to security is likely to help with financial decision making for security resources? Why or why not?	Principle 5
Q5. Does the use of a framework with all possible security categories included provide assurance to the process of securing an organisation? Why or why not?	Principle 2, Principle 3
Q6. After inputting an organisations security mechanisms into the framework, cell by cell, do you believe you could see the security gaps in an organisation and determine what else needs to be secured? Why or why not?	Principle 1, Principle 5
Q7. Would the analysis from a completed security framework help senior management or the CEO make security decisions or provide beneficial management information? Please give an exampl	Principle 1, Principle 4, Principle 5
Q8. What do you see as the benefits or features of the framework for an organisation using it?	Anecdotal free text from participants to encourage additional response not brought out by previous questions and focused on the positive use of the artifact
Q9. What are the problems or challenges of the framework for an organisation using it? Can they be solved?	Anecdotal free text from participants to encourage additional response not brought out by previous questions and focused on the challenges of the artifact

(*continued*)

Table 1. (*continued*)

Artifact survey questions	Design principle mapping/Question purpose
Q10. This framework is compliant to NIST and ISO27002 (international security industry standards). Does this information give you more confidence in the framework? Is the compliance important to you?	Principle 3
Q11. Is the framework easy to understand and use? Why or why not?	Usability and efficacy
Q12. Does it help to have the security categories broken down into organisational levels (the row perspectives)? Why or why not?	Principle 2, Principle 4
Q13. Have you found the framework and the background information educational? Please give an example	Anecdotal free text from participants to encourage additional response not brought out by previous questions and focused on education
Q14. Please provide any final thoughts on the theory, framework and supporting documentation?	Anecdotal free text from participants to encourage additional response not brought out by previous questions

In summary, the principles we identified provide the design foundation which supported the development of the security artifact to evaluate the design which is addressing the problem statement "will a holistic security model using EA provide security benefits in an organisation more effectively than a piecemeal approach".

3.2 The 36 Cells of the Artifact

Using the five principles identified in Sect. 4.1, the framework is developed based on the Zachman framework 2013 Version 3.0 [42] because it is the most complete, most referenced in our frameworks review, and historically the methodology that is chosen by others to base their frameworks on. Also the Zachman does not have a security view which increased the novelty of our approach. All 36 cells of the Zachman framework were explored and researched to determine exactly what the purpose of the cell was. All external research available was read thoroughly to provide a clear understanding of the framework and the purpose of each cell in it. This included identifying full definitions for each cell and the outer framework terms. Once EA and security were explored adequately and an expert level of knowledge was achieved, the outer edges of the proposed security version of the Zachman framework were identified. To ensure the integrity of the principles of EA, it was important to retain the organisational views (the rows) and the interrogatives (the columns). We methodically develop all 36 cells of the security instantiation by research and analysis of the 36 Zachman cells. The outcome is the ESA framework which is an exact matching overlay of the Zachman framework as a security instantiation.

Once the high level categories were defined for each cell, the detail needed to be developed to explain what each cell actually meant. Also whilst the high level definition provided the matching Zachman column/row reference for each security cell, the specific security ontological construct needed to be defined for organisational guidance when evaluating the framework. This resulted in four factors being defined. Those were:

1. Detailed explanation – what is the definition and purpose of the cell
2. Pictorial model – a pictorial description for ease of understanding to users
3. Artifact example – show the use of the cell using a real world example
4. Compliance mapping to ISO/IEC 27000 and NIST

Figure 2 is an example of the four factors defined for each cell.

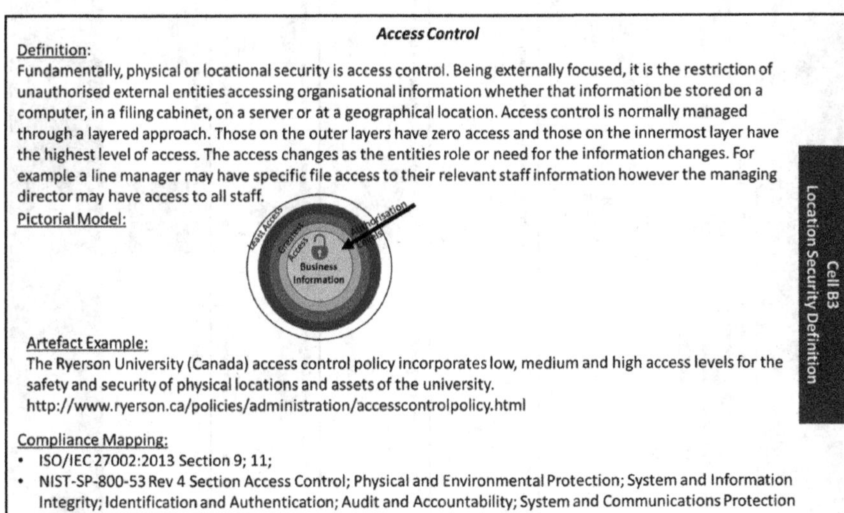

Fig. 2. Cell definition example.

3.3 The Security Architecture Framework for Enterprises Artifact

In summary, the notional artifact was completed and three layers of abstraction developed. The row/column categories, the detailed security definitions and the more detailed definitions (pictorial model, artifact example and compliance mapping) for use by organisation for understanding. The final framework is compliant with the 5 guiding design principles identified. Figure 3 is the completed Security Architecture Framework for Enterprises (SAFE) artifact.

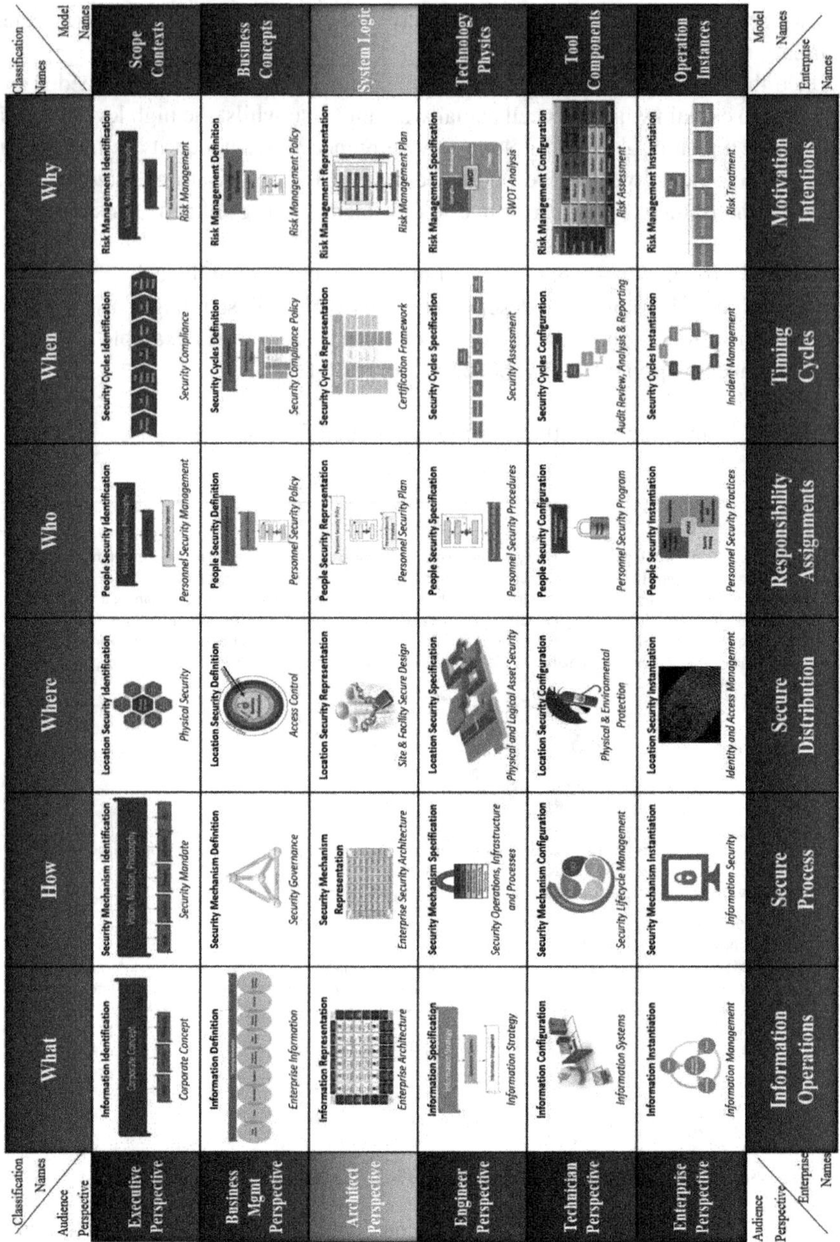

Fig. 3. The Security Architecture Framework for Enterprises (SAFE) artifact [23].

4 Artifact Assessment

To test the artifact design described previously, an Oppenheim structured expert evaluation survey was created to ask questions of a group of participants about the efficacy, validity, utility and quality of the artifact. We received 12 responses and an inductive grounded theory qualitative analysis was completed to derive the foremost themes indicated by the participants. The themes and anecdotal results are discussed and participant quotes are included.

4.1 Expert Evaluation Survey

We shared the artifact and supporting documentation for critique, to four categories of professionals – manager, security professional, IT professional and researcher. The participants are asked to review the framework and supporting documentation in the context of their own organisations and their expertise, carefully considering the utility of the design and its application in a working environment and compared to their current security situation. To test the utility, the participants work through each cell and determine if their organisation has a suitable security instance of the requirements indicated for that cell, using the provided explanatory notes. Just as an EA framework can build an organisation from its inception, so the security dimension we have created should functionally be able to build security into all aspects of the organisation. Theoretically a form of an organisational security ontology.

To gather the participant's inputs we designed a questionnaire using an Oppenheim [15] approach and following a successful rigorous ethical research approval process, distributed the questionnaire. The questionnaire is made up of five demographic questions - including security industry experience, job category, years of expertise; and 14 questions aimed at drawing out selected aspects of the initial research question and expert opinions of the design.

The survey responses were collated and grounded theory was used to draw out themes, through an inductive data collection that enables the participants to tell the story. Through the cyclical nature of the grounded theory methodology, each coding phase provided richer thematic results. For example the question "Have you found the framework and background information educational?" The first coding phase saw 13 raw responses from participants. However after a second, third and fourth coding phase, which essentially distilled the responses down to the key themes, the final three thematic responses became 1) Definitions, artifacts, models and references are a very strong tool; 2) Shows the full extent of issues involved in security - risk, difficulties and complexity; 3) Security policies and practices can be used to form a cohesive framework/security program.

4.2 Participant Demography

We received 12 participant questionnaires, of which 75% of participants were employed by a large company (200 + employees), 17% were from a small (1–19 employees) and 8% from a medium (20–199 employees). 42% had security industry experience and 75% had been in their current role for more than ten years and considered themselves experts in the field. The participants came from Industry (58%), Government (33%) or the Military (8%). Table 10 provides the demographic questions and the reason they were included in the survey.

4.3 Results of Design Principles Application

The following discussion describes the participant results to the 14 questions related to the Principles. The outcome shows an effective implementation of the Principles in guiding the design of the framework and the responses indicate that the artifact provides significant organisational security benefits more effectively than a piecemeal approach which successfully answers the research question. Table 1 provides the 14 questions and the design principle(s) they are mapped to or the purpose for inclusion.

Principle 1 – Security Mechanisms for all Organisational Assets. Survey questions two, three, six and seven were designed to test the principle that all organisational assets should be assessed for security mechanisms, noting that all security is risk based and therefore the answer can be that the organisation chooses not to secure the asset and accepts the risk, but the key is that all assets – people, process, technology and information, should be considered in the securing of an organisation. The participants indicated in Question two that the artifact was very comprehensive and there were no organisational assets missing from the artifact grid. To support this notion, the third question asked if a holistic approach – all assets, all departments, is likely to provide a more secure organisation. The responses were 100% in agreement with this question. One participant expanded further and explained how often media describes the extent an organisation will spend time and money on securing one part of an organisation, such as ICT, and the successful attack is in an area that was treated as less important or received less focus, such as physical security. The best security can be applied to a computer but if the attacker can simply walk away with the computer, then the organisational security has failed.

Questions six and seven focused on the potential gap analysis that is required to ensure all assets are secured and the executive buy in that is required to make those security decisions. Participants highlighted that the ontological nature of the grid – a list of security terms and the relationships between them, gives an organisation a complete list to work through to conduct the gap analysis and then bring the needs or risk choices to the executive to make a decision. The framework also demonstrates the interconnected system of security and the subsequent consequences of softening one aspect. It was also highlighted that the framework would provide an assurance to management that the recommendations they bring are based on a methodology.

Participant comments included "ensures all aspects of security are covered and assessed", "organises the complete security function", and "focuses organisations to include security elements not traditionally addressed".

Principle 2 – Ontological Phrases Are Used. Survey questions two, five and 12 were designed to test the principle of ontological security phrases rather than instances of a security mechanism. The ontological design principle provides flexibility to the users that the requirement for instances wouldn't. For example if an organisation is required to have a specified type of physical security such as a retina scanner for biometric screening of visitors to the building but the organisation is only ten people, it is unlikely that the organisation could afford or actually need such a large scale form of physical security. The use of the ontological phrase for physical security such as "identity and access management" from the artifact, emphasises to the organisation that physical security is required to be considered but the instance type is not mandated, allowing all organisation types, sizes and budgets to use the artifact. The responses from the participants indicated the categories allowed their subject matter experts, like physical security, to determine the best implementation for their organisation. It was also highlighted that the ontological phrases not being prescriptive allowed for flexibility and change when the organisational environment changed, such as growth or structure, or new threats emerged in the security environment. One participant mentioned that the categories were very encouraging to their small organisation and that they felt they were more likely to achieve a level of security assurance because categories were achievable but previous prescriptive instance-based frameworks they had tried to implement had been too costly, difficult and as a small organisation they did not have the expertise.

Participant comments included "provides better communication about security between all levels of the organisation", "provide an understanding of the gaps in security, the risks and remediation" and "provides good governance for security".

Principle 3 – Compliance to Security Industry Standard. Survey questions two, five and ten were designed to test the principle that it is important for the artifact to be in compliance with at least one security industry standard. In the Literature Review it was determined that the two most commonly used standards for security were ISO/IEC 27000 and NIST. The artifact was therefore designed to be in compliance with both of these standards and the survey questions were designed to understand the importance of compliance and assurance to organisations. Participants highlighted the two standards as best practice and therefore the framework, by association, would also be perceived as best practice and the use of a framework that was in compliance would aid in security audits as most audits now require compliance to pass. It was also noted that there is a level of credibility associated with standards, that it builds more confidence in the benefits and provenance of the framework, and this would lend a credibility to security programs and also to conversations with executive about security.

Participant comments include "compliance to NIST and ISO validates the framework in terms of academic rigor", "management are more likely to accept a model based on international standards", "compliance standards build more confidence in the benefits and provenance of the framework" and "tells me that the framework is based on best practice".

Principle 4 – Use of an Enterprise Architecture Reference. Survey questions four, seven, eight and 12 were designed to test the principle that the use of enterprise architecture as the primary model for the foundation of the framework is an effective choice. Enterprise architecture was chosen for two key reasons. The first is that EA was the most commonly used model for security frameworks when the review of 25 frameworks was done. Secondly EA adheres to and supports Principle 1 and Principle 5 directly and indirectly supports Principle 2 and 3 because EA is a model to build a complete organisation. In the same way, the research question and design principles were intended to develop a whole organisational security framework, not just for a department or a specific type of security.

The responses from participants discussed the importance of articulation of security mechanisms, including responsibilities for all levels of the organisation, the use of the architectural categories would provide the right information to the best people to understand it, and the rows and columns break up the complexity of security into identifiable chunks. The use of EA was also mentioned as helpful because large numbers of organisations are turning to EA to define the best use of their resources and having a security framework based on EA will complement, align and implement the organisations business models more effectively. Another response noted the use of a multi-faceted model like EA, aids understanding that security is also multi-faceted and that each department has something to contribute in the decision making and execution of security – fundamentally security is a whole-of-organisation responsibility.

Participant comments include "definitions, artifacts, models and references are a very strong tool", "fantastic concept that provides a single awareness for all security" and "would be used effectively and compliment the organisations existing enterprise architecture".

Principle 5 – Coverage is Organisationally Holistic. Survey questions three, four, six and seven were designed to test the principle that security should be considered in all departments of an organisation and not just individual departments like ICT and that all security in an organisation should be cohesively considered and managed not as separate departmental responsibilities or instances. The most frequent response to these survey questions was about the framework helping the organisations understand the other parts of the organisation that need security. By taking a holistic view, there was an educational factor involved, and security would be considered and implemented in areas that had not previously been considered. Mapping all of an organisations security in the single model would provide a view of security that had not previously been available. The consequences of this was exciting for some respondents mentioning better security coverage, strong gap analysis and therefore remediation, departmental responsibility definition and considered security decision making in understanding the organisation's risk exposure. Also a holistic view of an organisation's security is a balanced view of security for the purposes of resources and discussions where ICT or physical security is usually the focus.

Participant comments include "helps build trust because the right information is comprehensive and usable to the right audience", "security policies and practices can be used for a cohesive framework and security program" and "the structural configuration shows that security is a whole of organisation responsibility not just IT".

4.4 Theoretical Significance

The kernel knowledge for this research was the domain of enterprise architecture. As described in the Literature Review, EA is an established, comprehensive body of knowledge and models that are used to describe an organisation and its assets. Until this research and design study was conducted, security within EA had not been considered with the same depth as EA. There were other frameworks that used some of the principles of EA to describe security but none that strictly adhered to EA and all of its principles, and then used a fully researched process to create an artifact. This increased the novelty of the research and the outcome in both the artifact, the design and the evaluation, all indicate success to the extension of the kernel theory. There now exists a true enterprise security architecture framework and design principles to guide future users.

Similarly the security domain is also well established however there are very few models that address all forms of security within an organisation in a structured format that is fully compliant with industry standards. The collection of the security categories as a framework is also a form of ontology or categorisation system for organisational security. This research has extended the security domain body of knowledge by creating a design that provides both an ontology and a model for all organisations regardless of their size, budget or resources.

5 Related Work

The earliest enterprise security architecture (ESA) frameworks were developed in 1995 [43] and the few approaches available all agree that "the problem is that no standardised, comprehensive information security architecture currently exists" [5, 18]. Our analysis of existing work identified five surveys of ESA frameworks which gives a broad domain overview of the status of Enterprise Security Architecture as a discipline. The outcome of this review also led us to the need for a new, more comprehensive security framework survey which can be found in our past work [23] and includes 25 frameworks reviewed. The surveys are discussed below.

The Shariati et al. [44] 2010 Survey. Focuses on the importance of interoperability for organisational architectures, perceiving an organisation as a holistic seamless flow of information rather than compartments, which is a key requirement of enterprise architecture. The issue raised is that interoperability is a direct conflict with the principles of security. Such security principles as "need to know", physical defence of assets and confidentiality confirm the struggle. The paper's goal is to identify holistic security frameworks that support interoperability. The review provides a description of interoperability aspects and its importance in frameworks, specifically in the areas of technical, organisational and semantic. This inclusion helps the reader better understand the focus of the research. Furthermore, the holistic versus partial section provides a convincing discussion about the utility of holistic frameworks rather than partial frameworks which tend to have a limited domain specific use. The paper does not include a recommendation for a suggested framework that would incorporate interoperability. In contrast, this research has reviewed 25 frameworks and provides a recommended principles-based framework for an effective security framework.

The Oda et al. Review [45]. Aims to determine the effectiveness of ESA frameworks based on a number of criteria including business architecture, information architecture, technology architecture, security architecture, levels of abstraction and case studies. The review/survey looks at three frameworks, including the Zachman framework [46], which does not have a security element but is stated as being the foundation of all enterprise information architecture frameworks, and is included on that basis. The purpose is to determine the effectiveness of the architectures. The paper concludes with a case study of the enterprise information security architecture at the Oakland University in the U.S. The three architectures are explained and analysed in detail. However, the survey only considers two security architectures. Moreover, the Oakland University case study does not consider the introduced criteria.

The Da Veiga and Eloff Work [47]. Is centered on governance and, while not titled a review paper, does have a comprehensive "existing approaches" section and reviews four information security governance frameworks. The purpose of the paper is to derive a list of components (a principle or a security control or both) for the development of a new security governance framework. The review derives six components (leadership and governance; security management and organisation; security policies; security program management; user security management; technology protection and operations) placed into three categories: strategic, managerial or operational, and technical. It is not clear from the research why the particular frameworks were chosen because no selection criteria are given.

The 2006 Claycomb and Shin [48] Research. Is focused on enterprise security management architectures for mobile devices that use all of the aspects of organisational architectures. It reviews two related works using the criteria authentication, access control and audit. The review provides a detailed description of the suggested new security architecture including diagram specifications and a proof of concept implementation. Although the paper uses the phrase enterprise architecture, no reference to enterprise architecture principles is present and there are only two frameworks surveyed, which limits the analysis. The choice of criteria also indicates a technical focus, which would not provide a holistic security view of the organisation. In contrast, the chosen criteria in this research provides a complete view of an organisation choosing EA specifically because of the holistic aspect.

The Eloff and Eloff Survey [18]. Reviews five existing ESA frameworks from various fields including risk management and international standards. The survey then draws from the analysis to develop five principles for an ESA framework. The five principles are based on procedures, technology, and people, and are namely; holistic, security control synchronization, risk management, life-cycle implementation, and measures. The inherent challenge with this list of principles is its broadness, in that the scope of the principles is not defined and therefore it might be difficult to develop a comprehensive security architecture that meets all principles. The proposed principles in this work focus on security and enterprise architecture.

6 Conclusions

Security has never been more important to our connected world and to organisations, with the number of security breaches increasing every year and the high profile discussions of security issues in the media. A new approach to organisational security is a priority. In security, the whole is clearly greater than the sum of its parts and security maturity is not just technical but involves consideration of all parts of the organisation in a holistic manner. The benefits of a holistic approach require all aspects of security to be considered and risk managed based on the budget, size and mechanisms of the organisation, and provides a reduction in responsibility confusion and appropriate resourcing, would reduce security breaches and improve security factors in organisations. This research has designed a new holistic model for organisations to address security and the evaluation results indicate the research gap and practical organisational need have been achieved.

The research conducted a semi-systematic literature review of 25 organisational security structures demonstrated in our past work [23], to determine if a fully researched and holistic security methodology was available. The survey analysis showed that current security models lack research process and therefore lack case study analysis, replicability and research exploration. This was identified by a careful examination and review of the 25 security structures, their supporting documentation and the methodologies used. The result is very few structures met the holistic test and the most common construct to address an organisation holistically was Enterprise Architecture (EA). Furthermore, one of the important findings in the survey was the ontology gap. EA uses an ontology to describe the organisational classifications, simplifying structures for use. Organisational security does not currently have this classification structure. The development of an Enterprise Security Architecture (ESA) ontology is the first of its kind and provides an ESA language to articulate security in all its forms throughout an organisation. The structure can be used for compliance and assurance purposes, providing management with a tangible solution to the fiduciary and moral responsibilities of organisational security. The need for further research was highlighted.

Analysis identified the similarities and differences amongst the frameworks and proposed a set of design principles to guide the development of a security artifact. The design principles for the artifact were: 1) the securing of all assets, 2) the use of ontological phrases, 3) compliance to international security standards, 4) the use of EA as the reference model and 5) organisationally holistic in its implementation. The principles respect the key aspects of the two domains of security and enterprise architecture and provided a first step towards effectively combining them for the new artifact. The resulting research question was therefore:

Will a holistic security model, using Enterprise Architecture, provide security benefits to an organisation more effectively than a piecemeal approach?

The design of a holistic enterprise security architecture, highlights that security is not just technical but requires a focusing on all the organisational assets of people, technology, processes and information, which provides enterprise security management guidance to contemporary digitalised organisations of the 21st Century.

This research used the Design Science Research methodology due to the need for a designed and evaluated artifact. The qualitative analysis of an Oppenheim questionnaire given to expert evaluators to provide feedback for the artifact, was completed using the Grounded Theory Method, and the approach of the research was constructivist and inductive.

The designed and fully researched artifact produced in this work is the Security Architecture Framework for Enterprises (SAFE) (Fig. 3) and is based on the Zachman 2013 Version 3.0 EA construct which allows for the artifact to be used in conjunction with the Zachman EA or as a stand-alone organisational security model. SAFE is compliant with the five guiding design principles identified in the initial review and has been completed to three layers of abstraction. The completed artifact is a 6 × 6 framework and each cell was defined using 1) a detailed explanation, 2) pictorial model, 3) artifact example in the real world and 4) compliance mapping to ISO 27002 and NIST.

To determine the effectiveness of the framework in meeting security concerns and test the efficacy within real-world organisational environments, we shared the framework and supporting documentation with industry professionals together with a questionnaire for evaluation and asked them to consider the artifact in the context of their own organisations and expertise. The questionnaire was made up of five demographic questions about the participants and 14 questions about the artifact. The participants were made up of managers, security professionals, IT professionals and researchers. The questions about the artifact were mapped to the five design principles and the research question, and were designed to elicit meaningful responses to further guide the development and usability of the artifact. The responses were analysed using the qualitative analysis methodology, Grounded Theory.

The analysis of the questionnaire responses evaluating the security artifact, SAFE, indicates that the research gap has been bridged and that a holistic approach to organisational security, using EA, can provide security benefits more effectively than a piecemeal approach. The evaluation highlighted the usability of a holistic structure which demonstrates to the organisation, the interconnectedness and broad nature of security. Other benefits included reduction of security gaps, a categorisation framework for the entire security function, security governance structure, improved security program, compliance to best practice and a security nomenclature. Other opportunities include better financial decision making for the security function, improved organisational communication regarding security, and a strong educational tool for the organisation with the use of the provided definitions, framework, models and references. One challenge to non-security practitioners was the complexity of the artifact and a recommendation for a future improvement of the framework was a gap assessment workbook or user manual.

The theoretical significance of this research is the successful extension of the kernel theory, enterprise architecture, with a fully researched enterprise security architecture including all definitions and the five design principles successfully implemented. The security domain has benefited by the development of the first security categorisation system for organisations or an organisational security ontology.

To mature the concept further there would be benefit from future work such as a larger design study, a case study in an organisation or an organisational implementation study to explore further the ideas discussed in this research.

This work is important because organisational security has never been more necessary and the successful design and development of a security framework artifact that looks at all of the aspects of security throughout an organisation is an important step forward to achieve a comprehensive solution to a complex and challenging problem for our digital society. The success of this important security research provides an opportunity and a significant foundation for future ESA studies.

Acknowledgements. This work has been supported by an Australian Research Council Discovery Early Career Research Award (project number DE200101577).

References

1. ASD. Cyber Crime in Australia July to September 2019 (2020)
2. Patterson, T.: Holistic security: why doing more can cost you less and lower your risk. Comput. Fraud Secur. **6**, 13–15 (2003)
3. Roeleven, S., Broer, J.: Why Two Thirds of Enterprise Architecture Projects Fail. ARIS Expert Paper (2010)
4. Angelo, S.: Security Architecture Model Component Overview. Sans Security Essentials (2001)
5. Copeland, M.: Cyber Security on Azure. Apress, Berkeley (2017). https://doi.org/10.1007/978-1-4842-2740-4
6. Gorazo. Enterprise Architecture Literature Review (2014)
7. Anderson, R.: Security Engineering. John Wiley & Sons, New Jersey (2008)
8. Moulton, R., Coles, R.S.: Applying information security governance. Comput. Secur. **22**(7), 580–584 (2003)
9. Gregor, S., Hevner, A.R.: Positioning and presenting design science research for maximum impact. MIS Q. **37**(2), 337–355 (2013)
10. Hevner, A.R., et al.: Design science in information systems research. MIS Quarterly, pp. 75–105 (2004)
11. Nunamaker Jr., J.F., Chen, M., Purdin, T.D.: Systems development in information systems research. J. Manag. Inform. Syst. **7**(3), 89–106 (1990)
12. Venable, J., Pries-Heje, J., Baskerville, R.: FEDS: a framework for evaluation in design science research. Eur. J. Inform. Syst. **25**(1), 77–89 (2016)
13. Sein, M.K., et al.: Action design research. MIS Quarterly, pp. 37–56 (2011)
14. Peffers, K., et al.: The design science research process: a model for producing and presenting information systems research. In: Proceedings of the First International Conference on Design Science Research in Information Systems and Technology (DESRIST 2006). ME Sharpe, Inc. (2006)
15. Oppenheim, A.N.: Questionnaire Design, Interviewing and Attitude Measurement. Bloomsbury Publishing, London (2000)
16. Zachman, J.A.: A framework for information systems architecture. IBM Syst. J. **26**(3), 276–292 (1987)
17. EBI. E.B.I., Glossary (2015)
18. Eloff, J., Eloff, M.: Information security architecture. Comput. Fraud Secur. **11**, 10–16 (2005)
19. ITGI. I.G.I., Board briefing on IT governance. Information Systems Audit and Control Foundation (2001)
20. Anderson, R.: Why information security is hard-an economic perspective. In: Proceedings 17th Annual Computer Security Applications Conference, pp. 358–365. IEEE (2001)
21. ISACA: An Introduction to the Business Model for Information Security (2009)
22. Vaishnavi, V., Kuechler, W.: Design research in information systems (2004)
23. McClintock, M., et al.: Enterprise security architecture: mythology or methodology? In: International Conference on Enterprise Information Systems (2020)
24. Crotty, M.: The Foundations of Social Research: Meaning and Perspective in the Research Process. Sage, London (1998)
25. Hirschheim, R.: Information systems epistemology: an historical perspective. Res. Methods Inform. Syst. **9**, 13–35 (1985)
26. Fosnot, C.T.: Constructivism: Theory, Perspectives, and Practice. Teachers College Press, New York (2013)
27. Strauss, A., Corbin, J.: Basics of Qualitative Research Techniques. Sage Publications, New York (1998)

28. Mills, J., Bonner, A., Francis, K.: The development of constructivist grounded theory. Int. J. Qual. Methods **5**(1), 25–35 (2006)
29. Lee, A.S., Baskerville, R.L.: Generalizing generalizability in information systems research. Inform. Syst. Res. **14**(3), 221–243 (2003)
30. Williams, M.: Questionnaire design. Making Sense of Social Research, pp. 104–124 (2003)
31. Rattray, J., Jones, M.C.: Essential elements of questionnaire design and development. J. Clin. Nurs. **16**(2), 234–243 (2007)
32. Urquhart, C., Lehmann, H., Myers, M.D.: Putting the 'theory' back into grounded theory: guidelines for grounded theory studies in information systems. Inform. Syst. J. **20**(4), 357–381 (2010)
33. Starks, H., Brown Trinidad, S.: Choose your method: a comparison of phenomenology, discourse analysis, and grounded theory. Qual. Health Res. **17**(10), 1372–1380 (2007)
34. Strauss, A., Corbin, J.: Grounded theory methodology. Handb. Qual. Res. **17**, 273–85 (1994)
35. Martin, P.Y., Turner, B.A.: Grounded theory and organizational research. J. Appl. Behav. Sci. **22**(2), 141–157 (1986)
36. Siponen, M., Willison, R.: Information security management standards: problems and solutions. Inform. Manag. **46**(5), 267–270 (2009)
37. Bittler, R.S., Kreizman, G.: Gartner Enterprise Architecture Process: Evolution 2005. G00130849, Gartner, Stamford, CT, pp. 1–12 (2005)
38. Josey, A.: TOGAF Version 9.1 Enterprise Edition: An Introduction. The Open Group (2009)
39. USG. U.S.F.G., Introduction to the Federal Enterprise Architecture Framework V2 (2013)
40. DoD, C.: DoDAF Architecture Framework Version 2.02. Website, August 2010
41. ISO. I.S.O./I.E.C. 27000, 27001 and 27002 for information security management (2013)
42. Zachman, J.A.: The framework for enterprise architecture: background, description and utility. Zachman International (1996)
43. Sherwood, J., Clark, A., Lynas, D.: Enterprise security architecture. SABSA White Paper, vol. 2009 (1995)
44. Shariati, M., Bahmani, F., Shams, F.: Enterprise information security, a review of architectures and frameworks from interoperability perspective. Procedia Comput. Sci. **3**, 537–543 (2011)
45. Oda, S.M., Fu, H., Zhu, Y.: Enterprise information security architecture a review of frameworks, methodology, and case studies. In: ICCSIT 2009. IEEE (2009)
46. Zachman, J.P.: The Zachman Framework Evolution (2011)
47. Veiga, A.D., Eloff, J.H.: An information security governance framework. Inform. Syst. Manag. **24**(4), 361–372 (2007)
48. Claycomb, W., Shin, D.: Mobile-driven architecture for managing enterprise security policies. In: ACMSE 2006. ACM (2006)

GDPR Compliance Tools: Best Practice from RegTech

Paul Ryan[1,2(✉)], Martin Crane[1(✉)], and Rob Brennan[1(✉)]

[1] ADAPT Centre, School of Computing, Dublin City University, Glasnevin, Dublin 9, Ireland
paul.ryan76@mail.dcu.ie, {martin.crane,rob.brennan}@dcu.ie

[2] Uniphar PLC, Dublin, Ireland

Abstract. Organisations can be complex entities, performing heterogeneous processing on large volumes of diverse personal data, potentially using outsourced partners or subsidiaries in distributed geographical locations and jurisdictions. Many organisations appoint a Data Protection Officer (DPO) to assist them with their demonstration of compliance with the GDPR Principle of Accountability. The challenge for the DPO is to monitor these complex processing activities and to advise and inform the organisation with regard to the organisations demonstration of compliance with the Principle of Accountability. A review of GDPR compliance software solutions shows that organisations are being greatly challenged in meeting compliance obligations as set out under the GDPR, despite the myriad of software tools available to them. Many organisations continue to take a manual and informal approach to GDPR compliance. Our analysis shows significant gaps on the part of GDPR tools in their ability to demonstrate compliance in that they lack interoperability features, and they are not supported by published methodologies or evidence to support their validity or even utility. In contrast, RegTech has brought great success to financial compliance, using technological solutions to facilitate compliance with, and the monitoring of regulatory requirements. A review of the State of the Art identified the four success features of a RegTech system to be, strong data governance, automation through technology, interoperability of systems and a proactive regulatory framework. This paper outlines a set of requirements for GDPR compliance tools based on the RegTech experience and evaluate how these success features could be applied to improve GDPR compliance. A proof of concept prototype GDPR compliance tool was explored using the four success factors of RegTech, in which RegTech best practice was applied to regulator based self-assessment checklist to establish if the demonstration of GDPR compliance could be improved. The application of a RegTech success factors provides opportunities for demonstrable and validated GDPR compliance, notwithstanding the risk reductions and cost savings that RegTech can deliver and can facilitate organisations in meeting their GDPR compliance obligations.

Keywords: GDPR · Compliance · Accountability · Data protection officer · RegTech

J. Filipe et al. (Eds.): ICEIS 2020, LNBIP 417, pp. 905–929, 2021.
https://doi.org/10.1007/978-3-030-75418-1_41

1 Introduction

In May 2018, the European Union (EU) introduced the GDPR. This regulation brought about a high level of protection for data subjects, but also a high level of accountability for organisations [1]. The GDPR is a paradigm shift in data protection, towards greater choice and sovereignty for individuals, and more accountability for organisations [2]. The GDPR principle of accountability requires that a data controller must be able to demonstrate their compliance with the regulation (GDPR Recital 74). This requires an organisation "to act in a responsible manner, to implement appropriate actions, to explain and justify actions, provide assurance and confidence to internal and external stakeholders that the organisation is doing the right thing and to remedy failures to act properly" [3].

Organisations can be complex entities, performing heterogeneous processing on large volumes of diverse personal data, potentially using outsourced partners or subsidiaries in distributed geographical locations and jurisdictions. A challenge to complying with the accountability principle of the GDPR for organisations is demonstrating that these complex, dynamic activities and structures are meeting their regulatory obligations. The organisation must implement appropriate policies, procedures, tools and mechanisms to support their accountability practices [3].

Many organisations appoint a Data Protection Officer (DPO) to assist in this process. Bamberger describes the role as "the most important regulatory choice for institutionalising data protection" [4]. In practice the DPO is the early warning indicator of adverse events when processing personal data within the organisation [5]. The DPO must have "professional qualities and, in particular, expert knowledge of data protection law and practices" [GDPR Art 37]. This challenging role requires the DPO to monitor compliance and advise the organisation accordingly. The DPO acts independently of the organisation to assess and monitor the consistent application of the GDPR regulation and to ensure that the rights and freedoms of data subjects are not compromised [6]. The role of DPO encompasses a dynamic motion of policy generation, staff training, business process mapping and review, compliance record keeping, audit, data protection impact assessments, and compliance consultations [5]. The constant pace of business change allied with evolving legal interpretations require constant vigilance on the part of the DPO and create additional challenges for accountability. Fundamentally, it is the organisation, and not the DPO, that must be able to demonstrate that it is meeting the threshold that is the accountability principle as set out in the GDPR [3].

There are many solutions available to DPOs and organisations to help meet this challenge of demonstrating compliance to the accountability principle. We evaluate here the range of available tools, such as: privacy software solutions from private enterprise vendors, maturity models and regulator self- assessment tools. Despite the many GDPR compliance tools available, we highlight that the majority fail to meet the accountability principle. Most are not supported by published methodologies or evidence for their validity or even utility. They lack the ability to integrate or be integrated with other tools and the level of automation and innovation in this space has also been limited.

In contrast, RegTech has emerged as a framework for automating regulatory compliance in the Financial Industry. The "Global Financial Crisis (GFC)" of 2008 prompted financial regulators to introduce new compliance regulations [7], resulting in significant compliance challenges and compliance costs for organisations due to the complexity of these regulations. The digital transformation of data and the assimilation of regulatory compliance provisions into software code [8] to facilitate regulatory compliance has been enabled by developments such as process automation, the digitising of data, the use of semantic methods and machine learning algorithms. RegTech uses such tools to efficiently deliver compliance and risk reports in integrated toolchains. The evolution of RegTech has shown that IT can be used to support automated or semi-automated regulatory monitoring and reporting of compliance [9].

This paper proposes challenges for realising a RegTech approach to GDPR compliance whereby organisations leverage modern information technology to improve the organisational and external visibility of their GDPR compliance level. This approach requires automated data collection from relevant sources throughout the organisation and monitoring via GDPR compliance evaluation functions that could provide interoperable and machine-readable compliance metrics or reports for the organisation, suggested compliance actions and root cause analysis of compliance issues, using agreed data quality standards such as ISO8000. The role of monitoring, analysing and reporting the GDPR compliance status in an organisation is the task of the DPO. A RegTech approach to GDPR compliance could provide the DPO with the ability to track organisational compliance progress, identify areas of compliance weakness and benchmark their performance against other organisations. This would greatly enhance an organisation's ability to demonstrate and improve compliance and thus meet the GDPR accountability requirement.

This paper builds on "Design Challenges for GDPR RegTech", published in ICEIS 2020 [10], and seeks to identify the key success features of RegTech systems and evaluates how these success features could be applied to improve GDPR compliance. It features an updated and detailed review of GDPR compliance tools available to organisations and an enhanced State of the Art review of the key features of RegTech systems. Each of these key features are identified and evaluated to establish how they could be applied to GDPR Compliance.

The rest of this paper is structured as follows: Sect. 2 will discuss the accountability principle and what it means in practice to an organisation and the challenges they face to meet the accountability principle. The role of the DPO, and their part in compliance will be discussed in detail from the perspective of a practising DPO. Section 3 reviews the current approaches to GDPR compliance and critiques the many available offerings such as private enterprise software solutions, maturity models and self-assessment checklists. Section 4 examines RegTech literature to identify the key success features of RegTech systems to understand how RegTech is enhancing compliance using data driven solutions. Section 5 takes these success features and looks at them in a GDPR context to establish the challenges that organisations must face in developing the next generation of GDPR compliance tools. Section 6 will introduce a proof of concept compliance system where a Data Protection Regulators self- assessment checklist has been utilised based on RegTech best practice, to provide a simple efficient method to demonstrate GDPR compliance and meet the requirements of the accountability principle.

2 The GDPR Accountability Principle: A View from the DPO

In this section, this paper will discuss what the accountability principle of the GDPR means to organisations. The paper will look at the challenges that organisations are facing with demonstrating that they are meeting these obligations and it will discuss the role of the DPO in this process.

The Anglo-Saxon word "Accountability" has a broadly understood meaning of how responsibility is exercised and how it is made verifiable [11]. Accountability can be viewed to be an expression of how an organisation displays "a sense of responsibility—a willingness to act in a transparent, fair and equitable way" [12] and "the obligation to explain and justify conduct' [12]. The GDPR accountability principle requires a data controller "implement appropriate and effective measures to put into effect the principles and obligations of the GDPR and demonstrate on request" [11]. In 2018 the Centre for Information Policy Leadership (CIPL) [13] developed accountability-based data privacy and governance programs to encompass the key elements of accountability as described in Fig. 1.

Fig. 1. The accountability wheel–universal elements of accountability [13].

In practice, this can be viewed as "setting privacy protection goals based on criteria established in law, self-regulation and best practices and vesting the organisation with the responsibility to determine appropriate, effective measures to reach these goals" [13]. This is quite a challenging task for a data controller when challenged with a substantial legal text like the GDPR. There is a "lack of awareness of their obligations and duties in relation to personal data protection, it is urgent to define a methodology to be able to comply with the GDPR" [14].

In theory, the GDPR provides for certification methods in article 42 and 43 of the GDPR to assist a controller in demonstrating compliance. However, in practice this has proven to be a challenge for organisations as the European Union has not approved any

Certification body to certify compliance [15]. In fact, there are views being expressed that the GDPR certification process cannot be successful [15].

Many organisations appoint a DPO to assist with their GDPR compliance, however it is important to note that the demonstration of compliance obligations ultimately rests with the controller (organisation) and not the DPO. The role of DPO within the organisation covers a wide range of tasks as prescribed in Article 39 of the GDPR. The main tasks are to monitor, inform and advise the controller or processor regarding compliance with the GDPR, to provide advice such as data protection impact assessments, to provide training and awareness raising and to co-operate with and act as a contact point for the supervisory authority. The DPO must be at the heart of the accountability-based framework [5].

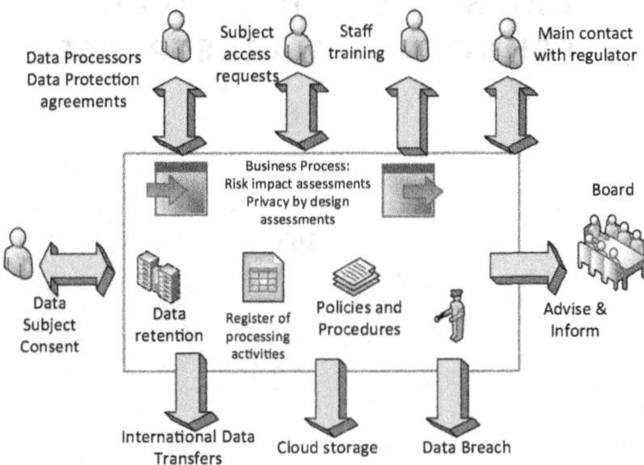

Fig. 2. The breadth and complexity of the role of Data Protection Officer [10].

The role of DPO requires a broad set of skills in GDPR legal compliance, and a detailed knowledge of business processes [5]. The DPO works with numerous stakeholders such as data subjects, employees, processors and regulators and provides consultancy and guidance on business processes. The role involves a broad spectrum of activities from maintaining a register of processing activities to dealing with data breaches, to completing data protection impact assessments. The DPO must have visibility of all activities and monitor and report compliance to the highest level in the organisation (see Fig. 2). The DPO is in essence "privacy on the ground" [16], in that the DPO is the early warning system for GDPR compliance within the organisation [5]. The challenge for the DPO is how to show that the organisation is accountable and how to demonstrate GDPR compliance.

3 An Analysis of Current Approaches to GDPR Compliance

This section discusses the broad range of tools and methods that are available to DPOs to demonstrate the GDPR compliance of their organisation. Each approach will be reviewed and critiqued.

3.1 Private Enterprise Software Solutions

There has been a call for tools and methods to assist organisations in meeting their GDPR compliance obligations [17]. This is being met by large financial investments by venture capital companies with over $500 million invested in privacy related start-ups around the world in 2017 [18]. This has continued into 2020 with significant funding continuing to go into privacy software vendors, such as One Trust $210 m, Ave Point $200 m, Privitar $80 m and BigID $50 m. There are over 304 vendors offering privacy software tools to organisations [19]. This has grown substantially from 2017 when there were 44 such vendors [20] (see Fig. 3).

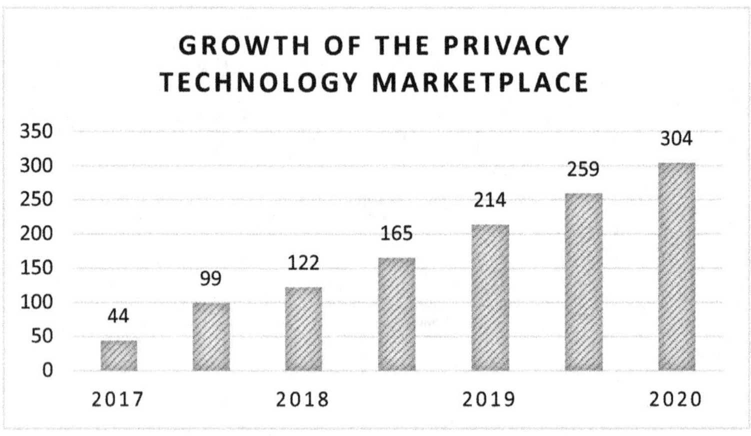

Fig. 3. Growth of the privacy technology marketplace [20].

These software solutions come in many forms ranging from simple questionnaires and templates to solutions that focus on individual aspects of compliance for GDPR such as statistical discovery tools for Data Subject Access Requests (DSAR's) where individuals exercise their right of access to information under Article 15 of the GDPR. The main categories of these privacy tools are as follows [20]:

- Activity Management - control and monitor access to personal data
- Assessment Managers - automate different functions of a privacy program, locating risk gaps, demonstrating compliance
- Consent managers - help organizations collect, track, demonstrate and manage users' consent.
- Data discovery - determine and identify personal data held
- Data mapping solutions - determine data flows throughout the enterprise.
- Data Subject Access Requests (DSAR's) - automation of process
- De-identification pseudonymisation tools
- Secure internal enterprise communications
- Data breach incident response solutions

- Privacy information managers - provide latest privacy laws around the world.
- Website scanning - catalogue cookies

Table 1. Privacy software tools, number of vendors per category [18, 19].

Privacy product category	No. of vendors offering this service 2019	No. of vendors offering this service 2020	Increase 2019 to 2020
Activity monitoring	86	93	+7
Assessment manager	102	118	+16
Consent manager	82	94	+12
Data discovery	91	107	+16
Data mapping	114	140	+26
Data subject requests	–	53	+53
De identification/Pseudonymity	45	55	+10
Enterprise communications	39	54	+15
Incident response	62	86	+24
Privacy information manager	72	96	+24
Website scanning	30	37	+7
Total	248	328	+80

Whilst there are a variety of privacy software solutions being offered by vendors, as displayed in Table 1 "there is no single vendor that will automatically make an organisation GDPR compliant" [18]. In fact, most solutions on offer from private enterprise solution providers cover 3 or less categories, see Fig. 4.

An accountability framework requires a comprehensive approach to compliance across the organisation. Whilst these software solutions go some way towards the demonstration of compliance, the author has identified several weaknesses in these private enterprise software solutions, as follows:

- they are not supported by published methodologies or evidence to support their validity or even utility
- many of these solutions are stand - alone in that they lack inter-operability with other GDPR compliance systems and hence cannot easily be assembled into toolchains providing comprehensive compliance reports and metrics, quality improvement processes or data analytics such as root cause analysis
- they focus on manual or semi-automated assessment approaches that are labour intensive, rely on domain experts and are not driven by quantitative operational data that is increasingly being generated by organisations

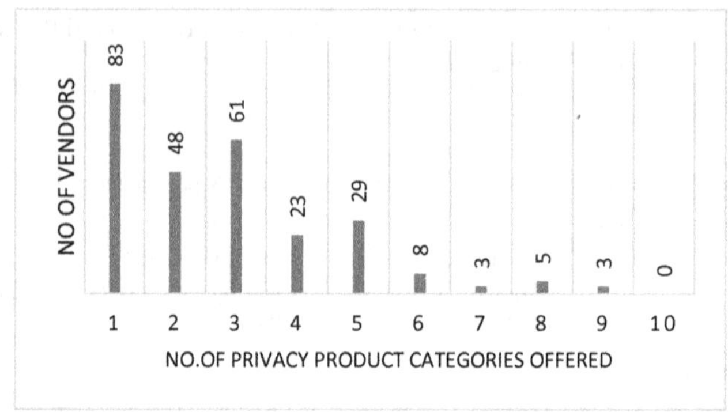

Fig. 4. No. of privacy product categories offered by no. of vendors [18].

- they are created by private enterprise and are based on an interpretation of the regulation, rather than being developed with the input of the regulator.

These solutions offer a starting point for GDPR compliance for an organisation however the lack of academic rigor or formal regulatory input and the inability to connect and build tool chains inhibits these solutions. The use of data driven inputs from heterogeneous sources and the mapping of business processes using agreed semantic standards would improve inputs to the evaluation tool. This would remove subjectivity and improve the quality of the outputs. GDPR compliance software must avoid the "pitfalls of a fragmented Tower of Babel approach." [21]. The 'best-of-breed' software point solution products could be used to feed a global evaluation tool to optimise and organise the outputs using agreed semantics.

3.2 Capability Maturity Models

Capability Maturity Models have been used for compliance monitoring for many years [22]. The American Institute of Certified Public Accountants privacy maturity model [23] was used to gain an understanding of an organisation's privacy compliance standing. It used a set of questions referred to as "generally accepted privacy principles" in the form of 73 measurable criteria. It gauged compliance along an axis of five maturity levels from ad hoc to optimized. The drawbacks of this methodology as a measure of compliance to the GDPR are that it predates the GDPR and would therefore need updating to reflect the new regulation. The more recent International Association of Privacy Professionals (IAPP) Maturity Framework [20] develops a series of checklists built through "collaboration between a team of highly experienced privacy and security professionals, lawyers and regulators." Both solutions provide visualisations of compliance on an axis and are an indicative measure of compliance. However, they do have several drawbacks:

- they are labour intensive and dependent on highly skilled labour/domain experts
- they are prone to human subjectivity, bias and errors
- they are infrequently updated

- the measures chosen utilise a selection of questions and checklists that require academic validation
- an unsuitability as part of an automated process and quality improvement toolchain

Whilst these maturity models are indicative of an organisations GDPR compliance position, the limitations outlined prevent these tools from developing any further without automation. Once automated, the lack of reporting and interoperability standards mentioned in the last section become relevant.

3.3 Accountability Toolkits and Self-assessment Checklists

Several data protection supervisory authorities have provided self-assessment checklists and accountability toolkits to assist organisations to prepare for GDPR. Their purpose is to support organisations to assess whether they have appropriate and effective internal policies, procedures and measures in place to ensure compliance with data protection regulations. These come in the form of a series of questions and checklists and are designed to assist the organisation in checking their compliance level. These toolkits are devised to provide broad coverage of all the principles of the GDPR. Just like maturity models these checklists provide an overview of compliance, however the main drawbacks of these tools for GDPR compliance are that they are fundamentally high-level self-assessments tools and are generic by nature and lack depth. Like maturity models, they rely on qualitative input of users, and they lack input or output interoperability with other solutions. However, the key benefit of these checklists and toolkits are that they have been developed by regulators, unlike maturity models and private enterprise software solutions, which have been developed independently.

3.4 Summary of Findings Regarding GDPR Compliance

This paper has demonstrated that there are several approaches that organisations are taking to GDPR compliance, each with their own weaknesses. The key questions are why is there such different approaches being taken by organisations to compliance? Why are some organisations choosing to rely on spreadsheets and templates, while others develop their own software and more again are investing in privacy tools? Do organisations understand the consequences of non-compliance with the GDPR? In recent analysis

Table 2. GDPR compliance of organisations 2019 [24].

GDPR compliance status	%
Fully compliant	9
Very compliant	36
Moderately compliant	42
Somewhat compliant	12
Not at all compliant	1

from the IAPP [24], the level of GDPR compliance by organisations is still considerably low (see Table 2).

Organisations are being greatly challenged in meeting their compliance obligations as set out under the GDPR, despite the myriad of software tools available to them [24]. Based on our personal experience and conversations with many DPOs and vendors, many organisations are choosing to use manual methods such as spreadsheets to manage their GDPR compliance. They continue to struggle with core GDPR compliance requirements such as Data Protection Impact Assessments (DPIA), Registers of Processing Activities (ROPA) and data inventory mapping (see Table 3). In fact, 76% do not use commercial software tools for carrying out any of these activities [25]. There are many solutions available to organisations to help meet their GDPR compliance, so why are organisations choosing not to implement these?

Table 3. Primary tools used to manage key GDPR compliance processes [25].

Primary tools used to manage key GDPR compliance processes	Data protection impact assessments	Data inventory and mapping	Register of processing activities
Manual /informal	47	45	44
Customised software	20	20	20
Internally developed systems	18	19	21
Governance, risk and compliance software (GRC)	8	10	9
Don't know	5	2	5
Outsourced	2	4	–
Not applicable	–	–	1

When it comes to another core element of the GDPR, data subject access requests (DSAR's), organisations are equally challenged. In 2019 DSAR's were the biggest cause of complaints to the Irish Data Protection Regulator [26]. An analysis of how DSAR's are addressed by organisations showed only 3% of DSAR's are automated, and 57% are completely manual (see Table 4) [25]. Despite there being 53 different software vendors (see Table 1) offering solutions to assist organisations, why are organisations failing to meet their compliance obligation in this area? Organisations are facing a steady stream of DSAR's and are spending significant human capital on satisfying requests [25].

In a similar analysis of the Data protection Impact Assessment (DPIA) personal data risk assessment process, most organisations were found to using manual / informal processes to complete this compliance document. This is also echoed with the Register of Processing Activities (ROPA) where organisations must maintain a recording of their personal data processing activities (see Table 3). These processes are part and parcel of an accountability framework, also mandated by article 30 of the GDPR [25]. Research

Table 4. How Data subject access requests are addressed by organisations [25].

Process used for SAR's	%
Fully automated	3
Partially automated	30
Entirely manual but mature	30
Entirely manual and ad-hoc	27
Still being designed	7
Don't know	2
Have not addressed yet	1

would indicate that organisations are using a variety of manual approaches to the maintenance of this critical record [27]. In fact, Table 3 shows a remarkably similar consistency on the part of the primary compliance tools used by organisations.

Organisations are greatly challenged to meet the accountability principle of the GDPR. The role of the DPO is a challenging one. They must monitor, advise and inform the organisation regarding their GDPR compliance. This is proving to be a significant challenge as many organisations continue to take a manual and informal approach to GDPR compliance [20].

4 An Analysis of the Key Features of RegTech Systems

In this section this paper will conduct a literature review to gain an understanding of what is Regtech, and to identify the key features of a Regtech system.

4.1 Methodology

A search of Google Scholar was conducted using the word "RegTech". The top 12 cited papers were identified, read and reviewed. Our conclusion was that the key success factors which have made RegTech successful are the adoption of developments in technology, the digital transformation of data, agreement on common standards to facilitate interoperability of systems, and the role of financial regulators as facilitators for the automation of regulation (see Table 5).

4.2 What is RegTech?

RegTech can be defined as "the use of technological solutions to facilitate compliance with, and the monitoring of regulatory requirements" [28]. RegTech has played an important role in making regulatory compliance more efficient and effective [21]. RegTech is information technology (IT) that (a) helps firms manage regulatory requirements and compliance imperatives by identifying the impacts of regulatory provisions on business models, products and services, functional activities, policies, operational procedures and controls; (b) enables compliant business systems and data; (c) helps control and manage regulatory, financial and non-financial risks; and (d) performs regulatory compliance reporting [21]. RegTech has evolved to address regulatory challenges in the financial system through innovative technology. It can support the technical handling of large amounts of data, sophisticated analysis and automated data processing within intermediaries as well as between intermediaries and supervisors [29].

Table 5. Literature review - the features of RegTech system.

Publication	Adoption of developments in technology	Digital transformation of data	Common standards and Agreed semantics, interoperability of systems	Role of supervisory authority as enabler/stakeholder
Fintech, RegTech and the reconceptualization of Financial Regulation [30]	✓	✓		✓
Fintech and RegTech Impact on regulators and banks [31]				✓
Fintech and regtech in a nutshell and the future in a sandbox [32]				✓
Adaptive financial regulation and RegTech a concept article on realistic protection for victims of bank failures [33]	✓			✓
Regtech compliance and the technology judgement rule [34]	✓			✓
The emergence of regtech 2.0 from know your customer to know your data [35]	✓	✓		✓

(*continued*)

Table 5. (*continued*)

Publication	Adoption of developments in technology	Digital transformation of data	Common standards and Agreed semantics, interoperability of systems	Role of supervisory authority as enabler/stakeholder
An innovative RegTech approach to financial risk monitoring & supervisory reporting [36]	✓	✓	✓	
RegTech is the new black – the growth of RegTech demand and investment [37]	✓			
Understanding regtech for digital regulatory compliance [21]	✓	✓	✓	✓
RegTech as a response to regulatory expansion in the financial sector [28]	✓	✓	✓	✓
Fintech and RegTech: enabling innovation while preserving financial stability [9]				✓

4.3 Adoption of New Technologies

The first key feature of Regtech is the adoption of new technologies by financial institutions. The Fintech revolution of 2008 has been a key contributor to the success of RegTech [35]. The evolution of Fintech has seen rapid growth and the creation of new opportunities through the application of Big Data, Internet of Things, Artificial Intelligence, machine learning, distributed ledger technology and blockchain, smart contracts and digital identity [29]. This revolution has brought about the creation of many start -up companies that have brought new technologies to help traditional financial entities to deal with the compliance burden placed on them. The advent of regulations such as Know-Your-Customer (KYC) and Anti-Money Laundering (AML) and EU directives such as Markets in Financial Instruments (MiFID and MiFID II), meant that organisations required an investment in technology to meet their compliance obligations.

These complex regulations required automated compliance monitoring and reporting. Organisation can no longer rely on humans to monitor and report complex areas like insider trading and money laundering. The use of technology has enabled the automatic detection, and reporting of irregular activities to a compliance officer, thus reducing the risk of human error in the form of an inattentive staff member. Similarly, tests, help organisations to determine what level of investment advice must be given to a customer based on the results of an automatically processed questionnaire [28]. Again, this solution helps an organisation to reduce errors and meet its legal obligations through process automation. These solutions remove the need for human intervention and make compliance less complex. RegTech tools are being used to leverage data from existing operational information systems and seek to provide agile solutions to improve compliance visibility, through the automation of mundane compliance tasks and reduction of risk to the organisation [28]. The foundation of compliance has been to prevent identify, respond to and remedy risk [38]. RegTech solutions are being developed to meet these regulatory requirements, but also to modernise compliance and generate a measurable value proposition to the organisation. RegTech solutions enhance the basics of compliance through enhanced data integration, the use of automation, predictive analytics and strategic process alignment [38].

4.4 The Digital Transformation of Data

The second key feature of RegTech systems has been the digital transformation of data. The KYC regulation requires an organisation to gather and validate information relating to their customers. Traditionally this was a complex manual process requiring the presentation of documents, recent photographs and proof of identity and address. The digitisation of KYC data has greatly reduced the risk of errors occurring. Similarly, the use of automation and AI to monitor AML regulations to ensure that all transactions undergo constant and scrutinous checks and reviews, has improved regulatory compliance with the prevailing AML guidelines. It entails accessing and evaluating relevant data from various sources, examining historical alerts, and searching for negative information on customers and their counterparties through public domain searches. The use of distributed ledger technology (DLT) has enabled the smooth and seamless sharing of information between financial organisations and regulators [36]. This has resulted in the faster delivery and verification of KYC data and significant savings on time spent on the manual reconciliation of documents.

In 2017, The Economist published an article entitled "The world's most valuable resource is no longer oil, but data" [39]. This works well when data governance is applied to the data, however upward of 80% of enterprise data today is in an unstructured form [40]. This can be a real challenge and source or risk for an organisation. Unstructured data is any data that resides in emails, files, PDFs or documents that does not have a pre-defined data model. If data is left unstructured it is a significant organisational problem, particularly if it contains sensitive personal data. Historically financial institutions held large amounts of paper-based records. The digital transformation of such data, has facilitated the adaption of new technologies and enabled the organisation to automate processes, gain efficiencies and optimise regulatory compliance.

4.5 Interoperability of Systems Through Common Standards and Agreed Semantics

The third key feature of RegTech systems is the application of semantic standards, through the ability to create machine-readable meta-models that enable data virtualisation across heterogeneous data stores [21]. Data held in siloed, heterogeneous databases can be virtualised, and ontologies and/or predictive analytics/machine learning algorithms and AI applied to automatically classify the data. It can then be used to monitor GDPR compliance, advise on risk and inform stakeholders. When such a knowledge-based model is linked and integrated with a regulatory knowledge base such as legal texts, then it is possible to enable automated regulatory compliance reporting of regulations as part of a business process. It enables an enhanced data-driven approach to the management risk and associated regulatory compliance reporting. "Semantic interoperability ensures that these exchanges make sense— that the requester and the provider have a common understanding of the "meanings" of the requested services and data" [21]. Regtech holds much promise for Regulators and organisation to solve a big problem through collaborative approaches to regulation with a small effort [21]. The full benefits of RegTech will only materialise if the pitfalls of a fragmented Tower of Babel approach are avoided. Semantic standards are the key to all this [21].

4.6 The Role of the Supervisory Authority as an Enabler

The fourth key feature of the success of Regtech has been the role that financial regulators have played, both as a beneficiary and an enabler of RegTech systems. In 2014 Andy Haldane, then Chief Economist at the Bank of England said "I have a dream. It is futuristic, but realistic. It involves a Star Trek chair and a bank of monitors. It would involve tracking the global flow of funds in close to real time in much the same way as happens with global weather systems and global internet traffic. Its centerpiece would be a global map of financial flows, charting spill-overs and correlations." [30].

The regulator now not only has access to periodic or real-time, fine-grained compliance reports, and the incremental improvements in compliance but also promote the design of a regulatory framework able to dynamically adapt to new rules and regulations [32]. The UK Financial Conduct Authority (FCA) has been a leader in this area. They sought to bring innovative technologies that had been successful in Fintech into the regulatory sphere, in a collaborative manner (see Fig. 5). The FCA identified areas of interest and called for tools. They advocated for technology accelerators where FinTech and financial firms were encouraged, in some cases, funding excellent innovations in the use of technology to deliver regulatory compliance and reporting. The regulator sought tool development in areas of interest such as real-time and system embedded compliance/risk evaluation tools, big data techniques, visualisation and automation tools, software integration tools and cloud technologies.

The FCA has hosted 7 Tech Sprints since 2016 to foster innovative and collaborative global financial markets, which brought together financial institutions, academics, regulators and technology companies to want to share experiences and to address key industry problems. An example of the success of these Tech Sprints is where the regulator brought

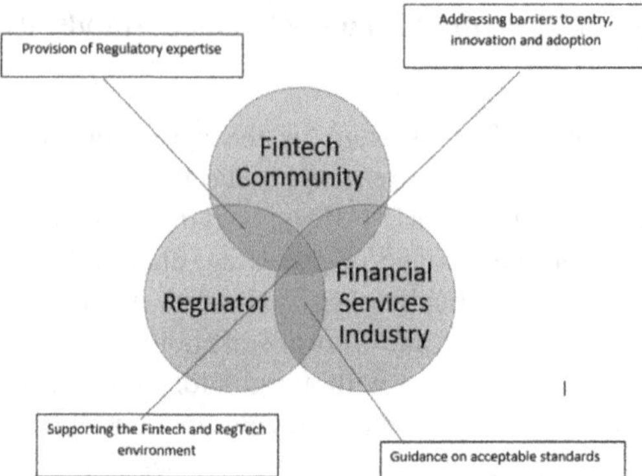

Fig. 5. A collaborative approach to RegTech fostered by the regulator [41].

the relevant stakeholders together to meet formulate an industry-wide open standards-based approach to regulatory compliance and reporting, which was required by both UK regulators and the financial industry [21]. The Tech Sprint was conducted to prove that standards based RegTech could help automate, and make more efficient and cost-effective, the task of regulatory reporting by financial institutions [21]. The traditional model of masses of documents stored in spreadsheets and PDF documents in servers was challenged. The Tech Sprint proposed solutions such as the use of Natural Language Processing, human and machine-readable rules, machine executable regulatory reports and linked knowledge bases.

4.7 Summary of RegTech Literature Review

The review of the literature identified the key success factors of RegTech systems. The role of the regulator as an enabler featured in almost all literature, and it was the primary success feature for RegTech. Without the regulator's facilitation to make regulations digital, RegTech could not have happened [21]. The adoption of new technologies features widely as a key success factor and is a key enabler to the success of RegTech. The role played by the digital transformation of data and the need for common standards and agreed semantics, Interoperability of systems is less pronounced in the literature, albeit that they are key elements in building RegTech systems. This may be that the literature is more financial in its nature than technology based.

5 RegTech Success Factors Applied to GDPR Compliance

In this section, we take the key success features of RegTech systems as described in Sect. 4 and evaluate how these success features could be applied to improve GDPR compliance. We have identified that the key success features of a RegTech system are the adaption of developments in technology, the digital transformation of data, agreement on common standards to facilitate interoperability of systems, and the role of regulators as facilitators for the automation of regulation. Each success factor will be discussed to establish how they could be applied in a GDPR context to facilitate GDPR compliance.

5.1 Enabling the Digital Transformation of Data

The first requirement for GDPR RegTech is the digital transformation of data. Whilst the Fintech revolution facilitated the digital transformation of data in the financial industry, the same cannot be said for personal data held by non-financial organisations.

Most organisations lack a data governance culture, with the necessary tools and organisational supports. Data governance builds on data management tools and capabilities within the organisation. To date, there is little coverage of data management practices in data protection regulatory compliance in the literature. It is critical that organisations develop and build a dedicated data management capability [2]. Organisations are facing challenges meeting their GDPR obligations due to a lack of common ground between legal and data management domains. Labadie proposes the need for a dedicated data management capability within organisations [2]. This would act as "an abstraction layer between the normative aspects of the regulation, i.e. the regulatory guidelines and compliance requirements (CR), and the concretized compliance requirements (CCR)" [2] to solve organisational problems, (see Fig. 6 for Labadie Capability Model).

System capabilities				
Define protected data scope	Identify data objects	Classify data attributes	Locate data records	
Manage consent	Implement consent items	Collect consent instances	Distribute consent	Enforce consent-based processing
Enable data processing rights	Delete data	Pseudonymize data	Transmit data in standardized form	
Organizational capabilities				
Orchestrate data protection activities	Assume data protection responsibilities	Oversee data protection activities	Control compliance of external processors	
Demonstrate compliant data processing	Maintain records of processing activities	Maintain documentation of system landscape	Supervise sensitive processing activities	
Disclose information	To individuals	To authorities		

Fig. 6. Capability model for GDPR data management [2].

The role that the data management domain plays in facilitating a RegTech approach is seen in Fig. 7. Developing concretized compliance requirements will facilitate compliance demonstration.

Organisations need to have clearly defined data principles, where data is viewed as an asset. The agreed uses of that data must be clearly defined, and the organisation must ensure that the use of data relates positively to the regulatory environment. They need to clearly set out as to what are the organisational behaviors for data quality, who will access the data, how data will be interpreted (metadata), and how long it will be retained [42] An organisation processing personal data must be able to locate and categorise all personal data held that they hold (data catalogs plus data classification). They need to understand the level of sensitivity of the data processed, where it is stored, who is the owner of the data, who has access to the data, how data evolves through a lifecycle, what lifecycle stage it is at and how long the data is retained for.

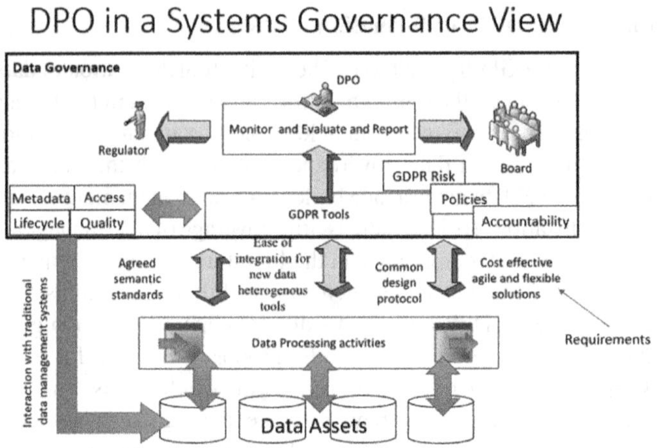

Fig. 7. Areas where IT will have to expand to support GDPR RegTech.

The GDPR is concerned with the protection of personal data that of a living individual. This type of data can be found in many guises within an organisation and can stretch to a substantial amount of data points. Such personal data can be found in many forms from HR files to CCTV images to internet browsing history. The challenge that organisations are facing about personal data is locating, classifying and cataloging this data i.e. creation of appropriate metadata to enable management of the personal data. The organisation must create management policies to enable machine-interpretation of data so that it is possible to automate business operations and enforcement. Effective digital policies require machine-readable models of the organisation, its people and resources, its data processing activities and so forth. Essentially this means having a "digital twin" (or data model) of the organisation and other relevant stakeholders like customers and the regulator. Building these models is not enough, organisations also need data governance, management and processing platforms that can interact to track data, processing activities and enforce policies. This would enable the automated creation of accountability logs with strong auditing capabilities. Once such systems are in place then value-added data governance and compliance tools can be built for areas such as risk management, data quality management, data value assessment, organizational change management and decision support tools.

5.2 Adoption of Developments in Technologies to GDPR Compliance

The second requirement for GDPR RegTech is the adaption of new technologies in the same manner that has been at the forefront of the successes of RegTech. A GDPR Regtech solution will require the same approach to new technology to facilitate efficient and effective compliance. The Fintech 3.0 revolution [43] which began in 2008 brought about the implementation of Big Data collection and analytics techniques, machine learning, Natural Language Processing (NLP), Artificial Intelligence (AI), cloud technology, devops (continuous development), Distributed Ledger tech, software integration tools and many other technologies in the financial industry. The cost of compliance and the need for agile solutions brought about the speedy and effective implementation of such new technologies. A RegTech approach to GDPR would require organisations to implement such technologies in a GDPR environment. The transformative nature of technology [30] enjoyed by RegTech can be achieved in the GDPR environment through by a new approach to technology at the nexus of data, digital identity, and regulation.

5.3 Agreement on Common Standards/Agreed Semantics for Personal Data Processing

The third requirement for GDPR RegTech is the need to make personal data interoperable between systems. Whilst the digital transformation of financial data in RegTech has facilitated the application of technology to this data, this may be more challenging in a GDPR environment where large amounts of data is held in manual forms (refer to Table 3 and 4). The semantic modelling of GDPR business processes would be a great benefit to an organisation and provide for machine-readable and interoperable representations of information that can be queried and verified based on open standards such as RDF, OWL, SPARQL, and SHACL [27, 44]. When these models are combined with legal knowledge bases, they become very useful for compliance evaluation and monitoring, which can help to harmonise and facilitate a common approach between legal departments and other stakeholders to the identification of feasible and compliant solutions around data protection and privacy regulations [2]. There has been progress to develop "Core Vocabularies', maintained by the Semantic Interoperability Community (SEMIC), that provide a simplified, reusable and extensible data model for capturing fundamental characteristics of an entity in a context-neutral fashion [45] in this area to foster interoperability. This work continues to be built on through the development of the W3C Data Privacy Vocabulary (DPV) and the PROV-O Ontology [45].

5.4 Data Protection Supervisory Authorities as an Enabler

The fourth requirement for GDPR RegTech is the need for proactive regulators, who will work with organisations to automate regulation and make compliance easier to achieve. The role of supervisory authority has been a major driving force in the success of RegTech, however Data Protection regulators are lagging behind their financial counterparts. There has been some effort by data protection regulators to make the business of compliance easier through the provision of guidance documents, self-assessment checklists and templates, however this remains far removed from the success of RegTech. Whilst each GDPR regulator must apply the GDPR consistently (GDPR recital 135) there have been very little in the form of a unified approach to technical solutions to facilitate GDPR compliance. In fact, an analysis of ROPA templates provided by regulators showed that there were inconsistencies in templates where some templates had as little as 12 fields whereas others had 43 fields. [27]. Perhaps there is a role for the European Data Protection Board (EDPB) as an independent European body, which contributes to the consistent application of data protection rules throughout the EU, and promotes cooperation between the EU's data protection authorities; this may be the way forward here. There have been a number of positive initiatives such as the development of the Internet Privacy Engineering network (IPEN), which promotes and advances the state-of-the-art of privacy engineering among regulators, academia, open source and business development, and other individuals who are committed to finding engineering solutions to privacy challenges. The over-all objective is to integrate data protection and privacy into all phases of the development process, from the requirements phase to production. The EDPB has also released open source software known as the "Website evidence collector" [46], which is a step in the right direction. There have been positive moves by the UK regulator (ICO) to develop sandboxes to work with innovators in privacy technology, however this is very much in the early days. For GDPR RegTech to be successful GDPR regulators need to move towards a symbiotic relationship with technology innovators and organisations processing personal data to develop open source compliance tools, digital regulations, sandboxes [9] and tech sprints. This would greatly accelerate the successes of GDPR RegTech solutions.

5.5 Summary

The key success features of RegTech systems are the adaption of developments in technology, the digital transformation of data, agreement on common standards to facilitate interoperability of systems, and the role of regulators as facilitators for the automation of regulation. If GDPR RegTech is to be realised it will require "a carefully tailored design of the technology, a joint effort of the regulators and the private sector and some shifts in corporate thinking [34]". Many Organisations have begun to invest or develop technology to facilitate compliance, whilst many others continue to rely on spreadsheets. The need for common ground between the legal and data management domain, would facilitate the digital transformation of data and be a major step towards digital compliance. Data Protection regulators have a major role to play as an enabler and facilitator of digital transformation of regulation and the formulation on common standards and agreed semantics for personal data processing.

6 Proof of Concept: An Evaluation Tool for GDPR Compliance

In this section we describe a prototype GDPR high-level evaluation tool that has been developed based upon the developments in RegTech, outlined in Sect. 5. The tool is an open-source high level GDPR compliance evaluation methodology that has been based on a self-assessment checklist created by a data protection regulator. It has been developed to measure the GDPR compliance level in an organisation. The evaluation tool was developed from the Irish Data Protection Commission self- assessment checklist [47] which segmented the GDPR into 8 regulatory sections and posed 54 questions in total. The tool is designed to be a layered information delivery system that provides information and insights so that the DPO can measure, monitor and manage business performance more effectively, and address accordingly [48].

The evaluation tool provides three layers of data as displayed in Fig. 8. The top level being a graphical overview of compliance for monitoring and reporting purposes, the second layer being the dimensional data that provides a view of each aspect of the GDPR and the final layer that being the detail of each GDPR compliance area.

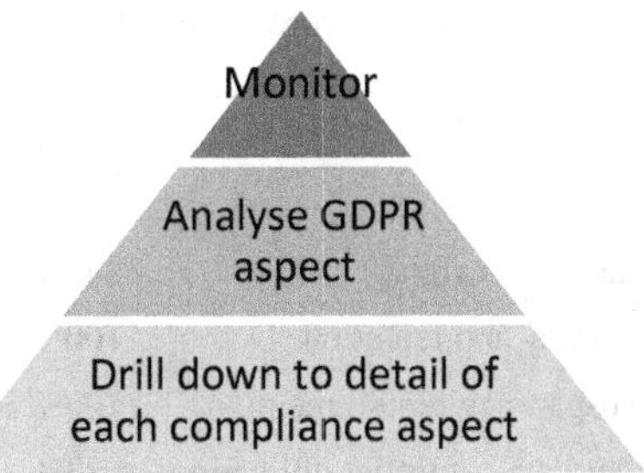

Fig. 8. GDPR evaluation dashboard overview [48].

It is planned that the tool will use the W3C Community group's data protection vocabulary [49] to describe the context using explicit semantics and the W3C Data Cube vocabulary to represent the time series of measurements across the different GDPR aspects or dimensions [50]. This development involved taking the self-assessment checklist and transforming it into an evaluation tool which was populated by a sample organisation each month for six months in total. The overall GDPR compliance monthly score for the organisation for each month is displayed in Fig. 9. This information gives the DPO a high-level view of compliance for the organisation.

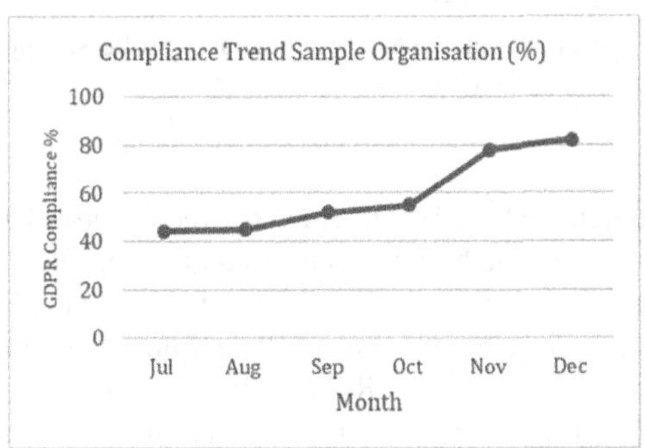

Fig. 9. Compliance trend for sample organisation [10].

The results from the evaluation tool can be also viewed by GDPR regulatory section to analyse how the organisation is performing in the various aspects of GDPR compliance, thus providing enhanced visibility to the DPO. In Table 6 the organisation is fully compliant in accuracy and retention but is only 50% compliant regarding data breaches. The data can be examined to another sub-level to provide the detail by GDPR aspect. Table 7 breaks out the Data Breach aspect and provides the granularity that a DPO needs to provide feedback to the controller to drive actions and improve the compliance of the organisation.

In the sample organisation the DPO can identify the non-compliant areas as identified in Table 3 and take the necessary actions to resolve.

Table 6. Compliance score per regulatory area [10].

GDPR section	Compliant %
Personal data	67%
Data subject rights	40%
Accuracy and retention	100%
Transparency requirements	100%
Other data controller obligations	83%
Data security	100%
Data breach	50%
International data transfers	100%
Total score	82%

Table 7. Non-compliance results for data breach [10].

Data breach section	Areas of failure
Are plans and procedures regularly reviewed?	Non-compliant
Are all data breaches fully documented?	Non-compliant
Are there cooperation procedures in place between data controllers, suppliers and other partners to deal with data breaches?	Non-compliant

This approach has demonstrated the use of a RegTech approach to GDPR compliance using a simple cost-effective method. It has utilised questions that have been created by regulatory authorities themselves so they could serve as a strong platform for the assessment of compliance. The evaluation tool meets the requirement of being comprehensive in that it covers the breadth of the GDPR and is informative in that it provides specific scores into GDPR areas requiring focus. The evaluation process is repeatable in that it can be run at intervals to generate compliance trends. The results yielded specific and relevant scores that can be used to drive corrective actions. The use of data driven inputs from heterogeneous sources and the mapping of business processes into the evaluation tool using agreed semantic standards would remove qualitative user inputs and would improve inputs to the evaluation tool. This would remove subjectivity and improve the quality of the outputs.

7 Conclusions

The accountability principle of the GDPR requires organisations be accountable for the demonstration of their compliance with the regulation. We have seen that the DPO is greatly challenged to monitor GDPR compliance as each of the available GDPR compliance tools, have their shortcomings. Our research question asked could a Regtech approach to the GDPR be successful? We reviewed RegTech literature and identified that the success features of a RegTech system are, the adoption of developments in technology, the digital transformation of data, agreement on common standards to facilitate interoperability of systems, and the role of regulators as facilitators for the automation of regulation. The application of these RegTech key success factors to the GDPR has already begun, in that organisations are moving towards technology as a solution to their GDPR compliance. Some of these organisations are acting independently and developing their own compliance software tools, whilst others are investing in privacy software or manually managing their compliance on spreadsheets. Organisations need a dedicated data management capability to build common ground between the legal and data management domains, to facilitate the digital transformation of data. GDPR Regulators have lacked the proactivity of financial regulators in the facilitation of automated digital compliance. This lack of leadership has resulted organisations facing the "pitfalls of a fragmented Tower of Babel approach" [21]. The role of the supervisory authority was identified as a key enabler and facilitator, for RegTech, however GDPR Regulators have been quite slow to take a similar role in comparison to financial regulators. We have shown that the use of technology to improve compliance monitoring and reporting can be achieved when flexible, agile, cost effective, extensible and informative tools are combined, particularly when the regulator facilitates digital compliance. The failure of GDPR regulators to actively promote digital regulatory compliance standards is resulting in slow progress towards digital compliance.

A proof of concept was conducted to demonstrate the application of some of these RegTech approaches to GDPR Compliance. A GDPR compliance tool was developed from a regulator supplied self-assessment tool to monitor and analyse organisational compliance that yielded a GDPR compliance output for an organisation. The compliance report that was generated from the evaluation tool can be used to identify GDPR areas where the organisation is not compliant, to trend their progress towards GDPR compliance over time and to benchmark performance versus other organisations. The DPO can use the results to direct resources to areas of non-compliance and improve their score, thus reducing the risk of GDPR fines. We have shown that a RegTech approach to GDPR can enable an organisation to meet its obligations to comply with the accountability principle.

Acknowledgements. This work is partially supported by Uniphar PLC., and the ADAPT Centre for Digital Content Technology which is funded under the SFI Research Centres Programme (Grant 13/RC/2106) and is co-funded under the European Regional Development Fund.

References

1. Buttarelli, G.: The EU GDPR as a clarion call for a new global digital gold standard. Int. Data Priv. Law, 77–78 (2016)
2. Labadie, C., Legner, C.: Understanding Data Protection Regulations from a Data Management Perspective: A Capability-Based Approach to EU-GDPR (2019)
3. Felici, M., Koulouris, T., Pearson, S.: Accountability for Data Governance in Cloud Ecosystems (2013)

4. Bamberger, K., Mulligan, D.: Privacy on the Ground: Driving Corporate Behaviour in United States and Europe, p. 204. MIT Press (2015)
5. Drewer, D., Miladinova, V.: The canary in the data mine. Comput. Law Secur. Rev. **34**, 806–815 (2018)
6. Eur. Union.: Charter of Fundamental Rights of The European Union. Official J. Article 8, (2012)
7. Johansson, E., Sutinen, K., Lassila, J., Lang, V., (eds.) Martikainen M., Lehner, OM.: RegTech-A Necessary Tool to Keep up with Compliance and Regulatory Changes? (2019)
8. Bamberger, K.: Technologies of compliance: Risk and regulation in a digital age. Texas Law Rev. (2009)
9. Arner, D.W., Zetzche, D.A., Buckley, R.F., Barberis, J.: Fintech and RegTech: enabling innovation while preserving financial stability. Georgetown J. Int. Affairs **18**, 47–58 (2017)
10. Ryan, P., Crane, M., Brennan, R.: Design Challenges for GDPR RegTech, ICEIS 92), pp. 787–795 (2020)
11. Article 29 Data Protection Working Party.: Opinion 3/2010 on the Principle of Accountability (2010)
12. Boven's, M.: Analysing and Assessing Accountability: A Conceptual Framework (2007)
13. Centre for Information Policy Leadership.: The Case for Accountability: How it Enables Effective Data Protection and Trust in the Digital Society (2018)
14. Da Conceicao Freitas, M., Silva M.: GDPR compliance in SME's: There is much to be done. J. Inf. Syst. Eng. Manage. (2018)
15. Lachaud, E.: Why the certification process defined in the General Data Protection regulation cannot be successful? Comput. Law and Secur. Rev. **32**, 814–826 (2016)
16. Heimes, R.: Top 10 operational impacts of the GDPR: Part 2 – The mandatory DPO (2016)
17. Piras, L., et al.: DEFeND architecture: a privacy by design platform for GDPR compliance. In: Gritzalis, S., Weippl, E.R., Katsikas, S.K., Anderst-Kotsis, G., Tjoa, A.M., Khalil, I. (eds.) TrustBus 2019. LNCS, vol. 11711, pp. 78–93. Springer, Cham (2019). https://doi.org/10.1007/978-3-030-27813-7_6
18. IAPP: 2018 Privacy Tech Vendor Report (2018)
19. IAPP: 2020 Privacy Tech Vendor Report (2020)
20. IAPP: 2019 The GDPR Maturity Framework. https://iapp.org/resources/article/the-gdpr-maturity-framework/. Accessed 31 July 2020
21. Butler, T., O'Brien, L.: Understanding RegTech for Digital Regulatory Compliance, Disrupting Finance (2019)
22. Humphrey, W.S.: Three process perspectives: Organizations, teams, and people. Ann. Softw. Eng. **4**, 39–72 (2002)
23. The American Institute of Certified Public Accountants (AICPA). Privacy Maturity Model (2011)
24. IAPP-EY annual privacy governance (2019)
25. IAPP – Trust Arc: Measuring Privacy Operations (2019)
26. Data Protection Commission. Annual Report 2020, p. 13 (2020)
27. Ryan, P., Pandit, H., Brennan, R.: Towards a Semantic Model of the GDPR Register of Processing Activities (2020)
28. Colaert, V.: RegTech as a response to regulatory expansion in the financial sector (2017)
29. Buckley, R.F., Arner, D.W., Zetzche, D.A., Weber, R.H.: The road to RegTech the astonishing example of the european union. J. Banking Regul. **21**, 36 (2020)
30. Arner, D., Barberis, J., Buckley, R.: FinTech, RegTech, and the Reconceptualization of Financial Regulation (2016)
31. Anagnostopoulos, I.: Fintech and RegTech impact on regulators and banks. J. Econ. Bus. **100**, 7–25 (2018)

32. Arner, D., Barberis, J., Buckley, R.: Fintech and RegTech in a Nutshell and the Future in a Sandbox. CFA Institute Research Foundation (2017)
33. Baxter, L.G., Duke, L.J.: Adaptive Financial Regulation and RegTech: A Concept Article on Realistic Protection for Victims of Bank Failures (2016)
34. Packin, N.G., Chi-Kent, L.: RegTech, Compliance and the Technology Judgement Rule (2018)
35. Arner, D.W., Barberis, J.N., Buckley, R.P.: The emergence of RegTech 2.0: From know your customer to know your data. J. Financ. Transform. **44**, 79 (2016). UNSW Law Research Paper No. 17-63. http://dx.doi.org/10.2139/ssrn.3044280
36. Kavassalis, P., Stieber, H., Breymann, W., Saxton, K., Gross, F.J.: An innovative RegTech approach to financial risk monitoring and supervisory reporting. J. Risk Finan. **19**(1), 39–55 (2018). https://doi.org/10.1108/JRF-07-2017-0111
37. Larsen, K., Gilani, S.: RegTech is the new black - the growth of RegTech demand and investment. J. Fin. Transform. Capco Inst. **45**, 22–29 (2017)
38. Deloitte.: Compliance Modernization is no Longer Optional https://www2.deloitte.com/lu/en/pages/strategy-operations/articles/compliance-modernization-no-longer-optional.html. Accessed 31 July 2020
39. The Economist: The World's Most Valuable Resource Is No Longer Oil, But Data (2017)
40. Vellingiri, J., Basser, K.K.: Data Management and Big Data Analytics: Data Management in Digital Economy (2019). https://doi.org/10.4018/978-1-5225-5137-9.ch001
41. Financial Conduct Authority: Call for Input: Supporting the development and adoption of RegTech (2015)
42. Khatri, V., Brown, C.V.: Designing Data Governance. Commun. ACM **53**(1) (2010)
43. Arner, D., Barberis, J., Buckley, R.: The Evolution of Fintech: A New Post-Crisis Paradigm? (2015)
44. GDPR Data Interoperability Model. Harshvardhan J. Pandit, Declan O'Sullivan, Dave Lewis (2018)
45. Pandit, H.J.: Representing Activities associated with Processing of Personal Data and Consent using Semantic Web for GDPR Compliance (2020)
46. https://joinup.ec.europa.eu/solution/website-evidence-collector/about
47. Data Protection Commission. https://www.dataprotection.ie/en/organisations/self-assessment-checklist
48. Eckerson, W.: Performance Dashboards: Measuring, Monitoring, and Managing Your Business, 2nd edition (2010)
49. Pandit, H., et al.: Creating a vocabulary for data privacy. In: Panetto, H., Debruyne, C., Hepp, M., Lewis, D., Ardagna, C.A., Meersman, R. (eds.) OTM 2019. LNCS, vol. 11877, pp. 714–730. Springer, Cham (2019). https://doi.org/10.1007/978-3-030-33246-4_44
50. Cyganiak, R.: The RDF data cube vocabulary. https://www.w3.org/TR/vocab-data-cube/. Accessed 31 July 2020

Correction to: A Modern Approach to Personalize Exergames for the Physical Rehabilitation of Children Suffering from Lumbar Spine

Cristian Gómez-Portes, Carmen Lacave, Ana I. Molina, David Vallejo, and Santiago Sánchez-Sobrino

Correction to: Chapter "A Modern Approach to Personalize Exergames for the Physical Rehabilitation of Children Suffering from Lumbar Spine" in: J. Filipe et al. (Eds.): *Enterprise Information Systems*, LNBIP 417, https://doi.org/10.1007/978-3-030-75418-1_35

In the originally published version of chapter 35, the special characters in the names of Cristian Gómez-Portes and Santiago Sánchez-Sobrino had been left out initially due to conversion errors. This has been corrected.

The updated version of this chapter can be found at https://doi.org/10.1007/978-3-030-75418-1_35

J. Filipe et al. (Eds.): ICEIS 2020, LNBIP 417, p. C1, 2021.
https://doi.org/10.1007/978-3-030-75418-1_42

Author Index

Printed in the United States
by Baker & Taylor Publisher Services